HANDBOOK of FLUIDIZATION and FLUID-PARTICLE SYSTEMS

CHEMICAL INDUSTRIES

A Series of Reference Books and Textbooks

Consulting Editor

HEINZ HEINEMANN
Berkeley, California

1. *Fluid Catalytic Cracking with Zeolite Catalysts*, Paul B. Venuto and E. Thomas Habib, Jr.
2. *Ethylene: Keystone to the Petrochemical Industry*, Ludwig Kniel, Olaf Winter, and Karl Stork
3. *The Chemistry and Technology of Petroleum*, James G. Speight
4. *The Desulfurization of Heavy Oils and Residua*, James G. Speight
5. *Catalysis of Organic Reactions*, edited by William R. Moser
6. *Acetylene-Based Chemicals from Coal and Other Natural Resources*, Robert J. Tedeschi
7. *Chemically Resistant Masonry*, Walter Lee Sheppard, Jr.
8. *Compressors and Expanders: Selection and Application for the Process Industry*, Heinz P. Bloch, Joseph A. Cameron, Frank M. Danowski, Jr., Ralph James, Jr., Judson S. Swearingen, and Marilyn E. Weightman
9. *Metering Pumps: Selection and Application*, James P. Poynton
10. *Hydrocarbons from Methanol*, Clarence D. Chang
11. *Form Flotation: Theory and Applications*, Ann N. Clarke and David J. Wilson
12. *The Chemistry and Technology of Coal*, James G. Speight
13. *Pneumatic and Hydraulic Conveying of Solids*, O. A. Williams
14. *Catalyst Manufacture: Laboratory and Commercial Preparations*, Alvin B. Stiles
15. *Characterization of Heterogeneous Catalysts*, edited by Francis Delannay
16. *BASIC Programs for Chemical Engineering Design*, James H. Weber
17. *Catalyst Poisoning*, L. Louis Hegedus and Robert W. McCabe
18. *Catalysis of Organic Reactions*, edited by John R. Kosak
19. *Adsorption Technology: A Step-by-Step Approach to Process Evaluation and Application*, edited by Frank L. Slejko
20. *Deactivation and Poisoning of Catalysts*, edited by Jacques Oudar and Henry Wise
21. *Catalysis and Surface Science: Developments in Chemicals from Methanol, Hydrotreating of Hydrocarbons, Catalyst Preparation, Monomers and Polymers, Photocatalysis and Photovoltaics*, edited by Heinz Heinemann and Gabor A. Somorjai
22. *Catalysis of Organic Reactions*, edited by Robert L. Augustine
23. *Modern Control Techniques for the Processing Industries*, T. H. Tsai, J. W. Lane, and C. S. Lin
24. *Temperature-Programmed Reduction for Solid Materials Characterization*, Alan Jones and Brian McNichol
25. *Catalytic Cracking: Catalysts, Chemistry, and Kinetics*, Bohdan W. Wojciechowski and Avelino Corma
26. *Chemical Reaction and Reactor Engineering*, edited by J. J. Carberry and A. Varma
27. *Filtration: Principles and Practices: Second Edition*, edited by Michael J. Matteson and Clyde Orr
28. *Corrosion Mechanisms*, edited by Florian Mansfeld
29. *Catalysis and Surface Properties of Liquid Metals and Alloys*, Yoshisada Ogino
30. *Catalyst Deactivation*, edited by Eugene E. Petersen and Alexis T. Bell
31. *Hydrogen Effects in Catalysis: Fundamentals and Practical Applications*, edited by Zoltán Paál and P. G. Menon

32. *Flow Management for Engineers and Scientists,* Nicholas P. Cheremisinoff and Paul N. Cheremisinoff
33. *Catalysis of Organic Reactions,* edited by Paul N. Rylander, Harold Greenfield, and Robert L. Augustine
34. *Powder and Bulk Solids Handling Processes: Instrumentation and Control,* Koichi Iinoya, Hiroaki Masuda, and Kinnosuke Watanabe
35. *Reverse Osmosis Technology: Applications for High-Purity-Water Production,* edited by Bipin S. Parekh
36. *Shape Selective Catalysis in Industrial Applications,* N. Y. Chen, William E. Garwood, and Frank G. Dwyer
37. *Alpha Olefins Applications Handbook,* edited by George R. Lappin and Joseph L. Sauer
38. *Process Modeling and Control in Chemical Industries,* edited by Kaddour Najim
39. *Clathrate Hydrates of Natural Gases,* E. Dendy Sloan, Jr.
40. *Catalysis of Organic Reactions,* edited by Dale W. Blackburn
41. *Fuel Science and Technology Handbook,* edited by James G. Speight
42. *Octane-Enhancing Zeolitic FCC Catalysts,* Julius Scherzer
43. Oxygen in Catalysis, Adam Bielanski and Jerzy Haber
44. *The Chemistry and Technology of Petroleum: Second Edition, Revised and Expanded,* James G. Speight
45. *Industrial Drying Equipment: Selection and Application,* C. M. van't Land
46. *Novel Production Methods for Ethylene, Light Hydrocarbons, and Aromatics,* edited by Lyle F. Albright, Billy L. Crynes, and Siegfried Nowak
47. *Catalysis of Organic Reactions,* edited by William E. Pascoe
48. *Synthetic Lubricants and High-Performance Functional Fluids,* edited by Ronald L. Shubkin
49. *Acetic Acid and Its Derivatives,* edited by Victor H. Agreda and Joseph R. Zoeller
50. *Properties and Applications of Perovskite-Type Oxides,* edited by L. G. Tejuca and J. L. G. Fierro
51. *Computer-Aided Design of Catalysts,* edited by E. Robert Becker and Carmo J. Pereira
52. *Models for Thermodynamic and Phase Equilibria Calculations,* edited by Stanley I. Sandler
53. *Catalysis of Organic Reactions,* edited by John R. Kosak and Thomas A. Johnson
54. *Composition and Analysis of Heavy Petroleum Fractions,* Klaus H. Altgelt and Mieczyslaw M. Boduszynski
55. *NMR Techniques in Catalysis,* edited by Alexis T. Bell and Alexander Pines
56. *Upgrading Petroleum Residues and Heavy Oils,* Murray R. Gray
57. *Methanol Production and Use,* edited by Wu-Hsun Cheng and Harold H. Kung
58. *Catalytic Hydroprocessing of Petroleum and Distillates,* edited by Michael C. Oballah and Stuart S. Shih
59. *The Chemistry and Technology of Coal: Second Edition, Revised and Expanded,* James G. Speight
60. *Lubricant Base Oil and Wax Processing,* Avilino Sequeira, Jr.
61. *Catalytic Naphtha Reforming: Science and Technology,* edited by George J. Antos, Abdullah M. Aitani, and José M. Parera
62. *Catalysis of Organic Reactions,* edited by Mike G. Scaros and Michael L. Prunier
63. *Catalyst Manufacture,* Alvin B. Stiles and Theodore A. Koch
64. *Handbook of Grignard Reagents,* edited by Gary S. Silverman and Philip E. Rakita
65. *Shape Selective Catalysis in Industrial Applications: Second Edition, Revised and Expanded,* N. Y. Chen, William E. Garwood, and Francis G. Dwyer
66. *Hydrocracking Science and Technology,* Julius Scherzer and A. J. Gruia
67. *Hydrotreating Technology for Pollution Control: Catalysts, Catalysis, and Processes,* edited by Mario L. Occelli and Russell Chianelli
68. *Catalysis of Organic Reactions,* edited by Russell E. Malz, Jr.
69. *Synthesis of Porous Materials: Zeolites, Clays, and Nanostructures,* edited by Mario L. Occelli and Henri Kessler
70. *Methane and Its Derivatives,* Sunggyu Lee
71. *Structured Catalysts and Reactors,* edited by Andrzej Cybulski and Jacob A. Moulijn
72. *Industrial Gases in Petrochemical Processing,* Harold Gunardson
73. *Clathrate Hydrates of Natural Gases: Second Edition, Revised and Expanded,* E. Dendy Sloan, Jr.
74. *Fluid Cracking Catalysts,* edited by Mario L. Occelli and Paul O'Connor
75. *Catalysis of Organic Reactions,* edited by Frank E. Herkes
76. *The Chemistry and Technology of Petroleum: Third Edition, Revised and Expanded,* James G. Speight
77. *Synthetic Lubricants and High-Performance Functional Fluids: Second Edition, Revised and Expanded,* Leslie R. Rudnick and Ronald L. Shubkin
78. *The Desulfurization of Heavy Oils and Residua, Second Edition, Revised and Expanded,* James G. Speight
79. *Reaction Kinetics and Reactor Design: Second Edition, Revised and Expanded,* John B. Butt
80. *Regulatory Chemicals Handbook,* Jennifer M. Spero, Bella Devito, and Louis Theodore

81. *Applied Parameter Estimation for Chemical Engineers*, Peter Englezos and Nicolas Kalogerakis
82. *Catalysis of Organic Reactions,* edited by Michael E. Ford
83. *The Chemical Process Industries Infrastructure: Function and Economics*, James R. Couper, O. Thomas Beasley, and W. Roy Penney
84. *Transport Phenomena Fundamentals*, Joel L. Plawsky
85. *Petroleum Refining Processes*, James G. Speight and Baki Özüm
86. *Health, Safety, and Accident Management in the Chemical Process Industries*, Ann Marie Flynn and Louis Theodore
87. *Plantwide Dynamic Simulators in Chemical Processing and Control*, William L. Luyben
88. *Chemicial Reactor Design*, Peter Harriott
89. *Catalysis of Organic Reactions*, edited by Dennis G. Morrell
90. *Lubricant Additives: Chemistry and Applications*, edited by Leslie R. Rudnick
91. *Handbook of Fluidization and Fluid-Particle Systems*, edited by Wen-Ching Yang
92. *Conservation Equations and Modeling of Chemical and Biochemical Processes*, Said S. E. H. Elnashaie and Parag Garhyan
93. *Batch Fermentation: Modeling, Monitoring, and Control*, Ali Çinar, Gülnur Birol, Satish J. Parulekar, and Cenk Ündey
94. *Industrial Solvents Handbook, Second Edition*, Nicholas P. Cheremisinoff

ADDITIONAL VOLUMES IN PREPARATION

Chemical Process Engineering: Design and Economics, Harry Silla

Petroleum and Gas Field Processing, H. K. Abdel-Aal, Mohamed Aggour,, M. A. Naim

Process Engineering Economics, James R. Couper

Thermodynamic Cycles: Computer-Aided Design and Optimization, Chih Wu

Re-Engineering the Chemical Processing Plant: Process Intensification, edited by Andrzej Stankiewicz and Jacob A. Moulijn

HANDBOOK of FLUIDIZATION and FLUID-PARTICLE SYSTEMS

edited by

Wen-Ching Yang

Siemens Westinghouse Power Corporation
Pittsburgh, Pennsylvania, U.S.A.

MARCEL DEKKER, INC.　　　　　　　　　　　NEW YORK · BASEL

Library of Congress Cataloging-in-Publication Data
A catalog record for this book is available from the Library of Congress.

ISBN: 0-8247-0259-X

This book is printed on acid-free paper.

Headquarters
Marcel Dekker, Inc.
270 Madison Avenue, New York, NY 10016
tel: 212-696-9000; fax: 212-685-4540

Eastern Hemisphere Distribution
Marcel Dekker AG
Hutgasse 4, Postfach 812, CH-4001 Basel, Switzerland
tel: 41-61-260-6300; fax: 41-61-260-6333

World Wide Web
http://www.dekker.com

The publisher offers discounts on this book when ordered in bulk quantities. For more information, write to Special Sales/ Professional Marketing at the headquarters address above.

Current printing (last digit):
10 9 8 7 6 5 4 3 2 1

PRINTED IN THE UNITED STATES OF AMERICA

Preface

Every chemical engineer, whether a student or practicing, has looked up technical information in *Perry's Chemical Engineering Handbook*. Its compilation was one of the most important contributions to the chemical engineering education and profession. After more than six decades, it remains one of the field's most useful general-purpose reference books. It was in this spirit of serving the profession that I undertook the task of compiling the *Handbook of Fluidization and Fluid-Particle Systems*. Through future revisions and additions, I sincerely hope that this handbook will become an archivable reference volume for every practitioner in this field, spanning the boundary of various disciplines. Fluidization and fluid-particle system engineering is being applied in industries as diverse as basic and specialty chemicals, mineral processing, coal and biomass gasification and combustion for power generation, environmental technologies, resource recovery, FCC petroleum refining, pharmaceuticals, biotechnology, cement, ceramics, and other solids handling and processing industries. The first focused handbook ever published in this extended field, it collects all relevant and important information in a single volume. Both fundamentals and applications are emphasized. Furthermore, all authors are internationally recognized practitioners in the area of fluidization and fluid-particle systems.

This handbook contains 28 chapters and is authored by 34 internationally recognized experts from seven countries; half of them are professors. Particle characterization and dynamics—important in all aspects of particle production, manufacturing, handling, processing, and applications—are discussed in Chapter 1. Chapter 2 presents the flow through fixed beds and summarizes packing characteristics of spherical and nonspherical particles, pressure-drop correlations for flow through fixed beds, and heat and mass transfer. Bubbling fluidized beds are presented in detail in Chapter 3, which covers all important aspects including jetting phenomena and particle segregation, topics not addressed extensively in other books on fluidization. Other important design considerations are treated in separate chapters: elutriation and entrainment in Chapter 4, effect of temperature and pressure in Chapter 5, gas distributor and plenum design in Chapter 6, effect of internal tubes and baffles in Chapter 7, attrition in Chapter 8, and modeling in Chapter 9. Heat transfer (Chapter 10) and mass transfer (Chapter 11) are also treated. The approaches for designing and scaling up fluidized bed reactors are elucidated in Chapter 12, "General Approaches to Reactor Design," and Chapter 13, "Fluidized Bed Scaleup."

Important industrial applications for fluidized bed reactors are also discussed, including fluid catalytic cracking (Chapter 14), gasifiers and combustors (Chapter 15), chemical production and processing (Chapter 16), coating and granulation (Chapter 17), and fluidized bed drying (Chapter 18).

The important variation of bubbling fluidized beds—the circulation fluidized beds—are discussed in detail in Chapter 19. Chapter 20 summarizes other nonconventional fluidized beds, including spouted beds, recirculating fluidized beds with a draft tube, jetting fluidized beds, and rotating fluidized beds. The solids handling, transport and circulating devices are described in Chapter 21, "Standpipe and Nonmechanical Valves," and Chapter 22,

"Cyclone Separators." Pneumatic transport is covered in Chapters 23 and 24. Instrumentation and measurement requirements are reviewed in Chapter 25.

The last three chapters examine the fluidized beds and fluid-particle systems involving liquid.

This handbook took more than four years to complete. Along the way, content was altered, format was changed, and chapters were revised to fit the page limitation. The final product is indeed one to be proud of by all who participated. A monumental endeavor such as this could not have been possible without the cooperation and dedication of all the authors, especially those who were asked to revise their chapters, sometimes several times. I am truly indebted to them all for taking the time out of their busy schedule and for their cooperation, dedication, and conscientious effort. The staff of the publisher, Marcel Dekker, Inc., also deserves credit for their patience and tenacity in shepherding the project to its eventual completion. Finally, I thank my family, especially my wife, Rae, for their continuous encouragement.

Wen-Ching Yang

Contents

Preface   *iii*
Contributors   *vii*

1   **Particle Characterization and Dynamics**   1
        Wen-Ching Yang

2   **Flow Through Fixed Beds**   29
        Wen-Ching Yang

3   **Bubbling Fluidized Beds**   53
        Wen-Ching Yang

4   **Elutriaton and Entrainment**   113
        Joachim Werther and Ernst-Ulrich Hartge

5   **Effect of Temperature and Pressure**   129
        J. G. Yates

6   **Gas Distributor and Plenum Design in Fluidized Beds**   155
        S. B. Reddy Karri and Joachim Werther

7   **Effect of Internal Tubes and Baffles**   171
        Yong Jin, Fei Wei, and Yao Wang

8   **Attrition**   201
        Joachim Werther and Jens Reppenhagen

9   **Modeling**   239
        Thomas C. Ho

10 **Heat Transfer**   257
        John C. Chen

11 **Mass Transfer**   287
        Thomas C. Ho

12 **General Approaches to Reactor Design**   309
        Peijun Jiang, Fei Wei, and Liang-Shih Fan

13   Fluidized Bed Scaleup   343
Leon R. Glicksman

14   Applications for Fluid Catalytic Cracking   379
Ye-Mon Chen

15   Applications for Gasifiers and Combustors   397
Richard A. Newby

16   Applications for Chemical Production and Processing   421
Behzad Jazayeri

17   Applications for Coating and Granulation   445
Gabriel I. Tardos and Paul R. Mort

18   Applications for Fluidized Bed Drying   469
Arun S. Mujumdar and Sakamon Devahastin

19   Circulating Fluidized Beds   485
John R. Grace, Hsiaotao Bi, and Mohammad Golriz

20   Other Nonconventional Fluidized Beds   545
Wen-Ching Yang

21   Standpipes and Nonmechanical Valves   571
T. M. Knowlton

22   Cyclone Separators   599
T. M. Knowlton

23   Dilute-Phase Pneumatic Conveying   619
George E. Klinzing

24   Electrostatics in Pneumatic Conveying   631
George E. Klinzing

25   Instrumentation and Measurements   643
Masayuki Horio, Rafal P. Kobylecki, and Mayumi Tsukada

26   Liquid–Solids Fluidization   705
Norman Epstein

27   Gas–Liquid–Solid Three-Phase Fluidization   765
Liang-Shih Fan and Guoqiang Yang

28   Liquid–Solids Separation   811
Shiao-Hung Chiang, Daxin He, and Yuru Feng

Index   *851*

Contributors

Hsiaotao Bi   Department of Chemical and Biological Engineering, University of British Columbia, Vancouver, British Columbia, Canada

John C. Chen   Department of Chemical Engineering, Lehigh University, Bethlehem, Pennsylvania, U.S.A.

Ye-Mon Chen   Shell Global Solutions US, Houston, Texas, U.S.A.

Shiao-Hung Chiang   Department of Chemical and Petroleum Engineering, University of Pittsburgh, Pittsburgh, Pennsylvania, U.S.A.

Sakamon Devahastin   Department of Food Engineering, King Mongkut's University of Technology Thonburi, Bangkok, Thailand

Norman Epstein   Department of Chemical and Biological Engineering, University of British Columbia, Vancouver, British Columbia, Canada

Liang-Shih Fan   Department of Chemical Engineering, The Ohio State University, Columbus, Ohio, U.S.A.

Yuru Feng   Department of Chemical and Petroleum Engineering, University of Pittsburgh, Pittsburgh, Pennsylvania, U.S.A.

Leon R. Glicksman   Departments of Architecture and Mechanical Engineering, Massachusetts Institute of Technology, Cambridge, Massachusetts, U.S.A.

Mohammad Golriz   Department of Applied Physics and Electronics, Umeå University, Umeå, Sweden

John R. Grace   Department of Chemical and Biological Engineering, University of British Columbia, Vancouver, British Columbia, Canada

Ernst-Ulrich Hartge   Technical University Hamburg-Harburg, Hamburg, Germany

Daxin He   Department of Chemical and Petroleum Engineering, University of Pittsburgh, Pittsburgh, Pennsylvania, U.S.A.

Thomas C. Ho   Department of Chemical Engineering, Lamar University, Beaumont, Texas, U.S.A.

Masayuki Horio   Department of Chemical Engineering, Tokyo University of Agriculture and Technology, Tokyo, Japan

Behzad Jazayeri   Fluor Daniel, Inc., Aliso Viejo, California, U.S.A.

Peijun Jiang   Department of Chemical Engineering, The Ohio State University, Columbus, Ohio, U.S.A.

Yong Jin   Department of Chemical Engineering, Tsinghua University, Beijing, People's Republic of China

S. B. Reddy Karri   Particulate Solid Research, Inc., Chicago, Illinois, U.S.A.

George E. Klinzing   Office of the Provost, University of Pittsburgh, Pittsburgh, Pennsylvania, U.S.A.

T. M. Knowlton   Particulate Solid Research, Inc., Chicago, Illinois, U.S.A.

Rafal P. Kobylecki[*]   Department of Chemical Engineering, Tokyo University of Agriculture and Technology, Tokyo, Japan.

Paul R. Mort   Procter & Gamble, Cincinnati, Ohio, U.S.A.

Arun S. Mujumdar   Department of Mechanical Engineering, National University of Singapore, Singapore

Richard A. Newby   Science & Technology Center, Siemens Westinghouse Power Corporation, Pittsburgh, Pennsylvania, U.S.A.

Jens Reppenhagen[†]   Technical University Hamburg-Harburg, Hamburg, Germany

Gabriel I. Tardos   Department of Chemical Engineering, The City College of the City University of New York, New York, U.S.A.

Mayumi Tsukada   Department of Chemical Engineering, Tokyo University of Agriculture and Technology, Tokyo, Japan

Yao Wang   Department of Chemical Engineering, Tsinghua University, Beijing, People's Republic of China

Fei Wei   Department of Chemical Engineering, Tsinghua University, Beijing, People's Republic of China

Joachim Werther   Technical University Hamburg-Harburg, Hamburg, Germany

Guoqiang Yang   Department of Chemical Engineering, The Ohio State University, Columbus, Ohio, U.S.A.

Wen-Ching Yang   Science & Technology Center, Siemens Westinghouse Power Corporation, Pittsburgh, Pennsylvania, U.S.A.

J. G. Yates   Department of Chemical Engineering, University College London, London, United Kingdom

[*] *Current affiliation*: Energy Engineering Department, Czestochowa Technical University, Czestochowa, Poland.
[†] *Current affiliation*: BMH Claudius Peters, Buxtehude, Germany.

HANDBOOK of
FLUIDIZATION and
FLUID-PARTICLE SYSTEMS

1

Particle Characterization and Dynamics

Wen-Ching Yang

Siemens Westinghouse Power Corporation, Pittsburgh, Pennsylvania, U.S.A.

1 INTRODUCTION

Particle characterization is important in all aspects of particle production, manufacturing, handling, processing, and applications. Characterization of particles is the first necessary task required in a process involving solid particles. The required characterization includes not only the intrinsic static parameters (such as size, density, shape, and morphology) but also their dynamic behavior in relation to fluid flow (such as drag coefficient and terminal velocity). In this chapter, the characterization of single particles with different available techniques is first introduced. The dynamic behavior of a single particle in the flow field at Stokes flow regime, the intermediate flow regime, and the Newton's law regime is then discussed. The chapter is concluded with coverage of multiparticle systems.

1.1 Characterization of Single Particles

The complete characterization of a single particle requires the measurement and definition of the particle characteristics such as size, density, shape, and surface morphology. Because the particles of interest are usually irregular in shape and different in surface morphology, there are many different ways and techniques to characterize the particles. Depending on the methods employed, the results may not be completely consistent. Some methods may be more appropriate than others for certain selected applications.

1.1.1 Definitions of Particle Size

The particle size is one or more linear dimensions appropriately defined to characterize an individual particle. For example, an ideal particle like a sphere is uniquely characterized by its diameter. Particles of regular shapes other than spherical can usually be characterized by two or three dimensions. Cubes can be uniquely defined by a single dimension, while cuboids require all three dimensions, length, width, and height. Two dimensions are required for regular isotropic particles such as cylinders, spheroids, and cones.

Irregular particles of practical interest, most often, cannot be uniquely defined. Their sizes are usually defined based on certain reference properties. The choice of any particular diameter for characterization of an irregular particle depends, in many cases, on the intended application. Unfortunately, in most cases, the correct choice of a representative diameter is uncertain. Many diameters have been defined to characterize the irregular particles. The more common ones are summarized below.

Volume Diameter
The volume diameter, d_v, is defined as the diameter of a sphere having the same volume as the particle and can be expressed mathematically as

$$d_v = \left(\frac{6V_p}{\pi}\right)^{1/3} \quad \text{where} \quad V_p = \text{volume of the particle} \quad (1)$$

Surface Diameter

The surface diameter, d_s, is defined as the diameter of a sphere having the same surface area of the particle. Mathematically it can be shown to be

$$d_s = \left(\frac{S_p}{\pi}\right)^{1/2} \quad \text{where} \quad S_p = \text{surface area of the particle} \quad (2)$$

Surface–Volume Diameter

The surface–volume diameter, d_{sv}, also known as the Sauter diameter, is defined as the diameter of a sphere having the same external-surface-area-to-volume ratio as the particle. This can be expressed as

$$d_{sv} = \frac{6V_p}{S_p} = \frac{d_v^3}{d_s^2} \quad (3)$$

Sieve Diameter

The sieve diameter, d_A, is defined as the width of the minimum square aperture in the sieve screen through which the particle will pass. The sieve diameter will be discussed in more detail when particle size analysis by sieving is discussed later in the chapter.

Stokes Diameter

The Stokes diameter, d_{st}, is the free-falling diameter of the particle in the Stokes law region and can be calculated from

$$d_{st} = \sqrt{\frac{18\mu U_t}{(\rho_p - \rho_f)g}} \quad \text{where} \quad U_t = \begin{array}{l} \text{terminal} \\ \text{velocity} \\ \text{of the} \\ \text{particle} \end{array} \quad (4)$$

Free-Falling Diameter

The free-falling diameter, d_f, is the diameter of a sphere having the same density and the same free-falling velocity (or terminal velocity) as the particle in a fluid of same density and viscosity. In the Stokes law region, the free-falling diameter is the Stokes diameter defined earlier.

Drag Diameter

The drag diameter, d_D, is defined as the diameter of a sphere having the same resistance to motion as the particle in a fluid of the same density and viscosity and moving at the same velocity.

Perimeter Diameter

The perimeter diameter, d_c, is the diameter of a circle having the same perimeter as the projected outline of the particle.

Projected Area Diameter

The projected area diameter, d_a, is defined as the diameter of a sphere having the same projected area as the particle viewed in a direction perpendicular to the plane of the greatest stability of the particle (see Fig. 1).

Feret Diameter

The Feret diameter, d_F, is a statistical diameter representing the mean value of the distances between pairs of parallel tangents to a projected outline of the particle, as shown in Fig. 1. The Feret diameter is usually used in particle characterization employing the optical imaging technique to be discussed later.

Martin Diameter

The Martin Diameter, d_M, is also a statistical diameter defined as the mean chord length of the projected outline of the particle, which appropriately bisects the area of the projected profile, as shown in Fig. 1. Like the Feret diameter, the Martin diameter is also employed most often during particle characterization using the optical imaging technique to be discussed later.

Only four of the above particle size definitions are of general interest for applications in packed beds and fluidized beds. They are the sieve diameter, d_A, the volume diameter, d_v, the surface diameter, d_s, and the surface–volume diameter, d_{sv}. The most relevant diameter for application in a fluidized bed is the surface–volume diameter, d_{sv}. For applications in catalytic reactors with different isometric catalyst shapes, Rase (1990) suggested that we use the equivalent diameters summarized in Table 1.

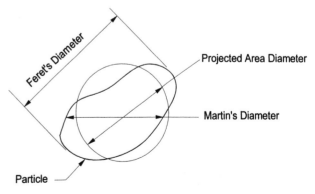

Figure 1 Illustration for projected area diameter, Feret diameter, and Martin diameter.

Table 1 Suggested Equivalent Particle Diameters for Catalysts in Catalytic Reactor Applications

Shape	Equivalent particle diameter, $d_\mathrm{p} = d_\mathrm{sv} = 6/a_\mathrm{s}$
Sphere	d_p = diameter of a sphere
Cylinder with length (l_y) equal to diameter (d_y)	$d_\mathrm{p} = d_\mathrm{y}$, the diameter of a cylinder
Cylinder, $d_\mathrm{y} \neq l_\mathrm{y}$	$d_\mathrm{p} = \dfrac{6d_\mathrm{y}}{4 + 2d_\mathrm{y}/l_\mathrm{y}}$
Ring with outside diameter of d_o, and inside diameter of d_i	$d_\mathrm{p} = 1.5(d_\mathrm{o} - d_\mathrm{i})$
Mixed sizes	$d_\mathrm{p} = \dfrac{1}{\sum\limits_i (x_i/d_{\mathrm{p}i})}$
Irregular shapes with $\phi = 0.5$ to 0.7	$d_\mathrm{p} = \phi d_\mathrm{v}; \; d_\mathrm{v} \approx d_\mathrm{A}$

Source: Rase (1990).

1.1.2 Definitions of Particle Shape

Natural and man-made solid particles occur in almost any imaginable shape, and most particles of practical interest are irregular in shape. A variety of empirical factors have been proposed to describe nonspherical shapes of particles. These empirical descriptions of particle shape are usually provided by identifying two characteristic parameters from the following four: (1) volume of the particle, (2) surface area of the particle, (3) projected area of the particle, and (4) projected perimeter of the particle. The projected area and perimeter must also be determined normal to some specified axis. For axisymmetric bodies, the reference direction is usually taken to be parallel or normal to the axis of symmetry.

All proposed shape factors to date are open to criticism, because a range of bodies with different shapes may have the same shape factor. This is really inevitable if complex shapes are to be described only by a single parameter. Thus in selecting a particular shape factor for application, care must be taken to assure its relevance.

Sphericity
Wadell (1933) proposed the "degree of true sphericity" be defined as

$$\phi = \frac{\text{Surface area of volume} - \text{equivalent sphere}}{\text{Surface area of particle}}$$

$$= \left(\frac{d_\mathrm{v}}{d_\mathrm{s}}\right)^2 = \frac{d_\mathrm{sv}}{d_\mathrm{v}} \tag{5}$$

For a true sphere, the sphericity is thus equal to 1. For nonspherical particles, the sphericity is always less than 1. The drawback of the sphericity is that it is difficult to obtain the surface area of an irregular particle and thus it is difficult to determine ϕ directly. Usually the more the aspect ratio departs from unity, the lower the sphericity. The sphericity, first introduced as a measure of particle shape, was subsequently claimed to be useful for correlating drag coefficient (Wadell, 1934). There is some theoretical justification for the use of sphericity as a correlating parameter for creeping flow past bodies whose geometric proportions resemble a sphere. But for other circumstances its use is purely empirical (Clift et al., 1978). Leva (1959) and Subramanian and Arunachalam (1980) suggested experimental methods using the Ergun equation for evaluation of the sphericity. This methodology will be discussed when Ergun equation is introduced in Chapter 2, "Flow through fixed beds." For regularly shaped solids, the sphericities can be calculated from Eq. (5), and they are presented in Table 2a. As for commonly occurring nonspherical particles, their sphericities are summarized in Table 2b.

Circularity
Wadell (1933) also introduced the "degree of circularity", defined as

$$\mathcal{C} = \frac{\substack{\text{Circumference of circle having same cross-}\\ \text{sectional area as the particle}}}{\text{Actual perimeter of the cross-section}}$$

$$\tag{6}$$

Table 2a   Sphericities of Regularly Shaped Solids

Shape	Relative Proportions	$\phi = d_{sv}/d_v$
Spheroid	1:1:2	0.93
	1:2:2	0.92
	1:1:4	0.78
	1:4:4	0.70
Ellipsoid	1:2:4	0.79
Cylinder	Height = diameter	0.87
	Height = 2 × diameter	0.83
	Height = 4 × diameter	0.73
	Height = $\frac{1}{2}$ × diameter	0.83
	Height = $\frac{1}{4}$ diameter	0.69
Rectangular parallelpiped	1:1:1	0.81
	1:1:2	0.77
	1:2:2	0.77
	1:1:4	0.68
	1:4:4	0.64
	1:2:4	0.68
Rectangular tetrahedron	—	0.67
Regular octahedron	—	0.83

Source: Adapted from Geldart (1986).

Unlike the sphericity, the circularity can be determined more easily experimentally from microscopic or photographic observation. For an axisymmetric particle projected parallel to its axis, $\not\subset$ is equal to unity. Use of $\not\subset$ is only justified on empirical grounds, but it has the potential advantage of allowing correlation of the

Table 2b   Sphericities of Commonly Occurring Nonspherical Particles

Material	Sphericity
Sand	
Round sand	0.86
Sharp sand	0.66
Crushed sandstone	0.8–0.9
Coal	
Pulverized coal	0.73
Crushed coal	0.63–0.75
Activated carbon	0.70–0.90
Mica flakes	0.28
Fischer–Tropsch catalyst	0.58
Common salt	0.84
Crushed glass	0.65
Silica gels	0.70–0.90
Tungsten powder	0.89
Sillimanite	0.75
Wheat	0.85

Source: Adapted from Geldart (1986).

dependence of flow behavior on particle orientation (Cliff et al., 1978).

Operational Sphericity and Circularity
Since the sphericity and circularity are so difficult to determine for irregular particles, Wadell (1933) proposed that ϕ and $\not\subset$ be approximated by "operational sphericity and circularity:"

$$\phi_{op} = \left(\frac{\text{Volume of particle}}{\text{Volume of the smallest circumscribing sphere}} \right)^{1/3}$$

$$(7)$$

$$\not\subset_{op} = \left(\frac{\text{Projected area of particle}}{\text{Area of the smallest circumscribing circle}} \right)^{1/2}$$

$$(8)$$

For the ellipsoids, the operational sphericity, ϕ_{op}, can be expressed as

$$\phi_{op} = (e_1 e_2)^{-1/3} \qquad (9)$$

For the rounded particles, it can be approximated by Eq. (9). The e_1 and e_2 are called the flatness ratio and the elongation ratio, respectively, and are defined as

$$e_1 = \frac{b}{t} \qquad \text{Flatness ratio} \qquad (10)$$

$$e_2 = \frac{l}{b} \qquad \text{Elongation ratio} \qquad (11)$$

where $t =$ thickness, the minimum distance between two parallel planes tangential to opposite surfaces. One of the two planes is the plane of the maximum stability. $b =$ breadth, the minimum distance between two parallel planes that are perpendicular to the planes defining the thickness and tangential to opposite surfaces. $l =$ length, projected on a plane normal to the planes defining t and b.

The three characteristic dimensions have an increasing order of magnitude $t < b < l$.

However, ϕ_{op} is not generally a good approximation to ϕ. Aschenbrenner (1956) showed that a better approximation to ϕ is given by a "working sphericity," ϕ_w, obtained from the flatness and elongation ratios:

$$\phi_w = \frac{12.8 \left(e_1 e_2^2 \right)^{1/3}}{1 + e_2(1 + e_1) + 6\sqrt{1 + e_2^2 \left(1 + e_1^2\right)}} \qquad (12)$$

The working sphericity has been found to correlate well with the settling behavior of naturally occurring mineral particles.

Wadell (1935) suggested that $\not\subset_{op}$ provides an estimate of ϕ based on a two-dimensional projection of a particle, and thus is sometimes called the "projection sphericity." The operation circularity, however, does not approximate ϕ for regular bodies and has virtually no correlation with settling behavior of natural irregular particles (Clift et al., 1978).

The Heywood Shape Factor
The Heywood shape factor is sometimes called the volumetric shape factor. Heywood (1962) proposed a widely used empirical parameter based on the projected profile of a particle as follows:

$$k = \frac{V_p}{d_a^3} \qquad (13)$$

where $d_a = \sqrt{4A_p/\pi}$.

The projected area diameter, d_a, is the diameter of the sphere with the same projected area as the particle. A number of methods have been suggested for obtaining an estimate for d_a. Even if d_a is available, the Heywood shape factor can only be evaluated if V_p is known. For naturally occurring particles or if a distribution of particle sizes or shapes is present, V_p may not be readily available. Automatic techniques for characterizing particle shape are under development. For a review, see Kaye (1973).

Heywood (1962) suggested that k may be estimated from the corresponding value, k_e, of an isometric particle of similar form by

$$k = \frac{k_e}{e_1 \sqrt{e_2}} \qquad (14)$$

Values of k_e for some regular shapes and approximate values for irregular shapes are given in Table 3.

Equation (14) is exact for regular shapes such as spheroids and cylinders. Heywood suggested that k be employed to correlate drag and terminal velocity, using d_a and the projected area to define Re and C_D, respectively. There are justifications for this approach because many natural particles have an oblate shape, with one dimension much smaller than the other two. Over a large Reynolds number range in the "intermediate" regime, such particles present their maximum area to the direction of motion, and this is the area characterized by d_a. There is also evidence that the shape of this projected area, which does not influence k, has little effect on drag (Clift et al., 1978).

A more modern approach uses fractal analysis, or Fourier transformation. The latter is a bit involved, requiring several coefficients for complex definition.

Table 3 Approximate Values of k_e and k

k_e for isometric irregular shapes	
Rounded	0.56
Subangular	0.51
Augular	
tending to a prismoidal	0.47
tending to a tetrahedron	0.38
k_e for selected natural particles	
Sand	0.26
Bituminous coal	0.23
Limestone	0.16
Gypsum	0.13
Talc	0.16
k for regular shape	
Sphere	0.524
Cube	0.696
Tetrahedron	0.328
Cylinder with an aspect ratio of 1	
viewing along axis	0.785
viewing normal to axis	0.547
Spheroids	
with an aspect ratio of 0.5	0.262
with an aspect ratio of 2	0.370

Source: Heywood (1962).

1.1.3 Definitions of Particle Density

There are several particle density definitions available. Depending on the application, one definition may be more suitable than the others. For nonporous particles, the definition of particle density is straightforward, i.e., the mass of the particle, M, divided by the volume of the particle, V_p, as shown in Eq. (15).

$$\rho_p = \frac{M_p}{V_p} = \frac{\text{mass of the particle}}{\substack{\text{volume that the particle would displace} \\ \text{if its surface were nonporous}}}$$

(15)

For porous particles with small pores, the particle volume in Eq. (15) should be replaced with the envelope volume of the particle as if the particles were nonporous as shown in Fig. 2. This would be more hydrodynamically correct if the particle behavior in the flow field is of interest or if the bulk volume of the particles is to be estimated. For total weight estimation, then the skeleton density should be known. The skeleton density is defined as the mass of the particle divided by the skeletal volume of the particle. In practice, the pore volume rather than the skeletal volume is measured through gas adsorption, gas or water displacement, and mercury porosimetry. These techniques will be discussed in more detail later. There are also porous particles with open and closed pores. The closed pores are not accessible to the gas, water or mercury and thus their volume cannot be measured. In this case, the calculated skeleton density would include the volume of closed pores as shown in Fig. 2. For nonporous particles, the particle density is exactly equal to the skeleton density. For porous particles, the skeleton density will be larger than the particle density.

When a porous particle is broken into smaller pieces, the particle density of the smaller pieces will usually be larger than the original particle density by virtue of the elimination of some pores. When the particle size becomes finer, the particle density will approach that of a nonporous particle because all the pores will be completely eliminated. In the process of handling porous particles, this tendency should be kept in mind, and the particle density should be carefully evaluated.

Knight et al. (1980) described a method of determining the density of finely divided but porous particles such as fluid bed cracking catalyst. They suggested to set a known mass and measurable volume of powder in resin and then measure the voidage of the sample sections of the resin. The results were satisfactory, though the method was quite tedious. Buczek and Geldart (1986) proposed to use very find dense powders as the pycnometric fluid to determine the density of porous particles. For a powder to be a suitable pycnometric fluid, it should be nonporous and free-flowing. Its density should be much larger than the porous particles to be measured, and its diameter should be much smaller than the porous particles and greater than the biggest pores of the porous particles. It should also be nonreactive toward the porous particles. Akapo et al. (1989) also proposed a novel method of determining particle density of porous aeratable powders. The method depends on measurement of bed expansion of a gas fluidized bed of the powder in the region between the minimum fluidization and the minimum bubbling. The Richardson and Zaki bed expansion equation was then applied to back out the particle density. The Richardson and Zaki bed expansion equation and the minimum fluidization and minimum bubbling regimes will be discussed in Chapter 3, "Bubbling Fluidized Beds."

2 PARTICLE CHARACTERIZATION TECHNIQUES

There are many techniques that can be employed to characterize particles, some simple and primitive and some complicated and sophisticated. Almost every technique is associated with intrinsic experimental errors and implicit assumptions. Thus care must be exercised to select proper techniques for your specific applications. The available techniques are reviewed in this section.

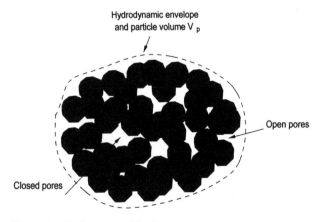

Figure 2 Skeleton particle density.

2.1 Methods for Direct Characterization of Particle Size and Shape

2.1.1 Sieve Analysis

The most commonly used method for classifying powders is to sieve the particles through a series of screens with standardized mesh size by sifting, swirling, shaking, or vibrating. Two standard mesh sizes, U.S. sieve size and Tyler sieve size, are usually used in the U.S. In the European practice, the British Standard and German DIN sieve sizes are also employed. The mesh number of a sieve refers to the number of parallel wires per inch in the weave of the screen. The American Society of Testing Materials (ASTM) specifications for standard mesh sizes for those systems are compared in Table 4. The mesh size was designed so that the aperture or opening of the alternating members in the series has a factor of $2^{1/2}$. For example, the U.S. 12 mesh with an opening of 1.68 mm is $2^{1/2}$ times the U.S. 16 mesh with an opening of 1.19 mm. The result of the sieve size analysis is commonly plotted in a logarithmic-scale graph expressing the cumulative weight percentage under size as the abscissa and the particle size as the ordinate, as shown in Fig. 3. There, the $2^{1/2}$ factor in the mesh size arrangement allows the size data points to be almost equally spaced on the logarithmic scale. Also, experimentally, many crushed materials yield a straight line if plotted as in Fig. 3.

The inaccuracies and uncertainties of sieve analysis stem from the discrete steps of the mesh size arranged at an approximate factor of $2^{1/2}$ between successive mesh sizes. Sieve analysis does not provide the information for the largest and the smallest particle sizes. The size cut provides an approximate value for the mean particle size within the cut. Sieve analysis also does not differentiate the particle shape. A needle-shaped particle can either pass through a mesh or be retained on the screen, depending on its orientation during sifting. The result of sieve analysis is also dependent on the time of sieving action, the particle loading on the sieve, and sieve blinding (also called pegging). Enlargement of aperture due to wire erosion of a sieve can cause discrepancy as well. For small particles, agglomeration due to static electricity or moisture can also occur.

The smallest mesh size for the Tyler Series is 400 mesh, equivalent to a 38 μm opening, while the smallest mesh size for the U.S. Series is 635 mesh, equivalent to a 20 μm opening. For particles finer than 20 μm, the surface and electrostatic forces become important, and particle classification by sieve analysis is not recom-mended. Detailed discussion of sieve analysis techniques can be found in standard textbook such as Allen (1975) and Kaye (1981).

2.1.2 Imaging Technique

Direct measurement of particle dimensions is also possible from enlarged photographic or electronic images of microscopes. There are three types of microscopes commonly employed, i.e., the optical microscope, the scanning electron microscope (SEM), and the transmission electron microscope (TEM). The optical microscope is employed for particles from 1 μm to about 150 μm. Both SEM and TEM make use of electron beams and can be used for particles from 5 μm down to as small as 0.01 μm. They are especially useful for revealing the surface morphology of extremely small particles.

Particles to be imaged in an optical microscope are usually dispersed in a drop of viscous fluid on a glass slide. Their images are then visually compared with a set of standard circles, geometric shapes, or linear grids to derive their actual sizes and shapes. The Martin and Feret diameters are also often used to characterize a particle. The Martin diameter is defined as the magnitude of the chord that divides the image into two equal areas with respect to a fixed direction (see Fig. 1). The Feret diameter of a particle image is defined as the length of the image as projected with respect to a specific reference direction (see Fig. 1). Both Martin and Feret diameters are intended to be statistical diameters, i.e., after characterization of many images. Thus any slight departure from the true randomness in the field of view of particle images can produce bias in the particle size characterization. Modern instruments couple television cameras interfaced with computers and sophisticated software can speed up the imaging analysis considerably.

Martin's statistical diameter is also referred to as the mean linear intercept or mean chord and has been shown to relate to the specific surface in the following manner (Herdan, 1960):

$$\text{Martin diameter} = \frac{4}{\rho S_v} \qquad (16)$$

where ρ = the density of packing and S_v = particle surface area per unit volume of particle.

Based on the experimental evidence, on the average, the Martin diameter is usually smaller than the mean projected diameter and the Feret diameter larger. Since both Martin and Feret diameters depend on the particle shape, the ratio of Feret diameter to Martin

Table 4 Summary of Various Types of Standard Sieve

U.S. mesh No.	Standard opening, mm	Tyler mesh No.	British mesh No.	Standard opening, mm	German DIN No.	Standard opening, mm
$3\frac{1}{2}$	5.66	$3\frac{1}{2}$	—	—	1	6.000
4	4.76	4	—	—	—	—
5	4.00	5	—	—	—	—
6	3.36	6	5	3.353	2	3.000
7	2.83	7	6	2.812	—	—
—	—	—	—	—	—	—
8	2.38	8	7	2.411	$2\frac{1}{2}$	2.400
10	2.00	9	8	2.057	3	2.000
12	1.68	10	10	1.676	4	1.500
—	—	—	—	—	—	—
14	1.41	12	12	1.405	—	—
—	—	—	—	—	—	—
16	1.19	14	14	1.204	5	1.200
—	—	—	—	—	—	—
18	1.00	16	16	1.003	6	1.020
20	0.84	20	18	0.853	—	—
—	—	—	—	—	8	0.750
25	0.71	24	22	0.699	—	—
—	—	—	—	—	—	—
30	0.59	28	25	0.599	10	0.600
—	—	—	—	—	11	0.540
35	0.50	32	30	0.500	12	0.490
40	0.42	35	36	0.422	14	0.430
45	0.35	42	44	0.353	16	0.385
—	—	—	—	—	—	—
50	0.297	48	52	0.295	20	0.300
60	0.250	60	60	0.251	24	0.250
70	0.210	65	72	0.211	30	0.200
80	0.177	80	85	0.178	—	—
100	0.149	100	100	0.152	40	0.150
—	—	—	—	—	—	—
120	0.125	115	120	0.124	50	0.120
140	0.105	150	150	0.104	60	0.102
170	0.088	170	170	0.089	70	0.088
—	—	—	—	—	—	—
200	0.074	200	200	0.076	80	0.075
230	0.062	250	240	0.066	100	0.060
270	0.053	270	300	0.053	—	—
325	0.044	325	—	—	—	—
400	0.038	400	—	—	—	—
635	0.020	—	—	—	—	—

diameter is usually fairly constant for the same material. For Portland cement, it is about 1.2; for ground quartz and ground glass, approximately 1.3 (Herdan, 1960).

The microscopic measurement technique is most suitable for particles relatively uniform in size and granular in shape, because a large number of particles, between 300 to 500, needs to be measured to minimize statistical error.

2.1.3 Gravity and Centrifugal Sedimentation

The falling speeds of particles in a viscous fluid under the influence of gravity are used to measure the particle

Particle Size Distribution

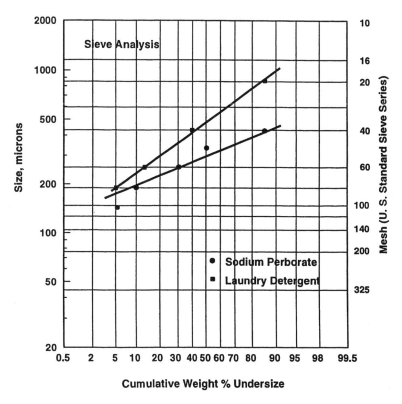

Figure 3   Graphic representation of sieve size analysis—logarithmic-scale graph expressing cumulative % oversize or undersize.

size in the gravity sedimentation technique. The measured speeds are then converted to Stokes diameters by applying the Stokes equation [Eq. (4)], assuming that the particles are all spherical in shape. Since the irregularly shaped particles fall with different orientations in the vertical direction and thus different settling velocities, similar irregularly shaped particles can have a range of Stoke diameters. Disturbances caused by the presence of other particles, the concentration effect, can also be important. Recommendations for the upper limit particle concentration to be used during the sedimentation analysis varies from 0.01 to 3.0% (Kaye, 1981). An upper limit of 0.05% was recommended by Kaye, and in certain practical cases up to 0.2% by volume is permissible. The probability of forming clusters increases with particle suspension concentration. The clusters tend to fall at a higher speed and thus introduce measurement error.

The hindering effect of the containing wall on the falling speed of the particles cannot be ignored either. For a spherical particle, the effect can be expressed by the Landenburg equation as

$$V_\infty = V_m\left(1 + 2.4\frac{d_p}{D}\right) \qquad (17)$$

According to Eq. (17), even if the column is 50 times the diameter of the particle, there is still a 5% reduction in the falling speed of the particle.

Two basic suspension systems, the line start system (or the two-layer sedimentation system) and the initially homogeneous system are employed in the gravity sedimentation technique. In the line start system, a thin layer of particles is placed at the top of the sedimentation column, and its settling behavior is analyzed by different techniques. In the initially homogeneous system, the column is homogenized first and its settling pattern is subsequently studied.

Classical techniques for measuring the sedimentation behavior include taking samples with a pipette, measurement of height of sediment layer at the bottom, and use of balance pan to measure the weight of settled particles. Modern sedimentometers make use of the diffraction pattern of a light beam, the power loss of an x-ray, or a Doppler shift of a laser beam. The modern techniques are primarily applied to monitor the sedimentation kinetics of an initially homogeneous suspension. The photosedimentometers measure the beam with a small-angle detector. This application is based on the Lambert–Beer law, which is

$$I_c = I_o e^{-\alpha CL} \tag{18}$$

where I_o = the original light beam intensity, I_c = the intensity of a light beam after passing through a particle suspension of concentration C, L = the length of the light beam path, and α = an empirical constant depending on equipment and particle characteristics.

In the x-ray sedimentometers, the x-ray absorption of a sedimentation suspension is used to measure the concentration gradient in the suspension. The law outlined by Olivier et al. (1970/1971) is employed:

$$\ln T = -\beta W_s \tag{19}$$

where T = the measured transmittance, X_c/X_o, X_o = the intensity of the x-ray beam passing through the sedimentation column filled with clear fluid, X_c = the intensity of the x-ray beam passing through a particle suspension of concentration C, W_s = the weight fraction of particles in the x-ray pass, and β = an empirical constant depending on equipment, particle, and suspending fluid characteristics.

Alternatively, centrifugal force instead of gravitational force can be created to enhance the sedimentation performance. Depending on the size of the centrifugal arm and the speed of rotation, many times the gravitational force can be applied.

The modern nonintrusive sedimentometers are efficient, can be adapted easily for hostile environments, and can provide information in situ and thus are ideal for quality control in manufacturing processes. The resolution and sensitivity allow measurement of Stokes diameters down to about $0.5\,\mu\text{m}$.

2.1.4 Characterization by Elutriation

Particle size characterization by elutriation makes use of the same kind of principles employed by the sedimentation. Instead of letting particles settle with gravity, the particles are actually carried out against gravity during elutriation. In vertical elutriators, the particles with terminal velocities less than the vertical fluid velocity will be elutriated out. By operating the elutriator at different flow conditions, the particle size distribution of the sample can be calculated. The flow in the elutriators is usually laminar flow, and the Stokes equation is used to estimate the Stokes diameter of the particle by assuming that the particle is spherical.

As in sedimentation, the concentration of the particles in the elutriators affects the results of measurement. Also the velocity in the elutriators tends to be parabolic rather than uniform and thus introduces errors in measurement. Centrifugal force can also be applied like that in the sedimentation to enhance the performance. Different designs, including horizontal elutriators, are available.

2.1.5 Cascade Impaction Technique

The cascade impaction technique is based on similar principles employed in the elutriation technique and is also based on the inertia of the particles. The particles with terminal velocities smaller than the flow velocity will be carried along by the gas flow. In addition, smaller particles will tend to follow the streamlines of the flow better than the larger particles owing to their smaller inertia. When the flow changes direction because of the presence of a plane surface, larger particles with larger inertia will impact on the plane surface and be collected. By successive decreases in flow velocity from stage to stage, the particles can be collected and classified into different particle size fractions. Earlier designs employed slides coated with adhesive to collect particles for microscopic analysis. Modern devices have many more variations. The cascade impaction technique is usually applied for particle sizes between 0.1 and $100\,\mu\text{m}$.

2.1.6 Resistivity and Optical Zone Sensing Techniques

The resistivity and optical zone sensing techniques measure the particle size by measuring the changes of resistivity or optical properties when the particles are passed through the sensing zone of the instruments. The well-known Coulter counter is a resistivity zone sensing instrument (see Fig. 4). The particles to be analyzed are first suspended and homogenized in an electrolyte and then are forced to pass through a cylindrical orifice placed between two electrodes. The passages of particles through the orifice generates voltage pulses that are amplified, recorded, and analyzed to produce a particle size distribution. The instrument is usually calibrated with standard particles such as latex spheres of known size. The data are analyzed by assuming that the sensing zone is isotropic (i.e., the exact location of the particles in the sensing zone is unimportant) and the pulse height is proportional to the volume of the particle. The results from this analysis are thus the equivalent spherical diameters of the particles. For the simple voltage-pulse-to-particle-

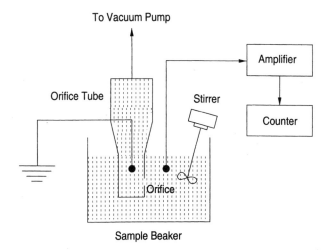

Figure 4 Electrical zone sensing (Coulter counter).

volume relationship to hold, the particle size to be analyzed needs to be less than about 40% of the orifice diameter. Particles smaller than 3% of the orifice diameter do not produce a reliable result. Thus for a sample of wide size distribution, changes of orifices of different diameters are usually necessary.

Possible errors during measurement arise from the possibility of more than one single particle occupying the sensing zone at the same time. This can easily be solved by diluting the suspension or making multiple measurements with increasingly dilute suspensions. If the particles substantially deviate from the spherical shape, the equivalent spherical diameters obtained with this technique may not have any physical significance. Another potential error stems from the fact that particles may not all pass through the axis of the orifice. Similar size particles will generate different voltage signals depending on whether they pass through the orifice at the axis or close to the cylindrical wall. Modern instruments are increasingly capable of editing the signals to reject those stray signals. The electrical zone sensing technique can be applied for particles ranging from 0.6 to 1200 μm.

The optical sensing technique measures the scattered light from a particle passing through a sensing volume illuminated by a light source, such as a white light or a laser. The intensity of the scattered light is then related to the size of the particle. In an ideal situation, a monotonic relationship exists between the intensity of the scattered light and the particle size and thus allows unique determination of the size of the particle. In reality, the light-scattering properties of a particle depend in a very complex way on its refractive index and shape, and on the wavelength of the light

used to illuminate the particle. Generally, there are two basic designs: the scattered light can either be collected in a narrow forward direction in the direction of the illuminating light or be collected in a wide angular range. The forward scattering systems are more suitable for sizing particles with light-absorbing properties, since for these particles there exists a monotonic relationship between the intensity and the size. For nonabsorbing particles, however, multiple values exist for particles larger than 1 μm. Thus most of the existing particle-sizing instruments make use of the second scattering geometry to collect the scattered light into a large angular range oriented either perpendicularly or axially with respect to the light beam direction, see Fig. 5. The instrument is usually calibrated with nonabsorbing spherical polystyrene latex particles, and the scattered light intensity is a monotonic function of the particle size. However, the presence of absorptivity can reduce substantially the sizing sensitivity of the instrument.

The light scattering theories employed for this technique are by Mie, Rayleigh, and Fraunhofer (Bohren and Huffman, 1986). When the dimensions of a particle are of the same order of magnitude of the wavelength of the incident light, Mie theory is used. When the particle is much smaller than the wavelength of the light, the appropriate light-scattering theory is that of Rayleigh. For particles much larger than the wavelength of the light, Fraunhofer diffraction is employed. For more detailed discussions of principles

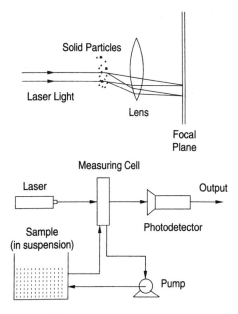

Figure 5 Laser-diffraction spectrometry.

behind this application, see Fan and Zhu (1998). An innovative instrumentation using three lasers to produce scattered light through an angular range from close to 0° up to 169° in one continuous pattern was described by Freud et al. (1993). Bonin and Queiroz (1991) also described an application in an industrial-scale pulverized coal-fired boiler measuring local particle velocity, size, and concentration.

Besides for particle sizing, the optical sensing technique can also be employed for measurement of the particle number concentration. Discussions mentioned earlier regarding the measurement errors for the resistivity sensing zone technique apply here as well. When more than one particle is present in the sensing zone, error can occur.

2.1.7 Techniques for Particle Surface Characterization

In many applications related to chemical reactions, the total surface area is more important than the particle size and shape. To measure the particle surface area, two commonly used techniques are the permeability technique and the gas adsorption technique. The permeability technique is a method for the determination of the power surface area by measuring the permeability of a powder bed. The permeability is defined in Chapter 2 in relation to Darcy's law. The derivation of pressure drop equations through power beds is quite involved and will be discussed in Chapter 2, for measuring the powder surface area the reader can refer to Allen (1975).

Another technique commonly used to measure the powder surface area and the pore size is the physical gas adsorption technique based on the well-known BET (Brunauer–Emmett–Teller) method on monolayer coverage of adsorptives such as nitrogen, krypton, and argon. The application is very well established, and detailed discussions are available in Allen (1975).

2.2 Effect of Particle Shape on Size Distribution Measured with Commercial Equipment

The effect of particle shape on particle size distribution was investigated by Naito et al. (1998) with commercial particle size analyzers based on five different measuring principles: electrical sensing zone, laser diffraction and scattering, x-ray sedimentation, photosedimentation, and light attenuation. The particles used are blocky aluminum oxide and barium titanate particles, flaky boron nitride particles, and rodlike silicon whisker ceramic powders. Altogether, four commer-

cially available models based on the electrical sensing zone principle, eight commercially available models based on the laser diffraction and scattering principle, three commercially available models based on the x-ray sedimentation principle, seven commercially available models based on the photosedimentation principle, and one commercially available model based on the light attenuation principle were employed in a round-robin evaluation. They found that the effect of anisotropic particles such as flaky and rodlike particles on the particle size measured by the commercial equipment is much larger than that of the isotropic block-shaped particles. In the x-ray sedimentation technique, the equivalent volume diameter is usually measured. The effect of particle shape is thus small, because the transmittance intensity of x-rays is independent of the particle shape and its orientation. Similarly, the effect of particle shape was also found to be small when the light attenuation technique was employed, because the particles were dispersed at random in the agitated sample suspension. The laser diffraction and scattering technique, on the contrary, resulted in a large range of size distributions, because the particles tended to orient along the shear flow, especially the anisotropic particles. The effect of particle shape also produced a large range of size distributions in the coarse-size range of the distribution in the photosedimentation technique owing to the turbulence of liquid created by the particle orientation, which resulted in intensity fluctuation of the transmitted light at the initial stage of gravitational sedimentation.

If the particle size distributions are characterized by the average diameters at the cumulative masses of 10, 50, and 90%, the data scattering measured by different techniques can be represented by the diameter ratios at the cumulative masses of 10, 50, and 90%, or DR_{10}, DR_{50}, and DR_{90}. Since the particle diameter measured by the electrical sensing zone technique is to be the equivalent volume diameter and is independent of the particle shape, its particle diameter is defined to be one. The particle diameters measured by other techniques are then ratioed with this diameter. The experimental results of DR_{10}, DR_{50}, and DR_{90} are summarized in Table 5. It can be seen that the results of particle analysis from different techniques can be quite different.

Because of differences in measuring techniques based on different properties of irregular particles, the particle size and size distribution obtained by different methods from the same sample are usually different. The size and distribution measured by sieve analysis will be different from that by laser diffract-

Table 5 Experimental Results of DR_{10}, DR_{50}, and DR_{90}

Sensing technique	Aluminum oxide	Barium titanate	Boron nitride	Silicon nitride whisker
DR_{10}				
Electrical sensing zone	1.00	1.00	1.00	1.00
Laser diffraction and scattering	0.61	0.77	0.94	0.61
X-ray sedimentation	0.82	1.10	—	0.69
Photosedimentation	0.82	0.95	0.58	0.75
Light attenuation	1.00	1.15	0.96	0.96
DR_{50}				
Electrical sensing zone	1.00	1.00	1.00	1.00
Laser diffraction and scattering	0.97	1.11	1.32	1.41
X-ray sedimentation	0.84	1.05	—	0.86
Photosedimentation	0.78	0.96	0.74	1.05
Light attenuation	1.33	1.20	1.11	1.04
DR_{90}				
Electrical sensing zone	1.00	1.00	1.00	1.00
Laser diffraction and scattering	1.15	1.33	1.56	2.32
X-ray sedimentation	0.90	1.14	—	1.01
Photosedimentation	1.01	1.17	1.53	2.43
Light attenuation	1.20	1.46	1.78	0.98

Source: Naito et al. (1998).

ometer, x-ray sedimentation, and so on. Conversion factors are available to convert particle size distributions obtained by one method to that measured by another technique. Austin (1998) suggested a technique to perform this correction and provided an equation for conversion between laser diffractometer and x-ray sedigraph measurements.

The applicable particle size ranges for various particle sizing techniques are summarized in Table 6.

2.3 Measurement of Mechanical Properties of Particles

2.3.1 Hardgrove Grindability Index (HGI)

The Hardgrove Grindability Index was originally developed by Babcock and Wilcox to measure the relative ease of pulverizing coal. This ASTM D409-71 method has since been employed for characterizing other particles as well. The method processes 50 g of air-dried particles screened to a size of 16×30 mesh ($1180 \times 600\,\mu m$) in a small ball-and-race mill for 60 revolutions. The amount of material (W_{200}) passing the 200 mesh ($75\,\mu m$) screen is then measured. The result is then compared with a standard index to arrive at the HGI or the HGI is calculated from the equation

$$HGI = 13 + 6.93 W_{200} \qquad (20)$$

The values of the HGI usually range from 15 to 140. The higher the HGI, the higher is the grindability of the material. The HGI has been found to correlate with the attrition characteristics of the particles in fluidized beds and in pneumatic transport lines (Davuluri and Knowlton, 1998). The HGI does not directly relate to hardness. For example, some materials such as plastics are difficult to grind.

2.3.2 Attrition Index

For application in fluidization and fluid–particle systems, the attrition index is probably the most important particle characteristic. The particle attrition can affect the entrainment and elutriation from a fluidized bed and thus subsequently dictate the design of downstream equipment. The attrition in a pneumatic transport line can change the particle size distribution of the feed material into a fluidized bed reactor and thus alter the reaction kinetics. Davuluri and Knowlton (1998) have developed standardized procedures to evaluate the Attrition Index employing two techniques, solids impaction on a plate and the Davison jet cup. The two test units used are shown in Figs. 6 and 7. They found that these two test techniques are versatile enough to be applicable for a wide range of materials, such as plastic, alumina, and lime-

Table 6 Applicable Particle Size Ranges for Various Particle Sizing Methods

Particle sizing methods	Applicable particle size range (µm)	Measured dimension
Sieving		
Dry	> 10	Sieve diameter
Wet	2–500	
Microscopic examination		
Optical	1.0–100	Length, projected area,
Electronic	0.01–500	statistical diameters
Zone sensing		
Resistivity	0.6–1200	Volume
Optical	1.0–800	
Elutriation		
Laminar flow	3–75	Stokes diameter
Cyclone	8–50	
Gravity sedimentation		
Pipette and hydrometer	1–100	
Photoextinction	0.5–100	Stokes diameter
X-ray	0.1–130	
Centrifugal sedimentation		
Mass accumulation	0.5–25	
Photoextinction	0.05–100	Stokes diameter
X-ray	0.1–5	
Centrifugal classification	0.5–50	
Gas permeability	0.1–40	
Gas adsorption	0.005–50	
Cascade impaction	0.05–30	

Source: Adapted from Pohl (1998) and Lloyd (1974).

stone. For a more recent review on attrition, see Werther and Reppenhagen (1999).

2.3.3 Abrasiveness Index

In pulverized coal combustion, the abrasiveness of the particles severely limits the life of the pulverizer grinding elements. The Abrasiveness Index in this application is usually determined by the Yancey–Geer Price apparatus (Babcock and Wilcox, 1992). In this test, four metal test coupons attached to a rotating shaft are rotated at 1440 rpm for a total of 12,000 revolutions in contact with a sample of 6350 µm × 0 (0.25 in. × 0) coal. The relative Abrasiveness Index is then calculated from the weight loss of the coupons. Babcock and Wilcox has estimated the wear of full-scale pulverizers on the basis of the Yancey–Geer Price Index.

2.3.4 Erosiveness Index

Babcock and Wilcox (1992) has also developed a method of quantifying the erosiveness of coal by subjecting a steel coupon to a stream of pulverized coal under controlled conditions. The weight loss of the coupon is an indication of the erosiveness of the particular coal and the potential damage to the processing and handling equipment, and other boiler components.

3 FLUID DYNAMICS OF A SINGLE PARTICLE

For a particle moving in a fluid, the force acting on the surface of a particle depends only on the flow of the fluid in its immediate vicinity. For the simplest case, let us consider a single particle moving at a velocity U_r relative to its immediate fluid around the particle. It is also assumed that the fluid is newtonian and that the U_r is constant. The fluid dynamic parameters can then be evaluated as follows.

3.1 Definition of Particle Drag Coefficient

The drag coefficient is defined as the ratio of the force on the particle and the fluid dynamic pressure caused by the fluid times the area projected by the particles, as shown in Eq. (21) and Fig. 8.

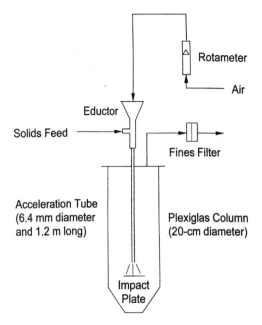

Figure 6 Schematic of attrition apparatus employing the solids impaction principle.

$$C_D = \frac{F}{(1/2)\rho_f U_r^2 A_p} \tag{21}$$

or

$$F = \frac{1}{2} C_D \rho_f U_r^2 A_p \tag{22}$$

The drag coefficient, C_D, is a function of particle's Reynolds number, $(Re)_p = U_r d_p \rho_f/\mu$, only. There are

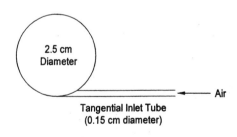

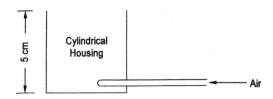

Figure 7 Schematic of attrition apparatus employing Davidson jet cup.

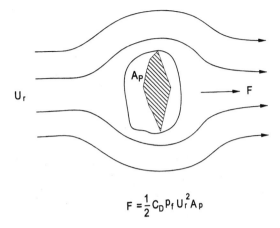

$$F = \frac{1}{2} C_D \rho_f U_r^2 A_p$$

Figure 8 Definition of particle drag coefficient.

three different regimes distinguishable depending on the magnitude of the particle's Reynolds number.

3.1.1 The Stokes Regime

$$C_D = \frac{24}{(Re)_p} \qquad \text{for} \qquad (Re)_p < 0.2 \tag{23}$$

The Stokes law regime is also commonly known as the creeping flow regime. In this regime, the viscosity of the fluid is dominating.

3.1.2 The Intermediate Regime

$$C_D = f\lfloor(Re)_p\rfloor \qquad \text{for} \qquad 0.2 < (Re)_p < 500 \tag{24}$$

In the intermediate regime, the drag coefficient is a function of the particle's Reynold number.

3.1.3 The Newton's Law Regime

$$C_D = 0.44 \qquad \text{for} \qquad (Re)_p > 500 \tag{25}$$

In the Newton's law regime, C_D is relatively constant, and the force F is largely due to the inertia of the fluid rather than to the viscosity of the fluid.

3.2. The Stokes Law Regime

Only in the Stokes law regime, $(Re)_p < 0.2$, have theoretical methods of evaluating C_D met with much success. The theoretical analysis starts with the viscous flow around a rigid sphere, which can be expressed as

$$F_s = 3\pi\mu d_p U_r \tag{26}$$

Two-thirds of the drag force is due to the viscous shear stresses acting on the particle surface (skin friction). The other one-third is due to the difference in pressure on its surface (form drag).

Equating Eqs. (22) and (26), we have

$$F_s = 3\pi\mu d_p U_r = \frac{1}{2} C_D \rho_f U_r^2 \frac{1}{4}\pi d_p^2 \qquad (27)$$

or

$$C_D = 24\left(\frac{\mu}{d_p U_r \rho_f}\right) = \frac{24}{(\mathrm{Re})_p} \qquad (28)$$

The linear relation between the F_s and U_r is an important property of the Stokes law. It enables considerable simplifications in many problems.

3.3   The Stokes Free Fall Velocity

Equating the drag force and the gravitational force for a single spherical particle, we have

$$F_s = \frac{\pi}{8} C_D \rho_f U_r^2 d_p^2 = \frac{\pi d_p^3}{6}(\rho_p - \rho_f)g \qquad (29)$$

or

$$C_D = \frac{4}{3}\frac{d_p(\rho_p - \rho_f)g}{\rho_f U_r^2} \qquad (30)$$

or

$$U_r = U_t = \frac{d_p^2(\rho_p - \rho_f)g}{18\mu} \qquad (31)$$

Equation (4) for the Stokes diameter can be derived from Eq. (31).

3.4   Wall Effect

When the fluid is of finite extent, there are two effects. The fluid streamlines around the particle impinge on the walls and reflect back on the particle, causing increasing drag. Also, since the fluid is stationary at a finite distance from the particle, this distorts the flow pattern and increases drag. The simple correction can be expressed as

$$F_s = 3\pi d_p \mu U_r\left(1 + k_c \frac{d_p}{L_w}\right) \qquad (32)$$

where L_w = the distance from the center of the particle to the walls, $k_c = 0.563$ (for a single wall), $k_c = 1.004$ (for two walls), and $k_c = 2.104$ (for a circular cylinder).

3.5   Corrections to the Stokes Approximation

In 1910, Oseen provided the first correction term to the Stokes approximation as well as a mathematically self-consistent first approximation, as shown in Eq. (33):

$$C_D = \frac{24}{(\mathrm{Re})_p}\left[1 + \frac{3(\mathrm{Re})_p}{16}\right] \qquad (33)$$

More recently, Proudman and Pearson (1957) provided the key to a more straightforward calculation of further correction terms. The results for the drag force on the sphere in accordance with Stokes', Oseen's, and Proudman and Pearson's calculations are given in Table 7. A very large and important collection of solutions of such problems can be found in the book by Happel and Brenner (1965).

3.6   Empirical Drag Coefficient Expression

The preferred correlations to be used in different ranges of particle Reynolds numbers were recommended by Clift et al. (1978) and are summarized in Table 8. In 1986, Turton and Levenspiel proposed a single correlation applicable for the complete range of Reynolds numbers and considerably simplified the calculation of the single-particle drag coefficient.

The Turton and Levenspiel (1986) equation is

$$C_D = \frac{24}{(\mathrm{Re})_p}\left[1 + 0.173(\mathrm{Re})_p^{0.657}\right]$$
$$+ \frac{0.413}{1 + 16.300(\mathrm{Re})_p^{-1.09}} \qquad (34)$$

Haider and Levenspiel (1989) subsequently improved the equation to cover the nonspherical particles and proposed

Table 7 Summary of Theoretical Expressions for the Drag Force $[(\mathrm{Re})_p \ll 1]$

Stokes (1851)
$$F_s = 3\pi d_p \mu U_r$$
Oseen (1910)
$$F = F_s\left[1 + \frac{3}{16}(\mathrm{Re})_p + O\left\{(\mathrm{Re})_p^2\right\}\right]$$
Proudman and Pearson (1957)
$$F = F_s\left[1 + \frac{3}{16}(\mathrm{Re})_p + \frac{9}{160}(\mathrm{Re})_p^2 \ln\left\{\frac{(\mathrm{Re})_p}{2}\right\} + O\left\{(\mathrm{Re})_p^2\right\}\right]$$

Table 8 Recommended Empirical Drag Correlations $w = \log_{10}(Re)_p$

$(Re)_p < 0.01$	$C_D = \dfrac{3}{16} + \dfrac{24}{(Re)_p}$
$0.01 < (Re)_p \leq 20$	$\log_{10}\left[\dfrac{C_D(Re)_p}{24} - 1\right] = -0.881 + 0.82w - 0.05w^2$
	$C_D = \dfrac{24}{(Re)_p}\left[1 + 0.1315(Re)_p^{(0.82-0.05w)}\right]$
$20 \leq (Re)_p \leq 260$	$\log_{10}\left[\dfrac{C_D(Re)_p}{24} - 1\right] = -0.7133 + 0.6305w$
	$C_D = \dfrac{24}{(Re)_p}\left[1 + 0.1935(Re)_p^{0.6305}\right]$
$260 \leq (Re)_p \leq 1500$	$\log_{10} C_D = 1.6435 - 1.1242w + 0.1558w^2$
$1.5 \times 10^3 \leq (Re)_p \leq 1.2 \times 10^4$	$\log_{10} C_D = -2.4571 + 2.5558w - 0.9295w^2 + 0.1049w^3$
$1.2 \times 10^4 \leq (Re)_p \leq 4.4 \times 10^4$	$\log_{10} C_D = -1.9181 + 0.6370w - 0.0636w^2$
$4.4 \times 10^4 \leq (Re)_p \leq 3.38 \times 10^5$	$\log_{10} C_D = -4.3390 + 1.5809w - 0.1546w^2$
$3.38 \times 10^5 \leq (Re)_p \leq 4 \times 10^5$	$C_D = 29.78 - 5.3w$
$4 \times 10^5 \leq (Re)_p \leq 10^6$	$C_D = 0.1w - 0.49$
$10^6 < (Re)_p$	$C_D = 0.19 - \dfrac{8 \times 10^4}{(Re)_p}$

Source: Adapted from Clift et al. (1978).

$$C_D = \frac{24}{(Re)_p}\left[1 + \left(8.1716e^{-4.0655\phi}\right)(Re)_p^{0.0964+0.5565\phi}\right]$$
$$+ \frac{73.69\left(e^{-5.0748\phi}\right)(Re)_p}{(Re)_p + 5.378e^{6.2122\phi}}$$

$$(35)$$

For spherical particles, Eq. (35) reduces to

$$C_D = \frac{24}{(Re)_p} + 3.3643(Re)_p^{0.3471} + \frac{0.4607(Re)_p}{(Re)_p + 2682.5}$$

$$(36)$$

3.7 Corrections for Nonspherical Particles

The book by Clift et al. (1978) contains an extensive review on this subject. The treatment is, however, mostly for the axisymmetric particles such as spheroids and cylinders and orthotropic particles such as rectangular parallelepipeds. For particles of arbitrary shape,

no fully satisfactory method is available for correlating the drag.

Pettyjohn and Christiansen (1948) determined the free-settling rates of isometric particles of the following shapes and sphericities: spheres ($\phi = 1$), cube octahedron ($\phi = 0.906$), octahedron ($\phi = 0.846$), cube ($\phi = 0.806$), and tetrahedron ($\phi = 0.670$). Their results suggest that the correction factor should be

$$K = 0.8431 \log\left(\frac{\phi}{0.065}\right) \qquad \begin{matrix} (Re)_t < 0.05, \\ 0.67 < \phi < 1 \end{matrix} \qquad (37)$$

where K is the ratio of the settling velocity of the volume-equivalent sphere to the settling velocity of the particle. Results on other isometric particles have also been published. These include the experimental results of Heiss and Coull (1952) using solid cylinders and rectangular parallelpipeds, Becker (1959) using prisms and cylinders, Christiansen and Barker (1965) using cylinders, prisms, and disks, Isaacs and Thodos (1967) using cylinders, and Hottovy and Sylvester (1979) using roundish irregularly shaped particles.

In the Stokes law regime, homogeneous symmetrical particles can take up any orientation during their settling in a fluid of infinite extent. Ellipsoids of uniform density and bodies of revolution with fore and aft symmetry can attain spin-free terminal states in all orientations. The terminal velocities of the ellipsoids will, however, depend on their orientation. A set of identical particles can therefore have a range of terminal velocities, though the range is generally fairly small. Asymmetric particles such as ellipsoids and discs do not generally fall vertically unless they are dropped with a principal axis of symmetry parallel to a gravity field. They tend to drift from side to side.

In the intermediate flow regime, particles adopt preferred orientations. Particles will usually align themselves with their maximum cross section normal to the direction of relative motion. There is no appreciable secondary motion in the intermediate flow regime, so results for flow past fixed objects of the same shape can be used if the orientation corresponds to the preferred orientation.

Secondary motion associated with wake shedding occurs in the Newton's law regime, and C_D is insensitive to $(Re)_p$. In this regime the density ratio, λ, plays an important role in determining the type of motion and the mean terminal velocity. For particles of arbitrary shapes,

$$C_D = \lambda^{-1/18}(5.96 - 5.51\phi) \quad \text{for} \quad 1.1 < \lambda < 8.6 \tag{38}$$

Pettyjohn and Christiansen (1948) also suggested a correction factor for nonspherical particles:

$$C_D = 5.31 - 4.88\phi \quad \begin{array}{l} 2 \cdot 10^3 < (Re)_t < 2 \cdot 10^5, \\ 0.67 < \phi < 1 \end{array} \tag{39}$$

C_D here is based on the cross-sectional area of the volume-equivalent sphere. The terminal velocity, U_t, can be calculated by

$$U_t = 0.49\lambda^{1/36}\left[\frac{g(\rho_p - \rho_f)d_v}{\rho_f(1.08 - \phi)}\right]^{\frac{1}{2}} \tag{40}$$

3.8 Drag Coefficient for Particles with Density Lighter than the Surrounding Fluid

All the above discussions are for particles with density larger than the surrounding fluid. For particles with density smaller than the surrounding fluid, it has long been assumed that the free rising velocity is governed by the same equations, the only difference being that of particle movement direction. However,

Karamanev and Nikolov (1992) and Karamanev (1994) showed recently that the relationship between C_D and $(Re)_p$ for particles with density much smaller than the fluid follows the standard drag curve only when $(Re)_p$ is less than 135. Otherwise the C_D is a constant equal to 0.95. Karamanev (1996) proceeded to suggest the following drag coefficient equations on the basis of the Archimedes number rather than the Reynolds number for calculation of the terminal falling or rising velocities.

For free falling spheres,

$$C_D = \frac{432}{Ar}\left(1 + 0.0470 Ar^{2/3}\right) + \frac{0.517}{1 + 154 Ar^{1/3}} \tag{41}$$

For free rising spheres,

$$C_D = \frac{432}{Ar}\left(1 + 0.0470 Ar^{2/3}\right) + \frac{0.517}{1 + 154 Ar^{1/3}} \tag{42}$$

$$\text{for} \quad Ar < 1.18 \cdot 10^6 d_p^2$$

$$C_D = 0.95 \quad \text{for} \quad Ar > 1.18 \cdot 10^6 d_p^2 \tag{43}$$

d_p being in meters.

4 CALCULATION OF TERMINAL VELOCITY FOR A SINGLE PARTICLE

The terminal velocity of a single particle is an intrinsic characteristic of the particle, and its calculation and measurement are as important as other intrinsic particle properties, such as particle size and density. Calculation of the terminal velocity of a single particle used to be an iterative process. More recent developments allow direct calculations without trial and error. These methods are introduced here.

4.1 Equation by Haider and Levenspiel (1989)

From Eq. (30), the terminal velocity for a single spherical particle can be obtained as

$$U_r = U_t = \sqrt{\frac{4}{3}\frac{d_p(\rho_p - \rho_f)g}{\rho_f C_D}} \tag{44}$$

By combining Eq. (44) and the recommended drag coefficient correlations in Table 8 or drag coefficient equations proposed by Turton and Levenspiel (1986), Eq. (34), and Haider and Levenspiel (1989), Eq. (35), the terminal velocity can be calculated. Haider and Levenspiel further suggested an approximate method for direct evaluation of the terminal velocity by defining a dimensionless particle size, d_p^*, and a dimensionless particle velocity, U^*, by

$$d_p^* = d_p \left[\frac{\rho_f (\rho_p - \rho_f) g}{\mu^2} \right]^{1/3} = Ar^{1/3} = \left[\frac{3}{4} C_D (Re)_p^2 \right]^{1/3}$$

$$\text{(45)}$$

$$U^* = U \left[\frac{\rho_f^2}{\mu (\rho_p - \rho_f) g} \right]^{1/3} = \frac{(Re)_p}{Ar^{1/3}} = \left[\frac{4}{3} \frac{(Re)_p}{C_D} \right]^{1/3}$$

$$\text{(46)}$$

The terminal velocity of the irregular particles can then be calculated by

$$U_t^* = \left[\frac{18}{(d_p^*)^2} + \frac{2.335 - 1.744\phi}{(d_p^*)^{0.5}} \right]^{-1} \qquad 0.5 < \phi < 1$$

$$\text{(47)}$$

For spherical particles, Eq. (44) reduces to

$$U_t^* = \left[\frac{18}{(d_p^*)^2} + \frac{0.591}{(d_p^*)^{0.5}} \right]^{-1} \qquad \phi = 1$$

$$\text{(48)}$$

4.2   Terminal Velocity by Polynomial Equations Fitted to Heywood Tables

Fouda and Capes (1976) proposed polynomial equations fitted to the Heywood (1962) tables for calculating large numbers of terminal velocities. The Heywood tables were recognized as being a simple and accurate method for calculating both the terminal velocity and the equivalent particle diameter, and they enjoyed wide

acceptance. The tables, however, are not very convenient to use, especially for applications using a computer. The tables are now fitted by Fouda and Capes (1976) in the form of

$$Y = \sum_{n=0}^{5} a_n X^n$$

$$\text{(49)}$$

$$\text{where} \quad Y = \log_{10}(Pd_t) \text{ or } \log_{10}\left(\frac{U_t}{Q} \right)$$

$$\text{(50)}$$

$$\text{and} \qquad X = \log_{10}\left(\frac{U_t}{Q} \right) \text{ or } \log_{10}(Pd_t)$$

$$\text{(51)}$$

The diameter d_t is the particle diameter of a sphere having the same terminal velocity as the particle. P, Q, U_t, and d_t are related in the following two dimensionless equations derived by Heywood:

$$C_D Re^2 = \frac{4}{3} g \frac{(\rho_p - \rho_f)}{\mu^2} \rho_f d_t^3 = P^3 d_t^3$$

$$\text{(52)}$$

$$\frac{Re}{C_D} = \frac{3}{4g} \frac{\rho_f^2}{(\rho_p - \rho_f)} \frac{U_t^3}{\mu} = \frac{U_t^3}{Q^3}$$

$$\text{(53)}$$

The resulting polynomial coefficients are summarized in Table 9.

The polynomial equations were reported to have an average deviation less than 0.15% and a standard deviation of less than 4.3%.

Similar types of equations were also proposed by Hartman et al. (1994) for nonspherical particles. They introduced the equations

Table 9   Polynomial Coefficients for Polynomial Equations Fitted to Heywood Tables [see Eq. (49)]

	Y	X	Polynomial coefficients
Terminal velocity calculation	$\log_{10}(U_t/Q)$	$\log_{10}(Pd_t)$	$a_0 = -1.37323$
			$a_1 = 2.06962$
			$a_2 = -0.453219$
			$a_3 = -0.334612 \times 10^{-1}$
			$a_4 = -0.745901 \times 10^{-2}$
			$a_5 = 0.249580 \times 10^{-2}$
Equivalent diameter calculation	$\log_{10}(Pd_t)$	$\log_{10}(U_t/Q)$	$a_0 = 0.785724$
			$a_1 = 0.684342$
			$a_2 = 0.168457$
			$a_3 = 0.103834$
			$a_4 = 0.20901 \times 10^{-1}$
			$a_5 = 0.57664 \times 10^{-2}$

$$\log(\mathrm{Re})_t(Y, \phi) = \log(\mathrm{Re})_t(Y, 1) + P(Y, \phi) \qquad (54)$$

and

$$\log(\mathrm{Re})_t(Y, 1) = 0.77481 - 0.56032 \log Y$$
$$+ 0.024246(\log Y)^2 - 0.0038056$$
$$(\log Y)^3 \qquad (55)$$

$$P(Y, \phi) = -0.10118(1 - \phi)\log Y + 0.092944$$
$$(1 - \phi)(\log Y)^2 - 0.0098356(1 - \phi)$$
$$(\log Y)^3 - 0.12666(1 - \phi)^2 \log Y$$
$$(56)$$

where

$$Y = \left(\frac{4}{3}\right)\left[\frac{g(\rho_p - \rho_f)\mu}{U_t^3 \rho_f^2}\right] \qquad (57)$$

These equations are good for $0.01 < (\mathrm{Re})_t < 16000$ and $0.67 < \phi < 1$. The equation fitted the experimental terminal velocities of limestone and lime particles to better than 20%

4.3 Calculation of Terminal Velocity of Porous Spheres

Calculation of the terminal velocity of a porous sphere is useful and important in applications in water treatment where settling velocities of a floc or an aggregate are estimated. It is also important in estimation of terminal velocities of clusters in fluidized bed applications. The terminal velocity of a porous sphere can be quite different from that of an impermeable sphere. Theoretical studies of settling velocity of porous spheres were conducted by Sutherland and Tan (1970), Ooms et al. (1970), Neale et al. (1973), Epstein and Neale (1974), and Matsumoto and Suganuma (1977). The terminal velocity of porous spheres was also experimentally measured by Masliyah and Polikar (1980). In the limiting case of a very low Reynolds number, Neale et al. (1973) arrived at the following equation for the ratio of the resistance experienced by a porous (or permeable) sphere to an equivalent impermeable sphere. An equivalent impermeable sphere is defined to be a sphere having the same diameter and bulk density of the permeable sphere.

$$\Omega = \frac{2\beta^2[1 - (\tanh \beta)/\beta]}{2\beta^2 + 3[1 - (\tanh \beta)/\beta]} \qquad (58)$$

where β is the normalized sphere radius expressed by

$$\beta = \frac{R}{\sqrt{k}} \qquad (59)$$

where k is the permeability and R is the radius of the sphere. The resistance ratio, Ω, is normally less than unity. For impermeable spheres, Ω approaches unity because β approaches infinity. For infinitely permeable spheres, β tends to zero and so does Ω.

Based on Eq. (58), the ratio of terminal velocities of permeable and impermeable spheres can be written as

$$\frac{(U_t)_p}{(U_t)_{ip}} = \frac{1}{\Omega} \qquad (60)$$

and the ratio of dimensionless drag coefficient can be expressed as

$$\frac{(C_D \mathrm{Re})_p}{(C_D \mathrm{Re})_{ip}} = \Omega \qquad (61)$$

where

$$(C_D \mathrm{Re})_p = \frac{4d_p^2 g[(1 - \varepsilon)(\rho_p - \rho_f)]}{3\mu(U_t)_p} \qquad (62)$$

and

$$(C_D \mathrm{Re})_{ip} = \frac{4d_p^2 g(\rho_p - \rho_f)}{3\mu(U_t)_{ip}} = 24 \qquad (63)$$

Equation (60) shows that a permeable sphere with the same diameter and bulk density as the impermeable sphere will have a higher terminal velocity than that of the impermeable sphere. At higher Reynolds number, the experiments by Masliyah and Polikar (1980) suggested the following equations:

For $15 < \beta < 33$

$$C_D = \frac{24\Omega}{(\mathrm{Re})_p}\left[1 + 0.1315(\mathrm{Re})_p^{(0.82 - 0.05w)}\right]$$
$$0.1 < (\mathrm{Re})_p \le 7 \qquad (64)$$

and

$$C_D = \frac{24\Omega}{(\mathrm{Re})_p}\left[1 + 0.0853(\mathrm{Re})_p^{(1.093 - 0.105w)}\right]$$
$$7 < (\mathrm{Re})_p < 120 \qquad (65)$$

where $w = \log_{10}(\mathrm{Re})_p$.

This confirms the theoretical study of Neale et al. (1973) that the drag experienced by a porous sphere at low Reynolds numbers is less than that for an impermeable sphere of similar diameter and bulky density. The effect of inertia at high Reynolds numbers is higher for a porous sphere than for an impermeable sphere of similar diameter and bulk density.

As far as the wall effect is concerned, the experiments show that, at low Reynolds numbers, the wall effect for a porous sphere is of the same order of magnitude as that for an impermeable sphere. At high Reynolds numbers, however, the wall effect become smaller and less significant for a porous sphere.

5 CHARACTERIZATION OF MULTIPARTICLE SYSTEMS

In most applications, the systems and processes contain large amounts of particles with size distribution; each size may also possess a distinguished shape. To describe properly these systems and processes for design and analysis, they need to be adequately characterized to reflect their physical and chemical potentials. In the following sections, different average particle diameter definitions are introduced along with statistical descriptions of particle size distribution. Depending on applications, one definition may be more suitable than others. Thus care must be exercised to select the proper characterization for each process.

5.1 Different Definitions of Average Particle Diameter

As mentioned earlier, naturally occurring particles are usually irregular in shape and also different in size. Different definitions of average particle diameter have been used to characterize an assemblage of particles. Some may be hypothetical while others may have physical significance. They are summarized here.

5.1.1 Arithmetic Mean

The arithmetic mean is defined as the sum of all diameters divided by the total number of particles

$$\bar{d}_{av} = \frac{\sum\limits_{i} n_i d_{pi}}{\sum\limits_{i} n_i} \qquad (66)$$

where n_i can be the number of particles or the weight percentage.

5.1.2 Surface Mean

The surface mean diameter is the diameter of a hypothetical spherical particle whose surface multiplied by the total number of particles in the assemblage would be equal to the total surface area of the assemblage, expressed mathematically as

$$\bar{d}_s = \sqrt{\frac{\sum\limits_{i} n_i d_{pi}^2}{\sum\limits_{i} n_i}} \qquad (67)$$

5.1.3 Volume Mean

The volume mean diameter is the diameter of a hypothetical spherical particle whose volume multiplied by the total number of particles in the assemblage would be equal to the total volume of the assemblage.

$$\bar{d}_v = \sqrt[3]{\frac{\sum\limits_{i} n_i d_{pi}^3}{\sum\limits_{i} n_i}} \qquad (68)$$

5.1.4 Volume–Surface Mean

The volume–surface mean is also known as the Sauter mean. This defines the average particle size based on the specific surface area per unit volume or per unit weight, as shown in Eq. (69):

$$\bar{d}_{vs} = \frac{\sum\limits_{i} n_i d_{pi}^3}{\sum\limits_{i} n_i d_{pi}^2} = \frac{1}{\sum\limits_{i} \dfrac{x_i}{d_{pi}}} \qquad (69)$$

5.1.5 Weight Mean

The weight mean is also known as the DeBroucker mean. This defines the average particle diameter based on the unit weight of the particles. The weight mean particle diameter is the diameter of a sphere whose surface area times the total number of particles equals the surface area per unit weight of the assemblage.

$$\bar{d}_w = \sum\limits_{i} v_i d_{pi} = \frac{\sum\limits_{i} n_i d_{pi}^4}{\sum\limits_{i} n_i d_{pi}^3} \qquad (70)$$

where

$$v_i = \frac{n_i d_{pi}^3}{\sum\limits_{i} n_i d_{pi}^3} \qquad (71)$$

5.1.6 Length Mean

The length mean is the average diameter obtained by dividing the total surface area of all particles with the summation of the diameters, as shown here:

$$\bar{d}_1 = \frac{\sum_i n_i d_{\mathrm{p}i}^2}{\sum_i n_i d_{\mathrm{p}i}} \tag{72}$$

5.1.7 Geometric Mean

The geometric mean is the logarithmic equivalent of the arithmetic mean. Being a logarithmic average, the geometric mean is always smaller than the arithmetic mean.

$$\bar{d}_{\mathrm{g}} = \sqrt[n]{d_{\mathrm{p}1}^{n_1} d_{\mathrm{p}2}^{n2} \cdots d_{\mathrm{p}n}^{n_n}} \tag{73}$$

$$\bar{d}_{\mathrm{g}} = \frac{\sum_i \left[n_i \log(d_{\mathrm{p}i}) \right]}{\sum_i n_i} \tag{74}$$

5.1.8 Harmonic Mean

This is an average diameter based on the summation of the reciprocals of the diameters of the individual components of an assemblage and can be expressed mathematically as

$$\bar{d}_{\mathrm{h}} = \frac{\sum_i n_i}{\sum_i \frac{n_i}{d_{\mathrm{p}i}}} = \frac{1}{\sum_i \frac{x_i}{d_{\mathrm{p}i}}} \tag{75}$$

The harmonic mean is actually related to the average spherical particle corresponding to the particle surface per unit weight. Mathematically, the harmonic mean is similar to the volume–surface mean or Sauter mean.

5.2 Statistical Characterization of Particles with a Size Distribution

Statistically, the particle size distribution can be characterized by three properties: mode, median, and mean. The mode is the value that occurs most frequently. It is a value seldom used for describing particle size distribution. The average or arithmetic mean diameter, $\bar{d}_{\mathrm{av}}$, is affected by all values actually observed and thus is influenced greatly by extreme values. The median particle size, $d_{1/2}$, is the size that divides the frequency distribution into two equal areas. In practical application, the size distribution of a typical dust is typically skewed to the right, i.e., skewed to the larger particle size. The central tendency of a skewed frequency distribution is more adequately represented by the median rather than by the mean (see Fig. 9). Mathematically, the relationships among the mean, median, and mode diameter can be expressed as

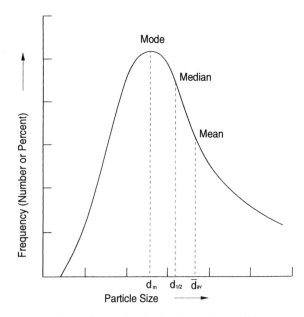

Figure 9 Skewed particle distribution of typical dust.

$$\bar{d}_{\mathrm{av}} = \log^{-1}\left(\log \bar{d}_{\mathrm{g}} + 1.151 \log^2 \sigma_{\mathrm{g}} \right) \tag{76}$$

$$d_{1/2} = \bar{d}_{\mathrm{g}} \tag{77}$$

$$d_{\mathrm{m}} = \log^{-1}\left(\log \bar{d}_{\mathrm{g}} - 2.303 \log^2 \sigma_{\mathrm{g}} \right) \tag{78}$$

Because of its mathematical properties, the standard deviation σ is almost exclusively used to measure the dispersion of the particle size distribution. When the skewed particle size distribution shown in Fig. 9 is replotted using the logarithm of the particle size, the skewed curve is transformed into a symmetrical bell-shaped curve as shown in Fig. 10. This transformation is of great significance and importance in that a symmetrical bell-shaped distribution is amenable to all the statistical procedures developed for the normal or gaussian distribution.

In the log-normal particle size distribution, the mean, median, and mode coincide and have an identical value. This single value is called the geometric median particle size, $\bar{d}_{\mathrm{g}}$, and the measure of dispersion, the geometric standard deviation, σ_{g}. Thus the log-normal particle size distribution can be described completely by these two characteristic values. To determine whether the particles have a distribution close to log-normal distribution, the particle cumulative frequency data can be plotted on a logarithmic probability graph paper. If the particle size distribution is log-normal, a straight line will result. The geometric median particle size is the 50% value of the distribution as shown in Fig. 11. The geometric standard deviation is equal to

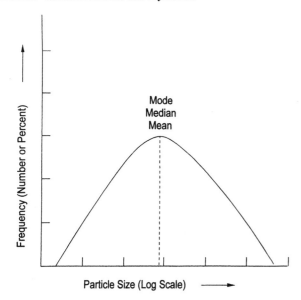

Figure 10 Log-normal particle size distribution.

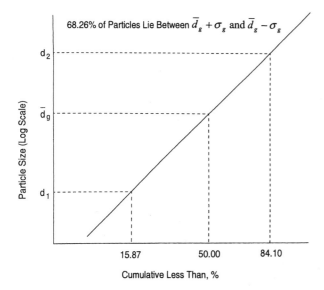

Figure 11 Log-normal distribution plotted on a logarithmic probability graph paper.

the ratio of 84.1% value divided by the 50% value or 50% value divided by the 15.9% value. Although 68.26% of the particles will lie in the particle size range between $\bar{d}_g + \sigma_g$ and $\bar{d}_g - \sigma_g$. If the particle size reduction is due to comminution such as crushing, milling, and grinding, the resulting particle size distribution very often tends to be log-normal distribution. Pulverized silica, granite, calcite, limestone, quartz,

soda, ash, sodium bicarbonate, alumina, and clay all were observed to fit the log-normal distribution.

Mathematically the log-normal distribution can be expressed as

$$f = \frac{1}{\sigma_g \sqrt{2\pi}} \exp\left[-\frac{(z - \bar{z})^2}{2\sigma_g^2}\right] \tag{79}$$

where

$$f = \frac{d\psi}{dz}, \qquad z = \ln(d_{pi}), \qquad \bar{z} = \frac{\sum_i z \, d\psi}{\sum_i d\psi} \tag{80}$$

and ψ can be number, surface, or weight.

Mathematically, the log-normal distribution can also be derived from the arithmetic distribution, by substituting x by $\ln x = z$ to give Eq. (81):

$$f = \frac{1}{\sigma \sqrt{2\pi}} \exp\left[-\frac{(x - \bar{x})^2}{2\sigma^2}\right] \tag{81}$$

$$\sigma^2 = \frac{\sum (x - \bar{x})^2}{n - 1} \tag{82}$$

An important characteristic of the log-normal distribution is that the transformation among the various particle size definitions and the statistical diameters can be performed analytically and graphically (Smith et al., 1929). The Hatch-Choate (1929) transformation equations are summarized here:

$$\log \bar{d}_{gm} = \log \bar{d}_{gc} + 6.908 \log^2 \sigma_g \tag{83}$$

$$\log \bar{d}_{av} = \log \bar{d}_{gc} + 1.151 \log^2 \sigma_g \tag{84}$$

$$\log \bar{d}_s = \log \bar{d}_{gc} + 2.303 \log^2 \sigma_g \tag{85}$$

$$\log \bar{d}_v = \log \bar{d}_{gc} + 3.454 \log^2 \sigma_g \tag{86}$$

$$\log \bar{d}_{vs} = \log \bar{d}_{gc} + 5.757 \log^2 \sigma_g \tag{87}$$

$$\log \bar{d}_w = \log \bar{d}_{gc} + 8.023 \log^2 \sigma_g \tag{88}$$

$$\log \bar{d}_{gc} = \log \bar{d}_{gm} - 6.908 \log^2 \sigma_g \tag{89}$$

$$\log \bar{d}_{av} = \log \bar{d}_{gm} - 5.757 \log^2 \sigma_g \tag{90}$$

$$\log \bar{d}_s = \log \bar{d}_{gm} - 4.605 \log^2 \sigma_g \tag{91}$$

$$\log \bar{d}_v = \log \bar{d}_{gm} - 3.454 \log^2 \sigma_g \tag{92}$$

$$\log \bar{d}_{vs} = \log \bar{d}_{gm} - 1.151 \log^2 \sigma_g \tag{93}$$

$$\log \bar{d}_w = \log \bar{d}_{gm} + 1.151 \log^2 \sigma_g \tag{94}$$

5.3 The Rosin–Rammler Distribution

The Rosin–Rammler (1933) distribution was first developed for broken coal, but it has since been found to be applicable to many other materials, such as cement, gypsum, magnetite, clay, dyestuffs, quartz, flint, glass, and ores. From probability considerations, the authors obtained

$$\frac{dW_{\mathrm{f}}(S_{\mathrm{a}})}{dS_{\mathrm{a}}} = 100\,nbS_{\mathrm{a}}^{n-1}\exp(-bS_{\mathrm{a}}^{n}) \tag{95}$$

where n and b are constants: b is a measure of the range of particle sizes present, and n is a characteristic of the substance. Integrating Eq. (95), we have

$$W_{\mathrm{f}} = 100\exp(-bS_{\mathrm{a}}^{n}) \tag{96}$$

or

$$\log\text{–}\log\left(\frac{100}{W_{\mathrm{f}}}\right) = \text{constant}+n\log(S_{\mathrm{a}}) \tag{97}$$

If $\log\text{–}\log(100/W_{\mathrm{f}})$ is plotted against $\log S_{\mathrm{a}}$, a straight line results. The peak of the distribution for $n = 1$ is at $100/e = 36.8\%$. This is used to characterize the degree of comminution of the material. In Eq. (97), W_{f} is the weight percent of material retained on the sieve of aperture S_{a}. If the mode of the particle size distribution curve is $(S_{\mathrm{a}})_{\mathrm{m}}$, Eq. (97) gives $b = 1/(S_{\mathrm{a}})_{\mathrm{m}}$. Since the slope of the line on the Rosin–Rammler graph depends on the particle size range, the ratio of $\tan^{-1}(n)$ and $(S_{\mathrm{a}})_{\mathrm{m}}$ is a form of variance. This treatment following Rosin and Rammler is useful for monitoring grinding operations for highly skewed distributions. It should be employed carefully, however, since taking the logarithm always reduces scatter, taking the logarithm twice as in this case, tends to obscure the actual scatter in the distribution.

5.4 Measurement of the Angle of Repose and the Angle of Internal Friction

Two important characteristics of powder rheology are the angle of repose and the angle of internal friction. Simple devices can be constructed to measure both. Figure 12 depicts a simple two-dimensional bed with transparent walls and a small orifice at the bottom of the bed. After filling the bed with the powder to be examined, the powder is allowed to flow out of the test device to the surface of the test stand. The angle of the powder-free surface measured from the flat surface of the test stand, the angle β in Fig. 12, is called the angle of repose. This angle is an intrinsic characteristic of the powder and should be close to a constant

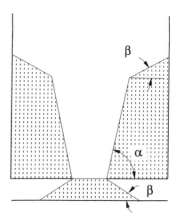

Figure 12 A two-dimensional device for management of angle or the repose and the angle of internal friction.

when the same powder is pulled on top of a flat surface. Normal powders have angles of repose of around 35°. The angles of repose for commonly occurring powders are summarized in Table 10.

The angle made by the free surface of the powder still remaining in the test device with the flat bottom of the test device, the angle α in Fig. 12, is called the angle of internal friction of the powder. The angle of internal friction is also an intrinsic property of the powder and can be considered as a shearing plane of the powder. In order for the powder to flow, the angle has to be higher than the angle of internal friction of the powder. Normal powders have angles of internal friction of around 70°. The angles of internal friction for commonly occurring powders are also listed in Table 10.

Both the angle of repose and the angle of internal friction can be modified and reduced by flow aids such as aeration, vibration, and the addition of other powders.

Table 10 Typical Angles of Repose and Angles of Internal Friction

Material	Angle of repose degree	Angle of internal friction degree
FCC catalyst	32	79
Sand	36	64
Resin	29	82
Wet ashes	50	—
Wheat	—	55
Oats	21	—

NOMENCLATURE

A_p = projected area of a particle

Ar = Archimedes number, $Ar = \dfrac{d_p^3 \rho_f (\rho_p - \rho_f) g}{\mu^2}$

a_s = exterior surface area/volume of catalyst pellet

b = breadth

C = concentration of particle suspension

C_D = drag coefficient

d_a = projected-area diameter

d_A = sieve diameter

$\bar{d}_{av}$ = arithmetic mean particle diameter

d_c = perimeter diameter

d_D = drag diameter

d_f = free-falling diameter

d_F = Feret diameter

$\bar{d}_g$ = geometric mean particle diameter or geometric median particle diameter

$\bar{d}_{gc}$ = geometric median diameter by count

$\bar{d}_{gm}$ = geometric median diameter by weight

$\bar{d}_h$ = harmonic mean particle diameter

d_i = inside diameter of a ring

$\bar{d}_l$ = length mean particle diameter

d_M = Martin diameter

d_m = mode particle diameter

d_o = outside diameter of a ring

d_p = equivalent particle diameter

d_{pi} = particle diameter of size i

d_p^* = dimensionless particle diameter

d_s = surface diameter

$\bar{d}_s$ = surface mean particle diameter

d_{st} = Stokes diameter

d_{sv} = surface–volume diameter

d_t = particle diameter of a sphere having the same terminal velocity as the particle

d_v = volume diameter

$\bar{d}_v$ = volume mean particle diameter

$\bar{d}_{vs}$ = volume–surface mean particle diameter

$\bar{d}_w$ = weight mean particle diameter

d_y = diameter of a cylinder

D = sedimentation column diameter

DR_{10} = particle diameter ratio at the cumulative mass of 10%

DR_{50} = particle diameter ratio at the cumulative mass of 50%

DR_{90} = particle diameter ratio at the cumulative mass of 90%

$d_{1/2}$ = median particle diameter

e_1 = flatness ratio

e_2 = elongation ratio

f = frequency of observation

F = force exerting on a particle

F_s = force exerting on a sphere

g = gravitational acceleration

I_c = intensity of light beam after passing through a particle suspension of concentration C

I_o = original light beam intensity

k = Heywood shape factor

k = permeability

K = ratio of the settling velocity of the volume-equivalent sphere to the settling velocity of the particle

k_c = a constant

k_e = Heywood shape factor for an isometric particle of similar form

l = length

l_y = length of a cylinder

L = length of light beam path

L_w = the distance from the center of a particle to the walls

M = mass of a particle

n_i = number of particles of size i or weight percent of particles of size i

R = radius of a sphere

$(Re)_p$ = Reynolds number based on the particle diameter

$(Re)_t$ = Reynolds number based on the terminal velocity of the particle

S_a = sieve aperture

$(S_a)_m$ = sieve aperture at mode of particle size distribution

S_p = surface area of a particle

S_v = particle surface per unit volume of particle

t = thickness

T = measured transmittance, X_c / X_o

U^* = dimensionless particle velocity

U_r = relative velocity between the particle and the fluid

U_t = terminal velocity of a single particle

$(U_t)_{ip}$ = terminal velocity of an impermeable particle

$(U_t)_p$ = terminal velocity of a permeable (or porous) particle

V_m = measured velocity in a column of diameter D

V_p = volume of a particle

V_∞ = falling speed of a sphere in an infinite fluid

W_f = weight fraction

W_s = weight fraction of particles in the x-ray pass

W_{200} = weight of material passing through 200 mesh screen in grams

x_i = weight fraction of particle size d_{pi}

X_c = intensity of x-ray beam passing through a particle suspension of concentration C

X_o = intensity of x-ray beam passing through the sedimentation column filled with clear fluid

μ = viscosity of fluid

ρ = packing density

ρ_f = density of fluid

ρ_p = density of particle

ϕ = sphericity

ϕ_{op} = operational sphericity

ϕ_w	=	working sphericity
$\not\subset$	=	circularity
$\not\subset_{op}$	=	operational circularity
α	=	an empirical constant depending on equipment and particle characteristics
α	=	angle of internal friction
β	=	an empirical constant depending on equipment, particle, and suspending fluid characteristics
β	=	normalized sphere radius
β	=	angle of repose
λ	=	density ratio
Ω	=	ratio of resistance experienced by a porous sphere to an equivalent impermeable sphere
Ψ	=	number, surface, or weight
σ	=	standard deviation
σ_g	=	geometric standard deviation

REFERENCES

S Akapo, TM Khong , CP Simpson , B Toussaint , JG Yates. A novel method for determining the particle density of porous aeratable powders. Powder Technol 58:237–242, 1989.

T Allen. Particle Size Measurement. London: Chapman and Hall, 1975.

B Aschenbrenner. A new method of expressing particle sphericity. J Sediment Petrol 26:15–31, 1956.

LG Austin. Conversion factors to convert particle size distributions measured by one method to those measured by another method. Part Part Syst Charact 15:108–111, 1998.

Babcock & Wilcox. Steam. 40th ed. Barberton: The Babcock & Wilcox Company, 1992, pp 12-8–12-9.

HA Becker. The effects of shape and Reynolds number on drag in the motion of a freely oriented body in an infinite fluid. Can J Chem Eng 37:85–91, 1959.

CF Bohren, RD Huffman. Absorption and Scattering of Light by Small Particles. New York: Wiley-Interscience, 1986.

MP Bonin, M Queiroz. Local particle velocity, size, and concentration measurements in an industrial-sale, pulverized coal-fired boiler. Comb and Flame 85:121–133, 1989.

B Buczek, D Geldart. Determination of the density of porous particles using very fine dense powders. Powder Technol 45:173–176, 1986.

EB Christiansen, DH Barker. The effect of shape and density on the free settling of particles at high Reynolds numbers. AIChE J 11:145–151, 1965.

R Clift, JR Grace, ME Weber. Bubbles, Drops, and Particles. New York: Academic Press, 1978.

RP Davuluri, TM Knowlton. Development of a standardized attrition test procedure. In: LS Fan, TM Knowlton, eds. Fluidization IX. New York: Engineering Foundation, 1998, pp 333–340.

N Epstein, G Neale. On the sedimentation of a swarm of permeable spheres. Chem Eng Sci 29:1841–1842, 1974.

LS Fan, C Zhu Principles of Gas–Solid Flows. Cambridge: Cambridge University Press, 1998, pp 13–16.

AE Fouda, CE Capes. Calculation of large numbers of terminal velocities or equivalent particle diameters using polynomial equations fitted to the Heywood tables. Powder Technol 13:291–293, 1976.

PJ Freud, MN Trainer, AH Clark, HN Frock. Unified scatter technique for full-range particle size measurement. Proc. Pittsburgh Conference, Atlanta GA, March 8–12, 1993.

A Haider, O Levenspiel. Drag coefficient and terminal velocity of spherical and non-spherical particles. Powder Technol 58:63–70, 1989.

J Happel, H Brenner. Low Reynolds Number Hydrodynamics. New Jersey: Prentice Hall, 1965.

M Hartman, O Trnka, K Svoboda. Free settling of non-spherical particles. Ind Eng Chem Res 33:1979–1983, 1994.

T Hatch, S Choate. Statistical description of the size properties of non-uniform particulate substances. J Franklin Inst 207:369–387, 1929.

JF Heiss, J Coull. Effect of orientation and shape. Chem Eng Prog 48:133–140, 1952.

G Herdan. Small Particle Statistics. 2nd ed. New York: Butterworths, 1960.

H Heywood. Uniform and non-uniform motion of particles in fluids. In: Proc Symp Interaction Fluids and Parts. London: Inst Chem Eng, 1962, pp 1–8.

JD Hottovy, ND Sylvester. Drag coefficients for irregularly shaped particles. Ind Eng Chem Process Des Devel 18:433–436, 1979.

JL Isaacs, G Thodos. The free-settling of solid cylindrical particles in the turbulent region. Can J Chem Eng 45:150–155, 1967.

DG Karamanev. On the rise of gas bubbles in quiescent liquids. AIChE J 40:1418–1421, 1994.

DG Karamanev. Equations for calculation of the terminal velocity and drag coefficient of solid spheres and gas bubbles. Chem Eng Comm 147:75–84, 1996.

DG Karamanev, LN Nikolov. Free rising spheres do not obey Newton's law for free settling. AIChE J 38:1843–1846, 1992.

BH Kaye. Automated decision-taking in fine particle science. Powder Technol 8:293–306, 1973.

BH Kaye. Direct Characterization of Fine Particles. New York: John Wiley, 1981.

MJ Knight, PN Rowe, HJ Macgillivray, DJ Cheesman. On measuring the density of finely divided but porous particles such as fluid bed cracking catalyst. Trans Instn Chem Engrs 58:203–207, 1980.

M Leva. Fluidization. New York: McGraw-Hill, 1959.

PJ Lloyd. Particle characterization. Chem Eng 120–122, April 29, 1974.

JB Masliyah, M Polikar. Terminal velocity of porous spheres. Can J Chem Eng 58:299–302, 1980.

K Matsumoto, A Suganuma. Settling velocity of a permeable model floc. Chem Eng Sci 32:445–447, 1977.

M Naito, O Hayakawa, K Nakahira, H Mori, J Tsubaki. Effect of particle shape on the particle size distribution measured with commercial equipment. Powder Technol 100:52–60, 1998.

G Neale, N Epstein, W Nader. Creeping flow relative to permeable spheres. Chem Eng Sci 28:1865–1874, 1973.

JP Olivier, GK Hickin, C Orr Jr. Rapid automatic particle size analysis in the sub-sieve range. Powder Technol 4:257–263, 1970/1971.

G Ooms, PF Mijnlieff, HL Beckers. Frictional forces exerted by a flowing fluid on a permeable particle, with particular reference to polymer coils. J Chem Phys 53:4123–4134, 1970.

CW Oseen. Über die Stokessche Formel und über die verwandte Aufgabe in der Hydrodynamik. Arkiv for Mathematik Astronomi och Fysik. Vol. 6, No. 29, 1910.

ES Pettyjohn, EB Christiansen. Effect of particle shape on free-settling rates of isometric particles. Chem Eng Prog 44:157–172, 1948.

M Pohl. Technology update: particle sizing moves from the lab to the process. Powder and Bulk Eng. 39–46, February 1998.

I Proudman, JRA Pearson. Expansions at small Reynolds numbers for the flow past a sphere and a circular cylinder. J Fluid Mech 2:237–262, 1957.

HF Rase. Fixed-Bed Reactor Design and Diagnostics. Boston: Butterworths, 1990, pp 118–119.

P Rosin, EJ Rammler. The laws governing the fineness of powdered coal. J Inst Fuel 7(31):29–36, 1933.

WO Smith, PD Foote, PF Busang. Capillary rise in sands of uniform spherical grains. Phys Rev 34:1271–1279, 1929.

P Subramanian, Vr Arunachalam. A simple device for the determination of sphericity factor. Ind Eng Chem Fundam 19:436–437, 1980.

DN Sutherland, CT Tan. Sedimentation of a porous sphere. Chem Eng Sci 25:1948–1950, 1970.

R Turton, NN Clark. An explicit relationship to predict spherical particle terminal velocity. Powder Technol 53:127–129, 1987.

R Turton, O Levenspiel. A short note on the drag correlation for spheres, Powder Technol 47:83–86, 1986.

H Wadell. Sphericity and roundness of rock particles. J Geol 41:310–331, 1933.

H Wadell. The coefficient of resistance as a function of Reynolds number for solids of various shapes. J Franklin Inst 217:459–490, 1934.

H Wadell. Volume, shape, and roundness of quartz particles. J Geol 43:250–280, 1935.

J Werther, J Reppenhagen. Attrition in fluidized beds and pneumatic conveying lines. In: WC Yang, ed. Fluidization, Solids Handlings, and Processing Westwood, NJ: Noyes Data Corporation, 1999, pp 435–491.

2

Flow Through Fixed Beds

Wen-Ching Yang

Siemens Westinghouse Power Corporation, Pittsburgh, Pennsylvania, U.S.A.

Fixed beds, also called packed beds, play a vital role in chemical processes. Their simplicity induced applications in many unit operations such as adsorption, drying, filtration, dust collection, and other catalytic and noncatalytic reactors. The primary operating cost for a fixed bed is the pressure drop through the packed bed of solids. Thus to understand the design and operation of a packed bed requires study of packing characteristics of particles and their effect on pressure drop through the bed.

1 PACKING CHARACTERISTICS OF MONOSIZED SPHERES

The simplest case of a packing system is a bed that consists of only uniform and regular monosized spherical particles. For this simple system, the voidage of an ordered arrangement of monosized spheres can be derived satisfactorily by mathematical considerations. Random packing of monosized spheres is more complicated and can be described mathematically through a coordination number. However, the voidage determination of random packing of monosized spheres should be done through experiments.

1.1 Regular Packing of Uniform-Sized Spheres

For convenience, the regular packing can be considered to be constructed from regular layers and rows. The two basic layers are the square layer with a 90° angle and the triangular or the simple rhombic layer of an angle of 60° (see Fig. 1). The vertical stacking of these layers yields six possible regular packings. For the six possible regular packings of spherical particles, the voidage is only a function of packing arrangement. The voidage is also independent of the particle size. The six arrangements are cubic, two orientations of orthorhombic, tetragonal-spheroidal, and two orientations of rhombohedral. They are graphically presented in Fig. 2. Formulae for calculating the voidage of these six arrangements have been developed and the packing characteristics of these ordered arrangements are summarized in Table 1 (Herdan, 1960). The closest packing is the rhombohedral with a voidage of 0.2595, and the loosest, the cubic with a voidage of 0.4764. The stability of the systems increases as the voidage decreases. The systems tend towards the orthorhombic state, especially if a mechanical disturbance such as vibration is applied to the systems. The voidage of a bed of monosized particles after prolonged shaking is usually about 0.395, approaching the characteristic of an eight-point contact packing of orthorhombic.

1.2 Random Packing of Uniform Monosize Spheres

Random packings of uniform monosize spheres are created by irregular and random arrangements of particles. According to Scott (1960), two reproducible states of random packing, dense random packing and loose random packing, can be experimentally created. The dense random packing can be developed by

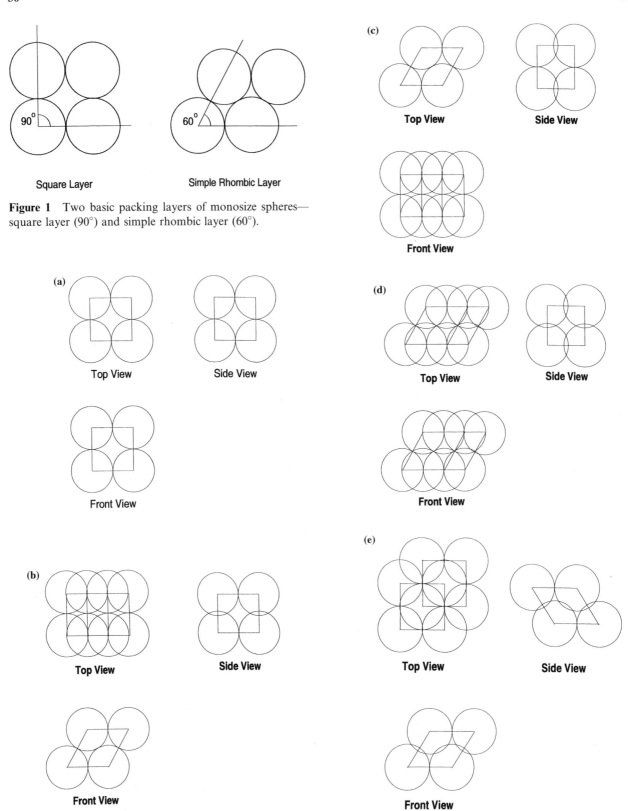

Figure 1 Two basic packing layers of monosize spheres—square layer (90°) and simple rhombic layer (60°).

Figure 2 Six possible arrangements of monosize spheres—one cubic (Figure 2a), two orthorhombic (Figures 2b and 2c), one tetragonal–spheroidal (Figure 2d), two rhombohedral (Figure 2e).

Table 1 Packing Characteristics of Ordered Uniform Monosized Spheres

Packing	Points of contact	Porosity, %	Surface of spheres per cm^3	Cross-sectional pore area per cm^2
Cubic	6	47.64	1.57/R	0.2146
Orthohombic				
(two orientations)	8	39.54	1.81/R	0.2146
Tetragonal-Spheroidal	10	30.19	2.10/R	0.093
Rhombohedral				
(two orientations)	12	25.95	2.22/R	—

pouring the spheres into cylindrical containers and then shaking for several minutes. The loose random packing is created by tipping the container horizontally and rotating slowly about its axis, and then returning slowly to its vertical position. The dense random packing density obtained by several authors ranged from 0.61 to 0.63, equivalent to a voidage of 0.39 and 0.37, corresponding closely to orthorhombic packing with a voidage of 0.3954. The loose random packing obtained has a voidage of 0.40 to 0.42, corresponding to a packing between the cubic packing (voidage = 0.4764) and orthorhombic packing. Haughey and Beveridge (1969) classified the packing into four different modes: very loose random packing, loose random packing, poured random packing, and close random packing. The loose random packing and poured random packing correspond to Scott's loose and dense random packings. The very loose random packing corresponds to the state where the bed is first fluidized and then the gas is slowly reduced until it is below the minimum fluidization. The bed so formed usually has a voidage of about 0.44. The close random packing is formed by vigorously shaking or vibrating the bed, and then the voidage is usually approaching 0.359 to 0.375.

1.3 Properties of Regular and Random Packing of Uniform Monosized Spheres

The coordination number, defined as the number of spheres in contact with any neighboring spheres, is used to characterize the voidage of packing. Table 2 lists the coordination numbers from 3 to 12 and its associated voidage. The coordination numbers of 4, 6, 8, 10, and 12 correspond to arrangements that are regular as shown in Table 1.

A model derived by Haughey and Beveridge (1966) relates the average coordination number with the mean bulk voidage as

$$n = 22.47 - 39.39\varepsilon \qquad 0.259 \leq \varepsilon \leq 0.5 \qquad (1)$$

A parameter called the layer spacing, βd_p, where d_p is the diameter of the sphere, has also been used to characterize the packing. For most common packings, β is between 0.707 and 1.0, the limits corresponding to the rhombohedral packing and cubic packing, respectively. Though the layer spacing has little physical meaning in random packing, the concept is useful for describing the packing. It is actually related to the bed voidage in the following way:

$$\beta = \sqrt{\frac{2}{3}} \left[\frac{\pi}{3\sqrt{2}(1-\varepsilon)} \right]^{1/3} \qquad (2)$$

1.4 Specific Surface Area of the Bed

The specific surface area of the bed is defined as the ratio of total particle surface area to the total bed volume. Since the number of spheres per unit volume of bed is

$$N_p = \frac{6(1-\varepsilon)}{\pi d_p^3} \qquad (3)$$

Table 2 Correspondence of Coordination Number and Voidage

Coordination Number	Voidage
3	0.7766
4	0.6599
5	0.5969
6	0.4764
7	0.4388
8	0.3955
9	0.3866
10	0.3019
11	0.2817
12	0.2595

The specific surface area can be evaluated as

$$S = \frac{6(1-\varepsilon)}{d_p} \qquad (4)$$

2 PACKING CHARACTERISTICS OF BINARY MIXTURES OF SPHERICAL AND NONSPHERICAL PARTICLES

The packing of binary mixtures of spheres is the simplest system and thus has received a lot of studies. The packing of binary mixtures of spheres depends on their diameter ratio and the percentage of large particles in the mixtures. Following McGeary's (1967) development, the minimum fraction of the coarse particles for closest packing is 73%, and the closest packing voidage is 0.14.

The dependence of maximum observed binary mechanical packing of spheres on the ratio of diameters between the large and small spheres shows that when the size ratio, or diameter ratio, decreases, the changes in voidage increase. With a diameter ratio larger than about 7 to 1, the change in voidage is only very slight. This ratio is very close to the critical diameter ratio of 6.5 for the smaller sphere to pass through the triangular opening formed by three larger spheres in close packing (see Fig. 7).

Abe et al. (1979) carried out theoretical analysis of a packed bed of binary mixture and proposed equations to estimate voidage at different degree of mixing of binary mixtures of spherical and nonspherical particles. If cohesive forces are absent, the binary mixture will yield a packed bed of minimum voidage at X_{min}, as shown in Fig. 3. In the region $X_b < X_{min}$, large particles are distributed randomly and evenly in the packed bed. This state is called complete mixing. In the region $X_b > X_{min}$, there are two different states possible. In the first state, the small particles are distributed in

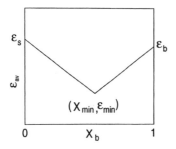

Figure 3 Packed bed of minimum voidage for binary particle mixtures.

the interstices of the big particles. When the fraction of the small particles decreases, segregation of small particles becomes inevitable, as shown in Fig. 4. Equations were proposed to calculate the average voidage of packed bed with binary mixtures.

For $0 \leq X_b < \bar{X}_b$,

$$\varepsilon_{av} = 1 - \frac{1-\varepsilon_s}{(1-X_b) + \alpha X_b(1-\varepsilon_s)} \qquad (5)$$

where

$$\alpha = 1 + f_c\left(\frac{d_s}{d_b}\right) \qquad (6)$$

$$f_c = 1 \qquad \text{for } \frac{d_s}{d_b} = 0.5 \qquad (7)$$

$$f_c = 1.2 \qquad \text{for } \frac{d_s}{d_b} = 0.25 \qquad (8)$$

$$f_c = 1.4 \qquad \text{for } \frac{d_s}{d_b} \leq 0.125 \qquad (9)$$

For $\bar{X}_b \leq X_b < X_{min}$,

$$\varepsilon_{av} = 1 - \frac{(1-\varepsilon_s)(X_b - \bar{X}_b)}{X_b\{1 - \bar{X}_b[1-(1-\varepsilon_s)\alpha]\}^2} \\ - \frac{\bar{X}_b(1-\varepsilon_s)}{X_b\{1 - \bar{X}_b[1-(1-\varepsilon_s)\alpha]\}} \qquad (10)$$

where

$$\bar{X}_b = \frac{a}{a + (1-\alpha a)(1-\varepsilon_s)} \qquad (11)$$

and

$$\alpha = \frac{(1-\varepsilon_b)}{[1 + 1.5(d_s/d_b)]^3} \qquad (12)$$

For $X_{min} \leq X_b \leq 1$,

$$\varepsilon_{av} = 1 - \frac{(1-\varepsilon_b)}{X_b(1+\beta_s\beta_c)^3} \qquad (13)$$

where

$$\beta_c = \sqrt[3]{\frac{1}{X_b}} - 1 \qquad (14)$$

and

$$\beta_s = \left(\frac{d_s}{d_b}\right)^n \qquad (15)$$

n equals 1/3 for crushed stones and 1/2 for round sand.

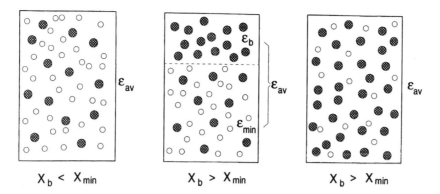

Figure 4 States of binary particle mixtures.

The intersection of Eqs. 10 and 13 gives the binary mixture of large and small particles where the voidage is the minimum, or X_{min} and ε_{min}:

ε_{av} = Average voidage of the binary mixture

ε_{min} = The minimum voidage of the binary mixture

ε_s = Voidage of a packed bed with small particles alone

ε_b = Voidage of a packed bed with large particles alone

X_b = Volume fraction of large particles

X_{min} = Volume fraction of large particles where the voidage of the binary mixture is the minimum

d_s = Diameter of the small particles

d_b = Diameter of the large particles

A common approach to predict the voidage of binary particle mixtures is to make use of the empirical correlation developed by Westman (1936) based on the total bed volume occupied by the specific volume of solid material $V = 1/(1 - \varepsilon)$. The equation was later modified by Yu et al. (1993) as shown in the following equation to apply to both spherical and non-spherical particles.

$$\left(\frac{V - V_b X_b}{V_s}\right)^2 + 2G\left(\frac{V - V_b X_b}{V_s}\right) \left(\frac{V - X_b - V_s X_s}{V_b - 1}\right) + \left(\frac{V - X_b - V_s X_s}{V_b - 1}\right)^2 = 1 \qquad (16)$$

The empirical constant G is independent of the composition of the mixtures but depends on the size ratio of the particles. Yu et al. (1993) gave the following values for G:

$$\frac{1}{G} = 1.355 r_p^{1.566} \qquad \text{for} \qquad r_p \leq 0.824 \qquad (17)$$

and

$$\frac{1}{G} = 1 \qquad \text{for} \qquad r_p \geq 0.824 \qquad (18)$$

where

$$r_p = \frac{d_s}{d_b} \qquad (19)$$

For the limiting case where $r_p = 1$, the voidage of the packed bed will not change through mixing of the binary particles. When r_p approaches 0, the interstices between large particles can be filled with the small particles as discussed earlier. The voidage and the specific solid volume of the packed bed become

$$\varepsilon = \varepsilon_b \varepsilon_s \qquad (20)$$

$$V = \frac{V_b V_s}{V_b + V_s - 1} \qquad (21)$$

$$X_b = \frac{1 - \varepsilon_b}{1 - \varepsilon_b \varepsilon_s} \qquad (22)$$

The effect of changing the particle size ratio on the packing of binary particles is summarized in Fig. 5.

For nonspherical particles, Yu et al. (1993) suggested to substitution of the particle diameter by the packing equivalent particle diameter calculated by

$$d_{pe} = \left(3.1781 - 3.6821\frac{1}{\phi} + 1.5040\frac{1}{\phi^2}\right)d_{ve} \qquad (23)$$

where d_{ve} is the volume equivalent particle diameter and ϕ is the Wadell's sphericity described in Chapter 1. This approach provides good agreement for packing characteristics of both spherical and nonspherical binary particles, though evaluation of the packing equivalent particle diameter is somewhat cumbersome.

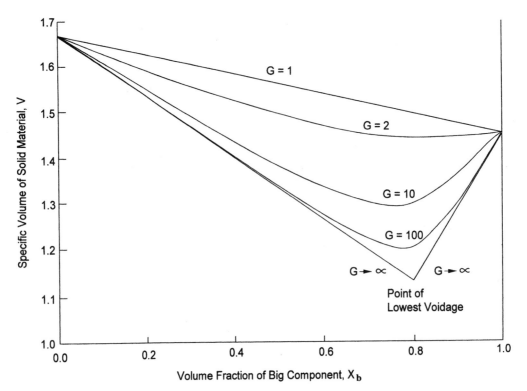

Figure 5 Effect of changing particle size ratio on the packing of binary particles. (Adapted from Finkers and Hoffmann, 1998).

An approach employing the structural ratio to predict the voidage of binary particle mixtures for both spherical and nonspherical particles is also proposed recently by Finkers and Hoffmann (1998). The structural ratio based on packing of spherical and non-spherical particles is defined as

$$r_{str} = \frac{\left(\frac{1}{\varepsilon_b} - 1\right)r_p^3}{1 - \varepsilon_s} \qquad (24)$$

The structural ratio is then used to calculate the empirical constant G for use in Eq. (16).

$$G = r_{str}^k + \left(1 - \varepsilon_b^{-k}\right) \qquad (25)$$

A value of -0.63 was suggested for k in Eq. (25). The proposed approach is good for both spherical and nonspherical particles. For binary particles with size distribution in each fraction, the approach did not fare as well. By changing the value k to -0.345, the proposed approach gave excellent results for data by Sohn and Moreland (1968). It was suggested that k relates to the particle size distribution in each fraction to provide an even more general equation of particle packing in a packed bed.

Packings of ternary and quaternary mixtures of solid particles, considerably more complex systems,

have also been studied by Ouchiyama and Tanaka (1989) and by Hoffmann and Finkers (1995). The packing characteristics of spheres of unequal sizes have also been investigated (Herdan, 1960). In rhombohedral packings of different spherical particle sizes, the voidage may be reduced to less than 0.15.

3 CRITICAL RATIO OF ENTRANCE AND CRITICAL RATIO OF OCCUPATION IN BINARY SYSTEMS

When the monosized spheres are arranged in normal loosest square or tightest rhombohedral packings, there are critical smaller spheres, which can pass through the openings formed by the larger monosized spheres, as shown in Figs. 6 and 7. These critical diameter ratios are called the "critical ratio of entrance."

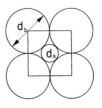

Figure 6 Critical ratio of entrance for square packing.

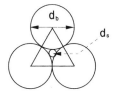

Figure 7 Critical ratio of entrance for rhombohedral packing.

For the loosest square packing, this critical diameter ratio can be expressed as

$$\frac{d_b}{d_S} = \frac{1}{\sqrt{2}-1} = 2.414 \qquad (26)$$

For the tightest rhombohedral packing, the critical ratio can be found from

$$\frac{d_b}{d_S} = \frac{1}{(2/\sqrt{3})-1} = 6.464 \qquad (27)$$

There also exists a critical size sphere that, though it cannot pass through the opening made by the mono-sized spheres, can occupy the volume enclosed by the monosized spheres without disturbing the basic packing. To make this possible, the smaller sphere has to be in the position already during packing of the larger spheres. This critical diameter ratio is named the "critical ratio of occupation." For the loosest square packing, the ratio can be calculated as (Cumberland and Crawford, 1987)

$$\frac{d_b}{d_M} = \frac{1}{0.732} = 1.366 \qquad (28)$$

For the tightest rhombohedral packing, there are two values:

$$\frac{d_b}{d_M} = \frac{1}{0.414} = 2.415 \qquad (29)$$

$$\frac{d_b}{d_M} = \frac{1}{0.225} = 4.444 \qquad (30)$$

where d_M is the particle diameter required during critical ratio of occupation.

4 PACKING OF NON-SPHERICAL PARTICLES

There has been very little theoretical and experimental work performed in this area owing to the complexity of such a system. Oman and Watson (1944) conducted experiments using particles of various shapes and found the voidage of the packing increased in the following order: cylinders, spheres, granules, Raschig

rings, and Berl saddles. Coulson (1949) studied packings of cubes, cylinders, and plates and found that the results depend on the height of fall of the particles. His results are shown in Table 3a. For random packed beds of uniform-sized particles, Brown (1966) suggested that the packing porosity depends on the sphericity of the particles and can be related as shown in Fig. 8 and Table 3b based on experimental findings. For small grains observable under the microscope, the sphericity can be obtained following Wadell (1935) and the average values obtained from the following equation after observing a sample of particles.

$$\phi = \frac{d_c}{D_c} \qquad (31)$$

where d_c = diameter of a circle equal in area to the projected area of the particle when resting on its larger face, D_c = diameter of the smallest circle circumscribing the projection of the particle.

When the packed bed is made up of a mixture of particles with different shapes, the sphericity is usually calculated from the arithmetic mean of the various sphericities.

Zou and Yu (1996) studied both loose and dense random packing of monosized nonspherical particles and found that the porosity was strongly dependent on both the particle shape and the packing method. The initial porosity of loose and dense random packings can be expressed as

For loose random packing:

$$\ln \varepsilon_{01,\text{cylinder}} = \phi^{5.58} \exp[5.89(1-\phi)] \ln 0.40 \qquad (32)$$

$$\ln \varepsilon_{01,\text{disk}} = \phi^{0.60} \exp[0.23(1-\phi)^{0.45}] \ln 0.40 \qquad (33)$$

For dense random packing,

$$\ln \varepsilon_{0d,\text{cylinder}} = \phi^{6.74} \exp[8.00(1-\phi)] \ln 0.36 \qquad (34)$$

$$\ln \varepsilon_{0d,\text{disk}} = \phi^{0.63} \exp[0.64(1-\phi)^{0.54}] \ln 0.36 \qquad (35)$$

The initial porosity of the nonspherical particles can be approximated by the proper use of the packing results of cylinders and disks shown in Eqs. (31) through (34) and expressed as

$$\varepsilon_0 = \frac{I_d}{I_c + I_d}\varepsilon_{0,\text{cylinder}} + \frac{I_c}{I_c + I_d}\varepsilon_{0,\text{disk}} \qquad (36)$$

where I_c is called the cylindrical index and I_d, the disk index. They are defined as follows.

$$I_c = |\phi - \phi_c| \qquad (37)$$

$$I_d = |\phi - \phi_d| \qquad (38)$$

Table 3a Rough Estimate of Bed Void Fraction

Particle of pellet	Normal charge	Dense packed	Multiplier for small tubes
Tablets	0.36	0.31	$1 + 0.43\, d_p/D$
Extrudates			
Short	0.40	0.33	$1 + 0.46\, d_p/D$
Long	0.46	0.40	$1 + 0.46\, d_p/D$
Spheres			
Uniform	0.40	0.36	$1 + 0.42\, d_p/D$
Mixed sizes	0.36	0.32	
Irregular	0.42 (average)	0.42 (average)	$1 + 0.3\, d_p/D$

Table 3b Voidage of Randomly Packed Beds with Uniformly Sized Particles Larger than 500 μm

	Voidage	
Sphericity	Loose packing	Dense packing
0.25	0.85	0.80
0.30	0.80	0.75
0.35	0.75	0.70
0.40	0.72	0.67
0.45	0.68	0.63
0.50	0.64	0.59
0.55	0.61	0.55
0.60	0.58	0.51
0.65	0.55	0.48
0.70	0.53	0.45
0.75	0.51	0.42
0.80	0.49	0.40
0.85	0.47	0.38
0.90	0.45	0.36
0.95	0.43	0.34
1.00	0.41	0.32

Source: Adapted from Brown, 1966.

where ϕ_c is an equivalent sphericity or aspect ratio defined as the ratio of the maximum length to the diameter of the circle having the same projected area normal to the maximum length, and ϕ_d, a sphericity defined as the ratio of the shortest length to the diameter of the circle having the same projected area.

4.1 Relationship Between the Hausner Ratio and the Sphericity

The Hausner ratio is defined as the ratio of tapped density to loose density, it is a measurement of the compressibility and cohesiveness of the powder.

Based on the work of Zou and Yu (1996), the Hausner ratio (HR) and the sphericity have the relationship

$$HR = 1.478 \cdot 10^{-0.136\phi} \tag{39}$$

5 FACTORS AFFECTING PACKING DENSITY IN PRACTICE

Particles, containers, and filling and handling methods contribute to packing density in practice. The important factors can be summarized as

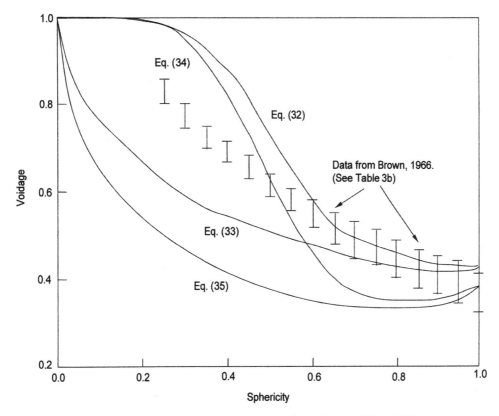

Figure 8   Relation between packing voidage and sphericity. (Adapted from Zou and Yu, 1996).

Particle shape, particle size, particle size distribution, particle coefficient of restitution, particle surface properties (friction)

Container shape, container size, container surface properties (friction)

Deposition method, deposition intensity, velocity of particle deposition

Vibratory compaction, pressure compaction

6   CORRELATIONS FOR FIXED BED BULK VOIDAGE

The voidage is the parameter most frequently employed to characterize the pressure drop in a fixed bed. For accurate determination, experimental technique such as the water displacement is usually used. In the application of catalytic reactors, the catalysis packing is important not only for pressure drop but also for heat and mass transfer. Furnas (1929) was probably the first to study the effect of wall on the packing of the particles. He proposed the following correlations for voidage close to the wall and at the core.

$$\varepsilon_{\mathrm{w}} = \left[1 + 0.6\left(\frac{d_{\mathrm{p}}}{D}\right)\right]\varepsilon - 0.6\left(\frac{d_{\mathrm{p}}}{D}\right) \qquad (40)$$

$$\varepsilon_{\mathrm{c}} = [\varepsilon + 0.3(1 - \varepsilon)]\left[1 + 0.6\left(\frac{d_{\mathrm{p}}}{D}\right)\right] - 0.6\left(\frac{d_{\mathrm{p}}}{D}\right) \qquad (41)$$

where ε is the average voidage experimentally measured or the ratio of void volume to the total volume of the packed bed. When the ratio of particle diameter to the bed diameter is less than 0.02, the correction for wall effect becomes negligible.

Figure 9 is a plot obtained by Benenati and Brosilow (1962) in a bed of spheres showing the radial oscillations of voidage away from the containing cylindrical wall. The observation was confirmed by Propster and Szekely (1977). Experiments performed by Goodling et al. (1983) also indicated that for uniformly sized spherical particles, the oscillations in voidage can be up to 5 particle diameters from the wall. The oscillations in voidage are down to 2 to 3 diameters from the wall, for a mixture of two spherical sizes, and down to 1 particle diameter from the wall for a mixture of three particle sizes.

Propster and Szekely (1977) also found that a marked local minimum in voidage existed in the inter-

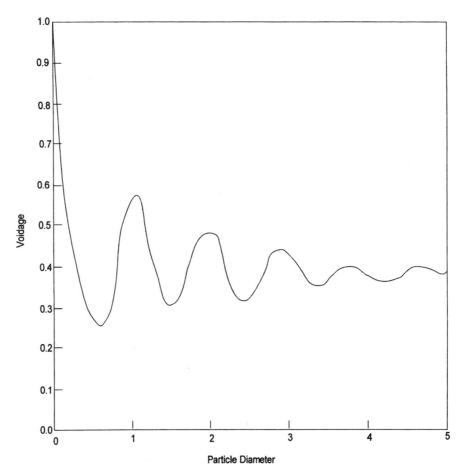

Figure 9 Schematic radial oscillations of voidage away from the containing cylindrical wall. (Adapted from Propster and Szekely, 1977.)

facial region when small particles were placed on top of a bed of larger particles because of penetration of small particles into the interstices of large particles. However, when the large particles were placed on top of the smaller particles, the local minimum in voidage was not as pronounced. When the particle ratio of the lower layer to the upper layer is less than 2, relatively little particle penetration was observed. Ready penetration of small particles could be observed when the ratio equals 6, corresponding to a critical ratio of entrance of 6.464 for the tightest rhombohedral packing (see Sec. 3). This observation will have practical application in the iron blast furnace or cupola where alternate layers of coke and iron ore of different particle sizes are charged into the reactors.

Fixed beds of very low tube-to–particle diameter ratio have also been proposed and studied. For these reactors, the effect of wall and particle shape on bulk voidage becomes important. It will be convenient to

have correlations for fixed bed bulk voidage for commonly used catalysts such as spheres, cylinders, and rings. The following correlations are those proposed by Dixon (1988). A more comprehensive review was conducted by Haughey and Beveridge (1969).

For spherical particles

$$\varepsilon = 0.4 + 0.05\left(\frac{d_{\mathrm{p}}}{D}\right) + 0.412\left(\frac{d_{\mathrm{p}}}{D}\right)^2 \qquad \frac{d_{\mathrm{p}}}{D} \leq 0.5 \quad (42)$$

$$\varepsilon = 0.528 + 2.464\left(\frac{d_{\mathrm{p}}}{D} - 0.5\right) \qquad 0.5 \leq \frac{d_{\mathrm{p}}}{D} \leq 0.536$$

$$(43)$$

$$\varepsilon = 1 - 0.667\left(\frac{d_{\mathrm{p}}}{D}\right)^3 \left[2\left(\frac{d_{\mathrm{p}}}{D}\right) - 1\right]^{-0.5} \qquad \frac{d_{\mathrm{p}}}{D} \geq 0.536$$

$$(44)$$

For full cylinders

$$\varepsilon = 0.36 + 0.10\left(\frac{d_{pv}}{D}\right) + 0.7\left(\frac{d_{pv}}{D}\right)^2 \qquad \frac{d_{pv}}{D} \leq 0.6$$

$$\text{(45)}$$

$$\varepsilon = 0.677 - 9\left(\frac{d_{pv}}{D} - 0.625\right)^2 \qquad 0.6 \leq \frac{d_{pv}}{D} \leq 0.7$$

$$\text{(46)}$$

$$\varepsilon = 1 - 0.763\left(\frac{d_{pv}}{D}\right)^2 \qquad \frac{d_{pv}}{D} \geq 0.7 \qquad \text{(47)}$$

where d_{pv} is the diameter of a sphere having the same volume as the cylinder.

For hollow cylinders

$$(1 - \varepsilon_{hc}) = \left[1 + 2\left(\frac{d_{yi}}{d_{yo}} - 0.5\right)^2\left(1.145 - \frac{d_{pv}}{D}\right)\right]$$
$$\left(1 - \frac{d_{yi}^2}{d_{yo}^2}\right)(1 - \varepsilon_{sc}) \qquad \frac{d_{yi}}{d_{yo}} \geq 0.5 \qquad \text{(48)}$$

where subscripts hc and sc denote the hollow and solid cylinders respectively, and d_{yi} and d_{yo} are the inside and outside diameter of the hollow cylinder, respectively.

7 FLOW THROUGH PACKED BEDS

There have been numerous investigations into flow through packed beds. Scheidegger (1961) critically reviewed the earlier models. A more recent review is by Molerus (1993). There were two main approaches, the discrete particle model and the pipe flow analogy. Both approaches gave reasonable predictions of pressure drop of flow through packed beds of spherical and near-spherical particles but were inadequate for beds with particles substantially different from the spheres. The pressure drop through the packed bed is due not only to the frictional resistance at the particle surface but also to the expansion and contraction of flow through the interstices among the particles

The most popular approach is the pipe flow analogy model, also called the capillary tube model or channel model, which approximates the flow through the packed bed by the flow through a bundle of straight capillaries of equal size. Further refinement produced the constricted tube model. In this model, an assembly of tortuous channels of varying cross sections simulates the varying dimensions and curvatures of pores in the packed bed. The major contributions following this approach include Blake (1922), Kozeny (1927),

Carman (1937), and Ergun (1952). The discrete particle model assumes that the packed bed consists of an assembly of discrete particles that possess their own boundary layer during the flow through the packed bed. The primary developments are due to Burke and Plummer (1928), Ranz (1952), Happel (1958), Galloway and Sage (1970), and Gauvin and Katta (1973). Conceptually, the discrete particle model is closer to the physical description of the flow through the packed bed, but the pipe flow analogy is historically more widely employed. The correlations developed through the pipe flow analogy are usually applicable for particles with sphericities larger than about 0.6. The correlations developed through the discrete particle model have wide applications, including particles of sphericities less than 0.6.

7.1 Darcy's Law

The theory of laminar flow through homogeneous porous media is based on a classical experiment originally performed by Darcy in 1856. Darcy's experiment is described in Fig. 10. The total volume of the fluid percolating through the fixed bed, Q, can be expressed in terms of the height of the bed and the bed area as

$$Q = -\frac{KA(h_2 - h_1)}{h} \qquad \text{(49)}$$

where K is a constant depending on the properties of the fluid and of the porous medium. The minus sign indicates that the flow is in the opposite direction of increasing height, h. Darcy's law can be restated in terms of the pressure P and the density ρ_f of the liquid. Assuming that the liquid density is constant, we have

$$Q = -\frac{K'A(P_2 - P_1 + \rho_f gh)}{h} \qquad \text{(50)}$$

and

$$P_1 = \rho_f g(h_1 - z_1) \qquad P_2 = \rho_f g(h_2 - z_2) \qquad \text{(51)}$$

Equations (49) and (50) are equivalent statements of Darcy's law. It is valid for a wide domain of flows. The flow of newtonian fluid at low Reynolds number is known to follow Darcy's law. Thus it is valid for arbitrary small pressure differentials for liquids. It has been used to measure flow rates by determining the pressure drop across a fixed porous bed. For liquids at high velocities and for gases at very low and very high velocities, Darcy's law becomes invalid.

Darcy's law in its original form is rather restricted in its usefulness. The physical significance of the constant K', known as the permeability constant and having

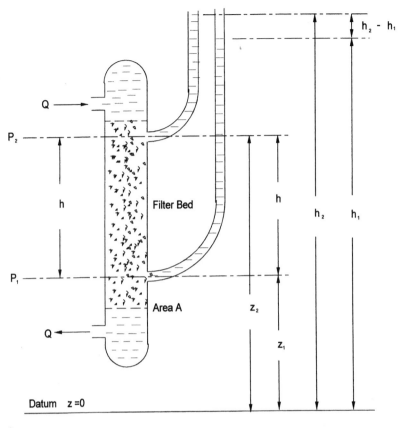

Figure 10 Darcy's experiment.

dimensions $M^{-1}L^3T^1$, has to be elucidated. The dependence of K' on the porous medium and on the liquid has to be separated for practical applications. Nutting (1930) proposed that

$$K' = \frac{k_s}{\mu} \tag{52}$$

where k_s is the specific permeability, having dimensions L^2, and the darcy is used as a unit for specific permeability (1 darcy $= 9.87 \times 10^{-9}\,\text{cm}^2$). By letting h becomes infinitesimal, Darcy's law can be written in the differential form

$$\vec{q} = \frac{Q}{A} = -\left(\frac{k_s}{\mu}\right)(\text{grad } P - \rho_f\,\vec{g}) \tag{53}$$

The differential form of Darcy's law is by itself not sufficient to determine the flow pattern in a porous medium for given boundary conditions, as it contains three unknowns, $\vec{q}$, P, and ρ_f. Two more equations are required to specify the problem completely. One is the dependence of the density on the pressure:

$$\rho_f = \rho_f(P) \tag{54}$$

and the other is the continuity equation

$$-\varepsilon\frac{\partial \rho_f}{\partial t} = \text{div}\left(\rho_f\,\vec{g}\right) \tag{55}$$

By eliminating all unknowns except the pressure, we have

$$\varepsilon\frac{\partial \rho_f}{\partial t} = \text{div}\left[\left(\frac{\rho_f k_s}{\mu}\right)(\text{grad } P - \rho_f\,\vec{g})\right] \tag{56}$$

7.2 Blake's Correlation

Blakes (1922) may be the first one to suggest using a modified dimensionless group incorporating the voidage, ε, in a particulate system. He proposed the use of the interstitial velocity, instead of the superficial velocity, and the reciprocal of the total particle surface area per unit volume, instead of the particle diameter, as the characteristic length. The Reynolds number and friction factor proposed by Blake are

$$(\text{Re})_B = \frac{Ud_p\rho_f}{\mu(1-\varepsilon)} \qquad (57)$$

$$f_B = \frac{\Delta P}{L}\frac{d_p}{\rho_f U^2 g}\frac{\varepsilon^3}{(1-\varepsilon)} \qquad (58)$$

These dimensionless groups have been used ever since as the basis for the pipe flow analogy in almost all packed bed correlations.

From dimensionless analysis, the pressure drop through a packed bed of particles can be obtained as

$$\frac{\Delta P}{L} \propto \mu^{2-n}\rho_f^{n-1}\left(\frac{U}{\varepsilon}\right)^n \underline{D}^{n-3} \qquad (59)$$

The interstitial fluid velocity in the average direction of fluid motion is used here. The $\underline{D}$ is a length analogous to the hydraulic radius of a conduit and is defined as

$$\underline{D} = \frac{\text{Mean cross-sectional area of flow}}{\text{Mean wetted perimeter of flow channels}} \qquad (60)$$

Multiplying the numerator and the denominator by L, the height of the bed, we have

$$\underline{D} = \frac{(\text{Total bed volume}) \times (\text{voidage})}{(\text{Total bed surface})} \qquad (61)$$

or

$$\underline{D} = \frac{\varepsilon}{S} = \frac{\varepsilon d_p}{6(1-\varepsilon)} \qquad (62)$$

where S is the total surface of solids per unit bed volume assuming spherical particles, it can be expressed as

$$S = \frac{6(1-\varepsilon)}{d_p} \qquad (63)$$

Equation (63) is essentially similar to Eq. (4). Substituting into Eq. (59), we obtain

$$\frac{\Delta P}{L} \propto \mu^{2-n}\rho_f^{n-1}U^n\frac{(1-\varepsilon)^{3-n}}{\varepsilon^3}d_p^{n-3} \qquad (64)$$

For conventional pressure drop through a packed bed of particles, $n = 2$. Equation (64) becomes

$$\frac{\Delta P}{L} = \frac{2f\rho_f U^2}{g d_p} \qquad (65)$$

The friction factor, f, is proportional to

$$f \propto \mu^{2-n}\rho_f^{n-2}d_p^{n-2}U^{n-2}\frac{(1-\varepsilon)^{3-n}}{\varepsilon^3} \qquad (66)$$

or

$$f \propto \frac{(1-\varepsilon)^{3-n}}{\varepsilon^3}\cdot\frac{1}{(\text{Re})_p^{2-n}} \qquad (67)$$

At low Reynolds numbers, where the pressure drop does not depend on the fluid density, $n = 1$, and

$$f \propto \frac{(1-\varepsilon)^2}{\varepsilon^3(\text{Re})_p} \qquad (68)$$

At high Reynolds numbers where the viscosity is not important, $n = 2$, and

$$f \propto \frac{(1-\varepsilon)}{\varepsilon^3} \qquad (69)$$

Because of the large dependence on the voidage, a 30-to-70-fold increase in f is predicted as the voidage changes from 0.3 to 0.7 in the turbulent and viscous ranges, respectively. The expression found above for f at low Reynolds numbers has been experimentally proved to be correct. At high Reynolds numbers, however, the expression seems to predict a somewhat stronger dependence of the friction factor on the voidage than is actually found experimentally. For a single particle, voidage becomes 1 by definition, and the Blake's Reynolds number approaches infinity. This is not surprising, since the Blake analogy based on a capillary flow analogy breaks down in this range, because such an analogy has no physical meaning.

7.3 The Brownell and Kats Correlation

Another correlation incorporating the voidage is that by Brownell and Katz (1947). They introduced a Reynolds number defined as

$$(\text{Re})_{p\varepsilon} = \frac{Ud_p\rho_f}{\mu\varepsilon^m} \qquad (70)$$

where the exponent m depends on the ratio of the sphericity to the porosity and ranges from 2 to 20. The use of dimensionless analysis in correlating the data is justified as long as the identified dimensionless groups represent true similarity. The Brownell and Kats correlation clearly does not represent true similarity, because it predicts that the Reynolds number decreases with increasing voidage, contradicting the experimental findings.

7.4 The Carman and Kozeny Correlations

Carman (1937) studied extensively the fluid flow through various packings in the viscous range and found that

$$f = \frac{90(1-\varepsilon)^2}{\varepsilon^3 (\mathrm{Re})_p} \quad \text{for} \quad 0.26 < \varepsilon < 0.89 \quad (71)$$

He also found that Eq. (71) can be applied to other regular shapes as long as their surfaces can be determined accurately and provided that d_p is expressed as $6V_p/S_p$. V_p is the particle volume, and S_p is the surface of the particle. For mixtures of various sizes and shapes, V_p and S_p should be taken as averages of all particles in the bed.

The Kozeny (1927) equation is usually expressed as, by combining Eqs. (65) and (71),

$$\frac{\Delta P}{L} = \frac{180(1-\varepsilon)^2 \mu U}{g \varepsilon^3 d_p^2} = \frac{5(1-\varepsilon)^2 \mu U}{g \varepsilon^3 (V_p/S_p)^2} \quad (72)$$

Kozeny derived the equation by assuming that a granular bed is equivalent to a group of parallel similar channels, commonly known as the pipe flow analogy. He started from the general equation for streamline flow through a uniform channel and assumed that the hydraulic radius R_h of the channel is ε/S. This is equivalent of assuming that the total internal surface and the total internal volume of the group of parallel similar channels are equal to the particle surface and the void volume of the bed. To find the ratio of the surface and volume of the particles, Carman proposed to measure the pressure drop through the bed of the irregular particles experimentally. The specific surface, S_p/V_p, can then be evaluated from

$$\frac{S_p}{V_p} = \sqrt{\frac{g \Delta P \varepsilon^3}{5(1-\varepsilon)^2 \mu U L}} \quad (73)$$

The pipe flow analogy breaks down beyond the creeping flow range. Thus beyond the creeping flow range, purely empirical models must be employed, the only remaining link with the pipe flow is the structure of the dimensionless groups.

7.5 The Ergun Correlation

There are many other correlations such as that by Oman and Watson (1944) and Happel's correlation (1958). The most widely used empirical correlation of this type is by Ergun (1952) employing Blake's definition of drag coefficient and the Reynolds number, as shown in Eqs. (57) and (58).

$$\frac{\Delta P}{L} \cdot \frac{g d_p \phi}{2 \rho_f U^2} \cdot \frac{\varepsilon^3}{(1-\varepsilon)} = 75 \frac{(1-\varepsilon)}{\phi(\mathrm{Re})_p} + 0.875 \quad (74)$$

For general applications, including irregular particles, the Ergun equation shown in Eq. (74) is expressed with sphericity by substituting ϕd_p for d_p, where d_p is the diameter of the irregular particle obtained by particle measurement techniques, such as sieving or the Coulter counter, described in Chapter 1. It can be seen that the Ergun equation reduces to the Blake–Kozeny–Carman equation at low Reynolds number, and at high Reynolds number, to the Burke–Plummer equation for turbulent flow. Many extensions and modifications of the Ergun equation have been proposed, such as that by Handley and Heggs (1968), Hicks (1970), Tallmadge (1970), Leva and coworkers (1947), and Rose and Rizk (1949). Among them the correlations suggested by Tallmadge and Leva et al. are for high Reynolds numbers, where the Ergun equation fails to fit the data well.

7.5.1 Use of Ergun Equation to Determine Sphericity Factor

Subramanian and Arunachalam (1980) suggested a technique making use of the Ergun equation to determine the sphericity of irregular particles. For very low flow rates, the viscous forces predominate, and the Ergun equation in Eq. (74) can be rearranged explicitly for the sphericity as

$$\phi = \left[\frac{150L(1-\varepsilon)^2 B\mu/(d_p^2 \varepsilon^3 \rho_f g)}{(1 - L/H_1)} \right]^{1/2} \quad (75)$$

with

$$B = \frac{\ln H_0 - \ln H_1}{t} \qquad H_0 > H_1 > L \quad (76)$$

By carrying out the experiment with a packed bed of irregular particles and by draining the liquid in a laminar flow from the height H_0 to H_1, the time, t, required can be determined. Substituting into Equations (75) and (76), the sphericity of the irregular particles can be obtained. A simple and accurate device was described in Subramanian and Arunachalam (1980).

7.6 Modified Ergun Equation

Gibilaro et al. (1985) modified the Ergun equation and proposed an alternative pressure drop equation on the basis of theoretical considerations:

$$\Delta P = \left(\frac{17.3}{(\text{Re})_p} + 0.336\right) \frac{\rho_f U^2 L}{d_p}(1 - \varepsilon)\varepsilon^{-4.8} \qquad (77)$$

The proposed equation compared well with published experimental data obtained from high-voidage fixed beds of spheres; it represented a significant improvement over that of Ergun. The equation is a combination of two equations, one for the laminar regime and the other for fully turbulent flow. The laminar flow regime equation is derived to match the Blake–Kozeny equation at ε equal to 0.4 and can be expressed as

$$\Delta P = \left(\frac{17.3}{(\text{Re})_p} + 0.336\right) \frac{\rho_f U^2 L}{d_p}(1 - \varepsilon)\varepsilon^{-4.8} \qquad (78)$$

Equation (78) gives an accurate prediction of fluidized bed expansion characteristics for the laminar regime and applies equally well to fixed and suspended particle systems.

The equation for the fully turbulent flow regime was also derived as

$$\Delta P = 0.336 \frac{\rho_f U^2 L}{d_p}(1 - \varepsilon)\varepsilon^{-4.8} \qquad (79)$$

The constant 0.336 is the result of matching $\varepsilon = 0.4$ in the Burke–Plummer (1928) equation, Eq. (80), which describes the normal packed bed pressure drop well:

$$\Delta P = 1.75 \frac{\rho_f U^2 L}{d_p} \frac{(1 - \varepsilon)}{\varepsilon^3} \qquad (80)$$

Equation (79) well represents the steady-state expansion characteristics of a turbulent regime in fluidized beds.

7.7 General Friction Factor Correlations

The pressure drop equations can be converted to general friction factor correlations as follows. From the Ergun equation, the friction factor correlations will be

$$f_e = \frac{\Delta P\, d_p \varepsilon^3}{\rho_f U^2 L(1 - \varepsilon)} = \frac{150(1 - \varepsilon)}{(\text{Re})_p} + 1.75 \qquad (81)$$

The friction factor correlation resulting from the equation by Gibilaro et al. (1985) becomes

$$f_p = \frac{\Delta P\, d_p \varepsilon^{4.8}}{\rho_f U^2 L(1 - \varepsilon)} = \frac{17.3}{(\text{Re})_p} + 0.336 \qquad (82)$$

Equation (82) gives a significantly better representation for data at higher Reynolds numbers, where Ergun equation consistently overestimates the observed friction factor. Wentz and Thodos (1963) also proposed a general equation for friction factor for packed and distended beds of spheres:

$$f_w = \frac{0.351}{\text{Re}^{0.05} - 1.2} \qquad (83)$$

Equation (83) is good for voidage between 0.354 and 0.882 and Reynolds numbers between 2,600 and 64,900.

7.8 Drag Coefficient for a Particle in an Array

The drag coefficient for a particle in an array was also derived by Gibilaro et al. (1985) as

$$C_{D1} = \frac{4}{3} f_p \varepsilon^{-3.8} \cong C_D \varepsilon^{-3.8} \qquad (84)$$

where C_D = drag coefficient for a particle in an infinite fluid.

Equation (84) provides accurate quantitative predictions of particulate fluidized bed expansion characteristics in both laminar and turbulent flow regimes. For the intermediate flow regime, only a qualitative trend was observed. Equation (84) is slightly different in dependence of voidage from the drag coefficient suggested by Wen and Yu (1966) as shown here:

$$C_{D1} = C_D \varepsilon^{-4.8} \qquad (85)$$

7.9 The General Correlation by Barnea and Mednick

A general correlation for the pressure drop through fixed beds of spherical particles, based on a discrete particle model corrected for particle interaction, was proposed by Barnea and Mednick (1978). They extended the standard C_D versus Re curve for single spheres to multiparticle systems by incorporating proper functions of the volumetric particle concentration. The modified drag coefficient and Reynolds number they suggested are

$$(\text{Re})_\phi = (\text{Re})_p \left\{ \frac{1}{\varepsilon \exp\left[\frac{5(1 - \varepsilon)}{3\varepsilon}\right]} \right\} \qquad (86)$$

$$(C_D)_\phi = \frac{8f\varepsilon^2}{3(1-\varepsilon)\left[1 + K(1-\varepsilon)^{1/3}\right]};$$

$$f = \frac{\Delta P}{L} \cdot \frac{g d_p}{2\rho_f U^2} \tag{87}$$

and

$$(C_D)_\phi = \left[0.63 + \frac{4.8}{(Re)_\phi}\right]^2 \tag{88}$$

where K is a constant

The correlation allows the prediction of pressure drop or drag in single-particle and multiparticle system with a single curve. The correlation provides good agreement with data in the creeping flow and intermediate regimes. Deviation is observed in the turbulent regime. For the data in the distended bed, the correlation also provides a good fit in the highly turbulent range. The range of applicability may be extended by the application of average particle size definitions, shape factors, and wall effect correlations.

7.10   The Concept of Stagnant Voidage

Happel (1958) introduced the interesting concept of a stagnant voidage, a part of the voidage in the packed bed that is occupied by the wake of the particles and is thus not available for the fluid flow. The concept is primarily employed in the discrete particle model but may be useful for other modeling effort.

Kusik and Happel (1962) derived this equation for estimating the stagnant voidage for the packing of spheres:

$$\varepsilon_s = 0.75(1-\varepsilon)(\varepsilon - 0.2) \tag{89}$$

Gauvin and Katta (1973) suggested a slightly different equation for a bed packed with spheres:

$$\varepsilon_s = 1.6(1-\varepsilon)(\varepsilon - 0.2) \tag{90}$$

They also proposed equations for packing of other isometric particles.

For packing of cylinders,

$$\varepsilon_s = 1.95 K_s(\varepsilon - 0.2)(1-\varepsilon) \tag{91}$$

For packing of ellipsoids,

$$\varepsilon_s = 2.5 K_s(\varepsilon - 0.2)(1-\varepsilon) \tag{92}$$

For packing of prisms and wafers,

$$\varepsilon_s = 0.93 K_d \varepsilon(1-\varepsilon) \tag{93}$$

For wood chips,

$$\varepsilon_s = 0.803\varepsilon(1-\varepsilon) \tag{94}$$

where K_s is the ratio of the mean projected area of a particle (sphere, cylinder, ellipsoid) to that of a sphere of the same volume, and K_d is the ratio of the mean projected area of a prism to that of a disc with the base having the same area as the larger face of the prism.

Galloway and Sage (1970) reported that the stagnant voidage, ε_s, varied from 0.172 to 0.157 when the Reynolds number was varied from 10,000 to 35,000 during their experiments.

7.11   Permeability of Packed Beds

Permeability of a packed bed can usually be estimated from the rearranged form of Kozeny–Carman equation:

$$\frac{R^2}{k} = \frac{75(1-\varepsilon)^2}{2\varepsilon^3} \tag{95}$$

where R is the radius of particle and k is the permeability.

Equation (95) has been found to provide a good estimation of permeability for packed beds of voidage between 0.26 and 0.80. Carman (1956) also found that if the hydraulic radius was used to replace the particle radius in Eq. (95), the equation was also good for mixtures of different particle sizes. The hydraulic radius for a bed of spherical particles can be calculated from

$$R_h = \frac{\varepsilon d_p}{6(1-\varepsilon)} \tag{96}$$

To account for wall effect, Mehta and Hawley (1969) modified the equation for hydraulic radius to give

$$R_h = \frac{\varepsilon d_p}{6(1-\varepsilon)M} \tag{97}$$

and

$$M = 1 + \frac{4 d_p}{6D(1-\varepsilon)} \tag{98}$$

where D is the diameter of packed column.

The permeability of a packed bed with polydisperse spherical or nonspherical particles can also be estimated using Eq. (95) with particle size calculated through the harmonic mean or the Sauter mean if the size distribution is not very broad. For wide size distributions, Li and Park (1998) have proposed equations for calculating the permeability for both spherical and nonspherical particles.

8 GAS VELOCITY DISTRIBUTION IN PACKED BEDS

Because of the uneven distribution of voidage across the packed bed created during the filling process, the radial gas velocity distribution in the packed bed is neither parabolic, as in an empty pipe, nor uniform. As pointed out earlier, the voidage close to the wall is usually higher, owing to the wall effect; the gas flow tends to be higher close to the wall. Schwarz and Smith (1953) measured the radial gas velocity profiles in a 2 in. pipe and a 4 in. pipe filled with 1/8 in. cylinders and found it indeed was the case (see Fig. 11). When the packed bed diameter becomes larger, the influence of the wall effect decreases. Thus for a large-diameter bed, the assumption of uniform radial velocity distribution can be a first approximation. Theoretical prediction of radial velocity distribution is not possible for a randomly packed bed. The radial gas distribution can also be profoundly affected by the design of gas inlet and outlet regions of the packed bed due to bypassing. Szekely and Poveromo (1975) employed the vectorial differential form of the Ergun equation to predict the flow maldistribution in a packed bed. The experimental measurements were found to be in reasonable agreement. They also found that a uniformly packed bed with a height/diameter ratio larger than one could also be used as a flow straightener

because it evened out the nonuniformities introduced upstream of the bed.

Cohen and Metzner (1981) indicated that for newtonian fluids, wall effect corrections could be neglected if the bed-to-particle-diameter ratio is larger than 30. For nonnewtonian fluids, it is 50. They also proposed a model dividing the bed into three regions—a wall region, a transition region, and a bulk region. The wall region extends a distance of one particle diameter from the wall. The transition can be up to six particle diameters from the wall where appreciable voidage oscillation occurs, as discussed in Sec. 6. In the bulk region, the remainder of the bed region, the voidage is essentially constant. For packed beds of small bed:-to:diameter ratios, the use of the single-region model based on the average voidage tends to over-predict the average mass flux. The triregional model proposed by Cohen and Metzner (1981) fitted the experimental data much better. For bed-to-diameter ratios less than 30, the transition region represent the largest fraction of the total bed cross-sectional area. The mass fluxes in the wall and transition regions were estimated to be larger than that of the bulk region by as much as 50% for fluids with power indexes of 0.25 and by 10% for fluids with power indexes of 1. Saunders and Ford (1940), employing a pitot tube, found the velocity in a ring about one particle diameter from the wall about 50% higher than the bulk gas velocity. Schwartz and Smith

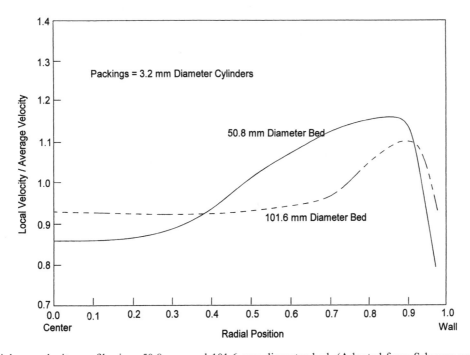

Figure 11 Radial gas velocity profiles in a 50.8 mm and 101.6 mm diameter bed. (Adapted from Schwarz and Smith, 1953.)

(1953), in a bed with a bed-to-particle-diameter ratio less than 30, also found a peak velocity at about one particle diameter away from the wall approximately 30% to 100% higher than the bulk velocity. For bed-to-particle-diameter ratio larger than 30, the bulk region has the largest cross-sectional area and thus has the largest contribution to the flow. The wall region is actually larger than the bulk region for packed beds with the bed to diameter ratios less than 18. This physical division is an important consideration for packed bed design in the laboratory and for attempting to use the laboratory data for scale-up. Since the voidage in the bulk region is constant and independent of the bed-to-particle-diameter ratio, its average flux decreases with increasing bed to a particle diameter ratio.

9   HEAT TRANSFER IN PACKED BEDS

As for the solid–fluid heat transfer coefficient for flow through a randomly packed bed, large variations of up to two- to five-fold have been observed (Barker, 1965). The primary difficulties have been the attempt to model the packed bed with simple average parameters that describe the complex local variations in packing voidage and the effects of particle shape, distribution, and velocity. Heat transfer in gas–solid packed bed systems was critically reviewed by Balakrishnan and Pei (1979). The overall heat transfer in the packed bed is quite complex and consists of the following mechanisms: (1) the conduction heat transfer between particles in both radial and axial directions, (2) the convective heat transfer between the bed particle and the flow gas, (3) the interaction of mechanisms (1) and (2), (4) heat transfer due to radiation between the bed particles, between the particles and the flowing gas, and between the flowing gas and the bed wall, and (5) heat transfer between the bed wall and bed particles. Most of the studies in the literature were directed toward developing correlations for the total heat transfer rate in terms of dimensionless parameters such as Reynolds number.

9.1   Rase Heat Transfer Correlations

Based on the recommendations by Rase (1990), the heat and mass transfer correlations to be used for packed bed calculations are summarized here.

9.1.1   Particle to Fluid Heat Transfer

Heat transferred to a single particle can be expressed as

$$q_p = h_s a_p (T_s - T) \tag{99}$$

The heat transfer coefficient can be evaluated from the equation

$$\frac{h_s d_p}{k_g} = 2 + 1.1 Pr^{1/3} (\text{Re})_p^{0.6} \qquad \text{accuracy to 25\%} \tag{100}$$

where $(\text{Re})_p = d_p U \rho_f / \mu$ and d_p is the diameter of a sphere or an equivalent sphere. U is the superficial velocity in the packed bed.

9.1.2   Heat Transfer Through Wall—One-Dimensional Model—Axial

For a cylindrical vessel with spherical particle packing,

$$q = h_i A_i (T - T_w) \tag{101}$$

where A_i = inside surface of the cylindrical vessel and T_w = wall temperature of the fluid at axial position of interest. For the homogeneous model, T of fluid and of bed are assumed identical.

$$\frac{h_i D}{k_g} = 2.26 (\text{Re})_p^{0.8} Pr^{0.33} \exp\left(-\frac{6 d_p}{D}\right)$$

$$20 \le (\text{Re})_p \le 7600 \text{ and } 0.05 \le \frac{d_p}{D} \le 0.3 \tag{102}$$

For a cylindrical vessel with cylindrical particles packing,

$$\frac{h_i D}{k_g} = 1.40 (\text{Re})_p^{0.95} Pr^{0.33} \exp\left(-\frac{6 d_p}{D}\right)$$

$$20 \le (\text{Re})_p \le 800 \text{ and } 0.03 \le \frac{d_p}{D} \le 0.2 \tag{103}$$

9.1.3   Wall Heat Transfer Coefficient—Two-Dimensional

$$q = h_w A_i (T_R - T_w) \tag{104}$$

were T_R = temperature at inside radius of the vessel. For spherical particle packing,

$$\frac{h_w d_p}{k_g} = 0.19 (\text{Re})_p^{0.79} Pr^{0.33}$$

$$20 \le (\text{Re})_p \le 7600 \text{ and } 0.05 \le \frac{d_p}{D} \le 0.3 \tag{105}$$

For cylindrical particle packing,

$$\frac{h_w d_p}{k_g} = 0.18(\text{Re})_p^{0.93} Pr^{0.33}$$

$$20 \leq (\text{Re})_p \leq 800 \quad \text{and} \quad 0.03 \leq \frac{d_p}{D} \leq 0.2 \tag{106}$$

9.1.4 Effective Radial Thermal Conductivity

$$\frac{q}{A_i} = k_g \left(\frac{\partial T}{\partial r}\right)_{r=R} \tag{107}$$

$$\frac{h_w d_p}{k_g} \frac{\varepsilon}{1-\varepsilon} = 0.27 \quad 500 < \frac{d_p U \rho_f}{\mu(1-\varepsilon)} < 6000$$

$$\text{and} \quad 0.05 < \frac{d_p}{D} < 0.15 \tag{108}$$

9.2 Heat Transfer Correlations Recommended by Molerus and Wirth (1997)

In a recent book by Molerus and Wirth (1997), the recommended heat transfer correlations for packed beds can be summarized as follows. For fully developed laminar flow, an approximation formula for the mean Nusselt number, derived from the pipe flow analogy, was proposed as

$$(\text{Nu})_D = \frac{hD_p}{k_g} = \sqrt[3]{3.66^3 + 1.61^3 (\text{Pe})_p \frac{D_p}{L_p}}$$

$$\text{for} \quad 0.1 < \frac{(\text{Pe})_p D_p}{L_p} < 10^4 \tag{109}$$

where $(\text{Pe})_p$ is the pipe flow Peclet number and L_p is the pipe length, equivalent to the bed depth. The D_p and $(\text{Pe})_p$ can be expressed as

$$D_p = \frac{2\varepsilon}{3(1-\varepsilon)} d_p \tag{110}$$

$$(\text{Pe})_p = \frac{\rho_f C_{pg} D_p U}{k_g} \tag{111}$$

where U is superficial flow velocity.

For heat transfer from a single particle inside a particle array,

$$\text{Nu} = \frac{h d_p}{k_g} = 5.49 \left(\frac{1-\varepsilon}{\varepsilon}\right) \tag{112}$$

For voidage values between 0.35 and 0.5, Eq. (112) gives a range of Nusselt numbers between 5 and 10.

For fluid heating by percolating through a particle array,

$$\frac{(\text{Pe})_d d_p}{L_p} = 7.83 \frac{(1-\varepsilon)^2}{\varepsilon} \tag{113}$$

where

$$(\text{Pe})_d = \frac{\rho_f C_{pg} d_p U}{k_g} \tag{114}$$

Equation (113) indicates that the heat transfer between the particles and the percolating fluid is very fast. Assuming $\varepsilon = 0.4$ and $L_p = n d_p$ and substituting into Eq. (113) we have

$$n = 0.14(\text{Pe})_d \tag{115}$$

For $(\text{Pe})_d < 10$, Eq. (115) implies that just one particle layer is enough to heat the percolating gas to the surface temperature of the particles.

9.3 Analytical Models for Heat Transfer with Immersed Surfaces

When the heat removal is entirely due to the flowing gas and there is no convective particle movement, Gabor (1970) proposed the following simple model. For heat transfer from a flat plate of length L_h immersed in the packed bed,

$$\frac{\partial T}{\partial z} = \frac{k_e}{C_{pg} G} \frac{\partial^2 T}{\partial y^2} \tag{116}$$

with the boundary conditions

$$T = T_w \quad \text{at} \quad y = 0 \quad \text{and} \quad 0 < z < L_h \tag{117}$$

$$T = T_o \quad \text{at} \quad y = \infty \tag{118}$$

$$T = T_o \quad \text{at} \quad z = 0 \tag{119}$$

For heat transfer from a cylindrical heater of length L_h and radius R_y immersed in the packed bed,

$$\frac{\partial T}{\partial z} = \frac{k_e}{C_{pg} G} \left(\frac{\partial^2 T}{\partial r^2} + \frac{1}{r} \frac{\partial T}{\partial r}\right) \tag{120}$$

with the boundary conditions

$$T = T_w \quad \text{at} \quad r = R_y \quad \text{and} \quad 0 < z < L_h \tag{121}$$

$$T = T_o \quad \text{at} \quad r = \infty \tag{122}$$

$$T = T_o \quad \text{at} \quad z = 0 \tag{123}$$

The solutions for the average heat transfer coefficients for the flat plate and cylinder cases were found by Gabor (1970).

For flat plate case,

$$h_{av} = \sqrt{\frac{4}{\pi} \frac{k_e C_{pg} G}{L_h}} \tag{124}$$

For the cylindrical heater case,

$$h_{av} = \sqrt{\frac{4}{\pi} \frac{k_e C_{pg} G}{L_h}} + \frac{1}{2} \frac{k_e}{R_y} \tag{125}$$

The effective thermal conductivity of the packed bed, k_e, can be estimated from

$$k_e = k_e^o + 0.1\left(C_{pg} d_p G\right) \tag{126}$$

The effective thermal conductivity of the packed bed expressed in Eq. (126) includes two terms, the conductivity term with no gas flow, k_e^o, and the convective term. The factor 0.1 was recommended by Yagi and Kunii (1957) for spherical particles. The conductivity term with no gas flow can be calculated below, following that suggested by Swift (1966) for orthorhombic particle packing with a voidage of 0.395.

$$k_e^o = 0.9065\left[\frac{2}{(1/k_g) - (1/k_s)}\right]\left[\frac{k_s}{k_s - k_g}\left(\ln\frac{k_s}{k_g}\right) - 1\right]$$
$$+ 0.0935 k_g \tag{127}$$

Botterill and Denloye (1978) conducted an even more detailed analysis of the heat transfer between an immersed cylindrical heater and the packed bed by dividing the heat transfer into two regions, a region of higher voidage within half a particle diameter from the immersed cylindrical surface, and the region of constant voidage outside the wall region. The radial transfer of heat for the wall region (region 1) is governed by

$$\rho_f U_1 C_{pg} \frac{\partial T_1}{\partial z} = k_{e1}\left(\frac{\partial^2 T_1}{\partial r^2} + \frac{1}{r}\frac{\partial^2 T_1}{\partial r^2}\right) \quad \text{at} \quad z > 0$$

$$\text{and} \quad R_t < r < \left(R_t + \frac{d_p}{2}\right) \tag{128}$$

In the region outside the wall region, region 2, the governing equation is

$$\rho_f U_2 C_{pg} \frac{\partial T_2}{\partial z} = k_{e2}\left(\frac{\partial^2 T_2}{\partial r^2} + \frac{1}{r}\frac{\partial^2 T_2}{\partial r^2}\right) \tag{129}$$

Heat transfer in region 1 was assumed to take place by three different mechanisms, neglecting the heat transfer through radiation. The three mechanisms are (1) heat transfer by turbulent diffusion in the radial direction, k_a, (2) heat transfer by molecular conduction in the fluid boundary layer, k_b, and (3) heat transfer through the thin film near particle contact points, k_c. Mechanisms (1) and (2) operate in series while mechanism (3) operates in parallel. The resulting effective bed conductivity can thus be expressed as

$$\frac{k_{e1}}{k_g} = \frac{k_c}{k_g} + \frac{1}{(k_g/k_a) + (k_g/k_b)} \tag{130}$$

$$\frac{k_a}{k_g} = 0.045(\text{Re})_1 \, \text{Pr} \tag{131}$$

where

$$(\text{Re})_1 = \frac{d_p U_1 \rho_f}{\mu} \quad \text{and} \quad \text{Pr} = \frac{C_{pg}\mu}{k_g}$$

The gas velocity in region 1, the wall region, is taken to be 50% larger than that in region 2, the bulk region.

$$\frac{h_2 d_p}{k_g} = 1.7(\text{Re})_1^{0.5} \tag{132}$$

$$\frac{k_c}{k_g} = \varepsilon_w + \frac{(1 - \varepsilon_w)}{2\phi_w + (2/3)(k_g/k_s)} \tag{133}$$

where

$$\phi_w = \frac{1}{4}\left\{\frac{[(K+1)/K]^2}{\ln K - (K-1)/K}\right\} - \frac{1}{3K} \quad K = \frac{k_s}{k_g}(134)$$

$$\varepsilon_w = 1 - \frac{(1 - \bar{\varepsilon})(0.7293 + 0.5139 Y)}{1 + Y} \tag{135}$$

where $Y = d_p/2R$.

The effective conductivity in region 2 can be evaluated as

$$\frac{k_{e2}}{k_g} = \frac{k_e^o}{k_g} + 0.075(Re)_2 Pr \qquad for \qquad (Re)_2 < 100$$

(137)

$$\frac{k_{e2}}{k_g} = \frac{K_e^o}{k_g} + 0.125(Re)_2 Pr \qquad for \qquad (Re)_2 > 100$$

(138)

Equations (137) and (138) need to be solved numerically. The agreement with experimental data was reported to be within 25%. The prediction of Gabor's model was consistently higher than the experimental data. The theoretical predictions show an almost linear dependence of the heat transfer coefficient on the particle Reynolds number. Because of its simplicity, Gabor's model may be used as a first approximation.

9.4 Mass Transfer in Packed Beds

Again based on recommendation by Rase (1990), the following correlations for a particle to fluid mass transfer have accuracy to within 25%

Mass transfer to a single particle can be calculated from

$$m_p = k_{sj} a_p (C_{js} - C_j)$$

(139)

The mass transfer coefficient can be obtained from

$$\frac{\mu}{\rho_f \vartheta_j} = 2 + 1.1 Sc^{1/3} Re^{0.6}$$

(140)

with accuracy to 25%.

The effective radial diffusivity is

$$N_{jr} = -\vartheta_r \left(\frac{\partial C_j}{\partial r}\right)_{z,r}$$

(141)

$$\frac{\varepsilon \vartheta_r}{U d_p} = \frac{1}{m} + \frac{0.38}{Re}$$

(142)

For $d_p/D > 0.1$,

$$m = 11 \qquad for \qquad Re > 400$$

(143)

$$m = 57.85 - 35.36 \log Re + 6.68(\log Re)^2$$

$$for \qquad 20 < Re < 400$$

(144)

For $d_p/D < 0.1$, divide ϑ_r calculated from the above equation by

$$\left[1 + 19.4\left(\frac{d_p}{D}\right)^2\right]$$

(145)

For more general equations in terms of Re, tortuosity, and ε, see Wen and Fan (1975).

NOMENCLATURE

A_i	= inside surface of the cylindrical vessel
a_p	= surface area of a particle
C_D	= drag coefficient of a particle in an infinite fluid
C_{D1}	= drag coefficient of a particle in an array
$(C_D)_\phi$	= drag coefficient for multiparticle systems
C_j	= concentration of component j in bulk region
C_{js}	= concentration of component j at particle surface
C_{pg}	= heat capacity of gas at constant pressure
D	= column diameter
D	= diameter of the cylindrical vessel
$\underline{D}$	= defined in Eq. (60)
d_b	= diameter of the large particles
d_c	= diameter of a circle equal in area to the projected area of the particle when resting on its larger face
D_c	= diameter of the smallest circle circumscribing the projection of the particle
d_M	= particle diameter required during critical ratio of occupation, see Sec. 3
d_p	= equivalent particle diameter
d_{pe}	= packing equivalent particle diameter
d_{pv}	= diameter of sphere with equivalent volume of the cylinder
d_s	= diameter of the small particles
d_{ve}	= volume equivalent particle diameter
d_{yi}	= inside diameter of the hollow cylinder
d_{yo}	= outside diameter of the hollow cylinder
f	= friction factor
f_B	= friction factor proposed by Blake, Eq. (58)
f_e	= general friction factor from Ergun (1952)
f_p	= general friction factor from Gibilaro et al. (1985)
f_w	= general friction factor from Wentz and Thodos (1963)
g	= gravitational acceleration
$\vec{g}$	= vector form of gravitational acceleration
G	= mass flow rate of gas
G	= an empirical constant
h_{av}	= average heat transfer coefficient, averaged over the length of the heater
h_i	= heat transfer coefficient at inside surface of the cylindrical vessel
h_s	= heat transfer coefficient to a single particle
h_w	= heat transfer coefficient at the wall
h_1	= fluid column height 1
h_1	= heat transfer coefficient in region 1

h_2 = fluid column height 2

h_2 = heat transfer coefficient in region 2

I_c = cylindrical index

I_d = disk index

k = permeability

k = an empirical constant

K = ratio of thermal conductivity of solid and gas, $K = k_e/k_g$

K = a constant

K' = permeability constant

k_s = heat transfer by turbulent diffusion in radial direction

k_b = heat transfer by molecular conduction in fluid boundary layer

k_c = heat transfer through the thin film near particle contact point

K_d = ratio of mean projected area of a prism to that of a disc with the base having the same area as the larger face of the prism

k_e = effective thermal conductivity of packed bed

k_{e1} = effective bed thermal conductivity in region 1

k_{e2} = effective bed thermal conductivity in region 2

k_e^o = effective thermal conductivity of the packed bed with zero gas flow

k_g = thermal conductivity of gas

k_s = specific permeability

k_s = thermal conductivity of solid

k_{sj} = mass transfer coefficient of component

K_s = ratio of mean projected area of a particle to that of a sphere of the same volume

L = bed height

L_h = length of heat transfer surface

L_p = pipe length

m = exponential coefficient

m_p = mass transfer rate

n = coordination number

n = an exponential coefficient

N_p = number of spheres per unit volume of bed

Nu = Nusselt number

$(Nu)_D$ = pipe flow Nusselt number

$(Pe)_d$ = pipe flow Peclet number defined in Eq. (114)

$(Pe)_p$ = pipe flow Peclet number defined in Eq. (111)

P = pressure

P_1 = pressure at point 1

P_2 = pressure at point 2

ΔP = pressure drop

Pe = Peclet number

q = heat transfer rate

$\vec{q}$ = vector quantity of volumetric flow rate per unit bed area

Q = volumetric fluid flow rate

q_p = heat transfer rate to a single particle

r = radial coordinate

r_p = diameter ratio of small particle to large particle

r_{str} = structural ratio defined in Eq. (24)

R = radius of sphere

R_t = radius of the heater

R_h = hydraulic radius

Re = Reynolds number, $= DU\rho_f/\mu$

$(Re)_B$ = Reynolds number proposed by Blake, Eq. (57)

$(Re)_p$ = Reynolds number based on particle diameter, $d_p U\rho_f/\mu$

$(Re)_{p\epsilon}$ = Reynolds number proposed by Brownell and Kats, and defined in Eq. (70)

$(Re)_\phi$ = Reynolds number for multiparticle systems

$(Re)_1$ = Reynolds number in region 1

$(Re)_2$ = Reynolds number in region 2

R_y = radius of cylindrical heater

S = specific particle surface area; total particle surface area per unit volume of bed

Sc = Schmidt number, $= \mu/\vartheta_j\rho_f$

S_p = particle surface

T = temperature

T_o = initial gas temperature

T_R = temperature at inside radius of the vessel

T_s = temperature of a single particle

T_w = temperature at the heater wall

T_w = temperature at wall

T_1 = temperature in region 1

T_2 = temperature in region 2

U = superficial fluid velocity

U_1 = superficial fluid velocity in region 1

U_2 = superficial fluid velocity in region 2

V = specific volume of solid material, $V = 1/(1-\varepsilon)$

V_b = volume of big particles

V_p = particle volume

V_s = volume of small particles

X_b = volume fraction of big particles

X_{min} = volume fraction of large particles where the voidage of the binary mixture is the minimum

X_s = volume fraction of small particles

y = horizontal coordinate

z = vertical coordinate

z = axial distance along the heat transfer surface

z_1 = vertical coordinate at point 1

z_2 = vertical coordinate at point 2

ϑ_j = diffusivity

β = layer spacing

μ = fluid viscosity

ρ_f = fluid density

ρ_p = particle density

ϕ = Wadell's sphericity, the ratio of the surface areas of a volume equivalent sphere and the actual particle

ϕ_c = an equivalent sphericity or aspect ratio defined as the ratio of the maximum length to the diameter of the circle having the same projected area normal to the maximum length

ϕ_d = a sphericity defined as the ratio of the shortest length to the diameter of the circle having the same projected area

ε = voidage of the fixed bed

ε_{av} = average voidage of the binary mixture

ε_b = voidage of a packed bed with large particles alone

ε_c = voidage at core region

ε_{hc} = voidage of the hollow cylinder packing

ε_{min} = the minimum voidage of the binary mixture

ε_o = initial packing voidage

$\varepsilon_{o,cylinder}$ = voidage of initial packing of cylinders

$\varepsilon_{o,disk}$ = voidage of initial packing of disks

$\varepsilon_{ol,cylinder}$ = voidage of initial loose packing of cylinders

$\varepsilon_{ol,disk}$ = voidage of initial loose packing of disks

$\varepsilon_{od,cylinder}$ = voidage of initial dense packing of cylinders

$\varepsilon_{od,disk}$ = voidage of initial dense packing of disks

ε_s = voidage of a packed bed with small particles alone

ε_s = stagnant voidage

ε_{sc} = voidage of the solid cylinder packing

ε_w = voidage at wall region

$\bar{\varepsilon}$ = average voidage

REFERENCES

E Abe, H Hirosue, A Yokota. Pressure drop through a packed bed of binary mixture. J Chem Eng Japan 12:302, 1979.

AR Balakrishnan, Pei DCT. Heat transfer in gas–solid packed bed systems, a critical review. Ind Eng Chem Process Des Dev 18:30–40, 1979.

JJ Barker. Heat transfer in packed beds. Ind Eng Chem 57(4):43–51, April 1965.

E Barnea, RL Mednick. A generalized approach to the fluid dynamics of particulate systems part III: general correlation for the pressure drop through fixed beds of spherical particles. Chem Eng J 15:215–227, 1978.

RF Benenati, CB Brosilow. Void fraction distribution in beds of spheres. AIChE J 8:359–361, 1962.

FC Blake. The resistance of packing to fluid flow. Trans Am Inst Chem Eng 14:415, 1922.

JSM Botterill, AOO Denloye. A theoretical model of heat transfer to a packed or quiescent fluidized bed. Chem Eng Sci 33:509–515, 1978.

GG Brown. Unit Operations. New York: John Wiley, 1966.

LE Brownell, DL Katz. Flow of fluids through porous media I. Single homogeneous fluids. Chem Eng Prog 43:537–548, October 1947.

SP Burke, WB Plummer. Gas flow through packed columns. Ind Eng Chem 20:1196–1200, 1928.

PC Carman. Fluid flow through granular beds. Trans Inst Chem Eng 15:150, 1937

PC Carman. The flow of Gases Through Porous Media. New York: Academic Press, 1956.

Y Cohen, AB Metzner. Wall effects in laminar flow of fluids through packed beds. AIChE J 27:705–715, 1981.

JM Coulson. The flow of fluids through granular beds: effect of particle shape and voids in streamline flow. Trans Inst Chem Eng 27:237–257, 1949.

DJ Cumberland, RJ Crawford. The Packing of Particles. Amsterdam: Elsevier, 1987, p 60.

H Darcy. Les Fontaines publiques de la ville de Dizon. Paris: Dalmont, 1856.

AG Dixon. Correlations for wall and particle shape effects on fixed bed bulk voidage. Can J Chem Eng 66:705–708, 1988.

S Ergun. Fluid flow through packed columns. Chem Eng Prog 48:89–94, February 1952.

HJ Finkers, AC Hoffmann. Structural ratio for predicting the voidage of binary particle mixtures. AIChE J 44:495–498.

CC Furnas. Flow of gases through beds of broken solids. Bull 307. US Bureau of Mines, 1929.

JD Gabor. Heat transfer to particle beds with gas flows less than or equal to that required for incipient fluidization. Chem Eng Sci 25:979–984, 1970.

TR Galloway, BH Sage. A model of the mechanism of transport in packed, distended, and fluidized beds. Chem Eng Sci 25:495–516, 1970.

WH Gauvin, S Katta. Momentum transfer through packed beds of various particles in the turbulent flow regime. AIChE J 19:775–783, 1973.

L Gibilaro, RDI Delice, SP Waldram, PU Foscolo. Generalized friction factor and drag coefficient correlations for fluid–particle interactions. Chem Eng Sci 40:1817–1823, 1985.

JS Goodling, RI Vachon, WS Stelpflug, SJ Ying, MS Khader. Radial porosity distribution in cylindrical beds packed with spheres. Powder Technol 35:23–29, 1983.

D Handley, PJ Heggs. Momentum and heat transfer mechanisms in regular shaped packings. Trans Inst Chem Eng 46:T251–T264, 1968.

J Happel. Viscous flow in multiparticle systems: slow motion of fluids relative to beds of spherical particles. AIChE J 4:197–201, 1958.

DP Haughey, GG Beveridge. Local voidage variation in a randomly packed bed of equal-sized spheres. Chem Eng Sci 21:905–916, 1966.

DP Haughey, GG Beveridge. Structural properties of packed beds—a review. Can J Chem Eng 47:130–140, 1969.

G Herdan. Small Particle Statistics. New York: Academic Press, 1960.

RE Hicks. Pressure drop in packed beds of spheres. Ind Eng Chem Fundam 9:500–502, 1970.

AC Hoffmann, HJ Finkers. A relation for the void fraction of randomly packed particle beds. Powder Technol 82:197–203, 1995.

J Kozeny. Über Kapillare Leitung des Wassers im Boden. Sitzungsber Akad Wiss Wien 136:271, 1927.

CL Kusik, J Happel. Boundary layer mass transfer with heterogeneous catalysis. AIChE J 8:163, 1962.

M Leva. Pressure drop through packed tubes: Part I. A general correlation. Chem Eng Prog 43:549–554, 1947.

Y Li, CW Park. Permeability of packed beds filled with polydisperse spherical particles. Ind Eng Chem Res 37:2005–2011, 1998.

RK McGeary. Mechanical packing of spherical particles. J Am Ceram Soc 44:513–522, 1961.

D Mehta, MC Hawley. Wall effects in packed columns. Ind Eng Chem Proc Design Dev 8:280–282, 1969.

O Molerus. Principles of Flow in Disperse Systems. London: Chapman and Hall, 1993.

O Molerus, KE Wirth. Heat Transfer in Fluidized Beds. London: Chapman and Hall, 1997.

PG Nutting. Bull Amer Assoc Petrol Geo 14:1337, 1930.

AO Oman, KM Watson. Pressure drops in granular beds. Refinery Mat Petrol Chem Tech 36:R795, 1944.

N Ouchiyama, T Tanaka. Predicting the densest packings of ternary and quaternary mixtures of solid particles. Ind Eng Chem Res 28:1530–1536, 1989.

M Propster, J Szekely. The porosity of systems consisting of layers of different particles. Powder Technol 17:123–138, 1977.

WE Ranz. Friction and transfer coefficients for single particles and packed beds. Chem Eng Prog 48:247–253, 1952.

HF Rase. Fixed-Bed Reactor Design and Diagnostics. Boston: Butterworths, 1990, pp 89–90.

HE Rose, MA Rizk. Further researches in fluid flow through beds of granular material. Proc Inst Mech Eng London 160:493–503, 1949.

OA Saunders, N Ford. Heat transfer in the flow of gas through a bed of solid particles. J Iron Steel Inst 141:291–316, 1940.

AE Scheidegger. The Physics of Flow Through Porous Media. New York: MacMillan, 1961.

CE Schwartz, JM Smith. Flow distribution in packed beds. Ind Eng Chem 45:1209–1218, 1953.

GD Scott. Packing of spheres Nature 188:908–911, 1960.

HY Sohn, C Moreland. The effect of particle size distribution on packing density. Can J Chem Eng 46:162–167, 1968.

P Subramanian, Vr Arunachalam. A simple device for determination of sphericity factor. Ind Eng Chem Fandam 19:436-437, 1980.

DL Swift. The thermal conductivity of spherical metal powders including the effect of an oxide coating. Int J Heat Mass Transfer 9:1061–1074, 1966.

J Szekely, JJ Poveromo. Flow maldistribution in packed beds: a comparison of measurements with predictions. AIChE J 21:769–775, 1975.

JA Tallmadge. Packed bed pressure drop—an extension to higher Reynolds numbers. AIChE J 16:1092–1093, 1970.

H Wadell. Volume, shape, and roundness of quartz particles. J Geol 43:250–280, 1935.

CY Wen, LT Fan. Models for Flow Systems and Chemical Reactors. New York: Marcel Dekker, 1975.

CY Wen, YH Yu. Mechanics of fluidization. Chem Eng Prog Symp Ser 62(62):100–111, 1966.

CA Wentz Jr, G Thodos. Pressure drops in the flow of gases through packed and distended beds of spherical particles. AIChE J 9:81–84, 1963.

AER Westman. The packing of particles: empirical equations for intermediate diameter ratios. J Am Ceram Soc 19:127–129, 1936.

S Yagi, D Kunii. Studies on effective thermal conductivities in packed beds. AIChE J 3:373–381, 1957.

AB Yu, N Standish, A McLean. Porosity calculation of binary mixtures of nonspherical particles. J Am Ceram Soc 76:2813–2816, 1993.

RP Zou, AB Yu. Evaluation of the packing characteristics of monosized nonspherical particles. Powder Technol 88:71–79, 1996.

3

Bubbling Fluidized Beds

Wen-Ching Yang

Siemens Westinghouse Power Corporation, Pittsburgh, Pennsylvania, U.S.A.

1 INTRODUCTION

For chemical reactor applications, the fixed (or packed) beds described in Chapter 2 have some major disadvantages. If the reactions are fast and highly exothermic or endothermic, hot or cold spots will form in the packed beds and render the reactor ineffective; or it increases the chance or unwanted byproduct production. By nature, the fixed bed operation has to be in batch mode, which is much less efficient than a continous operation. Sintering, plugging, and fluid maldistribution can also occur much more readily in packed beds. Comparing to fixed beds, fluidized beds have many advantages. Once the solids in the bed are fluidized, the solids inside the bed will behave just like liquid (Gelperin and Einstein, 1971; Davidson et al., 1977). The bed surface of a bubbling fluidized bed resembles that of a boiling liquid, and it can be stirred easily just like a liquid. Objects with a density lighter than the bulk density of the bed will float and those heavier will sink. If there is a hole on the side of a fluidizing bed, the solids will flow out like a liquid jet. The gas bubble size, shape, formation, rising velocity, and coalescence in the fluidized beds have quantitative similarity with those of gas bubbles in liquids.

The liquid like behavior of a fluidized bed thus allows the solids to be handled like a fluid, and continous feeding and withdrawal therefore become possible. The rigorous mixing in a fluidized bed results in a uniform temperature even for highly exothermic or endothermic reactions. This in turn provides an easier reactor control as well. The rigorous mixing also improves solids and fluid contacting, and it enhances heat and mass transfer. However, fluidized beds also possess some serious disadvantages. Rigorous solids mixing in the bed produces solid fines through attrition. Operating at high fludiization velocities, fines elutriation and entrainment can become a serious operational problem. Also, because of the rigorous mixing in the bed, a fluidized bed is essentially a continous stirred tank reactor with varying solids residence time distribution. These deficiencies may or may not be resolved through design. In practice, fluidization is still an empirical science (Yang, 1998a). Care must be exercised to select proper correlations for design and scale-up. A recent review of hydrodynamics of gas–solid fluidization can be found in Lim et al. (1995).

There are many different variations of fluidized beds in practice, which are covered in different chapters in this handbook. For conventional bubbling fluidized beds, the essential elements are depicted in Fig. 1. Depending on the applications, not all elements shown in Fig. 1 will be necessary. The design of the individual element can also vary substantially from one design to another design. For example, the cyclone can be arranged either internally or externally; the heat transfer tubes can be either vertical or horizontal, etc. These design aspects and various phenomena occurring in a bubbling fluidized bed will be discussed throughout this book.

2 POWDER CLASSIFICATIONS AND TYPES OF GAS FLUIDIZATION

The fluidization phenomena of gas–solids systems depend very much on the types of powders employed. There are several classifications available in the literature, all based on the original work by Geldart (1973).

2.1 Geldart Classification of Powders

Geldart (1973) was the first to classify the behavior of solids fluidized by gases into four clearly recognizable groups characterized by the density difference between the particles and the fluidizing medium, $(\rho_p - \rho_f)$ and by the mean particle size, d_p. Geldart's classification has since become the standard to demarcate the types of gas fluidization. The most easily recognizable features are

Group A	The bed particles exhibit dense phase expansion after minimum fluidization and before the beginning of bubbling. Gas bubbles appear at the minimum bubbling velocity.
Group B	Gas bubbles appear at the minimum fluidization velocity.
Group C	The bed particles are cohesive and difficult to fluidize.
Group D	Stable spouted beds can be easily formed in this group of powders.

Demarcation for Groups A and B Powders. For Group A,

$$\frac{U_{mb}}{U_{mf}} \geq 1$$

$$U_{mf} = \frac{8 \times 10^{-4} g d_p^2 (\rho_p - \rho_f)}{\mu} \tag{1}$$

$$d_p = \frac{1}{\left(\sum_i \frac{x_i}{d_{pi}}\right)} \tag{2}$$

$$U_{mb} = K d_p \tag{3}$$

Thus the powders are Group A powders if

$$\frac{8 \times 10^{-4} g d_p (\rho_p - \rho_f)}{K\mu} \leq 1 \tag{4}$$

For air at Room Temperature and Pressure, $K = 100$

For Group D Powders, $U_B \leq \dfrac{U_{mf}}{\varepsilon_{mf}}$,

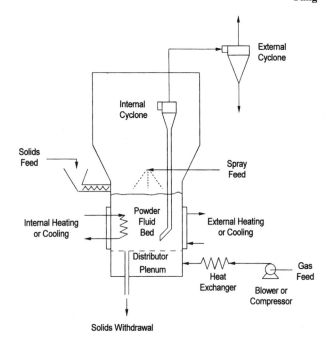

Figure 1   Essential elements of a bubbling fluidized bed.

$$U_B \leq \frac{8 \times 10^{-4} (\rho_p - \rho_f) g d_p^2}{\mu \varepsilon_{mf}} \tag{5}$$

Equation (5) is the demarcation between Group B and D powders. The powder classification diagram for the fluidization by air at ambient conditions was presented by Geldart, as shown in Fig. 2.

Minimum Bubbling Velocity. For gas fluidization of fine particles, the fluidization velocity at which the gas bubbles first appear is called the minimum bubbling velocity, U_{mb}. Geldart and Abrahamsen (1978) observed that

$$\frac{U_{mb}}{U_{mf}} = \frac{4.125 \times 10^4 \mu^{0.9} \rho_f^{0.1}}{(\rho_p - \rho_f) g d_p} \tag{6}$$

where the units are in kg, m, and s. Equation (6) gives $U_{mb} < U_{mf}$, which is generally true for Groups B and D powders. For fine particles (Group A), the ratio $U_{mb}/U_{mf} > 1$ was predicted and observed to increase with both temperature and pressure. The ratio also increases with smaller particles and lighter particles.

2.2 Molerus' Interpretation of Geldart's Classification of Powders

Molerus (1982) classified the powders by taking into account the interparticle cohesion forces. Free particle motion for Group C powder is suppressed by the

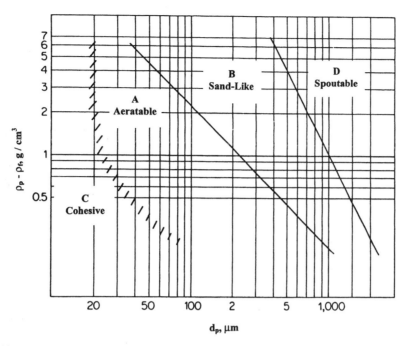

Figure 2 Geldart classification of powders.

dominant influence of cohesion forces, and its demarcation can be expressed by the equation

$$10 \times \frac{(\rho_p - \rho_f)d_p^3 g}{F_H} = K_1 \cong 10^{-2} \tag{7}$$

where F_H is the adhesion force transmitted in a single contact between two particles. For polypropylene powder, $F_H = 7.71 \times 10^{-7}$ newtons, and for glass beads and cracking catalyst, $F_H = 8.76 \times 10^{-8}$ newtons. K_1 is found experimentally.

For Group A and Group B transition, the following equation can be used.

$$\frac{\pi}{6}\frac{(\rho_p - \rho_f)d_p^3 g}{F_H} = K_2 \cong 0.16 \tag{8}$$

For Group B and Group D, the transition equation is

$$\frac{(\rho_p - \rho_f)d_p g}{\rho_f U_{mf}^2} \cong 15 \tag{9}$$

2.3 Particle Classification Boundaries Suggested by Grace

Based on additional data beyond that analyzed by Geldart, Grace (1986) suggested new boundaries between Groups A and B, and between Groups B and D of Geldart's classification. The new boundaries

are good also for gases other than air and for temperature and pressure other than ambient.

Boundary Between Group A and Group B

$$\text{Ar} = 1.03 \times 10^6 \left(\frac{\rho_p - \rho_f}{\rho_f}\right)^{-1.275} \tag{10}$$

For $(\rho_p - \rho_f)/\rho_f \cong 1000$ to 2000, Eq. (10) reduces to Ar $\cong$ 125 for air at atmospheric pressure. This compares to a value of Ar = 88.5 as suggested by Goossens (1998).

Boundary Between Group B and Group D

$$\text{Ar} = 1.45 \times 10^5 \tag{11}$$

This Group B/D boundary can be compared with Ar = 176,900, as suggested by Goossens (1998).

2.4 Goossen's Classification of Particles by Archimedes Number

Based on the hypothesis that the relative importance of laminar and turbulent phenomena governs the fluidization behavior, Goossen (1998) classified the powders on the basis of Archimedes number. The proposed classification is of general application, applying equally well in both liquid fluidization and gas fluidization. The four boundaries he suggested are as follows:

Group C boundary   Ar = 0.97   (12)
Group A/C boundary   Ar = 9.8   (13)
Group A/B boundary   Ar = 88.5   (14)
Group B/D boundary   Ar = 176,900   (15)

The demarcations between groups of powder compare well with that of Geldart's and Molerus' when air is used as the fluidization medium, except for the Group A/B boundary.

2.5 Powder Characterization by Bed Collapsing

The bed collapsing technique has been employed to study fluidization for various objectives (Rietema, 1967; Morooka et al., 1973; Geldart, 1986). Kwauk (1992) has instrumented the bed for automatic surface tracking and data processing to characterize powders. When the gas is abruptly turned off for an operating fluidized bed, the bed collapses in three distinct stages: (1) bubble escaping stage, (2) hindered settling stage, and (3) solids consolidation stage. The bed collapsing stage can be conveniently represented grahically in Fig. 3. Mathematically, the bed collapsing stages can be described as follows.

2.5.1 Bubble Escaping Stage, $0 < t < t_b$

During the bubble escaping stage, the change of bed surface is linearly dependent on the time, or

$$H_1 = H_0 - U_1 t \qquad (16)$$

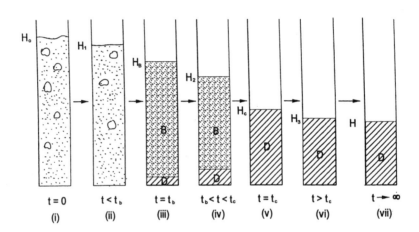

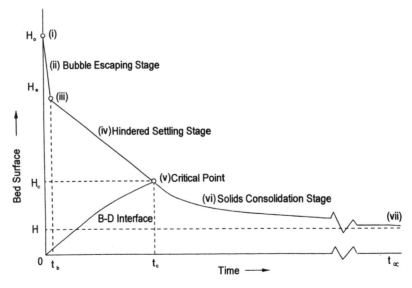

Figure 3 Bed collapsing stages during bed collapsing characterization of powders. (Adapted from Kwauk, 1992.)

where $U_1 = \dfrac{f_B(1 + f_w\varepsilon_e)U_B + \varepsilon_e(1 - f_B - F_w f_B)U_e}{(1 - f_B)(1 - \varepsilon_e)}$ (17)

and $f_B = \dfrac{H_0 - H_e}{H_0}$ (18)

2.5.2 Hindered Settling Stage, $t_b < t < t_c$

During the hindered settling stage, the bed surface settles at a constant velocity, and the settled layer at the bottom, layer D in Fig. 3, also increases in height. The change in bed height for both layers can be expressed as

$$H_2 = H_e - U_2 t \qquad (19)$$

$$
\begin{aligned}
H_D = &\left(\frac{H_\infty}{H_e - H_\infty}\right)U_2 t \\
&+ \left\{1 - \exp\left[-K_3\left(\frac{H_e - H_\infty}{H_e}\right)t\right]\right\} \\
&\frac{U_2 H_e}{(H_e - H_\infty)K_3}\left[\frac{1 - \varepsilon_e}{\varepsilon_e - \varepsilon_c} - \frac{H_e}{H_e - H_\infty}\right]
\end{aligned} \qquad (20)
$$

where K_3 is the rate constant for the increasing of height in layer D expressed as a first-order process:

$$-\frac{dH}{dt} = K_3(H_D - H_\infty) \qquad (21)$$

2.5.3 Solids Consolidation Stage, $t_c < t < t_\infty$

The consolidation stage is essentially the consolidation of layer D and thus the change in bed height can be expressed as

$$\frac{dH_3}{dt} = \frac{dH_D}{dt} = -K_3(H_D - H_\infty) \qquad (22)$$

or

$$H_3 = H_D = (H_c - H_\infty) \times \exp[-K_3(t - t_c)] + H_\infty \qquad (23)$$

Not all powders exhibit all three stages during bed collapsing experiments. Only Geldart's Group A powders exhibit all three stages. For Group B and D powders, the first bubble escaping stage and the second hindered settling stage are practically instantaneous. The transition between the hindered settling stage and the solids consolidation stage, the point (t_c, H_c) in Fig. 3, is called by Yang et al. (1985) "the critical point." The better to qualify the powders through the bed collapsing tests, Yang et al. (1985) also defined a dimensionless subsidence time of a powder, θ_s.

For small powders whose terminal velocity can be calculated by the Stokes law, θ_s can be expressed by

$$\theta_s = \frac{t_c}{d_p(\rho_p - \rho_f)H_\infty} \qquad (24)$$

The dimensionless subsidence time of a powder was found to be related to the ratio of minimum bubbling and minimum fluidization velocities

$$\ln\left(\frac{U_{mb}}{U_{mf}}\right) = 4\theta_s^{0.25} \qquad (25)$$

Thus for Group B and D powders, $U_{mb}/U_{mf} = 1$, the corresponding subsidence time is zero, and hence t_c is zero. Stages 1 and 2 occur almost instantaneously. For larger values of θ_s, the ratios of U_{mb} and U_{mf} become larger, and the bed exhibits the particulate fluidization more prominently.

Yang et al. (1985) have performed extensive experiments employing the bed collapsing technique to study the modification of fluidization behavior by addition of fines. They found that improvement in fluidization increases monotonically with increases in fines concentration in the bed. A more interesting study is the series for binary mixtures of Group B–C powders. They found that addition of cohesive Group C powder into Group B powder will improve the fluidization behavior of Group B powder. Conversely, addition of Group B powder into the cohesive Group C powder will also improve the fluidization quality of a cohesive powder. There is, however, some limitations on the effective amount to be added in the mixture. Addition beyond the maximum amount will adversely affect the fluidity and fluidization behavior of the mixture.

3 DIFFERENT REGIMES OF FLUIDIZATION

Fluidization regimes can be classified into two broad categories—particulate (smooth) and aggregative (bubbling) (Harrison et al., 1961). Most liquid fluidized beds under normal operation exhibit the particulate fluidization. In particulate fluidization, the solid particles usually disperse relatively uniformly in the fluidizing medium with no readily identifiable bubbles. Thus the particulate fluidization sometimes is also called homogeneous fluidization. In the heterogeneous or aggregative fluidization, voids containing no solids are usually observed. Those voids are called bubbles. Those voids can be well defined as in a bubbling fluidized bed or in a slugging bed, or they can be small voids where particle clusters dart to and from like in a turbulent bed or in a fast fluidized bed. There are a number of criteria available to determine whether a particular system will exhibit particulate or aggregative fluidization. Those criteria are summarized in Table 1.

It was suggested by Harrison et al. (1961) that aggregative fluidization may be expected if the ratio of maximum stable bubble size to particle diameter, $(D_B)_{max}/d_p$, is larger than 10, and particulate fluidization, if the ratio is less than or equal to unity. A transition region exists with the ratio between 1 and 10. They proposed to calculate the ratio as follows:

$$\frac{(D_B)_{max}}{d_p} = 71.3\left(\frac{\mu^2}{gd_p^3\rho_f^2}\right)$$
$$\left[\frac{(\rho_p/(\rho_p - \rho_f)) - \varepsilon_{mf}}{1 - \varepsilon_{mf}}\right]$$
$$\left[\left(1 + \frac{gd_p^3\rho_f}{54\mu^2}(\rho_p - \rho_f)\right)^{1/2} - 1\right]^2 \qquad (26)$$

For large particles, Eq. (26) reduces to

$$\frac{(D_B)_{max}}{d_p} = 1.32\left(\frac{\rho_p - \rho_f}{\rho_f}\right)\left[\frac{(\rho_p/(\rho_p - \rho_f)) - \varepsilon_{mf}}{1 - \varepsilon_{mf}}\right]$$
$$(27)$$

Eqs. (26) and (27) are only applicable when the maximum stable bubble size is much smaller than the bed diameter, i.e., without wall effect. Experimental

evidence of maximum bubble size was observed by Matsen (1973).

For gas–solid systems, there are at least five distinguishable regimes of fluidization observable experimentally: fixed bed, particulate fluidization, bubbling fluidization, slugging fluidization, and turbulent fluidization. When the operating velocity is higher than the transport velocity such that recycle of entrained particles is necessary to maintain a bed, additional fluidizing regimes are possible. The regimes of fluidization for circulating fluidized beds is discussed in more detail in Chapter 19, "Circulating Fluidized Beds." The different regimes for gas–solid fluidization are summarized in Table 2 and Figs. 4 and 5. Not all of these regimes can be observed in all systems, however, because some regimes are also dependent on the size of the equipment employed. For a recent review of fluidization regimes, see Bi and Grace (1995a).

3.1 Transition among Fixed Bed, Particulate Fluidization, and Bubbling Regime

For Geldart's Group B and Group D powders, the bed transfers from the fixed bed into a bubbling fluidized bed when the gas velocity is increased beyond the minimum fluidization velocity of the system. For Group A powders, no bubbles will be observed, instead the bed will expand homogeneously. The bubbles only appear when the gas velocity is increased beyond the minimum bubbling velocity. Thus the transition point from the

Table 1 Criteria for Transition Between Particulate and Aggregative Fluidization

Criteria for particulate fluidization	Reference
$\text{Fr} = \dfrac{U_{mf}^2}{gd_p} < 0.13$	Wilhelm and Kwauk (1948)
$\dfrac{(D_B)_{max}}{d_p} < 30; \quad (D_B)_{max} = \dfrac{2U_t^2}{g}$	Harrison et al. (1961)
$U_{mf} < 0.2$ cm/s	Rowe (1962)
$\dfrac{U_{mf}^3(\rho_p - \rho_f)H_{mf}}{g\mu D} < 100$	Romero and Johanson (1962)
$\dfrac{(\rho_p - \rho_f)}{\mu}\sqrt{gd_p^3} < C_1$ depending on ε and ε_{mf}	Verloop and Heertjes (1974)
$\left[\dfrac{gd_p^3\rho_f(\rho_p - \rho_f)}{\mu^2}\right]^m \left(\dfrac{\rho_{mf}}{\rho_f}\right)^{0.5} < C_2$	Doichev et al. (1975)

m and C_2 depend on range of operation

Table 2 Summary Description of Different Regimes of Fluidization

Velocity range	Fluidization regime	Fluidization features and appearance
$0 \leq U < U_{mf}$	Fixed bed	Particles are quiescent; gas flows through interstices
$U_{mf} \leq U < U_{mb}$	Particulate regime	Bed expands smoothly and homogeneously with small-scale particle motion; bed surface is well defined
$U_{mb} \leq U < U_{ms}$	Bubbling regime	Gas bubbles form above distributor, coalesce and grow; gas bubbles promote solids mixing during rise to surface and breakthrough
$U_{ms} \leq U < U_c$	Slug flow regime	Bubble size approaches bed cross section; bed surface rises and falls with regular frequency with corresponding pressure fluctuation
$U_c \leq U < U_k$	Transition to turbulent fluidization	Pressure fluctuations decrease gradually until turbulent fluidization regime is reached
$U_k \leq U < U_{tr}$	Turbulent regime	Small gas voids and particle clusters and streamers dart to and fro; bed surface is diffused and difficult to distinguish
$U > U_{tr}$	Fast fluidization	Particles are transported out of the bed and need to be replaced and recycled; normally has a dense phase region at bottom coexisting with a dilute phase region on top; no bed surface
$U \gg U_{tr}$	Pneumatic conveying	Usually a once-through operation; all particles fed are transported out in dilute phase with concentration varying along the column height; no bed surface

fixed bed to the bubbling regime is the minimum fluidization velocity for Group B and Group D powders while for Group A powders, it is the minimum bubbling velocity. The homogeneous expansion, also called particulate fluidization, occurs only in Group A powders for gas–solids systems. The particulate fluidization is especially important for liquid fluidized beds because where most of the occurrence is observed. Liquid–solids fluidization will be discussed in Chapter 26, "Liquid–Solids Fluidization."

3.2 Transition Between Bubbling and Slugging Regimes

A slugging regime occurs only in beds with bed height (H) over bed diameter ratio (D) larger than about 2. With large H/D ratios, the bed provides enough time for bubbles to coalesce into bigger ones. When the bubbles grow to approximately 2/3 of the bed diameter, the bed enters the slugging regime with periodic passing of large bubbles and regular large fluctuation of bed pressure drop corresponding to the bubble frequency. There are several correlations available to predict this transition, they are discussed in Sec. 10; "Slugging Beds."

3.3 Transition Between Bubbling and Turbulent Regimes

When the gas velocity is continuously increased, the bubbles grow bigger owing to coalescence, and the bubbling bed can transfer into a slugging bed if the bed diameter is small and the particle diameter is large, or into a turbulent bed if the bed diameter is large and the particle diameter is small. If the standard deviation of pressure fluctuation is measured and plotted against the superficial fluidization velocity, two characteristic velocities, U_c and U_k, first suggested by Yerushalmi and Cankurt (1979), can be identified. The velocity U_c corresponds to the bed operating conditions where the bubbles or slugs reach their maximum diameter and thus have the largest standard deviation of pressure fluctuation. Continuing increases beyond this velocity, large bubbles start to break up into smaller bubbles with smaller pressure fluctuation, and eventually the standard deviation of the pressure fluctuation reaches a steady state. This velocity is denoted as U_k, also a characterization velocity for transition from the bubbling regime to the turbulent regime. Subsequent study by Chehbouni et al. (1994), employing differential and absolute pressure transducers and a capacitance probe, concluded that the onset

Beds With Little Entrainment   **Transport & Circulating Fluidized Beds**
                                                                 (For transition velocities, see Bi & Grace, 1995)

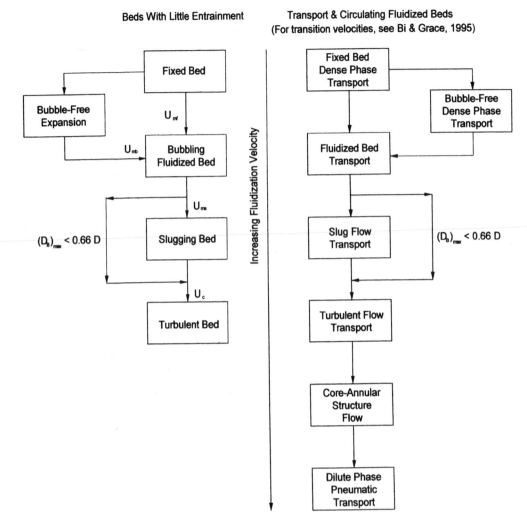

Figure 4   Regimes of fluidization for nontransporting systems.

of turbulent fluidization was at U_c, and the velocity U_k actually did not exist. The existence of U_k is an artifact due to the use of differential pressure transducers for experiments. They maintained that there is only one transition velocity, i.e., U_c. The turbulent fluidization starts at U_c and ends at the transport velocity, U_{tr}, the velocity capable of transporting all particles out of the reactor. There is still controversy in the literature on the actual transition boundary between the bubbling and the turbulent regimes.

In 1986, Horio proposed the following equations to calculate U_c and U_k (see also Horio, 1990).

$$(\mathrm{Re})_c = \frac{d_p \rho_f U_c}{\mu} = 0.936 \mathrm{Ar}^{0.472} \qquad (28)$$

$$(\mathrm{Re})_k = \frac{d_p \rho_f U_k}{\mu} = 1.46 \mathrm{Ar}^{0.472}$$

$$\text{for Canada et al. (1978) data} \qquad (29)$$

$$(\mathrm{Re})_k = \frac{d_p \rho_f U_k}{\mu} = 1.41 \mathrm{Ar}^{0.56}$$

$$\text{for Yerushalmi et al. (1978) data} \qquad (30)$$

The extensive literature data based on absolute pressure fluctuation and bed expansion measurements up to 1989 were correlated by Cai et al. (1989) to be

$$(\mathrm{Re})_c = \frac{d_p \rho_f U_c}{\mu} = 0.57 \mathrm{Ar}^{0.46} \qquad (31)$$

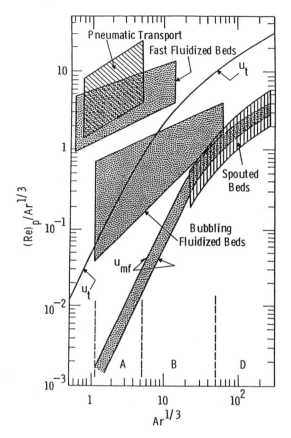

Figure 5 Regimes of fluidization for transporting and non-transporting systems.

Perales et al. (1990) also suggested equations to calculate U_k and U_{tr} as follows.

$$(\text{Re})_k = \frac{d_p \rho_f U_k}{\mu} = 1.95 \text{Ar}^{0.453} \tag{32}$$

$$(\text{Re})_{tr} = \frac{d_p \rho_f U_{tr}}{\mu} = 1.41 \text{Ar}^{0.483} \tag{33}$$

Since the U_k and U_{tr} are very similar, they suggest that the following equation alone may be used to calculate both U_k and U_{tr} for simplification:

$$(\text{Re})_{k,tr} = \frac{d_p \rho_f U_k}{\mu} = \frac{d_p \rho_f U_{tr}}{\mu} = 1.45 \text{Ar}^{0.484} \tag{34}$$

Bi and Fan (1992) affirmed the existence of turbulent regimes in gas–solid fluidization and suggested the following criteria for transition to turbulent fluidization.

$$(\text{Re})_k = \frac{d_p \rho_f U_k}{\mu} = 16.31 \text{Ar}^{0.136} \left(\frac{U_t}{\sqrt{gD}} \right)^{0.941}$$
$$\text{for } \text{Ar} \leq 125 \tag{35}$$

$$(\text{Re})_k = \frac{d_p \rho_f U_k}{\mu} = 2.274 \text{Ar}^{0.419} \left(\frac{U_t}{\sqrt{gD}} \right)^{0.0015}$$
$$\text{for } \text{Ar} \geq 125 \tag{36}$$

$$(\text{Re})_{tr} = \frac{d_p \rho_f U_{tr}}{\mu} = 2.28 \text{Ar}^{0.419} \tag{37}$$

Bi and Grace (1995b) correlated the literature data based on the differential pressure fluctuation and arrived at the following transition equation:

$$(\text{Re})_c = \frac{d_p \rho_f U_c}{\mu} = 1.24 \text{Ar}^{0.45}$$
$$\text{for } \quad 2 < \text{Ar} < 1 \times 10^8 \tag{38}$$

For gas fluidization of large particles, Catipovic et al. (1978) further subdivided the regimes into slow bubbles, fast bubbles, and rapidly growing bubble regimes. The transition equations are summarized here.

Regime Between Fast and Slow Bubbles

$$\frac{U_{mf}}{\varepsilon_{mf}} = U_B = \frac{\sqrt{gD_B}}{2} \tag{39}$$

Regime of Rapidly Growing Bubbles

In this regime, the bubble growth rate is of the same magnitude as the bubble rising velocity, and for shallow beds,

$$\frac{d(D_B)}{dt} = \frac{d(D_B)}{dh}\frac{dh}{dt} = U_B \tag{40}$$

or

$$\frac{d(D_B)}{dh} = 1 \tag{41}$$

3.4 Transition to Fast Fluidization

Continuing increases in operating velocity beyond that required at turbulent fluidization, a critical velocity, commonly called the transport velocity U_{tr}, will be reached where a significant particle entrainment occurs. Beyond this point, continuing operation of the bed will not be possible without recycle of the entrained solids. The bed is now said to be in the fast fluidization regime. The transition velocity has been correlated by Bi et al. (1995) as

$$U_{tr} = 1.53 \text{Ar}^{0.5} \quad \text{for} \quad 2 < \text{Ar} < 4 \times 10^{0.5} \tag{42}$$

For Group A and B particles, the transition velocity, U_{tr}, calculated from Eq. (42) is larger than the terminal velocity of the individual particles, while for Group D particles, the transition velocity equals essentially the terminal velocity of the individual particles.

However, a fluidized bed operating at a high gas velocity alone ($> U_{tr}$) does not make it a fast fluidized bed. A generally accepted definition of a fast fluidized bed is the coexistence of a dilute phase and dense phase regimes (see review by Rhodes and Wang, 1998).

4 THEORETICAL AND EMPIRICAL PREDICTIONS OF MINIMUM FLUIDIZATION VELOCITY

The phenomenon of fluidization can best be characterized by a $\Delta P / L$ versus U plot such as the one shown in Fig. 6. Below a characteristic gas velocity known as the minimum fluidization velocity, a packed bed of solid particles remains fixed, though a pressure drop across the bed can be measured. At the minimum fluidization velocity, all the particles are essentially supported by the gas stream. The pressure drop through the bed is then equal to the bed weight divided by the cross-sectional area of the bed, $\Delta P = W/A$. Further increases in gas velocity will usually not cause further increases in pressure drop. In actual practice, however, pressure drop at minimum fluidization velocity is actually less then W/A because a small percentage of the bed particles is supported by the wall owing to the less

than perfect design of the gas distributor, to the finite dimension of the containing vessel, and to the possibility of channeling. At the point of minimum fluidization, the voidage of the bed corresponds to the loosest packing of a packed bed. The loosest mode of packing for uniform spheres is cubic, as discussed in Chapter 2 or $\varepsilon_{mf} = (6 - \pi)/6 = 0.476$. Substituting into the original Carman equation, we obtain

$$\frac{\Delta P}{L} = \left[\frac{72}{\cos^2(\gamma)}\right] \cdot \frac{\mu U (1 - \varepsilon)^2}{d_p^2 \varepsilon^3} \tag{43}$$

and assuming

$$\frac{72}{\cos^2(\gamma)} = 180; \quad \gamma \text{ is usually from 48 to } 51° \tag{44}$$

we have

$$\frac{\Delta P}{L} = 459 \frac{\mu U_{mf}}{d_p^2} \tag{45}$$

At the point of minimum fluidization, the pressure drop is enough to support the weight of the particles and can be expressed as

$$\frac{\Delta P}{L} = (\rho_p - \rho_f)(1 - \varepsilon_{mf}) \tag{46}$$

Combining Eqs. (45) and (46) with the voidage at the minimum fluidization $\varepsilon_{mf} = 0.476$, we find

$$U_{mf} = 0.00114 \frac{g d_p^2 (\rho_p - \rho_f)}{\mu} \tag{47}$$

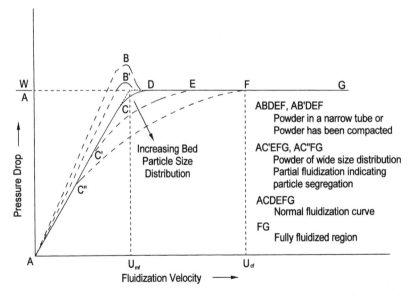

Figure 6 Pressure drop vs. fluidization velocity plot for determination of minimum fluidization velocity.

Leva (1959) employed the experimental values of voidage at minimum fluidization and arrived at the equations

$$U_{mf} = [0.0007(Re)_{mf}^{-0.063}]\frac{gd_p^2(\rho_p - \rho_f)}{\mu}$$

$$(Re)_{mf} = \frac{d_p U_{mf} \rho_f}{\mu} \qquad (48)$$

Equations (47) and (48) are comparable because $(Re)_{mf}^{-0.063}$ is on the order of unity for most fluidized systems where the Reynolds number at minimum fluidization, $(Re)_{mf}$, is generally between 10^{-2} to 10^2.

Rowe (1961) performed experiments on water flow through a regular array of spheres and found that the force on the single sphere in the array was 68.5 times the force on an isolated sphere at the same superficial velocity. If we assume that the same factor of 68.5 is also applicable at the minimum fluidization condition when the drag force of the fluid on the particles is just balanced by the net downward force on the particles, we have, for the low-Reynolds number case,

$$F = 3\pi\mu U d_p \qquad (49)$$

and

$$68.5 \times 3\pi\mu U_{mf}d_p = \frac{(\rho_p - \rho_f)\pi d_p^3 g}{6} \qquad (50)$$

or

$$U_{mf} = \frac{0.00081(\rho_p - \rho_f)gd_p^2}{\mu} \qquad (51)$$

Another widely employed approach is by simplification of the Ergun equation for the packed bed. By combining the original Ergun equation as shown in Eq. (52) and the pressure drop equation at minimum fluidization as shown in Eq. (46)

$$\frac{\Delta P}{L}\frac{g\phi d_p}{2\rho_f U^2}\frac{\varepsilon^3}{(1-\varepsilon)} = 75\frac{(1-\varepsilon)}{\phi(Re)_p} + 0.875 \qquad (52)$$

we obtain the equation for calculating the minimum fluidization velocity, including both the kinetic energy term and the viscous term, as

$$Ar = 150\frac{(1-\varepsilon_{mf})}{\phi^2\varepsilon_{mf}^3}(Re)_{mf} + 1.75\frac{1}{\phi\varepsilon_{mf}^3}(Re)_{mf}^2 \qquad (53)$$

Wen and Yu (1966a, 1966b) suggested a simplified form of the Ergun equation by assuming the followings based on experimental data:

$$\frac{1}{\phi\varepsilon_{mf}^3} \cong 14 \quad \text{and} \quad \frac{(1-\varepsilon_{mf})}{\phi^2\varepsilon_{mf}^3} \cong 11 \qquad (54)$$

In a simplified form, the Ergun equation, in terms of Reynolds number and Archimedes number, can be reduced to

$$(Re)_{mf} = \sqrt{C_1^2 + C_2 Ar} - C_1 \qquad (55)$$

There are many sets of values suggested for the constants C_1 and C_2 based on different databases used for correlation. A simplified set is shown in the table.

Reference	C_1	C_2
Wen and Yu (1966a)	33.7	0.0408
Richardson (1971)	25.7	0.0365
Saxena and Vogel (1977)	25.3	0.0571
Babu et al. (1978)	25.25	0.0651
Grace (1982)	27.2	0.0408
Chitester et al. (1984)	28.7	0.0494

According to Grace (1982), a change of C_1 from the 33.7 suggested by Wen and Yu to 27.2 improves the fit for fine particles. For the limiting cases for small particles where the kinetic energy term is not important, and for large particles where it is dominant, the simplified equations become
For $Ar < 10^3$,

$$(Re)_{mf} = 7.5 \times 10^{-4} \, Ar \quad \text{or}$$

$$U_{mf} = 0.00075\frac{(\rho_p - \rho_f)gd_p^2}{\mu} \qquad (56)$$

For $Ar > 10^7$,

$$(Re)_{mf} = 0.202\sqrt{Ar} \quad \text{or}$$

$$U_{mf} = 0.202\sqrt{\frac{(\rho_p - \rho_f)gd_p}{\rho_f}} \qquad (57)$$

It is worthwhile to note that at both small and large Archimedes numbers, the minimum fluidization velocity, U_{mf}, is directly proportional to the corresponding terminal velocity, U_t, of a single spherical particle of diameter d_p in an infinite medium. For small spheres where the Stokes law applies, U_t/U_{mf} approaches 74. For large spheres in the Newton's law regime, U_t/U_{mf} approaches 8.6.

Since in practical applications the particles are rarely in narrow distribution, the average particle size used in all equations dicussed so far for minimum fluidization velocity is recommended to be the surface-volume mean:

$$\bar{d}_{sv} = \frac{1}{\sum_i \dfrac{x_i}{d_{pi}}} \qquad (58)$$

where x_i is the weight fraction of particle size d_{pi}. All correlations available for calculating the minimum fluidization velocity have been reviewed by Babu et al. (1978). A method of estimating the minimum fluidization velocity at elevated temperatures and pressures was proposed by Yang (1998a). The effect of temperature and pressure on minimum fluidization velocity is discussed in more detail in Chapter 5, "Effect of Temperature and Pressure."

5 THE RATIO OF TERMINAL VELOCITY TO MINIMUM FLUIDIZING VELOCITY

For spherical particles, Bourgeois and Grenier (1968) proposed semitheoretical correlations relating the ratio of terminal velocity to minimum fluidizing velocity with the Archimedes number, Ar, as shown in Eqs. (59) through (64). Depending on whether the fluidized bed is fluidized by air or water, the velocity ratios are slightly different. The terminal velocity of a solid particle has been discussed in Sec. 4 of Chapter 1, "Particle Characterization and Dynamics."

For fluidization by air and $10^2 < Ar < 4 \times 10^4$,

$$R = \frac{U_t}{U_{mf}} = 135.7 - 45.0(\log Ar) + 4.1(\log Ar)^2$$
$$\sigma = 0.69 \quad \text{for} \quad 20 < R < 60 \qquad (59)$$

For fluidization by air and $4 \times 10^4 < Ar < 8 \times 10^6$,

$$R = \frac{U_t}{U_{mf}} = 26.6 - 2.3(\log Ar)$$
$$\sigma = 0.52 \quad \text{for} \quad 10 < R < 20 \qquad (60)$$

For fluidization by air and $Ar > 8 \times 10^6$,

$$R = \frac{U_t}{U_{mf}} = 10.8 \qquad \sigma = 0.5 \qquad (61)$$

For fluidization by water and $50 < Ar < 2 \times 10^4$,

$$R = \frac{U_t}{U_{mf}} = 132.8 - 47.1(\log Ar) + 4.6(\log Ar)^2$$
$$\sigma = 2.9 \quad \text{for} \quad 20 < R < 60 \qquad (62)$$

For fluidization by water and $2 \times 10^4 < Ar < 1 \times 10^6$,

$$R = \frac{U_t}{U_{mf}} = 26.0 - 2.7(\log Ar)$$
$$\sigma = 0.67 \quad \text{for} \quad 9 < R < 20 \qquad (63)$$

For fluidization by water and $Ar > 1 \times 10^6$,

$$R = \frac{U_t}{U_{mf}} = 90 \qquad \sigma = 0.3 \qquad (64)$$

The above equations are valid for spherical isometric particles with negligible wall effect. If the terminal velocity of a particle is available, the above equations can be used to estimate the minimum fluidization velocity as well.

6 TWO-PHASE THEORY OF FLUIDIZATION

The two-phase theory of fluidization was first proposed by Toomey and Johnstone (1952). The model assumed that the aggregative fluidization consists of two phases, i.e., the particulate (or emulsion) phase and the bubble phase. The flow rate through the emulsion phase is equal to the flow rate for minimum fluidization, and the voidage is essentially constant at ε_{mf}. Any flow in excess of that required for minimum fluidization appears as bubbles in the separate bubble phase. Mathematically, the two-phase theory can be expressed as

$$\frac{G_B}{A} = U - U_{mf} \qquad (65)$$

where G_B is the average visible volumetric bubble flow across a given cross section of the bed. For some systems, the two-phase theory is good and simple approximation, but there is also much experimental evidence that correction may be necessary, especially at high pressures. Experimental evidence indicated that the equation should be

$$\frac{G_B}{A} = U - K_4 U_{mf} \qquad \text{with} \qquad K_4 > 1.0 \qquad (66)$$

The reason for this deviation was suggested by Grace and Clift (1974) to be due primarily to two factors: an increase in interstitial gas velocity in the emulsion phase above that required for minimum fluidization and possibly through flow inside the bubbles. This led to their modified or n-type two-phase theory.

$$\frac{G_B}{A} = U - U_{mf}(1 + n\bar{\varepsilon}_B)$$
$$= U - [U_{mf}(1 - \bar{\varepsilon}_B) + (n+1)U_{mf}\bar{\varepsilon}_B] \qquad (67)$$

where $\bar{\varepsilon}_B$ is the average volume fraction of visible bubble phase in the bed. The flow through particulate phase is

$$\text{Flow through particulate phase} = U_{mf}(1 - \bar{\varepsilon}_B)A \qquad (68)$$

and the through flow relative to bubbles is

Through flow relative to bubbles $= (n+1)U_{mf}\bar{\varepsilon}_B A$ (69)

For the classical two-phase theory, $n = 0$. The model implicitly assumes that the superficial velocity in the emulsion phase remains at U_{mf}, while the velocity inside the rising voids in $(n+1)U_{mf}$ relative to the boundary of the bubbles. It was hoped by the authors that n would turn out to be a universal positive constant and thus the magnitude of n would become a measure of the overall through flow in freely bubbling beds. Experimental evidence available so far indicates that n is not universal. It is different from system to system and may be even different from observer to observer. Within a given system, n is also dependent on height and superficial gas velocity. The difficulty arises from the inability to measure the invisible flow components of the total flow, since the invisible flow component is substantial and its distribution remains unknown. The possibility of substantial through flow through the uneven distribution of bubbles was analyzed by Valenzuela and Glicksman (1981) theoretically.

7 VISCOSITY OF A FLUIDIZED BED

A number of investigators have measured the apparent viscosity of fluidized bed using methods available for ordinary liquids. For example, Matheson et al. (1949) and Furukawa and Ohmae (1958) employed a rotating paddle viscometer and found that a fully fluidized bed had a viscosity from 0.5 to 20 poises. The viscosity increases with the size of the particles in the bed. Other researchers, Kramers (1951), Dickman and Forsythe (1953), and Schugerl et al. (1961), used viscometers of slightly different designs and obtained practically similar results. For small shear stress, the fully fluidized beds behave as newtonian fluids. The viscosity of the fluidized bed is very high at close to minimum fluidization condition and decreases sharply with increases in gas flow.

As early as 1949, Matheson et al. reported that the addition of relatively small amounts of fines to a coarse bed will decrease the viscosity of the bed substantially. It remained for Trawinski (1953) to find an analytical explanation for this phenomenon. He assumed that the viscosity in a fluidized bed is primarily due to the rubbing of coarse particles among one another. The addition of fines will reduce the friction between coarse particles by coating a thin layer of fines on each coarse particle and thus decrease the viscosity of a fluidized

bed. The minimum effective amount of fines required to reduce the viscosity is thus that quantity necessary to coat each coarse particle with just a single layer of fine particles, as shown in Fig. 7.

In an actual fluidized bed, the bed is usually expanded during operation, and the voids occupied by the gas bubbles must be subtracted from the volume to give the weight fraction of the "film," which represents more closely the true minimum weight required. Assuming similar density for d_s and d_B, the weight fraction of fines can be calculated as

$$(W)_{d_s} = \frac{(1-\varepsilon)\left\{3\left[\left(\dfrac{d_S}{d_B}\right)+\left(\dfrac{d_S}{d_B}\right)^2\right]-\dfrac{(H-H_0)}{H_0}\right\}}{1+(1-\varepsilon)\left\{3\left[\left(\dfrac{d_S}{d_B}\right)+\left(\dfrac{d_S}{d_B}\right)^2\right]-\dfrac{(H-H_0)}{H_0}\right\}}$$

(70)

where H_0 and H represent the initial and the expanded bed height. Assuming the particles to be perfect spheres, the optimum two-component system with the maximum surface area has been computed to be the one in which the diameter of the fines is 22.5% that of the larger particles, and in which the smaller particles constitute 25% of the mixture by weight. A study by Geldart (1972) employing the surface–volume particle diameter seemed to confirm this as well. Some of the data by Matheson et al. (1949) and replotted by Geldart are presented in Fig. 8.

The importance of fines content in the fluidized bed reactors was also investigated by Grace and Sun (1990) from a different aspect. They found that fines spent a much longer time in the bubble phase and considerably enhanced the concentration inside the bubbles due to the through flow of fluidizing gas through the bubbles. This enhanced particles–gas contact and improved the reactor performance.

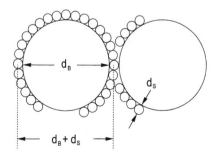

Figure 7 Schematic of coating of fine particles on coarse particles—Trawinski's model of reduction in fluidized bed viscosity upon addition of fines.

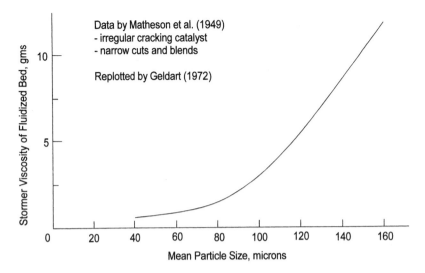

Figure 8 Matheson's data of fluidized bed viscosity. (Replotted by Geldart, 1972.)

Grace (1970) reviewed the literature available in this area in 1970 and concluded that the direct measurement techniques commonly used to determine the viscosity of newtonian liquids tended to alter the behavior of the bed, e.g., the regions underneath and above the probe tended to have high and low voidages. Thus the viscosity obtained by those methods may not be the true viscosity of a fluidized bed. He, in turn, proposed an indirect method based on the behavior of bubbles in fluidized beds.

The spherical-cap bubbles observed in fluidized beds can best be characterized by the included angle as shown in Fig. 9. The angle was found to be dependent only on the bubble Reynolds number, as shown in Eq. (71)

$$(Re)_e = \frac{D_e U_B \rho_f}{\mu} = 23e^{-0.0040} \qquad 200° < \theta < 260° \tag{71}$$

where D_e is the equivalent bubble diameter and can be calculated as

$$D_e = \left(\frac{6V_B}{\pi}\right)^{1/3} \tag{72}$$

The viscosity of the fluidized bed can then be estimated by the equation

$$\mu = \frac{D_e U_B \rho_p (1 - \varepsilon_{mf})}{(Re)_e} \tag{73}$$

The estimated range of the viscosity of fluidized beds is from 4 to 13 poises by using this method. These values are remarkably similar to that obtained by others using a rotating cylinder viscometer. For more discussion on the viscosity of fluidized beds, see that by Schugerl (1971).

The dense phase viscosity of fluidized beds at elevated pressure was studied by King et al. (1981) at pressures up to 20 bar by measuring the velocity of a falling sphere. They found that for particles less than about 100 µm, increases in pressure caused substantial decreases in viscosity. For larger particles, however, the viscosity is almost independent of pressure.

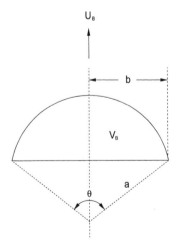

Figure 9 Schematic of spherical-cap bubbles for determination of fluidized bed viscosity.

8 BED EXPANSION

8.1 Expansion of an Aggregative Fluidized Bed

The bed expansion of an aggregative fluidized bed can be calculated using the equation

$$\frac{L}{L_{mf}} = \frac{(1 - \varepsilon_{mf})}{(1 - \varepsilon)} \tag{74}$$

Since the bed surface is oscillating because of constant bubble breaking through the bed surface, the bed height and the voidage are the time-averaged values.

8.1.1 Bed Expansion with a Constant Bubble Size

The bed height oscillation for a bed with a constant bubble size can be calculated from the absolute bubble velocity expressed as in Eq. (75), and the time it requires for the bubble to travel through the bed shown in Eq. (76) (Xavier et al., 1978).

$$U_A = (U - U_{mf}) + U_B \tag{75}$$

$$T = \frac{H_{max}}{U_A} \tag{76}$$

The bed expands from its minimum height, H_{mf}, to its maximum height, H_{max}, with a total bubble flow of $(U - U_{mf})A$, assuming that the two-phase theory holds, or

$$(H_{max} - H_{mf})A = (U - U_{mf})AT \tag{77}$$

The bed expansion equation is obtained by combining Eq. (75) through (77) to be

$$\frac{(H_{max} - H_{mf})}{H_{mf}} = \frac{(U - U_{mf})}{U_B} \tag{78}$$

If the bubble velocity, U_B, is known, the maximum bed expansion can be calculated from Eq. (78).

8.1.2 Bed Expansion with a Constant Bubble Size and an Array of Rods in the Bed

From volumetric balance, it can be written that

$$H_{max} - H_{mf} = (U - U_{mf})T \\ + H_{max}V_{Tmax} - H_{mf}V_{Tmf} \tag{79}$$

where V_{Tmax} = the fraction of bed occupied by rods when bed height is at the maximum, H_{max}, and V_{Tmf} = the fraction of bed occupied by rods when bed height is at the minimum, H_{mf}.

Eliminating the U_A and T among Eqs. (75), (76), and (79), we find

$$\frac{H_{max}(1 - V_{Tmax}) - H_{mf}(1 - V_{Tmf})}{H_{mf}(1 - V_{Tmf}) + H_{max}V_{Tmax}} = \frac{U - U_{mf}}{U_B} \tag{80}$$

8.1.3 Bed Expansion for the Case with Increases in Bubble Size Due to Coalescence

Xavier et al. (1978) employed the bubble coalescence correlation by Darton et al. (1977) and suggested to calculate the bed expansion from the following equation. The bubble coalescence will be discussed in more detail in a later section.

$$H = H_{mf} + \left(\frac{5b}{3}\right)\left[(H + B)^{0.6} - B^{0.6}\right] \\ - 5b^2\left[(H + B)^{0.2} - B^{0.2}\right] \tag{81} \\ + 5b^{2.5}\left\{\tan^{-1}\left[\frac{(H + B)^{0.2}}{b^{0.5}}\right] - \tan^{-1}\left[\frac{B^{0.2}}{b^{0.5}}\right]\right\}$$

where

$$b = 1.917\frac{(U - U_{mf})^{0.8}}{g^{0.4}} \quad \text{and} \quad B = 4\sqrt{A_o} \tag{82}$$

The A_o is the so-called catchment area for a bubble stream at the distributor plate. It is usually the area of distributor plate per orifice.

Geldart (1975) also developed an equation for bed expansion in freely bubbling beds with "sandlike" powders as follows:

$$\frac{H}{H_{mf}} - 1 = \frac{2}{d}\sqrt{\frac{2}{g}} \cdot Y(U - U_{mf})\left[\frac{(c + dH)^{1/2} - c^{1/2}}{H_{mf}}\right] \tag{83}$$

where $d = 0.027(U - U_{mf})^{0.94}$ and $c = 0.915\ (U - U_{mf})^{0.4}$ for porous distributor plates. Y is a correction for deviation from the two-phase theory. Quantitative values of Y are shown in Fig. 26.

For a distributor with N holes per cm^2, c can be calculated by

$$c = \frac{1.43}{g^{0.2}}\left(\frac{U - U_{mf}}{N}\right)^{0.4} \tag{84}$$

All units in Eqs. (83) and (84) are in cm and second.

9 BUBBLE PHASE IN FLUIDIZED BEDS

The bubbles in a fluidized bed have two basic properties. In general, they rise at a finite velocity, and they usually grow in size owing primarily to static pressure or to coalescence. It has been found that there is a striking similarity between the behavior of large gas bubbles in liquids and those in a fluidized bed (Davidson et al., 1977).

9.1 Bubbles in Liquids

In a liquid of small viscosity, the rate of rise of large bubbles depends primarily on inertial forces and surface tension. The viscous effect is negligible in comparison. The shape of the bubble will adjust itself to maintain the pressure inside the bubble constant. An approximate solution by Dumitrescu (1943) for a long bubble in a tube gives

$$U_B = 0.35\sqrt{gD} \tag{85}$$

Equation (85) was verified experimentally by Nicklin et al. (1962) for application to a finite bubble or to a slug rising in a tube. Davies and Taylor (1950) also provided a solution with a slightly different empirical constant.

$$U_B = 0.711\sqrt{gD_e} \tag{86}$$

where D_e is the diameter of the sphere having the same volume as the bubble and can be calculated from Eq. (72).

Uno and Kintner (1956) measured the rising velocity of gas bubbles in liquid contained in tubes of various diameters and found that the wall effect predominates when D_e is more than 1/3 of the bed diameter. The wall effect becomes negligible only when D_e is less than 0.1 of the bed diameter. The regime where the wall effect is dominant is generally called the slugging regime.

In the case of a stream of bubbles in a vertical tube generated continously by blowing air in at the bottom, the absolute upward rising velocity of each bubble is greater than the velocity of a similar size single bubble rising in a stagnant liquid. By making a simple material balance, it is possible to derive the absolute bubble velocity as

$$U_A = \frac{G}{A} + 0.35\sqrt{gD} \tag{87}$$

Nicklin et al. (1962) found experimentally that the absolute bubble velocity is

$$U_A = 1.2\frac{G}{A} + 0.35\sqrt{gD} \tag{88}$$

The factor 1.2 stems from the fact that the peak velocity at the middle of the tube is about 1.2 times the average velocity, owing to the nonuniform velocity profile. The bubbles evidently rise relative to the fastest moving liquid in the middle of the tube.

For a swarm of bubbles, the same thing applies:

$$U_A = U + U_B \tag{89}$$

From the continuity of gas flow it can be derived that

$$\frac{U_B}{U} = \frac{H_0}{(H - H_0)} \tag{90}$$

where H is the liquid height when gas velocity is U, and H_0 is the liquid height when gas velocity is zero. Equation (90) provides a relationship for estimating U_B from experimentally observable quantities. Equations (89) and (90) can be applied to a bubbling gas–solid fluidized bed by replacing U with $U - U_{mf}$, assuming that the two-phase theory applies.

9.2 Bubbles in Gas–Solid Fluidized Beds

Bubbles in gas–solid fluidized beds usually are spherical-capped as shown in Fig. 9 with the included angle equal to 240° found by experiments as compared to 120° derived theoretically. The bubbles in air–water systems have an angle of 100°.

The rising velocity of a single bubble in a quiescent bed has been found experimentally to be

$$U_B = 0.71g^{1/2}V_B^{1/6} \tag{91}$$

This compares with the experimental value of Davis and Taylor (1950) for bubbles in liquids shown in Eq. (86).

Equation (91) implies that the bubble rising velocity is independent of the type of bed materials used in the fluidized beds. This has been experimentally proven by Yasui and Johanson (1958).

A spherical-cap bubble with an included angle, θ, as shown in Fig. 9, has a volume of

$$V_B = \pi R_B^3 \left[\frac{2}{3} - \cos(\theta) + \frac{1}{3}\cos^3(\theta) \right] \tag{92}$$

If Eq. (91) is written in a general form, it can be expressed as

$$U_B = Cg^{1/2}V_B^{1/6} \tag{93}$$

Combining Eqs. (92) and (93), we have

$$C = \frac{2}{3}\pi^{-1/6}\left[\frac{2}{3} - \cos(\theta) + \frac{1}{3}\cos^3(\theta) \right]^{-1/6} \tag{94}$$

For a gas–solid fluidized bed where the bubble velocity can be expressed as in Eq. (91), i.e., $C = 0.71$, the included bubble angle is 120°.

When the bubbles are large enough to exceed approximately 25% of the column diameter, the bubble velocity is affected by the presence of the vessel wall. The bubble velocity can then be approximated by Eq. (85) for a long bubble in a tube containing liquid. The relative regime of application of Eqs. (85) and (91) for gas–solid bubbling fluidized beds is shown in Fig. 10.

9.3 Davidson's Isolated Bubble Model

In 1961, Davidson developed a simple theory that was capable of explaining a lot of phenomena relating to bubbles in fluidized beds observed experimentally. His development involves the following assumptions.

1. The particulate phase is an incompressible fluid with a bulk density similar to that of a fluidized bed at minimum fluidization. The continuity equation of the particles can thus be expressed as

$$\frac{\partial V_x}{\partial x} + \frac{\partial V_y}{\partial y} = 0 \tag{95}$$

2. The relative velocity between the particles and the fluidizing medium is assumed to be proportional to the pressure gradient within the fluid, and thus D'Arcy's law is applicable.

$$U_x = V_x - K\frac{\partial P}{\partial x} \quad \text{and} \quad U_y = V_y - K\frac{\partial P}{\partial y} \tag{96}$$

Since D'Arcy's law (see Chapter 2) is good only for low Reynolds numbers, the same limitation also applies here.

3. Fluidizing fluid is assumed to be incompressible, and thus the continuity equation can be written as

$$\frac{\partial U_x}{\partial x} + \frac{\partial U_y}{\partial y} = 0 \tag{97}$$

with the assumption that the voidage is constant at ε_{mf}.

Eliminating the velocities from Eqs. (95) through (97), we have

$$\frac{\partial^2 P}{\partial x^2} + \frac{\partial^2 P}{\partial y^2} = 0 \tag{98}$$

Equation (98) is exactly similar to that for a fixed bed, and thus the pressure distribution in a fluidized bed is unaffected by the motions of the particles in a fluidized bed.

4. The pressure throughout the bubble is constant.

5. The particulate phase behaves as an inviscid liquid.

6. The bubble has a circular cross section.

Davidson solved these equations in terms of particle motion, the pressure distribution within the fluidizing fluid, the absolute velocities of the fluidizing fluid, and the exchange between the bubble and the particulate phase. He derived the fluid stream function as

$$\psi_f = \left(U_B - \frac{U_{mf}}{\varepsilon_{mf}}\right)\left[1 - \left(\frac{R}{r}\right)^3\right]\frac{r^2\sin^2\theta}{2} \tag{99}$$

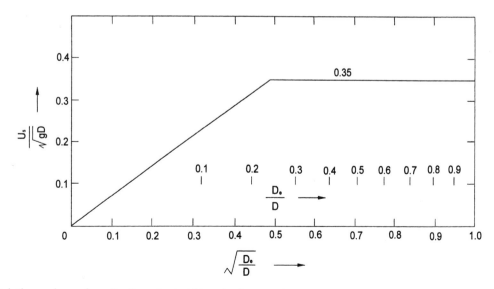

Figure 10 Relative regions of applications for bubble velocity equations.

$$\left(\frac{R}{R_{\mathrm{B}}}\right)^3 = \frac{U_{\mathrm{B}} + 2(U_{\mathrm{mf}}/\varepsilon_{\mathrm{mf}})}{U_{\mathrm{B}} - (U_{\mathrm{mf}}/\varepsilon_{\mathrm{mf}})} \qquad (100)$$

He found that the geometry of the stream function is crucially affected depending whether the bubble velocity U_{B} is larger or smaller than interstitial minimum fluidization velocity, $U_{\mathrm{mf}}/\varepsilon_{\mathrm{mf}}$.

9.3.1 Fast Bubbles Regime—When $U_{\mathrm{B}} > U_{\mathrm{mf}}/\varepsilon_{\mathrm{mf}}$

The fluidizing fluid in this case moves downward relative to the bubble motion. The fluid flows past the fictitious sphere of radius R with velocity $-(U_{\mathrm{B}}-U_{\mathrm{mf}}/\varepsilon_{\mathrm{mf}})$ at $r = \infty$. Inside the sphere of penetration of radius R, the fluid leaves the roof of the buble and recirculates back to the base of the bubble as shown in Fig. 11. The radius of the penetration for a 3-D bed can be calculated from Eq. (101).

$$R = \left[\frac{\dfrac{U_{\mathrm{B}}}{(U_{\mathrm{mf}}/\varepsilon_{\mathrm{mf}})} + 2}{\dfrac{U_{\mathrm{B}}}{(U_{\mathrm{mf}}/\varepsilon_{\mathrm{mf}})} - 1}\right]^{1/3} \cdot R_{\mathrm{B}} \qquad (101)$$

In this case, the bubbles in the fluidized bed are accompanied by a "cloud" while rising through the bed. For fast bubbles where the bubble velocity is large, or for fluidized beds of fine powders when the minimum fluidization velocity is small, the cloud is usually very thin. In most of the fluidized beds of pratical interests, this is the case. The theoretical findings here has been experimentally verified.

The stream function for the particles was obtained to be

$$\psi_{\mathrm{p}} = U_{\mathrm{B}}\left[1 - \left(\frac{R_{\mathrm{B}}}{r}\right)^3\right]\frac{r^2 \sin^2 \theta}{2} \qquad (102)$$

In this case, the particulate phase streams past a sphere of radius R_{B} as shown in Fig. 11. The relative velocity of the particulate flow to the void is $-U_{\mathrm{B}}$ at $r = \infty$.

9.3.2 Slow Bubbles Regime—When $U_{\mathrm{B}} < U_{\mathrm{mf}}/\varepsilon_{\mathrm{mf}}$

In this case, the fluidizing fluid moves upward relative to the bubble motion. This case is usual for beds of large particles and small bubbles. Since R is less than zero in this case, the majority of fluidizing fluid enters the void at the base and leaves from the roof. The fluidizing fluid, in essence, uses the bubble void as the shortcut. The fluidizing fluid penetrates the particulate phase freely from the bubble except for a small fraction of fluid in the shaded area, as shown in Fig. 12 in a circle of radius a', expressed in the following equation.

$$a' = \frac{|R|}{2^{1/3}} = 0.8|R| \qquad (103)$$

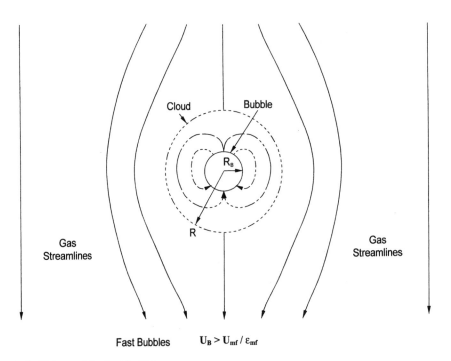

Figure 11 Davidson's bubble model—fast bubbles.

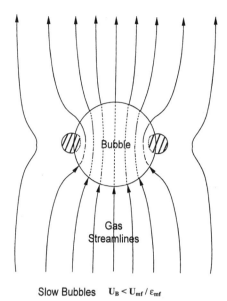

Slow Bubbles $U_B < U_{mf}/\varepsilon_{mf}$

Figure 12 Davidson's bubble model—slow bubbles.

The stream function for the particles in this case can be similarly expressed as in Eq. (102).

In both these cases, i.e., $U_B > U_{mf}/\varepsilon_{mf}$ and $U_B < U_{mf}/\varepsilon_{mf}$, the gas exchange between bubble and particulate phase was obtained by Davidson through an analogy to a fixed bed:

$$q = 3\pi R_B^2 U_{mf} \qquad \text{for a 3-D bed} \qquad (104)$$

The gas exchange rate q expressed in Eq. (104) is the total volumetric flow rate of gas passing through the bubble void. It can either pass through the bubble void or be recirculated back to the bottom of the bubble, depending on the relative magnitude of the bubble velocity and the minimum fluidizing velocity.

There are other models for the bubble motion in fluidized beds of more complexity, primarily those by Jackson (1963) and Murray (1965). They are more exact than the Davidson's model but give essentially the same results.

9.4 Bubble Formation in a Fluidized Bed

Bubble formation in a fluidized bed was found experimentally to be very similar to that in an inviscid liquid. At a very low gas flow rate, the frequency and size of the bubbles formed are primarily governed by a balance between the surface tension of the liquid and the buoyancy force of the bubble. The inertia of the liquid moved by the rising bubbles becomes more improtant than the surface tension at higher gas rates. It is in this

regime that the similarity between the formation of bubbles in a fluidized bed and that in an inviscid liquid is most applicable, because the surface tension in a fluidized bed is zero. At still higher gas rates, the momentum of gas issuing from the orifice can manifest as a jet before breakup into bubbles. The existence of a "jet" in a fluidized bed, however, has become a controversial issue in the literature that will be discussed in more detail later.

Using the analogy of bubble formation in an inviscid liquid, Davidson and Harrison (1963) derived equtions for both the bubble frequency and the bubble size (volume), assuming there is no gas leakage from the bubble to the emulsion phase.

$$t = \frac{1}{g^{0.6}}\left(\frac{6G}{\pi}\right)^{0.2} \qquad (105)$$

$$V_B = 1.138\frac{G^{1.2}}{g^{0.6}} \qquad (106)$$

At high gas flow rates where the bubble sizes are independent of the bed viscosity, the inviscid liquid theory can predict the bubble sizes satisfactorily. At low gas flow rates where the viscosity effects are quite pronounced, the inviscid liquid theory underestimates the bubble sizes. In this case, the following equation by Davidson and Schuler (1960) should be used.

$$V_B = 1.378\frac{G^{1.2}}{g^{0.6}} \qquad (107)$$

Equation (107) fits the data by Harrison and Leung (1961) and the data by Bloore and Botterill (1961) fairly well. There is, however, increasing evidence showing that the gas leakage from the bubble to the emulsion phase can be substantial, especially for large particles. Nguyen and Leung (1972) performed experiments in a 2-D bed with a fluidizing velocity of 1.2 times the minimum fluidizing velocity and found that the bubble volume could be better approximated by the following equation:

$$V_B = 0.53\frac{G}{f_n} \qquad f_n = \text{bubble frequency} \qquad (108)$$

From the computer enhanced video images of rising bubbles in fluidized beds, Yates et al. (1994) observed that the bubbles are surrounded by a region of emulsion phase in which the solids concentration is lower than that in the emulsion phase far from the bubbles. This region of increasing voidage was called the "shell" by Yates et al. The volumetric gas in the bubble and in the shell can be correlated as

$$V_s = 16 V_B^{0.42} \qquad (109)$$

Based on this equation, the volume of void following coalescence of two bubbles is about 28% larger than the volumes of the two constituent voids, because of the incorporation of gas from the shells.

9.5 Coalescence of Bubbles in Fluidized Beds

The bubbles in fluidized beds grow in size due primarily to three factors:

1. The effective hydrostatic pressure descreases toward the top of the fluidized bed.
2. Bubbles coalesce in the vertical direction with the trailing bubble catching up the leading bubble, and
3. Bubbles coalesce in the horizontal direction with the neighboring bubbles.

The effect of the hydrostatic pressure is usually small, and the bubbles grow in size owing largely to coalescence. There are a number of bubble coalescence models in the literature. A few of the more well-known ones are discussed here.

Geldart (1972) found that the fluidization behavior of Group B powders was independent both of the mean particle size and of particle size distribution. In particular, the mean bubble size was found to depend only on the type of the distributor, the distance above the distributor plate, and the excess gas velocity above that required at the minimum fluidization condition, $U - U_{mf}$. Mathematically, it can be expressed as

$$D_B = D_{Bo} + K H^n (U - U_{mf})^m \qquad (110)$$

According to the theory by Davidson and Harrison (1963), the size of a bubble issuing from a single orifice in a fluidized bed at the minimum fluidization condition can be calculated from Eq. (106). The same idea was extended by Geldart to multiorifice distributor plates by replacing G with $(U - U_{mf})/N_o$ and by replacing V_B with $(1 - f_w)\pi D_{Bo}^3 / 6$. By assuming the wake fraction f_w to be 0.25, we have

$$D_B = 1.43 \frac{\left(\dfrac{U - U_{mf}}{N_o}\right)^{0.4}}{g^{0.2}} + K H^n (U - U_{mf})^m \qquad (111)$$

Experimentally, it has been found that for the porous plates, the following equation applies.

$$D_B = 0.915 (U - U_{mf})^{0.4} + 0.027 H (U - U_{mf})^{0.94} \qquad (112)$$

For orifice plates, the following equation can be used:

$$D_B = 1.43 \frac{[(U - U_{mf})/N_o]^{0.4}}{g^{0.2}} + 0.027 H (U - U_{mf})^{0.94} \qquad (113)$$

The constant 0.027 is a dimensional constant with units of $(cm/s)^{-0.94}$. Equation (113) gives reasonable agreement with data from fluidized beds using industrial types of orifice distributor plates. Porous distributor plates, as expressed in Eq. (112), behave as though they contained approximately 1 hole per 10 cm^2 of bed area. The principal effect of adding fines to a fluidized bed of group B powders is the reduction of the mean particle size. At equal values of excess $(U - U_{mf})$, this results in increased bed expansion and solid circulation rates but produces no decrease in mean bubble size.

Mori and Wen (1975) assumed that all gas above the minimum fluidizing velocity went to form a single train of bubbles rising along the center line of the bed and calculated the diameter of bubble that would exist as

$$D_{BM} = 0.652 [A(U - U_{mf})]^{2/5} \qquad (114)$$

The bubble diameter at bed height H can then be estimated as

$$\frac{D_{BM} - D_B}{D_{BM} - D_{Bo}} = \exp\left(-0.3 \frac{H}{D}\right) \qquad (115)$$

For perforated plates, the initial bubble diameter D_{Bo} is expressed as

$$D_{Bo} = 0.347 \left[\frac{A(U - U_{mf})}{N_o}\right]^{2/5} \qquad (116)$$

where A is the area of the bed and N_o is the total number of orifices. For porous plates, the following expression should be used to estimate the initial bubble sizes.

$$D_{Bo} = 0.00376 (U - U_{mf})^2 \qquad (117)$$

The validity of the above equations has been tested within the ranges of the following parameters:

$30 < D < 130$ cm $\qquad\qquad$ $0.5 < U_{mf} < 20$ cm/s

$0.006 < d_p < 0.045$ cm $\qquad$ $U - U_{mf} < 48$ cm/s

Rowe (1976) suggested the following equation to estimate the bubble size in a fluidized bed:

$$D_B = \frac{(U - U_{mf})^{1/2} (H + h_o)^{3/4}}{g^{1/4}} \qquad (118)$$

Here h_o is an emperical constant and is a characteritics of the distributor plate. The h_o is effectively zero for a porous plate buy may be more than a meter for large tuyeres.

Darton et al. (1977) assumed that the bubbles are lined up as close together as possible, as shown in Fig. 13. They also defined a so-called "catchment area" for each particular bubble track. The bubble frequency can then be calculated by $U_B/2R_B$ with the bubble velocity $U_B = 0.711(gD_e)^{1/2}$. The bubble flow in each track can be calculated as follows assuming the two-phase theory.

$$(U - U_{mf}) = \left(\frac{\pi D_e^3}{6}\right)\left(\frac{U_B}{2R_B}\right)$$

$$= \left(\frac{\pi D_e^3}{12R_B}\right)(0.711\sqrt{gD_e}) \qquad (119)$$

If the bubbles are hemispheres, the volume of each individual bubble can be calculated by the equation

$$\left(\frac{2R_B^3}{3}\right) = \left[\frac{\pi D_e^3}{6}\right] \qquad (120)$$

Substituting Eq. (120) into Eq. (119), we have

$$D_{e(o)} = 1.63\left[\frac{(U - U_{mf})A_c}{\sqrt{g}}\right]^{2/5} ; \ A_c = \text{catchment area} \qquad (121)$$

The catchment area is defined as the area of distributor plate per hole. When two bubbles of equal volume from the nth stage coalesce to form a bubble of the $(n + 1)$th stage, we have

$$D_{e(n+1)}^3 = 2D_{e(n)}^3 \qquad (122)$$

If we further assume that the height of each bubble coalescence stage is proportional to the diameter of the catchment area, we have

$$A_c = \frac{\pi}{4}D_c^2 \qquad \text{and} \qquad h = \lambda D_c \qquad (123)$$

where D_c is the diameter of a circular catchment area for each bubble stream. We have from Eq. (123) that

$$h_n = \lambda D_{co} + \lambda D_{c1} + \cdots + \lambda D_{c(n-1)} \qquad (124)$$

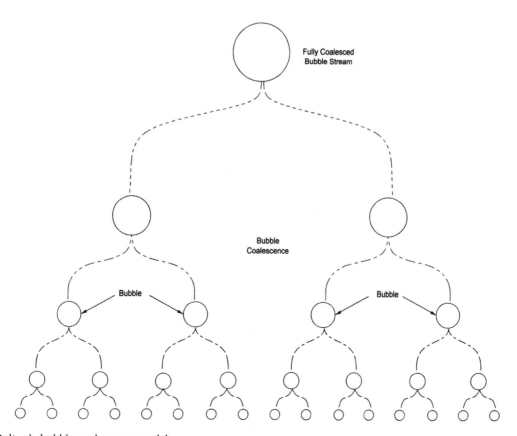

Figure 13 Dalton's bubble coalescence model.

Since $D_{e(n)} = 2^{n/3} D_{e(o)}$, the height of the nth stage is

$$h_n = \frac{0.62 g^{0.25} \lambda}{(U - U_{mf})^{0.5}} D_{e(o)}^{5/4} \sum_{n=0}^{n-1} 2^{5n/12} \qquad (125)$$

The total bed height can thus be expressed as

$$H = \frac{1.85 g^{0.25} \lambda}{(U - U_{mf})^{0.5}} (D_{e(n)}^{5/4} - D_{e(o)}^{5/4}) \qquad (126)$$

It was experimentally found by Darton et al. that $\lambda = 1.17$.

With the initial bubble diameter shown in Eq. (121), we have

$$D_{e(n)} = \frac{0.54(U - U_{mf})^{2/5}(H + 4.0\sqrt{A_c})^{4/5}}{g^{1/5}} \qquad (127)$$

If the bubbles grow to the size of the vessel diameter, the bed becomes a slugging bed. From this analysis, this occurs at the following conditions:

$$\frac{H}{D} > 3.5\left(1 - \frac{1}{\sqrt{N_o}}\right) \qquad (128)$$

Zenz (1977) assumed the bubble growth in the fluidized bed resembles the well-known Fibonacci series (Zenz, 1978) and propsoed the following equation for bubble growth.

$$\frac{D_B}{D_{Bo}} = 0.15\left(\frac{H}{D_{Bo}}\right) + 0.85 \qquad (129)$$

Bubble growth corresponding to the Fibonacci series is depicted in Fig. 14.

Choi et al. (1998) proposed a generalized bubble-growth model on mean bubble size and frequency for Geldart's Group A, B, and D particles. The model made use of empirical correlations for volumetric bubble flux and bubble splitting frequency. The proposed model correlated well with the extensive data reported in the literature on mean bubble size and frequency. They also found that the equilibrium bubble diameter increased linearly with the ratio of volumetric bubble flux to the splitting frequency of a bubble.

All the models cited for the growth of bubbles in fluidized beds have assumed some type of ordered progression in the bubble coalescence mechanism independent of solid movement in the beds. Whitehead (1979) pointed out that the pattern of bubble coalescence and solid circulation in large industrial fluidized beds depended on both the bed depth and the operating velocity. Figure 15 shows the patterns observed in large fluidized beds of Group B powders with industrial-type distributor plates (Whitehead and Young, 1967; Whitehead et al., 1977). Unless the solids move-

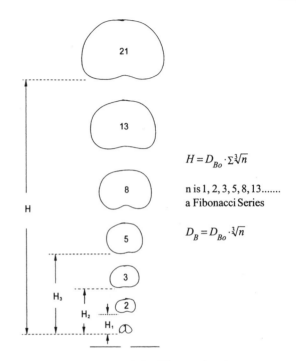

$$H = D_{Bo} \cdot \Sigma \sqrt[3]{n}$$

n is $1, 2, 3, 5, 8, 13\ldots\ldots$
a Fibonacci Series

$$D_B = D_{Bo} \cdot \sqrt[3]{n}$$

Figure 14 Bubble growth by Fibonacci series.

ment in the bed is also taken into account in developing the bubble coalescence model, the model may not accurately predict the actual phenomena occurring in industrial fluidized beds. Bubble growth in large fluidized beds was also studied by Werther (1967a).

9.5.1 Bubble Coalescence from Mutiple Entry Nozzles

Bubble coalescence from mutiple entry nozzles was studied by Yates et al. (1995) for a Group A powder by means of x-rays. The simple correlation expressed here was found to correlate the data well:

$$h_c = 39.6\left(\frac{U}{U_{mf}}\right)^{-1/3} l_s \qquad (130)$$

where h_c is the average height above the orifice at which coalescence was complete and l_s is the orifice separation distance. The volume of the bubble void and its associated gas shell following coalescence can be correlated as

$$V_s = 31 V_B^{0.42} \qquad (131)$$

This is approximately twice as large as the bubble volumn in a bubbling fluidized bed obtained by the same authors (Yates et al., 1994) as shown in Eq. (109).

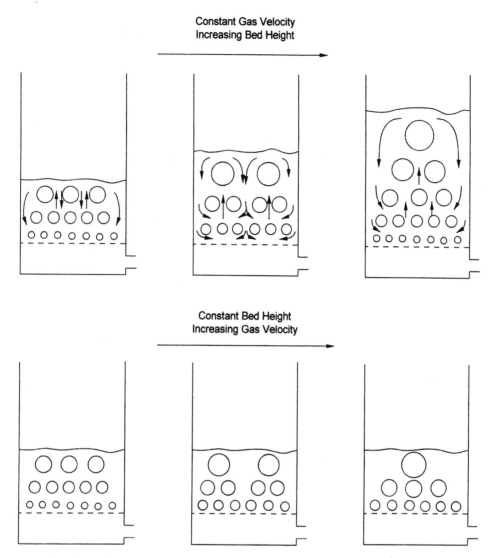

Constant Gas Velocity
Increasing Bed Height

Constant Bed Height
Increasing Gas Velocity

Figure 15 Patterns of bubble coalescence and solids circulation in large industrial fluidized beds. (Adapted from Whitehead, 1967.)

10 SLUGGING BEDS

The slugging phenomenon in fluidized beds has been extensively studied and is described in detail in Hovmand and Davidson (1971). A slugging bed is characterized by gas slugs of sizes close to reactor cross section that rise at regular intervals and divide the main part of the fluidized bed into alternate regions of dense and lean phase. The passage of these gas slugs produces large pressure fluctuations inside the fluidized bed. The occurrence of slugging is usually accompanied by deterioration in quality of bed mixing and gas–solid contacting. The slugging generally occurs in reactors of laboratory and pilot plant scale. There are basically two types of slugging fluidized beds. Type A slugging beds consists of axisymmetric round-nosed gas slugs, the solids flow past the gas slugs in an annular region close to the wall, as shown in Fig. 16. This type of slugging bed usually occurs with bed materials that fluidize easily, such as Group A and B powders. The type B slugging beds have slugs that are essentially square-nosed (see Fig. 16). The gas slugs occupy the complete bed cross section. The only way the solids can pass through the gas slug is by raining down through the slugs as solids streamers. For cohesive and particles of angular shape, this type of slugging bed is most prevalent. Slugging beds with combination of types A and B have also been observed, depending on the oper-

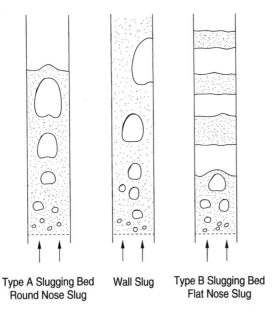

Type A Slugging Bed   **Wall Slug**   **Type B Slugging Bed**
Round Nose Slug   **Flat Nose Slug**

Figure 16 Type A and Type B slugging beds, and wall slugs.

ating gas velocity and the bed depth. The so-called wall slugs, which appear as half of a round-nosed slug, can also be classified as a variation of the type A slug.

10.1 Slugging Criteria

The Stewart and Davidson criterion on slugging (Stewart and Davidson, 1967) can be simply expressed as

$$U - U_{mf} > 0.2U_s = 0.2(0.35\sqrt{gD}) \quad \text{for slugging} \tag{132}$$

The criterion was developed based on a number of assumptions. The height of the slugs was assumed to be equivalent to the diameter of the bed. The distance between each gas slug was assumed to be two times the diameter of the bed and no coalescence of slugs occurred. The volume was assumed to be $\pi D^3/8$, and the slug rises at a velocity of $0.35(gD)^{0.5}$ relative to the surrounding solids by analogy to the gas–liquid systems. If the two-phase theory applies, we can write

$$(U - U_{mf})A = U_{SA}\varepsilon_S A \tag{133}$$

where U_{SA} is the absolute slug velocity and ε_S is the fraction of the bed occupied by the slugs, which can be expressed as

$$U_{SA} = (U - U_{mf}) + 0.35\sqrt{gD} \tag{134}$$

$$\varepsilon_S = \frac{\pi D^3/8}{(\pi D^2/4)3D} = \frac{1}{6} \tag{135}$$

Hence at the onset of slugging,

$$(U - U_{mf} = \left(\frac{1}{6}\right)\left[(U - U_{mf}) + 0.35\sqrt{gD}\right] \tag{136}$$

which reduces to Eq. (132).

In beds of height less than 30 cm, Baeyens and Geldart (1974) found that it was much more difficult to induce slugging. They proposed a modified slugging criterion as

$$(U - U_{mf}) > 0.07\sqrt{gD} + 1.6 \times 10^{-3}(60D^{0.175} - H_{mf})^2 \tag{137}$$

Equation (137) is essentially an empirical correlation, and c.g.s. units are to be used.

10.2 Slugging Bed Expansion

The bed surface of a slugging bed oscillates alternately between a maximum and a minimum following the eruption of each slug. The bed expansion equation was first derived by Matsen et al. (1969) following the two-phase theory:

$$\frac{(H_{max} - H_{mf})}{H_{mf}} = \frac{(U - U_{mf})}{0.35\sqrt{gD}} \tag{138}$$

The variation of the bed height of a slugging bed is shown in Fig. 17. Through same kind of argument, the total fluctuation of bed height in a slugging bed has been suggested by Kehoe and Davidson (1973a):

$$\frac{\Delta H}{D} = \left[\frac{(U - U_{mf})}{U_B}\right]N_T \tag{139}$$

where N_T is the number of bed diameters between the rear of a leading slug and the nose of a trailing one. Combining Eqs. (138) and (139) gives the minimum or collapsed bed height of a slugging bed:

$$\frac{H_{min}}{H_{mf}} = 1 + \left(1 - \frac{N_T D}{H_{mf}}\right)\left[\frac{(U - U_{mf})}{U_B}\right] \tag{140}$$

10.3 Slug Length and Slug Frequency

From the material balance of solids and gas, it can be calculated that the slug length is

$$L_S = \frac{[(U - U_{mf})N_T\sqrt{D}]}{0.35\sqrt{g}} \tag{141}$$

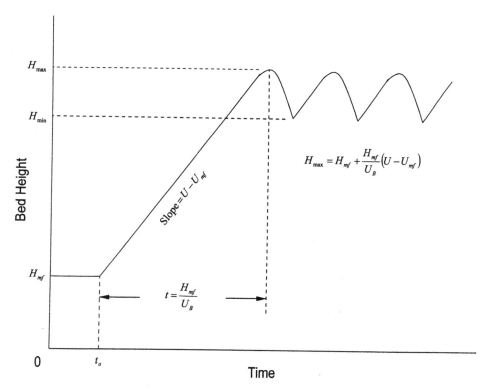

$$H_{\max} = H_{mf} + \frac{H_{mf}}{U_B}\left(U - U_{mf}\right)$$

$$t = \frac{H_{mf}}{U_B}$$

Figure 17 Slugging bed height.

Consequently, the slug frequency can be expressed as

$$f_S = \frac{(U - U_{mf})}{L_S} = \frac{0.35\sqrt{g}}{N_T\sqrt{D}} \tag{142}$$

This means that the slug frequency is independent of the operating velocity. The only unknown in Eq. (142) is the spacing between the slugs, which has to be determined experimentally. Values of N_T from 2 up to 8 have been observed experimentally (Kehoe and Davidson, 1973b).

Hovmand and Davidson (1968) derived an equation for calculating the slug length by assuming an interslug spacing to be two times the diameter of the bed.

$$\frac{L_S}{D} - 0.495\left(\frac{L_S}{D}\right)^{0.5}(1 + B) + 0.061 - 1.94B = 0 \tag{143}$$

The equation was later modified by Kehoe and Davidson (1973b) to read as

$$\frac{L_S}{D} - 0.495\left(\frac{L_S}{D}\right)^{0.5}(1 + B) + 0.061 - (T_S - 0.061)B = 0 \tag{144}$$

where

$$B = \frac{(U - U_{mf})}{U_B} > 0.2 \tag{145}$$

and T_S is called the stable slug spacing factor, the value of N_T at which a following slug ceases to be attracted to a leading one.

10.4 Minimum Bed Height Required for Slugging

From the analysis by Baeyens and Geldart (1974), there are three separate zones in a deep bed operating with an excess velocity. In Zone 1, there is a freely bubbling bed. There is a slugging bed in Zone 2, but the slug grows with continous coalescence. Only in Zone 3 is the slug coalescence complete and stable slug spacing is established. They suggest to calculate the minimum bed height for stable slugging using the following equation:

$$H_L = 60D^{0.175} \qquad D \text{ in cm} \tag{146}$$

The maximum bed height below which the bed will be freely bubbling can be calculated from

$$H_{fb} = \frac{(D - 2.51D^{0.2})}{0.13D^{0.47}} \qquad D \text{ in cm} \tag{147}$$

By using the bubble coalescence theory of Darton et al. (1977), it can be derived that the limiting minimum bed height required for slugging can be expressed as in Eq. (128).

10.5 Wall Slugs

Wall slugs usually form with large particles. The wall slugs move faster than the round-nosed slugs and its motion can be calculated as

$$U_S = 0.35\sqrt{2gD} \tag{148}$$

The wall slugs think they are in a bed of diameter $2D$.

10.6 Effect of Expanded Section on Slugging Bed Height

An effective way to suppress the slugging and to reduce the maximum slugging bed height is to expand the bed cross-sectional area toward the top of a fluidized bed. Yang and Keairns (1980b) developed a correlation to predict the effect of an expanded section on the slugging bed height. They divided the expansion of a slugging fluidized bed with an expanded section into three different stages, as shown in Fig. 18.

For $H_{mf} < h$, the maximum bed height can be calculated by the following two equations for Case I and Case II. For Case I, the gas leakage from the gas slug to the surrounding emulsion phase in the expanded section is assumed to be negligible. In Case II, the gas leakage from the gas slug to the surrounding emulsion phase in the expanded section is assumed to be instantaneous, so that the emulsion phase in the expanded section is always minimally fluidized.

For Case I where $H_{mf} < h$,

$$\frac{H_{max}}{H_{mf}} = \frac{h}{H_{mf}} + \left(\frac{D_1}{D_2}\right)^2 \left[1 + \left(\frac{D_1}{D_2}\right)^2 \cdot \frac{(U_1 - U_{mf})}{U_{B2}}\right]$$
$$\left\{1 - \frac{h}{H_{mf}} \cdot \frac{1}{\left[1 + \frac{(U_1 - U_{mf})}{U_{B1}}\right]}\right\} \tag{149}$$

For Case II where $H_{mf} < h$,

$$\frac{H_{max}}{H_{mf}} = \frac{h}{H_{mf}} + \frac{(U_2 - U_{mf})}{(U_1 - U_{mf})}\left[1 + \frac{(U_2 - U_{mf})}{U_{B2}}\right]$$
$$\left\{1 - \frac{h}{H_{mf}} \cdot \frac{1}{[1 + (U_1 - U_{mf})/U_{B1}]}\right\} \tag{150}$$

Similar equations were also developed for the design where $H_{mf} > h$, as is shown here:

For Case I where $H_{mf} > h$,

$$\frac{H_{max}}{H_{mf}} = 1 + \left(\frac{D_1}{D_2}\right)^2$$
$$\left[\frac{h}{H_{mf}} \cdot \frac{(U_1 - U_{mf})}{U_{B1}} + \left(1 - \frac{h}{H_{mf}}\right) \cdot \frac{(U_1 - U_{mf})}{U_{B2}}\right] \tag{151}$$

For Case II where $H_{mf} > h$,

$$\frac{H_{max}}{H_{mf}} = 1 + \left[\frac{h}{H_{mf}} \cdot \frac{(U_2 - U_{mf})}{U_{B1}}\right.$$
$$\left. + \left(1 - \frac{h}{H_{mf}}\right) \cdot \frac{(U_2 - U_{mf})}{U_{B2}}\right] \tag{152}$$

Experimental results from Yang and Keairns (1980b) indicated that the best results were obtained when the bubble velocities were assumed to be

$$U_{B1} = U_{B2} = 0.35\sqrt{gD_1} \tag{153}$$

This led to the conclusion that the gas leakage from the gas slug to the emulsion phase at the expanded section is very small, and the emulsion phase at the expanded section is actually less than minimally fluidized. By the time the bed surface expands into the expanded section, the bed tends to defluidize and cause bridge formation at the transition section, especially at high bed heights. On the basis of the experimental evidence, the expanded section will be effective in reducing slugging. Care must be exercised, however, in locating the expanded section at a proper distance from the gas distributor plate or to provide aeration at the transition to reduce the bridging tendency at the conical transition.

11 JETTING PHENOMENA IN FLUIDIZED BEDS

Jets in fluidized beds appear usually in the gas entry region of the bed, such as a gas distributor plate. The construction of a gas distributor plate can be variable and usually differs from application to application. The gas distributor plate designs for fluidized beds are discussed in Chapter 6, "Gas Distributor and Plenum Design." The gas distribution arrangement can be a simple perforated or orifice plate, a pipe sparger, or more complicated tuyeres and caps. When gas velocity through the orifices is low, the gas issued from the orifices appears as discrete bubbles. However, at a high orifice velocity, a standing jet with periodically truncated bubbles is formed. The

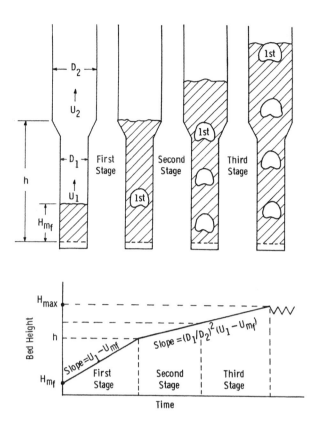

Figure 18 Effect of expansion sections on slugging bed height.

jets from a gas distributor plate are generally single phase, i.e., gas phase, and small in diameter, usually less than 15 mm. The study of jetting phenomena at the distributor plate is of interest because the intense gas–solid mixing, and the heat and mass transfer induced by the jets, are important for the overall performance of fluidized bed reactors. The jetting zone immediately above a grid is characterized by a time-averaged, sharp, vertical variation of bed density. A large fluctuation of bed density occurs in this zone, indicating extensive mixing and contacting of solids and gas. For chemical reactors where the reaction rate is fast, much of the conversion may occur in this jetting region. To achieve uniform gas distribution, a minimum pressure drop equivalent to at least 30% of the fluidized bed pressure drop is required. This necessitates a small orifice diameter in practice, generally less than 15 mm. Thus most of the existing jet correlations were developed on the basis of data generated from small jets.

When solids are fed pneumatically into operating fluidized beds, they generate jets as well. In this application, they are gas-solids two-phase jets. The gas and solids entrainment into the jets and the extent of the jet region (jet penetration depth) may characterize the performance of the reactor and sometimes dictate the reactor design. Because of the sizes of solids to be fed continously into the beds, the feeding nozzles and thus the jets are usually larger than 15 mm.

There is another category of jets in fluidized beds that tend to be substantially larger in size. For example, a large central jet is applied in a jetting fluidized bed, similar to a spouted bed, to induce solids mixing and circulation, to facilitate heat and mass transfer from one region of a fluidized bed to an adjacent region, and to promote granulation and agglomeration. This type of jetting fluidized beds were used for solids mixing and drying, coal gasification, and powder granulation. The study of large jets, substantially larger than 15 mm in diameter, is not as common (Yang, 1998b). The jetting fluidized beds are discussed in Chapter 20, "Other Nonconventional Fluidized Beds."

An ambiguity in studying the jetting phenomenon is the lack of consensus on what constitutes a jet (Rowe et al., 1979; Yang and Keairns, 1979). The gas issuing from an orifice might be in the form of bubbles, a pulsating jet (a periodic jet), or a permanent flamelike jet, depending on the relative properties of the gas, the bed material, and the operating conditions.

11.1 Different Jetting Regimes Observed in Fluidized Beds

Different jetting regimes were observed experimentally by different researchers using different bed configurations, different bed materials, and different operating conditions (see review by Massimilla, 1985).

When the gas velocity is low, the bed material is dense, and the particle size is small, the gas jet issuing from the orifice or the nozzle tends to be truncated into bubbles right at the orifice, a phenomenon very similar to that observed when gas is injected into a liquid medium. The chain of bubbles, at a frequency of 5 to 20 Hz, resembles a bubbling plum and can be called a bubbling jet although some authors, e.g., Rowe et al. (1979), prefer to reserve the word "jet" for the cases where a permanent flamelike void exists. The bubbling jet has been observed experimentally by Rowe et al. (1979) using x-rays in a three-dimensional bed and by Clift et al. (1976).

Markhevka et al. (1971) observed a jet in a fluidized bed located close to the wall formed elongated cavities, which were periodically truncated to become bubbles at the orifice. During the formation of the elongated voids, solids were observed to become entrained into the voids. Merry (1975) later proposed a correlation for jet penetration based on this observation. Tsukada and Horio (1990) studied the gas motion and bubble formation at the distributor of a semicircular fluidized bed with a transparenet front wall. They found that there were two different regions above the grid zone. One they called the "jet stem" region, where stable jets existed, and the other, the "bubble forming" region. The height of the bubble forming region is of the same order of magnitude as the initial bubble size. The fraction of the bubble forming region is larger when fine particles are used as bed material compared to the coarse particles.

With increasing gas density, such as operating under high pressure, and when the gas stream has a high momentum, such as delivering through a pneumatic transport line containing solids, the jet issuing into the fluidized bed exhibits itself as an oscillating jet plume, as shown in Figs. 19a and 19c. There is a high rate of coalescence immediately above the orifice or jet nozzle, and the truncation of the jet no longer occurs at the opening of the orifice, moreover, there is always a permanent void present at the minimum jet

penetration. At the maximum jet penetration depth, the jet plume is constructed through a series of bubbles with interconnecting roofs and bottoms. Solids were observed to penetrate through the bubble roof into the bottom of the bubble immediately above it. A bubble is formed at the end of the jet plume after the momentum of the jet dissipates. The oscillating jet plume was observed by Knowlton and Hirsan (1980) in a 1-foot diameter pressurized fluidized bed operated up to 53 atm; by Yang and Keairns (1978) using hollow epoxy spheres of density 210 kg/m^3 in a 1-foot diameter fluidized bed operated at atmospheric pressure; and by Yang and Keairns (1980a) and Kececioglu et al. (1984) in a jet delivered via a pneumatic transport line containing solids of different loadings.

Permanent flamelike jets have been observed primarily in two-dimensional beds, in fluidized beds with spoutable bed materials, and the emulsion phase around the jet is less than minimally fluidized. The permanent flamelike jet was observed by Kozin and Baskakov (1967), Zenz (1968), and Wen et al. (1977) in two-dimensional beds, and by Yang and Keairns (1980a) in a semicircular fluidized bed with a large spoutable bed material.

From the above classification, the word jet is used loosely to represent a region in the fluidized bed created mainly by the momentum of the gas stream or solids-carrying gas stream issuing from an orifice. The important parameters influencing the jetting modes are the characteristics of the bed material (particle size, particle density and perhaps particle size distribution), the gas density/solid density ratio, the gas velocity from the orifice, the solid loading in the gas stream, and the design configurations (such as nozzle or orifice; single or multiple orifices). A more objective technique for measurement of the jet penetration height by pressure signal analysis has recently being suggested by Vaccaro et al. (1997).

Grace and Lim (1987) proposed a simple criterion based on stable spouting correlation by Chandnani and Epstein (1986), as shown in Eq. (154), as a necessary but not sufficient condition for permanent flame jet formation in fluidized, spout-fluid, and spouted beds. They also observed that increasing temperature has a significant destabilizing influence on jet formation, and that a lower ratio of orifice and particle diameter is required for permanent jet formation at high temperatures. Larger auxiliary gas flow can also lead to unstable jets and absence of permanent flame jets. For jets other than in the vertical direction, the formation of permanent flame jets is less likely.

$$\frac{d_{or}}{d_p} \leq 25.4 \qquad (154)$$

In 1993, Roach proposed a critical Froude number, $(Fr)_c$, as a demarcation for jetting and bubbling. For systems above the critical Froude number, bubbles are formed; below it, jets are present. The critical Froude number is expressed as follows:

$$(Fr)_c = \frac{U}{\sqrt{gd_p}} = 520\beta^{-1/4}\left[\frac{(d_p/d_{or})}{(\rho_p/\rho_f)}\right]^{1/2} \qquad (155)$$

where U is the superficial fluidization velocity and β is the distributor porosity. The criterion thus relates the bubbling and jetting phenomena to the flow through multiple orifices on a distributor plate.

11.2 Characteristics of Jets in Fluidized Beds

The jets in fluidized beds have the following properties: the jet penetration depth, the jet expansion angle (or the jet half angle), gas and solids entrainment, initial bubble size, and frequency issuing from the jets. They will be discussed now.

11.2.1 Jet Momentum Dissipation

The jet momentum dissipation can be calculated from the following equation (Yang, 1998b):

$$\frac{U_j}{U_m} = 0.26\frac{x}{d_o} \quad \text{or} \quad \frac{U_m}{U_j} = 3.84\frac{d_o}{x} \qquad (156)$$

This compares to the velocity scale proposed for a homogeneous circular jet in an infinite medium as shown in Eq. (157).

$$\frac{U_j}{U_m} = 0.16\frac{x}{d_o} \quad \text{or} \quad \frac{U_m}{U_j} = 6.30\frac{d_o}{x} \qquad (157)$$

This means that the maximum jet velocity at the axis, U_m, for a jet in a fluidized bed, dissipates faster than that in a homogeneous medium by a factor of $6.3/3.84 = 1.64$. Equation (156) applies only in the region beyond the potential core.

11.2.2 Jet Potential Core

For small particles, like the cracking catalyst, experiments by Behie et al. (1970, 1975, 1976) indicated that the potential core, where the properties of the jet are essentially similar to that at the jet nozzle, is within about half of the nozzle diameter. Donsi et al. (1980) found that the potential core length depended on the particle characteristics and nozzle size in a very com-

plex way. Because of the lack of understanding, the potential core is usually neglected in the analysis of jets in fluidized beds.

11.2.3 Jet Penetration Depth

There are at least a dozen proposed correlations in the literature for calculating the jet penetration depth for a vertical jet, they are summarized in Table 3. The jet penetration correlations were also reviewed by Blake et al. (1984), Massimilla (1985), and more recently Kimura et al. (1994, 1995). The discrepancy between the prediction from different correlations sometimes can be more than an order of magnitude on the dependency of gas velocity or operating pressure. The primary reason is because of the complex jetting modes observed experimentally, which create conflicting definitions of jet penetration depth by different researchers. Those by Merry (1975), Knowlton and Hirsan (1980), and Kececioglu et al. (1984) are presented in Fig. 19.

The recommended approach to estimate the jet penetration depth is to make use of the two-phase Froude number defined as

$$(Fr)^{0.5} = \left[\frac{\rho_f}{(\rho_p - \rho_f)}\frac{U_j^2}{gd_o}\right]^{0.5} \qquad (158)$$

The two-phase Froude number was first suggested by Yang and Keairns (1978) to calculate the jet penetration depth in the equation

$$\frac{L_j}{d_o} = 6.5\left[\frac{\rho_f}{(\rho_p - \rho_f)}\frac{U_j^2}{gd_o}\right]^{0.5} = 6.5 \cdot (Fr)^{0.5} \qquad (159)$$

There was no corresponding theoretical basis proposed at the time. In fact, the dependence of jet penetration on the two-phase Froude number can be derived theoretically from the buoyancy theory of Turner (1973) as shown by Yang (1998a). The importance of the two-phase Froude number was also recently verified by Vaccaro (1997). Equation (159) was later modified for high temperature and high pressure applications (Yang, 1981).

$$\frac{L_j}{d_o} = 7.65\left[\frac{1}{R_{cf}}\frac{\rho_f}{(\rho_p - \rho_f)}\frac{U_j^2}{gd_o}\right]^{0.472} \qquad (160)$$

where

$$R_{cf} = \frac{(U_{mf})_p}{(U_{mf})_{atm}} \qquad (161)$$

Table 3 Summary of Jet Penetration Correlations

Jet penetration correlations	References

Vertical single jet

$$\frac{L_{\mathrm{j}}}{d_{\mathrm{o}}} = 6.5\left[\left(\frac{\rho_{\mathrm{f}}}{\rho_{\mathrm{p}} - \rho_{\mathrm{f}}}\right)\left(\frac{U_{\mathrm{j}}^2}{gd_{\mathrm{o}}}\right)\right]^{0.5}; \quad L_{\mathrm{j}} = \frac{(L_{\min} + L_{\max})}{2}$$

Yang and Keairns (1978); Yang (1998a)

$$\frac{L_{\mathrm{B}}}{d_{\mathrm{o}}} = 26.6\left(\frac{\rho_{\mathrm{f}}U_{\mathrm{j}}}{\rho_{\mathrm{p}}\sqrt{gd_{\mathrm{p}}}}\right)^{0.67}\left(\frac{U}{U_{\mathrm{cf}}}\right)^{-0.24}$$

Hirsan et al. (1980)

$$\frac{L_{\max}}{d_{\mathrm{o}}} = 19.3\left(\frac{\rho_{\mathrm{f}}U_{\mathrm{j}}}{\rho_{\mathrm{p}}\sqrt{gd_{\mathrm{p}}}}\right)^{0.83}\left(\frac{U}{U_{\mathrm{cf}}}\right)^{-0.54}$$

$$\frac{L_{\max}}{d_{\mathrm{o}}} = 7.65\left[\frac{(U_{\mathrm{cf}})_{\mathrm{atm}}}{(U_{\mathrm{cf}})_{\mathrm{p}}}\left(\frac{\rho_{\mathrm{f}}}{\rho_{\mathrm{p}} - \rho_{\mathrm{f}}}\right)\left(\frac{U_{\mathrm{j}}^2}{gd_{\mathrm{o}}}\right)\right]^{0.472} \quad \text{for} \quad U = U_{\mathrm{cf}}$$

Yang (1981); Yang (1998a)

$$\frac{L_{\max}}{d_{\mathrm{o}}} = C\left[\frac{(U_{\mathrm{cf}})_{\mathrm{atm}}}{(U_{\mathrm{cf}})_{\mathrm{p}}}\left(\frac{\rho_{\mathrm{f}}}{\rho_{\mathrm{p}} - \rho_{\mathrm{f}}}\right)\left(\frac{U_{\mathrm{j}}^2}{gd_{\mathrm{o}}}\right)\right]^{0.835} \quad \text{for} \quad U > U_{\mathrm{cf}}$$

Yang (1981); Yang (1998a)

$$1.81 \leq C \leq 4.21 \text{ depending on solid properties}$$

Vertical multiple distributor jets

$$\left(\frac{L_{\mathrm{j}}}{d_{\mathrm{o}}} + \frac{1}{2}\cot an\theta\right) = 13\left(\frac{\rho_{\mathrm{f}}U_{\mathrm{j}}}{\rho_{\mathrm{p}}\sqrt{gd_{\mathrm{p}}}}\right); \quad L_{\mathrm{j}} = \frac{(L_{\min} + L_{\max})}{2}$$

Shakhova (1968)

$$0.0144\frac{L_{\max}}{d_{\mathrm{o}}} + 1.3 = 0.5\log(\rho_{\mathrm{f}}U_{\mathrm{j}}^2); \quad \rho_{\mathrm{f}} \text{ in lb/ft}^3 \text{ and } U_{\mathrm{j}} \text{ in ft/s}$$

Zenz (1968)

$$\frac{L_{\max}}{d_{\mathrm{o}}} = \left(\frac{0.919d_{\mathrm{p}}}{0.007 + 0.566d_{\mathrm{p}}}\right)\frac{U_{\mathrm{j}}^{0.35}}{d_{\mathrm{o}}^{0.3}}; \quad \text{units in cgs system}$$

Basov et al. (1969)

$$\frac{L_{\max}}{d_{\mathrm{o}}} = 5.2\left(\frac{\rho_{\mathrm{f}}d_{\mathrm{o}}}{\rho_{\mathrm{p}}d_{\mathrm{p}}}\right)^{0.3}\left[1.3\left(\frac{U_{\mathrm{j}}^2}{gd_{\mathrm{o}}}\right)^{0.2} - 1\right]$$

Merry (1975)

$$\frac{L_{\mathrm{j}}}{d_{\mathrm{o}}} = 814.2\left(\frac{\rho_{\mathrm{p}}d_{\mathrm{p}}}{\rho_{\mathrm{f}}d_{\mathrm{o}}}\right)^{-0.585}\left(\frac{\rho_{\mathrm{f}}d_{\mathrm{o}}U_{\mathrm{j}}}{\mu}\right)^{-0.654}\left(\frac{U_{\mathrm{j}}^2}{gd_{\mathrm{o}}}\right)^{0.47}$$

Wen et al. (1977)

$$\frac{L_{\mathrm{j}}}{d_{\mathrm{o}}} = 15.0\left[\left(\frac{\rho_{\mathrm{f}}}{\rho_{\mathrm{p}} - \rho_{\mathrm{f}}}\right)\left(\frac{U_{\mathrm{j}}^2}{gd_{\mathrm{o}}}\right)\right]^{0.187}$$

Yang and Keairns (1979)

$$\frac{L_{\max}}{d_{\mathrm{o}}} = 1.3\left(\frac{U_{\mathrm{j}}^2}{gd_{\mathrm{p}}}\right)^{0.38}\left(\frac{\rho_{\mathrm{f}}d_{\mathrm{p}}U_{\mathrm{j}}}{\mu}\right)^{0.13}\left(\frac{\rho_{\mathrm{f}}}{\rho_{\mathrm{p}}}\right)^{0.56}\left(\frac{d_{\mathrm{o}}}{d_{\mathrm{p}}}\right)^{0.25}$$

Wen et al. (1982)

Table 3 Continued

Jet penetration correlations	References

Horizontal single and multiple jets

$$\frac{L_j}{d_o} = 24 \times 10^{-4} q \left(\frac{\rho_f}{\rho_p}\right)^{0.15} \left(\frac{g d_p^3 \rho_f^2}{\mu^2}\right)^{-0.30} \frac{U_j d_o \rho_f}{\mu}$$

Kozin and Baskakov (1967)

$0.5 \le q \le 1.5$; for a single jet $q = 0.5$; for multiple jets $q = 1.5$

$$\frac{L_j}{d_o} = 7.8 \left(\frac{\rho_f U_j}{\rho_p \sqrt{g d_p}}\right)$$

Shakhova (1968)

$$0.044 \frac{L_{max}}{d_o} + 1.48 = 0.5 \log(\rho_f U_j^2); \quad \rho_f \text{ in lb/ft}^3 \text{ and } U_j \text{ in ft/s}$$

Zenz (1968)

$$\frac{L_j}{d_o} + 4.5 = 5.25 \left[\frac{\rho_f U_j^2}{(1-\varepsilon)\rho_p g d_p}\right]^{0.4} \left(\frac{\rho_f}{\rho_p}\right)^{0.2} \left(\frac{d_p}{d_o}\right)^{0.2}$$

Merry (1971)

This fundamental relationship has been clouded over the years by different experimental arrangements, different definitions of jet penetration depth, and subjective observations of fluctuating jet penetration depth, which result in many jet penetration correlations that are inconsistent and of limited applicability. Equation (162) summarized the dependence of jet penetration depth on the design and operating parameters for the existing correlations. The wide range of dependence is not very comforting for practical applications in multi-million dollar projects.

$$L_j \propto \rho_f^{k_1} \rho_p^{k_2} d_o^{k_3} d_p^{k_4} U_j^{k_5} \mu^{k_6} g^{k_7} \left[\frac{(U_{mf})_{atm}}{(U_{mf})_p}\right]^{k_8} \tag{162}$$

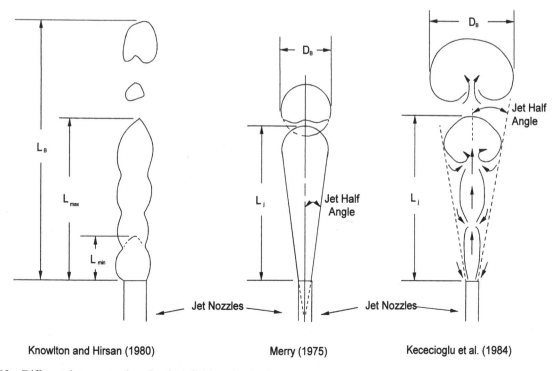

Knowlton and Hirsan (1980) Merry (1975) Kececioglu et al. (1984)

Figure 19 Different jet penetration depth definitions in the literature.

where

$k_1 = 0.30$ to 0.83	$k_2 = -0.3$ to -0.83
$k_3 = 0$ to 1.25	$k_4 = 0$ to -0.5
$k_5 = 0.4$ to 0.944	$k_6 = 0$ to -0.13
$k_7 = -0.2$ to -0.472	$k_8 = 0$ to 0.472

The proper advice for estimating jet penetration at high temperature and high pressure is that, if in doubt, carry out your experiments at atmospheric pressure and room temperature by simulating the gas/particle density ratio at high temperature and high pressure by using a low-density bed material.

The available correlations for horizontal jets are also summarized in Table 3. The correlation proposed by Zenz (1968, 1971) for horizontal jets was also claimed to be suitable for vertical downward jets.

11.2.4 Jet Half Angle

Merry (1975) correlated the experimental jet half angle reported in the literature and suggested the equation

$$\cot\theta = 10.4\left(\frac{\rho_p d_p}{\rho_f d_o}\right)^{-0.3} \tag{163}$$

Wu and Whiting (1988) recommended to change the coefficients to fit the available data better. Their proposed equation is

$$\cot\theta = 8.79\left(\frac{\rho_p d_p}{\rho_f d_o}\right)^{-0.236} \tag{164}$$

Based on the definition by Merry, the jet half angle can be calculated if the initial bubble size and the jet penetration depth are known.

$$\theta = \tan^{-1}\left[\frac{(D_B - d_o)}{2L_j}\right] \tag{165}$$

It is worthwhile to note that the jet half angle, in the most rigorous sense, only applies to the permanent flamelike jets. In other jetting modes, the jet half angle represents only the average behavior of the jetting boundary defined geometrically by Eq. (165). The existence of the jet half angle may not be generally accepted, but the concept considerably simplifies the analysis of the jetting region in fluidized beds. The same approach was also adapted for the bubbling jet created by injecting gas into a liquid medium where a jet half angle of about $10°$ was proposed (Anagbo, 1980).

A recent study by Vaccaro (1997) indicated that the jet half angle could be correlated with the two-phase Froude number quite well for $d_o/d_p > 7.5$.

Experimental jet half angle in the literature over a wide range of particle size (0.107×10^{-3} to 2.9×10^{-3} m), particle density (1117 to 11300 kg/m^3), jet diameter (0.5×10^{-3} to 17.5×10^{-3} m), operating pressure (1, 10, 15, and 20 bar), and operating temperature (20, 650, and 800°C) gave a value ranging from approximately 3.5 degree to close to 25 degree.

11.2.5 Gas Velocity Profiles in the Jet

The velocity profiles of the gas in the jet were usually obtained by pitot tubes. Depending on whether the gas momentum can be isolated, the results of the pitot tube measurement can either be translated into the velocity profiles or remain as the momentum distribution of the gas–solid mixture in the jet. Generally, the velocity profiles in the jet were found to be similar and could be approximated by the typical profiles of turbulent jets in homogeneous media (Abramovich, 1963). The Schlitchting type of similarity was obtained by Shakhova and Minaev (1972), De Michele et al. (1976), and Donadono et al. (1980); the Tollmien type of similarity was observed by Yang and Keairns (1980a), Yang et al. (1984a), and Yang (1998b). Both types of similarity are presented in Fig. 20. When the jet is not a permanent flamelike jet, the similarity represents the average properties at the particular axial cross section in question.

11.3 Initial Bubble Size and Frequency

When the momentum of the jets dissipates, bubbles are formed at the end of the jets. The initial bubble size was studied by Basov et al. (1969) and Merry (1975). They suggested the equations

$$\frac{D_B}{d_o} = 0.41\left(\frac{U_j^{0.375}}{d_o^{0.25}}\right) \qquad \text{units in cm and cm/s} \tag{166}$$

and

$$\frac{D_B}{d_o} = 0.33\left(\frac{U_j^{0.4}}{d_o^{0.2}}\right) \tag{167}$$

The experimental bubble frequency was found by Rowe et al. (1979), Sit and Grace (1981), and Kececioglu et al. (1984) to be around 5 to 10 per second, smaller than that predicted theoretically by Davidson and Harrison (1963). This discrepancy was believed to be due to gas leakage from the jet before forming visible bubbles. Large leakage has been reported by Nguyen and Leung (1972) and Buevich and Minaev (1976) with the corresponding equations

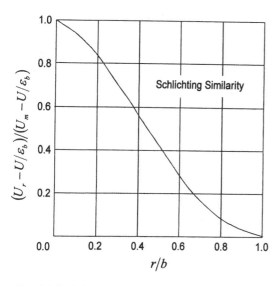

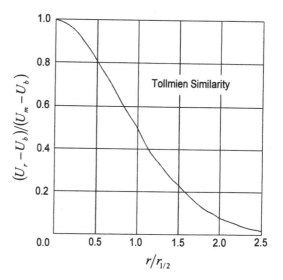

Figure 20 Schlitchting type and Tollmien type similarities for gas velocity profiles in jets in fluidized beds.

$$G = 0.53 V_B f_n \tag{168}$$

and

$$G = 0.37 V_B f_n \tag{169}$$

To describe the jet adequately, especially a large jet, the bubble size generated by the jet needs to be studied. A substantial amount of gas leaks from the bubble to the emulsion phase during the buble formation stage, particularly when the bed is less than minimally fluidized. A model developed on the basis of this mechanism predicted the experimental bubble diameter well when the experimental bubble frequency was used as an input (Yang et al., 1984b). The model is described in Yang (1999) and in Chapter 20, "Other Nonconventional Fluidized Beds."

11.4 Two-Phase Jets and Concentric Jets

The inlet jet velocity for a single-phase (gas) jet is calculated based on the cross-sectional area of the jet nozzle and the total volumetric flow rate of the jet. If the jet also carries solids, as in the case of feeding solids pneumatically into a fluidized bed, the inlet jet velocity should be calculated from the equation

$$\rho_f (U_j)_s^2 = \frac{M_g U_g + M_s U_s}{A_t} \tag{170}$$

Since the voidage in the inlet jet is close to 1, the gas velocity, U_g, can be calculated based on the cross-sectional area of the jet nozzle and the volumetric jet

flow rate. The solid particle velocity, U_s, can be calculated assuming the gas/solid slip velocity to be the terminal velocity of a single particle of the average size.

For a concentric jet with the inner jet carrying solid particles, the inlet jet velocity can be calculated based on the equation

$$\rho_f (U_j)_c^2 = \frac{(M_g)_i (U_g)_i + (M_s)_i (U_s)_i + (M_g)_o (U_g)_o}{A_t} \tag{171}$$

11.5 Gas and Solids Entrainment

The interchanges of gas between the jetting region and the outside emulsion phase can be studied by integration of the measured velocity profiles in the jet or by tracer gas analysis. The velocity profiles in a jet in a fluidized bed can be taken to be similar and can be approximated by either Schlitchting or Tollmien similarity. The turbulent jet equations already developed in the homogeneous medium can then be integrated to give the entrainment of gas into the jet region in a fluidized bed. It is generally agreed that gas is entrained from the emulsion region into the jet at close to the jet nozzle. This exchange reverses direction at larger distance from the jet nozzle (Donadono and Massimilla, 1978; Yang and Keairns, 1980a; Filla et al., 1981; Filla et al., 1983b; Gbordzoe et al., 1988, Yang, 1998b).

The rate of solids entrainment was studied by Donadono et al. (1980), Massimilla et al. (1981),

Yang and Keairns (1982a), Filla et al. (1981), and Filla et al. (1983b) by following solid tracer particles into the jet with movies. It was also measured by Donadono et al. (1980) and by Donsi et al. (1980) using an impact probe located inside the jetting region. The particle velocity and collision frequencies were obtained.

Fluid and particle entrainment into vertical jets in fluidized beds was also studied by Merry (1976) by using a two-dimensional bed of lead shot fluidized by water. Although the system employed is liquid–solid, the appearance of the jet and the motion of particles are very similar to those observed for gas–solid systems. The author considered that the results were equally applicable for gas–solid systems.

Merry (1976) also postulated a dividing streamline in the fluid. All the fluid inside this dividing streamline is entrained into the jet, while the fluid outside bypasses the jet. All fluid or particles intended to be entrained into the jet should be injected within this dividing streamline to be effective. A similar dividing streamline was also observed experimentally by Yang (1998b).

11.5.1 Model for Solid Entrainment into a Permanent Flamelike Jet

A mathematical model for solid entrainment into a permanent flamelike jet in a fluidized bed was proposed by Yang and Keairns (1982a). The model was supplemented by particle velocity data obtained by following movies frame by frame in a motion analyzer. The particle entrainment velocity into the jet was found to increase with increases in distance from the jet nozzle, to increase with increases in jet velocity, and to decrease with increases in solid loading in the gas–solid two-phase jet. High-speed movies indicated that the entrained particles tended to bounce back to the jet boundary more readily under high solid loading conditions. This may explain why the entrainment rate decreases with increases in solid loading in a two-phase jet. A ready analogy is the relative difficulty in merging into a rush-hour traffic as compared to merging into a light traffic.

The simple model by Yang and Keairns (1982a) for solid entrainment into a permanent flamelike jet shown in Fig. 21 resulted in the equation

$$W_j = 2\pi\rho_p(1 - \varepsilon_z)$$
$$\left[\frac{C_1\tan\theta}{3}L_j^3 + \frac{1}{2}\left(\frac{C_1 d_o}{2} + C_2\tan\theta\right)L_j^2 + \frac{C_2 d_o}{2}L_j\right]$$
(172)

The solids entrainment rate into a jet in a fluidized bed can be calculated from Eq. (172), if the empirical constants C_1 and C_2 and the jet half angle θ are known. For the first approximation, the jet half angle θ can be taken to be $10°$ as suggested by Anagbo (1980), a value very close to $7.5°$ obtained from solid particle tranjectories reported for a semicircular column. The jet half angle can also be calculated from Eqs. (163) and (164).

The model as formulated in this section cannot be used to predict a priori the solid entrainment rate into the jet because of the two empirical constants in Eq. (172). Lefroy and Davidson (1969) have developed a theoretical model based on a particle collision mechanism for entrainment of solid particles into a jet. The resulting equation for particle entrainment velocity is

$$V_{jz} \cong \frac{(1 - \varepsilon_j)}{(1 - \varepsilon_z)} \cdot \frac{(\varepsilon_j - \varepsilon_z)}{(1 - \varepsilon_z)} \cdot \frac{\pi^2 e(1 + e)d_p}{16r} \cdot V_j$$
(173)

Equation (173) predicts correctly the increase in solid entrainment into the jet with increases in jet velocity and the decrease with increases in solid loading in a two-phase jet.

11.6 Interacting Jets in Fluidized Beds

When there is more than one jet in a fluidized bed, jet interaction is to be expected in most cases. Thus jet interaction is an important phenomenon to study for operation and design. Unfortunately, not too much information is available in the literature. Wu and Whiting (1988) studied the interaction of two adjacent

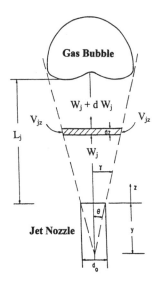

Figure 21 A model for solids entrainment into a jet in a fluidized bed.

jets and found that the jets behave like two isolated jets only at low nozzle velocities. At high nozzle velocities, the jets interact, and the jet penetration depth becomes constant. The jet interaction is divided into three different regions, as shown in Fig. 22.

Yang (unpublished data) also has studied the jet interaction for two and three jets in a two-dimensional beds. His observation of the jet interaction, shown in Fig. 23, is very much similar to that reported by Wu and Whiting. Yang also developed two correlations for the stagnant emulsion regions between the adjacent jets based on the pitch and jet diameter ratio. The height of the stagnant emulsion regions between the jets is inversely proportional to the interaction between the jets. The higher the jet velocity, the more intense the jet inaction, which in turn results in lower height of the stagnant emulsion regions between the jets, as evident in Fig. 23. The proposed equations are

$$\frac{Y_h}{P_j} = 1.1660 - 0.02876\left(\frac{U_j}{U_t}\right) \qquad \text{for two jets} \quad (174)$$

$$\frac{Y_h}{P_j} = 0.4445 - 0.01220\left(\frac{U_j}{U_t}\right) \qquad \text{for three jets} \quad (175)$$

For multiple-jet systems such as that above a gas distributor plate, Wu and Whiting (1988) suggested to make use of the jet half angle discussed earlier to derive an equation for jet penetration depth. Their proposed equation is

$$L_m = 8.79\left(\frac{\rho_p d_p}{\rho_f d_o}\right)^{-0.236}\left(\frac{P_j - d_o}{2}\right) \qquad (176)$$

Equation (176) predicts that the jet velocity does not affect the jet penetration depth in a multiple-jet system and infers that the jet region above the gas distributor has a constant height independent of the jet velocity.

Yang and Keairns (1979) correlated the jet penetration depth data for multiple-grid jets and proposed an equation to estimate the jet penetration depth employing the two-phase Froude number. The resulting equation, as shown below, did show a decreasing, but not negligible, effect of jet velocity on jet penetration depth for multiple-jet systems. This obviously is due to the jet interaction observed experimentally.

$$\frac{L_m}{d_o} = 15.0\left(\frac{\rho_f}{\rho_p - \rho_f} \cdot \frac{U_j^2}{g d_o}\right)^{0.187} \qquad (177)$$

12 PARTICLE MIXING AND SEGREGATION IN A GAS FLUIDIZED BED

Particle mixing and segregation phenomena are important in industrial fluidized beds, where particles of wide size distribution or particles of different densitites are usually handled. Studies have indicated that the fluidized bed reactors can be operated in different modes either to promote particle mixing or to enhance the

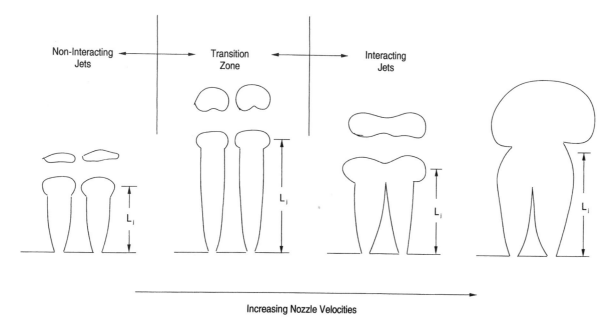

Figure 22 Schematic of jet interaction of multiple jets in fluidized beds. (Adapted from Wu and Whiting, 1988.)

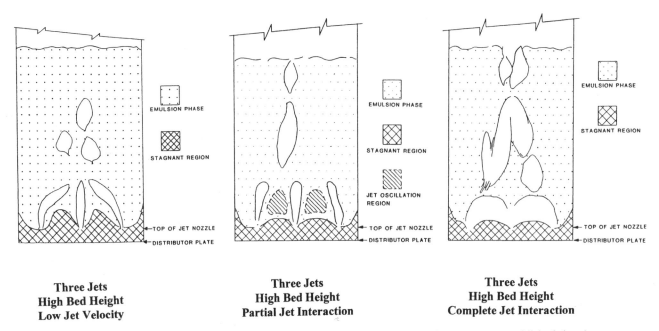

Figure 23 Schematic of jet interaction of multiple jets in fluidized beds. (Adapted from Yang, unpublished data.)

particle segregation. It is not unusual to have one part of the fluidized bed reactor operated in mixing mode while the other part of the same reactor is in segregation mode. It is important to point out that the segregation pattern or the particle distribution profile in the bed is set up by the dynamic equilibrium between the two competing mechanisms of solids mixing and particle segregation. Particle segregation can usually be prevented by operating a fluidized bed at a sufficiently high fluidizing velocity. On the other hand, a bed with particle size ratio between the largest and the smallest as small as 3 can be made to segregate by operating at a small enough fluidizing velocity (Wen and Yu, 1966b).

Past studies on particles in a gas fluidized bed have concentrated primarily on the mixing aspect of the phenomenon, notably those by Rowe and Nienow (1976) using two separate layers of flotsam and jetsam as a starting mixture. The flotsam is the lighter or smaller components; which tend to remain at the top layer of the bed upon fluidization, while the jetsam is those heavier or larger components, which tend to stay close to the bottom part of the fluidized bed. These words were coined originally by Rowe et al. (1972) and now have become widely accepted terminology. The works performed by Professor Rowe and his associates have been reviewed in Rowe and Nienow (1976). A quantitative analysis was proposed for the mixing of two segregating powders of different densi-

ties (Nienow et al., 1978a). In a separate study by Burgess et al. (1977), the initial condition of the bed was found to be important. The well-mixed initial condition (as compared to the unmixed initial condition of two separate layers of flotsam and jetsam) led to less segregation at all gas flow rates. Unfortunately, only limited experimental data are available with a well-mixed starting mixture (Chen and Keairns, 1975, 1978).

There are two primary objectives for investigating the particle segregation phenomenon in gas fluidized beds. In one respect, the fluidized beds are studied to determine the operating conditions required to promote bed mixing and eliminate or minimize particle segregation. A mixing index can generally be defined in this case to measure the closeness to perfect mixing. On the contrary, the other objective is to study the optimum conditions under which clean separation can be accomplished between different materials (or components) in the bed. For this case, both the degree and the rate of particle separation are important aspects of investigation.

It is generally recognized that both particle mixing and segregation in a gas fluidized bed require the presence of gas bubbles. No mixing or segregation of particles can occur in a fixed bed, because relative motion between the particles is not possible. The only way the segregation can happen in fixed beds is for one of the components to be sufficiently small, usually less than about 1/6 the diameter of the larger ones, to sieve through the interstices (Rowe et al., 192; see also Sec. 3 of Chapter 2). Both particle mixing and segregation occur simultaneously in a gas fluidized bed to maintain a dynamic equilibrium in particle distribution. Depending on the controlling mechanism under a specific operating condition, a gas fluidized bed can be operated under a wide range of characteristics, from a perfectly mixed to a completely segregated fluidized bed.

12.1 Particle Mixing in a Gas Fluidized Bed

The particle mixing in a gas fluidized bed is entirely induced by the passage of gas bubbles. Particles are picked up by the bubbles in the bubble wakes at close to the gas distributor plate and are carried up to the top of the fluidized bed. Along the way up, the bubble wakes exchange their content with the particles in the rest of the bed in a phenomenon commonly known as wake shedding. The gas bubbles also draw up a spout of the surrounding particles to produce an upward drift of particles. To balance this overall upward movement of the bed material, there is a mass circulation downward in the bubble-free region of the bed. The global circulation induced by the bubbles thus ensues, and the rapid mixing of bed material quickly follows. This mixing mechanism is qualitatively depicted in Fig. 24. The existence of bubble

wake (usually expressed as the volumetric ratio of wake volume and bubble volume, $\beta_w = V_w/V_B$), the phenomenon of drift (expressed as $\beta_d = V_d/V_B$), and wake shedding have been discussed in detail by Rowe et al. (1965) and reviewed by Rowe (1971). According to Rowe, the amount of solids induced upward by a single bubble is equal to about 0.6 times the bubble volume, of which about 60% is due to drift action (i.e., $\beta_w = 0.24$ and $\beta_d = 0.36$). These values agree reasonably well with the findings by Fane and Nghiem (1983), who found that the volume of solids set in motion by a single bubble is about 0.5 to 0.8 times the bubble volume and approximately 75% was due to drift. Using beds initially segregated into two pure horizontal layers, Burgess et al. (1977) found that the particle mixing between these two layers started at the moment the gas velocity exceeded the minimum fluidization velocity of the flotsam, $(U_{mf})_F$, for binary particle systems of equal density. The mixing was accomplished by gathering jetsam in the bubble wakes at the interface between the two layers. For binary systems of different densities, however, the mixing did not start until the gas velocity was above the minimum fluidization velocity of the jetsam, $(U_{mf})_J$. In this case, the bubbles must exist in the jetsam layer to induce mixing between the two layers.

Baeyens and Geldart (1973) also studied the solid circulation and found that for most Group B powders, the wake fraction, β_w, had an average of about 0.35

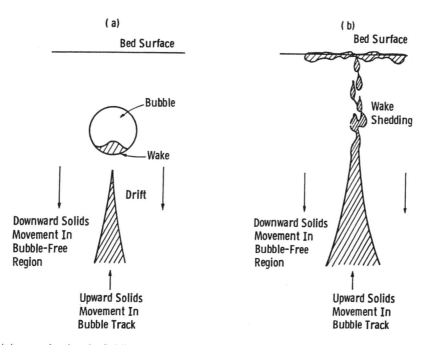

Figure 24 Solids mixing mechanism in fluidized beds—wake and drift.

For Group D powders, β_w is about 0.1. For Group A solids, it is 1.0. Geldart (1986) correlated the wake and drift fractions with the Archimedes number, as shown in Fig. 25.

12.1.1 Lateral Mixing of Solids in Gas–Solid Fluidized Beds

The lateral mixing of particles in gas–solid fluidized beds is induced by bubble movement through the bed, bubble bursting at the bed surface, and gross particle circulation inside the bed. Through the random walk theory, Kunii and Levenspiel (1969) have derived an equation for the lateral dispersion coefficient, D_{sr}, for lateral mixing of solids in gas–solid fluidized beds:

$$D_{sr} = \frac{3}{16}\left(\frac{\varepsilon_B}{1-\varepsilon_B}\right)\frac{U_{mf}D_B}{\varepsilon_{mf}} \qquad (178)$$

where ε_B is the volume fraction of bubble in the fluidized bed.

The lateral particle dispersion coefficient was also studied by Shi and Fan (1984) and Subbarao et al. (1985). A one-dimensional diffusion model was used by Shi and Fan (1984) to characterize lateral mixing of solids. Through dimensional analysis and nonlinear regression analysis of the literature data, they arrived at an equation for the lateral dispersion coefficient for general application.

$$\frac{D_{sr}}{(U-U_{mf})H_{mf}} = 0.46\left[\frac{(U-U_{mf})d_p\rho_f}{\mu}\right]^{-0.21}\left[\frac{h_{mf}}{d_p}\right]^{0.24}\left[\frac{\rho_p-\rho_f}{\rho_f}\right]^{-0.43} \qquad (179)$$

12.1.2 Convective Solids Transport and Mixing

The mixing in gas–solid fluidized beds is primarily due to two processes created by the passage of bubbles, bubble wake and bubble drift, as discussed earlier. The convective solids transport and mixing can be estimated based on the bubble properties, as suggested by Geldart (1986). The solid circulation flux in a fluidized bed can be calculated from the equation

$$J = \rho_p(1-\varepsilon_{mf})(U-U_{mf})Y(\beta_w+0.38\beta_d) \qquad (180)$$

where Y is a correction for deviation from the two-phase theory, and the factor 0.38 for drift flux is from the experimental evidence that the particles carried up in the drift travel on average at about 38% of the bubble velocity. The value of wake and drift fraction, β_w and β_d, can be obtained from Fig. 25, while the correction for deviation from the two-phase theory, Y, can be found from Fig. 26, also due to Geldart (1986). Values of β_w, β_d, and Y for common materials are summarized in Table 4.

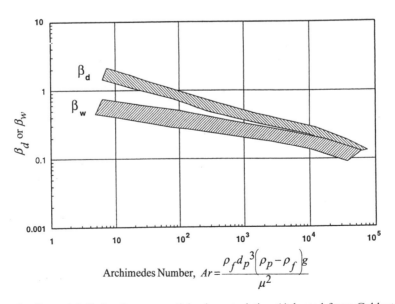

Figure 25 Dependence of wake and drift fraction on particle characteristics. (Adapted from Geldart, 1986.)

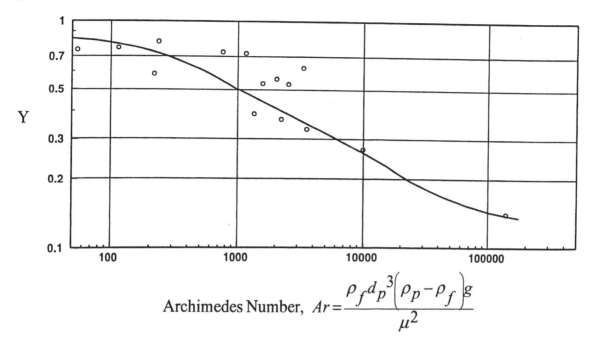

$$Ar = \frac{\rho_f d_p^3 \left(\rho_p - \rho_f \right) g}{\mu^2}$$

Archimedes Number,

Figure 26 Correction for deviation from the two-phase theory. (Adapted from Geldart, 1986.)

The bubble properties have also been measured by Werther (1976b) in large-diameter fluidized beds, 1 m in diameter; and an approach for calculating the convective transport of solids was proposed. The semi-empirical correlation suggested is

$$J = 0.67(1 - \varepsilon_{mf}) \rho_p \left(\frac{1}{\phi_B} - 1 \right)(U - U_{mf}) \qquad (181)$$

where ϕ_B is the bubble shape factor and can be calculated from

$$\phi_B = \{1 - 0.3 \exp[-8(U - U_{mf})]\} e^{-\omega h} \qquad (182)$$

and

$$\omega = 7.2(U - U_{mf}) \exp[-4.1(U - U_{mf})] \qquad (183)$$

The dimensions for h, the height above the distributor, and $(U - U_{mf})$ in Eqs. (182) and (183) are m and m/s. The suggested approach was said to be good over the following range of variables.

Solid density	$1115 \leq \rho_p \leq 2624$ kg/m³
Mean particle diameter	$100 \leq d_p \leq 635$ µm
Minimum fluidization velocity	$0.02 \leq U_{mf} \leq 0.24$ m/s
Excess gas velocity	$U - U_{mf} \leq 0.80$ m/s

12.1.3 Bed Turnover Time

In actual practice where particles are continuously being fed and withdrawn, the bed turnover time should be as short as possible compared to the particles' residence time in the bed. Geldart (1986) suggested a ratio of residence time to turnover time of about 5 to 10. With the known solid circulation rate, the bed turnover time can be calculated easily.

$$t_T = \frac{H_{mf} A \rho_p (1 - \varepsilon_{mf})}{JA} \qquad (184)$$

Substituting Eq. (180) into Eq. (184), we find

$$t_T = \frac{H_{mf}}{Y(U - U_{mf})(\beta_w + 0.38\beta_d)} \qquad (185)$$

Table 4 Wake Fraction, Drift Fraction, and Correction Factor for Two-Phase Theory for Some Common Materials

Powder	Size (µm)	β_w	β_d	Y
Catalyst	47	0.43	1.00	1.00
Angular sand	252	0.26	0.42	0.50
	470	0.20	0.28	0.25
Round sand	106	0.32	0.70	0.82
	195	0.30	0.52	0.65

Source: Baeyens and Geldart, 1973.

In reality, particles spend vastly different amounts of time in the bed, depending very much on the locations of feed and withdrawal nozzles, on flow and mixing patterns in the bed, and on many other factors. The particles' residence time is an even more important consideration in fluidized bed design.

12.1.4 Solids Residence Time Distribution in a Fluidized Bed

Particle residence time distribution in a fluidized bed is more close to that of a stirred tank reactor (CSTR) or ideal backmix reactor than that of a plug flow reactor (Yagi and Kunii, 1961). In a perfect plug flow reactor, all particles have the same residence time, which is equivalent to the mean residence time of particles and can be calculated by

$$\bar{t}_R = \frac{W}{F_o} \tag{186}$$

For a completely mixed stirred tank reactor, the residence time distribution can be expressed following ideal backmix reactor model.

$$R(t) = \left[\frac{1}{\bar{t}_R}\right]e^{-t/\bar{t}_R} \tag{187}$$

$R(t)\,dt$ is the fraction of solids staying in the bed for the time period between t and $t+dt$. Yagi and Kunii (1961) found experimentally that Eq. (187) represented the particle residence time in a fluidized bed quite well. The fraction of solids spending less than time t can then be calculated from

$$f = 1 - e^{-t/\bar{t}_R} \tag{188}$$

From Eq. (188), it can be seen that a significant amount of solids, about 18.2%, spends less than 20% of the average particle residence time in the bed.

The wide residence time distribution for an ideal backmix reactor is detrimental to high conversion of solids. However, the residence time distribution can be considerably narrowed by staging. The residence time distribution for a system with a N stages of fluidized beds of equal size can be estimated by

$$R(t) = \frac{1}{(N-1)!\bar{t}_i}\left(\frac{t}{\bar{t}_i}\right)^{N-1}e^{-t/\bar{t}_i}, \tag{189}$$

In some applications, a simple CSTR is not sufficient to describe the residence time distribution of particles. Habermann et al. (1998) described a metallurgical process where the fluidized bed showed a stagnant zone and a two-zone model with a CSTR and a stagnant dead-zone fitted the data much better. For the residence time distribution of systems with various combinations of series and parallel reactors, the reader can consult Levenspiel (1979).

12.2 Particle Segregation in a Gas Fluidized Bed

The mechanisms of particle segregation in a gas fluidized bed were first studied by Rowe et al. in a classical publication in 1972 using binary systems of near-spherical particles in both two-dimensional and three-dimensional cylindrical beds. The terminology of jetsam and flotsam, now widely accepted for the component that settles to the bottom and the component that floats to the top, respectively, was also first suggested in this paper. Three distinctly different mechanisms were found to be important in creating the relative movement of particles in the bed. The lifting of particles in the wake of a rising gas bubble, which was found to be the primary particle mixing mechanism discussed earlier, was also identified as the most important particle segregation mechanism. This is practically the only way the flotsam can be transported to the upper part of the bed. For the larger and denser particles known as jetsam, the migration down to the bottom of the bed involves two different mechanisms. The larger and denser particles usually descend by falling through the bubbles, while the smaller, denser particles percolate downward interstitially. The second mechanism, however, was found to be not an overly important one. It only occurs when the small, dense particles are sufficiently small. This interparticle percolation is restricted only to the regions recently disturbed by a passing bubble.

Tanimoto et al. (1980) investigated the movement of a small aggregate of jetsam induced by passage of a single gas bubble in a two-dimensional bed using cinema-photographic technique and found that the aggregate moved downward intermittently. The average descending distance of the aggregate was proportional to the bubble diameter and the distance from the bubble center. Chiba and Kobayashi (1982) proposed a segregation model based on this observation.

In a bed containing a small proportion of large particles in considerably smaller ones (with a size ratio of more than an order of magnitude), the sink and float behavior of the large particles depends on the ratio of absolute density of the large particles relative to the bulk density of the smaller ones (Nienow et al., 1978b). In practice, this is usually the case in fluidized bed combustion of coal, where the large coal particles constitute only a small proportion in a bed of predominantly smaller particles of ash or limestone. The particle shape was found to be not an important variable during separation unless the particles were platelike.

Nienow and Cheesman (1980) studied the mixing and segregation of a small proportion of large platelike particles in a gas fluidized bed of regular smaller particles using x-ray cinemaphotography. They concluded that the large flat particles with sphericity larger than 0.5 behaved essentially like regular particles with sphericity larger than 0.8. Extremely flat particles, those with sphericity less than 0.5, exhibited as jetsam over a small range of fluidization velocity, even when they should be flotsam by pure particle density consideration. They also discovered that the flat particles mixed better than the spheres owing to their "ability to 'fly' through the bed like a 'frisbee'."

12.2.1 Analogy to Gas–Liquid–Solid Phase Equilibrium

Kondukov and Sosna (1965) and Gelperin et al. (1967) (see also Gelperin and Einstein, 1971) were among the first to construct the phase equilibrium diagrams for binary and ternary systems in a gas fluidized bed. A gas fluidized bed behaves like a liquid in many ways. Its liquidlike behavior was reviewed by Gelperin and Einstein (1971) and by Davidson et al. (1977). A fluidized system goes through the phase transformations with changes in fluidizing velocity just like those experienced by a liquid during changes in temperature. The packed-fluidized-dilute phase states in a fluidized system correspond to the solid-liquid-gas phases in a liquid, with the minimum fluidization and the terminal velocities being the equivalent of the melting and boiling temperatures respectively.

The phase equilibrium diagrams are constructed with the beginning fluidization velocity, U_{bf}, the velocity at which the finer and lighter particles begin to fluidize, and the total fluidization velocity, U_{tf}, the velocity at which all particles including the large and heavy particles start to fluidize. Figure 27 shows the determination of U_{bf} and U_{tf} from the fluidization curve of a mixture suggested by those authors. Typical phase equilibrium diagrams constructed following this method are shown in Fig. 28 (Chen and Keairns, 1975). With increases in size ratio or density ratio of the components or decreases in operating pressure, the area between the curves of U_{bf} and U_{tf} increases as well, signifying the increase in segregation tendency of the particles in the bed.

Vaid and Sen Gupta (1978) proposed the only two equations for calculating U_{bf} and U_{tf} as follows:

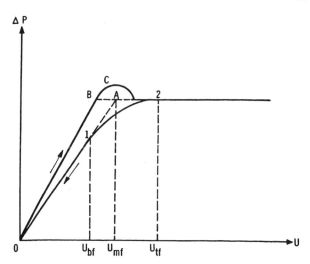

Figure 27 Determination of beginning fluidization velocity (U_{bf}) and total fluidization velocity (U_{tf}).

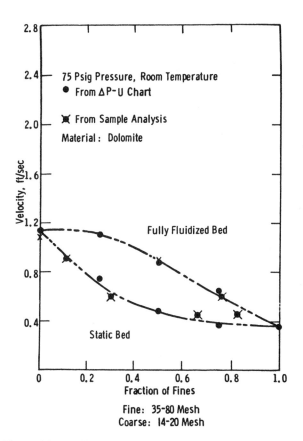

Figure 28 Typical phase equilibrium diagrams. (Adapted from Chen and Keairns, 1975.)

$$(Re)_{bf} = \frac{DU_{bf}\rho_f}{\mu} = [(18.1)^2 + 0.0192Ar]^{0.5} - 18.1$$

$$(190)$$

$$(Re)_{tf} = \frac{DU_{tf}\rho_f}{\mu} = [(24.0)^2 + 0.0546Ar]^{0.5} - 24.0$$

$$(191)$$

In a study on the phase equilibrium diagram of acrylic and dolomite particles, Yang and Keairns (1982b) concluded that the total fluidization velocity U_{tf} obtained from Fig. 27 or calculated from Eq. (191) had no physical significance in term of the state of particle mixing or separation. At U_{tf} the bed may be in "total fluidization," but the bed is far from perfectly mixed.

12.2.2 Classification of Flotsam and Jetsam

In a gas fluidized bed where the bed particles are of different densities, it will be beneficial to know which component will sink (jetsam) and which will float (flotsam). In most cases, especially the two-component systems, the classification of flotsam and jetsam is obvious. In some isolated cases, whether the particular component will behave as a flotsam or a jetsam will have to be determined experimentally. This is especially true for a bed of multicomponent mixture with a wide size and density distribution. For a two-component binary system, Chiba et al. (1980) suggested the following general rules:

One particular component may be a flotsam with respect to some components in the bed, while simultaneously it is also a jetsam relative to other components. Thus the distinction between a jetsam and a flotsam is less important in a multicomponent system with density differences. The component distribution at equilibrium is of primary concern. There are mathematical models available in the literature which enable the calculation of this equilibrium concentration profile in a gas fluidized bed, to be discussed later.

12.2.3 Minimum Fluidization Velocity of a Binary Mixture

The minimum fluidization velocity of a fluidized bed with a single component bed material, i.e., a bed material with particles of relatively narrow particle size distribution and of similar particle density, is well defined. For mixtures of particles of different sizes or densities, especially for those highly segregating system, the definition and determination of the minimum fluidization velocity are not as straightforward. Though the minimum fluidization velocity of a segregating mixture can still be defined conventionally following the procedure suggested for a single component system, the minimum fluidization velocity defined in this way loses its physical meaning. The particles in the bed are far from completely supported by the fluidizing gas at this velocity,

Case I	$d_B/d_S \leq 10$	
I_a	$\rho_B = \rho_S$	Jetsam = bigger component
I_b	$\rho_B \neq \rho_S$	Jetsam = heavier component
Case II	$d_B \gg d_S$ and bed material $\rightarrow$ 100% smaller component	
II_a	$\rho_B > (\rho_B)_S$	Jetsam = bigger component
II_b	$\rho_B < (\rho_B)_S$	Jetsam = smaller component
Case III	$d_B \gg d_S$ and bed material $\rightarrow$ 100% bigger component	
III_a	$\rho_B > \rho_S$	Jetsam = bigger component
III_b	$\rho_B < \rho_S$	Jetsam = either component may be jetsam
Case IV	The minor component is platelike with $\phi < 0.5$	
IV_a	Platelike particle is denser	Jetsam = platelike component
IV_b	Platelike particle is lighter	Jetsam = either component may be jetsam

In a gas fluidized bed of a mixture of wide size and density distribution, the distinction between a flotsam and a jetsam becomes less clear because the individual components are distributed axially into an equilibrium distribution governed primarily by hydrodynamics.

as was observed experimentally by Knowlton (1977) and by Chiba et al. (1979).

A study by Chiba et al. (1979) on the minimum fluidization velocity of binary particle mixtures indicated that the fluidization curve shown in Fig. 27 was not typical, and U_{bf} could not generally be determined with accuracy, especially for strong segregating systems. Since the segregating particle mixture will start to segregate when the gas velocity is higher than the minimum fluidization velocity of the mixture, the descending portion of the ΔP–U curve obtained with decreasing gas velocity is that of a partially segregated bed. Depending on the rate of particle separation and the time spent to obtain the complete fluidization curve, the descending portion will assume different paths.

For an ideal system where the particles are of small size difference and of equal density, both the ascending and descending portions of the fluidization curve will coincide, shown as curve (a) in Fig. 29. The conventional procedure will yield a minimum fluidization velocity, shown as $(U_{mf})_M$. For a highly segregating mixture where the particle separation rate is fast, the fluidization curve will follow curve (a) for the ascending portion but descend along curve (b). This is because the mixture will already have separated into two distinct layers of flotsam and jetsam when the gas velocity is higher than the minimum fluidization velocity of the jetsam. Curve (b) can be constructed a priori by adding together the fluidization curves of pure jetsam and pure flotsam, as shown in Fig. 29. The conventional procedure of determining the minimum fluidization velocity will give a velocity at $(U_{mf})_S$,

defined in Chiba et al. (1979) as the apparent minimum fluidization velocity. The descending portion of the fluidization curve of any real systems lying between (a) and (b), such as curve (c), depends on the rate of separation and the time spent on obtaining the fluidization curve. It is clear then that a unique U_{bf} may not be generally obtainable from the fluidization curve as shown in Fig. 27. The minimum fluidization velocity determined following the conventional procedure, $(U_{mf})_I$ shown in Fig. 29, is not unique. It should be noted that though the fluidized bed will be completely fluidized (i.e., no static region at the bottom) at U_{tf}, a value generally very close to the U_{mf} of the jetsam, $(U_{mf})_J$ in Fig. 29, the fluidized bed is far from completely mixed. As pointed out by Yang and Keairns (1982b) that U_{tf} does not adequately represent the equilibrium boundary determined experimentally for segregating systems.

The reported minimum fluidization velocities for binary systems in the literature are probably the velocities of either $(U_{mf})_M$ or $(U_{mf})_J$. For binary systems of small particle size and density difference, these two minimum fluidization velocities may be taken to be similar. Some of the equations proposed for calculating the minimum fluidization velocity of a binary mixture are summarized below.

Goossens et al. (1971) modified the equation by Wen and Yu (1966b) on the minimum fluidization velocity for single particle size systems by substituting with the mixture particle density, $\bar{\rho}_p$, and the mixture particle size, $\bar{d}_p$, of a binary mixture as follows:

$$\frac{\bar{d}_p U_{mf} \rho_f}{\mu} = \left[(33.7)^2 + 0.0408 \frac{\bar{d}_p^3 \rho_f (\bar{\rho}_p - \rho_f)}{\mu^2} \right]^{1/2} - 33.7$$

(192)

where

$$\frac{1}{\bar{\rho}_p} = \frac{\bar{x}_F}{\rho_F} + \frac{(1 - \bar{x}_F)}{\rho_J} \qquad \bar{d}_p = \left(\frac{R_o}{R} \right) (d_p)_F (d_p)_J$$

(193)

$$R = (1 - \bar{x}_F) \rho_F (d_p)_F + \bar{x}_F \rho_J (d_p)_J$$
$$R_o = (1 - \bar{x}_F) \rho_F + \bar{x}_F \rho_J$$

(194)

They found that the Wen and Yu equation as modified could be applied to binary mixtures different in both size and density for the minimum fluidization velocity, $(U_{mf})_M$.

Cheung et al. (1974) proposed an equation for binary systems with particles of similar density and with particle size ratios of less than 3. The totally empirical

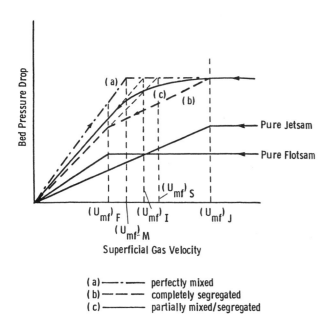

Figure 29 Determination of minimum fluidization velocity of a binary mixture.

equation for the minimum fluidization velocity of the mixture is expressed as

$$(U_{mf})_M = (U_{mf})_F \left[\frac{(U_{mf})_J}{(U_{mf})_F} \right]^{\bar{x}_J^2} \qquad (195)$$

The equation was later found to be also applicable for binary systems with particles of different densities (Chiba et al., 1979). For systems with different particle densities, the particle size ratio restriction of 3 still applies. Its extrapolation to multicomponent systems poses difficulties, however (Rowe and Nienow, 1975).

Chiba et al. (1979) also proposed two equations to estimate the $(U_{mf})_M$ for a completely mixed bed and the $(U_{mf})_S$ for a totally segregated mixture. By utilizing the Ergun equation and the constant voidage assumption, they proposed to estimate the $(U_{mf})_M$ for a completely mixed binary system as

$$(U_{mf})_M = (U_{mf})_F \frac{\bar{\rho}_p}{\rho_F} \left[\frac{\bar{d}_p}{(d_p)_F} \right]^2 \qquad (196)$$

The average mixture density and particle size are calculated from

$$\bar{\rho}_p = f_{VF}\rho_F + (1 - f_{VF})\rho_J \qquad (197)$$

$$\bar{d}_p = [f_{NF}(d_p)_F^3 + (1 - f_{NF})(d_p)_J^3]^{1/3} \qquad (198)$$

where f_{VF} is the volume fraction of flotsam and f_{NF} is the number fraction of the flotsam particles and can be evaluated as

$$f_{NF} = \left\{ 1 + \left(\frac{1}{f_{VF}} - 1 \right) \left[\frac{(d_p)_F}{(d_p)_J} \right]^3 \right\}^{-1} \qquad (199)$$

For a completely segregated bed, the following equation should be employed.

$$(U_{mf})_M = \frac{(U_{mf})_F}{[1 - (U_{mf})_F/(U_{mf})_J]\bar{x}_F + (U_{mf})_F/(U_{mf})_J} \qquad (200)$$

To make use of Eq. (196) requires prior knowledge of $(U_{mf})_F$ and Eq. (200), $(U_{mf})_F$ and $(U_{mf})_J$.

Uchida et al. (1983) modified the equation by Cheung et al. (1974), substituting the volume fraction for the weight fraction in the original equation and introducing an additional empirical constant, m, and suggested the following equations.

$$(U_{mf})_M = (U_{mf})_F \left[\frac{(U_{mf})_J}{(U_{mf})_M} \right]^{(1-f_F)^m} \qquad (201)$$

where

$$m = 0.17 \left[\left(\frac{d_J}{d_F} \right) \left(\frac{\rho_F}{\rho_J} \right) \right]^{0.437} \qquad (202)$$

The volume-mean particle diameter rather than the harmonic mean diameter should be used in Eq. (202). Using a wide variety of particles of different densities and sizes, these authors found that the minimum fluidization velocity depended more strongly on the volumetric fraction rather than on the mass fraction of the particles. The minimum fluidization velocity also is closely related to the mixing state of the mixture, confirming the observation by others in earlier investigations.

12.2.4 Minimum Fluidization Velocity of a Multicomponent Mixture

For multicomponent particles of equal density, Rowe and Nienow (1975) proposed to calculate the minimum fluidization velocity of the mixture using the equation

$$(U_{mf})_M = (U_{mf})_1 \left[\left(\frac{\varepsilon}{\varepsilon_1} \right)^3 \left(\frac{1 - \varepsilon_1}{1 - \varepsilon} \right)^{2-n} \right]^{1/n}$$
$$\times \left[\bar{x}_1 + \frac{d_{p1}}{d_{p2}} \bar{x}_2 + \frac{d_{p1}}{d_{p3}} \bar{x}_3 + \ldots \right]^{1-(3/n)} \qquad (203)$$

where n can be taken to be 1.053, an empirical value.

Kumar and Sen Gupta (1974) obtained the minimum fluidization velocity from the typical logarithmic plot of bed pressure drop versus fluidizing velocity for 4 single, 17 binary, 6 ternary, and 3 quaternary-component mixtures and found that the following simple empirical equation correlated the data well.

$$(Re)_{mf} = \frac{\bar{d}_p U_{mf} \rho_f}{\mu} = 0.0054(Ar)^{0.78} \qquad (204)$$

where

$$Ar = \frac{\bar{d}_p^3 \rho_f (\bar{\rho}_p - \rho_f)g}{\mu^2} \qquad (205)$$

$$\rho_p = \sum_{i=1}^{n} \rho_{pi} \qquad \text{arithmetic mean} \qquad (206)$$

and

$$\bar{d}_p = 1 / \sum_{i=1}^{n} \frac{x_i}{d_{pi}} \qquad \text{harmonic mean} \qquad (207)$$

Most of the literature correlations for calculating the minimum fluidization velocity of a multicomponent mixture are derived based on the assumption that the bed is completely mixed and homogeneous, corresponding to the $(U_{mf})_M$ shown in Fig. 29.

12.2.5 Degree of Particle Separation—Mixing and Segregating Indexes

To describe properly the state of particle distribution in a segregating gas fluidized bed, two indexes, the mixing index and the segregation index, have been defined. If the approach to the perfect mixedness is the primary interest, the mixing index is more appropriate. The mixing index for a binary system was first defined by Rowe et al. (1972) to be

$$M = \frac{(X_J)_u}{\overline{X}_J} \tag{208}$$

where $(X_J)_u$ is the fraction of jetsam in the upper part of the bed and $\overline{X}_J$ is the fraction of jetsam at the state of perfect mixing. Both $(X_J)_u$ and $\overline{X}_J$ are usually expressed in weight fraction. By this definition, the state of perfect mixing has a mixing index of $M = 1$, while a state of complete segregation, $M = 0$. The definition of mixing index was expanded by Naimer et al. (1982) to include cases where a unique value of $(X_J)_u$ cannot be obtained.

To describe the degree of particle separation in the bed, the segregation index defined by Chiba et al. (1982) may be more convenient. They defined the segregation index for a binary system, S, as

$$S = \frac{(X_F)_u}{\overline{X}_F} \tag{209}$$

where $(X_F)_u$ is the flotsam weight fraction in the upper part of the bed and $\overline{X}_F$ is the flotsam weight fraction at the state of perfect mixing. By this definition, $S = 1$ describes a state of perfect mixing and $S = 1/\overline{X}_F$

indicates a state of complete segregation. Since $\overline{X}_F = 1 - \overline{X}_J$ for a binary system, it can be readily derived that

$$M = \frac{1 - S\overline{X}_F}{1 - \overline{X}_F} = S + \frac{(1 - S)}{\overline{X}_J} \tag{210}$$

Because of the complexity of describing even a binary system, no general and useful mixing index has been suggested for multicomponent systems. Even the experimental studies are conspicuously lacking in the literature.

12.2.6 Effect of Particle Size, Density, Shape, and Gas Velocity

The segregation patterns for practical binary systems are shown in Fig. 30 for flotsam-rich systems and in Fig. 31 for jetsam-rich systems (Nienow et al., 1978a). Increasing operation velocity will drive the equilibrium toward a better mixed state. Similar segregation patterns can be drawn for binary systems with different density and size ratios. For systems with large density and size ratios, the segregation patterns are similar to those shown in Figs. 30a and 30b, and in Figs. 31a and 31b. The segregation patterns shown in Figs. 30c and 31c correspond to systems with smaller size and density ratios.

The effect of particle density, particle size, and gas velocity on the mixing of binary, ternary, and quaternary particle systems was studied by Nienow et al. (1978a, 1987) experimentally. They observed that significant improvement of mixing was obtained with perforated plate and standpipe distributors, compared to a porous distributor plate at similar superficial operating velocity. They also showed that the mixing index

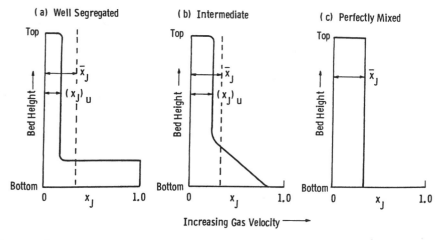

Figure 30 Practical states of equilibrium for flotsam-rich binary mixtures.

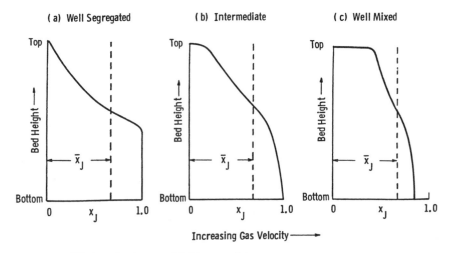

Figure 31 Practical states of equilibrium for jetsam-rich binary mixtures.

M could be correlated with the gas velocity by the following equation:

$$M = \frac{1}{(1 + e^{-Z})} \tag{211}$$

where

$$Z = \left[\frac{U - U_{\text{To}}}{U - (U_{\text{mf}})_{\text{F}}}\right] \cdot e^{U/U_{\text{To}}} \tag{212}$$

The takeover velocity U_{To}, where $dM/d(U - (U_{\text{mf}})_{\text{F}})$ is a maximum and $M = 0.5$, could be estimated from the following empirical equations:

$$\frac{U_{\text{To}}}{(U_{\text{mf}})_{\text{F}}} = \left[\frac{(U_{\text{mf}})_{\text{J}}}{(U_{\text{mf}})_{\text{F}}}\right]^{1.2}$$
$$+ 0.9(\rho_{\text{R}} - 1)^{1.1} d_{\text{ER}}^{0.7} - 2.2(\overline{x}_{\text{J}})^{0.5}(H^*)^{1.4} \tag{213}$$

The ρ_{R} in Eq. (213) is the density ratio; the d_{ER}, the size ratio; and the H^*, the reduced bed height. They are defined as

$$\rho_{\text{R}} = \frac{\rho_{\text{J}}}{\rho_{\text{F}}} \tag{214}$$

$$d_{\text{ER}} = \frac{\phi_{\text{J}} d_{\text{J}}}{\phi_{\text{F}} d_{\text{F}}} \tag{215}$$

$$H^* = 1 - \exp\left(-\frac{H}{D}\right) \tag{216}$$

Equation (213) gave reasonable prediction of U_{To} for binary systems of different particle densities $(\rho_{\text{J}}/\rho_{\text{F}} \neq 1)$. For binary systems with particles differing only in size, i.e., $\rho_{\text{J}}/\rho_{\text{F}} = 1$, Eq. (213) gave U_{To} values significantly larger than those measured experimentally. Equations (211) and (213) are only recommended for binary mixtures with volumetric jetsam concentration less than about 50%, and Eq. (213) is also only good for systems with a particle size ratio less than 3. Equation (213) cannot be applied to a bed with a high aspect ratio where slugging occurs.

The equations of Nienow et al. were modified later by Rice and Brainnovich (1986) and Peeler and Huang (1989). A more recent study on segregation by size difference was conducted by Wu and Baeyens (1998). They found that the excess gas flow rate required to prevent segregation in a fluidized bed with a wide size distribution of powder can be calculated from the mixing index expression shown in the equation

$$M = 1 - 0.0067 d_{\text{R}} 1.33\left(\frac{G_{\text{B}}}{A}\right)^{-0.75} \tag{217}$$

For $d_{\text{R}} \cong 2$, good mixing $(M \geq 0.9)$ can be achieved when the visible bubble flow (G_{B}/A) is larger than about 0.094 m/s. Since temperature has limited effect on the visible bubble flow rate, it is expected that temperature has very little effect on the particle separation due to difference in size alone.

Simple quantitative equation can only be formulated for binary systems of near spherical particles. For binary systems of granular particles, Rowe et al. (1972) suggested the following dependence for segregation tendency.

Segregation tendency

$$\propto [U - (U_{\text{mf}})_{\text{F}}]\left(\frac{\rho_{\text{J}}}{\rho_{\text{F}}}\right)^{-2.5}\left(\frac{d_{\text{B}}}{d_{\text{S}}}\right)^{-0.2} \tag{218}$$

According to Eq. (218), the effect of density difference is considerably more important than that of size difference. The main effect of the particle size is to alter the U_{mf} of the mixture. The exact form of the velocity effect is yet to be determined.

12.2.7   Effect of Pressure and Temperature

The effect of pressure on particle separation in a gas fluidized bed was studied by Chen and Keairns (1975) for systems of particles of both similar density (dolomite particle systems) and different densities (char-dolomite systems) up to a pressure of 690 KPa. They found that the tendency of particle segregation decreased with increases in pressure.

The pressure effect on solid mixing and segregation of binary mixtures was also investigated by Chiba et al. (1982) at pressures up to 0.8 Mpa. The total bed pressure drop and axial particle distribution for various systems using silica sand and coal char of different sizes were obtained. The U_{mf} prediction by Chenug et al. (1974), originally developed for binary mixtures of similar density and different particle sizes, was found to be applicable not only to particles of different sizes and densities at 2 atmospheric pressure but also to high pressures. The mixing between the binary components increased with increasing pressure. Through photographic measurement of bubble size, shape, and wake angle in a two-dimensional gas fluidized bed, they concluded that the increasing particle mixing (or the decreasing particle segregation tendency) at higher pressures was due primarily to the increases in wake fraction of the gas bubbles at higher pressures. The volumetric wake fraction was observed to increase from 0.02 at 0.1 Mpa to about 0.1 at 0.8 Mpa almost linearly.

12.2.8   Effect of Particle Segregation on Other Fluidization Phenomena

The effect of particle segregation on fine elutriation from gas fluidized beds was studied by Tanimoto et al. (1983) employing binary particle mixtures. They found that in segregated beds, where substantial amounts of fines tended to concentrate at the bed surface, the elutriation characteristics were very much different from those of the well-mixed beds. The elutriation rate decreased with increases in bed height rather than the opposite tendency observed in a well-mixed bed. The higher bed height increased the elutriation rate because of larger bubbles and because of more particles thrown into the freeboard by those larger bubbles. More fines tended to concentrate at

the bed surface with a lower bed height because there was less mixing provided by smaller bubbles existing in a shallower bed. The net effect was an increase in elutriation rate at lower bed heights. The concentration of fines at the bed surface, apparently, dominated the elutriation phenomenon in a segregated bed.

12.2.9   Rate of Particle Separation in a Gas Fluidized Bed

Past studies on particle separation have concentrated primarily on the mixing aspect of the phenomenon, notably those by Rowe, Nienow, Chiba, and their associates using two separate layers of flotsam and jetsam as a starting mixture. To design a fluidized-bed separator for a particle system, however, not only the degree of separation but also the rate of separation must be known. The rate of separation of different particle systems has not been studied systematically. Limited data were reported by Rowe et al. (1972) for coarse and fine ballotini particles. Chen and Keairns (1978) presented a limited amount of data on the rate of separation in a char–dolomite system with an unmixed starting mixture. The rate of separation of a wide size distribution of glass beads with the same particle density was investigated by Yoshida et al. (1980). A simple but effective apparatus was designed, and a comprehensive experimental program was carried out to study the rate of separation of different particle systems of different size and density ratios, by Yang and Keairns (1982b). These studies were concentrated, however, on batch separation. Continuous separation studies were also conducted by Yang et al. (1984b), Nienow and Naimer (1980), Iya and Geldart (1978), and Hussein et al. (1981), though only on a limited number of systems for certain specific applications. Those investigations are reviewed here.

Rate of Particle Separation in Batch Systems

The experiments conducted by Chen and Keairns (1978) were carried out using two-component mixtures of different sizes and/or different densities arranged with the heavy or large particles on top of the light or small particles. The particle separation information was obtained by sampling with a rotating turntable from two sampling ports located along the side of the bed at a rate of 1.5 to 3 seconds per sample when the bed was fluidized.

The study conducted by Yoshida et al. (1980) was also for binary systems but with particles of similar density. The bed of a selected binary system of glass beads was first fluidized at a velocity higher than the minimum fluidization velocity of the large component to promote the particle mixing. The velocity was then decreased to a preselected value between $(U_{mf})_F$ and $(U_{mf})_J$. The bed was slumped after a predetermined duration of operation, and the bed was removed layer by layer for analysis. The same procedure was repeated with longer operation duration until the steady state concentration profile was reached. It was found that a clean interface could usually be obtained between the flotsam-rich and the jetsam-rich beds when the particle size ratio of the binary components was larger than 3. For systems with size ratio less than 3, a transition region with changing composition existed between the two segregated beds.

A more comprehensive investigation was carried out by Yang and Keairns (1982b) employing acrylic particles ($\rho_p = 1110$ kg/m^3) as the flotsam and dolomite particles ($\rho_p = 2610$ kg/m^3) as the jetsam. Both types of particles have a relatively wide particle size distribution. Four mixtures of 20, 40, 60, and 80 weight percent of dolomite were studied at velocities ranging from the minimum fluidization velocity of the acrylic particles to slightly higher than that of the dolomite particles. The experiments were performed in a specially constructed bed with a main gas line leading to the bed and a gas bypass line, both controlled by individual but electrically interlocked solenoid valves. A known concentration of acrylic–dolomite mixture was first passed through a Riffle sampler at least four times to mix the mixture, and the mixture was then placed in the fluidized bed. Air was turned on to a desirable reading on the rotameter with the solenoid valve in the bypass line open and that in the main line closed. The needle valve in the bypass line was then adjusted to give a pressure drop in the line equivalent to that expected from the fluidized bed. An electrical switch interlocking both solenoid valves was then turned on to open the solenoid valve in the main line and to shut the solenoid valve in the bypass simultaneously. After a predetermined time, the switch was again turned off to reverse the flow. The fluidized-bed content was then vacuumed off layer by layer and the particle concentrations analyzed by screening. The procedure was then repeated for a different separation time duration, for a different separation gas velocity, or for a different particle mixture.

The transient particle concentration profiles for a 20 w/o and 60 w/o dolomite starting mixture are presented in Figs. 32 and 33. They are shown for 3, 5, 10, and 20 seconds of operation. When the jetsam concentration is low (e.g., 20 w/o dolomite mixture), both the top and bottom layers have relatively uniform concentration profiles throughout the transient time period at different operating velocities (see Fig. 32). For the 60 w/o dolomite mixture, however, the same is true only at lower operating velocities (Figs. 33a and 33b). At higher operating velocities, especially those higher than the minimum fluidization velocity of the dolomite (Fig. 33d), concentration gradients start to develop in both layers with a fuzzy transition between them. At equilibrium, the upper fluidized-bed layer and the bottom packed-bed layer usually have uniform particle concentrations if the operating fluidizing velocities are lower than the minimum fluidization velocity of the jetsam. There is also a small transition zone between the two layers. The jetsam in the bottom packed bed increases with an increase in operating velocity, while the upper fluidized bed almost consists of pure flotsam.

Visual observation of the experiments in this study indicated that the particle separation mechanisms were different at high and low fluidization velocities with the minimum fluidization velocity of dolomite, $(U_{mf})_J$, as the demarcation. At an operating velocity lower than $(U_{mf})_J$, there was a distinct packed bed at the bottom in equilibrium with a fluidized bed at the top with a short transition zone in between. Mixing between these two beds appeared to be minimum. When the operating velocity was increased to higher than $(U_{mf})_J$, the whole bed appeared to be fluidized, although a fuzzy particle interface was usually discernible. Both particle mixing and particle separation were apparently occurring along the bed height. At equilibrium the particle concentration profiles in both layers were essentially uniform. This was approximately true even when the operating velocity was higher than $(U_{mf})_J$. During the transient time period, this was still true except for the case with high jetsam concentration and high operating velocity (Fig. 33d). It appears, then, that considerable simplification in the mathematical model may be possible, at least for the highly segregating system of acrylic–dolomite.

A simple mathematical model was developed by Yang and Keairns (1982b) by assuming that the particle segregation is a fluidized bed could be simulated by two perfectly mixed fluidized beds in series with particle interchange between them. The short transition region observed experimentally was ignored in the model. The particle exchange was accomplished by bubble wakes from the bottom to the top fluidized bed and by bulk solids flow in the reverse direction. The resulting equation is

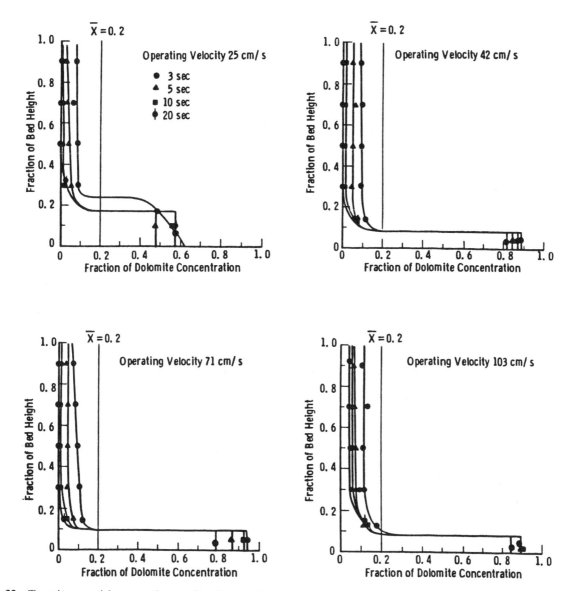

Figure 32 Transient particle separation profiles for acrylic–dolomite systems—20 w/o dolomite concentration.

$$\ln\left[\frac{V_1(1 - (C_F)_W) - V_{J1}}{V_1(1 - (C_F)_W) - V_{J1}^o}\right] =$$
$$-\frac{(U - U_{mf2}) \cdot A \cdot f_w \cdot (1 - \varepsilon_w)}{V_1} \cdot t \qquad (219)$$

If we assume further that $(C_F)_w = 1.0$, i.e., the bubble wake contains pure flotsam (acrylic particles), Eq. (219) can be written as

$$\ln\left(\frac{V_{J1}}{V_{J1}^o}\right) = -mt \qquad (220)$$

where

$$m = \frac{(U - U_{mf2}) \cdot A \cdot f_w \cdot (1 - \varepsilon_w)}{V_1} \qquad (221)$$

There is certainly some experimental evidence indicating that the assumption of $(C_F)_W = 1.0$ is reasonable at least for highly segregating systems (Yang and Keairns, 1982b).

Equation (220) was used to fit the experimental data with excellent results. The success of the correlation does suggest possible physical significance of the para-

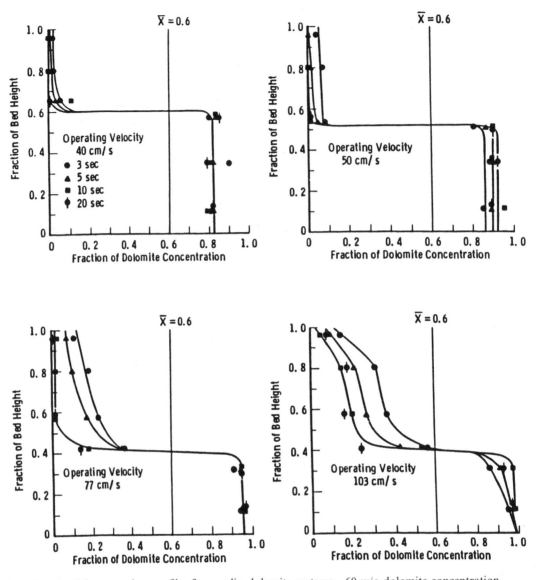

Figure 33 Transient particle separation profiles for acrylic–dolomite systems—60 w/o dolomite concentration.

meter, m, defined in Eq. (221). A constant m means a constant volumetric exchange rate between the two fluid beds during the transient period. A key assumption in Eq. (220) is that $(C_F)_W = 1.0$. This assumption is reasonable for a highly segregating system like the acrylic–dolomite system. The rate of particle separation depends critically on $(C_F)_W$ as shown in Eq. (219). Unfortunately, experimental values of $(C_F)_W$ for different systems are generally not available.

The mixing and segregation kinetics of mixtures of iron and glass particles, a strongly segregating bed, were also studied by Beeckmans and Stahl (1987) employing both initially fully mixed and initially com-

pletely segregated conditions. The data were analyzed based on a mixed bed of jetsam and flotsam residing on top of a pure jetsam bed. The interchange velocities of jetsam between the two beds were found to depend strongly on the excess fluidization velocity.

Continuous Operation and Industrial Applications

A good example of continuous particle separation is the industrial application in the ash-agglomerating fluidized bed gasifier. The ash remained from gasification of coal is agglomerated into bigger and denser agglomerates and is continuously removed from the fluidized bed gasifier via a fluidized bed separator. Cold flow simulation was conducted in a large scale cold model 3 meters in diameter using crushed acrylic particles ($\rho_p = 1100$ kg/m^3) to simulate the char and sand particles ($\rho_p = 2650$ kg/m^3) to simulate the ash agglomerates (Yang et al., 1986).

Continuous mixing experiments of two particulate species of different densities were also performed by Nienow and Naimer (1980) in a gas fluidized bed. They found that the continuous operation did not have any effect on the segregation pattern, and that the behavior of a continuously operated gas fluidized bed could generally be predicted from batch experiments. Similar mixing indices obtained from the batch experiments could be used under continuous operating conditions.

One promising application of the continuous particle separation operation is to use it as a solid-to-solid heat exchanger. The heat carriers, which are usually large and dense particles, are rained through a fluidized bed of fines. The application can be for drying as studied by Baskakov et al. (1975), for supplying the necessary heat to the endothermic reactions, as in the Union Carbide agglomerated-ash coal gasification process (Corder, 1973), or some other similar applications like the commercial Nitro-Top process (Drake, 1973).

The dynamics of coarse and dense particles raining through a gas fluidized bed of fine particles were studied by Baskakov et al. (1975), Iya and Geldart (1978), and Hussein et al. (1981) in fluidized beds of different configurations. Baskakov et al. (1975) found that the minimum fluidization velocity ratio between the jetsam and the flotsam had essentially no effect on the quality of particle separation for $(U_{mf})_J/(U_{mf})_F > 18$. When the ratio was less than 7, the degree of particle separation deteriorated noticeably. The bulk density and the mean size of the fines constituting the fluidized bed were found by Iya and Geldart (1978) to be the most important parameters affecting the rain-through velocity of the coarse particles. The size of the coarse particles did not appear to be important on rain-through velocity but it did affect the maximum circulation flux. Circulation flux of coarse particles up to 2.3×10^5 kg/h·m^2 was observed.

The later study by Hussein et al. (1981) attained only about 10% of the maximum circulation flux observed by Iya and Geldart (1978), probably because of the different sizes and densities of the particles employed. They also observed that the size of the descending balls and the fluidizing gas were important, contrary to that concluded by Iya and Geldart (1978). Obviously more studies are necessary in this

area of particle separation. Hussein et al. (1981) also investigated the heat transfer between the descending balls and the fluidized bed and determined the heat transfer coefficients to be in the range of 100 to 200 W/m^2·°K. This lies at the lower end of the heat transfer coefficients expected between immersed surfaces and a gas fluidized bed. A patent was granted for retorting and/or gasifying solid carbonaceous materials, such as coal, coke, shale, and tar sands, employing this concept (Mitchell and Sageman, 1979).

Continuous operation was also carried out by Beeckmans and Minh (1977) using the fluidized counter-current cascade principle with encouraging results. Separation can also be enhanced by the presence of an electrostatic field (Beeckmans et al., 1979) or baffles of various designs (Kan and Sen Gupta, 1978; Naveh and Resnick, 1974).

12.3 Mathematical Models for Prediction of Equilibrium Concentration Profiles

A gas fluidized bed is a complex reactor even if it contains only particles of similar size, shape and, density. The physical phenomena occurring in a gas fluidized bed depend not only on the particle characteristics, the operating pressure, and the temperature that will change the properties of the gas fluidizing medium but also on the physical size of the bed. New findings and surprises are still continuously being reported. When the bed material consists of particles of different sizes and densities, the mixing and segregation phenomena are much less understood. The equilibrium particle concentration profile established inside a gas fluidized bed at steady state is actually a dynamic equilibrium between the competing processes of mixing and segregation. Several mathematical models have been proposed to evaluate both the transient and the equilibrium particle concentration profiles. However, they are all restricted to systems of binary mixtures only.

The models proposed so far can be broadly classified into two categories. The first group of models is constructed on the basis of mechanistic phenomena observed in a bubbling gas fluidized bed. Thus they are usually similar in concept and only different in details. They all divide the bed into the wake (or bubble) phase and the bulk (or emulsion) phase. The assumptions made for the particle exchange between the two phases and the excursion of certain phenomena distinguish each individual model. The models in this group are due to Gibilaro and Rowe (1974), Burgess et al. (1977), and Yoshida et al. (1980). The second type of model is that suggested by Gelperin et al. (1977). He assumes that the particle separation process in a gas fluidized bed is a diffusion process and lumps all various effects into an effective particle diffusivity and a relative displacement rate. Readers can consult Yang (1986) or original articles for details.

12.4 Particle Separation Based on the Principle of Particle Terminal Velocity

All the discussion so far is on the particle separation in a gas fluidized bed based on velocities very close to the minimum fluidization velocity of the jetsam. Particle separation can also be accomplished by employing velocities close to that of the terminal velocity of the particles. The U-Gas coal gasification process developed by the Institute of Gas Technology utilized the principle of terminal velocity for separating ash agglomerates from char in the coal gasifier. A comprehensive study on this subject using a fluidized bed similar to that of the U-Gas process has been published by Leppin and Sahay (1980). Some experiments were also carried out by Chen and Keairns (1978).

NOMENCLATURE

A	=	cross-sectional area of fluidized bed
A_c	=	catchment area of a bubble stream
A_o	=	area of distributor plate per hole, $A_o =$ 0 for porous plate
Ar	=	Archimedes number
A_t	=	total cross-section area of a single jet or of concentric jets
b	=	jet half-thickness
C, C_1, C_2	=	constants
$(C_F)_w$	=	volumetric flotsam concentration (fraction) in the wake phase
$d_{p1}, d_{p2}----$	=	particle diameter of component 1, 2, and ---, respectively
d_B, d_S	=	particle diameter of big and small particles, respectively
d_F, d_J	=	particle diameter of flotsam and jetsam, respectively
d_o	=	diameter of a jet nozzle
d_{or}	=	orifice diameter
d_p	=	particle diameter of a single component system
$\bar{d}_p$	=	average particle size of a mixture
$(d_p)_F$	=	particle diameter of flotsam
$(d_p)_J$	=	particle diameter of jetsam
d_{pi}	=	particle diameter of ith component in a multicomponent mixture
d_R	=	particle diameter ratio, d_B/d_S

D	=	diameter of a fluidized bed
D_B	=	bubble diameter
D_{Bo}	=	initial bubble diameter
$(D_B)_{max}$	=	maximum bubble diameter
D_e	=	equivalent bubble diameter
D_{sr}	=	lateral dispersion coefficient of solids in fluidized beds
e	=	coefficient of restitution
F	=	force on a single particle; or total amount of gas leakage during bubble formation
f_F, f_J	=	volumetric fraction of flotsam and jetsam in the bed, respectively
F_H	=	adhesion force transmitted in a single contact between two particles
f_n	=	bubble frequency
f_{NF}	=	number fraction of flotsam
F_o	=	solids feed rate
Fr	=	Froude number
$(Fr)_c$	=	critical Froude number for bubbling and jetting demarcation
f_S	=	slug frequency
f_{VF}	=	volume fraction of flotsam
f_w	=	wake fraction, the volume of wake divided by the combined volume of bubble and wake
g	=	gravitational acceleration
G	=	gas flow rate
G_B	=	average visible volumetric bubble flow
G_j	=	gas flow rate through the jet nozzle
h	=	height above the distributor
H	=	bed height of the fluidized bed
H_1, H_2	=	bed height at stage 1 and stage 2, respectively, during bed collapsing tests
H_3, H_D	=	bed heights at stage 3 during bed collapsing tests
H_c	=	bed height at critical point in bed collapsing tests
H_e	=	bed height of emulsion phase
H_{max}	=	maximum bed height
H_{mf}	=	bed height at minimum fluidization
H_{min}	=	minimum bed height
H_o	=	initial bed height
H_∞	=	bed height at $t = \infty$
I	=	the uniformity index
J	=	solids circulation flux in fluidized beds
K, K_1, K_2, K_3, K_4	=	constants
L	=	bed depth; or bed width
L_j	=	jet penetration depth
L_m	=	jet penetration depth for multiple-jet systems
L_{mf}	=	bed depth at minimum fluidization
L_S	=	slug length
l_s	=	orifice separation distance or pitch
M	=	the mixing index

M_g	=	mass flow rate of gas
$(M_g)_i$	=	mass flow rate of gas in the inner jet
$(M_g)_o$	=	mass flow rate of gas in the outer jet
M_s	=	mass flow rate of solids
$(M_s)_i$	=	mass flow rate of solids in the inner jet
n	=	a constant
N_o	=	total number of orifices in a multiorifice distributor plate
N_T	=	number of bed diameters between the rear of a leading slug and the nose of a trailing one
P	=	pressure
P_j	=	pitch, distance between the centerline of two jet nozzles
ΔP	=	pressure drop
q	=	gas exchange rate between bubble and emulsion phase; or bulk/wake exchange rate per unit height
r	=	radial coordinate; or radial distance from the jet axis
$r_{1/2}$	=	radial position where jet velocity is $\frac{1}{2}$ of the maximum at jet axis
R	=	ratio of terminal velocity and minimum fluidization velocity; or radius of sphere of penetration (see Fig. 14)
R_B	=	radius of bubble
R_{cf}	=	ratio of minimum fluidization velocity at pressure P over that at atmospheric pressure
$(Re)_{bf}, (Re)_{mf}, (Re)_{tf}$	=	Reynolds' numbers based on the beginning fluidization velocity, U_{bf}, the minimum fluidization velocity, U_{mf}, and the total fluidization velocity, U_{tf}, respectively
$(Re)_p$	=	Reynolds' number based on the particle diameter
$(Re)_t$	=	Reynolds' number based on the terminal velocity
$R(t)$	=	particle resident time distribution in fluidized beds
S	=	the segregation index
t	=	time
t_b	=	time at the end of stage 1 in bed collapsing tests
t_c	=	time at the critical point in bed collapsing tests
$\bar{t}_i$	=	mean residence time of particles in ith stage
$\bar{t}_R$	=	mean residence time of particles
t_T	=	bed turnover time
U	=	superficial fluidization velocity
U_A	=	absolute bubble velocity
U_B	=	bubble velocity
U_b	=	gas velocity at jet boundary
U_{bf}	=	beginning fluidization velocity

U_c	=	a transition velocity between bubbling and turbulent fluidization regimes
U_{cf}	=	complete fluidization velocity
$(U_g)_i$	=	jet velocity in the inner jet
$(U_g)_o$	=	jet velocity in the outer jet
U_j	=	average jet velocity at the nozzle
$(U_J)_c$	=	jet velocity of a concentric jet
$(U_j)_s$	=	jet velocity of a simple two-phase jet
U_k	=	a transition velocity between bubbling and turbulent fluidization regimes
U_m	=	maximum velocity at the jet axis
U_{mb}	=	minimum bubbling velocity
U_{mf}	=	minimum fluidization velocity
$(U_{mf})_1$	=	minimum fluidization velocity of component 1 in a multicomponent mixture
U_{mf2}	=	minimum fluidization velocity of fluidized bed 2
$(U_{mf})_{atm}$	=	minimum fluidization velocity at atmospheric pressure
$(U_{mf})_F, (U_{mf})_J$	=	minimum fluidization velocities of flotsam and jetsam, respectively
$(U_{mf})_I, (U_{mf})_S$	=	minimum fluidization velocities of binary systems
$(U_{mf})_M$	=	minimum fluidization velocity of a mixture
$(U_{mf})_p$	=	minimum fluidization velocity at pressure p
U_r	=	jet velocity at radial position r
U_S	=	slug velocity; solid particle velocity; or relative displacement rate between jetsam and flotsam particles
U_{SA}	=	absolute slug velocity
$(U_s)_i$	=	solid particle velocity in the inner jet
U_t	=	terminal velocity of a single particle
U_{tf}	=	total fluidization velocity
U_{To}	=	the takeover velocity
U_{tr}	=	transport velocity, a transition velocity from the turbulent regime to the fast fluidization regime
V_1	=	volume of solids in fluid bed 1 excluding volume of the voids between particles
V_B	=	bubble volume
V_d	=	volume of bubble drift
V_j	=	mean particle velocity in the jet
V_{J1}	=	volume of jetsam in fluid bed 1 excluding the voids between particles
V_{J1}^0	=	volume of jetsam in fluid bed 1 at $t = 0$
V_S	=	slug volume; or volume of bubble shell
V_w	=	volume of bubble wake
W	=	bed weight
x	=	x coordinate; or distance from the jet nozzle
x_i	=	weight fraction of the ith component

$\bar{x}_F, \bar{x}_J$	=	average weight fraction of flotsam and jetsam, respectively
X_F, X_J	=	weight fraction of flotsam and jetsam, respectively
$\bar{X}_F, \bar{X}_J$	=	fraction of flotsam and jetsam, respectively, at the state of perfect mixing
$(X_F)_u, (X_J)_u$	=	weight fraction of flotsam and jetsam in the upper bed, respectively
y	=	y coordinate
Y	=	correction for deviation from the two-phase theory
Y_h	=	height of stagnant emulsion region between adjacent jets
z	=	axial coordinate
ρ_B, ρ_S	=	particle density of big particles and small particles, respectively
$(\rho_B)_S$	=	bulk density of the small particles
ρ_f	=	fluid density
ρ_F, ρ_J	=	particle density of flotsam and jetsam, respectively
ρ_p	=	particle density in a single component system
$\bar{\rho}_p$	=	average particle density of a mixture
ε	=	voidage
ε_1	=	voidage of bed consists of component 1
ε_b	=	bed voidage
ε_B	=	bubble fraction in the bed
$\bar{\varepsilon}_B$	=	average volume fraction of visible bubble phase
ε_c	=	voidage at critical bed height
ε_e	=	voidage of emulsion phase
ε_j	=	voidage inside the jet
ε_{mf}	=	voidage at minimum fluidization
ε_s	=	fraction of bed occupied by the slugs
ε_w	=	voidage in the wake
ε_z	=	voidage outside of jet in the emulsion phase at z
ψ_f	=	fluid stream function
ψ_p	=	particle stream function
ϕ	=	particle shape factor
ϕ_B	=	bubble shape factor
ϕ_F, ϕ_J	=	shape factor of flotsam and jetsam particles, respectively
α_b	=	angle of included angle in a bubble
β	=	distributor porosity
β_d	=	drift fraction, V_d/V_B
β_w	=	wake fraction, V_W/V_B
λ	=	the rate of competing effect between circulation and segregation, $= w/k$
σ	=	standard deviation
μ	=	fluid viscosity
θ	=	polar coordinate; included angle of a bubble; or jet half-angle
θ_s	=	dimensionless subsidence time of a powder

REFERENCES

Abramovich GN. The Theory of Turbulent Jets. Cambridge, MA: MIT Press, 1963.

Anagbo PE. Derivation of jet cone angle from bubble theory. Chem Eng Sci 35:1494–1495, 1980.

Babu SP, Shah B, Talwalkar A. Fluidization correlations for coal gasification materials—minimum fluidization velocity and fluidized bed expansion ratio. AIChE Symp Ser 74(176):176–186, 1978.

Baeyens J, Geldart D. Fluidization et ses Applications. Toulouse, 1973, p 182.

Baeyens J, Geldart D. An investigation into slugging fluidized beds. Chem Eng Sci 29:255–265, 1974.

Baskakov AP, Malykh GA, Shihko II. Separation of materials in equipment with a fluidized bed and with continuous charging and discharging. Int Chem Eng 15(2):286–289, 1975.

Basov VA, Markhevka VI, Melik-Akhnazanov TKh, Orochke DI. Investigation of the structure of a nonuniform fluidized bed. Intern Chem Eng 9:263–266, 1969.

Beeckmans JM, Minh T. Separation of mixed granular solids using the fluidized counter-current cascade principle. Can J Chem Eng 55:493–496, 1977.

Beeckmans JM, Stahl B. Mixing and segregation kinetics in a strongly segregated gas-fluidized bed. Powder Technol 53:31–38, 1987.

Beeckmans JM, Inculet II, Dumas G. Enhancement in segregation of a mixed powder in a fluidized bed in the presence of an electrostatic field. Powder Technol 24:267–269, 1979.

Behie LA, Bergougnou MA, Baker CGJ, Bulani W. Jet momentum dissipation at a grid of a large gas fluidized bed. Can J Chem Eng 48:158–161, 1970.

Behie LA, Bergougnou MA, Baker CGJ. Heat transfer from a grid jet in a large fluidized bed. Can J Chem Eng 53:25–30, 1975.

Behie LA, Bergougnou MA, Baker CGJ. Mass transfer from a grid jet in a large gas fluidized bed. In: Fluidization Technology. (Keairns DL, ed.). New York: McGraw-Hill, 1976, Vol. 1, pp 261–278.

Bi HT, Fan LS. Existence of turbulent regime in gas-solid fluidization. AIChE J 38:297–301, 1992.

Bi HT, Grace JR. Flow regime diagrams for gas-solid fluidization and upward transport. Int J Multiphase Flow 21:1229–1236, 1995a.

Bi HT, Grace JR. Effect of measurement method on velocities used to demarcate the onset of turbulent fluidization. Chem Eng J 57:261–271, 1995b.

Bi HT, Grace JR, Zhu JX. Regime transitions affecting gas-solids suspensions and fluidized beds. Chem Eng Res Des 73:154–161, 1995.

Blake TR, Wen CY, Ku CA. The correlation of jet penetration measurements in fluidized beds using nondimensional hydrodynamic parameters. AIChE Symp Series 80(234):42–51, 1984.

Bloore PD, Botterill JSM. Similarity in behavior between gas bubbles in liquid and fluidized solid systems. Nature 190:250, 1961.

Bourgeois P, Grenier P. The ratio of terminal velocity to minimum fluidizing velocity for spherical particles. Can J Chem Eng 46:325–328, 1968.

Buevich YA, Minaev GA. Mechanics of jet flows in granular layers, evolution of single jets and the nucleation mechanism. Inzh Fiz Zh 30(5):825–833, 1976.

Burgess JM, Fane AG, Fell CJD. Measurement and prediction of the mixing and segregation of solids in gas fluidized beds. Pac Chem Eng Congr 2:1405–1412, 1977.

Cai P, Chen SP, Jin Y, Yu ZQ, Wang ZW. Effect of operating temperature and pressure on the transition from bubbling to turbulent fluidization. AIChE Symp Series 85(270):37–43, 1989.

Canada GS, McLaughlin MH, Staub FW. Flow regimes and void fraction distribution in gas fluidization of large particles in beds without tube banks. AIChE Symp Ser 176(74):14–26, 1978.

Catipovic NM, Jovanovic GN, Fitzgerald TJ. Regimes of fluidization of large particles. AIChE J 24:543–547, 1978.

Chandnani PP, Epstein N. Spoutability and spout destabilization of fine particles with a gas. In: Fluidization (Ostergaard K, Sorensen A, eds.) New York: Engineering Foundation, 1986, pp 233–240.

Chehbouni A, Caouki J, Guy C, Klvana D. Characterization of the flow transition between bubbling and turbulent fluidization. Ind Eng Chem Res 33:1889–1896, 1994.

Chen JLP, Keairns DL. Particle segregation in a fluidized bed. Can J Chem Eng 53:395–402, 1975.

Chen JLP, Keairns DL. Particle separation from a fluidized mixture—simulation of Westinghouse coal-gasification combustor-gasifier operation. Ind Eng Chem Process Des Develop 17:135–141, 1978.

Cheung L, Nienow AW, Rowe PN. Minimum fluidization velocity of a binary mixture of different sized paticles. Chem Eng Sci 29:1301–1303, 1974.

Chiba T, Kobayashi H. Effect of bubbling on solid segregation in gas fluidized beds. In: Fluidization—Science and Technology (Kwauk M, Kunii, D. eds.). New York: Gordon and Breach, 1982, pp 79–90.

Chiba S, Chiba T, Nienow AW, Kobayashi H. The minimum fluidization velocity, bed expansion and pressure-drop profile of binary particle mixtures. Powder Technol 22:255–269, 1979.

Chiba S, Nienow AW, Chiba T, Kobayashi H. Fluidized binary mixtures in which the dense component may be flotsam. Powder Technol, 26:1–10, 1980.

Chiba S, Kawabata J, Yumiyama M, Tazaki Y, Honma S, Kitano K, Kobayashi H, Chiba T. Pressure effects on solid mixing and segregation in gas-fluidized beds of binary solid mixtures. In: Fluidization—Science and Technology (Kwauk M, Kunii, D. eds.). New York: Gordon and Breach, 1982, pp 69–78.

Chitester DC, Kornosky RM, Fan LS, Danko JP. Characteristics of fluidization at high pressure. Chem Eng Sci 39;253–261, 1984.

Choi JH, Son JE, Kim SD. Generalized model for bubble size and frequency in gas-fluidized beds. Ind Eng Chem Res 37:2559–2564, 1998.

Clift R, Filla M, Massimilla L. Gas and particle motion in jets in fluidized beds. Intern J Multiphase Flow 2:549–561, 1976.

Corder WC. Proc. 5th Synthetic Pipeline Gas Symp, Chicago, IL. 1973.

Darton RC, LaNauze RD, Davidson JF, Harrison D. Bubble growth due to coalescence in fluidized beds. Trans Inst Chem Eng 55:274–280, 1977.

Davidson JF. Symposium on fluidization—contribution to discussion. Trans Inst Chem Eng 39:230–232, 1961.

Davidson JF, Harrison D. Fluidized Particles. Cambridge: Cambridge University Press, 1963.

Davidson JF, Schuler BOG. Bubble formation at an orifice in an inviscid liquid. Trans Inst Chem Eng 38:335–342, 1960.

Davidson JF, Harrison D, Guedes de Carvalho JRF. On the liquidlike behavior of fluidized beds. Ann Rev Fluid Mech 9:55–86, 1977.

Davies RM, Taylor G. The mechanics of large bubbles rising through extended liquids and through liquids in tubes. Proc Roy Soc (London) A200:375–390, 1950.

De Michele G, Elia A, Massimilla L. The interaction between jets and fluidized beds. Ing Chim Ital 12:155–162, 1976.

Dickman R, Forsythe WL. Laboratory prediction of flow properties of fluidized solids. Ind Eng Chem 45:1174–1185, 1953.

Doichev K, Todorov S, Dimitrov V. Transition between particulate and aggregative fluidization at different state-of-flow of solids. Chen Eng Sci 30:419–424, 1975.

Donadono A, Massimilla L. Mechanisms of momentum and heat transfer between gas jets and fluidized beds. In: Fluidization. (Davidson JF, Keairns DL, eds.) Cambridge: Cambridge University Press, 1978, pp 375–380.

Donadono S, Maresca A, Massimilla L. Gas injection in shallow beds of fluidized, coarse solids. Ing Chim Ital 16:1–10, 1980.

Donsi G, Massimilla L, Colantuoni L. The dispersion of axisymmetric jets. In: Fluidization (Grace JR, Matsen JM, eds.). New York: Plenum Press, 1980, pp 297–304

Drake G. Proc Fert Soc No. 130, 1973.

Dumitrescu DT. Strömung an einer Luftblase im senkrechten Rohr. Z Angew Math Mech 23:139, 1943.

Ettehadieh B, Yang WC, Hadipur GB. Motion of solids, jetting and bubbling dynamics in a large jetting fluidized bed. Powder Technol 54:243–254, 1988.

Fane AG, Nghiem NP. Bubble induced mixing and segregation in a gas-fluidized bed. J Chinese Inst Chem Eng 24:215–224, 1983.

Filla M, Massimilla L, Vaccaro S. Solids entrainment in jets in fluidized beds: the influence of particle shape. Journées Européennes sur la Fluidisation, Toulouse, France, September 1981.

Filla M, Massimilla L, Vaccaro S. Gas jets in fluidized beds and spouts: a comparison of experimental behavior and models. Can J Chem Eng 61:370–376, 1983a.

Filla M, Massimilla L, Vaccaro S. Gas jets in fluidized beds: the influence of particle size, shape and density on gas and solids entrainment. Int J Multiphase Flow 9:259–267, 1983b.

Furukawa J, Ohmae T. Liquid-like properties of fluidized systems. Ind Eng Chem 50:821–828, 1958.

Gbordzoe EAM, Freychet N, Bergougnou MA. Gas transfer between a central jet and a large two-dimensional gas-fluidized bed. Powder Technol 55:207–222, 1988.

Geldart D. The effect of particle size and size distribution on the behavior of gas-fluidized beds. Powder Technol 6:201–205, 1972.

Geldart D. Types of gas fluidization. Powder Technol 7:285–292, 1973.

Geldart D. Predicting the expansion of gas fluidized beds. In: Fluidization Technology. (Kearins DL, ed.) Washington, DC: Hemisphere, 1975, pp 237–244.

Geldart D. Gas Fluidization Technology. Chichester: John Wiley, 1986.

Geldart D, Abrahamsen AR. Homogeneous fluidization of fine powders using various gases and pressures. Powder Technol 19:133–136, 1978.

Gelperin NI, Einstein VG. Heat transfer in fluidized beds. In: Fluidization. (Davidson JF, Harrison D, eds.) New York: Academic Press, 1971, pp 471–450.

Gelperin NI, Einstein VG, Nosov GA, Mamoshkina VV, Rebrova SK. Teor Osnovy Khim Tehnol 1:383, 1967.

Gelperin NI, Zakharenko VV, Ainshtein VG. Segregation of solid particles in a fluidized bed and the equilibrium distribution. Theor Foundations Chem Eng (Thor Osnovy Khim Tekh) 11:475–481, 1977.

Gibilaro LG, Rowe PN. A model for a segregating gas fluidized bed. Chem Eng Sci 29:1403–1412, 1974.

Goossens WRA. Classification of fluidized particles by Archimedes number. Powder Technol 98:48–53, 1998.

Goossens WRA, Dumont GL, Spaepen GL. Fluidization of binary mixtures in laminar flow region. Chem Eng Prog Symp Ser 67(116):38–45, 1971.

Grace JR. The viscosity of fluidized beds. Can J Chem Eng 48:30–33, 1970.

Grace JR. Fluidized-bed hydrodynamics. In: Handbook of Multiphase Systems. (Hetsroni G, ed.) Washington: Hemisphere, 1982, pp 8-5–8-64.

Grace JR. Contacting modes and behavior classification of gas–solid and other two-phase suspension. Can J Chem Eng 64:353–363, 1986.

Grace JR, Clift R. On the two-phase theory of fluidization. Chem Eng Sci 29:327–334, 1974.

Grace JR, Lim CT. Permanent jet formation in beds of particulate solids. Can J Chem Eng 65:160–162, 1987.

Grace JR, Sun G. Fines concentration in voids in fluidized beds. Powder Technol 62:203–205, 1990.

Habermann A, Winter, F. Hofbauer H, Gogolek PEG. Residence time distribution of particles in fluidized bed reactors for metallurgical processes. In; Fluidization (Fan LS, Knowlton TM, eds.). New York: Engineering Foundation, 1998, pp 117–124.

Harrison D, Leung LS. Bubble formation at an orifice in a fluidized bed. Trans Inst Chem Eng 39:409–413, 1961.

Harrison D, Davidson JF, de Kock JW. On the nature of aggregative and particulate fluidization. Trans Inst Chem Eng 39:202–211, 1961.

Hirsan I, Sishtla C, Knowlton TM. Presented at the 73rd AIChE Annual Meeting, Chicago, IL, November 1980.

Horio M. Hydrodynamics of circulating fluidization—present status and research needs. In: Circulating Fluidized Bed Technology III (Basu P, Horio M, Hasatani M, eds.). Oxford: Pergamon Press, 1990, pp 3–14.

Hovmand S, Davidson JF. Chemical conversion in a slugging fluidized bed. Trans Inst Chem Eng 46:190–203, 1968.

Hovmand S, Davidson JF. Pilot plant and laboratory scale fluidized reactors at high gas velocities: the relevance of slug flow. In: Fluidization. (Davidson JF, Harrison D, eds.) New York: Academic Press, 1971, pp 193–259.

Hussein FD, Paltra PP, Jackson R. Heat transfer to a fluidized bed using a particulate heat transfer medium. Ind Eng Chem Process Des Dev 20:511–519, 1981.

Iya SK, Geldart D. Flow of coarse particles through a fluidized bed of fines. Ind Eng Chem Process Des Dev 17:537–543, 1978.

Jackson R. The mechanics of fluidized beds: the motion of fully developed bubbles. Trans Inst Chem Eng 41:22–48, 1963.

Kan DD, Sen Gupta P. Relative performance of baffled and unbaffled fluidized bed classifiers. Indian J Tech 16:507–512, 1978.

Kececioglu I, Yang WC, Keairns DL. Fate of solids fed pneumatically through a jet into a fluidized bed. AIChE J 30:99–110, 1984.

Kehoe PWK, Davidson JF. The fluctuation of surface height in freely slugging fluidized beds. AIChE Symp Ser 69(128):41–48, 1973a.

Kehoe PWK, Davidson JF. Pressure fluctuations in slugging fluidized beds. AIChE Symp Ser 69(128):34–40, 1973b.

Kimura T, Matsuo H, Uemiya S, Kojima T. Measurement of jet shape and its dynamic change in three-dimensional jetting fluidized beds. J Chem Eng Japan 27:602–609, 1994.

Kimura T, Horiuchi K, Watanabe T, Matsukata M, Kojima T. Experimental study of gas and particle behavior in the grid zone of a jetting fluidized bed cold model. Powder Technol 82:135–143, 1995.

King DF, Mitchell RG, Harrison D. Dense phase viscosities of fluidized beds at elevated pressure. Powder Technol 28:55–58, 1981.

Knowlton TM. High-pressure fluidization characteristics of several particulate solids. AIChE Symp Ser 73(61):22–28, 1977.

Knowlton TM, Hirsan I. The effect of pressure on jet penetration in semi-cylindrical gas-fluidized beds. In: Fluidization (Grace JR, Matsen JM, eds.). New York: Plenum Press, 1980, pp 315–324.

Kondukov NB, Sosna MH. Khim Prom 6:402, 1965.

Kozin BE, Baskakov AP. Studies on the jet range in a bed of granular particles. Khim I Teckhnol Topliv I Masel 3:4–7, 1967.

Kramers H. On the "viscosity" of a bed of fluidized solids. Chem Eng Sci 1:35–37, 1951.

Kumar A, Sen Gupta P. Prediction of minimum fluidization velocity for multicomponent mixtures. Indian J Technol 12:225–227, May 1974.

Kunii D, Levenspiel, O. Fluidization Engineering. 1st ed. New York: John Wiley, 1969.

Kunii D, Levenspiel, O. Fluidization Engineering. 2nd ed. Boston: Butterworth-Heinemann, 1991.

Kwauk M. Fluidization: Idealized and Bubbleless, with Applications. New York: Ellis Horwood, 1992.

Lefroy GA, Davidson JF. The mechanics of spouted beds. Trans Inst Chem Eng 47:T120–T128, 1969.

Leppin D, Sahay GN. Cold-model studies of agglomerating gasifier discharge behavior. In: Fluidization (Grace JR, Matsen JM, eds.). New York: Plenum Press, 1980, pp 429–436.

Leva M. Fluidization. New York: McGraw-Hill, 1959.

Levenspiel O. The Chemical Reactor Omnibook. Corvallis: OSU Book Stores, 1979.

Lim KS, Zhu JX, Grace JR. Hydrodynamcis of gas–solid fluidization. Int J Multiphase Flow 21(Suppl):141–193, 1995.

Markhevka VI, Basov VA, Melik-Akhnazanov TKh, Orochko DI. The flow of a gas jet into a fluidized bed. Theor Foundations Chem Eng 5:80–85, 1971.

Massimilla L. Gas jets in fluidized beds. In: Fluidization. 2nd ed. (Davidson JF, Clift R, Harrison D., eds.). London: Academic Press, 1985, pp 133–172.

Massimilla L, Donsi G, Migliaccio N. The dispersion of gas jets in two-dimensional fluidized beds of coarse solids. AIChE Symp Ser 77(205):17–27, 1981.

Matheson GL, Herbst WA, Holt PH. Characteristics of fluid–solid systems. Ind Eng Chem 41:1099–1104, 1949.

Matsen JM. Evidence of maximum stable bubble size in a fluidized bed. AIChE Symp Ser 69(128):30–33, 1973.

Matsen JM, Hovmand S, Davidson JF. Expansion of fluidized beds in slug flow. Chem Eng Sci 24:1743–1754, 1969.

Merry JMD. Penetration of a horizontal gas jet into a fluidized bed. Trans Inst Chem Eng 49:189–195, 1971.

Merry JMD. Penetration of vertical jets into fluidized beds. AIChE J 21:507–510, 1975.

Merry JMD. Fluid and particle entrainment into vertical jets in fluidized beds. AIChE J 22:315–323, 1976.

Mitchell DS, Sageman DR. Countercurrent plug-like flow of two solids. U.S. Patent 4,157,245 (1979).

Molerus O. Interpretation of Geldart's type A, B, C and D powders by taking into account interparticle cohesion forces. Powder Technol 33:81–87, 1982.

Mori S, Wen CY. Estimation of bubble diameter in gaseous fluidized beds. AIChE J 21:109–115, 1975.

Morooka S, Nishinaka M, Kato Y. Sedimentation velocity and expansion ratio of the emulsion phase in a gas–solid fluidized bed. Kagaku Kogaku 37:485–493, 1973.

Murray JD. On the mathematics of fluidization: steady motion of fully developed bubbles. J Fluid Mech 21:465–493, 1965.

Naimer, NS, Chiba T, Nienow AW. Parameter estimation for a solids mixing/segregation model for gas fluidized beds. Chem Eng Sci 37:1047–1057, 1982.

Naveh E, Resnick W. Particle size segregation in baffled fluidized beds. Trans Inst Chem Eng 52:58–66, 1974.

Nguyen XT, Leung LS. A note on bubble formation at an orifice in a fluidized bed. Chem Eng Sci 27:1748–1750, 1972.

Nicklin DJ, Wilkes JO, Davidson JF. Two-phase flow in vertical tubes. Trans Inst Chem Eng 40:61–68, 1962.

Nienow AW, Cheesman, DJ. The effect of shape on the mixing and segregation of large particles in gas-fluidized bed of small ones. In: Fluidization. (Grace JR, Matsen JM, eds.) New York: Plenum Press, 1980, pp 373–380.

Nienow AW, Naimer NS. Continuous mixing of two particulate species of different density in a gas fluidized bed. Trans Inst Chem Eng 58:181–186, 1980.

Nienow AW, Rowe PN, Cheung LYL. A quantitative analysis of the mixing of two segregating powders of different density in a gas-fluidized bed. Powder Technol 20:89–97, 1978a.

Nienow AW, Rowe, PN, Chiba T. Mixing and segregation of a small proportion of large particles in gas fluidized beds of considerably smaller ones. AIChE Symp Ser 74(176):45–53, 1978b.

Nienow AW, Naimer NS, Chiba T. Studies of segregation/mixing in fluidized beds of different size paticles. Chem Eng commun 62:53–66, 1987.

Peeler JPK, Huang JR. Segregation of wide size range particle mixtures in fluidized beds. Chem Eng Sci 44:1113–1119, 1989.

Perales JF, Coll T, Llop MF, Puigjaner L, Arnaldos J, Cassal J. On the transition from bubbling to fast fluidization regimes. In: Circulating Fluidized Bed Technology III (Basu P, Horio M, Hasatani M, eds.). Oxford: Pergamon Press, 1990, pp 73–78.

Rhodes MJ, Wang XS. Defining fast fluidization. In: Fluidization. (Fan LS, Knowlton TM, eds.) New York: Engineering Foundation, 1998, pp 133–140.

Rice RW, Brainnovich Jr JF. Mixing/segregation in two- and three-dimensional fluidized beds: binary systems of equidensity spherical particles. AIChE J 32:7–16, 1986.

Richardson JF. Incipient fluidization and particulate systems. In: Fluidization. (Davidson JF, Harrison D, eds.) New York: Academic Press, 1971, pp 26–64.

Rietema K. The effect of interparticle forces on the expansion of a Ginigebeiys gas-fluidized bed. Proceedings Intern Symp Fluidization, Toulouse, 1967, pp 28–40.

Roach PE. Differentiation between jetting and bubbling in fluidized beds. Int J Multiphase Flow 19:1159–1161, 1993.

Romero JB, Johanson LN. Factors affecting fluidized bed quality. Chem Eng Prog Symp Ser 58(38):28–37, 1962.

Rowe PN. Drag forces in a hydraulic model of a fluidized bed, part II. Trans Inst Chem Eng London 39:175–180, 1961.

Rowe PN. The effect of bubbles on gas–solids contacting in fluidized beds. Chem Eng Prog Symp Ser 58(38):42–56, 1962.

Rowe PN. Experimental properties of bubbles. In: Fluidization. (Davidson JF, Harrison D, eds.) New York: Academic Press, 1971, pp 121–192.

Rowe PN. Prediction of bubble size in a gas fluidized bed. Chem Eng Sci 31:285–288, 1976.

Rowe PN, Nienow AW. Minimum fluidization velocity of multicomponent particle mixtures. Chem Eng Sci 30:1365–1369, 1975.

Rowe PN, Nienow AW. Particle mixing and segregation in gas fluidized beds. Power Tech 15:141–147, 1976.

Rowe PN, Partridge BA, Cheney AG, Henwood GA, Lyall E. The mechanisms of solids mixing in fluidized beds. Trans Inst Chem Eng 43:T271–T286, 1965.

Rowe PN, Nienow AW, Agbim AJ. A preliminary quantitative study of particle segregation in gas fluidized beds— binary systems of near spherical particles. Trans Inst Chem Eng 50:324–333, 1972.

Rowe PN, MacGillivray HJ, Cheesman DJ. Gas discharge from an orifice into a gas fluidized bed. Trans Inst Chem Eng 57:194–199, 1979.

Saxena and Vogel. The measurement of incipient fluidization velocities in a bed of corase dolomite at temperature and pressure. Trans Inst Chem Eng 55:184–189, 1977.

Schugerl K. Rheological behavior of fluidized systems. In: Fluidization. (Davidson JF, Harrison D, eds.) New York: Academic Press, 1971, pp. 261–292.

Schugerl K, Merz M, Fetting F. Rheologische Eigenschaften von gasdurchströmten Fliessbettsystemen. Chem Eng Sci 15:1–38, 1961.

Shakhova NA. Discharge of turbulent jets into a fluidized bed. Inzh Fiz Zh 14(1):61–69, 1968.

Shakhova NA, Minaev GA. Aerodynamics of jets discharged into fluidized beds. Heat Transfer Soviet Research 4(1):133–142, 1972.

Shi YF, Fan LT. Lateral mixing of solids in batch gas–solids fluidized beds. Ind Eng Chem Process Des Dev 23:337–341, 1984.

Sit SP, Grace JR. Hydrodynamics and mass transfer in the distributor zone of fluidized beds. Proc Second World Cong Chem Eng, Montreal, Canada, 1981, pp 81–84.

Stewart PSB, Davidson JF. Slug flow in fluidized beds. Powder Tech 1:61–80, 1967.

Subbarao D, Moghaddam E, Bannard JE. Lateral mixing of particles in fluidized beds. Chem Eng Sci 40:1988–1990, 1985.

Tanimoto H, Chiba S, Chiba T, Kobayashi H. Mechanism of solid segregation in gas fluidized beds: In: Fluidization. (Grace JR, Matsen JM, eds.) New York: Plenum Press, 1980, pp 381–388.

Tanimoto H, Chiba T, Kobayashi H. Effects of segregation of fine elutriation from gas-fluidized beds of binary solid mixture. J Chem Eng Japan 16:149–152, 1983.

Toomey RO, Johnstone HF. Gaseous fluidization of solid particles. Chem Eng Prog 48:220–226, 1952.

Trawinski H. Chem Ing Tech 25:201, 1953.

Tsukada M, Horio M. Gas motion and bubble formation at the distributor of a fluidized bed. Powder Technol 63:69–74, 1990.

Turner JS. Buoyancy Effects in Fluids. Cambridge: Cambridge University Press, 1973.

Uchida S, Yamada H, Tada I. Minimum fluidization velocity of binary mixtures. J Chinese Inst Chem Eng 14:257–264, 1983.

Uno S, Kintner RC. Effect of wall proximity on the rate of rise of simple air bubbles in a quiescent liquid. AIChE J 2:420–425, 1956.

Vaccaro S. Analysis of the variables controlling gas jet expansion angles in fluidized beds. Powder Technol 92:213–222, 1997.

Vaccaro S, Musmarra D, Petrecca M. A technique for measurement of the jet penetration height in fluidized beds by pressure signal analysis. Powder Technol 92:223–231, 1997.

Vaid RP, Sen Gupta P. Minimum fluidization velocities in beds of mixed solids. Can J Chem Eng 56:292–296, 1978.

Valenzuela JA, Glicksman LR. Gas flow distribution in a bubbling fluidized bed. Presented at AIChE 74th Annual Mtg, 1981.

Verloop J, Heertjes PM. On the origin of bubbles in gas-fluidized beds. Chem Eng Sci 29:1101–1107, 1974.

Wen CY, Yu YH. A generalized method for predicting the minimum fluidization velocity. AIChE J 12:610–612, 1966a.

Wen CY, Yu YH. Mechanics of fluidization. Chem Eng Prog Symp Ser 62(62):100–111, 1966b.

Wen CY, Horio M, Krishnan R, Kosravi R, Rengarajan P. Jetting phenomena and dead zone formation on fluidized bed distributors. Proc Second Pacific Chem Eng Cong, Vol. 2, 1182, 1977, Denver, Colorado.

Wen CY, Deole NR, Chen LH. A study of jets in a three-dimensional gas fluidized bed. Powder Technol 31:175–184, 1982.

Werther J. Bubble growth in large diameter fluidized beds. In: Fluidization Technology (Keairns DL, ed.). Washington: Hemisphere, 1976a, pp 215–235.

Werther J. Convective solids transport in large diameter gas fluidized beds. Powder Technol 15:155–167, 1976b.

Whitehead AB. Prediction of bubble size in a gas-fluidized bed. Chem Eng Sci 34:751–751, 1979.

Whitehead AB, Young AD. Fluidization performance in large scale equipment. In: Proc Intern Symp on Fluidization. Amsterdam: Netherlands University Press, 1967, pp 284–293.

Whitehead AB, Dent DC, McAdam JCH. Fluidization studies in large gas–solid systems part V. long and short term pressure instabilities. Powder Technol 18:231–237, 1977.

Wilhelm RH, Kwauk M. Fluidization of solid particles. Chem Eng Prog 44:201–217, 1948.

Wu CS, Whiting WB. Interacting jets in a fluidized bed. Chem Eng Comm 73:1–17, 1988.

Wu SY, Baeyens. J Segregation by size difference in gas fluidized beds. Powder Technol 98:139–150, 1998.

Xavier AM, Lewis DA, Davidson JF. The expansion of bubbling fluidized bed. Trans Inst Chem Eng 56:274–280, 1978.

Yagi S, Kunii D. Fluidized-solids reactors with continuous solids feed—I. Residence time of particles in fluidized beds. Chem Eng Sci 16:364–371, 1961.

Yang WC. Jet penetration in a pressurized fluidized bed. I&EC Fundamentals 20:297–300, 1981.

Yang WC. Particle segregation in gas-fluidized beds. In: Encyclopedia of Fluid Mechanics, Vol. 4. (Cheremisinoff NP, ed.) Houston: Gulf, 1986, pp 817–852.

Yang WC. 30 years of industrial research on fluidization— bridging the gap between theory and practice. In: Fluidization (Fan LS, Knowlton TM, eds.). New York: Engineering Foundation, 1998a, pp 31–43.

Yang WC. Comparison of jetting phenomena in 30-cm and 3-m diameter semicular fluidized beds. Powder Technol 100:147–160, 1998b.

Yang WC. Engineering and application of recirculating and jetting fluidized beds. In: Fluidization, Solids Handling, and Processing. (Yang WC, ed.) Westwood: Noyes, 1999, pp 236–330.

Yang WC, Keairns DL. Design and operating parameters for a fluidized bed agglomerating combustor/gasifier. In: Fluidization (Davidson JR, Keairns DL, eds.). Cambridge: Cambridge University Press, 1978, pp 208–214.

Yang WC, Keairns DL. Estimating the jet penetration depth of multiple vertical grid jets. I&EC Fundamentals 18:317–320, 1979.

Yang WC, Keairns DL. Momentum dissipation of and gas entrainment into a gas–solid two-phase jet in a fluidized bed. In: Fluidization (Grace JR, Matsen JM, eds.). New York: Plenum Press, 1980a, pp 305–314.

Yang WC, Keairns DL. The effect of an expanded section on slugging. AIChE J 26:144–148, 1980b.

Yang WC, Keairns DL. Solid entrainment rate into gas and gas–solid two-phase jets in a fluidized bed. Powder Technol 33:89–94, 1982a.

Yang WC, Keairns DL. Rate of particle separation in a fluidized bed. I&EC Fundamentals 21:228–235, 1982b.

Yang WC, Keairns DL, McLain DK. Gas mixing in a jetting fluidized bed. AIChE Symposium Series 80(234):32–41, 1984a.

Yang WC, Revay D, Anderson RG, Chelen EJ, Keairns DL, Cicero DC. Fluidization phenomena in a large-scale, cold-flow model. In: Fluidization. (Kunii D, Toei R, eds.) New York: Engineering Foundation, 1984b, pp. 77–84.

Yang WC, Ettehadieh B, Anestis TC, Gizzie RE, Haldipur GB. Fluidization phenomena in a large jetting fluidized bed. In: Fluidization V. (Ostergaard K, Sorensen A, eds.). New York: Engineering Foundation, 1986, pp 95–102.

Yang Z, Tung Y, Kwauk M. Characterizing fluidization by the bed collapsing method. Chem Eng Commun 39:217–232, 1985.

Yasui G, Johanson LN. Characteristics of gas pockets in fluidized beds. AIChE J 4:445–452, 1958.

Yates JG, Cheesman DJ, Segeev YA. Experimental observations of voidage distribution around bubbles in a fluidized bed. Chem Eng Sci 49:1885–1895, 1994.

Yates JG, Wu KT, Cheesman DJ. Bubble coalescence from multiple entry nozzles. In: Fluidization (Large JF, Laguerie C, eds.). New York: Engineering Foundation, 1995, pp 11–19.

Yerushalmi H, Cankurt NT. Further studies of the regimes of fluidization. Powder Technol 24:187–205, 1979.

Yerushalmi H, Cankurt NT, Geldart D, Liss B. Flow regimes in vertical gas–solid contact systems. AIChE Symp Ser 176(74):1–13, 1978.

Yoshida K, Kameyama H, Shimizu F. Mechanism of particle mixing and segregation. In: Fluidization (Grace JR, Matsen JM. eds.). New York: Plenum Press, 1980, pp 389–396.

Zenz FA. Bubble formation and grid design. Inst Chem Eng Symp Ser 30:136–145, 1968.

Zenz FA. Regimes of fluidized behavior. In: Fluidization. (Davidson JF, Harrison D, eds.) New York: Academic Press, 1971.

Zenz FA. How flow phenomana affect design of fluidized beds. Chem Eng 84(27):81–91, December 1977, pp 1–23.

Zenz FA. The fluid mechanics of bubbling beds. Fibonacci Quarterly 16(2):171–183, 1978.

4

Elutriation and Entrainment

Joachim Werther and Ernst-Ulrich Hartge

Technical University Hamburg-Harburg, Hamburg, Germany

1 INTRODUCTION

For most applications of fluidized beds, the process, e.g., the chemical reaction or the drying process, takes place in the dense fluidized bed either completely or at least in major part. Nevertheless, in many units the freeboard above the fluidized bed takes the larger volume of the whole unit compared to the fluidized bed. The task of the freeboard is mainly to prevent a too large amount of the bed material from being carried out of the unit by the gas stream. In order to design properly the freeboard, it is necessary to know not only the height, which is necessary to separate the solids as effectively as possible from the gas, but also the influences of the freeboard diameter, of particle properties, and of operating conditions on the entrainment flux. The entrainment flux has to be known for the design of gas/solid separators like cyclones or filters. On the other hand, the loss of bed material that is related to the entrainment may be important for the technical and also the economic success of a fluidized bed application.

While for most processes the transport of solids out of the fluidized bed reactor is a disadvantage, some processes take advantage of the loss of fines, examples being fluidized bed classifiers (Tasirin and Geldart, 2000) and the incineration of sewage sludge, where the fine ash is removed as fly ash entrained from the combustor.

In the present chapter we deal with the mechanisms underlying the entrainment, some experimental find-

ings, and the correlations to estimate the entrainment and the height necessary to minimize the entrainment.

2 DEFINITIONS

At the very beginning we define the terms that will be used throughout this chapter. Two terms are often used for the flow of solid out of fluidized beds. These are the terms entrainment and elutration. Entrainment means the flux of solids carried out of the fluidized bed by the gas in kg per unit cross-sectional area and second. Elutriation means the classifying effect of the fluidized bed entrainment—it characterizes the selective removal of particles of individual size from the fluidized bed.

The fluidized bed is usually divided into different vertical zones (Fig. 1). At the bottom there is the dense fluidized bed, above which the freeboard is located. The term freeboard denotes the space above the dense fluidized bed. The height between the dense bed surface and the gas outlet will be named here freeboard height H_{FB}. It could be argued whether it would be better to denote by "freeboard height" the total distance between the dense bed surface and the top of the vessel. However, most authors prefer the gas outlet as the upper limit of the freeboard, since this latter location determines the carryover of solids.

At higher gas velocities, the bed surface will start to fluctuate and will become more and more fuzzy. The sharp transition between the dense bed and the dilute

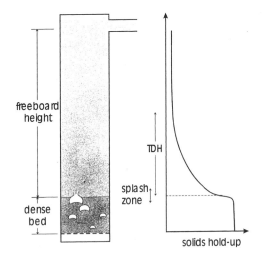

Figure 1 Zones in a fluidized bed vessel.

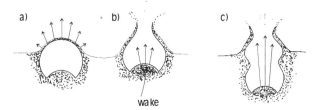

Figure 2 Bubbles bursting at the bed surface, which are ejecting solids into the freeboard (a) from the roof, (b) from the wake of a single bubble, and (c) from the wake of two coalescing bubbles.

freeboard will disappear, and there will be a zone characterized by a sharp decay of the solids concentration. This latter zone will be called the splash zone.

Experimental observations show that above the dense bed or the splash zone, respectively, the solids holdup gradually decays until it becomes constant or at least nearly constant. The distance between this point, where the solids concentration becomes (almost) constant, and the surface of the fluidized bed is called the Transport Disengaging Height (TDH).

3   EJECTION OF PARTICLES INTO THE FREEBOARD

In order to understand the process of entrainment we will have a look at the governing mechanisms first. The whole process may be divided into three subprocesses. At first the solids have to be transported from the bed into the freeboard; then they will rise, owing to their initial velocity, adjust their velocity to the gas velocity in the freeboard, and partly disengage; and—provided that the freeboard is sufficiently high—finally they will be transported by the upflowing gas to the gas outlet. In the following we will first deal with the mechanisms that cause the particles to be ejected from the bed into the freeboard.

The most important mechanism for the projection of particles from the fluidized bed surface into the freeboard is related to the eruption of bubbles (Levy et al., 1983). In Fig. 2 schematic drawings of bubbles bursting at the surface of a fluidized bed are shown. Different mechanisms are considered to be responsible

for the particle ejection, namely the ejection of particles from the bubble roof or nose (Fig. 2a) and the ejection of particles transported upwards in the wake of the bubble (Fig. 2b). While there is a general agreement on the important role of the bubbles, no such agreement exists with regard to the suitable mechanism. While Zenz and Weil (1958) and Chen and Saxena (1978) consider the nose mechanism to be the most important, George and Grace (1978) assumed the wake mechanism to be the governing one. Pemberton and Davidson (1986) postulated that particles might be ejected by both mechanisms, depending on the particle size, the fluidizing velocity, and the bed geometry. From their measurements they deduced that bubbles bursting one by one eject the solids from the roof while coalescing bubbles preferably throw the wake into the freeboard (cf. Fig. 2c). These findings are also confirmed by Levy et al. (1983) and by Peters et al. (1983), who took sequences of photographs to study the mechanism of particle projection. Pemberton and Davidson (1983) postulated that for particles belonging to the Geldart group A fluidized at high velocities, the ejection from the roof predominates, while for group B particles fluidized with lower velocities, ejection from the wake is governing the process. A difference between the two mechanisms lies in the amount of solids thrown into the freeboard and their size distribution. While the solids thrown into the freeboard from the bubble wake have nearly the same size distribution as the bed material, the solids thrown into the freeboard from the bubble roof are much finer. The bubble-roof model gives a much lower solids flux at the bed surface than the wake model, the difference between the fluxes being about one order of magnitude. Another difference between the two mechanisms is the dispersion of the solids ejected to the freeboard: while the solids from the wake are thrown into the freeboard as quite closely packed bulks of solids, the particles from the roof are well dispersed (Tanimoto et al., 1984, Hazlett and Bergougnou, 1992).

Starting velocities of the particles at the bed surface have been determined, e.g., by George and Grace (1978) and by Peters et al. (1983) and have been found to be significantly higher than the bubble rise velocity. George and Grace (1978) found that the starting velocity of the particles ejected into the freeboard is roughly twice the bubble rise velocity. These results have been confirmed by Hatano and Ishida (1983), who measured particle velocities by fiber-optical probes.

Another mechanism for the transport of particles into the freeboard, which may be particularly significant for coarse particle fluidized beds with a wide particle size distribution, is the interstitial velocity in the dense bed. Fine particles at the top of the dense bed will be transported to the bed surface and it will be blown out into the freeboard (Bachovchin et al., 1981). This latter transport mechanism is strongly classifying, but possible fluxes are low compared to the fluxes induced by the bursting bubbles.

4 GAS–SOLIDS FLOW IN THE FREEBOARD

After the solids have been ejected into the freeboard with a certain starting velocity, they will decelerate and either fall back to the fluidized bed or will be transported by the upflowing gas to the gas outlet. The most simple assumption for the modelling of this process is to assume that all particles behave independently of each other and to use a simple ballistic model (Zenz and Weil, 1958; Do et al., 1972). A comparison of trajectories calculated with this model and experimentally observed particle trajectories (Peters et al., 1983) is given in Fig. 3. There is a quite good agreement, but the small maximum heights between 1 and 2 cm should be noted.

The simple ballistic model implies that only particles with a terminal settling velocity u_t less than the superficial velocity U are entrained. But experimental observations show that also larger particles may be entrained (Geldart and Pope, 1983). Furthermore according to this model, there is no explanation for the experimental findings that entrainment decreases with increasing height even for fluidized beds of fines, which all are entrainable. A third contradiction of this model to experimental findings is that due to the decreasing solids velocities, the solids concentration above such a bed of fines would increase with height,

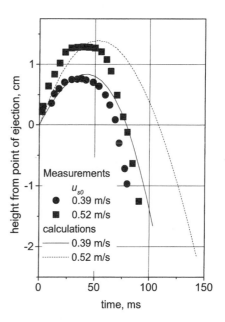

Figure 3 Measured and calculated particle trajectories. (After Peters et al., 1983.)

while all experimental observations show a decrease of the solids concentration with height.

In a bed with a wide solid size distribution, coarse particles with a settling velocity higher than the gas velocity are entrained, which can be explained by interactions between fine and coarse particles (Geldart et al., 1979; Geldart and Pope, 1983). The exchange of momentum may lower the settling velocity of the coarse particles, while the fines are decelerated. But this mechanism still gives no explanation of the decreasing entrainment of fines with increasing height.

Measurements of instantaneous solids concentrations (Hatano and Ishida, 1983; Yorquez-Ramirez and Duursma, 2001) show that particles are not moving individually through the freeboard; they form clusters, which allow slip velocities between gas and particles that are much higher than the single particle's terminal velocity. Such clusters will form also with very fine particles and thus provide an explanation of why the entrainment is decreasing with increasing height even for fine particle systems (Geldart and Pope, 1983; Kunii and Levenspiel, 1989).

The flow structure inside the freeboard becomes even more complicated when velocity profiles of the gas are taken into account. Such profiles are due to the laminar flow that prevails in many of the quite small test units used for entrainment experiments. For such a laminar profile, quite large particles may be entrained when rising in the center, while even very fine particles will move downward at the wall. For laminar velocity profiles this effect is much more pronounced than for turbulent velocity profiles, which prevail in larger units. This might also give an explanation for the observed dependence of the entrainment flux on the diameter of the freeboard. According to Tasirin and Geldart (1998a) there is a critical freeboard diameter above which the entrainment flux is nearly independent of the freeboard diameter. They explain the existence of this critical diameter by the transition from laminar flow to turbulent flow.

Besides the velocity profiles due to laminar or turbulent flow, there are also temporary fluctuations in the gas flow caused by the eruption of bubbles at the bed surface. A widely accepted model to explain the freeboard turbulence caused by the bubbles is the idea of "ghost bubbles," postulated by Pemberton and Davidson (1984). They assume that a bubble keeps its identity even after breaking through the bed surface and while traveling up through the freeboard. The ghost bubbles are characterized by a circulating flow, which continues the gas flow pattern inside and around the bubbles in the fluidized bed. Yorquez-Ramirez and Duursma (2000, 2001), however, state that vortex rings are generated by the erupting bubbles, which cause the freeboard turbulence. They identify the fluctuations of the level of the bed surface as another source of gas flow fluctuations.

5 EXPERIMENTAL INVESTIGATIONS OF ENTRAINMENT

After the description of the governing mechanisms responsible for the entrainment of particles from fluidized beds, a short overview about experimental investigations on entrainment and elutriation will be given. Since there is very large number of experimental investigations in the literature, we do not attempt to give a complete survey.

In most cases, experimental results are given either in terms of the total entrainment flux E_h in kg/(m^{2}s) at the height h above the distributor or in terms of an elutriation rate constant K_{ih}^*, defined as the ratio of the instantaneous rate of removal of solids of size d_{pi} based on the cross-sectional area A to the fraction of the mass of the bed material with size d_{pi}:

$$k_{ih}^* = \frac{E_{ih}(t)}{x_{Bi}(t)} \tag{1}$$

This can be rewritten as

$$K_{ih}^* = \frac{\dfrac{1}{A}\dfrac{d}{dt}[x_{Bi}(t) \cdot M_B(t)]}{x_{Bi}(t)} \tag{2}$$

E_{ih} denotes the carryover flux of a component of size d_{pi} at the height h above the distributor based on the cross-sectional area of the bed. x_{Bi} gives the mass fraction of particles with size d_{pi} in the bed, and M_B is the total mass of the bed material. For steady-state operation, i.e., when the carryover is continuously recycled to the bed, the fraction of particles with size d_{pi} is constant with time, and Eq. (1) can be rewritten as

$$K_{ih}^* = \frac{E_{ih}}{x_{Bi}} \tag{3}$$

For batch experiments, K_{ih}^* has to be calculated from the entrained mass $m_{i,t}$ of size d_{pi} accumulated during the time t. Therefore Eq. (1) has to be integrated. Provided the total mass of bed material does not change significantly, this gives

$$m_{i,t} = x_{Bi0} \cdot M_B\left[1 - \exp\left(-\frac{K_{ih}^* \cdot A}{M_B}\right)\right] \tag{4}$$

where x_{Bi0} is the initial mass fraction of particles with size d_{pi} in the bed mass. In order to obtain the total entrainment E_h, the individual entrained mass fluxes E_{ih} of particle sizes d_{pi} have to be accumulated:

$$E_h = \sum_i E_{ih} = \sum_i (x_{Bi} \cdot K_{ih}^*) \tag{5}$$

The mass fraction of the particles with size d_{pi} in the entrained flow can be calculated from

$$x_{i,E} = \frac{E_{ih}}{E_h} = \frac{x_{Bi} \cdot K_{ih}^*}{E_h} \tag{6}$$

Above a certain freeboard height, the entrainment flux and the elutriation rate constants become essentially independent of height. This height is called the transport disengaging height (TDH). The entrainment fluxes and elutriation rate constants above this height are given the index "∞", i.e. E_∞, $E_{i\infty}$. In the following, the influence of operating and design parameters will be discussed wherever possible in terms of the elutriation rate constants K_{ih}^* or $K_{i\infty}^*$.

5.1 Influence of Particle Size

For fine particles with a terminal velocity u_t below the gas velocity in the freeboard, the elutriation rate constant $K_{i\infty}^*$ increases with decreasing particles size. While there is a general consensus about this when group A powders are concerned there are some indications that there is a critical particle size, below which the elutriation rate constant $K_{i\infty}^*$ levels off or even decreases with decreasing size. Figure 4 shows experimental results measured by Smolders and Baeyens (1997) with limestone and α-alumina. The experimentally determined values for $K_{i\infty}^*$ are plotted against the particle size. In spite of some scatter, the general decrease in the elutriation rate constants below about 10 μm is obvious. Similar results were also presented by Tasirin and Geldart (1998b), Santana et al. (1999), and Tasirin et al. (2001).

The same trend was also observed for fluidized beds with binary particle mixtures—fines in a coarse particle fluidized bed—by Ma and Kato (1998). The reason for this effect is probably the formation of particle agglomerates in the case of very fine particles for which adhesion forces are large compared to the gravitational forces. Baeyens et al. (1992) suggest based on their measurements a correlation for the critical particle diameter $d_{p,crit}$, at which the inflection occurs:

$$d_{p,crit} \cdot \rho_s^{0.725} = 10325 \qquad (7)$$

$d_{p,crit}$ is to be inserted in μm and ρ_s in kg/m^3, respectively. Ma and Kato (1998) introduce a critical cohesion number

$$N_{coh}^* = \frac{C}{\rho_s \cdot d_{p,crit} \cdot g} = 4.5 \qquad (8)$$

as the criterion for the critical particle diameter $d_{p,crit}$, with $C = 0.455 \cdot \rho_s^{0.269}$ where ρ_s is to be inserted in kg/m^3. Experiments by Briens et al. (1992) also show the decrease of the elutriation rate constant with decreasing size for very fine particles. These latter authors showed furthermore that there is no influence of electrostatic effects on the critical particle size, even though the absolute values of the entrainment fluxes were significantly changed by electrostatic loading.

For the coarse particles with a terminal settling velocity u_t above the superficial gas velocity in the freeboard U, it is often assumed that they cannot be entrained if the freeboard is sufficiently high, but measurements by Geldart and Pope (1983) and Geldart et al. (1979) have shown that coarse particles may be entrained, too, if there is a sufficient entrainment flux

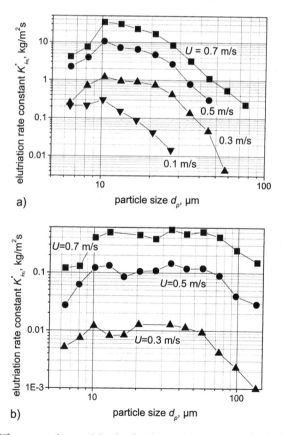

Figure 4 Elutriation rate constant $K_{i\infty}^*$ versus the particle size for fines and superfines for (a) limestone and (b) α-alumina. (Data from Smolders and Baeyens, 1997.)

of fines. The amount of entrained coarses particles is significantly influenced by the flux of fines.

While there is an influence of the entrainment of fines on the carryover of coarse material, no influence of the fines content on the elutriation rate constant of the fines can be found (Taneda et al., 1998). This is an assumption implicitly already made in Eq. (3).

5.2 Influence of Superficial Gas Velocity

The influence of the superficial gas velocity is quite obvious. Both the ejection of particles from the dense bed into the freeboard and the transport through the freeboard are affected by the superficial velocity. In Fig. 5, data measured by Choi et al. (1998) for sand are given as an example. In general, the elutriation rate increases proportionally with the gas velocity to a power of 2 to 4.

5.3 Effect of Freeboard Geometry

5.3.1 Influence of Freeboard Height

Close to the bed surface the carryover from the fluidized bed decreases significantly with increasing height. The decay in the entrainment becomes nearly constant, when the transport disengaging height (TDH) is exceeded (Fig. 6). Like the entrainment flux, the solids volume concentration in the freeboard decreases with increasing distance from the bed surface and becomes nearly constant above the TDH.

According to measurements by Nakagawa et al. (1994), the shape of the gas outlet has no significant influence on the solids holdup in the freeboard. Figure 7 shows their measurements in a 0.15×0.15 m^2 unit

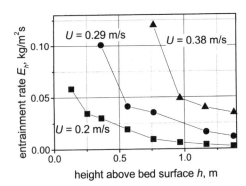

Figure 6 Influence of height above the dense bed surface on the entrainment flux for different gas velocities. ($D_t = 0.45$ m; $d_{p,50} = 260$ μm, quartz sand, data measured by Demmich, 1984.

with two different outlet configurations. As outlets, a $60°$ pyramidal with gas exit at the top, and a cubic cap with gas exit on one side, have been used.

5.3.2 Effect of Diameter

Only few studies of the influence of scale on the entrainment are available in the literature. Lewis et al., (1962) made a study on entrainment with units of 0.019 to 0.146 m diameter. As a result, they stated that entrainment is independent of size for units larger than 0.1 m in diameter. This result is supported by findings of Colakyan and Levenspiel (1984) and Tasirin and Geldart (1998c). The latter authors measured with units of 0.076 m and 0.152 m, respectively, and found the entrainment in the larger unit to be higher than in the smaller unit. Colakyan and Levenspiel (1984) found no significant influence of the diameter on the entrainment in beds of 0.1, 0.3, and 0.9 m diameter.

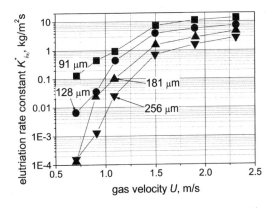

Figure 5 Influence of the superficial gas velocity on the elutriation rate, $D_t = 0.1$ m, air at ambient conditions. (Data measured by Choi et al., 1998.)

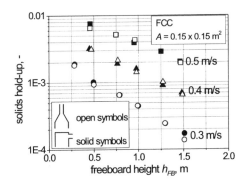

Figure 7 Axial particle holdup distribution obtained with different gas outlet configurations. (Nakagawa et al., 1994.)

While the above cited findings are all for units with constant diameter, Briens et al. (1990), Smolders and Baeyens (1997), and Tasirin and Geldart (1998a) studied the effect of an enlarged freeboard, i.e., of a freeboard diameter larger than the fluidized bed diameter. In this case the gas velocity is reduced by the square of the ratio of bed diameter D_{bed} to freeboard diameter D_{FB}, which leads to a significant reduction of the entrainment flux by increasing the freeboard diameter. Smolders and Baeyens (1997) found $E_\infty \propto (D_{bed}/D_{FB})^4$. Tasirin and Geldart (1998a) found that the elutriation rate constant $K^*_{i\infty}$ is only determined by the gas velocity of the expanded section, provided the expanded section is sufficiently high. If there is not sufficient length in the extended diameter section to form a uniform gas velocity profile across the whole cross-sectional area, the entrainment flux will be higher than expected for the averaged velocity in the expanded diameter section.

5.4 Influence of Bed Height and Internals in the Bed

Taking into account the role of bubbles for the transport of particles into the freeboard, an influence of bed height and also of internals in the fluidized bed could be expected. Baron et al. (1990) studied the influence of the bed height, using a fluidized bed column of 0.61 m diameter and silica sand with a surface mean diameter d_p of 65 μm. Their results for the entrainment flux E_∞ above the TDH and the entrainment flux E_0 just above the bed surface are shown in Figs. 8a and 8b, respectively. The results show a slight increase of the entrainment fluxes with increasing bed height, the influence being more pronounced for the entrainment flux E_0 at the bed surface and higher gas velocities. The reason for the increase of the entrainment flux is the increase of bubble sizes with height, which obviously dominates the decrease in bubble frequency and therefore in the probability of bubbles to coalesce near the bed surface. This indicates, that at least for the system studied by Baron et al., the "bubble nose" mechanism is the dominating mechanism for particle ejection into the freeboard. Choi et al. (1989), however, found no influence of bed height on entrainment during their study of entrainment from fluidized bed combustors. In this study they used two fluidized bed combustors with cross-sectional areas of 0.3×0.3 m^2 and 1.01×0.83 m^2, respectively. The smaller one was equipped with an extended freeboard of 0.45×0.45 m^2. In fact, the bed height effect in Fig. 8 is small compared, e.g., to the effect of the gas velocity.

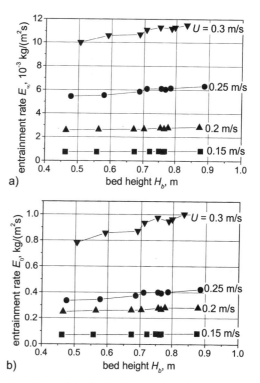

Figure 8 Effect of the bed height on the entrainment flux (a) above the TDH and (b) just above the bed surface. ($D_t = 0.61$ m, silica sand, $d_p = 61$ μm, data measured by Baron et al., 1990.)

Tweedle et al. (1970) showed by inserting screen packings into the fluidized bed that the entrainment was reduced due to the reduced bubble size.

5.5 Influence of Internals in the Freeboard

The picture of the influence of internals in the freeboard is not very clear. On the one hand, they may act as a kind of gas–solid separators, thus helping to reduce the TDH (Martini et al., 1976) and the entrainment flux (Baron et al., 1988). On the other hand, they may increase the entrainment due to the increased gas velocity in the freeboard (Tweedle et al., 1970), or they may not affect the entrainment at all (George and Grace, 1981). Specially designed baffles and inserts in the freeboard may help to reduce the entrainment flow significantly (Harrison et al., 1974).

5.6 Influence of Temperature and Pressure

Studies of the influence of pressure have been carried out by Chan and Knowlton (1984a, 1984b). They showed that the TDH increases linearly with pressure and that also the entrainment flux increases with pressure (Fig. 9). These effects may directly be related to the increasing density of the gas and thus decrease in single particle terminal velocity. Effects of the changing viscosity should not play any dominant role, since the influence of the pressure on viscosity is small in the pressure range investigated.

Another important operating parameter is the temperature, since most fluidized bed applications are operated at elevated temperatures, while most investigations on entrainment are performed at ambient conditions. Studies on the influence of temperature are reported by George and Grace (1981), Choi et al. (1997), and Wouters and Geldart (1998). George and Grace varied the temperatures in a quite narrow range using silica sand as bed material and did not find an influence of temperature on the entrainment. Choi et al. investigated a much wider range of temperatures and found a minimum of the entrainment for temperatures of about 450 to 700K in the case of sand (Fig. 10). Similar results were also obtained for emery ($\rho_s = 3891$ kg/m^3) and for cast iron ($\rho_s = 6158$ kg/m^3) particles as bed material. Wouters and Geldart measured a decrease of the elutriation rate constant with increasing temperature in the range between 270 and 670K for the fine FCC used. They did not find a minimum in their plot of the entrainment rate vs. the temperature.

6 MODELING

Considering the complex fluid mechanics in the freeboard and at the bed surface of the bubbling bed, it is not astonishing that up to now no rigorous model that

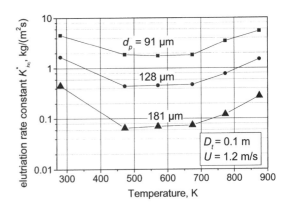

Figure 10 Influence of temperatures on the elutriation rate constant. (Sand, data measured by Choi et al., 1997.)

applies the methods of computational fluid dynamics (CFD) has been published in the literature. This would be quite a difficult task, particularly because a very wide range of solids volume concentrations—ranging from ε_{mf} in the suspension of the dense bed to nearly zero above the TDH—has to be covered. Models of the gas–solid flow in the freeboard are therefore based on either simplified physical descriptions or empirical correlations.

6.1 Entrainment for Freeboard Heights Exceeding TDH

As early as 1958 Zenz and Weil (1958) introduced the idea that the freeboard above the TDH behaves like a pneumatic transport line at choking conditions. This is the assumption on which most of the models up to now are based. The differences are mainly in the calculation of the choking load and of the particle size distribution of the entrained particles. Zenz and Weil (1958) calculated for each particle size class contained in the bed the choking load G_{si} (kg/m^2·s) for monosize particles of diameter d_{pi} with a correlation originally developed for pneumatic transport. The elutriation flux for these particles is then assumed to be the product of the mass fraction x_{Bi} of these particles in the bed and the choking load G_{si}:

$$E_{i\infty} = x_{Bi} \cdot G_{si} \tag{9}$$

This implies that, for all particle size classes, at least a mass flux equal to the elutriation flux $E_{i\infty}$ is ejected from the fluidized bed into the freeboard.

This choking load model has been modified by Gugnoni and Zenz (1980). They proceeded in two stages. First, they estimated the particle size distribution by using the Zenz and Weil model with a some-

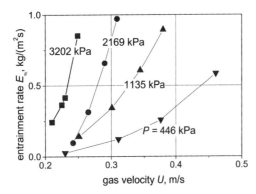

Figure 9 Influence of pressure on the total entrainment rates. ($D_t = 0.3$ m, sand, $d_p = 70$ µm, data measured by Chan and Knowlton, 1984a.)

what updated choking correlation. In a second step, the total flux of solids entrained above the TDH was then calculated with a correlation based on the calculated mean diameter of the entrained solids and the difference between the gas velocity and the minimum fluidizing velocity.

Briens and Bergougnou (1986) also used a two-step approach. They again assume that for each particle size the flow above the TDH is limited by choking, i.e., it cannot be larger than the choking load attributed to that specific particle size. Furthermore, the flux of each particle size class above the TDH cannot be larger than the flux of solids of this size ejected from the bed to the freeboard. While these assumptions are sufficient to calculate the entrainment from a monosize fluidized bed, there is one more assumption necessary to calculate the fraction of the choking load attributed to each specific particle size; for this purpose they assumed that the size distribution of the entrained material above the TDH will adjust itself so that the total entrainment flux reaches a maximum. To solve this model, correlations for the choking load and for the entrainment of particles from the bed into the freeboard are needed. The model then has to be solved iteratively.

Models that focus more on the entrainment of the particles from the dense bed to the freeboard are given by George and Grace (1978) and by Smolders and Baeyens (1997). Both models first calculate the entrainment flux E_{i0} at the bed surface based on such bubble properties as size, frequency, and velocity. For the flux above the TDH, it is simply assumed that only particles with a terminal velocity less than the gas velocity in the freeboard are entrained, the mass flux of these particles being the same as at the bed surface:

$$E_{i\infty} = \begin{cases} E_{i0} & \text{for} \quad U_{ti} < U_{FB} \\ 0 & \text{for} \quad U_{ti} \geq U_{FB} \end{cases} \quad (10)$$

These two models will give the same entrainment flux at the bed surface as above the TDH for fluidized beds operated with solids having all terminal velocities less than the superficial velocity. This latter finding is in contradiction with empirical findings which show a decay in the entrainment with height even for such fluidized beds of fines.

Besides the models described above there is quite a large number of empirical correlations for the elutriation rate constant $K_{i\infty}^*$ above the transport disengaging height. A selection of the most frequently cited and most recent correlations is given in Table 1. Comments with regard to special areas of application are also given in this table. All the correlations have

been converted so that the parameters are to be inserted in SI units, although the original correlations may have required the insertion of data in other units. In Fig. 11, numerical values for the elutriation constant $K_{i\infty}^*$ predicted by the different correlations are plotted vs. the gas velocity and compared with experimental data for sand of about 90 μm measured by Colakyan and Levenspiel (1984), Choi et al. (1998), and George and Grace (1981). It can be noticed that there are significant differences in magnitude and predicted tendencies, but also the experimental data show a considerable scattering. In Fig. 12, in addition, the elutriation rate constant is plotted vs. the particle diameter. Also in this figure, significant deviations between the correlations can be noticed even in the tendencies.

Both Fig. 11 and Fig. 12 could lead to the conclusion that it is rather hopeless to predict the entrainment from fluidized beds with a reasonable degree of accuracy. However, we see also that a group of correlations is not too far away from the experimental results. In a given situation one should always carefully select one or more correlations, which have been developed on a data basis, which includes the situation under consideration. For example, one should certainly not expect a correlation developed for fine particle systems to hold for a coarse particle system. The gas velocity, the type of solids, and particularly the particle size distribution should be for a given application within the range of tested parameters of the selected correlation. In any case, the results of such correlations should be treated with care: an uncertainty of plus/minus 100% should be taken into account.

6.2 Estimation of TDH

An important parameter for the design of a fluidized bed vessel is the TDH. To reduce the carryover from a fluidized bed, the freeboard should have, wherever possible, a height of at least the TDH. On the other hand, with respect to entrainment there is no additional advantage of increasing the height of the vessel beyond the TDH. As already mentioned, the definition of the TDH is not very sharp, since there is only a gradual decay of the entrainment flux vs. the height at the level of TDH. Again, there is no commonly accepted method for the calculation of the TDH, but just several empirical correlations. Some of these correlations are listed in Table 2. As can be seen from this table, the influencing parameters differ between the correlations. While some of them take only the superficial velocity into account, without any regard to the solids in the bed, other correlations depend on one or more of the

Table 1 Correlations for the Elutriation Constant $K_{i\infty}^*$

	U, m/s	D, m	$d_{p,bed}$ mm	Reference	Comments
1 $\dfrac{K_{i\infty}^* \cdot g \cdot d_{pi}^2}{\mu(U-u_{ti})^2} = 0.0015 \cdot Re_t^{0.6} + 0.01 \cdot Re_t^{1.2}$	0.3–1.0	0.07–1.0	0.1–1.6	Yagi and Aochi (1955) as cited by Wen and Chen (1982)	Cited by different authors in different ways, original publication inaccessible
2 $\dfrac{K_{i\infty}^*}{\rho_g \cdot U} = \begin{cases} 1.26 \cdot 10^7 \cdot \left(\dfrac{U^2}{g\, d_{pi}\rho_p^2}\right)^{1.88} & \text{for} \quad \dfrac{U^2}{g\, d_{pi}\rho_p^2} < 3.10 \\[2ex] 1.31 \cdot 10^4 \cdot \left(\dfrac{U^2}{g\, d_{pi}\rho_p^2}\right)^{1.18} & \text{for} \quad \dfrac{U^2}{g\, d_{pi}\rho_p^2} > 3.10 \end{cases}$	0.3–0.7	0.05 × 0.53	0.04–0.2	Zenz and Weil (1958)	Correlation aiming at FCC fluidized beds
3 $\dfrac{K_{i\infty}^*}{\rho_g(U-U_{ti})} = 1.52 \cdot 10^{-5} \left(\dfrac{(U-U_{ti})^2}{g\, d_{pi}}\right)^{0.5} \cdot Re_i^{0.725}$	0.6–1.0	0.102	0.7	Wen and Hashinger (1960)	
4 $\dfrac{K_{i\infty}^*}{\rho_g(U-U_{ti})} = 4.6 \cdot 10^{-2} \left(\dfrac{(U-U_{ti})^2}{g\, d_{pi}}\right)^{0.5} \cdot Re_t^{0.3} \cdot \left(\dfrac{\rho_s - \rho_g}{\rho_g}\right)^{0.15}$	0.9–2.8	0.031–0.067	0.7–1.9	Tanaka et al. (1972)	
5 $\dfrac{K_{i\infty}^*}{\rho_g \cdot U} = A + 130 \cdot \exp\left[-10.4\left(\dfrac{u_{ti}}{U}\right)^{0.5}\left(\dfrac{U_{mf}}{U-U_{mf}}\right)^{0.25}\right]$ with $A = 10^{-3} \ldots 10^{-4}$	0.6–2.4	0.91 × 0.91	0.06–1.0	Merrick and Highley (1974)	Correlation derived for bubbling fluidized bed combustors
6 $\dfrac{K_{i\infty}^*}{\rho_g \cdot U} = 23.7 \cdot \exp\left(-5.4\dfrac{u_{ti}}{U}\right)$	0.6–3.0	0.076 0.3	0.06–0.35 1.5	Geldart et al. (1979)	
7 $\dfrac{K_{i\infty}^*}{\rho_g \cdot U} = 9.43 \cdot 10^{-4} \left(\dfrac{U^2}{g\, d_p}\right)^{1.65}$	0.1–0.3	0.61 × 0.61	0–0.125	Lin et al. (1980)	

for $58 \le \left(\dfrac{U^2}{g\, d_{pi}}\right) \le 1000$; $0.1\,\text{m/s} \le U \le 0.3$ m/s;

$0 < d_p \le 74$ μm

No.	Equation				Reference	Remarks
8	$K_{i\infty}^* \left[\dfrac{\text{kg}}{\text{m}^2\text{s}}\right] = 0.011 \cdot \rho_s \left(1 - \dfrac{u_{ti}}{U}\right)^2$	0.9–3.7	0.92×0.92 0.3×0.3	0.3–1.0	Colakyan et al. (1981), Colakyan and Levenspiel (1984)	Focus on Geldart Group B and D particles
9	$K_{i\infty}^* \left[\dfrac{\text{kg}}{\text{m}^2\text{s}}\right] = 2.8 \cdot 10^{-2} \left(\dfrac{U - u_{ti}}{U}\right)^{1.6} \cdot \left(\dfrac{\rho_s - \rho_g}{\rho_s}\right)^{0.54 \cdot U_t^{2.1}} \cdot D_h$				Kato et al. (1985)	
10	$\dfrac{K_{i\infty}^*}{\rho_g \cdot U} = 1.6 \left(\dfrac{U}{u_{ti}}\right)\left(1 - \dfrac{u_{ti}}{U}\right)$ for $d_{pi} \leq \dfrac{10325}{\rho_s^{0.725}}$				Sciazko et al (1991)	
11	$K_{i\infty}^* \left[\dfrac{\text{kg}}{\text{m}^2\text{s}}\right] = 5.4 \cdot 10^{-5} \left(\dfrac{U}{0.2}\right)^{3.4} \left(1 - \dfrac{u_{ti}}{U}\right)^2$	0.2–0.7		0.03–0.78	Baeyens et al (1992)	Correlation takes cohesive forces into account and is focused on superfines in group A and C systems
12	$K_{i\infty}^* \left[\dfrac{\text{kg}}{\text{m}^2\text{s}}\right] = 0.35 \rho_s U (1 - \varepsilon)_H$ with $(1 - \varepsilon)_H = 7.41 \cdot 10^{-3} R^{1.87} A_t^{0.55} H_{FB}^{-0.64}$ and $R = \sum x_i \left(\dfrac{U - U_{ti}}{U_{ti}}\right)$ for $u_{ti} < U$	0.1–0.6	0.071 0.08×0.08 0.15×0.15	0.03–0.2	Nakagawa et al. (1994)	
13	$K_{i\infty}^* \left[\dfrac{\text{kg}}{\text{m}^2\text{s}}\right] = \begin{cases} 23.7 \cdot \rho_g \cdot U^{2.5} \exp\left(-5.4 \dfrac{u_{ti}}{U}\right) & \text{for } Re < 3000 \\ 14.5 \cdot \rho_g \cdot U^{2.5} \exp\left(-5.4 \dfrac{u_{ti}}{U}\right) & \text{for } Re > 3000 \end{cases}$ with $Re = \dfrac{D_t \cdot U}{\nu}$	0.2–0.8	0.076, 0.152	0.017–0.077	Tasirin and Geldart (1998c)	
14	$\dfrac{K_{i\infty}^* d_p}{\mu} = Ar^{0.5} \exp\left(6.92 - 2.11 F_g^{0.303} - \dfrac{13.1}{F_d^{0.902}}\right)$ with $F_g = g\, d_p (\rho_s - \rho_g)$ (gravity force per projection area) $F_d = C_d \dfrac{\rho_g U^2}{2}$ (drag force per projection area)	0.3–7.0	0.06–1.0	0.05–1.0	Choi et al. (1999)	Correlation based on a wide range of different units, materials, and operating conditions, e.g., temperature and pressure

All parameters to be inserted in SI units.

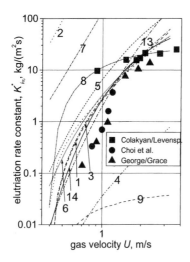

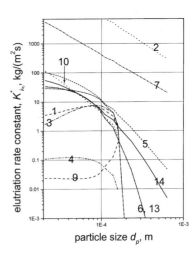

Figure 11 Comparison of predicted elutriation rate constants with measurements by Colakyan and Levenspiel (1984), Choi et al. (1998), and George and Grace (1981). Sand, $d_p = 90$ μm, for correlation 13 D = 0.6 m was assumed, and the terminal velocity was calculated with the correlation for the drag coefficient by Kaskas and Brauer as cited by Brauer (1971). The numbers denote the various correlations in Table 1.

Figure 12 Comparison of different correlations for the elutriation rate constant—influence of particle size. Sand, $U = 1$ m/s, for correlation 13 D = 0.6 m was assumed, the terminal velocity was calculated with the correlation for the drag coefficient by Kaskas and Brauer as cited by Brauer (1971). The numbers denote the various correlations in Table 1.

following parameters: the bubble diameter, the column diameter, and solids and gas properties.

For fluidized beds with FCC or comparable solids, Zenz and Othmer (1960) have given a graphical representation of the TDH vs. the fluidizing velocity U with the bed diameter as a parameter (Fig. 13). Here a significant influence of the bed diameter can be recognized. The Zenz–Othmer diagram is based on much industrial expertise and it is certainly helpful as a first guess.

6.3 Entrainment for Freeboard Heights Below TDH

To describe exactly the processes below the TDH, it is necessary to describe the formation and disintegration of clusters or strands, the acceleration and deceleration of clusters, the turbulent structures of the gas flow, and the interaction between the solids and the gas turbulence. All these effects are influenced by the processes inside or at the surface of the dense bed. The approaches published up to now reduce the complexity by reducing the flow either to three phases (Kunii and Levenspiel, 1989) or to two entrainment flows in par-

Table 2 Correlations for the Calculation of the Transport Disengaging Height (TDH)

	Equation	Reference
1	$\mathrm{TDH} = 0.85 U^{1.2}(7.33 - 1.2\log U)$	Chan and Knowlton (1984b)
2	$\mathrm{TDH} = 1000\dfrac{U^2}{g}$	Fournol et al. (1973)
3	$\mathrm{TDH} = 18.2 d_b$	George and Grace (1978)
4	$\mathrm{TDH} = 13.8 d_b$	Fung and Hamdullahpur (1993)
5	$\mathrm{TDH} = \dfrac{1500 H_b\,\mathrm{Re}_p}{\mathrm{Ar}}$	Sciazko et al. (1991)

All parameters to be inserted in SI units.

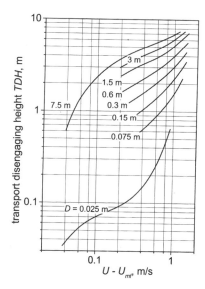

Figure 13 Transport disengaging height vs. excess gas velocity $U - U_{mf}$ for particles of the FCC catalyst type. The parameter is the bed diameter. The freeboard is assumed to have the same diameter as the bed. (After Zenz and Othmer, 1960.

or

$$E_{ih} = x_{Bi} \cdot K_{i\infty}^* + E_{i0} \cdot \exp(-a_i h) \qquad (12)$$

with E_{i0} being the flux of component i ejected from the bed to the surface.

To calculate the entrainment for freeboard heights below the TDH we need now in addition to the elutriation rate constant $K_{i\infty}^*$ the entrainment flux at the bed surface and the decay constant a_i. For the decay constant the value varies according to Wen and Chen (1982) between 3.5 and 6.4 m^{-1}. They recommend a constant value of $a = 4.0$ m^{-1} for systems for which no detailed information is available.

The entrainment at the bed surface E_{i0} can be recalculated from entrainment measurements at two different heights, if such data are available. If no experimental data are available a correlation for the entrainment at the bed surface has to be used. Some of the correlations published in the literature are given in Table 3.

Table 3 Correlations for the Calculation of the Entrainment Flux E_0 at the Bed Surface

	Equation	Reference
1	$E_0 = 3d_p \dfrac{(1 - \varepsilon_{mf})(U - U_{mf})}{d_b}$ bubble nose model	Pemberton and Davidson (1986)
	$E_0 = 0.1 \cdot \rho_s (1 - \varepsilon_{mf})(U - U_{mf})$ bubble wake model	
2	$E_0 = 3.07 \cdot 10^{-9} \dfrac{A \cdot d_b (U - U_{mf})^{2.5} \rho_g^{3.5} \cdot g^{0.5}}{\mu^{2.5}}$	Wen and Chen (1982)
3	$E_0 = 9.6 \cdot A \cdot (U - U_{mf})^{2.5} A \cdot d_b \cdot \left(\dfrac{298}{T}\right)^{3.5}$ (T in kelvins)	Choi et al. (1989)

All parameters to be inserted in SI units.

allel (Large et al., 1976). Since both models lead—at least from the practical point of view—to similar results, we will focus here on the latter one. Large et al. assume that the entrainment flux E_{ih} for a given component i consists of two fluxes in parallel, one flux $E_{i\infty}$ flowing continuously from the bed to the outlet and a second flux of clusters or strands ejected from the bed, which decreases exponentially with increasing height h above the bed surface:

$$E_{ih} = E_{i\infty} + E_{i0} \cdot \exp(-a_i h) \qquad (11)$$

NOMENCLATURE

a = Decay constant defined by Eq. (11), m^{-1}
A = Cross-sectional area of the bed, m^2
d_b = Diameter of a bubble, m
d_p = Particle size, m
D = Diameter of the fluidized bed column, m
D_{FB} = Diameter of the column in the freeboard region, m
D_h = Hydraulic diameter of the fluidized bed column (4A/circumference), m
E_∞ = Total entrainment flux above TDH, kg/(m^{2}s)

E_0 = Total entrainment flux at the surface of the fluidized bed, kg/(m²s)

E_h = Total entrainment flux at height h, kg/(m²s)

$E_{i\infty}$ = Entrainment flux of component i above TDH, kg/(m²s)

E_{i0} = Entrainment flux of component i at the surface of the fluidized bed, kg/(m²s)

E_{ih} = Entrainment flux of component i at height h, kg/(m²s)

h = Height above the distributor, m

$K_{i\infty}^*$ = Elutriation rate constant $E_{i\infty}x_{Bi}$ above TDH, kg/(m²s)

K_{ih}^* = Elutriation rate constant $E_{ih}X_{Bi}$ at height h, kg/(m²s)

M_b = Mass of solids in the fluidized bed, kg

Re = Reynolds number $\text{Re} = \frac{U \cdot D \cdot \rho_g}{\mu}$, -

Re_t = Particle Reynolds number defined by $\text{Re}_t = \frac{u_t \cdot d_p \cdot \rho_g}{\mu}$, -

t = Time, s

T = Temperature, K

U = Superficial velocity, m/s

U_{mf} = Minimum fluidizing velocity, m/s

u_t = Terminal velocity of a single particle, m/s

x_{Bi} = Mass fraction of particles of component i in the fluidized bed, -

Greek symbols

ρ_g = Gas density, kg/m³

ρ_s = Solids density, kg/m³

μ = Gas viscosity, Pa s

REFERENCES

Bachovchin DM, Beer JM, Sarofim AF. An investigation into the steady-state elutriation of fines from a fluidized bed. AIChE Symp. Ser. 77:76–85, 1981.

Baeyens J, Geldart D, Wu SY. Elutriation of fines from gas fluidized bed of Geldart A-type powders—effect of adding superfines. Powder Technology 71:71–80, 1992.

Baron T, Briens CL, Bergougnou MA. Reduction of particle entrainment from gas-fluidized beds with a screen of floating balls. AIChE Symp. Ser. 84:50–57, 1988.

Baron T, Briens CL, Galtier P, Bergougnou MA. Effect of bed height on particle entrainment from gas-fluidized beds. Powder Technology 63:149–156, 1990.

Brauer H. Grundlagen der Einphasen- und Mehrphasenströmungen. Frankfurt a. Main: Verlag Sauerländer, 1971, pp 196–200.

Briens CL, Bergougnou MA. A new model for entrainment from fluidized beds. AIChE Journal 32:233–238, 1986.

Briens CL, Bergougnou MA, Barton T. Reduction of particle entrainment from gas-fluidized beds. Prediction of the effect of disengagement zones. Powder Technology 62:135–138, 1990.

Briens CL, Bergougnou MA, Inculet II, Baron T, Hazlett JD. Size distribution of particles entrained from fluidized beds: electrostatic effects. Powder Technology 70:57–62, 1992.

Chan IH, Knowlton TM. The effect of pressure on entrainment from bubbling gas fluidized beds. In: Kunii D, Toei R, eds. Fluidization. New York: Engineering Foundation, 1984a pp 283–290.

Chan IH, Knowlton TM. The effect of system pressure on the transport disengaging height above bubbling fluidized beds. AIChE Symp. Ser. 80:24–33, 1984b.

Chen TP, Saxena SC. A theory of solids projection from a fluidized bed surface as a first step in the analysis of entrainment process. In: Davidson JF, Keairns DL, eds. Fluidization II. London: Cambridge University Press, 1978, pp 151–159.

Choi J-H, Son JE, Kim S-D. Solid entrainment in fluidized bed combustors. Journal of Chemical Engineering of Japan 22:597–606, 1989.

Choi J-H, Choi K-B, Kim P, Shun D-W, Kim S-D. The effect of temperature on the particle entrainment rate in a gas fluidized bed. Powder Technology 92:127, 1997.

Choi J-H, Ryu HJ, Shun D-W, Son JE, Kim S-D. Temperature effect on the particle entrainment rate in a gas fluidized bed. Ind. Eng. Chem. Res. 37:1130, 1998.

Choi J-H, Chang I-Y, Shun D-W, Yi C-K, Son J-E, Kim S-D. Correlation on the particle entrainment rate in gas fluidized beds. Ind. Eng. Chem. Res 38:2491–2496, 1999.

Colakyan M, Levenspiel O. Elutriation from fluidized beds. Powder Technology 38:223–232, 1984.

Colakyan M, Catipovic, N Jovanovic G, Fitzgerald T. AIChE Symp. Ser. 77:66–75, 1981.

Demmich J. Mechanisms of solids entrainment from fluidized beds. Ger. Chem. Eng. 7:386–394, 1984.

Do HT, Grace JR, Clift R. Particle ejection and entrainment from fluidised beds. Powder Technology 6:195–200, 1972.

Fournol AB, Bergougnou MA, Baker CGJ. Solids entrainment in a large gas fluidized beds. Canadian Journal of Chemical Engineering 51:401–404, 1973.

Fung AS, Hamdullahpur F. Effect of bubble coalescence on entrainment in gas fluidized beds. Powder Technology 77:251–265, 1993.

Geldart D, Pope D. Interaction of fine and coarse particles in the freeboard of a fluidized bed. Powder Technology 34:95–97, 1983.

Geldart D, Cullinan J, Georghiades S, Gilvray S, Pope DJ. The effect of fines on entrainment from gas fluidized beds. Trans. Inst. Chem. Eng. 57:269–275, 1979.

George SE, Grace JR. Entrainment of particles from aggregative fluidized beds. AIChE Symp. Ser. 74:67–74, 1978.

George SE, Grace JR. Entrainment of particles from a pilot scale fluidized bed. Canadian Journal of Chemical Engineering 59:279–284, 1981.

Gugnoni RJ, Zenz Fa. In: Grace JR, Matsen JN, eds. Fluidization. New York: Plenum Press. 1980, pp 501–508.

Harrison D, Aspinall PN, Elder J. Suppression of particle elutriation from a fluidised bed. Trans. Inst. Chem. Engrs. 52:213–216, 1974.

Hatano H, Ishida M. Study of the entrainment of FCC particles from a fluidized bed. Powder Technology 35:201–209, 1983.

Hazlett JD, Bergougnou MA. Influence of bubble size distribution at the bed surface on entrainment profile. Powder Technology 70:99–107, 1992.

Kato K, Tajima T, Mao M, Iwamoto H. In: Kwauk M, Kunii D, Zheng J, Hasatani M, eds. Fluidization '85—Science and Technology Amsterdam: Elsevier. 1985, pp 134–147.

Kunii D, Levenspiel O. Entrainment of solids from fluidized beds I. Hold-up of solids in the freeboard; II. Operation of Fast Fluidized Beds. Powder Technology 61:193–206, 1989.

Large JF, Martinie Y, Bergougnou MA. Interpretative model for entrainment in a large gas fluidized bed. International Powder Bulk Solids Handling and Processing Conference, Chicago, 1976.

Levy EK, Caram HS, Dille JC, Edelstein S. Mechanisms for solids-ejection from gas-fluidized beds. AIChE Journal 29:383–388, 1983.

Lewis WK, Gilliland ER, Lang PM. Entrainment from fluidized beds. Chem. Eng. Prog. Symp. Ser. 58:65–72, 1962.

Lin L, Sears JT, Wen CY. Elutriation and attrition of char from a large fluidized bed. Powder Technology 27:105–115, 1980.

Ma X, Kato K. Effect of interparticle adhesion forces on elutration of fine powders from a fluidized bed of binary particle mixture. Powder Technology 95:93–101, 1998.

Martini Y, Bergougnou Ma, Baker CGJ. Entrainment reduction by louvers in the dilute phase of a large gas fluidized bed. In: Keairns DL, ed. Fluidization Technology, Vol. II. 1976, pp 29–40.

Merrick D, Highley J. Particle size reduction and elutriation in a fluidized bed process. AIChE Symp. Ser. 70:366–378, 1974.

Nakagawa N, Arita S, Uchida S, Takamura H, Takarada T, Kato K. Particle hold-up and elutriation rate in the freeboard of fluid beds. J. Chem. Eng. Japan. 27:79–84, 1994.

Pemberton ST, Davidson JF. Elutration of fine particles from bubbling fluidized beds. In: Kunii D, Toei R, eds. Fluidization IV. New York: Engineering Foundation, 1983, pp 275–282.

Pemberton ST, Davidson JF. Turbulence in the freeboard of a gas fluidized bed, the significance of ghost bubbles. Chem. Eng. Sci 39:829–840, 1984.

Pemberton ST, Davidson JF. Elutriation from fluidized beds—I. Particle ejection from the dense phase into the freeboard. Chem. Eng. Sci. 41:243–251, 1986.

Peters MH, Fran L-S, Sweeney TL. Study of particle ejections in the freeboard region of a fluidized bed with an image carrying probe. Chem. Eng. Sci 38:485–487, 1983.

Santana D, Rodríguez JM, Macías-Machín. A. Modelling fluidized bed elutriation of fine particles. Powder Technology 106:110–118, 1999.

Sciazko M, Bandrowski J, Raczek J. On the entrainment of solid particles from a fluidized bed. Powder Technology 66:33–39, 1991.

Smolders K, Baeyens J. Elutriation of lines from gas fluidized beds: mechanisms of elutriation and effect of bed geometry. Powder Technology 92:35–46, 1997.

Tanaka I, Shinohara H, Hirose H, Tanaka Y. Elutriation of fines from fluidized bed. J. Chem. Eng. Japan 5:57–62, 1972.

Taneda D, Takahagi H, Aoshika S, Nakagawa N, Kato K. Effects of the powder properties upon the hold-up and the elutriation rate of fine particles in a powder–particle fluidized bed. In: Fan L-S, Knowlton TM, eds. Fluidization IX. New York: Engineering Foundation, 1998, pp 317–324.

Tanimoto H, Chiba T, Kobayashi H. The mechanism of particle ejection from the surface of a gas-fluidized bed. Int. Chem. Engng. 24:679–685, 1984.

Tasirin SM, Geldart D. Entrainment of fines from fluidized beds with expanded freeboard. Chem. Eng. Commun. 166:217–230, 1998.

Tasirin SM, Geldart D. The entrainment of fines and superfines from fluidized beds. Powder Handling and Processing 10:263–268, 1998b.

Tasirin SM, Geldart D. Entrainment of FCC from fluidized beds—a new correlation for the elutriation rate constants $K_{i\infty}^*$. Powder Technology 95:240–247, 1998c.

Tasirin SM, Geldart D. Separation performance of batch fluidized bed air classification. Powder Handling and Processing 12:39–46, 2000.

Tasirin SM, Anuar N, Rodzi F. Entrainment of fines (group C particles) from fluidized beds. In: Kwauk M, Li J, Yang WC, eds. Fluidization X. New York: Engineering Foundation, 2001, pp 445–452.

Tweedle TA, Capes CE, Osberg GL. Effect of screen packing on entrainment from fluidized beds. Ind. Eng. Chem. Process Des. Development 9:85–88, 1970.

Wen CY, Chen LH. Fluidized bed freeboard phenomena: entrainment and elutriation. AIChE Journal 28:117–128, 1982.

Wen CY, Hashinger RF. Elutriation of solid particles from a dense phase fluidized bed. AIChE Journal 6:220–226, 1960.

Wouters IM., Geldart D. Entrainment at high temperatures. In: Fan L-S, Knowlton TM, eds. Fluidization IX. New York: Engineering Foundation, 1998, pp 341–348.

Yagi S, Aochi T. Paper presented to the Society of Chemical Engineers of Japan, Spring Meeting, 1955.

Yorquez-Ramirez MJ, Duursma GR. Study of the flow pattern above an erupting bubble in an incipiently fluidized bed using image shifting. Chem. Eng. Sci 55:2055–2064, 2000.

Yorquez-Ramirez MJ, Duursma GR. Insights into the instantaneous freeboard flow above a bubbling fluidised bed. Powder Technology 116:76–84, 2001.

Zenz FA, Othmer DF. Fluidization and Fluid-Particle Systems. New York: Reinhold, 1960.

Zenz PA, Weil NA. A theoretical–empirical approach to the mechanism of particle entrainment from fluidized beds. AIChE Journal 4:472–479, 1958.

5

Effect of Temperature and Pressure

J. G. Yates

University College London, London, United Kingdom

1 INTRODUCTION

Practically all industrial gas–solid fluidized bed reactors operate at temperatures well above ambient, and some, such as those used in the production of polyolefins, also operate at elevated pressures. It is therefore important to know how fluidized beds behave under high temperatures and/or pressures and if possible to predict this behavior from observations made under ambient conditions. The emphasis here will be on those aspects of the subject that are of direct relevance to the design and operation of fluidized bed plant; a comprehensive review of the more academic aspects can be found in the review by Yates (1996).

As is made clear in other parts of this publication, solid particles in fluidized beds are held in suspension by the upward flow of gas. The velocity at which the particles first become suspended is the minimum fluidization velocity, u_{mf}, and as the gas velocity through the bed is increased, the particles pass through a number of flow regimes characterized as bubbling, turbulent, and fast. The transition velocities between one regime and another, as well as the terminal fall velocity of single particles, u_t, are all influenced by the physical properties of the gas, and these in turn are influenced by the temperature and pressure under which the bed is operated. Changes in the surface properties of the fluidized solids, particularly with increasing temperature, can also have an important influence on bed behavior. These and other factors will be considered in detail below.

The review will start by considering the two velocities at the extremes of the flow regimes, i.e. u_{mf} and u_t.

2 MINIMUM FLUIDIZATION VELOCITY

This may be found by measuring the pressure drop through a bed of particles as a function of the gas velocity. The frictional pressure drop, Δp, through a packed bed of monodispersed spheres, is given by the Ergun equation:

$$\frac{\Delta p}{L} = \frac{150(1-\varepsilon)^2}{\varepsilon^3}\frac{\mu u}{(\phi d_p)^2} + \frac{1.75(1-\varepsilon)}{\varepsilon^3}\frac{\rho_f u^2}{\phi d_p} \qquad (1)$$

where u is the fluid superficial velocity, L is the bed height, d_p is particle diameter, ϕ is the particle sphericity, μ is the fluid viscosity, ρ_f is the fluid density, and ε is the voidage of the bed. The first term on the right hand side of Eq. (1) represents the pressure loss through viscous effects and is the dominant term in the laminar flow region; the second term is the loss due to inertial forces and will be dominant at high Reynolds numbers.

At u_{mf} the weight of the bed is fully supported by the flow of gas, and the pressure drop through the bed is then equal to the bed weight per unit area:

$$\frac{\Delta p}{L} = (1 - \varepsilon_{mf})(\rho_p - \rho_f)g \qquad (2)$$

where ε_{mf} is the minimum fluidization voidage and ρ_p is the particle density. Combining (1) and (2) and multiplying through by $d_p^3 \rho_f / (1 - \varepsilon_{mf}) \mu^2$:

$$Ga = 150 \frac{(1 - \varepsilon_{mf})}{\phi^2 \varepsilon_{mf}^3} Re_{mf} + \frac{1.75}{\phi \varepsilon_{mf}^3} Re_{mf}^2 \qquad (3)$$

where

$$Ga = \frac{d_p^3 \rho_f (\rho_p - \rho_f) g}{\mu^2}; \qquad Re_{mf} = \frac{d_p u_{mf} \rho_f}{\mu} \qquad (4)$$

Wen and Yu (1966) showed that the voidage and shape factor terms in Eq. (3) could be correlated with much experimental data on the basis of the two approximations

$$\frac{(1 - \varepsilon_{mf})}{\phi^2 \varepsilon_{mf}^3} \approx 11 \qquad \frac{1}{\phi \varepsilon_{mf}^3} \approx 14 \qquad (5)$$

They lead to a modified form of Eq. (3):

$$Ga = 1650 \, Re_{mf} + 24.5 \, Re_{mf}^2 \qquad (6)$$

which may be rearranged to

$$Re_{mf} = \sqrt{(C_1)^2 + C_2 Ga} - C_1 \qquad (7)$$

The two constants C_1 and C_2 in the original Wen and Yu correlation are 33.7 and 0.0408, respectively, so that, knowing the effects of temperature and pressure on the gas density and viscosity terms contained in Eq. (4), u_{mf} can be calculated over any range of these parameters. However, Yang et al. (1985) pointed out that several other sets of constants have been proposed by other workers, making, Eq. (6) unreliable for predicting u_{mf} at elevated temperatures and pressures. Yang et al. (1985) also pointed out that the Wen and Yu correlation is nothing more than a relationship between Reynolds number, Re, and drag coefficient, C_D, since for a single spherical particle

$$C_D \frac{\pi d_p^2}{4} \rho_f \frac{u^2}{2} = \frac{\pi d_p^3}{6} (\rho_p - \rho_f) g \qquad (8)$$

or

$$C_D = \frac{4}{3} \frac{d_p (\rho_p - \rho_f) g}{\rho_f u^2} \qquad (9)$$

and

$$Ga = \frac{3}{4} Re^2 C_D \qquad (10)$$

The Wen and Yu correlation [Eq. (7)] can be rearranged to give

$$C_D \frac{(Re + C_1)^2 - C_1^2}{0.75 C_2 \, Re^2} \qquad (11)$$

showing the relationship between Re and C_D.

Yang et al. (1985) proposed using the methodology developed by Barnea and Mizrahi (1973) to obtain values for u_{mf} at elevated temperatures and pressures. These two workers proposed a general correlation for pressure drop through fixed beds of spherical particles based on a discrete particle model corrected for particle interaction. They incorporated a voidage term into a standard C_D vs. Re plot for a single sphere to give the following modified values for a multiparticle system:

$$(Re)_\varepsilon = Re \left\{ \frac{1}{\varepsilon \exp[5(1 - \varepsilon)/3\varepsilon]} \right\} \qquad (12)$$

$$(C_D)_\varepsilon = C_D \frac{\varepsilon^3}{1 + (1 - \varepsilon)^{1/3}} \qquad (13)$$

These correlations were extended by Barnea and Mednick (1975) to calculate u_{mf}. They suggested plotting the dimensionless diameter $(Re^2 \cdot C_D)_{mf}^{1/3}$ against the dimensionless velocity $(Re/C_D)_{mf}^{1/3}$ which at the point of minimum fluidization can be expressed as

$$\left(\frac{(Re)_\varepsilon}{(C_D)_\varepsilon} \right)_{mf}^{1/3} = \left(\frac{Re}{C_D} \right)_{mf}^{1/3} \left\{ \frac{\left[1 + \left(1 - \varepsilon_{mf}^{1/3} \right) \right]^{1/3}}{\varepsilon_{mf}^{4/3} \exp(5(1 - \varepsilon_{mf})/9\varepsilon_{mf})} \right\} \qquad (14)$$

Yang et al. (1985) combined Eq. (14) with the standard drag correlations recommended by Clift et al. (1978) (Table 1) and found good agreement with experimental values obtained at pressures of up to 63 bar and temperatures of up to 900°C.

A qualitative appreciation of the way u_{mf} is affected by elevated pressures was given by Rowe (1984), who rearranged Eq. (3) to express u_{mf} in terms of the operating variables as:

$$u_{mf} = \frac{\mu}{\rho_f d_p} 42.9(1 - \varepsilon_{mf})$$

$$\left\{ \left[1 + 3.0 \times 10^{-4} \frac{\varepsilon_{mf}^3}{(1 - \varepsilon_{mf})^2} Ga \right]^{1/2} - 1 \right\} \qquad (15)$$

Rowe (1984) used Eq. (15) to show how pressure would be expected to affect u_{mf} for particles with a density of 1250 kg/m³ and a range of sizes (Fig. 1). In these calculations the minimum fluidization voidage was set at 0.5 and the shape factor at unity; it is clear that for particles below about 100 μm in diameter

Table 1 Recommended Drag Correlations. $w = \log_{10} \text{Re}$

Range	Correlation
(A) $\text{Re} < 0.01$	$C_D = 3/16 + 24/\text{Re}$
(B) $0.01 < \text{Re} \le 20$	$\log_{10}\left[\dfrac{C_D \, \text{Re}}{24} - 1\right] = -0.881 + 0.82w - 0.05w^2$
	i.e., $C_D = \dfrac{24}{\text{Re}}\left[1 + 0.1315\,\text{Re}^{(0.82-0.05w)}\right]$
(C) $20 < \text{Re} \le 260$	$\log_{10}\left[\dfrac{C_D \, \text{Re}}{24} - 1\right] = -0.7133 + 0.6305w$
	i.e., $C_D = \dfrac{24}{\text{Re}}\left[1 + 0.1935\,\text{Re}^{0.6305}\right]$
(D) $260 \le \text{Re} \le 1500$	$\log_{10} C_D = 1.6435 - 1.1242w + 0.1558w^2$
(E) $1.5 \times 10^3 \le \text{Re} \le 1.2 \times 10^4$	$\log_{10} C_D = -2.4571 + 2.5558w - 0.9295w^2 + 0.1049w^3$
(F) $1.2 \times 10^4 < \text{Re} < 4.4 \times 10^4$	$\log_{10} C_D = -1.9181 + 0.6370w - 0.0636w^2$

Source: Clift et al., 1978.

[Group A in the Geldart (1973) classification], pressure is predicted to have little if any effect on u_{mf}. The reason for this is that the flow of gas around these small particles is laminar and so the fluid–particle interaction force is dominated by the gas viscosity (which is essentially independent of pressure in the range considered); the Galileo number, Ga, is a linear function of pressure. As particle size increases, however, inertial forces become more important and at $d_p > 500\,\mu m$ (Geldart Group B) they begin to dominate over the viscous forces; this causes u_{mf} to decrease

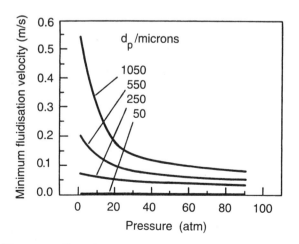

Figure 1 Effect of pressure on u_{mf}, based on Eq. (15). (From Rowe, 1984.)

sharply with pressure up to about 20 bar and more gradually thereafter. Similar conclusions had been reached by King and Harrison (1982) who showed that on the basis of Eq. (3) u_{mf} is independent of pressure for laminar flow ($\text{Re}_{mf} < 0.5$) while for turbulent flow ($\text{Re}_{mf} > 500$) u_{mf} is inversely proportional to the square root of gas density and hence pressure; pressure dependence would be expected to be weaker in the intermediate regime.

Olowson and Almstedt (1991) measured the u_{mf} of a range of particles in Geldart Groups B and D at pressures from 0.1 to 0.6 MPa and found a general decrease with increasing pressure. They found the effect of pressure to be well described by the Ergun equation and a number of simplified correlations derived from it, although the accuracy of these correlations differed significantly.

The effect of temperature on u_{mf} may also be seen by examination of Eq. (15). Since the density of a gas is inversely proportional to its absolute temperature, gas density will decrease with increasing temperature; the viscosity of a gas on the other hand increases with increasing temperature μ_g being proportional to T^n where n is usually between 0.5 and 1.0. The combined effect of changes in density and viscosity results in the Galileo number decreasing steadily with increasing temperature, but it is not immediately obvious from Eq. (15) how this will affect u_{mf}. Figure 2 shows Eq. (15) plotted with a square root dependence of gas

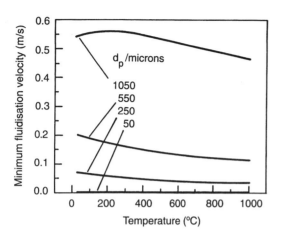

Figure 2 Effect of temperature on u_{mf}, based on Eq. (15). (From Rowe, 1984.)

viscosity on temperature, and it is clear that a rise in temperature causes the u_{mf} of Group A powders to decrease, although the effect is relatively modest. For Group B materials, the effect is first to raise the value through the gas density term and then to cause a decrease as the viscous term starts to dominate. Experimental verification of these trends comes from the work of Botterill and Teoman (1980), Svoboda and Hartman (1981), Botterill et al. (1982), Wu and Baeyens (1991) and Raso et al. (1992); all authors, however, stress the importance of choosing the correct value of the emulsion phase voidage, ε_{mf}, in using the Ergun equation for these predictions.

Recent work by Formisani et al. (1998) has shown the effect of increasing temperature on the voidage of fixed and fluidized beds, ε_0 and ε_{mf} respectively. They studied beds of silica sand, glass ballotini, salt, and fluidized cracking catalyst (FCC) and found for all an increase in voidage with increasing temperature, the dependence being expressed in the linear form by

$$\varepsilon_T = \varepsilon_{amb} + k(T - T_{amb}) \tag{16}$$

where the subscripts refer to temperature and ambient respectively and the parameter k is a function of particle properties. The authors conclude that the effect is due to the variation with temperature of interparticle forces, which runs parallel to the thermal variation of gas properties; the authors demonstrate that the predictive ability of the Ergun equation is only restored if the dependence of ε_{mf} is correctly accounted for. Yamazaki et al. (1995) studied the effect of chemisorbed water on the minimum fluidization voidage of silica particles at high temperatures and concluded that ε_{mf} decreases as humidity increases, the cause being the

reduction in van der Waals forces of wet surfaces due to the lower value of their Hamaker constant.

The effects of temperature and pressure on ε_{mf} have also been considered from a theoretical standpoint by Rowe (1989).

3 TERMINAL FALL VELOCITY OF SINGLE PARTICLES

The force acting on a single particle falling through a quiescent fluid is

$$F = M\frac{du}{dt} \tag{17}$$

$$= (\text{particle weight}) - (\text{drag force})$$

$$= \frac{\pi d_p^3}{6}(\rho_p - \rho_f)g - C_D\frac{\pi d_p^2}{4}\rho_f\frac{u^2}{2} \tag{18}$$

At u_t

$$\frac{du}{dt} = 0$$

Hence

$$u_t = \left[\frac{4d_p(\rho_p - \rho_f)g}{3\rho_f C_D}\right]^{1/2} \tag{19}$$

where C_D is, again, an empirical drag coefficient and a function of the Reynolds number (see Table 1). The form of this function has been a frequent subject of investigation, and work in this area has been reviewed by Khan and Richardson (1987) and by Hartman and Yates (1993). The relationship is complicated by such factors as the flow regime (laminar, turbulent, or intermediate) and the size, density, and shape of the particles involved. Haider and Levenspiel (1989) devised a method by which u_t can be calculated directly for nonspherical particles from their size, sphericity, etc. and from the physical properties of the system. The method uses a dimensionless particle size and a dimensionless gas velocity defined respectively by:

$$d_p^* = d_p\left[\frac{\rho_f(\rho_p - \rho_f)g}{\mu^2}\right]^{1/3} \quad u_t^* = u_t\left[\frac{\rho_f^2}{\mu(\rho_p - \rho_f)g}\right]^{1/3} \tag{20}$$

Then

$$u_t^* = \left[\frac{18}{(d_p^*)^2} + \frac{2.335 - 1.744\phi_s}{(d_p^*)^{0.5}}\right]^{-1} \tag{21}$$

for $0.5 < \phi_s < 1$ where ϕ_s is the particle sphericity.

Table 2 Particle Properties Used in u_t Calculations.

d_p/m	ρ_p/kg m^{-3}	ϕ	Group
0.0001	1500	0.8	A
0.0005	1500	0.8	B
0.001	2500	0.8	D

Terminal velocities can then be found directly from eqs (20) and (21). Based on this method, the terminal fall velocities in air of nonspherical particles in Groups A, B, and D of the Geldart classification have been calculated as functions of temperature and pressure (Table 2 and Figs. 3 and 4).

It can be shown that in the Stokes flow region ($Re_t < 0.1$):

$$u_t = \frac{d_p^2(\rho_p - \rho_f)g}{18\mu_f} \qquad (22)$$

so that apart from its negligibly small influence on the density difference term in Eq. (21), pressure will have no effect on u_t; temperature will of course have an influence, through its effect on gas viscosity. Figure 3 shows the effect of temperature and pressure on the terminal velocity of a Group A material (for which $Re_t < 1$), and it is clear that both cause a monotonic decrease under all superambient conditions. The pressure effect is transmitted through the gas density term, and the temperature effect through the gas viscosity term in Eq. (20). In the case of powders in Groups B and D (Fig. 4), increasing pressure has a stronger effect on u_t than in the case of Group A, but the trend is in the same direction.

Increasing the temperature on the other hand leads to an increase in u_t owing to the dominance of inertial forces over viscous forces in the case of these materials and to the effect of temperature in decreasing gas density.

4 BUBBLING BEDS

4.1 Pressure

It has long been recognized that pressure exerts a strong influence on the bubbling behavior of gas-fluidized beds. The literature up to 1993 was reviewed by Yates (1996), and this is reproduced and extended to 1998 in Table 3. The three principal groups of powders in Geldart's classification will be considered in turn.

4.1.1 Group A Powders

A general observation is that with these powders, while u_{mf} is unaffected (for the reasons given above), the region of bubble-free expansion between u_{mf} and the minimum bubbling velocity, u_{mb}, increases with increasing pressure; this is in accord with the theory of Foscolo and Gibilaro (1984). Furthermore, at the same values of volumetric flow rate, bubbles in beds of Group A materials generally become smaller as pressure increases. This could be due to (1) a greater proportion of gas flowing through the emulsion phase following an increase in emulsion-phase voidage or (2) a decrease in the stability of bubbles leading to their breakup into smaller voids. The question of there being a maximum stable bubble size was first explored by Harrison et al. (1961) and later developed

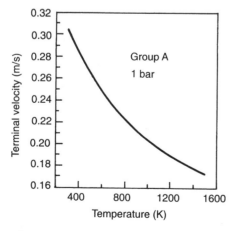
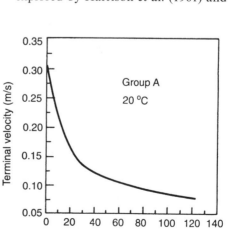

Figure 3 Effect of temperature and pressure on terminal fall velocity of Group A particles in air.

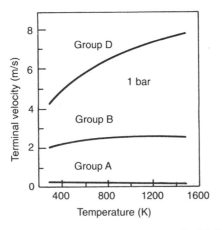 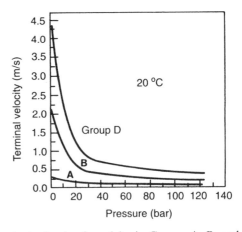

Figure 4   Effect of temperature and pressure on terminal fall velocity in air of particles in Groups A, B, and D.

by Davidson and Harrison (1963). The theory is based on the assumption that as a bubble rises in a fluid bed the shear force exerted by the particles moving downwards relative to the bubble sets up a circulation of gas within the bubble and that the velocity of this circulation, u_c, approximates to the bubble rise velocity, u_b. When through coalescence the bubble diameter increases to the point where $u_b > u_t$, the terminal fall velocity of the bed particles, the solids in the wake will be drawn up into the bubble, causing it to break up into smaller bubbles with lower rise velocities. The bubble would thus not be expected to exceed a certain size determined by the value of u_t, and beds of small, light particles should show "smoother" fluidization characteristics than those of coarser materials. Now u_t decreases as gas pressure increases (see above), and so if the foregoing theory is correct, Group A powders should show smoother behavior at elevated pressures; this effect has been widely observed (Hoffmann and Yates, 1986).

An alternative theory developed by Upson and Pyle (1973) and Clift et al. (1974) suggests that bubble breakup is caused by an instability in the bubble roof allowing particles to rain down through the void and so divide it into two. On the basis of hydrodynamic theory due to Taylor (1950), Clift et al. (1974) showed that the only factor determining the stability of the bubble roof is the apparent kinematic viscosity of the emulsion phase and that bubbles will become less stable as this viscosity decreases. A decrease in dense-phase viscosity would result from an increase in the voidage of that phase, and as we have already seen this is an increasing feature of beds of Group A powders as operating pressure is increased. King and Harrison (1980) reported results of a study of particles of Groups A and B at pressures of up to 25 bar. They

used x-rays to observe both slugs and bubbles and found that in all cases of Group A materials, breakup occurred by fingers of particles falling in from the roof and that the effect was more pronounced the higher the pressure.

In a separate study King et al. (1981) determined the viscosity of the emulsion phase of fluidized beds of glass ballotini of a range of sizes by measuring the rate of fall under gravity through the bed of a small metal sphere. The beds were fluidized by both carbon dioxide and nitrogen at pressures of up to 20 bar. They found that an increase in gas pressure led to a substantial decrease in the viscosity of the finest powder but that the viscosities of beds of powders larger than about 100 μm were almost independent of pressure.

4.1.2   Group B Powders

In the case of these materials King and Harrison (1980) reported no effect of pressure on bubble size up to 25 bar, but Hoffmann and Yates (1986) found that mean bubble diameters increased slightly up to 16 bar and decreased thereafter up to 60 bar; the extent of these effects was dependent on the vertical position in the bed at which the observations were made.

Hoffmann and Yates also measured changes in the value of the bubble velocity coefficient, K, with increasing pressure:

$$u_b = K\left(\frac{gd_b}{2}\right)^{1/2} \tag{23}$$

The work gave values of K for a freely bubbling bed of alumina powder at a point 40 cm above the distributor, decreasing from 0.9 at 1 bar to 0.7 at 20 bar; this was followed by a gradual increase to 1.2 at 60 bar. The observed decrease up to 20 bar is consistent with the

Table 3 Experimental Observation of Effects of Pressure on Gas-Fluidized Beds

Reference	Bed cross Section, cm	Particle properties material	ρ_p, gcm^{-3}	d_p, μm	Gas	Max. pressure, kPa	Observations with increasing pressure (P, kPa)
Godard and Richardson (1968)	diam. = 10.0	(i) diakon (ii) phenolic resin	1.18 0.24	125 186	air	1414	U_{mb}-U_{mf} increases.
Botterill and Desai (1972)	diam. = 11.4	(i) copper (ii) shot (iii) sand coal	— — —	150, 625, 160, 800, 2740 1430	air CO$_2$	1000	Quality of fluidization increases for large particles; heat transfer coefficients double from 1 to 10 bar for larger particles but not smaller ones.
Knowlton (1977)	diam. = 29.2	(i) coals cokes (ii) and chars siderite (FeCO$_3$)	1.12 -1.57 3.91	230–780 290	N$_2$	6900	(i) Fluidization becomes smoother at $P > 1035$. (ii) Bed entrainment increases significantly at $P > 1035$. (iii) No change in bed expansion or density.
Saxena and Vogel (1977)	diam. = 15.2	(i) dolomite (ii) sulphated dolomite	2.46 3.19	765 717	air	834	U_{mf} decreases and follows Ergun correlation.
Canada and MacLaughlin (1978)	30.5 square	glass beads	2.48 2.92	650 2600	air R-12	9000	Slugging less pronounced at high pressure; heat transfer coefficients increase.
Crowther and Whitehead (1978)	diam. = 2.7	(i) synclyst (ii) coals	0.9 1.3	63 19–63	Ar CF$_4$	6900	U_{mb}-U_{mf} increases; fully particulate fluidization occurs with finest particles.
Denloye and Botterill (1978)	diam. = 11.4	(i) copper shot (ii) sand (iii) soda glass	8750 2600 2450	160, 340, 620 160, 570, 1020, 2370 415	air, argon Co$_2$ and freon	1000	Bed-to-surface heat transfer coefficient increases.
Guedes de Carvalho et al. (1978)	diam. = 10	(i) ballotini (ii) sand	2.9 2.69	64 211 74	N$_2$ CO$_2$	2800	No significant effects with 74 μm sand and 21 μm ballotini; bubbles become smaller and break up from rear with 64 μm ballotini.

Table 3 Continued

Reference	Bed cross Section, cm	Particle properties material	ρ_p, gcm^{-3}	d_p, μm	Gas	Max. pressure, kPa	Observations with increasing pressure (P, kPa)
Subzwari et al. (1978)	2-D bed 46 × 15	FCC Powder	0.95	60	air	700	e_i increases; number and size of bubbles decrease: bubbles split by division from roof: maximum stable bubble size decreases.
Varadi and Grace (1978)	2-D bed 31 × 1.6	sand	2.65	250–295	air	2200	U_{mb}-U_{mf} increases; no increase in bubble splitting; bubbles split by division from roof.
Borodulya et al. (1980)	diam. 10 cm	sand	2.48 –2.60	126–1220	air	8000	Bed-to-surface heat transfer coefficient increases.
King and Harrison (1980)	diam. 10 cm	ballotini	2.9	64 101 475	air N$_2$ CO$_2$	2500	For 64 μm particles, bubbles become smaller and less stable; no effect of pressure on larger particles.
Knowlton and Hirsan (1980)	semicircle diam. 30.5 cm	(i) sand (ii) char (iii) siderite (FeCO$_3$)	2.63 1.16 3.99	438 419 421	Nz	5171	Jet penetration length increases; existing correlations underestimate jet lengths.
Rowe and MacGillivray (1980)	diam. = 20 cm	silicon carbide	3.19	58	air	400	Bubbles become smaller; bubble velocity increases; visible bubble flow diminishes.
Xavier et al.	diam. = 10 cm	(i) ballotini (ii) ballotini (iii) ballotini (iv) polymer	2.82 2.97 2.91 0.55	61 475 615 688	CO$_2$ N$_2$ N$_2$ N$_2$	2500	Bed-to-surface heat transfer for fine powders only slightly affected; for larger particles, heat transfer coefficient increases with pressure.
Kawabata et al. (1981)	2-D bed 30 × 1	sand	2.63 2.58 2.59	300 430 600	air	800	U_{mf} decreases; bubble sizes unchanged, but they become flatter; bubble velocity decreases.
Guedes de Carvallo et al. (1982)	diam. = 5	ballotini	2.9	64 101	N$_2$	500	Mass transfer between slug and the dense phase does not decrease as much as expected with pressure if gas diffusivity alone is the dominant process.

Reference	Diam.	Material	Density	Particle size	Gas		Observations
King and Harrison (1982)	diam. = 10	(i) ballotini (ii) sand (iii) polymer	2.82, −2.97 2.66 0.55	61–475 81, 288 688	N_2 CO_2	2500	For fine particles, U_{mf} and ε_{mf} are unaffected, but U_{mb} and ε_{mb} increase; for large particles, $U_{mf} = U_{mb}$ and decrease with increasing pressure. But $\varepsilon_{mf} = \varepsilon_{mb}$ and is unaffected.
Rowe et al. (1982)	diam. = 7.6	FCC powder	0.82	70, 82	Ar air CO_2	2000	$U_{mb}-U_{mf}$ increases
Sobreiro and Monteiro (1982)	diam. = 4.5	(i) ballotini (ii) alumina (iii) pyrrhotite	2.3 2.8 4.0	125, 177 250 88, 125 177 125	N_2	3500	U_{mf} is pressure independent in lamina flow regime but decreases with increasing pressure at higher Re values; ε_{mf} is pressure-independent; $U_{mb}-U_{mf}$ increases for fine powders.
Barreto et al. (1983)	diam. = 10	FCC powders	1.26 0.88	58–98	N_2	2000	Bubble size and velocity decrease; bubble frequency increases with pressure; more gas flows in bubble wakes and less in bubble voids.
Borodulya et al. (1983)	diam. = 10.5	(i) glass beads (ii) sand	2630 2630 2580 2700	3100 1250 1225 794	air	8100	Bed-to-surface heat transfer coefficient increases.
Rowe et al. (1984)	17.5 × 12.5	(i) alumina (ii) silicon carbide	1.42 3.19	450 262	N_2	8000	Geldart Group B behavior at low pressures changes to Group A behavior at high pressures; bubble velocity increases after a small initial decrease; bubble volume decreases after slight initial increase; bubbles flatten at high pressures.
Chan and Knowlton (1984)	diam. = 11.5	sand	2.60	20–300	N_2	3100	Solids entrainment increases sharply with increasing pressure.
Verkooijen et al. (1984)	diam. = 10	FCC powder	—	61	butene in N_2	1000	Number of interphase transfer units decreases from 1 to 2 bar and increases sharply above 5 bar.

Table 3 Continued

Reference	Bed cross Section, cm	Particle properties material	ρ_p, gcm^{-3}	d_p, μm	Gas	Max. pressure, kPa	Observations with increasing pressure (P, kPa)
Chiester et al. (1984)	diam. = 10.2	(i) coals	1.25	88–361	N$_2$	6485	ε_i increases; for coarse particles, bed height increases at a given gas flow rate; for fine particles it does not always do so. No evidence for maximum stable bubble size up to 810 kPa. Bubble splitting occurs from roof. Bed appears homogeneous at highest pressures.
		(ii) chars	1.12	157–376			
		(iii) ballotini	2.47	96–374			
	10.2 × 1.9	(i) coal	1.25	195			
		(ii) char	1.12	203			
Weimer and Quaderer (1985)	13	carbon	0.850	66	CO + H$_2$	8500	For 66 μm powder, ε_i increased from 0.53 to 0.74, u_i increases sevenfold, d_{bmax} decreases sixfold; pressure effects were similar but of lower magnitude.
		carbon	0.850	108			
		carbon	0.850	171			
Hoffmann and Yates (1986)	17.8 × 12.7	alumina	1.417	450	N$_2$	8000	d_{max} decreases but bubble velocity increases; bubble stability decreases; bubble flow in center of bed increases.
		alumina	1.488	695			
		silicon carbide	3.186	184			
Chan et al. (1987)	38.1	sand	2.565	100	N$_2$	3200	Bubble size decreases, bubble frequency increases.
		sand	2.565	200			
		sand	2.565	400			
		coke breeze	1.507	400			
		char	1.251	400			
Jacob and Weimer (1987)	9.7	carbon	0.85	44	CO + H$_2$	12,420	Particulate bed expansion is adequately described by Foscolo–Gibilaro theory.
		carbon		122			
Cai et al. (1989)	28.4	silica gel	0.834	476	air	800	Higher pressure causes smaller bubbles and lower transition velocity from bubbling to turbulent regimes.
		silica gel	0.706	280			
		silica gel	0.711	165			
		silica gel	0.844	1057			
		FCC	1.172	65			
		FCC	1.667	53			
		resin	1.330	566			
		sand	2.580	98			

Reference		Material	Density	Temperature	Gas	Pressure	Comments
Carsky et al. (1990)	13.5	sand fire clay	2.65 2.00	600 600	air	1300	Bubble size decreases up to 400 kPa and remains constant thereafter.
Olowson and Almstedt (1990)	20×30	sand	2.60	700	air	1600	Pierced bubble length decreases with increasing pressure and with decreasing u-u_{mf}.
Olsson et al. (1995)	20×30	sand	2.60	700	air	1600	Significant differences between beds operated with and without tubes. Initial increase with pressure in mean pierced bubble length, rise velocity, bubble volume fraction and visible bubble flow rate. These level off at pressures above about 0.5 MPa. As pressure increases, gas through-flow velocity decreases significantly.
Wiman et al. (1995)	20×30	sand	2.60	700	air	1600	Tube erosion decreases at high pressures. Bed-to-tube heat transfer coefficient increases significantly with increasing pressure.
Llop et al. (1996)	5	sand	2.65	213, 450 728, 1460	air	101	At reduced pressure, fluidization behavior is similar to that at atmospheric pressure. A modified Wen and Yu correlation is developed to calculate u_{mf}.
Wiman et al. (1997)	20×30	sand	2.60	450, 700	air	1600	Tube erosion in a sparse tube bank higher than in densely packed bank. Larger sand more erosive than smaller; strong coupling between tube erosion and bubble rise velocity.
Wiman et al. (1998)	20×30	sand	2.60	450	air	1600	A dimensionless drag force is a suitable scaling parameter in many cases. Increased gas–particle interaction at high pressures combined with turbulent fluctuations in the gas phase can account for increased instability of bubbles.

observations of Kawabata et al. (1981) on similar Group B powders. The increase in K above 20 bar coupled with the decrease in bubble diameter at these pressures means that although bubbles are becoming smaller, their rise velocities are increasing, the reverse of the tendency observed at ambient pressure. Hoffmann and Yates (1986) also reported an increase in bubble interaction leading to coalescence as pressure was increased, evidence that was further supported by the observed changes in the lateral distribution of bubbles; at atmospheric pressure bubbles were more or less evenly distributed across the bed, but at higher pressures the bubble flow was increasingly concentrated toward the vertical axis of the bed. This would seem to indicate an increase in bubble coalescence leading to the establishment of a central channel in which the resistance to gas flow is relatively low. Increased bubble coalescence should of course result in the formation of larger bubbles, but as mentioned above, bubble diameters tend to decrease as pressure rises above 20 bar, and it would seem likely that the explanation is that the stability of the coalesced bubbles is lower at the higher pressures, and that once formed they break up into ever smaller units. Indeed at the highest pressures studied, individual bubbles were hard to identify on the x-ray films, the whole bed having taken on the appearance of an ill-defined foaming mass of fluidized material.

The results of Hoffmann and Yates (1986) were corroborated by Olowson and Almstedt (1990) who used a combined capacitance and pressure probe to measure bubble characteristics in a bed of powder that was close to the Geldart B/D boundary ($d_p = 0.7$ mm, $\rho_p = 2600$ kg/m^3). They operated at pressures of up to 16 bar and at excess gas velocities from 0.1 to 0.6 m/s and found that the mean bubble frequency, the mean bubble rise velocity, the mean bubble volume fraction, and the visible bubble flow rate rose with both increasing pressure and excess gas velocity. The mean pierced length of bubbles was found to decrease, after an initial increase, with increasing pressure. An interesting observation was made with the pressure probe, which showed that the throughflow velocity of gas inside the bubbles decreased with increasing pressure, a fact that could also contribute to the growing instability of bubbles, since it is the drag force exerted by the throughflowing gas on particles in its roof that is responsible for bubble stability (Campos and Guedes de Carvalho, 1992); the drag force is proportional to the square of the fluid velocity [Eq. (8)]. This question was also considered by Olowson and Almstedt (1992).

Almstedt's group at Chalmers University has carried out a comprehensive series of studies on pressurized fluidized beds both with and without internal tube banks of different geometries (Olsson et al., 1995; Wiman et al., 1995; Wiman and Almstedt, 1997, 1998). The conclusions of these studies are summarised in Table 3. There has been some interest in operating fluidized beds at subambient pressures, and published work has been reviewed by Kusakabe et al. (1998) and Fletcher et al. (1993). Llop et al. (1996) studied the fluidization of coarse sand with a shape factor of 0.6 and more spherical particles of millet (0.9) over a range of pressures up to atmospheric. They found similar behavior at low pressures to that observed at atmospheric pressure and developed a modified Wen and Yu correlation to predict u_{mf} over the low-pressure range.

4.1.3 Group D Powders

These large, dense materials have been relatively little studied compared with those in Groups A and B. They are often difficult to fluidize smoothly, and there is a strong tendency towards spouting. Little is known about how pressure affects their bubbling behavior although Rowe (1984) has drawn attention to the inherent instability of fluidized Group D particles through the strong effect of pressure on their minimum fluidization velocity (Fig. 1). It is apparent from this that any small pressure changes that occur within the bed as a result of bubbles forming and passing through it will have a significant effect on u_{mf} that could lead to local defluidization and generally unstable behavior.

Some indication of the way high pressure operation could influence Group D fluidization comes from the study by King and Harrison (1980) of spouted beds. They measured the minimum spouting velocity of 1.1 mm diameter glass ballotini at pressures of up to 20 bar and found a marked decrease with increasing pressure. It would thus seem likely that Group D materials should follow the same trends as those shown by Group B powders but at higher pressures.

4.2 Temperature

Compared with the effects of high pressures referred to above, there has been relatively little work devoted to bubble dynamics at high temperatures. Furthermore, some of what has been published is contradictory. In an early study Mii et al. (1973) examined fluid beds of graphite particles (Group B) at temperatures of up to 800°C and found that both the frequency of bubble formation and the "fluidity" of the bed increased with increasing temperature; similar conclusions were reached by Yoshida et al. (1974) with beds of micro-spherical catalyst particles between 500 and 1000°C and by Otake et al. (1975) with beds of Group A cracking catalyst. Geldart and Kapoor (1976) studied bubble sizes in beds of spherical steel shot ($d_p = 0.118$ mm) at up to 300°C and found a reduction of up to 25% under equivalent conditions of bed height and excess gas velocity. This would be in keeping with the change in interstitial gas flow with rising temperature predicted by the theory of Foscolo and Gibilaro (1984). Kai and Furasaki (1985) also reported an improvement in the "quality" of fluidization in beds of FCC at up to 370°C.

Sittiphong et al. (1981) studied the eruption diameter of bubbles in beds of large particles ($d_p = 3$ mm) of a refractory material (Group D) and found a significant increase with temperature under equivalent conditions. This is contrary to what has been observed in small particle systems such as those mentioned above, but it is probably a result of the reduction in the value of minimum fluidization velocity for these large particles at high temperatures referred to earlier (Fig. 2); thus at the same value of the fluidization velocity more gas will flow as bubbles as u_{mf} decreases.

An important quantity in the design of bubbling-bed reactors containing internal heat transfer coils is the bed expansion ratio, δ. A model for δ in terms of a dimensionless drag force, F^*, was developed by Lofstrand et al. (1995) and compared with experimental work carried out with the pressurized unit at Chalmers University and an atmospheric pressure fluidized bed boiler fitted with bed internals. The bed expansion ratio is defined as

$$\delta = \frac{H_{fl} - H_{mf}}{H_{fl}} \tag{24}$$

and for a freely bubbling bed without internals the data correlated with

$$\delta = 0.11(F^* - 1)^{0.34} \tag{25}$$

where

$$F^* = \frac{Re}{\varepsilon_{mf}\phi^2 Ar}\left[150\frac{1-\varepsilon_{mf}}{\varepsilon_{mf}} + 1.75\phi Re\right] \tag{26}$$

For beds with staggered tube banks, the correlation was

$$\delta = 0.11(F^* - 1)^{0.34 ST} \tag{27}$$

where

$$S = \left(\frac{W}{W - N_h d}\right)^{0.89} \tag{28}$$

and

$$T = \left(\frac{H_{fl}}{H_{fl} - N_v d}\right)^{0.27} \tag{29}$$

In these expressions N_h and N_v are the average number of tubes in the horizontal and vertical directions, respectively, d is the tube diameter, and W is the bed width. The correlations apply adequately to a wide range of operating conditions with Group B and D powders and temperatures of up to 850°C.

5 JET PENETRATION

When gas first enters a fluidized bed from an orifice in the supporting grid it does so either in the form of discrete bubbles or as a flamelike jet that decays at some point above the grid into a bubble stream (Fig. 5). Whether one or the other forms seems to depend on the properties of the bed material. Thus Rowe et al. (1979) observed the point of entry of a gas using a nonintrusive x-ray technique and saw only bubbles forming with a well defined frequency. Observations in two-dimensional or semicylindrical beds frequently show the presence of jets that tend to be stabilized by the wall.

Grace and Lim (1987) proposed on the basis of much experimental evidence that the criterion for the formation of jets should be

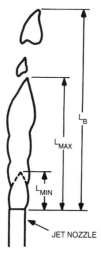

Figure 5 A jet entering a fluidized bed from an immersed nozzle. (From Knowlton and Hirsan, 1980.)

$$\frac{d_{or}}{d_p} \le 25.4 \tag{30}$$

where d_{or} and d_p are the diameters of orifice and particle, respectively. Under all other conditions, bubbles rather than jets would be expected to form.

A considerable number of studies have been reported on jet penetration, and several correlations have been developed giving the penetration length as a function of the physical properties of gas and particles and of operating conditions (Knowlton and Hirsan, 1980). However, most of these correlations were produced from experimental data obtained at ambient conditions, and they fail when applied at elevated temperatures and pressures. Hirsan et al. (1980) measured jet penetrations in beds of Group B powders at pressures of up to 50 bar and produced a correlation for L_{max} (Fig. 5) in terms of the Froude number and the ratio of fluid to particle density:

$$\frac{L_{max}}{d_0} = 26.6 \left(\frac{\rho_f}{\rho_p}\right)^{0.67} \left(\frac{u_o^2}{g d_p}\right)^{0.34} \left(\frac{u}{u_{cf}}\right)^{-0.24} \tag{31}$$

where u_o is the orifice gas velocity and u_{cf} is the superficial velocity necessary to completely fluidize the polydispersed powder. The correlation shows that the penetration length increases with pressure but decreases as the velocity of the fluidizing gas increases, results also obtained by Yates et al. (1986). Yang (1981) used the same data as Hirsan et al. (1980) and produced a slightly different correlation:

$$\frac{L_{max}}{d_0} = 7.65 \left[\left(\frac{1}{R_{cf}}\right)\left(\frac{\rho_f}{\rho_p - \rho_f}\right)\left(\frac{u_o^2}{g d_0}\right)\right]^{0.472} \tag{32}$$

where

$$R_{cf} = \frac{(u_{cf})_{pressure}}{(u_{cf})_{atmosphere}} \tag{33}$$

Yates and Cheesman (1987) measured jet penetrations in three-dimensional beds of two coarse powders at pressures of up to 20 bar at ambient temperature and at temperatures of up to 800°C at ambient pressure. The pressure results gave a correlation identical to Eq. (31) but the high temperature measurements had slightly different values of the coefficient and exponent; a correlation obtained from the combined results at temperature and pressure was

$$\frac{L_{max}}{d_0} = 9.77 \left[\left(\frac{1}{R_{cf}}\right)\left(\frac{\rho_f}{\rho_p - \rho_f}\right)\left(\frac{u_o^2}{g d_0}\right)\right]^{0.38} \tag{34}$$

This would appear to be the only correlation currently available which draws together results from measurements at both elevated temperatures and pressures.

6 HIGH VELOCITY OPERATION

6.1 General Characteristics

The overall structure of a fluidized bed changes as the fluidizing gas velocity is increased. At low velocities, the particles thrown by bursting bubbles into the freeboard region above the bed surface fall back after a short time, but as the gas velocity inceases and bubbling becomes more vigorous, the concentration of particles in the freeboard at any one time increases. The eruption of bubbles gives rise to pressure fluctuations within the bed, and these fluctuations increase with increasing velocity. The velocity at which the pressure fluctuations peak, U_c, has been assumed by some workers to mark the onset of a transition from bubbling to "turbulent" fluidization that is complete at the higher velocity, U_k, where the fluctuations level off. In this turbulent regime, at velocities in excess of U_k, the bed still has an upper surface, although it is much more diffuse than that present in the bubbling state, and considerable carryover of particles occurs.

As the gas velocity is increased beyond U_k a point is reached at which the particles are transported out of the bed altogether either in dense-phase or in dilute-phase pneumatic transport; the velocity at which this occurs is called the "transport velocity," U_{tr}. In order to maintain a constant inventory of particles in the bed at velocities in excess of U_{tr} it is necessary to recycle them via external cyclones and a standpipe, a geometry known as a "circulating fluidized bed." A comprehensive review of the literature on circulating systems has been given by Grace et al. (1997), and the subject is also dealt with in Chapter 19 of this handbook.

6.2 Temperature and Pressure Effects

The main interest here has been their effect on solids carryover and heat transfer (see below). Some work on transition velocities between flow regimes has been reported, however. Thus Cai et al. (1989) studied the effect of operating temperature (50 to 500°C) and pressure (0.1 to 0.8 MPa) on the transition from bubbling to turbulent fluidization of eight powders in Groups A and B of the Geldart classification fluidized in a column 150 mm in diameter and 3.8 m in height.

They found that the transition velocity, U_c, increased with increasing temperature but decreased with increasing pressure. They correlated their results with:

$$\frac{U_c}{\sqrt{gd_p}} = \left(\frac{\mu_{f20}}{\mu_f}\right)^{0.2} \left[K\left(\frac{\rho_{f20}}{\rho_f}\right)\left(\frac{\rho_p - \rho_f}{\rho_f}\right)\left(\frac{D_f}{d_p}\right)\right]^{0.27}$$

(35)

where D_f is the diameter of the fluidized bed, the subscript 20 refers to physical properties measured at 20°C and 1 bar, and

$$K = \left(\frac{0.211}{D_f^{0.27}} + \frac{2.42 \times 10^{-3}}{D_f^{1.27}}\right)^{1/0.27}$$

(36)

A study of a circulating bed at pressures of up to 50 bar was reported by Wirth (1992), but the type of flow investigated (i.e., transitional, turbulent, or fast) is not clear from the report.

Tsukada et al. (1993) determined the effect of pressures of up to 0.7 MPa on the three velocities U_c, U_k, and U_{tr} of an FCC powder. They found each velocity to decrease with increasing pressure raised to the power -0.3, a result not very different from that of Cai et al. (1989). Other conclusions were that the dilute phase volume fraction at U_{tr} was 0.8 at ambient pressure and decreased slightly with pressure, and that the dimensionless core diameter at U_{tr} was about 0.8 and was insensitive to pressure.

6.3 Entrainment and Elutriation

As has been demonstrated by a number of studies (e.g., Rhodes, 1989) there are strong links between the phenomena of turbulent and fast fluidization and the entrainment and elutriation of bed particles that occur at lower gas velocities. Entrainment occurs when gas bubbles burst at the bed surface and throw particles up into the freeboard region. At low gas velocities these particles quickly fall back into the bed and are retained, but as the fluidizing velocity is increased, more particles are transported to ever greater heights above the bed surface, and there exists a particle density gradient extending some distance above the surface. For sufficiently tall freeboards there will be a certain height at which the density gradient eventually falls to zero and above this height the entrainment flux will be constant. This height is called the "transport disengaging height" or TDH. If the bed solids have a wide size distribution and the gas velocity in the freeboard exceeds the terminal fall velocity of the smaller ones, then these will be carried out of the system or "elutriated."

There has been a large number of experimental studies of entrainment and elutriation, and as in turbulent and fast beds there are considerable areas of disagreement amongst them. The main reason for this is the often wide disparity between equipment scale (particularly bed diameter) and the size and size distribution of the particles investigated. Comprehensive reviews have been given by Kunii and Levenspiel (1991) and Tasirin and Geldart (1998).

It is conventional following Leva (1951) and Yagi and Kunii (1955) to consider elutriation to be a first-order process such that the rate of elutriation of particles within a particular size range, d_{pi}, is directly proportional to the mass fraction of that size range, x_i, in the bed. Thus

$$-\frac{1}{A_t}\frac{d}{dt}(x_i M) = \kappa_i^* x_i$$

(37)

where A_t is the bed cross-sectional area, M is the mass of particles in the bed and κ_i^* is the elutriation rate coefficient with units of $\mathrm{kg\,m^{-2}\,s^{-1}}$.

From Eq. (37) it may be seen that

$$x_i = x_{i0} \exp\left(-\frac{\kappa_i^* A_t t}{M}\right)$$

(38)

where x_{i0} is the initial mass fraction of the particles at time zero, so that the higher the value of κ_i^* the greater will be the rate of removal of particles from the bed. Many empirical correlations exist giving κ_i^* in terms of the physical properties of gas and particles (Kunii and Levenspiel, 1991), but all should be applied with caution outside the range of conditions under which they were established.

It is clear from an examination of these empirical correlations that the terminal fall velocity is an important factor in determining the value of the elutriation rate coefficient for any given size of particle. It is also apparent that the rate coefficient will increase as u_t decreases, so that increasing pressure would be expected to increase the rate of elutriation. Chan and Knowlton (1984) studied elutriation from a bed of Ottawa sand with a wide size distribution at pressures of up to 31 bar. They found the solids entrainment rate to increase significantly with increasing pressure and fluidizing gas velocity (Fig. 6).

The elutriation rate coefficient was found to be linearly proportional to the gas density up to an operating pressure of 20.7 bar but to increase rapidly and in a nonlinear manner at higher pressures. At pressures of up to 3.5 bar, the total entrainment rate was propor-

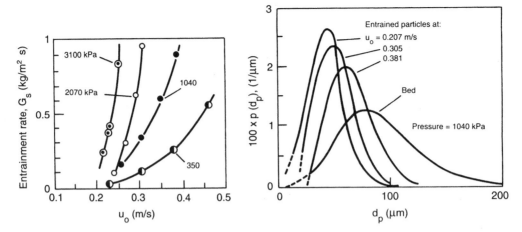

Figure 6 Effect of pressure and fluidizing gas velocity on solids entrainment based on the data of Chan and Knowlton (1984). (From Kunii and Levenspiel, 1991.)

tional to gas velocity to the power 4.1 but increased sharply to the 8.4 power of velocity at a pressure of 31 bar. Chan and Knowlton also observed that particles were entrained at fluidizing gas velocities lower than their terminal fall velocity and attributed this to momentum transfer from smaller particles to bigger ones, causing them to be carried over at velocities lower than the corresponding u_t. A similar observation had been made previously at ambient pressure by Geldart et al. (1979). Pemberton and Davidson (1983) also studied the effect of increased pressure on solids entrainment from bubbling beds and found a similar trend in the entrainment rate coefficient to that reported by Chan and Knowlton (1984). They attributed the increase in entrainment to the fact that at high pressures gas bubbles within a bed are generally smaller than for the same volumetric flowrate at ambient pressure; and that the entrainment flux produced by particle ejection from the roof of a bursting bubble is given by

$$E = \frac{3d_p\rho_p(1 - \varepsilon_{mf})(U - U_{mf})}{d_e} \qquad (39)$$

where d_e is the diameter of the sphere with the same volume as that of the bubble. Thus as d_e decreases, E increases. Alternatively, ejected particles could originate from the wakes of two bubbles that coalesce close to the surface. In this case the entrainment flux is independent of bubble diameter and so would be expected to be independent of pressure.

The effect on entrainment of increasing temperature was studied by Findlay and Knowlton (1985). They increased gas temperature while maintaining the

system pressure constant, so that the dominant effect was an increase in gas viscosity. This would also have the effect of decreasing terminal fall velocity and so increasing entrainment at a given fluidizing velocity; this is just what was observed, as may be seen from Fig. 7.

7 HEAT TRANSFER

7.1 Bubbling Beds

Their excellent heat transfer properties constitute one of the more attractive features of fluidized beds from a design point of view. Transfer between gas and parti-

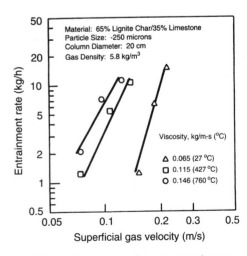

Figure 7 Effect of gas viscosity on entrainment. (From Findlay and Knowlton, 1985 and Knowlton, 1992.)

cles is normally very efficient, largely as a result of the high surface area of the particulate phase; a cubic metre of particles of diameter 100 μm has a surface area of the order of 30,000 m^2 (Botterill, 1986). Gas-to-particle transfer is thus rarely a limiting factor and will not be discussed further.

More interesting from a practical point of view is heat transfer between the bed material and immersed surfaces, and a number of studies of the effects of temperature and pressure on the mechanism of the process have been published. It is generally accepted that the overall heat transfer coefficient, h, between an immersed surface and a gas-fluidized bed can be expressed as the sum of three components:

$$h = h_{pc} + h_{gc} + h_r \qquad (40)$$

where h_{pc}, h_{gc}, and h_r are the particle convective, gas convective, and radiative transfer coefficients, respectively. The radiative component is of significance only above about 600°C in bubbling beds but becomes important at lower temperatures in the lean-flow risers of circulating systems (see below). The gas convective term is of importance only for beds of large Group B and Group D materials.

For Group A and small Group B powders h_{gc} is insignificant under normal operating conditions, so in the absence of radiation effects it is h_{pc} that dominates the heat transfer process. Mickley and Fairbanks (1955) first pointed out that bed-to-surface heat transfer is an unsteady state process in which "packets" of emulsion-phase material carry heat to or from the surface residing there for a short period of time before moving back into the bulk of the bed and being replaced by fresh material. On the basis of such a model the rate of heat transfer will be a maximum at the instant of contact but will decrease as the residence time of the packet at the surface increases and the local temperature gradient is reduced; when the surface is surrounded by a gas bubble, however, the transfer rate will fall rapidly to a very low value. The motion of the packets is of course directly related to the flow of gas bubbles through the bed, and their surface scouring action will increase as bubble flow increases up to a point where bubbles are the dominant phase; the rate of heat transfer by this mechanism would thus be expected to rise to a maximum value before falling away. The Mickley–Fairbanks model gives a value for the instantaneous heat transfer coefficient as

$$h_i = \left[\frac{(k_{mf}\rho_{mf}C_{mf})}{\pi\tau}\right]^{1/2} \qquad (41)$$

where k_{mf}, ρ_{mf}, and C_{mf} are the thermal conductivity, density, and heat capacity of the emulsion phase, respectively, and τ is the residence time of the packet at the wall. A time averaged heat transfer coefficient, h_p, may be obtained from Eq. (41) if the residence time distribution of the packets is known (Xavier and Davidson, 1985):

$$h_p = \int_0^\tau \frac{h_i\,dt}{\tau} \qquad (42)$$

$$= 2\left[\frac{k_{mf}\rho_{mf}C_{mf}(u - u_{mf})}{\pi L}\right]^{1/2} \qquad (43)$$

where for a bubbling bed L is half the equivalent bubble diameter and $\rho_{mf}C_{mf}$ may be assumed equal to $\rho_s(1 - \varepsilon_{mf})C_s$. In a bubbling bed, the presence of the bubbles must be allowed for in calculating h_{pc}:

$$h_{pc} = h_p(1 - \varepsilon_b) \qquad (44)$$

where ε_b is the volume fraction of bubbles in the bed.

The Mickley–Fairbanks model was found to over-predict heat transfer coefficients, and Baskakov (1964) suggested that this was due to an additional resistance, $1/h_f$, caused by the presence of a thin gas layer between the packet and the wall.

If the two resistances act independently and in series, then

$$h_{pc} = \frac{1}{[(1/h_p) + (1/h_f)]}(1 - \varepsilon_b) \qquad (45)$$

where, assuming the two-phase theory to apply

$$(1 - \varepsilon_b) = \frac{u_b}{(u - u_{mf} + u_b)} \qquad (46)$$

and

$$h_f = \frac{mk_g}{d_p} \qquad (47)$$

k_g being the gas thermal conductivity and u_b being the rise velocity of a single isolated bubble. Xavier and Davidson (1985) showed that literature values for m ranged from 4 to 10; they recommended a value of 6 for design purposes.

The effective conductivity of the emulsion phase may be expressed as

$$k_{mf} = k_e^0 + 0.1\rho_g C_g d_p u_{mf} \qquad (48)$$

k_e^0 being the conductivity of a fixed bed containing a stagnant gas.

The particle convective heat transfer coefficient may then be found from the physical properties of the system using Eqs. (48, 50–52), and (53) (Xavier and Davidson, 1985).

Examination of these relationships will show the expected effects of pressure and temperature.

7.1.1 Effect of Pressure

The main effect of increasing pressure will be to raise the gas density ρ_g, but this will have only a slight effect on the particle convective term through its influence on k_{mf} [Eq. (48)]. In the case of small particles, however, the suppression of bubbling caused by increasing pressure (see above) would be expected to increase the heat transfer coefficient by improving the "quality" of fluidization near the transfer surface. This was observed experimentally by Borodulya et al. (1980), who found an increase of 30% in the maximum heat transfer coefficient for 0.126 mm sand between 6 and 81 bar pressure.

For larger particles, the effect of pressure is to increase the gas convective component of the transfer coefficient, since in the case of these materials

$$h_{gc} \propto \mathrm{Re}_{mf} \tag{49}$$

Experimental confirmation of this trend comes from the work of Botterill and Desai (1972), Botterill and Denloye (1978a,b), Staub and Canada (1978), Canada and McLaughlin (1978), Borodulya et al. (1980), and Xavier et al. (1980).

The effect of pressure on bed-to-immersed-tube heat transfer was investigated by the Chalmers University group (Olssen and Almstedt, 1995), who found a significant increase in the bed-to-tube heat transfer coefficient with increasing pressure. Results from a previous study with the same bed showed a decrease in tube erosion with increasing pressure (Wiman et al., 1995). The work showed that while the convective heat transfer is most strongly coupled to the local bubble frequency, the erosion is most strongly coupled to the local bubble rise velocity and the local bubble flow rate.

7.1.2 Effect of Temperature

Increasing temperature has two effects (1) by decreasing gas density, the gas convective component of heat transfer is decreased slightly, and (2) by increasing the thermal conductivity of the gas, the effectivness of packets of emulsion phase in contact with the transfer surface is increased. The overall effect for Group A and B powders is to increase the convective transfer coefficient as was shown by Botterill and Teoman (1980). In the case of Group D powders where the gas convective component is dominant, an increase in temperature actually causes a decrease in the heat transfer coefficient. Botterill et al. (1981) showed that the radiant component of Eq. (40) begins to be important above 600°C, this was supported by the subsequent study of Ozkaynak et al. (1983).

7.2 Circulating Beds

In circulating fluidized bed (CFB) combustors, heat is transferred from the gas–solid mixture flowing through the riser to water-cooled surfaces in the wall. To model the convective heat transfer process, therefore, it is necessary to know, among other factors, the structure of the flowing suspension and its residence time at the wall. In a combustor operating at anything up to 1000°C, radiation heat transfer is also an important consideration and can in fact dominate over particle convection. The subject is thus complex, and we are still a long way from being able to describe the process in any detail. Reviews have been written by Grace (1986), Glicksman (1988), Leckner (1991), and Glicksman (1997).

The flow structure in CFB risers has been described above; it was shown to consist of a lean region in the center with a higher density annulus at the wall. The solid–gas suspension in the annulus is believed to move down the wall and to be responsible for transferring heat by convection. In CFB boilers, the risers are normally square in cross section with membrane–tube walls through which the heat transfer fluid circulates. In large units (> 50 MWth) the walls do not provide sufficient cooling surface, and it is necessary to immerse tube bundles in the riser (Leckner, 1991). In bubbling pressurized fluid bed combustors, heat is transferred to densely packed tubes that significantly alter the flow pattern compared to freely bubbling beds. There are few reports on this in the literature, but work by Olsson et al. (1995) adddresses the problem. Given the core-annular flow pattern in risers, it is to be expected that models developed for bed-to-surface heat transfer in bubbling beds can be applied in a suitably modified form to CFBs. Subarao and Basu (1986) used a cluster model similar to that of Mickley and Fairbanks (1955) and assumed that at any given time the fraction of surface covered by particle clusters is f and the fraction uncovered is $(1 - f)$. The average heat transfer coefficient can then be written as

$$h = h_c f + h_d (1 - f) \tag{50}$$

where h_c and h_d are the time-averaged heat transfer coefficients for clusters at the wall and in the dilute phase, respectively. Lints and Glicksman (1993) used the Baskakov model of heat transfer in bubbling beds [Eq. (45)] to derive an expression for h_c in a CFB as

$$h_c = \left[\frac{1}{h_f} + \frac{1}{h_e}\right]^{-1} = \left[\frac{\delta d_p}{k_g} + \sqrt{\frac{\pi\tau}{k_c c_{ps}\rho_s(1-\varepsilon_c)}}\right]^{-1} \tag{51}$$

where h_e is the transfer coefficient due to conduction within a cluster, δ is the ratio of film thickness to particle diameter, c_{ps} and ρ_s are, respectively, the heat capacity and density of the solid particles, ε_c is the cluster voidage, and k_c is the thermal conductivity of the cluster. If the thermal conductivity of the cluster is estimated using Eq. (48) then Eqs. (50) and (51) have five parameters that must be determined in order to predict the overall heat transfer coefficient: δ, τ, ε_c, f, and h_d. Lints and Glicksman (1993) discuss studies carried out to determine these parameters and show that experiment and model predictions agree quite well for laboratory scale CFBs. They sound a note of caution, however:

Although the same basic model of heat transfer should apply and the same hydrodynamic parameters should govern the process in larger beds the relation of those parameters within the wall layer to the average conditions across the entire bed can be expected to change substantially between beds of different size. This remains a critical area for further study.

The influence of the radiative component of the heat transfer coefficient was studied by Wu et al. (1989), using a refractory lined reactor column 7.32 m high and 152 × 152 mm in cross section. The circulating bed material was coarse sand, and the temperature range covered was 340–880°C. Average heat transfer coefficients to both a vertical tube and a membrane wall were found to increase almost linearly with suspension density and with temperature.

Radiation was found to play a significant role, especially at high temperatures and low suspension densities. The radiative component was estimated by treating the gas–solids suspension as a gray body such that

$$h_{rad} = \frac{\sigma(T_{susp}^4 - T_{surf}^4)}{(1/e_{susp} + 1/e_{surf} - 1)(T_{susp} - T_{surf})} \tag{52}$$

where T_{susp} and T_{surf} are the temperatures of the suspension and tube surface, respectively. The emissivity values, e_{susp} and e_{surf}, were set at 0.91 and the estimated

and experimental heat transfer coefficients were found to be in good agreement (Table 4).

A cross plot of the overall transfer coefficient against suspension temperature for the two transfer surfaces shows that the relative magnitude of increase of the radiation component is different for the two geometries owing, the authors conclude, to the better view factor of the vertical tube.

7.3 Application of Dimensional Analysis

The foregoing discussion of fluid bed heat transfer was based on theories of the hydrodynamic mechanism by which bed solids and gas come into contact with an immersed surface. An alternative approach to the problem is to arrange the relevant variables into dimensionless groups, thereby enabling behavior to be predicted over a wide range of operating conditions. An example of this is provided by the work of Molerus (1992a,b, 1993), who applied dimensional analysis to identify seven relevant groups based on the physical properties of the particulate material and the fluidizing gas:

$$\pi_1 = \frac{d_p^3 g(\rho_p - \rho_g)^2}{\mu^2}$$

$$\pi_2 = \frac{\rho_p}{\rho_g}$$

$$\pi_3 = \frac{C_g\mu}{k_g} \equiv Pr$$

$$\pi_4 = \frac{C_p\mu}{k_g} \tag{53}$$

$$\pi_5 = \frac{k_p}{k_g}$$

$$\pi_6 = \frac{hd_p}{k_g} \equiv Nu$$

$$\pi_7 = \sqrt[3]{\frac{\rho_p c_p}{k_g g}}(u - u_{mf})$$

where in addition to the normal symbols k represents thermal conductivity ($Wm^{-1}K^{-1}$). The maximum heat transfer coefficient in the laminar flow regime is then shown to be

$$\frac{h_{max}l_1}{k_g}\left(1 + \frac{k_g}{2C_p\mu}\right) = 0.09 \tag{54}$$

Table 4 Comparison of Experimental Data with Estimated Heat Transfer Coefficient for a Vertical Tube

kg/m³	°C			W/m² K			
Susp Dens	Avg susp temp	Tube surf temp	Gas conv comp	Particle conv comp	Rad comp	Total est h	Exp data h
15	701	83	13	19	68	100	106
	587	56	13	19	48	80	89
	343	45	14	19	22	55	62
60	701	83	13	76	68	157	166
	587	56	13	76	48	137	144
	343	45	14	76	22	112	110

Source: Wu et al., 1989

where l_1 is a laminar flow length scale defined by

$$l_1 = \left[\frac{\mu}{\sqrt{g}(\rho_p - \rho_g)}\right]^{2/3} \tag{55}$$

For Archimedes numbers in the range $10^5 < \text{Ar} < 10^8$ the method gives

$$\text{Nu}_{max} = 0.146\text{Pr}^{1/3} \tag{56}$$

and for $\text{Ar} > 10^8$,

$$\text{Nu}_{max} = 0.02469\text{Ar}^{0.4304}\text{Pr}^{1/3} \tag{57}$$

The predictions agree well with observations for both bubbling and circulating beds in the absence of radiation effects. For a more detailed account, the reader is referred to the paper by Molerus (1993).

8 SINTERING AND AGGLOMERATION

With certain fluidized bed materials, as the temperature of the bed is increased, a point is reached at which the particles begin to sinter by a softening of the surface and the formation of interparticle bonds. This temperature, which is often lower than the fusion temperature of the bulk material, is called the "minimum sintering temperature", T_s, and beds operated at temperatures higher than this can suffer catastrophic defluidization through large-scale agglomeration of particles. On the other hand, increasing the fluidizing velocity at bed temperatures in excess of T_s can prevent defluidization, as can increasing the initial size of the particles. Whether beds defluidize would seem to depend on a balance between the cohesiveness or "stickiness" of the particle surfaces and the kinetic

energy of the particles due to the fluid forces acting on them. Tardos et al. (1985a) stated that "basic knowledge in the area is so limited that industry relies mostly on empiricism to avoid defluidization of beds containing sticky particles," and despite the increased number of publications on the subject since that date, there is still no mechanistic model available that can reliably predict under what conditions defluidization will occur. A number of attempts at modeling the process have been made, however, and these will be reviewed following a brief description of sintering and ways of measuring it.

Sintering occurs as a result of the migration of lattice vacancies or the movement of atoms in the surface of particles (Siegell, 1984). If the rate of sintering is sufficiently high, then when particles of the same material come into contact, bonds will be formed between them leading to the formation of agglomerates. The temperature at which this occurs, T_s, can be measured with a dilatometer in which the relative expansion–contraction of a sample of granular material under load is measured as its temperature is increased (Compo et al., 1984). The change in length or dilation, ΔL, of a sample of glass beads divided by its initial length, L_0, is recorded and plotted as a function of temperature. An initial increase in the length ratio due to thermal expansion of the bulk material is observed, but at a higher temperature, as the surface begins to soften, and contraction counteracts expansion, the dilation ceases, and the curve shows a plateau. The temperature at which the curve turns sharply downward and at which the sample contracts at a rapid rate is the sintering temperature T_s.

Compo et al. (1987) combined dilatometer measurements with defluidization experiments for a variety of materials and concluded that the materials can be divided into two categories, (1) those such as glass beads, coal powder, and polyolefin granules that agglomerate quickly after T_s has been reached and the increase in gas velocity needed to keep the bed fluidized increases exponentially, and (2) the ionic solids such as sodium chloride and calcium chloride that form relatively weak agglomerates and for which only a slight increase in fluidizing velocity is necessary to restore fluidization once agglomeration has begun. None of the materials studied was found to defluidize below its minimum sintering temperature.

A system in which defluidization can present a particular problem is the combustion of low-rank coals in a circulating fluid bed (CFB) combustor. Some low-rank coals have a high content of sodium and sulphur, and at operating temperatures in the region of 850°C these are converted to molten alkali sulphates, which are deposited along with coal ash on particles of the bed material, usually sand or something similar, causing them to agglomerate and defluidize. The processes taking place are complex and poorly understood, although experimental studies by Manzoori and Agarwal (1993, 1994) have clarified some of the details.

Mechanistic models of the defluidization process have been explored by a number of authors and principally by Tardos and Pfeffer and their coworkers at City College New York. Thus Tardos et al. (1985a) set out to predict the limiting gas velocity U_s, necessary to break the largest agglomerate in the bed and thereby to keep a bed of sticky particles fluidized at temperatures above the minimum sintering temperature. They looked for conditions in which the dynamics of the bonding forces holding the agglomerates together and the forces leading to breakup (due to the motion of gas bubbles in the bed) are in equilibrium. For simplicity they assumed that the largest agglomerate occupied the entire cross section of the bed but conceded that in large beds defluidization would be likely to occur before this size was reached; in these cases the predicted limiting velocity, U_s, should be considered a lower bound for the fluidizing gas velocity. Further simplifying assumptions were that the agglomerates were cylindrical in shape and were fixed in the bed and not freely buoyant, and that the forces acting on them due to the flow of interstitial gas were negligible, the main force being shear due to the motion of bubbles. Based on an equation of Livshits et al. (1978) the maximum force, G_v, acting on the agglomerate was found in terms of the excess velocity $(u - u_{mf})$ and related to the pressure, q, acting on the bottom face of the cylinder by

$$q = \frac{4G_v}{\pi d_{ag}^2} \tag{58}$$

where d_{ag} is the diameter of the agglomerate. This pressure will cause failure of the structure if its value exceeds q_{max} where

$$q_{max} = \sigma_y \left(\frac{2h}{d_{ag}}\right)^2 A_1 \tag{59}$$

in which σ_y is the yield strength of the agglomerate and A_1 is a coefficient approximately equal to 2. From these equations a relationship is found between the excess velocity and the yield strength of the agglomerate, a quantity that was calculated on the basis of a technique devised by Rumpf (1977). The result is a rather complicated expressions, the evaluation of

which requires knowledge of two fundamental physical quantities, the surface viscosity η_s and the yield strength of a sinter neck σ_s, both as functions of temperature. The surface viscosity can be found from dilatometer data, and the yield strength can be either estimated using adhesion theory or calculated from experimental data (Tardos et al., 1985b).

An alternative approach based on the concept of the dissipation of the kinetic energy of particle collisions via surface viscosity has been studied by Ennis et al. (1991). It is particularly applicable to fluid bed granulation through the introduction of liquid binding agents but does appear to have some relevance for high temperature defluidization. The mechanism of coalescence is considered to be a function of a binder Stokes number St_v defined as

$$St_v = \frac{2Mu_0}{3\pi\mu a^2} \tag{60}$$

where M is the mass of a particle, u_0 is the initial relative velocity of two particles, a is the particle radius, and μ is the surface viscosity. For two colliding particles to rebound, the Stokes number must exceed a critical value given by

$$St_v^* = \left(1 + \frac{1}{e'}\right) \ln\left(\frac{h'}{h_a}\right) \tag{61}$$

where e' is the coefficient of restitution of the particles, h' is the thickness of the binder on the surface, and h_a is the characteristic length of surface asperities. Particles colliding with Stokes numbers less than this critical value are considered to have coalesced. The theory compared favorably with results of the defluidization experiments reported by Gluckman et al. (1975).

Seville et al. (1998) also identified two types of adhesion leading to agglomeration: viscoplastic which occurs with glassy materials, and a second type caused by the formation of an excessive amount of liquid on a surface resulting from melting or chemical reaction. Both types can lead to defluidization. The authors assumed that particles in a fluidized bed remain in close proximity in quiescent zones with little relative motion until they are disturbed by bubbles; the residence time, τ_b, of particles in these zones may be sufficiently prolonged to allow sintering to occur. They modeled the behavior of viscoplastic materials on the basis of τ_b and a second characteristic time, τ_s, which is that needed to form a strong sintered bond. Then equating the sintering time with the time required to turn the bed material over once, which is derived from two-phase theory as approximately

$$\tau_{b0} = \frac{H_{mf}}{(u - u_{mf})} \qquad (62)$$

which gives a relationship between the critical excess velocity required just to break up the agglomerates and the surface viscosity of the sintering particles:

$$u - u_{mf} = \frac{K_1 K_2}{\mu_0 \exp(E/RT)} \qquad (63)$$

Here K_1 is approximately equal to H_{mf} and K_2 is the critical size of the sinter neck. The denominator in Eq. (63) represents the temperature dependence of the surface viscosity. Favorable comparisons were drawn between the theory and defluidization experiments with polyethylene granules over a range of temperatures.

As was stated at the beginning of this chapter, practically all fluidized bed reactors operate at temperatures above the ambient, and yet the effect of temperature on the surface properties of fluidized solids is far from well understood. The work reported by the above authors has made a valuable contribution to our understanding, but the whole question of interparticle forces in fluidized beds and the effect on them of operating temperature and pressure are areas much in need of further study.

NOTATION

A, A_t = cross-sectional area, m^2
Ar = Archimedes number
C = heat capacity, $J\,kg^{-1}K^{-1}$
C_D = drag coefficient
D_f = bed diameter, m
D_g = gas diffusivity, $m^2\,Pa\,s^{-1}$
d = diameter, m
d_e = diameter of equivalent sphere, m
E = entrainment flux, $kg\,m^{-2}\,s^{-1}$
E_{mb} = modulus of elasticity, $N\,m^{-2}$
e = emissivity
e' = coefficient of restitution
Ga = Galileo number
G_v = force [Eq. (58)], N
g = acceleration due to gravity, $m\,s^{-2}$
h = heat transfer coefficient, $W\,m^{-2}\,K^{-1}$
h' = binder thickness, m
h_a = length of surface asperities, m
K = bubble velocity coefficient
k = thermal conductivity, $W\,m^{-1}\,K^{-1}$
k_c = chemical rate coefficient, $m\,s^{-1}$
L = height, m
M = mass, kg

m = constant in Eq. (47)
Nu = Nusselt number
Pr = Prandtl number
p = pressure, Pa
Re = Reynolds number
Sh = Sherwood number
St = Stokes number
T = temperature, K
t_c = particle burnout time, s
U, u_0 = superficial velocity, $m\,s^{-1}$
U_c = transition velocity, $m\,s^{-1}$
U_k = transition velocity, $m\,s^{-1}$
U_s = slip velocity, $m\,s^{-1}$
U_{tr} = transition velocity, $m\,s^{-1}$
u = velocity, $m\,s^{-1}$
u_t = terminal fall velocity, $m\,s^{-1}$
W = width, m
X = crossflow factor
x = mass fraction
z = vertical coordinate, m
z_i = location of point of inflection [Eq. (35)], m

Greek Symbols

α, γ = stoichiometric coefficients
δ = bed expansion ratio
ε = voidage
ϕ = sphericity
κ_i = elutriation rate coefficient, $kg\,m^{-2}\,s^{-1}$
μ = viscocity, $N\,s\,m^{-2}$
ρ = density, $kg\,m^{-3}$
σ = yield strength, $N\,m^{-2}$
τ = residence time, s

Subscripts

b = bubble
c = cluster
cf = complete fluidization
D = dense phase
f = fluid
fl = fluidized
g = gas
gc = gas convective
max = maximum
mb = minimum bubbling
m = minimum fluidization
o, or = orifice
p = particle
pc = particle convective
r = radiant
s = solids
t = terminal fall

REFERENCES

Barnea E, Mizrahi J. A generalised approach to the fluid dynamics of particulate systems: general correlation of fluidization and sedimentation in solid multiparticle systems. Chem Eng J 5:171, 1973.

Barnea E, Mednick RL. Correlation of minimum fluidization velocity. Trans I Chem E 53:278, 1975.

Barreto GF, Yates JG, Rowe PN. The effect of pressure on the flow of gas in fluidized beds of fine particles. Chem Eng Sci 38:1935–1945, 1983.

Baskakov AP. The mechanism of heat transfer between a fluidized bed and a surface. Int Chem Eng 4:320, 1964.

Borodulya VA, Ganzha VG, Podberezsky AI. Heat transfer in a fluidized bed at high pressure. In: Fluidization (JR Grace, JM Matsen, eds.). Plenum Press, New York, 1980, 201–207.

Borodulya VA, Ganzha VL, Podberezsky AI, Upadhyay SN, Saxena SE. High pressure heat transfer investigations for fluidized beds of large particles and immersed vertical tube bundles, Int J of Heat and Mass Transfer 26:1577–584, 1983.

Botterill JSM. Fluid bed heat transfer. In: Gas Fluidization Technology (D Geldart, ed.). John Wiley, Chichester, 1986.

Botterill JSM, Denloye AOO. Bed to surface heat transfer in a fluidized bed of large particles. Powder Tech 19:197–203, 1978a.

Botterill JSM, Denloye AOO. A theoretical model of heat transfer to a packed or quiescent fluidized bed. Chem Eng Sci 33:509–515, 1978b.

Botterill JSM, Desai M. Limiting factors in gas-fluidized bed heat transfer. Powder Tech 6:231–238, 1972.

Botterill JSM, Teoman Y. Fluid-bed behaviour at elevated temperatures. Fluidization 5 (JR Grace, JM Matsen, eds.). Plenum Press, New York, 1980, 93–100.

Botterill JSM, Teoman Y, Yuregir KR. Temperature effects on the heat transfer behaviour of fluidized beds. AIChE Symp Ser 77:(208)330–340, 1981.

Botterill JSM, Teoman Y, Yuregir KR. The effect of operating temperature on the velocity of minimum fluidization bed voidage and general behaviour. Powder Technology 31:101–110, 1982.

Cai P, Chen SP, Jin Y, Yu ZQ, Wang ZW. Effect of temperature and pressure on the transition from bubbling to turbulent fluidization. AIChE Symp Ser 85:(270)37–43, 1989.

Campos JBLM, Guedes de Carvalho JRF. Drag force on the particles at the upstream end of a packed bed and the stability of the roof of bubbles in fluidized beds. Chem Eng Sci 47:4057–4062, 1992.

Canada GS, MacLaughlin MH. Large particle fluidization and heat transfer at high pressure. AIChE Symp Ser 74:(176)27–37, 1978.

Carsky M, Hartman M, Ilyenko BK, Makhorin KE. The bubble frequency in a fluidized bed at elevated pressure. Powder Tech 61:251–254, 1990.

Chan IH, Knowlton TM. The effect of pressure on entrainment from bubbling gas fluidized beds. In: Fluidization (D Kunii, R Toei, eds.). Engineering Foundation, New York, 1984, pp 283–290.

Chan IH, Shishtla C, Knowlton TM. The effect of pressure on bubble parameters in gas fluidized beds. Powder Tech 53:217–235, 1987.

Chitester DC, Kornosky RM, Fan LS, Danko JP. Characteristics of fluidization at high pressure. Chem Eng Sci 39:253–261, 1984.

Clift R, Grace JR, Weber ME. Stability of bubbles in fluidized beds. I & E C Funds 13:45–51, 1974.

Clift R, Grace JR, Weber ME. Bubbles, drops and particles, Academic Press, New York, 1978.

Compo P, Pfeffer R, Tardos GI. Minimum sintering temperatures and defluidization characteristics of fluidizable particles. Powder Tech 51:85–101, 1987.

Compo P, Tardos GI, Mazzone D, Pfeffer R. Minimum sintering temperatures of fluidizable particles. Part Charact 1:171–177, 1984.

Crowther ME, Whitehead JC. Fluidization of fine particles at elevated pressure. In: Fluidization (JF Davidson, DL Kearins, eds.). Cambridge University Press, 1978, pp 65–70.

Davidson JF, Harrison D. Fluidised Particles, Cambridge University Press, 1963.

Denloye AOO, Botterill JSM. Bed to surface heat transfer in a fluidized bed of large particles. Powder Tech 19:197–203, 1978.

Ennis BJ, Tardos GI, Pfeffer R. A microlevel-based characterization of grannulation phenomena. Powder Tech 65:257–272, 1991.

Formisani B, Girimonte R, Mancuso L. Analysis of the fluidization process of particle beds at high temperature. Chem Eng Sci 53:951–962, 1998.

Findlay JG, Knowlton TM. Final Report for US Dept. of Energy, Project DE-AC21-83MC20314, 1985.

Fletcher JV, Deo MD, Hanson FV. Fluidization of a multi-sized Group B sand at reduced pressure. Powder Tech 76:141–147, 1993.

Foscolo PU, Gibilaro LG. A fully predictive criterion for the transition between particulate and aggregate fluidization. Chem Eng Sci 39:1667–1675, 1984.

Geldart D. Types of gas fluidization. Powder Tech 7:285–292, 1973.

Geldart D, Cullinan J, Georghiades S, Gilvray D, Pope DJ. The effect of fines on entrainment from gas fluidized beds. Trans I Chem E 57:269–275, 1979.

Geldart D, Kapoor DS. Bubble sizes in a fluidized bed at elevated temperatures. Chem Eng Sci 31:842–843, 1976.

Glicksman LR. Circulating fluidized bed heat transfer. In: Circulating Fluidized Bed Technology 2 (P Basu, JF Large eds.). Pergamon Press, Oxford, 1988.

Glicksman LR. In: JR Grace, AA Avidan, TW Knowlton, eds. Circulating Fluidized Beds. Blackie, London, 1997, pp 261–311.

Gluckman MJ, Yerushalmi J, Squires AM. Defluidization characteristics of sticky or agglomerating beds. In: Fluidization Technology (DL Kearins, ed.) Vol 2. Hemisphere, New York, 1975, pp 395–422,.

Goddard K, Richardson JF. The behaviour of bubble-free fluidized beds. I Chem E Symp Ser No. 30:126–135, 1968.

Grace JR, Avidan AA, Knowlton TM, (eds.). Circulating Fluidized Beds. Blackie, London, 1997.

Grace JR, Lim CT. Permanent jet formation in beds of particulate solids. Can J Chem Eng 65:160–162, 1987.

Grace JR. Heat transfer in circulating fluidized beds. In: Circulating Fluidized Bed Techonolgy (P Basu, ed.). Pergamon Press, Oxford, 1986.

Guedes de Carvalho JRF, King DF, Harrison D. Fluidization of fine particles under pressure. In: Fluidization (JF Davidson, DL Kearins, eds.). Cambridge University Press, 1978, pp 59–64.

Guedes de Carvalho JRF, King DF, Harrison D. Mass transfer from a slug in a fluidized bed at elevated pressures. Chem Eng Sci 37:1087–1094, 1982.

Haider A, Levenspiel O. Drag coefficient and terminal velocity of spherical and non-spherical particles. Powder Tech 58:63–70, 1989.

Harrison D, Davidson J, De Kock JW. On the nature of aggregative and particulate fluidization. Trans Inst Chem Engrs 39:202–211, 1961.

Hartman M, Yates JG. Free-fall of solid particles through fluids. Collect Czech Chem Commun 58:961–982, 1993.

Hirsan I, Shishtla C, Knowlton TM. The effect of bed and jet parameters on vertical jet penetration length in gas fluidized beds. Paper presented at 73rd Annual AIChE Meeting, Chicago, Illinois, Nov 16–20, 1980.

Hoffmann AC, Yates JG. Experimental observations of fluidized beds at elevated pressures. Chem Eng Commun 41:133–149, 1986.

Jacob KV, Weimer AW. High pressure particulate expansion and minimum bubbling of fine carbon powders. AIChE J 33:1698–1706, 1987

Kai T, Furusaki S. Behaviour of fluidized beds of small particles at elevated temperatures. J Chem Eng Japan 18:113–118, 1985.

Kawabata J, Yumiyama M, Tazaki Y, Honma S, Chiba T, Sumiya T, Ehdo K. Characteristics of gas fluidized beds under pressure. J Chem Eng Japan 14:85–89, 1981.

Khan AR, Richardson JF. Resistance to motion of a solid sphere in a fluid. Chem Eng Commun 62:135–150, 1987.

King DF, Harrison D. The bubble phase in high pressure fluidized beds. In: Fluidization (JR Grace, J M Matsen, eds.). Plenum Press, New York, 1980, 101–107.

King DF, Harrison D. The dense phase of a fluidized bed at elevated pressures. Trans I Chem E 60:26–30, 1982.

King DF, Mitchell FRG, Harrison D. Dense phase viscosities of fluidized beds at elevated pressures. Powder Tech 28:55–58, 1981.

Knowlton TM. High pressure fluidization characteristics of several particulate solids: primarily coal and coal-derived materials. AIChE Symp Ser 73:(161)22–28, 1977.

Knowlton TM, Hirsan I. The effect of pressure on jet penetration in semi-cylindrical gas-fluidized beds. In: Fluidization (JR Grace, JM Matesen, eds.). Plenum Press, New York, 1980, pp 315–324.

Kunii D, Levenspiel O. Fluidization Engineering. Butterworths, Boston, 1991.

Kusafabe K, Kuriyama T, Morooka S. Fluidization of fine particles at reduced pressure. Powder Tech 58:125–130, 1998.

Leckner B. Heat transfer in circulating fluidized bed boilers. Circulating Fluidised Technology 3 (P Basu, M Horio, M Hasatani, eds.). Pergamon Press, Oxford, 1991.

Leva M. Elutriation of fines from fluidized systems. Chem Eng Prog 47:(1)39–45, 1951.

Lints MC, Glicksman LR. Parameters governing particle-to-wall heat transfer in a circulating fluidized bed. Preprints 4th Int. Conf. on Circulating Fluidized Beds. AIChE J 350–355, 1993.

Livshits YY, Tamarin AI, Zabrodsky SS. Maximum forces acting on a body immersed in a fluidized bed. Fluid. Mech Sov Res 7:30, 1978.

Llop MF, Madrid F, Arnaldos J, Casal J. Fluidization at vacuum conditions. A generalized equation for the prediction of minimum fluidization velocity. Chem Eng Sci 51:5149–5158, 1996.

Lofstrand H, Almstedt AE. Andersson S. Dimensionless model for bubbling fluidized beds. Chem Eng Sci 50:245–253, 1995.

Manzoori AR, Agarwal PK. The role of inorganic matter in coal in the formation of agglomerates in circulating fluid bed combustors. Fuel 72:1069–1075, 1993.

Manzoori AR, Agarwal PK. Agglomeration and defluidization under simulated circulating fluidized-bed combustion conditions. Fuel 73(4):563–568, 1994.

Mickley HS, Fairbanks DF. Mechanism of heat transfer to fluidized beds. AIChE J 1:374–384, 1955.

Mii T, Yoshida K, Kunii D. Temperature effects on the characteristics of fluidized beds. J Chem Eng Japan 6:100–102, 1973.

Molerus O. Heat transfer in gas fluidized beds, Part 1. Powder Tech 70:1–14, 1992a.

Molerus O. Heat transfer in gas fluidized beds Part 2. Dependence of heat transfer on gas velocity. Powder Tech 70:15–20, 1992b.

Molerus O. Arguments on heat transfer in gas fluidized beds. Chem Eng Sci 48:761–770, 1993.

Olowson PA, Almstedt AE. Influence of pressure and fluidization velocity on the bubble behaviour and gas flow distribution in a fluidized bed. Chem Eng Sci 45:1733–1741, 1990.

Olowson PA, Almstedt AE. Influence of pressure on the minimum fluidization velocity. Chem Eng Sci 46:637–640, 1991.

Olowson PA, Almstedt AE. Hydrodynamics of a bubbling fluidized bed: influence of pressure and fluidization velocity in terms of drag force. Chem Eng Sci 47:357-366, 1992.

Olssen SE, Almstedt AE. Local instantaneous and time-averaged heat transfer in a pressurized fluidized bed with horizontal tubes: influence of pressure, fluidization velocity and tube-bank geometry. Chem Eng Sci 50:3231–3245, 1995.

Olssen SE, Wiman J, Almstedt AE. Hydrodynamics of a pressurized fluidized bed with horizontal tubes: influence of pressure fluidization velocity and tube bank geometry. Chem Eng Sci 50:581–592, 1995.

Otake T, Tone S, Kawashima M, Shibata T. Behaviour of rising bubbles in a gas-fluidized bed at elevated temperature. J Chem Eng Japan 8:388–392, 1975

Ozkaynak T, Chen JC, Frankenfield TR. An experimental investigation of radiation heat transfer in a high temperature fluidized bed. In: Fluidization (D Kunii, R Toei, eds.). Engineering Foundation, New York, 1983, pp 371–378.

Pemberton ST, Davidson JF. Elutriation of fine particles from bubbling fluidized beds. In: Fluidization (D Kunii, R Toei, eds.). Engineering Foundation, New York, 1983 pp 275–282.

Raso G, D'Amore M, Formisani B, Lignola PG. The influence of temperature on the properties of the particulate phase at incipient fluidization. Powder Tech 72:71–76, 1992.

Rhodes MJ. Upward flow of gas/solid suspensions, Part 2. A practical quantitative flow regime diagram for the upward flow of gas/solid suspensions. Chem Eng Res Des 67:30–37, 1989.

Rowe PN. The effect of pressure on minimum fluidization velocity. Chem Eng Sci 39:173–174, 1984.

Rowe PN. The dense phase voidage of fine powders fluidized by gas and its variation with temperature, pressure and particle size. In: Fluidization (JR Grace, LW Shemilt, MA Bergougnou, eds.). Engineering Foundation, New York, 1989, pp 195–202.

Rowe PN, Foscolo PU, Hoffmann AC, Yates JG. X-ray observation of gas fluidized beds under pressure. In: Fluidization (D Kunii, R Toei, eds.). Engineering Foundation, New York, 1983, pp 53–60.

Rowe PN, Foscolo PU, Hoffmann AC, Yates JG. Fine powders fluidized at low velocity at pressures up to 20 bar with gases of different viscosity. Chem Eng Sci 37:1115–1117, 1982.

Rowe PN, MacGillivray HJ. A preliminary x-ray study of the effect of pressure on a bubbling gas fluidized bed. Inst Energy Symp Ser 4:1–9, 1980.

Rowe PN, MacGillivray HJ, Cheesman DJ. Gas discharge from an orifice into a gas fluidized bed. Trans I Chem E 57:194–199, 1979.

Rumpf H. Particle Adhesion, in Agglomeration 1977 (KVS Shastry, ed.). AIME, 1997.

Saxena SC, Vogel GJ. The measurement of incipient fluidization velocities in a bed of coarse dolomite at temperature and pressure. Trans I Chem E 55:184–189, 1997.

Seville JPK, Siloman-Pflug H, Knight PC. Modelling of sintering in high temperature fluidized beds. Powder Tech 97:160–169, 1998.

Siegell JA. High temperature defluidization. Powder Tech 38:13–22, 1984.

Sittiphong N, George AH, Bushnell D. Bubble eruption diameter in a fluidized bed of large particles at elevated temperatures. Chem Eng Sci 36:1259–1260, 1981.

Sobreiro LEC, Monteiro JLF. The effect of pressure on fluidized bed behaviour. Powder Tech 33:95–100, 1982.

Staub FW, Canada GS. Effect of tube bank and gas density on flow behaviour and heat transfer in a fluidized bed. In: Fluidization (JF Davidson, DL Kearins, eds.). Cambridge University Press, 1978, pp 339–344.

Subarao D, Basu P. A model for heat transfer in circulating fluidized beds. Int J Heat Mass Transfer 29:487–489, 1986.

Subzwari MP, Clift R, Pyle DL. Bubbling behaviour of fluidized beds at elevated pressures. In: Fluidization (JF Davidson, DL Kearins, eds.). Cambridge University Press, 1978, pp 50–54.

Svoboda K, Hartman M. Influence of temperature on incipient fluidization of limestone, lime, coal ash and corundum. I and EC Proc Des Dev 20:319–326, 1981.

Tardos G, Mazzone D, Pfeffer R. Destabilization of fluidized beds due to agglomeration. Part 1: Theoretical model. Can J Chem Eng 63:377–383, 1985a.

Tardos G, Mazzone D, Pfeffer R. Destabilization of fluidized beds due to agglomeration. Part 2: Experimental verification. Can J Chem Eng 63:384–389, 1985b.

Tasirin SM, Geldart D. Entrainment of FCC from fluidized beds—a new correlation for elutriation rate constants. Powder Tech 95:240–247, 1998.

Taylor GI. The instability of liquid surfaces when accelerated in a direction perpendicular to their planes. Proc Roy Soc A201:192–196, 1950.

Tsukada M, Nakanishi D, Horio M. The effect of pressure on phase transition from bubbling to turbulent fluidization. Int J Multiphase Flow 19:27–34, 1993.

Upson PC, Pyle DL. The stability of bubbles in fluidized beds. In: Fluidization and Its Applications. Cépaduès-Éditions, Toulouse, 1973, pp 207–222.

Varadi T, Grace JR. High pressure fluidization in a two-dimensional bed. In: Fluidization (JF Davidson, DL Keairns, eds.). Cambridge University Press, 1978, pp. 55–58.

Verkooijen AHM, Rietema K, Thoenes D. The influence of pressure on the conversion in a fluidized bed. In: Fluidization (D Kunii, R Toei, eds.). Engineering Foundation, New York, 1983, pp 541–546.

Weimer AW, Quarderer GJ. On dense phase voidage and bubble size in high pressure fluidized beds of fine powders. AIChE J 31:1019–1028, 1985.

Wen CY, Yu YH. Mechanics of fluidization. Chem Eng Prog Symp. Ser 62:100–111, 1966.

Wiman J, Almstedt AE. Influence of pressure, fluidization velocity and particle size on the hydrodynamics of a freely bubbling fluidized bed. Chem Eng Sci 53:2167–2176, 1998.

Wiman J, Almstedt AE. Aerodynamics and heat transfer in a pressurized fluidized bed: Influence of pressure, fluidization velocity, particle size and tube geometry. Chem Eng Sci 52:2677–2696, 1997.

Wiman J, Mahpour B, Almstedt AE. Erosion of horizontal tubes in a pressurized fluidized bed: influence of pressure, fluidization velocity and tube bank geometry. Chem Eng Sci 50:3345–3356, 1995.

Wirth KE. Fluid mechanics of pressurized circulating fluidized beds. In: Fluidization (OE Potter, DJ Nicklin, eds.). Engineering Foundation, New York, 1992, pp 113–120.

Wu RL, Grace JR, Lim CJ, Brereton CMH. Suspension-to-surface heat transfer in a circulating fluidized bed combustor. AIChE J 35:1685–1691, 1989.

Wu SY, Baeyens J. Effect of the operating temperature on the minimum fluidization velocity. Powder Tech 67:217–220, 1991.

Xavier AM, Davidson JF. Heat transfer in fluidized beds. In: Fluidization (JF Davidson, R Clift, D Harrison, eds.). Academic Press, London, 1985.

Xavier AM, King DF, Davidson JF, Harrison D. Surface-bed heat transfer in a fluidized bed at high pressure. In: Fluidization (JR Grace, JM Matsen, eds.). Plenum Press, 1980, pp 209–216.

Yagi S, Kunii D. Studies on combustion of carbon particles in flames and fluidized beds. Proc. 5th Int. Symp., Combustion. Van Nostrand Reinhold, New York, 1955, pp 231–244.

Yamazaki R., Han NS, Sun ZF, Jimbo G. Effect of chemisorbed water on bed voidage of high temperature fluidized beds. Powder Tech 84:15–22, 1995.

Yang WC. Jet penetration in a pressurized fluidized bed. I and EC Fundams, 20:297–300, 1981.

Yang WC, Chitester DC, Kornosky RM, Keairns DL. A generalised methodology for estimating minimum fluidization velocity at elevated pressure and temperature. AIChE J 31:1086–1092, 1985.

Yates JG, Bejcek V, Cheesman DJ. Jet penetration into fluidized beds at elevated pressures. In: Fluidization (K Ostergaard, S Sorensen, eds.). Engineering Foundation, New York, 1986, pp 79–86.

Yates JG, Cheesman DJ. Unpublished results, 1987.

Yates JG. Effects of temperature and pressure on gas–solid fluidization. Chem Eng Sci 51:167–205, 1996.

Yoshida K, Ueno T, Kunii D. Mechanism of bed to wall heat transfer in a fluidized bed at high temperatures. Chem Eng Sci 29:77–82, 1974.

6

Gas Distributor and Plenum Design in Fluidized Beds

S. B. Reddy Karri

Particulate Solid Research, Inc., Chicago, Illinois, U.S.A.

Joachim Werther

Technical University Hamburg-Harburg, Hamburg, Germany

1 INTRODUCTION

The gas distributor (also called a grid) in a fluidized bed reactor is intended to induce a uniform and stable fluidization across the entire bed cross section, prevent nonfluidized regions on the grid, operate for long periods (years) without plugging or breaking, minimize weepage of solids into the plenum beneath the grid, minimize attrition of the bed material, and support the weight of the bed material during startup and shutdown. In practice, grids have taken a variety of forms, a few of which are discussed in subsequent pages. Whatever the physical form, all are fundamentally classifiable in terms of the direction of gas entry: upward, laterally, or downward. The choice depends on prevailing process conditions, mechanical feasibility, and cost. In the past, grid design has been more of an art than a science. However, more recent studies now allow grid design based on scientific principles.

2 TYPES OF GRIDS

2.1 Perforated Plates (Upwardly Directed Flow)

Main Advantages

Simple fabrication; most common; inexpensive; easy to modify hole size; easy to scale up or down; easy to

clean can be flat, concave, convex, or double dished; ports are easily shrouded.

Possible Disadvantages

Bed weepage to plenum; can be subject to buckling or thermal distortion; requires peripheral seal to vessel shell; requires support over long spans; high pressure drop required if weepage during operation is to be minimized

2.2 Bubble Caps and Nozzles (Laterally Directed Flow)

Main Advantages

Depending on the design, weeping is reduced or totally avoided; good turndown ratio; can incorporate caps as stiffening members; can support internals.

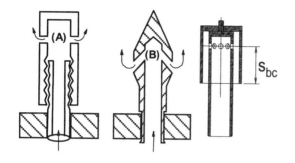

Possible Disadvantages

Expensive; difficult to avoid stagnant regions; more subject to immediate bubble merger; difficult to clean; difficult to modify; not advisable for sticky solids; requires peripheral seal; ports not easily shrouded.

Details of some nozzles that are currently used in circulating fluidized beds (CFB) combustors are shown in Fig. 1 (VGB, 1994). There are significant differences between bubble caps (No. 7 in Fig. 1) and nozzles (No. 1 in Fig. 1) with respect to the prevention of solids back flow: in the case of nozzles, the high velocity of the gas jet prevents the solids from flowing back into the wind box. On the other hand, in the case of the bubble cap design, the gas flowing out of the bubble cap into the bed has a rather low velocity. In this case, the backflow of solids is avoided by letting the gas flow downward from the holes in the inner tube to the lower

edge of the cap. The separation distance s_{bc} is responsible for the sealing effect of the bubble cap.

2.3 Sparger (Laterally or Downwardly Directed Flow)

Main Advantages

Can minimize weeping; good turndown ratio; low pressure drop; can support internals; can undergo thermal expansion without damage; ports are easily shrouded; well suited to multilevel fluid injection; solids can flow from above the grid to below.

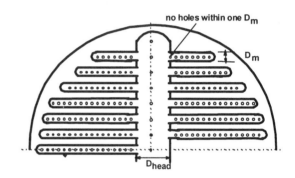

Possible Disadvantages

Defluidized solids beneath the grid; can be a less forgiving mechanical design.

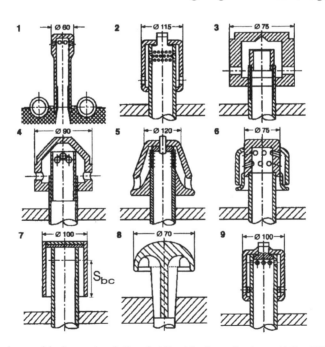

Figure 1 Distributors and nozzles used in large circulating fluidized bed combustors. (After VGB, 1994.)

2.4 Conical Grids (Laterally Directed Flow)

Main Advantages

Promotes solid mixing; prevents stagnant solids buildup; minimizes solids segregation. Facilitates the easy discharge of solids.

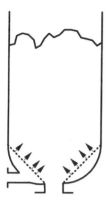

Possible Disadvantages

Difficult to construct; requires careful design to ensure good gas distribution; requires high pressure drop for good gas distribution.

2.5 Pierced Sheet Grids (Laterally Directed Flow)

Produced by punching holes in a relatively thin plate. Holes are of a semielliptical shape with slanting, strongly conical openings in the direction of entry. It is primarily used in fluid bed drying applications. Holes can be oriented in such a way to promote certain mixing patterns or drive the solids toward discharge nozzle.

Main Advantages

Promotes solid mixing; prevents stagnant solids buildup. Facilitates discharge of most of the solids. The holes are angled so that the grids can be non-weeping for coarse solids.

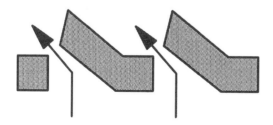

Possible Disadvantages

Difficult to construct, facilitates only small hole sizes, requires reinforcement underneath the sheet to support the bed.

Among the foregoing advantages and limitations, the designer must select those most pertinent or critical to the process application. There are, for example, instances in which solids below the grid level are tolerable, where grid thermal expansion is significant, where bed solids are very friable, where pressure drop, and therefore the cost of compressive horsepower, is critical, where solids are "sticky" and must be kept in motion throughout, where internal impellers or stirrers must be provided, or where grids are expected to have a short life due to corrosion. These and many other specifics have dictated a host of design variations, some of which are illustrated below. It should be emphasized that each application requires thoughtful engineering consideration before final design selection.

3 GRID DESIGN CRITERIA

3.1 Jet Penetration

Gas flowing from the grid holes can take the form of either a series of bubbles or a permanent jet, depending on system parameters and operating conditions. However, a permanent jet prevails for most industrial conditions. Jet penetration is one of the most important design parameters since it helps in

1. Determining how far to keep the bed internals, such as feed nozzles, heat exchanger tubes, etc., away from the grid to minimize erosion of internals.
2. Deciding on grid design parameters such as hole size and the gas jet velocity required to achieve a certain jetting region.
3. Minimizing or maximizing particle attrition at grids.

Knowlton and Hirsan (1980) reported that the jet penetration for upwardly directed jets fluctuated greatly. Karri (1990) noted that jet penetration can vary as much as 30% for upwardly directed jets. However, the jet emanating from a downwardly directed grid hole is stable, and its penetration length does not significantly fluctuate with time. Figure 2 indicates jet penetration configurations for jets oriented upwardly, horizontally, and downwardly. According

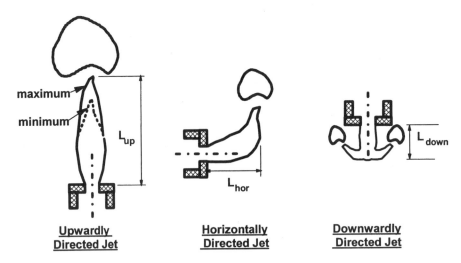

Figure 2 Jet penetrations at grid holes for different orientations.

to Karri, the jet penetrations for various orientations can be approximately related by:

$$L_{up} \approx 2L_{hor} \approx 3L_{down} \tag{1}$$

There are numerous jet penetration correlations (Zenz, 1969; Shakhova, 1968; Merry, 1971; Yang and Keairns, 1979; Knowlton and Hirsan, 1980; Yates et al., 1986; Blake et al., 1990; Roach, 1993) in the literature. Massimilla (1985) and Karri (1990) have shown that the jet penetrations predicted by these correlations can vary by a factor of 100 or more. Among them, Merry's correlation for horizontal jets was shown (Karri, 1990; Chen and Weinstein, 1993; Roach, 1993) to give reliable predictions, although this correlation was derived for horizontal jets issuing into an incipiently fluidized bed, which is not exactly the same situation as for a grid jet. Merry's correlation to calculate the penetration of horizontal jets is

$$\frac{L_{hor}}{d_h} = 5.25 \left(\frac{\rho_{g,h} U_h^2}{\rho_p (1 - \varepsilon_{mf}) g d_p} \right)^{0.4} \left(\frac{\rho_{g,b}}{\rho_p} \right)^{0.2} \left(\frac{d_p}{d_h} \right)^{0.2} \tag{2}$$

The jet penetration lengths for upwardly and downwardly directed jets can be calculated from Eq. (1). These equations take into account the effects of pressure and temperature on jet penetration. Knowlton and Hirsan (1980) and Yates et al. (1986) found that the jet penetration increases significantly with system pressure. In addition, Findlay and Knowlton (1985) found that the jet penetration decreases with increasing system temperature. Bed internals should not be placed

in the jetting zone near the grid, otherwise the internals could be severely eroded.

3.2 Grid Pressure Drop Criteria

For a grid, achieving equal distribution of gas flow through many parallel paths requires equal resistances and sufficient resistance to equal or exceed the maximum value of any unsteady state pressure fluctuation. It has been determined experimentally that the "head" of solids in some fluidized beds above an upwardly directed grid port can vary momentarily by as much as 30%. This is due to large fluctuations in the jet penetration for an upwardly directed jet, as discussed in the previous section. The equivalent variation downstream of a downwardly directed port is less than 10%. Thus as a rule of thumb, the criteria for good gas distribution based on the direction of gas entry are (Karri, 1990):

1. For upwardly and laterally directed flow:

$$\Delta P_{grid} \geq 0.3 \Delta P_{bed} \tag{3}$$

2. For downwardly directed flow

$$\Delta P_{grid} \geq 0.1 \Delta P_{bed} \tag{4}$$

and

3. Under no circumstances should the pressure drop across a large-scale commercial grid be less than 2500 Pa, i.e.,

$$\Delta P_{grid} \geq 2,500 \, Pa \tag{5}$$

Several investigators (Hiby, 1964; Zuiderweg, 1967; Whitehead, 1971; Siegel, 1976; Mori and Moriyama, 1978) have found the ratio of distributor pressure drop to bed pressure drop to be in the range of 0.015 to 0.4.

If turndown is desired, the grid pressure drop criteria (Eqs. 3 and 4) should apply at the minimum gas flow rate. This can be a problem for circulating fluidized bed combustors, since this means that under full load the grid pressure drop will be unacceptably high. Also, if the grid is curved, i.e., concave, convex, or conical, the criterion must apply with respect to the lowest hole on the grid. Take an example of a fluid bed with curved grid, as shown in Fig. 3.

A pressure balance across the curved grid can be written as

$$\Delta P_h \text{ (Highest hole)} = \Delta P_h \text{ (Lowest hole)} \\ + \rho_B g (H_{high} - H_{low}) \tag{6}$$

i.e.,

$$\Delta P_h \text{ (Highest hole)} = \Delta P_h \text{ (Lowest hole)} \\ + 480 \times 9.8 \times 0.9 \\ = \Delta P_h \text{ (Lowest hole)} + 4235\,\text{Pa} \tag{7}$$

Therefore the lowest grid hole has the lowest pressure drop, and hence the pressure drop criterion must apply with respect to the lowest hole on the grid.

3.3 Design Equations

The following equations can be used to design perforated plates, spargers, and bubble cap types of grids:

Pressure drop across the grid:

$$\Delta P_{grid} = K g \rho_B L_B \qquad \Delta P_{grid} \geq 2,500\,\text{Pa} \tag{8}$$

where $K = 0.3$ for upward and lateral gas entry and 0.1 for downward gas entry.

The gas velocity through the grid hole (orifice equation):

$$U_h = C_d \sqrt{\frac{2 \Delta P_{grid}}{\rho_{g,h}}} \tag{9}$$

The orifice discharge coefficient, C_d, is typically about 0.6 for gas flowing through an orifice in a pipe (for a ratio of orifice diameter to pipe diameter in the range of 0 to 0.2). This value of the orifice coefficient is for a sharp-edged orifice. However, grids are not sharp edged, and the orifice coefficient is then greater than 0.6. A typical value of C_d for a grid hole is about 0.8. Actually, the value of C_d depends on the grid plate thickness and the hole pitch. It can be calculated from Fig. 4 (Karri, 1991).

Volumetric flow rate of gas:

$$Q = N \frac{\pi d_h^2}{4} U_h \tag{10}$$

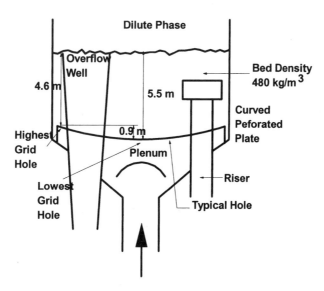

Figure 3 A typical fluid bed showing a curved perforated plate.

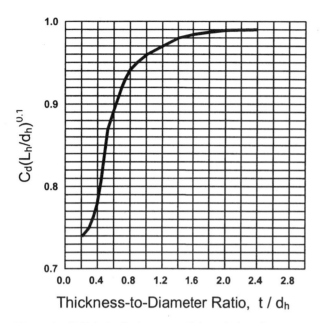

Figure 4 Grid hole discharge coefficient design chart.

3.3.1 Hole Size

To increase the gas residence time in the bed, it is desirable to introduce the greatest number of small gas bubbles as possible into the bed. This can be achieved by maximizing N at the expense of d_h in Eq. (10) (within the limits of mechanical, cost, and scaleup constraints). To minimize stagnant zones, the

number of grid holes per m^2 should be ≥ 10. In practice, the number of grid holes per square meter should be greater than 20.

3.3.2 Hole Layout

To increase the uniformity of fluidization, it is common to lay out the holes in triangular or square pitch, as shown in Fig. 5. All the holes in a grid with triangular pitch are equidistant. This is not the case for a grid with square pitch. Triangular pitch will also result in more holes per unit area.

The relationship between the grid hole pitch, L_h, and the number hole density (holes per unit area of the bed), N_d, depends on whether the holes are laid out in triangular or square pitch.

3.4 Additional Criteria for Sparger Grids

Additional distribution criteria are used for sparger grids. To keep the pipe header pressure drop down to acceptable levels and to ensure good gas distribution, the following criteria (Karri, 1990) should be met:

1. The manifold should be sized based on the following equation:

$$\left(\frac{D_m^2}{N_h d_h^2}\right)^2 > 5 \qquad (11)$$

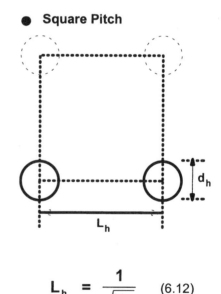

Figure 5 The relationship between hole density and grid hole pitch for both triangular and square pitch.

The parameters in Eq. (11) are defined in Fig. 6. Eq. (11) ensures that the pressure drop in the manifold is negligible compared to the pressure drop in the holes, which are determining the grid pressure drop.

Similarly, the main header pipe should be sized based on the equation

$$\left(\frac{D_{\text{head}}^2}{N_{\text{m}}D_{\text{m}}^2}\right)^2 > 5 \qquad (12)$$

2. In some instances, two to three different hole size are used on a given manifold to get better gas distribution.
3. The gas velocity in the header/manifold pipe should be $< 25\,\text{m/s}$ for best distribution.
4. Holes should not be located closer than one D_{m} from any sharp bend or tee in the header/manifold to prevent solids from being sucked into the manifolds due to the vena contracta effect.

3.5 Port Shrouding or Nozzle Sizing

Shrouds are generally placed around grid holes to reduce the velocity at the gas–solid interface and reduce particle attrition (de Vries et al., 1972). Shrouds simply consist of short pipes centered over the smaller grid holes that have been selected in size and number to operate at a hole velocity defined by Eq. (9).

To be effective, shrouds must be long enough to "contain" the expanding (11° included angle) gas jet leaving the grid orifice (Karri, 1991).

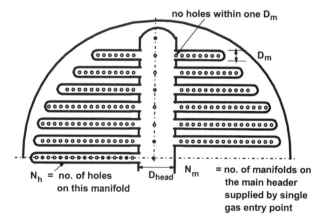

Figure 6 Manifold sparger grid showing the definitions of various parameters.

As can be seen from Fig. 7, the minimum shroud length should be:

$$L_{\text{min}} = \frac{D_{\text{s}} - d_{\text{h}}}{2\tan 5.5°} \qquad (13)$$

In practice, it is prudent to increase L_{min} by a factor of 50 to 100%. A shroud length less than L_{min} causes significantly more erosion and attrition than no shroud at all. Significant attrition can also occur if the shroud is not centered over the smaller hole.

The nozzle or shroud details inside a sparger pipe grid are illustrated in Fig. 8.

If properly sized and installed, particle attrition is reduced by a factor (Karri, 1990) calculated from

$$\frac{\text{particle attrition without shrouds}}{\text{particle attrition with shrouds}} = \left(\frac{D_{\text{s}}}{d_{\text{h}}}\right)^{1.6} \qquad (14)$$

4 PARTICLE ATTRITION AT GRIDS

Solids immediately surrounding the gas jets issuing from the grid are ingested into the jets. These particles are accelerated and collide with the particles near the tip of the jet. Figure 9 depicts how the particles are picked up and slammed into a fluidized, yielding bed for an upwardly directed jet. However, downward-pointing jets generally issue into a nonfluidized area of particles. Therefore particles picked up by downwardly directed jets, issuing into a nonyielding unaerated bed, result in a greater degree of particle attrition than those for upwardly directed jets. Karri (1990) reported that downwardly directed jets have approximately twice the steady-state attrition rate as that of upwardly directed jets. The attrition rates for upwardly and laterally directed jets are essentially the same. Grid jet attrition is discussed in greater detail in Chapter 8 of this book.

5 EROSION

5.1 Erosion at Bed Walls and Internals

Erosion in the grid region is primarily due to high-velocity submerged jets impinging on distributor parts, bed walls, or bed internals. Therefore one should estimate the jet penetration heights for a given grid design and check for the following:

1. Bed internals should not be placed in the jetting zone near the grid, otherwise the internals could be severely eroded.

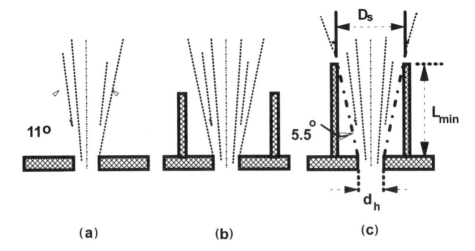

Figure 7 (a) Diverging free jet; (b) shroud too short to contain the jet; (c) minimum shroud length required to contain jet.

2. Nozzles should not be located any closer than half the jet penetration height from the bed wall.

The basic equation for erosion rate is of the form (Karri, 1990)

$$\text{Erosion} \propto \frac{K_e \rho_{g,h}^2 U_h^3 d_h^2 d_p^3 \rho_p^2}{\varphi} \qquad (15)$$

5.2 Erosion at Distributor Nozzles

Erosion in the nozzle or orifices is often associated with weepage of solids. This can be avoided by carefully designing a grid with the proper pressure drop criteria, as presented in Sec. 3.2. Poorly designed bubble caps tend to have erosion problems due to the secondary circulation of solids. Therefore bubble caps should be designed to minimize secondary circulation of solids.

Erosion has often been experienced at the nozzles used in CFB combustors (Fig. 1). A dominant mechanism leading to erosion is the pressure-induced gas flow reversal that will be discussed below in Sec. 6. Solids which have entered into the nozzle during a period of flow reversal are entrained out once the gas flows at high velocity in the outward direction again. The entrained solids in the high-velocity flow in the nozzle hole may cause severe erosion of the wall of the hole. A

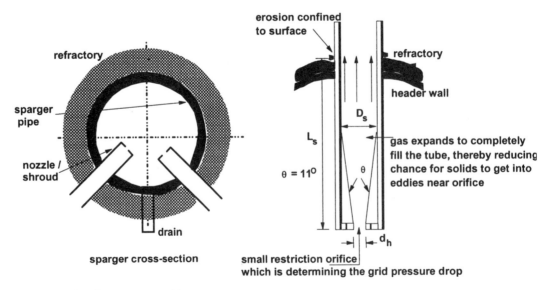

Figure 8 Shroud design for a sparger grid.

Figure 9 The mechanism of particle attrition at a submerged jet.

Figure 10 Erosion marks around the gas outlet holes of the nozzle with 5 mm diameter holes. (From Hartge and Werther, 1998.)

second mechanism of erosion was observed by high-speed video in cold models (Hartge and Werther, 1998): even when the gas was flowing out of the hole into the bed, a region near the mouth of the orifice could be observed where the gas jet entrained particles into the hole. These entrained particles caused erosion at the outer edge of the hole. Figure 10 shows the photography of a nozzle that had been painted in black before the experiment. After 60 hours of operation, the erosion marks were clearly visible. They were particularly obvious at the lower edges of the holes, which is due to the fact that the jet issuing from a horizontal bore tends to bend into the upward direction (see Fig. 2), which gives more surface area to entrain solids at the lower edge of the hole.

6 WEEPAGE OF SOLIDS

Solids weepage has been a major problem during the development of circulating fluidized bed (CFB) combustors in the past two decades. Seemingly well-designed nozzle grids experienced weepage to such an extent that CFB boilers had to be shut down after several days or weeks of operation, because most of the bed inventory had wept through the grid. More recent investigations (Hartge and Werther, 1998; Karri, 1991) have revealed that pressure fluctuations in the dense bottom bed of the CFB riser may cause the backflow of solids through the grid hole. Figure 11

shows measurements of the pressure drop between the wind box and at a height of 0.3 m above the distributor in a circulating fluidized bed. The average pressure drop of gas distributor and bed was about 40 mbar during these measurements. As can be seen, sometimes negative pressure drops occurred, i.e., the pressure at a height of 0.3 m above the distributor was higher than

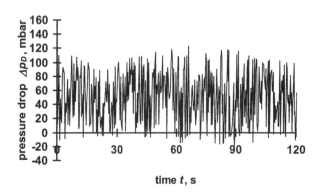

Figure 11 Pressure flucutations measured in a circulating fluidized bed between the windbox and at a height of 0.3 m above the distributor (riser diameter, 0.4 m; superficial gas velocity, 3 m/s; solids mass flux, 22 kg/m²s).

the pressure below the distributor. In such cases, gas flow reversal occurs, which results in weepage of solids into the nozzles (Fig. 12).

In order to prevent such a flow reversal, the design pressure drop of the nozzle in the example of Fig. 11 should be roughly 20 mbar larger, i.e., the pressure drop of the grid has to take into account the largest possible pressure fluctuations. These pressure drop fluctuations are significantly higher for Group B particles with a wide particle size distribution which are typical of CFB combustion. Another reason that makes it difficult to keep the grid pressure drop always higher than the largest pressure fluctuations in the bed is the necessity of frequent turndown operations for CFB combustors. Reducing the load to 50% means a reduction of the grid pressure drop to 25% of its value at full load. In CFB combustion, the bubble cap design of Fig. 1 (case 7) has shown to be more effective. Here the solids have to be transported upwards in the annulus between the bubble cap and central tube against gravity during the gas flow reversal. The solids are prevented from entering the central tube if the separation length s_{bc} is kept between 70 and 100 mm. Figure 13 shows solids inflow during the period of gas flow reversal.

7 EFFECTS OF TEMPERATURE AND PRESSURE

System temperature and pressure affect the momentum of grid jets via the gas density. The momentum of the gas jets is proportional to $\rho_{g,h} U_h$. When the tempera-

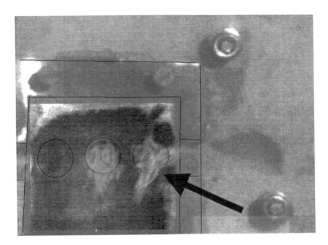

Figure 12 Solids flowing into the nozzle shaft during gas flow reversal.

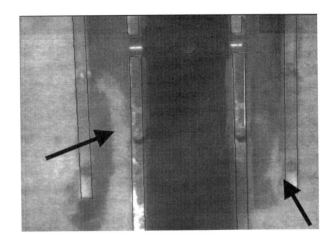

Figure 13 Solids flowing into the cap during gas flow reversal.

ture is increased, the gas density decreases. For the same gas jet velocity, this decreases the momentum of the jets and therefore decreases the jet penetration and the attrition at the grid. Similarly, when system pressure is increased, gas density increases, gas jet momentum increases, and therefore the jet penetration and the attrition at the grid are increased.

8 PLENUM DESIGN

The plenum, or windbox, is the chamber immediately below the grid. If the bed-pressure-drop-to-grid-pressure-drop ratio is high enough, the plenum design will probably not be too important. However, for the case where this ratio is low, the plenum design may determine whether the bed will operate satisfactorily.

The typical plenum designs showing various configurations for introducing gas into the plenum are illustrated in Fig. 14. Common sense dictates that certain plenum designs be preferred over others. If the gas enters the plenum from the bottom it is preferable that the plenum have a large enough distance between the outlet of the supply pipe and the grid to prevent the gas from preferentially passing through the middle of the grid. When gas enters a plenum from the side, it is preferable to rout the gas to the middle of the plenum (Fig. 14c) rather than have the supply pipe end at the wall of the plenum. In addition, horizontal-to-vertical down gas entry (Fig. 14c) is preferable over the horizontal-to-vertical up gas entry (Fig. 14b).

If the gas–solid or gas–liquid suspension needs to be introduced into the plenum, as for example in a polyethylene reactor and some FCC regenerators, it is

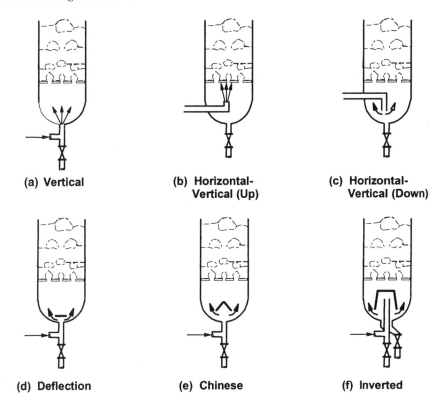

Figure 14   Different plenum configurations.

preferable to introduce the suspension at the lowest point of the plenum (Fig. 14a,d,e) to minimize the accumulation of solids or liquids in the regions inaccessible to reentrainment For two-phase systems, it is preferable to have some sort of deflection device (Fig. 14d,e,f) between the outlet of the supply pipe and the grid to prevent the solids from preferentially passing through the middle of the grid due to their high momentum. This preferential bypassing of solids causes maldistribution of gas. In addition, the configuration of Figs. 14e and 14f are preferable over the configurations of Figs. 14a,d.

9   POWER CONSUMPTION

Since the grid contributed a considerable fraction of total pressure drop across a given fluid bed system, it is always important to estimate the power consumption of the blower that drives the gas through this system. Suppose a stream of gas is to be compressed from an initial pressure of P_1 to a higher pressure of P_2 to pump it through the entire fluid-bed system. Using thermodynamics for adiabatic reversible compression with negligible kinetic and potential energy effects, the ideal shaft work to compress each kilogram of gas is given by

$$-W_{s,ideal} = \int_{P_1}^{P_2} \frac{dP}{\rho_g} \tag{16}$$

If an ideal gas behavior is assumed, then Eq. (16) transforms into

$$-W_{s,ideal} = \frac{\gamma}{\gamma - 1} P_1 Q_1 \left[\left(\frac{P_2}{P_1} \right)^{(\gamma-1)/\gamma} - 1 \right] \tag{17}$$

or

$$-W_{s,ideal} = \frac{\gamma}{\gamma - 1} P_2 Q_2 \left[1 - \left(\frac{P_1}{P_2} \right)^{(\gamma-1)/\gamma} \right] \tag{18}$$

Due to heat of compression, the raise in temperature can be calculated from

$$T_2 = T_1 \left(\frac{P_2}{P_1} \right)^{(\gamma-1)/\gamma} \tag{19}$$

where γ = ratio of specific heats of gas $\cong$ 1.67, 1.4, and 1.33 for monatomic, diatomic, and triatomic gases, respectively.

However, for real operations with its frictional losses, the actual work required is always greater than the ideal and is given by

$$W_{s,actual} = \frac{W_{s,actual}}{\eta} \tag{20}$$

where η is the blower efficiency, approximately given by

$\eta = 0.55\text{--}0.75$ for a turboblower
$ = 0.6\text{--}0.8$ for a roots blower
$ = 0.8\text{--}0.9$ for an axial blower or a two-stage reciprocating compressor

Equation (20) can be used not only for power consumption but also to size the correct horse power motor to drive the blower.

The actual temperature of gas leaving a well-insulated (adiabatic) but not 100% efficient compressor is then calculated from

$$T_2 = T_1 + \frac{T_1}{\eta_1}\left[\left(\frac{P_2}{P_1}\right)^{(\gamma-1)/\gamma} - 1\right] \tag{21}$$

10 DESIGN EXAMPLES

10.1 FCC Grid Design

Example 1. A 13-m-ID bed of FCC catalyst ($d_p = 60\,\mu$m) 3 m deep is to operate at a superficial gas velocity of 0.6 m/s. The bed density is 480 kg/m^3. the density of the gas entering the bed is 0.64 kg/m^3. Design the following grid types: (1) a flat perforated plate, and (2) concentric-ring type downflow sparger. Assume the grid thickness to be 0.025 m.

Solution. Perforated Plate Design

Determine ΔP_{bed} and ΔP_{grid}:

$$\Delta P_{bed} = g\rho_B L_B = 9.8 \times 480 \times 3 = 14,112\,\text{Pa}$$

Choose ΔP_{grid} to be 30% of ΔP_{bed}

$$\Delta P_{grid} = 0.3 \quad \Delta P_{bed} = 4,234\,\text{Pa}$$

Determine the gas velocity through the grid holes (assume a typical value for $c_d \cong 0.77$):

$$U_h = C_d\sqrt{\frac{2\Delta P_{grid}}{\rho_{g,h}}} = 0.77\sqrt{\frac{2 \times 4234}{0.64}} = 88.6\,\text{m/s}$$

Determine the volumetric gas flow rate at the conditions below the grid. For this example, assume that the temperature of the gas below the grid is the same as in the bed. This may not be the case in an actual plant.

$$Q = U_{sup}\frac{\pi D^2}{4} = 0.6\frac{\pi(13)^2}{4} = 79.6\,\text{m}^3/\text{s}$$

Determine the number of grid holes required:

Since $Q = N\dfrac{d_h^2}{4}U_h$

Therefore

$$N = \frac{Q}{U_h}\frac{1}{\pi d_h^2/4} = \frac{79.6}{88.6}\frac{1}{\pi d_h^2/4} = \frac{1.14}{d_h^2}$$

The hole density is

$$N_d = \frac{N}{(\pi/4)D^2} = \frac{1.14}{d_h^2}\frac{1}{(\pi/4)(13)^2} = \frac{0.0086}{d_h^2}$$

Determine the hole pitch for a triangular arrangement:

$$L_h = \frac{1}{\sqrt{N_d \sin 60°}} = 11.59 \cdot d_h$$

Downwardly Directed Gas Sparger Design

Choose ΔP_{grid} to be 10% of ΔP_{bed}

$$\Delta P_{grid} = 0.1\Delta P_{bed} = 1,411\,\text{Pa} = 14.4\,\text{cm}\,\text{H}_2\text{O}$$

1411 Pa is less than the minimum of 2500 Pa ΔP required for a grid. Therefore use $\Delta P_{grid} = 2500\,\text{Pa}$.

$$U_h = 0.77\sqrt{\frac{2 \times 2500}{0.64}} = 68\,\text{m/s}$$

$$N = \frac{79.6}{68}\frac{1}{\pi d_h^2/4} = \frac{1.5}{d_h^2}$$

Various combinations of N and d_h satisfy the pressure drop requirements for the two grid type as shown in the table:

d_h	Number of holes (N)	
m	perforated plate	downflow sparger
0.005	45,600	60,000
0.01	11,400	15,000
0.025	1,824	2,400
0.05	456	600

To proceed with the design, it is necessary to select a hole size (judgment call). For the purpose of this example, a hole size of 0.025 m will be chosen to compare the different grid types. This hole diameter does not result in an excessive number of holes for both types of grids.

Check the value for C_d for the perforated plate:

$$N_d = 13.8 \text{ holes/m}^2 \quad \text{and} \quad L_h = 0.29 \text{ m}$$

$$\frac{t}{d_h} = \frac{0.025}{0.025} = 1$$

From Fig. 3

$$C_d \left(\frac{L_h}{d_h}\right)^{0.1} = 0.96$$

$$C_d = 0.96 \left(\frac{0.025}{0.29}\right)^{0.1}$$

$$= 0.75 \text{ vs. } 0.77 \text{ (initial guess)}$$

There is a fairly good agreement between the initial and calculated values for C_d. If not, one must repeat the calculations using the calculated C_d until both values agree.

Therefore, for a 0.025 m hole diameter, the *perforated plate* has 1,824 holes arranged in a triangular pitch of 0.29 m. The hole density is 13.8 hole/m^2.

For the sparger grids, it remains to determine the sparger configuration and pipe-header size. Pipe headers can be laid out in various configurations. The design calculations will depend on the configuration one chooses.

Concentric-Ring Sparger. Consider for example, a configuration of *four* concentric rings of 0.4 m diameter supplied by a number of gas entry points. This design results in 2,401 holes.

Determine the hole pitch:

$$L_h = \frac{97.13}{2401} = 0.04 \text{ m}$$

To determine the header-pipe size, first determine the maximum number of holes in ring section supplied by a single *effective* entry of gas. If outermost ring is supplied by four gas entry points, then the number of effective gas entry points is 8, and the number of holes in each section of ring No. 4 would be $N_h = 978/8 = 122$. Then Eq. 11 gives

$$\left(\frac{D_{head}^2}{N_h d_h^2}\right)^2 > 5$$

$$\left(\frac{D_{head}^2}{122 \times 0.025^2}\right)^2 > 5 \quad \text{or} \quad D_{head} > 0.41 \text{ m}$$

Summary. For an orifice diameter of 0.025 m, the downwardly directed *concentric-ring sparger* has 2,401 nozzles placed on four concentric rings. The pitch is 0.04 m. Sometimes the holes are staggered on the sparger pipe. Also it is a common practice to place two nozzles at a given cross section as shown in Fig. 8.

Example 2. For the conditions of Example 1 of perforated plate design, estimate the submerged jet height in the fluidized bed.

Solution. Perforated Plate

$$U_h = 88.6 \text{ m/s} \quad \rho_{g,h} = 0.64 \text{ kg/m}^3$$

$$\rho_{g,b} = 0.5 \text{ kg/m}^3 \quad d_h = 0.025 \text{ m} \quad N = 1,824$$

$$d_p = 60 \text{ μm} \quad \rho_p = 1440 \text{ kg/m}^3 \quad \varepsilon_{mf} = 0.42$$

Gas jet penetration depth using Merry's correlation (Eq. 3.4.2) for horizontal jets

$$\frac{L_{hor}}{d_h} = 5.25 \left(\frac{\rho_{g,h} U_h^2}{\rho_p (1 - \varepsilon_{mf}) g d_p}\right)^{0.4} \left(\frac{\rho_{g,b}}{\rho_p}\right)^{0.2} \left(\frac{d_p}{d_h}\right)^{0.2}$$

Sparger Grid, Concentric Ring Type

Ring no. (i)	Radius of each ring (r$_i$), m	Length of each ring (L$_i$) 2πr$_i$, m	% of total length	Number of holes on each ring (N$_i$)
1	1.43	8.98	9.24	222
2	3.05	19.16	19.73	474
3	4.68	29.41	30.28	727
4	6.30	39.58	40.75	978
Total =	—	97.13	—	2,401

$$L_{hor} = 5.25\left(\frac{0.64 \times 88.6^2}{1440(1-0.42)9.8 \times 65 \times 10^{-6}}\right)^{0.4}$$

$$\left(\frac{0.5}{1440}\right)^{0.2}\left(\frac{65 \times 10^{-6}}{0.025}\right)^{0.2} \times 0.025 = 0.32\,m$$

From Eq. (1),

$$L_{up} \approx 2L_{hor} \approx 2 \times 0.32 \approx 0.64\,m$$

Example 3. For the conditions and the perforated plate defined in Example 1, design a shroud having an ID twice that of the grid hole, i.e., $D_S = 2d_h = 0.05\,m$

Solution. Perforated Plate

The minimum length of the shroud should be

$$L_{min} = \frac{0.05 - 0.025}{2\tan 5.5°} = 0.13\,m$$

The gas jet velocity emanating from the shroud is

$$U_{h,s} = U_h\left(\frac{d_h}{D_s}\right)^2 = 88.6\left(\frac{0.025}{0.05}\right)^2 = 22.2\,m/s$$

Particle attritions rate will be reduced by a factor calculated from Eq. (14):

$$\frac{\text{particle attrition without shrouds}}{\text{particle attrition with shrouds}} = \left(\frac{D_s}{D_h}\right)^{1.6}$$

$$= \left(\frac{0.05}{0.025}\right)^{1.6} = 3.0$$

Thus adding a shroud to the grid reduces the attrition rate to 67% of the rate without a shroud.

10.2 Polyethylene Reactor Grid Design

Example 4. Design a flat perforated-plate grid for the polyethylene reactor schematically shown in Fig. 15 and calculate the gas jet penetration depth. Use a triangular pitch. System parameters are

$$U_{sup} = 0.5\,m/s \qquad \rho_{g,h} = 19.2\,kg/m^3$$
$$\rho_{g,b} = 17\,kg/m^3 \qquad \rho_p = 641\,kg/m^3$$
$$\rho_B = 272\,kg/m^3 \qquad \Delta P_{grid} = 0.4\Delta P_{bed}$$
$$d_h = 0.01\,m \qquad d_p = 508\,\mu m \qquad \varepsilon_{mf} = 0.45$$
$$t = 0.019\,m$$

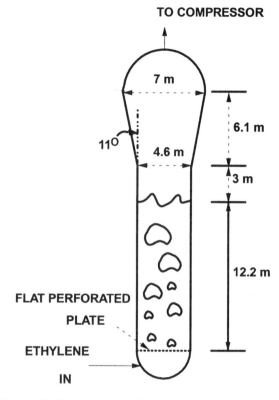

Figure 15 Schematic of polyethylene reactor.

Solution

Determine ΔP_{bed} and ΔP_{grid}

$$\Delta P_{bed} = g\rho_B L_B = 9.8 \times 272 \times 12.2 = 32,520\,Pa$$
$$\Delta P_{grid} = 0.4\Delta P_{bed} = 13,008\,Pa$$

Determine the gas velocity through the grid hole (*trial and error*). Assume $C_d = 0.8$.

$$U_h = C_d\sqrt{\frac{2\Delta P_{grid}}{\rho_{g,h}}} = 0.8\sqrt{\frac{2 \times 13008}{19.2}} = 29.5\,m/s$$

Determine the volumetric flow rate of gas

$$Q = U_{sup}\frac{\pi D^2}{4} = 0.5\frac{\pi(4.6)^2}{4} = 8.3\,m^3/s$$

Determine the number of grid holes required:

$$N = \frac{Q}{U_h}\frac{1}{\pi d_h^2/4} = \frac{8.3}{29.5}\frac{1}{(\pi/4)(0.01)^2} = 3582$$

Hole density:

$$N_d = \frac{3582}{(\pi/4)(4.6)^2} = 215\,holes/m^2$$

Determine the hole pitch:

$$L_h = \frac{1}{\sqrt{N_d \sin 60°}} = \frac{1}{\sqrt{215 \sin 60°}} = 0.073 \, \text{m}$$

Check the value for C_d:

$$\frac{t}{d_h} = \frac{0.019}{0.01} = 1.9$$

From Figure 3.4.4,

$$C_d \left(\frac{L_h}{d_h} \right)^{0.1} = 0.98$$

therefore

$$C_d = 0.98 \left(\frac{0.01}{0.073} \right)^{0.1} = 0.803 \approx 0.80 \, \text{(great guess)}$$

Gas jet penetration depth using Merry's correlation [Eq. (2)] for horizontal jets:

$$\frac{L_{hor}}{d_h} = 5.25 \left(\frac{\rho_{g,h} U_h^2}{\rho_p (1 - \varepsilon_{mf}) g d_p} \right)^{0.4} \left(\frac{\rho_{g,b}}{\rho_p} \right)^{0.2} \left(\frac{d_p}{d_h} \right)^{0.2}$$

$$L_{hor} = 5.25 \left(\frac{19.2 \times 29.5^2}{641(1 - 0.45) 9.8 \times 508 \times 10^{-6}} \right)^{0.4}$$

$$\left(\frac{17}{641} \right)^{0.2} \left(\frac{508 \times 10^{-6}}{0.1} \right)^{0.2} \times 0.01 = 0.55 \, \text{m}$$

From Eq. (1),

$$L_{up} \approx 2 L_{hor} \approx 2 \times 0.55 \approx 1.1 \, \text{m}$$

Coalescence factor:

$$\lambda = \frac{L_h}{L_{up}/2} = \frac{0.073}{1.1/2} = 0.13 < 1$$

Therefore Jets coalesce. The low value of λ indicates that the bed of solids is probably suspended above the coalesced jets. Therefore the solids rarely come into contact with the grid. This type of design reduces the chances of grid pluggage due to "sticky" polyethylene solids.

Summary: The perforated plate has 3,582 holes, each of 0.01 m diameter, arranged in a triangular pitch of 0.073 m. The hole density is 215 holes/m².

10.3 Power Consumption

Example 5. Determine the compressor power to pass reactant gas into the plenum of the fluid bed system. Also calculate the temperature rise due to heat of compression. The system parameters are

$\Delta P_{grid} = 6 \, \text{kPa};$ $\Delta P_{bed} = 15 \, \text{kPa};$ $\Delta P_{cyclones+filters} = 12 \, \text{kPa};$ pressure at the exit of the filters = 350 kPa.

Gas entering the compressor: $T_1 = 20°\text{C};$ $P_1 = 101 \, \text{kPaq};$ $Q_1 = 10 \, \text{m}^3/\text{s}.$

Use $\eta = 0.85;$ $\gamma = 1.4$

Solution

Determine compressor discharge pressure, P_2:

$$P_2 = P_{exit} + \Delta P_{cyclones+filters} + \Delta P_{bed} + \Delta P_{grid}$$

$$= 350 + 12 + 15 + 6 = 388 \, \text{kPa}$$

Determine ideal power consumption, $W_{s,ideal}$

$$-W_{s,ideal} = \frac{\gamma}{\gamma - 1} P_1 Q_1 \left[\left(\frac{P_2}{P_1} \right)^{(\gamma-1)/\gamma} - 1 \right]$$

$$-W_{s,ideal} = \frac{1.4}{1.4 - 1} 101 \times 10 \left[\left(\frac{383}{101} \right)^{(1.4-1)/1.4} - 1 \right]$$

$$= 1638 \, \text{kW}$$

Determine actual power consumption, $W_{s,actual}$:

$$W_{s,actual} = \frac{W_{s,ideal}}{\eta} = \frac{1638}{0.85}$$

$$= 1927 \, \text{kW (or 2587 hp)}$$

Determine the temperature rise, T_2:

$$T_2 = T_1 + \frac{T_1}{\eta_1} \left[\left(\frac{P_2}{P_1} \right)^{(\gamma-1)/\gamma} - 1 \right]$$

$$T_2 = 293 + \frac{293}{0.85} \left[\left(\frac{383}{101} \right)^{(1.4-1)/1.4} - 1 \right]$$

$$= 453 \, \text{(or 180°C)}$$

NOMENCLATURE

C_d	=	discharge coefficient; see Fig. 4
d_h	=	grid hole diameter, m
d_p	=	Sauter mean particle size, m
D	=	diameter of fluid bed, m
D_{head}	=	diameter of the main header pipe, m
D_m	=	diameter of the manifold pipe, m
D_s	=	shroud or nozzle diameter, m
g	=	gravitational acceleration 9.8 m/s²
H_{high}	=	elevation of highest grid hole for curved grid, m
H_{low}	=	elevation of lowest grid hole for curved grid, m

K	=	grid pressure-drop coefficient; see Eq. (8)
		0.3 for upward gas entry;
		0.1 for lateral and downward gas entry
K_e	=	erosion constant, Eq. (15)
L_B	=	operating bed depth, m
L_{down}	=	jet penetration for downwardly directed jet, m
L_h	=	grid hole pitch, cm
L_{hor}	=	jet penetration for horizontally directed jet, m
L_{min}	=	minimum shroud or nozzle length, m
L_s	=	shroud or nozzle length, m
L_{up}	=	jet penetration for upwardly directed jet, m
N	=	number of grid holes
N_d	=	number of hole density (holes per unit area of the bed), holes/m^2
N_h	=	maximum number of holes per manifold pipe section supplied by gas entry
N_m	=	number of manifolds on the main header supplied by single gas entry point
P_1	=	pressure of gas entering the blower, Pa
P_2	=	pressure of gas leaving the blower, Pa
Q	=	total volumetric gas flow entering the grid, m^3/s
Q_1	=	total volumetric gas flow entering the blower, m^3/s
Q_2	=	total volumetric gas flow leaving the blower, m^3/s
t	=	grid thickness, m
T_1	=	temperature of gas entering the blower, $^{\circ}K$
T_2	=	temperature of gas leaving the blower, $^{\circ}K$
U_h	=	velocity of gas through the grid hole, m/s
U_{sup}	=	superficial gas velocity, m/s
$W_{s,actual}$	=	actual power consumption due to shaft work, W
$W_{s,ideal}$	=	ideal power consumption due to shaft work, W
α	=	energy efficiency factor
γ	=	ratio of specific heats of gas
ρ_B	=	operating bed density, kg/m^3
$\rho_{g,b}$	=	density of gas at bed operating conditions, kg/m^3
$\rho_{g,h}$	=	density of gas entering the grid hole (plenum conditions), kg/m^3
ρ_p	=	particle density, kg/m^3
ε_{mf}	=	voidage at minimum fluidizing conditions
θ	=	included angle of gas jet, degrees
ΔP_{bed}	=	pressure drop across the dense bed, Pa
ΔP_{grid}	=	pressure drop across the grid, Pa

ΔP_h	=	pressure drop across the grid hole, Pa
φ	=	particle shape factor
η	=	compressor efficiency

REFERENCES

Blake TR, Webb H, Sunderland PB. Chem Eng Sci 45:365, 1990

Chen L, Weinstein H. AIChE J 39(12):1901, 1993.

Findlay J, Knowlton TM. Final Report for US Dept of Energy, Project, DE-AC21-83MC20314, 1985.

Hartge E-U, Werther J. In: Fan LS, Knowlton TM, eds. Fluidization IX. Engineering Foundation, 1998, p 213.

Hiby JW. Chem-Ing Techn 36:228, 1964.

Karri SBR. PSRI Research Report No. 60, 1990.

Karri SBR. Grid Design Chapter, PSRI Design Manual, 1991

Knowlton TM, Hirsan I. In: Grace JR, Matsen JM, eds. Fluidization. Plenum Press, 1980, p 315.

Massimilla L. In: Davidson JF et al., eds. Fluidization. Academic Press, 1985, p 133.

Merry JMD. Trans Instn Chem Engrs 49:189, 1971.

Mori S, Moriyama A. Inst Chem Eng 18:245, 1978.

Roach PT. Fluid Dyn Res 11:197, 1993.

Shakhova NA, Inzh Fiz Zh 14(1):61, 1968.

Siegel R. AIChE J 25:590, 1976.

Sishtla C, Findlay J, Chan I, Knowlton TM. In: Grace JR, Shemilt LW, Bergougnou MA, eds. Fluidization VI. Engineering Foundation, 1989, p 581.

VGB Powertech, Design Report M218H, VGB Kraftwerkstechnik, 1984.

de Vries RJ, van Swaaij WPM, Mantovani C, Heijkoop A. Proc 5th Europ Symp Chem Reaction Engng, Amsterdam, 1972, p 39–59.

Whitehead AB. In Davidson JF, Harrison D, eds. Fluidization. Academic Press, 1971, p 781.

Yang WC, Keairns DL. Ind Eng Chem Fundam 18:317, 1979.

Yates JG, Bejcek V, Cheesman DJ. In: Ostergaard K, Sorensen A, eds. Fluidization V. Engineering Foundation, 1986, p 79.

Zenz FA, Othmer DF. Fluidization and Fluid-Particle Systems. Reinhold, 1960, p 171.

Zenz FA. Inst Chem Eng Symp 30:136, 1968.

Zuiderweg. In: Drinkenburg, ed. Proc Int Symp on Fluidization. Netherlands University Press, 1967, p 739.

7

Effect of Internal Tubes and Baffles

Yong Jin, Fei Wei, and Yao Wang

Tsinghua University, Beijing, People's Republic of China

In the past sixty years, worldwide research has led to an essential understanding of the most important properties of fluidized beds. In particular, it has been found that the flow structures of G–S, L–S, and G–L–S fluidization mainly vary with increasing fluid velocity. For example, several flow patterns or regimes have been identified in gas–solid fluidized beds (Grace, 1986), such as particulate fluidization, bubbling/slugging fluidization, turbulent fluidization, fast fluidization, and pneumatic conveying regimes (Fig. 1).

The complexity of gas–solid flow structure of the above regimes is as follows:

1. There are quite different flow behaviors on an equipment scale, such as the intensity of gas–solid overall movement and backmixing, gas–solid dispersion, and the residence time distribution (RTD) of both gas and solid in beds.
2. The bed parameters on a micro scale vary with the radial/axial bed locations and time.
3. There are many substructures in beds, such as bubbles, clouds, wakes, channels, and clusters; their size distribution in time and space, as well as the frequency of their presence and disappearance, are normally quite complicated.
4. The effect of particle size and size distribution on flow regime is complex.

The influence of complex flow structures on reactor performance is complicated, even in scaled up reactors. Internals are usually introduced to modify the above gas–solid flow structures, in an effort to form a more uniform and active gas–solid flow to enhance heat and mass transfer so as to improve the overall performance of fluidized bed reactors, especially to make the scale-up easier. Therefore the study and improvement of internals are very important for raising the productivity of fluidized bed reactors and extending the usefulness of fluidization processes.

In this chapter the mechanism of the effect of internals to improve the performance of fluidized beds is discussed, and highly efficient internal tubes and baffles used in gas–solid fluidized beds are presented. First, the inherent structures of two-phase flow without internals are introduced in Sec. 1 so that the importance and necessity of internals for gas–solid fluidized beds can be explored. Second, various internal tubes and baffles are collected from the literature; they are classified and discussed in Sec. 2. Recent progress and some simple criteria for designing in this area are also covered. Section 3 focuses on explaining the effects of internals on two-phase flow.

1 INHERENT FLOW STRUCTURES OF GAS–SOLID FLOW

Bubbles, clusters and nonuniform flow structures are inherent characteristics of the gas–solid fluidization systems, which greatly influence the performance of fluidized beds in many applications. Thus an understanding of the inherent flow structures is essential for introducing internal tubes and baffles with clear

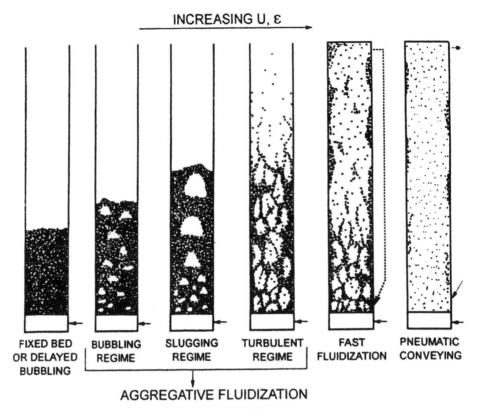

INCREASING U, ε

Figure 1   Flow regime of gas-solid fluidization. (From Grace, 1986.)

purpose, and judging the improvement effect of internals objectively.

1.1   Flow Structure of Bubbling Fluidization

When the gas velocity approaches and goes over the minimum fluidization velocity, U_{mf}, all fine bed particles that belong to Group A in the Geldart particle classification scheme are suspended by the gas. The whole bed expands uniformly, and the gas passes the bed through the particles. This is called the particulate fluidized regime. As the gas velocity increases to bubbling velocity, U_b, bubbles begin to appear in the bed. But for the Group B particles, the U_{mf} and U_b are the same value.

One of the most characteristic phenomena of gas–solid fluidized beds is the formation of gas bubbles, which dominate the behavior of fluidized beds in relatively low-velocity regimes. In analyzing the behavior of bubbling fluidized beds, it is essential to distinguish between the bubble phase and the emulsion phase, the latter consisting of particles fluidized by interstitial gas. A bubbling bed can conveniently be defined as a bed in

which the bubble phase is dispersed and the emulsion phase is continuous.

For a gas–solid fluidized bed, in which most gas passes through the bed as bubbles, the operational region of gas velocity is greatly extended by the presence of bubbles. This is one of the main reasons that G–S fluidized beds are more widely used in industries than L–S fluidized beds.

In addition, the rising bubbles cause the motion of particles, which obviously intensifies the solids mixing on the macro scale and leads to the temperature uniformity and high bed/surface heat transfer characteristic. This is another attractive advantage of G–S fluidization.

Consider the case of gas forming a bubble within a mass of particles. The only way in which the solids can be moved out of the way to form a bubble is that there must be a force being exerted by the gas flowing past the solids. Most of the gas then finds its way back into the bubble to complete its passage through the bed. The resulting flow into and out of the emulsion phase enhances the mass transfer. In the case where the bubble eruption and collapse occur at the bed surface, there is also intensely enhanced mixing due to convection effects compared with normal diffusion effects.

However, when a bed does not contain internals of any sort, the movement of the bubbles in the bed is unrestricted. As bubbles rise, they gradually increase in size and tend to move horizontally toward the center of the bed (Fig. 2). Much of the gas is "short-circuited" through the bubbles, which greatly limits the conversion and selectivity of a chemical reaction, especially for the Group B particles at high superficial gas velocities.

Bubble size is very important for a G–S fluidized bed. The gas in a bubble can directly contact the bed particles only in the clouds around the bubbles. The performance of a fluidized bed reactor can be improved by decreasing the bubble size and renewing the bubble surface for interchanging the gas between bubbles and the interstitial gas in the emulsion phase. The equilibrium bubble is one that is constantly splitting and coalescing. In a small pilot plant or lab facility, bubbles will not grow to very large size. However, as the system is scaled up, the bubble size increases and with it the bubble velocity. At higher superficial gas velocities, especially for the group B particles, the maximum bubble diameter appears to be as wide as the bed diameter. Such large bubbles would violently shake the unit as tons of catalyst is splashed when they come out of the bed surface. In addition, mass transfer from such an enormous bubble would be so poor that it would limit the ability of reaction. In order to prevent either of these phenomena from taking place, we must find out a mechanism that can limit the maximum stable bubble size.

Therefore the main effect of internals in a bubbling fluidized bed is to break up and renew bubbles, thus enhancing the interchange of gas between the bubbles and the emulsion phase. Chemical reactions and mass transfer can be improved only when bubbles are small and evenly distributed throughout the bed volume.

1.2 Flow Structure of Turbulent Fluidization

As the gas velocity of a bubbling fluidized bed is slowly raised, the heterogeneous two-phase character of the bed first peaks and then gradually gives way to a condition of increasing uniformity in a turbulent state in which large discrete bubbles or voids are wholly absent. A turbulent fluidized bed is considerably more diffuse than a bubbling fluidized bed because of the larger renewal frequency of bubbles and the greater freeboard activity at higher gas velocities. On the whole, a turbulent fluidization has found a more wide application field than a bubbling fluidization.

The most attractive feature of a turbulent fluidized bed is its uniform temperature field. Thus the conversion can be obviously improved at fully developed turbulent fluidization for most isothermal catalytic fluidized bed reactors compared with bubbling fluidization. At the same time, the solids backmixing is also violent in turbulent fluidization, which in turn intensifies the gas backmixing. A wide gas RTD is very harmful to the selectivity of a chemical process with side reactions.

As the renewal frequency of bubbles is already fairly large in a turbulent fluidized bed, introducing internals into such a bed is mainly not for breaking bubbles,

(a) U_g=0.0417m/s　　　(b) U_g=0.395m/s　　　(c) U_g=0.51m/s

Figure 2 Bubble behavior in a free fluidized bed.

eliminating gas "short-circuits" or enhancing gas inter-change between the bubble phase and the emulsion phase, but for limiting the gas axial backmixing and promoting the radial movement of the gas and solids.

The transition to a turbulent behavior is related to particle size and density. The larger the particle size and density, the higher the transition gas velocity, U_c. It has been found that the transition from a bubbling to a turbulent regime occurs at a smaller U_c when suitable internals are applied, which is especially important for Group B particles. Detailed information is given in Sec. 3.4.

1.3 Flow Structure of Fast Fluidization

Transition from turbulent to fast fluidization occurs at the transport velocity, U_{tr}, where significant numbers of particles are carried out from the top of the column. At the same time, continuous and smooth feeding of solids into the bottom of the riser should be maintained in order to keep the stability of a fast fluidization. Thus a fast-fluidization regime is connected with a circulating fluidized bed (CFB).

Nonuniform flow structure in space is the characteristic of a typical CFB core–annulus structure in a CFB along the radial direction. In the center of the bed, gas and particle velocities are higher and solids concentration lower, while in the wall region gas and particle velocities are much lower and solids concentration is much higher (Fig. 3). When at high solids circulating rate, downflow of gas–solid will exist in the wall region, which does not exist in bubbling and turbulent fluidized beds.

The axial solids distribution is obviously affected by the end configuration of the bed, the total solids inventory, and the gas velocity. A typical axial solids distribution is dense in the riser bottom and dilute in the top section (Fig. 4). When the inlet restriction is strong, the axial solids concentration decreases gradually from the bottom to the top section, presenting an exponential distribution. Reducing the solids inlet restriction results in a higher solids circulation rate and higher solids holdup in the riser bottom region, and an axial solids profile like "S" comes into being. This is attributed to pressure buildup in the solid circulation system. The inflexion point of the S, where the solids density turns from high to low, rises with the increase of solids inventory obviously.

This nonuniform flow structure in space leads to two main problems: (1) inefficient gas–solid contact. On the one hand, most gas only comes into contact with dilute particles in the center of the bed, while in

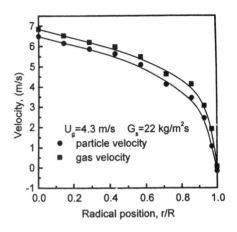

(a) gas/solids velocity distribution

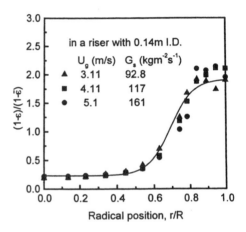

(b) solids concentration distribution

Figure 3 Radial distribution of gas/solids velocity and solids concentration in a CFB.

the wall region, where there are large amounts of clusters, only a little gas passes by. On the other hand, the radial gas exchange and diffusion rate between the center and the wall region is rather low, which results in a large difference of chemical reaction rate between them. (2) Significant gas and solids backmixing is due to the nonuniform radial gas–solid flow structure, especially when there is downflow in the wall region.

Thus for a fast fluidization, internals are used to redistribute the axial and radial gas–solid flow structure, that is, to improve the uniformity of gas–solid flow structure in space and to change the core–annulus flow structure, so as to promote radial gas–solid exchange.

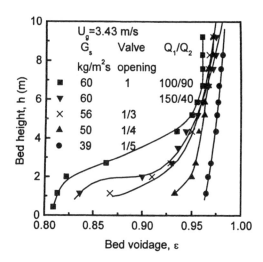

Figure 4 Axial distribution of solids concentration in a CFB.

1.4 Flow Structure of Pneumatic Conveying

When a continuous particle feeding of a CFB is cut off, particles become dilute and well distributed in the bed. Because of the low solids holdup and short residence time, this bed can only be used in pneumatic conveying, or in particular chemical reactors, where the reaction rate is very high or extra active catalyst is used. All internals used in risers can be applied in this situation, so it will not be discussed here.

In general, nonuniform structures, in both time and space, is widespread in bubbling, turbulent, and fast fluidization regimes. On the one hand, such nonuniformity can enhance the mass and heat transfer of a bed. On the other hand, it decreases the contact efficiency of gas and solids and makes the scale-up rather difficult. Internals are usually introduced not to eliminate the nonuniform flow structure completely but to control its effect on chemical reactions. The function of internals varies in different fluidization regimes, as do the types and parameters of internals. Taking these purposes into consideration, internals may be successfully applied to catalytic reactors with high conversion and selectivity, and some other physical processes.

If you have one of the following four requirements, please do not hesitate to choose a suitable type of internals:

1. To modify the nonuniform flow structure of gas–solids, that is, to avoid the overall nonuniformity on equipment-scale, and to make the nonuniformity in time much smaller than the characteristic time of a chemical reaction;

for example, to reduce the average bubble size in a bubbling fluidized bed.
2. To enhance the gas–solid contact and subsequently the mass and heat transfer, so as to eliminate them as rate limiting factors to a chemical reaction.
3. To improve the gas–solid movements and mixing along the radial direction, and to decrease the axial gas–solid backmixing to a certain extent.
4. To control the holdup of solids and to reduce the carryover of particles.

2 CLASSIFICATION OF VARIOUS INTERNALS

Internals include tubes, baffles, and other obstacles inside a fluidized bed. In some cases, for example, with a strongly exothermic or endothermic reaction in which substantial amounts of heat need to be removed or supplied from the bed, immersed tubes may be a necessity as surfaces for heat exchange. In other cases, baffles or other obstacles can act as internals to improve the quality of fluidization or to divide a bed into a number of stages in parallel or in series to promote smooth fluidized bed operation. It should be recognized that cyclone diplegs, downflow standpipes, feeding nozzles, and detective probes only come into contact with a very small part of the bed, and their effects on overall bed behavior are likely to be negligible. Although the end configurations of the inlet and the outlet of a fluidized bed (i.e., of a CFB) can also affect the two-phase flow structure, they will not be discussed here. Therefore, in this section, only baffles, tubes, and some novel geometric structures, which are introduced to significantly strengthen the contact efficiency of gas–solids and to improve the fluidization performance, are treated. Internals have different structural shapes and arrangements in a bed, hence each has its own performance characteristics. Internals reported in the literature are classified into five groups as baffles, tubes, packings, inserted bodies, and other configurations, which are listed in Table 1 in detail.

2.1 Baffles

Baffles can prevent bubbles from growing continuously, redistribute bubbles across the cross section of the bed, strengthen the heat and mass exchange

Table 1 Various Types of Internals

Category	Type	Configuration	Main features	References	Aspect studied	Application field
	Wire mesh		1. Increase bubble splitting 2. Low erosion	Dutta and Suciu (1992)	Effectiveness of baffles	Chemical reaction process (synthetic reaction of ethyl acetate)
	Perforated plate		1. Improve radial solids distribution 2. Increase bubble frequency	Gelperin (1971) Kono (1984) Zheng (1990) Dutta and Suciu (1992) Zhao (1992) Hartholt (1997)	Classification of a binary mixture of coal particles Bubble frequency, dense bed height Radial distribution of solids Effectiveness of baffles Solids backmixing Mixing and segregation of solids	Turbulent fluidized bed Gas–solid fluidized bed Circulating fluidized bed
Baffles	Single- or multiple-turn plate		1. Enhance gas and solids exchange 2. Decrease elutriation 3. Improve radial bubble distribution			Pilot chemical reactors (syntheses of phthalic anhydride and butadiene)
	Louver plate		1. Enhance gas and solids exchange 2. Decrease elutriation 3. Improve radial bubble distribution	Jin (1982)	Bubble behavior	Pilot chemical reactors
	Ring		1. Improve radial voidage distribution 2. Enhance radial gas and solids mixing 3. Improve gas-solids contact efficiency 4. Suppress axial solids mixing 5. Increase conversion of ozone decomposition reaction	Zhu et al. (1997) Zheng et al. (1991b,1992) Jiang et al. (1991)	Radial and axial voidage distributions Chemical reaction (ozone decomposition)	Fast fluidized bed Catalytic circulating fluidized bed area

Category	Type	Diagram	Effects	References	Characteristics	Application
Tubes	Horizontal banks		1. Split bubbles 2. Improve the uniformity of gas distribution	Olsson (1995) Löfstrand (1995) Olowson (1994) Bayat (1990) Yates (1987) Levy (1986) Jodra (1983, 1979) Glass and Harrison (1964)	Tube bundles design for pressure fluidized bed Bed expansion Bubble behavior and gas flow distribution Tube bundles design with the consideration of tube erosion Bubble size, tube erosion, tube bundle design Bubble eruption, particle elutriation process Prediction of the bubble size distribution Bubble size modification, tube bundles design Visual observation	Two dimensional fluidized bed bubble fluidized bed Pressure fluidized bed Cylindrical reactor (butanol and ammonia)
	Vertical banks		1. High heat-exchange coefficient 2. Low erosion phenomenon 3. Low scale-up effect	Volk (1962)	Scale up of chemical reactor	Turbulent fluidized bed
Packings	Station, irregular		1. Keep bubbles small and uniformly distributed 2. Reduce carryover of particles 3. Increase bed expansion 4. Increase chemical conversion 5. Impede solids motion	Huang (1997)	Pressure drop Solids holdup	Gas–solid–solid fluidized bed reactor
	Station, regular		1. Increase pressure drop 2. Increase solids holdup 3. Improve gas–solid contact efficiency	van der Ham et al. (1991, 1993, 1994)	Hydrodynamics and mass transfer	Circulating fluidized bed (CFB) Small scale CFB Pilot-plant scale CFB
	Floating		1. Binary particles with different size and density 2. Larger particles float freely in the bed	Goikhman (1969)		
Inserted bodies	Pagoda-shaped bodies	(See Fig. 8)	1. Break up bubbles 2. Enhance gas–solid contact	Jin et al. (1982)	Bubble behavior Fluid mechanics Pilot-plant trials	Turbulent fluidized bed reactor

Table 1 Continued

Category	Type	Configuration	Main features	References	Aspects studied	Application field
	Ridge-shaped bodies	(See Fig. 9)	1. Control bubble size 2. Improve the quality of fluidization 3. Extend the range of turbulent fluidization	Jin et al. (1986)	Bed expansion Bed homogeneity Change of flow region Emulsion phase behavior	Bubble fluidized bed Turbulent fluidized bed
	Inverse cone		1. Improve radial solids distribution 2. Decrease solids holdup	Zheng et al. (1990)	Radial solids distribution	Circulating fluidized bed
Inserted bodies (cont.)	Bluff bodies		1. Increase pressure drop 2. Decrease solids holdup 3. Increase gas and particle velocities 4. Improve radial voidage distribution 5. Enhance gas–solid contact efficiency 6. Reverse radial flow pattern	Gan et al. (1990)	Concentration profiles	Fast fluidized bed
	Spiral flow pates		1. Enhance gas–solid contact 2. Lengthen solids residence time	Li (1997) Cui (1996)	Vortex strength Cocurrent and countercurrent flow patterns Bed pressure drop Average density	Fast fluidized bed
	Swages		1. Decrease pressure drop	Davies and Graham (1988)	Pressure drop	Vertical pneumatic conveying tubes
Other configur-ations	Center circulating tube	(See Fig. 12)	1. Very high solid circulation rate 2. Uniform residence time distribution 3. Simple structure	Milne et al. (1992, 1994) Liu et al. (1993) Wang et al. (1993) Fusey et al. (1986)	Stable operation Temperature distribution Axial solid distribution	Circulating fluidized bed

between gas and solids, and decrease the rate of solids elutriation.

Moreover, baffles are useful for continuous processes since they divide a fluidized bed into an arbitrary number of stages in series. Unbaffled beds are notable for remarkable temperature uniformity, rapid solids mixing (approximate a well-mixed stage), and substantial amounts of gas backmixing, except for columns with large height/diameter ratios. These characteristics can be altered considerably by horizontal partitions. The RTD of gas and solids can be effectively narrowed in a baffled fluidized bed (Overcashier, 1959), though not to the extent corresponding to an equivalent number of perfect-mixing stages in series, and it is possible to establish temperature gradients between successive stages of the bed.

At the same time, the vibration of the fluidized bed can be alleviated by eliminating large bubbles and realizing uniform distribution of bubble size. With the decrease of the bubble size and velocity, a bed with internals has a higher bed surface than a free bed at the same gas velocity. However, the baffles tend to impede solids movement so that surface/bed heat transfer coefficients decrease; particle segregation can also occur, and then it is difficult to have a full fluidization in all compartments simultaneously. In addition, the total pressure drop across the bed will be slightly increased by the horizontal baffles.

As shown in Table 1, there are many kinds of baffles used in practice, such as wire meshes, perforated plates, single- or multiple-turn plates, ring baffles, and so on. Dutta and Suciu (1992) initiated an experimental investigation to study systematically the capacity of baffles of various designs for breaking bubbles in a medium-sized cold bed. The experimental conditions chosen and the baffle designs are believed to approach the hydrodynamic conditions of many commercial bubbling beds more closely than did earlier studies, such as those by Jin et al. (1982), Jodra and Aragon (1983), and Yates and Ruiz-Matinez (1987). Dutta and Suciu's results indicate that the effectiveness of a baffle may be directly correlated with the percentage of the opening area and the number of openings per unit cross section of the baffle.

When compared with free fluidized beds, beds containing horizontal wire meshes or perforated plates usually have smaller bubbles and fluidize more smoothly. In the 1970s, mesh baffles were used in the synthetic reaction process of ethyl acetate, where the active carbon was used as carrier.

The single- and multiple-turn plates are very effective in breaking bubbles. Although this kind of baffle is much more complex in structure, it is not easily eroded and can be used without maintenance for several years. Single- and multiple-turn plates have been successfully used to improve the synthetic yield and selectivity of phthalic anhydride and butadiene in pilot-scale fluidized bed chemical reactors.

The function of louver plates is quite similar to that of single- and multiple-turn plates. The only difference between them is that a louver plate cannot lead to gas swirling. It is also reported that louver plates have been used in pilot reactors, for example, for the synthesis of aniline.

When gas velocity is high enough, for instance, the reverse flow of solids through the baffle slots is inhibited by the upflow gas. Thus the expansion ratio of the bed increases very much, and a dilute phase of particles forms under each baffle at the same time, which is not good for chemical reactions. The flow of solids between the stages and the radial distribution of gas are of special importance in the design of multistage fluidized beds. A type of baffle plates with underflow downcomers has been developed that substantially improves the performance of multistage fluidized beds (Kono and Hwang, 1983)

The influence of ring baffles on the reduction of nonuniformity of a fast fluidization was investigated by Zheng et al. (1991). Jiang et al. (1991) also installed ring baffles in a riser 0.1 m in diameter to test their effect on chemical conversion. Three ring baffles with opening area ratios of 70%, 90%, and 95% were installed in a riser 7.6 cm in diameter and 3 m in height by Zhu et al. (1997). They found that the ring baffles could reduce radial nonuniformity and evenly redistribute solids in the radial direction. A detailed study on the radial solids distribution around the ring baffle shows the formation of a more dense region above the ring with a 70% opening ratio as compared with that above those rings with opening ratios of 90% and 95%, and the formation of a more dilute region below all three rings. Results of the axial voidage distribution with the presence of rings show the formation of a zigzag type axial profile instead of the regular S-shaped profile.

Although baffles are recommended to help reactor performance, their effectiveness will depend on the baffle design and the hydrodynamic conditions in the reactor, controlled by the following four factors:

2.1.1 Suitable Particle Systems

For group B particles for which bubble growth is significant, it is possible to keep the bubbles small by using baffles. However, for group A particles, the bubbles are usually small enough and will not be radically affected by the baffles unless the baffles are spaced sufficiently close to interfere with the overall solids circulation. Therefore baffles are usually applied to processes using fluidizing particles much different (say larger and/or heavier) from the conventional FCC-type catalysts, as has been demonstrated by Dutta and Suciu (1992)

2.1.2 Reasonable Design

A careful design should consider the choice of a baffle, which is effective in narrowing the RTD of solids, and the right configuration of it.

Perforated plates and ring baffles are adaptable to many solids processing operations which have a low reaction rate and demand higher solid-phase conversion.

The general criteria for the selection of optimum ring openings are to obtain uniform radial solids distribution and to prevent solids accumulation (dead zone).

For the selection of baffles, besides the requirements for good solids distribution, one should also reduce the dead zone and erosion. For example, rings with an inclined surface on top would encourage solids to slide into the core region.

After a proper type of baffle is selected, the baffle diameter and spacing should be decided. Usually the baffle diameter is smaller than that of the reactor. Thus there is an annular gap near the reactor wall, which contributes to axial solid backmixing. A certain degree of solid backmixing can overcome the shortcomings of solid segregation and the obvious axial temperature gradient in a baffled fluidized bed. However, too big a gap will lead to serious gas shortcuts that can significantly decrease the baffle efficiency. So a suitable annular gap should be selected in the range of 10 to 50 mm. To allow solids to be exchanged from stage to stage, the openings in the baffles should be made sufficiently large that particles can pass through. Or alternatively, downcomers that are similar to those used in distillation columns may be provided.

The baffle spacing is another important design parameter. We know that there are few rules for selecting baffle spacing. Many practices indicate that proper baffle spacing approximately equals the bed diameter. However, most pilot plant beds in China of 1 to 3 meters in diameter have baffle spacing of between 400 and 600 mm.

The scale-up of a fluidized bed with baffles is not straightforward, since the effect of baffles on the gas and solid flow is complex and is also likely dependent on the bed diameter. In Zhu's study (1997), the 90% ring can be considered to be the optimum ring opening for the 76 mm riser. This may not be true for a riser with a larger diameter.

2.1.3 Appropriate Operating Condition

The operating conditions are so extremely important that baffles may or may not improve the fluidized bed performance. For example, with a ring baffle, the improvement on the radial flow structure of a CFB is achieved by properly coupling the ring size and the operating conditions (Zhu et al., 1997). Determining the optimum operating conditions requires a trial and error approach.

2.1.4 Correct Installation

If transverse baffles, of suitable types and correct design are not installed quite horizontally, or if they move while the bed is being operated, instability of operation and other serious consequences like dead spots and channeling may occur.

2.2 Tubes

Although the improvement of gas–solid contact is usually the main objective when various baffles are selected, some concomitant improvement, when a bank of heat exchanger tubes must be used, is also welcome. The fluidized beds with tubes immersed in the bed have a number of important applications in industry.

To take advantage of the excellent heat transfer characteristics of gas fluidized beds, they are frequently packed with heating or cooling tubes through which a suitable heat transfer fluid is circulated. Supply or removal of heat in this way serves to control the temperature of the bed, which is a major factor in the design of chemical reactors whose product selectivity is an important consideration.

Heat exchanging tubes have an important effect on the performance of the bed even when they are located relatively far below the free surface of the bed. A vast majority of investigations on beds containing tubes has been devoted to attempts to improve the heat transfer between the bed and the immersed tubes. However, the same important effects that tubes have on fluidized bed performance have not been stressed in detail.

Major problems associated with the design of a tube bank are related to the orientation arrangement of tubes, the optimal tube diameter, the tube spacing, and the distance between the tube bank and the distributor. All of them control or are related to the flow of solids in the tube bank, and they consequently determine the heat transfer coefficient between the bed and the tube surface and the rate of mass transfer in the bed; finally they determine the rate of chemical reaction, if the bed is used as a reactor.

Heat transfer tubes may be positioned either vertically or horizontally within the bed, but whatever orientation they take; they will inevitably exert an influence on the pattern of gas flow through the bed.

Horizontal tubes are commonly used in fluidized bed combustors. The first qualitative information on the flow patterns near horizontal tubes was obtained by visual observation of the phenomena in two-dimensional beds (Glass and Harrison, 1964). Sitnai and Whitehead (1985) summarized the results of different experimental work on hydrodynamics in fluidized beds containing in-bed horizontal tubes, but most of the investigations were made under atmospheric conditions. Furthermore, Almstedt (1985), Almstedt and Ljungström (1987), and Almstedt and Zakkay (1990) presented measurements of in-bed hydrodynamics from pressurized fluidized bed combustors (PFBC) burning coal. While these beds contain horizontal cooling tubes, the possibility of varying operating conditions during combustion is limited. More systematic studies of the influence of pressure, fluidization velocity, and tube bank geometry on the hydrodynamics have been carried out in model beds operating at room temperature, i.e., by Staub et al. (1980) and recently by Olowson (1994) and Olsson et al. (1995).

Provided particle transport conditions are not approached, a thin cushion of air will form at the underside of an object immersed in an air fluidized bed, while a defluidized region of solid particles will rest on the topside. Especially when a tube diameter is fairly large, because of these effects, horizontal tubes or cylindrical obstacles do not have particularly good particle–surface contact except at the obstacle surface near the ends of the horizontal diameter of the obstacle. The relatively poor heat transfer coefficients at the top and bottom of horizontal tubes are partly compensated for by good heat exchange at the tube sides where bubbles may sometimes be formed. At fluidizing flow rates close to incipient fluidization, horizontal tubes may be at least as satisfactory for heat transfer purposes as vertical tubes, but at higher flow rates the vertical orientation is normally preferable. Horizontal tubes may cause rising bubbles to split, but this is only really effective when the tubes are much smaller than the bubbles and when "direct hits" occur. In any case, bubble coalescence usually takes place within a short distance above the horizontal obstacle, so the overall effect of horizontal tubes on the mean bubble size in the bed is seldom large, unless, of course, the tubes form an array, that fills the bed.

In industrial application of fluidization, horizontal heat transfer tubes are used to improve the uniformity of gas distribution at the base of a bed equipped with a low-pressure drop distributor. Except where horizontal tubes are used to aid gas distribution in this manner, or where a bed has a very small height/diameter ratio, fluidization considerations suggest that it is generally preferable to arrange tubes vertically in fluidized beds rather than horizontally.

Optimum tube spacing of a horizontal tube tank depends on the size of the solids used. The choice of horizontal spacing determines the extent of the vertical solids movement. The distance of separation between horizontal tubes is a major factor controlling the bubble breakage in fluidized beds. Reducing the distance of separation leads to increased bubble splitting to the extent that the average volume of the daughter bubbles is reduced to about 25% of that of the parent bubble (Yates, 1987).

Various triangular arrangements are used; the most typical, in order to increase tube packing density, are (1) triangular with vertical separation between rows equal to center-to-center horizontal tube spacing, (2) equilateral, and (3) "Rivesville" arrangement, in which face-to-face tube spacing between vertically staggered tubes are half of the horizontal face-to-face tube spacing (Cherrington and Golan, 1978). The limits of horizontal spacing (center to center) generally used are between two and four tube diameters, with three-tube-diameter spacing being favorable in experimental studies. For commercial design, where high surface area of tubes per unit bed volume is required, close spacing and small diameter tubes are preferable. The larger the ratio of bubble diameter to tube diameter, the more likely are the bubbles to split. The tube spacing is not critical at high fluidizing velocities; in that case, the radial mixing of solids is intensive. Closer tube spacing usually results in a slight decrease in heat transfer coefficient, but the penalty is compensated by a higher heat exchange area per unit bed volume (Sitnai, 1985). The horizontal arrangement affects the heat transfer of a tube bank much more than the vertical arrangement (Zabrodsky et al., 1981). In some cases, with a triangular array of tubes that occupied 8% of the bed volume, the bubble size is roughly 1.5 times the tube spacing (Glicksman et al., 1987). It seems possible, by matching the geometrical arrangement of the tubes with the operating conditions of the bed (i.e., gas velocity, particle size, U_{mf}, etc.), to exert control over the size of the bubbles.

A vertical tube bank is usually used for its high heat-exchanging ability and low erosion wastage, although it is not as effective as a horizontal tube bank in breaking bubbles. However, it has been observed that the heat transfer coefficient will decrease along vertical tubes in the upward direction in a circulating fluidized bed boiler, because many particles move along the tubes, keeping close to the wall and with little radial exchange. This can be prevented by putting up cornices along the tube bundle 200 mm apart (Fig. 5), which may effectively improve the heat exchanging efficiency.

Rising bubbles, which may be as large as several feet in diameter, set up forces that can cause vigorous transverse vibrations of immersed vertical tubes. There is evidence that these vibrations improve heat transfer rates and bed uniformity, so that it is not usually desirable to damp out these vibrations completely, whereas the vibrations set a minimum on the thickness of the vertical tubes and need to be considered when specifying the size, shape, and material of construction of the tubes. Moreover, vertical tubes

divide the bed into a number of parallel longitudinal compartments standing side by side. It has been asserted that scale-up should be based on the equivalent diameter (4 times the effective diameter of the fluidized bed over the effective saturation of the bed), which goes a long way toward overcoming the problems of the scale-up effect. This is especially true for beds in turbulent fluidization regimes of Geldart A particles. The reason is that for fine particles, the biggest stable bubble size is rather small and is sufficient for a vertical tube bank to prevent bubbles from centralizing while they are rising. But for coarse particles, the narrow spaces between the vertical tubes make it easy for channeling and gushing to occur because of the inability to break bubbles. These phenomena tend to have very adverse effects at higher gas velocities. For this reason, vertical tube banks are widely used when the weight fraction of fines in the catalyst is relatively high. Volk et al. (1962) used rods 4.8 cm in diameter as vertical inserts in a fluidized bed, and scale-up was made on that basis; favorable chemical conversions in large-scale beds were obtained.

It is reported that a large pilot fluidized bed reactor with vertical tubes (shown as Fig. 6), whose equivalent diameter is 10–20 cm, can achieve the same conversion and yield as a small reactor 10–20 cm in diameter (Ye, 1987). This result has been verified in the iron reduction process and in the synthesis process of carbon monoxide hydrogenation.

On the other hand, slugging and channeling are easily present in parallel compartments between vertical tubes, which are often regarded as harmful to the performance of fluidized beds. The gas–solid exchange and heat transfer rate may significantly decrease when slugging and channeling take place. This can be prevented by choosing a proper tube diameter and spacing. When the ratio of the tube diameter to the average bubble size is larger than 1/3, gushing flow occurs in the bed. When this ratio is smaller than 1/5, some bubbles adhere to the vertical tube, and the fluidization performance will be improved by decreasing the bubble velocity and the bubble coalescence (Ye, 1987). The gap width between any part of adjacent surfaces should be at least 30 particle diameters. If smaller gaps exist, channels tend to appear and lead to a general deterioration in the performance of the fluidized bed.

2.3 Packings

Early at the beginning of the development of fluidization (in the 1950s), Rasching rings and Berl saddle packings were packed in fluidized beds to enhance

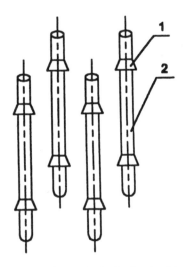

Figure 5 Structure of a vertical tube bank with cornices. 1 – cornice; 2 – vertical tube.

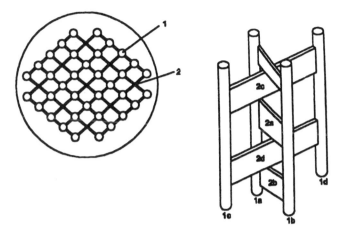

Figure 6 Vertical internals used in the iron reduction process. 1 – vertical tube; 2 – crosspiece.

the efficiency of gas–solid contact. But they prevent the particle movement so seriously that the bed would easily be defluidized.

Up to the 1960s, packings in fluidized beds received considerable attention in North America, and beds filled with packings have fallen into two groups (Grace and Jarrison 1970):

1. Screen-packed beds in which the packings consist of small open-ended wire mesh screen cylinders
2. Packed fluidized beds in which small particles are fluidized in the interstices of much larger unfluidized particles

Compared with other types of packings, screen cylinder packings occupy only 5% of the bed volume and thus attract special attention. There were many investigations of it on gas–solid contact, bed expansion, bubble size, solids entrainment, fluidization quality, and effect on chemical reactions. It was found that the screen cylinder type tends to keep bubbles small and uniformly distributed so that slugging is suppressed, carryover of particles is reduced, bed expansion is increased, and chemical conversions are increased. Although metal screens have many advantages, they are easy to transfigure in fluidization beds for poor strength, and cannot be operated for a long period of time. Moreover, solids motion is impeded so that heat transfer rates are less favorable than in corresponding unpacked systems, and segregation of particles may be significant. Dead spots and channeling also tend to occur, and it may be difficult to dump the fluidized particles from a bed containing fixed packings. But till now, chemical reactors packed with metal screens are still applied in the acrylonitrile indus-

try, and they have effectively improved the synthesis yield of acrylonitrile.

In the 1990s, a large research effort was put into the second group and some equipment has been developed. For example, the gas–solid–solid circulating fluidized bed reactor (GSSCFR) is such a novel reactor that can carry out, in addition to a primary reaction, a heat transfer, a mass transfer, or a momentum transfer operation or even another reaction simultaneously. In a GSSCFR, a packed bed and a fluidized one are interconnected as shown in Fig. 7. Large catalyst particles are fixed in the packed bed. Another amount of fine particles is fluidized, which is often used as an adsorbent or a heat carrier, circulating between the packed bed and the fluidized bed. For reversible reactions, the powder adsorbent can selectively adsorb and remove the desired product while it flows through the packed bed, so the reactor can shift the equilibrium in favor of the desired product. For strongly exothermic or endothermic reactions, the powder can remove or supply the heat of reaction and make a uniform temperature distribution in the reaction zone.

In a GSSCFR, the behavior of a gas–solid two-phase concurrent flow through the packed bed greatly affects the mass and heat transfer. The gas and powder flow concurrently through a packed bed to intensify the gas–solid contact (Yamaoka, 1986; Shibata et al., 1991; Song et al., 1995; Van der Ham et al., 1996). Huang et al. (1996a,b) extended the research started by Song et al. (1995), and examined the axial distribution of powder holdup in the packed bed; they developed a mathematical model to predict the powder mass flux and the mean dynamic powder holdup. They found that the powder mass flux is influenced by the powder inventory, the gas velocities, and the void of

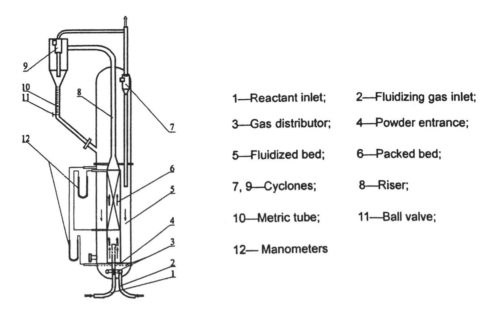

Figure 7   Structure of a gas–solid–solid circulating fluidized reactor (GSSCFR).

1—Reactant inlet;   2—Fluidizing gas inlet;

3—Gas distributor;   4—Powder entrance;

5—Fluidized bed;   6—Packed bed;

7, 9—Cyclones;   8—Riser;

10—Metric tube;   11—Ball valve;

12— Manometers

the packed bed. More powder inventory, less gas velocities, and larger voids of the packed beds are of benefit for increasing the powder mass flux and hence for enhancing the heat and mass transfer.

A GSSCFR exhibits good operating stability and simple construction and can be operated at a wider range of superficial gas velocity than a gas–solid–solid trickle flow reactor (Yamaoka, 1986; Shibata et al., 1991; Song et al., 1995). The experimental results of oxidation–dehydrogenation of butene to butadiene show that the C_4H_8 conversion and C_4H_6 selectivity can be significantly increased in a GSSCFR (Huang et al., 1998).

Some early Soviet work (Goikhman et al., 1968) suggests that packings (which are much larger than the fluidized particles) may be more effective if they are light enough to float freely in the bed instead of being essentially fixed in position as described above. This appears to be a peculiar approach, but more work is necessary before floating packings can be evaluated with confidence.

Despite the considerable amount of research work that has been carried out in the laboratory on packings in fluidized beds, the technique has not yet found a place in large-scale industrial applications because of its fatal weakness of preventing solids movement.

It is well known that, compared with a packed bed reactor, the most outstanding advantage of a fluidized bed reactor is its strong ability for heat exchange. That is why it is especially suitable for processes with strong

heat effects. The favorable bed/surface heat transfer rates and solids mixing characteristics of gas fluidized beds result from the motion of particles, motion caused by rising bubbles; therefore it is important not to impede the particle movement greatly. However, most packings in a fluidized bed seriously prevent the movement of particles, and they are harmful to heat transfer too. Though the effective thermal conductivity of a packed fluidized bed can be about 75 times that of a normal packed bed (Ziegler, 1963), it is still lower than that of a fluidized bed. It seems that packed fluidized beds are only useful when it is not practicable to divide the solid material into particles fine enough to permit fluidization at reasonable fluid flow rates, or when a highly exothermic reaction takes place on the surface of the large fixed particles.

2.4   Insert Bodies

It has been suggested that a combination of horizontal and vertical internals has an important effect on improving fluidization quality. Thus various kinds of insert bodies come into being, such as: pagoda-shaped bodies by Jin et al. (1982), ridge-shaped bodies by Jin et al. (1986b), inverse cones by Zheng et al. (1990), half oval bluff-bodies by Gan et al. (1990), and spiral flow plates by Li (1997).

On the basis of an empirical investigation on the effects of the behavior of the bubbles in fluidized beds and different types of internals on the motions of the bubbles and the solid particles, Jin et al. (1982) have designed a type of pagoda-shaped body (Fig. 8). The pagoda-shaped bodies are made up of three separate parts welded together: a perforated pipe with square cross section, an inclined sieve plate inside the pipe, and a square cornice. The vertical perforated pipe is like the normal type of vertical tube, its purpose being to restrict the bubble size and to guard against any tendency of the bubbles to congregate in the center of the bed as they move upwards. Most of the bubbles are forced into the perforated pipes through the sieve openings and the larger holes, only to be redispersed through the holes further up the pipes in the form of strings of small bubbles. However, owing to the presence of four larger holes and numerous sieve holes on the wall of the pipe, radial movement of the solid particles is quite vigorous, which is advantageous as far as heat and mass transfer are concerned.

The effect of the inclined sieve plates inside the pipe and the cornice outside the pipe is to break up and renew the bubbles. At higher gas velocities, especially, the effect is to force the bubble phase to pass through the sieve holes or around the pipe. During this process, the bubbles are continuously broken up and renewed, the long-distance backmixing of solids and gas in the emulsion phase is cut off, and relatively high gas–solid exchange coefficients between the bubble phase and the emulsion phase are obtainable. In addition, as the perforated pipe is empty, two-phase flow occurs both inside and outside the pipe, and the effective space inside the bed is not reduced. Also, the inclined

sieve plates inside and the cornices outside adjacent pipes are arranged in an overlapping fashion, thereby making the spatial distribution of solids more uniform and avoiding the formation of air "cushions" in the bed. Vertical and radial circulation of the particles is thus improved.

From the viewpoint of mechanical design, the pagoda-shaped bodies are simple to construct, easy to fabricate, and mechanically strong. During startup and shutdown, they do not experience any violent impingement of the catalyst, and during operation, they do not readily suffer from thermal deformation, and can remain in working condition for a long period of time. Since the orientation of the tubes is the same as that of heat exchanger tubes, their installation, maintenance, and overhaul are quite easy. Through experiments in a laboratory reactor (30 cm in I.D.), a medium-sized reactor (1 m in I.D.), and a pilot reactor (2 m in I.D.), pagoda-shaped bodies have proved to be very effective at high gas velocity, catalyst loading rate, and reaction yield.

The ridge-shaped bodies (Jin, 1986b), which are said to be capable of controlling the size of the bubbles, improving the quality of fluidization, and extending the range of turbulent fluidization, are composed of vertical heat exchange tubes and ridge baffles (Fig. 9). Figure 10 shows the structure of an oxidizing reactor with ridge-shaped bodies.

The pagoda- and ridge-shaped bodies are used in turbulent fluidized beds. However, for a CFB, a more simple construction is adopted, such as inverse

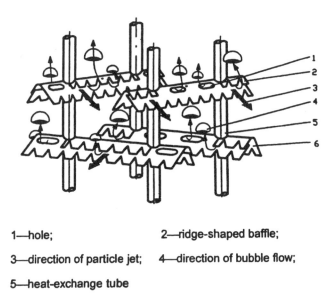

1—hole;   2—ridge-shaped baffle;

3—direction of particle jet;   4—direction of bubble flow;

5—heat-exchange tube

Figure 8   Construction of the pagoda-shaped bodies.

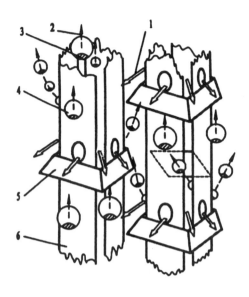

1—direction of particle jet;   2—direction of bubble flow;

3—bubble wake;   4—bubble;

5—cornice;   6—perforated pipe

Figure 9 Construction of the ridge-shaped bodies.

cones and bluff bodies. According to Gan et al. (1990), by forcing gas and solids to flow upward in the narrow ring section near the wall, half oval bluff bodies can reverse the radial flow distribution in a circulating fluidized bed and cause higher solids concentration and lower gas and solids velocities in the core, and lower solids concentration and higher gas and solids velocities in the wall region. Gas and solids are forced to flow upward in the narrow annular region near the wall. At the same time, the gas–solid mixing along the radial direction could be also greatly intensified.

Besides the single- or multiple-turn plate, the spiral flow plate is another kind of internal that may lead to gas and solid swirling. The main functions of a spiral flow plate are lengthening the residence time of solids and enhancing the contact between gas and solids. There are many kinds of spiral flow plates, with different structures and opening area ratios, and even with different arrangements, which are applied in drying processes (Fig. 11). The vortex strength is a common index to evaluate them. Li (1997) found that a reversed conic plate is better than a conic spiral flow panel in PVC drying processes.

Generally, the introduction of solid obstacles into a bed further complicates the system, and there is as yet no relevant theoretical analysis available.

2.5 Other Configurations

There are also other elements in fluidized beds, not installed as internals, but necessary for some purposes.

One example is that by Davies and Graham (1988). They swaged indentations 3.5 mm deep and 16 mm wide onto the wall of a riser 0.152 m in diameter to test their effect on pressure drop in the riser. They found that swages evenly spaced along the circular riser wall would scrape the dense downflow solids away from the wall and cause higher bed voidage and lower pressure drop in the fully developed region than those in a bare tube.

Another example is the strange "nose" in a circulating fluidized boiler. As we know, the circulating coal ash usually feeds into a boiler from one side of the bed wall, which easily results in annular flow and decreases the burning efficiency. Thus a protruding object like a "nose" is set at the other side in order to prevent the nonuniform flow.

The third example, fast fluidization, can also be achieved in a concentric circulating fluidized bed. One option is to have solids flow up in the center and down in the annular shell. This has been called an internally circulating fluidized bed. Alternatively, one can have solids flow up in the annular shell and

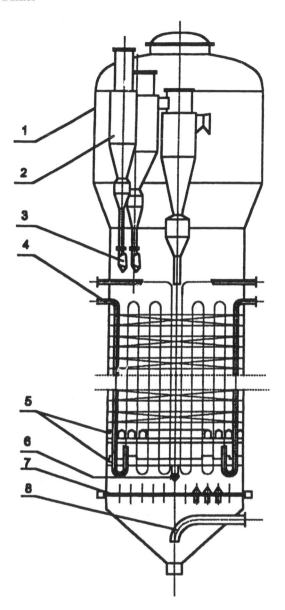

1—main body;   2—cyclone seperator;   3—flutter valve;   4—heat-exchange tube;

5—ridge-shaped body;   6—cone end cap;   7—distributor;   8—inlet duct

Figure 10 Structure of an oxidizing reactor with ridge-shaped bodies.

flow down in the center, leading to what has been called the integral circulating fluidized bed. These two options are shown in Fig. 12. Since both have been abbreviated to ICFB, the former will be identified as i-ICFB and the latter as o-ICFB with the i and o indicating the inside and outside location of the riser.

The i-ICFB originated from the spouted bed with a draft tube. Preliminary hydrodynamic studies by Milne

et al. (1992a,b) and by earlier workers (Fusey et al., 1986) show stable operation of this system. One advantage is that a very high solids circulation rate (near $600 \, kg/m^2.s$) can be realized in this system (Milne et al., 1994). Milne et al. (1994) have shown that a stable, isothermal operation at high temperature can also be achieved in an i-ICFB unit and have suggested that it can be ideal for short contact time reactions where

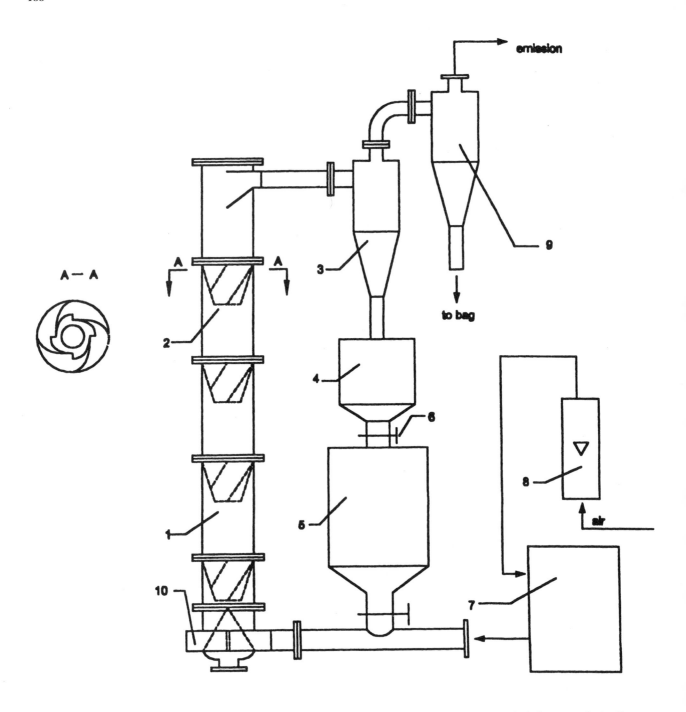

Figure 11 Diagrams of multistage dryer with spiral flow plates. 1—spiral flow bed; 2—spiral flow panel; 3—first-stage cyclone; 4—measure tank; 5—storage tank; 6—butterfly value; 7—electric heater; 8—flowmeter; 9—second-stage cyclone; 10—inlet.

uniform residence time distribution and stable high temperature operation are needed, such as in pyrolysis.

o-ICFB has been developed as an alternative to the regular CFB systems to address the concerns of non-

uniform solids feeding at the riser bottom and the energy requirements of a large amount of external solids separation and recirculation. There are several advantages: (1) simple structure; (2) high solids circu-

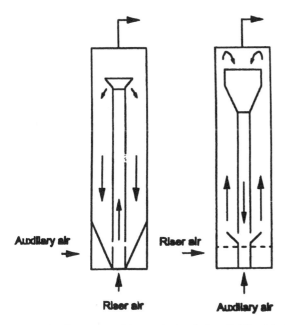

Figure 12 Structure of integral circulating fluidized bed.

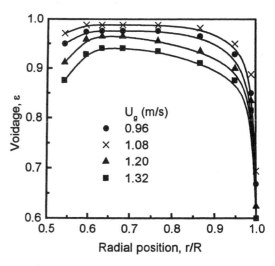

Figure 13 Typical radial solids concentration profiles in an o-CFB.

lation rate; (3) uniform solids feeding. As for i-ICFB, the solids circulation rate is not independently controlled but varies with riser and auxiliary gas velocities. Liu et al. (1993a,b) and Wang et al. (1993) found that axial solids distributions in the riser of o-ICFB are very similar to those in regular CFB risers. The cross-sectional average solids holdup is somewhat higher than that in regular CFB risers operated under similar conditions (Wang et al., 1993). Typical radial solids concentration profiles measured by Wang et al. (1993) show that a "core-annulus" structure still exists inside the annular riser, with the dense layer at the outer wall slightly thicker than that on the inner wall (Fig. 13). Measuring at two vertical planes in the annular riser on the opposite sides of the central standpipe, Wang et al. (1993) found that the radial voidage distributions are symmetrical.

It has already been stressed that it is impossible to specify a single type of internal that is best for all applications. This is because each successful application of fluidization depends upon different properties of fluidized beds to varying degrees, and a type of internal that helps one process may hinder another. For example, particle segregation is an adverse effect in many applications, but not, of course, if a fluidized bed is used as a particle classifier; then an internal member that promotes segregation (i.e., fixed packings, horizontal screens) will be appropriate. In many continuous operations, horizontal baffles may be of considerable benefit. Where heat transfer is of para-

mount importance or where an existing system does not perform as well as expected, vertical tubes warrant serious consideration. Clearly, combinations of various kinds of internals are likely to be useful in certain applications.

3   EFFECT OF INTERNAL TUBES AND BAFFLES

Many references in the literature mention that the improved fluidization results from the use of tubes and baffles as internals. Harrison and Grace (1971) and Sitnai and Whitehead (1985) have reviewed the influence of internals on the behavior of fluidized beds for reactor design purposes. For example, on the one hand, most transverse baffles in a fluidized bed effectively narrow the gas–solid residence time distribution, prevent rushing at the surface of the bed, and reduce the entrainment of solid particles. On the other hand, a disadvantage of the use of baffles in a bed is the presence, underneath each baffle, of a region of lower density where the solid particles are more dilute. Another disadvantage is the axial size-classification of particles. Furthermore, increasing the diameter of the fluidized bed magnifies these problems.

In this section, the effect of internals on a fluidized bed is analyzed from the following four aspects.

3.1   Bubble Behavior

Bubble size is a very important parameter in the design and simulation of fluidization. We can accept that

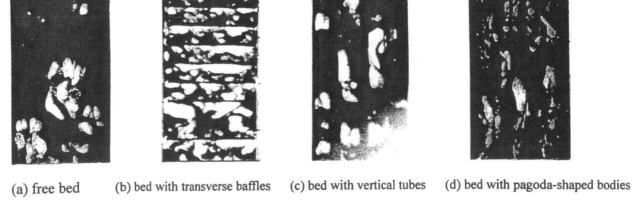

(a) free bed   (b) bed with transverse baffles   (c) bed with vertical tubes   (d) bed with pagoda-shaped bodies

Figure 14 Bubble behavior in fluidized beds with different internals.

most of the effects that internals produce in a fluidized bed are derived from the same primary effect: control of the bubble size and its distribution. The movement of bubbles in a bed containing no internals has already been discussed in Sec. 1.1.

Jin et al. (1982) have investigated the ability of internals to break up bubbles in a two-dimensional fluidized bed by the photographical method (Fig. 14). For comparison, Fig. 14a shows the bubbles in a free fluidized bed. When an internal module is introduced, such as a baffle, a tube or a pagoda-shaped body, the situation is modified to different degrees according to the module type, as shown in Figs. 14b,c,d.

Figure 15 shows the effect of transverse baffles (louver plates) on the behavior of bubbles at different gas velocities. The baffles do not visibly influence the bubble behavior (coalescence or rise velocity) at low gas velocities (Fig. 15a). Small bubbles push upward

past each layer of baffles, only to collapse when they reach the surface of the bed. With the increase of gas velocity, a small, nonstationary, particle-free section is observed underneath a baffle with bubbles splitting above it. At much higher gas velocities, the effects of baffles become apparent—breakup, regeneration, and uniform distribution of bubbles. When the superficial gas velocity further increases, stationary particle-free sections are observed under each baffle and smaller bubbles are formed above them. The particles, which follow the bubbles upwards, mainly circulate through the bed in one of two ways: through the spaces around the baffles, or directly past the louvers in the baffles. Figure 15b illustrates the rain of particles, which is observed coming off the louver plates. Under these conditions, contact between the gas and solids is excellent.

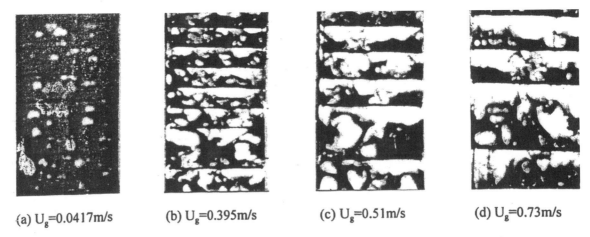

(a) U_g=0.0417m/s   (b) U_g=0.395m/s   (c) U_g=0.51m/s   (d) U_g=0.73m/s

Figure 15 Effect of transverse baffles (louver plates) on the behavior of bubbles at different gas velocities.

As the superficial gas velocity is increased still further, bubbles aggregate into large cushions underneath the baffles. The fraction of the volume of the bed which is occupied by the cushions of coalesced bubbles keeps on increasing with the gas velocities (see Fig. 15c). At higher gas velocities (Fig. 15d), a very large part of the bed is occupied by these air cushions, giving a very large void ratio and expansion ratio. Since a too-high gas velocity affects the vertical circulation of the particles, there is a gradual lessening and eventual disappearance of the "rain" of particles coming off the louvers of the baffles. The particles tend to become classified according to their fineness. A thick air "cushion" is also detrimental to heat exchange.

Figure 16 shows the effect of vertical tube banks on the behavior of the bubbles. The main effect of the presence of pipes inside a bed is to even up the spatial distribution of the bubbles within the bed and to reduce their tendency to congregate near the center of the bed. Such pipes also tend to have relatively little effect on bubble breakup.

To sum up, within a definite range of gas velocities, transverse baffles as well as vertical tubes are able to control, break up and renew the bubbles and thereby substantially improve the gas–solid contact within the bed. At higher gas velocities, however, such baffles and tubes will have a detrimental effect on fluidization performance.

The presence of horizontal tubes decreases the dense-phase fluidity, which prevents bubbles from accelerating and consequent coalescence. Horizontal tubes tend to cause the bubbles to split, resulting in smaller bubbles in the region just above the tubes. Processes of bubble splitting are followed by coalescence. This reaches an equilibrium with a relatively small bubble, similar in size to the tube spacing.

Under some circumstances, horizontal tubes will cause splitted bubbles to recombine, thus nullifying the beneficial effects of splitting (Yates, 1987). Large bubbles can rise over horizontal rods without breaking up (Xavier et al., 1978), and the bed expansion is not much influenced by the rods (Newby and Keairns, 1978).

Figure 17 shows the effect of pagoda-shaped bodies on the behavior of the bubbles in a two-dimensional bed at different gas velocities. When the gas velocity is very low, a large number of small bubbles adhere to the surface of the perforated pipe, passing back and forth through the holes during the course of the upward movement, with the inserts exerting a definite effect on the bubble velocities (Fig. 17a). As the gas velocity is raised, there is a slight increase in the bubble diameter, but the presence of the inserts prevents the bubbles from growing too much; some of the bubbles move away from the wall of the pipe (Fig. 17b). At higher superficial gas velocity, it is possible to observe each individual bubble being constrained and broken up by the pagoda-shaped bodies (Fig. 17c). At still higher superficial gas velocities, it becomes more and more difficult to discern the behavior of individual bubbles (Fig. 17d). Under these conditions, the bed could be classified into two regimes: a bubble phase, in which groups of bubbles are of no definite shape, and in which the solid particles are sparsely scattered and move upward more or less in train with the bubbles; and an emulsion phase, in which the solid particles are present in greater concentration and moving downward. The equivalent diameter of the bubbles is basically constant over a rather wide range of superficial gas velocities, irrespective of the height of the bed. Raising the superficial gas velocity under these conditions increases the volume of the bubble phase

(a) U_g=0.0417m/s   (b) U_g=0.395m/s   (c) U_g=0.51m/s

Figure 16 Effect of vertical tube banks on the behavior of the bubbles.

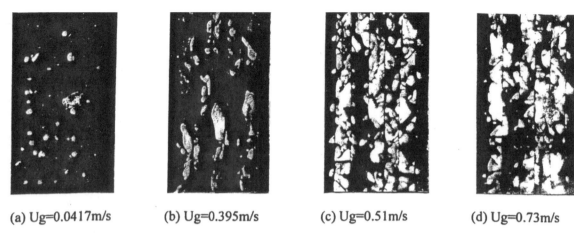

(a) Ug=0.0417m/s   (b) Ug=0.395m/s   (c) Ug=0.51m/s   (d) Ug=0.73m/s

Figure 17 Effect of pagoda-shaped bodies on the behavior of the bubbles.

and intensifies turbulence. Also, there is an increasingly rapid exchange between the bubble and emulsion phases throughout the bed.

In general, by catching and breaking up the bubbles, the internals have the effect of reducing their size. On one hand, when a bubble is split by an internal, the sum of the volumes of the daughter bubbles is less than the volume of the parent one, indicating that some of the original bubble gas leaks into the emulsion phase of the bed during the splitting process. Furthermore, the amount of leakage increases as the number of daughter bubbles increases. These facts could be of importance in the design of fluidized bed reactors. With the reduction of the mean bubble size, the expansion ratio of a bed changes. Figure 18 shows quite clearly the effects that different types of internals have on the expansion ratio in a two-dimensional fluidized bed. On the other hand, small bubbles rise relatively slowly in a fluidized bed, and so the result is to increase the total concentration of bubbles in the bed. A thick mist of small bubbles has a beneficial effect on the conversion of a chemical reaction.

Two different methods have been developed for prediction of the bubble size in a bed with internals formed by horizontal rows of tubes (Jodra and Aragon, 1983). One, based on a new correlation, makes it possible to find the size of a bubble above the tube as a function of the operating conditions and characteristics of the tube arrangement. The other is based on the general theory of bubble formation in submerged orifices. Both methods can be used to estimate bubble sizes, which are acceptably close to the experimental values, and can be applied to other types of internals, such as perforated plates, meshes and arrangements of horizontal bars of noncircular cross sections.

Besides splitting bubbles, internals alter the eruption mechanism for single bubbles at the bed surface, causing a significant fraction of the single bubbles to generate large wake eruptions. Levy et al. (1986) have found that without any internals in the bed, when single bubbles erupt, the bulge material of a bubble always reached a greater height above the bed free surface than the wake material. With horizontal tubes present, up to 60 percent of the single bubble eruptions are characterized by the upward projection of wake material, reaching a height above the free surface which far exceeds those attained by the bulge material. The wake-dominated eruption of the single bubble appears to be due to the formation of elongated vertical cavities in the region between the internals and the free surface. A quantitative understanding of the effects of internals on bubble eruptions may lead to

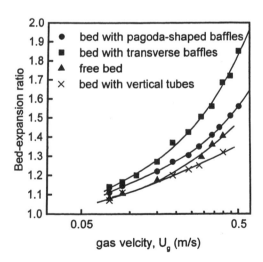

Figure 18 Effects of internals on the expansion ratio in a two-dimensional fluidized bed.

techniques for designing the internals to minimize solids elutriation.

Recently, it has been observed that the bubble behavior is affected by pressure. Olsson (1995) has found that there is an initial increase in the mean pierced length, the mean bubble rise velocity, the mean bubble volume fraction, and the mean visible bubble flow rate as the pressure is increased. With a further increase in pressure, these visible bubble flow parameters start to level off or even decrease. Generally speaking, this decrease is more pronounced for the high excess gas velocity than for low velocity. This behavior is governed by a balance between bubble coalescence and bubble splitting. At low pressures, bubble coalescence dominates over splitting, while, as the pressure is increased, the splitting process becomes predominant.

Moreover, the distribution of bubbles is also related to pressure (Olsson, 1995). Experiments show that at low excess gas velocity, there is a redistribution of the visible bubble flow toward the center of the bed as the pressure is increased. At high excess gas velocity, however, the influence of pressure is not the same. For lower pressures, there is still a redistribution of the bubbles toward the center of the bed with increasing pressure, but for higher pressures the distribution becomes somewhat more uniform again as the pressure is increased further. Olsson's explanation for the bubble distribution becoming more uniform at combinations of high excess gas velocity and pressure was that the bed was developing a more turbulent behavior, with increased bubble splitting and consequently a more dispersed bubble flow.

3.2 Flow Distribution

As discussed in Sec. 1, nonuniform flow is an inherent characteristic of gas–solid fluidization systems. A typical radial solids distribution in a riser without any inserts is not uniform: the solids concentration is low in the center and high in the wall region. In many circumstances, internals are aimed at redistributing gas and solids flow, in an effort to form a more uniform flow structure to improve interphase contact efficiency so as to increase the overall performance of reactors.

Ring baffles and swages scrape the downflowing solids away from the wall towards the center, leading to a more uniform radial solids distribution. Using an optical fiber concentration probe, Zheng et al. (1991, 1992) have showed that the radial voidage distribution is much more uniform with ring baffles present (Fig. 19). Jiang et al. (1991) have confirmed these findings by

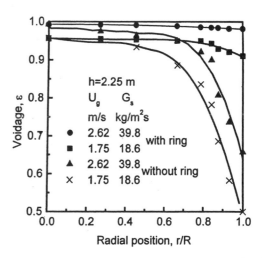

Figure 19 Radial voidage distribution in a ring-baffled fluidized bed.

showing that the conversion of the ozone decomposition reaction is improved with the addition of ring baffles. Arrangement of a narrow seam near the wall further improves the performance. The gas preferably flows upward in the annular region so that the radial distribution becomes more uniform. Similarly, a half-oval bluff body forces the gas and solids to flow upwards in the narrow annular region (Fig. 20). Hence the dilute core and dense annulus flow pattern are disrupted (Gan et al., 1990). As a consequence, gas–solid contacting is greatly improved.

However, not all internals provide favorable gas and solids redistribution. Salah et al. (1996) have found that ring baffles with smaller opening areas can make the radial voidage distribution less uniform under low gas velocity. With a ring of 44% open area, Zheng et al. (1990) reported increased nonuniformity under certain operating conditions. For risers operated at high gas velocity and high solids flux, e.g., the riser in FCC units, particles mainly flow upwards at the wall so that the installation of ring type baffles may actually induce particle downflow, leading to more radial flow segregation. Fortunately, Zheng et al. (1990) have found that a plate with larger orifices and more opening area near the wall is very effective in flow redistribution. Therefore, properly designed internals can help to redistribute the gas and solids flow.

At the same time, internals can also change the average voidage. The results of Zheng et al. (1992) indicate that ring baffles significantly reduce the average solids concentration. Gan et al. (1990) also found that their bluff body baffle reduced the average solids concentration by scraping off superfluous particles

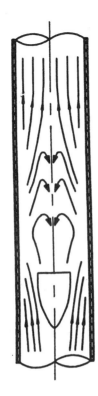

Figure 20 Flow pattern in a riser with a bluff body.

near the wall. However, Jiang et al. (1991) reported a higher local solids concentration in the presence of ring baffles. With the packed bar internals, which take up a large cross-sectional area of the bed (50%), van der Ham et al. (1993) measured 1.6 to 5 times higher solids holdup compared with an empty column. Although part of this increase was due to particles settling on top of the bars, at least some was due to the bars. It would appear that baffles with less flow restriction (e.g., less reduction in cross-sectional area) mainly increase the actual gas velocity, scraping the dense downflow solids away from the wall, leading to higher bed voidage and lower pressure drop as observed by Davies and Graham (1988) and Zheng et al. (1991, 1992), while baffles with a significant reduction of cross-sectional area constrict the solids flow, leading to lower bed voidage (e.g., Jiang et al., 1991; Van der Ham et al., 1993). The bluff body seems to be an exception. It occupies 50% of the riser cross section, significantly increases the gas and particle velocities, and causes a higher pressure drop; it does not increase the solids holdup. This may be attributed to the smooth parabolic shape facing the gas and solids flow.

Besides the gas and solids distributions, the installation of internal baffles can significantly affect the pressure distribution in a fluidized bed.

Pressure fluctuations in the fluidized bed have been used successfully as a measure of bubbling intensity, and the method is found particularly useful for the study of behavior of beds with internals (Kang et al., 1967; Newby and Keairns, 1978; Staub and Canada, 1978). Jiang et al. (1991) measured an increase in pressure drop when four ring baffles with an opening area of 56% were installed in their riser 0.1 m in I.D. Gan et al. (1990) experienced a localized high-pressure reduction across their bluff body, accompanied by significant solids acceleration.

3.3 Gas and Solids Mixing

The degree of axial mixing of gas and particles in fluidized beds is important for many continuous as well as batch processes, and control thereof is desirable. It is very important to limit backmixing and maintain good gas–solid contact while using fluidization for solid or gas–solid processing, as these operations need high solid-phase conversion and gas-phase utilization.

First, gas backmixing is substantially reduced by immersed obstacles. The higher the gas velocity the smaller the gas backmixing (Ye, 1987). There is little gas backmixing between the top and the bottom of each horizontal baffle. The axial gas-mixing coefficient is lowered only in beds of small particles or at low gas velocities. The larger the particles and the higher the gas velocity, the less effective is the horizontal baffles.

Horizontal tubes completely suppress gas backmixing in a large range of velocities, but vertical tubes are less effective at higher velocities. A noteworthy result with vertical tubes is the development of marked nonuniformity of radial gas mixing.

Secondly, most internals are likely to suppress axial solids backmixing. Subdividing the bed with horizontal baffles is a common and successful practice to improve gas–solid contact and to limit solids backmixing. Many configurations of baffles not only can constrain bubble growth but also can narrow the RTD of gas and solids. For example, a 1 m I.D. bed with baffle spaces of 100 to 200 mm, with a perforation ratio smaller than 30% and a small space between the plate and the wall, can reduce axial solids backmixing and lead to a large temperature gradient along the axis. Zheng et al. (1992) have found that the axial solids dispersion of a ring baffle is only half of that in an empty column.

The baffled fluidized bed, when used to carry out a reaction, is a kind of multistage fluidized reactor, which can be modeled by treating the overall mixing of solids as a series of perfectly mixed stages with solids backmixing between them. Consequently, the determination of the solids backmixing rate between stages is important.

Experimental results show that, for a given bed material, the main factors influencing the solids backmixing rate are gas velocity, baffle free area, and baffle spacing. Based on a theory of particle entrainment, a mathematical model has been developed to predict the solids backmixing rate in smooth bubbling fluidization (Zhao et al., 1992).

Modeling of solids mixing in larger scale beds with horizontal tubes has been reported by several workers. Solids tracers having similar properties to the bed material are usually used in mixing experiments. One of the simpler models (Sitnai et al., 1981) postulates one upflow and two downflow regions. From the interpretation of mixing data, a high exchange rate of solids between the main downflowing dense phase and the upward moving drift phase is inferred. At fluidizing velocities of up to 1 m/s with particles about $700\,\mu m$, the axial solids dispersion coefficient has little effect on the overall solids transport, and the solids exchange coefficient between regions remains constant.

Solids mixing in beds with an assembly of vertical tubes has not been widely investigated. Ramamoorthy and Subramanian (1981), applying a one-dimensional diffusion model, determined an empirical correlation to predict the axial diffusion coefficient at given fluidizing velocity and packing density of vertical rods. The diffusivity is substantially decreased in the presence of internals and further decreased with a reduction in the spacing of internals.

Most internals appear to improve the radial gas and solids mixing. Using FCC particles with a mean diameter of $89\,\mu m$ impregnated with ferric oxide as catalysts for ozone decomposition, Jiang et al. (1991) observed a more uniform radial distribution of ozone concentration in the riser with ring baffles (Fig. 21) and attributed this to higher radial gas and solids mixing. Owing to the enhanced gas–solid contact efficiency, increased gas and solids radial mixing and suppressed axial mixing, CFB reactor performance is expected to improve when internal baffles are present. Jiang et al. (1991) showed that ozone conversion was significantly higher when ring baffles were installed in a CFB reactor.

In summary, the addition of baffles, tubes, and other obstacles restricts large-scale solids movement, reduces the gas and solids backmixing, improves the radial gas and solids mixing, and causes segregation of particles of different sizes.

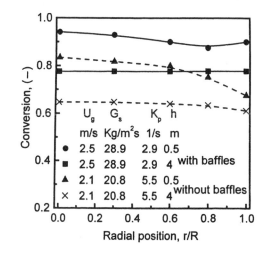

Figure 21 Radial distribution of ozone concentration in a riser with ring baffles.

3.4 Transition to Turbulent Fluidization

Flow patterns in fluidized bed chemical reactors have a strong influence on the performance of reactors. Most commercial catalytic reactors are run successfully in the turbulent fluidized region. Numerous investigators have used the method of analyzing the pressure fluctuations in the bed to characterize the transition from bubbling to turbulent fluidization; see, e.g., Canada et al. (1978), Yerushalmi and Cankurt (1979), Sadasivan et al. (1980), Satija and Fan (1985), Rhodes and Geldart (1986), Noordergraaf et al. (1987), Andersson et al. (1989), Cai et al. (1989), and Olowson (1991). These fluctuations are caused mainly by the activity of the bubbles in the bed, e.g., by the bursting of bubbles at the surface and by the coalescence and splitting of bubbles in the bed.

The transition to a more turbulent behavior is associated with many factors, such as fluidization pressure, particle properties, and the geometric conditions of the bed. It is also influenced by the internals in the bed. The effect of internals on the transition velocity U_c has been systematically studied by Jin et al. (1986a). It can be seen from Fig. 22 that with internals in beds, the transition from a bubbling to a turbulent regime occurs at smaller U_c. Johnsson and Andersson (1990) have also suggested that tubes in a bed can cause a transition to a turbulent behavior. Similar result has been obtained by Olsson et al. (1995). They found that the tube banks caused the bubbles to break up at lower pressure and gas velocity, resulting in a more turbulent behavior than for the bed without tubes.

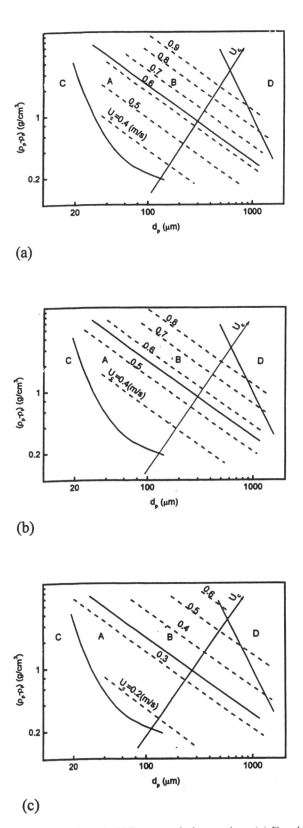

(a)

(b)

(c)

Figure 22 Effect of internals on the transition from bubbling to turbulent regime. (a) Free bed, (b) bed with vertical tubes, (c) bed with pagoda-shaped bodies.

Experiment results indicate that different types of internals have effects on U_c to different extents. Jin et al. (1986a) have proposed a correlation to predict U_c as follows:

$$\frac{U_c}{\sqrt{gd_p}} = \left[\frac{kD_f}{d_p} \frac{\rho_p - \rho_j}{\rho_f}\right]^n$$

where $n = 0.27$ and D_f is a parameter called performance diameter which has length dimension. The KD_f is found to characterize the geometric structure of the beds, and its value is obtained as follows:

$KD_f = 3.67 * 10^{-3}$ for free bed ($280 * 280$ mm)
$KD_f = 2.32 * 10^{-3}$ for bed with vertical tubes
$KD_f = 3.42 * 10^{-4}$ for bed with pagoda-shaped bodies

The average error of above formula is about 6.58%.

REFERENCE

Almsedt AE. Measurements of bubble behavior in a pressurized fluidized bed burning coal, using capacitance probes—Part II. Proceedings of the 8th International Conference on Fluidized Bed Combustion. Houston, 1985, pp 865–877.

Almstedt AE, Ljungström EB. Measurements of the bubble behavior and the oxygen distribution in a pilot scale pressurized fluidized bed burning coal. Proceedings of the 9th International Conference on Fluidized Bed Combustion, Boston, 1987, pp 575–585.

Almstedt AE, Zakkay V. An investigation of fluidized-bed scaling-capacitance probe measurements in a pressurized fluidized bed combustor and a cold model bed. Chem Eng Sci 45:1071–1078, 1990.

Andersson S, Johnsson F, Leckner B. Fluidization regimes in non-slugging fluidized beds. Proceedings of the 10th International Conference on Fluidized Bed Combustion. San Francisco, 1989, pp 239–247.

Cai P, Chen SP, Yu ZQ, Jin Y. Effect of operating temperature, pressure and internal baffles on the average voidage of gas–solid fluidized beds. Petro Chem Eng 18(9):610–615, 1989.

Canada GS, McLaughlin MH, Staub FW. Flow regimes and void fraction distribution in gas fluidization of large particles in beds without tube banks. AIChE Symp Ser No 176, 74:14–26, 1978.

Cherrington DC, Golan LP. Am. Pet. Inst. Proc. Div. Refining. 43rd. 57:164–182, 1978.

Davies CE, Graham KH. Pressure drop reduction by wall baffles in vertical pneumatic conveying tubes. In CHEMECA '88, Australia's Bicentennial International Conference for the Process Industries, Institute of Engineers, Australia, Sydney, Australia, August 1988, Vol. 2, pp 644–651.

Dutta S, Suciu GD. An experimental study of the effectiveness of baffles and internals in breaking bubbles in fluid beds. J Chem Eng (Japan.) 25(3):345–348, 1992.

Gan N, Jiang DZ, Bai DR, Jin Y, Yu ZQ. Concentration profiles in fast-fluidized bed with bluff-body (in Chinese). J Chem Eng Chinese Univ 3(4):273–277, 1990.

Glass DH, Harrison D. Chem Eng Sci 19:1001–1002, 1964.

Glicksman LR, Lord WK, Sakagami M. Bubble properties in large-particle fluidized beds. Chem Eng Sci 42(3):479–491, 1987.

Goikhman ID, Oigenblik AA, Genin LS, Filippova LA. Effect of floating spherical packing on the residence-time distribution of gas in a fluidized bed (in Russian). Khim Technol Topl Masel 13(10):36–38, 1968.

Grace JR. Contacting modes and behaviour classification of gas-solid and other two-phase suspensions. Can J Chem Eng 64:353–363, 1986.

Grace JR, Jarrison D. Design of fluidised beds with internal baffles. Chem Proc Eng 6:127–130, 1970.

Harrison D, Grace JR. In: Davidson JF, Jarrison D, eds. Fluidization. New York: Academic Press, 1971, pp 599–626.

Huang SY, Wang ZW, Jin Y, Yu ZQ. Solids mass flux and mean dynamic solids hold-up in the packed section of the gas-solid-solid circulating fluidized bed (in Chinese). Chem Reaction Eng Tech 12:369–376, 1996a.

Huang SY, Wang ZW, Jin Y, Yu Z. Axial pressure drop and powders hold-up distribution in packed section of gas-solid-solid circulating fluidized bed (in Chinese). J Chem Eng Chinese University 10:377–383, 1996b.

Huang SY, Wang ZW, Jin Y. Studies on gas-solid-solid circulating fluidized bed reactors, Chem Eng Sci 54(13–14):2067–2075, 1999.

Jiang P, Bi HT, Jean RH, Fan LS. Baffle effects on performance of catalytic circulating fluidized bed reactor. AIChE J 37:1392–1340, 1991.

Jin Y, Yu ZQ, Zhang L, Shen JZ, Wang ZW. Pagoda-shaped internal baffles for fluidized bed reactors. Intern Chem Eng 22(2):269–279, 1982.

Jin Y, Yu ZQ, Wang ZW, Cai P. A criterion for transition from bubbling to turbulent fluidization. In: Östergaard K, Sörensen A, eds. Fluidization V. Elsinore, Deamarh, May 18–23, 1986a, pp 289–296.

Jin Y, Yu ZQ, Zhang L, Yao WH, Cai P. Ridge type internal baffle for fluidized bed reactor. Petro Chem Eng (China) 15(5):269–277, 1986b.

Jodra LG, Aragon JM. Prediction of the bubble-size distribution in fluidized beds with internal baffles. Inter Chem Eng 23(11):18–30, 1983.

Johnsson F, Andersson S. Expansion of a bubbling fluidized bed with internals. Report A. Department of Energy Conversion, Chalmers University of Technology, Göteborg, Sweden, 1990, pp 90–184.

Kang WK, Sutherland JP, Osberg GL. Pressure fluctuations in a fluidized bed with and without screen cylindrical packings. Ind Eng Chem Fundam 6(4):499–504, 1967.

Kono HO, Hwang JT. The staged underflow fluidized bed. AIChE Symposium Series 222, Vol. 78, pp 37–46, 1983.

Levy EK, Freeman RA, Caram HS. Effect of horizontal tubes on bubble eruption behavior in a gas fluidized bed. In: K Östergaard, A Sörensen, eds. Fluidization V. Elsinore, Denmark, May 18–23, 1986, pp 135–142.

Li LX. Experimental study on spiral flow and design and application of a new combined drying system. MS thesis, Tsinghua University. 1997.

Liu DJ, Zheng CG, Li H, Kwauk M. A preliminary investigation on integral circulating fluidized bed (in Chinese). Eng Chem Metal 64–67, 1993a.

Liu DJ, Zheng CG, Li H, Kwauk M. Preliminary studies on integral circulating fluidized bed (in Chinese). In: Proceedings 6th National Conf. on Fluidization. Wuhan, China, October 1993b, pp 63–66.

Milne BJ, Berruti F, Behie LA. Solids circulation in an internally circulating fluidized bed (ICFB) reactor. In: Fluidization VII (OE Potter, DJ Nicklin, eds.). Engineering Foundation, New York, 1992a, pp 235–242.

Milne BJ, Berruti F, Behie LA, de Bruijn TJW. The internally circulating fluidized bed (ICFB): a novel solution to gas bypassing in spouted beds. Can J Chem Eng 70:910–915, 1992b.

Milne BJ, Berruti F, Behie LA, de Bruijn, TJW. The hydrodynamics of the internally circulating fluidized bed at high temperature. In: Avidan AA (ed.), Circulating Fluidized Bed Technology IV. AIChE, New York, 1994, pp 28–31.

Newby RA, Keairns DL. Fluidized bed heat transport between parallel, horizontal tube-bundles. In: Davidson JF, Keairns DL, eds. Fluidization. London: Cambridge University Press, 1978, pp 320–326.

Noordergraaf IW, van Kijk A, van der Bleek CM. Fluidization and slugging in large-particle systems. Powder Technol 52:54–68, 1987.

Olowson PA. Hydrodynamics of a fluidized bed-influence of pressure and fluidization velocity. PhD thesis, Chalmers University of Technolgy, Göteborg, 1991.

Olowson PA. Influence of pressure and fluidization velocity on the hydrodynamics of a fluidized bed containing horizontal tubes. Chem Eng Sci 49(15):2437–2446, 1994.

Olsson SE, Wiman J, Almstedt AE. Hydrodynamics of a pressurized fluidized bed with horizontal tubes: influence of pressure, fluidization velocity and tube-bank geometry. Chem Eng Sci 50(4):581–592, 1995.

Overcashier RH. Some effects of baffles on a fluidized system. AIChE J 5(1):54–60, 1959.

Ramamoorthy S, Subramanian N. Axial solids mixing and bubble characteristics in gas-fluidized beds with vertical internals. Chem Eng J (Lausanne) 22(3):237–242, 1981.

Rhodes MJ, Geldart D. Transition to turbulence? In: Östergaard K, Sörensen A, eds. Fluidization V. New York: Elsinore, Deamarh, 1986, pp 281–288.

Sadasivan N, Barreteau K, Laguerie C. Studies on frequency and magnitude of fluctuations pressure drop in gas-solid fluidized beds. Powder Technol 26:67–74, 1980.

Salah H, Zhu JX, Zhou YM, Wei F. Effect of internals on the hydrodynamic circulating fluidized beds. Chinese Society of Particuology, Beijing, China, pp Eq 1.1–1.6, May 1996.

Satifa S, Fan LS. Characteristics of slugging regime and transition to turbulent regime for fluidized beds of large coarse particles. AIChE. J, 31:1554–1562, 1985.

Shibata K, Shimizu M, Inaba S, Takahashi R, Yagi J. One-dimensional flow characteristics of gas-powder two phase flow in packed beds (in Japanese). Iron Steel 77:236–243, 1991.

Sitnai O, Dent DC, Whitehead AB. Bubble measurement in gas-solid fluidized beds. Chem Eng Sci 36(9):1583, 1981.

Sitnai O, Whitehead AB. Immersed tubes and other internals. In: Davidson JF, Clift R, Harrison D, eds. Fluidization, 2nd ed. Academic Press, London, 1985, pp 473–493.

Song X, Wang Z, Jin Y, Tanaka Z. Powder Technology 83:127–131, 1995.

Staub FW, Canada GS. Effect of tube bank and gas density on flow behavior and heat transfer in fluidized beds. Fluid Proc Eng Found Conf 2nd, London: Cambridge Univ Press: 1978, pp 339–344.

Staub FW, Wood RT, Canada GS, Mclaughlin MH. Two-phase flow and heat transfer in fluidized beds. EPRI Report CS-1456, RP525-1, 1980.

Van der Ham AGJ, Prins W, Van Swaaij WPM. Hydrodynamics of a pilot-plant scale regularly packed circulating fluidized bed. AIChE Symp Ser 89(296):53–72, 1993.

Van der Ham AGJ, Heesink ABM, Prins W, Van Swaaij WPM. Proposal for a regenerative high-temperature process for coal gas cleanup with calcined limestone. Ind Eng Chem Res 35(5):1487–1495, 1996.

Volk W, Johnson CA, Stotler HH. Effect of reactor internals on quality of fluidization. Chem Eng Progr 58(3):44–47, 1962.

Wang ZL, Yao JZ, Liu SJ, Li HZ, Kwauk M. Studies on the voidage distribution in an internally circulating fluidized bed (in Chinese). In: Proceedings 6th National Conf. on Fluidization, Wuhan, China, October 1993, pp 150–154.

Xavier AM, Lewis DA, Davidson JF. Trans Inst Chem Eng 56:274–280, 1978.

Yamaoka H. Iron Steel (in Japanese) 72:403–410, 1986.

Yates JG, Ruiz-Martinez RS. Interaction between horizontal tubes and gas bubbles in a fluidized bed. Chem Eng Comm 62:67–78, 1987.

Ye YH. Internals for fluidized beds. In: Wang ZX, Ye YH. eds. Chemical Engineering Handbook (section 20, Fluidization). Beijing: Chemical Industry Press, 1987, pp 138–177.

Yerushalmi J, Cankurt NT. Further studies of the regimes of fluidization. Powder Technol 24:187–205, 1979.

Zabrodsky SS, Epanov YG, Galershtein DM, Saxena SC, Kolar AK. Int. J. Heat Mass Transfer 24(4):571–579, 1981.

Zhao J, Zhong X, Xu HQ. A model of solid backmixing between stages in a gas-fluidized bed with perforated baffles. Powder Technol 73:37–41, 1992.

Zheng CG, Tung YG, Zhang WN, Zhang JG. Impact of internals on radial distribution of solids in a circulating fluidized bed (in Chinese). Eng Chem Metallurgy 11(4):296–302, 1990.

Zheng CG, Tung YK, Li HZ, Kwauk M. Characteristics of fast fluidized beds with internals. In: Potter OE, Nicklin DJ, eds. Fluidization VII. Engineering Foundation, New York, 1992, pp 275–284.

Zheng CG, Tung YK, Xia YS, Hun B, Kwauk M. Voidage redistribution by ring internals in fast fluidization. In: Dwauk M, Hasatani M, eds. Fluidization '91. Science and Technology. Beijing: Science Press, 1991, pp 168–177.

Zhu JX, Salah M, Zhou YM. Radial and axial voidage distributions in circulating fluidized bed with ring-type internals. J Chem Eng Jpn 30(5):928–937, 1997.

Ziegler EN, Brazalton WT. Ind Eng Chem (Proc Des Dev), 2:276, 1963.

8

Attrition

Joachim Werther and Jens Reppenhagen*

Technical University Hamburg-Harburg, Hamburg, Germany

1 INTRODUCTION

In any fluidized bed process—be it a combustor, a dryer, or a chemical reactor—the bed material is in vigorous motion and thus inevitably subjected to mechanical stress due to interparticle collisions and bed-to-wall impacts. This mechanical stress leads to a gradual degradation of the individual bed particles, which is quite often an unwanted phenomenon that is therefore termed "attrition."

The main consequence of attrition in fluidized bed processes is the generation of fines that cannot be kept inside the system. Hence attrition creates an additional burden on the filtration systems, i.e., the collection systems must be larger (Merrick and Highley, 1974; Vaux, 1978), and most importantly there is a loss of valuable material. In a combustion process, for example, the attrition-induced production of fines will lead to a loss of both unburned fuel and sorbent, causing a decrease in the efficiency of combustion and sulfur retention (Pis et al., 1991; Ray et al., 1987b). Other examples are the heterogeneously catalyzed chemical reactors, where the compensation for the lost material by the addition of fresh makeup catalyst can amount to very high operating expenses due to which a given process may even become uneconomical (Contractor et al., 1989; Patience and Mills, 1994).

Simultaneously with that unwanted production of fines, there is a further significant effect of attrition,

namely its influence on the bed particle size distribution: it causes in any case a gradual shrinking of the so-called mother particles from which the fines are drawn until they are finally lost from the system as well (Kunii and Levenspiel, 1969; Ray et al., 1987b). In addition, there might be even a fragmentation of bed particles, resulting in a significant increase in the number of particles and a corresponding decrease in the mean particle size (Blinichev et al., 1968, Pis et al., 1991). Whether with a batch process or a continuous one, attrition will thus always lead to differences in the bed particle size distribution in comparison to that of the feed material. Some authors (such as Kraft et al., 1956; de Vries et al., 1972) report, for example, on the effect that the bed material in a considered FCC process got too coarse by the addition of makeup catalyst in compensation for the lost mass of fines. Braca and Fried (1956), however, observed the opposite effect, where the bed material of the FCC process got too fine since the attrition-produced fines were mostly kept inside the system.

Since the particle population determines almost all relevant mechanisms in a fluidized bed system, attrition may thus strongly affect the performance of a fluidized bed process. For instance, the elutriation and entrainment effects (cf. Chapter 4 of this book), the heat transfer from bed to inserts (Molerus, 1992) or the conversion and selectivity of reactions (Werther, 1992) are affected either directly or through the bed hydrodynamics by the particle size distribution. Therefore some authors (such as Ray et al., 1987a) even claim that a fluidized bed with attritable materials

*Current affiliation: BMH Claudius Peters, Buxtehude, Germany.

cannot be designed and simulated appropriately unless the attrition activity is described.

Naturally, the primary interest in the field of attrition is to reduce its extent to a minimum. An obvious approach is to use an optimum attrition resistant bed material. Hence the majority of work deals with the particles' degradation in standardized bulk attrition tests to determine the relative tendency of materials to attrit (Davuluri and Knowlton, 1998). These so-called friability tests are quite useful in the quick and first assessment of a candidate bed material. But they solely focus on one factor affecting attrition, namely the materials' properties, whereas the prevailing stress mechanisms may differ considerably from those in the process. For that reason, the pure tests cannot predict the quantitative extent of attrition that will occur with the tested materials in a technical fluidized bed process. They cannot even guarantee that the ratio of the attrition rates of two materials will be the same in the test and in the process (Werther and Reppenhagen, 1999a; Boerefijn et al., 2000).

Hence the pure attrition tests are not sufficient when a quantitative prediction of the attrition-induced material loss or a prediction of the effect on the bed particle size distribution is required. This is for instance the case in the design procedure of a process, where the capacity of the dust collection system and the lifetime of the bed material must be evaluated (Vaux and Fellers, 1981; Zenz, 1974) and where—above all—the hydrodynamics and thus the bed particle size distribution must be clear (Ray et al., 1987a; Zenz, 1971). Another example is when a new generation of catalyst is to be developed for an existing fluidized bed reactor: here it might not be sufficient to know the relative hardness in comparison to the previous catalyst generation. For a comprehensive cost–benefit analysis, it is rather necessary to predict the attrition-induced loss rate and to be sure about the hydrodynamics and thus the particle population of the bed material. The same is valid when a fuel or a sorbent is to be changed in a combustion process.

In order to achieve such a comprehensive insight into the attrition phenomena of a given process, it is necessary to consider meticulously each individual effect of the various influencing factors related to the material properties, the system design, and the operating conditions. The present chapter attempts to summarize systematically the findings available in the open literature and thus to draw up a possible approach in the attrition modeling.

The first key point in dealing with attrition seems to be a distinction between the different ways of the particles' degradation, such as fragmentation or abrasion (e.g., Blinichev et al., 1968; Pell 1990), which are based on quite different mechanisms. For this reason Sec. 2 will deal with the various modes of attrition and the factors affecting them. Furthermore, there should be an unambiguous definition of the way in which the attrition extent and/or its impact on the particle size distribution is measured and quantified. This definition must suit both the prevailing attrition modes and the attrition-induced effect that is to be considered. Section 3 will therefore deal with the respective assessment procedures. This provides a basis on which the influences of both the material properties and the process conditions, i.e., system design and operating conditions, can be described.

Considering the influence of the material properties, it must be acknowledged that attrition is a statistical effect, i.e., there are differences in the attrition susceptibility of the individual bed particles, and the attrition stress is acting randomly on them. As a consequence, there is usually not the influence of one particular property evaluated; rather there is an evaluation of the materials' statistical attritability as a whole. Section 4 summarizes the relevant attrition test procedures. The results of these attrition tests must then be transferred to the actual process by means of a physically sound description of the process conditions (Werther and Reppenhagen, 1999; Boerefijn et al., 2000). This, however, requires a distinction between different regions of the system that apply different types of stress to the solids. For this reason, Sec. 5 first summarizes the attrition mechanisms prevailing in the relevant sources of a fluidized bed system, and Sec. 6 then finally deals with a description of attrition in the entire process.

2   THE MODES OF ATTRITION AND THE FACTORS AFFECTING THEM

Both the mechanism and the extent of attrition depend not only on the properties of the solids but also on the process design and to a large extent on the process conditions. Clift (1996) has stated that attrition is a triple-level problem, i.e., one must deal with phenomena on three different length and time scales: of the processing equipment, of the individual particles, and of the subparticle phenomena such as fracture, which leads to the formation of fines. The appearance of attrition, therefore, can differ between the various processes and even between individual regions in a given process (cf. Sec. 5).

In general, the mode of attrition may vary from pure abrasion to total fragmentation of the particles (Blinichev et al., 1968; Pell, 1990). Abrasion implies that exclusively asperities are removed from the particles' surfaces. It thus produces a lot of elutriable fines, whereas the particle size distribution of the mother particles is hardly changed. In contrast to this, fragmentation is a process of particle breakage into similarly sized pieces. As a result, the number of particles increases, and the particle size distribution becomes significantly broader with a distinctly smaller mean diameter compared to the original one. The two modes and their different effects on the particle size distribution are sketched in Fig. 1.

There are various attrition modes that can be regarded as lying somewhere in between these two extremes, as for example the initial breakage of fresh catalyst particles from which irregularities and corners are knocked off. All these modes are based on different mechanisms and must therefore be described separately (Pis et al., 1991).

The necessity of distinguishing different modes of attrition is illustrated by the work of Reppenhagen and Werther (1999a), for example, dealing with catalyst attrition in cyclones. Focusing on the mode of pure abrasion, they derived a physically sound equation for the description of the attrition-induced production of fines and thus for the attrition-induced material loss through the cyclone overflow (cf. Sec. 5.3). This model approach is confirmed for various types of catalysts in a wide range of operating conditions. But when a specific threshold velocity at the cyclone inlet is exceeded, a so-called surface fragmentation adds to the production of fines and leads to a much higher loss rate as predicted from the model approach (Fig. 19). Reppenhagen and Werther (1999a) identified the surface fragmentation by scanning electron microscope (SEM) images (cf. Fig. 5) and could thus determine the boundaries for pure abrasion conditions and thus for their cyclone attrition model, which deals solely with pure abrasion.

As it is obvious from the above example, the individual modes do usually not occur separately. They are rather combined in varying proportions, which makes a description of the entire attrition process rather complicated. The only attrition mode that can occur separately is abrasion, since it has the lowest threshold energy. The other attrition modes, the threshold energy of which is much higher, are at least combined with abrasion. In general, the extent of abrasion in relation to fragmentation depends on various factors that will be discussed below.

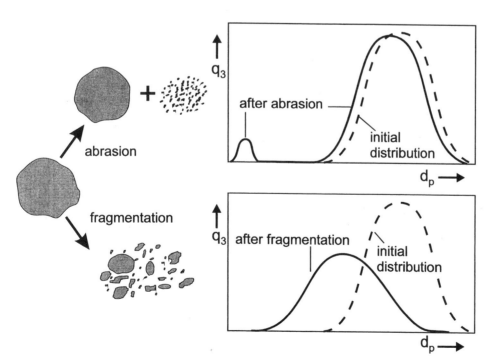

Figure 1 Attrition modes and their effects on the particle size distribution (q_3 = mass density distribution of particle sizes d_p).

Several rules have been suggested to identify the dominant mode in a given case. They all use the change in the particle size distribution as a criterion. For instance, Dessalces et al. (1994) assessed the degradation behavior of various industrial FCC catalysts by the change in the ratio of the 90% and 10% values of the cumulative mass distribution of the particle sizes. They got high values of this ratio ($>$ 100) if abrasion was dominant and smaller values ($<$ 10) in the case of fragmentation.

In the following, the various factors are summarized that have been identified to affect attrition in its extent or in its mechanisms. Basically, they can be classified into two major groups, namely the various factors related to material properties on the one hand and the factors related to the process conditions on the other.

2.1 Material Properties

2.1.1 Particle Structure

On the part of the material properties, the particle structure has the most fundamental influence on the catalysts' attrition behavior. The extent of degradation as well as its mode will strongly depend on whether a particle has a crystallized or an amorphous structure or is an agglomerate (British Materials Handling Board, 1987).

As an example, catalysts that are used in fluidized bed reactors are usually produced by spray drying. This type of catalyst is on the one hand rather susceptible to attrition, but on the other hand it tends to abrasion rather than to fragmentation. Hence it is often assumed that abrasion is the only or at least the governing attrition mode in fluidized bed reactors (e.g., Werther and Xi, 1993).

Arena et al. (1983) and Pis et al. (1991) studied the attrition of amorphous materials such as coal or limestone and found that the size distributions of the attrited materials were independent of the initial particle size and of most operating parameters. Ray et al. (1987a) assumed that unlike crystalline materials, amorphous materials may have some kind of "natural grain size" to which the degradation finally leads.

2.1.2 Particle Size

The initial particle size is of primary interest with respect to particle breakage, because smaller particles tend to contain fewer faults in form of microcracks or imperfections and are thus more difficult to break than larger ones. However, the mechanisms of breakage will not be further discussed in the present chapter. instead we refer to a comprehensive survey given by the British Materials Handling Board (1987).

2.1.3 Particle Size Distribution

Besides the breakage probability of the particles, the size distribution determines another important parameter with respect to attrition, namely the volume specific surface area of the bed material. This parameter is particularly relevant with respect to abrasion, where it leads to two contrary effects.

On the one hand, it determines the total surface area that is exposed to abrasion. On the other hand, at given process conditions, the specific surface area determines the specific surface energy available for attrition or abrasion. In other words, with a finer bed material there may be an increase in total surface exposed, but the energy that is acting on a unit surface is decreasing. This is particularly relevant with respect to mixtures. As an example, Arena et al. (1983) investigated coal attrition in a mixture with sand under hot but inert conditions. As they increased the sand particle size while keeping its mass in the bed constant, they observed an increase in the coal attrition rate. They interpreted their results by assuming that the abrasion energy is shared out on the entire material surface. On the same basis Ray et al. (1987a) developed their "attrition rate distribution model" for abrasion in a fluidized bed.

These findings may also explain the observed influence of the fines content on the attrition propensity of a given material: Forsythe and Hertwig (1949) already noticed a reduction of the degradation of FCC catalysts in jet attrition tests due to the presence of fines. They themselves supposed some kind of cushioning effect that limits the force of collision impact and thus limits the degradation of the coarse particles. The effect of fine particles is of strong interest because they are produced by attrition, so attrition inhibits itself if the fine particles remain in the system.

More recently, Reppenhagen and Werther (1999a) and Werther and Reppenhagen (1999) found in their catalyst attrition tests that the abrasion-induced production of fines increases linearly with the surface mean diameter of the tested material.

2.1.4 Particle Shape and Surface Structure

The particle shape is a relevant parameter, because irregular and angular particles are inclined to have their corners knocked off in collisions and thus become rounder and naturally smaller with time. A macroscopically smooth surface is therefore less prone to breakage, but it may still undergo abrasion. With respect to the latter, the microscopic surface structure is of interest. Surface asperities may be chipped off and lead to abrasion.

In their investigation on cyclone attrition, Reppenhagen and Werther (1999a) compared, for example, the attrition resistance of two different catalysts, the chemical composition of which is identical but their shape and surface structure is different (cf. Fig. 2). Whereas the catalyst termed 97-G(fresh) is irregularly shaped and has a rough surface, the catalyst termed 97-R(fresh) is almost spherical, and its surface shows only few asperities. Reppenhagen and Werther (1999a) attributed these morphological differences to their finding that the 97-G(fresh) catalyst was almost twice more attritable than the 97-R(fresh) material.

Both the particle shape and the surface structure will inevitably change with time inside a process. This will in any case happen as a gradual effect, which is discussed in Sec. 2.2.3. But it may also happen instantaneously due to excessive stress, for example the surface fragmentation of catalyst particles that has been observed by Reppenhagen and Werther (1999a) (cf. Figs. 19, 4, and 5).

2.1.5 Pretreatment and Preparation or Processing History

The unsteady-state attrition rate of a given material will normally depend on the pretreatment to which it was subjected. The higher the intensity and the longer the duration of the foregoing stress, the more probable it is that the weakest points of the particles have already been attrited.

The pretreatment may already occur in the processing route of a given material. If we consider for example the spray-drying process of catalysts, the weakest agglomerates will immediately break down during the production and handling of the material. Otherwise, there is a pretreatment at earlier times in the process itself. A quite often quoted example is the difference between a fresh catalyst material and the corresponding equilibrium catalyst taken from the process (Boerefijn et al., 1998). In their cyclone attrition tests, Reppenhagen and Werther (1999a) compared the attrition resistance of an equilibrium FCC catalyst obtained from a refinery with that of the fresh material. They found the fresh material to be almost five times more attritable than the equilibrium catalyst and concluded that some of the material properties must have been changed inside the FCC process. These properties may be chemical ones with resulting changes in the material strength, but in any case Reppenhagen and Werther (1999a) observed morphological changes in the SEM images shown in Fig. 3. The surfaces of the fresh particles are rougher than the ones of the equilibrium particles. Consequently, there are more asperities that can be chipped off and thus contribute to the attrition extent. Furthermore, the equilibrium particles are a bit more spherical than the fresh ones, which means according to Sec. 2.1.4 an increase in the attrition resistance.

Another example might be the sulfation of limestone sorbents in combustion processes (Anthony

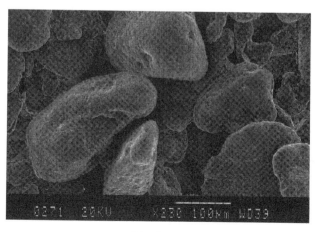

97-G(fresh)

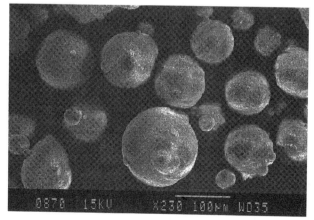

97-R(fresh)

Figure 2 SEM images of two catalyst materials. The surface mean diameter was 112 microns in the case of the "97-R(fresh)" and 103 microns in the case of the "97-G(fresh)." (From Reppenhagen and Werther, 1999a.)

Figure 3 SEM images of an equilibrium catalyst and the corresponding fresh material. (Reppenhagen and Werther, 1999a.)

and Granatstein, 2001). In any case, these are gradually acting phenomena, that are strongly related to the influence of the solids residence time, the effect of which is discussed in Sec. 2.2. But there may be also singular phenomena that affect the materials attritability. A broken steam pipe in the stripping unit of an FCC plant may for example cause a steam jet that generates extra stress on the catalyst particles. Such a singular event would not only cause a higher extent of attrition during its occurrence but also it would most probably have a long-term effect. The reason for this is a serious damage of the morphology of the particles. Reppenhagen and Werther (1999a) observed such a phenomenon during their cyclone attrition tests as they exceeded, for several passes through the cyclone, a certain threshold energy. The result of this experiment is shown in Fig. 4.

A batch of catalyst particles was passed in a first phase through a cyclone under conditions of pure abrasion, then in a second phase under conditions where additional surface fragmentation takes place, and finally in a third phase again under conditions of the first phase of pure abrasion. It is quite obvious that the surface fragmentation leads to an increased susceptibility to abrasion, i.e., at the end of the experiment the abrasion rate is distinctly higher than at the beginning, even though the operating conditions are identical. This can be explained by comparing SEM images taken from the catalyst particles before and after the subjection to the excessive stress (cf. Fig. 5). It can be seen that parts of the particles' surfaces have been chipped off, leading to a new surface that is distinctly rougher than the old one and therefore more prone to abrasion. However, with an increasing number of

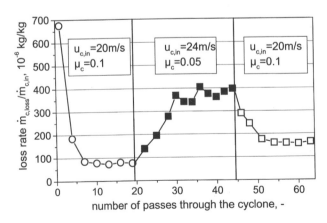

Figure 4 Influence of surface fragmentation on the cyclone loss rate at conditions of pure abrasion ($u_{c,in} = 20$ m/s, $\mu_c = 0.1$). (From Reppenhagen and Werther, 1999a.)

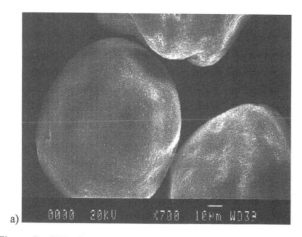

Figure 5 SEM images of a given catalyst material (a) after subjecting it to pure abrasion, and (b) after subjecting it to abrasion with additional surface fragmentation. (*Note*: The bridges between the particles in image (b) are not due to sintering but to an excess gold layer resulting from the SEM preparation step.) (Reppenhagen and Werther, 1999a.)

passes under conditions of pure abrasion, the rough surface, which is responsible for the increased loss rate, would become smoother again and the cyclone loss rate would then decrease to its initial value.

2.2 Process Conditions

The process conditions, which are resulting from the system design and the operating conditions, will primarily influence the attrition of the bed material by generating the stress on the individual particles. In general, the stress leading to attrition of a given bulk material may be a mechanical one due to compression, impact, or shear, a thermal one owing to evaporation of moisture or temperature shock, or a chemical one by molecular volume change or partial conversion of the solid into the gas phase.

2.2.1 Gas and Solids Velocity

The gas velocity is usually directly related to the particle velocity, which is the most important factor in generating the mechanical stress by interparticle collisions or by particle–wall impacts (British Materials Handling Board, 1987). The forces involved in the degradation process may be generated by high-speed collisions resulting preferably in breakage. Alternatively, the energy may be transmitted through a matrix of comparatively slowly moving particles resulting mostly in abrasion. Particularly in the distributor region of a fluidized bed, where grid jets are issuing into the bed of particles, the impact velocities

between particles can be extremely high and result in significant particle breakage.

2.2.2 Wall-Hardness

One can assume that the particle degradation increases with the hardness of the vessel wall. This effect will increase with increasing ratio of particle-to-vessel diameter and may thus be only relevant inside a cyclone, or in the vicinity of bed inserts.

2.2.3 Solids Residence Time

The relationship between the solids residence time and the amount of material that is produced by attrition is generally nonlinear. As an example, Fig. 6 shows the typical time dependence of the attrition of a fresh catalyst that is subjected to attrition in one of the test devices described in Sec. 4.

The elutriated mass is defined to be the attrition product, and consequently the attrition rate is defined as elutriated mass per unit time. It is clearly seen that the rate of attrition is decreasing with time. The reason is that at the beginning the fresh catalyst particles have very irregular shapes and contain many faults. This results in a high rate of initial particle degradation, during which the particles break and their edges and asperities are knocked off. With progressing time, the particles become smaller, rounder, and smoother, and the number of their weak points decreases. The elutriation rate therefore decreases continuously with time and tends to a more or less constant value, which can be interpreted as a kind of steady-state level where only abrasion takes place.

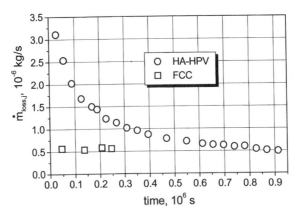

Figure 6 Time dependence of catalyst attrition in a submerged jet test. $D_t = 0.05\,\mathrm{m}$, $u_{or} = 100\,\mathrm{ms}^{-1}$, $d_{or} = 2\,\mathrm{mm}$, and HA-HPV is a fresh catalyst, whereas FCC is a spent catalyst from a commercial unit). (From Werther and Xi, 1993.)

The shape of the curve in Fig. 6 is typical for all particle degradation processes. However, one has to distinguish between batch and continuous processes. In batch processes the whole bed material is always at the same state of attrition, which changes during the operating time according to Fig. 6. On the other hand, there is a particle residence time distribution in continuous processes. The attrition rate of the entire material is therefore constant, although the state of attrition of each particle again changes according to Fig. 6.

Many attempts have been made to describe the time dependence of the attrition rate in batch fluidized bed processes. Gwyn (1969) studied the degradation of catalysts in a small-scale test apparatus and defined the elutriated particles as the only attrition product. He described the increase of the elutriated mass m_{loss} with time t based on the initial solid bed mass $m_{b,0}$ by the now widely known Gwyn equation,

$$\frac{m_{loss}}{m_{b,0}} = K_a \cdot t^b \qquad (1)$$

For several batches of the same catalyst material with quite different mass median diameters, Gwyn found the exponent b to be constant, whereas the attrition constant K_a was found to decrease with mean particle size. Equation (1) is therefore valid for a particular size distribution only. Other empirical correlations for the time dependence of the mass ratio $m_{loss}/m_{b,0}$ have been suggested by e.g. Pis et al. (1991) and Dessalces et al. (1994).

2.2.4 Temperature

There are three conceivable temperature effects that may influence the particle degradation in either a direct

or an indirect way, i.e., thermal shock, changes in particle properties, changes in the gas density.

The heating of fresh cold particles fed into a hot process can cause various phenomena that may lead to particle degradation. These are thermal stress, decrepitation, evaporation of moisture, hydrate decomposition, and impurity transformation. On the other hand, particle properties such as strength, hardness, and elasticity may also be affected by the temperature. With respect to the resistance to degradation, there is an optimum temperature range for any specific type of material. At lower temperatures, particles become brittle, while at higher temperatures they may soften, agglomerate, or melt and lose discrete particulate properties. Consequently, it is important that the particle friability, which is the major factor for the attrition propensity of a given material, is assessed under conditions that are similar to those found in the process where the attrition is occurring. Moreover, the temperature can have a strong effect on the gas density, which affects the fluidization state and with it the particle motion and the stress the particles are subjected to.

2.2.5 Pressure

The absolute pressure is unlikely to have a direct effect on attrition unless it affects the amount of adsorbed surface layers. But there is again an effect on the gas density that is similar to the effect of temperature mentioned above. Moreover, the rate of pressure change may have more influence.

2.2.6 Chemical Reaction

On the one hand, a chemical reaction of the particulate material may generate stress within the particles leading to fracture. In the case of gas–solid reactions, the particle degradation is also desired, because it accelerates the reaction by extending the reactive surface. A commercially relevant example is the particle degradation of solid fuels in combustion processes. This latter topic has been extensively studied by Massimilla and coworkers. The reader is referred for further details to a comprehensive review given by Chirone et al. (1991).

On the other hand, a chemical reaction of the particulate material may affect the mechanical attrition resistance of the particles, as for example the sulfation of limestone sorbents in combustion processes (Anthony and Granatstein, 2001) or the different degrees of oxidation of a catalyst (Xi, 1993).

3 ASSESSMENT OF ATTRITION

As attrition basically affects everything from a single particle to the entire process, there are various criteria on the basis of which attrition may be qualitatively and quantitatively characterized. One may base this assessment on the observation of an individual particle. Alternatively, the fate of a group of particles may be examined, or the effect of attrition on the bulk properties of the given material may be taken for this assessment. The British Materials Handling Board (1987) and Bemrose and Bridgwater (1987) give examples for the different methods.

The chosen type of assessment procedure basically depends on whether the attrition propensity of a given material, i.e., its material strength, or attrition in a given process is to be considered. Furthermore, the choice of the assessment procedure depends on the attrition effect that is of interest. When the entire process is considered, it is usually either the assessment of the attrition-induced material loss or the assessment of the attrition-induced changes in the particle size distribution of the bed material.

3.1 Assessment of the Material's Strength

The assessment of the material's strength is commonly based on so-called friability tests or attrition tests (cf. Sec. 4), in which a single particle or a bulk sample of the material is for a certain extent subjected to a specific stress. Data from these tests are usually presented as single numbers called friability or attrition indices. Most of these numbers are derived from a comparison of the material's content of a previously defined fraction of fines before and after the test (British Materials Handling Board, 1987; Davuluri and Knowlton, 1998).

A frequently quoted example is the Hardgrove Grindability Index, which was originally developed to assess the grindability of coal. It measures the generation of material less than 74 microns produced in a ball-ring pulverizer during 60 revolutions on a sample with a size range between 590 and 1190 microns (cf. ASTM D409-71).

Such indices may give a ranking of the material's friability but cannot be directly related to the process attrition, which would be the most appropriate assessment. This means, if at given process conditions a material A yields compared to a material B twice as high attrition, the ratio of the respective attrition indices of the materials A and B should be 2.

Since the attrition propensity of a given material strongly depends on the type of stress to which it is subjected (cf. Sec. 4), the requirement of a direct relation to the process will be met only by attrition tests that simulate the process stress. The results from those tests are usually expressed by a so-called attrition rate [cf. Eqs. (2) and (3)], which is also taken for the assessment of process attrition. Hence for each considered process there is an empirical or even physically sound equation available that describes the attrition rate in dependence on the chosen operating conditions. The material's strength is then simply expressed as a proportionality constant that is independent of the operating conditions, the so-called attrition rate constant [e.g., K_j in Eq. (7)].

3.2 Assessment of the Attrition-Induced Material Loss

In fluidized bed experiments, most authors consider the attrition-induced material loss only, i.e., they assume that all attrition products are elutriated. Consequently, they measure either the decrease in bed mass (Kono, 1981; Kokkoris and Turton, 1991, 1995) or the elutriated mass (Seville et al., 1992; Werther and Xi, 1993). The results are commonly expressed by a so-called total or overall attrition rate r_{tot}, which is accordingly defined by the relative change of bed weight m_b with time,

$$r_{tot} = -\frac{1}{m_b}\frac{dm_b}{dt} \qquad (2)$$

or, alternatively, by the relative change in elutriated mass m_{loss} with time,

$$r_{tot} = \frac{1}{m_b}\frac{dm_{loss}}{dt} \qquad (3)$$

Whether Eq. (2) or Eq. (3) is chosen will normally depend on the type of equipment used. When comparing different authors' works, attention should be paid to the definition of m_b. For batch processes it may be either the initial bed mass or the instantaneous inventory of the bed. Special attention has to be paid to a definition of attrition rates in the case of continuous processes where fresh material must be added to compensate for reacted or withdrawn material and for the attrition losses. The fresh material may contain elutriable fines which add to the measurable elutriation rate thus leading to an apparently higher attrition rate.

Furthermore, it should be noted that a system's material loss is also affected by the efficiency and the cut size of the gas/solids separation unit. Differences in this type of system must be taken into account when different authors' works are compared, or when a test result is scaled-up to a full-scale process. Section 4.3, which contains a description of the various test equipment and procedures, will deal with this matter in more detail.

Some authors (Vaux and Keairns, 1980, Sishtla et al., 1989; Pis et al., 1991) define the time integral of the attrition rate as the attrition extent $E_{tot}(t)$,

$$E_{tot}(t) = \int_0^t r_{tot}(\tau) \cdot d\tau \qquad (4)$$

This definition, however, includes effects of initial breakage and of initial fines content and will therefore normally not give an unambiguous assessment of attrition.

3.3 Assessment of Changes in the Particle Size Distribution

The above definition of the attrition rate considers the bed material as a whole and quantifies solely the production of elutriable material without taking all breakage events or the shrinking of the so-called mother particles into account. More insights into the attrition mechanisms can be obtained from the observation of the change in the particle size distribution as demonstrated by Zenz and Kelleher (1980) and by Lin et al. (1980). An example of one of the results obtained by Zenz and Kelleher (1980) is shown in Fig. 7.

Such a description of the effects of attrition gives good information about changes in the content of particles in a particular size class. But it can only describe the fate of a batch of particles where the extent of the changes depends on the length of the considered time interval during which the catalyst is subject to attrition. Furthermore, this type of assessment includes effects of initial breakage and the sifting of initial fines. As a consequence of the above limitations, this type of assessment is able to describe only the results of attrition and not the dynamics of the attrition mechanisms themselves. This can be best explained on the basis of the mass balance of a given size interval i,

$$\Delta m_i = m_{feed,i} - m_{loss,i} + \sum_{h=i+1}^{n} m_{h,i} - \sum_{k=1}^{i-1} m_{i,k} \qquad (5)$$

The mass content in the interval i may thus be increased by a mass $m_{feed,i}$ that is fed to the system

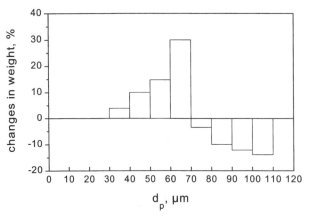

Figure 7 Changes in the particle size distribution of a FCC catalyst material that was subjected to cyclone attrition. (Zenz and Kelleher, 1980.) The respective experimental procedure is described in Sec. 5.3.

and may be decreased by a mass $m_{loss\,i}$ that is lost from the system in a considered time interval. Furthermore, there is an internal mass transfer between the individual size intervals. The interval i receives attrited particles from coarser size intervals h and loses attrited particles to size intervals with finer material, k, which is in Eq. (5) expressed by the transfer masses $m_{h,i}$ and $m_{i,k}$, respectively.

The assessment on the basis of changes in the particle size distribution considers the left-hand side of Eq. (5), i.e., the absolute change in mass Δm_i is measured. But the individual contributions that led to these changes, i.e., the terms on the right-hand side of Eq. (5), are not considered by this type of assessment. It may be possible to take the contributions of feed and system loss by means of measurements into account, but it is not possible at all to unravel the internal transfer masses $m_{h,i}$ and $m_{i,k}$ between the size intervals just by observing the changes in the particle size distribution. For this reason, it is not possible to extrapolate the description to a different duration of the attrition process or even to transfer it to other initial particle size distributions. For this purpose a description via particle population balances is needed.

A description of the attrition process that is based on a particle population balance requires a detailed knowledge about the kinetics of the attrition mechanisms of the individual particles. In particular, the influence of the individual particle size on the attrition extent, and the size distribution of the corresponding attrition products, must be known. With regard to comminution processes it is therefore very common to divide the bulk material in several size ranges to which both a probability of degradation and an individual size distribution of attrition products are assigned. The entire description is then realized by a concept of two separate functions, namely the so-called selection function or probability function $S(x)$ and the breakage function $B(x, y)$ (Austin, 1972, Schubert, 1975). The selection function describes the mass fraction of a differential size interval $(x, x + dx)$ which is subjected to the particle size reduction and therefore lost to smaller size intervals. The assignment of this lost mass fraction to the particular size intervals is realized by the breakage function. It describes the mass fraction of breakage products of size y originating from a particle of size x.

$S(x)$ and $B(x, y)$ can be defined as vectors and matrices where the elements are applied to discrete particle size ranges. In detail, the vector elements s_i describe the rate of material loss of a particular size interval of mean diameter x_i and the various matrix

elements $b_{i,k}$ describe the distribution of attrition products from the intervals i into the intervals k of smaller particle sizes. With a vector of the feed size distribution they can be combined in a matrix equation that yields a size distribution vector of the product. Since these vector and matrix elements must be usually obtained experimentally, the use of selection and breakage functions is much more cumbersome than working with a single index number. This concept has therefore been used almost exclusively for the description of commercially interesting comminution processes but very seldom for the description of catalyst attrition (Wei et al., 1977).

However, if attrition exclusively occurs by the mode of abrasion, the respective particle balance appears to be less complex. There are several mathematical models of particle population balances suggested in the literature that deal with that particular mode (Levenspiel et al., 1969; Overturf and Reklaitis, 1983; Ray et al., 1987b; Werner et al., 1995; Reppenhagen, 1999; Reppenhagen and Werther, 2001) for fluidized bed systems, or Hansen and Ottino (1997) for a general description of the abrasion process. A comparison with the above-discussed concept of selection and breakage functions shows that all these approaches assume that the particles of a given size fraction are either entirely subjected to abrasion or left undamaged. This means that the selection function is either completely omitted, since all particle size ranges are assumed to be attritable (Levenspiel et al., 1969; Overturf and Reklaitis, 1983; Werner et al., 1995), or is reduced to a step function from zero where no abrasion is assumed to one at a particle size above which all particles are assumed to be attritable (Ray et al., 1987b; Hansen and Ottino, 1997). The respective breakage functions are also comparatively simply organized in the above approaches. The particle-size dependence of the attrition extent and thus the mass of attrition-produced fines per size interval and unit time is either described by a first-order kinetics (Levenspiel et al., 1969) or by a potential dependence (Hansen and Ottino, 1997). The entire mass of fines is then either simply assumed to be lost (Levenspiel et al., 1969) or is assigned to a particular size interval with a given constant size distribution (Ray et al., 1987b; Hansen and Ottino, 1997).

In most of the above approaches the balances are solved analytically and the abrasion-induced shrinking of the mother particles is thus taken into account by simply shifting the size distribution in accordance with the respective material losses. In contrast to this, Reppenhagen and Werther (2001) recently developed a discretized particle population balance, which describes the mass transfer between neighboring size intervals on the basis of the natural coupling with the fractional fines production, i.e., the mass of produced fines originating per unit time from a size interval K_i corresponds to the material loss of the initial parent particles of this fraction.

4   ATTRITION TESTS

The term "attrition test" generally covers all experiments in which the effects of attrition are considered and assessed. This might vary from the observation of the degradation of only a single particle (Boerefijn et al., 2000) to the consideration of an entire process (Werther and Reppenhagen, 1999). However, analogously to the division of the assessment procedures, the large number of experiments can be divided into two major fields of application, namely the tests of material friability and the experiments to study attrition phenomena. They will be separately discussed in the following sections. The relevant test devices will be discussed afterwards.

4.1   Friability Tests

Friability tests are often used for a comparison of different types of materials to select the most attrition-resistant one (Vaux and Fellers, 1981; Davuluri and Knowlton, 1998). A field where friability tests are of particular importance is catalyst development (Dart, 1974). As an example, Contractor et al. (1989) used a submerged-jet attrition test (described below) in their development of a new generation of fluidized bed VPO-catalyst.

Several attempts have been made to develop a standardized attrition test procedure to determine the relative tendency of materials to attrit. A candidate procedure had to meet at least the following two criteria: the amount of sample material required should be small and the test time should be relatively short (Davuluri and Knowlton, 1998). Unfortunately, the determination of a material's friability is not as simple. The crucial point is the interaction of the material properties with the type of stress to which a material is subjected. In this context Pell (1990) gave a frequently quoted thought experiment to illustrate the difficulties: "If we took a batch of rubbers and a batch of diamonds, and rubbed them on abrasive paper, we would conclude that the diamonds were more attrition resistant. If we instead struck the parti-

cles with a hammer we would conclude that the rubber were more attrition resistant."

In this context Contractor et al. (1989) conclude that the relative attrition rate depends on the attrition test method used. Knight and Bridgwater (1985) subjected spray-dried powders to a compression test, a shear test, and a test in a spiral classifier. They found that each test gave a different ranking of the materials. Werther and Reppenhagen (1999a) observed this phenomenon as they subjected various types of fluidized bed catalysts to both a cyclone attrition and a jet attrition test, each simulating one of the three major attrition sources in fluidized bed systems (cf. Sec. 5).

Obviously, there can be no universal procedure for the measurement of a material's propensity to attrition. The attrition resistance is relative and depends on both the material and the stress. An appropriate attrition test should therefore duplicate at least the dominant stress occurring in the considered fluidized bed process, in order to ensure that the ranking of the materials in the test will be identical to that in the process.

However, besides the simple ranking there is quite often even a quantitative prediction of the process attrition requested. This requires both an attrition model with a precise description of the process stress and—as an input parameter to the model—precise information on the material's attritability under this specific type of stress. This calls for attrition/friability tests that duplicate the process stress entirely. As will be elucidated in Sec. 5, the stress in a given fluidized bed system will be generated from at least three sources, i.e., the grid jets, the bubbling bed, and the cyclones. For each there is a corresponding friability test procedure.

4.2   Experiments to Study Attrition Mechanisms

Attrition can normally not be investigated directly in a large-scale process. It is, for example, impossible to analyze the entire bulk of material, and it is nearly impossible or at least very expensive to perform a parameter analysis in a running industrial process. For this reason, attrition has to be investigated in small-scale experiments. The results of these experiments require a model or at least an idea of the governing attrition mechanisms to be applied to the large-scale process. In principle, there are two different "philosophies" of attrition modeling:

The most commonly used philosophy is to design a bulk test facility that simulates the process stress.

Examples of such test facilities are the various jet attrition test devices (such as Forsythe and Hertwig, 1949; Gwyn, 1969; Werther and Xi, 1993), which simulate the stress generated in the vicinity of a gas distributor. The results can either be directly extrapolated to a large-scale process or reveal relationships that may serve as guidelines in the design of this process. It should be noted that most of these tests are batch tests, but most full-scale processes are continuous. Hence the transfer of the results can be difficult owing to the time dependence of the degradation.

Furthermore, it must be taken into account that such a bulk test can only give results that are based on statistics, i.e., they reveal the average effect on all particles but cannot indicate a specific mechanism of a single particle breakage. In any case attention must be paid when the properties of the bulk sample tested differ from those in the real process. As an example, Werther and coworkers (Werther and Xi, 1993; Reppenhagen and Werther, 1999a) used previously screened catalysts in their investigation of the mechanisms of jet and cyclone attrition but also evaluated the effect of the particle size distribution to take the shift in the particle size into account (Werther and Reppenhagen, 1999).

In the second attrition-modeling philosophy (Yuregir et al., 1987; Boerefijn et al., 1998, 2000), it is believed that the mechanics of particle interaction in process test devices and in the large-scale processes are not sufficiently understood to maintain the necessary dynamic similarity and that the analysis of the experimental data for relating the trend to the large-scale operation is therefore not straightforward (Boerefijn et al., 2000). Instead, test devices have to be used where the material is subjected to a well-defined stress like compression, impact, or shear. The chosen test should preferably apply the type of stress that is considered as the main source of attrition in the large-scale process. The complete process attrition can than be described by coupling the test results with a model that describes the motion of fluid and particles and with it the frequency and extent of the stress. Unfortunately, the practical use of this latter approach is at the limit of our current state of knowledge. First of all, it is not easy to define the relevant stress of a process, i.e., the stress is usually composed of more than one of the above mentioned pure types of stress, and it is usually changing in dependence on various parameters. As an example, Boerefijn et al. (2000) reported a gradual change of the grid-jet-induced stress from shear to impact stress with increasing orifice-to-particle size ratio. Moreover, the necessary models of the hydrodynamics in the various regions are not yet fully developed (Boerefijn et al., 2000). Nevertheless, an attempt has been made in one case: Ghadiri and coworkers (Ghadiri et al., 1992a, 1994, 1995; Boerefijn et al., 1998, 2000) have applied this concept to grid-jet-induced attrition using a single particle impact test.

4.3 Test Equipment and Procedures

Detailed reviews of the various test procedures are for example given by Bemrose and Bridgwater (1987) and the British Material Handling Board (1987). The present section is therefore confined to the tests that are most relevant to the subject of fluidized bed attrition.

4.3.1 Tests Applying Well-Defined Stress

As mentioned earlier, one can distinguish three pure and well-defined mechanical stresses on bulk solids material, namely compression, impact, and shear. There are numerous tests that are based on compression and shear, e.g., Paramanathan and Bridgwater (1983), Neil and Bridgwater et al. (1994), Shipway and Hutchings (1993), but they are not further discussed in this chapter because these stresses are usually not relevant to fluidized beds. On the other hand, impact stress occurs whenever particles hit walls or other particles. Attrition caused by impacts can thus be observed, e.g., in grid jets, in the wake of bubbles, in cyclones, or due to free fall. Consequently, there is a great variety of impact tests that try to simulate these particular stresses.

There are, for example, various drop shatter tests in which the material falls under gravity onto a hard surface or a fixed bed. Such a test was carried out by Zenz and Kelleher (1980), who considered catalyst attrition due to free fall in a CFB downcomer. However, the probably most relevant impact tests are those where pneumatically accelerated particles are impacted onto a target. Yuregir et al. (1986, 1987) pioneered this type of test in their work on NaCl salt. In the meantime, such test devices have found broad industrial application as friability tests. For example, Fig. 8 shows the setup used by Davuluri and Knowlton (1998). It requires approximately 100 grams of material to conduct a test. The velocity at which the solids strike the impact plate was varied from 46 to approximately 144 m/s, but in the upper range a material-specific threshold velocity exists, above which the particles completely shatter. This is for most materials a velocity greater than 76 m/s. If velocities above the threshold

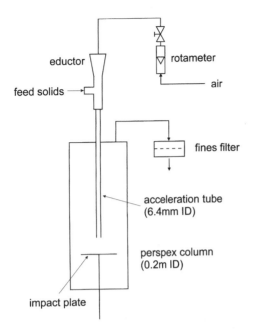

Figure 8 Schematic drawing of impact test facility used by Davuluri and Knowlton (1998).

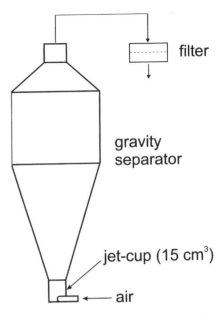

Figure 9 Schematic drawing of the Grace–Davison jet-cup attrition test. (After Weeks and Dumbill, 1990.)

velocity are used, then no relative attritability information can be obtained.

In addition to its use as a pure friability test, this type of test has also been used in the above-mentioned grid-jet modeling approach of Ghadiri and coworkers (Ghadiri et al., 1992a, 1994, 1995; Boerefijn et al., 1998, 2000). In this approach the particles were fed in a single array to the eductor.

4.3.2 Grace–Davison Jet-Cup Attrition Test

The Grace–Davison jet-cup attrition test is often used to test the friability of catalysts (Weeks and Dumbill, 1990; Dessalces et al., 1994). The jet-cup apparatus is sketched in Fig. 9.

The catalyst sample is confined to a small cup, into which gas is tangentially added at a high velocity (about 150 m/s). After a test run over a period of an hour, the so-called Davison Index (DI) is determined by measuring the increase in the weight fraction of particles below 20 microns. The increase is determined from both the fraction of elutriated fine material and a particle size analysis of the remaining fraction in the cup.

Some authors (Dessalces et al., 1994) assume that the stress in the jet-cup is similar to that prevailing in gas cyclones. With respect to fine catalysts, this type of test works as good as the impact test described above, but its applicability is limited to smaller sizes because

larger particles tend to slug in the small cylinder. However, in the catalyst development, where at first only a little batch of catalyst is available, this apparatus is an important friability test, because it requires only a small amount of material (approximately 5 to 10 g).

4.3.3 Fluidized Bed Tests

Fluidized bed tests may be used for both purposes, the determination of the catalysts' friability (Forsythe and Hertwig, 1949; Gwyn, 1969) and the investigation of attrition mechanisms (Werther and Xi, 1993). Most fluidized bed tests are currently carried out as so-called submerged-jet tests, where high-velocity gas jets submerged in a fluidized bed produce high attrition rates in a well-defined short period of time. The majority of these tests are based on the device suggested by Forsythe and Hertwig (1949) (Fig. 10a).

The setup consists of a 0.0254 m ID and 1.52 m long glass pipe, which bears a canvas filter at its upper end, and which is sealed by an orifice plate at the bottom. This latter plate contains a single 0.4 mm ID orifice in its center. The apparatus is operated in such a way that the jet gas velocity approaches the speed of sound in the orifice. The filter keeps all material inside the system. To assess the degradation extent, one should screen the material by wet sieving through −325 mesh (44 µm). The attrition rate is defined as the ratio of the

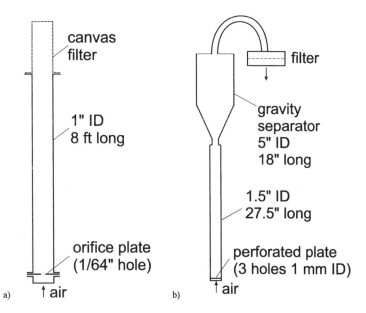

Figure 10 Comparison of two different designs of a submerged jet test (a) after Forsythe and Hertwig (1949); (b) after Gwyn (1969).

increase in weight percent of −325 mesh material and the weight percent age of +325 mesh particles in the initial material. Using this test procedure, one has to take into account that the attrition-produced fines that are kept in the system may affect the material's attritability. Moreover, there are difficulties in using this test as a friability test, because the results are time integrated and have to be assessed with the help of the particle size distribution. The materials that are to be compared should therefore have a similar initial particle size distribution and an identical pretreatment.

These difficulties are avoided in the design suggested by Gwyn (1969), shown in Fig. 10b. Here, the attrition products are not kept inside the system; it is rather assumed that they are elutriated. In the enlarged diameter top section gravity separation defines the limiting diameter of the elutriable particles. The attrition rate is assumed to be given by the elutriation rate. The steady-state elutriation rate can therefore be used as a friability index.

It should be noted here that the quantitative results obtained in a Gwyn-type attrition apparatus will in general depend not only on the cut size of the gravity separator but also on the entrainment and elutriation conditions in the main column. Werther and Xi (1993) compared, for example, attrition test results of the same catalyst obtained from three differently sized Gwyn-type units, one with column A having 50 mm ID and 500 mm height, one with column B having 50 m ID and 1000 mm height, and one with column C having 100 mm ID and 1110 mm height. Although in all columns the same stress was applied to the material and the same gas velocity in the enlarged top sections was adjusted, column A led to significantly higher results in the measured elutriated mass. From a comparison of the particle size distribution of the elutriated material, it could afterwards be concluded that the height of column A had not exceeded the transport disengaging height (TDH), and that coarser particles were allowed to reach the gas outlet. From this observation, Werther and Xi (1993) suggested that as a design rule for this type of attrition test, the height between the bed surface and the gas outlet should at least exceed the TDH.

Moreover, it should be noted that, when one compares different materials' test results, the density of the particulate materials must be taken into account. If the gas mass flow and the temperature are kept constant, then a variation in the solids density will result in a shift of the cut size and thus in the amount of material collected as "attrition product." Another point concerns the particle size distribution of the bed material. If the original solid sample is prepared by sieving e.g. or sifting so that the smallest size is significantly larger than the cut size of the gravity separator (Werther and Xi, 1993), one can be fairly sure that the elutriated material is indeed due to attrition. On the other hand, if the cut size of the gravity separator is located somewhere inside the original particle size distribution, at steady state the elutriated material will inevitably consist of both attrited debris and mother particles, which have shrunk due to attrition to a size below the cut size. It is clear that in this latter case the attrition rate will differ from the one obtained with the previous test. However, it must be generally noted that the Gwyn-type test procedure is particularly suited to measuring and investigating the attrition mode of pure abrasion. It solely measures the produced and elutriated fines but not the rest of the size reduction, i.e., neither the shrinking of most of the mother particles from which the fines are withdrawn nor the breakage into debris larger than the cut size.

Both devices described above were developed in order to test the friability of FCC catalysts. Nowadays the application of these or similar tests is a common procedure in the development of fluidized bed catalysts. Contractor et al. (1989), for example, used a submerged-jet test to compare the attrition resistance of newly developed VPO catalysts. In fact, such tests can be applied to any type of fluidized bed processes. Sometimes they have to be slightly modified to adapt them to the process under consideration. The drilled plate may, for example, be substituted by a porous plate if only attrition in the bed is of interest. Even temperature and pressure can be adapted. Vaux and Fellers (1981) investigated for example the friability of limestone sorbent that is used for fluidized bed combustion. By surrounding a Gwyn-type test facility with a heating system, they took thermal shock and reaction into account.

4.3.4 Cyclone Tests

Even though cyclones are often regarded as the main attrition source in a fluidized bed system (Pell, 1990), the number of publications on cyclone attrition test experiments in the open literature is very small. It may be that experiences with the performance of cyclones are often considered proprietary. However, this lack of information can also be explained by the difficulty of making a distinction between collection efficiency and attrition. The cyclone is a separator designed to keep as much material as possible inside the fluidized bed system. The material that cannot be kept inside the system, i.e., the loss or elutriation, is usually defined as attrition. But in fact, the elutriation rate originates from both the sifted original fines entering the cyclone (from the original feed particle size distribution or produced by attrition in other parts of the fluidized bed system) and the fines that are actually

produced by attrition in the cyclone. Moreover, not all attrition products will be directly elutriated from the cyclone. Instead, a part will be collected by strands of the material and will be transported via the solids return line into the fluidized bed. In subsequent passes through the cyclone the accumulated attrition-produced fines will be elutriated owing to the sifting effect of the cyclone. The same is valid when instead of the attrition-induced elutriation rate the shift in the particle size distribution is considered: there is always a combined effect of attrition and gas solids separation, and above all there is the influence of the other sources of attrition in the solids circulation loop that feeds the cyclone. However, the influence of other attrition sources can be eliminated by a test setup in which a cyclone is operated in isolation and the material handling system does not create additional attrition stress on the material.

Zenz (1974) and Zenz and Kelleher (1980) considered attrition-induced changes in the particle size distribution of FCC catalysts in isolated cyclones of various sizes (0.1 m to 0.76 m ID). The experimental setup simply consisted of the cyclones themselves, a gas–solids separation unit attached to the cyclone overflow consisting of gravity separator and filter and a subsequent suction fan. In the experiments a given batch of catalysts was sucked from a bucket through the respective cyclone. The underflow of the cyclone was directly collected in another bucket, and the particles leaving the cyclone through the overflow were collected in the gas–solids separation unit. After each pass the cyclone's overflow and underflow were mixed and once again sucked through the cyclone. The reason for the recombination of over- and underflow after each pass was to eliminate the interaction with the collection efficiency of the cyclone and thus to assign the measured effects to attrition only. However, this procedure does not entirely correspond to the operating conditions of a cyclone in a real process, where the cyclone overflow is usually lost and thus is not able to reenter the cyclone. Therefore this procedure might lead to results that do not entirely fit with process conditions, since attrition is also affected by the particle size distribution and particularly by the fines content of the material, i.e., there is some kind of cushioning effect of the fines that reduces the entire attrition of the material (cf. Sec. 2.1.3).

Reppenhagen and Werther (1999a) developed an alternative method of studying the attrition mechanisms inside the cyclones independently of the collection efficiency. The basic idea is also to pass a batch of particles several times through an isolated cyclone but to consider the adjusting steady-state loss rate instead of the particle size distribution of the material, i.e., cyclone catch and cyclone loss are not mixed. The experimental setup is shown in Fig. 11.

The cyclone is operated in a suction mode that makes it possible to introduce the solids via a vibrating feeder into a tube attached to the cyclone inlet. The separate feeding of the solids allows an independent variation of gas inlet velocity $u_{c,in}$ and solids loading μ_c. After each pass through the cyclone, the collected material can be used as a feed material for the next run by exchanging the underflow and feeder hoppers. The cyclone overflow is connected to a sieve and a filter in series. The appropriately sized sieve is intended to collect accidentally elutriated mother particles, whereas the filter is intended to collect the abrasion-produced fine particles. Since a part of these abrasion-produced fines will not reach the filter but stick to the sieve, after each pass the sieve is cleaned. For this purpose the connection between sieve and filter is detached and a nozzle is mounted onto the top of the filter housing. With this vacuum-cleaner-like arrangement the sieve is then "cleaned" from its bottom side, by moving the sieve relative to the nozzle. All particles smaller than its mesh size are then sucked through the sieve and are collected on the filter. Hence, the entire amount of elutriated abrasion-produced fines can be measured by the increase in weight of the filter.

Figure 12 gives an example of the experimental results obtained with a previously screened spent FCC catalyst, i.e., the smallest particles were sufficiently larger than the particle size at which the separation efficiency of the cyclones becomes unity. The cyclone loss rate per single pass is plotted against the number of passes for given operating conditions (gas velocity at the cyclone inlet $u_{c,in}$, solids-to-gas loading ratio μ_c). After each pass the cyclone loss was determined from the increase of the filter weight, and the loss rate was evaluated by relating the solids mass in the loss to the respective solids input mass.

In analogy to Fig. 6, a large initial cyclone loss rate is observed, which rapidly decreases and after 10 to 20 passes reaches a steady-state value. The high value of the loss rate during the first couple of passes is due to the added effects of attrition and the sifting of fines that could not be separated by the previous sieving. They may have stuck to the surface of coarser particles and are only removed inside the cyclone due to interparticle and particle–wall collisions. After 34 passes, the operating conditions were changed to a higher inlet velocity of 20 m/s, which due to a constant solids feeding rate led to a corresponding reduction in the solids loading. The measurements show a slow increase of the cyclone loss rate, which after another 15 passes leads to a steady state again. This slow transition is probably caused by abrasion-produced fines sticking to the larger particles that are sent into the catch of the cyclone.

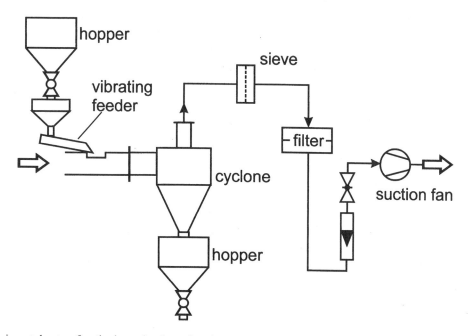

Figure 11 Experimental setup for the investigation of cyclone attrition. (After Reppenhagen and Werther, 1999a.)

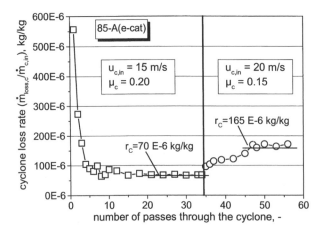

Figure 12 Loss rate obtained with a screened spent FCC catalyst as a function of the number of passes through the cyclone ($u_{c,in}$: gas velocity at the cyclone inlet; μ_c: solids to gas loading ratio, 90 mm ID cyclone). (After Reppenhagen and Werther, 1999a.)

Hence it can be concluded that for each operating condition the fines concentration in the material tends to a characteristic value of which the accumulation of fines is balanced by the release of fines. When this characteristic concentration is reached, the loss rate is at steady state, i.e., it is equal to the production rate of fines. Reppenhagen and Werther (1999a) suggested that we take this steady-state value as a characteristic value for both the assessment of a material's attritability and the study of cyclone attrition mechanisms in dependence on the various influencing parameters. For this purpose, they defined the cyclone attrition rate as

$$r_c = \frac{\dot{m}_{loss,c,steady\text{-}state}}{\dot{m}_{c,in}} = \frac{\dot{m}_{a,c,fines}}{\dot{m}_{c,in}} \qquad (6)$$

where $\dot{m}_{loss,c,steady\text{-}state}$ is the cyclone loss flow at steady state, $\dot{m}_{a,cfines}$ is the production rate of fines due to attrition inside the cyclone, and $\dot{m}_{c,in}$ is the mass flow into the cyclone.

It should be noted that Reppenhagen and Werther (1999a) have exclusively concentrated on the attrition of FCC and fluidized bed catalysts, the abrasion-produced fines of which were throughout all the experiments entirely elutriable. In other words, independently of the cyclone, the operating conditions, and the catalyst's material density, the produced fines were sufficiently small to be in a particle size range in which the grade efficiency of the cyclone is smaller than unity. However, for a transfer of the described test concept to other investigations, the same attention must be paid as in the case of the Gwyn-type tests, i.e., when comparing results with different materials the density of the particulate material and its particle size distribution

must be taken into account. As for the latter parameter, Reppenhagen and Werther (1999a) have given a simple example. They compared the test results of a previously screened fraction with those of an unscreened fraction of the same material. By taking the differences in the mean particle size into account (cf. Sec. 5.3), they observed a distinctly higher loss rate (approximately 25% higher) in the case of the unscreened fraction owing to the additional loss of sufficiently shrunk mother particles.

In general, it must be recognized that the greater effort, and the large amount of material required, for the currently available cyclone attrition test methods will generally prevent them from being used as a pure friability test, i.e., a test for a sole ranking of various materials without any reference to a particular process. However, the tests are useful for both the study of general attrition mechanisms in cyclones and the evaluation of a given material's susceptibility to the particular attrition stress inside cyclones, which may be needed as an input parameter in a cyclone attrition model.

5 SOURCES OF ATTRITION IN A FLUIDIZED BED SYSTEM

A first approach to finding attrition in fluid beds was made by Zenz (1971). He pointed out that there are various regions in a fluidized bed reactor system in which the stress acting on the bed particles and the corresponding attrition mechanisms are quite different. In the subsequent works (Zenz, 1974; Vaux and

Keairns, 1980), attempts have been made to identify these individual sources. Later on, Zenz and Kelleher (1980) revealed the necessity of studying each of these sources in isolation in order to get detailed information about the particular attrition mechanisms. They suggested that we investigate the vicinity of a multihole gas distributor, the bubbling fluidized bed, the cyclones, the conveying lines, and the solids feeding devices as separate regions where the attrition mechanisms are different. However, usually the gas distributor, the bubbling bed, and the cyclones are regarded as the most relevant attrition sources (Pell, 1990). The present section is therefore also confined to these sources.

The investigation of the gas distributor and the bubbling bed attrition were both almost exclusively carried out in various derivations from the Gwyn-type test facility. Hence the material loss is the only attrition result considered, and the derived model approaches thus exclusively deal with the production of elutriable fines. There is consequently a lack of direct information on the role of attrition in the adjustment of the bed particle size distribution.

5.1 Grid Jet as a Source of Attrition

Gas distributors of fluidized beds are often designed as perforated or nozzle plates. Since a minimum pressure drop is required to obtain a uniform gas distribution over the bed's cross-sectional area, the open surface area is rather small, and the gas jets issuing from the distributor holes are at high velocity. Particles are entrained by these jets, accelerated to high velocities, and impacted onto the fluidized bed suspension at the end of the jets, resulting in particle degradation similar to that in jet grinding processes (Kutyavina and Baskakov, 1972).

One peculiarity of the jet-induced attrition is that the jets affect only a limited bed volume above the distributor, which is defined by the jet length. Hence as soon as the jets are fully submerged, their contribution to attrition remains constant, with further increasing bed height. Figure 13 shows some experimental results obtained by Werther and Xi (1993).

The jet penetration length can be estimated by various correlations, such as those given by Zenz (1968), Merry (1975), Yates et al. (1986), or Blake et al. (1990). However, in most commercial fluidized bed processes the bed is much higher than the jet penetration length. Hence jet-induced attrition cannot be investigated in isolation, because there is always some additional attrition of the bubbling bed. For this reason many authors

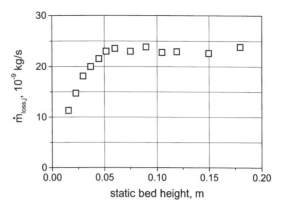

Figure 13 Influence of the static bed height on jet attrition of spent FCC catalyst in a submerged jet test facility ($D_t = 0.05$ m, $u_{or} = 100$ m/s, $d_{or} = 2$ mm). (Werther and Xi, 1993.)

(Blinichev et al., 1968; Kutyavina and Baskakov, 1972; Arastoopour and Chen, 1983; Contractor et al., 1989) considered the overall attrition rate resulting from both attrition sources. However, in order to get direct insights into the mechanisms of jet attrition it is necessary to separate the jet contribution from the measured overall attrition rate. This can be done in two different ways.

Seville et al. (1992) and Ghadiri et al. (1992b) measured the attrition rates at various static bed heights. Assuming a linear increase of the attrition rate with the bed height above the jetting region, they extrapolated the measured attrition rates back to the jet length calculated from one of the available correlations. The extrapolated value was then taken as the attrition rate of the jetting region. Unfortunately, no comprehensive study of the attrition mechanisms were carried out on the basis of this strategy. Only a few parameter effects were tested, which did not lead to an unambiguous description of the jet attrition mechanisms.

An alternative strategy to study the jet-induced attrition is suggested by Werther and Xi (1993). They used a Gwyn-type test apparatus with a particular distributor design in which a separately fed nozzle was integrated into a porous plate (Fig. 14). At first the bed was only aerated via the porous plate. In this way the contribution of the bubbling bed attrition could be measured without any additional attrition sources. In a second step, the apparatus was operated with a chosen jet gas velocity. To maintain the cut size of the gravity separator above the bed at some prescribed level, they kept the superficial gas velocity constant by supplying auxiliary air through the porous plate. The resulting loss rate of the system originates

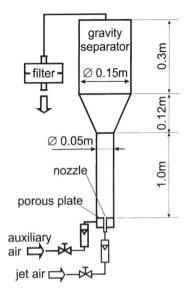

Figure 14 Gwyn-type test apparatus suggested by Werther and Xi (1993) to investigate the jet-induced attrition separately from the bubble-induced attrition.

from both the jets and the bubbling bed. And the jet attrition rate can be now calculated by subtracting the bubble attrition rate measured before from the overall attrition rate. This is a certain oversimplification, since it assumes that the bubble attrition will always be the same regardless of the ratio of jet air mass flow to auxiliary mass flow. However, this method holds fairly well for low jet gas velocities, where the contribution of the jet gas flow to the total gas flow is rather small. For higher jet gas velocities, the jet attrition is so high that the contribution of bubble attrition may be neglected.

Using the above-described experimental setup, Werther and Xi (1993) carried out a comprehensive experimental program to study grid-jet attrition. They found that at steady-state attrition conditions (cf. Fig. 6), the jet-induced attrition exclusively occurs in the mode of pure abrasion. On the basis of this result they suggested and validated a model that considers the energy utilization of this abrasion process by relating the surface energy created to the kinetic energy that has been spent to produce this surface area. Assuming that there is no interaction between the individual jets of a given distributor, they derived the following relationship for the grid-jet-induced generation of fines:

$$\dot{m}_{abr.,fines,j} = n_{or} \cdot K_j \cdot \rho_f \cdot d_{or}^2 \cdot u_{or}^3 \qquad (7)$$

where n_{or} is the number of orifices in the distributor, ρ_f is the density of the jet gas, d_{or} is the diameter of the

orifices, and u_{or} is the jet gas velocity inside an orifice. K_j is a constant that characterizes the solids' susceptibility to abrasion. It turns out that this parameter can be expressed as the product of the surface mean diameter d_{pb} of the bed solids and a particle-size-independent material specific constant C_j, i.e., the solids susceptibility to abrasion increases linearly with its surface mean diameter

$$K_j = C_j \cdot d_{pb} \qquad (8)$$

A substitution of Eq. (8) into Eq. (7) yields

$$\dot{m}_{abr.,fines,j} = n_{or} \cdot C_j \cdot d_{pb} \cdot \rho_f \cdot d_{or}^2 \cdot u_{or}^3 \qquad (9)$$

With regard to the distributor design, Werther and Xi (1993) have shown that the attrition effect of a horizontal jet equals that of an upward facing jet, i.e., the same equation and an identical attrition rate constant can be used for the modeling. In contrast to this, downward issuing jets were found to generate significantly higher attrition rates because of the impact of entrained particles onto the fixed bed of particles at the nose of the jet, i.e., there is probably a different attrition mechanism involved requiring a different attrition model or at least a different attrition rate constant for a given material.

Another issue that should be considered with respect to distributor design is the influence of its open surface area A_o. For such a consideration, Eq. (9) can be complemented by the following dependencies:

$$A_o \propto n_{or} \cdot d_{or}^2 \qquad (10)$$

$$u_{or} = \frac{\dot{V}}{A_o} \qquad (11)$$

leading to the expression

$$\dot{m}_{abr.,fines,j} = \propto \frac{\dot{V}^3}{A_o^2} \qquad (12)$$

According to Eq. (12), for a given gas volumetric flow rate, the decisive quantity for the distributor attrition rate is the open surface area A_o, which suggests that it is unimportant with respect to attrition whether A_o originates from a few large or from many small orifices. But this conclusion is somehow in contrast to the findings of Boerefijn et al. (2000). They observed a change in the attrition stress with increasing orifice-to-particle size ratio from pure shear to impact stress and thus a change in the predominant attrition mode from pure abrasion to the fracture of microspheroids.

5.2 Bubble-Induced Attrition

Bubble-induced attrition originates from low-velocity interparticle collisions. Since the bubble rise velocity is of the order of 1 m/s, the energy is generally not high enough to shatter the bed particles into fragments. For this reason, most laboratory experiments have shown the bubbles to be a minor source of attrition. However, in a deep fluidized bed, with several meters of height, the contribution of bubble-induced attrition may be a significant factor.

Again, as in the case of jet attrition, attention must be paid to the isolation of that part of the attrition that is due to bubbles. There are basically two ways to do this. One is to use a porous plate distributor in order to avoid any grid jets. The other is the above described procedure suggested by Seville et al. (1992) and Ghadiri et al. (1992b): the measurement of the production rate of fines at different values of the static bed height permits us to eliminate the grid jet effects (cf. Sec. 5.1).

Even though most of the attrition tests presented in the literature deal with bubble-induced attrition, the respective attrition mechanisms are not quite clear yet. There are various theoretical and empirical approaches that can in accordance with Eq. (3) be summarized in the following definition of a bubble-induced steady-state attrition rate:

$$r_b = \frac{\dot{m}_{loss,b,steady-state}}{m_b} = \frac{\dot{m}_{attr.,fines,b}}{m_b}$$
$$= K_b \cdot m_b^n \cdot (u - u_{min})^z \qquad (13)$$

where $\dot{m}_{loss,b,steady-state}$ is the bubble-attrition-induced loss rate from a system under steady-state conditions, which can be assumed to be identical to the bubble-attrition-induced production rate of elutriable material, $\dot{m}_{attr.,fines,b}$ at this state. m_b is the bed mass, the exponent of which varied between 0 and 1. In the case of $n = 0$, it is implied that the attrition stress does not change with the bed height; there is simply a linear increase with the amount of treated material. In contrast to this, an exponent $n > 0$ implies an increase of the attrition stress with the bed height. The velocity u_{min} is regarded as a threshold velocity above which the bubble-induced attrition occurs. Its value varied from $u_{min} = u_{mf}$ to $u_{min} \gg u_{mf}$. The exponent z is in the vast majority of studies set to a value of one, but in some cases it is even set to a value of 3, i.e., a linear dependence of the attrition extent on the input of kinetic energy with the fluidizing gas is assumed. In the following,

the various theoretical and experimental findings are briefly summarized.

Merrick and Highley (1974) have modeled bubble-induced attrition as a comminution process. According to Rittinger's law of size reduction (cf. Perry, 1973), the rate of creation by abrasion of new surface area $\Delta S/\Delta t$ is proportional to the rate of energy input $\Delta E/\Delta t$:

$$\frac{\Delta S}{\Delta t} \propto \frac{\Delta E}{\Delta t} \qquad (14)$$

Since the size distribution of the fines produced by abrasion is approximately constant, the rate of production of new surface can be taken to be proportional to the mass rate of production of fines,

$$\frac{\Delta S}{\Delta t} \propto \dot{m}_{attr.,fines,b} = r_b \cdot m_b \qquad (15)$$

The total rate of input of energy to the fluidized bed is given by the product of the volumetric flow rate of gas ($u \cdot A_t$) and the pressure drop, which may be expressed as weight of the bed divided by the bed's cross-sectional area A_t. However, only part of the input energy is available for bubble formation and thus for comminution. The input energy ($u_{mf} \cdot m_b \cdot g$) is required per unit time for keeping the particles in suspension. That part of the rate of input of energy that remains for bubble formation, and thus for attrition, is then given by $((u - u_{mf}) \cdot m_b \cdot g)$. Insertion into Eq. (15) yields

$$r_b = \frac{\dot{m}_{attr.,fines,b}}{m_b} = K_b \cdot (u - u_{mf}) \qquad (16)$$

where K_b is an abrasion rate constant. With a similar approach, Ray et al. (1987a) arrived at the same result. This approach supports the idea of a linear dependence of the bubble-induced attrition rate on the bed mass, i.e., the attrition stress does not change with the bed height, and $n = 0$ in Eq. (13). In contrast to this Vaux (1978), Ulerich et al. (1980), and Vaux and Schruben (1983) proposed a mechanical model based on the kinetic energy of particles agitated by the bubble motion, where they took the increase of the bubble velocity with bed height into account. The authors conclude that the bubble-induced attrition rate is proportional to the product of excess gas velocity and bed height or bed mass, respectively:

$$r_b = \frac{\dot{m}_{attr.,fines,b}}{m_b} = K_b' \cdot m_b \cdot (u - u_{mf}) \qquad (17)$$

This inconsistency with respect to the influence on the bed height is also reflected by the available experimental data on bubble-induced attrition: Merrick and Highley (1974) and Pis et al. (1991) have found the attrition rate r_b to be independent of bed height. Kono (1981) found this value to be proportional to the static bed height with the exponent 0.78, and Ulerich et al. (1980) and Xi (1993) found r_b to be proportional to the bed height.

Obviously the role of bed height is not yet fully understood. Ray et al. (1987a) have explained these discrepancies by a consideration of bubble growth with height above the distributor. They argue that as long as the bubble size increases with height, the efficiency of the transformation of kinetic energy to free surface energy might increase, thus leading to an attrition rate that increases with bed height. However, as the bed height reaches the limits of slugging in small-diameter columns or maximum attainable bubble size in a large diameter bed, an extra bed height will not vary the conditions of bubbling and thus will result in r_b becoming independent of bed height. However, further experiments are certainly needed in this area.

Although the dependence on the gas velocity appears to be reasonably explained by the above described model approaches, the experimental data available in the literature are giving even in this respect an inconsistent picture: Merrick and Highley (1974), Arena et al. (1983), and Pis et al. (1991) also found the linear dependence on the excess gas velocity $(u - u_{mf})$ to be valid. As an example, Fig. 15 shows the results of Pis et al. (1991), which were obtained in a fluidized bed column of 0.14 m in diameter. The distributor had orifices of 1 mm in diameter on a 5 mm square pitch. Unfortunately, no distinction was made between the measured attrition rate and the influence

of the grid jets. However, their influence might be negligible in the present case due to the relatively small jet velocity.

Ray et al. (1987a) obtained a fairly different result as they considered the attrition of narrow fractions of limestone with particle sizes between 1.09 and 0.77 mm in a 0.1 m diameter bed, which was equipped with a porous plate (Fig. 16). Similar results were obtained by Xi (1993), who investigated the attrition of fine catalyst particles with a minimum fluidization velocity u_{mf} of 0.002 m/s (Fig. 17). As is obvious from Figs 16 and 17, the attrition rate extrapolates to zero at a fluidizing velocity u_{min} which is significantly larger than u_{mf}. This means that a minimum kinetic energy or a minimum extent of bubbling is necessary to cause attrition in these cases.

More recently, Werther and Reppenhagen (1999) correlated their bubbling bed attrition test results obtained under pure abrasion conditions in a 200 mm ID Gwyn-type test plant even to the excess gas velocity raised to a power of 3 (Fig. 18), which indicates that the bubbling bed attrition is linearly increasing with the excess kinetic energy supplied to the system:

$$r_b = \frac{\dot{m}_{loss,b,steady\text{-}state}}{m_b} = \frac{\dot{m}_{abr.,fines,b}}{m_b} \qquad (18)$$
$$= K_b^* \cdot (u - u_{mf})^3$$

5.3 Cyclones as Attrition Sources

For the reasons explained above there is only a limited amount of work published in the open literature on cyclone attrition. In fact there are results from only

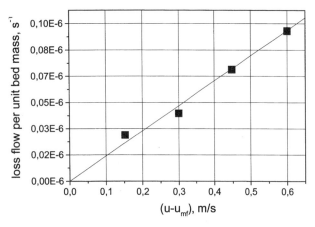

Figure 15 Variation of the steady-state attrition-induced loss rate of coal ash (0.2 to 0.315 mm) with the excess gas velocity $(u\text{-}u_{mf})$. (Pis et al., 1991.)

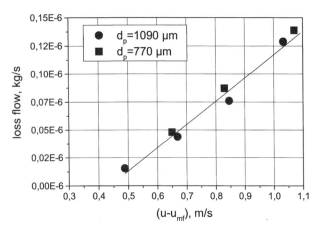

Figure 16 Variation of the steady-state attrition-induced loss flow of limestone fractions with $(u\text{-}u_{mf})$. (Ray et al., 1987a.)

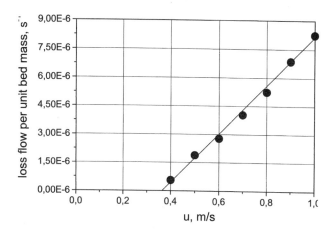

Figure 17 Variation of the steady-state attrition-induced loss flow per unit bed mass of a catalyst with superficial gas velocity (AVN 802, $u_{mf} = 0.002\,\text{m/s}$). (Xi, 1993.)

two groups (Zenz, 1974; Zenz and Kelleher, 1980; Reppenhagen and Werther, 1998, 1999a; Werther and Reppenhagen, 1999). The experimental procedures of both groups have already been explained in Sec. 4.3.4.

As a result of their experiments with FCC catalyst material in isolated cyclones, Zenz (1974) and Zenz and Kelleher (1980) observed even after a few passes a significant change in the particle size distribution (cf. Fig. 7). The content of coarser particles was decreased in comparison to their initial one, whereas the content of smaller particles was increased. From many experiments of this kind, Zenz and Kelleher

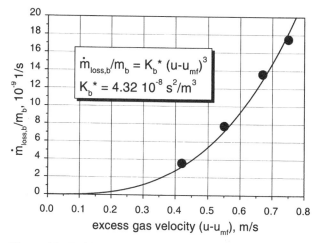

Figure 18 Bubbling bed attrition test results obtained by Werther and Reppenhagen (1999) in a 200 mm diameter Gwyn-type test plant under pure abrasion conditions with fresh FCC catalyst.

(1980) derived design correlations for the estimation of cyclone attrition for fluidized bed systems. Unfortunately the details of these estimations remained proprietary.

Hence the only detailed reports yet available in the open literature on attrition in cyclones are those of Reppenhagen and Werther (1998, 1999a), who exclusively concentrated on the attrition of fluidized bed catalyst particles. Assuming that the catalysts attrit in a cyclone exclusively in the mode of pure abrasion, they suggested a model that regards cyclone attrition under steady-state conditions as a comminution process: it considers the efficiency of such a process by relating the surface energy created by comminution to the kinetic energy, which has been spent to create this new surface area. According to the model, the above defined cyclone attrition rate r_c [cf. Eq. (6)] can be calculated from

$$r_c = \frac{\dot{m}_{\text{loss,c,steady-state}}}{\dot{m}_{c,\text{in}}} = \frac{\dot{m}_{\text{abr.,fines,c}}}{\dot{m}_{c,\text{in}}} = K_c \cdot \mu_c^n \cdot u_{c,\text{in}}^2 \tag{19}$$

where K_c is the cyclone attrition rate constant, which summarizes all particle properties that are relevant to the abrasion process, and n is an exponent assuming a power law dependence of the attrition rate on the solids-to-gas loading ratio μ_c at the entrance of the cyclone. According to the procedure described in Sec. 4.3.4, the authors carried out comprehensive attrition tests with a spent FCC catalyst and a 90 mm ID cyclone in order to scrutinize this model approach. In these experiments the gas velocity $u_{c,\text{in}}$ at the entrance of the cyclone was varied between 8 and 24 m/s with the solids loading μ_c ranging from 0.05 to 1. In Fig. 19 the measured cyclone attrition rates r_c obtained under steady-state conditions are plotted against the cyclone inlet velocity on a double logarithmic grid.

The relationship between r_c and $u_{c,\text{in}}$ predicted by Eq. (19) is seen to be confirmed over a wide range, which is indicated by the solid lines. However, as indicated by the dashed lines, there are some distinct deviations when the gas velocity exceeds a certain threshold. The threshold on its part seems to be dependent on the solids loading, i.e., the smaller the solids loading the smaller the gas velocity from which the deviation occurs. Reppenhagen and Werther (1999a) explain this deviation by the occurrence of another attrition mechanism in addition to abrasion, namely the so-called surface fragmentation of the catalyst particles, which results from the combination of increasing kinetic energy at increasing velocities and increasing

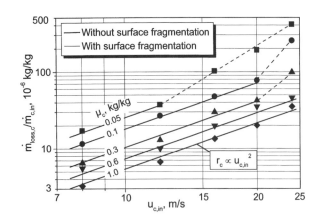

Figure 19 Influence of the cyclone inlet velocity $u_{c,in}$ on the cyclone attrition rate r_c at different solids loadings measured by Reppenhagen and Werther (1999a) in a 90 mm ID cyclone. Material: spent FCC catalyst; $u_{c,in}$: cyclone inlet velocity; μ_c: solids-to-gas loading ratio.

single-particle/wall interactions with decreasing solids loading. However, the authors could clearly identify the threshold for the onset of this additional attrition mode and thus the range of applicability of Eq. (19) by means of scanning electron microscope images (cf. Sec. 2.1.5, especially Figs. 4 and 5). Furthermore, it should be noted here that the operating conditions leading to fragmentation are rarely encountered in industrial fluidized bed applications since high cyclone inlet velocities are normally avoided in order to keep the cyclone pressure drop at a reasonable level.

After identification of the thresholds, Reppenhagen and Werther (1999a) derived the value of the exponent n in Eq. (19) from all measurements taken under conditions of pure abrasion (straight lines drawn in Fig. 19) to $n = -0.5$. They explained the negative value of n by some kind of "cushioning" effect, i.e., the chance for a given particle to impact on the wall decreases with increasing solids concentration in the flow. With $n = -0.5$, the model equation can now finally be written as

$$r_c = K_c \cdot \frac{u_{c,in}^2}{\sqrt{\mu_c}} \tag{20}$$

In further experiments, where the additional attrition mode of surface fragmentation was avoided by keeping the inlet gas velocity below 20 m/s and the gas solids loading above 0.1, this model equation could be validated for various types of catalyst and differently designed and sized cyclones with tangential inlet. As an example, the test results for five different catalysts

are shown in Fig. 20, where the cyclone attrition rate is plotted as a function of $u_{c,in}^2/\mu^{0.5}$.

In complete agreement with Eq. (20), the attrition rate for each material is proportional to $u_{c,in}^2/\mu^{0.5}$. However, the absolute value of the attrition rate depends on the properties of the individual material, which are summarized in the rate constant K_c. Moreover, the authors found that this constant depends not only on the type of material but also on the surface mean diameter d_{pc} of the solids that enter the cyclone. Hence they suggested that we subdivide the attrition rate constant K_c into

$$K_c = C_c \cdot d_{p,c} \tag{21}$$

where a particle-size-independent attrition rate constant C_c characterizes now solely the material properties as, for example, strength, shape, and surface roughness. Substituting Eq. (21) into Eq. (20) finally leads to

$$r_c = C_c \cdot d_{pc} \cdot \frac{u_{c,in}^2}{\sqrt{\mu_c}} \tag{22}$$

6   ATTRITION IN THE OVERALL FLUIDIZED BED SYSTEM

In conclusion from the sections above, two main distinctions must be made when attrition in an overall fluidized bed system is considered: Primarily it must be recognized that there are several attrition sources in a fluidized bed system with distinctly different attrition mechanisms, which must be described separately.

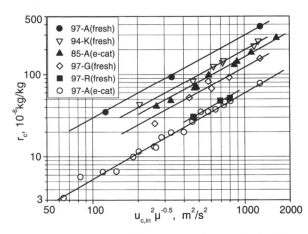

Figure 20 Cyclone attrition test results obtained with various types of catalysts in a 90 mm ID cyclone. (Reppenhagen and Werther, 1999a.)

Furthermore, a distinction must be made between the various attrition modes that may even occur at a single attrition source depending on the energy level applied.

Although this concept is widely known, up to now it has rather seldom been applied. Even most of those publications that are just focusing on a Gwyn-type test facility do not distinguish between grid-jet attrition and bubble-induced attrition. This may be the main reason for the strong discrepancies in the previously published literature where experimental findings are compared, which are obtained from different systems and different solids under quite different operating conditions. Ghadiri et al. (1992a) gave a simple example of such a discrepancy by comparing the published findings for the influence of the superficial gas velocity on the attrition-induced loss flow from fluidized bed systems: Some authors (Seville et al., 1992) relate their experimental results to $u - u_{mf}$, while others assume the attrition rate to be proportional to u^n, where the exponent varies from $n = 1$ (Patel et al., 1986) to $n = 5.8$ (Blinichev et al., 1968).

However, more recently the present authors (Werther and Reppenhagen, 1999; Reppenhagen and Werther, 1999b) have demonstrated that a strict observance of the above concept dramatically helps to overcome these discrepancies. They considered the attrition-induced loss flow of catalyst material from the cold model fluidized bed unit with external solids recirculation that is schematically shown in Fig. 21. The cyclone overflow is connected to a filter that collects the elutriated material. Under conditions of steady-state attrition, the loss rate was then obtained from the increase in the filter weight per unit time. As bed material a fresh FCC catalyst material was used. In order to ensure that no original particle but only attrition-produced debris is collected on the filter, the catalyst was previously sieved to remove its fraction of elutriable fines.

The results of two different test series are shown in Fig. 22, where they are plotted as a function of the superficial gas velocity in a double logarithmic grid. The first series of measurements was taken by using a porous plate as a gas distributor, which was substituted by a perforated plate (640 orifices of 0.7 mm ID) in the second series. For both series a strong sensitivity against the gas velocity can be observed. In the case of the porous plate distributor, a 10% increase in the superficial gas velocity leads almost to a doubling of the loss flow. In case of the perforated plate, the sensitivity against the gas velocity is lower, but the absolute values of the loss flows are higher.

In order to describe such parameter dependencies, the conventional approach would be simply to correlate the measured overall loss flows in the form of a power law. For the data shown, this would lead to a dependence on the superficial gas velocity raised to a power of 4.2 or even 7, respectively. A physically sound explanation for these exponents is certainly difficult, and it is obvious that such a correlation obtained with the one system design is not directly transferable to the other even though only the distributor was exchanged, which reveals that such a model approach

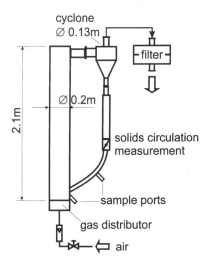

Figure 21 Experimental setup for the investigation of the attrition-induced loss flow of fresh FCC catalyst from a fluidized bed system. (After Werther and Reppenhagen, 1999; Reppenhagen and Werther, 1999b.)

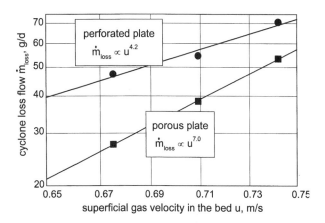

Figure 22 Measured loss flows from the fluidized bed system shown in Fig. 21 operated with previously screened fresh FCC catalyst. (After Werther and Reppenhagen, 1999; Reppenhagen and Werther, 1999b.)

results in exactly those discrepancies described by Ghadiri et al. (1992a). In contrast to such a simple correlation of the overall loss flow, Werther and Reppenhagen (1999) have chosen to consider the relevant mechanisms with respect to the attrition-induced loss flow of the system individually. In the following it will be briefly described:

In a first step, they identified three regions as main attrition sources, namely the grid jets, the bubbling bed itself, and the cyclone section. Accordingly, they described the quasi-stationary loss rate of the overall system as the sum of their individual contributions: Since the stress conditions throughout all the experiments could be regarded to be under pure abrasion conditions, the model equations Eqs. (9), (18), and (22) could be applied to describe the generation of elutriable fines; the following reasoning was made:

Since the abrasion-produced fines are rather small, i.e., typically smaller than about 1 to 3 microns, the respective grade efficiency of the cyclones is distinctly smaller than unity. Therefore it can be assumed that the fines are immediately lost after their production, even if they are produced in the jetting region or in the bubbling bed. In contrast to this, it can be assumed that after an initial phase where the original elutriable material is sifted off, the mother particles are entirely kept inside the system. Due to the mode of pure abrasion the change in their particle size is negligible, and so is their contribution to the loss rate. Hence some kind of steady state can be assumed where the production of fines, which originate from the almost unchanged mother particles, is balanced by the loss rate of fines.

Hence we can write:

$$\dot{m}_{\text{loss,tot,steady-state}} = \dot{m}_{\text{loss,j,steady-state}} + \dot{m}_{\text{loss,b,steady-state}}$$
$$+ \dot{m}_{\text{loss,c,steady-state}}$$
$$= \dot{m}_{\text{abr.,fines,j}} + \dot{m}_{\text{abr.,fines,b}}$$
$$+ \dot{m}_{\text{abr.,fines,c}}$$
$$= n_{\text{or}} \cdot C_{\text{j}} \cdot d_{\text{pb}} \cdot \rho_{\text{f}} \cdot d_{\text{or}}^2 \cdot u_{\text{or}}^3$$
$$+ K_{\text{b}}^* \cdot m_{\text{b}} \cdot (u - u_{\text{mf}})^3$$
$$+ \dot{m}_{\text{c,in}} \cdot C_{\text{c}} \cdot d_{\text{pc}} \cdot \frac{u_{\text{c,in}}^2}{\sqrt{\mu_{\text{c}}}}$$

$$(23)$$

However, for a consistent description of the overall process it is reasonable to substitute the orifice velocity u_{or}, the solids loading μ_{c} and the cyclone inlet velocity $u_{\text{c,in}}$ by

$$u_{\text{or}} = u \cdot \frac{D_{\text{t}}^2}{n_{\text{or}} \cdot d_{\text{or}}^2} \qquad (24)$$

with D_{t} being the diameter of the fluidized bed column,

$$\mu_{\text{c}} = \frac{\dot{m}_{\text{c,in}}}{\rho_{\text{f}} \cdot u_{\text{c,in}} \cdot A_{\text{c,in}}} \qquad (25)$$

and

$$u_{\text{c,in}} = u \cdot \frac{A_{\text{t}}}{A_{\text{c,in}}} \qquad (26)$$

which leads to the following completed model equation for the overall process:

$$\dot{m}_{\text{loss,tot}} = C_{\text{j}} \cdot d_{\text{pb}} \cdot \rho_{\text{f}} \cdot \frac{D_{\text{t}}^6}{d_{\text{or}}^4 \cdot n_{\text{or}}^2} \cdot u^3$$
$$+ K_{\text{b}}^* \cdot m_{\text{b}} \cdot (u - u_{\text{mf}})^3$$
$$+ C_{\text{c}} \cdot d_{\text{pc}} \cdot \sqrt{\dot{m}_{\text{c,in}}} \cdot \sqrt{\rho_{\text{f}}} \cdot \frac{A_t^{2.5}}{A_{\text{c,in}}^2} \cdot u^{2.5} \quad (27)$$

From this model equation (27), it is obvious that the attrition effects in the overall system depend not only on the prevailing attrition mechanisms but also on the solids transport in the freeboard which determines the solids flow into the cyclone and thus the amount and the particle size distribution of the material that is subjected to cyclone attrition. Hence in a second step of the overall modeling, these transport effects must be taken into account, which has in a first approach been simply done by measuring the necessary data.

Figure 23 shows a comparison of the experimental data depicted in Fig. 22 with the calculation from the model equation (27). The required attrition rate constants C_{j}, K_{b}^*, and C_{c} that describe the materials susceptibility to attrit in the respective regions have been determined by the corresponding attrition tests as described in Sec. 4.3. C_{j} has been determined from exactly that Gwyn-type test facility that is shown in Fig. 14 and was set to zero in the case of the porous plate distributor; K_{b}^* has been measured in a 200 mm ID Gwyn-type test apparatus, and C_{c} has been determined from exactly that cyclone attrition-test procedure that is described in Sec. 4.3.4 using the equipment sketched in Fig. 11. The parameters $\dot{m}_{\text{c,in}}$ and d_{pc} were measured in the apparatus sketched in Fig. 21 under the assumption that $\dot{m}_{\text{c,in}}$ may be

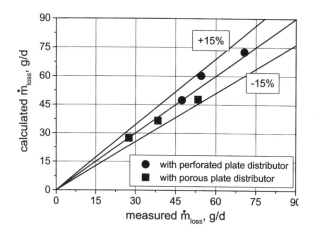

Figure 23 Comparison of measured and calculated loss flows for the two different system designs. The input data for the calculation that are affected by entrainment had been measured. (After Werther and Reppenhagen, 1999; Reppenhagen and Werther, 1999b.)

approximated by the solids circulation rate measured in the return line. Figure 23 shows a very good agreement between the measured and the calculated overall loss flows even though two different process designs are considered.

However, as mentioned above, Eq. (27) requires information on the solids transport effects. Hence for an a priori modeling, the implementation of an entrainment/elutriation model is required. Werther and Reppenhagen (1999) have given an example of such a modeling approach: the bed particle size distribution and thus d_{pb} were simply assumed to be constant, and $m_{c,in}$ and d_{pc} were substituted by

$$\dot{m}_{c,in} = A_t \cdot G_s = A_t \cdot \sum_i G_{si}$$
$$= A_t \cdot \sum_i w_i \cdot K_i^* \qquad (28)$$

$$d_{pc}(u) = \frac{G_s}{\sum_i \frac{G_{si}}{x_i}} = \frac{G_s}{\sum_i \frac{w_i \cdot K_i^*}{x_i}} \qquad (29)$$

where G_s is the entire solids elutriation rate from the bed, x_i is the mean diameter of the size interval i, G_{si} is the fractional elutriation rate for the size interval i, w_i is the weight fraction of the size interval i in the bed material, and K_i^* is the respective elutriation rate constant, which was calculated from a correlation suggested by Tasirin and Geldart (1998):

$$K_i^* = 14.5 \cdot \rho_f \cdot u^{2.5} \cdot \exp\left(-5.4 \frac{u_{ti}}{u}\right) \qquad (30)$$

where u_{ti} is the terminal settling velocity of particles in the size interval i. As a result, Eq. (27) can be written as

$$\dot{m}_{loss,tot} = C_j \cdot d_{pb} \cdot \rho_f \cdot \frac{D_t^6}{d_{or}^4 \cdot n_{or}^2} \cdot u^3$$
$$+ K_b^* \cdot m_b \cdot (u - u_{mf})^3$$
$$+ 3.81 \cdot C_c \cdot d_{pc}(u) \cdot \rho_f \cdot \frac{A_t^3}{A_{c,in}^2} \qquad (31)$$
$$\cdot \sqrt{\sum_i w_i \cdot \exp\left(-5.4 \frac{u_{ti}}{u}\right)} \cdot u^{3.75}$$

On the basis of Eq. (31), the total loss rate and the contributions of the individual sources of a given system can be a priori calculated in dependence on the superficial gas velocity. Using the same values for the material's attrition rate constants and the system design as in the above considerations, this is demonstrated in Fig. 24.

Obviously, there is a strong sensitivity of cyclone attrition inside a fluidized bed system against the superficial gas velocity. It is significantly higher than the sensitivities of jet- and bubble-induced attrition, respectively. Furthermore, it is obvious that owing to these different dependencies, the role of the main attrition source changes with the gas velocity. In the lower velocity range—in this particular example below 0.55 m/s—the gas distributor is the main attrition source, whereas at higher velocities cyclone attrition is dominant.

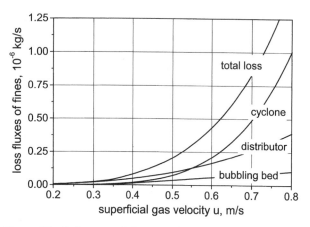

Figure 24 Influence of the superficial gas velocity on the extent of attrition in the individual regions and in the overall fluidized bed system (fluidized bed facility from Fig. 21, screened FCC catalyst, bed mass 5.5 kg). (After Werther and Reppenhagen, 1999.)

From the above-summarized work of Werther and Reppenhagen (1999), it must be concluded that attrition phenomena in an overall fluidized bed system can be only modeled when the relevant attrition sources are identified and separately described. Each individual attrition source model should take into account the specific design of the system, the operating conditions, and the relevant material properties of the solids. The latter can be obtained from attrition tests using for each source a particular designed experimental setup that simulates the relevant process stress. Finally, the solids transport and segregation effects must be taken into account when combining the individual model results again to a description of the overall process.

6.1 Continuous Operation

Up to now, attrition has been considered in batch processes only. But many industrial fluidized bed reactors are operated in a continuous mode. This means that the cyclone loss as well as the reacted material (e.g., in all noncatalytic gas–solid reactions) is compensated for by the addition of freshly fed material. Furthermore, in some processes there is an additional withdrawal of bed material, which is also compensated for by fresh solids in order to keep the bed inventory at a required level. As a consequence, there is a residence time distribution of the solids, and the time dependence of attrition has to be taken into account. According to Sec. 2.2, there will be both high initial attrition of freshly fed material and steady-state attrition of "old" particles. Zenz and coworkers (1971, 1972, 1980) were among the first to suggest calculation procedures for the content of fines, which is attained at equilibrium as a result of attrition and the addition of new catalyst in a fluidized bed system. Levenspiel et al. (1969) and Kunii and Levenspiel (1969) suggested the consideration of particle balances in the system. Newby et al. (1983) proposed a simplified mass balance for the entire system and distinguished between continuous and instantaneous attrition, which was connected to the feed rate. Fuertes et al. (1991) coupled a description of time dependence derived from batch processes with a residence time distribution function of a continuous stirred tank reactor. Ray et al. (1987b) and Werner et al. (1995) presented population balance models to describe process attrition. In all these models, however, the attrition mechanism is again treated in a rather superficial way without making a distinction between the different mechanisms prevailing in the respective parts of the system.

6.2 Changes in the Bed Particle Size Distribution

As mentioned in the introduction, the effect of attrition on the particle size distribution is quite often as relevant as the attrition-induced loss is. The reason is quite obvious: it is the strong dependence of the process performance on the bed particle size distribution. In the chemical industry, for example, the content of fines, i.e., the mass of particles below 44 microns, has often been observed to have a strong effect on the fluidized bed reactor performance. de Vries et al. (1972) reported an increase in the conversion of gaseous hydrogen chloride in the Shell chlorine process from 91 to 95.7% with an increase of the fines content in the bed material from 7 to 20%. The same effect was observed by Pell and Jordan (1988) with respect to the propylene conversion during the synthesis of acrylonitrile. They reported on an increase of the conversion from 94.6 to 99.2% as the fines content was changed from 23 to 44%.

Dealing with FCC processes, Zenz (1971) has given the following statement:

> Ideally it should be possible to predict simply from the fresh feed catalyst size analysis and a specific reactor and cyclone geometry how the bed analysis and reactor losses will change with time and how these will converge to an equilibrium. If attrition were not a significant factor it is obvious that the addition of fresh make-up catalyst coarser than the losses would cause the bed size distribution to become continually coarser until theoretically losses would be reduced to zero.

Despite its practical relevance the number of publications dealing with this topic is rather small. In fact, the few publications that focus on an overall system with solids recirculation (such as Zenz, 1971, 1974; Ewell et al., 1981; Gierse, 1991) describe attrition only by an undefined term of particle degradation, to which neither a particular mechanism nor a particular region is assigned. However, more recently the present authors (Reppenhagen and Werther, 2001) suggested a particle population balance for a fluidized bed system that allows a description of the fate of the individual particles in terms of attrition and transport effects. It thus provides a description of the dynamic adjustment in the steady-state particle size distribution and the solids loss rate of a given system. In the following this work is briefly summarized.

As in the authors' previous work (cf., e.g., Sec. 6) they considered a process where the attrition mode is pure abrasion. Focusing on a discretized particle size distribution they sketched the particular effect of this attrition mode in Fig. 25.

From each size interval K_i there is a mass flow $\dot{m}_{\text{loss},i}$ of fines into the smallest size interval K_1, which is the fraction of the abrasion-produced fines. It can be assumed that all particles of the size interval K_i are involved in the generation of these fines. As a consequence, all particles shrink, and some of them become smaller than the lower boundary of their size interval. They must thus be assigned to the smaller size interval K_{i-1}. This mass transfer between neighboring intervals is denoted as a mass flow $\dot{m}_{i,i-1}$. On the other hand, the rest of the material remains in its original size interval, even though the particles are also reduced in size. These phenomena of mass transfer can be summarized in a set of mass balances for the individual size fractions,

$$\frac{dm_i}{dt} = -\dot{m}_{\text{loss},i} - \dot{m}_{i,i-1} + \dot{m}_{i+1i} \qquad i = 2 \cdots n-1 \tag{32}$$

when n is the number of size intervals. An exception is made for both the interval of the finest particles K_1 and the interval of the coarsest particles K_n. The interval K_1 receives all abrasion-produced fines originating from the other size intervals and the shrunk particles from the neighboring size interval K_2. But there is no material loss due to a further attrition of the particles within the interval. Even if there were a further particle degradation, the attrition products would remain within the size interval K_1. On the other hand, the particles within the size interval K_n undergo abrasion. This results in both the loss of fine material and the loss of shrunk

particles. But there is no coarser size fraction from which shrunk particles could be received.

$$\frac{dm_1}{dt} = \sum_{i=2}^{n} \dot{m}_{\text{loss},i} + \dot{m}_{2,1} \tag{33}$$

$$\frac{dm_n}{dt} = -\dot{m}_{\text{loss},n} - \dot{m}_{n,n-1} \tag{34}$$

In order to solve the set of mass balances, Eqs. (32), (33), and (34), both the fractional loss of fines $\dot{m}_{\text{loss},i}$ and the resulting mass transfer between the neighboring size intervals $\dot{m}_{i,i-1}$ must be known for each size interval. Hence the influence of the individual particle sizes must be taken into account. Based on the description of the abrasion-induced loss flow from an overall system that is summarized in Eq. (27) this can be done as follows.

Assuming that the particle size dependent rate constant K_b^* for bubble-induced attrition can—in analogy to the other sources—be written as the product of a particle size independent constant C_b and the surface mean diameter of the bed material d_{pb}:

$$\dot{m}_{\text{loss}^*} = C^* \cdot d_{\text{p}^*} \cdot \xi^* \quad \begin{aligned} &|\xi^* = \xi_{\text{b}}, \quad \xi_{\text{j}}, \quad \xi_{\text{c}} \qquad C^* = C_{\text{b}}, C_{\text{j}}, C_{\text{c}} \\ &|d_{\text{p}^*} = d_{\text{pb}}, d_{\text{pc}}, \quad d_{\text{pj}} \qquad \text{with } d_{\text{pj}} = d_{\text{pb}} \end{aligned} \tag{35}$$

with $\xi_{\text{b}} = m_{\text{b}} \cdot (u - u_{\text{mf}})^3$; $\xi_{\text{j}} = \rho_{\text{f}} \cdot d_{\text{or}}^2 \cdot u_{\text{or}}^3$; $\xi_{\text{c}} = \sqrt{\dot{m}_{\text{c,in}} \cdot \rho_{\text{f}} \cdot A_{\text{c,in}}} \cdot u_{\text{c,in}}^{2.5}$ representing the respective influences of geometry and operating parameters. With the definition of the surface mean diameter, Eq. (35) can be written as

$$\dot{m}_{\text{loss}^*} = \sum \dot{m}_{\text{loss}^*\text{j}} = C^* \cdot \xi^* \cdot \sum \bar{x}_i \cdot \Delta Q_{2,i} \tag{36}$$

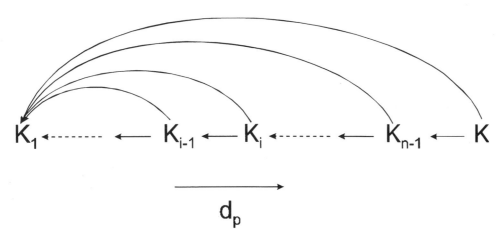

Figure 25 Abrasion-induced mass transfer between particle size intervals. (From Reppenhagen and Werther, 2001.)

Q_2 is the cumulative surface area distribution of the particle sizes. From this the material loss of a size fraction can be deduced as

$$\dot{m}_{\text{loss}^*j} = C^* \cdot \xi^* \cdot \bar{x}_i \cdot \Delta Q_{2,i} \tag{37}$$

According to Eq. (37) for a given size interval the material loss due to fines generation increases with its geometric mean particle size $\bar{x}_i$ and its fraction of the entire particle surface, i.e., $\Delta Q_{2,i}$.

The mass transfer between the neighboring size intervals, $\dot{m}_{i,i-1}$, can be derived from the natural coupling with the fractional fines production: the mass of produced fines originating per unit time from a size interval K_i corresponds to the material loss of the initial parent particles of this fraction. Provided that the loss is evenly contributed by all particles of this size interval, and moreover assuming that the particles are spherical, the mass loss per unit time of a single particle can be derived by

$$\dot{m}_{\text{loss},i,p} = \frac{\dot{m}_{\text{loss},i}}{N_{p,i}} = \dot{m}_{\text{loss},i} \cdot \frac{\rho_{s,a} \cdot \pi \cdot \bar{x}_i^3}{\Delta Q_{3,i} \cdot m_{\text{tot}} \cdot 6} \tag{38}$$

with $N_{p,i}$ being the number of particles in the size interval K_i. $\rho_{s,a}$ is the apparent density of the solids, and m_{tot} is the total mass of solids. Provided that $\dot{m}_{\text{loss},i}$ is constant in the time interval Δt, the mass balance for a single particle can be written as

$$m_{i,p,t} = m_{i,p,t+\Delta t} + \dot{m}_{\text{loss},i,p} \cdot \Delta t \tag{39}$$

This can be transferred to

$$\bar{x}_{i,t} = 3\sqrt{\frac{6}{\pi} \cdot \left(\frac{\pi}{6} \bar{x}_{i,t+\Delta t}^3 + \frac{\dot{m}_{\text{loss},i,p} \cdot \Delta t}{\rho_{s,a}} \right)} \tag{40}$$

A rearrangement of Eq. (40) and insertion of Eq. (38) yields

$$\bar{x}_{i,t} = \bar{x}_{i,t+\Delta t} \cdot 3\sqrt{1 + \frac{\dot{m}_{\text{loss},i} \cdot \Delta t}{\Delta Q_{3,i} \cdot m_{\text{tot}}}} \tag{41}$$

which can now be transferred to an abrasion-induced reduction in the particle diameter Δd_i:

$$\Delta d_i = \bar{x}_{i,t} - \bar{x}_{i,t+\Delta t} = \bar{x}_{i,t+\Delta t}$$
$$\cdot \left(3\sqrt{1 + \frac{\dot{m}_{\text{loss},i} \cdot \Delta t}{\Delta Q_{3,i} \cdot m_{\text{tot}}}} - 1 \right) \tag{42}$$

From this a critical particle size x_i^* might be derived, which characterizes the upper bound of the particular particle fraction within the size interval K_i that shrinks during the time interval Δt into the smaller size interval K_{i-1}:

$$x_i^* = x_i + \Delta d_i \tag{43}$$

with x_i being the smallest particle size in the size interval K_i. According to Fig. 26, the transferring mass fraction can then be calculated as

$$m_{i,i-1} = \left(m_i - m_{\text{loss},i} \right) \cdot \frac{x_i^* - x_i}{x_{i+1} - x_i} \tag{44}$$

However, as shown in Sec. 6, in addition to these local attrition phenomena, the movement of the solids must be described. The flow sheet in Fig. 27 summarizes the mechanisms that are taken into account in the authors' work.

In the fluidized bed itself ideal mixing of the solids is assumed. Both freshly fed particles and reentering particles from the return line are thus evenly distributed in the bed. The entrainment from the fluidized bed into the cyclone section is described by the correlation suggested by Tasirin and Geldart (1998). The cyclone is modeled as a series connection of an attrition unit and a subsequent gas–solids separator. Since in typical fluidized bed processes the solids loading is distinctly higher than its critical value, the gas–solids separation is described according to the critical load hypothesis suggested by e.g. Trefz and Muschelknautz (1993), which again divides the cyclone separation into a series of a spontaneous separations of the surplus mass at the cyclone inlet and a subsequent so-called inner separation of the remaining critical mass inside the vortex. The German standard design procedure (VDI Heat Atlas, 1993) assumes an empirically found distribution of the particles that are subjected to separation in the vortex. This procedure cannot, however, be used in connection with population balancing. The authors have therefore considered two different extreme approaches: one assumes a particle-size-sensitive

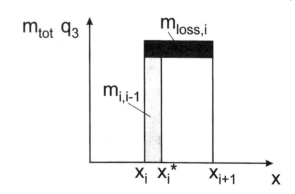

Figure 26 Approach to derive the mass transfer between neighboring size intervals. (From Reppenhagen and Werther, 2001.)

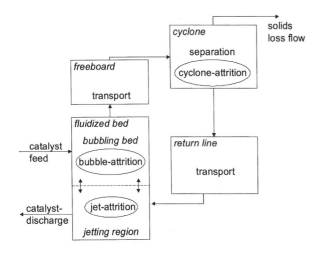

Figure 27 The particle population balance model of a fluidized bed system. (Reppenhagen and Werther, 2001.)

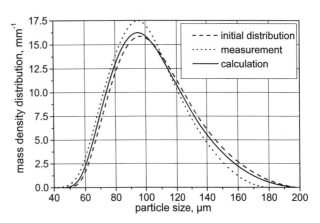

Figure 28 Measured and calculated shift in the size distribution when a batch of catalyst was fed 600 times to a cyclone. (Reppenhagen and Werther, 2001.)

separation at the inlet that leaves the fine particles in the critical mass flow entering the vortex

$$\dot{m}_{crit} = \dot{m}_{fluid} \cdot \mu_G = \dot{m}_{c,in} \cdot \int_0^{x_{limit}} q_3(x)\,dx \qquad (45)$$

and causes the coarser particles as the surplus mass flow to form the strand. This separation is assumed to be ideal. The alternative approach assumes that the particles entering the vortex have the same size distribution as the material entering the cyclone.

The part of the entering solids mass flow that exceeds the critical load is completely attributed to the strands flowing directly into the catch of the cyclone section. The catch feeds the standpipe, where a certain amount of material is stored. It passes in the downward direction, until it is finally refed to the bed.

From a previous work (Reppenhagen and Werther, 1999a), the results of a long-term cyclone attrition experiment are available for comparison with the above-derived approach for the abrasion-induced shift in the PSD. In this experiment, a batch of catalyst has been fed 600 times to an isolated 90 mm ID cyclone operated at an inlet gas velocity of $u_{c,in} = 18$ m/s and a solids loading of $\mu_c c = 0.3$. Figure 28 shows the measured and the calculated PSD after the experiment in comparison to the initial one.

It is obvious that the developed description of the particle shrinking is basically suitable to describe the abrasion effect on the particle size distribustion qualitatively, i.e., the shift and the narrowing of the PSD with the resulting increase of its modal value are predicted. But it can be seen that the shift is underestimated by the model. In particular, the experimentally

observed strong shift in the coarse particle size range is not sufficiently described. One reason for this underestimation might be the common problem of discretized particle balances, that either the mass or the number of particles cannot be conserved (e.g., Hill and Ng, 1995). In the above approach, this leads to a wrong increase in the number of particles: prior to its transfer a mass, $m_{i,i-1}$, is assigned to the mean particle diameter $\bar{x}_i$, and afterwards the same mass is assigned to the smaller mean particle diameter $\bar{x}_{i-1}$. This can only be compensated by a wrong increase in the number of particles. According to Eq. (38), the increased number of particles results in a wrong decrease in the material loss per single particle and with it to an underestimation of the shrinking.

In the experiment shown above, the loss rate was measured for each pass of the material through the cyclone. In Fig. 29, this data set is now compared

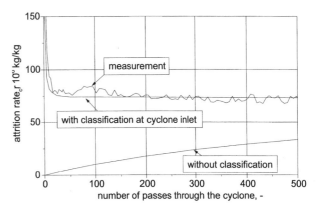

Figure 29 Measured cyclone loss rate in comparison with the simulation. (Reppenhagen and Werther, 2001.)

with numerical predictions of the separation process in the cyclone. As is obvious from this figure, the assumption of a particle-size-sensitive separation at the cyclone inlet gives an excellent representation of the measurements, whereas the nonclassifying mechanism is far from being suitable.

As there were no experimental data available to evaluate the entire above-derived particle population balance, a plausibility check with a fictitious industrial scale system has been made instead. The chosen data of the system are given in the list here:

Bed design and inventory
  Column diameter $D_t = 4.0$ m
  Column height higher than TDH
  Bed mass $m_b = 18,000$ kg
  Mass in return line $m_{return} = 500$ kg
Distributor design
  Type: perforated plate
  Number of orifices $n_{or} = 5,000$
  Orifice diameter $d_{or} = 0.005$ m
Design of the cyclone section
  Single stage
  Number of cyclones in parallel $N_c = 1$
  Outer diameter $D = 1.35$ m
  $\Delta p = 2000$ Pa at operating conditions
Operating conditions
  Fluid: air at $T = 293$K and $p = 1.0 \cdot 10^5$ Pa
  Superficial gas velocity $u = 0.61$ m/s
  No catalyst discharge
Initial catalyst material
  PSD of the initial material shown in Fig. 31
  Surface mean diameter of initial material
  $d_p = 59.10^{-6}$ m
  Jet attrition constant $C_j = 9.5 \cdot 10^{-6}$ s^2/m^3
  Bubble attrition constant $C_b = 0.4 \cdot 10^{-3}$ m^{-2}
  Cyclone attrition constant $C_c = 1.2 \cdot 10^{-3}$ s^2/m^3

For this system, the particle population balance is solved in discretized time steps Δt. At the beginning of each time step, except the very first one, the fresh material and the material from the return line are fed and mixed with the bed material. It follows the calculation of jet-induced and bubble-induced attrition. Afterwards the particular material mass fractions are determined that are entrained from the bed in the course of this time step. These material fractions are then subjected to the combination of attrition and gas–solid separation inside the cyclone section. Finally, the catch of the cyclone is added to the return line, which in the next time step feeds its excess mass to the bed. At the starting time $t = 0$, the particle size distributions in both the bed and the return line are identical to that of

the initial material. As computational time step, a value of $\Delta t = 10$ s was determined by numerical experiments. As a first result, Fig. 30 shows the calculated loss flow of the system during the initial 10 days.

After a very high initial value of approximately 3 t/d, the loss flow rapidly decreases and approaches asymptotically a significantly smaller value. The high initial loss rate can be explained by a strong sifting of the bed material due to the elutriation of the initial fines. However, the stepped shape of the graph is much more noticeable than this initial value. It mainly results from the combination of the discretized description of the particle size distribution and the cyclone model with its idea of an absolutely particle-size-sensitive inlet separation: Owing to the continuous sifting in the cyclone there is a depletion of the fine particles in the inventory, and the threshold diameter x_{limit} from Eq. (45) thus increases with time. This results in an increasing content of larger particles with higher grade efficiency in the vortex, and the system's loss rate decreases accordingly. Since there is only one grade efficiency assigned to one size interval, a step in the loss rate occurs whenever x_{limit} is moving, into a larger size interval. However, with decreasing loss rate and the thus slower change in the size distribution, the time between these steps becomes longer until x_{limit} stays within a size interval. From this time on, the changes in the size distribution are even more slight and the system tends to a steady state wherein the elutriated particles are balanced by both the freshly fed and the attrition-produced particles.

Figure 31 shows a comparison of the calculated size distribution at steady state with that of the initial material. As can be seen, the major changes occurred in the finer particle size range, i.e. the sifting effect of the cyclone dominates the attrition-induced shrinking.

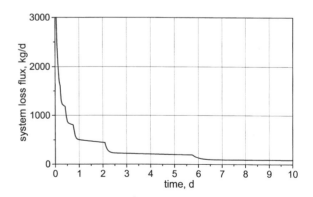

Figure 30 Calculated loss flow of the system during the initial 10 days. (Reppenhagen and Werther, 2001.)

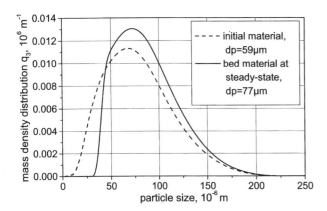

Figure 31 Simulation of the bed particle size distribution in a fluidized bed system. (Reppenhagen and Werther, 2001.)

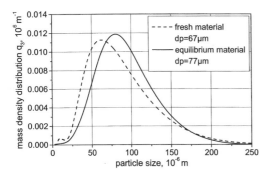

Figure 32 FCC process in a refinery. (Reppenhagen and Werther, 2001.)

However, despite the underestimation of the particle shrinking by the present approach (see above), the dominating effect of the gas–solid separation is fairly well confirmed by industrial findings. Figure 32 shows a comparison of the initial and the equilibrium catalyst material of an industrial FCC process. As in the simulation, there are no changes in the coarse particle size range but only in the small particle size range. However, the sifting of the finer particles is not entirely that indicated by the simulation, which can again be explained by the assumption of a single and comparatively large cyclone in the simulation.

NOMENCLATURE

$A_{c,in}$ = cross section of the cyclone inlet area, m^2
A_o = open surface area of a perforated plate distributor, m^2
A_t = cross sectional area of the fluidized bed column, m^2
b = exponent in the Gwyn Equation, Eq. (1)

$B(x, y)$ = breakage function
C_b = particle size independent rate constant of bubble-induced attrition, s^2/m^3
C_c = particle size independent rate constant of cyclone attrition, defined by Eq. (21), s^2/m^3
C_j = particle size independent rate constant of jet-induced attrition, defined by Eq. (8), s^2/m^3
C^* = particle size independent rate constant, defined by Eq. (35), s^2/m^3
Δd_i = abrasion-induced reduction in the particle diamter in interval i, defined by Eq. (42), m
d_{or} = diameter of an orifice in a multihole gas distributor, m
d_p = surface mean diameter, m
d_p^* = surface mean diameter, defined by Eq. (35), m
d_{pc} = surface mean diameter of the material that enters the cyclone, m
$d_{pc}(u)$ = surface mean diameter of the elutriated material, defined by Eq. (29), m
d_{pb} = surface mean diameter of the bed material, m
d_{pj} = surface mean diameter for jet-induced attrition, defined by Eq. (35), m
D_t = diameter of the fluidized bed column, m
E_{tot} = total attrition extent of a given system, defined by Eq. (4)
G_s = total rate of solids elutriation from the bed, $kg/(m^2 \cdot s)$
G_{si} = fractional elutriation rate for the size interval i, $kg/(m^2 \cdot s)$
K_a = attrition rate constant in the Gwyn equation, Eq. (1), s^{-b}
K_b = rate constant of bubble-induced attrition, defined by Eq. (13), $m^{-1} \cdot kg^{-n}$
K_b' = rate constant of bubble-induced attrition, defined by Eq. (17), $m^{-1} \cdot kg^{-n}$
K_b^* = rate constant of bubble-induced attrition, defined by Eq. (18), s^3/m^3
K_c = particle size dependent rate constant of cyclone attrition, defined by Eq. (19), s^2/m^2
K_i^* = elutriation rate constant of the size interval i, $kg/(m^2 \cdot s)$
K_j = particle size dependent rate constant of jet-induced attrition, defined by Eq. (7), s^2/m^2
$\dot{m}_{abr.,fines}$ = mass of abrasion-produced fines per unit time, kg/s
m_b = bed mass, kg
$m_{b,0}$ = initial bed mass, kg
$\dot{m}_{c,in}$ = solids mass flow rate into the cyclone, kg/s
$\dot{m}_{crit}$ = critical solids mass flow rate, defined by Eq. (45), kg/s
$\dot{m}_{fluid}$ = mass flow of the fluid, kg/s
m_i = material mass in the size interval i, kg
$\dot{m}_{i,i-1}$ = mass transfer flow rate due to particle shrinking from the size interval i to $i-1$, kg/s

m_{loss} = mass lost from a given system, kg

$\dot{m}_{loss,c}$ = produced mass of fines per unit time by cyclone attrition, kg/s

$\dot{m}_{loss,i}$ = lost mass of fines per unit time from the size interval i, kg/s

$\dot{m}_{loss,i,p}$ = lost mass of a single particle per unit time, kg/s

$\dot{m}_{loss,j}$ = produced mass of fines per unit time by jet-induced attrition, kg/s

$\dot{m}_{loss,tot}$ = attrition-induced loss flow rate of the entire fluidized bed system, kg/s

m_{return} = mass in return line, kg

m_{tot} = total mass of solids in a given system, kg

N_c = number of primary cyclones in parallel within a fluidized bed system

n_{or} = number of orifices in a multihole gas distributor

$N_{p,i}$ = number of particles in the size interval K_i

p = pressure, Pa

Δp = pressure drop, Pa

$Q_2(x)$ = cumulative particle size distribution in particle surface

$\Delta Q_{2,i}$ = fraction of the size interval i on the entire particle surface

$q_3(x)$ = mass density particle size distribution, m^{-1}

$Q_3(x)$ = cumulative particle size distribution in mass

$\Delta Q_{3,i}$ = fraction of the size interval i on the entire material mass

r_b = bubble-induced attrition rate, defined by Eq. (13)

r_c = cyclone attrition rate, defined by Eq. (6)

r_{tot} = overall attrition rate, defined by Eqs. (2) and (3), 1/s

$S(x)$ = selection function

t = time, s

u = superficial gas velocity, m/s

$u_{c,in}$ = gas velocity at the cyclone inlet, m/s

u_{min} = threshold velocity for bubble-induced attrition, defined by Eq. (13), m/s

u_{mf} = minimum fluidization velocity, m/s

u_{or} = gas velocity in the orifice of a multihole gas distributor, m/s

u_{ti} = terminal velocity of the size interval i, m/s

V = volumetric flow rate, m^3/s

w_i = weight fraction of the size interval i in the bed material

x = particle size, m

$\bar{x}_i$ = geometric mean particle size of the size interval i, m

x_{limit} = critical particle size in the absolutely particle-size-sensitive inlet separation, defined by Eq. (45), m

Greek Symbols

ξ = parameter defined by Eq. (35)

μ_c = solids loading at the cyclone inlet, defined by Eq. (19)

μ_G = critical solids loading for the carrying capacity of the gas

$\rho_{s,a}$ = apparent density of the catalyst material, kg/m^3

ρ_f = density of the fluid, kg/m^3

τ = time, s

Subscripts and Indices

0 = value at initial state

abr = caused by abrasion

attr. = caused by attrition

b = bed

c = cyclone

feed = related to the feed

fines = related to fines

h = index to number a certain size interval

i = index to number a certain size interval

j = jet

k = index to number a certain size interval

loss = lost material

or = orifice

steady-state = under steady state conditions

tot = related to the entire system

REFERENCES

Anthony EJ, Granatstein DL. Sulfadation phenomena in fluidized bed combustion systems. Progress in Energy and Combustion Science 27:215–236, 2001.

Arastoopour H, Chen C-Y. Attrition of char agglomerates. Powder Technol 36:99–106, 1983.

Arena U, D'Amore M, Massimilla L. Carbon attrition during the fluidized combustion of coal. AIChE J 29:40–48, 1983.

Austin LG. A review—introduction to the mathematical of grinding as a rate process. Powder Technol 5:1–17, 1972.

Bemrose CR, Bridgwater J. A review of attrition and attrition test methods. Powder Technol 49:97–126, 1987.

Blake TR, Webb H, Sunderland PB. The nondimensionalization of equations describing fluidization with application to the correlation of jet penetration height. Chem Eng Sci 45:365–371, 1990.

Blinichev VN, Streltsov VV, Lebedeva ES. An investigation of the size reduction of granular materials during their processing in fluidized beds. Int Chem Eng 8(4):615–618, 1968

Boerefijn R, Zhang SH, Ghadiri M. Analysis of ISO fluidized bed test for attrition of fluid cracking catalyst particles. In: Fan LS, Knowlton TM, eds. Fluidization IX. New York: Engineering Foundation, 1998, pp 325–332.

Boerefijn R, Gudde NJ, Ghadiri M. A review of attrition of fluid cracking catalyst particles. Adv Powder Technol 11(2):145–174, 2000.

Braca RM, Fried AA. Operation of fluidization processes. In: Othmer DF, ed. Fluidization. New York: Reinhold, 1956, pp 117–138.

British Materials Handling Board. Particle Attrition. Trans. Tech. Publications Series on Bulk Materials Handling 5, 1987.

Chirone R, Massimilla L, Salatino P. Comminution of carbons in fluidized bed combustion. Progress in Energy and Combustion Science 17:297–326, 1991.

Clift R. Powder technology and particle science. Powder Technol 88:335–339, 1996.

Contractor RM, Bergna HE, Chowdhry U, Sleight AW. Attrition resistant catalysts for fluidized bed-systems. In: Grace JR, Shemilt LW, Bergougnou MA, eds. Fluidization VI. New York: Engineering Foundation, 1989, pp 589–596.

Dart JC. Mechanical tests to determine strength and abrasion resistance of catalysts. AIChE Symposium Series 143(70):5–9, 1974.

Davuluri RP, Knowlton TM. Development of a standardized attrition test procedure. In: Fan LS, Knowlton TM, eds. Fluidization IX. New York: Engineering Foundation, 1998, pp 333–340.

Dessalces G, Kolenda F, Reymond JP. Attrition evaluation for catalysts used in fluidized or circulating fluidized bed reactors. Preprints of the First International Particle Technology Forum, Part II, Denver, CO, 1994, pp 190–196.

de Vries RJ, van Swaaij WPM, Mantovani C, Heijkoop A. Design criteria and performance of the commercial reactor for the shell chorine process. 5th Europ Symp Chem Reac. Eng, Amsterdam, 1972, (B9) pp 59–69.

Ewell RB, Gadmer G, Turk WJ. FCC catalyst management. Hydrocarbon Processing 60:103–112, 1981.

Forsythe WL, Hertwig WR. Attrition characteristics of fluid cracking catalysts. Ind Eng. Chem 41:1200–1206, 1949.

Fuertes AB, Pis JJ, Garcia JC, Rubiera F, Artos V. Prediction of attrition in a continous fluid-bed system. Powder Technol 67:291–293, 1991.

Geldart D. Gas Fluidization Technology. Chichester: John Wiley, 1986.

Ghadiri M, Cleaver JAS, Tuponogov VG. Modelling attrition rates in the jetting region of a fluidized bed. Preprints of the Symposium Attrition and Wear, Utrecht, 1992a, pp 79–88.

Ghadiri M, Cleaver JAS, Yuregir KR. Attrition of sodium chloride crystals in a fluidized bed. In: Potter OE, Nicklin DJ, eds. Fluidization VII. New York: Engineering Foundation, 1992b, pp 604–610.

Ghadiri M, Cleaver JAS, Tuponogov VG, Werther J. Attrition of FCC powder in the jetting region of a fluidized bed. Powder Technol 80:175–178, 1994.

Ghadiri M, Cleaver JAS, Tuponogov VG. Influence of distributor orifice size on attrition in the jetting region of fluidized beds. In: Preprints Fluidization VIII(2). Tours, 1995, pp 799–806.

Gierse M. Kornhaushalt in Zirkulierenden Wirbelschichten. Brennstoff-Wärme-Kraft 43:459–462, 1991.

Gwyn JE. On the particle size distribution function and the attrition of cracking catalysts. AIChE J 15:35–38, 1969.

Hansen S, Ottino JM. Fragmentation with abrasion and cleavage: analytical results. Powder Technol 93:177–184, 1997.

Hilligardt K, Werther J. The influence of temperature and solids properties on the size and growth of bubbles in gas fluidized beds. Chem Eng Technol 10:272–280, 1987.

Hill J, Ng KM. New discretization procedure for the breakage equation. AIChE J 41:1204–1216, 1995.

Hirschberg B. Solids mixing and segregation in a circulating fluidized bed. PhD dissertation, Technical University Hamburg-Harburg, 1997.

Knight PC, Bridgwater J. Comparison of methods for assessing powder attrition. Powder Technol 44:99–102, 1985.

Kokkoris A, Turton R. The reduction of attrition in fluidized beds by the addition of solid lubricants. AIChE Symp Ser. 87:20–31, 1991.

Kokkoris A, Turton R. A phenomenological model predicting the attrition and the reduction of attrition due to the addition of solid lubricants in slugging beds. Powder Technol 84:39–47, 1995.

Kono H. Attrition rates of relatively coarse solid particles in various types of fluidized beds. AIChE Symp Ser 77:96–106, 1981.

Kraft WW, Ullrich W, O'Connor W. The significance of details in fluid catalytic cracking units: engineering design, instrumentation and operation. In: Othmer DF, ed. Fluidization. New York: Reinhold, 1956, pp 184–211.

Krambrock W. Die Berechnung des Zyklonabscheiders und praktische Gesichtspunkte zur Auslegung. Aufbereitungstechnik 12(7):391–401, 1971.

Kunii D, Levenspiel O. Fluidization engineering. New York: Robert E. Krieger, 1969.

Kutyavina TA, Baskakov AP. Grinding of fine granular material with fluidization. Chem Techn Fuels Oils 8:210–213, 1972

Levenspiel O, Kunii D, Fitzgerald T. The processing of solids of changing size in bubbling fluidized beds. Powder Technol 2:87–96, 1969.

Lin L, Sears JT, Wen CY. Elutriation and attrition of char from a large fluidized bed. Powder Technol 27:105–115, 1980.

Merrick D, Highley J. Particle size reduction and elutriation in a fluidized bed process. AIChE Symp Ser 137(70):367–378, 1974.

Merry JMD. Penetration of vertical jets into fluidized beds. AIChE J 21:507–510, 1975.

Molerus O. Heat transfer in gas fluidized beds. Powder Technol 70:1–20, 1992.

Mothes H, Löffler F. Zur Berechnung der Partikelabscheidung in Zyklonen. Chem Eng Process 18:323–331, 1984.

Muschelknautz E. Auslegung von Zyklonabscheidern in der technischen Praxis. Staub-Reinhaltung der Luft 30:187–195, 1970.

Neil AU, Bridgwater J. Attrition of particulate solids under shear. Powder Technol 80:207–219, 1994.

Newby RA, Vaux WG, Keairns DL. Particle attrition in fluidized-bed systems. In: Kunii D, Toei R, eds. Fluidization IV. New York: Engineering Foundation, 1983, pp 233–239.

Overturf BW, Reklaitis GV. Fluidized-bed reactor model with generalized particle balances. AIChE J 5:813–829, 1983.

Patel K, Nienow AW, Milne IP. Attrition of urea in a gas-fluidised bed. Powder Technol 47:257–262, 1986.

Pell M, Jordan SP. Effects of fines and velocity on fluidized bed reactor performance. AIChE Symp Ser. 84:68–73, 1988.

Pell M. Gas fluidization. In: Handbook of Powder Technology. Amsterdam: Elsevier, 1990, p 8, pp 97–106.

Pis JJ, Fuertes AB, Artos V, Suarez A, Rubiera F. Attrition of coal and ash particles in a fluidized bed. Powder Technol 66:41–46, 1991.

Paramanathan BK, Bridgwater J. Attrition of solids—II. Chem Eng Sci 38:207–224, 1983.

Patience GS, Mills PL. Modelling of propylene oxidation in a circulating fluidized-bed reactor. New Developments in Selective Oxidation II, 1994, pp 1–18.

Ray YC, Jiang TS, Wen CY. Particle attrition phenomena in a fluidized bed. Powder Technol 49:193–206, 1987a.

Ray YC, Jiang TS, Jiang TL. Particle population model for a fluidized bed with attrition. Powder Technol 52:35–48, 1987b.

Reppenhagen, J. Catalyst attrition in fluidized bed systems. PhD dissertation, Technical University Hamburg-Harburg, Aachen: Shaker, 1999.

Reppenhagen J, Werther J. Particle attrition influences cyclone performance. Preprints of the 4th European Symposium Separation of Particles from Gases, Nürnberg, 1998, pp 63–72.

Reppenhagen J, Werther J. Catalyst attrition in cyclones. Powder Technol 113:55–69, 1999a.

Reppenhagen J, Werther J. Catalyst attrition in fluidized bed systems with external solids circulation. In: Werther J, ed. Circulating Fluidized Bed Technology VI. Frankfurt a. M.: DECHEMA e.V.,1999b, pp 971–976.

Reppenhagen J, Werther J. The role of catalyst attrition in the adjustment of the steady-state particle sizes distribution in fluidized bed systems. In: Kwauk M, Li J, Yang WC, eds. Fluidization X. New York: Engineering Foundation, 2001, pp 69–76.

Schubert H. Aufbereitung fester mineralischer Rohstoffe. Band I, 3. Aufl. Leipzig: VEB Deutscher Verlag für Grundstoffindustrie, 1975.

Seville JPK, Mullier MA, Adams MJ. Attrition of agglomerates in fluidized beds. In: Potter OE, Nicklin DJ, eds.

Fluidization VII. New York: Engineering Foundation, 1992, pp 587–594.

Shipway PH, Hutchings IM. Attrition of brittle spheres by fracture under compression and impact loading. Powder Technol 76:23–30, 1993.

Sishtla C, Findlay J, Chan I, Knowlton TM. The effect of temperature and gas velocity on fines generation in non-reactive fluidized beds of coal char. In: Grace JR., Shemilt LW, Bergougnou MA, eds. Fluidization VI. Alberta, 1989, pp 581–588.

Tasirin SM, Geldart D. Entrainment of FCC from fluidized beds—a new correlation for the elutriation rate constants. Powder Technol 95:240–247, 1998.

Trefz M, Muschelknautz E. Extended cyclone theory for gas flows with high solids concentration. Chem Eng Technol 16:153–160, 1993.

Ulerich NH, Vaux WG, Newby RA, Keairns L. Experimental/engineering support for EPA's FBC program: Final report. Vol. 1. Sulfur Oxide Control. EPA-600/7-80-015a, 1980.

Vaux WG. Attrition of particles in the bubbling zone of a fluidized bed. Proc Am Power Conf 40:793–802, 1978.

Vaux WG, Fellers AW. Measurement of attrition tendency in fluidization. AIChE Symp Ser 77:107–115, 1981.

Vaux WG, Keairns DL. Particle attrition in fluid-bed processes. In: Grace JR, Matsen JM, eds. Fluidization III. New York: Engineering Foundation, 1980, pp 437–444.

Vaux WG, Schruben JS. Kinetics of attrition in the bubbling zone of a fluidized bed. AIChE Symp Ser 222(79):97–102, 1983.

VDI-Heat Atlas. Fullarton JW. 1st English ed. Düsseldorf: VDI-Verlag, 1993.

Weeks SA, Dumbill P. Method speeds FCC catalyst attrition resistance determinations. Oil Gas J 88:38–40, 1990.

Wei J, Lee W, Krambeck FJ. Catalyst attrition and deactivation in fluid catalytic cracking systems. Chem Eng Sci 32:1211–1218, 1977.

Werner A, Haider M, Linzer W. Modelling of particle population in fluidized beds of particles differing in size and physico-chemical behaviour. In: Preprint Fluidization VIII(1), Tours, 1995, pp 557–564.

Werther J. Scale-up modeling for fluidized bed reactors. Chem Eng Sci 47:2457–2462, 1992.

Werther J, Reppenhagen J. Catalyst attrition in fluidized-bed systems. AIChE J 45:2001–2010, 1999.

Werther J, Xi W. Jet attrition of catalyst particles in gas fluidized beds. Powder Technol 76(1):39–46, 1993.

Xi W. Katalysatorabrieb in Wirbelschichtreaktoren. PhD dissertation. Technical University Hamburg-Harburg, 1993.

Yates JG, Bejcek V, Cheesman DJ. Jet penetration into fluidized beds at elevated pressures. In: Ostergaard K, Sorensen A, eds. Fluidization V. New York: Engineering Foundation, 1986, pp 79–86.

Yuregir KR, Ghadiri M, Clift R. Observation of impact attrition of granular solids. Powder Technol 49:53–57, 1986.

Yuregir KR, Ghadiri M, Clift R. Impact attrition of sodium chloride crystals. Chem Eng Sci 42:843–853, 1987.

Zenz FA. Bubble formation and grid design. Int Chem Eng Symp Ser 30:136–139, 1968.

Zenz FA. Find attrition in fluid beds. Hydrocarbon Processing 50:103–105, 1971.

Zenz FA. Help from project E-A-R-L. Hydrocarbon Processing 53:119–, 1974.

Zenz FA, Kelleher GH. Studies of attrition rates in fluid–particle systems via free fall, grid jets, and cyclone impact. J Powder Bulk Tech 4:13–20, 1980.

Zenz FA, Smith R. When are fines at equilibrium. Hydrocarbon Processing 51:104–106, 1972.

9

Modeling

Thomas C. Ho

Lamar University, Beaumont, Texas, U.S.A.

1 INTRODUCTION

The conversion in gas–solids fluidized bed reactors has been observed to vary from plug flow, to well below mixed flow, mainly depending on reaction and fluidization properties (Levenspiel, 1972). Historically, two classes of models have been proposed to describe the performance of fluidized bed reactors; one is based on a pseudohomogeneous approach and the other on a two-phase approach. The pseudohomogeneous approach, where the existence of more than one phase is not taken into account, proposes that we use the conventional multiphase flow models for the fluidized bed reactors. These conventional models may include ideal flow models, dispersion models, residence time distribution models, and contact time distribution models. The two-phase approach, however, considers the fluidized bed reactors to consist of at least two phases, a bubble and an emulsion, and proposes a separate governing equation for each phase with a term in each equation describing mass interchange between the two phases. Among the two-phase models, the bubbling bed model proposed by Kunii and Levenspiel (1969) and the bubble assemblage model proposed by Kato and Wen (1969) have received the most attention. Figure 1 illustrates the development and evolution of these various flow models for fluidized bed reactors.

2 PSEUDOHOMOGENEOUS MODELS

The first attempt at modeling gas–solids fluidized bed reactors employed ideal or simple one-parameter models, i.e., plug flow, complete-mixed, dispersion, and tank-in-series models. However, the observed conversion of sometimes well below mixed flow in the reactors could not be accounted for by these models, and the approach was dropped by most researchers. The next attempt considered residence time distribution (RTD) model in which all the gas in the bed is considered equal in terms of gas residence time. However, since an operating fluidized bed consists of a bubble phase and an emulsion phase of completely different gas contacting hydrodynamics, the approach is inadequate and was also dropped.

Gilliland and Kundsen (1970) then modified the RTD models and assumed that the faster gas stayed mainly in the bubble phase and the slower in the emulsion. Their approach was to distinguish the effect of the two classes of gas with different effective rate constant. The overall conversion equation in their proposed model is

$$\frac{C_A}{C_{A_0}} = \int_0^\infty e^{-kt} E \, dt \qquad (1)$$

and the following equation was proposed to describe the effective rate constant, i.e.,

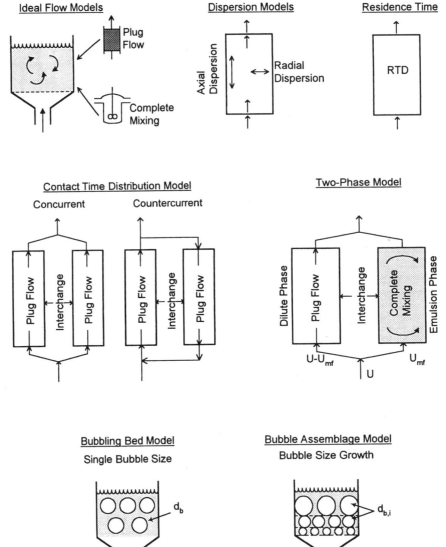

Figure 1 Development and evolution of fluidized bed models.

$$k = k_0 t^m \tag{2}$$

where m is a fitted parameter. Its value is small for short staying gas and high for long staying gas. Inserting Eq. (2) into (1) yields

$$\frac{C_A}{C_{A_0}} = \int_0^\infty \exp\left(-k_0 t^{m+1}\right) E \, dt \tag{3}$$

The above equation describes the conversion and is referred to as the contact time distribution (CTD) model. Although improvement was made over the RTD models, the problem with this approach involves obtaining a meaningful E function to use in Eq. (3) from a measured C curve obtained at the exit of a bed. Questions remain as to how the measured C curve represents the necessary E function in the calculation due to the considerable backmixing between the faster and slower gas occurring in an operating fluidized bed.

3 TWO-PHASE MODELS

The discouraging result with the previous approach has then led to the development of a sequence of models based on the two-phase theory of fluidization originally proposed by Toomey and Johnstone (1952). The two-phase theory states that all gas in excess of that necessary just to fluidize the bed passes through the bed in the form of bubbles. The term two-phase model, however, represents a broad range of models with various basic assumptions that may or may not directly follow the original two-phase theory. For example, some models consider the wakes and clouds, while others do not; some models propose the use of single-size bubbles, while others allow for bubble growth, and some models use the two-phase flow distribution following the two-phase theory, while others neglect the percolation of gas through the emulsion. In addition, different models may propose different interphase mass transfer mechanisms.

In the following subsections, the general two-phase models are briefly reviewed, and then we give a more detailed description of two of the more popular two-phase models, namely the bubbling bed model proposed by Kunii and Levenspiel (1969) and the bubble assemblage model proposed by Kato and Wen (1969).

3.1 General Two-Phase Models

In most two-phase models, a fluidized bed is considered to consist of two distinct phases, i.e., a bubble phase and an emulsion phase. Each phase is represented by a separate governing equation with a term in each equation describing mass interchange between the two phases. A general expression of the two-phase models, therefore, consists of the following two equations (from Wen and Fan, 1975). For the bubble phase we have

$$F\left(\frac{\partial C_{A,b}}{\partial t}\right) - FD_b\left(\frac{\partial^2 C_{A,b}}{\partial h^2}\right) + FU\left(\frac{\partial C_{A,b}}{\partial h}\right) \\ + F_0(C_{A,b} - C_{A,e}) + F_s k C_{A,b} = 0 \tag{4}$$

and for the emulsion phase, we have

$$f\left(\frac{\partial C_{A,e}}{\partial t}\right) - fD_e\left(\frac{\partial^2 C_{A,e}}{\partial h^2}\right) + fU\left(\frac{\partial C_{A,e}}{\partial h}\right) \\ + F_0(C_{A,e} - C_{A,b}) + f_s k C_{A,e} = 0 \tag{5}$$

where the term FU represents the rise velocity of gas in the bubble phase and the fU represents the rise velocity of gas in the emulsion phase ($= U_e$).

In the development of the two-phase models, most investigators used a simplified form of the two-phase model by either assuming or estimating some of the

terms in the above two equations. Table 1 summarizes experimental investigations of model parameters associated with the two-phase model. As indicated in Table 1, most of the studies assumed steady state ($\partial C_{A,b}/\partial t = 0$ and $\partial C_{A,e}/\partial t = 0$) with $D_e = 0$ (plug flow in emulsion phase) or $D_e = \infty$ (completely mixed in emulsion phase); also $D_b = 0$ (plug flow in bubble phase) and $U_e = U_{mf}$. The parameters investigated include gas interchange coefficient (F_0), particle fraction in bubble phase (γ_b), and reaction rate constant (k). Among the investigations, several authors reported that the flow patterns in the emulsion phase, whether it is assumed to be plug flow or completely mixed, do not significantly affect the model prediction (Lewis et al., 1959; Muchi, 1965). However, Chavarie and Grace (1975) reported that the two assumptions do affect their predictions based on the model of Davidson and Harrison (1963).

A number of theoretical studies on the two-phase model were also carried out and reported in the literature. They are listed in Table 2. The ones proposed by Davidson and Harrison (1963) and Partridge and Rowe (1966) are briefly reviewed below.

3.1.1 Model of Davidson and Harrison (1963)

One of the representative two-phase models is the one proposed by Davidson and Harrison (1963). This model follows the two-phase theory of Toomey and Johnstone (1952) and has the following assumptions:

1. All gas flow in excess of that required for incipient fluidization passes through the bed as bubbles.
2. Bubbles are of uniform size throughout the bed.
3. Reaction takes place only in the emulsion phase with first-order kinetics.
4. Interphase mass transfer occurs by a combined process of molecular diffusion and gas throughflow.
5. Emulsion phase (dense phase) is either perfectly mixed (DPPM) or in plug flow (DPPF).

Note that the key parameter in their model is the equivalent bubble diameter, which was assumed to be a constant.

Chavarie and Grace (1975) reviewed the model and stated that the DPPM model is too conservative in its estimate of overall conversion and that the concentration profiles are in no way similar to their experimental counterparts. Regarding the DPPF model, they stated that while the DPPF model offers a fair dense phase

Table 1 Experimental Investigation of Two-Phase Mode Parameters

Authors	Model	Experimental conditions	Experimental results
Shen & Johnstone (1955)	$\gamma_b = 0$ $U_e = U_{mf}$ $D_e = 0$ or ∞	decomposition of nitrous oxide $d_t = 11.4$ cm $L_{mf} = 26 \sim 32$ cm $d_p = 60 \sim 200$ mesh	parameter F_o $k = 0.06 \sim 0.05$ (1/s)
Mathis & Watson (1956)	$D_e = 0$ $U_e = U_{mf}$	decomposition of cumene $d_t = 5 \sim 10.2$ cm $L_{mf} = 10 \sim 31$ cm $d_p = 100 \sim 200$ mesh	parameter: F_o, γ_b $k = 0.64$ (1/s)
Lewis et al. (1959)	$U_e = 0$ $D_e = 0$ or ∞	hydrogenation of ethylene $d_t = 5.2$ cm $L_{mf} = 11 \sim 53$ cm $d_p = 0.001 \sim 0.003$ cm	parameter: F_o, γ_b $k = 1.1 \sim 15.6$ (1/s) $\gamma_b = 0.05 \sim 0.18$, $F = 0.4 \sim 0.8$
Gomezplata & Shuster (1960)	$\gamma_b = 0$ $U_e = U_{mf}$ $D_e = 0$	decomposition of cumene $d_t = 5 \sim 10.2$ cm $L_{mf} = 3.8 \sim 20$ cm $d_p = 100 \sim 200$ mesh	parameter: F_o, γ_b $k = 0.75$ (1/s)
Massimila & Johnstone (1961)	$\gamma_b = 0$ $U_e = U_{mf}$ $D_e = 0$	oxidation of NH_3 $d_t = 11.4$ cm $L_{mf} = 26 \sim 54$ cm $d_p = 100 \sim 325$ mesh	parameter: F_o $k = 0.071$ (1/s)
Orcutt et al. (1962)	$\gamma_b = 0$ $U_e = 0$	decomposition of ozone $d_t = 10 \sim 15$ cm $L_{mf} = 30 \sim 60$ cm $d_p = 0.001 \sim 0.003$ cm	$k = 0.1 \sim 3.0$ (1/s)
Kobayashi et al. (1966a)	$U_e = U_{mf}$ $D_e = 0$	decomposition of ozone $d_t = 8.3$ cm $L_{mf} = 10 \sim 100$ cm $d_p = 60 \sim 80$ mesh	parameter: γ_b $k = 0.1 \sim 0.8$ 1/s $\gamma_b = 15 (L/L_{mf} - 1)$
Kobayashi et al. (1966b)	$U_e = U_{mf}$ $D_e = 0$	decomposition of ozone $d_t = 20$ cm $L_{mf} = 10 \sim 100$ cm $d_p = 60 \sim 80$ mesh	parameter: γ_b $k = 0.2 \sim 3.5$ 1/s $\gamma_b = 0.1 \sim 0.3$
Kato (1967)	$U_e = U_{mf}$ $D_e = 0$	packed fluidized bed hydrogenation of ethylene $d_t = 8.7$ cm $L_{mf} = 10 \sim 30$ cm $d_p = 100 \sim 200$ mesh, $d_p = 1 \sim 3$ cm	parameters: γ_b $k = 1.1 \sim 3.3$ 1/s $\gamma_b = 0.35 \sim 0.45$
Kobayashi et al. (1967)	$D_e = 0$ $U_e = U_{mf}$	residence-time curve gas: air tracer: He particle: silica, gel $d_t = 8.4$ cm	$F_o = 11/d_b$
De Grout (1967)	$D_e = 0$ $U_e = U_{mf}$	residence time curve gas: air tracer: He particle: silica $d_t = 10 \sim 150$ cm	$H_K = 0.67\, d_t^{0.25}\, L^{0.5}$ where L = bed height (m) $H_K = U/F_o$; U[m/s]
Kato et al. (1967)	$U_e = 0$ $D_e = 0.68 \left(\dfrac{U - U_{mf}}{U_{mf}} \right) d_p \varepsilon_p$	residence-time curve gas: air, H_2, N_2 tracer: H_2, C_2H_4, C_3H_5 particle: silica-alumina, glass $d_t = 10$ cm, $d_p = 1 \sim 3$ cm	$F_o = 5 \sim 3$ 1/s for $U/U_{mf} = 2 \sim 30$ $M = 0.4 \sim 0.2$ 1/s for $U/U_{mf} = 2 \sim 30$

Table 2 Theoretical Study of the Two-Phase Model

Authors	Parameter assumed	Method	γ_b or F_0	Remarks
Van & Deemter (1961)	$\gamma_b = 0$ $U_e = 0$ $D_e = D_s$	A steady state analysis of gas backmixing and residence time curve and first-order reaction by two-phase model	$H_k = \dfrac{F_0 L}{U}$ $H_k = 0.5 \sim 2.5$, $\gamma_b = 0$	Parameter F_0 is not related to the bubble movement in the bed
Davidson & Harrison (1963)	$D_e = 0$ or ∞ $\gamma_b = 0$	Estimation of conversions for a first-order reaction	$F_0 = \dfrac{5.85 D^{1/2} g^{1/4}}{d_b^{5/4}} + \dfrac{4.5 U_{mf}}{d_b}$	Parameter d_b model does not account for bubble growth in the bed
Muchi (1965)	$U_e = U_{mf}$ $0 < D_e < \infty$	A study of effect of F_0, γ_b, D_e, U_e on conversion of a first-order reaction		No relation between bubble movement and parameter
Mamuro & Muchi (1965)	$U_e = U_{mf}$ $\gamma_b = 0$	Analysis of a first-order reaction based on the two-phase cell model	$F_0/\phi = 0.05$ ϕ = shape factor of bubble	
Kobayashi & Arai (1965)	$U_e = 0$ $D_e = 0$	A study of the effect of k, γ_b, D_e, and F_0 on conversion of a first-order reaction		Parameters γ_b, F_0, D_e are not related to the bubble movement
Partridge & Rowe (1966)	$D_e = 0$ $\gamma_b \neq 0$	Consider bubble-cloud phase reaction in bubble-cloud phase with non-first-order kinetics	$Sh_c = 2 + 0.69 Sc^{0.33} Re_c^{0.5}$ $Sh_c = \dfrac{K_{g,c} d_c}{D}$	Allow the variation of bubble population with height
Van & Deemter (1967)	$D_e = 0$	Analysis of backmixing residence time curve of tracer gas and the first-order reactions	$F_0 = 0.4 \sim 1.2$ (1/s)	Parameters γ_b, F_0, U_e, are not related to the bubble growth in the bed

concentration profile, predicted dense phase concentrations are far too high and too close to the bubble phase concentrations. They concluded that the mass transfer rate in the model is too high.

3.1.2 Model of Partridge and Rowe (1966)

Another representative two-phase model is the one proposed by Partridge and Rowe (1966). In this model, the two-phase theory of Toomey and Johnstone (1952) is still used to estimate the visible gas flow, as in the model of Davidson and Harrison (1963). However, this model considers the gas interchange to occur at the cloud-emulsion interphase, i.e., the bubble and the cloud phase are considered to be well-mixed, the result being called bubble-cloud phase. The model thus interprets the flow distribution in terms of the bubble-cloud phase and the emulsion phase. With the inclusion of the clouds, the model also allows reactions to take place in the bubble-cloud phase. The rate of interphase mass transfer proposed in the model, however, considers the diffusive mechanism only (i.e., without throughflow) and is much lower than that used in the model of Davidson and Harrison (1963).

Chavarie and Grace (1975) also reviewed the model and found that the model has a tendency to overestimate the cloud size which deprives the concentration profiles and related features obtained from the model of any physical meaning, owing to the inherent physical incompatibility. A similar problem regarding the model was also reported by Ellis et al. (1968).

3.1.3 Modifications and Applications

In recent years, several modified versions of the two-phase model were proposed for modeling fluidized bed reactors. They include a model proposed by Werther (1980) for catalytic oxidation of ammonia, in which the mass transfer process is expressed in terms of film theory, as described in Danckwerts (1970); a model proposed by Werther and Schoessler (1986) for catalytic reactions; a model proposed by Borodulya et al. (1995) for the combustion of low-grade fuels; a model proposed by Arnaldos et al. (1998) for vacuum drying; and a model proposed by Srinivasan et al. (1998) for combustion of gases. The modifications include the consideration of axial mass transfer profile, the inclusion of a wake phase in addition to the bubble and emulsion phases, and the consideration of the growth of bubbles in the bubble phase.

3.2 Bubbling Bed Model

The bubbling bed model proposed by Kunii and Levenspiel (1969) can be considered a modified version of the two-phase model where, in addition to the bubble and the emulsion phases, a cloud-wake phase is also considered. The model represents a group of models often referred to as backmixing or dense phase flow reversal models (see also Van Deemter, 1961; Latham et al., 1968; Fryer and Potter, 1972). A key difference between this model and the rest of the two-phase models is that the interphase mass transfer considers two distinct resistances, one from the bubble phase to the cloud-wake phase, and the other from the cloud-wake phase to the emulsion phase.

The derivation of the model involves the following background theory and observations reported by Davidson and Harrison (1963) and Rowe and Partridge (1962), i.e.,

1. Bubble gas stays with the bubble, recirculating very much like smoke rising and only penetrating a small distance into the emulsion. This zone of penetration is called the cloud since it envelops the rising bubble.
2. All related quantities such as the velocity of the rise, the cloud thickness, and the recirculation rate, are simple functions of the size of rising bubble.
3. Each bubble of gas drags a substantial wake of solids up the bed.

3.2.1 Derivation of the Model

Based on the above observations, the bubbling bed model assumes that

1. Bubbles are of one size and are evenly distributed in the bed.
2. The flow of gas in the vicinity of rising bubbles follows the Davidson model (Davidson and Harrison, 1963).
3. Each bubble drags along with it a wake of solids, creating a circulation of solids in the bed, with upflow behind bubbles and download in the rest of the emulsion.
4. The emulsion stays at minimum fluidizing conditions; thus the relative velocity of gas and solid remains unchanged.

With the above assumptions, material balances for solids and for gas give in turn

(Upflow of solids with bubble)

$\quad$ = (Downflow of solids in emulsion) $\qquad$ (6)

(Total throughflow of gas)

$\quad$ = (Upflow in bubble) + (Upflow in emulsion)

$\hfill$ (7)

Letting

$$u_{br} = 0.711(gd_b)^{0.5} \qquad (8)$$

the above material balances give (1) The rise velocity of bubbles, clouds, and wakes, u_b:

$$u_b = U - U_{mf} + u_{br} = U - U_{mf} + 0.711(gd_b)^{0.5}$$
$$\hfill (9)$$

(2) the bed fraction in bubbles, δ:

$$\delta = \frac{(U - [1 - \delta - \alpha\delta]U_{mf})}{u_b} \cong \frac{U - U_{mf}}{u_b} \qquad (10)$$

(3) the bed fraction in clouds, β:

$$\beta = \frac{3\delta(U_{mf}/\varepsilon_{mf})}{u_{br} - (U_{mf}/\varepsilon_{mf})} \qquad (11)$$

(4) the bed fraction in wakes, ω:

$$\omega = \alpha\delta \qquad (12)$$

(5) the bed fraction in downflowing emulsion including clouds, $\bar{\omega}$:

$$\bar{\omega} = 1 - \delta - \alpha\delta \qquad (13)$$

(6) the downflow velocity of emulsion solids, u_s:

$$u_s = \frac{\alpha\delta u_b}{1 - \delta - \alpha\delta} \qquad (14)$$

(7) the rise velocity of emulsion gas, u_e:

$$u_e = \frac{U_{mf}}{\varepsilon_{mf}} - u_s \qquad (15)$$

Using Davidson's theoretical expression for bubble-cloud circulation and the Higbie (1935) theory for cloud-emulsion diffusion, the interchange of gas between bubble and cloud is then found to be

$$K_{bc} = 4.5\frac{U_{mf}}{d_b} + 5.85\frac{\mathcal{D}^{0.5}g^{0.25}}{d_b^{1.25}} \qquad (16)$$

and between cloud and emulsion

$$K_{ce} = 6.78\left(\frac{(\varepsilon_{mf}\mathcal{D}u_b)}{d_b^3}\right)^{0.5} \qquad (17)$$

The above expressions indicate that if ε_{mf}, α, U_{mf}, and U are known or measured, then all flow properties and regional volumes can be determined in terms of one parameter, the bubble size. The application of this model to chemical conversion is described below.

3.2.2 Model Expression for First-Order Kinetics

For a first-order catalytic reaction occurring in a gas–solid fluidized bed with $\varepsilon_A = 0$, the rate equation may be expressed as

$$-r_{A,s} = \frac{1}{V_s}\left(\frac{dN_A}{dt}\right) = kC_A \qquad (18)$$

If the bed is assumed to be operated at a fairly high gas flow rate with vigorous bubbling of large rising bubbles, then both the gas flow in the emulsion and the cloud volume become so small that we can ignore the throughflow of gas in these regions. Consequently, as an approximation, flow through the bed occurs only in the bubble phase. The disappearance of A in rising bubble phase, therefore, can be formulated as (see Fig. 2):

(Disappearance from bubble phase) = (Reaction in

$\quad$ bubble) + (Transfer to cloud and wake)

(Transfer to cloud and wake) = (Reaction in cloud

$\quad$ cloud and wake) + (Transfer to emulsion)

(Transfer to emulsion) = (Reaction in emulsion)

$\hfill$ (19)

In symbols, the above expressions become

$$-r_{A,b} = -\frac{1}{V_b}\left(\frac{dN_A}{dt}\right) = \gamma_b kC_{A,b} +$$
$$K_{bc}(C_{A,b} - C_{A,c})$$
$$K_{bc}(C_{A,b} - C_{A,c}) = \gamma_c kC_{A,c} + K_{ce}(C_{A,c} - C_{A,e})$$
$$K_{ce}(C_{A,c} - C_{A,e}) = \gamma_e kC_{A,e} \qquad (20)$$

Experiment has shown that

$$\gamma_b = 0.001 \sim 0.01 \qquad (21)$$

and

$$\alpha = 0.25 \sim 1.0 \qquad (22)$$

Also, by the material balance expressions, Eqs. 6 to 15, it can be shown that

$$\gamma_c = (1 - \varepsilon_{mf})\left[3\frac{\left(\frac{U_{mf}}{\varepsilon_{mf}}\right)}{\left(u_{br} - \frac{u_{mf}}{\varepsilon_{mf}}\right)} + \alpha\right] \qquad (23)$$

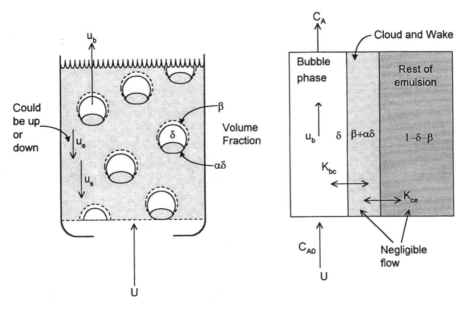

Figure 2 Bubbling bed model. (From Kunii and Levenspiel, 1969.)

and

$$\gamma_e = \frac{(1 - \varepsilon_{mf})(1 - \delta)}{\delta} - \gamma_b - \gamma_c \qquad (24)$$

On eliminating all intermediate concentration in Eqs. 20a, b, and c, we find that

$$-r_{A,b} = \left(\gamma_b k + \frac{1}{\chi_1}\right) C_A \qquad (25)$$

where

$$\chi_1 = \frac{1}{K_{bc}} + \frac{1}{\chi_2} \qquad (26)$$

$$\chi_2 = \gamma_c k + \frac{1}{\chi_3} \qquad (27)$$

and

$$\chi_3 = \frac{1}{K_{ce}} + \frac{1}{(\gamma_e k)} \qquad (28)$$

Inserting for plug flow of gas through the bed yields the desired performance expression, or

$$\ln\left(\frac{C_{A_0}}{C_A}\right) = \left(\gamma_b k + \frac{1}{\chi_1}\right)\frac{L_{fluidized}}{u_b} \qquad (29)$$

where approximately

$$\frac{L_{fluidized}}{u_b} = \frac{1 - \varepsilon_{packed}}{1 - \varepsilon_{mf}}\frac{L_{packed}}{u_{br}} \qquad (30)$$

Note that since the bubble size is the only quantity that governs all the rate quantities with the exception of k,

the performance of a fluidized bed is a function of the bubble size. For small bubbles, the results from the bubbling bed model may range between a plug flow model and a mixed flow model; for large bubbles, however, they may be well below those predicted by a mixed flow model.

3.2.3 Examples and Model Applications

Examples illustrating the model and additional discussions on the model are found in Levenspiel (1972) and Kunii and Levenspiel (1991). Applications of the model were reported in several recent studies, including scale-up studies of catalytic reactors by Botton (1983) and Dutta and Suciu (1989), a gasification study of coal carried out by Matusi et al. (1983), and a study of fast fluidized bed reactors by Kunii and Levenspiel (1998).

3.3 Bubble Assemblage Model

Since the development of the bubbling bed model, various other hydrodynamic models have been proposed using other combinations of assumptions such as changing bubble size with height in the bed, negligible bubble-cloud resistance, negligible cloud-emulsion resistance, and nonspherical bubbles. Among them, the bubble assemblage model, proposed by Kato and Wen (1969), considers changing bubble size with height in the bed. The model has the following assumptions:

1. A fluidized bed may be represented by n compartments in a series. The height of each compartment is equal to the size of each bubble at the corresponding bed height.
2. Each compartment is considered to consist of a bubble phase and an emulsion phase. The gas flows through the bubble phase, and the emulsion phase is considered to be completely mixed within the phase.
3. The void space within the emulsion phase is considered to be equal to that of the bed at the incipient fluidizing conditions. The upward velocity of the gas in the emulsion phase is U_e.
4. The bubble phase is assumed to consist of spherical bubbles surrounded by spherical clouds. The voidage within the cloud is assumed to be the same as that in the emulsion phase, and the diameter of the bubbles and that of clouds is given by Davidson (1961) as

$$\left(\frac{d_c}{d_b}\right)^3 = \frac{u_{br} + 2(U_{mf}/\varepsilon_{mf})}{u_{br} - (U_{mf}/\varepsilon_{mf})} \tag{31}$$

$$\text{for} \quad u_{br} \geq \frac{U_{mf}}{\varepsilon_{mf}}$$

where

$$u_{br} = 0.711(g d_b)^{0.5} \tag{32}$$

Note that the calculation proposed above would not be applicable for large particles where u_{br} may be less than U_{mf}/ε_{mf}.

5. The total volume of the gas bubbles within the bed may be expressed as $(L - L_{mf})S$.
6. Gas interchange takes place between the two phases. The overall mass interchange coefficient per unit volume of gas bubbles is given by

$$F_d = F_o + K'M \tag{33}$$

7. The bubbles are considered to grow continuously while passing through the bed until they reach the maximum stable size or reach the diameter of the bed column. The maximum stable bubble size, $d_{b,t}$, can be calculated by (Harrison et al., 1961)

$$d_{b,t} = \left(\frac{u_t}{0.711}\right)^2 \frac{1}{g} \tag{34}$$

8. The bed is assumed to be operating under isothermal conditions since the effective thermal diffusivity and the heat transfer coefficient are large.

3.3.1 Key Equations in the Bubble Assemblage Model

In addition to the above assumptions, the model has the following key equations to estimate various bubbling properties:

1. Bubble size, based on Cooke et al. (1968):

$$d_b = 0.14\rho_p d_p \left(\frac{U}{U_{mf}}\right) h + d_o \tag{35}$$

where

$$d_o = 0.025 \frac{[6(U - U_{mf})/(n_o\pi)]^{0.4}}{g^{0.2}} \tag{36}$$

2. Bubble velocity, following Davidson and Harrison (1963):

$$u_b = (U - U_{mf}) + u_{br} = (U - U_{mf}) + 0.711(g\, d_b)^{0.5} \tag{37}$$

3. Bed expansion, based on Assumption 5 and Eqs. 35 through 37,

$$\frac{L - L_{mf}}{L_{mf}} = \frac{U - U_{mf}}{0.711(g\, d_{b,a})^{0.5}} \tag{38}$$

where $d_{b,a}$ is the average bubble diameter of the bed given by

$$d_{b,a} = 0.14\rho_p d_p \frac{U}{U_{mf}} \frac{L_{mf}}{2} + d_o \tag{39}$$

4. Voidage of the bed, ε
   (a) For $h \leq L_{mf}$,

   $$1 - \varepsilon = \frac{L_{mf}}{L}(1 - \varepsilon_{mf}) \tag{40}$$

   (b) for $L_{mf} \leq h \leq L_{mf} + 2(L - L_{mf})$,

   $$1 - \varepsilon = \frac{L_{mf}}{L}(1 - \varepsilon_{mf})$$
   $$- 0.5\frac{L_{mf}(1 - \varepsilon_{mf})(h - L_{mf})}{2L(L - L_{mf})} \tag{41}$$

5. Superficial gas velocity in emulsion phase, U_e

$$\frac{U_e}{U_{mf}} = 1 - \frac{\varepsilon_{mf}\alpha'\theta u_b}{U_{mf}(1 - \theta - \alpha'\theta)} \tag{42}$$

where

$$\theta = \frac{L - L_{mf}}{L} \tag{43}$$

and α' is the ratio of the volume of emulsion transported upward behind a bubble (volume of wake) to the volume of a bubble. The value of α' is approximately $0.2 \sim 0.3$ according to the experimental study of Rowe and Partridge (1965). Therefore under normal experimental conditions, Eq. (42) yields $U_e/U_{mf} = 0.5$ for $U/U_{mf} = 3$, and $U_e/U_{mf} = 0$ for $U/U_{mf} = 5 \sim 6$. However, in the model, U_e was assumed to be zero based on the experimental findings of Latham (1968) and the argument presented by Kunii and Levenspiel (1969).

6. Interchange coefficient, F_d, based on Eq. (33) Since no experimental data are available for the particle interchange rate, M, or the adsorption equilibrium constant for the reacting gas on particle surfaces, K', the model neglects gas interchange due to adsorbed gas on interchanging particles. Equation (33) therefore can be reduced to

$$F_d = F_o \tag{44}$$

where the following equation based on the experimental work of Kobayashi et al. (1967) was proposed to describe F_o:

$$F_o = \frac{0.11}{d_b} \tag{45}$$

3.3.2 Calculation Procedure Based on Bubble Assemblage Model

Let the height of the nth compartment be h_n, where $n = 1, 2, 3,$ to n (see Fig. 3). Based on an arithmetic average of the bubble size, the height of the initial compartment immediately above the distributor becomes

$$\Delta h_1 = \frac{d_o + (\psi \Delta h_1 + d_o)}{2} \tag{46}$$

or

$$\Delta h_1 = \frac{2d_o}{2 - \psi} \tag{47}$$

where

$$\psi = 0.14 \rho_p d_p \frac{U}{U_{mf}} \tag{48}$$

which is a proportionality constant relating the bubble diameter for a given operating condition. The height of the second compartment then becomes

$$\Delta h_2 = 2d_o \frac{2 + \psi}{(2 - \psi)^2} \tag{49}$$

and that of nth compartment becomes

$$\Delta h_n = 2d_o \frac{(2 + \psi)^{n-1}}{(2 - \psi)^n} \tag{50}$$

The number of bubbles in the nth compartment becomes

$$N = \frac{6S(\varepsilon - \varepsilon_{mf})}{\pi(\Delta h_n)^2(1 - \varepsilon_{mf})} \tag{51}$$

The volume of cloud in the nth compartment can be computed from Eq. (31) as

$$V_{cn} = \frac{N\pi(\Delta h_n)^3}{6} \frac{3(U_{mf}/\varepsilon_{mf})}{u_{br} - U_{mf}/\varepsilon_{mf}} \tag{52}$$

where

$$u_{br} = 0.711[g(\Delta h_n)]^{0.5} \tag{53}$$

The total volume of the bubble phase (bubble and cloud) and that of the emulsion phase in the nth compartment are, respectively,

$$V_{bn} = \frac{N\pi(\Delta h_n)^3}{6} \frac{u_{br} + 2(U_{mf}/\varepsilon_{mf})}{u_{br} - (U_{mf}/\varepsilon_{mf})} \tag{54}$$

and

$$V_{en} = S \Delta h_n - V_{bn} \tag{55}$$

The distance from the distributor to the nth compartment is then

$$h_n = \Delta h_1 + \Delta h_2 + \Delta h_3 + \cdots + \Delta h_n \tag{56}$$

The gas interchange coefficient based on unit volume of bubble phase (bubble and cloud) can be shown as

$$F'_{on} = F_{on} \frac{u_{br} - (U_{mf}/\varepsilon_{mf})}{u_{br} + 2(U_{mf}/\varepsilon_{mf})} \tag{57}$$

Hence, the material balance for the gaseous reactant around the nth compartment becomes, for the bubble phase,

$$(SUC_{A,b})_{n-1} = \{F'_{on}V_b(C_{A,b} - C_{A,e})\}_n + (r_{A,c}V_c)_n + (SUC_{A,b})_n \tag{58}$$

and for the emulsion phase,

$$\{F'_{on}V_b(C_{A,b} - C_{A,e})\}_n = (r_{A,e}V_e)_n \tag{59}$$

where $r_{A,c}$ and $r_{A,e}$ are the reaction rates per unit volume of the cloud and the emulsion phase, respectively. Note that for the first-order reaction, the expressions for $r_{A,c}$ and $r_{A,e}$ are $r_{A,c} = k_r C_{A,b}$ and $r_{A,e} = k_r C_{A,e}$, respectively.

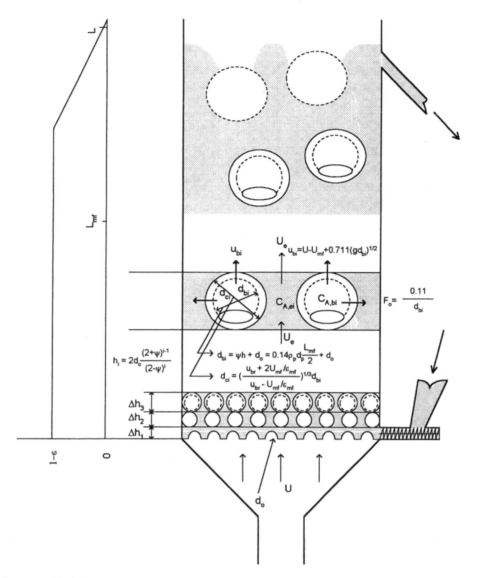

The figure contains the following labeled equations:

$$u_{bi} = U - U_{mf} + 0.711(gd_{bi})^{1/2}$$

$$F_o = \frac{0.11}{d_{bi}}$$

$$h_i = 2d_o \frac{(2+\psi)^{i-1}}{(2-\psi)^i}$$

$$d_{bi} = \psi h + d_o = 0.14\rho_p d_p \frac{L_{mf}}{2} + d_o$$

$$d_{ci} = \left(\frac{u_{br} + 2U_{mf}/\varepsilon_{mf}}{u_{br} - U_{mf}/\varepsilon_{mf}}\right)^{1/3} d_{bi}$$

Figure 3 Main features of bubble assemblage model. (From Kato and Wen, 1969.)

3.3.3 Computational Procedure for Conversion

The computational procedure for conversion and concentration profile in a fluidized bed reactor is given below. The following operating conditions are needed, i.e., particle size (d_p), particle density (ρ_p), minimum fluidization velocity (U_{mf}), gas superficial velocity (U), distributor arrangement (n_o), column diameter (d_t), incipient bed height (L_{mf}), reaction rate constant (k_r), and order of reaction. It should be noted that the model requires no adjustable parameters.

First, Eq. (38) is used to calculate the expanded bed height, L. Next, Eq. (50) is used to compute the size of the nth compartment. Using Eqs. (51) through (55), the volumes of the cloud, the bubble phase, and the emul-

sion phase for the nth compartment are then calculated. The nth compartment concentrations, $C_{A,bn}$ and $C_{A,en}$, are computed from $(C_{A,b})_{n-1}$ and $(C_{A,e})_{n-1}$ using Eqs. (58) and (59). The calculations are repeated from the distributor until the bed height equivalent to L_{mf} is reached. For bed height above L_{mf}, the voidage is adjusted by Eq. (41), and V_{cn}, V_{bn}, and V_{en} are obtained using the same procedure as that shown for the height smaller than L_{mf}. The calculation is repeated until the bed height reaches $L_{mf} + 2(L - L_{mf})$.

3.3.4 Discussion of the Bubble Assemblage Model

The computation using the bubble assemblage model indicates that for most of the experimental conditions tested, the number of compartments is usually greater than 10. This means, in terms of the flow pattern, that the gas passing through the bubble phase is close to plugflow. In the emulsion phase, because $U_e = 0$ is used, the flow pattern is essentially considered as a dead space interchanging gas with the bubble phase similar to that in the bubbling bed model. The flow pattern in the emulsion phase, however, does not expect to affect significantly the reactor behavior according to Lewis (1959) and Muchi (1965). It is worth pointing out that when the reaction is slow, any model, either a plugflow, a complete mixing, a bubbling bed, or a bubble assemblage model will represent the experimental data well. However, when the reaction is fast, a correct flow model is needed to represent the data.

3.3.5 Examples and Applications

More detailed discussion regarding the model performance is given by Wen and Fan (1975) and Mori and Wen (1975). The literature-reported applications of this model include a combustion study of coal with limestone injection by Horio and Wen (1975), a coal gasification study by Mori et al. (1983), a study on catalytic oxidation of benzene by Jaffres et al. (1983), a catalytic ammoxidation of propylent by Stergiou and Laguerie (1983), a silane decomposition study by Li et al. (1989), a catalytic oxidation of methane by Mleczko et al. (1992), and a study on chlorination of rutile by Zhou and Sohn (1996).

3.4 Comparison of Models

In an attempt to compare the effectiveness of various two-phase models, Chavarie and Grace (1975) carried out a study of the catalytic decomposition of ozone in a two-dimensional fluidized bed where the reactor performance and the ozone concentration profiles in both phases were measured. They compared the experimental data with those predicted from various two-phase models reviewed above and reported that

1. The model of Davidson and Harrison (1963), which assumes perfect mixing in the dense phase (DPPM), underestimates seriously the overall conversion for the reaction studied. While the counterpart model that assumes piston flow in the dense phase (DPPF) gives much better predictions of overall conversion, still the predicted concentration profiles in the individual phases are in poor agreement with the observed profiles.

2. The model of Partridge and Rowe (1966) makes allowance for variable bubble sizes and velocities and for the presence of clouds. Unfortunately, for the conditions of their work, the overestimation of visible bubble flow by the two-phase theory of Toomey and Johnstone (1952) led to incompatibility between predicted cloud areas and the total bed cross section. This mechanical incompatibility prevented direct application of these models to the reaction data obtained in their work.

3. The bubbling bed model of Kunii and Levenspiel (1969) provides the best overall representation of the experimental data in this study. Predicted bubble phase profiles tend gently to traverse the measured profiles, while dense phase profiles are in reasonable agreement over most of the bed depth. Overall conversions are well predicted. The success of the model can be mainly attributable to (1) the moderate global interphase mass transfer, (2) the negligible percolation rate in the dense phase, (3) the occurrence of reaction within the clouds and wakes assumed by the model, and (4) the use of average bubble properties to simulate the entire bed.

4. The Kato and Wen (1969) bubble assemblage model, though better suited to represent complex hydrodynamics due to allowance for variable bubble properties, fails to account for observed end effects in the reactor. While this model was found to give the best fit for the bubble phase profile, dense phase profiles and outlet reactant concentrations were seriously overpredicted.

It is worth pointing out that a general observation made by Chavarie and Grace (1975) was that none of the models (tested correctly) accounted for the considerable end effects observed at both the inlet and the outlet, i.e., near the distributor (grid region) and in the space above the bed surface (freeboard region).

4 MULTIPLE-REGION MODELS

The grid region near the distributor and the freeboard space above the bed surface have bed hydrodynamics significantly different from the main body of the bubbling bed reactor. A number of authors have observed the abnormally high rate of reaction near the distributor, apparently due to the very high rate of interphase mass transfer. Similarly, many observations have been made of temperature increase in the freeboard region, indicating additional reaction in the region. A realistic approach for modeling a fluidized bed reactor would therefore require the consideration of three consecutive regions in the bed: the grid region near the bottom, the bed region in the middle, and the freeboard region above the bed surface.

4.1 Models for the Grid Region

The grid region plays an important role in determining the reaction conversion of fluidized bed reactors, especially for fast reactions where the mass transfer operation is the controlling mechanism. Experimental studies have indicated that changing from one distributor to another, all other conditions remaining fixed, can cause major changes in conversion (Cooke et al., 1968; Behie and Kehoe, 1973; Bauer and Werther, 1981). It is generally observed that, in the grid region, additional mass transfer can take place owing to the convective flow of gas through the interphase of the forming bubbles. The flow of gas through the forming bubbles into the dense phase, and then returning to the bubble phase higher in the bed, represents a net exchange between the two phases. The experimental work of Behie and Kehoe (1973) indicated that the mass transfer coefficient in the grid region, k_{je}, can be 40 to 60 times that in the bubble region, i.e., k_{be}.

Behie and Kehoe (1973) and Grace and De Lasa (1978) proposed similar sets of equations to describe fluidized bed reactors considering both the grid and bed regions. The model of Grace and De Lasa (1978) contains the following three equations: (1) for the jet phase in the grid region ($0 \leq h \leq J$),

$$U\left(\frac{dC_{A,j}}{dh}\right) + k_{je}a_j(C_{Aj} - C_{A,e}) = 0 \qquad (60)$$

(2) for the bubble phase in the bed region ($J \leq h \leq L$),

$$\beta' U\left(\frac{dC_{A,b}}{dh}\right) + k_{be}a_b(C_{A,b} - C_{A,e}) = 0 \qquad (61)$$

and (3) for the emulsion phase in both the grid and bed regions ($0 \leq h \leq L$),

$$U(1 - \beta')(C_{A,e} - C_{A,jJ}) + \int_0^J k_{je}a_j(C_{A,e} - C_{A,j})dh$$
$$+ \int_J^L k_{be}a_b(C_{A,e} - C_{A,b})\,dh + k_r C_{A,e} L_{mf} = 0 \qquad (62)$$

It should be noted that the difference between this model and the corresponding model for the bed region only is the replacement of Eq. (61) by Eq. (60) in the grid region ($0 \leq h \leq L_j$). The model therefore predicts a much higher conversion in the grid region because the mass transfer coefficient in the region can be 40 to 60 times greater than in the bed region, as reported by Behie and Kehoe (1973). Among the grid region studies, Ho et al. (1987) reported dynamic simulation results of a shallow jetting fluidized bed coal combustor using the grid region model.

In another attempt, Sit and Grace (1986) measured time-averaged concentrations of methane in the entry region of beds of 120 to 310 μm particles contained in a 152 mm column with a central orifice of diameter 6.4 mm and auxiliary tracer-free gas. The following equations were proposed for the particle–gas mass transfer in this region:

$$V_b\left(\frac{dC_{A,j}}{dt}\right) = Q_{or}C_{A,or} + k_{be1}S_{ex,b}C_{A,e} \\ - (Q_{or} + k_{be1}S_{ex,b})C_{A,j} \qquad (63)$$

where Q_{or} represents the convective mechanism defined as

$$Q_{or} = u_{or}\left(\frac{\pi}{4}\right)d_{or}^2 \qquad (64)$$

and k_{be1} represents the diffusive mechanism that can be predicted by the penetration theory expression, i.e.,

$$k_{be1} = \left(\frac{4\mathcal{D}\varepsilon_{mf}}{\pi t_f}\right)^{0.5} \qquad (65)$$

Their results indicated that the convective mechanism has a greater effect on the grid region mass transfer. Since bubbles are also smaller near the grid, they concluded that favorable gas–solid contacting occurs in this region primarily due to convective outflow followed by recapture. Note that Eq. (63) can be used to replace Eq. (60) for the grid region modeling.

4.2 Models for the Freeboard Region

When bubbles burst and release their gases at the surface of a fluidized bed, it is generally observed that particles are ejected into the freeboard space above the surface. These particles can originate either from the dense phase just ahead of the bubble at the moment of eruption (Do et al., 1972) or from those solids that travel in the wake of the rising bubbles (Basov et al., 1969; Leva and Wen, 1971; George and Grace, 1978).

As for studies of freeboard region hydrodynamics and reactions, Yates and Rowe (1977) developed a simple model of a catalytic reaction based on the assumption that the freeboard contained perfectly mixed, equally dispersed particles derived from bubble wakes. The fraction of wake particles ejected, f', was a model parameter. The governing equation of the model was proposed as

$$-\frac{dC_{A,cell}}{dt} = \frac{k_g A_p}{V_{cell}}\left(C_{Ah} - C_{Ap}\right) \qquad (66)$$

where V_{cell} can be determined by

$$V_{cell} = \frac{3V_p}{f'(1-\varepsilon_{mf})}\frac{U - U_{mf}}{U - u_t} \qquad (67)$$

and k_g can be evaluated from the expression of Rowe et al. (1965), i.e.,

$$\mathrm{Sh} = \frac{k_g d_p}{\mathcal{D}} = 2 + 0.69\,\mathrm{Sc}^{0.33}\mathrm{Re}_t^{0.5} \qquad (68)$$

with the Schmidt number, Sc, defined as

$$\mathrm{Sc} = \frac{\mu}{\rho_g \mathcal{D}} \qquad (69)$$

and the terminal Reynolds number, Re_t, defined as

$$\mathrm{Re}_t = \frac{u_t d_p \rho_g}{\mu} \qquad (70)$$

The model equations were solved for various combinations of the parameters, and it was reported that in many cases freeboard reactions lead to greater conversion than that achieved by the fluid bed itself. The details of the results are also reviewed by Yates (1983).

5 CONCLUDING REMARKS

Numerous theoretical and experimental studies have been carried out in the past five decades in an attempt to model gas–solids fluidized bed reactors. However, the modeling of such reactors remains an art rather than a science. Models that work well for certain reaction processes may not work for others. In other words, there has not been any single model universally applicable to all processes carried out in such reactors. However, all these modeling attempts do generate valuable insights regarding reactor behavior, which leads to improved design and operation. Nevertheless, additional studies are needed to further our knowledge in the modeling of such reactors.

NOTATION

A_p	=	surface of freeboard particles appearing in Eq. (66), m^2
a_b	=	transfer area of bubbles to emulsion per unit volume of reactor, m^2/m^3
a_j	=	transfer area of jets to emulsion per unit volume of reactor, m^2/m^3
C_A	=	concentration of A, kg mol/m^3
C_{Ah}	=	concentration of A at periphery of gas cell appearing in Eq. (66), kg mol/m^3
C_{Ao}	=	initial concentration of A appearing in Eqs. (1) and (3), kg mol/m^3
C_{Ap}	=	concentration of A at particle surface appearing in Eq. (66), kg mol/m^3
$C_{A,b}$	=	concentration of A in bubble phase, kg mol/m^3
$C_{A,bn}$	=	concentration of A in bubble phase at nth compartment, kg mol/m^3
$C_{A,c}$	=	concentration of A in cloud phase, kg mol/m^3
$C_{A,cell}$	=	concentration of A in gas cell appearing in Eq. (66), kg mol/m^3
$C_{A,e}$	=	concentration of A in emulsion phase, kg mol/m^3
$C_{A,en}$	=	concentration of A in emulsion phase at nth compartment, kg mol/m^3
$C_{A,j}$	=	concentration of A in jet phase, kg mol/m^3
C_{AjJ}	=	concentration of A at the tip of jets, kg mol/m^3
$C_{A,or}$	=	concentration of A at the distributor orifice, kg mol/m^3
D_b	=	axial dispersion coefficient of reactant in bubble phase, m^2/s
D_e	=	axial dispersion coefficient of reactant in emulsion phase, m^2/s
d_b	=	bubble diameter, m
$d_{b,a}$	=	average bubble diameter, m
$d_{b,t}$	=	maximum stable bubble diameter, m
d_c	=	diameter of cloud, m
d_0	=	initial bubble diameter at the distributor, m
d_{or}	=	diameter of distributor orifice, m
d_p	=	particle diameter, m
d_t	=	diameter of bed column, m
E	=	probability density function
F	=	volumetric fraction of gas in the bubble phase
F_d	=	overall gas interchange coefficient per unit volume of gas bubble, 1/s
F_o	=	gas interchange coefficient per unit volume of gas bubble, 1/s
F_{on}	=	gas interchange coefficient at the nth compartment per unit bubble volume, 1/s
F'_{on}	=	F_{on} per unit volume of bubble phase (bubble and cloud), 1/s
F_s	=	volume fraction of solids in bubble phase
f	=	volumetric fraction of gas in the emulsion phase

f' = fraction of wake particles ejected appearing in Eq. (66)

f_s = volumetric fraction of solids in emulsion phase

g = gravitational acceleration, m/s^2

h = distance from the distributor, m

h_1 = distance between the distributor and the top of the first compartment, m

Δh_1 = length of the first compartment, m

h_n = distance between the distributor and the top of the nth compartment, m

Δh_n = length of the nth compartment, m

J = height of jet phase, m

K' = adsorption equilibrium constant

K_{bc} = gas interchange coefficient between the bubble and cloud phases, 1/s

K_{ce} = gas interchange coefficient between the cloud and emulsion phases, 1/s

k = reaction rate constant, 1/s

k_{be} = bubble to emulsion mass transfer coefficient, m/s

k_{bel} = bubble to emulsion mass transfer coefficient during bubble formation, m/s

k_g = mass transfer coefficient, m/s

k_{je} = jet to emulsion mass transfer coefficient, m/s

k_0 = reaction rate constant defined in Eq. (2), 1/s

k_r = first-order reaction rate constant based on volume of emulsion or cloud, 1/s

L = bed height, m

$L_{fluidized}$ = bed height under fluidization condition, m

L_{mf} = bed height at minimum fluidization, m

L_{packed} = bed height under packed condition, m

M = Solid interchange coefficient between bubble and emulsion phases per unit volume of bubble, 1/s

m = fitted parameter appearing in Eqs. (2) and (3)

N = number of bubbles in nth compartment

N_A = moles of A, kg mol

n_0 = number of distributor holes

Q_{or} = gas flow through orifice, m^3/s

$r_{A,b}$ = rate of reaction per unit bubble volume, kg mol/(m^3 s)

$r_{A,c}$ = rate of reaction per unit cloud volume, kg mol/(m^3 s)

$r_{A,e}$ = rate of reaction per unit emulsion volume, kg mol/(m^3 s)

$r_{A,s}$ = rate of reaction per unit solid volume, kg mol/(m^3 s)

Re_t = Reynolds number of particle at terminal velocity

S = cross-sectional area of the bed, m^2

$S_{ex,b}$ = surface area of bubbles, m^2

Sc = Schmidt number, defined as $Sc = \mu/(\rho_g \mathcal{D})$

Sh = Sherwood number, defined as $Sh = k_g d_p / \mathcal{D}$

t = time, s

t_f = bubble formation time, s

U = superficial gas velocity, m/s

U_e = superficial gas velocity in the emulsion phase, m/s

U_{mf} = minimum fluidization velocity, m/s

u_b = rise velocity of bubbles in a fluidized bed, m/s

u_{br} = rise velocity of a single bubble in a fluidized bed, m/s

u_e = rise velocity of emulsion gas, m/s

u_{or} = velocity of gas through orifice, m/s

u_s = downflow velocity of emulsion solids, m/s

u_t = terminal velocity of particle, m/s

V_b = volume of bubble phase, m^3

V_{bn} = volume of bubble phase at the nth compartment, m^3

V_c = volume of cloud phase, m^3

V_{cell} = volume of a freeboard gas cell expressed in Eq. (67), m^3

V_{cn} = volume of cloud phase at the nth compartment, m^3

V_e = volume of emulsion phase, m^3

V_{en} = volume of emulsion phase at the nth compartment, m^3

V_p = volume of freeboard particles appearing in Eq. (67), m^3

V_s = volume of solids, m^3

α = ratio of cloud volume to bubble volume

α' = ratio of wake volume to bubble volume

β = bed fraction in clouds

β' = fraction of gas flow in the bubble or jet phase

γ_b = volume fraction of solids in bubbles in a fluidized bed

γ_c = volume fraction of solids in clouds in a fluidized bed, -

γ_e = volume fraction of solids in emulsion phase in a fluidized bed,

$\mathcal{D}$ = diffusivity of reactant gas, m^2/s

δ = bed fraction in bubbles

ε = void fraction

ε_A = fractional volume change on complete conversion of A

ε_{mf} = void fraction at minimum fluidization

ε_{packed} = void fraction under packed bed condition

θ = bed expansion fraction defined in Eq. (43)

π = constant, ≈ 3.1416

ρ_g = gas density, kg/m^3

ρ_p = density of particle, kg/m^3

μ = gas viscosity, kg/(m s)

χ_1 = expression defined in Eq. (26)

χ_2 = expression defined in Eq. (27)

χ_3 = expression defined in Eq. (28)

ω = bed fraction in wake, appearing in Eq. (12)

$\bar{\omega}$ = bed fraction in downflowing emulsion, appearing in Eq. (13)

ψ = expression defined in Eq. (48)

REFERENCES

Arnaldos J, Kozanoglu B, Casal J. Vacuum fluidization: application to drying. In: Fan LS, Knowlton TM, eds. Fluidization IX, Durango, Colorado. New York: Engineering Foundation, 1998, pp 709–716.

Basov VA, Markhevkd VI, Melik-Akhnazarov TK, Orochko DI. Investigation of the structure of a nonuniform fluidized bed. Int Chem Eng 9:263–273, 1969.

Bauer W, Werther J. Proc 2nd World Congress Chem Eng 3:69, 1981.

Behie LA, Kehoe P. The grid region in a fluidized bed reactor. AIChE J 19:1070–1072, 1973.

Borodulya VA, Dikalenko VI, Palchonok GI, Stanchits LK. Fluidization and combustion of solid biological waste and low-grade fuels: experiment and modelling. In: Large JF, Laguerie C, eds. Fluidization VIII, Tours, France. New York: Engineering Foundation, 1995, pp 351–358.

Botton R, Vergnes F, Bergougnou MA. Validation by means of industrial data of Kunii-Levenspiel type bubble models which can be used in the scale-up to commercial size of fluidized bed reactors. In: Kunii D, Toei R, eds. Fluidization IV, Kashikojima, Japan. New York: Engineering Foundation, 1983, pp 575–582.

Chavarie C, Grace JR. Performance analysis of a fluidized bed reactor. II. Observed reactor behavior compared with simple two-phase models. Ind Eng Chem Fund 14:79–86, 1975.

Cooke MJ, Harris W, Highley J, Williams DF. Intn Chem Engrs Symp Ser 30:21, 1968.

Danckwerts PV. Gas Liquid Reaction. London: McGraw-Hill, 1970.

Davidson JF, Harrison D. Fluidized Particles. Cambridge: Cambridge University Press, 1963.

De Groot JH. Proceedings of the International Symposium on Fluidization, 1967, pp 348–361.

Do H, Grace JR, Clift R. Particle ejection and entrainment from fluidized beds. Powder Technology 6:195–200, 1972.

Dutta S, Suciu GD. Unified model applied to the scale-up of catalytic fluid bed reactors of commercial importance. In: Grace JR, Shemilt LW, Bergougnou MA, eds. Fluidization VI, Banff, Canada. New York: Engineering Foundation, 1989, pp 311–318.

Ellis JE, Partridge BA, Lloyd DI. Proc Tripartite Chem Eng Conf, Montreal, 1968, pp 43.

Fryer C, Potter OE. Bubble size variation in two-phase models of fluidized bed reactors. Powder Technology 6:317–322, 1972.

George SE, Grace JR. Entrainment of particles from aggregative fluidized beds. AIChE Symp Ser 74(176):67–74, 1978.

Gilliland ER, Knudsen CW. Paper 16d, AIChE Annual Meeting, Chicago, 1970.

Gomezplata A, Shuster WW. Effect of uniformity of fluidization on catalytic cracking of cumene. AIChE J 6:454–459, 1960.

Grace JR, De Lasa HI. Reaction near the grid in fluidized beds. AIChE J 24:364–366, 1978.

Harrison D, Davidson JF, De Kock JW. On the nature of aggregative and particulate fluidisation. Trans Inst Chem Engrs 39:202–211, 1961.

Higbie R. Trans Am Inst Chem Eng 31:365, 1935.

Ho TC, Ko KN, Chang CC, Hopper JR. Dynamic simulation of a shallow jetting fluidized bed coal combustor. Powder Technology 53:247–256, 1987.

Horio M, Wen CY. Analysis of fluidized-bed combustion of coal with limestone injection. In: Keairns DL, ed. Fluidization Technology, Vol 2. Washington, DC: Hemisphere, 1976, pp 289–320.

Ishii T, Osberg GL. Effect of packing on the catalytic isomerization of cyclopropane in fixed and fluidized beds. AIChE J 11:279–287, 1965.

Jaffres JL, Chavarie C, Patterson I, Perrier M, Casalegno L, Laguerie C. Conversion and selectivity modeling of the oxidation of benzene to maleic anhydride in a fluidized bed reactor. In: Kunii D, Toei R, eds. Fluidization IV, Kashikojima, Japan. New York: Engineering Foundation, 1983, pp 565–573.

Kato K. PhD dissertation. Tokyo Institute of Technology, Tokyo, Japan, 1967.

Kato K, Imafuku K, Kubota H. Chemical Engineering (Japan) 31:967, 1967.

Kato K, Wen CY. Bubble assemblage model for fluidized bed catalytic reactors. Chem Eng Sci 24:1351–1369, 1969.

Kobayashi H, Arai F. Chemical Engineering (Japan) 29:885, 1965.

Kobayashi H, Arai F, Isawa S, Sunagawa T, Miya K. Chemical Engineering (Japan) 30:656, 1966a.

Kobayashi H, Arai F, Tanaka T, Sakaguchi Y, Sagawa N, Sunagawa T, Shiba T, Takahashi K. The 6th Reaction Engineering Symposium (Japan), 1966b, pp 13.

Kobayashi H, Arai F, Sunagawa T. Chemical Engineering (Japan) 31:239, 1967.

Kunii D, Levenspiel O. Bubbling bed model—model for flow of gas through a fluidized bed. Ind Eng Chem Fund 7:446–452, 1968.

Kunii D, Levenspiel O. Fluidization Engineering. New York: John Wiley, 1969.

Kunii D, Levenspiel O. Fluidization Engineering, 2nd ed. Boston: Butterworth-Heinemann, 1991.

Kunii D, Levenspiel O. Conversion expressions for FF reactors. In: Fan LS, Knowlton TM, eds. Fluidization IX, Durango, Colorado. New York: Engineering Foundation, 1998, pp 677–684.

Latham R, Hamilton C, Potter OE. Back-mixing and chemical reaction in fluidised beds. Brit Chem Eng 13:666–671, 1968.

Leva M, Wen CY. Elutriation. In: Davidson JF, Harrison D, eds. Fluidization, Chap 14. London: Academic Press, 1971, pp 627–649.

Levenspiel O. Chemical Reaction Engineering, 2nd ed. New York: John Wiley, 1972.

Lewis WK, Gilliland ER, Glass W. Solid-catalyzed reaction in a fluidized bed. AIChE J 5:419–426, 1959.

Li KY, Peng SH, Ho TC. Prediction of silicon powder elutriation in a fluidized bed reactor for the silane decomposition reaction. AIChE Symp Ser 85(270):77–82, 1989.

Mamuro T, Muchi I. J of Ind Chem (Japan) 68:126, 1965.

Massimilla L, Johnstone HF. Reaction kinetics in fluidized beds. Chem Eng Sci 16:105–112, 1961.

Mathis JF, Watson CC. Effect of fluidization on catalytic cuemene dealkylation. AIChE J 2:518–528, 1956.

Matsui I, Kojima T, Furusawa T, Kunii D. Gasification of coal char by steam in a continuous fluidized bed reactor. In: Kunii D, Toei R, eds. Fluidization IV, Kashikojima, Japan. New York: Engineering Foundation, 1983, pp 655–662.

Mleczko L, Rothaemel M, Andorf R, Baerns M. Fluidized bed reactor performance for the catalytic oxidative coupling of methane to C_{2+} hydrocarbons. In: Potter OE, Nicklin DJ, eds. Fluidization VII, Brisbane, Australia, 1992, pp 487–494.

Mori S, Wen CY. Simulation of fluidized bed reactor performance by modified bubble assemblage model. In: Keairns DL, ed. Fluidization Technology, Vol 1. Washington DC: Hemisphere, 1976, pp 179–203.

Mori S, Nomura S, Kurita M, Hiraoka S, Yamada I. Modeling for the fluidized bed coal gasifier. In: Kunii D, Toei R, eds. Fluidization IV, Kashikojima, Japan. New York: Engineering Foundation, 1983, pp 639–646.

Muchi I. Memoirs of the Faculty of Engineering, Nagoya University (Japan) 17(1), 1965.

Orcutt JC, Davidson JF, Pigford RL. Reaction time distributions in fluidized catalytic reactors. Chem Eng Prog Sym Series 58(38):1–15, 1962.

Partridge BA, Rowe PN. Chemical reaction in a bubbling gas-fluidised bed. Trans Inst Chem Engrs 44:T335–T348, 1966.

Rowe PN, Partridge BA. Proc Symp on Interaction Between Fluids and Particles, Inst Chem Eng London, 1962, pp 135.

Rowe PN, Partridge BA. An x-ray study of bubbles in fluidized beds. Trans Inst Chem Eng 43:157–175, 1965.

Rowe PN, Claxton KT, Lewis JB. Heat and mass transfer from a single sphere in an extensive flowing fluid. Trans Inst Chem Eng 43:14–31, 1965.

Shen CY, Johnstone HF. Gas–solid contact in fluidized beds. AIChE J 1:349–354, 1955.

Sit SP, Grace JR. Interphase mass transfer during bubble formation in fluidized beds. In: Ostergaard K, Soorensen A, eds. Fluidization V, Elsionore, Denmark. New York: Engineering Foundation, 1986, pp 39–46.

Srinivasan RA, Sriramulu S, Agarwal PK. Mathematical modeling of the combustion of gases in bubbling fluidized beds. In: Fan LS, Knowlton TM, eds. Fluidization IX, Durango, Colorado. New York: Engineering Foundation, 1998, pp 693–700.

Stergiou L, Laguerie C. An experimental evaluation of fluidized bed reactor models. In: Kunii D, Toei R, eds. Fluidization IV, Kashikojima, Japan. New York: Engineering Foundation, 1983, pp 557–564.

Toomey RD, Johnstone HP. Gaseous fluidization of solid particles. Chem Eng Progress 48:220–226, 1952.

Van Deemter JJ. Mixing and contacting in gas–solid fluidized beds. Chem Eng Science 13:143–154, 1961.

Van Deemter JJ. Proceedings of the International Symposium on Fluidization, Eindhoven, 1967, pp 322.

Werther J. Modeling and scale-up of industrial fluidized bed reactors. Chem Eng Science 35:372–379, 1980.

Werther J, Schoessler M. Modeling catalytic reactions in bubbling fluidized beds of fine particles. In: Van Swaaij WPM, ed. Heat and Mass Transfer in Fixed and Fluidized Beds. New York: Hemisphere, 1986, pp 355–370.

Wen CY, Fan LT. Models for Flow Systems and Chemical Reactors. New York: Marcel Dekker, 1975.

Yates JG. Fundamentals of Fluidized-Bed Chemical Processes. London: Butterworths, 1983.

Yates JG, Rowe PN. A model for chemical reaction in the freeboard region above a fluidized bed. Trans Inst Chem Eng 55:137–142, 1977.

Zhou L, Sohn HY. Mathematical modeling of fluidized-bed chlorination of rutile. AIChE J 42:3102–3112, 1996.

10

Heat Transfer

John C. Chen

Lehigh University, Bethlehem, Pennsylvania, U.S.A.

1 INTRODUCTION

In many gas fluidized bed applications, heat transfer takes place either naturally or by design. The heat transfer could be between the solid and the gas phase, between the two-phase mixture and a solid surface, or both. An example is the fluidized combustion of coal, wherein coal particles are fluidized by air. The exothermic oxidation of carbon at particle surface causes an increase in particle temperature, and this naturally leads to heat transfer from the hot particles to the fluidizing air. To maintain an overall energy balance of the bed, it is then necessary to transfer heat from the particle–gas medium to some cooling surface, i.e., heat exchanger tubes. In this example, both types of heat transfer (between gas and particles, and between fluidized medium and submerged surfaces) occur. This chapter deals with these two types of heat transfer in gas fluidized beds. Whether a heat transfer process occurs naturally or by design is not of consequence; the engineer often needs to determine the rate of transfer in either case. In the above example, rapid heat transfer between particle and gas reduces the temperature of the burning particle and affects such aspects as ash agglomeration in the bed. The rate of heat transfer from the hot bed material to heat exchanger determines the amount of heat exchanger surface that must be provided. Because heat transfer between particle and gas phases tends to be rapid, it is often of less concern and will be treated in less detail. In contrast, the rate of heat transfer between the flui-

dized bed (particles and gas mixture) and submerged surfaces is often of engineering interest and will be treated in greater detail.

Gas fluidized beds operate in a number of different regimes, as described in Chapter 3 of this book. The mechanisms of heat transfer are significantly different for different fluidization regimes. Consequently, the correlations and models for prediction of heat transfer coefficients are regime specific and should only be used within their range of applicability. This chapter deals with the two regimes of gas fluidization most commonly encountered in industrial applications:

Dense bubbling fluidized beds (BFBs)
Fast circulating fluidized beds (FFBs)

For each regime, first a description is given of the general hydrodynamic characteristics. Then heat transfer characteristics are described and correlations/models useful for thermal design are presented. In many instances, the subject is still under research and no single design method has won general acceptance. For this reason, several alternative approaches are presented, for both BFBs and FFBs.

To use the information given in this chapter, one must first determine the applicable regime of fluidization. Kunii and Levenspiel (1991) and Grace (1986a) have presented excellent reviews of the available information, including graphs showing the operational boundaries for various fluidization regimes. The information presented by Grace can be replotted in a generalized map using a dimensionless particle diameter d_p^* and dimensionless velocity U^*, as shown in Fig. 1. These variables are discussed below and defined in Eqs. (4) and (7), respectively. Knowing the physical properties of gas and particles, one can then set the limits of superficial gas velocity for either bubbling or fast fluidization using the regime map of Fig. 1.

2 DENSE BUBBLING FLUIDIZED BEDS (BFBs)

2.1 Hydrodynamics

Consider the upward flow of a gas through a bed of packed particles. At some superficial velocity, the upward drag force exerted by the gas on the particles balances the downward body force of gravity. This is the condition of minimum fluidization, marking the transition between packed beds and fluidized beds. For particles with diameters in the range of 50 to 500 microns and densities in range of 0.2 to 5,000 kg/m³, fluidization usually can be achieved smoothly with increasing gas velocity. As denoted by Geldart (1973), such particles are classified as type A or B

and include the majority of particles encountered in fluidized beds applications. For such particles, gas velocities above the minimum fluidization velocity result in the occurrence of gas bubbles in the bed, wherein some fraction of the gas flows through the suspension of particles as a continuum phase, while the remaining fraction flows as discrete bubbles rising through the suspension. This is the regime commonly called dense bubbling fluidization. The upper limit of gas velocity for this regime is related to terminal velocity of the particles, beyond which interfacial drag becomes sufficient to entrain the particles out of the bed. Lines (a) and (b) in Fig. 1 indicate the usual limits of gas velocity for the operation of BFBs.

To establish the appropriate fluidization regime for any given application, one needs to calculate the minimum fluidization velocity and the terminal velocity of the bed particles. The superficial velocity of the gas for minimum fluidization (U_{mf}) can be calculated by solving the following equation for Re_{mf}:

$$Ar = 1.75\frac{(Re_{mf})^2}{\phi_s(\varepsilon_{mf})^3} + 150\frac{(1-\varepsilon_{mf})(Re_{mf})}{(\phi_s)^2(\varepsilon_{mf})^3} \qquad (1)$$

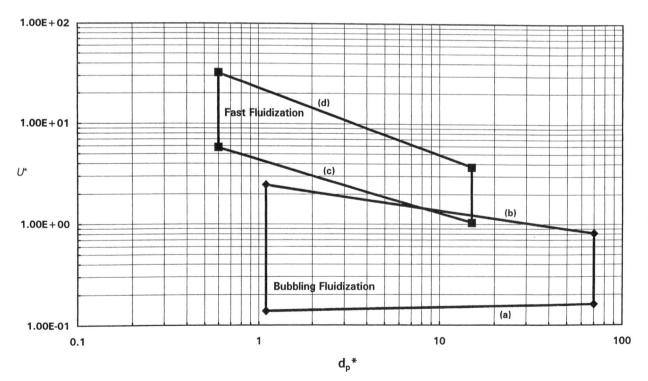

Figure 1 Map of bubbling and flat fluidization regimes.

where $\mathrm{Re_{mf}}$ is the Reynolds number at U_{mf},

$$\mathrm{Re_{mf}} = \frac{U_{mf} d_p \rho_g}{\mu_g} \tag{2}$$

where ϕ_s is the sphericity of the particles and ε_{mf} is the void fraction of the bed at minimum fluidization.

Ar is the Archimedes number, defined as

$$\mathrm{Ar} = \frac{g\rho_g(\rho_s - \rho_g)}{\mu_g^2} d_p^3 \tag{3}$$

Note that a dimensionless particle diameter can be defined as

$$d_p^* \equiv \mathrm{Ar}^{1/3} \tag{4}$$

If ϕ_s and ε_{mf} are not known for Eq. (1), an estimate of the minimum fluidization velocity can be obtained by the equation of Wen and Yu (1966):

$$\mathrm{Re_{mf}} = [1140 + 0.0408\mathrm{Ar}]^{1/2} - 33.7 \tag{5}$$

The terminal velocity (U_t) is given by the expression

$$U_t = \sqrt{\frac{4g d_p(\rho_s - \rho_g)}{3\rho_g C_D}} \tag{6}$$

where C_D is the drag coefficient for a single particle. In the case of near-spherical particles, over the range $1 < \mathrm{Re_t} < 1,000$, C_D is given by the relationship

$$C_D = 18.5\mathrm{Re_t}^{-3/5} \tag{6a}$$

where the Reynolds number at terminal velocity ($\mathrm{Re_t}$) is defined by Eq. (52). Substituting (6a) into Eq. (6), an explicit equation for the terminal velocity is obtained,

$$U_t = \left[0.072g \frac{d_p^{8/5}(\rho_s - \rho_g)}{\rho_g^{2/5} \mu_g^{3/5}} \right]^{5/7} \tag{6b}$$

A dimensionless gas velocity can be defined as

$$U^* \equiv \frac{U_g - U_{mf}}{U_t - U_{mf}} \tag{7}$$

with values of 0 and 1 at U_{mf} and U_t, respectively. The two dimensionless parameter d_p^* and U^* are used to delineate regimes of fluidization in Fig. 1. As seen from Fig. 1, some bubbling beds are operated at gas velocities slightly above the terminal velocity (i.e., $U^* > 1$), relying on various detrainment devices to reduce elutriation of particles out of the bed.

2.2 Heat Transfer

2.2.1 Particle Gas Transfer

Heat transfer between particles and gas in a fluidized bed may be compared to gas convection from a single fixed particle, and to gas convection from a packed bed of fixed particles. A common definition of the heat transfer coefficient can be used for all three cases, based on the surface area of a single particle (a_p),

$$h_p \equiv \frac{q}{a_p(T_p - T_g)} \tag{8}$$

where for spherical particles, $a_p = \pi d_p^2$. The mechanism for heat transfer becomes increasingly more complex as the process changes from single particles, to fixed beds, to fluidized beds. In the case of a single particle, the heat transfer mechanism is single-phase convection, governed by the boundary layer at the particle surface. In fixed beds, one has the added complication of particle packing and the complex gas flow pattern that results. In bubbling fluidized beds, the process is further complicated by the motion of the suspended particles and by macromixing caused by gas bubbles. In all three situations, the heat transfer coefficient is found to increase with increasing velocity of the gas relative to the particles. As may be expected, the heat transfer coefficient also increases with increasing thermal conductivity, increasing density, and decreasing viscosity of the gas.

The magnitude of the heat transfer coefficient between particles and gas in a bubbling fluidized bed is generally not large. Sample values of h_p for common applications are of approximate order 1 to 100 W/m^2-K. Nevertheless, the rate of heat transfer between particles and gas per unit bed volume is extremely high, due to the large interfacial surface area. Consequently, it is common to find that thermal equilibrium between particles and gas is reached quickly, within a short distance from the point of gas injection. In many designs, the isothermal condition is assumed throughout the particle–gas mixture. Concern with the actual rate of heat transfer between particles and gas arise primarily in situations where one phase or the other is an intense heat source, such as burning particles in a fluidized combustor.

In those situations wherein one is concerned with the rate of heat transfer between particles and gas, correlations for the effective heat transfer coefficients are required. While Eq. (8) gives an exact definition of the heat transfer coefficient h_p, its application is subject to some ambiguity. The problem arises from the model utilized to specify the particle and gas temperatures (T_p, T_g) within the fluidized bed. Recognizing that particle diffusivity is high in bubbling beds, it is common to treat the particles as being isothermal (well mixed) within the bed. Choices for the gas range from the perfectly mixed model to the plug-flow model. Values of h_p derived from any specific experiment, and the resulting correlation, would be different depending on the gas model selected. Application of the correlations must be consistent to the derivations and utilize the same model for gas flow. Choice of a gas-flow model is somewhat arbitrary, since the well-mixed and the plug-flow models both represent idealized limiting cases, and neither is an accurate representation for the bubbling fluidized bed. Kunii and Levenspiel (1969) pointed out that results based on the plug-flow model show less scatter and have the advantage of allowing analogous comparison to mass transfer between particles and fluid. Based on this reasoning, the data and correlations given below are all based on this model, with particles taken as isothermal within the bed and gas passing through the bed in perfect plug flow.

Experimental measurements of h_p have been made by various investigators, using both steady-state and transient techniques. Kunii and Levenspiel (1969, 1991) collected findings of 22 studies and presented the results in terms of a particle Nusselt number and the particle Reynolds number, respectively defined as

$$\text{Nu}_p \equiv \frac{h_p d_p}{k_g} \tag{9}$$

$$\text{Re}_p \equiv \frac{U_g d_p \rho_g}{\mu_g} \tag{10}$$

All of these measurements were obtained with gas of Prandtl number $\text{Pr} \cong 0.7$. The data are shown in Fig. 2, bounded between the two dashed curves. For comparison, Ranz's correlation (1952) for convection from a single sphere is also plotted on Fig. 2, as the dot–dash curve. The equation for Ranz's correlation is

$$\text{Nu}_p = 2 + 0.6\text{Re}_p^{0.5}\text{Pr}_g^{0.33} \tag{11}$$

It is seen that the Nusselt numbers for BFBs fall below those for convection from a single sphere, for Reynolds numbers less than 20. In fact, the magnitude of Nu_p for fluidized beds drops below the value of 2.0, which represents the lower limit of conduction heat transfer. The cause of this is the bubbling phenomenon. Low Reynolds numbers correspond to beds of fine particles (small d_p and U_g), wherein bubbles tend to be clouded with entrained particles. This diminishes the efficiency of particle–gas contact below that represented by idealized plug flow, resulting in reduced values of Nu_p. As particle diameter increases (coarse particle beds), bubbles are relatively cloudless and gas–particle contact improves. This is shown in Fig. 2 where the Nusselt numbers of fluidized beds are seen to increase with

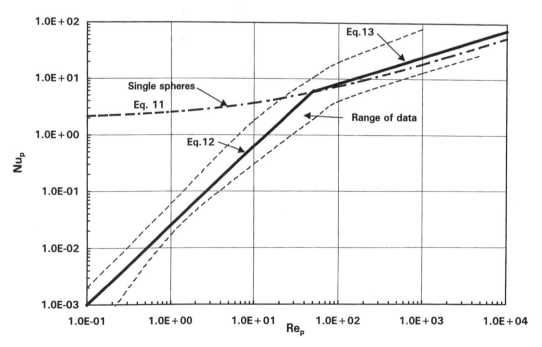

Figure 2 Nusselt numbers for particle–gas heat transfer in dense bubbling beds for $\text{Pr}_g = 0.7$.

increasing Reynolds number, approaching and slightly exceeding that of single particles at $Re_p > 60$.

The experimental data shown in Fig. 2 can be approximated by the following equations, representing the two straight lines in the figure:

$$Nu_p = 0.0282 Re_p^{1.4} Pr_g^{0.33} \quad \text{for} \quad 0.1 \le Re_p \le 50 \tag{12}$$

$$Nu_p = 1.01 Re_p^{0.48} Pr_g^{0.33} \quad \text{for} \quad 50 < Re_p \le 10^4 \tag{13}$$

It should be noted that the Prandtl number term in these two equations is somewhat speculative, since the experimental data did not cover a sufficient range for good correlation. Equations (12) and (13) include Pr to the 0.33 power, based on the expectation that dependence is similar to that for single spheres, as shown in Eq. (11). In applying heat transfer coefficients calculated by Eqs. (12) and (13), it is important to use a model that considers the particles to be well mixed and the gas to be in plug flow, in order to be consistent with the definition of h_p as discussed above.

2.2.2 Bed–Surface Transfer

It is sometimes necessary to cool or heat dense bubbling fluidized beds, and this is usually accomplished by the insertion of heat transfer tubes carrying cooling or heating fluids into the bed. Heat transfer occurs between the fluidized particle/gas medium (often referred to as the "bed") and the submerged tube surfaces (often referred to as "walls"). For this situation, one requires a heat transfer coefficient based on the surface area of the submerged wall:

$$h_w \equiv \frac{q}{a_w(T_b - T_w)} \tag{14}$$

where a_w is the submerged surface area, T_w is the temperature of the submerged surface, and T_b is the temperature of the particle/gas medium in the bed. For this analysis, any small temperature difference between the particles and the gas is neglected.

Due to its engineering importance, this bed-to-surface heat transfer coefficient has been measured by many investigators for various geometries and operating conditions (see Mickley and Trilling, 1949; Wender and Cooper, 1958; Vreedenberg, 1960; Botterill and Desai, 1972; Chen, 1976; Ozkaynak and Chen, 1980). Typical characteristics of h_w are:

 h_w is several times greater than the heat transfer coefficient for single-phase gas convection.

 h_w increases steeply as gas velocity exceeds the minimum fluidization velocity.

 h_w attains a maximum value at some specific velocity that depends on the particle size.

 Beyond the maximum point, h_w decreases slightly with further increase of gas velocity.

 h_w decreases with increasing particle size.

Figure 3 shows data of Chandran et al (1980) for the average heat transfer coefficients around a horizontal tube submerged in fluidized beds of spherical glass particles with various mean diameters. The coefficients are plotted against the dimensionless velocity defined by Eq. (7). It can be seen that all the characteristics noted above are displayed by these data.

With bubbling dense beds, it is recognized that mechanisms contributing to heat transfer at submerged surfaces include gaseous convection during times of bubble contact, particle conduction/convection during times of particle contact, and radiation in the case of high-temperature operation. The effective heat transfer coefficient is often represented as the sum of the coefficients for various contributing mechanisms:

$$h = f_l h_l + (1 - f_l)h_d + h_r \tag{15}$$

where

h_l = heat transfer coefficient for lean gas phase contact

h_d = heat transfer coefficient for dense particle phase contact

h_r = heat transfer coefficient for radiation

f_l = time fraction of contact by lean phase

This additive representation implies that the various mechanisms can be superimposed, weighted by the time fraction of lean or dense phase contacts at the heat transfer surface. While not exact, this is a useful approach.

Convective Heat Transfer. Most researchers concur that for bubbling fluidized beds, the gaseous convection is significantly enhanced by the presence of solid particles (Ozkaynak and Chen, 1980; Kunii and Levenspiel, 1991). Models often have focused on the dense/particle phase contribution, or on the total convective contribution,

$$h_c = f_l h_l + (1 - f_l)h_d \tag{16}$$

Several different approaches have been proposed for the estimation of h_c. Each approach has its advantages and disadvantages, and all involve some degree of empiricism. The following paragraphs summarize the various approaches. Attention is centered on vertical

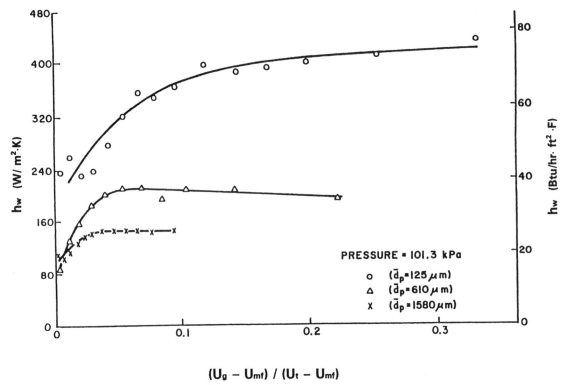

Figure 3 Heat transfer coefficients for horizontal tube in bubbling beds of glass spheres, fluidized by air at atmospheric pressure. (Data of Chandran et al., 1980.)

and horizontal tubes, since these are the geometries of most practical interest.

The most common approach assigns thermal resistance to a gaseous boundary layer at the heat transfer surface. The enhancement of heat transfer is then attributed to the scouring action of solid particles on the gas film, decreasing the effective film thickness. The works of Leva et al. (1949, 1952), Dow and Jakob (1951), Levenspiel and Walton (1954), Vreedenberg (1958), and Andeen and Glicksman (1976) utilized this approach. Correlations following this approach generally present a heat transfer Nusselt number in terms of the fluid Prandtl number and a modified Reynolds number, with either the particle diameter or the tube diameter as the characteristic length scale. Some useful examples of such correlations are given here.

Leva's correlation (1952) for vertical surfaces, for larger particles, is

$$\mathrm{Nu_c} \equiv \frac{h_c d_p}{k_g} = 0.525(\mathrm{Re_p})^{0.75} \tag{17}$$

Wender and Cooper's correlation (1958) for vertical tubes (SI units) is

$$\frac{h_c d_p}{k_g} = 3.51 \times 10^{-4} C_R (1 - \varepsilon) \mathrm{Re_p^{0.23}} \left(\frac{c_{pg}\rho_g}{k_g} \right)^{0.43}$$

$$\left(\frac{c_{ps}}{c_{pg}} \right)^{0.8} \left(\frac{\rho_s}{\rho_g} \right)^{0.66}$$

$$\text{for} \quad 10^{-2} \le \mathrm{Re_p} \le 10^2 \tag{18}$$

where C_R is a correction factor for tubes located away from the bed axis.

An empirical fit to the data of Vreedenberg (1952) gives the following function for C_R:

$$C_R = 1.07 + 3.04\left(\frac{r}{R_b} \right) - 3.29\left(\frac{r}{R_b} \right)^2 \tag{19}$$

where r is the radial position of the heat transfer tube and R_b is the radius of the bed.

Vreedenberg's correlations (1958) for horizontal tubes are

$$\frac{h_c D_t}{k_g} = 420 \left(\frac{\rho_s}{\rho_g} \text{Pr}_g \frac{\mu_g^2}{g \rho_s^2 d_p^3} \right)^{0.3} \text{Re}_D^{0.3} \tag{20a}$$

$$\text{for} \quad \frac{\rho_s}{\rho_g} \text{Re}_p \geq 2550$$

and

$$\frac{h_c D_t}{k_g} = 0.66 \text{Pr}_g^{0.3} \left(\frac{\rho_s (1 - \varepsilon)}{\rho_g \varepsilon} \right)^{0.44} \text{Re}_D^{0.44} \tag{20b}$$

$$\text{for} \quad \frac{\rho_s}{\rho_g} \text{Re}_p \leq 2050$$

where $\text{Re}_D \equiv D_t \rho_g U_g / \mu_g$ and $\text{Re}_p \equiv d_p \rho_g U_g / \mu_g$. In the intermediate range, Vreedenberg recommends using the average of the values calculated by Eqs. (20a) and (20b).

Andeen and Glicksman (1976) slightly modified Vreedenberg's correlation for large particles to account for different void fractions, resulting in the equation

$$\frac{h_c D_t}{k_g} = 900(1 - \varepsilon) \left(\frac{\rho_s}{\rho_g} \text{Pr}_g \left(\frac{\mu_g^2}{g \rho_s^2 d_p^3} \right) \right)^{0.3} \text{Re}_D^{0.3} \tag{21}$$

$$\text{for} \quad \frac{\rho_s}{\rho_g} \text{Re}_p \geq 2550$$

which gives the same result as Eq. (20a) for a void fraction of $\varepsilon = 0.53$. In this and other correlations that require a value for void fraction ε, it would have to be either known a priori or estimated from hydrodynamic models (see Chapter 3 of this book). For example, if the volume fraction of bubbles (δ) is determined from hydrodynamic models, the average bed void fraction (ε) is approximated by

$$\varepsilon \cong \delta + (1 - \delta)\varepsilon_{mf} \tag{22}$$

Equations (20a,b) and (21) calculate average values of the heat transfer coefficient around the circumferences of horizontal tubes. The experiments of Chandran et al. (1980) showed that the local heat transfer coefficients are actually significantly different at different circumferential positions. For small particles, the maximum heat transfer coefficient was found at the bottom and sides of the tube. The heat transfer coefficient at the top of the tube was noticeably smaller, implying the existence of a stagnant cap of solid particles that insulate that portion of the heat transfer surface. For the larger particles, a fairly uniform heat transfer coefficient is obtained, albeit with smaller magnitude than

the coefficient for small particles. Similar observations have been reported by Berg and Baskakov (1974) and Saxena et al. (1978). The detailed prediction of local heat transfer coefficients around horizontal tubes is beyond the scope of this book. The reader is referred to the article of Chandran and Chen (1985b), which presents a method for such calculations.

A second approach to modeling of the convective heat transfer coefficient considers combined gaseous and particle convection. Molerus et al. (1995) developed a semiempirical correlation of Nu_c, which is given by the following equation, for spherical (or nearly spherical) particles:

$$\frac{h_c l}{k_g} = \frac{0.125 \varepsilon_{s,mf}}{B_1 [1 + B_2 (k_g / 2 c_{ps} \mu_g)]} + 0.165 \text{Pr}_g^{1/3}$$
$$\left(\frac{\rho_g}{\rho_s - \rho_g} \right)^{1/3} \left(\frac{1}{B_3} \right) \tag{23}$$

where

$$l = \left(\frac{\mu_g}{\rho_s - \rho_g} \right)^{2/3} \left(\frac{1}{g} \right)^{1/3} \tag{24}$$

$$B_1 \equiv 1 + 33.3 \left[\left(\frac{U_e \rho_s c_{ps}}{U_{mf} g k_g} \right)^{1/3} U_e \right]^{-1} \tag{25}$$

$$B_2 \equiv 1 + 0.28 \varepsilon_{s,mf}^2 U_e U_{mf} \left(\frac{\rho_g}{\rho_s - \rho_g} \right)^{1/2} \left(\frac{\rho_s c_{ps}}{g k_g} \right)^{2/3} \tag{26}$$

$$B_3 \equiv 1 + 0.05 \left(\frac{U_{mf}}{U_e} \right) \tag{27}$$

$$U_e \equiv U_g - U_{mf} \tag{28}$$

the excess gas velocity above the minimum fluidization velocity.

Borodulya et al. (1991) developed a different correlation, which emphasizes the parametric effects of bed pressure and temperature as reflected by changes of gas and particle properties. Their equation is

$$\frac{h_c d_p}{k_g} = 0.74 \text{Ar}^{0.1} \left(\frac{\rho_s}{\rho_g} \right)^{0.14} \left(\frac{c_{ps}}{c_{pg}} \right)^{0.24} \varepsilon_s^{2/3}$$
$$+ 0.46 \text{Re}_p \text{Pr}_g \frac{\varepsilon_s^{2/3}}{\varepsilon} \tag{29}$$

where the first term represents particle convection and the second term represents gas convection. Equation (29) does not differentiate between vertical and horizontal surfaces and is claimed to be good for the following ranges of parameters: $0.1 < d_p < 4\,\text{mm}$; $0.1 < P < 10\,\text{MPa}$; $140 < \text{Ar} < 1.1 \times 10^7$.

A third approach is the so-called packet theory approach, which considers the heat transfer surface to be contacted alternately by gas bubbles and packets of closely packed particles. This leads to a surface renewal process wherein heat transfer occurs primarily by transient conduction between the heat transfer surface and the particle packets during their time of residence on the surface. Mickley and Fairbanks (1955) treated the particle packet as a pseudo-homogeneous medium with solid volume fraction $(1 - \varepsilon)_{pa}$ and thermal conductivity k_{pa}, obtaining the following expression for the average heat transfer coefficient due to particle packets:

$$h_{pa} = 2\left[\frac{k_{pa}\rho_s c_{ps}(1-\varepsilon)_{pa}}{\pi\tau_{pa}}\right]^{1/2} \quad (30)$$

where τ_{pa} is the root-square-average residence time for packets on the heat transfer surface,

$$\tau_{pa}\left[\frac{\sum \tau_i}{\sum (\tau_i)^{1/2}}\right]^2 \quad (31)$$

summed over a statistical number of packets. For small or medium size particles, this packet renewal mechanism accounts for the major portion of convective transfer so that one can take $h_c \cong h_{pa}$ in Eq. (16).

To utilize Eq. (30), one needs effective values of the packet properties in the vicinity of the heat transfer surface. In recognition of the higher voidage in the first layer of particles next to the surface, Chandran and Chen (1985) rewrote Eq. (30) as

$$h_{pa} = 2C\left[\frac{k_{pab}\rho_s c_{ps}(1-\varepsilon)_{pab}}{\pi\tau_{pa}}\right]^{1/2} \quad (32)$$

where k_{pab} and $(1 - \varepsilon)_{pab}$ are thermal conductivity and volume fraction in the bulk packet (far from the wall), respectively. C is a correction factor to account for the variation in packet voidage within the transient conduction zone near the wall. It takes the form

$$C = \exp\left[\frac{-a_1}{\text{Fo}^{a_2+a_3\ln\text{Fo}}}\right] \quad (33)$$

where

$$\text{Fo} = \text{Fourier modulus} = \frac{k_{pab}\tau_{pa}}{c_{ps}\rho_s(1-\varepsilon)_{pab}d_p^2}$$

$$a_1 = 0.213 + 0.117w + 0.041w^2$$
$$a_2 = 0.398 - 0.049w \qquad\qquad (33a)$$
$$a_3 = 0.022 - 0.003w$$
$$w = \ln(k_{pab}/k_g)$$

If specific information on void fraction in the bulk packet is lacking, $(1 - \varepsilon)_{pab}$ may be approximated as $(1 - \varepsilon)_{mf}$. The thermal conductivity of the bulk packet then can be estimated by the general model of Kunii and Smith (1960). The following equation is an empirical representation of Kunii and Smith's results, appropriate in the case of gas fluidized beds, with packets of loosely packed particles:

$$k_{pab} = k_g\left[\varepsilon_{pab} + \frac{(1-\varepsilon)_{pab}}{\phi_k + 2k_g/3k_s}\right] \quad (34)$$

where ϕ_k is given by

$$\phi_k = 0.305\left(\frac{k_s}{k_g}\right)^{-0.25} \quad \text{for} \quad 1 \le \frac{k_s}{k_g} \le 1,000 \quad (35)$$

Ozkaynak and Chen (1980) showed that this modified packet model could predict the measured convective heat transfer coefficient h_c with good success, when packet residence time τ_{pa} is known. When τ_{pa} is not known a priori, Kunii and Levespiel (1991) suggest that it can be estimated from bubble frequency (f_b) and the volume fraction of bubbles at the surface (δ_w),

$$\tau_{pa} = \frac{1 - \delta_w}{f_b} \quad (36)$$

See Chapter 3 of this book for methods of estimating bubble frequency and volume fraction. It especially needs to be kept in mind in the case of horizontal heat transfer tubes that bubbles that encounter the tube tend to be deflected to one side of the tube or the other. In such a situation, a position on the tube surface would be exposed to bubbles at approximately half of the frequency calculated for an unobstructed bed.

A fourth and different approach is that of Martin (1984), who utilized an analogy between particle motion in fluidized beds and kinetic motion of molecules in gases. From the kinetic theory of gases, Martin developed a model to account for thermal energy transport by particle motion across the boundary layer at surfaces. The resulting Nusselt number for convection was obtained as

$$\frac{h_d d_p}{k_g} = Z(1 - \varepsilon_b)\left(1 - e^{-N/C_c Z}\right) \tag{37}$$

where

$$Z \equiv \frac{\rho_s c_{ps} d_p w_p}{6k_g} \tag{38}$$

N = Nusselt number for heat transfer upon collision of particle with wall

C_c = dimensionless parameter inversely proportional to contact time of a particle on wall

w_p = average random particle velocity.

For the common situation of $k_s \gg k_g$,

$$N \cong 4(1 - \text{Kn})\ln\left(1 + \frac{1}{\text{Kn}}\right) - 4 \tag{39}$$

where Kn is the Knudsen number for the gas in the gap between the particle and the wall. From kinetic theory, for gases with accommodation constant approximately unity,

$$Kn = \frac{4k_g(2\pi RT/M_g)^{1/2}}{Pd_p(2c_{pg} - R/M_g)} \tag{40}$$

Assuming that the random particle velocity can be described by Maxwell's distribution, the average particle random velocity is derived as

$$w_p = \sqrt{\frac{g d_p(\varepsilon_b - \varepsilon_{mf})}{5(1 - \varepsilon_b)(1 - \varepsilon_{mf})}} \tag{41}$$

The dimensionless parameter C_c was left as an empirical parameter. From comparison with experimental data, Martin (1984b) suggested a value of ~ 2.6 for C_c. Using Martin's model, the convective coefficient h_c would thus be calculated by Eqs. (37)–(41).

All the design relationships given above, whatever the approach, treat the heat transfer wall as an isolated surface in the fluidized bed. In practice, the geometry of most interest is a bundle of tubes, composed of rows of cylindrical tubes placed either horizontally or vertically inside the fluidized bed. Wood et al. (1978) showed that the presence of a horizontal tube bundle can suppress slugging by breaking up the bubbles,

resulting in a decrease of heat transfer coefficients of up to 20%, at gas velocities that correspond to the slugging regime. At higher gas flow rates, this effect is reduced, owing to the uniform nature of the bubbling with or without tube banks. Tests made at atmospheric pressure further showed that heat transfer coefficients were independent of location within the tube bundle. For vertically placed tube bundles, Noe and Knudsen (1968) reported no difference in measured heat transfer coefficients for single or multiple tubes. In a tube bundle, the tubes are usually arranged in either a triangular or a square array, with sufficient spacing between neighboring tubes (pitch) to permit the mixing of the fluidized medium and the passage of gas bubbles. Following the information summarized by Pell (1990), it is suggested that the pitch-to-diameter ratio for tubes in a bundle be in the range of 2 to 4 or greater, in order to achieve good mixing within the fluidized bed.

Radiative Heat Transfer. For many years, there were contradictory opinions regarding the significance of radiative heat transfer in fluidized beds. Estimates of the radiative contribution to total heat transfer ranged from 2% to 30% for high temperature beds. Improved experimental measurements since 1980 have resolved much of this uncertainty. With direct measurements of both radiative heat flux and total heat flux in bubbling beds, Ozkaynak et al. (1983) showed that the radiant contribution to total heat transfer was less than 15% at bed temperatures below 500°C. At higher temperatures, radiant contribution increased linearly with bed temperature, becoming greater than 35% when bed temperature exceeded 800°C. This increase was not more rapid, i.e., proportional to the fourth power of absolute temperature, because convective/conductive heat transfer also increased with temperature as a result of the changes in the gas's physical properties.

The radiative heat transfer coefficient in Eq. (15) is defined as

$$h_r \equiv \frac{q_r}{a_w(T_b - T_w)} \tag{42}$$

where q_r is the radiative heat transfer rate on the wall surface.

Models to calculate q_r or h_r can be grouped into two approaches. The first and simpler approach is to use the Stefan–Boltzman equation for radiant exchange between opaque gray bodies. Treating the wall and bed medium as two opposing parallel surfaces with respective emissivities of e_w and e_b, the radiative heat transfer rate is

$$q_r = a_w e_{bw} \sigma (T_b^4 - T_w^4) \qquad (43)$$

where σ is the Stefan–Boltzman constant. The effective emissivity for the bed–wall combination (e_{bw}) is given by the expression

$$e_{bw} = \frac{1}{1/e_b + 1/e_w - 1} \qquad (44)$$

The resulting expression for the radiative heat transfer coefficient is then

$$h_r = \left(\frac{e_b e_w}{e_b + e_w - e_b e_w} \right) \frac{\sigma (T_b^4 - T_w^4)}{(T_b - T_w)} \qquad (45)$$

Equation (45) is applicable for tubes in a bundle with large pitch-to-diameter ratio, where particle/gas medium separates adjacent tubes. In using this equation, values of wall and bed emissivity (e_b, e_w) are required. Assuming that e_w is known for specific heat transfer surfaces, the need is then to estimate e_b. For BFBs of normal dimensions, e_b would be greater than the emissivity of individual particles and approach unity for almost all types of particles. The investigation of Ozkaynak et al. (1983), with experimental measurements of radiant heat flux, provides an indication of the effective bed/wall emissivity. For beds with particles similar to sand particles, e_{bw} was found to be in the range of 0.8 to 1.0 at bed temperatures greater than 700°C, where radiation is significant. It was also found that this effective emissivity is fairly insensitive to the superficial gas velocity. Thus a simple, approximate model for radiant heat transfer in bubbling fluidized beds is Eqs. (43)–(45), with e_b taken as approximately equal to 0.9.

The second type of approach takes a more mechanistic representation of the radiant transport process. These models recognize that radiant photons are emitted, absorbed, and scattered by the solid particles in the fluidized bed. Bhattacharya and Harrison (1976) utilized this approach to account for radiation exchange from one layer of particles with 25 neighboring layers. The particles were treated as an absorbing and emitting medium so that radiation was attenuated exponentially with distance. A more rigorous model was presented by Chen and Chen (1981), whereby the Mickley–Fairbanks packet model was modified to include simultaneous radiative and conductive heat transfer during alternating contact of the heat transfer surface by gas bubbles and particle packets. The gas phase was taken to be transparent to thermal radiation, while the particle packet was treated as a radiatively participative medium with absorption, emission, and scattering of photons. During bubble contact, radiation was directly exchanged between parallel surfaces representing the heat transfer wall and the boundary of the bubble. During packet contact, Hamaker's two-flux formulation of radiant transport was used to describe the absorption, scattering, and emission process within the packet:

$$\frac{dI}{dy} = -(A + S)I + SJ + A\sigma T^4$$
$$\frac{dJ}{dy} = (A + S)J - SI + A\sigma T^4 \qquad (46)$$

where

I, J = forward and backward radiant fluxes, respectively
A = volumetric absorption coefficient
S = volumetric scattering coefficient

The one-dimensional transient energy equation completed the system of equations

$$k_{pa} \frac{\partial^2 T}{\partial y^2} + A(I + J)2A\sigma T^4 = \rho_s c_{ps}(1 - \varepsilon)_{pa} \frac{\partial T}{\partial t} \qquad (47)$$

Radiant heat flux from the bed into the wall surface is then

$$\frac{q_r}{a_w} = J - I \qquad (48)$$

The radiation cross sections of the particle packets (A, S) are required. These would have to be measured, as done by Cimini and Chen (1987), or estimated from particle emissivity using the approach of Brewster and Tien (1982). Numerical solutions of this model have been shown to be in good agreement with experimental data (see Chen and Chen, 1981). While this approach is more rigorous, it requires significant effort and would be appropriate only for situations wherein radiative heat transfer is of special importance. In most design applications, the first approach, using Eqs. (43)–(45), is much easier and would be sufficient for the purpose.

Freeboard Heat Transfer. In some applications, e.g., fluidized combustors, heat exchanger tubes are located in the freeboard space above the bed, as well as within the bed. The coefficients h_w, h_c, and h_r, defined by Eqs. (15, 16, and 42), are also appropriate to represent the heat transfer process at the surface of such freeboard tubes. If the tube is placed so that it sees mostly bed material (not the vessel walls), Eqs. (44) and (45) may be used to estimate the radiative contribution. Evaluation of the convective coefficient is more difficult, since this must account for the phenomenon of particles splashing from the fluidized bed into the freeboard space, some to fall back into the bed and the rest being entrained out by exiting gas. The concentration of particles in the freeboard space decreases with increasing elevation above the bed, increasing particle size, and decreasing gas velocity. The presence of these particles enhances gaseous convection so that the magnitude of h_c varies in a similar manner.

Experimental data and models for the heat transfer coefficients in freeboards space are limited. Biyikli et al. (1983) measured heat transfer coefficients on horizontal tubes in freeboards of fluidized beds with particle mean diameters ranging from 275 to 850 μm. These investigators discovered that data for particles of different sizes and materials could be unified into a family of curves by plotting the normalized ratios of heat transfer coefficients against dimensionless velocity:

$$h_c^* \equiv \frac{h_c - h_g}{h_{cb} - h_g} \quad \text{versus} \quad U^* \equiv \frac{U_g - U_{mf}}{U_t - U_{mf}} \quad (49)$$

where

h_c = the convective coefficient in freeboard
h_g = the single-phase gas convective coefficient at velocity U

h_{cb} = the convective coefficient for a surface submerged in the bed

Standard correlations may be used to obtain the single-phase gas convective coefficient, h_g. For the common case of horizontal tubes, the correlation of Douglas and Churchill (1956) is recommended:

$$\frac{h_g D_t}{k_g} = 0.46\left(\frac{D_t \rho_g U_g}{\mu_g}\right)^{1/2} + 0.00128\left(\frac{D_t \rho_g U_g}{\mu_g}\right)$$

$$\text{for} \quad \left(\frac{D_t \rho_g U_g}{\mu_g}\right) \geq 50 \quad (50)$$

In Eq. (50), gas properties should be evaluated at film temperature (average of bed and wall temperatures) in cases where the temperature difference is large.

Figure 4 reproduces the results of Biyikli et al (1983), showing that h_c^* is independent of particle material and size. h_c^* is dependent only on U^* and the elevation H in freeboard, as measured from the surface level of the collapsed (unfluidized) bed. Because the particle splashing process is affected by many parameters (e.g., distribution of particle sizes, bubble coalescence, etc.), it is difficult to find a general correlation of h_c for the freeboard region. As a first estimate, the following empirical correlation, based

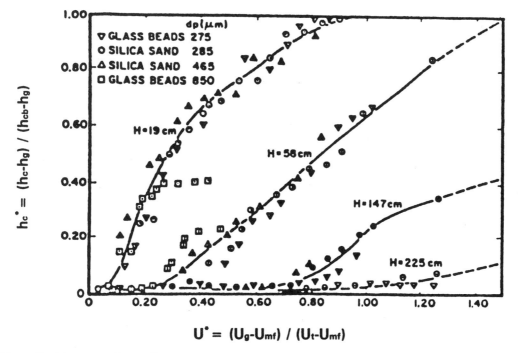

Figure 4 Normalized heat transfer coefficients for horizontal tube in freeboard of bubbling beds. H measured from level of collapsed (packed) bed. (Data of Biyikli et al., 1983.)

on data of Biyikli et al. (1983), may be utilized for horizontal tubes in freeboard:

$$h_c^* = 1.3e^{-1.53H/U^*} \qquad (51)$$

with the constraint that $0 \leq h_c^* \leq 1.0$. Note that the constant 1.53 has the dimension of (1/m), with H measured in meters from the level of the collapsed (unfluidized) bed. Equation (51) can be used also for tube bundles if the bundle has open spacing, i.e., on a pitch-to-diameter ratio greater than 3.

3 FAST CIRCULATING FLUIDIZED BEDS (FFBs)

3.1 Hydrodynamics

When the gas velocity in a fluidized bed exceeds the terminal velocity of the bed particles, upward entrainment of particles out of the bed occurs. To maintain solid concentration in the fluidized bed, an equal flux of solid particles must be injected at the bottom of the bed as makeup. Operation in this regime, with balanced injection and entrainment of particles, is termed fast or circulating fluidization. Figure 1 presents a generalized map of this fast fluidization regime, in terms of the dimensionless particle diameter and gas velocity, d_p^* and U^*. Lines (c) and (d) indicate the lower and upper bounds of gas velocity, respectively, for the usual FFB operation. It is seen that for Geldart types A and B particles, fast fluidization permits superficial gas velocities an order of magnitude greater than that for bubbling fluidization. In the case of catalyst particles (density of ~ 1800 kg/m^3), superficial gas velocities approaching 10 m/s may be encountered. Heat transfer in fast fluidized beds is strongly affected by the volumetric concentration of solid particles as they are transported by the gas through the bed. For this reason, we start with a review of the hydrodynamic characteristics of FFBs.

In fast fluidized beds, solid particles are fed into the bed at the bottom, to be entrained by the upward flowing gas. The solid and gas phases pass through the bed in cocurrent two-phase flow. Since the particles enter with little vertical velocity, they are accelerated by shear drag of the gas, gaining velocity by momentum transfer from the high-speed gas. This hydrodynamic development results in the volumetric concentration of solids decreasing with axial elevation along the length of the bed. Measurements obtained by Herb et al. (1989) and Hartige et al. (1986) indicate that many meters of axial length may be required to approach fully developed flow conditions (i.e., constant solids concentration). The axial height required for this hydrodynamic development has been found to increase with increasing particle size, solid mass flux, and bed diameter. Cross-sectional-averaged solid concentrations typically range from ~ 15 volume % at bed bottom to less than 3% at bed top. This is significantly less than the solids concentrations of 30–40% commonly encountered in bubbling fluidized beds.

In addition to the axial variation of solid concentration, fast fluidized beds have the complication of significant radial (horizontal) variation in solid concentration. Radial profiles of solid volume fraction measured by Beaude and Louge (1995) indicate there is increasing nonuniformity of solid concentration with increasing solid mass flux. At the higher mass fluxes, local solid volume fraction adjacent to the bed wall approach a magnitude of 30%, in contrast to solid fractions of 1–3% near the bed centerline. Herb et al. (1989b) and Werther (1993) noted that such radial distributions could be normalized as a general similarity profile for various operating conditions by utilizing the dimensionless ratio of local solid concentration to cross-sectional-averaged solid concentration.

Solid mass flow flux and velocity also vary across the radius of fast fluidized beds. Experimental measurements obtained by Herb et al. (1992) show that while local solid fluxes are positive upward in the core of the bed, they can become negative downward in the region near the bed wall. The difference between core and wall regions becomes increasingly greater as total solid mass flux increases. The downward net flow of solid in the region near the bed wall has significance for heat transfer at the wall.

Another hydrodynamic complication found in fast fluidized beds is the tendency for particles to aggregate into strands or clusters, as reported by Horio et al. (1988) and Chen (1996). The concentration of solid particles in such clusters is significantly greater than in the bed itself, and it increases with increasing radial position and with increasing total solid flux (see Soong et al., 1993). This characteristic also directly affects heat transfer at walls.

The above is a very condensed discussion of the hydrodynamic characteristics of fast fluidized beds. As presented below, heat transfer is strongly dependent on the time-averaged local concentration of solid particles and is therefore influenced by these hydrodynamic characteristics. Almost all the heat transfer models require information on solid concentration. The reader is referred to Chapter 19 of this book for more detailed discussion of the hydrodynamics, and for models for predicting solid concentrations in fast circulating fluidized beds.

3.2 Heat Transfer

3.2.1 Particle Gas Transfer

There is little information on the process of heat transfer between particles and gas in fast fluidized beds. One reason for this situation is the expectation that the heat transfer rates are high and therefore of little concern. The smaller concentration of particles (in comparison to bubbling dense beds) implies less intense heat source concentration in the cases of exothermic reactions, and this also alleviates potential concern. At the time of this writing, there is no established model or correlation for estimation of the heat transfer coefficient between particles and gas during fast fluidization.

Given this situation, one possible approach is to treat the particles as if they were individually suspended by the upward flowing gas. With this simplification, the relative velocity between particle and gas would be equal to the terminal velocity U_t, as given by Eq. (6). The heat transfer coefficient based on particle surface area, as defined by Eq. (8), would be given by Ranz's correlation (Eq. 11). In applying this correlation for suspended particles, the particle Reynolds number should be based on the terminal velocity rather than the superficial gas velocity,

$$\text{Re}_t \equiv \frac{d_p \rho_g}{\mu_g} U_t \qquad (52)$$

By using U_t rather than U_g, the effective Reynolds number is reduced so that the calculated Nusselt numbers tend toward the conservative conduction limit of 2. Obviously this approach greatly simplifies the actual phenomena, and disregards the complications of particle acceleration, clustering, and downflow at walls. Estimates of the particle/gas heat transfer coefficient (h_p) from Eqs. (11) and (52) should be considered only as approximate.

3.2.2 Bed–Surface Transfer

Similar to bubbling beds, fast fluidized beds often require cooling or heating in order to maintain a steady thermal state. Unlike bubbling beds, the use of submerged heat exchanger tubes is not common practice in FFBs, because of concern about tube erosion by the faster gas/particle flows. The usual practice is to use walls of the FFB as the heat exchange sur-

faces, by mounting cooling/heating tubes in the walls. Additionally, "membrane walls" of vertical tubes may be placed within the fast fluidized bed, to increase total heat transfer area. In either case, engineering design is concerned with the heat transfer coefficient for vertical surfaces in contact with the gas/particle medium in the fast fluidized bed, h_w as defined by Eq. (14).

Experimental measurements of this heat transfer coefficient have been obtained by many investigators, including Kiang et al., 1976; Furchi et al., 1988; Dou et al., 1991, 1993; Ebert et al., 1993; and Wirth, 1995. Parameters that have been found to affect h_w include gas velocity, solids flow rate (mass flux), mean particle diameter, axial location of the heat transfer surface along the height of the fluidized bed, radial location of the heat transfer surface across the bed, and physical dimension of the heat transfer surface. General characteristics noted from the collected data include

- h_w is higher than that for gas convection at the same velocity but lower than that found in dense bubbling beds.
- h_w decreases with increasing particle diameter.
- h_w increases with increasing solid mass flux.
- h_w increases with increasing radial position toward the bed wall.
- h_w decreases with increasing elevation along the bed.
- At low elevations, h_w decreases with increasing gas velocity.
- At higher elevations, h_w increases with increasing gas velocity.
- h_w is greater for small heat transfer surfaces, than that for large surfaces.

Most of these parametric effects parallel those for solid concentration and therefore imply a strong dependence of h_w on local concentration of solid particles.

Many different models and correlations have been proposed for the prediction of the heat transfer coefficient at vertical surfaces in FFBs. At time of this writing, no single correlation or model has won general acceptance. The following discussion presents a summary of some potentially useful approaches. It is helpful to consider the total heat transfer coefficient as composed of convective contributions from the lean-gas phase and the dense-particle phase plus thermal radiation, as defined by Eqs. (15) and (16). All correlations based on ambient temperature data, where thermal radiation is negligible, should be considered to represent only the convective heat transfer coefficient h_c.

Convective Heat Transfer. The simplest correlations for the convective heat transfer coefficient empirically relate h_c to the solids concentration, expressed in terms of the mixture density for the gas/solid medium in the bed (ρ_b),

$$\rho_b \equiv \rho_s(1 - \varepsilon)_b + \rho_g \varepsilon_b = \rho_s \varepsilon_{s,b} + \rho_g \varepsilon_b \quad (53)$$

Since gas density is usually negligible compared to solid density, ρ_b is essentially equal to the mass concentration of solid per unit mixture volume. Some examples of this type of correlation, for convective heat transfer at vertical walls of fast fluidized beds, are given below. Due to accumulating evidence that the vertical size of the heat transfer surface affects the average h_c, the correlations are divided with regard to short or long surfaces.

For short surfaces of $L_h \leq 0.5\,\text{m}$, the correlation of Wen and Miller (1961) is

$$\frac{h_c d_p}{k_g} = \left(\frac{c_{ps}}{c_{pg}}\right)\left(\frac{\rho_b}{\rho_s}\right)^{0.3}\left(\frac{U_t}{g d_p}\right)^{0.21}\text{Pr}_g \quad (54)$$

and the correlation of Divilio and Boyd (1993) is

$$h_c = 23.2(\rho_b)^{0.55} \quad (55)$$

For long surfaces of $L_h > 0.5\,\text{m}$, the correlation of Werdmann and Werther (1993) is

$$\frac{h_c d_p}{k_g} = 7.46 \times 10^{-4}\left(\frac{D_b \rho_g U_g}{\mu_g}\right)^{0.757}\left(\frac{\rho_b}{\rho_s}\right)^{0.562} \quad (56)$$

and the correlation of Fraley et al. (1983) is

$$h_c = 2.87(\rho_b)^{0.9} \quad (57)$$

To use any of these correlations, one needs the value of the cross-sectional-averaged bed density ρ_b. Methods for estimating this parameter are discussed in Chapter 19 of this book. As a precautionary note, Dou et al. (1991) have shown the heat transfer coefficient to be dependent on local solid concentration in the vicinity of the surface (ρ_m), rather than the cross-sectional-averaged concentration (ρ_b). Therefore the above correlations for convective coefficient at the wall of FFBs can be justified only if ρ_b and ρ_m are related by a generic similar profile. The existence of such a similar profile has been suggested by Herb et al. (1989b) and Werther (1993). Examination of Eqs. (54) and (57) also indicates significant differences regarding the parameters appropriate for the correlation of h_c. Particle size (d_p), bed diameter (D_b), particle terminal velocity (U_t), heat capacity of solid (c_{ps}), and gas thermal properties (μ_g, k_g, ρ_g, c_{pg}) are incorporated into some correlations but not in others. Given this situation, the

user is cautioned to refer to the original references and apply a given correlation only within the range of parameters covered by that correlation's experimental data base.

In contrast to the purely empirical correlations given above, an alternate modeling approach is suggested by Molerus (1993), Wirth (1993, 1995), and Molerus and Wirth (1997). These researchers reasoned that the wall-to-bed heat transfer in FFBs is governed by the fluid flow immediately near the wall. They assumed existence of a thin layer near the wall with low solids concentration, providing significant thermal resistance between the wall and the falling strands of relatively high solids concentrations. Wirth and Molerus report that a dimensionless pressure-drop number that measures bed-averaged solids concentration and the Archimedes number are sufficient to characterize this gas–particle flow and to correlate the convective heat transfer coefficient. At low Archimedes numbers, heat transfer occurs primarily by gas conduction. At higher Archimedes numbers, gaseous conduction and convection both contribute to heat transfer. The correlation suggested by Wirth (1995) simply assumed that conduction and convection were additive, which results in the equation

$$\frac{h_c d_p}{k_g} = 2.85\left[\frac{\Delta P/\Delta L_b}{g(\rho_s - \rho_g)(\varepsilon_{s,mf})}\right]^{0.5} + 3.28 \times 10^{-3}\text{Re}_{pa}\text{Pr}_g \quad (58)$$

where Re_{pa} = Reynolds number for falling cluster packets = $\rho_g d_p V_{pa}/\mu_g$ and V_{pa} = velocity of falling cluster packets. Based on physical reasoning, Wirth (1990) obtained the following relationship for the Reynolds number of falling clusters:

$$\text{Re}_{pa} = \frac{2(1 - \phi)^2\text{Re}_{mf} + \phi\text{Re}_t}{2[\phi + (1 - \phi)\varepsilon_{mf}]}\sqrt{1 - 4\left(\frac{F_1}{F_2}\right)} \quad (59)$$

where

$$F_1 \equiv \lfloor(1 - \phi)^4\text{Re}_{mf}^2 + \phi(1 - \phi)^2\text{Re}_{mf}\text{Re}_t - 1891(1 - \varepsilon_{mf})\phi^3(1 - \phi)\text{Ar}\rfloor \quad (60)$$

$$F_2 \equiv [2(1 - \phi)^2\text{Re}_{mf} + \phi\text{Re}_t]^2 \quad (61)$$

$$\phi \equiv 1 - \frac{2.3\Delta P/\Delta L_b}{g(1 - \varepsilon_{mf})(\rho_s - \rho_g)} \quad (62)$$

In contrast to the above correlations for total convective coefficient h_c, some models estimate the separate heat transfer contributions from lean and dense phases as represented by h_l and h_d in Eq. (15). It is generally believed that dense phase convection is the dominant mechanism, and lean phase convection is a relatively unimportant mechanism. This simplification is valid where local solids concentration is greater than 5% by volume, as at the bottom regions of FFBs, close to vertical walls, or in operations with high mass flux of solid flow. Measurements of Ebert et al. (1993) indicate that the lean phase convection can become significant, contributing up to 20% of total convective heat transfer, when the average solids concentration becomes less than 3%, as is often found in upper regions of fast fluidized beds. Rather than neglecting lean phase convection, it is better to include an estimate of h_l in total convection. A conservative estimate for h_l is the Dittus–Boelter correlation for single phase gas convection,

$$\frac{h_l D_b}{k_g} = 0.023 \left(\frac{D_b \rho_g U_g}{\mu_g}\right)^{0.8} (\text{Pr}_g)^{0.33} \tag{63}$$

There is evidence that this approach underestimates the lean phase convection, since gaseous convection is enhanced by suspended particles. Lints (1992) suggested that this enhancement could be partially taken into account by increasing the gas thermal conductivity in Eq. (63) by a factor of 1.1.

To estimate the contribution of dense phase convection, one approach considers the heat transfer surface to be contacted periodically by clusters or packets of closely packed particles. Experimental evidence supporting this concept is reported by Dou et al. (1993). Their dynamic measurements at the wall of a fast fluidized bed found significant correlation between instantaneous solids concentration and heat transfer coefficient, indicating a direct dependence of the dense phase heat transfer on cluster contact. This approach leads to a surface renewal model whereby the heat transfer to/from the dense phase occurs by transient conduction between the heat transfer surface and the particle clusters during their time of residence at the surface. This is analogous to the packet theory for heat transfer in bubbling beds, and so Eqs. (30)–(35) can be utilized to calculate the dense phase convective coefficient by taking $h_d = h_{pa}$. Application of this cluster model requires information on solids concentration in the cluster ($\varepsilon_{s,pab}$) and the effective thermal conductivity of the cluster (k_{pa}). Unlike the emulsion phase in BFBs, which have solids fractions close to that at minimum fluidization ($\varepsilon_{s,mf}$), clusters in FFBs have solids fractions significantly less than $\varepsilon_{s,mf}$. Experimental data for $\varepsilon_{s,pab}$ are very sparse. Some lim-

ited data may be found in the articles of Wu, 1989; Louge et al., 1990; Lints, 1992; and Soong et al., 1993. From the available data, Gu and Chen (1998) developed the following correlation to estimate the solids concentration in clusters ($\varepsilon_{s,pab}$) as a function of the local solids concentration in the bed medium ($\varepsilon_{s,m}$),

$$\varepsilon_{s,pab} = 0.57 \left[1 - \left(1 - \frac{\varepsilon_{s,m}}{0.57}\right)^{3.4}\right] \tag{64}$$

Once the solids concentration in clusters is determined, the thermal conductivity of clusters (k_{pa}) can be estimated by the expression of Chandran and Chen (1985) as

$$k_{pa} = k_{pab} C^2 \tag{65}$$

where k_{pab} is given by Eq. (34) with

$$\varepsilon_{pab} = (1 - \varepsilon_{s,pab}) \tag{66}$$

and C as given by Eq. (33). The result is same as Eq. (32), written for the dense-phase convective coefficient in FFB,

$$h_d = 2C \left[\frac{k_{pab} \rho_s c_{ps} \varepsilon_{s,pab}}{\pi \tau_{pa}}\right]^{1/2} \tag{67}$$

This surface renewal model also requires the average residence time of clusters on the heat transfer surface (τ_{pa}). Unfortunately, there is insufficient information to correlate this residence time directly with operating parameters (e.g. U_g, D_p, ρ_s, G_s, L_b). If the average cluster falls down along the heat transfer surface with a velocity V_{pa}, over a contact length L_c, the residence time for that cluster would be the ratio L_c/V_{pa}. This would be the effective mean value of τ_{pa} on a large heat transfer surface. Cluster velocities measured by Hartge et al. (1988), Horio et al. (1988), Lints (1992), and Soong et al. (1995) range from 0.3 to 3 m/s. In absence of specific information, it is suggested that Wirth's correlation, Eq. (59), be used to calculate the Reynolds number for falling clusters (Re_{pa}), from which V_{pa} can be obtained from the definition of Re_{pa}.

Measurements of contact length L_c are lacking, but an inference can be drawn from the experiments of Burki et al. (1993) in fast fluidized beds of sand and catalyst particles. They found that local heat transfer coefficients on a vertical wall decreased with distance down the heat transfer surface, over an "entry length" of approximately 0.3–0.5 m. This result implies that L_c is approximately 0.4 m, for their operating conditions. At this stage of development, we can only take an approximate estimate for the cluster contact time:

$$\tau_{pa} = \frac{L_c}{V_{pa}} = \frac{L_c}{Re_{pa}}\left(\frac{\rho_s d_p}{\mu_g}\right) \qquad (68)$$

with $L_c \cong 0.4\,m$ and Re_{pa} from Eq. (59). Clearly this will require additional research. It should be noted that in cases where the vertical length of the heat transfer surface is less than the contact length ($L_b < L_c$), L_h should be substituted for L_c in Eq. (68).

Finally, to utilize this approach in the prediction of the total convective heat transfer coefficient h_c, one needs to determine the average fraction of wall surface covered by the dense phase (clusters). By definition this is equal to the fraction of time that a spot on the heat transfer surface is covered by the dense phase, and denoted as $(1 - f_l)$ in Eq. (16). Measurements of $(1 - f_l)$ have been obtained by Wu et al. (1989), Louge et al. (1990), Rhodes et al. (1992), and Dou et al. (1993). Figure 5 plots $(1 - f_l)$ against the cross-sectional-averaged bed concentration of solids ($\varepsilon_{s,b}$), showing the range of experimental data. The following equation gives an empirical fit to these data and may be used to estimate $(1 - f_l)$ for vertical walls in fast fluidized beds.

$$1 - f_l = 1 - 0.018\left(\frac{1}{\varepsilon_{s,b}}\right)^{0.6} \qquad (69)$$

$$\text{for} \qquad 0.002 \le \varepsilon_{s,b} \le 0.3$$

In summary, the various variables required for the estimation of the convective heat transfer coefficient (h_c) by the surface renewal model, for vertical walls in FFBs, may be determined by the equations listed in following tabulation:

Variable	Equation No.
h_c	16
h_l	63
h_d	67
k_{pab}	34
$\varepsilon_{s,pab}$	64
τ_{pa}	68
f_l	69

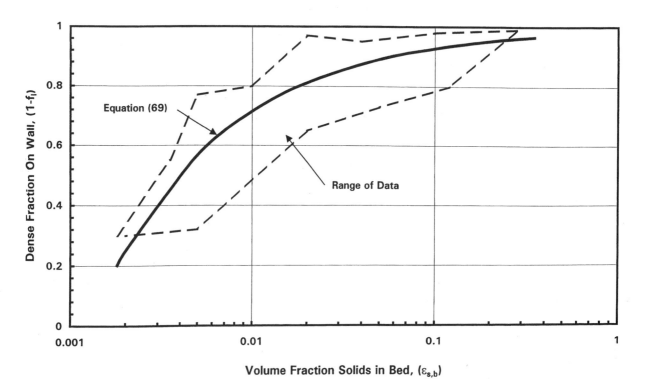

Figure 5 Fraction of wall surface in FFBs contacted by dense phase.

Radiative Heat Transfer. Models for estimating the radiative heat transfer in fast fluidized beds parallel those proposed for dense bubbling beds, as reviewed above. From a phenomenological point of view, the suspension of particles in gas acts as a radiatively participative medium, with absorption, scattering, and emission of radiant energy. Only a few researchers have attempted to model this physics in FFBs. Chen et al. (1988) used the two-flux radiative model of Eqs. (46)–(48) to analyze simultaneous convection and radiation in high temperature FFBs. Turbulent convection was combined with discrete radiative fluxes in the three transport equations, including a term for volumetric heat generation. Sample results for fluidized combustor conditions predicted a significant interdependence of radiation and convection. This conclusion was verified by the experimental measurements of Han (1992), who showed that the contribution of radiative heat transfer was strongly effected by bed-averaged solid concentration, decreasing with increasing solid mass flux and decreasing gas velocity. While this participative media model is mechanistically sound, it requires information on the absorption and scattering coefficients for the particle/gas medium and is relatively difficult to apply. Its use is recommended only for cases where detailed estimates of local radiative transfer is desired; in such cases the reader is referred to the papers of Chen et al. (1988) and Radauer et al. (1996) for the necessary equations and methods of solution.

The more common approach for engineering design is to treat the particle–gas suspension as an equivalent gray surface parallel to the heat transfer surface; see, for example, Palchonok et al. (1995), Anderson and Lechner (1993), and Mahalingam and Kolar (1991). In this approach, Eqs. (42)–(45) are used to calculate the radiative heat flux (q_r) and the coefficient (h_r). It is necessary first to estimate the effective bed emissivity (e_b) in FFBs. Grace (1986b) suggests that the emissivity of the particle–gas suspension can be approximated as

$$e_b = 0.5(1 + e_s) \qquad (70)$$

where e_s is the surface emissivity of the solid particles. This is a reasonable approximation for large FFBs, where the mean free path of radiation photons is short compared to bed dimensions, i.e., the gas/particle medium is optically "thick." Experimental measurements reported by Han et al. (1996) and Han (1992) indicate that when the particle/gas medium is not optically thick, the bed emissivity is dependent on the solid

concentration and bed dimension, as well as on particle emissivity. From the theory of radiant transmission through absorbing media, the effective emissivity of the bed would be expected to be of functional form

$$e_b = 1 - \exp\left(\frac{-C_e \varepsilon_{s,b} e_s D_b}{d_p}\right) \qquad (71)$$

where $\varepsilon_{s,b}$ is the solid volume fraction in the bed, D_b is the bed diameter between opposing heat transfer surfaces, and C_e is a dimensionless parameter. Figure 6 plots the range of data measured by Han (1992) for sand particles of $e_s = 0.7$, in an FFB of $D_b = 5\,\text{cm}$. A value of 0.95 for C_e in Eq. (71) is seen to fit the data and is recommended for design calculations.

One cautionary note should be kept in mind when using Eqs. (42)–(45) and (71) to calculate radiative heat transfer in FFB. The bed's absolute temperature T_b is normally assumed to be uniform across the bed and is used as the source or sink temperature in Eqs. (42) and (43). This assumption may be inappropriate in those cases in which a dense annular region of particles shields the FFB wall from the bulk bed. In such situations, it is the average temperature of the particles in the annular layer that should be taken as the source/sink temperature for calculation of radiant heat flux to/from the wall. This requires a mass and heat balance analysis for the material flowing in the annulus, and the reader is referred to Chapter 19 for necessary hydrodynamic models.

4 EXAMPLES

4.1 Bubbling Fluidized Bed

Consider a bed of diameter 0.203 m and loaded with spherical glass beads of mean diameter 125 µm. We wish to fluidize the particles with ambient air, at a superficial gas velocity of $U_g = 0.16\,\text{m/s}$. Properties of the two phases are

	Gas	Particles
Density, ρ (kg/m^3)	1.18	2,480
Heat capacity, c_p (kJ/kg K)	1.00	0.753
Thermal conductivity, k (W/m K)	.0262	0.890
Viscosity, μ (kg/m s)	1.85×10^{-5}	
Pr	0.71	

According to Geldart's classification, these particles are of type B. The void fraction at minimum fluidization is found to be $\varepsilon_{mf} = 0.42$.

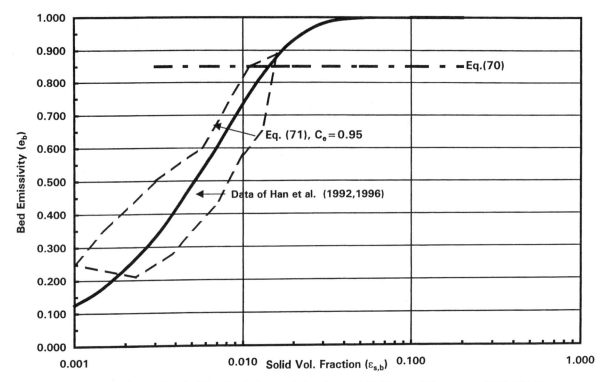

Figure 6 Effective emissivities for fast fluidized beds for particles of $\varepsilon_s = 0.7$. (Data of Han et al., 1992, 1996.)

4.1.1 Fluidization Regime

At the intended operating condition, would the bed fluidize in the bubbling regime?

The answer is found by spotting the operating point on the map of fluidization regimes plotted in Fig. 1. Calculating the necessary parameters, we find the Archimedes number from Eq. (3):

$$\mathrm{Ar} = \frac{g\rho_g(\rho_s - \rho_g)}{\mu_g^2}d_p^3$$

$$= \frac{(9.81)(1.18)(2480 - 1.18)(125 \times 10^{-6})^3}{(1.85 \times 10^{-5})^2} = 164$$

The Reynolds number at minimum fluidization is found by solving Eq. (1):

$$\mathrm{Ar} = 1.75\frac{(\mathrm{Re}_{mf})^2}{\phi_s(\varepsilon_{mf})^3} + 150\frac{(1 - \varepsilon_{mf})(\mathrm{Re}_{mf})}{(\phi_s)^2(\varepsilon_{mf})^3}$$

$$164 = 1.75\frac{(\mathrm{Re}_{mf})^2}{(1.0)(0.42)^3} + 150\frac{(1 - 0.42)(\mathrm{Re}_{mf})}{(1.0)^2(0.42)^3}$$

$$\mathrm{Re}_{mf} = 0.139$$

from which we obtain the minimum fluidization velocity by Eq. (2),

$$U_{mf} = \mathrm{Re}_{mf}\left(\frac{\mu_g}{\rho_g d_p}\right) = 0.139\left(\frac{1.85 \times 10^{-5}}{1.18(125 \times 10^{-6})}\right)$$

$$= 0.0175\,\mathrm{m/s}$$

The terminal velocity is obtained from Eq. (6b),

$$U_t = \left[0.072g\frac{d_p^{8/5}(\rho_s - \rho_g)}{\rho_g^{2/5}\mu_g^{3/5}}\right]^{5/7}$$

$$= \left[0.072 \times 9.81\frac{(125 \times 10^{-6})^{8/5}(2480 - 1.18)}{(1.18)^{2/5}(1.85 \times 10^{-5})^{3/5}}\right]^{5/7}$$

$$= 0.730\,\mathrm{m/s}$$

and the dimensionless velocity is, by definition, from Eq. (7),

$$U^* \equiv \frac{U_g U_{mf}}{U_t - U_{mf}} = \frac{0.160 - 0.0175}{0.730 - 0.0175} = 0.20$$

The dimensionless particle diameter is, from Eq. (4),

$$d_p^* \equiv \mathrm{Ar}^{1/3} = 164^{1/3} = 5.46$$

This locus is within the bubbling regime on the map of Fig. 1, so the proposed operation will be a dense bubbling fluidized bed.

4.1.2 Particle–Gas Heat Transfer

If the gas enters the bed 30K cooler than the temperature of the particles in the bed, how long a distance must the gas travel within the bed before that difference is reduced to 0.5K? At the superficial gas velocity of $U_g = 0.16$ m/s, take the average void fraction in the bubbling bed to be $\varepsilon_b = 0.585$, and the average solid fraction to be $\varepsilon_{s,b} = 0.415$.

The heat transfer coefficient between particle and gas is given by Eq. (12) or Eq. (13), depending on the particle Reynolds number, which is defined by Eq. (10),

$$Re_p \equiv \frac{U_g d_p \rho_g}{\mu_g} = \frac{0.16 \times 125 \times 10^{-6} \times 1.18}{1.85 \times 10^{-5}} = 1.28$$

For this small Re_p Eq. (12) is appropriate

$$h_p = \frac{k_g}{d_p} Nu_p = \frac{0.0262}{125 \times 10^{-6}} \left(0.0282 Re_p^{1.4} Pr_g^{0.33}\right)$$

$$= \frac{0.0262}{125 \times 10^{-6}} \left(0.0282 \times 1.28^{1.4} \times 0.71^{0.33}\right)$$

$$= 7.46 \, \text{W/m}^2\text{K}$$

Using the plug-flow model for gas, and assuming the particles to be well mixed, the energy balance for the gas gives

$$\rho_g c_{pg} U_g dT_g = h_p a_p n_p (T_p - T_g) \, dL_b$$

where

a_p = surface area of single particle = πd_p^2
n_p = number of particles per unit bed volume = $6(1 - \varepsilon)/(\pi d_p^3)$
$a_p n_p = 6(1 - \varepsilon)/d_p = 6(1 - 0.585)/(125 \times 10^{-6})$
 $= 1.99 \times 10^4$ particles/m^3.

Integration of the energy balance equation results in the following expression for length (L_b) required to change gas temperature from T_1 to T_2:

$$L_b = \frac{\rho_g c_{pg} U_g}{a_p n_p h_p} \ln\left(\frac{T_p - T_1}{T_p - T_2}\right) = \frac{1.18 \times 1,000 \times 0.16}{1.99 \times 10^4 \times 7.42}$$

$$\ln\frac{30}{0.5} = 5.23 \times 10^{-3} \, \text{m}$$

Thus in 5.2 mm, the gas reaches within $\frac{1}{2}$ degree K of the particle temperature. For all practical purposes, the bed of gas and particles can be considered to be isothermal.

4.1.3 Bed-to-Tube Heat Transfer

If a heat transfer tube of a 2.9 cm diameter is submerged horizontally within the fluidized medium at the above operating conditions, what would be the surface heat transfer coefficient estimated by various empirical correlations?

Since the bed temperature is low, radiative heat transfer may be neglected; thus only the convective coefficient needs to be calculated. For a horizontal tube, the correlations of Vreedenberg (1958), Borodulya (1991), and Molerus and Schweinzer (1989) can be utilized. For Vreedenberg's correlation, we first need to calculate the quantity.

$$\left(\frac{\rho_s}{\rho_g} Re_p\right) = \frac{2480}{1.18} \times 1.28 = 2,690 \geq 2550$$

At this magnitude, Eq. (20a) is appropriate.

$$\frac{h_c D_t}{k_g} = 420 \left(\frac{\rho_s}{\rho_g} Pr_g \left(\frac{\mu_g^2}{g \rho_s^2 d_p^3}\right)\right)^{0.3} Re_D^{0.3}$$

$$= 420 \left(\frac{2480}{1.18} \times 0.71\right.$$

$$\left. \times \frac{(1.85 \times 10^{-5})^2}{9.81 \times 2480^2 \times (125 \times 10^{-6})^3}\right)^{0.3} Re_D^{0.3}$$

$$= 82.1 \times Re_D^{0.3}$$

where

$$Re_D \equiv \frac{D_t \rho_g U_g}{\mu_g} = \frac{0.029 \times 1.18 \times 0.16}{1.85 \times 10^{-5}} = 296$$

and the heat transfer coefficient is obtained as

$$h_c = \frac{k_g}{D_t}\left(82.1 Re_D^{0.3}\right) = \frac{0.0262}{0.029} \times 82.1 \times 296^{0.3}$$

$$= 409 \, \text{W/m}^2 \, \text{K}$$

The correlation of Borodulya et al. (1991), Eq. (29), is also appropriate for this case:

$$\frac{h_c d_p}{k_g} = 0.74 Ar^{0.1} \left(\frac{\rho_s}{\rho_g}\right)^{0.14} \left(\frac{c_{ps}}{c_{pg}}\right)^{0.24} \varepsilon_s^{2/3}$$

$$+ 0.46 Re_p Pr_g \frac{\varepsilon_s^{2/3}}{\varepsilon}$$

$$= 0.74 \times 164^{0.1} \times \left(\frac{2480}{1.18}\right)^{0.14} \left(\frac{0.753}{1}\right)^{.24} 0.415^{2/3}$$

$$+ 0.46 \times 1.28 \times 0.71 \frac{0.415^{2/3}}{0.585}$$

$$= 1.87 + 0.398 + 2.27$$

which shows that particle convection (first term) dominates over gas convection (second term).

The resulting heat transfer coefficient is then calculated as

$$h_c = 2.27 \times \frac{k_g}{d_p} = 2.27 \times \frac{0.0262}{125 \times 10^{-6}} = 476 \, \text{W/m}^2 \, \text{K}$$

The correlation of Molerus et al. (1995) calls for Eq. (23). The necessary parameters are given by Eqs. (24)–(28):

$$U_e \equiv U_g - U_{mf} = 0.160 - 0.0175 = 0.143 \, \text{m/s}$$

$$l = \left(\frac{\mu_g}{\rho_s - \rho_g}\right)^{2/3} \left(\frac{1}{g}\right)^{1/3} = \left(\frac{1.85 \times 10^{-5}}{2480 - 1.18}\right)^{2/3}$$

$$\left(\frac{1}{9.81}\right)^{1/3} = 1.77 \times 10^{-6} \, \text{m}$$

$$B_1 \equiv 1 + 33.3 \left[\left(\frac{U_e \rho_s c_{ps}}{U_{mf} g k_g}\right)^{1/3} U_e\right]^{-1} = 1 + 33.3$$

$$\left[\left(\frac{0.143 \times 2480 \times 753}{0.0175 \times 9.81 \times 0.0262}\right)^{1/3} 0.143\right]^{-1} = 1.60$$

$$B_2 \equiv 1 + 0.28 \varepsilon_{s,mf}^2 U_e U_{mf} \left(\frac{\rho_g}{\rho_s - \rho_g}\right)^{1/2} \left(\frac{\rho_s c_{ps}}{g k_g}\right)^{2/3}$$

$$= 1 + 0.28 \times 0.58^2 \times 0.143 \times 0.0175$$

$$\left(\frac{1.18}{2480 - 1.18}\right)^{1/2} \left(\frac{2480 \times 753}{9.81 \times 0.0262}\right)^{2/3} = 1.194$$

$$B_3 \equiv 1 + 0.05 \left(\frac{U_{mf}}{U_e}\right) = 1 + 0.05 \left(\frac{0.0157}{0.143}\right) = 1.005$$

and the Nusselt number is obtained from Eq. (23),

$$\frac{h_c l}{k_g} = \frac{0.125 \varepsilon_{s,mf}}{B_1 [1 + B_2 (k_g / 2 c_{ps} \mu_g)]} + 0.165 \text{Pr}_g^{1/3}$$

$$\left(\frac{\rho_g}{\rho_s - \rho_g}\right)^{1/3} \left(\frac{1}{B_3}\right)$$

$$\frac{h_c l}{k_g} = \frac{0.125 \times 0.58}{1.60 [1 + 1.194 (0.0262/2 \times 753 \times 1.85 \times 10^{-5})]}$$

$$+ 0.165 \times 0.71^{1/3} \left(\frac{1.18}{2480 - 1.18}\right)^{1/3} \left(\frac{1}{1.005}\right)$$

$$= 0.0328$$

which gives the value for the heat transfer coefficient

$$h_c = 0.0328 \frac{k_g}{l} = 0.0328 \frac{0.0262}{1.77 \times 10^{-6}} = 486 \, \text{W/m}^2 \, \text{K}$$

4.1.4 Heat Transfer Coefficient from Packet Model

Use the packet model to estimate the convective heat transfer coefficient for the above case of a horizontal tube in a bubbling bed. Additionally, the bed has a porous plate distributor, and the center line of the heat transfer tube is located of $L_t = 0.19 \, \text{m}$ above the distributor.

To apply the packet model, we first need to calculate some bubble characteristics, utilizing hydrodynamic theory. The necessary equations can be obtained from Chapter 3, or from the book of Kunii and Levenspiel (1991). First, to estimate the bubble diameter (d_b) at the elevation of the tube,

$$d_b = d_{bm} - \frac{d_{bm} - d_{bo}}{e^{0.3 L_t / D_b}}$$

where

$$d_{bo} = 2.78 (U_g - U_{mf})^2$$

$$g = 2.78 (0.16 - 0.0175)^2$$

$$9.81 = 0.00575 \, \text{m} = 0.575 \, \text{cm}$$

$$d_{bm} = 0.65 \left[\frac{\pi D_b^2 (U_g - U_{mf})}{4}\right]^{0.4}$$

$$= 0.65 \left[\frac{\pi \times 20.3^2 (16 - 1.75)}{4}\right]^{0.4} = 19.0 \, \text{cm}$$

so that

$$d_b = d_{bm} - \frac{d_{bm} - d_{bo}}{e^{0.3 L_t / D_b}} = 19.0 - \frac{19.0 - 0.575}{e^{0.3(0.19/0.203)}}$$

$$= 5.09 \, \text{cm} = 0.0509 \, \text{m}$$

The rise velocity of the bubble, relative to the gas/ particle emulsion, is

$$u_{br} = 0.853\sqrt{gd_b}\left(e^{-1.49d_b/D_b}\right)$$

$$= 0.853\sqrt{9.81 \times 0.0509}\left(e^{-1.49 \times 0.0509/0.203}\right)$$

$$= 0.415\,\text{m/s}$$

and the actual velocity of the bubble approaching the tube is

$$u_b = U_g - U_{mf} + u_{br} = 0.160 - 0.0175 + 0.415$$

$$= 0.558\,\text{m/s}$$

Since $U_g \gg U_{mf}$, the bed is vigorously bubbling and the volume fraction of the bubbles in the bed is

$$\delta_w = \frac{U_g}{u_b} = \frac{0.160}{0.558} = 0.287$$

The frequency of bubbles at moderate-to-high elevations in a bubbling bed is ~ 2 per second. Since bubbles are equally likely to divert to one or the other side of a horizontal tube, the frequency of bubble contact by any spot on the tube would be approximately 2/2, or 1 per second. Equation (36) then estimates the average residence time of particle packets at the tube surface as

$$\tau_{pa} = \frac{1 - \delta_w}{f_b} = \frac{1 - 0.285}{1.0} = 0.715\,\text{s}$$

Next we calculate the effective thermal conductivity of the bulk packet by Eqs. (35) and (34):

$$\phi_k = 0.305\left(\frac{k_s}{k_g}\right)^{-0.25} = 0.305\left(\frac{0.89}{0.0262}\right)^{-0.25} = 0.126$$

$$k_{pab} = k_g\left[\varepsilon_{pab} + \frac{(1-\varepsilon)_{pab}}{\phi_k + 2k_g/3k_s}\right] = 0.0262$$

$$\left[0.42 + \frac{1 - 0.42}{0.126 + 2 \times 0.0262/3 \times 0.89}\right]$$

$$= 0.115\,\text{W/ms}$$

Knowing the residence time, the Fourier number for transient conduction into the packet is calculated by Eq. (33a),

$$\text{Fo} = \frac{k_{pab}\tau_{pa}}{c_{ps}\rho_s(1-\varepsilon)_{pab}d_p^2}$$

$$= \frac{0.115 \times 0.715}{0.753 \times 2480(1 - 0.42)(125 \times 10^{-6})^2} = 4860$$

For this large Fourier number, Eq. (33) shows $C \cong 1.0$. The packet model, Eq. (32), then gives a value for the convective heat transfer coefficient:

$$h_c \cong h_{pa} = 2C\left[\frac{k_{pab}\rho_s c_{ps}(1-\varepsilon)_{pab}}{\pi\tau_{pa}}\right]^{1/2}$$

$$= 2 \times 1 \times \left[\frac{0.115 \times 2480 \times 753 \times 0.58}{\pi \times 0.715}\right]^{1/2}$$

$$= 466\,\text{W/m}^2\text{K}$$

4.1.5 Heat Transfer Coefficient from Kinetic Model

Martin's (1984) model uses an analogy to the kinetic theory of gases to calculate the convective heat transfer coefficient for the particle phase in fluidized beds. Using this model, predict h_c for above case.

First, we calculate the Knudsen number for a gas in the gap between particle and wall, using Eq. (40),

$$Kn = \frac{4k_g(2\pi RT/M_g)^{1/2}}{Pd_p(2c_{pg} - R/M_g)}$$

where, for ambient air,

$$
\begin{aligned}
M_g &= 0.029\,\text{kg/mol} \\
T &= 293\text{K} \\
P &= 1.01 \times 10^5\,\text{Pa}
\end{aligned}
$$

and the gas content is $R = 8.31\,\text{J mol}^{-1}\,\text{K}^{-1}$.
Hence

$$\text{Kn} = \frac{4k_g(2\pi RT/M_g)^{1/2}}{Pd_p(2c_{pg} - R/M_g)}$$

$$= \frac{4 \times 0.0262\sqrt{2\pi(8.31 \times 293/0.029)}}{1.01 \times 10^5 \times 125 \times 10^{-6}(2 \times 753 - 8.31/0.029)}$$

$$= 0.00495$$

The Nusselt number for particle/wall collision is obtained from Eq. (39):

$$N \cong 4(1 - \text{Kn})\ln\left(1 + \frac{1}{\text{Kn}}\right) - 4 = 4(1 - 0.00495)$$

$$\ln\left(1 + \frac{1}{0.00495}\right) - 4 = 17.2$$

and the random particle velocity is obtained from Eq. (41):

$$w_p = \sqrt{\frac{gd_p(\varepsilon_b - \varepsilon_{mf})}{5(1 - \varepsilon_b)(1 - \varepsilon_{mf})}}$$

$$= \sqrt{\frac{9.81 \times 125 \times 10^{-6}(0.585 - 0.42)}{5(0.415)(0.58)}}$$

$$= 0.013\,\text{m/s}$$

The value for Z is calculated by Eq. (38):

$$Z = \frac{\rho_s c_{ps} d_p w_p}{6 k_g} = \frac{2200 \times 753 \times 125 \times 10^{-6} \times 0.013}{6 \times 0.0262}$$

$$= 17.1$$

and taking a value of 2.6 for the empirical constant C_c, Eq. (37) gives the convective coefficient as

$$h_d = \left(\frac{k_g}{d_p}\right) Z (1 - \varepsilon_b)(1 - e^{-N/C_c Z}) = \left(\frac{0.0262}{125 \times 10^{-6}}\right)$$

$$\times 17.1 \times (0.415)(1 - e^{-17.2/2.6 \times 17.1})$$

$$= 477 \, \mathrm{W m^2 \, K}$$

4.1.6   Comparison of Calculated Coefficients with Data

The specifications given above represent the bubbling fluidized bed of Chandran et al. (1980). Their experimental measurement of heat transfer coefficients on the specified horizontal tube is plotted in Fig. 3. Compare the values of h_c calculated by the various models to the experimental data of Chandran et al.

At the specified gas velocity of $U_g = 0.16 \, \mathrm{m/s}$, the experimental value of the surface heat transfer coefficient is, $(h_c)_{exp} = 395 \, \mathrm{W/m^2 \, K}$. The following tabulation shows the corresponding values calculated from the various models, and the percent deviation from experimental value:

Model	$(h_c)_{calc'd}$ (W/m^2 K)	Deviation (%)
Vreedenberg (1958)	409	+3.5
Borodulya et al. (1991)	476	+20.5
Molerus and Schweinzer (1989)	486	+23.0
Packet model	466	+18.0
Martin (1984)	477	+20.8
Average deviation		17.2

It is seen that for this single case, the older Vreedenberg correlation came closest to the measured value. Needless to say, a single case such as this should not be taken to be a general assessment of models. While there are differences in the calculated coefficients, the average deviation of 17% is reasonable and is representative of current predictive capability.

4.1.7   Freeboard Heat Transfer

For the BFB operating at the above conditions, what would be the heat transfer coefficient (average) around a horizontal tube of diameter 0.029 m, located in the freeboard with its centerline 0.2 m above the level of the collapsed (packed) bed?

Equation (51) describes the exponential decrease of the heat transfer coefficient with increasing elevation in the freeboard of bubbling beds, for horizontal tubes. For the present case, $H = 0.2 \, \mathrm{m}$, and $U^* = 0.2$. Hence

$$h_c^* = 1.3 e^{-1.53 H/U^*} = 1.3 e^{-1.53 \times 0.2/0.2^*} = 0.281$$

From the definition of h_c^*, Eq. (49),

$$h_c^* \equiv \frac{h_c - h_g}{h_{cb} - h_g}$$

where

h_{cb} = the convective coeffient for tube submerged in the bubbling bed.

     = 395 W/m^2 K from Chandran's data (see example in Sec. 4.1.6)

h_g = the gas convective coefficient, obtained from correlation of Douglas and Churchill, Eq. (50):

$$\frac{h_g D_t}{k_g} = 0.46 \left(\frac{D_t \rho_g U_g}{\mu_g}\right)^{1/2} + 0.00128 \left(\frac{D_t \rho_g U_g}{\mu_g}\right)$$

$$= 0.46 \left(\frac{0.029 \times 0.160 \times 1.18}{1.85 \times 10^{-5}}\right)^{1/2}$$

$$+ 0.00128 \left(\frac{0.029 \times 1.18 \times 0.160}{1.85 \times 10^{-5}}\right) = 8.29$$

$$h_g = 8.29 \frac{k_g}{D_t} = 8.29 \frac{0.0262}{0.029} = 7.49 \, \mathrm{W/m^2 \, K}$$

Hence the heat transfer coefficient for the freeboard tube is

$$h_c = h_c^* (h_{cb} - h_g) + h_g = 0.281(395 - 7.49)$$

$$+ 7.49 = 116 \, \mathrm{W/m^2 \, K}$$

Note that this is 15 times greater than the single-phase gas convection coefficient, the increase being attributable to particles splashing into the freeboard region.

4.2   Fast Fluidized Bed

Consider a small-scale FFB, with sand particles of diameter 125 µm mean, fluidized by ambient air. The diameter of the bed is $D_b = 0.15 \, \mathrm{m}$, with a vertical length of 11 m. Properties of the air and particles are as in the table:

	Gas	Particles
Density, ρ (kg/m^3)	1.18	2,200
Heat capacity, c_p (kJ/kg K)	1.00	0.800
Thermal conductivity, k (W/m K)	0.0262	0.900
Viscosity, μ (kg/m s)	1.85×10^{-5}	
Pr	0.71	
Sphericity, ϕ_s		0.95

According to Geldart's classification, these particles are of type B. Minimum fluidization with ambient air occurs with void fraction of $\varepsilon_{mf} = 0.45$. In examples below, we consider the operation of this bed with a superficial gas velocity of $U_g = 5.0$ m/s and a particle mass flux of $G_s = 40$ or 60 kg/m^2 s.

4.2.1 Fluidization Regime

At the intended operating condition, would the bed fluidize in the fast regime?

The generalized regime map of Fig. 1 can be used to determine the operational characteristic of this case. The Archimedes number is found from Eq. (3),

$$Ar = \frac{g\rho_g(\rho_s - \rho_g)}{\mu_g^2} d_p^3$$

$$= \frac{(9.81)(1.18)(2200 - 1.18)(125 \times 10^{-6})^3}{(1.85 \times 10^{-5})^2} = 145$$

The Reynolds number at minimum fluidization is found by solving Eq. (1),

$$Ar = 1.75 \frac{(\text{Re}_{mf})^2}{\phi_s(\varepsilon_{mf})^3} + 150 \frac{(1 - \varepsilon_{mf})(\text{Re}_{mf})}{(\phi_s)^2(\varepsilon_{mf})^3}$$

$$145 = 1.75 \frac{(\text{Re}_{mf})^2}{(0.95)(0.45)^3} + 150 \frac{(1 - 0.45)(\text{Re}_{mf})}{(0.95)^2(0.45)^3}$$

$$\text{Re}_{mf} = 0.144$$

from which we obtain the minimum fluidization velocity by Eq. (2),

$$U_{mf} = \text{Re}_{mf}\left(\frac{\mu_g}{\rho_g d_p}\right) = 0.144\left(\frac{1.85 \times 10^{-5}}{1.18(125 \times 10^{-6})}\right)$$

$$= 0.0182 \text{ m/s}$$

The terminal velocity is obtained from Eq. (6b),

$$U_t = \left[0.072g \frac{d_p^{8/5}(\rho_s - \rho_g)}{\rho_g^{2.5}\mu_g^{3/5}}\right]^{5/7}$$

$$= \left[0.072 \times 9.81 \frac{(125 \times 10^{-6})^{8/5}(2200 - 1.18)}{(1.18)^{2/5}(1.85 \times 10^{-5})^{3/5}}\right]^{5/7}$$

$$= 0.67 \text{ m/s}$$

and the dimensionless velocity is by definition,

$$U^* \equiv \frac{U_g - U_{mf}}{U_t - U_{mf}} = \frac{5.0 - 0.0182}{0.671 - 0.0182} = 7.63$$

The dimensionless particle diameter is, from Eq. (4),

$$d_p^* \equiv Ar^{1/3} = 145^{1/3} = 5.25$$

This locus is within the fast regime, near its upper boundary, on the map of Fig. 1. Therefore the proposed operation will be a fast fluidized bed.

4.2.2 Convective Heat Transfer at the Wall

Dou et al. (1993) measured the convective heat transfer coefficient at the wall of the FFB described above. They used a small heat transfer surface of $L_h = 3.2$ cm, located 5.5 m above the air inlet port. These investigators also obtained measurements of the cross-sectional bed density (ρ_b) at this elevation. Their data, for two mass fluxes of particles, are:

G_s (kg/m^2 s)	ρ_b (kg/m^3)	h_c (W/m^2 K)
40	12	80
60	24	120

Use models that give empirical correlations of h_c as functions of ρ_b for short surfaces to calculate the values of the convective heat transfer coefficients, and compare them with the above experimental data.

The correlation of Wen and Miller (1961), Eq. (54), gives

$$h_c = \frac{k_g}{d_p}\left(\frac{c_{ps}}{c_{pg}}\right)\left(\frac{\rho_b}{\rho_s}\right)^{0.3}\left(\frac{U_t}{gd_p}\right)^{0.21} \text{Pr}_g$$

$$h_c = \frac{0.0262}{125 \times 10^{-6}}\left(\frac{0.8}{1.0}\right)\left(\frac{12}{2200}\right)^{0.3}$$

$$\left(\frac{0.671}{9.81 \times 125 \times 10^{-6}}\right)^{0.21} 0.71 = 93.7 \text{ W/m}^2 \text{ K}$$

$$\text{for} \qquad G_s = 40 \text{ kg/m}^2 \text{ s}$$

$$= \left(\frac{24}{12}\right)^{0.3} \times 93.7 = 115\,\mathrm{W/m^2\,K}$$

$$\text{for} \qquad G_\mathrm{s} = 60\,\mathrm{kg/m^2\,s}$$

The correlation of Divilio and Boyd (1993), Eq. (55), gives

$$h_\mathrm{c} = 23.2(\rho_\mathrm{b})^{0.55} = 23.2(12)^{0.55} = 91.0\,\mathrm{W/m^2\,K}$$

$$\text{for} \qquad G_\mathrm{s} = 40\,\mathrm{kg/m^2\,s}$$

$$= \left(\frac{24}{12}\right)^{0.55} \times 91.0 = 133\,\mathrm{W/m^2\,K}$$

$$\text{for} \qquad G_\mathrm{s} = 60\,\mathrm{kg/m^2\,K}$$

The table at foot of this page compares the calculated values to the experimental data.

While these results are quite satisfactory, the reader is cautioned that comparison for a single case should not be taken as a definitive assessment. As for all empirical correlations, application should be limited to conditions within the range of the correlations' original database.

4.2.3 Convective Coefficient Calculated from Pressure Drop

The model of Molerus and Wirth (1993, 1995) correlates the convective heat transfer coefficient with bed pressure drop. Calculate h_c for the above case by this model and compare with the data of Dou et al. (1993) at $G_\mathrm{s} = 60\,\mathrm{kg/m^2\,s}$.

Since the heat transfer surface of Dou is 5.5 m above the gas and particle inlets, acceleration and wall friction effects would be negligible and bed pressure drop related directly to bed density:

$$\frac{\Delta P}{\Delta L_\mathrm{b}} \cong g\rho_\mathrm{b} = 9.81 \times 24 = 235\,\mathrm{Pa/m}$$

$$\text{for} \quad G_\mathrm{s} = 60\,\mathrm{kg/m^2\,s}$$

Several other required quantities have been calculated in example 2.1:

$$\mathrm{Ar} = 144$$

$$\mathrm{Re_{mf}} = 0.145$$

$$U_\mathrm{t} = 0.671\,\mathrm{m/s}$$

$$\mathrm{Re_t} \equiv \frac{\rho_\mathrm{g} U_\mathrm{t} d_\mathrm{p}}{\mu_\mathrm{g}} = \frac{1.18 \times 0.671 \times 125 \times 10^{-6}}{1.85 \times 10^{-5}} = 5.35$$

Additional required quantities are obtained from Eqs. (59)–(62):

$$\phi \equiv 1 - \frac{2.3\Delta P/\Delta L_\mathrm{b}}{g(1 - \varepsilon_\mathrm{mf})(\rho_\mathrm{s} - \rho_\mathrm{g})}$$

$$= 1 - \frac{2.3 \times 235}{9.81 \times (1 - 0.45)(2200 - 1.18)} = 0.0456$$

$$F_1 \equiv \lfloor (1 - \phi)^4\,\mathrm{Re_{mf}^2} + \phi(1 - \phi)^2\,\mathrm{Re_{mf}\,Re_t} - 189$$

$$(1 - \varepsilon_\mathrm{mf})\phi^3(1 - \phi)\mathrm{Ar}\rfloor$$

$$= \lfloor (1 - 0.0456)^4\,0.144^2$$

$$+ 0.0456(1 - 0.0456)^2\,0.145$$

$$\times 5.35 - 189(1 - 0.45)0.0456^3$$

$$(1 - 0.0456)145\rfloor$$

$$= -1.31$$

$$F_2 \equiv \left[2(1 - \phi)^2\,\mathrm{Re_{mf}} + \phi\mathrm{Re_t}\right]^2$$

$$= \left[2(1 - 0.0456)^2\,0.144 + 0.0456 \times 5.35\right]^2$$

$$= 0.258$$

$$\mathrm{Re_{pa}} = \frac{2(1 - \phi)^2\mathrm{Re_{mf}} + \phi\mathrm{Re_t}}{2[\phi + 1(1 - \phi)\varepsilon_\mathrm{mf}]}\sqrt{1 - 4\left(\frac{F_1}{F_2}\right)}$$

$$= \frac{2(1 - 0.0456)^2\,0.145 + 0.0456 \times 5.35}{2[0.0456 + (1 - 0.0456)0.45]}$$

$$\sqrt{1 - 4\left(\frac{-1.31}{0.258}\right)} = 2.48$$

	h_c W/m^2 K % diff. from exp.	
Source	$G_\mathrm{s} = 40\,\mathrm{kg/m^2\,s}$	$G_\mathrm{s} = 60\,\mathrm{kg/m^2\,s}$
Experimental, Dou et al. (1993)	80	120
Calc'd by Eq. (54), Wen and Miller (1961)	94 (+17%)	115 (−4%)
Calc'd by Eq. (55), Davilio and Boyd (1993)	91 (+14%)	133 (+11%)

The convective Nusselt number is given by Eq. (58) as

$$\frac{h_c - d_p}{k_g} = 2.85 \left[\frac{\Delta P / \Delta L_b}{g(\rho_s - \rho_g)(\varepsilon_{s,mf})} \right]^{0.5} + 3.28$$

$$\times \, 10^{-3} \, \mathrm{Re}_{pa} \mathrm{Pr}_g$$

$$= 2.85 \left[\frac{235}{9.81(2200 - 1.18)(0.55)} \right]^{0.5} + 3.28$$

$$\times \, 10^{-3} \times 2.48 \times 0.71$$

$$= 0.401 + 0.0058 = 0.407$$

from which the heat transfer coefficient is obtained as

$$h_c = \left(\frac{k_g}{d_p} \right) \times 0.407 = \left(\frac{0.0262}{125 \times 10^{-6}} \right) \times 0.407$$

$$= 85.3 \, \mathrm{W/m^2 \, K}$$

which is 29% below the experimental value of $120 \, \mathrm{W/m^2 \, K}$.

4.2.4 Convective Coefficient from Surface Renewal Model

The surface renewal model accounts for convective heat transfer by the dense-packet phase (clusters) and by the lean-gas phase. Calculate h_c for the above case by this model and compare with the data of Dou et al. (1993) at $G_s = 60 \, \mathrm{kg/m^2 s}$

To calculate the dense phase coefficient (h_d), we need to determine several parameters of the packet/cluster model. From example 2.3, we have $\mathrm{Re}_{pa} = 2.48$, whence the velocity of falling clusters is obtained as

$$V_{pa} = \mathrm{Re}_{pa} \frac{\mu_g}{\rho_g d_p} = 2.48 \frac{1.85 \times 10^{-5}}{1.18 \times 125 \times 10^{-6}}$$

$$= 0.311 \, \mathrm{m/s}$$

For the case of interest, the length of heat transfer surface (L_h) is smaller than the cluster contact length (L_c), so we use the heater length in Eq. (68) to obtain the cluster contact time,

$$\tau_{pa} = \frac{L_h}{V_{pa}} \cong \frac{0.032}{0.311} = 0.103 \, \mathrm{s}$$

The average solid fraction is obtained from the average bed density,

$$\varepsilon_{s,m} \cong \frac{\rho_b}{\rho_s} = \frac{24}{2200} = .0110$$

Gu and Chen's correlation, Eq. (64), gives the solids concentration in clusters,

$$\varepsilon_{s,pab} = 0.57 \left[1 - \left(1 - \frac{\varepsilon_{s,m}}{0.57} \right)^{3.4} \right] = 0.57$$

$$\left[1 - \left(1 - \frac{0.0110}{0.57} \right)^{3.4} \right] = 0.0365$$

Next, Eqs. (35) and (34) calculate the thermal conductivity for clusters of this concentration,

$$\phi_k = 0.305 \left(\frac{k_s}{k_g} \right)^{-0.25} = 0.305 \left(\frac{0.900}{0.0262} \right)^{-0.25} = 0.126$$

$$k_{pab} = k_g \left[\varepsilon_{pab} + \frac{(1-\varepsilon)_{pab}}{\phi_k + 2k_g/3k_s} \right] = 0.0262$$

$$\left[(1 - 0.0365) + \frac{0.0365}{0.126 + 2 \times 0.0262/3 \times 0.90} \right]$$

$$= 0.0318 \, \mathrm{W/mK}$$

Assuming that the Fourier number is sufficiently large to take $C \cong 1.0$ from Eq. (33), the coefficient for dense phase convection is then obtained from Eq. (67) as

$$h_d = 2 \left[\frac{k_{pa} \rho_s C_{ps} \varepsilon_{s,pa}}{\pi \tau_{pa}} \right]^{1/2}$$

$$= 2 \left[\frac{0.0318 \times 2200 \times 800 \times 0.0365}{\pi \times 0.103} \right]^{1/2}$$

$$= 159 \, \mathrm{W/m^2 \, K}$$

Equation (63) is used for an estimate of the lean phase convective coefficient,

$$h_t = 0.023 \left(\frac{k_g}{D_b} \right) \left(\frac{D_b \rho_g U_g}{\mu_g} \right)^{0.8} (\mathrm{Pr}_g)^{0.33}$$

$$= 0.023 \left(\frac{0.0262}{0.15} \right) \left(\frac{0.15 \times 1.18 \times 5}{1.85 \times 10^{-5}} \right)^{0.8} (0.71)^{0.33}$$

$$= 19.9 \, \mathrm{W/m^2 \, K}$$

The weighting factor (f_l) is obtained from Eq. (69),

$$1 - f_l = 1 - 0.018 \left(\frac{1}{0.0110} \right)^{0.6} = 0.731$$

and the combined convective heat transfer coefficient is calculated by Eq. (16),

$$h_c = f_l h_l + (1 - f_l) h_d = 0.269 \times 19.9 + 0.731 \times 159$$

$$= 122 \, \mathrm{K/m^2 \, K}$$

which is within 2% of experimental value.

4.2.5 Heat Transfer in a High-Temperature Bed

Consider the sand particles described above fluidized in a combustor of diameter $D_b = 0.5$ m at atmospheric pressure and 1150K temperature, with superficial gas velocity of 5 m/s and solid mass flux of 60 kg/m^{2}s. Assume that ε_{mf} remains at 0.45. If large areas of the bed wall were to be used as a heat transfer surface, maintained at 550K, what would be the effective heat flux into that surface?

We first require the physical properties of the bed material at the operating conditions. Properties of the particles are little changed by temperature, and so will be taken at the values given above. Gas properties will be taken to be those of air at 1 atm and 1150K:

	Gas	Particles
Density, ρ (kg/m^3)	0.307	2,200
Heat capacity, c_p (kJ/kg K)	1.17	0.800
Thermal conductivity, k (W/m K)	0.0743	0.900
Viscosity, μ (kg/m s)	4.51×10^{-5}	
Pr	0.71	
Sphericity, ϕ_s		0.95

We first check the fluidization regime of the intended operation.

Using Eqs. (3), (1), (2), (6b), and (4), the operating parameters are determined as

$$\text{Ar} = \frac{g\rho_g(\rho_s - \rho_g)}{\mu_g^2}d_p^3$$

$$= \frac{(9.81)(0.307)(2200 - 0.307)(125 \times 10^{-6})^3}{(4.51 \times 10^{-5})^2}$$

$$= 636$$

$$\text{Re}_{mf} \cong \frac{\text{Ar}(\phi_s)^2(\varepsilon_{mf})^3}{150(1 - \varepsilon_{mf})} = \frac{6.36(0.95)^2(0.45)^3}{150(1 - 0.45)}$$

$$= 6.34 \times 10^{-3}$$

$$U_{mf} = \text{Re}_{mf}\left(\frac{\mu_g}{\rho_g d_p}\right) = 0.00634\left(\frac{4.51 \times 10^{-5}}{0.307(125 \times 10^{-6})}\right)$$

$$= 0.00745\,\text{m/s}$$

and the dimensionless velocity is by definition, Eq. (7),

$$U_t = \left[0.072g\frac{d_p^{8/5}(\rho_s - \rho_g)}{\rho_g^{2/5}\mu_g^{3/5}}\right]^{5/7}$$

$$= \left[0.072 \times 9.81\frac{(125 \times 10^{-6})^{8/5}(2200 - 0.307)}{(0.307)^{2/5}(4.51 \times 10^{-5})^{3/5}}\right]^{5/7}$$

$$= 0.673\,\text{m/s}$$

The dimensionless velocity and particle diameter are then obtained from Eqs. (7) and (4),

$$U^* = \frac{U_g - U_{mf}}{U_t - U_{mf}} = \frac{5.0 - 0.00745}{0.673 - 0.00745} = 7.50$$

$$d_p^* \equiv \text{Ar}^{1/3} = 6.36^{1/3} = 1.85$$

This locus is within the fast regime, near its center, on the map of Fig. 1. Therefore the proposed operation will be a fast fluidized bed. In actual operation, the bed average density, at elevation of the heat transfer surface, is found to be $\rho_b = 45$ kg/m^3. The values for volume fractions of solid and gas phases are then

$$\varepsilon_{s,b} = \frac{\rho_b}{\rho_s} = \frac{45}{2200} = 0.0205$$

$$\varepsilon_b = 1 - \varepsilon_{s,b} = 1 - 0.0205 = 0.980$$

We now have a choice of several models for the estimation of the heat transfer coefficient. The operating conditions are close to those used by Werdmann and Werther (1993) in developing their correlation, which is particularly appropriate for large heat transfer surfaces. Thus from Eq. (56),

$$h_c = 7.46 \times 10^{-4}\left(\frac{k_g}{d_p}\right)\left(\frac{D_b\rho_g U_g}{\mu_g}\right)^{0.757}\left(\frac{\rho_b}{\rho_s}\right)^{0.562}$$

$$= 7.46 \times 10^{-4}\left(\frac{0.0743}{125 \times 10^{-6}}\right)$$

$$\left(\frac{0.5 \times 0.307 \times 5.0}{4.51 \times 10^{-5}}\right)^{0.757}\left(\frac{45}{2200}\right)^{0.562}$$

$$= 79.4\,\text{W/m}^2\,\text{K}$$

Due to high temperatures, we also need to calculate radiative heat transfer. The effective emissivity of the bed is obtained from Eq. (71) using the recommended value of 0.95 for C_e and taking the particle emissivity to be $e_s = 0.7$:

$$e_b = 1 - \exp\left(-\frac{C_e \varepsilon_{s,b} e_s D_b}{d_p}\right)$$

$$= 1 - \exp\left(-\frac{0.95 \times 0.0205 \times 0.7 \times 0.5}{125 \times 10^{-6}}\right) = 1.00$$

Assuming the bed wall to be oxidized steel, a reasonable value for wall emissivity is $e_w \cong 0.8$. The bed–wall combined emissivity is calculated by Eq. (44) as

$$e_{bw} = \frac{1}{1/e_b + 1/e_w - 1} = \frac{1}{1/1.00 + 1/0.8} = 0.80$$

The radiative heat transfer coefficient is obtained from Eq. (45),

$$h_r = \frac{e_{bw}\sigma(T_b^4 - T_w^4)}{T_b - T_w}$$

$$= \frac{0.80 \times 5.67 \times 10^{-8}(1150^4 - 550^4)}{1150 - 550} = 125\,\text{W/m}^2\,\text{K}$$

The total heat transfer coefficient and heat flux are, respectively,

$$h_w = h_c + h_r = 79.4 + 125 = 204\,\text{W/m}^2\text{K}$$

$$\frac{q_w}{a_w} = h_w(T_b - T_w) = 204(1150 - 550)$$

$$= 1.22 \times 10^5\,\text{W/m}^2$$

Note that this high heat flux is representative of operations for FFB combustors. Note also that 61% of the heat transfer is due to thermal radiation.

NOMENCLATURE

a = interfacial surface area; m^2

A = radiation absorption coefficient, Eq. (41); 1/m

Ar = Archimedes number, Eq. (3); dimensionless

c_p = specific heat capacity; J/kg K

C = correction factor for conductivity, Eqs. (32 and 33); dimensionless

C_c = contact parameter in Martin's model, Eq. (64); dimensionless

C_e = emissivity parameter, Eq. (70); dimensionless

C_D = coefficient of drag, Eq. (6); dimensionless

C_R = correction factor for radial position, Eq. (18); dimensionless

d_b = bubble diameter, cm or m

d_{bm} = maximum bubble diameter, cm or m

d_{bo} = initial bubble diameter at distributor, cm or m

d_p = particle diameter, m

d_p^* = dimensionless particle diameter, Eq. (4)

D = diameter of bed or of tube; m

e = eradiative emissivity; dimensionless

f = time fraction of contact by lean or dense phase, Eq. (16); dimensionless

f_b = frequency of bubbles; 1/s

F_v = view factor for radiation, Eq. (38); dimensionless

F_o = Fourier modulus, Eq. (33); dimensionless

g = gravitational acceleration; 9.81 m/s^2

h = heat transfer coefficient; W/m^2 K

h_c^* = dimensionless heat transfer coefficient, Eq. (44)

h_p = heat transfer coefficient based on particle surface area, Eq. (8); W/m^2 K

h_w = heat transfer coefficient based on submerged wall area, Eq. (14); W/m^2 K

H = height in freeboard, measured from surface level of collapsed bed; m

I = forward radiative flux, Eq. (41); W/m^2

J = backward radiative flux, Eq. (41); W/m^2

k = thermal conductivity; W/m K

Kn = Knudsen number for gas in particle–wall gap, Eq. (67); dimensionless

l = length scale in Eqs (23) and (24); m

L_b = axial elevation in FFB; m

L_c = contact length of clusters on wall, Eq. (62); m

L_h = vertical length of heat transfer surface in FFB; m

M_g = molecular weight of gas; kg/mol

Nu = Nusselt number; dimensionless

Nu_p = Nusselt number based on particle diameter, Eq. (9); dimensionless

P = pressure; Pa

Pr = Prandtl number ($c_p\mu/k$), dimensionless

q = heat transfer rate; W

r = radial position within bed; m

R = universal gas constant

R_b = radius of bed; m

Re_D = Reynolds number based on tube diameter, Eq. (20); dimensionless

Re_{mf} = Reynolds number at minimum fluidization velocity, Eq. (2), dimensionless

Re_p = Reynolds number based on particle diameter, Eq. (10); dimensionless

Re_t = Reynolds number at terminal velocity, Eq. (46); dimensionless

S = radiation scattering coefficient, Eq. (41); 1/m

t = time, s

T = temperature; K

U = superficial gas velocity; m/s

U_e = excess superficial velocity over minimum fluidization, Eq. (28); m/s

U^* = dimensionless velocity, Eq. (7)

V_{pa} = velocity of particle packet or cluster; m/s

w_p = average random particle velocity, Eq. (65); m/s

y = coordinate distance away from surface; m

Greek Symbols

δ_w = bubble volume fraction at wall, Eq. (36); dimensionless

ε = volume fraction gas (void fraction); dimensionless
ε_s = volume fraction solids, equal to $(1 - \varepsilon)$; dimensionless
ϕ_s = sphericity of particles, Eq. (1); dimensionless
μ = viscosity; kg/(m s)
ρ = density; kg/m^3
ρ_b = density of particle–gas medium, Eq. (47); kg/m^3
σ = Stefan–Boltzmann constant; 5.67×10^{-8} W/m^2 K^4
τ = contact (residence) time; s
τ_{pa} = residence time of packets, Eq. (31); s

Subscripts

b = bed, or bubble
c = convective
d = dense phase
g = gas
l = lean phase
m = local gas–particle medium
mf = minimum fluidization
p = particle
pa = packet or cluster
pab = bulk packet
w = wall or submerged surface
r = radiative
s = solid
t = terminal, or tube

REFERENCES

Andeen BR, Glicksman LR. Heat transfer to horizontal tube in shallow fluidized beds. National Heat Transfer Conference, St Louis, MO, paper no. 76-HT-67, 1976.

Anderson BA, Lechner B. Local lateral distribution of heat transfer on tube surface of membrane walls in CFB boiler. In: Avidan AA, ed. Circulating Fluidized Bed Technology IV. New York: AIChE, 1993, pp 311–318.

Battacharya, Harrison D. Heat transfer in high temperature fluidized beds. Proc of Europ Conf Particle Tech, 1976, p. K7.

Beaude F, Louge M. Similarity of radial profiles of solid volume fraction in a circulating fluidized bed. In: Fluidization VIII. New York: Engineering Foundation, 1995, pp 245–253.

Berg BV, Baskakov AP. Investigation of local heat transfer between a fixed horizontal cylinder and a fluidized bed. Intern Chem Eng 14, no. 3:440–443, 1974.

Biyikli S, Tuzla K, Chen JC. Heat transfer around a horizontal tube in freeboard region of fluidized beds. AIChE J 29, no. 5:712–716, 1983.

Biyikli S, Tuzla K, Chen JC. Freeboard heat transfer in high-temperature fluidized beds. Powder Tech 53:187–194, 1987.

Borodulya VA, Teplitsky YS, Sorokin AP, Markevich II, Hassan AF, Yeryomenko TP. Heat transfer between a

surface and a fluidized bed: consideration of pressure and temperature effects. Int J Heat Mass Transf 34, no. 1:47–53, 1991.

Botterill JSM, Desai M. Limiting factors in gas–fluidized bed heat transfer. Powder Tech 6, no. 4:231-238, 1972.

Brewster MO, Tien CL. Radiative heat transfer in packed fluidized beds: dependent versus independent scattering. J Heat Transf 104, no. 4:574–580, 1982.

Burki V, Hirschberg B, Tuzla K, Chen JC. Thermal development for heat transfer in circulating fluidized beds. AIChE Annual Meeting, St. Louis, MO, 1993.

Chandran R, Chen JC. Influence of the wall on transient conduction into packed media. AIChE J 31, no. 1:168–170, 1985a.

Chandran R, Chen JC. A heat transfer model for tubes immersed in gas fluidized beds. AIChE J 31, no. 2:244–252, 1985b.

Chandran R, Chen JC, Staub FW. Local heat transfer coefficients around horizontal tubes in fluidized beds. AIChE J 102, no. 2:152–157, 1980.

Chen JC. Heat transfer to tubes in fluidized beds. Award lecture. National Heat Transfer Conference, St. Louis, MO, paper no. 76-HT-75, 1976.

Chen, JC. Thomas Baron Plenary Lecture—Clusters. AIChE Symp Series, 92, No. 313:1–6, 1996.

Chen JC, Chen KL. Analysis of simultaneous radiative and conductive heat transfer in fluidized beds. Chem Eng Commun 9:255–271, 1981.

Chen JC, Cimini RJ, Dou SH. A theoretical model for simultaneous convective and radiative heat transfer in circulating fluidized beds. In: Basu P, Large JF, eds. Circulating Fluidized Bed Technology II. Oxford: Pergamon Press, 1988, pp 255–262.

Cimini RJ, Chen JC. Experimental measurement of radiant transmission through packed and fluidized media. Exp Heat Transfer 1:45–56, 1987.

Divilio RJ, Boyd TJ. Practical implications of the effect of solids suspension density on heat transfer in large-scale CFB boilers. In: Circulating Fluidized Bed Technology IV. New York: Engineering Foundation, 1993, pp 334–339.

Dou S, Herb B, Tuzla K, Chen JC. Heat transfer coefficients for tubes submerged in circulating fluidized bed. Experimental Heat Transfer 4:343–353, 1991.

Dou S, Herb B, Tuzla K, Chen JC. Dynamic variation of solid concentration and heat transfer coefficient at wall of circulating fluidized bed. In: Potter OE, Nicklin DJ, ed. Fluidization VII. New York: Engineering Foundation, 1993, pp 793–801.

Douglas WYM, Churchill SW. Recorrelation of data for convective heat transfer between gases and single cylinders with large temperature differences. Chem Eng Prog Symp Ser 51, no. 17:57, 1956.

Dow WM, Jacob M. Heat transfer between a vertical tube and a fluidized air–solid mixture. Chem Eng Progr 47, no. 12:637–648, 1951.

Ebert T, Glicksman L, Lints M. Determination of particle and gas convective heat transfer components in circulating fluidized bed. Chem Eng Sci 48:2179–2188, 1993.

Fraley LD, Lin YY, Hsiao KM, Solbakken A. ASME Paper 83-HT-92, 1983.

Furchi JCL, Goldstein L Jr, Lombardi G, Mohseni M. Experimental local heat transfer in a circulating fluidized bed. In: Basu P, Large JF, eds. Cirulating Fluidized Bed Technology II. Oxford, UK: Pergamon Press, 1988, pp 263–270.

Geldart D. Types of gas fluidization. Powder Tech 7, no. 5:285–292, 1973.

Glicksman LR. Heat transfer in circulating fluidized beds. In: Grace JR, Avidan AA, Knowlton TM, eds. Circulating Fluidized Beds. London: Chapman and Hall, 1997, pp 261–311.

Grace JR. Can J Chem Eng 64:353, 1986a.

Grace J. Heat transfer in circulating fluidized beds. In: Circulating Fluidized Bed Technology. New York: Engineering Foundation, 1986b, pp 63–81.

Gu WK, Chen JC. A model for solid concentration in circulating fluidized beds. In: Fan LS, Knowlton T, eds. Fluidization IX. New York: Engineering Foundation, 1998, pp 501–508.

Han GY. Experimental study of radiative and particle convective heat transfer in fast fluidized beds. PhD dissertation, Dept. of Chem. Eng., Lehigh University, Bethlehem, PA, 1992.

Han GY, Tuzla K, Chen JC. Performance of an entrained particle heat exchanger. Heat Transfer Eng 17, no. 4:64–71, 1996.

Hartige EU, Li Y, Werther J. Flow structures in fast fluidized beds. In: Fluidization V. New York, Engineering Foundation, 1986, pp 345–352..

Herb BE, Dou S, Tuzla K, Chen JC. Axial solid concentration in CFBs: experimental measurements and model predictions. AIChE Annual Meeting, 1989a.

Herb B, Tuzla K, Chen JC. Distribution of solid concentrations in circulating fluidized bed. In: Fluidization VI. New York, Engineering Foundation, 1989b, pp 65–72.

Herb BE, Dou S, Tuzla K, Chen JC. Solid mass fluxes in circulating fluidized beds. Powder Tech 70, no. 3: 197–205, 1992.

Horio M, Morshita K, Tachibana O, Murata M. Solid distribution and movement in circulating fluidized beds. In: Basu P, Large JF, eds. Circulating Fluidized Bed Technology II. Oxford, UK: Pergamon Press, 1988, pp 147–154.

Kiang KD, Lin KT, Nack H, Oxley JH. Heat transfer in fast fluidized beds. In Fluidization Technology. New York: Engineering Foundation, 1976, pp 471–483.

Kunii D, Smith JM. Heat transfer characteristics of porous rocks. AIChE J, no. 1:71–78, 1960.

Kunii D , Levenspiel O. Fluidization Engineering. 1st ed. New York: John Wiley, 1969.

Kunii D , Levenspiel O. Fluidization Engineering. 2nd ed. Boston: Butterworth-Heinemann, 1991.

Leva M, Grummer M. A correlation of solids turnovers in fluidized systems. Chem Eng Progr 48, no. 6:307–313, 1952.

Leva M, Weintrub M, Grummer M. Heat transmission through fluidized beds of fine particles. Chem Eng Progr 45, no. 9:563–572, 1949.

Levenspiel O, Walton JS. Chem Eng Progr Symp Series 50, no. 9:1, 1954.

Lints M. Particle to wall heat transfer in circulating fluidized beds. PhD dissertation, Mass Inst Tech Cambridge, MA, 1992.

Lints MC, Glicksman LR. The structure of particle clusters near wall of a circulating fluidized bed. AIChE Symp Series, 89, no. 296:35–47, 1993.

Louge M, Lischer J, Chang H. Measurements of voidage near the wall of a circulating fluidized bed riser. Powder Tech 62:269–276, 1990.

Mahalingam M, Kolar AK. Heat transfer model for membrane wall of a high temperature circulating fluidized bed. In: Basu P, Horio M, Hasatani M, ed. Circulating Fluidized Bed Technology III. Toronto: Pergamon Press, 1991, pp 239–246.

Martin H. Heat transfer between gas fluidized beds of solid particles and the surfaces of immersed heat exchanger elements, part I. Chem Eng Processes 18:157–169, 1984a.

Martin H. Heat transfer between gas fluidized beds of solid particles and the surfaces of immersed heat exchanger elements, part II. Chem Eng Processes 18:199–223, 1984b.

Mickley HS, Trilling CA. Heat transfer characteristics of fluidized beds. Ind Eng Chem 41, no. 6:1135–1147, 1949.

Mickley HS, Fairbanks DF. Mechanism of heat transfer to fluidized beds. AIChE J 1, no. 3:374–384, 1955.

Molerus O. Fluid dynamics and its relevance for basic features of heat transfer in circulating fluidized beds. In: Avidan AA, ed. Circulating Fluidized Bed IV. New York: AIChE, 1993, pp 285-290.

Molerus O, Burschka A, Dietz S. Particle migration at solid surfaces and heat transfer in bubbling fluidized beds II, Prediction of heat transfer in bubbling fluidized beds. Chem Eng Sci 50, no. 5:879–885, 1995.

Molerus O, Schweinzer J. Prediction of gas convective part of the heat transfer to fluidized beds. In Fluidization IV. New York: Engineering Foundation, 1989, pp 685–693.

Molerus O, Wirth KE. Heat Transfer in Fluidized Beds. London: Chapman and Hall, 1997.

Noe AR, Knudsen JG. Local and average heat transfer coefficients in a fluidized bed heat exchanger. Chem Eng Progr Symp Series 64, no. 82:202–211, 1968.

Ozkaynak TF, Chen JC. Emulsion phase residence time and its use in heat transfer models in fluidized bed. AIChE J 26, no. 4:544–550, 1980.

Ozkaynak TF, Chen JC, Frankenfield TR. An experimental investigation of radiant heat transfer in high temperature fluidized bed. Fluidization IV. New York, Engineering Foundation, 1983, pp 371–378.

Palchonok GI, Breitholz C, Anderson BA, Lechner B. Heat transfer in the boundary layer of a circulating fluidized bed boiler. In: Large JF, Laguerie C, eds. Fluidization VIII. New York: Engineering Foundation, pp 291–299, 1995.

Pell M. Gas Fluidization. Amsterdam: Elsevier, 1990.

Radauer HG, Glatzer A, Linzer W. A model combining convective and radiative heat transfer in CFB boilers. In: Circulating Fluididized Bed Technology V. New York: Engineering Foundation, 1996, p 9.

Ranz WE. Friction and transfer coefficients for single particle and packed beds. Chem Eng Progress 48, no. 5:247–253, 1952.

Rhodes MJ, Mineo H, Hirama T. Particle motion at the wall of a circulating fluidized bed. Powder Tech 70:207–214, 1992.

Saxena SC, Grewal NS, Gabor JD, Zabrodsky SS, Galershtein DM. Heat transfer between a gas fluidized bed and immersed tubes. Adv in Heat Transfer 14:145–247, 1978.

Soong CH, Tuzla K, Chen JC. Identification of particle clusters in circulating fluidized bed. In: Avidan AA, ed. Circulating Fluidized Bed Technology IV. New York: AIChE, 1993, pp 809–814.

Soong CH, Tuzla K, Chen JC. Experimental determination of cluster size and velocity in circulating fluidized beds. In: Large JF, Laguerie C, ed. Fluidization VIII. New York: Engineering Foundation, 1995, pp 219–227.

Vreedenberg HA. J Appl Chem 2 (suppl 1):s26, 1952.

Vreedenberg HA. Heat transfer between a fluidized bed and a horizontal tube. Chem Eng Sci 9, no. 1:52–60, 1958.

Vreedenberg HA. Heat transfer between a fluidized bed and a vertical tube. Chem Eng Sci 11, no. 4:274–285, 1960.

Wen CY, Miller EN. Heat transfer in solids–gas transport lines. Ind. Eng. Chem. 53, no. 1: 51–53, 1961.

Wen CY, Yu YH. A generalized method for predicting the minimum fluidization velocity. AIChE J 12, no. 3:610–612, 1966.

Wender L, Cooper GT. Heat transfer between fluidized-solids bed and boundary surfaces—correlation of data. AIChE J 4, no. 1:15–23, 1958.

Werdermann CC, Werther J. Solids flow pattern and heat transfer in an industrial scale fluidized bed heat exchanger. Proc 12th Intern Conf on Fluid Bed Combustion 2:985–990, 1993.

Werther J. Fluid mechanics of large-scale CFB units. In: Avidan AA, ed. Circulating Fluidized Bed Technology IV. New York: AIChE, 1993, pp 1–14.

Wirth KE. Zirkulierende Wirbelschichten-strömungsmechanische Grundlagen. Anwendung in der Feuerungstechnik, 1990.

Wirth KE. Prediction of heat transfer in circulating fluidized beds. In: Avidan A, ed. Circulating Fluidized Bed Technology IV. New York: AIChE 1993, pp 291–296.

Wirth KE. Heat transfer in circulating fluidized beds. Chem Eng Sci 50, no. 13:2137–2151, 1995.

Wood RT, Staub FW, Canada GS, McLaughlin MH. Two-phase flow and heat transfer. Tech Rep RP 525-1, General Electric Co, Schenectady, NY, 1978.

Wu R, Grace J, Lim C, Brereton C. Suspension to surface heat transfer in a circulating fluidized bed combustor. AIChE J 35:1685–1691,1989.

11

Mass Transfer

Thomas C. Ho

Lamar University, Beaumont, Texas, U.S.A.

1 INTRODUCTION

Gas–solid fluidized bed reactors are widely used by industry. Examples include fluid catalytic cracking (FCC), combustion and gasification, uranium processing, roasting of sulfide ore, reduction of iron ores, and drying (Zenz, 1978). Of all the phenomena affecting the performance of bubbling fluidized beds as chemical reactors, interphase mass transfer is probably of primary importance in most cases (Chavarie and Grace, 1996a,b).

The particle–gas mass transfer in a gas–solid fluidized bed has been an essential subject of investigation since the invention of the fluidized bed technology. Historically, there have been two approaches in modeling the rate of mass transfer in fluidized bed reactors. One approach, called the homogeneous bed approach, considers the fluidized bed reactor to behave like a fixed bed and correlates the fluidized bed mass transfer coefficient in a manner similar to that in a fixed bed based on a plug-flow model. The other approach, called the bubbling bed approach, considers the fluidized bed to consist of two phases, a bubble and an emulsion, and the gas interchange between the two phases constitutes the rate of mass transfer. The objective of this chapter is to review the two approaches.

It is essential to point out that the mechanism of mass transfer, as will be discussed in this chapter, represents only one of the potential rate controlling steps governing reactor performance in the system.

Depending on the types of gas–solid reactions, the rate controlling mechanisms may include (Levenspiel, 1972):

1. Particle–gas mass transfer (gas film diffusion) control
2. Pore diffusion (ash layer diffusion) control
3. Surface phenomenon control

This chapter is devoted to the first mechanism listed above, i.e., mass transfer between the gas and the bed particles at the gas–particle interphase in a bubbling fluidized bed.

2 HOMOGENEOUS BED APPROACH

The homogeneous bed approach correlates the mass transfer coefficient in a fluidized bed in a manner similar to that in a fixed bed based on a plug-flow model. The transfer equation in this approach for a fixed or a fluidized bed closely resembles that for a single sphere suspended in a gas stream. However, the magnitude of the mass transfer coefficient for these various bed particles may be significantly different depending on particle size and operating conditions. The following subsections review separately the particle–gas mass transfer for various bed particles based on this homogeneous bed approach.

2.1 Transfer Between Single Spheres and Surrounding Gas

The rate of mass transfer between well-dispersed single spheres and the surrounding gas can be described by the following mass transfer equation (Kunii and Levenspiel, 1991):

$$\frac{dN_A}{dt} = k_{g,single} S_{ex,single} \left(C_A^i - C_A \right) \qquad (1)$$

where dN_A/dt represents the transfer rate of A from the particle surface to the gas stream, $k_{g,single}$ represents the mass transfer coefficient of the particle, $S_{ex,single}$ is the exterior surface of the particle, C_A^i is the concentration of A at the gas–particle interphase, and C_A is the concentration of A in the gas stream. The mass transfer coefficient $k_{g,single}$ under this single-sphere condition has been well established, both theoretically and experimentally (Froessling, 1938), as

$$Sh_{single} = \frac{k_{g,single} d_{sph} y}{\mathcal{D}} = 2 + 0.6(Re_{sph})^{0.5}(Sc)^{0.333} \qquad (2)$$

In the above equation, Re_{sph} is the particle Reynolds number defined as

$$Re_{sph} = \frac{\rho_g u d_{sph}}{\mu} \qquad (3)$$

Sc is the Schmidt number defined as

$$Sc = \frac{\mu}{\rho_g \mathcal{D}} \qquad (4)$$

y is the logarithmic mean mole fraction of the inert or nondiffusing component, and $\mathcal{D}$ is the gas phase diffusion coefficient. Equation 2 implies that the Shewood number has a theoretical minimum at Sh = 2.0 even when the particle is exposed to a stationary gas, i.e., u (or Reynolds number) equals zero. The equation also indicates that the mass transfer coefficient ($k_{g,single}$) is proportional to the diffusion coefficient of the gas ($\mathcal{D}$), and is inversely proportional to the particle diameter (d_{sph}) and the mole fraction of the inert component (y). Note that for nonspherical particles, d_{sph} in Eqs. (2) and (3) can be replaced by screen size d_p as suggested by Kunii and Levenspiel (1991), and the two equations become Eqs. (5) and (6), respectively:

$$Sh_{single} = \frac{k_{g,single} d_p y}{\mathcal{D}} = 2 + 0.6(Re_p)^{0.5}(Sc)^{0.333} \qquad (5)$$

$$Re_p = \frac{\rho_g u d_p}{\mu} \qquad (6)$$

It is essential to note that Eq. (2) [or (5)] is valid only for a single sphere, or single spheres well dispersed in a gas stream. For packed bed or fluidized bed particles, the particle–gas mass transfer coefficient may be higher or lower than that estimated from Eqs (2) [or (5)], depending on the particle Reynolds number.

2.2 Transfer Between Fixed Bed Particles and Flowing Gas

The rate of particle–gas mass transfer from a differential bed segment of a fixed bed in the axial direction can be represented by an equation resembling Eq. (1) based on a plug-flow model, i.e.,

$$\frac{dN_A}{dt} = k_{g,bed} S_{ex,particles} \left(C_A^i - C_A \right) \qquad (7)$$

However, in this equation, dN_A/dt represents the combined mass transfer rate from all the particles in the segment, $k_{g,bed}$ is the average mass transfer coefficient of the particles, and $S_{ex,particles}$ is the total exterior surfaces of all individual particles in the bed segment.

It is worth pointing out that the average mass transfer coefficient for fixed bed particles may be higher or lower than that for single spheres at an identical superficial gas velocity. On one hand, the low fixed bed voidage may enhance the mass transfer coefficient owing to the creation of increased true gas velocities at the gas–particle interphase for fixed bed particles. On the other hand, the actual particle surfaces involved in the particle–gas mass transfer are smaller for fixed bed particles than those for well dispersed particles owing to particle blockage in the fixed bed arrangement. The use of total exterior surfaces of all individual particles in Eq. 7, therefore, overestimates the active surfaces involved in the mass transfer process. The net effect of this overestimation is to decrease the average mass transfer coefficient evaluated based on Eq. (7).

There have been several reports on the average mass transfer coefficient for fixed bed particles. In general, for particle Reynolds numbers greater than 80 (relatively large particles), the average mass transfer coefficient for fixed bed particles is found to be higher than that estimated by Eq. (2) [or (5)] for well-dispersed single spheres. On the other hand, for particle Reynolds numbers less than 80 (relatively smaller particles), the trend is reversed, i.e., the average mass transfer coefficient for fixed bed particles is much smaller than that estimated by Eq. (2). It is apparent that the increase in mass transfer coefficient for relatively large particles in the fixed bed setup is due to the effect of the increase in true gas velocity at the gas–particle interface. However, the decrease in average mass transfer coefficient for smaller particles is due to the effect of particle blockage resulting in the overestimation of active surfaces in Eq. (7), as discussed previously.

2.2.1 Average Mass Transfer Coefficient, $k_{g,bed}$

For particle Reynolds numbers greater than 80, Ranz (1952) reported that the following equation describes the particle–gas mass transfer coefficient for fixed bed particles:

$$\text{Sh}_{bed} = \frac{k_{g,bed} d_p y}{\mathcal{D}} = 2 + 1.8(\text{Re}_p)^{0.5}(\text{Sc})^{0.333} \tag{8}$$

$$\text{for } \text{Re}_p > 80$$

In the above equation, Re_p and Sc are similarly defined as in Eqs. (6) and (4), but the superficial gas velocity is used in the calculation of Re_p, i.e.,

$$\text{Re}_p = \frac{\rho_g U d_p}{\mu} \tag{9}$$

For smaller particles, both Hurt (1943) and Resnick and White (1949) reported that the particle–gas mass transfer coefficient for fixed bed particles can be lower than that estimated from Eq. (2) for single spheres. Note that their original results were reported in the form of j_D-factors; those results, however, can be easily converted to the Sherwood number. The converted Sherwood number from the above two studies along with those from Eqs. (2) (for single spheres) and (8) (for large fixed bed particles) are plotted against particle Reynolds numbers in Fig. 1. As indicated in the figure, the Sherwood number for fixed bed particles can be higher or lower than that estimated by Eq. (2) for well-dispersed single spheres. For smaller particles, the Sherwood number can be much lower than the theoretical minimum of the Sherwood number for single spheres, i.e., $\text{Sh}_{single} = 2.0$.

2.2.2 Total Particle Exterior Surfaces, $S_{ex,particles}$.

Besides the average mass transfer coefficient, another essential aspect in the application of Eq. (7) for predicting the rate of mass transfer is the estimation of the total exterior surfaces of the particles in the bed segment. Two methods are discussed below for estimating $S_{ex,particles}$, one by Kunii and Levenspiel (1991) and the other by Carmon (1941).

Method of Kunii and Levenspiel (1991). The method proposed by Kunii and Levenspiel (1991) is based on the equation

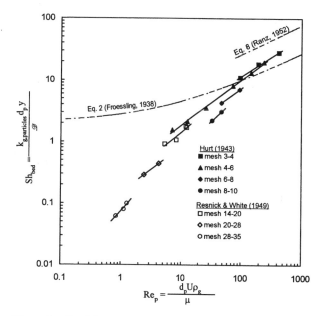

Figure 1 Particle–gas mass transfer coefficient for fixed bed particles.

$$S_{ex,particles} = V_{seg}(1 - \varepsilon_{bed})a' \tag{10}$$

where V_{seg} is the volume of the bed segment, ε_{bed} is the bed voidage, and a' is the ratio of particle surface to particle volume defined as:

$$a' = (\text{surface of a particle/volume of a particle})$$

$$= \frac{\pi d_{sph}^2 \phi_s}{\pi d_{sph}^3/6} \tag{11}$$

$$= \frac{6}{d_{sph}\phi_s}$$

where ϕ_s is the sphericity of bed particles ($\phi_s = 1.0$ for spherical particles and $0 < \phi_s < 1.0$ for all other particles), and d_{sph} is the diameter of spherical particles having the same volume as the bed particles. Equations (10) and (11) can be combined to yield

$$S_{ex,particles} = V_{seg} \frac{6(1 - \varepsilon_{bed})}{\phi_s d_{sph}} \tag{12}$$

The sphericity for different solids has been calculated and is given in Table 1. Note that the term $(\phi_s d_{sph})$ appearing in Eq (12) is called the effective particle diameter (d_{eff}) and can be estimated according to the following approximations:

(a) For large particles (> 1 mm), d_{sph} can be calculated from the measurement of the size of particles by calipers or a micrometer if the

Table 1 Sphericity of Particles

Type of particles	Sphericity ϕ_s	Source
Sphere	1.00	(a)
Cube	0.81	(a)
Cylinder		
$h = d$	0.87	(a)
$h = 5d$	0.70	(a)
$h = 10d$	0.58	(a)
Disks		
$h = d/3$	0.76	(a)
$h = d/6$	0.60	(a)
$h = d/10$	0.47	(a)
Activated carbon and silica gels	0.70–0.90	(b)
Broken solids	0.63	(c)
Coal		
anthracite	0.63	(e)
bituminous	0.63	(e)
natural dust	0.65	(d)
pulverized	0.73	(d)
Cork	0.69	(d)
Glass, crushed, jagged	0.65	(d)
Magnetite, Fischer–Tropsch catalyst	0.58	(e)
Mica flakes	0.28	(d)
Sand		
round	0.86	(e)
sharp	0.66	(e)
old beach	as high as 0.86	(f)
young river	as low as 0.53	(f)
Tungsten powder	0.89	(f)
Wheat	0.85	(d)

(a) From geometric considerations. (b) From Leva (1959). (c) From Uchida and Fujita (1934). (d) From Carmån (1937). (e) From Leva et al. (1948, 1949). (f) From Brown et al. (1949).
Source: Kunii and Levenspiel, 1991.

particles are regular in shape, or by weighing a certain number of particles if their density is known, or by fluid displacement if the particles are nonporous. The sphericity ϕ_s can be measured directly, estimated from Table 1, or evaluated by the pressure drop experiments as outlined in Kunii and Levenpiel (1991).

(b) For intermediate particles ($40\,\mu m < d_p < 1\,mm$), screen analysis is the most convenient way to measure particle size. The d_{eff} is then estimated based on the following suggestions:

1. For irregular particles with no seemingly longer or shorter dimension (hence isotropic in shape),

$$d_{eff} = \phi_s d_{sph} \cong \phi_s d_p \qquad (13)$$

2. For irregular particles with one somewhat longer dimension, but with a length ratio not greater than 2 : 1 (eggs, for example),

$$d_{eff} = \phi_s d_{sph} \cong d_p \qquad (14)$$

3. For irregular particles with one somewhat shorter dimension, but with a length ratio not greater than 1 : 2, then roughly,

$$d_{eff} = \phi_s d_{sph} \cong \phi_s^2 d_p \qquad (15)$$

4. For very flat or needlelike particles, estimate the relationship between d_p and d_{eff} from the ϕ_s values for the corresponding disks and cylinders. An alternative method for these

particles is to conduct the pressure drop experiments to determine the effective particle size.

(c) For very small particles ($< 40 \, \mu$m), there are two methods to determine the effective particle size. One is to use the scanning of magnified photographs of particles and the other is to conduct sedimentation experiments of particles in a known fluid; the terminal velocity of these particles will give the diameter of the equivalent sphere.

Method by Carmon (1941). The method proposed by Carmon (1941) for the total exterior areas of fixed bed particles is based on the equation

$$S_{\text{ex,particles}} = (S)(L)(A) \tag{16}$$

where

$$S = S_{\text{a}}(1 - X) \tag{17}$$

and

$$S_{\text{a}} = 14 \left[\frac{X^3}{K_{\text{t}}(1 - X)^2} \right]^{0.5} \tag{18}$$

In this equation, K_{t} is estimated by

$$K_{\text{t}} = \frac{\mu L Q}{A(1 - \Delta P_{\text{p}})} \tag{19}$$

Table 2 summarizes the various properties of particles in the investigation carried out by Resnick and White (1949). They reported that the following equation roughly holds for a quick estimation of S appearing in Eq. (16) for the particles involved in their experiments, i.e.,

$$S = \frac{S_{\text{d}}}{d_{\text{p}}} \tag{20}$$

where S_{d} was reported to have an average value of 4.476. The S_{d} values for various size particles in their experiments are also summarized in Table 2.

2.3 Transfer Between Fluidized Bed Particles and Fluidizing Gas

The homogeneous bed approach proposes the same governing equation for a fluidized as that for a fixed bed. In this approach, the rate of particle–gas mass transfer in a differential segment of the bed is again expressed in the following mass transfer equation:

$$\frac{dN_{\text{A}}}{dt} = k_{\text{g,bed}} S_{\text{ex,particles}} (C_{\text{A}}^{\text{i}} - C_{\text{A}})$$

In the above equation, the term $S_{\text{ex,particles}}$ represents the total exterior surfaces of the fluidized particles in the segment of the bed, and $k_{\text{g,bed}}$ represents the average mass transfer coefficient associated with these fluidized particles. As will be shown below, for the same group of particles, $k_{\text{g,bed}}$ is always higher under fluidized bed operations than that under fixed bed operations.

It should be pointed out that, in this homogeneous bed approach, the mass transfer coefficient between fluidized particles and fluidizing gas is difficult to evaluate and that the evaluated coefficient is essentially empirical in nature. This is especially due to the complex bubbling behavior and the highly nonuniform hydrodynamic phenomena within a fluidized bed, which make the accurate determination of the terms in Eq. (7) almost impossible. Nevertheless, several groups of workers have measured the average mass transfer coefficient based on the homogeneous bed approach. Table 3 briefly summarizes several of the studies reviewed by Kunii and Levenspiel (1991). Additional review of the studies has been provided

Table 2 Properties of Naphthalene Particles

Mesh size	ρ (g/cm^3)	X (—)	S_{a} (cm^2/cm^3)	S (cm^2/cm^3)	d_{p} (cm)	S_{d} (—)
14–20	0.646	0.391	88.1	53.65	0.089	4.775
20–28	0.626	0.411	106.5	62.73	0.073	4.579
28–35	0.615	0.422	167.7	96.93	0.045	4.362
35–48	0.651	0.389	187.8	114.75	0.039	4.475
48–65	0.607	0.430	262.5	149.63	0.028	4.189
Average	0.629	0.409				4.476

Source: Resnick and White, 1949.

Table 3 Mass Transfer Experiments Included in Figure 2

Investigators	Gas	Process*	Particles	$d_{p,m}$ (µm)	d_t (cm)	L_m (cm)
Resnick and White (1949)	Air H$_2$ CO$_2$	s. naphthalene	Naphthalene	210–1700	2.2 4.4	1.3–2.5
Kettenring et al. (1950)	Air	v. water	Silica gel Alumina	360–1000	5.9	10–15
Chu et al. (1953)	Air	s. naphthalene	Glass Lead Rape seed	710–1980	10.2	0.3–9.2
Richardson and Szekely (1961)	Air H$_2$	a. CCl$_4$ a. water	Active carbon Silica gel	88–2580	3.0	$5d_p$
Thodos et al. (1961, 69, 72)	Air	s. dichlorobenzene v. water v. nitrobenzene v. n-decane	Alumina Celite	1800–3100	3.8 9.5 11.3	0.6–7.0

* s. = sublimation, v. = vaporization, a. = adsorption.
Source: Kunii and Levenspiel, 1991.

by Krell et al. (1990). Typical results from these past studies are summarized in the following subsections.

2.3.1 Average Mass Transfer Coefficient, $k_{g,bed}$

The reported average mass transfer coefficients for fluidized bed particles are summarized in Fig. 2. As indicated, the average mass transfer coefficient for fluidized bed particles, again, can be higher or lower than that estimated from Eq. (2) for well-dispersed single spheres. The Sherwood number for fluidized bed particles is generally lower than that for single spheres at lower Reynolds numbers; however, it can be higher at higher Reynold numbers (Re$_p$ > 80). Similar to that for fixed bed particles, the Sherwood number for fluidized particles can be well below the theoretical minimum of the Sherwood number for single spheres, i.e., Sh$_{single}$ = 2.0. This is due to the potential overestimation of active surfaces involved in the mass transfer operation. In a bubbling fluidized bed, most of the particles are expected to stay in the emulsion phase with C_A^i at equilibrium with C_A in the emulsion gas. Many of these particles may thus be considered as inerts from the mass-transfer point of view because they do not contribute to significant amount of mass transfer to the bubbling gas. The fact that all the particle surfaces are included in the mass transfer calculation as described by Eq. (7), therefore, overestimates $S_{ex,particles}$, which in turn decreases the average mass transfer coefficient to a value even below the theoretical minimum for single spheres.

An attempt was made to correlate the average mass transfer coefficient reported by Resnick and White (1949) for small size fluidized particles, and the following equations are obtained (for air system, Sc = 2.35):

1. For particles with size between mesh 14 and 20 (d_p = 1000 µm),
$$\text{Sh}_{bed} = 0.200\,\text{Re}_p^{0.937} \quad \text{for } 30 < \text{Re}_p < 90 \tag{21}$$

2. For particles with size between mesh 20 and 28 (d_p = 711 µm),
$$\text{Sh}_{bed} = 0.274\,\text{Re}_p^{0.709} \quad \text{for } 15 < \text{Re}_p < 80 \tag{22}$$

3. For particles with size between mesh 28 and 35 (d_p = 570 µm),
$$\text{Sh}_{bed} = 0.773\,\text{Re}_p^{1.107} \quad \text{for } 8 < \text{Re}_p < 60 \tag{23}$$

4. For particles with size between mesh 35 and 48 (d_p = 410 µm),
$$\text{Sh}_{bed} = 0.071\,\text{Re}_p^{0.926} \quad \text{for } 6 < \text{Re}_p < 40 \tag{24}$$

5. For particles with size between mesh 48 and 65 (d_p = 275 µm),
$$\text{Sh}_{bed} = 0.041\,\text{Re}_p^{1.036} \quad \text{for } 4 < \text{Re}_p < 15 \tag{25}$$

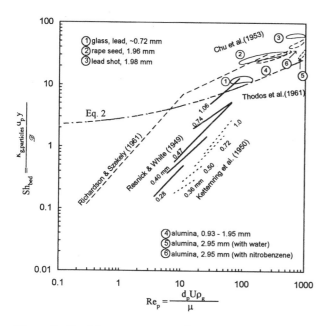

Figure 2 Particle–gas mass transfer coefficient for fluidized bed particles. (From Kunii and Levenspiel, 1991.)

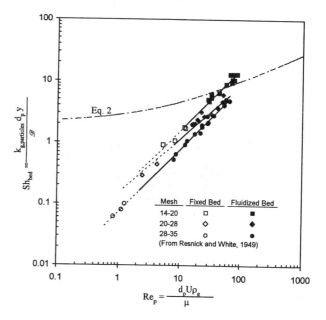

Figure 3 Comparison of mass transfer coefficient in fixed and fluidized beds.

2.3.2 Total Particle Exterior Surfaces, $S_{ex,particles}$

The procedure for determining the total particle exterior surfaces in a bed segment of a fluidized bed is essentially identical to that in a fixed bed, as described previously, i.e., using Eqs. (10) through (15). The ε_{bed} appearing in Eqs. (10) and (12), however, represents bed voidage in a fluidized bed (or ε_f) in this application.

2.3.3 Comparison of $K_{g,bed}$ between Fixed Bed and Fluidized Bed Particles

The experimental study of Resnick and White (1949) included the interesting observation that, although the bed hydrodynamics is significantly different, the measured mass transfer coefficients for the fixed and fluidized beds can be correlated with Re_p by the same set of constants as shown in Fig. 3. In the figure, the dashed lines represent fixed bed operations and the solid lines represent fluidized bed operations. As indicated, the Sherwood number (or mass transfer coefficient) for the same group of particles continues to increase with the Re_p, even during the transition of the operation from a fixed bed to a fluidized bed. It appears that Eqs. (21) through (25), although derived for fluidized bed operations, can be extrapolated to fixed bed operations as well.

2.4 Transfer Between Immersed Isolated Spheres and a Fluidized Bed

Another type of mass transfer encountered in a fluidized bed involves the transfer between an immersed active particle and a fluidized bed. Fluidized bed combustors are a typical example where active coal particles burn in the matrix of chemically inert particles of sand and ash which constitute 97 to 99% of the whole bed mass. The mass transfer between the immersed active particles and the fluidizing gas is expected to be different from that in a bed consisting of only active particles, because the inert particles may serve as an impermeable barrier for mass flow.

The subject of mass transfer involving isolated active particles and the fluidizing gas has been studied by several groups of researchers. Hsiung and Thodos (1977) measured the mass transfer coefficients of a batch of naphthalene particles fluidized in a matrix of inert particles similar in size and density. Vandershuren and Delvossalle (1980) experimentally studied the drying process of moist particles in a fluidized bed of similar dry particles. Palchonok and Tamarin (1983, 1985) measured the mass transfer coefficients for an isolated active particle of various sizes ($d_{p,a} = 3.5–30\,mm$) freely moving within a fluidized bed of large inert particles ($d_{p,i} = 1.2–6.3\,mm$). Kok et al. (1986) measured the particle–gas mass transfer coefficient in a 30 cm ID air fluidized bed consisting of inert glass beads and a varying mass fraction of subliming naphthalene beads. The above studies were comprehensively reviewed by Agarwal and La Nauze (1989) and Linjewile and Agarwal (1990).

In an an attempt to provide a better correlation for the mass transfer coefficient, Palchonok et al. (1992) derived an interpolation formula that couples an equation derived by Palchonok and Tamarin (1984), i.e.,

$$Sh_a = 0.117 Ar^{0.39} Sc^{0.33} \tag{26}$$

with an expression suggested by Avedesian (1972) for the limiting Sherwood number, i.e.,

$$Sh_{limit} = \frac{k_{g,a} d_{p} y}{\mathcal{D}} = 2\varepsilon_{mf} \tag{27}$$

to become

$$Sh_a = 2\varepsilon_{mf} + 0.117 Ar^{0.39} Sc^{0.33} \tag{28}$$

where Ar is the Archimedes number, defined as

$$Ar = g d_{p}^{3} \frac{\rho_{p} - \rho_{g}}{v^2 \rho_{g}} \tag{29}$$

The authors reported that Eq. (28) describes the experimental results of their own and those from other research groups well. The Sherwood number estimated from Eq. (28) is plotted against Re_p in Fig. 4

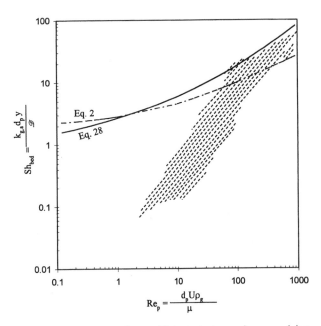

Figure 4 Mass transfer coefficients between immersed isolated active particles and a fluidized bed of inert particles. (Shaded area: range of particle–gas mass transfer coefficient measured in a fluidized bed of only active particles).

along with that estimated from the equation for single spheres, i.e., Eq. 2. The figure indicates that, at higher Reynolds numbers, the Sherwood number for active particles in a fluidized bed of inert particles is higher than that for single spheres suspended in a gas stream. However, at low Reynolds numbers, the trend is reversed, i.e., the Sherwood number for active particles in a fluidized bed is seen to be slightly lower than that for single spheres well-dispersed in a gas stream. This is possibly due to the resistance by inert particles at low Reynolds numbers. Again, the shaded area in Fig. 4 represents the range of particle–gas mass transfer coefficient measured in a fluidized bed as reviewed earlier. Additional correlation equations for the immersed active particles may be found elsewhere (see, e.g., Cobbinah et al., 1987; Hemati et al., 1992).

2.5 Summary on Homogeneous Bed Approach

The homogeneous bed approach correlates the particle–gas mass transfer coefficients in a fluidized bed in a manner similar to those in a fixed bed based on a plug-flow model. The experimentally measured average mass transfer coefficients for the bed particles under either the fixed- or the fluidized-bed operation can be higher or lower than those estimated from the theoretically derived and experimentally confirmed correlation equation for well-dispersed single spheres. For fine particles, the measured average mass transfer coefficients can be well below the theoretical minimum for single spheres. These low values of average mass transfer coefficients, however, should not be interpreted as the violation of theoretical laws. They are caused by the overestimation of the transfer potentials and/or the active surfaces involved in the mass transfer. The measured mass transfer coefficients in this approach, therefore, should be treated as empirical in nature. Figure 5 illustrates comments made by Kunii and Levenspiel (1991) regarding the homogeneous bed approach.

3   BUBBLING BED APPROACH

The bubbling bed approach takes into consideration the heterogeneous nature of a fluidized bed, in particular, the coexistence of bubble and emulsion phases. In this approach, the bubble phase is assumed to consist of spherical bubbles surrounded by spherical clouds. Historically, there have been three classes of models proposed for describing the mass transfer process in a fluidized bed based on the bubbling bed

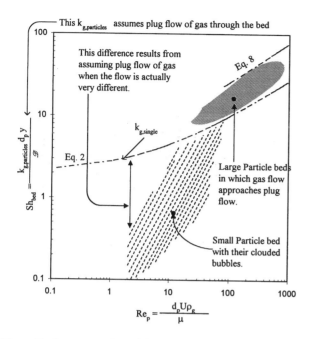

Figure 5 Summary of particle–gas mass transfer coefficient. (From Kunii and Levenspiel, 1991.)

approach. One class is based on bubble–emulsion transfer as represented by Kunii and Levenspiel (1969, 1991); the second class is based on cloud–emulsion transfer as represented by Partridge and Rowe (1966); the third class is based on empirically correlated bubble–emulsion transfer as represented by Chavarie and Grace (1976a). The three classes of models are described below.

3.1 Model of Kunii and Levenspiel (1991)

Kunii and Levenspiel (1969, 1991) considered the vaporization or sublimation of A from all particles in the bed. They assumed that fresh gas enters the bed only as bubbles, and that at steady state the measure of sublimation of A is given by the increase in C_A with height in the bubble phase. They further assumed that the equilibrium is rapidly established between C_A at the gas–particle interphase and its surroundings. The above assumptions lead to a mass transfer equation in terms of a bubble–emulsion mass transfer coefficient, K_{GB}:

$$\frac{dN_A}{dt} = u_b V_{bubble}\left(\frac{dC_{A,b}}{dz}\right) = K_{GB}V_{bubble}(C_{A,c} - C_{A,b}) \tag{30}$$

It is worth pointing out that, in this approach, the cloud phase and the emulsion phase are assumed to be perfectly mixed, which leads to $C_A^i = C_{A,e} = C_{A,c}$. Figure 6a defines the term K_{GB} graphically.

3.1.1 Relation Between K_{GB} and $k_{g,bed}$

The above equation proposed by Kunii and Levenspiel (1991), i.e., Eq. (30), can be derived from the traditional mass transfer expression described previously based on the homogeneous bed approach, i.e., Eq. (7):

$$\frac{dN_A}{dt} = k_{g,bed}S_{ex,particles}(C_A^i - C_A) \tag{7}$$

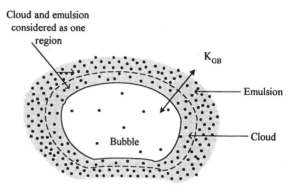

(a)

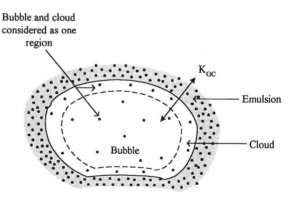

(b)

Figure 6 Graphical description of K_{GB} and K_{GC}. (a) Bubble-cloud interchange (from Kunii and Levenspiel, 1991), (b) cloud-emulsion interchange (from Partridge and Rowe, 1966.)

With the assumption that fresh gas enters the bed only as bubbles, Eq. (7) can be rewritten as

$$\frac{dN_A}{dt} = k_{g,bed} S_{ex,particles} (C_A^i - C_{A,b}) \quad (31)$$

where $C_{A,b}$ is the concentration of A in the bubble phase. Equation (31) can then be rewritten as

$$\frac{d(V_{bubble} C_{A,b})}{dt} = k_{g,bed} S_{ex,particles} (C_A^i - C_{A,b}) \quad (32)$$

The above equation can be rearranged to become

$$V_{bubble} \left(\frac{dC_A}{dt} \right) = k_{g,bed} S_{ex,particles} (C_A^i - C_{A,b}) \quad (33)$$

or

$$\frac{dC_{A,b}}{dt} = k_{g,bed} \frac{S_{ex,particles}}{V_{bubble}} (C_A^i - C_{A,b}) \quad (34)$$

Note that Eq. (32) implies that

$$\frac{dN_A}{dt} = V_{bubble} \frac{dC_{A,b}}{dt} \quad (35)$$

or

$$\frac{dN_A/dt}{V_{bubble}} = \frac{dC_{A,b}}{dt} \quad (36)$$

Then, with the nomenclature given in Fig. 7, for a segment of bed of height dz, we can write

$$dt = \frac{dz}{u_b} \quad (37)$$

Inserting Eq. (37) into Eq. (34) yields

$$\frac{dC_{A,b}}{dt} = u_b \left(\frac{dC_{A,b}}{dz} \right) = k_{g,bed} \left(\frac{S_{ex,particles}}{V_{bubble}} \right)_{segment}$$
$$(C_A^i - C_{A,b}) \quad (38)$$

where the terms $S_{ex,particles}$ and V_{bubble} in the bed segment can be formulated respectively by

$$S_{ex,particles} = A \, dz (1 - \varepsilon_f) a' \quad (39)$$

and

$$V_{bubble} = A \, dz \, \delta \quad (40)$$

In the above equations, a' is the ratio of particle surface to particle volume defined previously as

$$a' = \left(\frac{\text{surface of a particle}}{\text{volume of a particle}} \right) = \frac{\pi d_p^2 \phi_s}{\pi d_p^3 / 6}$$
$$= \frac{6}{d_p \phi_s} \quad (11)$$

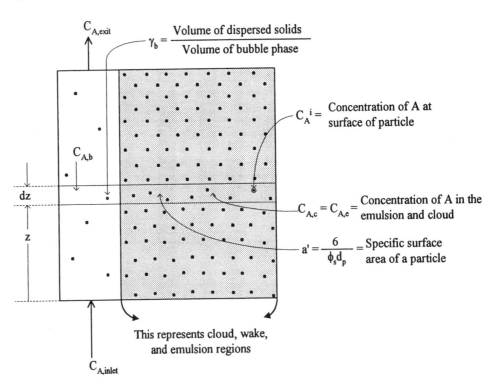

Figure 7 Nomenclature used in developing the model of Kunii and Levenspiel. (From Kunii and Levenspiel, 1991.)

ε_f is the void fraction in a fluidized bed as a whole, and δ is the bubble fraction in the bed segment. Equations (39) and (40) can be combined to yield

$$\left(\frac{S_{\text{ex,particles}}}{V_{\text{bubble}}}\right)_{\text{segment}} = \frac{(1-\varepsilon_f)a'}{\delta} \tag{41}$$

Inserting Eq. (41) into Eq. (38) yields

$$\frac{dC_{A,b}}{dt} = u_b\left(\frac{dC_{A,b}}{dz}\right) = \frac{k_{g,\text{bed}}(1-\varepsilon_f)a'}{\delta}(C_A^i - C_{A,b}) \tag{42}$$

and upon combining this with Eq. (36), the equation becomes

$$\frac{dN/dt}{V_{\text{bubble}}} = u_b\frac{dC_{A,b}}{dz} = \frac{k_{g,\text{bed}}(1-\varepsilon_f)a'}{\delta}(C_A^i - C_{A,b}) \tag{43}$$

With the assumption that equilibrium is rapidly established between C_A at the gas–particle interphase and its surroundings, i.e., $C_A^i = C_{A,e} = C_{A,c}$, Eq. (43) can be expressed as

$$\frac{dN_A}{dt} = u_b V_{\text{bubble}}\left(\frac{dC_{A,b}}{dz}\right) = \frac{k_{g,\text{bed}}(1-\varepsilon_f)a'}{\delta} \\ V_{\text{bubble}}(C_{A,c} - C_{A,b}) \tag{44}$$

A comparison between Eqs. (44) and (30) yields

$$K_{GB} = \frac{k_{g,\text{bed}}(1-\varepsilon_f)a'}{\delta} \tag{45}$$

Note that Eq. (45) implies that

$$k_{g,\text{bed}} = \frac{\delta}{(1-\varepsilon_f)a'}K_{GB} \tag{46}$$

Combining with Eq. (11) the equation yields

$$\text{Sh}_{\text{bed}} = \frac{k_{g,\text{bed}}d_p y}{\mathcal{D}} = \frac{\delta/[(1-\varepsilon_f)a']K_{GB}d_p y}{\mathcal{D}} \\ = \frac{y\phi_s d_p^2 \delta}{6\mathcal{D}(1-\varepsilon_f)}K_{GB} \tag{47}$$

It is worth mentioning that, according to Kunii and Levenspiel (1991), K_{GB} represents the transfer of A from the particle phase to the bubble phase via two sources, i.e., transfer from particles dispersed in the bubble phase and transfer of gas across the bubble–cloud boundary. These are discussed below.

3.1.2 K_{GB} for Nonporous and Nonadsorbing Particles

For a fluidized bed with nonporous and nonadsorbing particles, the particles dispersed in the bubble phase will not contribute to any additional mass transfer; the transfer of gas across the bubble–cloud boundary therefore is the only source of mass transfer, i.e.,

$$K_{GB} = K_{bc} \tag{48}$$

where K_{bc} is the bubble-cloud interchange coefficient derived by Davidson and Harrison (1963), i.e.,

$$K_{bc} = 4.5\frac{U_{\text{mf}}}{d_b} + 5.85\frac{\mathcal{D}^{0.5}g^{0.25}}{d_b^{1.25}} \tag{49}$$

Note that the above expression for K_{bc} considers the combined effects of through-flow of gas (q) and the mass transfer coefficient between bubble and cloud (k_{bc}), i.e.,

$$K_{bc} = q + k_{bc}S_{\text{ex,bubble}} \tag{50}$$

where

$$q = \frac{3\pi}{4}U_{\text{mf}}d_b^2 \tag{51}$$

and

$$k_{bc} = 0.975\mathcal{D}^{0.5}\left(\frac{g}{d_b}\right)^{0.25} \tag{52}$$

Inserting Eq. (48) into Eq. (47) yields

$$\text{Sh}_{\text{bed}} = \frac{y\phi_s d_p^2 \delta}{6\mathcal{D}(1-\varepsilon_f)}K_{bc} \tag{53}$$

The above equation provides a correlation for estimating Sh_{bed} (or $k_{g,\text{bed}}$) based on particle and fluidization properties, which then can be used to determine the rate of mass transfer employing Eq. (7) based on the homogeneous bed approach. In the application, the following equations provided by Kunii and Levenspiel (1991) are needed for calculating fluidization properties appearing in Eq. (53)

$$\delta = \frac{U - U_{\text{mf}}}{u_b} \tag{54}$$

$$u_b = (U - U_{\text{mf}}) + u_{br} \tag{55}$$

$$u_{br} = 0.711(g\,d_b)^{0.5} \tag{56}$$

$$1 - \varepsilon_f = (1 - \varepsilon_{\text{mf}})(1 - \delta) \tag{57}$$

The above equations indicate that, besides gas and particle properties, the fluidization properties are a function of bubble size only.

3.1.3 K_{GB} for Highly Adsorbing Particles

For highly adsorbing or sublimable particles, both the particles dispersed in the bubble phase and the bubble–cloud gas interchange can contribute to the particle–gas mass transfer, and the expression for K_{GB} takes the form

$$K_{GB} = \gamma_b a' k_{g,single} + K_{bc} = \gamma_b \frac{6}{d_p \phi_s} k_{g,single} + K_{bc} \tag{58}$$

Since $k_{g,single}$ can be related to Sh_{single} by

$$k_{g,single} = \frac{Sh_{single} \mathcal{D}}{d_p y} \tag{59}$$

Equation (58) can be rearranged to become

$$K_{GB} = \gamma_b \left[\frac{6}{(d_p \phi_s)} \frac{Sh_{single} \mathcal{D}}{d_p y} \right] + K_{bc}$$

$$= \gamma_b \frac{6 Sh_{single} \mathcal{D}}{\phi_s d_p^2 y} + K_{bc} \tag{60}$$

Inserting Eq. (60) in Eq. (47) yields:

$$Sh_{bed} = \frac{y \phi_s d_p^2 \delta}{6 \mathcal{D}(1 - \varepsilon_f)} \gamma_b \frac{6 Sh_{single} \mathcal{D}}{\phi_s d_p^2 y} + K_{bc}$$

$$= \frac{\delta}{(1 - \varepsilon_f)} \left\{ \gamma_b Sh_{single} + \frac{\phi_s d_p^2 y}{6 \mathcal{D}} K_{bc} \right\} \tag{61}$$

Like Eq. (53) this equation provides a correlation for estimating Sh_{bed} (or $k_{g,bed}$) based on particle and fluidization properties, which then can be used to determine the rate of mass transfer employing Eq. (7). Eqs. (54) through (57) are also needed for calculating fluidization properties appearing in Eq. (61). Again, all these equations indicate that, besides gas and particle properties, the fluidization properties are a function of bubble size only. Note that, for a given bed of solids and constant bubble size, Eq. (61) reduces to the form (Kunii and Levenspiel, 1991):

$$Sh_{bed} = a Re_p + b \tag{62}$$

The equation indicates that Sh_{bed} is linearly related to the particle Reynolds number with a slope a and an intercept b. This conclusion is similar to but not in complete agreement with that generated empirically based on the curve fitting of experimental results, where an exponential relationship was derived as expressed in Eqs. (21) through (25). It may be that the bubbles do not have a constant size over a wide range of Re_p. However, for a short range of Re_p, this conclusion seems reasonable.

3.1.4 K_{GB} for Porous or Partially Adsorbing Particles

For porous but nonadsorbing or partially adsorbing particles, Kunii and Levenspiel (1991) derived the following equation involving η_d:

$$Sh_{bed} = \frac{\delta}{(1 - \varepsilon_f)} \left[\gamma_b Sh_{single} \eta_d + \frac{\phi_s d_p^2 y}{6 \mathcal{D}} K_{bc} \right] \tag{63}$$

where

$$\eta_d = \frac{1}{1 + \alpha/m} \tag{64}$$

and

$$\alpha = \frac{6 k_{g,single} t_{p,mean}}{\phi_s d_p} \tag{65}$$

Note that, in Eq. (64), m is the adsorption equilibrium constant defined as

$$C_{As} = m C_A^i \tag{66}$$

where C_{As} is the concentration of tracer A within the particle in equilibrium with the concentration C_A^i of tracer gas at the gas–particle interphase. Note that for nonporous and nonadsorbing particles, $m = 0$ and $\eta_d = 0$, and Eq. (63) reduces to Eq. (53). For highly adsorbing particles, m is on the order of thousands, in which case $\eta_d \to 1.0$ and Eq. (63) becomes Eq. (61). For porous but nonadsorbing particles, $m = \varepsilon_p$, and Eq. (63) can be used to estimate Sh_{bed}.

In demonstrating the effect of the particle adsorbing property on the mass interchange coefficient, Wakabayashi and Kunii (1971) reported that the K_{GB} can be greatly enhanced even with a small value of ($\gamma_b \eta_d$). Their reported values are shown in Table 4. Rietema and Hoebink (1975) also reported that their measured gas interchange coefficients increase in bubble size and commented that this was just the reverse of the prediction of Eq. (49) for K_{bc}. According to Kunii and Levenspiel (1991), this is because the first term of Eq. (63) dominates for large bubbles, caused by more vigorous splitting and coalescing of bubbles.

3.2 Model of Partridge and Rowe (1966)

In this class of models, the cloud surrounding the bubble is considered the primary mass transfer boundary, and the bubble and the cloud phases are considered a perfectly mixed single phase, as indicated in Fig. 6(b). The mass transfer equation then has the form

Table 4 Effect of $\gamma_b \eta_d$ on K_{GB}

$d_{b,m}$ (m)	0.04	0.06	0.08	0.11
$K_{GB,measured}$ (s^{-1})	11–16	9–18	8–14	7–11
calculated, with $\gamma_b \eta_d = 0$	5	3	2	1.6
K_{GB}				
calculated, with $\gamma_b \eta_d = 4 \times 10^{-4}$	13	12	12	11

Source: Kunii and Levenspiel, 1991

$$\frac{dN_A}{dt} = V_{cloud}\left(\frac{dC_{A,c}}{dt}\right) = k_{gc} S_{ex,cloud}(C_{A,e} - C_{A,c}) \tag{67}$$

where V_{cloud} is the volume of gas in the bubble–cloud phase, k_{gc} the cloud emulsion mass transfer coefficient, and $S_{ex,cloud}$ the cloud exterior surface area. By analogy with the transfer from a drop of one immiscible liquid rising through another, Partridge and Rowe (1966) proposed the mass transfer coefficient to be correlated by the Sherwood number equation

$$Sh_c = \frac{k_{gc}d_c}{\mathcal{D}} = 2 + 0.69\,Sc^{0.33}Re_c^{0.5} \tag{68}$$

where Sc is the Schmidt number, defined previously as

$$Sc = \frac{\mu}{\rho_g \mathcal{D}} \tag{4}$$

and Re_c is a Reynolds number defined in terms of the relative velocity, u_R, between rising cloud and emulsion ($u_R = u_c u_e$), defined as

$$Re_c = \frac{u_R d_c \rho_g}{\mu} \tag{69}$$

In this equation, d_c is the diameter of the sphere with the same volume as the cloud, while $\mathcal{D}$, μ, and ρ_g represent diffusivity, viscosity, and density of the fluidizing gas, respectively.

The mass transfer equation given in Eq. (67) can also be expressed in terms of cloud-to-emulsion gas interchange coefficient, K_{GC}, as follows:

$$\frac{dN_A}{dt} = V_{cloud}\left(\frac{dC_{A,c}}{dt}\right) = K_{GC} V_{cloud}(C_{A,e} - C_{A,c}) \tag{70}$$

where K_{GC} can be related to k_{gc} by Partridge and Rowe, 1966)

$$\begin{aligned} K_{GC} &= \frac{k_{gc}\pi d_c^2 \varepsilon_{mf}}{V_{cloud}} \\ &= \frac{3.9\varepsilon_{mf}\mathcal{D}Sh_c}{(V_{cloud})^{0.67}} \end{aligned} \tag{71}$$

Equations (70) and (71) were used in the models of Rowe (1963) and Partridge and Rowe (1966). It is essential to note that the equation assumes a purely diffusive mechanism and does not consider gas through-flow between the cloud and the emulsion phases.

3.3 Model of Chavarie and Grace (1976a)

Chavarie and Grace (1976a) measured mass transfer rates for bubbles containing ozone injected into an air-fluidized two-dimensional bed and proposed the following empirical equation for their results:

$$\frac{dN_A}{dt} = V_{bubble}\left(\frac{dC_{A,b}}{dt}\right) = k_{gt} S_{ex,bubble}(C_{A,e} - C_{A,b}) \tag{72}$$

They compared the measured k_{gt} with the predictions of various models available in the literature and found that those which assumed a combination of through-flow and diffusion tend to overestimate the rate of transfer, while those which assume a purely diffusive mechanism underestimate it. Table 5 illustrates the results from various models. They concluded that the through-flow equation proposed by Murray (1965) best fit their data, which lead to

$$k_{gt} = \frac{U_{mf}}{\pi} \tag{73}$$

In a follow-up attempt, Sit and Grace (1978) derived the following equations to include the diffusive mechanism in the expression. For a circular two-dimensional bubble, the equation has the form

$$k_{gt} = \frac{U_{mf}}{\pi} + \left(\frac{4\mathcal{D}\varepsilon_{mf}u_b}{\pi d_b}\right)^{0.5} \tag{74}$$

Table 5 Interphase Mass Transfer Models and Coefficients for Ozone Exchange in an Air-Fluidized Two-Dimensional Bed

Model	Reference	Coefficient	k_{gt} equivalent	Calculated k_{gt} (m/s)
1. Penetration	Chavarie (1973)	$k_{gc} = \dfrac{4\mathcal{D}\varepsilon_{mf}u_b}{\pi d_c}$	k_{gc}	0.0051
2. Cloud two film	Walker (1970)	$k_{gc} = \dfrac{0.93\mathcal{D}\varepsilon_{mf}U_{mf}\sqrt{\alpha-1}}{d_c}$	$\dfrac{2}{3}k_{gc}$	0.0017
3. Partridge and Rowe	Partridge and Rowe (1966)	$k_{gc} = \dfrac{0.26\mathcal{D}\mathrm{Re}_c^{0.6}\mathrm{Sc}^{0.33}}{d_c}$	$\varepsilon_{mf}k_{gc}$	0.0020
4. Chiba and Kobayashi	Chiba and Kobayashi (1970)	$k_{gc} = \sqrt{\dfrac{4\mathcal{D}\varepsilon_{mf}^2 u_b(\alpha-1)}{\pi d_c \alpha}}$	k_{gc}	0.0030
5. Kunnii and Levenspiel	Kunii and Levenspiel (1969)	$k_{gb} = 0.6\mathcal{D}^{1/2}(g/d_b)^{1/4}$ $q = 2U_{mf}d_b w$ $Q = q + k_{gb}S_b$ $k_{gc} = \dfrac{4\mathcal{D}\varepsilon_{mf}u_b}{\pi d_c}$	$\dfrac{k_{gc}Q}{k_{gc}S_b+Q}$	0.0051
6. Toei et al.	Toei et al. (1969)	$k_{gc} = \dfrac{1.02\varepsilon_{mf}}{1+2\varepsilon_{mf}/(\alpha-1)}\sqrt{\dfrac{u_b\mathcal{D}(\alpha-1)}{d_b\alpha}}\sqrt{\dfrac{\alpha-1}{\alpha+1}}$	$k_{gc}+k_{gs}$	0.0056
7. Murray through-flow, no diffusion	Murray (1965)	$k_{gb} = 0$ $q = U_{mf}d_b w$	q/S_b	0.0160
Empirical value (from Charvarie and Grace, 1976a)				0.0160

Source: Chavarie and Grace, 1976a.

and for a spherical three-dimensional bubble, it has the form

$$k_{gt} = \frac{U_{mf}}{4} + \left(\frac{4\mathcal{D}\varepsilon_{mf}u_b}{\pi d_b}\right)^{0.5} \tag{75}$$

The above equation can be converted to K_{GT}, the mass interchange coefficient per unit volume of bubble, based on the appropriate surface-to-volume ratio, i.e.,

$$K_{GT} = \frac{6}{d_b}k_{gt}$$
$$= \frac{1.5U_{mf}}{d_b} + \frac{12}{d_b^{1.5}}\left(\frac{\mathcal{D}\varepsilon_{mf}u_b}{\pi d_b}\right)^{0.5} \tag{76}$$

The corresponding mass transfer equation for K_{GT} appearing in Eq. (76) is

$$\left(\frac{dN_A}{dt}\right) = V_{bubble}\left(\frac{dC_{A,b}}{dt}\right) = K_{GT}V_{bubble} \tag{77}$$
$$(C_{A,e} - C_{A,b})$$

Note that this equation is essentially in the same form as that of Eq. (30) proposed by Kunii and Levenspiel (1991), i.e., K_{GT} reported by Sit and Grace (1978) is directly comparable to K_{GB} described previously by Kunii and Levenspiel (1991).

In a similar attempt, Hatano and Ishida (1986) studied the particle–gas mass transfer coefficient in a three-dimensional fluidized bed of nonadsorbing glass beads with $d_p = 0.18\,\text{mm}$ using optical fiber probes. Tracer gas concentrations in and around rising single bubbles was measured continuously by the penetrative probes, while the bubble boundary and the zone with the prominent particle movement were detected by reflective probes. The measurements provided necessary data for determining the interphase mass transfer coefficient. The following equation resulted from their study:

$$k_{gt} = 0.127 \mathcal{D}^{0.33} g^{0.33} \left(\frac{u_b \varepsilon_{mf}}{U} \right)^{0.22} \tag{78}$$

where u_b can be calculated from Eq. (55). Note that, in addition to single-bubble experiments, they also measured the interphase mass transfer coefficient associated with paired bubbles. However, they found that the mass transfer coefficient remains unchanged in both single-bubble and paired-bubble conditions. This finding apparently is different from that reported in the literature from other researchers, as reviewed below.

3.3.1 Correlation for Multiple-Bubble Beds

The correlation equations described previously for K_{GC} and K_{GT} are all based on single-bubble experiments. Although they provide valuable insight and lead to useful comparisons, these correlation equations may not be directly applicable to freely bubbling three-dimensional beds. In such beds, complex bubbling phenomena such as growth, splitting, coalescence, and wake shedding occur constantly along the bed, and the interphase mass transfer mechanism is expected to be much more complex than those in single-bubble beds. Sit and Grace (1981) used a noninterfering technique to measure the concentration of ozone in pairs of bubbles injected into a bed of 390 μm glass beads fluidized by ozone-free air. They found that the transfer of ozone tracer from the bubble phase to the emulsion phase is enhanced over the transfer from isolated bubbles in the same particles and the same column. Bubble growth is also greater for the case where pairs of bubbles are introduced than when bubbles are present in isolation. They also found that enhancement of interphase mass transfer for interacting bubbles increases with particle size and can be explained in terms of enhancement of the through-flow component of transfer, while the diffusive component remains unaltered.

Their study has resulted in the following two modified forms of Eqs. (74) and (75) for a freely bubbling, two- or three-dimensional fluidized bed. For two-dimensional beds, the modified equation has the form

$$k_{gt,m} = 0.4 U_{mf} + \left(\frac{4 \mathcal{D} \varepsilon_{mf} u_{b,m}}{\pi d_{b,m}} \right)^{0.5} \tag{79}$$

and for three-dimensional beds, the modified equation is

$$k_{gt,m} = \frac{U_{mf}}{3} + \left(\frac{4 \mathcal{D} \varepsilon_{mf} u_{b,m}}{\pi d_{b,m}} \right)^{0.5} \tag{80}$$

Again, as with Eq. (76) for three-dimensional beds, $k_{gt,m}$ can be converted to $K_{GT,m}$ by the expression

$$K_{GT,m} = \left(\frac{6}{d_{b,m}} \right) k_{gt,m}$$
$$= \frac{2.0 U_{mf}}{d_{b,m}} + \frac{12}{d_{b,m}^{1.5}} \left(\frac{\mathcal{D} \varepsilon_{mf} u_{b,m}}{\pi d_{b,m}} \right)^{0.5} \tag{81}$$

According to Yates (1983), Eqs. (80) and (81) would seem to give the best available values of interphase mass coefficients in three-dimensional beds of nonadsorbing particles. Note again that $K_{GT,m}$ appearing in Eq. (81) is directly comparable to K_{GB} appearing in Eq. (30) proposed by Kunii and Levenspiel (1991).

3.4 Mass Transfer in the Grid Region

The grid region plays an important role in determining the reaction conversion of fluidized bed reactors, especially for fast reactions where the mass transfer operation is the controlling mechanism. Experimental studies have indicated that changing from one distributor to another, all other conditions remaining fixed, can cause major changes in conversion (Cooke et al., 1968; Behie and Kehoe, 1973; Bauer and Werther, 1981). It is generally observed that, in the grid region, additional mass transfer can take place because of the convective flow of gas through the interphase of the forming bubbles. The flow of gas through the forming bubbles into the dense phase and then returning to the bubble phase higher in the bed represents a net exchange between the two phases. The experimental work of Behie and Kehoe (1973) indicated that the mass transfer coefficient in the grid region, k_{je}, can be 40 to 60 times that in the bubble region, i.e., k_{be}.

Sit and Grace (1986) measured time-averaged concentrations of methane in the entry region of beds of 120 to 310 μm particles contained in a 152 mm column with a central orifice of diameter 6.4 mm and auxiliary tracer-free gas. The following equations were proposed for the particle–gas mass transfer in this region:

$$V_{bubble} \left(\frac{dC_{A,b}}{dt} \right) = Q_{or} C_{A,or} + k_{be1} S_{ex,bubble} C_{A,e}$$
$$- (Q_{or} + k_{be1} S_{ex,bubble}) C_{A,b} \tag{82}$$

where Q_{or} represents the convective mechanism defined as

$$Q_{or} = u_{or}\left(\frac{\pi}{4}\right)d_{or}^2 \tag{83}$$

and k_{be1} represents the diffusive mechanism, which can be predicted by the penetration theory expression, i.e.,

$$k_{be1} = \left(\frac{4\mathcal{D}\varepsilon_{mf}}{\pi t_f}\right)^{0.5} \tag{84}$$

Their results indicated that the convective mechanism has a greater effect on the grid region mass transfer. Since bubbles are also smaller near the grid, they concluded that favorable gas–solid contacting occurs in this region primarily because of convective outflow followed by recapture.

3.5 Effect of Particle Adsorption

As pointed out by Kunii and Levenspiel (1991), the mass transfer rate in a fluidized bed can be substantially enhanced for highly adsorbing particles even though the amount of particles dispersed in the bubble phase may be small. Similar observations have also been reported in the literature, see, e.g., Yoshida and Kunii (1968), Drinkenburg and Rietema (1973), Yates and Constans (1973), Chiba and Kobayashi (1970), Nguyen and Porter (1976), and Gupalo et al. (1978). Among these investigators, Chiba and Kobayashi (1970) derived a theoretical expression for the ratio of interphase mass transfer coefficients in the presence (K'_{be}) and absence (K_{be}) of adsorption effects:

$$\frac{K'_{be}}{K_{be}} = \left(1 + 0.67m\frac{1 - \varepsilon_{mf}}{\varepsilon_{mf}}\left(1.5 + \frac{\alpha'}{\alpha' + 1}\right)\right)^{0.5} \tag{85}$$

where m is the adsorption equilibrium constant defined in Eq. (66) and α' is defined as

$$\alpha' = \frac{u_b \varepsilon_{mf}}{U_{mf}} \tag{86}$$

Equation (85) was later confirmed experimentally by Bohle and van Swaaij (1978), who measured mass transfer coefficients for a number of adsorbing (e.g., propane) and nonadsorbing (e.g., helium) gases in a fluidized bed of silica-alumina. A typical comparison is shown in Fig. 8, where the effect of adsorption is obvious with the enhancement being as high as 100%.

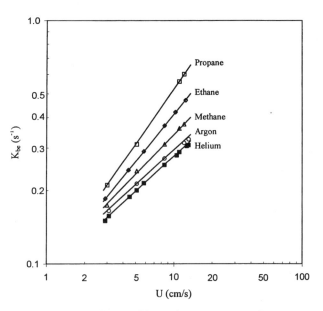

Figure 8 Dependence of interphase mass transfer rate on has adsorptivity. (From Bohle and van Swaaij, 1978.)

4 RELATION BETWEEN MASS AND HEAT TRANSFER COEFFICIENTS

The correlation equations for particle–gas mass and heat transfer coefficients are closely related when the Sherwood number (Sh) is equivalent to the Nusselt number (Nu), and the Schmidt number (Sc) is equivalent to the Prandtl number (Pr). All the particle–gas correlation equations for the Sherwood number and the Nusselt number are therefore interchangable. For example, the correlation equation for estimating the Nusselt number for particles in a fixed bed is expressed as

$$Nu_{bed} = \frac{hd_p}{k_g} = 2 + 1.8(Re_p)^{0.5}(Pr)^{0.333} \tag{87}$$

$$\text{for } Re_p > 80$$

whereas the Prandtl number (Pr) is defined as

$$Pr = \frac{C_p \mu}{k_g} \tag{88}$$

Note that the corresponding equation for the Sherwood number has the form

$$Sh_{bed} = \frac{k_{g,bed}d_p y}{\mathcal{D}} = 2 + 1.8(Re_p)^{0.5}(Sc)^{0.333} \tag{89}$$

$$\text{for } Re_p > 80$$

The above discussion implies that the correlation equations derived for Nu_{bed} based on heat transfer experiments can also be converted to the Sherwood number for estimating $k_{g,particles}$.

5 CONCLUDING REMARKS FROM KUNII AND LEVENSPIEL (1991)

It is worth bringing up the concluding remarks of Kunii and Levenspiel (1991) at this point to conclude the chapter regarding particle–gas mass transfer in a fluidized bed. Three points were made: first, particles dispersed in bubbles should be taken into account when kinetic processes, such as mass transfer, are carried out in fluidized beds; second, when dealing with a gaseous component that is adsorbed or somewhat captured by the bed solids (or else desorbed), K_{bc} should be used carefully to represent the movement of these adsorbed gaseous components; third, the mass transfer coefficient measured for the bed as a whole, $k_{g,bed}$, is model dependent. For large-particle cloudless bubble beds, the plug flow model closely matches the flow conditions in the bed, and the $k_{g,bed}$ should match the single particle coefficient, i.e., $k_{g,single}$. However, for fine-particle clouded bubble beds, the $k_{g,bed}$ may be well below that of $k_{g,single}$.

6 EXAMPLE CALCULATIONS

Two examples are given here to demonstrate the effects of d_b and γ_b on the prediction of the particle–gas mass transfer coefficient based on Eq. (61). The calculated results are then compared with the experimental data of Resnick and White (1949).

6.1 Example 1 Effect of d_b on Sh_{bed} based on Eq. (61)

Data. $\rho_s = 1.06\,g/cm^3$, $\varepsilon_{mf} = 0.5$, $\phi_s = 0.4$, $\gamma_b = 0.005$, $\rho_g = 1.18 \times 10^{-3}\,g/cm^3$, $\mu = 1.8 \times 10^{-4}$ $g/(cm\,s)$, $\mathcal{D} = 0.065\,cm^2/s$, $Sc = 2.35$, $\eta_d = 1.0$, $y = 1.0$, $U_{mf} = 1.21\,cm/s$, $u_t = 69\,cm/s$.

Determine. The relation between Sh_{bed} and Re_p based on Eq. (61) with $d_b = 0.1$, 0.2, 0.37, 0.5, and 1 cm.

Solution. Equation (61) has the form

$$Sh_{bed} = \frac{\delta}{1 - \varepsilon_f}\left(\gamma_b Sh_{single} + \frac{\phi_s d_p^2 y}{6\mathcal{D}} K_{bc}\right)$$

From Eq. (6)

$$Re_p = \frac{\rho_g u d_p}{\mu}$$

At $u_t = 69\,cm/s$,

$$Re_{p,t} = \frac{\rho_g u_t d_p}{\mu} = \frac{(0.00118)(69)(0.028)}{0.00018} = 12.66$$

Then Eq. (2) gives

$$Sh_{single} = 2 + 0.6(12.66)^{0.5}(2.35)^{0.33} = 4.84 \qquad (A)$$

From Eq. (49) (using $d_b = 0.37\,cm$),

$$
\begin{aligned}
K_{bc} &= 4.5\frac{U_{mf}}{d_b} + 5.85\frac{\mathcal{D}^{0.5} g^{0.25}}{d_b^{1.25}} \\
&= 4.5\left(\frac{1.21}{0.37}\right) + 5.85\frac{(0.065)^{0.5}(980)^{0.25}}{(0.37)^{1.25}} \\
&= 43.63\,(s^{-1})
\end{aligned}
\qquad (B)
$$

Combining Eqs. (54) through (57), i.e.,

$$\delta = \frac{U - U_{mf}}{u_b}$$

$$u_b = U - U_{mf} + u_{br}$$

$$u_{br} = 0.711(g d_b)^{0.5}$$

$$(1 - \varepsilon_f) = (1 - \varepsilon_{mf})(1 - \delta)$$

we obtain

$$
\begin{aligned}
\frac{\delta}{1 - \varepsilon_f} &= \frac{U - U_{mf}}{u_{br}(1 - \varepsilon_{mf})} \\
&= \frac{U - 1.21}{0.711(980 \times 0.37)^{0.5}(1 - 0.5)} \\
&= \frac{U - 1.21}{6.77}
\end{aligned}
\qquad (C)
$$

Also

$$\frac{\phi_s d_p^2 y}{6\mathcal{D}} = \frac{(0.4)(0.028)^2(1)}{6 \times 0.065)} = 8.04 \times 10^{-4} \qquad (D)$$

Combining (A) through (D) yields

$$
\begin{aligned}
Sh_{bed} &= \frac{\delta}{1 - \varepsilon_f}\gamma_b Sh_{single} + \frac{\phi_s d_p^2 y}{6\mathcal{D}} K_{bc} \\
&= \frac{U - 1.21}{6.77}[(0.005)(4.84) + 0.000804(43.63)] \\
&= 0.0088U - 0.011
\end{aligned}
\qquad (E)
$$

304

Ho

Then, since

$$\mathrm{Re}_p = \frac{\rho_g U d_p}{\mu} = \frac{(0.028)U(0.00118)}{0.00018} = 0.184U \quad (F)$$

Combining expressions (E) and (F) gives

$$\mathrm{Sh}_{bed} = 0.048\,\mathrm{Re}_p - 0.011 \quad (G)$$

Note that the above equation is generated based on $d_b = 0.37$ cm. This calculation procedure can be easily repeated for different bubble sizes at 0.1, 0.2, 0.5, and 1.0 cm. The generated equations for these different bubble sizes are plotted in Fig. 9 along with the experimentally observed data of Resnick and White (1949). The results indicate that the correlation based on $d_b = 0.37$ cm describes the experimental data best. The results in Fig. 9, however, indicate that the Sherwood number is sensitive to the bubble size selected; a smaller bubble size generates a higher Sherwood number, i.e., a higher mass transfer coefficient.

6.2 Example 2. Effect of γ_b on Sh_{bed} based on Eq. (61)

Data. Same as Example 1; but, with $d_b = 0.37$ cm.

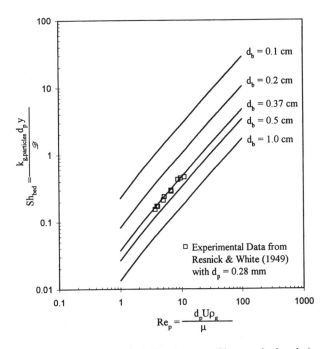

Figure 9 Effect of bubble size on Sh_{bed} calculated in Example 1.

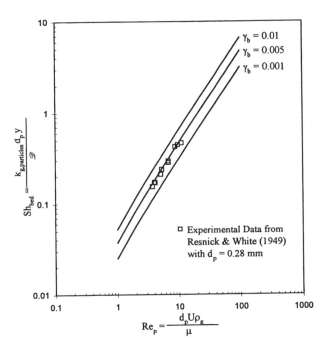

Figure 10 Effect of γ_b based on Eq. (61) with $d_b = 0.37$ cm.

Determine. The relation between Sh_{bed} and Re_p based on Eq. (61) with $d_b = 0.37$ cm and $\gamma_b = 0.001$, 0.005, and 0.01.

Solution. The identical procedure as is demonstrated in Example 1 can be used to evaluate Sh_{bed} at different γ_b. The results are shown in Fig. 10 where, as expected, a higher γ_b will generate a higher Sherwood number. The results in Fig. 10 clearly indicate that, in addition to bubble size, the selection of γ_b is also essential in the estimation of the particle–gas mass transfer coeffficient based on Eq. (61) proposed by Kunii and Levenspiel (1991).

NOMENCLATURE

A	= cross-sectional area of bed, m^2
Ar	= Archimedes number defined in Eq. (29), Ar $= g d_p^3 (\rho_p - \rho_g)/(v^2 \rho_g)$
a'	= surface areas of solid per volume of solid, m^{-1}
C_A	= concentration of A in bulk gas phase, kg-mol/m^3
C_A^i	= concentration of A at gas–particle interphase, kg-mol/m^3
$C_{A,b}$	= concentration of A in bubble phase, kg-mol/m^3

$C_{A,c}$ = concentration of A in cloud phase, kg-mol/m^3

$C_{A,e}$ = concentration of A in emulsion phase, kg-mol/m^3

C_{As} = concentration of A within a particle defined in Eq. (65), kg-mol/m^3

C_p = heat capacity at constant pressure,

d = diameter, m

d_b = effective bubble diameter, m

$d_{b,m}$ = mean effective bubble diameter defined in Eqs. (79) and (80), m

d_c = cloud diameter appearing in Eq. (68), m

d_{eff} = effective particle diameter defined in Eqs. (13)–(15), m

d_{or} = orifice diameter, m

d_p = particle diameter, m

d_p = particle diameter defined in Eq. (20), cm

$d_{p,a}$ = particle diameter of active particles, m

d_{pi} = particle diameter of inert particles, m

$d_{p,m}$ = mean particle diameter appearing in Table 3, cm

d_{sph} = equivalent spherical diameter of a particle, m

d_t = bed diameter appearing in Table 3, cm

g = gravitational acceleration, 9.8 m/s^2

h = height or thickness, m

h = particle–gas heat transfer coefficient, Wm^{-2} K^{-1}

K_{bc} = coefficient of gas interchange between bubble and cloud phase, m^3/s

K'_{be} = K_{be} associated with adsorbing particles, m^3/s

K_{GB} = interchange coefficient between bubble and cloud–emulsion, s^{-1}

K_{GC} = interchange coefficient between bubble-cloud and emulsion, s^{-1}

K_{GT} = K_{GB} appearing in Eqs. (76) and (77), s^{-1}

$K_{GT,m}$ = mean K_{GB} appearing in Eq. (81), s^{-1}

K_t = permeability defined in Eq. (19)

k_{bc} = mass transfer coefficient between bubble and cloud, m/s

k_{be} = bubble-to-emulsion mass transfer coefficient, m/s

k_{be1} = bubble-to-emulsion mass transfer coefficient during bubble formation, m/s

k_g = thermal conductivity of gas, Wm^{-1} K^{-1}

$k_{g,a}$ = mass transfer coefficient of active particles immersed in a fluidized bed, m/s

$k_{g,bed}$ = average mass transfer coefficient of bed particles, m/s

$k_{g,single}$ = mass transfer coefficient of well-dispersed single spheres, m/s

k_{gc} = mass transfer coefficient defined in Eq. (66) based on cloud surface, m/s

k_{gt} = mass transfer coefficient defined in Eq. (72) based on bubble surface, m/s

$k_{gt,m}$ = mean mass transfer coefficient defined in Eqs. (79) and (80), m/s

k_{je} = jet to emulsion mass transfer coefficient, m/s

L_m = bed height appearing in Table 2, cm

L = bed height defined in Eq. (16), cm

m = Adsorption equilibrium constant, defined in Eq. (65)

N_A = mole of A, kg-mol

Nu_{bed} = Nusselt number of bed particles, $Nu_{bed} = (hd_p)/k_g$

ΔP_p = pressure drop through porous beds appearing in Eq. (19), g/cm^2

Q = air flow rate appearing in Eq. (19), cm^3/s

Q_{or} = gas flow through orifice, m^3/s

q = through-flow of gas defined in Eq. (51), m^3/s

Re_c = Reynolds number based on cloud properties, $Re_c = (\rho_g u_c d_c)/\mu$

Re_p = particle Reynolds number, $Re_p = (\rho_g U d_p)/\mu$

Re_{sph} = particle Reynolds number based on d_{sph}, $Re_p = (\rho_g U d_{sph})/\mu$

S = surface area per unit bulk volume of bed defined in Eqs. (16) and (17), cm^{-1}

S_a = specific surface area per unit volume of solid as defined in Eqs. (17) and (18), cm^{-1}

Sc = Schmidt number, $Sc = \mu/(\rho_g \mathcal{D})$

S_d = $S_d = (Sd_p)$ as defined in Eq. (20)

$S_{ex,single}$ = exterior surface of a single sphere or well-dispersed single spheres, m^2

$S_{ex,particles}$ = total exterior surface of bed particles, m^2

$S_{ex,bubble}$ = exterior surface of bubbles, m^2

$S_{ex,cloud}$ = exterior surface of clouds, m^2

Sh = Sherwood number, $Sh = (k_g d_p y)/\mathcal{D}$

Sh_a = Sherwood number of isolated active particles, $Sh_a = (k_{g,a} d_p y)/\mathcal{D}$

Sh_{bed} = average Sherwood number of bed particles, $Sh_{bed} = (k_{g,bed} d_p y)/\mathcal{D}$

Sh_{limit} = limit Sherwood number defined in Eq. (27)

Sh_{single} = Sherwood number of single spheres, $Sh_{single} = (k_{g,single} d_p y)/\mathcal{D}$

t = time, s

t_f = bubble formation time, s

$t_{p,mean}$ = mean particle residence time staying in bubble phase, s

U = superficial gas velocity, m/s

U_{mf} = minimum fluidization velocity, m/s

u = gas velocity, m/s

u_b = bubble rise velocity, m/s

$u_{b,m}$ = mean bubble rise velocity defined in Eqs. (79) and (80), m/s

u_{br} = rise velocity of a bubble with respect to the emulsion phase, m/s

u_c = cloud (= bubble) rise velocity, m/s

u_e = velocity of the emulsion gas, m/s

u_{or} = velocity of gas through orifice, m/s

u_R = relative velocity, defined as $u_R = u_c - u_e$

u_t = particle terminal velocity, m/s

V_{bubble} = volume occupied by bubble phase, m^3

V_{cloud} = volume occupied by bubble–cloud phase, m^3

V_{seg} = volume of a bed segment, m^3

X	=	void fraction of bed appearing in Eqs. (17) and (18)
y	=	mole fraction of inert component
z	=	distance above the distributor, m
α	=	parameter of Eq. (64), defined in Eq. (65)
α'	=	parameter of Eq. (85), defined in Eq. (86)
γ_b	=	volume of solids dispersed in bubbles per unit bubble volume
$\mathcal{D}$	=	gas diffusion coefficient, m^2/s
δ	=	bubble fraction in a fluidized bed
ε_{bed}	=	void fraction in a fixed bed
ε_f	=	void fraction in a fluidized bed as a whole
ε_{mf}	=	void fraction in a bed at minimum fluidization velocity
ε_p	=	void fraction in a fluidized bed particle
μ	=	gas viscosity, $kg\ m^{-1}\ s^{-1}$
ν	=	kinematic viscosity, m^2/s
ρ_g	=	density of gas, kg/m^3
ρ_p	=	density of particle, kg/m^3
η_d	=	dispersion parameter defined in Eq. (64)
ϕ_s	=	sphericity of a particle
π	=	constant, ≈ 3.1416

REFERENCES

Agarwal PK, La Nauze RD. Transfer processes local to the coal particle: a review of drying, devolatilization and mass transfer in fluidized bed combustion. Chem Eng Res Des 67:457–480, 1989.

Avedesian MM. Combustion of char in fluidized beds. PhD dissertation: University of Cambridge, 1972.

Bauer W, Werther J. Proceedings of 2nd World Congress Chemical Engineering, 1981, p 69.

Behie LA, Kehoe P. The grid region in a fluidized bed reactor. AIChE J 19:1070–1072, 1973.

Behie LA, Bergougnou MA, Baker CGJ. Mass transfer from a grid jet in a large gas fluidized bed. In: Keairns DL, ed. Fluidization Technology, Vol 1. Washington DC: Hemisphere, 1976, pp 261–278.

Bohle H, Van Swaaij WPM. In: Davidson JF, Keairns DL, eds. Fluidization. London: Cambridge University Press, 1978, p 167.

Brown GG. Unit Operations. New York: John Wiley, 1950.

Carmon PC. Fluid flow through granular beds. Trans Inst Chem Eng 15:150–166, 1937.

Carmon PC. Symposium on New Methods for Particle Size Determination in the Subsieve Range. Amer Soc Testing Materials, 1941, p 24.

Chavarie C. Chemical reaction and interphase mass-transfer in gas fluidized beds. PhD dissertation, McGill University, 1973.

Chavarie C, Grace JR. Interphase mass transfer in a fluidized bed. In: Keairns DL, ed. Fluidization Technology, Vol 1. Washington DC: Hemisphere, 1976a, pp 251–259.

Chavarie C, Grace JR. Interphase mass transfer in a gas-fluidized bed. Chem Eng Sci 31:741–749, 1976b.

Chiba T, Kobayashi H. Gas exchange between the bubble and emulsion phases in gas–solid fluidized beds. Chem Eng Sci 25:1375–1385, 1970.

Chu JC, Kalil J, Wetteroth WA. Mass transfer in a fluidized bed. Chem Eng Prog 49:141–149, 1953.

Cobbinah S, Laguerie C, Gilbert H. Simultaneous heat and mass transfer between a fluidized bed of fine particles and immersed coarse porous particles. Int Heat and Mass Transfer 30:395–400, 1987.

Cooke MJ, Harris W, Highley J, Williams DF. Int Chem Engrs Symp Ser 30:21, 1968.

Davidson JF, Harrison D. Fluidised Particles. London: Cambridge University Press, 1963.

Froessling N. The evaporation of falling drops. Gerland Beitr Geophys 52:170, 1938.

Gupalo YP, Ryazantsez YS, Sergeez YA. In: Davidson JF, Keairns DL, eds. Fluidization. London: Cambridge University Press, 1978, p 162.

Hatano H, Ishida M. In: Ostergaard K, Soorensen A, eds. Fluidization V, Elsionore, Denmark. New York: Engineering Foundation, 1986, pp 119–127.

Hemati M, Mourad M, Steinmetz D, Laguerie C. Continuous and intermittent drying of maize in a flotation fluidized bed. In: Potter OE, Nicklin DJ, eds. Fluidization VII, Brisbane, Australia. New York: Engineering Foundation, 1992, pp 831–840.

Hsiung TH, Thodos G. Mass transfer in gas-fluidized beds: measurement of actual driving forces. Chem Eng Sci 32:581–592, 1977.

Hurt DM. Principles of reactor design: gas-solid interface reactions. Ind Eng Chem 35:522–528, 1943.

Kettenring KN, Manderfield EL, Smith JM. Heat and mass transfer in fluidized systems. Chem Eng Prog 46:139–145, 1950.

Kok J, Stark NL, van Swaaij WPM. In: Ostergaard K, Soorensen A, eds. Fluidization V, Elsionore, Denmark. New York: Engineering Foundation, 1986, pp 433–441.

Krell L, Kunne HJ, Morl L. Flow regimes and heat and mass transfer in gas-fluidized beds of solids. Internat Chem Eng 30:45–56, 1990.

Kunii D, Levenspiel O. Fluidization Engineering. New York: John Wiley, 1969.

Kunii D, Levenspiel O. Fluidization Engineering, 2nd ed. Boston: Butterworth-Heinemann, 1991.

Leva M. Fluidization. New York: McGraw-Hill, 1959.

Leva M, Grummer M, Weintraub M, Pollchick M. Introduction to fluidization. Chem Eng Prog 44:511–520, 1948.

Leva M, Grummer M, Weintraub M, Storch HH. A study of fluidization of an iron fischer–tropsch catalyst. Chem Eng Prog 44:707–716, 1948.

Leva M, Weintraub M, Grummer M, Pollchick M. Fluidization of an anthracite coal. Ind Eng Chem 41:1206–1212, 1949.

Levenspiel O. Chemical Reaction Engineering, 2nd ed. New York: John Wiley, 1972.

Linjewile TM, Agarwal PK. The 23rd Symposium on Combustion. Combustion Inst, 1990, p 917.

Murray JD. On the mathematics of fluidization: Part 1. Fundamental equations and wave propagation. J Fluid Mech 21:465–493, 1965.

Nguyen HV, Potter OE. Adsorption effects in fluidized beds. In: Keairns DL, ed. Fluidization Technology, Vol 2. Washington DC: Hemisphere, 1976, pp 193–200.

Palchonok GI, Tamarin AI. Mass transfer at a moving particle in a fluidized bed of coarse material. J Eng Phys 47:916–922, 1984.

Palchonok GI, Dolidovich AF, Andersson S, Leckner B. Calculation of true heat and mass transfer coefficients between particles and a fluidized bed. In: Potter OE, Nicklin DJ, eds. Fluidization VII, Brisbane, Australia. New York: Engineering Foundation, 1992, pp 913–920.

Partridge BA, Rowe PN. Chemical reaction in a bubbling gas-fluidised bed. Trans Inst Chem Engrs 44:335–348, 1966.

Ranz WE. Friction and transfer coefficients for single particles and pack beds. Chem Eng Prog 48:247–253, 1952.

Resnick WE, White RR. Mass transfer in systems of gas and fluidized solids. Chem Eng Prog 45:377–390, 1949.

Richardson JF, Szekely J. Mass transfer in a fluidised bed. Trans Inst Chem Engrs 39:212–222, 1961.

Rietema K, Hoebink J. In: Keairns DL, ed. Fluidization Technology, Vol 1. Washington DC: Hemisphere, 1976, pp 279–288.

Sit SP, Grace JR. Interphase mass transfer in an aggregative fluidized bed. Chem Eng Sci 33:1115–1122, 1978.

Sit SP, Grace JR. Effect of bubble interaction on interphase mass transfer in gas fluidized beds. Chem Eng Sci 36:327–335, 1981.

Sit SP, Grace JR. Interphase mass transfer during bubble formation in fluidized beds. In: Ostergaard K, Soorensen A, eds. Fluidization V, Elsionore, Denmark. New York: Engineering Foundation, 1986, pp 39–46.

Thodos G, Riccetti RE. Mass transfer in the flow of gases through fluidized beds. AIChE J 7:442–444, 1961.

Thodos G, Wilkin GS. In: Drinkenburg AAH, ed. Proc Int Symp on Fluidization. Amsterdam: Netherlands University Press, 1967, pp 586–594.

Thodos G, Wilkin GS. Mass transfer driving forces in packed and fluidized beds. AIChE J 15:47–50, 1969.

Thodos G, Yoon P. Mass transfer in the flow of gases through shallow fluidized beds. Chem Eng Sci 27:1549–1554, 1972.

Toei R, Matsuno R, Miyigawa H, Nishitani K, Komagawa Y. Gas transfer between a bubble and the continuous phase in a gas-solid fluidized bed. Internat Chem Eng 9:358–364, 1969.

Uchida S, Fujita S. J Chem Soc, Ind Eng Section (Japan) 37:1578 (1583, 1589, 1707), 1934.

Vandershuren I, Delvossalle C. Particle-to-particle heat transfer in fluidized bed drying. Chem Eng Sci 35:1741–1748, 1980.

Wakabayashi T, Kunii D. Contribution of solids dispersed in bubbles to mass transfer in fluidized bed. J Chem Eng (Japan) 4:226–230, 1971.

Walker BV. Gas-solid contacting in bubbling fluidised beds. PhD dissertation, Cambridge University, 1970.

Yates JG. Fundamentals of Fluidized-Bed Chemical Processes. London: Butterworths, 1983.

Yates JG, Constans JAP. Residence time distributions in a fluidized bed in which gas adsorption occurs: stimulus–response experiments. Chem Eng Sci 28:1341–1347, 1973.

Yoshida K, Kunii D. Stimulus and response of gas concentration in bubbling fluidized beds. J Chem Eng (Japan) 1:11–16, 1968.

Zenz FA. Encyclopedia of Chemical Technology, Chap 10, 3rd ed. Kirk Olhmer, 1978, p 548.

12

General Approaches to Reactor Design

Peijun Jiang, Fei Wei,* and Liang-Shih Fan

The Ohio State University, Columbus, Ohio, U.S.A.

1 INTRODUCTION

Fluidized bed reactors have a number of characteristics ideal for industrial applications and hold advantages over fixed bed reactors (e.g., Kunii and Levenspiel, 1991; Fan and Zhu, 1998). Among these characteristics are good particle mixing, good temperature control, and adaptability to high-pressure and high-temperature operations. Fluidized solid particles behave like liquids, allowing online particle addition/removal to adjust the catalyst activity and reducing downtime for catalyst replacement. Furthermore, this behavior permits easy transport of particles between reactors, adaptation of continuous operation of solids, more precise and automatic control, and greater flexibility in system configuration. Fluidized bed reactors can be operated over a wide range of gas flow rates, thereby allowing selection of the optimum contact time or residence time of gas and solid particles. Due to simple geometric configurations, fluidized bed reactors are suitable for large-scale operations. Other features of fluidized bed reactors are

The rapid mixing of solids, due to bubbles and strong turbulent flow, leads to isothermal and hot-spot-free operations.

A bed of well-mixed solids represents a large thermal flywheel and thus responds slowly to abrupt changes in operating conditions.

A flexible process temperature control enables optimization of other process variables to increase product yields and minimize wastes and by products.

The circulations of solids between two fluidized beds allows continuous catalyst regeneration.

The use of small particles gives rise to effectiveness factors close to unity

A high gas-to-particle mass and heat transfer rates can be established.

Some cohesive particles can be processed due to strong particle motion and interactions.

Disadvantages of fluidized bed reactors for industrial applications include

The gas bypass, in the form of bubbles, reduces the gas–solid contact efficiency in bubbling beds.

Poor gas–solid contact results from particle accumulation near the wall in risers.

A complicated bubble hydrodynamics and mixing patterns require special development efforts for process scaleup.

Nonuniform products are produced owing to wide distributions of gas and solids residence time in the bubbling regime.

The strong backmixing for both particles and gas results in low conversion and poor selectivity.

The strong solids motion leads to erosion of internal parts.

The attrition of catalyst particles leads to catalyst loss in cyclones.

*Current affiliation: Tsinghua University, Beijing, People's Republic of China

Particle entrainment leads to loss of particles and air pollution.

1.1 Features and Commercial Processes of Fluidized Bed Reactors

Fluidized beds have been applied to physical, chemical, metallurgical, mineral, and other operations. Examples of these applications are given in Table 1.

Fluidized bed reactors are often considered for implementation because of the solids mixing requirement for temperature uniformity, highly endothermic or exothermic reactions, small size of particles, continuous solids flow, and/or short gas–solid contact time. Catalysts are commonly the fluidizing particles in the fluidized reactor for chemical synthesis. Solid reactants are commonly utilized in metallurgical and mineral processing. Operating conditions required for each application primarily depend on the process rate derived from the transport and kinetics of the reactions concerned. For fast reactions, high-velocity fluidization or a fast fluidization regime is used. For slow reactions, a low-velocity fluidization or bubbling fluidization regime is used. In the bubbling beds, gas flows through the interstitial space between particles and

predominantly bypasses the bed in the form of bubbles without effectively contacting solid particles. On the other hand, bubbles act as agitators to enhance gas–solid mixing. Fluidized bed reactors typically consist of a vertical column (reactor), a gas distributor, cyclones, heat exchangers, an expanded section, and baffles, as shown in Fig. 1a. The gas distributor provides desired distributions of fluidizing gas and support for particles in the bed. Generally, a minimum pressure drop across the distributor is required to ensure the uniformity of gas distribution in the bed. Gas distributors can be designed in many ways, examples being sandwiched, stagger perforated, dished, grated, porous, tuyere, and cap types (Karri and Knowlton, 1999). In the bubbling regime, the distributor geometry strongly influences the initial bubble size from the distributor, which may then affect the bubble size in the bed. The cyclone separates solid particles from the outlet gas and returns solid particles back into the dense bed through the dipleg. Several cyclones may be combined to form a multistage cyclone system that resides either inside or outside the bed. Although cyclones are the most widely adopted gas–solid separators in fluidized beds, other types of gas–solid separators, such as fast separator and filter, are also employed. The expanded

Table 1 Examples of Industrial Applications of Fluidized Bed Reactors

Physical operations	Chemical syntheses	Metallurgical and mineral processes	Other
Heat exchange, catalyst cooler	Phthalic anhydride synthesis	Uranium processing	Coal combustion
Solids blending	Propylene ammoxidation to acrylonitrile	Reduction of iron oxide	Coal gasification
Particle coating	Maleic anhydride synthesis	Metal heat treatment	Fluidized catalytic cracking
Drying (PVC, MBS, ABS)	Ethylene dichloride synthesis	Roasting of sulfide ores	Incineration of solids waste
Adsorption	Methanol to gasoline and olefin processes	Crystalline silicon production	Cement clinker production
Granulation, drying of yeast	Syngas to gasoline	Titanium dioxide production	Microorganism cultivation
Binding	Dimethylbenzene (m) ammoxidation to dinitrilebenzene (m)	Calcination of $Al(OH)_3$	Coal plasma pyrolysis to acetylene
Particle separation	Vinyl acetate synthesis Nitrobenzene hydrogenation to aminobenzene Olefin polymerization	Pyrolysis of oil shale Silicon chloride to $SiCl_3H$	Biomass pyrolysis

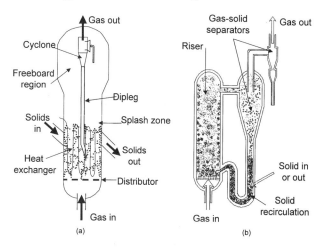

Figure 1 Typical fluidized bed reactors. (a) Low-velocity fluidized bed, and (b) high-velocity fluidized bed.

section on the top of the fluidized bed reduces the linear gas velocity in the freeboard, thereby enhancing particle settling. The expanded section may not always be necessary and depends on the operating conditions and design of the gas–solid separator. The heat exchanger removes generated heat or adds required heat to the fluidized bed by a cooling or heating fluid. Heat exchangers can be either immersed in the dense bed or placed both in the dense bed and the freeboard. Heat exchangers are also placed along the wall as in fluidized bed coal combustors. The proper design of baffles or other types of internal structure reorganize flow, enhance the breakup of bubbles, promote gas–solid contact, and reduce particle entrainment. Commonly, baffles refer to any internal structures other than diplegs and immersed heat exchangers, although the latter may also perform the baffle function. Baffles can be designed with a number of variations: horizontal and vertical grates, fins of different sizes and directions, mesh, or even pagoda-like shapes (Jin et al., 1980). The benefits of baffles are more distinct for coarse particle (Groups B and D) beds than for fine particle beds (Group A), because the bubbles in the former case are larger.

In the high-velocity fluidized bed, the bed can be operated in turbulent, fast fluidization, dilute transport, and downer flow regimes. In the turbulent regime, the bubble/void phase gradually becomes indistinguishable from the emulsion phase, and the particle entrainment rate increases significantly with increasing gas velocity. Upon further increase in gas velocity, the bubble/void phase eventually disappears and the gas evolves into a continuous phase in the fast

fluidization regime. As the gas flow rate increases beyond the point corresponding to the disappearance of the bubble/void phase, a drastic increase in the entrainment rate of the particles occurs so that a continuous feeding of particles into the fluidized bed is required to maintain a steady operation. As a result, high-velocity operations of fluidized beds require a solids recycle loop. The most widely used fluidized bed is a solids recycle loop in a riser or a circulating fluidized bed (CFB). The riser (or downer), gas–solid separator, standpipe, and solids flow control device are the four integral parts of a CFB loop (Fig. 1b) with the riser (or downer) acting as the main component of the system. In the riser, gas and solids flow concurrently upward. During this operation, the fluidizing gas is introduced at the bottom of the riser, where solid particles are fed via a control device from a standpipe and carried upward in the riser. Particles exit at the top of the riser into gas–solid separators, after which the separated particles flow to the standpipe and return to the riser.

In a CFB, particle separation is typically achieved by cyclones. The efficiency of the solids separator can affect the particle size distribution and solids circulation rate in the system. The standpipe provides holding volume and a static pressure head for particles recycling to the riser. The standpipe may be connected to a heat exchanger or spent catalyst regenerator. In this situation, the standpipe is a transport line providing direct solids passage from the riser. The entrance and exit geometries of the riser have significant effects on the gas and solids flow behavior in the riser.

Gas velocity and solids circulation rate are two operating variables for circulating fluidized bed reactors. The key to smooth operation of a CFB system is effective control of the solids circulation rate to the riser. The solids flow control device serves two major functions: sealing riser gas flow to the standpipe and controlling the solids circulation rate. Both mechanical and nonmechanical valves are used to perform these functions. Typical mechanical valves are rotary, screw, butterfly, and sliding valves. Nonmechanical valves include L-valves, J-valves, V-valves, seal pots, and other variations. Blowers and compressors are commonly used as gas suppliers. Operating characteristics of these gas suppliers are directly associated with the dynamics and instability of riser operation and must be taken into consideration.

Both the low and high velocity operations of fluidized beds can be conducted in a variety of configurations as exemplified in Fig. 2. Each configuration has advantages and disadvantages. Figure 2a shows

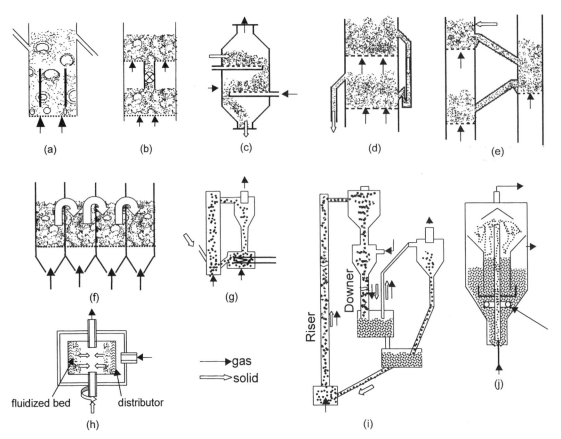

Figure 2 Variations of fluidized bed reactor configurations. (a) Baffled bed, (b) multi-stage bed, (c) multi-stage with counter-current solid flow, (d) multi-stage with external solid circulation, (e) fluidized bed with online solid exchange, (f) multi-cell bubbling bed, (g) CFB, (h) centrifugal fluidized bed, (i) downer, (j) spouted bed.

the fluidized bed with vertical baffles that reduce the bubble size, yielding more uniform flow structures in the fluidized bed. For drying processes involving sticky particles or particles with high moisture contents, shallow bed operation is essential to smooth fluidization. Shallow beds are also operated with small bubble sizes, providing efficient gas–solid contact. Figures 2b and c show the configurations of multistage shallow bed operations. For multistage drying operation, particles are introduced from the top stage and dried through multistage fluidized beds while hot gas enters at the bottom stage. This countercurrent gas–solid contact improves the efficiency of the fluidized bed operation. Particle exchange between fluidized beds or between regions in a fluidized bed can be accomplished through external or internal solids circulation, as shown in Figs. 2d and e. The external or internal solids circulation allows the coupled reaction and regeneration scheme to be implemented for the solid particles, thereby

prolonging particle on-stream time. Reactors in series and parallel are common arrangements for fluidized bed operations (Figs. 2e and f). These systems can be used in conjunction with multistep reaction processes. The crosscurrent multicell type of fluidized bed (Fig. 2f) allows recovery of rapidly released gas products from solid particles where the particle residence time is to be long. A fluidized bed can be also operated in the presence of an external force field other than the gravitational field, such as centrifugal and magnetic fields. The configuration shown in Fig. 2h is a typical centrifugal fluidized bed. For fast reactions and processes requiring on-line catalyst regeneration, the circulating fluidized bed is usually used as shown in Figs. 2g and i. The downer reactor offers advantages for ultrafast reactions and processes with good product selectivity. Through gas spouting, spouted beds (Fig. 2j) offer means to fluidize large particles (Group D) with well-structured internal particle circulation.

1.2 General Procedure for the Development of Fluidized Bed Reaction Processes

Many factors affect optimum fluidized bed reactor performance, including hydrodynamics, heat and mass transfer of interparticles and intraparticles, and complexities of reaction kinetics. The design of fluidized bed reactor processes follows the general approach for multiphase reactor processes. Krishna (1994) and Jazayeri (1995) outlined the general procedure for this process development. The design of the processes can be described by considering various factors as illustrated in Fig. 3.

The fluidized bed reactor design requires understanding the reaction chemistry. The essential knowledge for a design engineer may include the reaction kinetics, conversion or yield, and selectivity, thermodynamics, and process parameters (e.g., operating temperature and pressure as well as heat of reaction) affecting the reaction.

Information on intrinsic kinetics is an essential element in the analysis of reactor performance. Kinetic data should be obtained over a range of temperatures spanning the entire operation conditions. It is also essential to obtain the kinetic data under much higher conversion conditions than those anticipated from the fluidized bed reactor, due to the two-phase flow nature (see Chapter 3), as most reaction occurs in the emulsion/dense phase where the gas reactant is highly converted. Often the reaction kinetics employed is obtained from a slugging bed reactor or a fluidized bed reactor in which transport properties are embedded in the kinetic information; thus such information cannot represent the intrinsic kinetics. For example, Squires (1994) indicated the effects of transient solids mixing patterns on the reaction kinetics in bubbling fluidized bed reactors. Gas bubbles act as agitators in the bubbling bed and lead to rapid mixing and global circulation of solid particles. The bed also undergoes strong exchange of gas between the emulsion phase and the bubble phase by means of gas flow through the bubble and bubble coalescence and breakup. As a consequence, the solid particles are exposed to gas of highly fluctuating chemical compositions. Thus the reaction kinetics of a fluidized bed reactor could be different from the intrinsic kinetics commonly obtained in a TGA (thermogravimetric analyzer) or a differential reactor with steady variations in the concentration of chemical compounds. Ideally, the reaction kinetics should be examined in a differential reactor under a chemical environment similar to that experienced in a fluidized bed reactor and reexamined in a pilot fluidized bed reactor.

Also, in many situations, the pseudo-kinetics expression is used owing to the lack of detailed kinetics information or for simplicity. Useful pseudo-kinetics can only be obtained in an environment as close as possible to that of the proposed commercial operations. For instance, the catalyst decay in a FCC (fluid catalytic cracking) system is mainly caused by

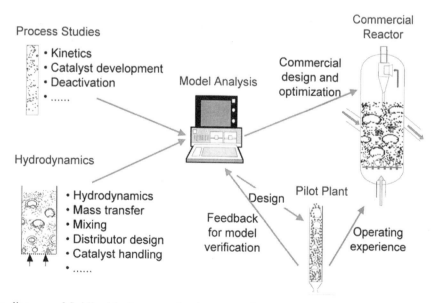

Figure 3 Schematic diagram of fluidized bed reactor developments. (From Krishna, 1994; Jazayeri, 1995.)

coking. The means and extent of coke formation on the catalysts depend on the reaction environment and catalyst residence time. Refiners commonly employ a microactivity test (MAT) unit to establish the activity of catalysts for particular feedstock. The MAT is based on the concept of continuously contacting a hydrocarbon feedstock with a catalyst sample of approximately 1 gram during 75 to 100 seconds of residence time. In the MAT apparatus, the catalyst/oil ratio used depends on the catalyst residence time. The detailed procedure is defined in ASTM (D3907-80). This procedure would yield a significant discrepancy from the conventional riser reactor operation in which the catalyst flow and the hydrocarbon flow are set for a given operating condition and the catalyst/oil ratio would not be a function of catalyst residence time. Another significant discrepancy between the MAT and the riser operation is in contact times. In a conventional riser reactor, the catalyst and the hydrocarbon stay in intimate contact for about 2–5 seconds before being separated in cyclones. In the MAT unit, however, the catalyst reacts with hydrocarbons for as long as 75–100 seconds. In this sense, the MAT technique only allows one to establish a relative performance indicator of catalytic materials and is of limited use in its application to catalytic riser reactors. Thus the kinetic models derived from the data obtained by MAT are of little practical usage for effectively simulating riser reactors and scaling up.

A small hot unit is always used for examining the feasibility of the fluidized bed reactor operation for a new reaction. The hot unit provides information on conversion, yield, and selectivity as well as by-product distributions for a given fluidization regime. Squires (1994) developed a vibrated bed microreactor capable of varying the Peclet number and of back-to-front particle mixing independently. With such a reactor, the sensitivity of reaction to Peclet number and macroscale solids mixing could be expediently examined.

As shown in the process development diagram (Fig. 3), hydrodynamic studies are essential. Such studies would provide information on the basic flow patterns, mixing, particle attrition behavior, and mass and heat transfer for a specific design of a fluidized bed. Operational stability is also examined. Like the reaction kinetic studies, hydrodynamic information is obtained under operating conditions similar to those intended for commercial reactors. In general, fluidized bed reactor design involves studying specific reactions at progressively increased reactor scales from bench to demonstration, in conjunction with analysis based on

models that incorporate the kinetics and the transport processes.

A pilot unit integrates the reactor with all process-related components, such as feed, catalyst recovery, product separation and purification, downstream, and recycles. The pilot unit should be operated in the same flow regime as the commercial unit to ensure that the gas–solid contact scheme and mixing patterns are similar. For a deep bed with a moderate fluidizing velocity, a bench or small pilot unit usually operates in the slugging regime. However, operation in the slugging regime is not possible in a large commercial unit. Therefore, care should be exercised when extrapolating data from pilot units to large-scale commercial units. Operational constraints such as impurities of feed, length of operating period, and equipment reliability should be taken into account during pilot testing. Impurity in the product stream may lead to reduction of product yield due to catalyst deactivation or poisoning.

A pilot reactor needs to be as small as possible to maintain its flexibility of small-scale operation and to provide a means for readily testing alternative designs. However, it needs to be large enough to provide scale-up information for designing the commercial reactor. Mathematical modeling is commonly used in the process of sizing the pilot scale unit. The mathematical models are also used as tools or guides for the successful scale-up to commercial scale. Model development has become an integrated approach in a new process development and in optimal reactor design. The model can be empirical, semiempirical, or mechanistic. The mechanistic model takes into account the interplay between the reaction kinetics and the transport process. Due to the complicated nature of fluidized bed flow, proper assumptions are important for the simplification of model equations. Figure 4 shows a typical flow diagram for describing the mathematical models that provide predictions of product rate, product yield, conversion, species concentration, and temperature profiles in a new process development. Experimental verification of the model parameter using pilot or commercial plant data is necessary.

Two basic approaches are often used for fluidized bed reactor modeling. One approach is based on computational fluid dynamics developed on the basis of the mass, momentum, and energy balance or the first principle coupled with reaction kinetics (see Chapter 9). Another approach is based on phenomenological models that capture the main features of the flow with simplifications by assumption. The flow patterns of plug flow, CSTR (continuous-stirred tank reactor),

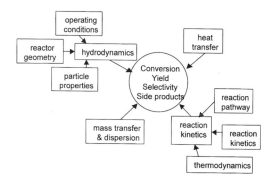

Figure 4 Flow diagram of reactor model developments.

and dispersion are commonly assumed in the model formulation. Most of the phenomenological models developed are derived using only mass balance without considering momentum and/or energy balance. Additional relationships are obtained through empirical correlations. The two-phase theory is a typical example of a phenomenological model. The success of model prediction often relies on the accuracy of empirical correlations established for accounting for transport properties. New commercial reactors are commonly developed through examining incremental scale-up units.

2 KEY STEPS IN FLUIDIZED BED REACTOR DESIGN

Important factors to consider in selecting gas–solid reactors include gas–solid contact schemes, nature of the reactions, temperature, and pressure among many other factors. Of particular pertinence in determining desirable reactor performance can be proper selection of particles, flow regime, and system configurations.

2.1 Particle Selection and Catalyst Development

Particles are the bed material employed in fluidized bed reactors and can be reactants (e.g., coal and limestone), products (e.g., polyethylene), catalysts, or inert. The choice of particle size, in general, affects the hydrodynamics, transport processes, and hence the extent of reactor conversion. Particles experience particle–particle collisions, friction between particles and walls or internals, and cyclones. In some cases, the catalyst material is inherently susceptible to attrition, and special preparation to enhance the attrition resistance is required. For example, the vanadium phosphate metal oxide (VPO) catalysts developed for butane oxidation

to maleic anhydride in a circulating fluidized bed reactor (Contractor et al., 1987) have an attrition problem that can be treated by spray-drying active catalyst microspheres together with a polysilicic acid hydrogel under conditions that allow the silica (SiC or SiO_2) to migrate to the outer surface and form a strong porous layer on the outer surface (Bergna, 1988). This porous shell allows the reactant and product molecules to diffuse in and out of the catalyst particles without significantly affecting the maleic anhydride selectivity.

For catalytic reactions, most active sites of the catalysts are dispersed on the surface of supported particles. Once appropriate active compositions are identified, catalyst development can focus on the selection of support materials. Most reactions in gas-fluidized bed reactors are carried out under high-temperature conditions to overcome high activation energies and ensure the presence of a gas phase. Similarly, elevated pressures are typically utilized to enhance the concentration of reactants on solid surfaces. The catalyst supports should be chemically and physically stable under high temperature, because the structure must remain unchanged for a long period of time. Supports provide an enhanced thermal and mechanical stability to the catalyst material as well as reduce the usage of expensive metals. Selection of inert support can significantly affect the fluidized bed operation. Inorganic materials generally satisfy the rigid constraint. Only inorganic lattices have been found to possess a static and stable three-dimensional backbone capable of sustaining a porous structure under severe reaction conditions. The commonly used catalysts in fluidized bed reactors include zeolites, amorphous silica, and alumina or other metal oxides. Aluminum oxides are commercially available in different forms. The major phase is γ-alumina, which has small pores, a surface area up to $200\,m^2/g$ and surface hydroxyl groups. Amorphous silica can be prepared with high surface area (up to $500\,m^2/g$). Zeolites are aluminosilicates of well-defined crystal structure and regular pore size so they can be engineered to catalyze reactions with shape selectivity. Each of these materials can serve as either a catalyst or a support. The activity and selectivity of catalysts are dictated by the nature of active sites and may be modified by the presence of promoter or doping agents. For fluidized bed reactors, the supporting particles are tailored to achieve optimum hydrodynamics for a specific reaction. Catalyst support particle sizes and pore structures are generally chosen to ensure minimum intraparticle mass transfer resistance. Another reason to avoid intraparticle diffusion resistance may result from selectivity considerations. Minimizing internal transport resistance ensures that desired products rapidly diffuse out of the catalyst particle and avoid further undesirable reactions. For a consecutive reaction scheme, high diffusion resistance decreases intermediate compositions, while final product conversion increases. The specific surface area and pore structures are also very important, and the effects are reaction specific. The reaction of ammoxidation of propylene to acrylonitrile preferably uses catalysts with specific surface areas below $100\,m^2/g$ and a pore radius greater than $50\,\text{Å}$. The catalysts are designed to provide limited oxygen access to the reactant, and a higher surface area would only enhance the extent of oxidation. Thus the formation of desired products while restricting further oxidation is possible. Conventional processes for catalytic cracking of heavy hydrocarbon feedstock to gasoline and distillate fractions typically use a large-pore molecular sieve, such as zeolite Y, as the primary cracking component. Upon the addition of a medium-pore zeolite, such as ZSM-5, to the cracking catalyst composition, an increase in the octane number of the gasoline fraction can be obtained. Conventional ZSM-5 cracking additives are implemented with a crystal size in excess of 0.2 microns, since smaller crystal materials reduce hydrothermal stability and hence rapidly lose activity when exposed to the high-temperature steam generated during FCC regeneration.

Particles in gas–solid reactions in fluidized bed reactors require appropriate preparation (e.g., grinding, surface treatment) to achieve optimum efficiency. Coal must be ground to a particle size range suitable for operation in a fluidized bed combustor. Sorbet powders are synthesized to obtain desirable pore structure. Limestone is extensively used in the in situ removal of acid gas species, such as SO_2 from fluidized bed coal combustors, and H_2S from advanced dual-cycle gasification systems. Removal of SO_2 involves the injection or fluidization of dry calcium-based powders in the high-temperature environment (800–1150°C) of the combustor, calcination of the sorbent to produce CaO, and further reaction with SO_2 to form higher molecular volume $CaSO_4$.

$$CaCO_3(s) \leftrightarrow CaO(s) + CO_2(g)$$

$$CaO(s) + SO_2(g) + \frac{1}{2O_2(g)} \leftrightarrow CaSO_4(s)$$

The optimum particle size and temperature range of operation differ slightly for each of the above steps. The initial reactivity of the sorbent and the ultimate sulfur capture are strongly influenced by available surface area, pore size, and volume characteristics. Commercially mined limestone powder used in SO_2 capture suffers from a very low surface area (less than $3\,m^2/g$) and negligible porosity. As a result, a significant loss of internal pore volume is observed due to the pore filling/pore plugging by the large molecule, $CaSO_4$. Specifically, Ghosh-Dastidar et al. (1996) have shown that the pores generated by the calcination of limestone are typically less than $50\,\text{Å}$ and are very susceptible to pore blockage and plugging. This leads to premature termination of sulfation and incomplete utilization of sorbent. Ghosh-Dastidar et al. (1996) reported that carbonate powders with high-surface-area and porosity can exhibit very high reactivity and conversion when compared to high-surface-area hydrates. Fan et al. (1998) have developed a novel sorbent based on optimization of the surface area and pore size distribution of the calcium carbonate powder. Such optimized calcium carbonate powders lead to the generation of CaO with pores in the 50–200 Å size range. This powder is generated by a precipitation process in a three-phase reactor system by bubbling CO_2 through a $Ca(OH)_2$ suspension in the presence of anionic surfactants.

On an individual particle level, particle size and density play a dominant role in dictating the heat and mass transfer rates and hydrodynamics. Usually, particle sizes are chosen so that resistance of mass and heat transfer between particles and surrounding flow is negligible. Mass transfer between the gas and a single sphere can be estimated by (Froessling, 1938):

$$Sh = 2.0 + 0.6Sc^{1/3}Re_p^{1/2} \qquad (1)$$

For catalytic reactions, particles used in fluidized bed processes are usually in the range of 40 to $100\,\mu m$ in mean diameter. Similarly, particle-to-gas heat transfer coefficients in dense phase fluidized beds can be estimated by (Kunii and Levenspiel, 1991):

$$\frac{h_{gp}d_p}{\lambda} \approx 2 + (0.6 - 1.8)Re_p^{1/2}Pr^{1/3} \qquad Re_p = \frac{Ud_p\rho_g}{\mu_g}$$
$$(2)$$

For high-activity catalysts, the required heating time of a particle has to be relatively short. This is especially true for systems in which the products and selectivity are sensitive to temperatures. For high-velocity operations, solid particles are recycled internally or externally back into the reactors. The temperature of the particles in the reactor is usually different from that from recycle. It is imperative that particles of two different temperatures rapidly mix so that the temperature variations in the entrance zones can be minimized. Small particle size can often eliminate these concerns.

Selection of particle size and density is further complicated by hydrodynamics, which strongly depends on particle physical properties. From a hydrodynamic point of view, particle selection must consider the interplay among flow regime, mixing characteristics, and particle residence time. Particle design provides the most effective tools to improve reactor performance for existing plants through modifications of hydrodynamic behavior. Approaches for particle design include

Optimum particle size and density
Mixture of two groups of particles (e.g., add fine particles to a coarse particle bed and vice versa)
Modification of particle size distributions
Use of fine particles (with interparticle forces to be smaller than the hydrodynamic forces)

These approaches are discussed in conjunction with the flow regime behavior, which is given in the following section.

The particle size distribution (PSD) significantly impacts the reactant conversion in a fluidized bed reactor. Sun and Grace (1990) examined the three different particle size distributions, wide, narrow, and bimodal, on the performance of a catalytic fluidized bed reactor using the ozone decomposition reaction. They found that a fluidized bed with particles of wide size distribution yields the highest reactant conversion. Of further interest, the property of particle entrainment and elutriation is a function of particle size, density, and shape. Both entrainment and elutriation rates increase with decreasing particle size and density. Particle separation efficiency may serve as another measure for particle selection.

2.2 Flow Regime Relevancy

The factors considered for choosing a fluidization/flow regime include

1. Gas–solid contact pattern and interphase mass transfer
2. Backmixing characteristics for both gas and particle phases
3. Reaction kinetics
4. Heat transfer
5. Residence time and reactor height/diameter
6. Ability of solid handling in large quantity
7. Productivity or throughput

Contact schemes of gas–solid systems in fluidized beds are classified by the state of gas and solids motion. For noncirculating systems, the gas at low velocity merely percolates through the voids between packed particles, while the particles remain motionless in the fixed bed state. With an increase in gas velocity, particles move apart and become suspended; the bed has entered the fluidization state. Further increase of gas velocity subjects the flow to a series of transitions from a bubbling fluidization regime at low velocities to a dilute transport regime at high velocities accompanied by significant variations in gas–solid contact behavior. Some gas–solid flow structures associated with given flow regimes are presented in Fig. 5.

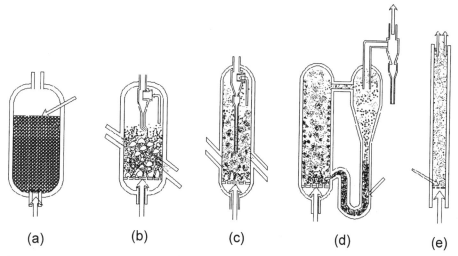

Figure 5 Flow regime diagram of fluidized bed reactors. (a) Fixed bed, (b) bubbling fluidized bed, (c) turbulent fluidized bed, (d) circulating fluidized bed, and (e) dilute transport bed.

2.2.1 Characteristics of Flow Regimes

The bubbling regime is characterized by the coexistence of a bubble phase and a dense/emulsion phase as shown in Fig. 5b. The majority of the fluidizing gas is present in the form of bubbles, and as a result, the gas velocity through the dense phase is very low (on the order of 0.01 m/s for Group A particles). The dense phase occupies about 40–60% of the bed, depending on the particles and operating conditions. Figure 6 shows typical values for dense phase fraction, bubble fraction, and overall voidage for FCC particles under different gas velocities. Once formed, bubbles rise, grow, coalesce, break up, and finally erupt on the bed surface. Bubbles induce the drift effect on surrounding particles and entrain particles in the wake region, thereby inducing vigorous local and global circulation of solid particles in the dense phase. Although bubbles in fluidized beds resemble those in gas–liquid systems, the interface of bubbles is permeable to gas flow. The exchanges of gas between bubble and emulsion phases are accomplished by through-flow, bubble coalescence and breakup, and diffusion. A small fraction of particles may be entrained into the bubble phase depending on the fine content and operating conditions. The solid particles in the bubble phase play an important role in the reactant conversion when fast reactions and/or slow interphase exchange are encountered.

The turbulent regime is often regarded as a transition regime between the bubbling and fast fluidization regime. Bubbles or voids are still present, although they are less distinguishable in the dense suspension. With an increase in gas velocity, the dense phase structure varies gradually, and local nonuniformity of solids concentration emerges. In this regime, interactions between gas voids and the dense/emulsion phase are vigorous and provide an effective gas–solid contact.

The coexistence of a bottom dense region and an upper dilute region characterizes the solids flow in a riser. The solids holdup within the dense phase zone ranges from 0.1 to 0.3, while the holdup profiles in the dilute region can be approximated by an exponential decay. A core-annular flow structure along the radial direction develops with a dense particle layer in the wall region and a dilute core region. Particle exchanges occur between the dilute core and the dense wall region, as temporal and spatial accumulations of particles in the wall region form a transient dense particle layer or streamers. Nonuniform particle distributions or the presence of localized dense zones usually results from the existence of particle clusters and greatly influences the hydrodynamic characteristics.

The hydrodynamics of a circulating fluidized bed is further complicated by the existence of significant variations in solids concentration and velocity in the radial direction. A more uniform distribution can be achieved at conditions of lower solids concentrations under higher gas flow conditions. In the dilute transport regime, the solids concentration is very low and both gas and solids have short residence times.

A concurrent downward flow circulating fluidized bed, or a downer, is a new alternative flow arrangement for a high-velocity system. A downer reactor system has similar system configurations to a riser reactor system except that both the gas and the solid particles flow downward. Concurrent downward flow of particles and gas reduces the residence time of solid particles because the downward flow is in the same direction as gravity. More uniform radial gas and solids flow than those in a riser can be achieved. The downer leads to more uniform contact time between the gas and solids. With these advantages, downer reactors have been proposed for processes such as fluid catalytic cracking, which requires short contact time and uniform gas and solids residence time distributions.

As discussed, gas–solid fluidized bed behavior varies not only with flow regime but also with other factors such as particle properties, operating pressures, and temperatures. Grace (1986a) summarized the effects of particle properties and operating conditions on fluidization behavior and prepared a flow regime diagram.

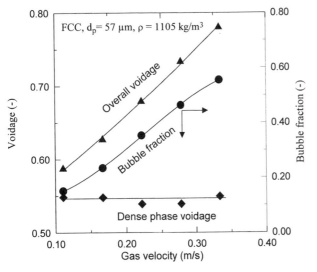

Figure 6 Effects of gas velocity on the voidage of dense and dilute phases in a fluidized bed. (From Yang et al., 1997.)

The flow regime diagram was further modified by Kunii and Levenspiel (1997) and is shown in Fig. 7. The flow diagram is presented as a function of non-dimensional particle diameter and gas velocity defined as

$$d_p^* = d_p\left[\frac{\rho_g(\rho_p - \rho_s)g}{\mu_g^2}\right]^{1/3} \quad (3)$$

and

$$U^* = U\left[\frac{\rho_g^2}{\mu_g g(\rho_p - \rho_g)}\right]^{1/3} \quad (4)$$

For given particles and operating velocity, the gas–solid contact pattern can be determined using this diagram. Likewise, for a given flow regime, this diagram could provide available combinations of particle properties and gas velocity.

2.2.2 Gas–Solid Contact and Interphase Mass Transfer

Gas in the dense and bubble phases plays different roles in a bubbling bed reactor. When gas enters the fluidized bed reactor, the gas in the bed flows to the dense and bubble phases. The gas reactant reacts in the dense phase upon contact with the particles. Interphase mass transfer allows gas reactant and product transfer between the bubble phase and the dense phase. As a bubble rises through a dense or emulsion phase region

in a bubbling bed, gas in the surrounding dense phase could either flow through the bubble phase or circulate between the bubble phase and surrounding dense phase, depending on the ratio of bubble rise velocity to interstitial gas velocity in the dense phase. As depicted in Fig. 8a, when the bubble rise velocity is greater than the interstitial gas velocity in the dense phase, a "clouded" bubble forms in which the circulatory flow of gas takes place between the bubble phase and the surrounding clouded region. The cloud region size decreases as the bubble rise velocity increases. On the other hand, when the interstitial gas velocity in the dense phase is greater than the bubble rise velocity, gas flows through the bubble phase yielding a cloudless bubble, as shown in Fig. 8b. Bubbles in reactors with coarse particles, e.g., Group D, are typically cloudless bubbles, while those with fine particles, e.g., Group A, are typically cloud bubbles. As shown in Fig. 8a, for gas to transfer from the bubble to the emulsion phase, resistances exist at the bubble–cloud interface and the cloud–emulsion phase interface. As indicated in Fig. 9, product formation may take place in bubble, cloud, or dense phase. Davidson and Harrison (1963) assumed that the gas exchange rate is made up of through-flow and gas diffusion. The through-flow velocity is on the order of minimum fluidization velocity, U_{mf}. The volumetric mass transfer coefficient in the bubbling regime, defined based on per unit volume of the bubble phase, can be estimated by (Sit and Grace, 1981):

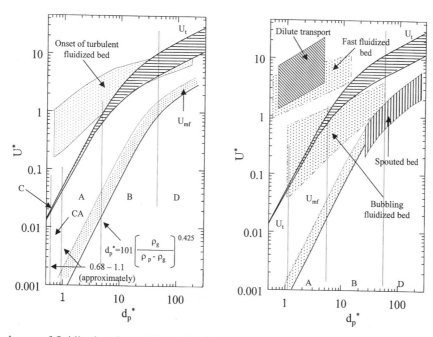

Figure 7 Generalized map of fluidized regimes. (From Kunii and Levenspiel, 1997.)

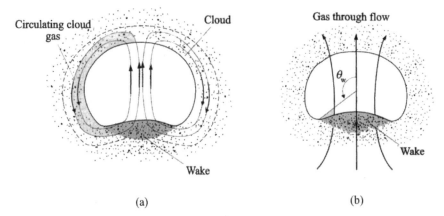

(a)   (b)

Figure 8 Bubble configurations and flow patterns around a fast bubble and a slow bubble. (a) Fast bubble ($U_b > V_d$), (b) slow bubble ($U_b < V_d$).

$$k_q = \frac{U_{mf}}{3} + \left(\frac{4D_m \varepsilon_{mf} U_b}{\pi d_b}\right)^{1/2} \qquad (5)$$

The effectiveness of gas–solid contact is reflected in the reactant conversions. Figure 10 shows the extents of reactant conversions in a bubbling bed reactor for various reactions as a function of dimensionless rate coefficients along with model predictions based on both the plug flow and the CSTR flow patterns for bubbling bed reactors (Kunii and Levenspiel, 1991). The conversions are typically lower than those predicted by either model, and this is an indication of inefficient gas–solid contact.

In the turbulent regime, interactions between gas voids and the dense phase are strong, yielding extensive global and local solids mixing, thereby enhancing gas–solid contact. In high-velocity operations of a fluidized bed reactor, contact resistance of the bulk gas with particles in clusters or wall layers is present. Particle accumulation in the wall region also imposes a hindrance on gas–solid contact owing to low gas flow near the wall. Since significant amounts of particles are in the wall layer, this layer plays an important role in mass and heat transfer. The extent of solid accumulation in the wall region can be analyzed using the core–annular model, and the mass balance of solid particles can be expressed as

$$\frac{\pi}{4}D^2 G_s = \frac{\pi}{4}(D - 2\delta_w)^2 V_{pc}(1 - \bar{\varepsilon})_c \rho_p + \frac{\pi}{4}$$
$$(4D\delta_w - 4\delta_w^2)V_{pw}(1 - \bar{\varepsilon}_w)\rho_p \qquad (6)$$

The cross-sectional averaged voidage is

$$\frac{\pi}{4}D^2(1 - \bar{\varepsilon}) = \frac{\pi}{4}(D - 2\delta_w)^2(1 - \bar{\varepsilon}_c) + \frac{\pi}{4}$$
$$(4D\delta_w - 4\delta_w^2)(1 - \bar{\varepsilon}_w) \qquad (7)$$

Upon rearrangement of Eqs. (6) and (7), the ratio of solid particles in the wall region to the overall particle content can be expressed as

$$\frac{(1 - \bar{\varepsilon}_w)4h(1 - h)}{(1 - \bar{\varepsilon})} = 1 - \frac{(1 - 2h)^2(1 - \bar{\varepsilon}_c)}{(1 - \bar{\varepsilon})}$$
$$= \frac{V_{pc} - G_s/[\rho_p(1 - \bar{\varepsilon})]}{V_{pc} - V_{pw}} \qquad (8)$$
$$h = \frac{\delta_w}{D}$$

Using the correlation of radial voidage profiles developed by Zhang et al. (1991),

$$\varepsilon(r) = \bar{\varepsilon}^{(0.191 + \phi^{2.5} + 3\phi^{11})} \qquad \phi = \frac{r}{R} \qquad (9)$$

A ratio of 0.51 is obtained for typical conditions ($\bar{\varepsilon} = 0.88$ and $h = 0.05$).

Like Fig. 10 for the bubbling bed reactor, the effects of gas–solid contact in a riser reactor can be described in terms of the extent of reactant conversion for various reactions along with predictions as given in Fig 11

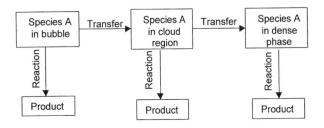

Figure 9 Gas transfer pathway in the bubbling regime.

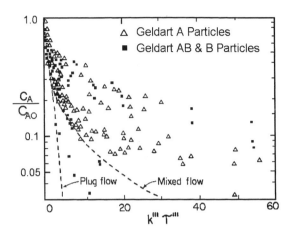

Figure 10 Fraction of unconverted reactant A as a function of dimensionless reaction rate coefficient in bubbling bed reactors. (From Kunii and Levenspiel, 1991.)

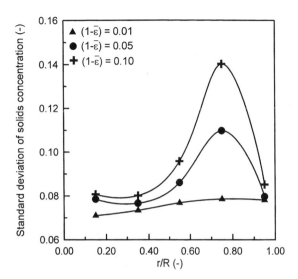

Figure 12 Radial profiles of the standard deviation of local solids concentration with frequency lower than 12.5 Hz.

by the plug-flow and CSTR models. It is seen that an ineffective gas–solid contact induces lower conversions than predicted from either the plug-flow or the CSTR flow pattern and results from the passage of a large fraction of gas through the dilute central core region of the riser. Gas mixing is governed by the combined effects of gas diffusion and turbulent convection. Turbulent intensity is indirectly reflected by the variations of solids concentration fluctuations, as shown in Fig. 12. High gas entrainment in the core region

diminishes the agitation or eddy turbulence in the lateral direction. The dense wall layer is of low turbulent intensity. The strong turbulent actions on the interface between core and wall regions are not sufficient to sweep accumulated particles away from the wall under most operating conditions.

2.2.3 Solids Mixing

Bubbles entrain particles during their rise in a bubbling fluidized bed, and the entrained particles are carried up to the bed surface and erupted to the freeboard. These particles then move downward in the bed, creating a solids circulation pattern. On the local scale, rising bubbles displace surrounding dense phase solid particles through wakes and drift effects. Furthermore, irregular motion, coalescence, and breakup of bubbles cause local particle mixing. Solids mixing has a significant influence on (1) gas–solid contact, (2) temperature profiles, (3) heat transfer, and (4) design of particle feeding and discharge. In the bubbling regime, the solid flow approaches a completely mixed flow pattern, and the particle residence time distribution can be expressed as

$$f(t_p) = e^{-t_p/\tau_p} \qquad \tau_p = \frac{\text{total solid in the bed}}{\text{particle feeding rate}} \quad (10)$$

In the turbulent regime, strong interactions between gas flow and particles lead to a well solids mixing state. Little, however, has been done to quantify such a state of mixing. Particle segregation occurs in the radial direction of a fast fluidized bed. Particles move down-

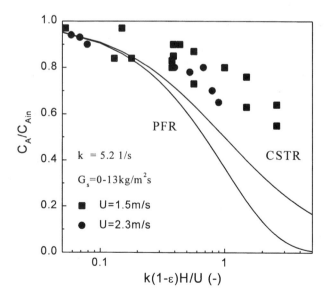

Figure 11 Fraction of unconverted reactant A as a function of dimensionless reaction rate coefficient in a riser reactor. (From Jiang et al., 1991.)

ward along the reactor wall under some operating conditions. Downward particle flow results in local particle recirculation, leading to the flow deviating from plug flow with poor gas–solid contact. On the other hand, downward solids flow extends particle residence time and its distribution. Particle residence time affects the extent of solids reactions or catalytic reactions. For catalytic reactions, as the reactants are absorbed on the catalyst surface, the actual reaction time will depend on the rate processes of reaction and absorption. For an operation in a circulating fluidized bed or pneumatic transport reactor, both the gas reactant residence time and the particle residence time in the reactors are important. For reactions sensitive to catalyst activity, nonuniform products and low selectivity might be a result of solids backflow due to the prolonged residence of low activity catalysts.

2.2.4 Gas Mixing

Fluid flow without any backmixing represents an ideal plug flow, while that with infinite backmixing represents an ideal CSTR. Gas dispersion is a function of bubble size and reactor scale. As the gas velocity increases, bubble motion becomes more vigorous. At the critical gas velocity, U_c, the gas dispersion reaches a maximum as the bed undergoes transition from the bubbling to the turbulent regime; trends in the gas dispersion coefficient, as shown in Fig. 13, reflect the change of the heterogeneous nature of the bed. The gas dispersion coefficient presented is approximately four orders of magnitude higher than the molecular diffu-

sion due to turbulence-dominated gas mixing. For high-velocity operations in the riser, the backmixing in the core region is negligible, while the backmixing near the wall is extensive.

Tracer techniques are commonly used to determine the gas dispersion coefficients in fluidized bed reactors. The tracer concentration measured at the outlet in response to a pulse or step input of the tracer at the inlet can be used to calculate the dispersion coefficient based on the dispersion models in a form similar to Eq. (11), i.e.,

$$\frac{\partial C}{\partial t} + \bar{u}\frac{\partial(C)}{\partial z} = D_{ax}\frac{\partial^2 C}{\partial z^2} + \frac{D_r}{r}\frac{\partial}{\partial r}\left(r\frac{\partial C}{\partial r}\right) \quad (11)$$

Empirical correlations for the dispersion coefficient in bubbling fluidized bed reactors are available in Wen and Fan (1975).

2.2.5 Heat Transfer

Three major components contributing to surface heat transfer are radiation, particle convection, and gas convection. Particles seldom directly contact heat transfer surfaces and are usually separated by a thin gas layer through which particle heat convection occurs. In gas–solid fluidized beds, radiation may be neglected when the bed temperature is below 400°C. The relative effect of particle convection to gas convection on heat transfer depends appreciably on the types of particles used in fluidization. Particle convection is the dominant mechanism for small particles ($d_p < 400\,\mu m$), such as Group A particles. Gas convection becomes dominant for large particles ($d_p > 1500\,\mu m$), such as Group D particles (Maskaev and Baskakov, 1974) and for high-pressure or high-velocity fluidization. For Group B particles, both particle and gas convection are significant. Molerus (1992) studied the effects of system properties on heat transfer and the data plotted in Fig. 14a shows the general trend of heat transfer for three groups of particles. The increasing and decreasing behavior is a result of interplay between the particle and gas convective heat transfer. Figure 14b shows the variations of the wall-to-bed heat transfer coefficient with local solids concentration in a circulating fluidized bed. Heat transfer coefficients are influenced by other factors such as particle size and distribution, solids concentration, and geometry of the heat transfer surface; a detailed account of heat transfer is given in Chapter 10. In general, heat transfer coefficients increase with the solids concentration for fine particles. Thus heat

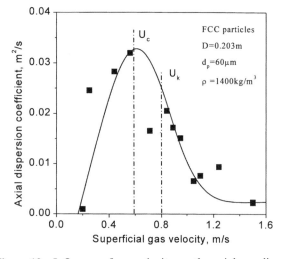

Figure 13 Influence of gas velocity on the axial gas dispersion in a turbulent fluidized bed. (From Wei et al., 2000a.)

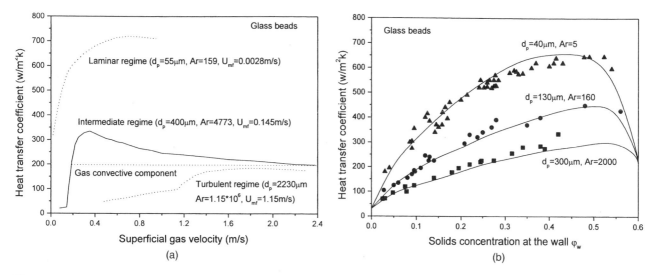

Figure 14 Heat transfer variations with gas velocity and solids concentration. (a) Dependence of heat transfer coefficient on gas velocity for three types of particles. (From Molerus, 1992), (b) dependence of heat transfer coefficient on solids concentration in a CFB. (From Shi et al., 1998.)

transfer is generally higher in a bubbling bed than that in a CFB.

2.2.6 Flow Regime Selection

Each flow regime has its unique flow characteristics, which are illustrated in Table 2 using Group A particles as an example. Knowing these flow characteristics can aid in the selection of a flow regime for a given reaction. The solids holdup in a packed state is typically in the range of 0.45–0.65 and decreases with increasing gas velocity. Typical axial profiles of solids holdup for Groups A and B particles are given in Fig. 15, while radial profiles for the fast fluidization and dilute transport regimes are given in Fig. 16. The turbulent fluidized bed regime provides improved gas–solid contact efficiency with heat transfer and gas backmixing comparable to or slightly higher than those in the bubbling bed regime while maintaining a relatively high solids concentration. This unique regime has been applied to many process operations such as the MTG process developed by Mobil, butane oxidation, and acrylonitrile production. When the gas velocity to the fast fluidization regime is increased, the core–annular flow structure occurs. The relative dimension of the annular region to the core region is not proportional to the size of the riser, which makes riser scale-up a challenging issue. To minimize the segregated flow between the core and wall regions, baffles and other internal structures (Jiang et al., 1991; Ran et al., 1999) can be used. Manipulations of particle size distribution have also

been suggested as a means for improving the gas–solid contact. The flow in a dilute transport reactor is close to plug flow. At high operating gas velocities the solids holdup is very low. This flow regime is applicable for fast reactions with short residence time requirements. As noted, fluidized bed reactors are often used due to nearly isothermal operation and intensive mixing. Clearly, this is the case for coal combustion and catalyst regeneration in FCC systems. Cracking catalysts deactivate rapidly by coking, and the regeneration process is dominated by highly exothermic coke combustion. The residence time for catalyst regeneration determines the regenerator size. Operation in the turbulent regime ensures efficient contact between oxygen and catalysts while avoiding the formation of hot spots and preventing catalysts deactivation by sintering and steaming in catalyst regeneration.

For selective reactions requiring isothermal operation with desired conversion sensitive to backmixing, such as butane oxidation, partial oxidation of natural gas, and acrylonitrile production, a compromise between isothermal operation and the plug–flow requirement must be made in regime selection. The turbulent regime can accommodate this compromise. However, regime selection also requires consideration of catalyst characteristics and system configurations. Oxidation of hydrocarbons is routinely carried out in fixed-bed bundle reactors and fluidized bed reactors with simultaneous presence of oxygen and hydrocarbon at the active sites of the catalyst. Sze and Gelbein (1976) proposed a reactor–regenerator process for oxi-

Table 2 Comparisons of Fluidization Properties of Bubbling, Turbulent, Fast Fluidization, and Dilute Transport Beds with Group A Particles

	Bubbling bed	Turbulent bed	Fast fluidization	Dilute transport
Flow nature	Bubbles and emulsion phase	Dispersed dilute phase and dense phase	Core-annular flow, particle accumulation near wall and particle clustering	Dispersed dilute flow with a thin particle layer on the wall
Solids holdup	0.5, decrease with gas velocity	0.3–0.5, decrease with gas velocity	Bottom dense region: 0.05–0.4, Upper dilute region: < 0.05, decrease with gas velocity and increase with solids flow rate	< 0.05, decrease with gas velocity and increase with solids flow rate
Axial profiles of solids holdup	Fig. 16	Fig. 16	Fig. 16	Fig. 16
Radial profiles of solids fraction	Uniform	Slightly dense in wall region	Large variations	Uniform except in the thin wall layer
Gas backmixing (dispersion)	High	$Pe_a = 0.1$–0.4	$Pe_a = 5$–8, close to plug flow in central zone; strong backmixing near the wall	Plug flow
Radial gas mixing	$Pe_r = 2$–20	$Pe_r = 2$–20	$Pe_r = 100$–1000	$Pe_r = 100$–200
Axial solids mixing	Bubble agitation	Very well	Poor, $Pe_a = 5$–10	Poor
Radial solids mixing	Well	Very well	Core–annular structure poor, $Pe_a = 100$–1000	Poor
Gas–solid contact	Poor	Well	Poor in wall region	Well
Temperature profile	Uniform	May have slight axial gradient	Slight axial gradient	Large axial gradient
Heat transfer	Very high	Very high	Low	Poor
Solids flow	Entrainment	High entrainment	Solids circulation	Solids transport
Gas velocity (throughput)	Low	Higher	Highest	Highest
Residence time distribution	Long/wide	Long/wide	Short/relatively wide	Short/narrow

dative synthesis of aromatic nitriles. The same type of process with a circulating fluidized bed reactor is used for selective oxidation of *n*-butane to maleic anhydride (Contractor and Sleight, 1987). In this process, *n*-butane is oxidized in the riser by the oxygen-loaded catalyst, which in turn is reoxidized in a separate regenerator operated in the bubbling bed regime. The separated reaction–regeneration configuration allows for optimum flow regime to be employed for each of the reactors associated with different types of reaction.

Catalyst regeneration also provides a means of maintaining overall catalyst activity within systems with rapid catalyst deactivation. Deactivated catalyst particles can be continuously removed from a reaction zone while regenerated catalyst from a regeneration zone is continuously recycled back to the reaction zone in a flow scheme similar to that shown in Fig.

17. The averaged activity of a fluidized bed can be maintained even though each catalyst particle undergoes rapid deactivation–regeneration. However, this process does not apply to situations in which the product and selectivity are sensitive to catalyst activity, as deactivated catalysts may lead to undesirable products.

Advances in catalyst development have led to a shift in industrial focus from the case of low-throughput bubbling bed reactors with low catalyst activities to high-throughput turbulent bed reactors or risers with high catalyst activities. Catalytic cracking of petroleum feedstock to light oils, gasoline, and solvents can be carried out in bubbling beds or riser reactors. In bubbling beds, catalyst particles are well mixed and hot spot formation is avoided. With the development of new zeolite catalysts with high activities in the 1960s, riser reactors were subsequently implemented and pro-

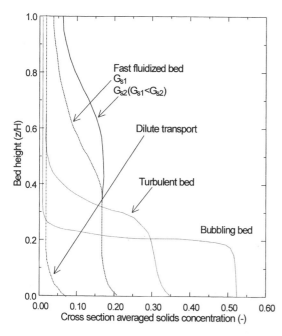

Figure 15 Typical axial profiles of solids concentration in various fluidization regimes.

duced an increase of more than 10% gasoline yield over conventional bubbling bed reactors. With new catalyst development, FCCs can also be operated in a downer reactor due to the short residence time often associated with a downer. Rapid feed heating is required and may be attained, at least in part, by hot catalyst from the regenerator returning to the reactor through the feed stream. The particle temperature may be significantly higher than the average cracking temperatures, and rapid mixing is necessary to avoid

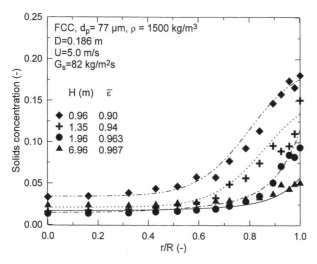

Figure 16 Radial profiles of solids concentration in a CFB.

undue thermal cracking. Furthermore, for short residence time reactors, the initial gas and solids flow development is important in order to control the reaction selectivity and product distribution. As a result, the downer distributor design has a significant effect on the gas–solid flow structure in the downer, especially in the entrance section. A good feed nozzle design that provides excellent gas–solid mixing and uniform distributions of gas and solids over the entrance cross section would enhance the gas and solid flow development.

Increasing gas velocity or gas throughput in a reactor eventually leads to a high production rate with transition of flow regimes to turbulent or riser fluidization. Some designs and modifications have been implemented on bubbling bed reactors to operate as turbulent bed reactors:

1. Increasing operating gas velocity to greater than 0.3 m/s
2. Decreasing particle sizes
3. Increasing fine content ($d_p < 45 \, \mu m$)
4. Increasing bed length-to-diameter ratio
5. Placing internal baffles, bubble suppressor for staging operation

Circulating fluidized bed (CFB) risers are particularly advantageous for the following reactor process characteristics:

1. High rates of reaction
2. Reactions with reversible deactivation requiring continuous catalyst regeneration and high selectivity
3. Varying feed and product requirements with staged reactant inlets that require independent control of gas and solids retention time and

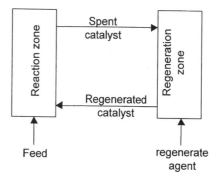

Figure 17 Schematic diagram of separated reaction and regeneration system.

where heat removal or supply through solids circulation is essential

In practice, regime selection cannot be accomplished without considering other factors. For instance, velocity selection must be incorporated into the distributor design. In the bubbling regime, the gas velocity through a distributor has to be high enough to overcome orifice resistance and prevent weeping. For reactions taking place near the particle melting and sticking temperature, strong agitation is necessary to prevent the particles from agglomeration. Furthermore, CFB riser operating conditions affect the efficiency of downstream equipment such as cyclones, filters, and standpipes.

2.3 Process Requirements

Successful process developments rely on the understanding of process requirements and constraints. Fluidized bed reactors have been used in many chemical processes. The typical reactions can be grouped as catalytic reactions, gas–solid reactions, noncatalytic gas reactions, and polymerization. The following describes process requirements for these systems.

2.3.1 Catalytic Reactions

Catalyst selection should be based on catalyst reactivity, reaction selectivity, and physical properties such as particle size, density, and resistance to attrition. For process development, heat and mass transfer phenomena together with reactivity and physical properties of catalysts must be taken into account. The catalytic process begins with gas reactant transferring to the catalyst outer surface and subsequent intraparticle diffusion of the reactant through the pores of the catalyst. Reactants then absorb onto the catalyst surface and react to form product. These products desorb from the surface, and, through intraparticle diffusion, the products exit from the pores and outer catalyst surface. Consider the example of the ammoxidation of propylene to produce acrylonotrile over multicomponent molybdenum/bismuth catalysts:

$$C_3H_6 + NH_3 + \tfrac{1}{2}O_2 \rightarrow CH_2 = CH - CN + 3H_2O$$

$$\Delta H_{R,298K} = -515 \, kJ/mol$$

The reaction network of propylene ammoxidation is not fully understood, but a kinetic model can be approximated by the network shown in Fig. 18. For decades, the Sohio acrylonitrile process has utilized turbulent fluidized beds for propylene ammoxidation

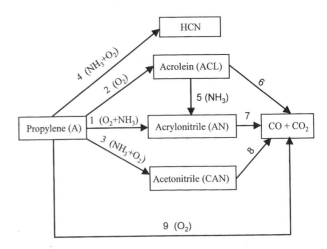

Figure 18 Reaction network for acrylonitrile by catalytic ammoxidation of propylene. (From Wei et al., 1996.)

to synthesize acrylonitrile. The industrial significance of this process was designated by the ACS as a National Historic Landmark process in 1996. The Sohio acrylonitrile process is an innovative, single-step catalytic reaction over metal oxide catalysts and is implemented, today, in more than 90% of acrylonitrile production throughout the world. The gas mixture, consisting of propylene and ammonia, is preheated to 150°C and fed into a fluidized bed catalytic reactor that is operated at 420–460°C and 1.3–2 bar. The molar feed ratio of propylene/ammonia/air is 1 : 1.15 : 10 and gives a minimum excess ammonia over propylene with about 10% stoichiometric excess air in respect to propylene. Catalyst particles range from 50 to 70 µm in mean diameter, and the superficial gas velocity ranges from 0.4 to 0.8 m/s with a residence time between 8 and 10 seconds. As shown in Fig. 18, this reaction is dominated by the competition between partial and completed oxidation. The catalyst is regenerated in situ for this single region and single reactor design.

Internal heat exchangers are used to remove heat and control the reaction temperature of the reactor within ±5°C. Heat transfer tubes are usually arranged vertically and must be capable of controlling bed temperature for the worst possible scenario including reaction runaway. Heat exchangers can also act as internals to prevent the formation of large bubbles in the fluidized bed, thereby reducing overoxidation, which is directly associated with gas backmixing. Internal structures also allow fluidized beds to be operated in the turbulent regime at lower gas velocities. The structure of the internals has a significant influence on the performance of a turbulent fluidized bed reactor and is discussed further in Chapter 7.

Instead of in situ catalyst activation, a catalyst can be regenerated in a fluidized bed reactor with a separated reaction zone and regeneration zone. The unique features of this type of reactor are reflected by the gas distributor design. These distributors consist of an air distributor and a propylene/ammonia distributor that plays three roles: (1) distributing the three reactant gases uniformly in the reactor, (2) forming a separated regeneration and reaction zone within the reactor to facilitate reactions of the redox mechanism, (3) flammable controls for the reactant mixtures. The air distributor in the bottom of the fluidized bed reactor is usually a perforated plate with uniformly distributed holes. Direct contact of catalyst particle within a high-velocity air jet from the distributor may cause severe catalyst attrition. Thus holes on the distributor must be sufficiently large (e.g., 20–35 mm I.D.) to reduce the gas velocity lower than the jetting velocity but not excessively large to yield low pressure drops (caused by low gas velocity) across the distributor, which would lead to nonuniform air distribution.

To avoid spontaneous ignition of the reactant mixture, propylene/ammonia is fed into the reactor through a second distributor placed above the air distributor, separated from the air feed. In this way, propylene and air are not mixed prior to contact with the catalysts. This second distributor is usually a tubular type with a horizontal O-ring or branch tubular arrangement designed to create a uniform distribution across the plate while minimizing downward dispersion of propylene and ammonia into the regeneration zone. In order to avoid possible explosion during mixing of feed components and in case of catalyst deactivation or absence of catalyst, a proper distance (e.g., 0.3–0.6 m) between the two distributors must be maintained. Creation of an oxygen-rich regeneration zone between the two distributors provides catalyst contact with oxygen to reoxidize the reduced catalyst from the reaction zone. Oxygen-loaded catalysts are then returned to the reaction zone where the catalysts come into contact with propylene and ammonia to release crystal oxygen and produce the desired acrylonitrile. The spent catalysts must once again be reoxidized.

The drawbacks of using a turbulent bed for partial oxidization reactions are backmixing and low throughput. A wide distribution in solid residence time also makes controlling the reaction difficult. With new developments in catalysts, synthesis of acrylonitrile can be completed in a few seconds and suggests that the turbulent fluidized beds may be replaceable with high-density riser type reactors (Wei et al., 1994a). High gas velocity and high solids circulation rates in high-density risers can reduce backmixing significantly and improved gas–solid contact efficiency. The small hot model and pilot plant tests show that the average reaction rate in a riser can be as high as 0.20–0.35 kg propylene/kg catalyst/hour, which is three to five times higher than that in a turbulent fluidized bed. However, more efforts are needed to commercialize this concept as experimental observations indicate that nonuniform distributions of solids flow in the radial direction worsens under high-density conditions (Wei et al., 1994b, 1997, 1998). The gas and solids mixing in the radial direction decreases significantly with increasing solids concentration and thus implies that results obtained in small units may not correlate well with large units and pose scale-up challenges.

2.3.2 Gas–Solid Reactions

Unlike the catalytic reaction discussed above, gas–solid reactions involve the solid particle as well as the gas in the reaction. Typical examples of industrial applications include spent FCC catalyst regeneration, calcination, coal combustion, gasification, and silicon chlorination. Owing to the solid particle involvement in the reaction, significant changes in the chemical compositions and physical properties of the particles occur during the reaction. Particles reduce in size and/or increase in porosity in some reactions like coal combustion, whereas particles increase in size and/or decrease in porosity in other reactions such as limestone sulfation. As a result, the particle properties vary unlike those particle properties in catalytic reactions. However, as with catalytic reactions, gas–solid reactions take place on the particle surface as gas reactant adsorbs to the surface.

Consider the example of spent FCC catalyst regeneration where C_mH_n is present as spent catalyst:

$$C_mH_n + O_2 \rightarrow CO_2 + CO + H_2O$$

For decades, regeneration has utilized bubbling fluidized beds. Air is fed into a fluidized bed regenerator which is operated at 550–800°C and 1.3–2 bar. The spent FCC catalyst particles range from 50 to 70 μm in mean diameter, and the superficial gas velocity ranges from 0.3 to 1.8 m/s with a solids residence time between 5 and 30 minutes. This reaction is dominated by reaction kinetics or mass transfer depending on the fluidization regimes and temperature. The unique features of the regenerator are reflected by the various fluidization regimes used for the reactor design to improve efficiency. For kinetic-control reactions, high solids concentration and high temperature are the primary choice of the reactor conditions; the bubbling fluidized bed is the usual choice for high solids concentration and extensive heat transfer. The poor mass transfer between bubble phase and emulsion phase yields low regenerator efficiency and requires the unit to be one of the largest among the FCC processes. On the other hand, the extensive heat transfer with high solids concentration allows for easy heat removal and hence easy reaction control.

With the use of zeolite catalyst in FCC, the requirement of complete catalyst regeneration becomes important in improving the selectivity and yield of the gasoline product (Avidan et al., 1990). Therefore the one-stage well-mixed bubbling or turbulent fluidized bed regenerator cannot achieve high efficiency with more than 95% of the conversion. Moreover, the steam generated by burning hydrogen in coke is detrimental to the stability of catalyst owing to the high temperatures produced. A two-stage fluidized bed reactor with different reactor combinations has been developed to meet the process requirements. Among various combinations, the turbulent-circulating fluidized bed combination achieves great success. This combination uses the turbulent bed for the first stage and the circulating fluidized bed for the second stage. The first stage quickly burns most of the hydrogen and 70% of the carbon with high density and low temperatures. Then, at high temperatures, half-regenerated catalyst is separated from flue gas primarily containing steam and is transported into the circulating fluidized bed. Once it is within the circulating fluidized bed, fresh air is supplied to burn out the rest of the carbon at high temperature, high oxygen concentration, and nearly plug-flow condition. Thus the catalyst can be completely regenerated with high efficiency.

2.3.3 Gas-Phase Olefin Polymerization

Olefin polymerization through heterogeneous catalysis is one of the most important processes to produce polyolefin resins. In this process, small catalyst particles are continuously fed into reactors operated under controlled temperature, pressure, and chemical composition. Gas-phase olefin polymerization can be carried out in fluidized bed reactors and in vertical and horizontal stirred bed reactors. To compete with other process types and maintain a high level of efficiency, the monomer mixture must be operated close to its dew

point without condensing to obtain high monomer concentration and high yields.

Olefin polymerization with metallocene catalysts involves initiation, chain propagation, formation of a dead chain with a saturated chain end through the chain transfer agent, β-hydride elimination to form a dead chain with a vinyl terminal double bond, insertion of a macromer with a vinyl end group, and catalyst deactivation. Assuming that all the reactions associated with each step are first order, the reaction processes can be expressed as

Initiation (activation)	$A + M \xrightarrow{k_a} P_{1,0}$
Propagation	$P_{i,b} + M \xrightarrow{k_p} P_{i+1,b}$
Long chain branching	$P_{i,b} + R_{m,n}^{=} \xrightarrow{k_b} P_{i+m,b+n+1}$
Chain transfer	$P_{i,b} + CTA \xrightarrow{k_c} R_{i,b} + A$
β-hydride elimination	$P_{i,b} \xrightarrow{k_\beta} R_{i,b}^{=} + A$
Deactivation	$P_{i,b} \xrightarrow{k_d} R_{i,b}$

where M is monomer, P is the living (growing) polymer, R is terminated polymer, CTA is chain transfer agent, and A is catalyst. Typical consumption for modern catalysts is on the order of 10,000–50,000 gram of polymer per gram of catalyst. The polymerization reaction is exothermic on the order of 20 cal/mol.

The UNIPOL process, developed by Union Carbide in 1968, is a gas phase fluidized bed polymerization process and initially developed to make high-density polyethylene with Ziegler-Natta catalysts. Along with the development of these catalysts, enhancements to process efficiency and extension of the method scheme to produce linear low-density polyethylene, polypropylene, elastomers, and other ethylene-α olefin copolymers have been developed (Burdett et al., 1998). A flow diagram of the UNIPOL process is shown in Fig. 19. Four primary operations make up this process: monomer purification, reaction, resin degassing, and resin pelleting. The operating gas velocity is relatively high (≈ 0.6 m/s) with a conversion of only about 2% ($< 5\%$) per pass. Upon exiting the reactor, the unreacted/inert gas mixture is cooled, compressed, and then recycled back into the reactor until the overall conversion reaches nearly 100%. Polyethylene polymerization is typically carried out at a pressure of 20–30 atm and a temperature of 75–105°C. In order to avoid agglomeration and sheeting, the operating temperature is limited by the particle softening temperature. The fluidized bed reactor starts with an initial charge of polymer particles and establishes steady operation by supplying inert gas or a mixture of inert gas and monomers. After the fluidized bed reaches a certain temperature, catalyst particles are injected into the bed continuously and initiate poly-

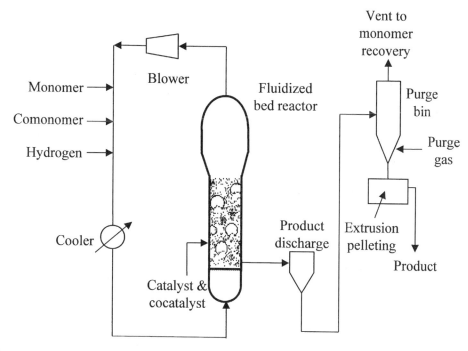

Figure 19 Schematic diagram of fluidized bed reactor for gas-phase polymerization. (From Burdett et al., 1998.)

merization to form a broad size distribution of polymer particles. In polymerization, reactant diffuses from the gas phase, through the boundary layer surrounding the particles, and into the catalyst pores to the active sites where polymerization through a coordination-insertion mechanism occurs to form a solid polymer. The catalyst particles are composed of small metal fragments and explode into a large number of smaller fragments during the reaction. In the early stage of polymerization, these smaller particles are quickly encapsulated by polymer molecules and grow to large polymer particles. For polyethylene, the reaction begins with small particles (20–200 μm in diameter) and finishes, 3–5 hours later, with a mean particle diameter of 200–2000 μm. From gas-phase monomer to solid-state polymer, ethylene experiences a dramatic physicochemical transition. Similarly, the particle morphology undergoes a dramatic change, and the developmental stages are shown in Fig. 20. Implementation of fluidized beds for gas-phase polymerization also provides a typical example in which particles grow in size during a short reaction time.

Key factors to fluidized bed polymerization processes are localized/overall temperature control and handling of sticky particles due to process conditions operated near the resin melting temperature. Without proper cooling, the reactor temperature would increase until the catalysts become inactive or particles fuse to the reactor wall. Often, the production rate of the gas-phase polymerization in a fluidized bed reactor is limited by the maximum heat removal rate. The preferred methods for removing heat from a fluidized bed reactor involving gas-phase polymerization are

1. Recycling unreacted gas through a heat exchanger before returning the gas to the fluidized bed aids in cooling the reactor. The heat removal rate depends on the external heat exchanger capacity and gas flow rate. In order to maintain this cooling sink, monomer consumed in the polymerization reaction is replaced by adding excess monomer gas to the recycle stream.

2. Liquid, with a boiling point lower than the reaction temperature, is intermittently injected into the bed with a spray pattern and immediately evaporates, thereby removing heat. The injected liquid can be monomer to be polymerized or inert liquid hydrocarbon.

3. Fluidized bed reactors are connected in series as a staging operation. The fluidizing gas from the prior reactor is cooled off before entering the next fluidized bed.

4. Part of the gaseous stream leaving the top of the reactor may be condensed by means of an external heat exchanger and then reinjected into the fluidized bed in liquid form. The boiling point of condensable liquid has to be lower than the operating temperature of polymerization. Thus the reaction heat can be removed

owing to the large latent heat of vaporization and high heat transfer rates of condensable liquids. The condensable liquid can be monomers and inert liquids (e.g., pentane, isopentane, butane, and hexane).

Particle temperature control is another issue associated with fluidized bed polymerization as high particle temperature will cause localized melting and fouling. High temperature may also cause catalyst deactivation and alter the monomer compositions in the growing polymer particles. For example, the monomer composition becomes more ethylene rich at higher temperature in polyethylene polymerization. As discussed earlier, particle temperature varies depending on the reaction rate (heat generation) and heat transfer rate between the particle surface and surrounding bulk flow. Hutchinson and Ray (1987) developed a single-particle model to predict the temperature profile as a function of process conditions, reaction kinetics, and particle growth. The mass balance equation for a growing particle is:

$$\frac{\pi}{6}d_p^3 r_p = \pi d_p^2 k_s (M_\infty - M_p) \tag{12}$$

The heat balance for a single particle in the condensation model of operation can be written as

$$(-\Delta H_p)\frac{\pi}{6}d_p^3 r_p = \pi d_p^2 h_{gp}(T_p - T_\infty) + q\pi d_p^2(-\Delta H_c) \tag{13}$$

Model analysis has revealed the existence of multiple unstable operational conditions (Hutchinson and Ray,

1987) and that a uniform concentration of active catalyst helps reduce high initial catalyst activities, thereby reducing the likelihood of polymer melting.

Uniform temperature distribution across the bed is achieved through strong particle mixing and thus favors high-velocity operation. Also, particle size distributions range widely from 10 to 3,000 µm in fluidized bed polymerization systems, and this wide size distribution enables smooth fluidization. Resin particles are often characterized by rough surfaces and irregular shapes (see SEM photo in Fig. 20). Polymers are dielectric materials with low electrical dissipation rates, and as a result they carry strong electrostatic charges. Therefore special attention to reactor design and to the handling of these charged particles is required. For some processes, carbon particles may be added to reduce the handling problems of these particles.

The polymerization reactor can be operated under condensing mode conditions in which the recycled gas stream from the reaction is partially condensed prior to its reintroduction to the reactor. The condensing mode of operation allows the latent heat of vaporization to be used to absorb the substantial heat of reactions generated from polymerization reactions, thereby enhancing production rates of the reactor. Under the condensing mode of operation, the state of fluidization varies with the extent of gas condensation as illustrated in Fig. 21. For low liquid injection rates or low liquid contents, the evaporative liquid vapor is well dispersed in the fluidizing gas yielding a gaseous stream of higher

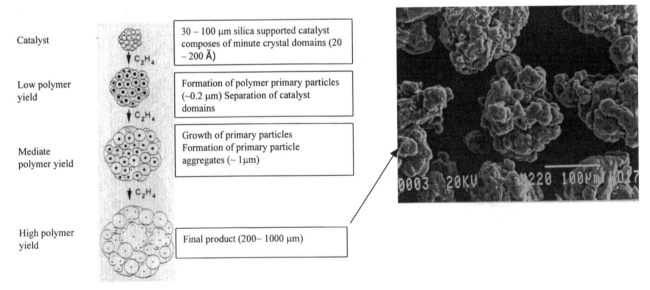

Figure 20 Morphology developments and particle growth of polyethylene particles during gas-phase polymerization. (From Xie et al., 1994).

density and viscosity and hence higher drag forces with the solid particles. Under this state of liquid evaporation, the bed expansion in the fluidized bed would be higher than that of a fluidized bed without liquid injection (Fig. 21a) at a given gas velocity. The bed expansion will not be higher, however, with any further increase in liquid injection rates; particle agglomeration is seen (Fig. 21b), although smooth fluidization can be maintained. With continued increase in liquid contents, a segregation of the agglomerates is observed along the axial direction, even though the bed remains in a fluidized state (Fig. 21c). In the medium range of liquid contents, channeling in the bottom accompanied by smooth fluidization in the upper region appears in the bed (Fig. 21d). Higher liquid content leads to strong binding forces between particles, and channeling occurs throughout the entire bed (Fig. 21e). As a result, a nonuniform distribution of liquid content is often observed in the bed, which leads to localized high liquid content that can cause severe particle agglomeration and local defluidization. With full liquid injections, the bed becomes a slurry bed (Fig. 21f). Bottom liquid injections may cause liquid saturation and channeling in the bottom zone, sometimes even at low overall liquid contents. Liquid injection plays a key role in enhancing liquid content but may require injection nozzles to atomize the liquid rapidly for rapid evaporation and a uniform mix of droplets with particles. The fluidizing gas must be able to pass through the interstitial voids among particles and exert sufficient drag force on the particle to overcome capillary forces, to achieve fluidization without severe particle agglomeration. The addition of fine particles can improve fluidization by absorbing the surface moisture of polymer particles into the interstitial voids of these fine particles that attach to the polymer particle surface. This would create a layer of fine particles on the granular surface and reduce the capillary forces of the granules.

Fluidization quality in a condensing mode of operation can be monitored by pressure drop measurements. As shown in Fig. 22, the fluidized bed with dry polyethylene powders exhibits a typical pressure drop curve of a normal fluidized bed. With moderate liquid contents, the pressure drop decreases with an increase in gas velocity before the bed reaches the fluidized state due to partial channeling. In the fluidization state, the presence of large agglomerates and intermittent collapse of local channeling cause large fluctuations in pressure drops. The pressure drop curve of dry particles can be used as a baseline for monitoring fluidization quality. Pressure drop can also be coupled with bed expansion to monitor further the fluidization quality. During fluidization, the bed expands, and a low bed expansion is a characteristic of high moisture contents in the gas stream (Fig. 23). The low bed expansion, however, may also reveal the existence of dead zones, channeling, and/or defluidization.

Evaporative liquid injection is a common practice in the feed nozzle area of the FCC riser. Similar fluidization phenomena can be observed in the fluidized bed processes like particle drying and particle gradulation (see Sec. 2.3.4). Proper atomization of injected liquid and dispersion of the flow to enhance mixing is the key to smooth fluidized bed operation with high liquid injection rates and low liquid vaporization time.

2.3.4 Physical Operations

Physical operations of fluidized bed technology are very diverse. A brief description of processes and requirements for selected examples is given as follows.

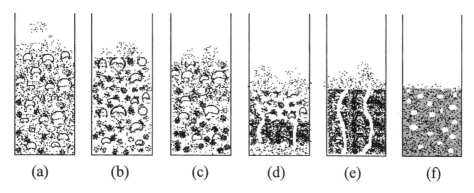

Figure 21 Particle flow patterns under various liquid contents in a fluidized bed reactor. (a) Dry particle fluidization, (b) particle agglomeration, (c) agglomerate segregation, (d) bottom channeling, (e) whole bed channeling, (f) paste or slurry bed.

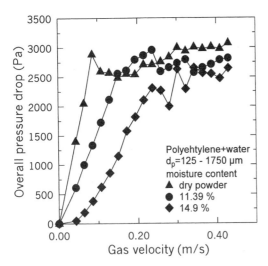

Figure 22 Effects of moisture content on the overall pressure drop in a fluidized bed.

Fluidized beds have been used for coating operations for many years in the production of pharmaceuticals, fertilizers, salts and the solidification of radioactive waste.

In the coating process, a film is formed around a particle by atomizing the coating agent into a fluidized bed. The process consists of spraying a liquid solution into the bed of fluidized particles, coating particles with liquid solution, and then evaporating the solvent from the particle surface. The solute forms a layer or film around individual particles causing the particle size to increase. If the liquid spraying rate is high enough, the coated particles may collide before the solution completely dries and cause particles to agglomerate. This can be minimized by using high fluidization velocity and large particles. However, at lower coating flow rates and higher fluidizing velocities, the film on particles may undergo an "onion ring" growth pattern. By taking into account the hydrodynamics and coating kinetics, optimized conditions can be achieved.

Fluidized beds are often implemented in industry for drying granular materials. For each individual particle, the drying process involves moisture migration from the inner core regions to the particle surface. The surface moisture is then vaporized or evaporated into the fluidizing medium. Meanwhile, heat has to be transferred from the fluidizing medium to the inside of each particle. Diffusion, capillary flow, or internal pressure increase by vaporization or decrease by drying could be mechanisms of moisture migration. For drying, in the constant drying rate period, the moisture on a particle surface is over saturated, and the drying rate is controlled by fluidization characteristics. In the varying drying rate period, the heat conduction and moisture diffusion to the particle surface are rate-controlling factors. The drying course varies with particle temperature and heating rate, since drying is often accompanied by particle shrinkage and deformation. Optimized fluidized bed dryer designs must emphasize the difference in drying kinetics. The strategy is to enhance particle mixing and heat transfer in the constant rate drying period. When materials with a high internal resistance against moisture migration are dried, the proper heating rate and system temperature are essential. Heat can be supplied by the fluidization gas in a fluidized bed dryer, but the gas flow need not be the only source. Heat may be effectively introduced by heating surfaces such as panels or tubes immersed in the fluidized bed. The system temperature can therefore be adjusted by controlling the external heating source. Uniform temperature distribution is another requirement for heat-sensitive materials to avoid overheating. Special attention to the requirements of fluidized bed design must be made when high moisture or sticky materials are involved. Using stirred fluidized beds and mixing dried material with fresh feeds are effective ways to prevent overheating and sticking.

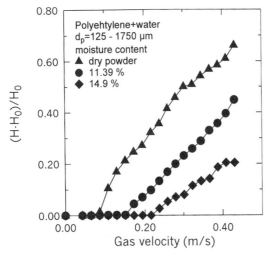

Figure 23 Effects of moisture content on the bed expansion in a fluidized bed.

3 REACTOR MODELING AND ESTIMATION OF DESIGN PARAMETERS

As indicated in Fig. 4, the hydrodynamics and reaction kinetic models, both phenomenological modeling

(Kunii and Levenspiel, 1990) and computational fluid dynamics (CFD) modeling (Gidaspow, 1994), have been extensively studied in the literature. Considerable effort remains to be made in regard to CFD modeling as a predictive tool for fluidized bed performance in terms of conversion and selectivity.

Phenomenological and empirical models are routinely utilized for predictive purposes in commercial fluidized bed reactors in industry today.

3.1 Bubbling Fluidized Bed Reactors

Bubbles are the key features of the bubbling fluidized bed reactor. Two-phase theory, which considers the bed to be composed of a bubble phase and an emulsion phase, is often the basis in reactor model formulation. For the cases of fine particle fluidization, bubbles are surrounded by a cloud region, as implied in Fig. 24; depending on model consideration, the cloud region can be treated separately or lumped with either the bubble or the emulsion. Many models have been developed with variations in the degree of gas backmixing in each of the two phases and the interphase gas exchange. Grace (1986b) provided a thorough description of reactor models. Modeling is also described in detail in Chapter 9.

The simplest case of two-phase models is shown in Fig. 25a, in which the cloud region is merged with the dense phase. Based on the two-phase model, mass balance equations in each phase for a given species can be formulated, and the reactor outlet concentration can be predicted when the hydrodynamic variables and reaction kinetics are known. Assuming a particle-free bubble phase and plug flow of gas bubbles, the mass balance for species A can be expressed as:

$$U_b \frac{dC_{ba}}{dz} - k_q a_b \varepsilon_b (C_{dA} - C_{bA}) = 0 \qquad (14)$$

with a boundary condition

$$C_{bA} = C_{Ain} \quad \text{at} \quad z = 0 \qquad (15)$$

Assuming complete mixing in the emulsion phase with respect to gas and solid particles, the mass balance for species A in the emulsion phase can be written as

$$U_d(C_{Ain} - C_{dA}) + \int_0^{H_f} k_q a_b \varepsilon_b (C_{bA} - C_{dA})\, dz \\ - (1 - \varepsilon_b)\varepsilon_{ds} H_f r_A = 0 \qquad (16)$$

where r_A is the reaction rate per unit volume of catalysts. In obtaining Eq. (16), the mass transfer resistance between the catalyst particles and the surrounding gas flow is neglected. The concentration of Species A in the emulsion phase is constant. The concentration profiles of species A in the bubble phase can be obtained by integration of Eq. (14) as

$$C_{bA} = C_{dA} + (C_{Ain} - C_{dA})\exp\left(-\frac{k_q a_b \varepsilon_b}{U_b}z\right) \qquad (17)$$

The concentration of species A at the bed surface can be obtained from the mass balance at bed height H_f:

$$UC_{Aout} = U_b C_{bA}\big|_{z=H_f} + U_d C_{dA} \qquad (18)$$

The analytical solutions of Eqs. (14)–(16) for different types of reaction kinetics were given by Grace (1986b). This simple model captures the key features of bubbling fluidized bed reactor and is convenient to use.

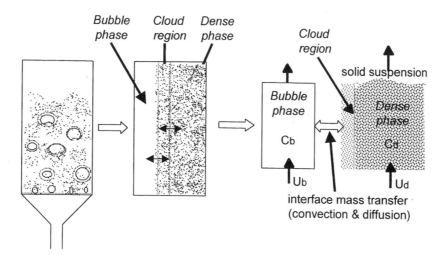

Figure 24 Configurations of a two-phase model for bubbling fluidized bed reactors.

As discussed previously, the addition of fine particles is commonly practiced in the operation of fluidized bed reactors. In some cases, the addition of fine particles is necessary in order to provide smooth bed operations. In systems with fine particles, the assumption of a solids-free bubble phase is no longer valid. Thus the two-phase model needs to be modified to account for the fine particle presence and hence reaction in the bubble phase. The mass balance of species A in the bubble phase can now be expressed by

$$U_b \frac{dC_{bA}}{dz} + k_q a_b \varepsilon_b (C_{dA} - C_{bA}) - \varepsilon_b \varepsilon_{bs} r_A = 0 \quad (19)$$

The mass balance equation for species A in the emulsion phase remains essentially the same as Eq. (16). Grace (1986b) further simplified the model by assuming that no net gas flows through the emulsion phase, i.e., $U_d = 0$. Under this assumption, Eqs. (16) and (19) can be solved analytically for simple reaction kinetics. Effects of fine particle solids concentration in the bubble phase on the reactor conversion can also be examined analytically.

Two-phase models could not be directly applied to account for the entrance effects of the reactor. For shallow beds or reactors with strong jetting, the distributor effects dominate the hydrodynamic behavior of the bed and thus have to be considered differently from that of the in-bed region. The two-phase model that applies to the in-bed region could be modified to describe the entrance effects, as conceptually depicted in Fig. 25b. Detailed descriptions of distributor effects on hydrodynamics are given in Chapter 6.

3.2 Turbulent Fluidized Bed Reactors

In the turbulent fluidized bed reactor, the two-phase flow nature is still distinguishable though significantly less distinct than that in the bubbling regime. Thus the models developed based on the two-phase concept for the bubbling regime can be extended for the turbulent regime. Due to an increased presence of particles in the bubble phase, the bubble phase is referred to as the dilute phase as shown in Fig. 25a. For catalytic reactions in a turbulent fluidized bed reactor with an assumption of plug flow in the dilute phase, the concentration of species A can be expressed as

$$\frac{d(\varepsilon_b V_b C_{bA})}{dz} + k_q a_b \varepsilon_b (C_{bA} - C_{dA}) - \varepsilon_b \varepsilon_{bs} r_A = 0 \quad (20)$$

Likewise, the concentration of species A in the dense phase can be written as

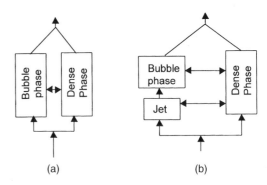

Figure 25 Schematic representation of two-phase model developed for bubbling fluidized bed. (a) Two-phase model, (b) two-phase model with jetting.

$$\frac{d[(1-\varepsilon_b)(1-\varepsilon_{ds})V_d C_{dA}]}{dz} - \frac{d}{dz}$$
$$\left[(1-\varepsilon_b)(1-\varepsilon_{ds})D_{ax}\frac{dC_{dA}}{dz}\right]$$
$$+ k_q a_b \varepsilon_b (C_{dA} - C_{bA}) - (1-\varepsilon_b)\varepsilon_{ds} r_A = 0 \quad (21)$$

When the flow properties remain unchanged along the axial direction in the dense phase, Eq. (21) can be rearranged as

$$\frac{d^2 C_{dA}}{dz^2} - \frac{V_d}{D_{ax}}\frac{dC_{dA}}{dz} - \frac{k_q a_b}{D_{ax}}\frac{\varepsilon_b}{(1-\varepsilon_b)(1-\varepsilon_{ds})}C_{dA}$$
$$+ \frac{\varepsilon_{ds}}{(1-\varepsilon_{ds})}\frac{r_A}{D_{ax}}$$
$$+ \frac{k_q a_b}{D_{ax}}\frac{\varepsilon_b}{(1-\varepsilon_b)(1-\varepsilon_{ds})}C_{bA} = 0 \quad (22)$$

The term V_b/D_{ax} is associated with the Peclet number, $\mathrm{Pe}_a = HV_b/D_{ax}$. For the first order reaction, $r_A = kC_A$, and the reaction rate term is associated with the Damkholer number, $\mathrm{Da} = kH(1-\varepsilon)/D_{ax}$.

High-velocity operation in the turbulent bed yields significant particle entrainment to the upper dilute region or freeboard. The one-dimensional plug flow model can be used to describe the reaction in this region. The mass balance for species A can be expressed as

$$\frac{d(UC_A)}{dz} - \varepsilon_{fs} r_A = 0 \quad (23)$$

where ε_{fs} is the solids volume fraction in the freeboard. The total gas (reactants, products, and inerts) volumetric flow rate Q, can be expressed as

$$Q = AU = \frac{n_g R_m T}{P} \tag{24}$$

where n_g is the total molar flow rate for all gas species at a given axial position, z. The total gas flow rate varies due to molar flow rate changes of the individual species through reactions as described by Eqs. (20) through (23). For convenience, an averaged value can be assumed, and therefore the molar flow rate for species A, n_{gA}, can be expressed by

$$U C_A A = n_{gA} \tag{25}$$

In the upper dilute region,

$$\frac{1}{A} \frac{dn_g}{dz} = \varepsilon_{bs} \sum r_i \tag{26}$$

The inlet boundary condition for the dense and dilute phase equations in a fluidized bed can be given respectively as

$$V_d C_{A0} = V_d C_{dA}\big|_{z=0^+} - D_{ax}\left(\frac{dC_{dA}}{dz}\right)\bigg|_{z=0} \quad \text{at} \quad z = 0 \tag{27}$$

$$C_{bA} = C_{A0} \quad \text{at} \quad z = 0 \tag{28}$$

where C_{A0} is the inlet concentration of species A. The outlet boundary conditions for the dense phase equation can be written as

$$\left(\frac{dC_A}{dz}\right)\bigg|_{z=H} = 0 \quad \text{at} \quad z = H \tag{29}$$

To exemplify the approach, the butane oxidation to maleic anhydride process is presented. The required information on hydrodynamics and mass transfer can be obtained from the equations given in Table 3. These equations have been verified for turbulent fluidized bed reactor applications of butane oxidation. Different correlations are required for other applications.

Figure 26 describes a triangular reaction network for butane oxidation (Buchanan and Sundaresan, 1986). The reactions for each pathway can be expressed as

r_1: $2CH_3 - CH_2 - CH_2 - CH_3 + 7O_2 \rightarrow 2C_4H_2O_3$
$\qquad + 8H_2O$

r_2: $C_4H_{10} + mO_2 \rightarrow (13-2m)CO + (2m-9)CO_2$
$\qquad + 5H_2O$

r_3: $C_4H_2O_3 + bO_2 \rightarrow (6-2b)CO + (2b-2)CO_2$
$\qquad + H_2O$

where b and m are stoichiometric coefficients. The reaction rate for each pathway is written as

$$r_1 = \frac{k_1 C_{bta}}{1 + K_B(C_{bta}/C_{O_2}) + K_{MA}(C_{ma}/C_{O_2})} \tag{30}$$

$$r_2 = \frac{k_2 C_{bta}}{1 + K_B(C_{bta}/C_{O_2}) + K_{MA}(C_{ma}/C_{O_2})} \tag{31}$$

$$r_3 = \frac{k_3 C_{ma}}{1 + K_B(C_{bta}/C_{O_2}) + K_{MA}(C_{ma}/C_{O_2})} \tag{32}$$

where k_1, k_2, k_3, K_B, and K_{MA} are rate constant parameters. The underlying assumptions made in the model equations are as follows:

The fluidized bed reactor is operated under isothermal conditions.
Mass transfer resistance on the catalyst particle in the dense phase is negligible.
Catalyst activity is uniform and remains unchanged throughout the fluidized bed operation.
Reaction takes place in the dense and dilute phases of the bed as well as in the freeboard.

To summarize the model calculation, the concentrations of each species (reactant and products) in the bed are calculated using Eqs. (20) and (21). Equation (23) is used to calculate the concentration profiles of each species in the upper dilute region. The total gas flow rate can be calculated by addition of the molar flows of each species at any given axial location within the bed. The total flow rate can be converted to the volumetric gas flow rate, and the superficial gas velocity may be determined using Eqs. (24) and (25). The values for hydrodynamic parameters are then calculated for the obtained gas velocity and the set of equations is solved numerically. Figure 27 shows the axial profiles of butane and maleic anhydride concentrations calculated based on the model equations. The effects of fine particle ($d_p < 45\,\mu m$) contents on butane conversion are also predicted as shown in Fig. 28. As surmised from the figure, the content of fine particles affects fluidization properties and reactor conversion.

3.3 Riser and Downer Reactors

The hydrodynamic model development for a circulating fluidized bed follows the same approach as bubbling and turbulent beds. In the macroscale, the gas–solid flow is characterized by a coexistence of a bottom dense region and an upper dilute region. The flow in the radial direction can be described by a core-annular structure with a dense particle region close to

Table 3 Correlations Used for Model Simulation of Butane Oxidation in a Turbulent Fluidized Bed Reactor

Parameters	Correlations	Sources
U_d	$U_d^2 + 85.7\varepsilon_{ds}\dfrac{\mu_g}{d_p\rho_g}U_d - 0.571(1-\varepsilon_{ds})^3\dfrac{\rho_p d_p g}{\rho_g} = 0$	This chapter
V_d	$V_d = U_d/(1-\varepsilon_{ds})$	This chapter
V_b	$V_b\varepsilon_b = U - U_d = U - V_d(1-\varepsilon_b)(1-\varepsilon_{ds})$	This chapter
U_c	$\dfrac{U_c}{\sqrt{gd_p}} = \left(\dfrac{\mu_{20}}{\mu}\right)^{0.20}\left[\dfrac{\rho_{g20}}{\rho_g}\dfrac{\rho_p - \rho_g}{\rho_g}\dfrac{D}{d_p}\right]^{0.27}\left(\dfrac{0.211}{D^{0.27}} + \dfrac{2.42\times10^{-3}}{D^{1.27}}\right)$	Cai et al. (1989)
ε	$\varepsilon = 3.0\left(\dfrac{U+1}{U+2}\right)\mathrm{Ar}^{-0.055}\left(\dfrac{\rho_p - \rho_g}{\rho_g}\right)^{-0.132}\left(\dfrac{d_p}{D}\right)^{-0.02}\exp(0.157F_{45})$	Jiang et al. (1999)
ε_b	$\varepsilon_b = \dfrac{\varepsilon + \varepsilon_{ds} - 1}{\varepsilon_{ds}}$ $\quad$ assume $\quad$ $\varepsilon_s \approx 0$	This chapter
ε_{mb}	$\varepsilon_{mb} = \dfrac{0.321\exp(0.157F_{45})\rho_g^{0.042}\mu_g^{0.081}}{g^{0.205}d_{sv}^{0.176}}$	Xie et al. (1995)
ε_{ds}	$(1-\varepsilon_{ds}) = \varepsilon_{mb} + 0.02\mathrm{Ar}^{0.13}\left(\dfrac{U}{U_c}\right)^3$	This chapter
ε_{bs}	$\varepsilon_{bs} = 0.07\left(\dfrac{U}{U_t}\right)^{0.3}\exp(0.25F_{45})$	This chapter
ε_{fs}	$\varepsilon_{fs} - \varepsilon_s^* = (\varepsilon_{bf} - \varepsilon_s^*)\exp[-a(z-H_f)]$ $\quad$ $\varepsilon_{bf} = (1-\varepsilon)$	Kunii and Levenspiel (1997)
W_{tf}	$\dfrac{W_{tf}}{A\rho_p} = \displaystyle\int_{H_f}^{H}\varepsilon_{fs}dz = \varepsilon_s^*(H-H_f) + (\varepsilon_{bf} - \varepsilon_s^*)\dfrac{1-\exp[-a(H-H_f)]}{a}$	This chapter
H_f	$W_t = \rho_p A(1-\varepsilon)H_f + W_{tf}$	Jiang et al. (1999)
$k_q a_b$	Experiments	Jiang et al. (1999)
D_{ax}	Experiments	Jiang et al. (1999)

the wall and a central dilute region. In the mesoscale, the flow is characterized by the presence of particle clusters moving upward in the core and moving downward near the wall. Phenomenological models attempt to capture these main flow features and express the flow behavior in simplified mathematical form to describable levels. Cluster concepts evolved from the observations of large slip velocity between gas and particles. This type of model describes the motion of clusters and can be combined with reaction models to predict reactor performance when cluster properties are known.

Various core-annular models have been developed to describe the gas–solid flow (Horio et al., 1988; Bai et al., 1995; Bolton and Davidson, 1998). The main difference among these models lie in the degree of complexity and in the assumptions associated with simplifications. Core-annular flow structures become dominant in the upper dilute region. Thus, when the dilute flow is predominantly present in the riser, models based on core-annular structures can be applied to reactor models. Kunii and Levenspiel (1990) extended the conventional fluidized bed model (a dense lower region coupled with a freeboard upper region) to cir-

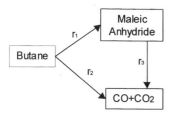

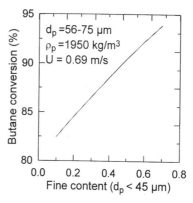

Figure 26 Kinetic network for butane oxidation to maleic anhydride. (From Buchanan and Sundaresan, 1986.)

Figure 28 Effects of fine particle content on the butane conversion in a turbulent fluidized bed reactor. (From Jiang and Fan, 1999.)

culating fluidized beds. The two-phase theory was used to describe the lower dense region, while the entrainment model proposed by Wen and Chen (1982) was adopted to account for the upper dilute region. When the radial solids concentration variation is not prevalent, the entrainment model can provide a sound account of the macroscale hydrodynamic behavior. Detailed information on hydrodynamic modeling is available in Chapter 19.

Although the flow structures have been well recognized, quantitative information on the parameters describing the structure is far from sufficient. As experimental results show continuous profiles for the

solids concentration distributions and gas dispersion profiles in both radial and axial directions, no clear, definitive demarcation can define the core and annular regions, cluster and surrounding suspension, and the lower dense and upper dilute regions. Since distributions of the flow properties in the radial and axial direction are nonuniform in the riser flow, a general two-dimensional dispersion model can be used to describe this flow. Considering a reaction under isothermal conditions and steady-state operations, the general form of the mass balance equation in the 2D cylindrical coordinate for a species A can be expressed as

$$\frac{\partial}{\partial z}\left(V_{gz}\varepsilon C_A\right) + \frac{1}{r}\frac{\partial}{\partial r}\left(rV_{gr}\varepsilon C_A\right) = \frac{\partial}{\partial z}\left(\varepsilon D_{ax}\frac{\partial C_A}{\partial z}\right)$$

$$+ \frac{1}{r}\frac{\partial}{\partial r}\left(r\varepsilon D_r\frac{\partial C_A}{\partial r}\right) + (1-\varepsilon)r_A \quad (33)$$

Note that D_{ax}, D_r, ε, and V_g vary with radial and axial positions. For an axial symmetric flow, $V_{gr} = 0$ and $V_{gz} = f(r)$. With the assumption that D_{ax}, D_r, ε, and V_g remain unchanged within a defined computational cell and may vary from element to element, the material balance for a species A in a given element, Eq. (33), can be further simplified to

$$\frac{\partial C_A}{\partial z} = \frac{H}{Pe_a}\frac{\partial^2 C_A}{\partial z^2} + \frac{D}{Pe_r}\frac{1}{r}\frac{\partial}{\partial r}\left(r\frac{\partial C_A}{\partial r}\right) + \frac{(1-\varepsilon)}{\varepsilon V_{gz}}r_A$$

$$(34)$$

Noting that $\varepsilon V_{gz} = U$, the boundary conditions for Eq. (34) can be written as

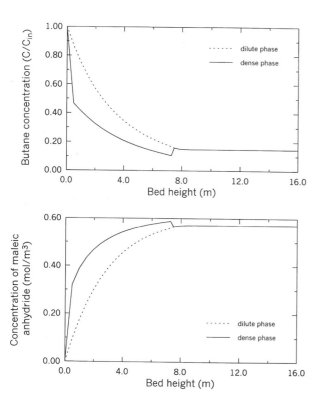

Figure 27 Model predictions of concentration in a turbulent fluidized bed. (From Jiang and Fan, 1999.)

$$C_{A0} = C_A|_{z=0^+} - \frac{H}{\text{Pe}_a}\left(\frac{dC_A}{dz}\right)\bigg|_{z=0^+} \qquad \text{at} \qquad z = 0 \tag{35}$$

$$\frac{\partial C_A}{\partial r} = 0 \qquad \text{at} \qquad r = 0 \quad \text{and} \quad r = \frac{D}{2} \tag{36}$$

$$\frac{\partial C_A}{\partial z} = 0 \qquad \text{at} \qquad z = H \tag{37}$$

The mass balance equations for a given species in the reaction system need to be solved in conjunction with a hydrodynamic model in order to simulate the reactor performance.

For illustrative purposes, the following hydrodynamic model and correlations are used in the model calculation of catalytic reactions of propylene ammoxidation to acrylonitrile. These hydrodynamic models and correlations are appropriate for a high-density riser under propylene ammoxidation conditions. Specifically, the axial profiles of cross-sectional averaged solids concentration was obtained by fitting the cluster-based model proposed by Li and Kwauk (1980) with the experimental data from Wei et al. (1998). The axial profile of solids concentration obtained can be expressed as:

$$\ln\left(\frac{\bar{\varepsilon} - \varepsilon_a}{\varepsilon^* - \bar{\varepsilon}}\right) = -\frac{1}{Z_0}(z - Z_i) \tag{38}$$

where ε_a, ε^*, and Z_0 are empirical constants that can be calculated using the correlations

$$\varepsilon_a = A\left[\frac{18\text{Re}_s + 2.7\text{Re}_s^{1.687}}{\text{Ar}}\right]^{0.0741}$$

$$\text{Re}_s = \frac{d_p \rho_g}{\mu_g}\left(\frac{U}{1 - \varepsilon_a} - \frac{G_s}{\varepsilon_a \rho_p}\right) \tag{39}$$

$$\varepsilon^* = B\left[\frac{18\text{Re}_s + 2.7\text{Re}_s^{1.687}}{\text{Ar}}\right]^{0.02857} \tag{40}$$

$$Z_0 = 500\exp[-69(\varepsilon^* - \varepsilon_a)] \tag{41}$$

where, for acrylonitrile catalyst, $A = 0.484$ and $B = 0.95$. Z_i can be obtained by the overall solids balance in the circulating loop.

The radial profiles of solids concentration are calculated using the correlation (Wei et al., 1998)

$$\frac{1 - \varepsilon(z, r)}{1 - \bar{\varepsilon}} = 2.3 - \frac{2.1}{1 + \exp[20(r/D) - 7.665]} \tag{42}$$

Experimental data for gas velocity profiles in high-density risers are very limited. Radial profiles of the gas

velocity are expected to follow a trend similar to that of the particle velocity (Wei et al., 1998). Using the form of the equation for the radial particle velocity profile, Wei et al. (1998) suggested that the gas velocity profile could be obtained by fitting the radial solids concentration profiles and considering the mass balance of the gas flow. Specifically, the mass balance of gas flow at any cross section of the riser is expressed as

$$\int_0^{D/2} 2r V_{gz}(r)\varepsilon(r)\,dr = \frac{D^2 U}{4} \tag{43}$$

The radial profile of gas velocity can be obtained by substituting the correlation for radial profiles of solids concentration, Eq. (42), in Eq. (43):

$$\frac{V_{gz}(r)}{U} = \frac{2.6}{1 + \exp[20(r/D) - 10q_0]} \tag{44}$$

where q_0 is an empirical constant that can be calculated from

$$q_0 = 2.5 - \frac{2}{1 + \exp[-1.7149(1 - \bar{\varepsilon}) - 1.263]} \tag{45}$$

Based on the ammoxidation of propylene to the acrylonitrile reaction kinetic network shown in Fig. 18, a prediction for the reactor performance of a high-density rise is made. The axial profiles of the main compounds for typical high-density riser conditions are shown in Fig. 29 along with the experimental data obtained in a pilot-scale reactor. Axial Peclet numbers for gas and solids in the circulating fluidized bed range from 4 to 8 (Liu et al., 1999) and vary slightly with operating conditions. Due to the axial gas backmixing, the changes in concentration along the axial position are less than those in a reactor with plug flow. The reaction is first order with respect to propylene and zeroth order with respect to O_2 and NH_3. Therefore a much higher reaction rate at the inlet region exists than in a well-mixed reactor. Extensive gas backmixing in a reactor results in overoxidization of acrylonitrile and a yield reduction.

Models can also be used for parameter sensitivity analysis. Due to the complexity of reaction networks and hydrodynamics, the effects of various factors on the reactor performance are complex. Model analysis provides a guiding tool for process development. The effects of Pe_a on the yield of acrylonitrile are shown in Fig. 30. As shown, when Pe_a is less than 0.05, the yield of acrylonitrile changes marginally with Pe_a, and the reactor can be considered well mixed. When Pe_a is greater than 10, the yield of acrylonitrile is almost the same as that in a plug-flow reactor. Model simulation also reveal the existence of a Pe_a-sensitive range

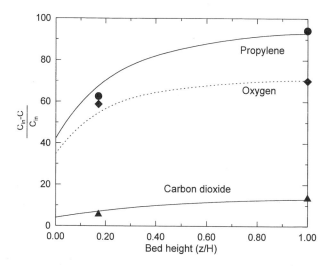

Figure 29 Axial profiles of concentrations for propylene, oxygen, and carbon dioxide in the process of ammoxidation of propylene. (From Wei et al., 2000a.)

from 0.02 to 10, in which the yield of acrylonitrile increases significantly with an increase in Pe_a.

Figure 31 shows the model analysis of the effects of radial gas dispersion coefficient on radial profiles of propylene concentration. The radial mass transfer has a significant effect on the conversion and yield. When the radial Peclet number decreases from 1400 to 200, the conversion of propylene increases by over 10%, and the yield of acrylonitrile increases by about 7%. Since the reaction is first order with respect to propylene, risers are operated under dilute conditions at $Pe_r = 200$, so the radial concentration distribution of propylene is uniform and radial mass transfer is not

the limiting factor to the reaction. Nonuniform distributions of catalyst in a radial direction develop as the particle density of the riser increases. The low turbulent intensity and low gas velocity in the wall region result in poor mass transfer, and the low mass transfer and high catalyst density yield a reactant-deficient zone in the wall region. Since the reaction network of propylene ammoxidation is represented by a combination of parallel and series reactions with the intermediate product being the desired product as shown in Fig. 18, poor mass transfer favors complete oxidation instead of acrylonitrile formation.

4 CONCLUSION

Fluidized bed reactor systems have been proved useful, reliable, and cost-effective, suitable to many industrial applications. The general approaches for designing a fluidized bed reactor are described in this chapter with specific attention given to the knowledge required for the selection of flow regimes and to the interplay between transport properties and kinetics. Examples of several industrial fluidized bed reactor processes are given to illustrate the operational principles. No single universal approach for designing a fluidized bed reactor system exists. At present, fluidized bed reactor system design relies heavily on correlations, engineering models, and plant observations. The computational fluid dynamic approach may offer viable and attractive options over the traditional approaches; however, challenges remain in considering reactive

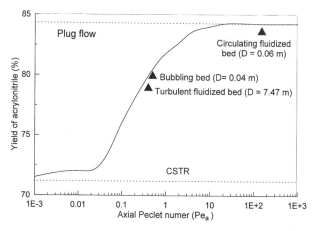

Figure 30 Effects of axial Peclet number on acrylonitrile yield in a high-density riser. (From Wei et al., 1997.)

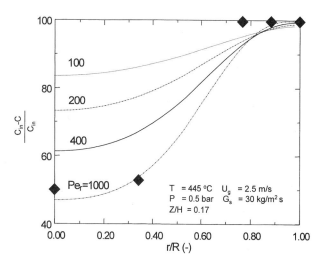

Figure 31 Effects of radial Peclet number on radial profiles of propylene concentration in a high-density riser. (From Wei et al., 2000a.)

flow that involves the incorporation of kinetics, heat, and mass transfer properties into fluid dynamic calculation. Continued research efforts are necessary toward the development of fluidized bed processes for new industrial applications. More measurements on the microscopic properties of the fluidized bed reactor such as velocities, concentrations, and temperatures for a given flow regime and reaction kinetics are necessary to allow further refinement of the current reactor models and their predictions.

NOTATION

A	=	cross-sectional area of a fluidized bed
Ar	=	Archimedes number ($= \rho_g(\rho_s - \rho_g)g d_p^3/\mu_g^2$)
a_b	=	interfacial surface area of bubble per unit bubble volume ($a_b = 6/d_b$)
C	=	concentration
C_A	=	concentration of species A
C_{A0}	=	inlet concentration of species A
C_b	=	reactant concentration in the dilute phase
C_d	=	reactant concentration in the dense phase
d_b	=	mean bubble diameter based on equivalent spherical volume
d_p	=	particle diameter
d_p^*	=	nondimensional particle size ($= d_p[\rho_p(\rho_p - \rho_g)g/\mu_g^2]^{1/3}$)
d_{sv}	=	surface–volume equivalent particle diameter ($= 1/\Sigma x_i/d_p$)
D	=	column diameter
Da	=	Damkoler number
D_{ax}	=	gas dispersion coefficient in axial direction
D_m	=	molecular diffusion coefficient
D_r	=	gas dispersion coefficient in radial direction
F_{45}	=	percentage of fine contents
g	=	gravitational acceleration constant
G_s	=	solids circulation rate
H	=	fluidized bed height
H_0	=	static bed height
H_f	=	dense-phase bed height
ΔH_c	=	latent heat
ΔH_p	=	heat of polymerization reaction
$\Delta H_{R,298K}$	=	heat of reaction at 298K
h_{gp}	=	particle-to-gas heat transfer coefficient
k	=	reaction rate constant
k'''	=	first-order catalytic rate constant defined by Kunii and Levenspiel (1991) in Fig. 10
k_d	=	decay coefficient of catalyst activity
k_q	=	interfacial mass transfer coefficient (volumetric transfer rate per unit bubble surface area)
k_s	=	mass transfer coefficient
K_1, K_2, K_3	=	rate constant parameters
K_b, K_{MA}	=	rate constant parameters

m	=	reaction order
n_g	=	molar gas flow rate
M_p	=	monomer concentration at particle surface
M_∞	=	monomer concentration in surrounding medium
P	=	bed pressure
ΔP	=	pressure drop
Pe_a	=	Peclet number based on axial dispersion coefficient ($= UH/D_{ax}$)
Pe_r	=	Peclet number defined based on radial dispersion coefficient ($= UD/D_r$)
Pr	=	Prandtl number ($= \nu/\alpha$)
q	=	condensated vaporization rate
q_0	=	empirical constant defined in Eqs. (44) and (45)
Q	=	total gas volumetric flow rate
Re_p	=	Reynolds number based on particle diameter ($= \rho_g U D_p/\mu_g$)
Re_s	=	Reynolds number defined in Eq. (39)
R	=	column radius
r	=	radial coordinate in cylindrical coordinate system
r_i	=	reaction rate per unit volume of catalyst for species i
r_A	=	reaction rate per unit volume of catalyst for species A
R_m	=	gas constant
r_p	=	polymerization rate per unit volume of particle
Sc	=	Schmidt number ($= \mu_g/\rho_g D_m$)
Sh	=	Sherwood number ($= k_s d_p/D_m$)
t	=	time
T	=	temperature
T_∞	=	bulk (or surrounding medium) temperature
T_p	=	particle temperature
t_p	=	particle residence time
u	=	velocity
$\bar{u}$	=	average velocity
U	=	superficial gas velocity [$= \varepsilon_b V_b + (1 - \varepsilon_b)V_d$]
U^*	=	nondimensional gas velocity ($= U[\rho_g^2/\mu_g(\rho_p - \rho_g)g]^{1/3} = Re/Ar^{1/3}$)
U_b	=	superficial bubble rising velocity
U_c	=	transition velocity from bubbling to turbulent regimes
U_d	=	superficial gas velocity in the dense phase
U_{mf}	=	minimum fluidization velocity
U_t	=	terminal velocity
V_b	=	interstitial (linear) gas velocity in the bubble phase
V_d	=	interstitial (linear) gas velocity in the dense phase
V_g	=	linear gas velocity
V_{gz}	=	linear gas velocity in axial direction
V_{gr}	=	linear gas velocity in the radial direction
V_{pc}	=	particle velocity in the dilute core region

V_{pw}	=	particle velocity in the dense wall region
W_t	=	weight of catalyst in the whole bed
W_{tf}	=	weight of catalyst in free board
z	=	vertical (axial) coordinate
Z_0	=	empirical constant in Eq. (41)
Z_i	=	empirical constant in Eq. (38)

Greek Letters

α	=	thermal diffusivity
δ_w	=	thickness of the dense wall region
ε	=	overall or local voidage
$\bar{\varepsilon}$	=	cross-sectional averaged voidage
ε_a	=	empirical constant in Eq. (39)
ε_b	=	volume fraction of the dilute (or bubble) phase (= bubble volume/bed volume)
$\bar{\varepsilon}_c$	=	averaged voidage in the dilute core region
$\bar{\varepsilon}_w$	=	average voidage in the dense wall region
ε_d	=	volume fraction of the dense (or emulsion) phase (= $1 - \varepsilon_b$)
ε_{ds}	=	volume fraction of solid particles in the dense phase (per unit volume of dense-fraction)
ε_{bs}	=	volume fraction of solid particles in the dilute phase (per unit volume of dilute phase
ε_{fs}	=	volume fraction of solid particles in the free board region
ε_{mb}	=	voidage at incipient bubbling
ε_{mf}	=	voidage at incipient fluidization
ε^*	=	empirical constant in Eq. (40)
λ	=	thermal conductivity
μ_g	=	gas viscosity
ν	=	kinematic viscosity
ρ_p	=	particle density
ρ_g	=	gas density
τ'''	=	contact time defined by Kunii and Levenspiel (1991) in Fig. 10
τ_p	=	mean particle residence time

Subscripts

A	=	species A in the reaction
b	=	bubble phase in bubbling bed or dilute phase in turbulent bed and riser
bta	=	butane in the reaction of butane oxidation
d	=	dense phase
in	=	inlet
ma	=	maleic anhydride in the reaction of butane oxidation
O_2	=	oxygen in the reaction of butane oxidation
out	=	exit
s	=	solid (or particle) phase

REFERENCES

Avidan, AA, Shinnar R. Development of catalytic cracking technology, a lesson in chemical reactor design. Ind Eng Chem Res 29:931–942, 1990.

Bai DR, Zhu JX, Jin Y, Yu ZQ. Internal recirculation flow structure in vertical upward flowing gas–solid suspensions I, a core/annular model. Powder Technol 85:171–178, 1995.

Bergna HE. Method of making maleic anhydride. U.S. Patent, 4,769,477, 1988.

Bolton LW, Davidson JF. Recirculation of particles in fast fluidized risers. In: Basu P, Large JF, eds. Circulating Fluidized Bed Technology II. Toronto: Pergamon Press, 1988.

Buchanan JS, Sundaresan S. Kinetics of redox properties of vanadium phosphate catalysts for butane oxidation. Applied Catalysis 26:211–226, 1986.

Burdett ID, Eisinger RS, Cai P, Lee KH. Recent developments in fluidized bed process for olefin polymerization. AIChE Annual Meeting, Miami, FL, 1998.

Cai P, Chen SP, Jin Y, Yu ZQ, Wang ZW. Effect of operating temperature and pressure on the transition from bubbling to turbulent fluidization. AIChE Symp Ser 85(270):37–43, 1989.

Contractor RM, Sleight AW. Maleic anhydride from C-4 feedstock using fluidized bed reactor. Catalyst Today 1:587–607, 1987.

Davidson JF, Harrison D. Fluidized Particles. New York: Cambridge University Press, 1963.

Fan LS, Zhu C. Principles of Gas–Solids Flows. New York: Cambridge University Press, 1998.

Fan LS, Ghosh-Dastidar A, Mahuli S. Calcium carbonate sorbent and methods of making and using same. US Patent 5, 779,464, 1998.

Froessling N. The evaporation of falling drops. Gerlands Beitr. Geophys 52:170, 1938.

Ghosh-Dastidar A, Mahuli S, Agnihotri R, Fan LS. Investigation of high-reactivity calcium carbonate sorbent for enhanced SO_2 capture. Ind Eng Chem Res 35:598–606, 1996.

Gidaspow D. Multiphase Flow and Fluidization-Continuum and Kinetic Theory Descriptions. San Diego: Academic Press, 1994.

Grace JR. Contacting modes and behavior classification of gas–solid and other two-phase suspension. Can J Chem Eng 64:353–363, 1986a.

Grace JR. Fluidized beds as chemical reactors. In: Geldart D, ed. Gas Fluidization Technology. New York: John Wiley, 1986b, pp 285–339.

Horio M, Morishita K, Tachibana O, Murata N. Solids distribution and movement in circulating fluidized beds. In: Basu P, Large JF, eds. Circulating Fluidized Bed Technology II. Toronto: Pergamon Press, 1988.

Hutchinson RA, Ray WH. Polymerization of olefins through heterogeneous catalysis, VII. Particle ignition and extinction phenomena. J Appl Polym Sci 34:657–676, 1987.

Jazayeri B. Successfully scale up catalytic gas-fluidized beds. Chem Eng Prog 91(4):26–31, 1995.

Jiang P, Fan LS. Characteristics of turbulent fluidized beds with fine particles at high temperatures and pressures. AICHE Annual Meeting, Dallas, 1999.

Jiang P, Bi H, Jean RH, Fan LS. Baffle effects on performance of catalytic circulating fluidized bed reactor. AIChE J 37:1392–1400, 1991.

Jin Y, Yu ZQ, Shen JZ, Zhang, L. Pagoda-type of vertical internal baffles in gas fluidized beds. Int Chem Eng 20:191–196, 1980.

Karri SBR, Knowlton TM. Gas distributor and plenum design in fluidized beds. In: Yang WC, ed. Fluidization Solids Handling and Processing. Westwood, NJ: Noyes, 1999.

Krishna R. A systems approach to multiphase reactor selection. Advances Chem Eng 19:201–249, 1994.

Kunii D, Levenspiel O. Entrainment of solid from fluidized beds, II. Operation of fast fluidized beds. Powder Technol 61:193–206, 1990.

Kunii D, Levenspiel O. Fluidization Engineering. 2d ed. Boston: Butterworth-Heinemann, 1991.

Kunii D, Levenspiel O. Circulating fluidized-bed reactors. Chem Eng Sci 52:2471–2482, 1997.

Li Y, Kwauk M. The dynamics of fast fluidization. In: Grace JR, Masten, JM, eds. Fluidization. New York: Plenum Press, pp 537–544.

Liu JZ, Grace JR, Bi H, Morikawa H, Zhu J. Gas dispersion in fast fluidization and dense suspension upflow. Chem Eng Sci 54:5441–5449, 1999.

Maskaev VK, Baskakov AP. Features of external heat transfer in a fluidized bed of coarse particles. Int Chem Eng 14:80–83, 1974.

Molerus O. Heat transfer in gas fluidized beds, Part 2. Dependence of heat transfer on gas velocity. Powder Technol 70:15–20, 1992.

Ran X, Wang ZW, Wei F, Jin Y. Particle velocity and solids fraction profiles of a swirling flow in a riser. Proceedings of the Sixth International Conference on Circulating Fluidized Bed, Würzburg, Germany, Aug. 1999, pp 131–139.

Shi D, Nicolai R, Reh L. Wall-to-bed heat transfer in circulating fluidized beds. Chem Eng Proc 37:287–293, 1998.

Sit SP, Grace JR. Effect of bubble interaction on the interphase mass transfer in gas fluidized beds. Chem Eng Sci 36:327–335, 1981.

Squires A. Origins of the fast fluid bed. Adv in Chem Eng 20:1–37, 1994.

Sun G, Grace JR. The effect of particle-size distribution on the performance of a catalytic fluidized-bed reactor. Chem Eng Sci 45(8):2187–2194, 1990.

Sze MC, Gelbein AP. Make aromatic nitriles this way. Hydrocarbon Process 3:103–106, 1976.

Wei F, Lu FB, Jin Y, Yu ZQ. Petrochemical Engineering (in Chinese) 23:631–636, 1994a.

Wei F, Jin Y, Yu ZQ. Macro visualization of gas solids flow structure in high density circulating fluidized beds. In: Avidan AA, ed. Circulating Fluidized Bed Technology IV. New York: AIChE, 1994b, pp 588–596.

Wei F, Lai ZP, Jin Y, Yu ZQ. A CFB reactor model for the synthesis of acrylonitrile. Proceedings of Asian-Pacific Conference on Chemical Reaction Engineering, Beijing, China, 1996, pp 625–630.

Wei F, Lu FB, Jin Y, Yu ZQ. Mass flux profiles in a high density circulating fluidized bed. Powder Technol 91:189–195, 1997.

Wei F, Lin HF, Cheng Y, Wang ZW, Jin Y. Profiles of particle velocity and solids fraction in a high-density riser. Powder Technol 100:183–189, 1998.

Wei F, Du B, Fan LS. Gas mixing in a turbulent fluidized bed under high temperature/pressure conditions. AIChE Annual Meeting, Los Angeles, CA, 2000a.

Wei F, Wan XT, Hu YQ, Wang ZG, Yang YH, Jin Y. A pilot plant study and 2-D dispersion-reactor model for a high-density riser reactor. Chem Eng Sci, in press, 2000b.

Wen CY, Chen LH. Fluidized bed freeboard phenomena: entrainment and elutriation. AIChE J 28(1):117–128, 1982.

Wen CY, Fan LT. Models for Flow Systems and Chemical Reactors. New York: Marcel Dekker, 1975.

Xie HY, Geldart D. Fluidization of FCC powders in the bubble-free regime: effect of types of gases and temperature. Powder Technol 82:269–277, 1995.

Xie T, McAuley KB, Hsu JCC, Bacon DW. Gas phase ethylene polymerization: production processes, polymer properties, and reactor modeling. Ind Eng Chem Res 33:449–479, 1994.

Yang G, Wei F, Jin Y, Yu ZQ, Wang Y. Unique properties of 30-mm particles as the catalyst of fluidized-bed reactors. AIChE J 43:1190–1193, 1997.

Zhang W, Tung Y, Johnsson F. Radial voidage profiles in fast fluidized beds of different diameters. Chem Eng Sci 46(12):3045–3052, 1991.

13

Fluidized Bed Scaleup

Leon R. Glicksman

Massachusetts Institute of Technology, Cambridge, Massachusetts, U.S.A.

1 INTRODUCTION

Typically, the development of a new commercial fluidized bed process involves a laboratory bench scale unit, a larger pilot plant, and a still larger demonstration unit. Many of the important operating characteristics change between the different size units. The critical problem of scale-up is how accurately to account for the performance changes with plant size to optimize the performance of a full-size commercial unit. In addition, it would be helpful if the link between smaller units could be understood, so that the smaller unit could be used to improve the commercial plant performance or solve existing problems.

One discouraging problem is the decrease in reactor or combustor performance that is found when some pilot plants are scaled up to larger commercial plants. These problems can be related to poor gas flow patterns, undesirable solid mixing patterns, and physical operating problems; see Matsen (1985). In most instances, it is difficult to make any observations or measurements directly within the commercial fluidized bed. Thus problem solving is hindered by a lack of understanding of the underlying cause of poor behavior. In the Synthol CFB reactors constructed in South Africa, first scale-up from the pilot plant increased the gas throughput by a factor of 500. Shingles and McDonald (1988) describe the severe problems initially encountered and their resolution.

In some scaled-up fluidized bed combustors, the lower combustion zone has been formed into a narrow rectangular cross section. Sometimes the lower section is divided into two separate subsections; this is referred to sometimes as a pant leg design, to provide better mixing of fuel and sorbent in a smaller effective cross section, and to reduce the potential maldistribution problems in the scaled-up plant.

Matsen (1985) pointed out a number of additional problem areas in scale-up such as consideration of particle size balances, which change over time due to reaction, attrition, and agglomeration. Erosion of cyclones, slide valves, and other components owing to abrasive particles are important design considerations for commercial units that may not show up in pilot plants.

As the diameter of a fluidized bed is increased, the bed hydrodynamics may not remain similar. In some instances, the flow regime may change between small and large beds even when using the same particles, superficial gas velocity, and particle circulation rate per unit area. The issue of scale-up involves an understanding of these hydrodynamic changes and their influence on chemical reaction rates, conversion efficiency, and thermal conditions through variations in gas distribution, gas–solid contact, residence time, and solid circulation and mixing.

There are several avenues open to deal with scale-up. There are a large number of empirical and semiempirical correlations that exist in the fluidized bed literature to predict fluid dynamic or overall process behavior. In addition, there are probably a large number of proprietary correlations used by individual companies. The danger lies in extrapolating these relations to new geometric configurations of the riser or inlet, which lead to flow conditions outside the range of previous data, or to beds that are much larger. Avidan and coauthors in a 1990 review of FCC summed up the state of the art: "basic understanding of complex fluidization phenomena is almost completely lacking. While many FCC licensors and operators have a large body of in-house proprietary data and correlations, some of these are not adequate, and fail when extrapolated beyond their data base"; see Avidan et al. (1990).

As an example, consider the influence of mean particle size. In the early work on bubbling fluidized bed combustors, attempts were made to use relations from the classic fluidization literature that had been developed from experiments with smaller particles, such as FCC applications. In many cases, the relationships for small particles gave erroneous results for combustors with much larger particles. For example, the two-phase theory equating the excess gas velocity above minimum fluidization to the visible bubble flow was substantially in error for large particle systems. Jones and Glicksman (1986) showed that the visible bubble flow in a bubbling bed combustor was less than one-fifth of u_0-u_{mf}. In other cases, even the trends of the parametric behavior with particle size were changed. Heat transfer to immersed surfaces in fine particle bubbling beds increases strongly with a decrease in the mean particle size. For large particle beds, the heat transfer decreases with decreasing particle diameter in some instances.

Numerical models have been developed that are more closely based on fundamental principles. The models range from simple one-dimensional calculations to complex multidimensional computational fluid dynamics solutions. There is no doubt that such numerical models are a great aid in synthesizing test data and guiding the development of rational correlations. In a recent model evaluation, modelers were given the geometry and operating parameters for several different circulating beds and asked to predict the hydrodynamic characteristics without prior knowledge of the test results; see Knowlton et al. (1995). None of the analytical or numerical models could reliably predict all of the test conditions. Few of the models could come close to predicting the correct vertical distribution of solid density in the riser, and none could do it for all of the test cases. Although it is tempting to think that these problems can be solved with the "next generation of computers," until there is better understanding and thorough verification of the fundamental physical models and equations used to describe the

hydrodynamics, the numerical models will not stand alone as reliable scale-up tools.

Another approach to scale up is the use of simplified models with key parameters or overall coefficients found by experiments in large beds. For example, May (1959) used a large-scale cold reactor model during the scale-up of the fluid hydroforming process. This technique must be used with care. A large cold model may not directly simulate the hydrodynamics of a real process that operates at elevated pressure and temperature unless care is taken in the selection of flow rates, particle sizes, and particle densities for the cold tests.

Johnsson et al. (1987) have shown examples of verification of a model for a 2.13 m diameter industrial phthalic anhydride reactor. Several bubbling bed models gave good overall prediction of conversion and selectivity when proper reaction kinetics were used. The results were shown to be quite sensitive to the bubble diameter. The comparison was a good check of the models for the reaction kinetics, but the reactor model required accurate bubble size estimates obtained from measurements of overall bed density in the reactor.

As Matsen expresses it, after over a half a century of scale-up activity in the chemical process industry, "such scale-up is still not an exact science but is rather a mix of physics, mathematics, witchcraft, history and common sense which we call engineering"; see Matsen (1997).

Since changes in the bed size primarily influence scale-up through changes in the bed hydrodynamics, one focus of this chapter will be on experimental results and models that deal explicitly with the influence of bed diameter on hydrodynamic performance for both bubbling and circulating fluidized beds. The changes in the bed dynamics will, in turn, impact the overall chemical conversion or combustion efficiency through changes in the particle-to-gas mass transfer and the heat transfer from the bed to immersed surfaces or the bed wall. Several examples of this influence will be presented.

The second focus of this chapter will be on the design rules for small-scale experimental models that permit the direct simulation of the hydrodynamics of a hot, possibly pressurized, pilot plant or commercial bed. By the use of this modeling technique, beds of different diameters, as well as different geometries and operating conditions, can be simulated in the laboratory. To date, this technique has been successfully applied to fluidized bed combustors and gasifiers. Rules for the use of this experimental modeling technique for fluidized bed operations with different particle sizes, as well as for the simulation of bed-to-solid-surface heat transfer, will also be given. A brief review of verification experiments and comparisons to large-scale commercial systems will be shown.

2 REACTOR MODELING: BED DIAMETER INFLUENCE

Representative results provide a prospective of reactor modeling techniques that deal with bed size. There are limited results in the open literature, and in all probability there are additional unpublished proprietary materials in this area. Early studies of fluidized reactors recognized the influence of bed diameter on conversion due to less efficient gas–solid contacting. In bubbling beds that are larger in diameter and deeper, bubbles can grow to larger size. The larger bubbles allow more gas to bypass the dense particle-laden phase. This becomes particularly striking as the bubble size approaches the bed depth and at large volume fractions of bubbles within the bed. Smaller bubbles provide improved gas–solid contacting and more uniform solids mixing.

Experimental studies have been used to predict reactor performance. Frye et al. (1958) used a substitute reaction of ozone decomposition to study hydrocarbon synthesis. The ozone decomposition can be run at low pressures and temperatures and can be rate controlled in the same way and by the same catalyst as the reaction under development. Frye and coworkers used three beds, 2, 8, and 30 inches in diameter to study the size influence. We should interject a caution that the use of pressures and temperatures different from those of the actual reaction may mean that the hydrodynamics of the substitute reaction model will differ from the actual application; this will be illustrated later in the chapter. Figure 1 shows the apparent reaction rate constant for the different bed diameters at two different bed heights, with the other parameters held constant. Note that the rate constant decreased by roughly a factor of 3 between the 2 inch and 30 inch beds.

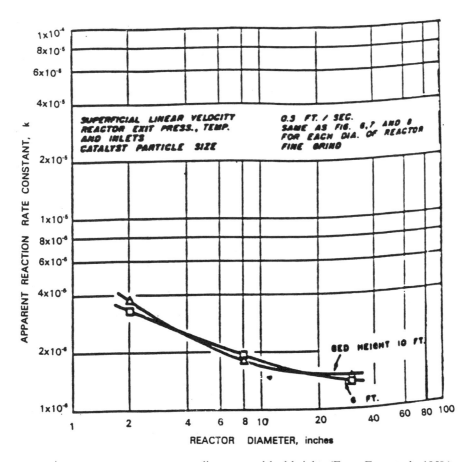

Figure 1 Apparent reaction-rate constant vs. reactor diameter and bed height. (From Frye et al., 1958.)

May (1959) reports results of tests done in cold models used to simulate the flow through large reactors whose performance had been found to be inferior to that of smaller pilot units. The importance of this problem can be appreciated from the scale of the equipment used, a 5 foot diameter unit used for the scale-up tests. This unit was fluidized with compressed air at 27 to 38°C (80 to 100°F) and pressures up to 689 kPa (100 psi). Gas residence time in the bed was determined by the use of tracer gas. Radioactive solid tracers were introduced into the bed to determine solid mixing. The data obtained in the larger units are much more erratic, with evidence of large-scale mixing patterns. Figure 2 shows the axial mixing coefficients obtained in experiments with different size beds. Mixing in the larger diameter bed is an order of magnitude larger than that in a small laboratory unit. The measured hydrodynamic behavior of the gas and solid was combined with a reaction model to predict the reactor behavior. Here again, there should be concern about the accuracy with the air experiments done at ambient temperature. Use of identical bed geometry and bed solid material does not guarantee identical hydrodynamics. The shift in gas properties from the cold model to the hot reactor may cause a marked difference in behavior. Additional scaling parameters must be maintained constant between the reactor and the cold model to insure identical hydrodynamics, and in some cases just to guarantee identical flow regimes!

Volk et al. (1962) show the effect of bed diameter on the conversion of CO in the "Hydrocol" reaction in which hydrogen and carbon dioxide are converted over an iron catalyst to hydrocarbons and oxygenated hydrocarbons in a bubbling or possibly slugging bed. Figure 3 shows the CO conversion. It is seen that the conversion rate is reduced as the reactor diameter increases. This is probably due to larger bubbles in the larger diameter bed. In general, smaller bubbles give more uniform solid mixing and a better gas–solid constant. Larger bubbles can allow more gas by passing through the bed. Volk used vertical tubes within the reactor to reduce the equivalent diameter of the system, equal to the hydraulic diameter, defined as four times the free cross-sectional area divided by the wetted perimeters of all surfaces in the cross section. The performance correlated with the equivalent diameter. It was also found that bed expansion was correlated with bed diameter. The internals serve to suppress the bubble size, although in some cases bubbles will preferentially stay near the vertical tubes and affect bed performance. In their process, larger beds were built with internals that kept the equivalent diameter the same as that of smaller units. The large units with internals appeared to give comparable gas-to-solid contacting. The use of vertical internals may not be feasible for a number of reasons, such as tube erosion. The use of the equivalent diameter approach may not be universally valid.

Van Swaaij and Zuiderwag (1972) used the ozone decomposition reaction to study the conversion characteristics in a bubbling bed. Studies were made with beds of 5, 10, 23, 30, and 60 cm diameter, respectively and up to 300 cm bed heights. The results were com-

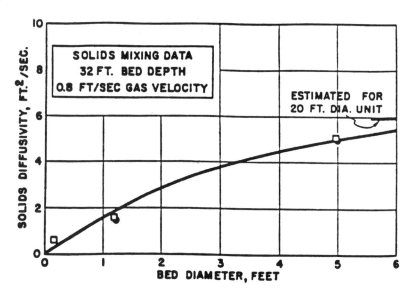

Figure 2 Solid diffusivity in axial direction for large units. (From May, 1959.)

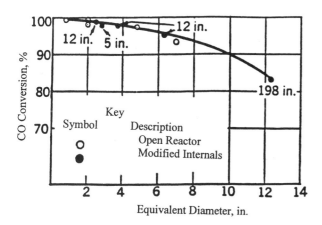

Figure 3 CO conversion in Hydrocol reaction for several reactor diameters. (From Volk et al., 1962.)

pared with predictions using a two-phase flow model with the mass transfer coefficient between the bubble and dense phase derived from residence time distribution results of gas tracer tests using pulse response. The results from the ozone conversion and the residence time distribution interpreted by the two-phase model gave reasonably similar results. In these cases, the mass transfer between phases is the limiting resistance for the reaction. Note that for larger bed diameters the mass transfer coefficient decreases. Van Swaaij and Zuiderweg (1973) showed that the inclusion of vertical tubes in a bed gave bubble-to-dense-phase mass transfer results that were roughly equivalent to those of a smaller open bed with the same hydraulic diameter. The solids' axial mixing was higher than that predicted using the hydraulic diameter.

Bauer et al. (1981) measured the influence of bed diameter on the catalytic decomposition of ozone. Figure 4 shows the decrease of the conversion with bed diameter for Bauer's data. This figure also shows the influence of distributor design on conversion. In many small-scale experiments, a porous plate is used that will give better performance than the distributors used in large shallow commercial bed designs.

Avidan and Edwards (1986) successfully scaled up from bench scale to demonstration plant from 0.04 m to 0.6 m diameter while maintaining nearly 100% conversion for a fluid bed methanol-to-gasoline process. In this case they ran at a high superficial gas velocity; the bed was in the turbulent flow regime suppressing bubbles. By this technique, they eliminated the losses associated with gas bypassing in bubbles.

3 INFLUENCE OF BED DIAMETER ON HYDRODYNAMICS

3.1 Bubbling Beds

In the studies mentioned above, the major objective was the experimental determination of conversion as a function of overall design parameters and particle properties. There have also been studies of the influence of bed diameter on the hydrodynamics in an effort to understand the cause of the conversion loss with bed size increase. These studies have aided in the development of physical models of reactor performance. De Groot (1967) measured gas residence time, bed expansion, and solid axial mixing in a series of beds at different diameters fluidized with air at ambient conditions. He used a narrow size range and a broad size range of crushed silica with sizes below 250 μm. Beds with diameters of 0.1, 0.15, 0.3, 0.6, and 1.5 m were used in the tests. There was a substantial decrease respectively in bed expansion and bubble fraction for narrow size range particles at large bed diameters, indicating the possibility of gas bypassing in bubble channels, Fig. 5. The axial diffusivity also increased with bed diameter and was a strong function of particle size distribution.

Werther (1974) measured the bubble characteristics in cylindrical beds of diameters 100, 200, 450, and 1000 mm for fine particles with a mean diameter of 83 μm. He showed that for bed diameters commonly used for laboratory experiments, 200 mm or smaller, the bed diameter had a strong effect on the bed hydrodynamics. There was a zone of preferred bubble flow near the bed walls at lower elevations, Fig. 6. The

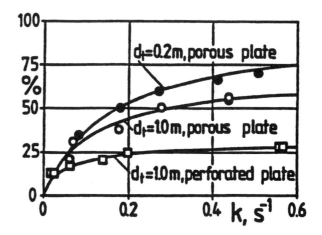

Figure 4 Conversion catalytic decomposition of ozone for different bed diameters and distributors. (From Werther, 1992.)

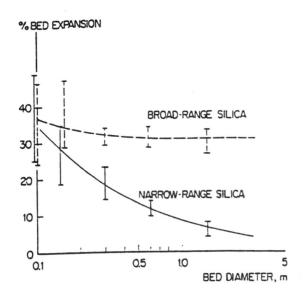

Figure 5 Bed expansion as a function of bed diameter at a fluidization velocity of ≈ 0.20 m/s. (From DeGroot, 1967.)

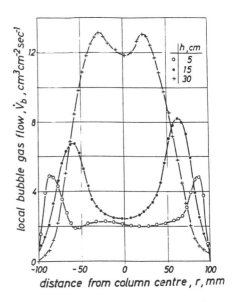

Figure 6 Bubble gas flow V_b as a function of the distance r from the vessel center line in a height of 30 cm above the distributor in beds of different diameters D_B ($u_0 = 9$ cm/s, $H = 50$ cm. (From Werther, 1974.)

bubbles grew in size and moved toward the centerline, presumably by coalescence, higher up in the bed. The transition to slugging occurred higher up in the larger bed at the same superficial gas velocity. The bubble velocity increased with height until slug flow was approached, after which the velocity decreased. For the only case shown, the 100 mm bed, the maximum velocity occurred when the bubble dimension was about one third of the bed diameter. In larger beds, the bubble rise velocity was higher for the same bubble volume, Fig. 7. Hovmand and Davidson (1971) reviewed data on bubble rise velocity and concluded that when the bubble diameter exceeded one-third of the bed diameter, the bed diameter rather than the bubble diameter governed the bubble rise velocity. Note that Werther's results at different superficial velocities are well correlated by the drift flux form,

$$u_b = \Phi(u_0 - u_{mf}) + c\sqrt{gd_v} \qquad (1)$$

It is curious that the influence of bed size appeared to hold even when the bubble was much smaller than the bed diameter. This may be tied to the local concentration of bubbles within certain sections of the bed. The increased local bubble flow led to higher coalescence rates and higher local bubble velocities. There was a distinction in bubble velocity between Geldart group A and B powder while the ratio of visible bubble flow to u-u_{mf} seems to be independent of the

Geldart group. Werther also found that the visible bubble flow, the product of the number of bubbles per unit time crossing a given surface and their respective volumes was considerably less than u-u_{mf}. This was especially true in the lower regions of the smaller bed and throughout the large diameter beds. The residence time of bubbles was significantly higher for small beds, with diameters of 200 mm or less, than it was for large diameter beds. Werther concludes that the smallest bed diameter that appears

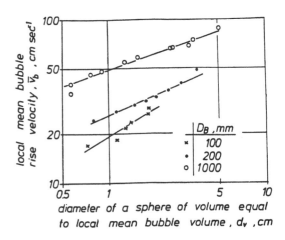

Figure 7 Relationship between local mean bubble size and local mean bubble rise velocity in beds of different diameters ($u_0 = 9$ cm/s). (From Werther, 1974.)

suitable for obtaining good scale-up results is 500 mm. It should be pointed out that this criterion would probably vary with flow conditions, bed depth, and particle size. Also, Werther's experiments were carried out with air at ambient conditions. Although the trends and physical picture may be similar for beds at elevated temperature and/or pressure, there will probably be some changes in the numerical limits for this criterion.

Whitehead (1967) found patterns of bubble tracks in a large 1.2 m square bed similar to the patterns observed by Werther: preferred bubble tracks near the walls and corners of a shallow open bed and merging of bubbles toward the bed center at higher elevations. Nguyen et al. (1979) also found that a horizontal tube bank in the large bed caused smaller bubbles that appeared in more random locations across the upper surface of the bed. This work was carried out using fine solids at low superficial velocity, 15 cm/s, and modest bed depths.

Geldart (1970) showed a substantial distinction between bubble sizes in two-dimensional and three-dimensional beds. He used 128 µm river sand in a 30.8 cm round bed and a 68 × 1.27 cm rectangular cross section bed. The bubbles in the three-dimensional round bed were larger. There were differences in the visible bubble flow rate at the same superficial velocity. Geldart ascribes the differences in bubble diameter to differences in visible bubble flow rate as well as to out-of-line bubble coalescences in the three-dimensional bed.

Glicksman and McAndrews (1985) determined the effect of bed width on the hydrodynamics of large particle bubbling beds. Sand particles with a mean diameter of 1 mm were fluidized by air at ambient conditions. A rectangular-cross-section bubbling bed was used. The bed width ranged from 7.6 to 122 cm, while the other cross-sectional dimension was held constant at 122 cm. Most experiments were carried out with an open bed. The bubble rise velocity increased with the bed width; when the bubble velocity was represented as the drift flux plus the rise velocity of an isolated bubble,

$$u_b = u_0 - u_{mf} + \phi\sqrt{gd_B} \tag{2}$$

φ varied from 0.4 in a two-dimensional bed to 0.6 in the three-dimensional bed. The mean vertical chord length of bubbles decreased with bed width; see Fig. 8. The visible bubble flow decreased dramatically with an increase of bed width at a fixed superficial velocity; see Fig. 9. Representing the total gas flow as the sum of the visible bubble displacement, the flow through the dense bed at minimum fluidizing conditions and the gas throughflow in the bubbles can be written as,

$$u_0 = Q_b + (1 - \delta)u_{mf} + m\delta u_{mf} \tag{3}$$

The gas throughflow coefficient m increased from a mean value of 3.6 for the two-dimensional bed to a mean value of 11.7 for the largest, 1.22 × 1.22 m bed cross section. The bed depths at minimum fluidization were 46 and 76 cm in the tests. For these rather shallow beds there were no observable preferred bubble tracks, and the location of erupting bubbles was random across the bed surface. For the cases observed, with u_0/u_{mf} varying from 1.3 to 1.8, the influence of the wall was absent when the bubble diameters were roughly one-fifth of the bed width or less. For deeper beds or higher gas velocity, the ratio of bubble diameter to bed width is expected to be the best criterion for determining when wall effects will be negligible. For one test series, five staggered rows of horizontal tubes with a horizontal center-to-center spacing of 15.2 cm were placed in the 1.22 m square bed. The tube bed results are shown in Figs. 8 and 9 with a T symbol. The behavior of bubbles in the bed with the tube bank resembled that of an open bed with a smaller width. Different tube arrangements and spacing yield different effective bed widths. Glicksman and Yule (1991) showed that the bed expansion had different rates of change with superficial velocity when different horizontal tube arrangements were tested in the large cold bed. The differences are due to varying maximum bubble sizes and the possibility of vertical bubble alignments to augment gas bypass through the dense bed substantially.

3.2 Mixing

Van Deemter (1980) surveyed data on solid mixing in fluidized beds of different diameters. Many of the experiments in large beds were considered inconclusive. The compiled data for longitudinal fluid dispersion, M_L, gas back mixing, M_B, and longitudinal dispersion of solids, M_S are shown in Fig. 10. Note the strong influence of bed diameter on mixing. The scatter in the data was attributed to differences in gas velocity and range of particle sizes. It may also be set up by different measurement techniques. Large differences between small and large bed diameters may be due to flow regime transitions.

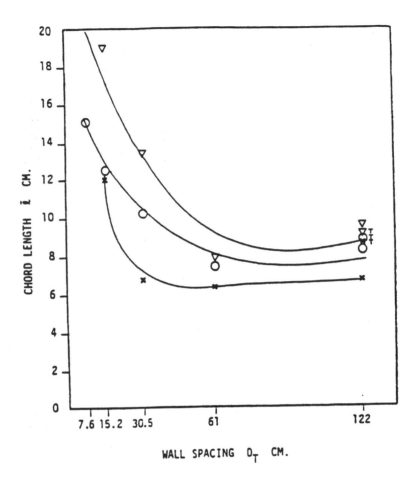

Figure 8 Variation of main chord length with wall spacing. $I_m = 76$ cm, probe height $= 46$ cm, X, 1.3 U_{mf}; O; 1.5U_{mf}; ∇; 1.8 U_{mf}; T denotes with tubes. (From Glicksman and McAndrews, 1985.)

The differences in behavior between small laboratory beds and larger demonstration units can in part be attributed to a switch from porous plate distributors in the small bed to discrete hole or bubble caps in the larger beds. The porous plates give a better quality of fluidization, e.g., smaller bubbles, for shallow beds and beds of moderate depth; see Rowe and Stapleton (1961).

Yerushalmi and Avidan (1985) suggest that the axial dispersion coefficient of solids in slugging and turbulent flow varies approximately linearly with the bed diameter, as with Thiel and Potter (1978).

3.3 Influence of Bed Diameter on Circulating or Fast Fluidized Beds

Arena et al. (1988) measured the hydrodynamic behavior of two circulating fluidized beds with riser diameters of 0.041 and 0.12 m ID of roughly the same height. At the same superficial gas velocity and solid recirculation rate, the larger diameter column had a higher solids fraction. The average slip velocities derived from these data are also higher for the large diameter riser; see Hartge et al. (1985). Yerushalmi and Avidan (1985) found a similar trend when comparing a 15.2 and 7.6 cm column. Noymer et al. (1995) also compared two columns of 5.08 and 7.68 cm diameter of the same height, which were used to simulate larger pressurized fluidized bed combustors. They found higher solids loading for the larger diameter riser at equal gas velocity and solid recirculation. In addition, the fraction of the wall covered by clusters was higher for the larger diameter column when the two beds had equal solids flow and when the two beds had equal cross section averaged solids concentration. The increased wall coverage should lead to higher heat transfer rates from the bed to the wall.

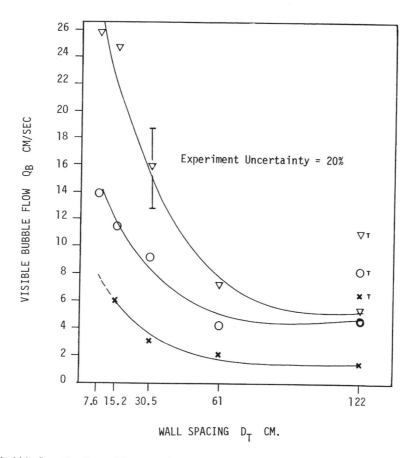

Figure 9 Visible bubble flow Q_b, $I_{mf} = 76$ cm, probe height $= 46$ cm. X, $1.3U_{mf}$; $\bigcirc$; $1.5U_{mf}$; ∇; $1.8U_{mf}$; T denotes with tubes. Experiment uncertainty $= 20\%$. (From Glicksman and McAndrews, 1985.)

Rhodes et al. (1992) compared the solids flux profiles across the cross sections of a 0.152 and a 0.305 m diameter circulating bed riser. They found a region where the solid profile, given by the ratio of local flux to average flux, had a similar variation over the cross section and was insensitive to the level of solid flux. The variation of the local solids flux over the radius was a function of the gas velocity and the riser diameter. In the larger riser, the profiles were somewhat flatter and the thickness of the downflowing region relative to the bed radius was smaller. The comparisons were not exact since the cross sections compared for the two beds were at different heights.

Zhang et al. (1991) carried out investigations with three different fast bed systems with diameters of 32, 90, and 300 mm. They found that the radial voidage distribution, given as a ratio to the cross-sectional average, was independent of bed diameter and solids recycle rate. The similarity does not hold at transition

to the turbulent regime. The results are for the center of the riser excluding the entrance and exit regions. It would be interesting to determine if the similar voidage profiles hold for larger diameter risers.

The thickness of the downflowing layers at the wall of the CFB is typically defined as the distance from the wall to the position of zero average vertical solid flux. Zhang et al. (1995) made measurements of the layer thickness on a 12 MWth and a 165 MW CFB boiler. They found that the thickness increased for the larger bed. They related data from many different beds, Fig. 11, with the equivalent bed diameter, taken as the hydraulic diameter, using the form

$$\delta = 0.05 \mathrm{De}^{0.74} \tag{4}$$

The thickness, δ, was found to be insensitive to particle concentration, gas velocity, and height within the furnace. This suggests that the thickness results from a balance of solids internal circulation that is generally

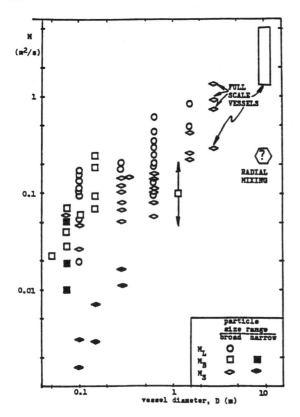

Figure 10 Mixing coefficients for different vessel diameters; M_L, longitudinal dispersion of fluid; M_B, gas backmixing; M_S: longitudinal solid dispersion. (From Van Deemter, 1980.)

much higher than net throughflow. If the local solids flux profile, as a ratio of the cross-sectional average, is roughly invariant over the cross section for these larger

beds, the thickness of the wall layer follows from a mass balance. The upflow is proportional to the core area, $\pi D^2/4$, and the downflow to the product of perimeter and layer thickness, proportional to $\pi\delta D$. Thus the thickness, δ, should vary as the ratio of cross-sectional area to perimeter, i.e., proportional to the hydrodynamic diameter.

Patience et al. (1992) developed a dimensionless correlation for the mean slip factor between gas and solid by using solid suspension data from various small laboratory beds. The proposed correlation relates the slip to the Froude number based on the bed diameter. It remains to be seen if the correlation will hold at Froude numbers typical of large beds and if other dimensionless factors are important for large beds.

Johnsson and Leckner (1995) observed the flow patterns in a large laboratory circulating bed, $1.7 \times 1.4\,\mathrm{m}$ in cross section. The lower zone was characterized by a dense bottom similar to a bubbling bed with exploding bubbles. The bed porosity was 0.7 or less. These observations are in contrast to those made by other investigators in much smaller circulating beds.

3.4 Flow Transition

Hovmand and Davidson (1971) review Stewart's criterion for the transition from bubbling to slug flow,

$$\frac{u - u_{\mathrm{mf}}}{0.35(gD)^{1/2}} = 0.2 \tag{5}$$

and show it gives good agreement with most experiments. When $u - u_{\mathrm{mf}}$ is larger than the value found from Eq. (5), the bed will be in slug bed. Thus a small

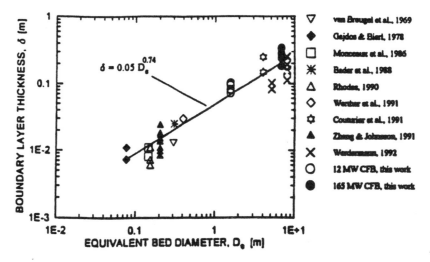

Figure 11 Empirical correlation and experimental data of thickness of downflowing layer at the wall of a CFB as a function of the equivalent bed diameter. (From Zhang et al., 1995.)

diameter laboratory bed may exhibit behavior that is far different from that of a scaled-up pilot plant. The small bed may be well into slug flow whereas the larger bed may be in bubbling bed. There is some data that suggests the ratio of bed depth to bed diameter may also influence the bubble-to-slug transition.

Thiel and Potter (1977) carried out slugging flow experiments in beds of three different diameters, the largest being 0.22 m in diameter. They found that the bed aspect ratio influenced the onset of slug formation. They also found that the transition to turbulent flow occurred at much lower superficial gas velocity in the 0.22 m diameter bed than in the 0.1 m bed. With fluid cracking catalyst, the transition velocity decreased from 20 cm/s to 2.5 cm/s. Yerushalmi and Avidan (1985) assert that in large shallow beds where slugging does not occur, transition from bubbling to turbulent flow should be independent of bed size.

4 EXPERIMENTAL MEANS TO ACCOUNT FOR SCALE-UP: USE OF SCALE MODELS

Since large pilot plants are costly, it may not be feasible to undertake an extensive scale-up program. Furthermore, it is difficult to observe or measure detailed behavior within large hot beds. Thus many test results from beds at elevated temperatures and pressures are confined to overall operating characteristics, leaving the designers to speculate on the cause of shifts in performance with bed size. Full-scale cold models are also expensive and require lengthy construction to modify the bed geometry. As we will see below, the full-scale cold test may not correctly simulate the hydrodynamics of the actual process at elevated temperature and pressure. Indeed, a familiar occurrence in atmospheric bubbling fluidized bed combustors is the marked difference in flow behavior between a bed fluidized with cold air and the same bed, using the same particles, run at normal operating conditions.

A technique, that can assist in the scale-up of commercial plant designs is the use of scale models. A scale model is an experimental model that is smaller than the hot commercial bed but that has identical hydrodynamic behavior. Usually the scale model is fluidized with air at ambient conditions and requires particles of a different size and density than those used in the commercial bed. The scale model relies on the theory of similitude, sometimes through the use of Buckingham's pi theorem, to design a model that gives identical hydrodynamic behavior to the commercial bed. Such a method is used in the wind tunnel

testing of small model aircraft or in the towing tank studies of naval vessels.

Once a technique has been established to design a small cold model that simulates the hydrodynamics of a hot (possibly pressurized) fluidized bed, then a series of differently sized models can be used to determine the influence of bed size on the performance of commercial beds; see Fig. 12. Model A′ simulates the behavior of commercial bed A, model B′ simulates a larger commercial bed B, and so forth. Then by comparing models A′, B′, with C′ we can determine the expected changes in operating characteristics when commercial bed A is replaced by larger beds B and C.

Designing a model fluidized bed that simulates the hydrodynamics of a commercial bed requires accounting for all of the important mechanical forces in the system. In some instances convective heat transfer can also be scaled, but at present, proper scaling relationships for chemical reactions or hydromechanical effects, such as particle attrition or the rate of tube erosion, have not been established.

4.1 Development of Scaling Parameters

There are several approaches to developing the correct scaling relationships. Probably the most straightforward is the nondimensionalization of the governing

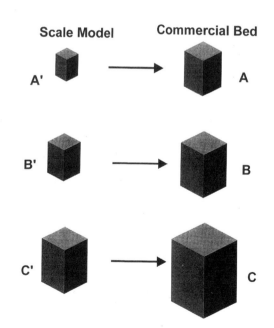

Figure 12 Use of scale models with different bed diameters to simulate the influence of diameter on the hydrodynamics of a hot commercial reactor.

equations. If we can write the proper equations governing fluid and particle dynamic behavior, we can develop the proper scaling relationships even if we cannot solve the equations (at present we cannot). In essence, if a model is designed that follows the exact same equations and boundary conditions as the commercial bed, with the same values of the coefficients, then the model and the commercial bed should have identical behavior. The trick is to nondimensionalize all of the terms of the equations so that the model and the commercial beds have identical nondimensionalized values of the parameters. The complete development of this is given in Glicksman et al. (1994) and Glicksman (1997).

An alternate approach is to design a scale model so that the ratio of all of the important forces is the same in the model as it is in the full-scale bed, and a the scale model uses geometry similar to that of the larger fluidized bed. If the ratio of forces acting on a particle in the scale model bed have the same ratio as they do for a particle in the full bed, then the trajectory of the particle motion should be the same for both beds. The same follows for the motion of the gas and the motion of clusters of particles or particles in a dense phase. From this point of view, we must identify all the important forces in the system. Table 1 lists these forces.

To characterize the gas flow in the fluidized bed, important forces are viscous and inertia as well as particle-to-gas forces. For the particles, the important forces include gravity, particle inertia, gas interaction with the particles such as drag, collisional forces between particles and between particles and wall, and particle surface forces such as electrostatic and adhesion forces. For larger particles, in Geldart groups B or D, the particle surface forces can be neglected. The question is complicated for smaller particles because the surface forces are difficult to quantify. We must also include the distributions of particle sizes, since they can influence the relative magnitude of, say, gravity to drag forces.

For this work we will concentrate on particles of sufficient size and mass so to minimize interparticle surface forces relative to other forces. Limited results from Litka and Glicksman (1985), Chang and Louge (1992), and Glicksman et al. (1991a) suggest that, within the normal range of particle and wall coefficient of restitution and particle-to-particle sliding friction, these parameters have a modest influence on the bed dynamics. The particle-to-particle and particle-to-wall collisions may still play an important role in the bed dynamics. This only assumes that the changes in the coefficients of restitution and friction are small enough to justify their omission in the list of important dimensionless parameters that can be controlled or modified.

To construct a scale model of a commercial bed, first the geometry must be similar. The linear dimensions of all components in the commercial bed, bed diameter, height, internal tube diameter and spacing, etc., must be reduced by the same factor in the model. The location coordinates of a particle within the bed are also reduced in the same fashion.

When all of the force ratios given in Table 1 are set equal between a scale model and a commercial bed, then the scale model should show identical dynamic behavior. This behavior refers to the flow regime, particle and gas velocity (nondimensionalized as a ratio to the superficial velocity), and particle location (the coordinates nondimensionalized as a ratio to the bed diameter or bed height). Thus at a given location X/D_m, y/D_m in the model, where D_m is the model diameter, the velocity u/u_{0m} is the same as u/u_{0c} at XD_c, y/D_c in the commercial bed.

The requirement of equal force ratios for all of the terms given in Table 1 is referred to as the full set of scaling relations. They can be written as

Table 1 Important ratio of forces acting in a fluidized bed.

$\rho_s U_0 d_p/\mu$	Particle inertia/gas viscous force
$\rho_f U_0 L/\mu$	Gas inertia/Gas viscous force
U_0^2/gL	Inertia/gravity force
ρ_s/ρ_f	Solid inertia/Gas inertia force
	Surface Forces—Collision, Adhesion; Form is Debatable
	Additional Important Ratios
$G_s/\rho_s U_0$	Solid recycle volumetric flow/Gas volumetric flow rate
L/D	Bed height/Bed diameter
PSD	Particle size distribution, χ_1 (mass fraction of d_i) for d_i/d_p (mean solid diameter) for all d_i's
φ	Particle sphericity

$$\frac{u_0^2}{gL}, \frac{\rho_s u_0 d_p}{\mu}, \frac{\rho_f u_0 L}{\mu}, \frac{\rho_s}{\rho_f}, \frac{G_s}{\rho_s u_0}, \frac{L}{D},$$

$$\text{bed geometry, } \varphi, \text{ PSD} \qquad (6)$$

Where φ is the particle sphericity and PSD is the dimensionless particle size distribution.

Alternatively, the terms can be rewritten by multiplying terms together, although the number of independent dimensionless parameters stays the same. Thus, Eq. (6) can be rewritten, for example, as

$$\frac{\rho_f \rho_s d_p^3 g}{\mu^2}, \frac{\rho_s}{\rho_f}, \frac{u_0^2}{gL}, \frac{\rho_f u_0 L}{\mu}, \frac{G_s}{\rho_s u_0}, \frac{L}{D},$$

$$\text{bed geometry, } \varphi, \text{ PSD} \qquad (7)$$

He et al. (1995) have shown that for spouting beds, cohesive factors are important, and this group must be augmented by including the internal friction angle and the loose packed voidage to achieve similar scale models. Since interparticle friction can occur in slugging beds, these additional parameters should be included to scale slugging beds properly.

5 SIMPLIFIED SCALING RELATIONSHIPS

For scaling to hold with the full set of scaling relationships, all of the dimensionless parameters given in Eqs. (6) or (7) must be identical in the scale model and the commercial bed under study. If the small scale model is fluidized with air at ambient conditions, then the fluid density and viscosity are fixed and there is only one unique modeling condition that will allow complete similarity. In some cases, this requires a model that is too large and unwieldy or will not permit simulation of a very large bed.

In most situations, one would expect that not all of the parameters are of first order importance. By reducing the number of parameters that must be maintained in the model it will be possible to model larger commercial beds with small scale models. This will involve simplifications of the interparticle drag at the extreme of small and large Reynolds numbers based on particle diameter. If the same simplification can be shown to hold in both of these limits, it is reasonable to consider application of the simplification over the entire range of conditions.

By representing the particle to air drag force by the Erqun equation or single particle drag, it can be shown, Glicksman et al. (1994), that in the limit of small particle Reynolds numbers the governing parameters can be reduced to

$$\frac{u_0^2}{g^2}, \frac{\rho_s}{\rho_f}, \frac{u_0}{u_{mf}}, \frac{L_1}{L_2}, \frac{G_s}{\rho_s u_0}, \text{ bed geometry, } \phi, \text{ PSD} \qquad (8)$$

At the other extreme, at large particle Reynolds numbers it can be shown that the same set of governing parameters, Eq. (8), applies.

Since the same simplified set of dimensionless parameters holds exactly at both high and low Reynolds numbers, it is reasonable to expect that they hold, at least approximately, over the entire range of conditions for which the drag coefficient can be determined by the Erqun equation or an equation of similar form.

For the more general case, Fig. 13 shows the value of β the dimensionless drag coefficient relative to β at low Re over a range of conditions when u_0/u_{mf} is 10 and 3, respectively, and Fr and ϕ_s remain constant. When u_0/u_{mf} and the slip velocity are high there is a larger variation of dimensionless drag coefficient with Reynolds number. Note that β does not vary with particle Reynolds number when the Reynolds number remains above about 10^3 or below about 10. When u_0/u_{mf} is 1000, a condition is approached with very fine particle bubbling beds or circulating beds. The use of the Erqun relationship is questionable except for the dense lower part of the bed.

5.1 Range of Validity of Simplified Scaling

To determine the validity of the simplified scaling laws over a wide range of conditions, the simplified scaling laws have been used, Eq. (8), to design hypothetical models whose linear dimensions are one-fourth and one-sixteenth, of the linear dimensions of a model designed using the full set of scaling laws, Eq. (6). To determine the validity of the smaller, simplified models, the dimensionless drag coefficient $\beta L/\rho_s u_0$ will be compared between the simplified models and the model using the full set of scaling laws. Figure 14 shows a comparison of the exact model and the simplified models for a pressurized fluidized combustor. Using the full set of scaling laws, the exact model, fluidized by ambient air, is approximately the same size as the combustor. The simplified models are reduced in size by their respective assumed length scale. The other parameters of the simplified model are then calculated to match the simplified parameters. For example, when the length scale is reduced to one-fourth that of the exact model, the velocity is reduced by one-half to keep the Froude number constant. The particle diameter is then reduced appropriately to keep the ratio of u_0/u_{mf} con-

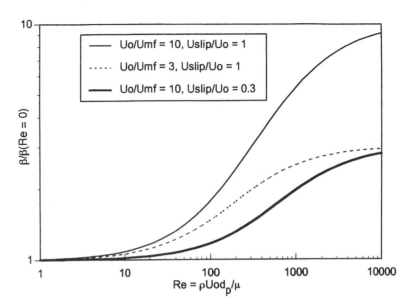

Figure 13 Ratio of drag coefficient to low Reynolds number drag coefficient, $u_0/u_{mf} = 10$ and 3. (From Glicksman et al., 1993b.)

stant. These calculations were carried out over a range of particle Reynolds numbers, Re_{pE}, based on the full scaling law, or exact, model. In the simplified scaling relationships, the Reynolds number is not maintained constant. The concern is how much the drag coefficient is impacted by the shift in Reynolds number. It was found that the particle Reynolds number for the one-fourth scale simplified model remained roughly equal to 0.34 Re_{pE} over a wide range of values for Re_{pE}, whereas the particle Reynolds number for the one-sixteenth scale model was roughly 0.12 Re_{pE}.

Using these Reynolds number scale factors, the error in the dimensionless drag coefficient $\beta L/\rho_s u_0$ using the simplified scaling models is shown on Fig. 15 for u_0/u_{mf} of 10, plotted as a function of Re_{pE} based on parameters for the exact scaled bed. For a particle Reynolds number of 1000 or less, which corresponds to pressurized beds with particles of 1 mm or less, the error in the drag coefficient with the simplified scaling laws is 20% or less for a one-quarter length scale model. The error is 40% or less for a one-sixteenth length scale model. At u_0/u_{mf} of 1000 and u_{slip}/u_{mf} of one-fiftieth the errors for the one-sixteenth scale model are 20% or less for Re_{pE} less than 10^3. For particles of 0.2 mm or less, corresponding to a Reynolds number of 100 or less, the errors in drag coefficient are minimal. When the Ergun equation applies for the drag coefficient, a one-quarter scale model based on the simplified scaling laws should be valid for any conditions. A one-sixteenth scale model

should be valid for particle diameters of about 0.2 mm or less for a pressurized bubbling bed with u_0/u_{mf} of 10 and u_{slip}/U_0 of 0.3. At u_0/u_{mf} of 1000 and u_{slip}/U_0 of one-fiftieth, the one-sixteenth scale model should be valid for pressurized beds with particles up to 1 mm in diameter. These conclusions apply when the particle-to-fluid drag term is given by the Ergun equation or similar relationships and the scaled particles are not so small that interparticle surface forces come into play.

5.2 Clusters

In the freeboard of a bubbling bed or in the upper portion of a circulating bed where particles generally are considered to act in clusters or groups, a similar examination of scaling of the gas-to-solid drag can be made. Consider all the particles grouped into clusters with an effective diameter d_c and the clusters occupying a volume fraction $_c$ of the bed volume. The cluster-to-gas drag will be represented by the drag coefficient for a solid sphere of diameter d_c, C_D.

If the reduced scale models faithfully reproduce the dynamics of the exact case, the cluster dimensions should scale directly with the linear dimensions of the bed. Thus, a one-quarter linear scale model that has a velocity one-half that of the exact case will have a cluster Reynolds number (Re_{dc}) one-eighth that of the exact bed. From the relationship of C_D with Re the change of C_D with model scale at a given Reynolds number of the exact bed can be determined. Figure 16

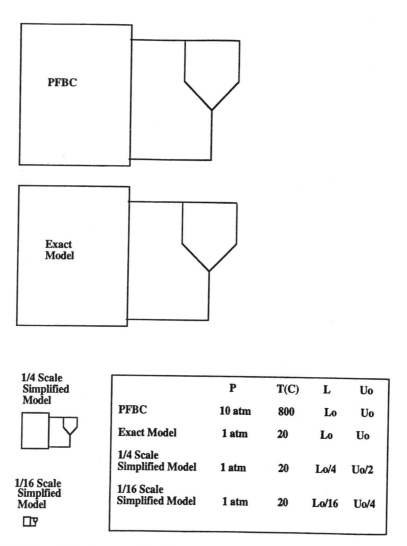

Figure 14 Exact and simplified models of a pressurized fluidized bed combustor. (From Glicksman et al., 1993b.)

shows the shift in C_D using the C_D relationship of White (1974) for length scales of one-fourth, one-eighth, and one-sixteenth of the exact bed length as a function of the cluster Reynolds number of the exact bed. Also shown in the figure is the typical Reynolds number of an atmospheric combustor with a 0.3 m cluster diameter, approximately 1.5×10^4. In a bubbling bed, the cluster diameter in the freeboard should be at least equal in size to the diameter of bubbles erupting at the bed surface. For beds with horizontal tubes, the bubble diameter will be equal to or larger than the horizontal tube spacing. In a bubbling bed without tubes, the bubbles and clusters can be much larger. In an open circulating bed, the cluster diameter is more difficult to determine. It is reasonable to assume that its diameter is proportional to the bed diameter, equal in magnitude to the bed diameter or one order of magnitude smaller. From these considerations, the Reynolds number based on the cluster diameter should be 10^4 or larger in an atmospheric combustor with a cluster diameter of 0.2 m. The cluster Reynolds number should be 10^5 or larger in a pressurized combustor. From Fig. 16 it can be seen that a one-quarter scale or an eighth scale model should have drag coefficients similar to those of the exact bed. For pressurized beds, the drag coefficients should be very close in magnitude.

Considering the drag acting on a single particle in a dispersed flow, the validity of the simplified relationships can be examined by comparing the terminal velocity predicted by the simplified relationship and the exact value.

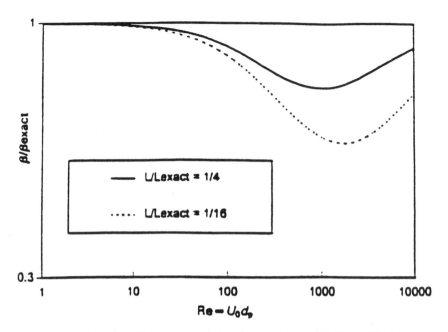

Figure 15 Error in drag coefficient for simplified scaling relationships, $u_0/u_{mf} = 10$. (From Glicksman et al., 1993b.)

The errors in u_t/u_{mf} is shown in Fig. 17 for simplified scale models at two different linear dimensions. Scaling a combustor with comparatively small particles, 0.2 mm or less, gives good agreement for u_t/u_{mf} even at one-sixteenth linear scale, while for large particles a linear scale of one-fourth gives fair agreement for u_t/u_{mf}. Since u_0/u_{mf} is held constant in the simplified scaling laws, close agreement of u_t/u_{mf} also results in close agreement of u_t/u_0.

6 FURTHER SIMPLIFICATIONS IN THE SCALING RELATIONSHIP

6.1 Viscous Limit

Glicksman (1984) showed that the list of controlling dimensionless parameters could be reduced if the fluid–particle drag is primarily viscous or primarily inertial. For very low particle Reynolds numbers the scaling relationships can be further simplified to

$$\frac{\rho_s u_0 d_p^2}{\mu D}, \phi_s, \frac{gD}{u_0^2}, \frac{D}{L}, \frac{G_s}{\rho_s u_0}, \text{ bed geometry, PSD} \qquad (9)$$

or alternatively, this can be written as

$$\frac{u_{mf}}{u_0}, \phi_s, \frac{gD}{u_0^2}, \frac{D}{L}, \frac{G_s}{\rho_s u_0}, \text{ bed geometry, PSD} \qquad (10)$$

Note that this is a subset of the simplified scaling laws presented above with the solid-to-gas density ratio removed.

Recent results indicate that in most scaling interactions, elimination of the ratio of solid-to-gas density ratio from the set of scaling parameters leads to discrepancies between the scale models and the larger beds being simulated; see Glicksman et al. (1998).

7 DESIGN OF SCALE MODELS

7.1 Full Set of Scaling Relationships

We will first consider the steps to design a model that is similar to another bed based on the full set of scaling parameters; see Eqs. (6) or (7).

To construct a model that will give behavior similar to another bed, for example, a commercial bed, all the dimensionless parameters listed in Eqs. (6) or (7) must have the same value for the two beds. The requirements of similar bed geometry is met by the use of geometrically similar beds; the ratio of all linear bed dimensions to a reference dimension such as the bed diameter must be the same for the model and the commercial bed. This includes the dimensions of the bed internals. The dimensions of elements external to the bed, such as the particle return loop, do not have to be matched as long as the return loop is designed to provide the proper external solids flow rate and size dis-

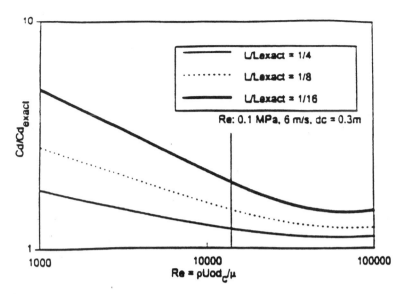

Figure 16 Error in cluster drag coefficient for fixed u_0/u_{mf} using C_d for a solid sphere. (From Glicksman et al., 1993b.)

tribution, and solid or gas flow fluctuations in the return loop do not influence the riser behavior; see Rhodes and Laussman (1992).

Proper conditions must be chosen to design a scale model to match the dimensionless parameters of the commercial bed. To model a gas fluidized commercial bed, a scale model using air at standard conditions is most convenient, although several investigators have used other gases; see Fitzgerald and Crane (1980), Fitzgerald et al. (1984), Chang and Louge (1992) or pressurized scale models, Almstedt and Zakkay (1990), Di Felice et al. (1992 a,b). The gas chosen for

the model, along with the gas pressure and temperature, determines the values of ρ_f and μ. The particle density for the model is chosen to match the density ratio, so that

$$\left(\frac{\rho_f}{\rho_s}\right)_m = \left(\frac{\rho_f}{\rho_p}\right)_c \quad (11)$$

where the subscript m is for the model and c is for the commercial bed. For the remaining parameters, the form of Eq. (6) will be chosen for the dimensionless parameters. Combining the Reynolds number based

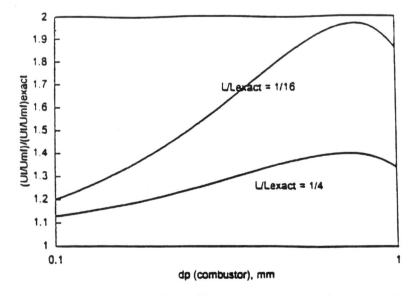

Figure 17 Error in terminal velocity using simplified scaling at 1013 kPa and 800°C. (From Glicksman et al., 1994.)

on bed diameter and the square root of the Froude number, and rearranging, we obtain

$$\frac{\rho_f u_0 D}{\mu_f} \frac{\sqrt{gD}}{u_0} = \left(\frac{D^{3/2}\sqrt{g}}{v_f}\right)_m = \left(\frac{D^{3/2}\sqrt{g}}{v_f}\right)_c \quad (12)$$

$$\left(\frac{D_m}{D_c}\right) = \left(\frac{(v_f)_m}{(v_f)_c}\right)^{2/3} \quad (13)$$

All of the linear dimensions of the model are scaled to the corresponding dimensions of the commercial bed by the ratio of the kinematic viscosities of the gas raised to the two-thirds power. By taking the ratio of the Reynolds number based on the particle diameter to the Reynolds number based on the bed diameter,

$$\frac{\rho_f u_0 D}{\mu_f} \cdot \frac{\mu_f}{\rho_f u_0 d_p} = \left(\frac{D}{d_p}\right)_m = \left(\frac{D}{d_p}\right)_c \quad (14)$$

The particle diameters in the model scale by the same factor as the bed diameter, by the ratio of the kinematic viscosities to the two-thirds power.

Equating the Froude number and rearranging, we find

$$\frac{u_{0m}}{u_{0c}} = \left(\frac{v_{fm}}{v_{fc}}\right)^{1/3} = \left(\frac{D_m}{D_c}\right)^{1/2} \quad (15)$$

Thus the velocity scales are the square root of the linear dimension scale.

By satisfying both Eq. (13) and Eq. (15), the Reynolds number and the Froude numbers are kept identical between the model and the commercial bed.

Combining G_s/ρ_s U_0 and the product of the Reynolds and Froude numbers along with Eq. (13) it can be shown that

$$\frac{(G_s/\rho)_m}{(G_s/\rho_s)_c} = \left(\frac{v_{fm}}{v_{fc}}\right)^{1/3} \quad (16)$$

so that the ratio of solids flow to solids density scales as the ratio of the cube root of the kinematic viscosity.

Once the model fluid and its pressure and temperature are chosen, which sets the gas density and viscosity, there is only one unique set of parameters for the model that gives similarity when using the full set of dimensionless parameters. The dependent variables, as nondimensionalized as X/D, y/D, u/u_0 are the same in the respective dimensionless time and spatial coordinates of the model as the commercial bed. The spatial variables are nondimensionalized by the bed diameter, so that the dimensional and spatial coordinates of the model are proportional to the two-thirds power of the kinematic viscosity, as given by Eq. (13),

$$\frac{x_m}{x_c} = \left(\frac{v_m}{v_c}\right)^{2/3} \quad (17)$$

Since the velocity scales with $v^{1/3}$, the ratio of time scales can be expressed as

$$\frac{t_m}{t_c} = \left(\frac{v_{fm}}{v_{fc}}\right)^{1/3} \quad (18)$$

Similarly, it can be shown that the frequency scales as

$$\frac{f_M}{f_c} = \left(\frac{v_{fc}}{v_{fm}}\right)^{1/3} \quad (19)$$

Table 2 gives the values of design and operating parameters of a scale model fluidized with air at ambient conditions which simulates the dynamics of an atmospheric fluidized bed combustor operating at 850°C. Fortunately, the linear dimensions of the model are much smaller, roughly one-quarter those of the combustor. The particle density in the model must be much higher than the particle density in the combustor to maintain a constant value of the gas-to-solid density ratio. Note that the superficial velocity of the model differs from that of the combustor along with the spatial and temporal variables.

When modeling a pressurized hot bed (Table 3), the ambient temperature model fluidized with air has dimensions very close to those of the pressurized combustor. If another gas is used in the model, particularly a gas with a higher density, the model can be made much smaller than the pressurized combustor (Table 4). Care must be taken to select a safe modeling gas and one that yields a solid density for the model that is available.

7.2 Design of Scale Models Using the Simplified Set of Scaling Relationships

The simplified scaling relationships, Eq. (8) offer some flexibility in the model design, since fewer parameters must be matched than with the full set of scaling relationships. When the fluidizing gas, and the pressure and temperature of the scale model, are chosen, the gas density and viscosity for the scale model is set. The model must still be geometrically similar to the commercial bed. There is still one free parameter. Generally, this will be the linear scale of the model. For the simplified scaling relationships, the gas-to-solid density ratio must be maintained constant,

$$\left(\frac{\rho_f}{\rho_s}\right)_m = \left(\frac{\rho_f}{\rho_p}\right)_c \quad (20)$$

Table 2 Atmospheric combustor modeled by a bed fluidized with air at ambient conditions

Given	Commercial bed	Scale model, full scaling laws
Temperature (°C)	850	25
Gas viscosity (10^{-5} kg/ms)	4.45	1.81
Density (kg/m^3)	0.314	1.20
Derived from scaling laws		
Solid density	ρ_{sc}	$3.82\rho_{sc}$
Bed diameter, length, etc.	D_c	$0.225D_c$
Particle diameter	d_{pc}	$0.225d_{pc}$
Superficial velocity	u_{0c}	$0.47u_{0c}$
Volumetric solid flux	$(G_s/\rho_s)_c$	$0.47(G_s/\rho_s)_c$
Time	t_c	$0.47t_c$
Frequency	f_c	$2.13f_c$

With ρ_f of the model set by the fluidized gas and its state, the solid density in the model follows from Eq. (20). Choosing the length coordinate of the model, D_m, the new free parameter, the superficial velocity in the model is determined so that the Froude number remains the same,

$$\left(\frac{u_0^2}{D}\right)_m = \left(\frac{u_0^2}{gD}\right)_c \tag{21}$$

so that

$$\frac{u_{cm}}{u_{0c}} = \left(\frac{D_m}{D_c}\right)^{1/2} \tag{22}$$

Note that in the simplified case the velocity scaling is not uniquely tied to just the gas properties as it is in the full scaling relationship. With U_0 and ρ_s set the solids recycle rate can be determined by

$$\left(\frac{G_s}{\rho_s u_o}\right)_m = \left(\frac{G_s}{\rho_s u_o}\right)_c \tag{23}$$

$$\frac{G_m}{G_{sc}} = \left(\frac{\rho_{sm}}{\rho_{sc}}\right)\left(\frac{u_{0m}}{u_{0c}}\right) = \left(\frac{\rho_{fm}}{\rho_{fc}}\right)\left(\frac{D_m}{D_c}\right) \tag{24}$$

Finally, the mean particle size for the model as well as the sphericity and particle size distribution must be determined. The particle size is determined by the need for equal values of u_0/u_{mf} between the model and the commercial bed:

$$\left(\frac{u_0}{u_{mf}}\right)_m = \left(\frac{u_0}{u_{mf}}\right)_c \tag{25}$$

Table 3 Pressurized combustor modeled by a bed fluidized with air at ambient conditions

Given	Commercial bed	Scale model, full scaling laws
Temperature (°C)	850	25
Gas viscosity (10^{-5} kg/ms)	4.45	1.81
Density (kg/m^3)	3.14	1.20
Pressure (bar)	10	1
Derived from scaling laws		
Solid density	ρ_{sc}	$0.382\rho_{sc}$
Bed diameter, length, etc.	D_c	$1.05D_c$
Particle diameter	d_{pc}	$1.05d_{pc}$
Superficial velocity	u_{0c}	$1.01u_{0c}$
Volumetric solid flux	$(G_s/\rho_s)_c$	$1.01(G_s/\rho_s)_c$
Time	t_c	$1.01t_c$
Frequency	f_c	$0.98f_c$

Table 4 Pressurized combustor modeled by a bed fluidized with refrigerant vapor 134a at ambient conditions

Given	Commercial bed	Scale model, full scaling laws
Temperature (°C)	850	20
Gas viscosity (10^{-5} kg/ms)	4.45	1.19
Density (kg/m^3)	3.14	4.34
Pressure (bar)	10	1
Derived from scaling laws		
Solid density	ρ_{sc}	$1.38\rho_{sc}$
Bed diameter, length, etc.	D_c	$.334D_c$
Particle diameter	d_{pc}	$.334d_{pc}$
Superficial velocity	u_{0c}	$.58u_{0c}$
Volumetric solid flux	$(G_s/\rho_s)_c$	$.58(G_s/\rho_s)_c$
Time	t_c	$.58t_c$
Frequency	f_c	$1.7f_c$

$$(u_{mf})_m = (u_{mf})_c \left(\frac{u_{cm}}{u_{0c}}\right) = (u_{mf})_c \left(\frac{D_m}{D_c}\right)^{1/2} \qquad (26)$$

In general, u_{mf} is a function of the particle diameter and gas properties, as well as ϕ and ε_{mf}. Once the fluidizing gas and the length of scale of the model are chosen, the proper particle diameter is that which gives the value of u_{mf} needed in Eq. (26).

If both the model and commercial bed are in the region where the respective Reynolds number based on particle diameter and gas density is very low, then a single algebraic relationship can be developed. In that region,

$$u_{mf} \sim \frac{\rho_s d_p^2}{\mu} \qquad (27)$$

$$\frac{d_{pm}}{d_{pc}} = \left(\frac{\rho_{fc}\mu_m}{\rho_{fc}\mu_m}\right)^{1/2} = \left(\frac{D_m}{D_c}\right)^{1/4} \qquad (28)$$

when both $(Re_{dp})_m$ and $(Re_{dp})_c < 20$.
When the Reynolds number of the model and commercial bed are both very large,

$$U_{mf}^2 \sim d_p \frac{\rho_s}{\rho_f} \qquad (29)$$

Combining Eq. (29) with Eqs. (22) and (25), we find

$$\frac{u_{mf_m}^2}{u_{mf_c}^2} = \frac{d_{pm}}{d_{pc}} \frac{(\rho_s/\rho_f)_m}{(\rho_s/\rho_f)_c} = \frac{u_{0m}^2}{u_{0c}^2} = \frac{D_m}{D_c} \qquad (30)$$

Since the gas-to-solid density ratio of the model and the commercial beds must be the same to satisfy the simplified scaling relationships Eq. (30) becomes

$$\frac{d_{pm}}{d_{pc}} = \left(\frac{D_m}{D_c}\right)^{1/4} \qquad (31)$$

when both $(Re_{dp})_m$ and $(Re_{dp})_c > 1000$.

Tables 5 and 6 show the values of the mean particle diameter for models of an atmospheric and pressurized commercial bed, for different selected linear scale ratios between the model and the commercial bed.

By the use of the simplified scaling parameters, the linear scale factor can be changed as shown in Tables 5 and 6. Note that as the linear scale factors are changed the particle diameter changes much more slowly. The scale model of the 10 atmosphere bed has a mean particle diameter that is quite close to the mean particle diameter of the commercial bed. The model particles have a substantially lower density in this case.

It is not clear where cohesive forces will become important. The use of very dense particles (for the models of the one atmosphere bed) will cause a shift of the boundary of cohesive influence as given, for example, by Geldart's classification. However, adequate experimental data are still lacking with such dense fine particles to set the limits of cohesive influence.

Note that for completeness the nondimensional particle size distribution, the sphericity, and the internal angle of friction (for slugging and spouting beds) should also be matched between the two beds.

7.3 Hydrodynamic Scaling of Bubbling Beds

Experiments using scaled models of bubbling beds have been carried out since 1980 using the scaling relationships presented in previous sections. The earlier

Table 5 Scale models of atmospheric commercial hot bed using the simplified scaling relationship

Commercial beds		Particle diameter of model with bed linear scale factor	
d_p	u_{mf}	$Dm/Dc = 1/4$	1/9
40 μm	7.45×10^{-4} m/s	10 μm	8 μm
60	1.68×10^{-3}	15	12
100	4.66×10^{-3}	24	20
200	1.86×10^{-2}	49	40
400	7.42×10^{-2}	98	80
1000	0.441	245	198

Commercial bed $t = 800°C$, $P = 1$ atm, $\rho_s = 2500$ kg/m³ gas:air. Model bed $\rho_s = 8960$ kg/m³ gas:air at STP.

work was concerned with experimental testing and verification of the scaling relationships. Hot bed behavior was compared to cold scaled models in several studies. In others, cold beds of different sizes were compared to each other. For bubbling beds there is the sense of what constitutes sufficient verification of the scaling relationships. Since bubbles are the prime motive agents for both gas and solids displacement, a detailed verification should involve comparison of bubble properties through the large bed and the scale model. This is preferable to comparison of overall performance parameters for a bed based on input and exhaust measurements.

7.4 Verification of Scaling Relationships for Bubbling and Slugging Beds

Most early experiments devoted to verifying the scaling relationships have dealt with the full set of scaling relationships. Recent experiments have dealt with the simplified set of dimensionless parameters. In some experiments, additional scaling parameters were unintentionally matched.

Fitzgerald and Crane (1980) were among of the first to evaluate the full set of hydrodynamic scaling parameters. They compared the hydrodynamics of two scaled beds using pressure fluctuation measurements and movies. In one bed cork particles were fluidized with air; the other bed used sand fluidized with pressurized refrigerant 12 vapor. Movies showed qualitative agreement between bubble growth and the solids flow in the beds.

Fitzgerald et al. (1984) measured pressure fluctuations in an atmospheric fluidized bed combustor and a quarter-scale cold model. The full set of scaling parameters was matched between the beds. The autocorrelation function of the pressure fluctuations was similar for the two beds but not within the 95% confidence levels they had anticipated. The amplitude of the autocorrelation function and the experimentally determined time-scaling factor differed from the expected value. They suggested that the differences could be due to electrostatic effects.

Table 6 Scale model of 10 atm commercial hot bed using the simplified scaling relationships

Commercial beds		Particle diameter of model with bed linear scale factor	
d_p	u_{mf}	$Dm/Dc = 1/4$	1/9
40 μm	7.44×10^{-4} m/s	31 μm	25 μm
60	1.66×10^{-3}	46	38
100	4.65×10^{-3}	78	63
200	1.85×10^{-2}	155	126
400	7.18×10^{-2}	310	250
1000	0.329	610	550

Commercial bed $t = 800°C$, $P = 1$ atm, $\rho_s = 2500$ kg/m³ gas:air. Model bed $\rho_s = 896$ kg/m³ gas:air at STP.

364

Glicksman

Nicastro and Glicksman (1984) experimentally verified the full set of scaling laws for bubbling fluidized beds. They compared the time-resolved differential pressure measurements from a bubbling fluidized bed combustor and a scaled cold model. Good agreement was obtained between the measurements that were proportional to the bubble size and frequency. Figure 18 presents the comparisons. They concluded that hydrodynamic similarity had been achieved between the hot combustor and the cold model. When actual hot bed material was used in the cold model, which was a violation of the scaling laws, the model's behavior was very different from that of the hot bed.

Horio et al. (1986) used three geometrically similar bubbling beds, fluidized with ambient air, to verify their proposed scaling laws. The solid-to-gas density ratio was maintained constant in the experiments, although it was not one of the proposed scaling parameters. By maintaining a constant density ratio, they

in essence used the simplified set of scaling parameters, Eq. (8). Video analysis of bubble eruptions at the bed surface were used to determine that similarity was achieved for the cross-sectional average bubble diameter, bubble diameter distribution, and radial distribution of superficial bubble velocity. Horio et al. (1989) verified the bubbling bed scaling relations for solid mixing and segregation. Sand was used as a bed material in straight and tapered bed geometries. A bed sectioning technique was used to measure the transient radial dispersion coefficient and the distribution of float tracers. They concluded that bed mixing and the behavior of floating bodies obey the scaling laws in both straight and tapered beds. The solid-to-gas density ratio was again held constant in the tests, satisfying the simplified set of scaling laws.

Newby and Keairns (1986) made bubbling bed scaling comparisons between two cold models using the full set of scaling laws. One bed was fluidized with two different 200 μm glass powders using ambient air. The second bed, which was a half-scale model of the first, used pressurized air to fluidize 100 μm steel powder. High-speed movies showed good agreement between the nondimensional bubble frequencies in the two beds. Figure 19 is a plot of the nondimensional bubble frequencies as a function of bed Froude number.

Zhang and Yang (1987) carried out scaling comparisons between two two-dimensional beds with u_0^2/gD and u_0/u_{mf} matched between them. They also inadvertently kept the solid-to-gas density ratio constant; thus they matched the simplified scaling parameters. They found through photographs that the beds appeared qualitatively similar. The beds also had similar dimensionless freeboard entrainment rates and dimensionless bed heights over a range of u_0/u_{mf}.

Roy and Davidson (1989) considered the validity of the full and viscous limit scaling laws at elevated pressures and temperatures. The nondimensional dominant frequency and amplitude of the pressure drop fluctuations were used as the basis of the comparison. They concluded that when the full set of scaling parameters is matched, similarity is achieved.

Di Felice et al. (1992a) investigated the validity of the full set of scaling laws for bubbling and slugging fluidized beds. They used an experimental facility that permitted the pressurization of different diameter test sections to match the scaling parameters. Minimum fluidization measurements, video measurements of bed expansion, and pressure fluctuation data were used to compare the similarity of five different bed configurations. Three of the beds were scaled properly,

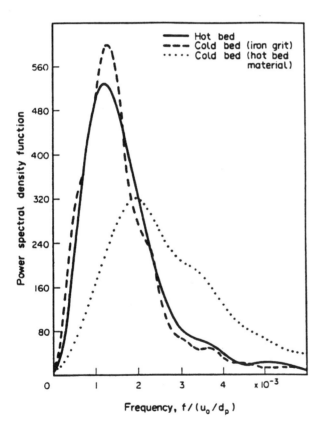

Figure 18 Comparison of dimensionless power spectra of differential pressure fluctuations. Double probe across levels 2 and 3; $x/L = 0.0$, coal burning bubbling bed combustor. Full set of scaling laws with iron grit in cold bed; hot bed material in cold bed violates scaling laws. (From Nicastro and Glicksman, 1984.)

the fourth had a mismatched particle sphericity, and the fifth was purposefully misscaled relative to the others. In the bubbling regime, good agreement in the nondimensional bed expansion measurements was obtained for all but the bed with the misscaled particle sphericity. The pressure fluctuations for the three properly scaled beds in the bubbling regime showed good agreement, while the misscaled beds exhibited poor agreement with the other three. Figure 20 is a plot of the dimensionless dynamic pressure variance for the five beds in the bubbling regime. The two sets of data that deviate from the other three correspond to the misscaled beds.

In the slugging regime, Di Felice et al. (1992a) found that the bed expansion characteristics were similar to those in the bubbling regime, but the pressure fluctuation characteristics for all five beds were in poor agreement with each other. They attributed this to the importance of particle material properties and some particle–particle interaction effects that are not accounted for in the full set of scaling laws. This is discussed further below in the light of recent results for spouting beds.

Di Felice et al. (1992b) evaluated the full set of scaling laws for three different Geldart powder categories (A, B, and D) in the bubbling and slugging fluidization regimes. Pressure fluctuations were used as the basis for the scaling comparisons. In the bub-

bling regime, the RMS and dominant frequencies of the pressure fluctuations showed good agreement for all three powder categories. In the slugging regime, Geldart groups B and D exhibited fair agreement in the RMS of their pressure fluctuations, but their dominant frequencies disagreed. They found that the full set of scaling laws are valid for bubbling beds fluidizing powders in Geldart groups A, B, and D. They also concluded that the full set of scaling relationships is not sufficient for slugging beds where particle–particle interactions are also thought to be important.

7.5 Verification of Scaling Laws for Spouting Beds

He et al. (1997) extended the scaling considerations to spouting beds. They showed that for spouting beds the full set of scaling relationships, Eq. (5), must be augmented with two new parameters, the internal friction angle and the loose packed voidage. By systematic tests in differently sized cold beds as well as comparisons between hot and cold beds, they showed excellent agreement when the full set of scaling parameters, augmented with the two spouting bed parameters, were held constant. Close agreement was found for spout diameter, fountain height, longitudinal pressure profiles, and dead zone boundary; see Fig. 21. When the

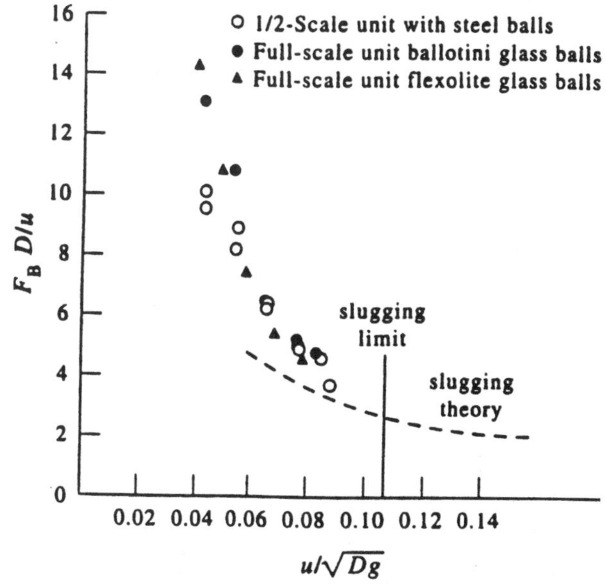

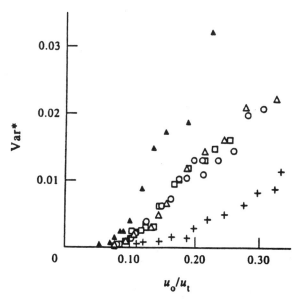

Figure 19 Comparison of nondimensional bubble frequencies from two cold scaled models. (From Newby and Keairns, 1986.)

Figure 20 Comparison of dynamic pressure variance for three properly scaled beds and two misscaled beds in bubbling regime. Properly scaled: □, laposorb; △, sand; ○, bronze. Intentionally misscaled: +, iron; ▲, sand. (From DiFelice et al., 1992a)

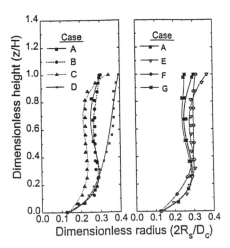

Figure 21 CDimensionless spout diameters as a function of dimensionless height for small columns. Case A, test case; Case B, all dimensionless parameters matched, bed diameter halved; Case C, particle Reynolds number mismatched; Case D, Froude number mismatched; Case E, density ratio, Reynolds number mismatched; Case F, bed Reynolds number mismatched; Case G, internal friction angle, loose packed voidage mismatched. (From He et al. 1995.)

internal friction angle and sphericity were mismatched, there was a large disagreement in fountain height.

The internal friction angle is also important for slugging beds; see Zenz and Othmer (1960). DiFelice et al. (1992 a,b) did not report their values; it could be that the disagreement they found in their slugging bed tests was due to mismatches of the internal friction angle.

7.6 Verification of Scaling Relationships for Pressurized Bubbling Beds

Almstedt and Zakkay (1990) made scaling comparisons between a hot PFBC with horizontal tubes and a pressurized cold scale model using the full set of scaling laws. The cold model had linear dimensions one-half those of the hot bed. A capacitance probe was used to measure the mean values of the bubble frequency, pierced length, bubble rise velocity, and bubble volume fraction. Scaling comparisons were made using the dimensionless form of these dependent hydrodynamic parameters. Three different bed materials were used in the cold bed: olivine sand and two different size distributions of the hot bed material, one properly scaled and one out of scale. Almstead and Zakkay concluded that behavior that is hydrodynamically similar to that of a pressurized fluidized bed

combustor can be achieved using a properly scaled cold model.

Glicksman and Farrell (1995) constructed a scale model of the Tidd 70 MW$_c$ pressurized bubbling bed combustor. The scale model was fluidized with air at atmospheric pressure and temperature. They used the simplified set of scaling relationships to construct a one-quarter length scale model of a section of the Tidd combustor. Low-density polyurethane beads were used to obtain a close fit with the solid-to-gas density ratio for the combustor as well as the particle sphericity and particle size distribution. Differential pressure measurements were made between several vertical elevations within the bed. The solid fraction profiles were obtained from the vertical pressure profile with a hydrostatic assumption. The cold model solid fraction profile showed very close agreement with data taken from pressure taps in two different locations within the combustor; see Fig. 22. The probability density functions of the cold model and combustor give very close agreement. The power spectral density of the combustor exhibited several distinct peaks at increasingly higher frequencies. All but the first peak were not seen in the cold model. The peaks could be due to tube vibrations in the hot bed, fluctuations upstream or downstream of the bed, or hydromechanical interactions between the bed and the internals.

Farrell (1996) experimentally evaluated the importance of the solid-to-gas density ratio (ρ_s/ρ_f) for scaling the hydrodynamics of bubbling and slugging fluidized beds. Two bed materials, polyethylene plastic ($\rho_s = 918 \, \text{kg/m}^3$) and a dolomite/limestone sorbent mixture ($\rho_s = 2670 \, \text{kg/m}^3$), were used to create a mismatch in the density ratio. The size of the particles was chosen so that the remaining simplified scaling parameters were matched. The internal angle of friction was similar between the two materials.

In one case, the solid fraction of the sorbent material was less than the plastic in the lower regions of the bubbling bed, with good agreement in the upper section of the bed. However, for the same conditions, the dimensionless standard deviation of the time-varying pressure drop showed the best agreement in the bottom of the bed with a large discrepancy in the upper portion of the bed. The plastic bed material has a much broader transition region between its fully bubbling and fully slugging regimes than the sorbent material, and the nature of this transition is different in the two materials. Therefore the solid-to-gas density ratio influences both the hydrodynamics in the bubbling regime and the boundary at which the transition to slugging occurs. This is consistent with the conclusion

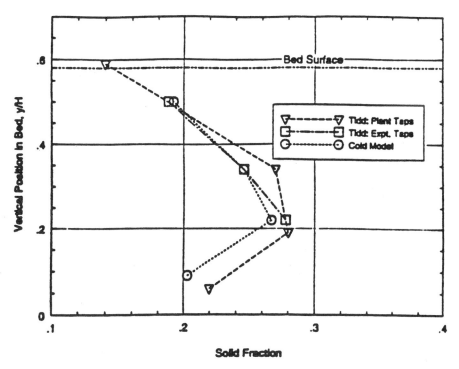

Figure 22 Solid fraction profile comparisons for Tidd PFBC and cold model based on simplified scaling laws (From Glicksman and Farrell, 1995.)

of Glicksman et al. (1993b), who found it essential to match the density ratio when scaling circulating fluidized bed hydrodynamics.

8 APPLICATIONS OF SCALING TO COMMERCIAL BUBBLING FLUIDIZED BED UNITS

A substantial number of experimental investigations have demonstrated the validity of scaling. This has increased awareness of the concept and confidence in its application. Scaling has many useful applications. The dynamic characteristics of different bed designs can be quickly compared. The influence of bed diameter on hydrodynamic behavior can be studied by the use of several different models. The models allow easy experimental examination of existing operating characteristics. The beds also can be used to confirm quickly the influence of proposed modifications in operating parameters and bed geometry. Since the models usually operate at ambient conditions, it is possible to instrument them to observe detailed behavior. This allows a better understanding of the fundamental physics as well as the identification of hydrodynamic factors needed for proper correlation of performance.

The earliest scaling studies were directed at atmospheric bubbling bed combustors. To date, a rich variety of questions have been addressed.

Jones and Glicksman (1986) constructed a model of the 20 MW bubbling bed pilot plant jointly sponsored by the Tennessee Valley Authority and the Electric Power Research Institute (EPRI) at Paducah, Kentucky. Figure 23 shows a photograph of the model of the in-bed tubes installed in the scale model. The model, which is roughly 100 by 120 cm in cross section, simulates two-thirds of the entire 20 MW pilot plant. Care was taken to match the pilot plant tube bundle geometry and distributor design. Steel grit particles with the same dimensionless size distribution and sphericity as the hot bed material were used. The full set of scaling parameters was matched in the model and the combustor. The largest discrepancy was in the solid-to-gas density ratio, which was 18% smaller in the model than in the pilot plant.

The measured bubble velocity for an actively bubbling bed was found to agree closely with the drift flux form proposed by Davidson and Harrison (1963). In contrast, the volumetric flow rate of the bubbles was found to be far less than that predicted by the two-phase hypothesis; see Fig. 24.

Figure 23 Model of 20 MW bubbling fluidized bed combustor showing tube arrangement. (From Jones and Glicksman, 1986.)

The larger particles were thrown high in the splash zone, higher than predicted by a ballistic trajectory using the bubble rise velocity as the initial velocity and neglecting any air drag. Later observations of this model showed that when bubbles erupt at the surface, the accompanying gas flow has a velocity much higher than the bubble rise velocity; see Glicksman and Piper (1987). This sets up a substantial gas bypass from the distributor to the surface of a relatively shallow bed. The observations from the model led to a mechanistic model for gas throughflow aided by the low resistance of the bubble cavity; see Yule and Glicksman (1988), and an accurate prediction of bubble volume flow rate and bed expansion; see Glicksman and Yule (1991).

A major question in the design of a commercial sized bubbling bed is the need to identify part load operating techniques. While reducing the total combustion rate, it is desirable to keep the bed operating temperature constant. This requires a reduction of the heat transfer to the water filled tubes within the bed. One technique uses the contraction of the bed that accompanies a decrease in superficial velocity. As the bed contracts, some of the tube rows are uncovered, reducing the net heat transfer. The scale model allowed many different tube arrangements to be tested. The validity of the scaling technique was confirmed by a comparison of the bed expansion measured for the pilot plant and that found in the model equipped with the same tube bank geometry; see Fig. 25.

A second method of reducing load while maintaining constant bed temperature is to reduce the superficial velocity of a portion of the bed distributor to a

value below u_{mf}. In this design, the bed does not contain vertical partitions above the distributor. The scale model was used to determine the rate of growth of the fixed bed in the defluidized zone along with the heat transfer to tubes in that region. Figure 26 shows a typical pattern of particle accumulation in a slumped zone adjacent to an actively fluidized zone. Heat transfer coefficients are also shown. Note that tubes near the upper surface of the solids that experience a downflow of solids have a very high heat transfer rate.

The heat transfer from tubes in the freeboard was also measured for the 20 MW model. The variation of measured overall heat transfer coefficient in the 20 MW pilot plant versus velocity agreed closely with that predicted from the scale model test. When the bed height is lowered, uncovering some tubes, the heat transfer is reduced because there are fewer particles contacting the tube surface.

Ackeskog et al. (1993) made the first heat transfer measurements in a scale model of a pressurized bubbling bed combustor. These results shed light on the influence of particle sizes, density, and pressure levels on the fundamental mechanism of heat transfer, e.g., the increased importance of the gas convective component with increased pressure.

A multisolid bed contains a mixture of large solids that are contained in a dense region at the bottom of the bed and finer particles that recirculate through the bed and external cyclone. Ake and Glicksman (1989) used a cold scale model of a multisolids combustor to determine the dense bed expansion. The measured expansion in a properly scaled quarter-scale model using steel pellets to simulate the coarse particles and to satisfy the solid-to-gas density ratio gave good agreement with field data; see Fig 27. It was also demonstrated that an improperly scaled cold model, using the same coarse material as the hot bed, had an incorrect gas-to-particle density ratio and substantially overpredicted the bed expansion.

Tube erosion has been observed in both atmospheric and pressurized bed combustors. The scaling analysis presented earlier can be used to construct an accurate hydrodynamic simulation of the commercial bed. This can be used to investigate qualitatively the factors related to tube wear such as the location of highest wear around the circumference of an individual tube and the location within the bed of highest wear. Quantitative wear rates cannot be obtained from model tests unless the parameters governing both the hydrodynamics and the wear phenomena are matched between the model and the commercial bed. The inclusion of more large particles causes an increase in the

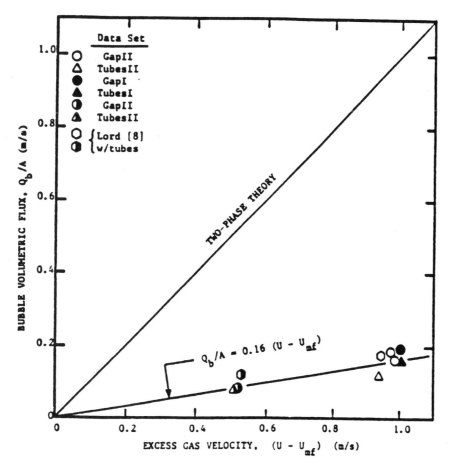

Figure 24 Deviation from two-phase theory for model of bubbling fluidized bed combustor. (From Jones and Glicksman, 1986.)

wear rate. The second row of tubes is shielded by the first row, reducing the incidence of strikes by large high-velocity bubbles. In the interface between an actively fluidized and a slumped bed, there is a flow of particles down the slumped surface, which causes higher wear. The relative changes in wear rate with tube location as well as the circumferential location with highest wear agreed closely between the cold model and the 20 MW combustor.

Other boiler manufacturers have used scale models to aid in the design of large bubbling bed combustors. Yang et al. (1995) used atmospheric cold models 30 cm and 3 m in diameter to simulate a large jetting fluidized bed gasifier to be operated at about 12 atm. They used the full set of scaling relations that required the model to be about the same dimensions as the hot bed. A half-round bed with a transparent wall was used to photograph the jet behavior. They measured and developed correlations for jet penetra-

tion, bubble frequency, vertical bubble size, jet half-angle, and others. They also found considerable gas leakage from the bubbles to the emulsion phase. This data covers much larger bed sizes than had been examined before.

9 HYDRODYNAMIC SCALING OF CIRCULATING BEDS

Given the success in scaling bubbling beds, research has progressed to scaling of circulating fast beds. The initial research has focused on the verification of both the full and the simplified scaling relationships for circulating beds. The verification using data from combustors is complicated by the difficulty in accurately measuring the recycle rate of solids, an important parameter. Figure 28 presents the range of scaling parameters for experimental studies undertaken for

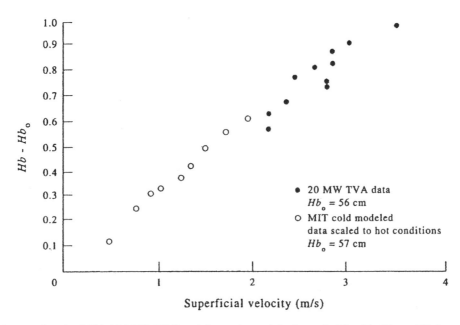

Figure 25 Bed expansion for TVA 20 MW$_e$ FBC and for scale model, shown in Fig. 23. (From Glicksman et al., 1989.)

circulating in terms of the Froude number and the Reynolds number based on particle diameter and the superficial gas velocity.

Horio et al. (1989) experimentally verified their proposed circulating fluidized bed scaling laws. The solid-to-gas density ratio was not varied in the tests; thus they effectively verified the simplified set of scaling laws. Two cold scaled CFBs, fluidized using ambient air, were used in the verification. Good agreement in the axial solid fraction profiles was obtained for most of the conditions tested. An optical probe was used to verify similarity in the annular flow structures and the cluster velocities.

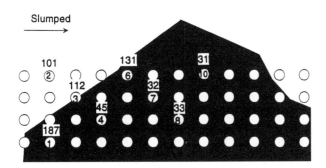

Figure 26 Particle accumulation in slumped zone adjacent to active bed: air velocity through active region = 151 cm/s; air velocity through slumped region = 7.3 cm/s; numbers above heater are heat transfer coefficient in W/m$^{2\circ}$C after 15–30 min. (From Glicksman et al., 1992.)

Ishii and Murakami (1991) evaluated Horio et al. (1989) CFB scaling relationships using two cold CFB models. Solids flux, pressure drop, and optical probe measurements were used to measure a large number of hydrodynamic parameters to serve as the basis for the comparison. Fair to good similarity was obtained between the beds.

Tsukada et al. (1991) applied Horio et al. (1989) CFB scaling laws at several different elevated pressures (viscous limit scaling laws). A single bed and bed material were used in the study. A pressure vessel was used to vary the gas pressure. They found that as the pressure was increased the axial solid fraction profile changed, indicating a change in the hydrodynamics. In this study, the gas-to-solid density ratio changes with pressure level. It is likely based on the recent results of Glicksman et al. (1993b) that the change in gas-to-solid density ratio led to the influence of the pressure level on the bed hydrodynamics. This points out a deficiency of the viscous limit, Horio's (1989) scaling relationship.

Glicksman et al. (1991a) made scaling comparisons between an experimental circulating fluidized bed combustor and a scaled cold model based on the full set of scaling laws. Due to uncertainties in the hot bed solid circulation measurements, the cold bed solids flux was adjusted until the average bed solid fraction matched that of the hot bed. Good agreement was obtained between the vertical solid fraction profiles except near the top of the beds. Disagreement may have

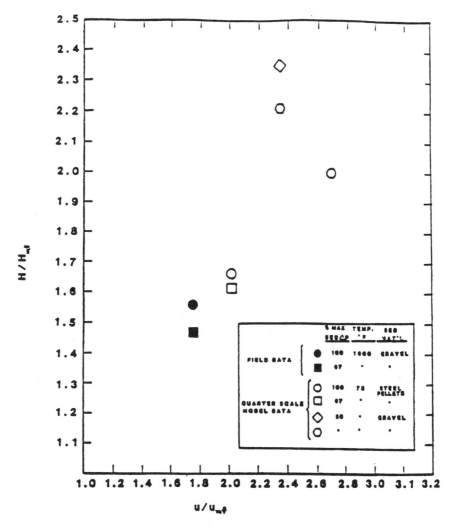

Figure 27 Expanded bed height model data compared to field data for multisolids fluidized bed for properly scaled steel pellets and misscaled gravel particles. (From Ake and Glicksman, 1989.)

been due to the differences between the gas-to-solid density ratio in the cold model and the hot bed. Good agreement was also obtained in the comparison of the probability density distribution and the Fourier transform of the pressure fluctuations.

Glicksman et al. (1993a) evaluated the full set of scaling laws for circulating fluidized beds. Solid fraction data were obtained from the 2.5 MW_{th} Studsvik atmospheric CFB prototype. The full set of scaling laws was evaluated through solid fraction profile comparisons between Studsvik and a one-quarter scale cold model. Fairly good agreement was obtained; the profiles most closely matched in the top of the beds. Differences between the profiles were attributed to uncertainty in the hot bed solid flux measurements and to the mismatch in the solid-to-gas density ratio.

A modified set of the simplified scaling relationships were also evaluated by Glicksman et al. (1993b) in a series of comparison tests using circulating beds in which the gas-to-solid density ratio was not held constant. The average solid fraction profiles, Fig. 29, solid fraction probability density functions, and power spectral densities were all in poor agreement. It is believed the beds were operating near the point of incipient choking condition as predicted by the Yang (1983) correlation. Because this correlation indicates that choking is a strong function of the solid-to-gas density ratio, it was concluded that this parameter must be matched to model bed hydrodynamics near the boundary between different flow regimes.

The simplified scaling relationships were used by Glicksman et al. (1993b) to compare two geometrically

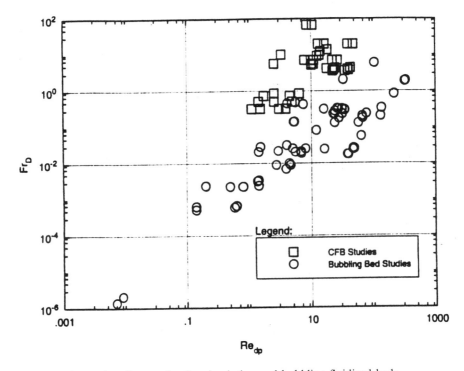

Figure 28 Range of experimental scaling studies for circulating and bubbling fluidized beds.

similar beds, one having linear dimensions four times larger than the other. In one series of tests, properly sized plastic particles were used in both beds; in another test series, glass particles were used in the two beds. The average solid fraction profiles showed excellent agreement; see Fig. 30. The probability density functions and power spectral densities also agreed well.

Glicksman et al. (1993b) verified the simplified scaling laws for hot beds by comparing the solid fraction profiles for the Studsvik bed, the one-quarter scale cold

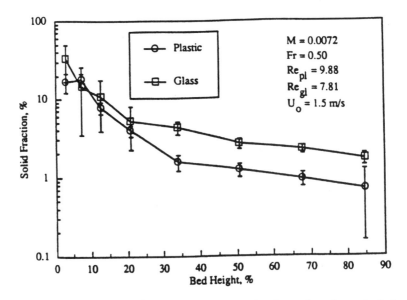

Figure 29 Solid fraction profiles, glass/plastic viscous limit scaling; density ratio mismatched: low velocity case. (From Glicksman et al., 1993a.)

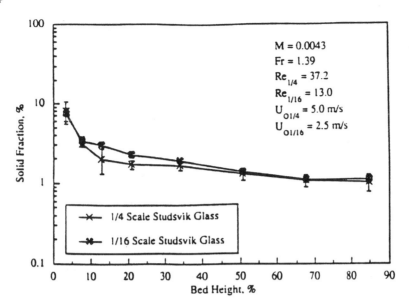

Figure 30 Solid fraction profiles, glass simplified scaling, high velocity case. (From Glicksman et al., 1993a.)

model, and a one-sixteenth scale cold model. The one-sixteenth scale model had a cross sectional area of 16 cm^2 to simulate a 2.5 MW combustor! The average solid fraction profiles were in good agreement for most of the conditions tested. The agreement was excellent between the one-quarter scale cold model, which used the full set of scaling laws, and the one-sixteenth scale model that used the simplified set of scaling laws. Thus any disagreement between the Studsvik bed and the one-sixteenth scale model is not due to the simplifica-

tions of the full set of scaling laws. The density ratio was not matched exactly between the hot bed and the two cold beds, which may have affected the agreement. Figure 31 provides a typical comparison of the solid fraction profiles in the three beds. The authors concluded that the simplified set of scaling laws, which includes the solid-to-gas density ratio, gives acceptable results over a wide range of particle densities and bed sizes, even when the length ratio is as small as one-sixteenth for an atmospheric combustor.

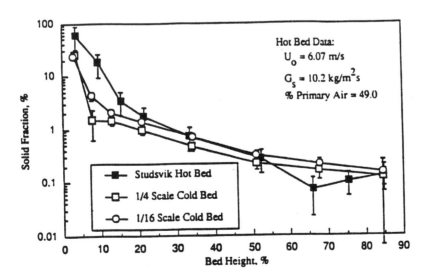

Figure 31 Solid fraction profiles, hot bed scaling with simplified scaling laws, low velocity, 49% primary air. (From Glicksman et al., 1993a.)

Glicksman et al. (1995) used the simplified scaling parameters to construct a one-half linear scale model of a Foster Wheeler circulating bed combustor pressurized to 14 bar. The combustor has a 20.3 cm inner diameter with an overall height of 8.3 m with both a primary and a secondary air supply. The solids recycle rate was accurately determined by a calorimetric balance of a fluidized bed heat exchanger in the return loop of the circulating bed. The cold model, one-half scale, used polyethylene plastic particles to match the dimensionless particle size distribution as well as the gas-to-solid density ratio.

The time averaged vertical pressure difference was used to determine the solid fraction distribution in the combustor of the cold model. Figure 32 shows the close agreement between the combustor and the cold model. Three test cases had similar solid fraction versus height profiles. The fourth, operated at low gas velocity and solids recycle rate, had a more abrupt decrease of solid fraction with bed height. The cold model reflected the same behavior. The probability density functions also were in agreement except for some discrepancy near the secondary air inlets. This might be due to fluctuations set up by the bubbling bed heat exchanger upstream of the secondary air

inlet to the combustor. The bubbling bed was not duplicated in the cold scale model.

Subsequent experiments verified that a 1/6.5 linear scale model of another Foster Wheeler circulating bed combustor, designed with the simplified scaling relationships, exhibited very close agreement with the full scale hot combustor. Figure 33 shows a comparison of the average solid fraction versus bed height between the pressurized hot combustor and the 1/6.5 linear scale model, using plastic particles fluidized by ambient air. Figure 34 shows a comparison of the Fourier transform versus frequency, based on an equivalent hot frequency.

Further verification of the scaling laws are needed in terms of radial solids distribution and solids diffusivity. These have not been compared between hot and cold beds. In addition, work needs to be carried out to see the lower limit of solids diameter for which the present set of scaling parameters holds. At some point, surface forces will have an important influence on bed dynamics. This will require additional scaling parameters that include these effects. The range of validity of the present set of scaling parameters for fluidized beds for FCC operations and for cyclone separators remains uncertain at the present time.

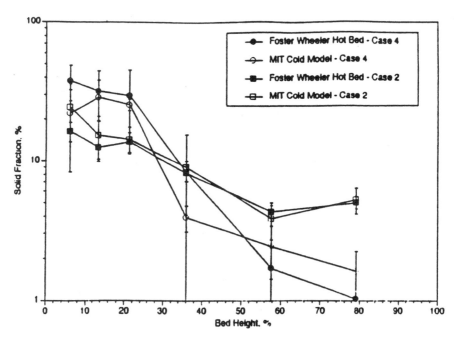

Figure 32 Solid fraction profile comparison: between pressurized circulating fluidized bed combustor and one-half size scale model based on simplified scaling. Two different operation conditions. (From Glicksman et al., 1995.)

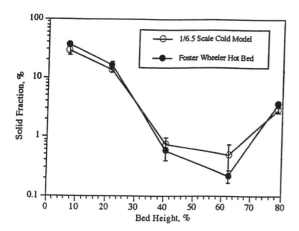

Figure 33 Average solid fraction: 11.5 m high CFB combustor at 7.5 bar compared to 1/6.5 scale cold model.

10 CONCLUSIONS

As fluidized beds are scaled up from bench scale to commercial plant size, the hydrodynamic behavior of the bed changes, resulting, in many cases, in a loss of performance. Although there have been some studies of the influence of bed diameter on overall performance as well as detailed behavior such as solids mixing and bubble characteristics, generalized rules to guide scale-up are not available. The influence of bed diameter on performance will differ for different flow regimes of fluidization.

Small, properly scaled laboratory models operated at ambient conditions have been shown accurately to simulate the dynamics of large hot bubbling and circulating beds operating at atmospheric and elevated pressures. These models should shed light on the overall

operating characteristics and the influence of hydrodynamics factors such as bubble distribution and trajectories. A series of differently sized scale models can be used to simulate changes in commercial bed behavior with bed size.

The scale models must be carefully designed. Failure to match one or more important dimensionless parameters will lead to erroneous simulation results. Modeling can be extended to particle convective heat transfer. Wear or erosion of in-bed surfaces can be qualitatively studied, although quantitative assessment requires the identification and simulation of additional wear-related parameters.

Most of the simulation effort has been applied to fluidized bed combustors that use relatively large particles. Simulation can also be used for other fluidization processes in the petrochemical industry. Research should be undertaken to identify the proper scaling parameters for beds fluidized with smaller particles. Similar simulations may also apply to components such as cyclones.

ACKNOWLEDGMENTS

The author would like to acknowledge the contributions of many present and former M.I.T. students who carried out scaling studies of fluidized beds. Much of the M.I.T. research mentioned in this chapter has been sponsored by the Electric Power Research Institute, the National Science Foundation, the U.S. Department of Energy, and the Tennessee Valley Authority.

NOTATION

C_D	=	Drag coefficient of sphere
D	=	Bed diameter
d_B	=	Bubble diameter
d_c	=	Cluster diameter
d_p	=	Particle diameter
d_V	=	Equivalent bubble diameter
G_S	=	Solids recycle rate per unit area
g	=	Acceleration of gravity
L	=	Typical bed dimensions
Q_B	=	Bubble volume flow rate
t	=	Time
u	=	Gas velocity
u_b	=	Bubble rise velocity
u_{mf}	=	Minimum fluidization velocity
u_0	=	Superficial gas velocity
u_f	=	Terminal velocity
x, y	=	Coordinates

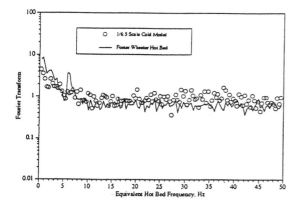

Figure 34 Frequency response: 11.5 m high CFB combustor at 7.5 bar compared to 1/6.5 scale cold model.

Dimensionless

PSD = Dimensionless particle size distribution
Re = Reynolds number

Greek

β = Drag coefficient
δ = Bubble volume fraction
δ = Boundary layer thickness in a fast bed
Δ = Distributor plate voidage
ε = Void fraction
ε_c = Cluster void fraction
ε_{mf} = Void fraction at minimum fluidization
ρ_f = Fluid density
ρ_S = Solid density
μ = Fluid viscosity
ν = Fluid kinematic viscosity
φ = Particle sphericity

Subscripts

$()_c$ = Commercial bed
$()_m$ = Model bed

Superscripts

$()'$ = Dimensionless
$(-)$ = Vector

REFERENCES

Ackeskog HBR, Almstedt AE, Zakkay V. An investigation of fluidized-bed scaling: heat transfer measurements in a pressurized fluidized-bed combustor and a cold model bed. Chem Eng Sci 48:1459, 1993.

Ake TR, Glicksman LR. Scale model and full scale test results of a circulating fluidized bed combustor. Proceedings of 1988 Seminar on Fluidized Bed Combustion Technology for Utility Appl. EPRI, 1989, pp 1–24.

Almstedt AE, Zakkay V. An investigation of fluidized-bed scaling-capacitance probe measurements in a pressurized fluidized-bed combustor and a cold model bed. Chem Eng Sci 45(4):1071, 1990.

Arena U, Cammarota A, Massimilla L, Pirozzi D. The hydrodynamic behavior of two circulating fluidized bed units of different size. In: Basu P, Large JF, eds. Circulating Fluidized Bed Technol II. Oxford: Pergamon Press, 1988.

Avidan A, Edwards M. Modeling and scale up of Mobil's fluid bed MTG process. In: Ostergaard K, Sorensen A, eds. Fluidization V. New York: Engineering Found, 1986.

Avidan A, Edwards M, Owen H. 50 years of catalytic cracking. Oil Gas Journal 33, 1990.

Bauer W, Werther J, Emig G. Influence of gas distributor design on the performance of fluidized bed reactor. Ger Chem Eng 4:291, 1981.

Chang H, Louge M. Fluid dynamic similarity of circulating fluidized beds. Powder Technol 70:259, 1992.

Davidson JF, Harrison D. Fluidized Particles. Cambridge: Cambridge University Press, 1963.

DeGroot JH. Scaling-up of gas-fluidized bed reactors. In: Proceedings of the International Symposium on Fluidization. Drinkenburg AAH, ed. Amsterdam: Netherlands University Press, 1967.

DiFelice R, Rapagna S, Foscolo PU. Dynamic similarity rules: validity check for bubbling and slugging fluidized beds. Powder Technol 71:281, 1992a.

DiFelice R, Rapagna S, Foscolo PU, Gibilaro LG. Cold modelling studies of fluidised bed reactors. Chem Eng Sci 47:223, 1992b.

Farrell PA. Hydrodynamic scaling and solids mixing in pressurized fluidized bed combustors. PhD dissertation, Massachusetts Institute of Technology, Cambridge, MA, 1996.

Fitzgerald TJ, Crane SD. Cold fluidized bed modeling. Proceedings of 6th International Conference on Fluidized Bed Combustion, III, 1980, p 815.

Fitzgerald TJ, Bushnell D, Crane S, Shieh Y. Testing of cold scaled bed modeling for fluidized-bed combustors. Powder Technol 38:107, 1984.

Frye CG, Lake WC, Eckstrom HC. Gas-solid contacting with ozone decomposition reaction. AIChE J 4(4):403, 1958.

Geldart D. The size and frequency of bubbles in two- and three-dimensional gas-fluidised beds. Powder Technol 4:41, 1970.

Glicksman LR. Scaling relationships for fluidized beds. Chem Eng Sci 39:1373, 1984.

Glicksman LR. Scaling relationships for fluidized beds. Chem Eng Sci 43:1419, 1988.

Glicksman LR. Heat transfer in circulating fluidized beds. In: Grace JR, Avidan A, Knowlton TM, eds. Circulating Fluidized Beds. London: Chapman and Hall, 1997, pp 261–311.

Glicksman LR, Farrell P. Verification of simplified hydrodynamic scaling laws for pressurized fluidized beds: part I. Bubbling fluidized beds. Proc. 13th Int. Conference for Fluidized Bed Combustion, 1995, p 981.

Glicksman LR, McAndrews G. The effect of bed width on the hydrodynamics of large particle fluidized beds. Powder Technol 42:159, 1985.

Glicksman LR, Piper GA. Particle density distribution in a freeboard of a fluidized bed. Powder Technol 53:179, 1987.

Glicksman LR, Yule T. Prediction of the particle flow conditions in the freeboard of a freely bubbling fluidized bed. AIChE J, 1991c.

Glicksman LR, Yule T, Dyrness A, Carson R. Scaling the hydrodynamics of fluidized bed combustors with cold models: experimental confirmation. Proc. 9th Int. Conf. on Fluidized Bed Combustors, 1987, p 511.

Glicksman LR, Mullens G, Yule TW. Tube wear tests in the MIT scaled fluidized bed. Proceedings of EPRI Workshop, Nov, 1987b.

Glicksman LR, Yule T, Carson R, Vincent R. Comparison of results from TVA 20 MW fluidized bed combustor with MIT cold scale model. Proc. 1988 Seminar on Fluidized Bed Combustion Technology for Utility Applications, EPRI GS-6118, 1-20-1, 1989.

Glicksman LR, Westphalen D, Brereton C, Grace J. Verification of the scaling laws for circulating fluidized beds. In: Basu P, Horio M, Hasatani M, eds. Circulating Fluidized Bed Technology. Oxford: Pergamon Press, 1991a.

Glicksman LR, Yule T, Dyrness A. Prediction of the expansion of fluidized beds containing tubes. Chem Eng Sci 46:1561, 1991b.

Glicksman LR, Yule T, Arencibia A, Pangan A. Heat transfer to horizontal tubes in a slumped bed adjacent to a fluidized bed. Proceedings of 7th Eng. Foundation Conf. on Fluidization. New York: Engineering Foundation, 1992, p 813.

Glicksman LR, Hyre MR, Westphalen D. Verification of scaling relations for circulating fluidized beds. Proc. 12th Int. Conf. on Fluidized Bed Combustion, 1993a, p 69.

Glicksman LR, Hyre MR, Woloshun K. Simplified scaling relationships for fluidized beds. Powder Technol 77:177, 1993b.

Glicksman LR, Hyre M, Farrell P. Dynamic similarity in fluidization. Intern J Multiphase Flow, 20(suppl): 331, 1994.

Glicksman LR, Hyre M, Torpey M, Wheeldon J. Verification of simplified hydrodynamic scaling laws for pressurized fluidized beds: Part II. Circulating fluidized beds. Proc. 13th Int. Conference for Fluidized Bed Combustion, 1995, p 991.

Glicksman LR, Hyre M, Farrell PA. Importance of the solid-to-gas density ratio for scaling fluidized bed hydrodynamics. Proc of the 9th International Conference on Fluidization, Fluidization IX Conference, 1998.

Hartge E-U, Li Y, Werther J. Flow structures in fast fluidized beds. Proc. 5th Engineering Foundation Conf. on Fluidization, Elsinore Denmark, 1985, p 345.

He Y-L, Lim CJ, Grace JR. Scale-up studies of spouted beds. Chem Eng Sci 52:329–339, 1997.

Horio M, Nonaka A, Sawa Y, Muchi I. A new similarity rule for fluidized bed scale-up. AIChE J 32:1466, 1986.

Horio M, Ishii H, Kobukai Y, Yamanishi N. A scaling law for circulating fluidized beds. J Chem Eng Japan 22:587, 1989.

Hovmand S, Davidson JF. Slug flow reactors. In: Davidson JF, Harrison D, eds. Fluidization. London: Academic Press, 1971.

Ishii H, Murakami I. Evaluation of the scaling law of circulating fluidized beds in regard to cluster behaviors. In: Basu P, Horio M, Hasatani M, eds. Circulating Fluidized Bed Technology III. Oxford: Pergamon Press, 1991.

Johnsson F, Leckner B. Vertical distribution of solids in a circulating fluidized bed furnace. Proceedings 13th International Conference on Fluidized Bed Combustions. American Society of Mechanical Engineers, 1995, pp 671–702.

Johnsson JE, Grace JR, Graham JJ. Fluidized bed reactor model verification on a reactor of industrial scale. AIChE J 33(4):619, 1987.

Jones L, Glicksman LR. An experimental investigation of gas flow in a scale model of a fluidized-bed combustor. Powder Technol 45:201, 1986.

Knowlton TM, Geldart D, Matsen J. Oral presentation. Fluidization VIII. International Symposium of the Engineering Foundation, Tours, France, 1995, pp 14–19.

Litka T, Glicksman LR. The influence of particle mechanical properties on bubble characteristics and solid mixing in fluidized beds. Powder Technol 42:231, 1985.

Matsen JM. Fluidized beds in scaleup of chemical processes: conversion from laboratory scale tests to successful commercial size design. In: Bisio A, Kabel RL, eds. Scaleup in Chemical Processes. New York: John Wiley, 1985, p 347.

Matsen JM. Design and scale-up of CFB catalytic reactors. In: Grace JR, Avidan AA, Knowlton TM, eds. Circulating Fluidized Beds. London: Chapman and Hall, 1997, pp 489–503.

May WG. Fluidized-bed reactor studies. Chem Eng Progress 55 (12):49, 1959.

Newby RA, Keairns DL. Test of the scaling relationships for fluid-bed dynamics. In: Ostergaard K, Sorensen A, eds. Fluidization V. New York: Engineering Foundation, 1986.

Nguyen HV, Potter OE, Whitehead AB. Bubble distribution and eruption diameter in a fluidized bed with a horizontal tube bundle. Chem Eng Sci 34:1163, 1979.

Nicastro MT, Glicksman LR. Experimental verification of scaling relationships for fluidized bed. Chem Eng Sci 39:1381, 1984.

Noymer PD, Hyre MR, Glicksman LR. The influence of bed diameter on hydrodynamics and heat transfer in circulating fluidized beds. Fluidization and Fluid-Particle Systems, AIChE, 1995, pp 86–90.

Patience GS, Chaouki J, Berruti F, Wong R. Scaling considerations for circulating fluidized bed risers. Powder Technol 72:31, 1992.

Rhodes MJ, Laussman P. A study of the pressure balance around the loop of a circulating fluidized bed. Can J Chem Eng 70:625, 1992.

Rhodes MJ, Wang XS, Cheng H, Hirama T, Gibbs BM. Similar profiles of solids flux in circulating fluidized bed risers. Chem Eng Sci 47(7):1635, 1992.

Rowe PN, Stapleton WM. The behavior of 12-inch diameter fast fluidized beds. Trans Instn Chem Engrs 39:181, 1961.

Roy R, Davidson JF. Similarity between gas-fluidized beds at elevated temperature and pressure. Fluidization VI. New York: Engineering Foundation, 1989.

Shingles T, McDonald AF. Commercial experience with synthol CFB reactors. In: Basu P, Large BP, eds. Circulating Fluidized Bed Technology I. Oxford: Pergamen Press, 1988.

Thiel WJ, Potter OE. Slugging in fluidized beds. Ind Eng Chem Fundam 16(2):242, 1977.

Thiel WJ, Potter OE. The mixing of solids in slugging gas fluidized beds. AIChE J 24:561, 1978.

Tsukada M, Nakanishi D, Takei Y, Ishii H, Horio M. Hydrodynamic similarity of circulating fluidized bed under different pressure conditions. Proc. 11th International Conference Fluidized Bed Combustion, 1991, p 829.

Van Deemter JJ. Mixing patterns in large-scale fluidized beds. In: Grace JR, Matsen JM, eds. Fluidization. New York: Plenum Press, 1980.

Van Swaaij WPM, Zuiderweg FJ. Investigation of ozone decomposition in fluidized beds on the basis of a two-phase model. Chemical Reaction Eng. Proc. of the 5th European/2nd Int Symp. Chem. Reaction Eng. Amsterdam/London/New York: Elsevier, 1972.

Van Swaaij WPM, Zuiderweg FJ. The design of gas-solids fluidized beds—prediction of chemical conversion. Proc. International Symposium on Fluidization and its Appl. Ste Chimie Industrielle, Toulouse, 1973.

Volk W, Johnson CA, Stotler HH. Effect of reactor internals on quality of fluidization. Chem Eng Prog 58:44, 1962.

Werther J. Influence of bed diameter on the hydrodynamics of gas fluidized beds. AIChE Symposium Ser 70(141):53, 1974.

White FM. Viscous Fluid Flow. New York: McGraw-Hill, 1974.

Whitehead AB, Young AD. Fluidisation performance in large scale equipment. Proc. Intern. Symp. on Fluidisation, Eindhoven, 1967, p 294.

Yang WC. Criteria for choking in vertical pneumatic conveying lines. Powder Technol 35:143, 1983.

Yang WC, Newby RA, Keairns DL. Large-scale fluidized bed physical model: methodology and results. Powder Technol 82:331, 1995.

Yerushalmi Y, Avidan A. High velocity fluidization. In: Davidson JF, Clift R, Harrison D, eds. Fluidization. 2d ed. New York: Academic Press, 1985.

Yule T, Glicksman LR. Gas flow through erupting bubbles in fluidized beds. AIChE Symp Ser 262(84):1, 1988.

Zenz FA, Othmer DF. Fluidization and Fluid-Particle Systems. New York: Reinhold, 1960.

Zhang MC, Yang RYK. On the scaling laws for bubbling gas-fluidized bed dynamics. Powder Technol 51:159, 1987.

Zhang W, Tung Y, Johnsson F. Radial voidage profiles in fast fluidized beds of different diameters. Chem Eng Sci 46(12):3045, 1991.

Zhang W, Johnsson F, Leckner B. Fluid dynamic boundary layers in CFB boilers. Chem Eng Sci 50(2):201, 1995.

14

Applications for Fluid Catalytic Cracking

Ye-Mon Chen

Shell Global Solutions US, Houston, Texas, U.S.A.

1 INTRODUCTION

1.1 What Is FCC?

Fluid catalytic cracking (FCC) is the primary conversion unit in most refineries in the United States. It converts low-value heavy components of crude oil into a variety of high-value lighter products. A modern FCC unit consists of three major sections: the reactor/regenerator section, the main fractionation/gas plant section, and the flue gas cleaning/power recovery section. This chapter focuses mainly on the reactor/regenerator section of the FCC process, paying special attention to its relationship with fluidization and fluid–particle systems. Readers who are interested in broader aspects of the FCC process can refer to several recent publications, such as Sadeghbeigi (1995) and Wilson (1997).

In the reactor/regenerator section of the FCC process, liquid hydrocarbon feedstock is preheated, mixed with steam, and injected into the riser reactor through feed nozzles. Hot regenerated catalyst is drawn from the regenerator and contacts the hydrocarbon feed. The two are thoroughly mixed in the lower riser by fine atomization through the feed nozzles. The temperature of regenerated catalyst drops rapidly as the feed vaporizes and reacts, converting the feedstock into lighter products while a coke layer deposits on the spent catalyst, temporarily deactivating the FCC catalyst. Product vapor and spent catalyst are separated by riser termination devices, such as cyclone systems, at

the end of the riser. The product vapor goes to the main fractionation and gas plants for product separation. The spent catalyst enters the stripper where steam is introduced to further recover entrained/adsorbed hydrocarbon products. The spent catalyst then passes through a spent catalyst transfer line and enters the regenerator. As the catalyst activity is restored by burning off the coke layer on spent catalyst, a large amount of heat is released, which heats up the regenerated catalyst. The flue gas from the regenerator goes to flue gas cleaning and the power recovery system. The regenerated catalyst then returns to the reactor riser. This completes the catalyst circulation loop.

1.2 How the FCC Unit Fits in a Typical US Refinery

The FCC unit usually has the second highest throughput after the crude distillation unit in most U.S. refineries. Crude oil contains hydrocarbons of different boiling points ranging from light gas to residue. Typically, crude oil is first distilled in an atmospheric, or crude, distillation unit to produce a wide range of hydrocarbon products according to the different boiling points in the original crude. The residue from the atmospheric distillation unit, the atmospheric residue, is then fed to the vacuum distillation unit, which produces vacuum gas oil and vacuum residue. The vacuum gas oil from the vacuum distillation unit is the traditional FCC feedstock with typical boiling points ranging from 650 to 1050°F. Depending on the type of crude and the operating conditions of the vacuum distillation unit, the portion of vacuum gas oil can range from 25 to 35% of the basic crude. Many refineries use delayed cokers to take the vacuum residue from the vacuum distillation unit as feedstock. Liquid products from cokers, the coker gas oils, are another common feedstock for the FCC unit.

Over the years, FCC has evolved into a very flexible process that can take a wide variety of other feedstocks, including hydrotreated gas oils, deasphalted oils, slop oils, and lube extracts. In addition, many modern FCC units are capable of the direct processing of atmospheric residue or vacuum residue, which is called residual or resid fluid catalytic cracking (RFCC).

The FCC process also plays a key role as a feed preparation unit for downstream refining processes. For example, light petroleum gas (LPG) produced by the FCC unit is a common feedstock to the alkylation unit.

1.3 Why is FCC Important?

The FCC unit has several important functions in refining and petrochemical manufacture. First of all, the FCC unit is a gasoline machine. About 45% of worldwide gasoline production comes either directly from FCC units or indirectly from FCC downstream units, such as alkylation units.

The second function of the FCC unit is to reduce the amount of residue in crude oil. In the U.S. market, the product slate produced directly from the distillation units contains too little transportation fuel (gasoline and diesel) and too much residue. The FCC unit enables refineries to convert most of the residue into lighter products, thus fully utilizing the crude oil to meet the market demands.

The third function of the FCC unit is to provide flexibility to the refining operation. The FCC unit can adjust operating conditions to maximize the production of gasoline, middle distillate (LCO), or light petroleum gas (LPG). This flexibility enables the refinery to adjust its product slate to meet seasonal demands in the market. Refineries in the U.S. also take in different crude diets according to changes in the crude market. The flexibility of the FCC unit enables the refinery to process different crude diets as the crude market changes.

The fourth function of the FCC unit is to produce light olefins for downstream refining processes, such as alkylation and other petrochemical processes, as in MTBE, PE, and PP plants.

1.4 Why Is Particulate Technology Important to FCC?

FCC is by far the most important petrochemical process that involves the broadest scope of particulate technology. Although the chemistry of the FCC process is rather complex, many of the operational problems are in fact associated with handling fluid–particle systems and related mechanical issues, such as erosion. In particular, the reactor/regenerator section of the FCC unit includes many aspects of fluid–particle systems that are critical to FCC design and operation.

For instance, the lower riser is typically operated in the fast fluidization regime, whereas the upper riser is in the dilute transport regime. In between, the feed injection section is a complicated transition that involves rapid mixing, heat transfer, and reaction between the FCC catalyst and the feedstock. The challenge is to achieve fast mixing and vaporization of the liquid feedstock, which plays a key role in achieving the overall riser performance. In the middle section of the riser, the phenomena of catalyst segregation and backmixing are related to cluster formation of the catalyst. This presents another challenge to the riser reactor design. At the end of the riser are riser termination and reactor cyclone systems, which handle gas–solid separation. The challenge is to achieve a fast and clean separation that minimizes the degradation of the hydrocarbon products by minimizing post riser cracking.

The stripper is typically operated in the bubbling regime utilizing various internals to improve steam–catalyst contacting efficiency. The challenge is to maximize the contacting efficiency in order to recover the most hydrocarbons by proper design of stripper internals and steam injection systems. The regenerator is typically the largest vessel of the FCC process operated in the turbulent to fast fluidization regime. The challenge is to maximize catalyst regeneration by proper distribution of spent catalyst and air.

Both the reactor and the regenerator have cyclone systems in the dilute phase to separate catalyst and to return it back to the system via cyclone diplegs. Prediction of the transport disengaging height (TDH) is critical to determining the locations of cyclone inlets. Prediction of bed entrainment is critical to estimating the catalyst loading to the cyclones. Designs of cyclones and diplegs are critical to FCC operation in maintaining proper particle size distribution of catalyst in the system. Last, but certainly not least, the standpipe is the main driving force for catalyst circulation between the regenerator and the reactor, which drives the FCC process. The challenge is to maximize pressure buildup and flow stability in the standpipe by proper design of the standpipe inlet and aeration.

2 FUNDAMENTALS OF THE FCC PROCESS

2.1 FCC Catalyst

The FCC catalyst is the heart and soul of the process. Both chemical and physical properties of the catalyst determine how the FCC unit is designed and operated. Since fresh catalyst is added to the FCC unit regularly, and catalyst is also withdrawn and lost through cyclone systems, the most important catalyst properties to FCC operation are those of the equilibrium catalyst.

The physical form of a modern FCC catalyst is a fine powder with a broad particle size distribution (PSD), mostly in the range between 10 to 150 microns. The average particle size is around 70 to 80 microns. It is a typical group A particle according to the Geldard classification (1973). The apparent bulk density of FCC catalyst is about 50 to 60 lb/ft^3. In addition to average particle size and density, one parameter critical to FCC operation is the content of fines in the range of 0 to 40 microns in the equilibrium catalyst. If the content of the fines drops, the fluidization property of the equilibrium catalyst will deteriorate rapidly even though the increase in the average particle size is relatively small. This typically leads to catalyst circulation problems. The loss of fines is also an indication that the cyclone systems are not functioning properly.

A modern FCC catalyst includes four major components. They are zeolite, matrix, binder, and filler. The first component, zeolite, is the primary active ingredient of the FCC catalyst, which can vary in the range of 15 to 50 wt% of the catalyst. It is a molecular sieve with a well-defined lattice structure, which provides the selectivity to allow only a certain size range of hydrocarbon molecules to enter the catalyst lattice structure. The acidic sites on the zeolite provide most of the activity of the FCC catalyst.

The basic building blocks of zeolite are silica and alumina tetrahedra. Each tetrahedron of the basic block has either a silicon or an aluminum atom at the center and four oxygen atoms at the corners. The pore diameter of the zeolite lattice is approximately 8 angstroms. The unit cell size (UCS) is the distance between the two repeating cells in the zeolite structure. Fresh FCC catalyst has a UCS in the range of 24.5 to

24.75 angstroms. In general, zeolite does not allow hydrocarbon molecules with a molecular diameter greater than 8 to 10 angstroms to enter the structure. The acidic sites of zeolite catalyst come from the alumina tetrahedron. The aluminum at the center of the tetrahedron is at a +3 oxidation state surrounded by four oxygen atom at the corners, which are shared with neighboring cells. Thus each alumina tetrahedron carries a net charge of −1, which has to be balanced by a positive ion. Sodium ion is commonly used in catalyst production for this purpose, which is later replaced by an ammonium ion. Upon drying of the catalyst, ammonia is vaporized, resulting in the formations of acidic sites of both the Bronsted and Lewis types. The Bronsted acidic sites can be further exchanged with rare earth material, such as cerium and lantheium.

The second component, matrix, plays an important role for cracking large hydrocarbon molecules, thus contributing significantly to the overall performance of FCC catalyst. Matrix also contains alumina but mostly in amorphous form, which is another source of the catalytic activity. The function of matrix is to provide active sites in larger pores in the amorphous form of alumina, which allow larger hydrocarbon molecules to diffuse in and crack into smaller molecules. This precracking function of the matrix enables FCC catalyst to process heavier feedstock with large hydrocarbon molecules, which are otherwise unable to enter the zeolite structure. The last two components, the filler and the binder, provide the physical integrity and mechanical strength of the FCC catalyst.

Metal contamination has detrimental effects on FCC catalyst performance and should be closely monitored. Sodium, which is present in fresh FCC catalyst, can also be present in the feedstock. Other metals, which include nickel, vanadium, iron, and copper, originate mostly from the heavy ends of the hydrocarbon feedstock. As catalyst comes in contact with feed in the unit, the metal components deposit on the catalyst and stay there. Sodium is known to neutralize the acidic sites and causes the zeolite structure to collapse. Nickel, vanadium, iron, and copper are promoters for dehydrogenation reactions. Vanadium is also known to destroy zeolite activity. FCC units, which process heavy feedstock, typically have higher metal contamination on the equilibrium catalyst.

2.2 Catalytic Cracking Reactions

Cracking of a hydrocarbon molecule means breaking the long carbon-to-carbon chain into shorter hydrocarbon molecules. There are two parallel mechanisms of cracking reactions. One is thermal cracking, which is a slower process, and the other is catalytic cracking.

The reaction kinetics of catalytic cracking goes mainly through the intermediate of a carbenium ion. The first step of catalytic cracking begins with the reaction of a hydrocarbon molecule at the acidic sites of the FCC catalyst, forming a carbenium ion ($R-CH_2^+$). This can occur from either adding a H^+ charge to an olefin molecule at the Bronsted acidic site, or removing a H^- charge from a paraffin molecule at a Lewis acidic site. The three key catalytic reactions following the formation of the carbenium ion are isomerization, beta scission, and hydrogen transfer.

The carbenium ion has the tendency to isomerize, because the primary ions are the least stable and the tertiary ions are the most stable. Thus, regardless of the initial forms of carbenium ions, and hence of the original hydrocarbon molecules in the feedstock, they tend to rearrange themselves preferentially toward secondary and tertiary ions. This has a significant impact on the FCC product properties. Because the tertiary ion is the preferred carbenium ion formation, the products of catalytic cracking reactions contain highly branched molecules.

The actual cracking step of the catalytic reaction is the beta scission of the carbenium ions. The cracking step occurs at the beta position because the carbon-to-carbon bond beta to the charged carbon on the carbenium ions is the weakest bond of the entire hydrocarbon chain and the easiest to break. The beta scission of a carbenium ion produces an olefin and a new carbenium ion. The new carbenium ion produced by the beta scission can repeat the beta scission or undergo first isomerization and then beta scission. The combination of beta scission and the fact that the primary carbenium ion is the least stable means that the hydrocarbon molecule formed by catalytic cracking contains at least three carbon atoms. This results in high production of C3 and C4 in LPG from the catalytic cracking reactions. As the hydrocarbon chain length of the carbenium ion becomes shorter, the reaction rate of the beta scission becomes lower. Thus the rate of catalytic cracking decreases as conversion gets higher in the upper riser.

Hydrogen transfer is another important mechanism in FCC catalytic reaction kinetics. Unlike beta scission and isomerization, which involve only a single molecule, hydrogen transfer is a bimolecular reaction. In order for hydrogen transfer to occur, two hydrocarbon molecules have to be adsorbed on two active sits on the catalyst, and the two active sites have to be close

enough to allow the two adjacent molecules to interact. One important outcome of hydrogen transfer is that it enables a shorter, less reactive carbenium ion to transfer its charge to a longer, more reactive molecules, thus further propagating the catalytic cracking reactions. Another important outcome is the redistribution of hydrogen atoms in the FCC products. For example, hydrogen transfer between an olefin and a naphthene produces a paraffin and a cyclo-olefin. Subsequent hydrogen transfer can further convert the cyclo-olefin to an aromatic molecule. A rare-earth-exchanged zeolite catalyst increases the tendency to hydrogen transfer. This is because the rare earth forms bridges between acidic sites on the catalyst. Since hydrogen transfer reaction requires adjacent acidic sites to occur, bridging these sites with rare earth promotes hydrogen transfer. The end results of hydrogen transfer are that it reduces the reactivity of the gasoline produced by FCC, which improves gasoline stability. This is because olefins are the most reactive species in the gasoline that are subject to secondary reactions, and hydrogen transfer reduces olefins in the gasoline. However, hydrogen transfer also has major drawbacks to the quality of the FCC products. It lowers light olefins in the LPG, lowers the octane number of the gasoline, and produces more aromatics in gasoline and light cycle oil (LCO).

Dehydrogenation is an undesirable catalytic reaction, which is promoted by metal contamination on equilibrium catalyst. The dehydrogenation reaction extracts hydrogen molecules from hydrocarbon molecules, thus increasing the production of highly aromatic products and coke. Nickel contamination is known to promote dehydrogenation. Other metals, such as copper and iron, are also known to increase hydrogen production. Metal contamination comes mainly from the heavy ends of the hydrocarbon feedstock, particularly the residues, as discussed previously.

2.3 Thermal Cracking Reactions

Thermal cracking reactions are undesirable but inevitable parallel reactions to the catalytic cracking reactions under normal FCC reactor conditions. The hydrocarbon molecules will crack owing to high temperature alone in absence of active FCC catalyst. Before the development of the FCC process, thermal cracking used to be the primary conversion unit. Today, refineries still utilize thermal cracking processes, such as delayed coking and visbreaking for thermally cracking heavy residues.

The reaction kinetics of thermal cracking goes through a different intermediate of free radicals. Two free radicals are formed when a hydrocarbon chain splits at any carbon-to-carbon or hydrogen-to-carbon bond. This is the rate-determining step of thermal cracking reactions. Splitting a hydrocarbon molecule produces two uncharged molecules with each molecule having an unpaired electron. Since more energy is required to split a hydrogen-to-carbon bond, this is less likely to occur. For splitting primary, secondary, or tertiary carbon-to-carbon bonds, there is little difference in energy requirements. Thus the formation of free radicals is nonselective. Methyl and ethyl free radicals are as likely to form as other longer free radicals.

Once free radicals are formed, they are extremely reactive. They can undergo alpha and beta scissions, hydrogen transfer, and polymerization. The scissions of free radicals are the actual thermal cracking reactions that produce smaller hydrocarbon molecules. When a scission reaction of a free radical occurs, it produces an olefin and a new free radical. However, one important difference from the catalytic reactions is that free radicals cannot undergo isomerization. Thus the end product from thermal cracking has fewer branched hydrocarbons. Another important difference is that the beta scission in the catalytic cracking reaction produces hydrocarbons longer than C3, whereas alpha and beta scissions of free radicals produce molecules and radicals including C1 and C2. Thus, thermal cracking leads to a product that is rich in C1 and C2 light gases.

Hydrogen transfer occurs when a free radical extracts a hydrogen atom from a hydrocarbon molecules, which ends the free radical and creates a new free radical. The hydrogen-to-primary-carbon bond is the strongest and the hydrogen-to-tertiary-carbon bond is the weakest within a hydrocarbon chain. Furthermore, the rate of hydrogen transfer decreases as the hydrocarbon chain of the free radical increases. Thus a methyl radical produced by the alpha scission tends to extract a hydrogen atom from a neutral hydrocarbon molecule to form a methane molecule and a new secondary or tertiary free radical. The same mechanism applies to the formation of ethane through beta scission followed by hydrogen transfer.

Two free radicals can recombine into a single molecule and terminate the chain reactions of thermal cracking. The olefins formed initially by scission reactions can also undergo polymerization and condensation to form coke. All in all, thermal cracking produces more C1/C2 light gases as well as more coke than catalytic cracking. The challenge in the FCC process is to minimize thermal cracking while maximizing catalytic cracking.

2.4 Heat Balance

Like every system operating under steady-state conditions, an FCC unit must stay in heat balance at all times. However, because catalyst circulation between the regenerator and the reactor serves the dual purposes of providing reaction activity and the heat requirement to the reactor, the heat balance in the FCC process has special purposes that other processes do not have. Thus, heat balance is the key to better understanding of how different variables interact with one another in the FCC unit.

Most FCC units only have a few independent variables. Typically, these independent variables are the feed rate, feed preheat temperature, reactor/riser temperature, air flow rate to the regenerator, and catalyst activity. The feed rate and air flow rate to the regenerator are set by flow controllers. The feed temperature is set by the feed temperature controller. Catalyst activity is set by catalyst selection and fresh catalyst addition rate. Reactor temperature is controlled by the regenerator slide valve that regulates the catalyst circulation rate. The catalyst circulation rate is not directly measured or controlled. Instead, the unit relies on the heat balance to estimate the catalyst circulation rate. Except for these independent variables, other variables, such as regenerator temperature, degree of conversion, and carbon-on-catalyst, etc., will vary accordingly to keep the FCC unit in heat balance. These variables are dependent variables.

The overall heat balance requires that the FCC unit constantly adjust itself to produce and burn just the right amount of coke in the regenerator to supply the heat requirements for the entire unit. The major heat requirements are heating and vaporizing the hydrocarbon feed, heat of reaction, heating air and steam to system temperature, and heat losses of the unit. Combustion of coke in the regenerator is the single source of heat supply to run the FCC unit. The heat generated in the generator is carried to the rest of the unit by catalyst circulation, as described by the following equation:

$$\frac{\text{Heat transfer}}{\text{lb of feed}} = \frac{\text{catalyst}}{\text{oil}}(\text{catalyst heat capacity})$$
$$(T_{\text{regen}} - T_{\text{react}})$$

The first term on the right-hand side is the catalyst-to-oil ratio, which has the dimension of pounds of catalyst circulation per pound of feed processed. The last term is the temperature difference between the regenerator and the reactor.

The material balance requires that coke burned in the regenerator be equal to coke formed in the reactor, which is called the coke yield. Coke yield has the dimension of pounds of coke formed per pound of feed processed. Since catalyst circulation is the carrier of the coke on catalyst, the coke yield can be expressed as

$$\text{Coke yield} = (\text{delta coke}) \frac{\text{catalyst}}{\text{oil}}$$

The first term on the right-hand side is the delta coke, which is the difference in coke concentration (lb coke/ lb catalyst) between the spent and the regenerated catalyst. The second term on the right-hand side is the catalyst-to-oil ratio.

Through these relationships of heat and material balances, a change in one independent variable will trigger a domino effect on other variables in an FCC unit. For instance, if the reactor/riser temperature is increased at a given feed temperature by opening the regenerator slide valve, it will trigger the following chain of responses (on a per-pound-of-feed-processed basis):

Catalyst circulation will increase owing to the opening of the regenerator slide valve.

Reaction conversion will increase because of higher reaction rate at higher catalyst-to-oil ratio and reactor temperature. This is the most common reason for increasing the reactor temperature.

Coke yield will increase because of higher conversion. This also provides additional heat for higher reaction conversion and other heating requirements.

Delta coke will decrease because there is more catalyst circulating to the reactor to receive the coke formation.

More air to the regenerator is required to burn more coke. The increase in air requires an independent adjustment to the flow controller because air flow rate to the regenerator is not a dependent variable.

Regenerator temperature will most likely increase because of higher coke combustion rate. However, the increase in regenerator temperature is typically much less than the increase in reactor temperature.

2.5 Pressure Balance and Catalyst Circulation

Catalyst circulation between the reactor and the regenerator has two critical functions in the FCC process. One is to maintain the activity of the regenerated catalyst by burning and removing coke on the spent catalyst. The other is to keep the unit in heat balance by continuously removing heat from the regenerator and adding it to the reactor and the rest of the unit.

Since FCC catalyst is kept above minimum fluidization conditions everywhere in this catalyst circulation loop, the fluidized catalyst is free to flow from one place to another. Thus the catalyst circulation is driven by the overall pressure balance of the unit, and the circulation rate is regulated by the two slide valves, i.e., the stripper slide valve and the regenerator slide valve. A minimum pressure drop across each slide valve is set in the control system to guard against flow reversal, which is a very serious safety issue. For instance, a reverse flow of hydrocarbon vapor from the reactor to the oxygen-rich regenerator can lead to a sudden increase in combustion reaction and regenerator temperature. In the extreme case, a catastrophic explosion could occur. The overall pressure balance of the unit determines the pressure drops available for the slide valve control and hence the maximum catalyst circulation rate.

The first step in the overall pressure balance is to determine the hydraulic heads in different parts of the FCC unit. Typical apparent density in the riser is 4 to 6 lb/ft³. For side entry feed nozzle configuration, the lower riser section prior to feed injection has a much higher density than the remaining of the riser, which depends on the preacceleration fluidization velocity. Typical standpipe density is 35 to 40 lb/ft³. An inclined standpipe is known to have a slightly lower density than a vertical standpipe. Typical stripper bed density is about 35 to 40 lb/ft³, whereas the regenerator bed has a lower density, in the range of 25 to 30 lb/ft³. The dilute phase in both the reactor and the regenerator has a density below 1 lb/ft³.

The next step is to estimate transition pressure losses, which include riser liftpot, riser termination, cyclone, spent catalyst distributor, and other flow transitions. By subtracting these transition pressure losses from the hydraulic pressure, the remainder would be the pressure drop available for slide valve control. While other parameters in pressure balance seldom change much at all, it is not uncommon to have the standpipe apparent density lower than the typical value. Because standpipes are long columns, a lower standpipe apparent density means a substantial loss in hydraulic heads and hence lower available pressure drops for slide valve control, leading to a lower achievable catalyst circulation rate. Commercial experience has demonstrated that a proper design and operation of the standpipe, particularly the standpipe inlet and fluidization, can increase the catalyst circulation rate by up to 30 to 40% (Chen, 2001).

2.6 Hydrogen Balance

Every chemical reaction must satisfy an overall material balance as well as individual atomic balances. It is important to discuss the hydrogen balance in the FCC process, because hydrogen in the feedstock is the limiting component of the cracking reactions.

Typical FCC feedstock contains about 12 wt% of hydrogen. The average hydrogen content of lighter cracking products is about 13.5 wt% whereas the average hydrogen content of the heavier liquid cracking products is about 9 wt%. Since typical hydrogen content in coke is in the range of 5 to 7 wt% and coke yield is about 5 wt% on feed, a simple hydrogen balance shows that typical conversion of an FCC unit should be about 70%. Any deviation from these typical values will shift the hydrogen balance accordingly. For instance, an increase in feedstock hydrogen content will allow the unit to have higher conversion, leading to higher production of cracking products, coke and less heavier liquid cracking products.

Hydrogen balance also provides other insights into the FCC process. For example, it is known that thermal cracking produces more C1/C2 dry gas. Because these shortest hydrocarbon molecules have the highest hydrogen contents, more coke and heavy cracked oils are also produced by thermal cracking to satisfy the hydrogen balance. Thus thermal cracking is undesirable. Dehydrogenation reaction, which is promoted by metal contamination, has similar effects to those of thermal cracking in increasing the production of coke and heavy cracked oils based on the hydrogen balance. Another example is to look at the effect of adding residuals to the feedstock. As the percentage of residual in feed increases, the hydrogen content in the overall feedstock will drop. Thus hydrogen balance requires that a residual cracking unit operate at a much lower conversion and have a higher coke yield than a typical FCC unit.

3 MODERN FCC DESIGN

Several engineering companies such as UOP, Stone and Webster, Kellogg Brown and Root, and ABB/Lummus as well as major oil companies such as Exxon/Mobil, Shell, and Total/ElfFina all have their own FCC designs (Meyers, 1997). Although each FCC design is different, the basic design philosophies are actually quite similar. The following discussion will use mainly the Shell FCC design, shown in Fig. 1, as an example to elucidate major components in a typical modern FCC process. As shown in the figure, the FCC unit is a two-vessel, side-by-side design with the regenerator to the left and reactor/stripper to the right. The two vessels are connected to each other with catalyst transfer lines. Each major component of this FCC design is discussed in detail in the following sections.

3.1 Feed Injection System

The feed injection system is the most critical component of the modern FCC riser reactor design. Several factors have made the feed injection system increasingly important over the years. The most important factor is that the FCC reaction time is getting shorter and shorter. This was not the case for earlier FCC units when the catalyst was not as active and the reaction time was relatively long, i.e., on the order of minutes. Thus, mixing time and evaporation time of the feed induced by the feed injection system were not as critical. However, due to the development of the highly active zeolite catalyst, typical reaction time has been shortened to a few seconds in the riser reactor. Some of the most recent FCC reactor designs have gone to

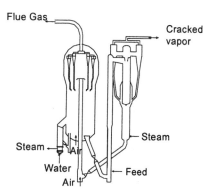

Figure 1 Modern shell FCC design.

below 1 second (Kauff et al., 1996). Since catalytic cracking reactions can only occur in the vapor phase after liquid hydrocarbon feedstock is vaporized, mixing and feed vaporization must take place as quickly as possible in order for catalytic cracking reactions to complete in a few seconds or less. Otherwise, only thermal cracking can take place. Furthermore, liquid hydrocarbon, which is not vaporized, is burned as coke in the regenerator.

The second factor is that the regenerator temperature is getting higher, which is also a contributing factor to the shorter reaction time as discussed previously. Because of the higher regenerator temperature, control of thermal cracking in the reactor becomes more critical. Typical riser top temperature of modern FCC units is controlled in the range of 950 to 1050°F. But typical regenerated catalyst temperature is much higher, in the range of 1250 to 1350°F. Thus both catalytic and thermal cracking reaction rates are much higher in the lower riser section because the catalyst is fresh and active, and the temperature is high. The feed injection system plays the key role of controlling thermal cracking by cooling off the lower riser section with fast mixing and vaporization of the liquid feed.

The third factor is that the FCC feedstock is getting heavier. As the feed gets heavier, the boiling point increases, which makes feed vaporization more difficult. At the same time, the viscosity of the feed also increases, which makes feed atomization more difficult. Thus better feed atomization must be achieved to disperse the feed into even finer droplets in order to vaporize the heavier feed in time for catalytic cracking reactions to take place in the riser reactor within seconds.

The feed injection system in FCC consists of two key components: the feed nozzle itself and the preacceleration region, which is the lower riser transition preparing the catalyst flow prior to feed injection. These two design components must work in concert in order to achieve desirable riser performance. There are many different FCC feed nozzle designs in the patent literature (Chen and Dewitz, 1998; Chen et al., 1999; Bedaw et al., 1993; Haruch, 1995). All these modern FCC feed nozzles belong to the same category of twin-fluid atomizers that utilize pressurized gas to assist the atomization of liquid feed. Figure 2 shows the feed nozzle design of Chen and Dewitz (1998). In this design, the atomization gas flows along the center pipe, passes through multiple holes at the end of the gas cap, and mixes with the liquid feed from the annulus prior to exiting the nozzle slits.

Figure 3 shows fine atomization and feed distribution of the fan spray pattern generated by the feed nozzle of Chen and Dewitz (1998) during a pilot plant test.

The feed nozzle has three major functions: feed atomization, feed distribution, and mixing with catalyst.

1. Feed atomization. The objective of feed atomization is to generate a large surface area for fine droplets to vaporize as quickly as possible. A feed nozzle produces a spray with a certain droplet size distribution, both in time and space. The challenge of a successful feed nozzle design is to produce the finest, most stable, and uniform feed atomization using the least amount of energy. The quality of feed atomization is typically measured by the Sauter mean diameter of the droplet size distribution. However, the mean droplet size is only one of several important measurements. It is desirable to produce a droplet size distribution as narrow as possible to limit the number of large droplets. Another key performance parameter of a feed nozzle is the energy efficiency, which is measured by the amount of atomization steam used and the feed pressure required to achieve desirable atomization. Commercial experience has demonstrated that a successful feed nozzle design can achieve fine atomization with atomization steam in a range between 1 to 3 wt% on feed and with a pressure drop through the nozzle of no more than 25 to 60 psi.

2. Feed distribution. Feed distribution is measured by the uniformity of liquid flux across the entire spray pattern generated by the feed nozzle, both in time and in space. The objective is to achieve uniform coverage of feedstock across the riser. Commercial experience has shown that the best spray pattern to achieve such an objective is a wide-angle fan spray such as the one shown in Fig. 3.

3. Mixing with catalyst. The objective is to achieve a flat radial riser temperature profile as quickly as possible. In doing so, the regenerated catalyst is uniformly cooled down by the vaporization of hydrocarbon feedstock, thus minimizing thermal cracking reactions in the lower riser section. In addition to feed atomization and feed distribution, commercial experience has shown that the feed injection angle, defined by the angle between the riser

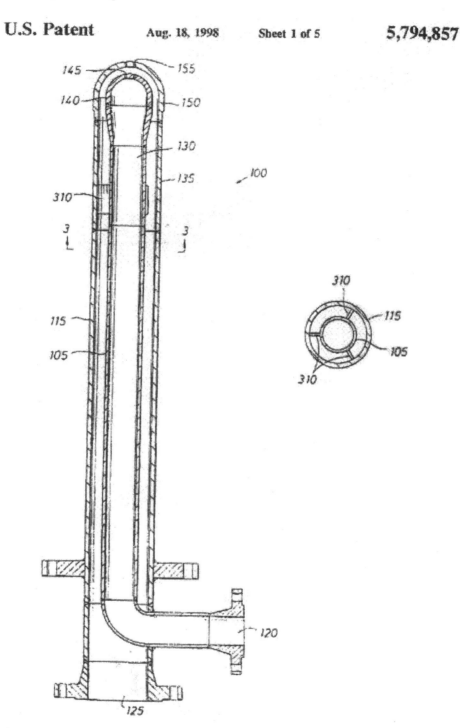

Figure 2   CCU feed nozzle design. (From Chen and Dewitz, 1998.)

Figure 3 Atomization and spray pattern of the feed nozzle shown in Fig. 2. (From Chen and Dewitz, 1998.)

axis and the feed injection, plays a significant role in mixing the feed with catalyst and hence achieving a desirable temperature profile in the riser. Most FCC units have feed nozzles installed through riser shrouds at a fixed angle. A new feed nozzle design by Chen et al. (1999) as shown in Fig. 4, enables an FCC unit to adjust the feed injection angle while using the existing riser shrouds. This enables an FCC unit to optimize the mixing of feed with catalyst by adjusting the feed injection angle to achieve the best performance of the unit.

Commercial experience has confirmed that using better feed nozzles can substantially reduce dry gas and increase gasoline yield. These results are in line with the expectation that thermal cracking reactions, which are the primary source for dry gas, are reduced. As a result, catalytic cracking reactions are maximized, and more desirable cracking products are produced. Depending on unit constraints and market demands, other steps can be taken, such as increasing riser temperature, increasing catalyst activity, or adding ZSM-5 catalyst, in order to take full advantage of the dry gas reduction.

The second component of the feed injection system is the design of the low riser transition prior to feed injection. There are two common configurations of the lower riser to prepare for feed injection; one is bottom entry and the other is side entry configura-

tion. Most of the modern FCC units have the side entry configuration, as shown in Fig. 1, in which regenerated catalyst is lifted from the riser bottom to the feed injection point by fluidizing gas, normally steam. The section between the riser bottom and the feed injection point is called the catalyst preacceleration zone. A typical lift steam velocity for the preacceleration is in the range between 3.0 to 10.0 ft/s, which is in the turbulent to fast fluidization regime. The length for the preacceleration zone is in the range from 6 to 20 feet. At the end of preacceleration, multiple feed nozzles located on the circumference of the riser are used to inject the feedstock. With proper designs of the preacceleration and feed nozzles, backmixing of catalyst in the lower riser is minimized with the side entry configuration.

Most of the older FCC units have the bottom entry configuration. In this case, a single or multiple nozzles located at the bottom of the riser are used to introduce the feedstock directly into the region where hot regenerated catalyst enters the riser. The catalyst condition in the riser bottom is denser, chaotic, and highly erosive. Earlier feed nozzles with the bottom entry configuration emphasize mechanical robustness, but with primitive feed atomization and poor performance. Because of this disadvantage many bottom entry FCC units have been converted to side entry units in recent years. However, recent commercial experience has demonstrated that, with proper feed nozzle design (Chen, 2002), a bottom entry configuration can achieve feed contacting and riser performance similar to the

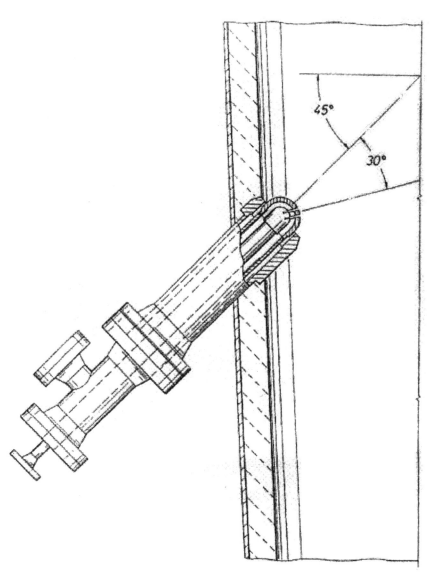

Figure 4 CCU feed nozzle design. (From Chen et al., 1999.)

side entry configuration. In some cases, the bottom entry configuration can even be the preferred option because it does not require the preacceleration zone, which can take up substantial riser volume. In addition, bottom entry configuration has lower pressure drop in the riser, which has a positive impact on catalyst circulation.

3.2 Riser/Reactor

Modern FCC units have vertical risers, which provide the proper residence time for the hydrocarbon feedstock to vaporize, react, and crack in presence of catalyst. Due to the highly active modern FCC catalyst, the cracking reactions are essentially completed at the end of the vertical riser. Thus the reactor vessel in the modern FCC unit downstream of the riser is actually a containing vessel for spent catalyst/product vapor separation. In some older FCC units, however, the reactor vessel is designed to provide additional residence time for the cracking reactions to complete downstream of the riser while using a less active catalyst.

Due to the cracking reactions, the molar flow rate and the volumetric flow rate of the hydrocarbon vapor increase as it moves up the riser. In a typical FCC unit, the volumetric expansion is in the range of 3 to 4 times of the vaporized feedstock. Thus the riser diameter may be increased once or twice after feed injection to keep the vapor velocity within the range of 40 to 80 feet/s. Some modern FCC risers could further include boundary layer strip rings, e.g., Gwyn (1993), to reduce catalyst backmixing along the riser wall and to achieve performance closer to an ideal plug flow reactor.

A modern FCC riser design also includes a separation device at the end of the riser. The objective is to achieve a quick and clean separation of spent catalyst and hydrocarbon vapor in order to minimize post riser cracking. Post riser cracking is undesirable because the riser is designed to achieve optimum conversion, and additional cracking reactions after the riser will push the conversion outside the optimum range. One riser termination design, called the prestripping cyclone, uses a dead end tee at the riser top followed by a special cyclone, as described in Parker et al. (1987). Figure 5 shows further details of this design in which the upper part is a rough-cut cyclone for catalyst/vapor separation and the lower part is a prestripper bed where steam is injected to remove entrained vapor. Catalyst not captured in the prestripping cyclone is separated in downstream secondary cyclones. In most modern FCC designs, the secondary cyclones are close-coupled with the primary cyclones, such as examples shown in Wilson (1997), to minimize post riser cracking. The reactor cyclones can be designed to operate either inside or outside the reactor vessel, as shown in Fig. 1.

3.3 Stripper

After the spent catalyst/product vapor separation from a riser termination system, it is inevitable that some hydrocarbon will be entrained by the spent catalyst from the exit of the riser termination system. Part of the entrained hydrocarbon resides in the vapor space between catalyst particles, and the remainder is physically adsorbed on the surface and inside the pores of the catalyst. The primary objective

of a stripper is to recover the entrained and adsorbed hydrocarbon as much as possible with a limited amount of steam. The other objective is to reduce hydrocarbon carryunder into the regenerator, which will be burned as coke.

Most FCC strippers are designed with a countercurrent contacting of descending spent catalyst and ascending stripping steam, operating in the bubble flow regime. Any hydrocarbon not stripped by steam will enter the regenerator and burned as coke. Thus the overall coke entering the regenerator includes the coke on the catalyst and the entrained hydrocarbon. Since the hydrocarbon vapor removable by stripping is relatively rich in hydrogen compared to coke deposition on the spent catalyst, a direct measurement of the stripper performance is the level of hydrogen content in the overall coke entering the regenerator. Another measurement of the stripper performance is the energy efficiency expressed in terms of the amount of stripping steam used. A properly designed stripper should be able to reach a hydrogen-in-coke level of 5 to 7 wt% using 2 to 4 pounds of steam per 1000 pounds of catalyst circulation.

It is important to note that the hydrocarbon vapor continues to react and crack under typical FCC stripper conditions. Thus early dissociation of the hydrocarbon vapor from catalyst is critical to minimize overcracking of high-value hydrocarbon products into low-value light gas and coke. This can be achieved by a proper design of two-stage stripping. In the first stage, a relatively high flow of stripping steam is used to displace quickly the easily desorbed and entrained hydrocarbon vapor. This is typically done by placing a steam ring in the upper part of the stripper or by using a prestripping cyclone, as shown in Fig. 5. In second stage stripping, a moderate flow rate of the stripping steam combined with an adequate residence time removes crackable fragments of the more strongly adsorbed heavy hydrocarbons.

3.4 Regenerator

The primary objective of the regenerator is to burn off coke deposition on the spent catalyst to restore catalyst activity. However, other important aspects should also be taken into consideration as well. The challenge of a successful regenerator design is to achieve the following objectives all at the same time, to

Achieve low coke on regenerated catalyst (CRC) to restore catalyst activity.
Burn more coke at a given amount of blower air.

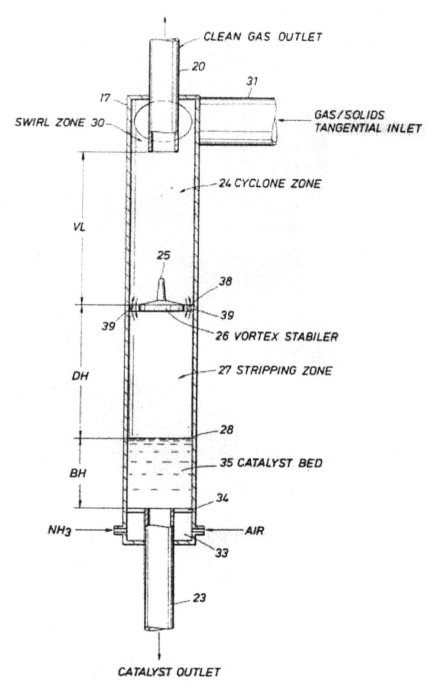

Figure 5 CCU stripper cyclone. (From Parker et al., 1987.)

Minimize catalyst deactivation.

Minimize afterburn, which will be discussed later in this section.

There are two types of regenerators in FCC operations; one operates in the partial combustion mode and the other in the total combustion mode. In partial combustion mode, a less-than-theoretical, or stoichiometric, amount of air is provided to the regenerator. Only part of the carbon in coke is reacted to carbon dioxide, and the remainder of the carbon is reacted to carbon monoxide. Ideally, all oxygen should be consumed and no oxygen should be present in the flue gas.

One important parameter to monitor in partial combustion regeneration is the CO/CO_2 ratio in the flue gas, which is typically operated in the range from 0.5 to 2.0.

In the total combustion mode, excess air is provided to the regenerator. Ideally, all the carbon component in the coke should be reacted to carbon dioxide, and no carbon monoxide should be present in the flue gas.

One important parameter to monitor in total combustion regeneration is the oxygen content in the flue gas, which is typically operated in the range from 1.0 to 3.0 percent on a dry basis.

Coke on regenerated catalyst (CRC) is a key performance measurement for both partial and total combustion regenerators. Units with total combustion regenerators can achieve a typical CRC level of about 0.05 wt% or lower. Units with partial combustion regenerators have higher CRC levels, typically about 0.1 wt% or higher.

A partial combustion regenerator is typically designed to operate in a single-stage countercurrent flow with fluidizing air supplied close to the bottom of the vessel and spent catalyst distributed close to the top, as shown in Fig. 1. Typical fluidization velocity is in the range of 2 to 9 feet/s in turbulent to fast fluidization regimes. Partial combustion regenerators have several advantages over total combustion regenerators. One is that more coke can be burned at a given amount of blower air because partial combustion of carbon requires less than the stoichiometric amount of air. Another benefit is that catalyst deactivation is minimized by maintaining moderate regenerator temperatures. A potential drawback of the partial combustion regenerator is higher coke on regenerated catalyst (CRC).

The phenomenon of afterburn in the partial combustion regenerator is due mostly to the escape of oxygen through the regenerator bed. This is likely to

happen when the spent catalyst is poorly distributed and some areas of the regenerator have little coke to burn. As a result, oxygen in the fluidizing air in that particular area is not completely consumed in the regenerator bed. When oxygen reaches the freeboard and reacts with carbon monoxide from other parts of the regenerator to produce carbon dioxide, a large amount of heat combustion is released. Since only a little catalyst is present in the freeboard, the heat capacity is low and the temperature escalates quickly, which is called afterburn. Severe afterburn, sometimes as high as 150°F, can cause severe mechanical damage to the regenerator cyclone system.

Both high CRC and afterburn in partial combustion regenerators can be overcome by proper design of the spent catalyst distributor and the air grid. Figure 6 shows one example of a modern spent catalyst distributor designed by Khouw et al. (1994), which distributes spent catalyst laterally through several horizontal arms.

The most significant advantage of total combustion regenerators is that the CRC is low and catalyst

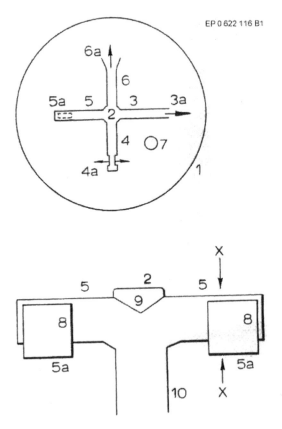

Figure 6 CCU spent catalyst distributor design. (From Khouw et al., 1994.)

activity is higher. Potential drawbacks of total combustion regenerators include higher catalyst deactivation owing to higher regenerator temperature, lower coke burning capacity owing to the requirement of excess air, and higher heat release owing to total combustion reaction. Catalyst deactivation can be reduced by using two-stage regeneration; see, e.g., Herbst et al. (1989). In the two-stage regeneration, the first stage is operated at a moderate temperature to burn off the hydrogen component on the coke, which has a higher reaction rate, and some of the carbon component. The second stage uses excess air to burn off the carbon component on the coke at higher temperature to carbon dioxide. Because of the absence of water vapor in the second stage regenerator, catalyst deactivation at high temperature can be minimized. Two-stage regeneration is more expensive and more complicated to operate. The coke burning issue is typically addressed by having a large air blower, which adds to higher capital and operational costs. The third issue in total combustion regeneration is the heat balance, because more heat is generated in the regenerator at the same coke burning rate when carbon is reacted to carbon dioxide instead of carbon monoxide. If more heat is generated in the regenerator than what is required for the reactor, a common solution is to add a catalyst cooler to the regenerator to remove the excess heat, as shown in Fig. 1. The catalyst cooler is a heat exchanger that produces steam while removing heat from regenerator. This also adds to capital costs.

The phenomenon of afterburn also occurs in total combustion regenerators. This is most likely to happen when the spent catalyst is poorly distributed and a particular area of the regenerator has too much coke. Although, on average, air is always in excess in the total combustion mode, local areas could have insufficient air to burn all the carbon to carbon dioxide if spent catalyst is not well distributed. As a result, some carbon monoxide will escape through the regenerator bed from these areas. Excess oxygen is always present in the freeboard to react with carbon monoxide and to produce carbon dioxide, causing an afterburn problem. Afterburn in total combustion regenerators can be overcome by using higher excess air, promoter, or a better design of the spent catalyst distributor, e.g., Fig. 6.

Residual cracking poses another challenge to regenerator design and operation. For a given conversion, coke yield is much higher in residual cracking compared to the conventional FCC because of heavier feedstocks and higher metal contents. Catalyst coolers

are required, regardless whether the regenerator operates in partial or total combustion, in order to keep the unit in heat balance. In addition, metal contamination on equilibrium catalyst is high, which has detrimental effects on the catalytic reactions. There are several ways of dealing with this problem. These include selection of catalyst with metal trapping capability, off-line removal of metal contamination, and management of equilibrium catalyst to control the maximum concentration of metals.

3.5 Third-Stage Separator

A modern FCC unit may further include a power recovery system to recover energy from the regenerator flue gas, which is a high volumetric gas flow at an elevated temperature and a moderate pressure. It becomes critical to control particulates in the flue gas in order to protect the blades of an expander in the power recovery system. However, most regenerator cyclone systems have a limited capability of removing particles around 10 microns.

A special design called the third-stage separator (TSS) is shown in Fig. 7 (Dries et al., 2000) and can be used for this purpose. It uses swirl tubes to generate a very high centrifugal force and removes practically all particles of 10 microns and larger from the hot flue gas. In fact, the TSS design has recently been improved so that it has a compatible capability for catalyst removal as an electro static precipitator (ESP) from the emission control viewpoint. A TSS can be installed either inside or outside the regenerator vessel.

3.6 Standpipe and Standpipe Inlet

Standpipe flow is the main driving force behind catalyst circulation between the regenerator and the reactor, which drives the entire FCC process. The objective of a standpipe is rather straightforward. It is to build a hydraulic head by holding a column of fluidized catalyst. This enables the catalyst to move in the direction against pressure gradient, from low pressure at the top to high pressure at the bottom of the standpipe. When standpipes do not function properly, the FCC unit cannot circulate catalyst at the design rate and the unit is forced to cut feed rate. Even with such an important role in the FCC process, standpipe flow remains poorly understood. The existing theories on standpipe flow, e.g., Leung (1977) and Chen et al. (1984), are overly simplified.

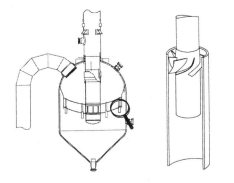

TSS-version	emission
original, <1985	~ 100 mg/Nm^3
post-1985	~ 80 mg/Nm^3
per 1997	~ 50 mg/Nm^3
per 1998	< 50 mg/Nm^3

(all subject to incoming particle size)

Figure 7 Third stage separator, blow-up of swirl tube, emissions improvements.

Designing a standpipe is simple from mechanical viewpoint, and yet very complex from hydrodynamic viewpoint. A standpipe is typically a vertical, or nearly vertical, column with a constant diameter, which is the simplest mechanical structure of the entire FCC unit. There are only two design elements that influence the standpipe flow besides the slide valve, which controls the outlet end of the standpipe. The first design element is aeration along the standpipe. As pressure increases down the standpipe, the gas phase within the fluidized catalyst is compressed and must be made up by additional fluidizing gas. Otherwise, a long standpipe can become defluidized at the lower end and restrict the catalyst flow. The complex nature of the aeration design is that an overaerated standpipe has a hydrodynamic behavior similar to that of an underaerated standpipe, i.e., it cannot build proper pressure, and the catalyst flow is unstable. In order to distinguish the two cases, one must look beyond the superficial symptoms. In the overaerated case the instability is caused by the formation and release of large bubbles, whereas in the underaerated case it is caused by defluidization leading to a stick-slip flow.

The second design element is the standpipe inlet. Because both stripper and regenerator fluidized beds are operated at relatively high gas flow rates, excessive gas can be drawn into the standpipe through the inlet, which is highly undesirable. To address the gas entrainment issue, the conventional design is to add an inlet hopper at the top of the standpipe. The typical inlet hopper size is about 2 to 2.5 times the standpipe diameter. The concept of the inlet hopper is to provide enough residence time for small entrained bubbles to coalesce into larger ones. Since larger bubbles have a

higher rising velocity, these large bubbles have a better chance to escape from the inlet hopper, thus minimizing gas entrainment into the standpipe. The complex nature of the design is that a poorly designed inlet hopper can either entrain too much gas or cause defluidization. The overall hydrodynamic behaviors of the two cases are again very similar, i.e., the standpipe cannot build proper pressure and the catalyst flow is unstable. If catalyst defluidizes in the inlet hopper, putting a local aeration into the hopper can solve the problem (Chen, 1986). However, if the hopper draws too much gas into the standpipe, any additional aeration will only worsen the situation. One particular problem with the conventional inlet hopper is that the amount of gas entrainment cannot be controlled. Thus a standpipe can suddenly become unstable as the catalyst circulation rate is increased or decreased beyond a certain rate.

A new standpipe inlet design has recently been developed by Chen and Brosten (2001), as shown in Fig. 8. Instead of using an inlet hopper, this new inlet design uses a disk positioned directly below the standpipe inlet. The concept is to trap fluidizing gas from below, causing a local partial defluidization above the disk and forming a dense bed region near the proximity of the standpipe inlet. A small amount of fluidization gas can be introduced above the disk to control the fluidization condition of the standpipe inlet region independently of regenerator and stripper fluidization conditions, which are set by process requirements. Commercial experience has demonstrated that both catalyst circulation rate and stability can be significantly improved by simply replacing the conventional standpipe inlet with the better inlet design of Fig. 8.

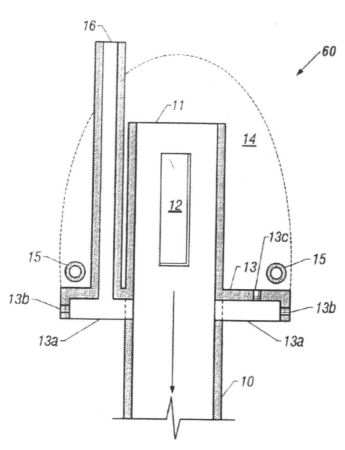

Figure 8   CCU standpipe inlet design. (From Chen and Brosten, 2001.)

REFERENCES

Bedaw RE. Atomizing spray nozzle for mixing a liquid with a gas. U.S. Patent 5,240,183, 1993.

Chen Y. A theoretical investigation of an aerated hopper flow. J Chinese Inst Chem Eng 17:195, 1986.

Chen Y. Feed nozzle. U.S. Patent 6,387,247, 2002.

Chen Y, Brosten D. Standpipe inlet. U.S. Patent, 6,228,328, 2001.

Chen Y, Dewitz TS. Feed nozzle. U.S. Patent 5,794,857, 1998.

Chen Y, Rangachari R, Jackson R. Theoretical and experimental investigation of fluid and particle flow in a vertical standpipe. Ind Eng Chem Fundam 23:354, 1984.

Chen Y, Brosten D, Nielson JW. Feed nozzle. U.S. Patent 5,979,799, 1999.

de Kruijff GT, van Hattem A. Apparatus for solids–fluid separation. U.S. Patent 5,1998,005, 1993.

Dries H, Patel M, van Dijk N. New advances in third-stage separators. World Refining 30–34, Oct. 2000.

Geldart D. Types of gas fluidization. Power Technol 7:285–292, 1973.

Gwyn JE. Entrance, exit and wall effects on gas/particulate solids flow regimes. Proceedings of 4th International Conf on Circulating Fluid Beds, Somerset, Pennsylvania, 1993, pp 679–684.

Harauch J. Nozzle for use in fluidized catalytic cracking. European Patent EP 0-717-095-A2, 1995.

Herbst JA. Fluid catalytic cracking process and apparatus for more effective regeneration of zeolite catalyst. U.S. Patent 4,814,068, 1989.

Kauff DA, Bartholic CA, Steves CA, Keim MR. Successful application of the MSCC process. Paper AM-96-27, NPRA meeting, 1996.

Khouw FHH, van Poelje WM, van der Honing G. Process and apparatus for distributing fluids in a container. E.P. Patent 0622116, 1994.

Leung LS. Design of fluidized gas–solids flow in standpipes. Powder Technol 16:1, 1977.

Meyers RA. Handbook of Petroleum Refining Processes. 2d ed. 1997, pp 3.3–3.99.

Parker WA, Dewitz TS, Hinds GP, Gwyn JE, Bilgic AH, Hardesty DE. Apparatus for the separation of fluid cracking catalyst particles from gaseous hydrocarbons. U.S. Patent 4,692,311, 1987.

Sadeghbeigi R. Fluid Catalytic Cracking Handbook. 1995.

Wilson JW. Fluid catalytic cracking—technology and operation. 1997.

15

Applications for Gasifiers and Combustors

Richard A. Newby

Siemens Westinghouse Power Corporation, Pittsburgh, Pennsylvania, U.S.A.

1 INTRODUCTION

Carbonaceous fuels are used throughout the world as major energy sources. They exist primarily in solid forms, or heavy liquid forms, containing significant process and environmental contaminants compared to cleaner fuels such as natural gas. Coals and biomass fuels are two of the many carbonaceous fuel types that can be effectively processed in fluidized bed reactors, and they are currently of great industrial interest for fluidized bed gasification and combustion.

Fluidized bed gasifiers and fluidized bed combustors for carbonaceous fuels are relatively simple chemical reactors that can be utilized for several industrial applications that include chemical synthesis, process heat supply, steam generation, and power generation. Only relatively recently have carbonaceous fuel fluidized bed combustors and gasifiers reached or approached commercial status, with fluidized bed combustors gaining commercial acceptance earlier than fluidized bed gasifiers.

These fluidized bed technologies compete directly with commercially mature entrained-particle and moving-bed gasifiers and combustors (pulverized-fuel-fired boilers, and stoker-fired boilers) not only with respect to plant cost and efficiency but also with respect to plant availability. The fluidized bed combustor and gasifier competitive positions relative to these more mature combustor and gasifier technologies are still being established and depend, to a great extent, on the specific application and the properties of the carbonaceous fuel being processed.

Fluidized bed gasifiers differ from fluidized bed combustors in their basic reaction mechanisms, functions, features, and operating conditions. They have in common, though, several characteristics. Fluidized bed gasifiers and combustors operate at relatively low temperatures, below the fuel ash "agglomeration" temperature, and they have relatively uniform temperature throughout their volume. Their designs are tailored to the properties and behavior of the carbonaceous fuels to be processed. Finally, they must overcome operational and performance difficulties associated with the phenomena of nonuniform fluidized bed gas mixing and bypassing, nonuniform fuel feeding and mixing, bed overtemperature zones, bed agglomeration, ash deposition, particle erosion, materials corrosion, and residual carbon conversion.

The many variations in design configurations and features of the key types of fluidized bed gasifiers and combustors have been developed from fluidization principles, chemical reactor principles, test observations, and empirical rules, coupled with conventional engineering practice. Coals and biomass fuels represent two contrasting extremes in fuel properties and behavior, as well as in industrial application capacity, and have thus evolved fluidized bed processing technologies having somewhat differing configurations and features. Other carbonaceous fuels, having their own unique set of properties and behavior, have preferred fluidized bed processing configurations and features that, to some extent, may be drawn from these two carbonaceous fuel types by analogy.

2 FLUIDIZED BED GASIFIERS

Fluidized bed gasifiers for coals and biomass fuels can, in principle, generate reducing gas, liquid, and char products for a variety of applications. Several fluidized bed gasifier configurations have been proposed that apply well-known principles of bubbling bed and circulating bed fluidization technologies. The fluidized bed gasifier configurations nearing, or at, commercial demonstration primarily generate a fuel gas product and operate in fluidization regimes and with maximum feasible fuel capacities that differ considerably. The design considerations applied to these fluidized bed gasifier configurations reflect these differences, as well as differences in their design fuel properties.

2.1 Fluidized Bed Gasifier Principles

The principles of design and operation of fluidized bed gasifiers for carbonaceous fuels are relatively simple. The challenge is to design the fluidized bed gasifier for reliable operation, meeting its performance requirements while avoiding the consequences of the undesirable phenomena that typify fluidized bed gasification.

2.1.1 Gasifier Products and Applications

Fluidized bed gasification might be applied to carbonaceous fuels to produce steam or process heat, chemicals, or electric power, or combinations of these. The fluidized bed gasifier combines "processed" carbonaceous fuel (dried and sized) with oxidant, steam, and/or an external heat source to produce a "raw reducing gas" stream, a "raw ash" stream, or a "raw char" stream, and possibly a "raw liquid products" stream. Inert gases, such as nitrogen and recycled product gas will also be fed to the gasifier, functioning as solids transport, aeration, and purge gases. These inert gases can contribute significantly to the makeup of the raw reducing gas. Sorbent particles or inert bed makeup particles may also be fed to the gasifier.

The fluidized bed gasifier raw reducing gas stream may be processed further to generate a "syngas" used to produce chemicals, or a "fuel gas" used to generate process steam, process heat, or electric power (Ghate and Longanbach, 1988; Schobert et al., 1998). In some gasification processes, liquid products (tars and oils) might also be extracted from the raw reducing gas as

another chemical synthesis feed stock. The gasifier raw char might be applied for the synthesis of chemicals, or combusted for steam, process heat, or electric power generation. Raw ash may be processed further to extract remaining fuel energy content (e.g., residual carbon) or sensible heat, and to prepare it for disposal or byproduct use.

"Gasification" is used in this chapter as a generic term representing several alternative methods for converting a carbonaceous fuel into a primarily gaseous form having significant heating value (steam reforming, hydrogasification, hydropyrolysis, partial oxidation, carbonization, thermal pyrolysis). Specific gasification alternatives considered in this chapter are designated as "partial oxidation," "carbonization," and "pyrolysis." Fuel gasification based on combining substoichiometric oxidant with carbonaceous fuel to combust partially the char and volatiles is termed either "partial oxidation" or "carbonization," depending on the extent of fixed-carbon conversion that occurs. Gasification under substoichiometric oxidation conditions, intended to produce essentially no char, is "partial oxidation," under conditions intended to produce char as a product it is designated "carbonation."

2.1.2 Functions and Requirements

The functional requirements of the fluidized bed gasifier are to convert efficiently and reliably the carbonaceous fuel into a raw reducing gas stream and a raw char, or raw ash stream meeting all of the product stream compositional, temperature, pressure, and flow rate specifications that are imposed on the gasifier. The fluidized bed gasifier is designed to promote a reaction environment having good gas–particle contacting, good particle–particle mixing, and relatively uniform temperature conditions, and to avoid operational difficulties resulting from the agglomeration, deposition, erosion, and corrosion characteristics associated with the gasification of carbonaceous fuels. The design engineer is challenged to select fuel and sorbent/inert particle feed sizes, fluidization conditions, temperature environment, and gasifier reactant feed stream proportions that avoid the nonidealities common to fluidization (for example, defluidization, gas bypassing, excessive particle attrition and elutriation, particle segregation, generation of hot spots in the dense bed and freeboard, and nozzle and drain plugging).

The "efficiency" of the gasifier conversion might be defined in a variety of ways, based on the energy content of the raw reducing gas, raw liquid products, raw char, or raw ash streams, and the energy content of the carbonaceous fuel feed stream. It is most common to determine the "cold gas efficiency," defined as the ratio of the energy content of the raw reducing gas at 25°C to the carbonaceous fuel input energy and the fractional fuel carbon conversion as measures of the gasifier performance. In some cases, there may also be a specification on the maximum acceptable carbon content of the raw ash stream.

Availability is a measure of the time that the gasifier is able to operate within performance specifications relative to the time that the overall process plant desires to operate. It is a critical factor in the commercial feasibility of the fluidized bed gasifier technology.

2.1.3 Reaction Environment

Fluidized bed carbonaceous fuel gasification features relatively low-temperature, generally less than 1050°C, nonslagging gasification. It is also characterized by nearly uniform gasifier temperature throughout. While fluidized bed gasification can operate effectively with a wide range of carbonaceous fuels, the gasifier operating conditions and design features must be specifically selected to optimize performance with each carbonaceous fuel type.

The simplified, overall reaction steps (some nonstoichiometric relationships) are

Partial combustion of carbonaceous fuel char and volatiles
Carbonaceous fuel {C, H, O, N, S,..., mineral matter} + oxidant {O_2, N_2} → raw reducing gas {CO, CO_2, H_2, H_2O, CH_4, N_2, higher hydrocarbons} + raw reducing gas contaminants {nitrogen compounds (NH_3, HCN), sulfur compounds (H_2S, COS), halogen compounds (HCl), alkali metal compounds (NaCl, NaOH, KCl, KOH)} + liquids {oils and tars} + ash/char {mineral matter, carbon,...}
Heterogeneous char reforming
Char {C, H, O, N, S, ..., mineral matter} + H_2O/CO_2 → CO, H_2, H_2S, ..., ash
Homogeneous gas-phase reactions
$CO + H_2O \rightarrow CO_2 + H_2$
$C + 2H_2 \rightarrow CH_4$
Boudouard reaction
$2CO \rightarrow C(solid) + CO_2$

Steam may be fed to the gasifier in some cases to provide a reaction environment that will promote fixed-carbon reforming and minimize the production of fine carbon particles by the Boudouard reaction. The exothermic partial combustion reactions produce the reaction heat required to heat the input reactant streams and perform the endothermic volatilization and char reforming reactions. The extent of conversion of char into raw reducing gas depends on the operating conditions in the fluidized bed (fuel-to-oxidant feed ratio, fuel-to-steam feed ratio, bed temperature and pressure), and the char particle residence time in the fluidized bed. It has been observed that limestone particles, added as sulfur sorbents, may also catalyze char reforming reactions as well as induce tar cracking (Simell et al., 1992; Agrawal and Haldipur, 1988).

Alternatively, fluidized bed gasifiers may be operated without oxidant supply, having only injected "heat" and steam to "pyrolyze" the carbonaceous fuel, driving off its volatile content and partially reforming the char. The simplified overall reactions are in this case,

Pyrolysis of the carbonaceous fuel
Carbonaceous fuel {C, H, O, N, S, ..., mineral matter} + "heat" → volatiles + tars/oils + char
Heterogeneous char reforming reactions
Char {C, H, O, N, S, ..., mineral matter} + H_2O and H_2 → CO, H_2, CH_4, H_2S, and ash
Homogeneous gas-phase reactions
Volatiles {C, H, N, S, ... + H_2O and CO_2 → CO, H_2, ..., $CO + H_2O \rightarrow CO_2 + H_2$
Boudouard reaction
$2CO \rightarrow C(solid) + CO_2$

The heat input may be "direct" heat input, for example, injected hot solids or hot recycled product fuel gas; or indirect heat input, as by high-temperature heat transfer surfaces located in the fluid bed.

With respect to fluidization phenomena, the fluidized bed will consist of a distribution of char and ash particles, with their relative amounts depending on the carbonaceous fuel reactivity and the gasifier operating conditions. Sorbent or inert particles, if fed to the gasifier, may occupy the greatest portion of the bed. The size distribution of particles in the gasifier bed, and the bed density profile in the gasifier, depend on a number of factors, including the carbonaceous fuel and sorbent/inert particle feed size and density distribution, the attrition properties of the char and sorbent/inert particles after exposure to the gasifier environment, the gasifier axial velocity profile, the existence of high attrition zones (e.g., high-velocity jets) in the bed, and the use of cyclones to recycle elutriated particles.

The gasification reaction phenomena in fluidized beds involve fuel devolatilization followed by the partial combustion of volatiles and char separately with oxygen, and char reforming by H_2O and CO_2 in both the bubble phase and the emulsion phase of the fluidized bed. This multiphase mass transfer and reaction aspect of fluidized bed gasification is an inherent feature of all fluidized bed reactor systems. Carbonaceous fuel devolatilization and combustion have similar, characteristically high initial reaction rates and are thus strongly influenced by rate-limiting fluidized bed mixing and heat transfer phenomena. Char reforming, though, is characteristically orders of magnitude slower than devolatilization and combustion, and it is a controlling reaction step in efficient fluidized bed gasification (Wen and Tonen, 1978).

The fluidized bed may be designed and operated to behave as a bubbling bed with a level of turbulence depending on the fluidization velocity and the particle size and density distribution in the bed, or as a circulating bed having fast fluidization behavior. Recirculation of fines collected by cyclones will generally be used even with bubbling bed gasifiers.

2.2 Status of Fluidized Bed Gasifier Technology

Gasification of carbonaceous fuels is a relatively old technology. During the twentieth century, many fluidized bed gasifiers have been proposed to gasify coals, biomass fuels, petroleum products, and other carbonaceous fuels for the purpose of producing liquid fuels, synthetic natural gas, syngases, chars, and fuel gases. Some low-pressure fluidized bed biomass fuel gasifiers are considered commercial.

Hebden and Stroud (1981) have reviewed many of the fluidized bed coal gasifiers under development during the 1970s. Similarly, Seglin and Bresler (1981) review fluidized bed low-temperature coal pyrolysis technology of the 1970s. Low-temperature pyrolysis in fluidized bed reactors for the purpose of tar/liquid and coke generation has not achieved commercial status, and only limited activities are currently progressing (Ghate and Longanbach, 1988).

The coal-fueled fluidized bed gasifier types nearing demonstration and commercialization, and their applications, are characterized in Table 1. These represent both operating facilities and projects in design or construction. The three types of gasification are represented: partial oxidation, carbonization, and pyrolysis, and bubbling bed fluidization is the prevalent regime.

Table 2 presents a similar characterization of biomass-fueled gasifiers nearing demonstration and commercialization. Partial oxidation at both elevated pressure and low pressure are included, and both bubbling bed and circulating bed fluidization are feasible regimes. As with coal, the biomass gasifiers are primarily fuel gas applications for steam or combustion turbine power.

Figure 1 illustrates the general schemes representative of the fluidized bed coal gasification processes of significance today that are listed above, and Fig. 2 represents the key biomass fuel gasification processes. The process schemes shown are oversimplified and show only the major streams, gasifier components, and process interfaces relating to the primary fluidized bed conversions. Gasifier stream "interfaces" are identified in the figures to emphasize the critical need for satisfying all interface requirements. The figures also indicate "optional" streams that are used depending on the specific fuel properties and the application needs. Nonmechanical valve locations for solids recycle streams are identified, without aeration gas being shown. In general, feed streams carrying solids also require an "inert" transport gas (for example, nitrogen or recycled product gas), and these transport gases are not included in either Fig. 2 or 3. Inert gases are also used at various places within the gasifiers for purging, aeration, local fluidization, and stripping, and these also are not shown in the figures.

Three schemes are shown in Fig. 1: (A) pressurized bubbling bed partial oxidation of coal; (B) pressurized bubbling bed carbonization of coal; and (C) low-pressure bubbling bed pyrolysis of coal. Figure 1A is representative of pressurized bubbling bed partial oxidation of coal. The sketch indicates that the bubbling bed fluidized bed gasifier vessel cross-section is expanded in the freeboard to minimize particle elutriation, and this is generally true of all bubbling bed coal gasifiers. Elutriated fines recycling is used in general, with nonmechanical valves to seal the recycle system and to control the reinjection of fines into the fluidized bed gasifier. The raw fuel gas will have a low heating value of about 3.3–5.6 MJ/Nm^3 (LHV) for air-blown operation, and a medium heating value of about 8.2–11.2 MJ/Nm^3 (LHV) for oxygen-blown operation. The vessel will drain a low-carbon ash product.

The gasifier must interface with the pressurized oxidant delivery system (air or oxygen-blown gasifiers being feasible for coals); with the pressurized coal and sorbent preparation and feed systems; with the raw fuel gas pressurized processing system; and with the pressurized ash processing system. Coals may be

Table 1   Leading Coal Fueled Gasifier Systems

Gasifier type/Fluidization type/Pressure mode: Partial oxidation/Bubbling bed/Elevated

Developer	Applications	Description	References
Kellogg KRW gasifier	Fuel gas for combustion turbine power	20,000 MJ/h development unit testing completed on wide range of coals; 800,000 MJ/h coal demonstration plant startup initiated 1998; air-blow operation; 2100–2400 kPa operating pressure	Sierra Pacific Power Co., 1994; Demuth and Smith, 1998
IGT U-gas gasifier	Coke oven fuel gas	105,000 MJ/h commercial plant operational; air-blow operation; 160–220 kPa operating pressure	Bryan and Hoppe, 1998
Rheinbraum AG HTW gasifier	Fuel gas for combustion turbine power; syngas for methanol	Lignite operation only; 450,000 MJ/h demonstration project completed; air and oxygen-blown operations	Renzenbrink et al. 1998

Gasifier type/Fluidization type/Pressure mode: Carbonization/Bubbling bed/Elevated

Developer	Applications	Description	References
Foster Wheeler Development Corp.	Fuel gas and char for combustion turbine power	10,000 MJ/h development unit testing completed on range of coals; 50,000 MJ/h pilot plant construction completed 1998; air-blow operation	Robertson, 1995
British Coal	Fuel gas for turbine power; char for steam generation	40,000 MJ/h pilot plant operation completed 1992; air-blow operation; integrated gas cleaning and turbine	Dawes et al., 1997

Gasifier type/Fluidization type/Pressure mode: Pyrolyzer/Bubbling bed/Low

Developer	Applications	Description	References
Tsinghua University	Fuel gas for town gas; char for pyrolysis heat and steam	5,000 MJ/h development unit operated on Chinese coals; air-blown operation; 100,000 MJ/h demonstration plant in design	Lu et al., 1996

fed in either dry or paste form; the paste form is claimed to improve feed reliability and cost. If the application is to provide a pressurized fuel gas to fire a combustion turbine, the raw fuel gas must be well cleaned of particulate and alkali species to protect the combustion turbine. If the pressurized fuel gas is used to fire a process heater, such high levels of cleaning may not be needed. Coals will generally result in a raw fuel gas that needs to be desulfurized, and the fuel gas ammonia content may be a concern for stack gas NOx emissions.

Figure 1B is representative of pressurized bubbling bed carbonization of coal. Here, cyclone fines recycling is not used, and the carbonizer vessel and cyclone will drain the raw char product. The gasifier must interface with the pressurized air delivery system; with the pressurized coal and sorbent preparation and feed systems; with the raw fuel gas pressurized processing system; and with the pressurized char processing system. Coals may be fed in either dry or paste forms. The raw fuel gas will have a low heating value of about 3.3–5.6 MJ/Nm^3 (LHV), similar to the raw fuel gas issued from the pressurized air blown bubbling bed partial oxidation gasifier.

The third scheme shown, Fig. 1C, is a low-pressure bubbling bed coal pyrolysis process based on a parallel circulating bed char combustor that provides hot circulating solids to the gasifier. The bubbling bed pyro-

Table 2 Leading Biomass-fueled Gasifier Systems

Gasifier type/Fluidization type/Pressure mode: Partial oxidation/Bubbling bed/Elevated

Developer	Applications	Description	References
Carbona Corp (IGT Renugas gasifier)	Fuel gas for combustion turbine power	100,000 MJ/h pilot unit operated on range of biomass fuels; integrated gas cleaning; 750,000 MJ/h alfalfa-stem plantation unit in design	DeLong et al., 1995
Cratech gasifier	Fuel gas for small combustion turbines	10,000 MJ/h pilot unit operating on cotton gin mill waste; integrated gas cleaning; 100 kW turbine operation planned	Craig and Purvis, 1998

Gasifier type/Fluidization type/Pressure mode: Partial oxidation/Circulating bed/Elevated

Developer	Applications	Description	References
Foster Wheeler Energia Oy Bioflow gasifier	Fuel gas for combustion turbine power	70,000 MJ/h demonstration testing ongoing since 1995; integrated with 6 MWe combustion turbine operation; wood chip fuel	Stahl and Neergaard, 1998

Gasifier type/Fluidization type/Pressure mode: Partial oxidation/Circulating bed/Low

Developer	Applications	Description	References
Lurgi gasifier	Fuel gas for process heat; fuel gas for combustion turbine power	350,000 MJ/h unit for cement kiln fuel gas in operation; 160,000 MJ/h unit for combustion turbine in design; air-blown operation	Hirschfelder and Vierrath, 1998; DeLange and Barbucci, 1998
Foster Wheeler Energia Oy	Fuel gas for process heat and steam	55,000–230,000 MJ/h commercial units in operation; lime kiln process heat and steam; air-blown operation	Nieminen and Kivila, 1998
Termiska Processor AB (TPS) gasifier	Fuel gas for process heat and steam; fuel gas for combustion turbine power	Two, 50,000 MJ/h units in commercial operation for steam generation; 300,000 MJ/h eucalyptus unit designed for combustion turbine plant in Brazil; 100,000 MJ/h wood chip unit for combustion turbine in UK in construction	Barducci et al., 1995; Pitcher et al., 1998

Gasifier type/Fluidization type/Pressure mode: Pyrolysis/Circulating bed/Low

Developer	Applications	Description	References
Battelle gasifier	Fuel gas for process heat and steam; fuel gas for combustion turbine power	145,000 MJ/h wood chip in operation for steam generation; future conversion for combustion turbine operation planned	Farris et al., 1998

lyzer is fluidized by steam, and its design and operation must be closely integrated with the circulating bed combustor. The pyrolysis gasifier must also interface with low-pressure coal and sorbent preparation and feed systems. The low-pressure raw fuel gas can be used for process heat or for steam generation, and it must be processed to meet the requirements of those applications. The raw fuel gas may also be cooled, scrubbed and compressed to act as a combustion turbine fuel gas. The product fuel gas, after cooling and steam condensation, will have a medium heating value of about 12–16 MJ/Nm^3 (LHV). The fluidized bed

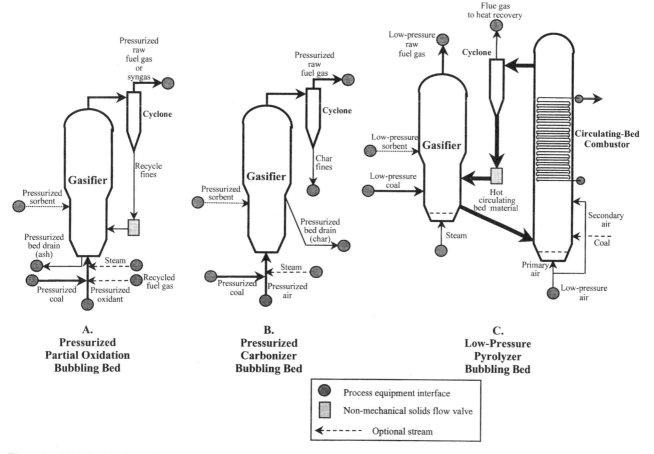

Figure 1 Fluidized bed, coal gasification schemes nearing demonstration.

combustor produces substantial steam for export, and a separate coal stream may be fed to the fluidized bed combustor to raise additional steam.

Figure 2 shows representations of biomass fuel fluidized bed gasifiers. Four types are illustrated: (A) pressurized bubbling bed partial oxidation of biomass fuels; (B) pressurized circulating bed partial oxidation of biomass fuels; (C) low-pressure, circulating bed partial oxidation of biomass fuels; and (D) low-pressure circulating bed pyrolysis. For all of these, proper biomass fuel drying and sizing is critical to their successful operation.

Figure 2A is representative of pressurized bubbling bed air blown partial oxidation of biomass fuel. The sketch indicates that fines recycling might be used, with nonmechanical valves to control the reinjection of fines into the fluidized bed gasifier. The vessel will drain a low-carbon ash product. The sketch also suggests that overbed air injection might be used for fuel gas partial oxidation as a means for fuel gas tar destruction. The

gasifier must interface with the pressurized air delivery system (only air being considered feasible with biomass fuel gasification applications); with the pressurized fuel and inert feed systems; with the raw fuel gas pressurized processing system; and with the pressurized ash processing system. Biomass fuels will be fed only in dry form.

The application is to provide a pressurized fuel gas to fire a combustion turbine, and the raw fuel gas must be well cleaned of particulate and alkali metal species to protect the combustion turbine. Here, it is expected that tar generation will be small, and special provision for tar cracking or removal will not be needed. The fuel gas must be cooled to the extent needed to achieve this cleaning, probably to a maximum temperature of 650°C. Fuel gas ammonia content may be a particular concern for NOx stack emissions when using some biomass fuels. Biomass fuels generally will not require fuel gas desulfurization, but some agricultural plantation crops do have significant sulfur content and may

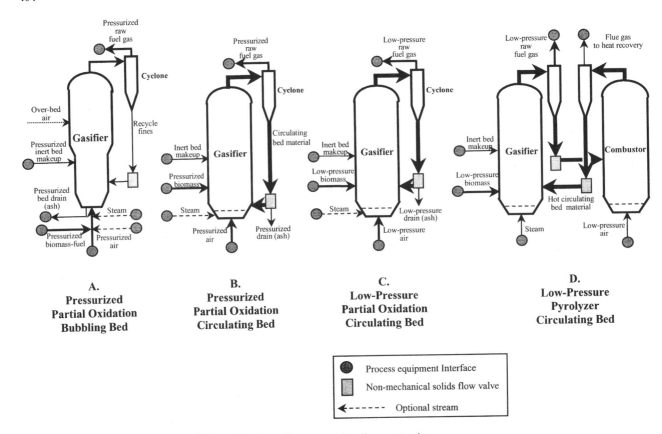

Figure 2 Fluidized bed, biomass fuel gasification schemes nearing demonstration.

require gas desulfurization. The raw fuel gas will have a low heating value similar to those raw fuel gases produced in pressurized air blown coal gasifiers.

Figure 2B is a pressurized circulating bed air blown partial oxidation of biomass fuel. Here biomass fuel is fed as a dry solid into a circulating bed of inert particles. While circulating bed gasification may not be practical for most coals, the high reactivity of biomass fuels makes circulating bed gasification an effective process. The raw fuel gas must be processed so that it can serve as a combustion turbine fuel gas. Again, it is expected that tar generation will be small, and special provision for tar cracking or removal will not be needed. Large bed circulation rates are used to maintain vigorous gas–particle mixing and nearly uniform temperatures throughout the gasifier. The raw fuel gas heating value and carbon utilization will be in the same range as the other air blown gasifiers.

The third scheme, Fig. 2C, is low-pressure (near atmospheric pressure) circulating bed air blown partial oxidation of biomass fuel. Here low-pressure biomass fuel is fed as a dry solid into a circulating bed of inert particles fluidized by low-pressure air and steam. The

process challenges of pressurizing and feeding biomass fuels at low pressure are much less than in the two previous high-pressure schemes. The low-pressure raw fuel gas may be cooled to reduce hot gas piping expense and used to fire a conventional boiler–steam generator with little gas cleaning. Alternatively, the raw fuel gas might be cooled and scrubbed so that it can be compressed to provide a combustion turbine fuel gas. It is expected again that tar generation will be small in most cases.

The fourth scheme, Fig. 2D, is a low-pressure biomass pyrolysis process based on parallel circulating beds, one a pyrolyzer and the other a char combustor. The high reactivity of biomass fuels make this process feasible in relatively compact vessels. The raw fuel gas will have a medium heating value of about 13.0–16.7 MJ/Nm^3 (LHV). The low-pressure raw fuel gas can be used to fire a conventional boiler after cooling, with little fuel gas cleaning; or it may be cooled, scrubbed, and compressed to act as a combustion turbine fuel gas. The concentrated pyrolysis fuel gas will generally contain relatively high tar content, and tar cracking, or recovery, will need to be incorporated into the raw fuel

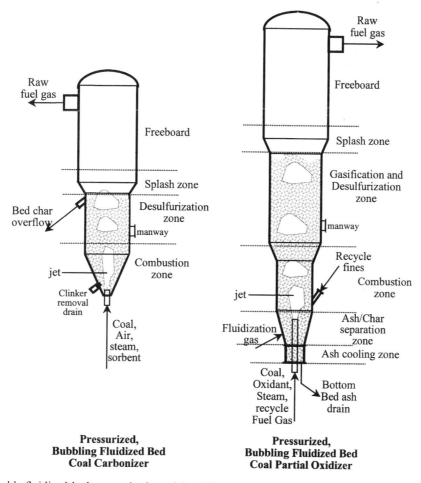

Figure 3 Comparable fluidized bed pressurized partial oxidizer and carbonizer.

gas processing system for the combustion turbine application. Fuel gas tar content may also be a process concern for the boiler fuel gas application.

New techniques for pyrolysis of biomass fuels by indirect heat input are currently at various stages of development and demonstration for small-capacity fuel gas supply applications: the BrightStar Technologies low-pressure steam fluidized pyrolyzer using fuel firing to heat the windbox (Chau and Rovner, 1995); the MTCI low-pressure steam fluidized pulse combustion pyrolyzer with immersed heat transfer surfaces in the bed (Monsour et al., 1995); and the Iowa State University cyclic (gasification–oxidation), steam fluidized pyrolyzer with high-temperature phase change materials contained in tubes immersed in the bed (Stobbe et al., 1996). These new pyrolyzer configurations all have significant materials, scale-up, and process integration challenges to overcome.

Generally, the bubbling bed gasifiers have mean bed particles characteristic of Geldart group B classification (Geldart, 1973), with no overlap into the A or D groups. The pressurized bubbling bed gasifier appears to approach the turbulent fluidization regime much more closely than does low-pressure bubbling bed gasification, indicating the potential for more vigorous fluidization, but also more extensive particle elutriation. The low-pressure and pressurized circulating bed gasifiers operate well above the terminal velocity, although the low-pressure operating region indicates that low-pressure circulating bed gasifiers may operate very near the transition between turbulent and fast fluidization.

The single gasifier module maximum fuel capacity is sometimes applied as an indicator of the feasibility of competing configurations. A key factor influencing the single gasifier module fuel capacity of fluidized bed gasification technologies is the cross-sectional area of the fluid bed gasifier. The major parameters controlling the cross-sectional area of the fluid bed gasifier are

The oxidant type (air or oxygen), or pyrolysis heat source type
Operating pressure: low-pressure or pressurized
Gas velocity in the freeboard
Fuel properties

Fluidized bed gasifiers are generally cylindrical vessels, even for low-pressure operation. A 5 m vessel diameter is about the limit for shop fabrication and road/rail shipment of pressure vessels. The fluidized bed gasifier vessels require either field fabrication or the use of multiple parallel gasifiers, as their fuel thermal input capacity exceeds limiting capacity.

The pressurized circulating bed biomass fuel partial oxidation gasifier is the most compact of the gasifiers; and the low-pressure circulating bed biomass fuel pyrolysis gasifier is relatively compact despite its low operating pressure. The pressurized, bubbling bed air blown coal partial oxidation gasifiers are fairly compact, with the oxygen blown gasifier having very large capacity in a single module. The low-pressure bubbling bed coal pyrolysis gasifier is very large in diameter and will have only limited coal capacities.

2.3   Fluidized Bed Gasifier Design Considerations

The design of fluidized bed gasifiers requires that transport models be developed that apply the conservation principles for mass, energy, and momentum to some degree of fundamental detail. The mass and energy balances are closely coupled, and their solutions, applying empirical reaction kinetic and multiple phase mixing models, provide estimates of solids and gas composition and temperature profiles and input and output stream conditions. The momentum balances provide the gasifier pressure profile and total pressure drop when applied with appropriate fluidized bed phase density models. These empirically based models provide the means needed to size the gasifier equipment and scale pilot plant data.

The design of fluidized bed gasifiers involves the selection of several interrelating design, operating, and performance parameters and requires the consideration of performance and cost trade-offs for any specific application. The type of fluidized bed gasifier, the fuel properties, and the application requirements strongly influence the fluidization parameter values selected for the design. To make such designs properly, for all of the types of gasifiers, fuels, and applications described in this chapter, it is currently necessary to apply highly empirical fluidized bed reactor models that are closely calibrated with pilot plant test data, thermodynamic estimates, and cold flow modeling observations (Yang et al., 1995). Iterative design codes are needed to consider all of the design parameters, fuel properties, process performance requirements, and trade-offs.

2.3.1   Gasification of Coal

Figure 3 represents a pressurized coal fueled bubbling bed partial oxidizer and a carbonizer of comparable fuel thermal input capacity, showing their major components and features, and relating to the respective process schemes shown in Figs. A and B. The gasifier vessels, like all fluidized bed gasifier vessels, are uncooled refractory lined pressure vessels having appropriately located nozzles for gas and solids inlet and outlet flows and manways for inspection and maintenance.

The bubbling bed partial oxidizer is separated axially into several perceived functional zones: the combustion zone, the gasification zone, the ash/char separation zone, and the freeboard zone. Each of these zones is provided with a diameter resulting in acceptable fluidization and mixing at the local flow conditions, and a height sufficient to perform the zone functions effectively.

When gasifying caking coals, the coal is introduced into the combustion zone via a vertical jet, operating at 25–50 m/s nozzle velocity, that intensely mixes the feed streams of oxidant, coal, recycled fuel gas, and steam with internally circulating char, ash, and sorbent particles entrained into the jet. This results in the complete and rapid consumption of oxygen, the pyrolysis and partial combustion of char and volatiles, and the generation of a hot zone in the jet that allows ash agglomeration, if it is desirable with the coal being gasified. Noncaking coals may be fed more simply by mechanical or pneumatic means at the vessel wall, since coal agglomeration is not a concern in this case, but high reactivity coals might still be fed through a central jet to avoid nonuniform volatiles release in the bed. The oxidant may also be injected as jets from nozzles distributed around the vessel wall but must always be injected with vigorous mixing and solids entrainment to avoid local hot spots and excessively large ash agglomerates or deposits.

In the ash/char separation zone, if it is desirable to incorporate such a zone, controlled fluidization results in efficient segregation of the larger ash agglomerates from the char particles, acting as a mechanism to maintain a high char content in the bed but a low carbon content in the drained ash. An alternative ash–char separation method is pneumatic stripping of char from the draining ash agglomerates. Bottom bed drainage is performed with countercurrent cooling of the draining ash and heating of injected gases. Bed drainage must be controlled to provide and maintain sufficient gasifier bed depth. The feeding of limestone to the gasifier bed catalyzes the steam/carbon dioxide–char reforming reactions and makes char–ash separation less important, with the carbon content in the bed being lowered. Coals having high reactivity, such as many subbituminous coals and lignites, may not require ash agglomeration or special ash–char separation features to be used efficiently, and may be gasified at lower bed temperatures.

The gasification zone is an expanded lower velocity zone above the combustion zone where the slower char–steam/carbon dioxide reforming reactions occur that result in efficient utilization of the coal char. The reaction conversion of this zone is greatly influenced by the large bubbles issued from the combustion zone jet. The freeboard is both a disengaging zone for the very large bubbles that break the bubbling bed surface and a zone for homogeneous fuel gas conversions to occur. Tar formation is generally relatively small, but various ash-related deposits can form within the gasifier vessel.

The recycle cyclone, if this is used to reduce carbon losses, must be placed at a sufficient elevation above the relatively dense fluidized bed to provide a sufficiently long standleg to balance the circulation loop pressure profile. The bed elutriation will tend to result in recycled fine particles having low bulk density compared to the dense fluidized bed. Fine recycled particles must be injected into the combustion zone effectively to provide additional carbon conversion.

Figure 3 shows that the equivalent pressurized air blown bubbling bed coal carbonizer vessel might be about half the total height of, and slightly smaller in diameter than, the coal partial oxidation vessel due to its primary objective of generating a char product rather than a low-carbon ash. Similar central jet feeding of coal may be used into a combustion zone, but segregated char–ash draining is not required. Simpler top bed overflow drainage can be used to control the bed height. Generally there is also no need to recycle overhead fines back into the carbonizer vessel, this stream being primarily product char. The fluidized bed carbonizer may operate at a lower temperature than the fluidized bed partial oxidizer. Thus the bubbling bed coal carbonizer design is significantly simpler than the bubbling bed coal partial oxidizer design.

Several design decisions and estimates that must be addressed to generate a reliable and efficient pressurized bubbling bed coal partial oxidizer or carbonizer are

Fluidized bed temperature—this represents a trade-off between carbon utilization, oxidant and steam consumption, raw fuel gas heating value, and the undesirable possibilities of deposits, agglomerates, nozzle and drain plugging, and excessive tar and ammonia generation.

Coal and sorbent feed size distributions.

Oxidant/coal, steam/coal, recycle fuel gas/coal feed ratios

Gasification and combustion zone fluidization velocity

Freeboard zone velocity and height to minimize particle carryover—a trade-off influencing equipment cost, particle recycle system capacity, and gasifier performance.

Oxidant injection design.

Steam and recycle fuel gas distribution for local fluidization to avoid dead zones in the bed and to minimize local hot spots resulting in uncontrolled ash agglomeration.

Coal feed location(s) and transport method.

Recycled fine particles feed location and transport method.

Combustion zone height and gasification zone height required for effective partial oxidation, carbonization, and desulfurization.

Special features for limiting bed drainage carbon losses, if needed to meet performance requirements.

Overall gasifier and recycle loop pressure drop profiles and heat losses.

Effective turndown and control methods.

Design to accommodate multiple fuel types.

Matching gasifier input and output conditions and requirements with compatible capabilities at interfacing equipment and systems, and achieving overall gasifier–process integration.

Low-pressure bubbling bed pyrolysis of coal, as shown schematically in Fig. 1C, must apply many of the same types of design considerations as those listed above for pressurized bubbling bed coal partial oxidation, but it differs in its characteristics, performance, and behavior expectations in several ways.

The performance of the low-pressure bubbling bed pyrolysis gasifier will be extremely sensitive to the volatile content of the coal and its char–steam reforming reactivity.

The coal feed size distribution may need to be restricted in its top and bottom sizes, for example, limited to 0.5–3.2 mm, to avoid fine particle losses and to obtain good pyrolysis conversion.

No oxidant supply is required, and fluidization is provided by steam, as well as by the evolution of fuel gas from the pyrolyzing coal particles—a major trade-off exists between factors relating to steam consumption and fluidization velocity.

The pyrolysis gasifier temperature (greater than 800°C for most coals) is controlled by the char fueled circulating bed combustor temperature (greater than 900°C) and the solids circulation rate, and the gasifier bed temperature may be limited to lower than optimum values by ash agglomeration limitations in the circulating bed combustor.

The circulating bed media may be coal ash particles or limestone particles, depending on the process desulfurization needs—the limestone reaction conversions in this cyclic oxidizing–reducing environment may have reaction heat effects that influence the process performance.

The low-pressure bubbling bed pyrolysis unit will be much shallower than the pressurized bubbling bed partial oxidation gasifier to meet its pyrolysis conversion and pressure drop constraints.

Rapid mixing of hot circulating solids, fed above the pyrolysis bed, with the coal feed, is required, as well as good distribution of coal to the pyrolysis bed so that nonuniform volatile release to the freeboard does not occur.

Greater tar and ammonia content in the bubbling bed pyrolysis fuel gas is expected than in the pressurized bubbling bed partial oxidation coal gasifier, although this may be moderated by the large steam content.

The oxidizing–reducing interfaces between the gasifier and combustor must be properly sealed and purged.

2.3.2 Gasification of Biomass Fuels

A pressurized bubbling bed partial oxidizer for biomass fuel, shown schematically in Fig. 2A, will be designed much like the pressurized bubbling bed coal partial oxidizer in Fig. 3. The higher reactivity of the biomass fuel relative to coal, and the probable elimination of in-bed desulfurization with biomass fuels, might result in shorter combustion and char gasification zones if the gasifier operating temperature is not limited to low values due to bed agglomeration potential. The high reactivity of biomass fuels makes relatively uniform fuel feed distribution even more important than it is with coals. Reactive biomass fuels may not require the use of fine particle recycling or char–ash separation features. Secondary air might be injected into the freeboard of the biomass fuel gasifier to promote tar destruction through partial fuel gas combustion. The biomass fuel bubbling bed would consist largely of inert sandlike particles containing small amounts of biomass fuel char and ash.

The same design considerations must be applied with biomass fuels as with coals. Biomass fuel gasification is relatively simple due to the generally highly reactive nature of biomass fuels. The greatest risk in biomass fuel gasification is associated with ash, or ash-inert particle agglomeration within the bed that may lead to defluidization and plugging of nozzles and drains. With the tendency for biomass fuel bed agglomeration, relatively low partial oxidation temperatures may need to be selected that result in relatively high carbon losses. The tendency for greater formation of tars in the raw fuel gas, and increased gasifier drain material carbon content, as the temperature is decreased, must be weighed against the tendency for greater bed agglomeration, and possibly greater alkali vapor release, as the gasifier temperature is increased. The reliable, controllable preparation, pressurization, and feeding of biomass fuels are significant challenges to be fulfilled to make the technology a commercial success.

Pressurized circulating bed air blown partial oxidation of biomass fuels, shown schematically in Fig. 2B, is feasible because of the high reactively of biomass fuels. Proper biomass fuel drying and fuel and inert bed material particle size control (size reduction or pelletization) are important to the gasifier performance. The biomass fuel is fed several meters above the air distributor to a relatively low-density zone of the fast fluidized bed. The initial pyrolysis of the biomass fuel, with some partial combustion of the volatiles and char, will occur rapidly near the fuel injection point, and the vessel diameter may be expanded at this point owing to the large evolution of biomass fuel volatiles. Steam reforming of char particles will continue throughout the fast bed riser, with the conversion becoming less effective at the lower density higher elevations of the fast bed. The recirculating solids, containing unconverted char and sand, are injected into the turbulent, dense fluidized bed at the base of the vessel, near the air distributor, so that unconverted char can be burned.

Low-pressure circulating bed partial oxidation of biomass fuels, shown schematically in Fig. 2C, is conceptually very similar in design to the pressurized circulating bed partial oxidizer. It faces similar performance constraints and design considerations to those for the pressurized circulating bed partial oxidizer. The low-pressure partial oxidizer fluidization and reaction kinetics are greatly influenced by its lower pressure operation, but it can operate effectively with fuel and inert material feed particle sizes, velocities, temperatures, and bed depths that are selected to provide acceptable performance.

Low-pressure circulating bed pyrolysis of biomass fuels, shown schematically in Fig. 2D, differs significantly from coal pyrolysis in that the volumetric evolution of fuel gas from the pyrolyzing biomass fuel particles is sufficient to fluidize the inert bed media in the gasifier vessel. Only local fluidization with steam, or nitrogen, is needed to avoid defluidized, nonmixing locations in the vessel. The circulating bed pyrolyzer and the char fueled circulating bed combustor are both adiabatic reactors. A balance is required between the carbon conversion in the pyrolyzer at its selected operating temperature, the carbon feed rate to the circulating bed combustor and its operating temperature, and the solids circulation rate between the vessels. The relative complexity of this thermal and chemical balance scheme, and its high potential for raw fuel gas, are design issues. Proper sealing and stripping at the reducing–oxidizing interfaces is required.

3 FLUIDIZED BED COMBUSTORS

Fluidized bed combustors for coals and biomass fuels are applied for steam generation, process heat supply, and electric power generation. They apply the principles of bubbling fluidized beds, and circulating fluidized beds, and they operate at both atmospheric pressure (AFBC) and elevated pressures (PFBC). AFBC configurations, both bubbling bed and circulating bed, have achieved commercial status, and PFBC is in an early commercial scale-up phase. The design considerations and the characteristics of fluidized bed combustors for coals and biomass fuels reflect the significant differences between these two types of fuels.

3.1 Fluidized Bed Combustor Principles

Fluidized bed combustors burn carbonaceous fuels to produce steam, process heat, or electric power. The major distinction of fluidized bed combustors compared to other combustor types is that carbonaceous fuel combustion occurs at a relatively low temperature of 760–930°C, within an almost uniform temperature reactor environment. Carbonaceous fuel is fed into a fluidized bed consisting primarily of sulfur-sorbent (limestone) particles, or inert particles, and containing smaller portions of fuel char and ash particles. The bed is fluidized by air, fed in excess of that needed for complete fuel combustion, to perform combustion within the relatively well-mixed, relatively uniform temperature of the fluidized bed. The fluidized bed combustor may operate at near-atmospheric pressure for steam generation, process heat production, or Rankine cycle electric power generation; or at elevated pressure for combined cycle power generation.

The fluidized bed temperature, high enough to perform efficient combustion and sulfur removal, but low enough to avoid bed agglomeration, is maintained by rapidly removing heat to generate steam through heat transfer surfaces immersed in the fluidized bed. A "raw combustion gas" is produced that is processed, primarily removing entrained particles, so that the "product combustion gas" can have further energy extracted from it. This gas processing is considerably less complex than the gas processing associated with fluidized bed gasification. Bed drain solids, a mixture of the carbonaceous fuel ash and sulfur-sorbent products, are produced that may require further processing (heat removal, chemical/physical processing) before disposal or by-product use.

The functional requirements for fluidized bed combustors are generally more complex than those for fluidized bed gasifiers: they must efficiently and reliably combust the carbonaceous fuel, achieve specified sulfur removal performance with economic consumption of sorbent, generate a raw combustion gas meeting flow, composition (SO_2, NO_x, CO, O_2, particulate), and temperature specifications, generate steam meeting flow, pressure, temperature, and quality requirements, and generate a raw ash stream meeting its flow, composition, and temperature constraints. The fluidized bed combustor's overall thermal efficiency is the ratio of the recoverable thermal energy content of the output streams (steam, raw combustion gas, raw ash) to the carbonaceous fuel input energy. This thermal efficiency is strongly influenced by the carbonaceous fuel properties and the nature of the process application and is sensitive to unburned carbon losses. The combustion inefficiency is the fuel value of the waste solids streams (primarily based on their carbon content) over the carbonaceous fuel input.

While fluidized bed combustion may be performed in a staged-combustion arrangement, having both reducing and oxidizing zones, the overall reaction steps are representative of oxidizing conditions (non-stoichiometric relationships):

Combustion of carbonaceous fuel

Carbonaceous fuel {C, H, O, N, S, ..., mineral matter} + air {O_2, N_2} → raw Gas {CO_2, H_2O, N_2, excess O_2} + raw gas contaminants {CO, SO_2, SO_3, NO_x, halogen compounds [HCl], alkali metal compounds (NaCl, Na_2SO_4, KCl, K_2SO_4)} + ash {mineral matter, residual carbon}

Limestone sulfation

Limestone {($CaCO_3$, $MgCO_3$, mineral matter } + SO_2 + O_2 → $CaSO_4$, CaO, $CaCO_3$, MgO, CO_2

The fluidized bed combustion phenomenon involves carbonaceous fuel devolatilization followed by the combustion of volatiles and char separately with oxygen in the dense bed bubble and emulsion phases, and within the dilute phases of the fluidized bed (Leckner, 1998). In general, both reducing and oxidizing regions exist within the fluidized bed combustor even with vigorous mixing, especially in the vicinity of carbonaceous fuel feed points. Similarly, limestone sulfation involves calcination and contacting of limestone particles with sulfur oxides and oxygen in the bed emulsion and bubble phases.

Fluidized bed combustors may be designed and operated as either bubbling fluidized beds or as circulating fluidized beds. They may also be operated near atmospheric pressure (AFBC) or at elevated pressures (PFBC) where the product combustion gas is expanded through a turbine to generate electric power.

Bubbling bed combustors normally operate with Geldard group B particles that provide generally "slow bubble" behavior with good bubble-to-emulsion phase gas mixing. Circulating bed combustors, with finer particle sizes and higher velocities, operate in the fast fluidization regime, providing excellent gas-particle mixing due to the large slip velocity between gas and particle clusters characteristic of fast fluidization.

The pressurized bubbling bed combustors (bubbling PFBC) appear to approach the turbulent fluidization regime much more closely than do the low-pressure bubbling bed combustors (bubbling AFBC). The low-pressure and pressurized circulating bed combustors (circulating AFBC and circulating PFBC) operate well above the terminal velocity, although circulating AFBC may operate just above the transition between turbulent and fast fluidization.

The low-pressure and pressurized fluidized bed combustors operate in similar regimes, but with different fluidization characteristics with regard to mixing and particle elutriation and entrainment. Changes in operating pressure also result in significant changes in combustion and sulfation reaction kinetics. Devolatilization and combustion, being kinetically rapid even at the relatively low temperature of fluidized bed combustors, are generally limited by large-scale mixing phenomena in the dense bed and dilute regions (gas-gas, and gas-particle mixing). Combustion of CO and residual carbon represent the limiting reaction steps.

Figure 4 illustrates the bubbling bed and circulating bed configurations and is representative of both AFBC and PFBC. The process schemes shown are oversimplified and show only the process steps relating to the fluidized bed conversions. Bubbling bed combustors are depicted in the figure to be larger in cross section but shorter in height than circulating bed combustors. This results from several factors: bubbling bed combustors operate at much lower fluidization velocities, and with much denser beds than circulating bed combustors; bubbling bed combustors normally pass all of the combustion air through the bubbling bed, while circulating bed combustors use staged combustion, where only about half of the combustion air is used as primary air in a primary turbulent bubbling bed, and the remaining air is injected into the fast bed zone as secondary air.

Bubbling bed combustors conduct almost all of their combustion and, simultaneously, considerable heat transfer within the relatively dense bubbling bed. The freeboard is primarily used to complete CO and volatiles combustion and to disengage solids from the raw combustion gas before the gas reaches the convective heat transfer surface. Elutriated particles that are captured by cyclones may be recycled, at relatively low temperature, to the fluidized bed combustor.

In contrast, circulating bed combustors perform partial combustion and devolatilization in the primary, reducing bed zone, and complete combustion in the relatively dilute, but well mixed, fast fluidized bed zone. Heat transfer surface is distributed throughout the secondary fast fluidized bed combustor and the solids recycle leg (water-cooled cyclone and external heat exchanger), requiring relatively large furnace heights similar to those of pulverized fuel-fired combustors. Solids circulation rates are very large, so that little temperature difference occurs across the full height of the circulating bed.

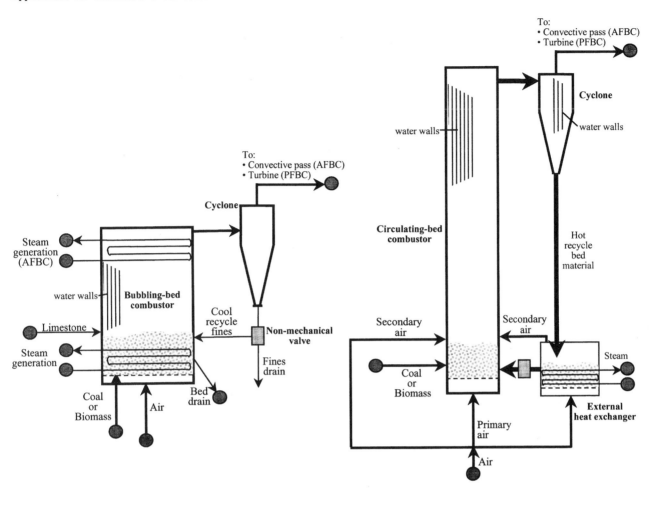

Bubbling-Bed FBC   **Circulating-Bed FBC**

Figure 4 Fluidized bed combustion process schemes.

In AFBC applications, the convective heat transfer surface is arranged so that maximum steam generation is accomplished by cooling the product combustion gas down to the lowest permissible stack temperature. In PFBC operation, the product combustion gas is maintained at its highest permissible temperature, that of the fluidized bed combustor, so that maximum turbine power and efficiency can be obtained. PFBC operates at bed temperatures (815–870°C) much lower than conventional combustion turbine inlet temperatures (1150–1430°C), so the turbine performance in PFBC is relatively poor, but it still contributes significantly to the power plant generating output and efficiency.

Another form of PFBC, topped-PFBC, has been devised to maximize the gas turbine performance. In topped-PFBC, a fluidized bed carbonizer (see Sec. 3), operated at 815–930°C, generates a low–heating value fuel gas and a char product. The char is combusted in a pressurized fluidized bed combustor at 870–930°C, producing a hot combustion product gas stream containing excess oxygen. Both fluid bed reactors use in-situ sulfur removal with limestone. The carbonizer fuel gas and the fluidized bed combustor product gas combine in a low-NO$_x$ topping combustor, resulting in an expansion gas temperature characteristic of modern heavy-duty gas turbines. The two parallel gas streams from the carbonizer and combustor are cleaned at, or near, their generation temperatures so that the combined streams meet both the gas turbine protection requirements and the power plant environmental emission standards. The hot gas cleaning system performs particle removal and alkali removal functions, if alkali metal content is a concern.

3.2 Status of Fluidized Bed Combustors

AFBC is a commercial technology that is widespread across the world. Coal-fueled AFBC competes with conventional stoker-fired boilers, with practical capacity of 20,000 to 300,000 MJ/h fuel energy input, all the way up to the lower end of base-load pulverized fuel–fired boilers, having practical capacities of 250,000 to 9,000,000 MJ/h. Among the many claimed practical advantages of AFBC are the following:

Feed coarse particle sizes of "crushed" carbonaceous fuel compared to pulverized fuel–fired combustors, thus not requiring expensive, power-consuming fuel pulverization.

Feed broad particle size ranges of carbonaceous fuels, thus not requiring the elimination of fine particles as set by stokers.

Operate at low temperatures, eliminating ash slagging, reducing ash deposition, and allowing a greater variety of carbonaceous fuels to be effectively combusted.

Can accomplish in-combustor desulfurization with limestone, generating a dry solid waste product and produce low NOx emissions without special controls.

Characterized by high fluidized bed-to-surface heat transfer coefficients that result in compact heat exchanger surface even with the lower temperature driving forces relative to stokers and pulverized fuel–fired combustors.

Combustion efficiency higher than stokers and comparable to pulverized fuel–fired combustors.

Biomass fueled AFBC is commercial at small capacities commensurate with biomass fuel supplies but is not as widespread as coal-fired AFBC. PFBC with coals is a relatively new commercial power generation technology still being scaled up to base-load electrical utility capacities. No activity on PFBC with biomass fuels is underway, as it is not appropriate for the limited biomass fuel supplies available, and given the practical problems of pressurized biomass fuel operation (fuel feeding, limited bed temperature to avoid agglomeration, high alkali metal vapor potential).

The key technology suppliers for large-capacity fluidized bed combustors are primarily conventional stoker and PC boiler manufacturers (for example, Asea Brown Boveri, Babcock and Wilcox, Foster Wheeler Pyropower, Kvaerner Pulping), while a variety of vendors are involved in smaller capacity AFBC. A key factor influencing the single fluidized bed combustor commercial capacity of these technologies is the cross-sectional area of the combustor vessel. The major parameters influencing the cross-sectional area of the combustor vessel are

Fluidization velocity (characteristic of circulating and bubbling bed operations)

Operating pressure: atmospheric pressure or pressurized

Excess air level

Fuel properties

AFBC combustor vessels are normally of rectangular cross section, with flat, water-walled, shop- or field-fabricated construction. PFBC combustor vessels may be cylindrical pressure vessels having refractory lining and wall-walled construction. Alternatively, the PFBC combustor vessel may have rectangular water-walled construction, similar to AFBC, and must then be contained within a larger external pressure vessel.

The two key parameter values are fluidization velocity and pressure. Clearly, the scale of some of the fluidized bed combustor vessels, particularly for AFBC, has grown to enormous levels. PFBC vessels having relatively large capacities can be shop fabricated pressure vessels:

500,000–800,000 MJ/h for bubbling bed PFBC

2,400,000–3,5000,000 MJ/h for circulating bed PFBC

Outstanding engineering innovations in the area of fluidized bed design to achieve economical vessel arrangements while maintaining acceptable fluidization, distribution and mixing of fuel and air, steam generation, and turndown capability has been required for the fluidized bed combustor technologies to reach large commercial scales.

3.3 Fluidized Bed Combustor Design Considerations

The design of fluidized bed combustors, as with the design of fluidized bed gasifiers, requires that transport models be developed that apply the conservation principles for mass, energy, and momentum to some degree of fundamental detail. The fluidized bed combustor mass and energy balances are closely coupled, and their solutions, applying empirical reaction kinetic and multiple phase mixing models, provide estimates of solids and gas composition and temperature profiles and input and output stream conditions. The momentum balances provide fluidized bed combustor pressure profiles when applied with appropriate fluidized bed phase density models.

These empirically based models provide the means needed to size the combustor equipment and scale pilot plant data. A significant aspect of fluidized bed combustor modeling and design is associated with heat transfer from the fluidized bed to immersed tube surfaces for steam generation (LaFanechere et al., 1998). The general arrangement of fluidized bed combustion systems, especially the mechanical arrangements of immersed steam generating heat transfer surfaces and convective passages, have been strongly influenced by commercial stoker-fired boiler and pulverized fuel–fired boiler designs.

The design of fluidized bed combustors involves the selection of several interrelating design, operating, and performance parameters and requires the consideration of performance and cost trade-offs for any specific application. The type of fluidized bed combustor and the application requirements strongly influence the fluidization parameter values selected for the design. The fuel properties (heating value, moisture content, ash content, volatile/fixed carbon ratio, sulfur content, nitrogen content, chlorine content, ash alkalinity, alkali metal content, and ash fusibility) represent the major parameters that the designer must apply, and these properties may limit the potential performance (thermal efficiency, environmental, operating cost, availability) that can be achieved by the fluidized combustor (Makansi, 1990).

The design of a fluidized bed combustor must result in acceptable unit availability, accounting for the most significant operating problems that arise (Makansi, 1997; Jones, 1995):

Tube failures
Refractory damage
Plugging and erosion of nozzles and drains
Deposits and blockages of circulating solids leg seal-
    valve
Wear and plugging of coal and limestone feed lines
Coal handing and feeding system
Ash handling system

Many of these phenomena are common to almost all fluidized bed reactors, and fluidized bed combustor design improvements are evolving as experience increases.

The nature of the fluidized bed combustor application and the fuel properties lead to unique design configurations and features in coal fueled AFBC, biomass fueled AFBC, and coal fueled PFBC.

3.3.1   Coal Fueled AFBC

The features of bubbling bed and circulating bed AFBC relating to fluidization technology are illustrated in Fig. 5. The AFBC configuration and operating conditions are influenced by the coal properties, the steam conditions (superheated temperature and pressure, feed water temperature, need for reheating), and the unit capacity and turndown requirements (Bernstein et al., 1995). A requirement to operate with several coals may also exist, which means the design must be based on the "worst" coal properties.

Bubbling bed AFBC with coals has been determined most suitable for smaller capacity steam generators because of its low fluidization velocity (1.5–4 m/s), resulting in a large fluidized bed cross-sectional area compared to circulating bed AFBC (Gaglia and Hall, 1987). Nonetheless, several large-capacity bubbling bed AFBC units have been constructed and are successfully operating (Anderson et al., 1997; Takahashi et al., 1995). Larger scale bubbling bed AFBC units (greater than 50,000 MJ/h thermal input) generally must be laid out in modularized fluid bed compartments, either as a single level, "ranch" style design, or as a "stacked" compartment design, with differing steam generation functions placed in separate combustion compartments (Manaker et al., 1982).

Figure 5 shows bubbling bed AFBC with two fluidized compartments on a single level. The air inlet plenum, or windbox, is compartmentalized so that individual bed sections can be defluidized for control, and these bed sections may have physical boundaries between them. The key features are the dense bubbling bed with its splash zone and freeboard zone, the air distributor, the coal and limestone feed arrangements, the heat transfer surface arrangements, and the fines recycling.

Bubbling bed AFBC may use underbed feeding of coal or overbed feeding, and coal is fed in dry form. Underbed feeding requires a multitude of pneumatic feed points, as many as one per 1–2 m^2 of bed cross section to achieve acceptable coal combustion efficiency (96–98%). Overbed feeding can use a more economical spreader feed system accommodating relatively large surface areas of bed. Overbed feed, though, requires a coarser coal top size (< 25 mm with removal of coal fines) than underbed feeding (< 12 mm × 0 to 6 mm × 0) and generally achieves a combustion efficiency about 1% lower than underbed feeding. Fluidized bed particle and gas mixing, particle segregation phenomena, splash-zone behavior, and particle attrition and elutriation phenomena control

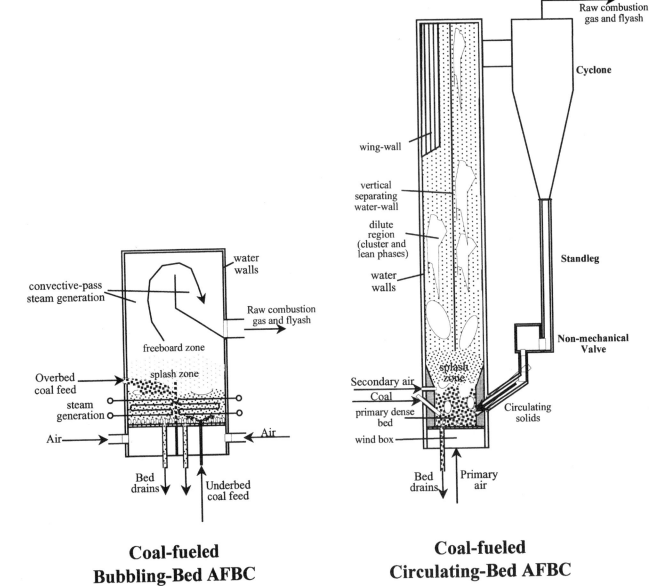

**Coal-fueled
Bubbling-Bed AFBC**

**Coal-fueled
Circulating-Bed AFBC**

Figure 5 Coal fueled AFBC configurations.

the performance of both the underbed and the overbed feed systems. Nonuniform char–bed mixing and volatiles–oxygen mixing may result in bed material agglomeration, high carbon losses, volatiles breakthrough with above-bed burning, reduced sulfur removal, and other undesirable results.

Typical bubbling bed temperatures are 815–870°C, and excess air levels range from 20 to 25%, with lower reactivity coals requiring higher excess air, higher temperatures, and fewer coal feed points than higher reactivity coals. Bubbling bed AFBC places

horizontal heat transfer surface within the dense fluidized bed, achieving high heat transfer coefficients (250–400 W/m²°C) but also facing the possibility of steam tube erosion. Heat transfer surface is also placed above the dilute freeboard zone where convective steam generation occurs. Bubbling bed combustor dense beds are generally about 1–1.5 m deep, allowing them to achieve acceptable pressure drops (15–25 kPa), while the splash zone and freeboard height is about 10 m from the top of the dense bed. The combustion gas velocity above the transport disengaging height is usually accelerated up to normal furnace convective pass velocities.

Smaller scale bubbling bed AFBC units can apply innovations not feasible at large capacities, such as internal fluidized bed recirculation schemes. Turndown can be achieved by a number of methods, including shutting down bed compartments, reducing fluidization velocity, reducing fluidized bed depth, and using flue gas recirculation.

Circulating bed AFBC uses a more complex combustor design than bubbling bed AFBC, as is illustrated in Fig. 5. It uses no heat transfer in the dense, reducing, primary bubbling bed zone, and it places all heat transfer surfaces exposed to high bed velocity in a vertical orientation to minimize the possibility of erosion. Refractory covering is also used to protect the metal heat transfer surfaces in some locations. The dense, primary, bubbling bed zone is usually refractory lined to minimize heat transfer surface erosion. The dense bed is a turbulent bed fluidized by substoichiometric primary air (about half of the total air) at a fluidization velocity of about 2–4 m/s. The dense bed is supported by an air distributor whose primary function is to distribute air uniformly to induce particle mixing between injected coal and recycle solids, and carry out partial coal oxidation. The distributor must operate within specified pressure drop limits, without plugging, erosion damage, or buckling. A variety of designs are in use, including refractory insulated bubble nozzles, water-cooled bubble nozzles, and pipe distributors. Allowance must also be made for bed material drainage from the primary bed, with the drainage ports integrated with the air distributor, but most drainage is taken from the low-carbon solids in the recycle leg.

Coal is fed to the dense bed by a small number of simple slide chutes fed by rotary valves and located along one wall of the unit. A single coal feed point may service 9–30 m^2 of bed cross section (Brereton, 1997). Alternatively, coal may be fed at a single point into the solids recirculation leg. Limestone (less than 6 mm in diameter) is normally fed separately above the bed.

Recirculating bed media are fed hot into the primary bed zone at a ratio of 50 to 100 times the coal feed rate. Various types of reinjection, nonmechanical valves and seals are used, such as seal pots, L-valves, J-valves, and others. These are simple refractory lined ducts having solids holding volumes that provide a loop seal, and appropriate aeration nozzles that induce and control solids flow. Cyclones are normally used to separate recirculating solids from the over-head gas stream, with solid fluxes on the order of 10 kg/m^2 s (Brereton, 1997). Alternative devices may also be used to remove entrained particulates for recycling, such as impactors (Belin and Flynn, 1991). Nonmechanical valves are discussed in Chapter 21, "Standpipes and Nonmechanical Valves." Cyclones are discussed in Chapter 22, "Cyclone Separators."

Secondary air is injected from nozzles located at the vessel wall at an elevation close to the top of the dense bed zone, resulting in increased fluidization velocity to about 4–9 m/s with high particle entrainment from the dense bed, completion of char and volatile combustion, and vigorous circulating bed, gas–particle mixing due to the high slip velocity and solids backflow that exists. Additional higher levels of secondary air injection may also be used. Circulating bed AFBC can operate with excess air of 10-20%, slightly lower than that of bubbling bed AFBC. The secondary air jets must penetrate sufficiently far into the fast fluid bed to promote effective mixing, and this jet penetration length limits the allowable furnace depth to about 9–11 m (Makansi, 1993). Even though coal is not uniformly mixed in the dense primary bed zone, the vigorous circulating bed furnace provides ample mixing and residence time to achieve high levels of char and volatiles combustion and high sulfur removal efficiency.

The density of the circulating bed falls rapidly with increased elevation above the dense bed and reaches density levels as small as about 10–40 kg/m^3 near the top of the circulating bed furnace. The bed density profile may be influenced by the design of the solids–gas exit region (Horio, 1997). The furnace height must be sufficient to provide residence time for efficient combustion and sulfur removal, and must also accommodate the required heat transfer surface for steam generation while resulting in acceptable gas pressure drop. Typically, the furnace height is limited to about 40 m.

The bed-to-surface heat transfer coefficient (about 85–280 W/m^2 °C) by solids convection and radiation is lower in circulating bed AFBC than in bubbling bed AFBC due to the lower bed density and vertical heat transfer surface orientation, and the heat transfer coefficient decreases with increased height. The dilute zone of the furnace is water-walled, with the heat transfer surfaces placed at the perimeter of the rectangular vessel enclosure. Additional vertical heat transfer surface walls or "wing walls" may be hung within the vessel to increase its heat removal capacity. Another means to increase the heat transfer surface is to place an "external heat exchanger" in the circulating solids loop (see Fig. 4). This external heat exchanger is a relatively low-velocity (1.5–3 m/s) bubbling fluidized bed containing horizontal heat transfer surface; it is fluidized by secondary air and has no in-bed reducing zones.

Scaling to larger capacity single circulating bed AFBC units requires proper engineering of the geometry of the rectangular furnace and the use of the several heat transfer surface configuration options, and it should provide for acceptable coal feed distribution and secondary air injection (Lee, 1997). Models of the integrated fluidized bed combustor and solids circulation system are required to design the equipment and control scheme properly. Circulating bed AFBC turndown techniques differ significantly from those utilized by bubbling bed AFBC. These have been reviewed by Brereton (1997).

3.3.2 Biomass Fueled AFBC

The relatively small capacity of biomass fueled AFBC applications (generally less than 300,000 MJ/h fuel thermal input), and the highly reactive nature of biomass fuels make bubbling bed AFBC a highly compact and effective combustion technology. Many aspects of biomass fueled bubbling bed AFBC are similar to coal fueled bubbling bed AFBC:

The fluidization velocity is about 3 m/s or less to minimize bed material elutriation.

The bed temperature is 760–870°C depending on the biomass fuel ash agglomeration tendencies.

The total excess air is about 20–30%.

The turndown is achieved by compartmentalizing the bed and by flue gas recirculation.

The bubbling bed depth is about 1 m with in-bed heat transfer surface (finned tubes).

The differences between coal-fired bubbling bed AFBC and biomass fueled bubbling bed AFBC, though, are significant (Douglas and Morrison, 1997; Hanson, 1991).

Coarse size biomass fuel feeding is used (100% less than 90–150 mm maximum particle dimension).

An inert bed of sand or limestone particles is used.

Above-bed biomass fuel feeding through chutes is used.

The dense bed is operated substoichiometrically, with primary air ranging from 40 to 80% of the total air.

At the top of the freeboard, the gas is accelerated (velocity as high as 5 m/s) and secondary air is injected to complete volatiles combustion, result-

ing in a combustion gas temperature as much as 200°C above the dense bed temperature.

Special features may be used to provide online removal of oversized tramp materials brought in with the biomass fuel.

Agglomeration in the fluidized bed, due to alkali metal (Na, K) reactions with the inert bed sand particles is common and may require periodic renewal of the bed material. Convective surface fouling can also be extensive and may require the use of screens before the superheater surfaces.

In larger installations, biomass fueled circulating bed AFBC can also be used, with designs very similar to those used in coal fueled, circulating bed AFBC (Skoglund, 1997). Circulating bed AFBC with biomass fuels has little or no advantage over bubbling bed AFBC with biomass fuels.

3.3.3 Coal Fueled PFBC

As with bubbling bed and circulating bed AFBC technology evolution, bubbling bed PFBC has led the development of PFBC over the still-to-be-demonstrated circulating bed PFBC. Bubbling bed and circulating bed PFBC both appear to be attractive options for advanced high-efficiency power generation systems for large electric generating applications.

The basic features of bubbling bed and circulating bed PFBC are analogous to those of bubbling bed and circulating bed AFBC, except for the following points:

The elevated pressure of PFBC (1000–1500 kPa) means that even the bubbling bed combustor cross-sectional area is reasonable for shop fabrication at relatively large capacities.

The elevated pressure of PFBC means that the dense bubbling bed can be relatively deep (3–5 m) and still meet pressure drop limits, so that the distribution of coal feed points is not as demanding as it is in bubbling bed AFBC.

The bubbling bed PFBC combustor vessel is typically tapered near the base to provide higher fluidization velocities and more intense mixing in the regions where coal is fed.

Coal may be fed by either dry lock hopper methods or by more reliable water–paste methods.

The PFBC vessel, cyclones, solids circulation equipment, and the external heat exchanger may be constructed as independent cylindrical pressure vessels. Alternatively, flat water-walled construction, like that used in AFBC may be placed, along with cyclones and many other pieces of process equipment designed for low-pressure operation, within the pressurized environment of a single large external pressure vessel, allowing the combustor to retain many of its AFBC features.

Figure 6 illustrates the PFBC power plant configuration, showing the main power plant components. The figure is representative of bubbling bed PFBC with the currently commercial process configuration: multiple stages of cyclones are used to remove particulate from the combustion gas, and the combustion gas is expanded through a ruggedized turbine expander, an expander designed to tolerate large particulate content in the gas. The stack gas must be further cleaned of particulates in commercial equipment (electrostatic precipitator or fabric bag filter) to meet environmental standards. The PFBC unit and several other components are located in a large external pressure vessel that is pressurized and cooled by the warm compressor air. This arrangement separates the thermal boundary from the pressure boundary and provides several practical equipment and process advantages over the use of separate pressure vessels for each component (McDonald et al., 1987; Huryn et al., 1987).

Bubbling PFBC uses dense bed depth adjustment as a means for load control, exposing normally immersed heat transfer surfaces to the freeboard zone; a vessel for storage and transfer of bed ash is used for this purpose and is housed in the external pressure vessel. The external heat exchanger used with circulating PFBC represents a relatively large fluidized bed that is also housed in this pressure vessel (Walter et al., 1997). In scaling up PFBC, the size of the external pressure vessel is a major consideration. For example, the Cottbus PFBC plant has a fuel thermal input (lignite fed dry) of about 700,000 MJ/h and operates at a pressure ratio of about 13 : 1. Its external pressure vessel is about 32 m high and 11 m in diameter (Walter et al., 1997). The Karita PFBC unit has about five times the fuel thermal input as the Cottbus and feeds coal as a water paste (Jeffs, 1997). With its higher operating pressure, the Karita external pressure vessel might be about 50 m high, with a diameter of about 15–16 m, a formidable vessel to fabricate, ship, and install.

Topped PFBC applies both fluidized bed gasification and combustion technologies in concert. Topped

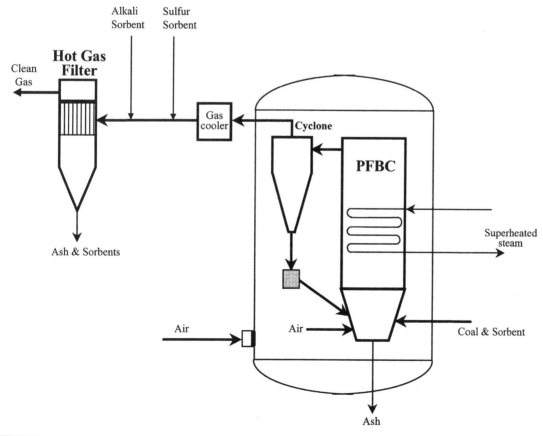

Figure 6 PFBC power plant schematic.

PFBC will operate with conventional heavy-duty combustion turbines having higher firing temperatures and higher pressure ratios than those used in PFBC. Hot gas filters must be used to remove particulate from both the combustion gas and the carbonizer fuel gas to protect the turbine expander from erosion and deposition damage. The additional challenge of topped PFBC is to integrate and control the combined operation of the carbonizer and combustor and their interfacing components.

The design of the fluidized bed carbonizer and combustor unit for topped PFBC change little from those design factors already reviewed. Because a higher turbine inlet temperature is used than in PFBC, the operating pressure of the fluidized bed combustor would be higher, resulting in smaller bed diameter and smaller external pressure vessel diameter than for a comparable PFBC. The fluidized bed combustor is fed warm char from the fluidized bed carbonizer at appropriate char feed distribution points, and the combustor must be sized to burn efficiently the low-volatile char and convert the CaS from the fluidized bed carbonizer into $CaSO_4$. Topped PFBC also has the option of feeding coal to the fluidized bed combustor as well as to the fluidized bed carbonizer, making much greater power plant flexibility possible.

4 THE FUTURE OF FLUIDIZED BED GASIFIERS AND COMBUSTORS

The recent practical advancements of fluidized bed gasifiers and combustors have occurred with little dependence on the fundamental understanding of fluidization phenomena and associated reaction phenomena. Fluidized bed gasifiers and combustors are designed following highly empirical principles based on the extrapolation of pilot plant test data. The likelihood of further improvements in the competitiveness of fluidized bed gasifiers and combustors depends in part on the health of the commercial markets for their applications and the relative economic and environmental pressures on the carbonaceous fuels that they utilize (coals, biomass fuels, and others).

Improved understanding of fluidization fundamentals related to gasification and combustion phenomena may eventually guide commercial equipment design and selection of special features. More immediate impacts may result, though, from engineering innovations based on operating plant observations that improve performance, reliability, operability, and cost. Experimentation (laboratory scale conceptual testing, cold flow simulations, pilot plant parametric testing, demonstration plant experience) will continue to be an important avenue for such engineering innovations.

REFERENCES

Agrawal RK, Haldipur GB. Advanced gasifier-desulfurizer process development for SNG application. Final Report by KRW Energy Systems, Inc., to Gas Research Institute, GRI-89/0011, 1988.

Anderson KD, Manaker AM, Stephans EA Jr. Operating experience of the Tennessee Valley Authority's 160-MW atmospheric fluidized bed combustion demonstration unit. Proceedings of the 14th International Conference on Fluidized Bed Combustion, ASME, 1997, pp 39–45.

Barducci G, Daddi P, Polzinetti G, Ulivieri P. Thermic and electric power production and use from gasification of biomass and RDF: Experience at CFBG plant at Greve in Chianti. Proceedings of the Second Biomass Conference of the Americas, Portland, 1995, NREL/CP-200-8098, pp 565–575.

Belin F, Flynn TJ. Circulating fluidized bed boiler solids system with in-furnace particle separator. Proceedings of the 1991 Fluidized Bed Combustion Conference, ASME, 1991, pp 287–294.

Bernstein N, Goidich S, Li S, Phalen J. Influence of fuel type and steam cycle on CFB boiler configuration. Proceedings of the 13th Fluidized Bed Combustion Conference, ASME, 1995, pp 27–37.

Brereton C. Combustion performance. In: Grace JR, Avidan AA, Knowlton T, eds. Circulating Fluidized Beds. New York: Blackie, 1997, pp 376–416.

Bryan BG, Hoppe JA. Status of the Shanghai Coking and Chemical Company's U-gas coal gasification plant. Proceedings of the Fifteenth Annual International Pittsburgh Coal Conference, Pittsburgh (CD-ROM), 1998.

Chau LL, Rovner JM. Wood gasification for combined cycle power applications. Paper presented at the AIChE Summer National Meeting, July, 1995.

Craig JD, Purvis CR. A small scale biomass fueled gas turbine engine. Paper presented at the ASME, International Gas Turbine and Aeroengine Congress and Exhibition, Stockholm, Sweden, June 1998.

Dawes SG, Mordecai M, Welford GB. Recent design and cost studies for air blown gasification. Proceedings of the 14th International Conference on Fluidized Bed Combustion, ASME, 1997, pp 347–358.

DeLong MM, Oelke EA, Onischak M, Schmid MR, Wiant BC. Sustainable biomass energy production and rural economic development using alfalfa as feedstock. Proceedings of the Second Biomass Conference of the Americas, Portland, 1995, NREL/CP-200-8098, pp 1582–1591.

Demuth JE, Smith HG. Pinon Pine Project gasifier startup. Presented at the Advanced Coal-Based Power and Environmental Systems '98 Conference, Federal Energy Technology Center, Morgantown, 1998.

Douglas MA, Morrison SA. The design of an advanced BFB steam generator for biomass. Proceedings of the 14th International Conference on Fluidized Bed Combustion, ASME, 1997, pp 151–157.

Farris M, Paisley MA, Irving J, Overend RP. The biomass gasification process by Battelle/FERCO: design, engineering, construction, and startup. Proceedings of the EPRI 1998 Gasification Technologies Conference, San Francisco, 1998.

Gaglia BN, Hall A. Comparison of bubbling and circulating fluidized bed industrial steam generation. Proceedings of the 1987 International Conference on Fluidized Bed Combustion, ASME, 1987, pp 18–25.

Geldart D. Powder Technology 7:285, 1973.

Ghate M, Longanbach J. Coal gasification—technology status report. DOE/METC-88/0262, January 1988.

Hanson JL. Agricultural waste fired fluid bed combustor, Delano, California. Proceedings of the 1991 International Conference on Fluidized Bed Combustion, ASME, 1991, pp 1105–1109.

Hebden D, Stroud HJF. Coal gasification processes. In: Elliott MA, ed. Chemistry of Coal Utilization. Second supplementary volume. New York: John Wiley, 1981, pp 1599–1752.

Hirschfelder H, Vierrath H. CFB biomass gasification for energy and industry operational results. Proceedings of the EPRI 1998 Gasification Technologies Conference, San Francisco, 1998.

Huryn JB, Wickstrom B, Rederstorff B. Design concept for a PFBC commercial plant. Proceedings of the 1987 International Conference on Fluidized Bed Combustion, ASME, 1987, pp 272–277.

Jeffs E. Karita: a quantum leap for PFBC. Turbomachinery International, March 1997, pp 16–17.

Jones C. O&M experience underscores maturity of CFB technology. Power, May 1995, pp 46–55.

LaFanechere L, Basu B, Jestin L. Use of an expert system to study the effect of steam parameters on the size and configuration of circulating fluidized bed boilers. Journal of Engineering for Gas Turbines and Power—Transactions of the ASME 120(4):813–819, 1998.

Leckner B. Fluidized bed combustion: mixing and pollutant limitation. Progress in Energy and Combustion Science 24(1):31–61, 1998.

Lee YY. Design considerations for CFB boilers. In: Grace JR, Avidan AA, Knowlton T, eds. Circulating Fluidized Beds. New York: Blackie, 1997, pp 417–440.

Lu Z, Li D, Shen Y, Xu X, Cao B. An industrial demonstration project on tri-generation process for gas, heat and power using CFB technology. Presented at the China–Korea Technology Cooperation Conf. On Coal Applications, Tsinghua University, China, 1996.

Makansi J. Fuel type, preparation emerge as critical to FBC design. Power, January 1990, pp 41–44.

Makansi J. Can fluid bed take on P-C units in the 250–400 MW range? Power, September 1993, pp 45–50.

Makansi J. Designers, users air concerns on CFBs. Power, March/April 1997, pp 4–5.

Manaker AM, Dunn GC, Cruikshank KA, Wyatt JN. TVA's technical assessment and cost comparison of final conceptual designs by Babcock Contractors, Inc. (BCI), Babcock and Wilcox Company (B&W), and Combustion Engineering, Inc. (CE), of a 200-MW Atmospheric Fluidized Bed Combustion (AFBC) Demonstration Plant. Proceedings of the 7th Conference on Fluidized Bed Combustion, 1982, pp 691–700.

McDonald DK, Weitzel PS, Thornblad P. PFBC design integration—ASEA Babcock PFBC's approach. Proceedings of the 1987 International Conference on Fluidized Bed Combustion, ASME, 1987, pp 200–207.

Monsour N, Durai-Swamy K, Voelker G. MTCI/thermochem steam reforming process for biomass. Proceedings of the Second Biomass Conference of the Americas, Portland, Oregon, 1995, NREL/CP-200-8098, pp 543–552.

Nieminen J, Kivela M. Biomass CFG gasifier connected to a 350 MW_{th} steam boiler fired with coal and natural gas—thermie demonstration project in Lahti in Finland. Biomass and Bioenergy 15(3):251–257, 1998.

Pitcher K, Hilton B, Lundberg H. The ARBRE project: progress achieved. Biomass and Bioenergy 15(3):213–218, 1998.

Renzenbrink W, Wischnewski R, Engelhard J, Mittelstadt A. High temperature Winkler (HTW) coal gasification—a fully developed process for methanol and electricity production. Proceedings of the EPRI 1998 Gasification Technologies Conference, San Francisco, 1998.

Robertson A. Second generation PFBC systems research and development—pilot plant becomes demonstration plant design. Presented at the Advanced Coal-Fired Power Systems '95 Review Meeting, Federal Energy Technology Center, Morgantown, 1995.

Seglin L, Bresler SA. Low-temperature pyrolysis technology. In: Elliott MA, ed. Chemistry of Coal Utilization. Second Supplementary Volume. New York: John Wiley, 1981, pp 785–846.

Schobert HH, Rusinko F, Mathews JP. W(h)ither the coal industry?—the long-term view. Proceedings of the Fifteenth Annual International Pittsburgh Coal Conference, Pittsburgh (CD-ROM), 1998.

Sierra Pacific Power Company. Tracy Power Station—Unit No. 4, Pinon Pine Power Project. Public Design Report, 1994, NTIS, DOE/MC/29309—4056.

Simell P, Kurkela E, Stahlberg P. Formation and catalytic decomposition of tars from fluidized-bed gasification. Presentation at the International Conference on Advances in Thermochemical Biomass Conversion, Interlaken, Switzerland, 1992.

Skoglund B. Six years of experience with Sweden's largest CFB boiler. Proceedings of the 14th International Conference on Fluidized Bed Combustion, ASME, 1997, pp 47–56.

Stahl K, Neergaard M. IGCC power plant for biomass utilization. Biomass and Bioenergy 15(3):205–211, 1998.

Stobbe S, Oatley J, Brown RC. Indirectly-heated biomass gasification using latent-heat ballasting of a fluidized reactor. Proceedings of the IECEC 96 Conference, Institute of Electrical and Electronics Engineers, 1996. pp 1724–1729.

Takahashi M, Nakabayashi Y, Kimura N. The 350 MWe Takehara plant. VGB Kraftwerkstechnik 75:427–434, 1995.

Walter E, Krartz HT, Almhen P. Lignite fired combined cycle heat and power plant using pressurized fluidized bed combustion. Proceedings of the 14th International Conference on Fluidized Bed Combustion, ASME, 1997, pp 359–369.

Wen CY, Tone S. Coal conversion reaction engineering. In: Chemical Reaction Engineering Reviews, Houston (Luss D, Weekman VW, eds.). ACS Symposium Series 72, 1978, pp 85–109.

Yang WC, Newby RA, Keairns DL. Large-scale fluidized bed physical model: methodology and results. Powder Technology 82:333–381, 1995.

16

Applications for Chemical Production and Processing

Behzad Jazayeri

Fluor Daniel, Inc., Aliso Viejo, California, U.S.A.

This chapter provides an overview of some of the more important commercial applications of fluidized bed technology not covered in the other sections. Recent developments with high potentials of commercialization in the near future are also discussed.

1 CHEMICAL SYNTHESIS

1.1 Acrylonitrile

The Sohio Company (now BP Chemicals) introduced in 1950s what is now considered to be the most successful application of fluidized beds for chemical synthesis (Fig. 1). In this application, air, propane, and ammonia are reacted together in a turbulent fluid bed of Geldart group A (Chapter 3) catalyst to produce acrylonitrile by the reaction:

$$C_3H_8 + NH_3 + \frac{3}{2}O_2 \rightarrow C_2H_3CN + 3H_2O \qquad (1)$$

The features of the original process as reported by Kunii and Levenspiel (1991a) and some recent developments follow. Reactor operating conditions are 400 to 500°C at about 1.7 atmosphere, with a reactor contact time of 5 to 20 seconds. In addition to acrylonitrile, carbon monoxide, carbon dioxide and water, small quantities of hydrogen cyanide, acetonitrile, and acrolein are also produced. The original plants were designed to operate at around 52 cm/s superficial velocity. Modern plants are now operating with fines

content (10 to 45 micron) of around 25–45% at about 66 cm/s (Pell and Jordan, 1987), and possibly as high as 100 cm/s. BP has introduced several new catalysts with increasing selectivity to acrylonitrile. With these changes, propylene conversion is now approaching 100%, and per pass reactor yield to propylene is about 85%. The reaction is highly exothermic, releasing about 670–750 kJ/mol. The heat of reaction is removed by direct generation of high pressure steam in serpentine coils located inside the reactor. The reactor hydraulic diameter (Volk et al., 1962) is reported as 100 to 150 cm.

Sohio initially tested the process in both a small bench scale unit and a circa 61 cm pilot unit before building the first commercial plant (circa 300 cm). Over 51 units have been built, representing over 95% of the world's acrylonitrile capacity. The largest units now utilize reactors with estimated diameters of 800 to possibly 1000 cm.

In the process, air is introduced uniformly into the bottom of the reactor via a distributor plate. The propylene and ammonia are introduced into the fluidized bed above the distributor plate via a separate sparger. The design of these gas distribution systems has evolved over the years improving the reactor yield (Ohta and Yokura, 1996.)

The original units used several parallel sets of three-stage cyclones for catalyst recovery. Recently, one cyclone vendor has announced the use of multiple cyclones made of a single stage followed by two second stages.

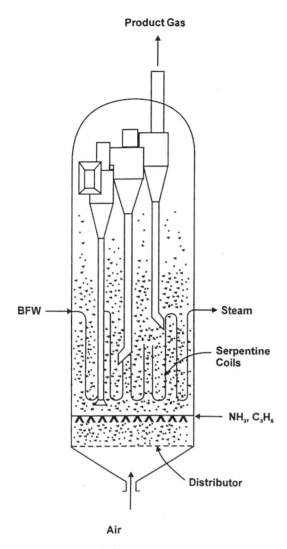

Figure 1 Sohio acrylonitrile reactor.

pane or propylene as feedstock.) Based on a review of literature and patents issued to BP and Mitsubishi, it is the author's opinion that the propane-based process will operate with a low per pass propane conversion. If so, such a process is likely to use recycle, which may increase the process complexity and cost. Use of recycle may also introduce additional safety issues as the recycle gas will contain unconverted propane and possibly some oxygen. Additionally, it is the author's opinion that the capacity of a given reactor will be lower with the propane-based catalyst. In such a situation, revamping an existing propylene based plant to propane can result in loss of production. Since in a revamp situation, the reactor volumetric throughput cannot be increased beyond the capacity of the downstream recovery equipment, one possible solution to this issue may be to use enriched air, or possibly oxygen. Use of enriched air or oxygen is likely to increase the operating cost. It remains to be seen how these challenges will be resolved and how, in the ultimate analysis, the economics of the propane-based process compare with the more mature propylene-based process. In late 1998, BP announced that it has successfully completed the pilot plant program and will move ahead with a demonstration scale plant. It therefore appears that BP has resolved these issues and that this new process may indeed offer an economic advantage.

1.2 Maleic Anhydride

Production of maleic anhydride by the vapor phase partial oxidation of C4s is an extremely exothermic reaction (1410 KJ/mol) and is therefore well suited to fluidized bed application. The first commercial fluidized bed application was by Mitsubishi Chemical Industries in 1970. This process converted a mixture of butadiene and butene to maleic anhydride in a fluidized bed of 6 m diameter using a vanadium phosphorous oxide (VPO) catalyst (Kunii and Levenspiel, 1991a.) In the early 1980s, three new fluid bed processes were developed for conversion of *n*-butane to maleic anhydride. The first, by Badger, was taken through the demonstration stage but was never commercialized because of the global depression of the early 1980s. The second process was developed by BP Chemicals and has been licensed once. The third process was commercialized by Lummus at about the same time as BP's. Nine units have been licensed to date. The Lummus process is now licensed by Lonza.

The process developed by these four companies uses different formulations of VPO catalyst in a conventional fluidized bed similar in design to the Sohio acry-

The Monsanto Company has also commercialized a fluidized bed for the production of acrylonitrile for its own use. The Monsanto reactor is expected to have similar features to that of BP but uses a different catalyst. In early 1990s, Monsanto decided to license this process and has licensed one unit in Korea.

BP Chemicals recently announced the piloting of a propane-based fluid bed process for acrylonitrile production (1997 annual report). Mitsubishi has also received patents on a similar propane-based process. Use of propane may offer production cost benefits as propane is typically cheaper than propylene. The propane-based catalyst is believed to be different from the propylene based catalyst, ruling out the possibility of a swing operation depending on the market prices (a single catalyst producing acrylonitrile using either pro-

lonitrile reactor (Fig. 1). Air, flowing through a grid plate, is used to fluidize the catalyst. Vaporized C4s is sparged into the fluidized bed above the grid plate. The reaction heat is removed by direct generation of steam inside serpentine cooling coils placed in the fluidized bed. The fluidized bed reactor can operate within the flammability envelope with much higher concentrations of *n*-butane in the total feed compared with the fixed bed process. Reactor operating conditions are 410 to 420°C at about 1 to 5 atmospheres. Main reaction by-products are acetic and acrylic acids. The primary reaction is

$$C_4H_{10} + O_2 \rightarrow C_4H_2O_3 \rightarrow H_2O + CO + CO_2 \quad (2)$$

An evaluation of patents and literature in this field has led this author to conclude the following:

Fixed bed reactors are tubular in design and have features similar to that of a shell and tube heat exchanger with fixed tube sheets. Molten salt flowing on the shell side is used to remove the heat of reaction. Mechanical considerations limit the number of tubes in such reactors. As a result, the author estimates the fixed bed reactor capacity to be limited to circa 15,000–20,000 MT/year. Multiple reactors are therefore required for higher capacities. The fluidized bed, on the other hand, can be designed as a single reactor unit with capacities of circa 70,000 MT/year and possibly more.

Fixed bed reactors are limited by flammability considerations to about 1.2 to 1.8 mol% butane in the feed, as the butane and air are premixed outside the reactor. The fluidized bed can operate with up to possibly 5 mol% butane in the feed because butane is sparged separately into the bed and the mass of catalyst acts as a heat sink. The higher butane-to-air feed ratio reduces the size of the downstream gas handling section of the process, reducing the capital cost.

The sensible heat of the purge gas vented to atmosphere from the fluidized bed process can be as low as 30% of that in the fixed bed process because of lower gas volumetric flow. Therefore for a given reactor capacity more of the reaction heat is recovered in generating steam.

The fluidized bed maleic anhydride process, however, has also several apparent disadvantages:

The catalyst volume required in the fluidized process is several times higher than that of the fixed bed process. Since VPO catalyst is very expensive, the additional initial cost of the catalyst may negate any savings in the downstream equipment.

Reaction selectivity to maleic anhydride tends to be lower in the fluidized bed process.

These competing characteristics tend to favor the fluidized bed when a larger capacity unit is required and there is a high demand for steam. The author estimates a rough break point to be around 10,000–20,000 MT/year capacity.

The maleic anhydride fluidized bed reactor also offers several design and operational problems that must be carefully addressed:

The VPO catalyst used is somewhat dense. Therefore the particle size distribution of the catalyst, the design of the reactor internals, and the operating velocity must be carefully selected. Failure to do so will result in gas bypassing, which will affect reactor conversion and increased oxygen slip.

The VPO catalyst is susceptible to loss of selectivity when exposed to reducing environment. To prevent catalyst overreduction, a threshold minimum oxygen concentration is needed in the reactor offgas. Additionally, the reactor must operate at higher velocities to enhance gas/solid contact.

To maintain suitable oxidation environment for the catalyst, butane conversions must be kept within a certain range.

Meeting the above constraints results in operation with a reactor offgas that is close to autoignition and is therefore susceptible to afterburn. Engineering solutions have been developed to address this and permit the operation of these units in a safe manner.

Little information has been published on the above four processes, but some educated guesses can be made. The catalyst is Geldart group A with possibly 25 to 55% fines (10 to 45 micron) to counter its higher density. Operation within the turbulent regime is needed to enhance gas/solid contact and promote uniformity between the emulsion and void gas and therefore minimize catalyst overreduction. Operating velocities of about 40 and possibly as much as 100 cm/s are therefore required.

In an effort to resolve the design issues discussed above, DuPont has developed and commercialized a circulating fluidized bed for conversion of butane to maleic anhydride (Contractor, 1985). The process is depicted conceptually in Fig. 2. The VPO-based catalyst is structurally different from the conventional flui-

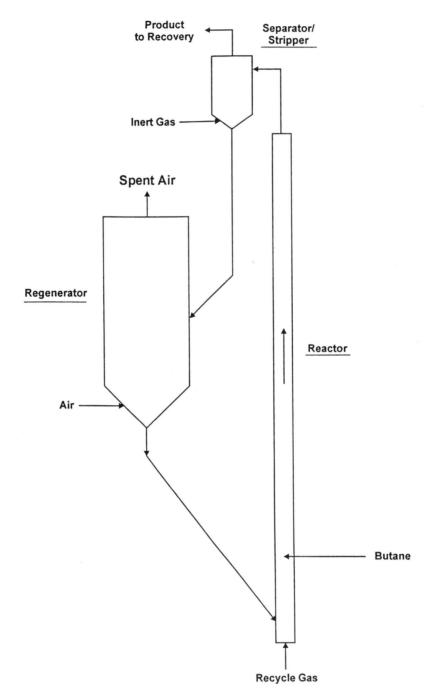

Figure 2 DuPont maleic anhydride reactor.

dized bed catalyst discussed above. Details of the catalyst structure are provided by Contractor et al. (1978). A significant departure from conventional fluidized bed design is the use of the catalyst as oxygen carrier. The catalyst is oxidized in a separate regeneration vessel by contact with air. It is then circulated to a riser reactor where it is reduced by reaction with butane. The reactor operates at high velocities with a relatively short contact time. Per pass conversion of butane is therefore low, requiring recycling of unconverted butane. Butane feed concentration can be high, possibly over 20 mol%. To avoid catalyst overreduction in such a harsh environment, the reactor is designed to minimize catalyst residence time and operates at relatively low temperatures, possibly as low as 360°C. This mode of operation helps maintain reaction selectively even though butane concentrations are very high.

The reduced catalyst is separated from the reactor offgas, stripped to minimize butane and maleic anhydride losses, regenerated by contact with air in the regenerator, and then returned back to the reactor. DuPont has indicated that a further stripping step may be used between the regenerator and reactor. DuPont claims that by carrying out the oxidation and reduction cycles in two separate zones, each zone can be designed to operate close to its optimal point, thus increasing reaction selectivity and catalyst productivity. Afterburn of the reactor offgas is not expected, as the gas contains little or no oxygen.

DuPont bench tested this process in mid-1980s. A 15 cm diameter by 27.4 m high riser was operated for over one year to prove the process concept (Contractor et al., 1993). The table shows the conditions for the demonstration unit.

Parameter	Range tested
Reaction temperature, °C	360–420
Reaction pressure, atmosphere	< 5
Catalyst flux, kg/cm^2 s	250–1100
Gas velocity, m/s	4–10
Butane concentration, mol%	< 25

The riser solid volume concentration approached 0.2. The first commercial facility was started in the mid-1990s.

The DuPont process incorporates design features of fluid catalytic cracking (Chapter 14) in addition to many features that are unique to chemical synthesis reactors.

The Mitsubishi process described above has since been converted to *n*-butane feedstock. Mitsubishi just recently announced piloting an enhanced process with butane recycle and catalyst regeneration using a new non-VPO catalyst formulation, which is claimed to reduce cost.

1.3 Oxychlorination of Ethylene

Ethylene is reacted with dry hydrogen chloride (HCl) and either air or pure oxygen over a supported cupric chloride catalyst in a fluidized bed to produce ethylene dichloride (EDC) and water by the vapor phase reaction

$$C_2H_4 + 2HCl + \frac{1}{2}O_2 \rightarrow C_2H_4Cl_2 + H_2O \qquad (3)$$

The oxychlorination reactor is part of a balanced vinyl chloride monomer complex (Fig. 3). The EDC is the intermediate product and is thermally cracked to VCM and HCl. Operating conditions of the oxychlorination reactor are 220 to 245°C at 2.5 to 6 atmospheres. For the air-based process, ethylene and air are fed in slight excess of stoichiometry to ensure high conversions of hydrogen chloride and minimize losses of excess ethylene in the vent gas. Under these conditions, 94 to 99% conversion of ethylene and 98 to 99.5% conversion of hydrogen chloride are obtained at a reported selectivity to EDC of 94 to 97% (Kirk-Othmer, 1997.) The oxygen-based process operates at lower temperatures and has a higher yield and provides a drastic reduction in the volume of vent gas. It is operated with a much higher excess of ethylene. BF Goodrich (now Geon) is the dominant licensor of the fluidized bed process. The first commercial plant was started in 1960s. A total of thirty-six plants have been licensed. Other licensors are listed by Naworski and Velez, 1983.

The cupric chloride catalyst used in the fluidized bed process is a group A powder with a fines content (10–45 micron) of about 30 wt% (Naworski and Velez, 1983.) The catalyst is subject to stickiness under certain (usually upset) conditions that promote formation of dendritic growths of cupric chloride on the catalyst surface. Particle agglomeration can occur with loss of fluidity, and, in severe cases, collapse of the bed.

The oxychlorination reactor is a vertical cylindrical shell made of carbon steel with a support grid/air sparger system and internal cooling coils. Internal or external cyclones are used to minimize catalyst carryover. The reactor internal parts are made from corrosion-resistant alloy. The reactor has many design features depicted in Fig. 1.

1.4 Phthalic Anhydride

Coal tar naphthalene is oxidized with air over a vanadia catalyst in a fluidized bed according to the following highly exothermic partial oxidation reaction:

$$C_{10}H_8 + 4SO_2 \rightarrow C_8H_4O_3 + 2H_2O + 2CO_2 \qquad (4)$$

Reaction yields are about 100 kg of phthalic anhydride per 100 kg of naphthalene with desulfurized feed.

Desulfurized molten naphthalene is introduced into the bed above the distributor plate through multiple spray nozzles, while air is admitted from below. Contact of reactants occurs inside the fluidized bed,

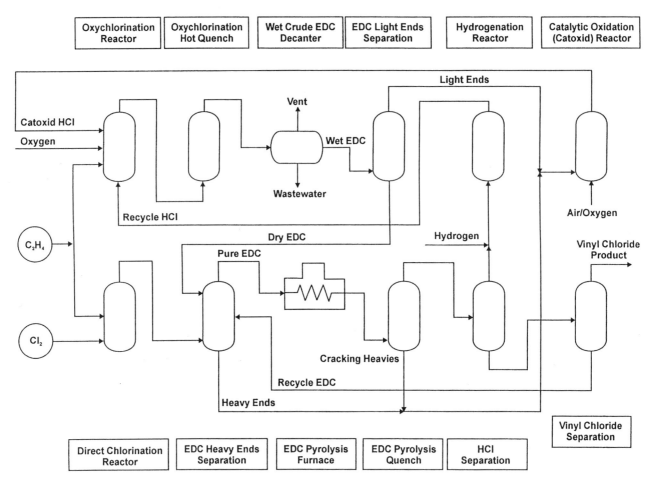

Figure 3 Balanced oxychlorination process.

permitting reactor operation within the flammability envelope (Graham, 1970). The reactor operates at around 2.7 atmosphere and between 345 and 385°C, with a velocity of between 30 and 60 cm/s and a contact time of 10 to 20 seconds (Graham and Way, 1962). The catalyst is Geldart group A with fines content (10 to 45 micron) of 28 % (Johnsson et al., 1987). The reactor is believed to operate within the turbulent regime. Besides carbon monoxide, carbon dioxide, and water, small quantities of maleic anhydride, naphthoquinone, and benzoic acid are also formed as by-products.

The fluidized bed process was commercialized by Sherwin Williams in the mid-1940s but is now known as the Badger phthalic anhydride process. Over 20 reactors have been built, with the largest having a capacity of 100,000 metric tons per year. The last reactor was built in China in the mid-1980s. The process is now considered noncompetitive with o-xylene-based

fixed bed plants due to limited availability of coal tar naphthalene. Attempts to develop an o-xylene-based fluid bed process have been hampered by the unavailability of a suitable attrition resistance catalyst support (Bolthrunis, 1989).

Excellent accounts of the development, operation, and safety aspect of this process have been provided elsewhere (Miseralis et al., 1991; Graham, 1970) and are summarized here.

The original Sherwin Williams first-generation reactor used an external shell-and-tube catalyst cooler and external ceramic filter elements (Fig. 4a). Reaction heat was removed by circulating catalyst through the external cooler via a standpipe. The cooled catalyst was returned back to the reactor through the distributor plate. This operation was never satisfactory and suffered from both mechanical and operational problems. Reactor throughput was often limited due to insufficient heat removal capability. In the mid-1950s, Badger

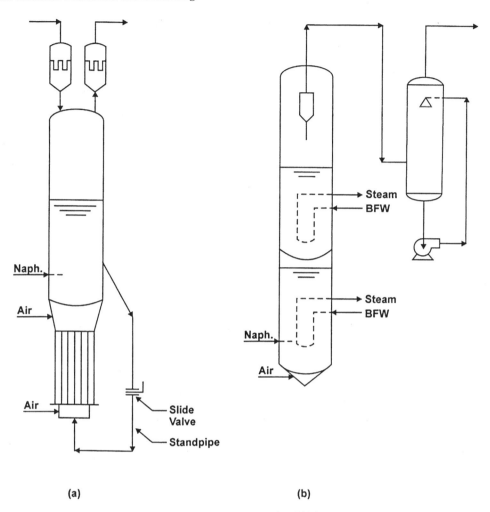

(a)   (b)

Figure 4 Badger phthalic anhydride reactors. (From Miseralis et al., 1991.)

suggested experimenting with internal serpentine cooling coils as a way to increase the heat removal capability of the reactor. This experiment proved to be very successful. Heat removal was increased so much that the external cooler was no longer needed. The behavior of the fluidized bed was also markedly improved with the addition of the coils. The second generation reactors were designed with internal cooling coils only. Medium to high pressure steam was generated directly in the coils.

Entrained catalyst leaving the bed was originally removed by externally mounted ceramic filter elements. The hot reactor offgas is susceptible to afterburn (Graham and Way, 1962). Long runs of pipes between the reactor and filter housings were used to cool the gas to about 260°C before the catalyst fines were removed. The recovered dust was blown back to the reactor with heated air to prevent thermal shock of the filter elements. Catalyst was purged periodically from the filters to prevent buildup of very fine dust (below 10 micron) in the reactor. The fine dust is produced by attrition, and at high enough levels it can affect the fluidity of the catalyst. Ceramic filter elements are susceptible to cracking from thermal shock and mechanical stress. They were eventually replaced by filter elements constructed of glass-wool-wound steel pipe cores and finally micrometalic filters with marked improvement in operating reliability. With the improvement in cyclone design and operation, the filters were eventually replaced by internal multistage cyclones. Figure 4b shows a third-generation phthalic anhydride reactor. Two fluidized beds are stacked one on top of the other. The bottom reactor is used for reaction. The top reactor is a quench zone. This quench zone is necessary to prevent afterburn of the reactor offgas after the entrained catalyst is removed by the cyclones.

A venturi scrubber is also provided to remove the small quantity of catalyst carryover from the cyclones and to protect downstream switch condensers in the event that the cyclones malfunction.

2 GAS TO LIQUIDS

2.1 Sasol

Synthesis of hydrocarbons from H_2 and CO gases by the Fischer–Tropsch reaction is strongly exothermic and occurs at about 200 or 340°C and 20 to 50 atmosphere by the reaction:

$$nCO + 2nH_2 \rightarrow (CH_2)_n + nH_2O \qquad (5)$$

The higher temperature operation produces gasoline range products and is used exclusively with fluidized beds. The lower temperature operation is used for wax production in either fixed bed tubular or slurry bed reactors. Sasol of South Africa commercialized the first fluidized bed Fischer–Tropsch process in 1955.

The synthol fluidized bed process produces hydrocarbons in the gasoline boiling range using an iron catalyst. The scale-up and reactor design development are depicted in Fig. 5 (Jones, 1991). Sasol 1A and B used a circulating fluidized bed design based on the original 10 cm pilot plant work of the WM Kellogg company in the 1940s. These first commercial reactors have a diameter of 2.3 m and operate at about 2 m/s. They represent a 500-fold increase in gas throughput compared to the pilot plant. Catalyst is circulated between the upflow leg of the reactor and the downflow leg of the disengaging hopper and standpipe. The reactors were originally built with two banks of fixed tube sheets for heat removal. The reacting gas and catalyst mixture flowed through the tubes and was cooled by oil circulating on the outside. Operating problems plagued the startup and successful operation of the Sasol 1A and B reactors (Shingles and Jones, 1986). The disengaging hopper cone was too shallow. Funnel flow occurred, causing bridging of the catalyst. The standpipe was found to operate in stick–slip flow (Chapter 21) and was subject to bridging and loss of circulation. The

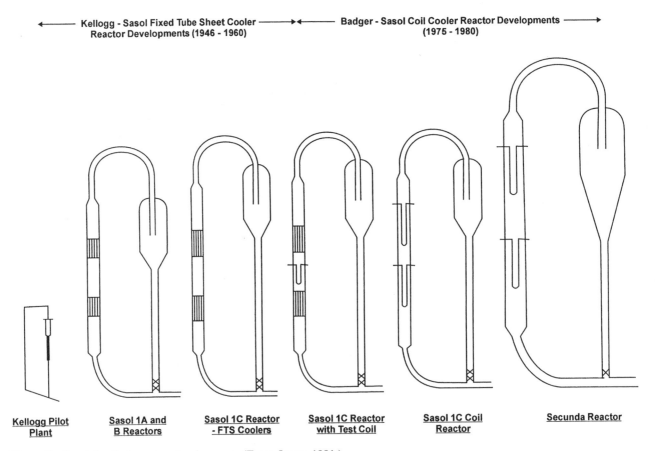

Figure 5 Sasol Synthol reactor development. (From Jones, 1991.)

reactors were designed originally to operate with continuous catalyst addition and withdrawal to maintain catalyst activity. This was never practiced in these first generation reactors as the valve technology capable of reliable operation was not available at that time. The most serious problem was however with the tube sheets. Localized pressure fluctuations resulted in preferential gas flow through some tubes, leaving the others with a dense solid pocket that quickly plugged due to formation of waxy compounds. The increased gas flow through the open tubes then resulted in tube erosion.

It took Sasol two years to learn to operate these first reactors with acceptable run times between shutdowns. In 1960s, a third reactor, Sasol 1C, was added; it used longer tubes for increased heat removal and a taller standpipe to improve the pressure recovery. In 1974, Sasol, reacting to the world oil crisis, decided to build the much larger Sasol 2 complex. A cooperative program was started with Badger to review the Synthol reactor operation and determine the best options to eliminate the operating problems. Excellent accounts of this cooperative development have been provided by others (Shingles and Jones, 1986; Jones, 1991). A brief review follows.

Sasol 2 was built using the Synthol concept and was deemed a major success. The disengaging hopper cone angle was increased to effect mass flow, eliminating previous bridging problems. The standpipes were designed with smaller diameters, eliminating stick–slip flow. However, the most significant design improvement was the replacement of the fixed tube sheets with serpentine cooling coils. By allowing the gas/solid mixture to flow on the outside of the coils, the instability problems associated with previous fixed tube sheet design were eliminated, allowing reliable continuous operation. According to Jones (1991), a test coil bank was installed in Sasol 1C to demonstrate feasibility and measure erosion. The test proved successful. Very high heat transfer coefficients were measured, and the mechanical integrity of the design was proven. The 10 Synthol reactors of Sasol 2 (Jones, 1991) have now operated for over 20 years with no reported problems with the coils. In the early 1980s, the Sasol 3 complex, essentially a copy of Sasol 2, was added. The second generation Synthol technology was licensed for use in the Mosselbay project that started in 1991. The Sasol 2 reactors are apparently still operated batchwise. As the catalyst ages, the density is reduced, and the mass flow through the reactor must be increased to maintain constant holdup (Shingles and McDonald, 1988).

The fractional solid holdup within the reactor ranges from 0.12 at the start of the run, to 0.17 at the end of the run.

In 1974, as part of the original Sasol–Badger development program, Badger tested and confirmed the possibility of using a conventional fixed fluidized bed (FFB) with the iron catalyst. At that time, this new reactor concept was given a lower priority due to schedule constraints of the Sasol 2 project and the need for demonstrating the new concept. The new reactor concept was bench tested in 1983 by Sasol in a 100 bbl/day unit. In 1989, Sasol returned to this idea. By the early 1990s, a 3500 bbl/day demonstration unit, designed by Badger, was brought on-line and proved to be an instant success. Following this, an 11000 bbl/day unit was started in 1995. Sasol has announced plans to replace all of the Synthol reactors with the third-generation FFB reactors. Figure 6 shows the comparison of the Synthol and FFB reactors. Some details of the FFB design and development have been published elsewhere (Silverman et al., 1986; Jaeger et al., 1990). The FFB reactor is believed to operate in the turbulent regime.

In the early 1980s, Sasol bench tested and then piloted in a 5 cm unit, a new slurry phase reactor concept for conversion of syngas to waxy liquids using an iron catalyst. The wax is upgraded by conventional technology to naphtha and diesel. In 1990, Sasol, in cooperation with Badger, demonstrated this new concept in a 100 bbl/day, 1 m diameter unit. A commercial unit of 2500 bbl/day capacity (circa 5 m reactor) was started in 1993. The slurry reactor is reported to operate at $240°C$ and about 20 atmospheres and will eventually replace the older fixed bed Arge reactors used at Sasol since 1955. The slurry reactor is shown in Fig. 7. Syngas is bubbled into the reactor via a support grid/distributor plate and rises through the bed. Reaction between the syngas and the liquid occurs on the surface of the suspended catalyst. Gaseous products and unreacted syngas exit the bed via the freeboard, while liquid product is withdrawn from the bed. Internal serpentine cooling coils are used to remove the heat of reaction by direct steam generation.

The slurry phase reactor provides much better and more flexible temperature control compared to the Arge reactor. It can operate at higher temperatures and with more active catalyst without formation of coke or catalyst breakup. The key operating issues are catalyst/liquid separation and catalyst attrition. Sasol has apparently resolved these design issues successfully.

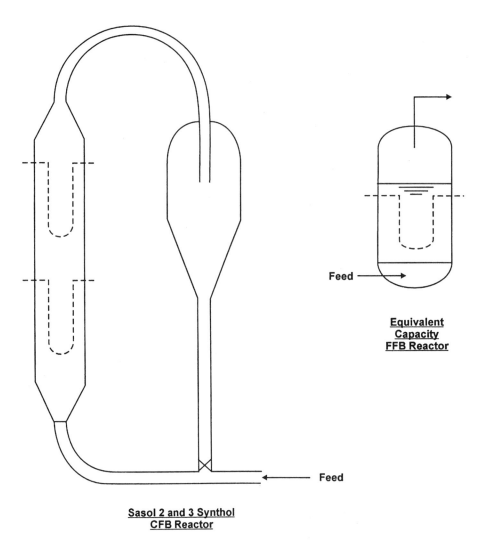

Sasol 2 and 3 Synthol
CFB Reactor

Figure 6 Sasol Synthol and FFB reactors. (From Jones, 1991.)

2.2 Exxon

Exxon (now ExxonMobil) has developed a two-step process for conversion of natural gas to liquid hydrocarbons. The process, referred to as AGC-21, utilizes a fluidized bed for catalytic partial oxidation/steam reforming of natural gas to syn gas, and a slurry bed reactor for conversion of the syn gas to hydrocarbons. The hydrocarbons are processed further to produce liquid products that are used as refinery or chemical plant feedstock.

The feed gas to the syn gas reactor is first treated for removal of H_2S and sulfur compounds. The treated gas, together with oxygen and steam, is then converted to syn gas in a high-temperature high-pressure fluidized bed reactor (Goetsch et al., 1989). An alpha alumina (95 wt% minimum) supported catalyst containing 0.5 to 2.5 wt% nickel and substantially no silica is used. The catalyst density is very high, between 2.4 and 3.9 g/cm^3. Exxon claims that catalyst fines leaving the reactor cyclone system can lay down in the overhead piping and will promote the reverse reaction of syn gas to methane, increasing methane slip. The above catalyst characteristics are considered essential by Exxon for minimizing fines generation and thus achieving high methane conversion. The preferred reactor temperature is between 980 and 1000°C. Operating pressure can be as high as 40 atmosphere, but the preferred range is 20 to 30 atmospheres. The fluidized bed syngas generator is shown in Fig. 8. Natural gas/steam and oxygen enter the bed via separate spargers. The reactor offgas is separated from entrained solids using

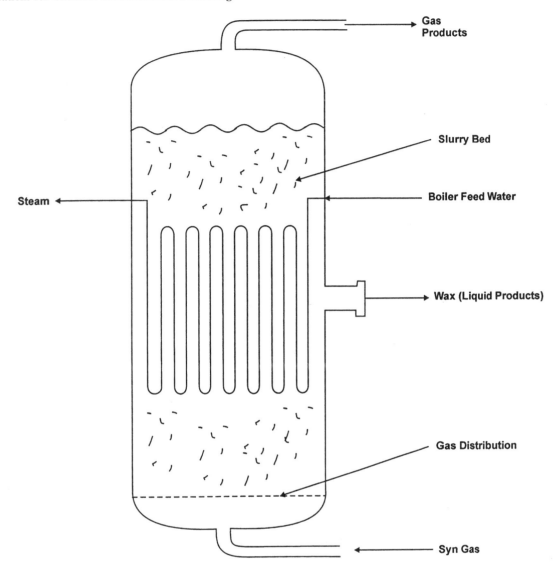

Figure 7   Sasol slurry bed reactor.

external cyclones and is then rapidly quenched to prevent the reverse reaction to methane and carbon formation by the Boudard reaction. Exxon claims that a minimum cooling rate of 150°C and preferably 200°C is required. A circa 1.5 meter pilot plant was operated in early 1990s to prove the syngas reactor concept.

Conversion of syngas to hydrocarbons is by the Fischer–Tropsch synthesis. A multiphase slurry bed reactor is used. Information on the slurry reactor is limited, but the general design is expected to be similar to that in Fig. 7. Exxon claims that the addition of an inert solid enhances bed expansion and reactor performance.

2.3   Syntroleum

The Syntroleum company has announced the development of an air-based gas conversion process. Syngas is produced using an air-blown authothermal reformer. The syngas and diluent nitrogen is then passed on a once-through basis through the conversion reactor. Syntroleum is developing fixed, fluidized, and slurry bed conversion reactors using cobalt catalyst. The company claims favorable economics at capacities as low as 5,000 barrels per day and has joined up with ARCO and Texaco to develop slurry bed conversion reactors.

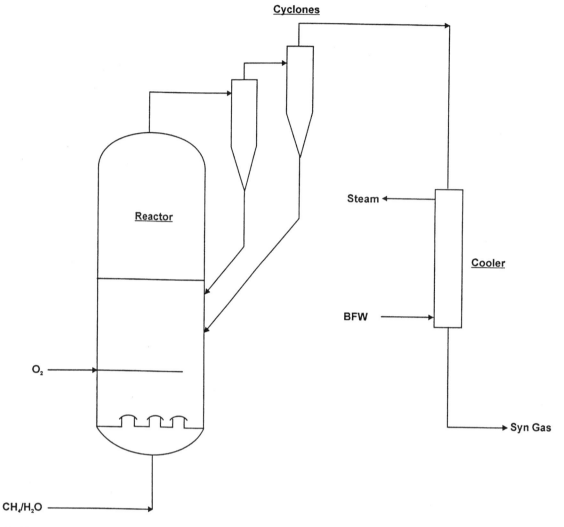

Figure 8 Exxon syn gas reactor.

3 POLYMERIZATION OF OLEFINS

The UNIPOL gas phase fluidized bed reactor for the production of polymers was commercialized by Union Carbide (now Dow Chemical) in 1968. This reactor produced high-density polyethylene (HDPE). The UNIPOL process was extended in 1975 to the production of linear low-density polyethylene (LLDPE) and in 1985 to polypropylene production. In the late 1980s, BP Chemicals began licensing its own gas phase Innovene fluidized bed process in competition with Union Carbide. The UNIPOL process currently holds the lion's share of the market with over 120 reaction lines sold or under construction. The reactor design is similar for all types of polymers and is shown conceptually in Fig. 9. A similar reactor is used by BP Chemicals, which has licensed 27 units.

Monomer, comonomer, hydrogen, and unconverted monomer are fed through a distributor plate into the fluidized bed, while catalyst and cocatalyst are fed into the bed above the distributor plate. Polymerization occurs on the catalyst surface creating particles in the 250 to 1000 micron range. The reaction is highly exothermic and has a low per pass conversion of the monomer. Chromium, Ziegler, or the new generation of metallocene catalysts are used in the process. Reactor operating conditions range from 75 to 105°C and from 20 to 25 atmosphere depending on the type of polymer being produced and the catalyst. The bed aspect ratio has been reported at 2.7 to 4.7.

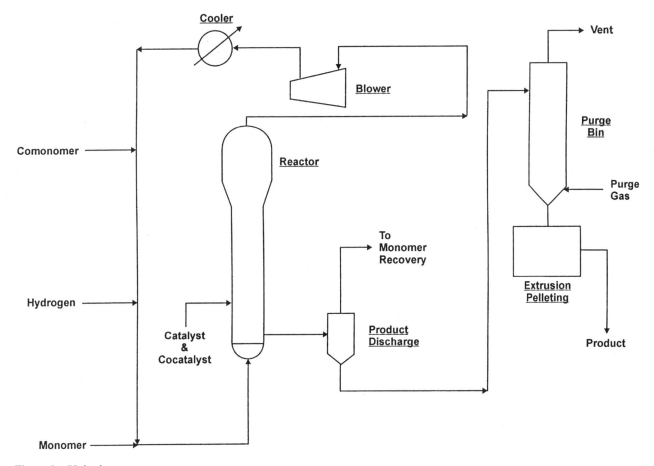

Figure 9   Unipol process.

Key reactor design issues are catalyst addition point, polymer withdrawal point, distributor plate design, aspect ratio, operating velocity, and temperature control. Temperature uniformity is critical as the reactor operates close to the polymer melting point. Maldistribution can result in polymer sheet formation or bed collapse; under severe conditions, there can be complete solidification of the reactor.

The upper section of the reactor is expanded to reduce velocity and minimize polymer entrainment. In the UNIPOL process, the gas and the small amount of entrained polymer are cooled and recycled back to the process. In the BP design, a cyclone is used first to recover the polymer before the gas is cooled and recycled.

Reactor capacity is limited by the system heat removal capability. Internal cooling coils cannot be used, as these will be quickly rendered useless by polymer film formation. Temperature control in the first generation reactors was accomplished by cooling of the recycle gas. Capacity of the newer plants have

been increased by as much as 200% by operating in the condensing mode. The condensing technology is offered by Union Carbide, BP, and Exxon. The recycle gas is cooled below its dew point to form some liquid. The latent heat of the liquid is then used to increase the reactor capacity. Union Carbide and Exxon have joined forces to offer supercondensing operation with as much as 50% liquid in the total feed. BP claims operation with up to circa 40% liquid. In the Union Carbide and Exxon process, the gas and liquid are fed together through the distributor plate. In the BP process, the gas and liquid are separated, and the liquid is fed above the distributor plate (Newton et al., 1995).

Swing reactor operation, where the operating conditions are changed to make different polymers, have been considered theoretically but do not appear to have been used in practice. Dual reactors in series have been used for impact polypropylene copolymer production (Burdett, 1992).

The polymer product and unreacted catalyst is withdrawn from the reactor and degassed with nitrogen in a dual lock hopper system to drive off interstitial hydrocarbon gases (Burdett, 1992). The amount of unreacted catalyst withdrawn with the polymer is extremely small owing to the very high productivity of the catalyst. As a result, the unreacted catalyst is left in the polymer.

4   HEAVY OIL UPGRADING

In addition to gasification (Chapter 15), several fluidized bed processes are available for upgrading of heavy oil. These upgrading processes fall into two groups. The first group upgrades heavy oil by carbon rejection. These include resid cracking (see Chapter 14), fluid coking, and flexicoking. The second group upgrades heavy oil by hydrogen addition. These

include H-Oil, LC Fining, Combi-Cracking, HDH, and U-can.

4.1   Fluid Coking

This continuous fluidized bed process was developed by Exxon (now ExxonMobil) in the 1950s. The process was piloted in a 100 BPD unit. The first commercial plant was built in 1955 with a capacity of 3,800 BPD, representing a 38-fold scale-up. A total of thirteen units have been built, the largest having a capacity of 72,000 BPD. The process is shown in Fig. 10.

Heavy oil is sprayed into a reactor of fine coke particles with an average diameter of 150 to 200 micron, fluidized by steam. The reactor operates at 500 to 600°C. The reaction products are coke, light ends, and distillates. The vapor products leave the reactor

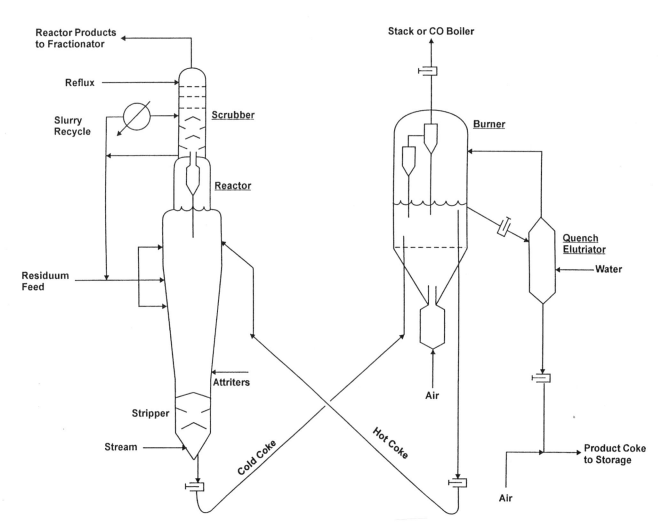

Figure 10   Exxon Fluid Coking process.

as gas bubbles, pass through cyclones and into a scrubber, where they are quenched before being fractionated. The coke particles from the reactor are first steam stripped for product recovery and then flow to the burner vessel, where a portion of the coke is burnt with air. Hot coke is then returned back to the reactor to supply the heat of reaction.

4.2 Flexicoking

This process uses the same configuration as the fluid coker but has an additional coke gasification step (Fig. 11), where excess coke is gasified to refinery fuel. The process is commercially proven and is licensed by Exxon.

4.3 H-Oil and LC Fining

These processes use an ebullated reactor. The H-Oil process is licensed by HRI (now Axens). The LC

fining is licensed by Lummus. Both processes have been proven commercially. The H-Oil reactor is shown in Fig. 12. Oil and hydrogen are fed into the bottom of the reactor, containing an ebullated bed of catalyst, where the oil is hydrogenated in the liquid phase. The reactor is operated in back-mixed mode by recycling the liquid. Catalyst activity is maintained by on-line addition and withdrawal. Five H-Oil units have been built ranging in capacity from 2,500 to 35,000 BPSD. The first unit was commercialized in 1963.

4.4 Combi-Cracking

The heavy oil is first hydrocracked in a liquid phase fluidized bed and is then hydrotreated in a fixed bed reactor. The upgraded product is very low in sulfur. The process was developed by Veba. The liquid phase reactor has been demonstrated in a 3,000 BPD unit.

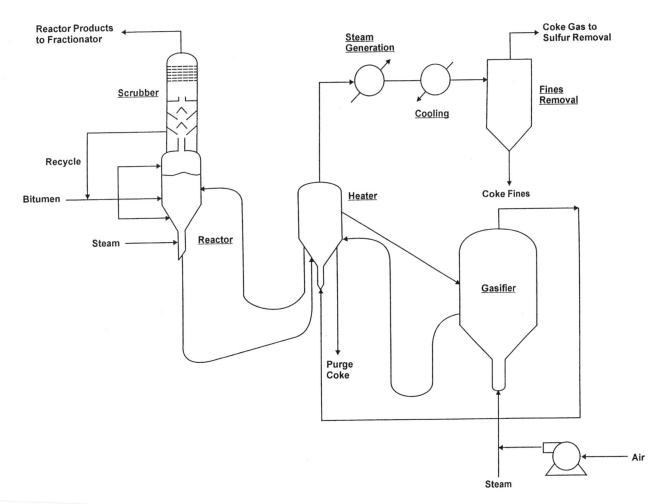

Figure 11   Exxon Flexicoking process.

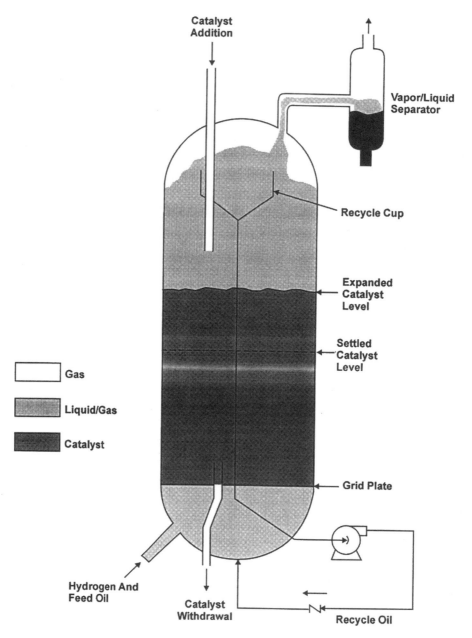

Figure 12 H-Oil ebullated bed reactor.

4.5 HDH

This process, developed by Intevep, has been only piloted. It is very similar to the first step in Combi-Cracking. A slurry bed is used to hydrocrack the heavy oil.

4.6 U-CAN

A bubble slurry phase reactor is used to upgrade the heavy oil. The process was developed by SCN-Lavalin. The reactor was demonstrated in a 5,000 BPD unit.

5 SEMICONDUCTOR SILICON

MEMC Electronic Materials uses a fluidized bed reactor originally developed by the Ethyl Corporation to produce high-purity silicon for use in the semiconductor industry. The process was piloted (Ibrahim and Johnston, 1990) using several pilot plant scale units of increasing size before being demonstrated in a 125 tons per year market development unit. A commercial plant with 1250 tons per year capacity was started in 1987. The high purity silicon is produced in the fluidized bed reactor by decomposition of silane in a hydrogen atmosphere by the reaction

$$SiH_4 \rightarrow Si + 2H_2 \tag{6}$$

This reaction is exothermic, but additional heat must be provided to balance thermal losses. Recycled purified hydrogen and silane are introduced continuously in the bottom of the reactor (Fig. 13), while high-purity silicon seed particles are added and the product withdrawn on a periodic basis. The product is cooled and then collected in storage hoppers. An inert gas is used to displace any entrained process gas. The product silicon is classified to remove fines before being transferred to shipping containers. The hydrogen-rich reactor offgas is cooled, cleaned of fines, compressed, purified, and returned to the reactor

To maintain high purity, equipment and piping made of high-purity silicon are used to transfer and handle product and seed silicon. Storage and shipping containers are stainless steel with fluorocarbon lining. The reactor is constructed with an internal lining coated with silicon. Additional silicon deposition during the course of normal operation of the reactor provides the ultimate high-purity containment of the reactor. The authors indicate it is undesirable for silane to decompose in the reactor above the fluidized bed, as this produces fines that deposit on the upper parts of

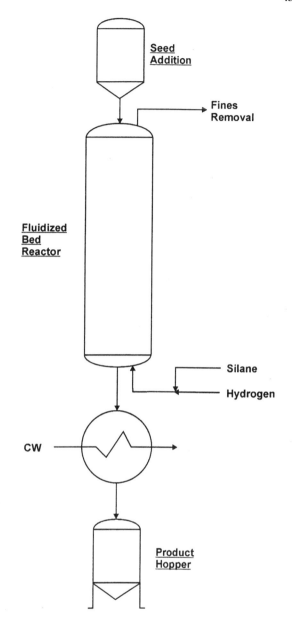

Figure 13   MEMC high-purity silicon process.

the reactor. Use of a quench zone in this section of the reactor is further suggested as a way to prevent this problem.

The chemistry of the process is complex. Heterogeneous decomposition results in particle size growth by direct deposition of silicon onto the particles as the silane decomposes. Homogenous decomposition occurs as the silane decomposes in the bubble or gas stream and results in silicon dust ranging from submicron size to 10 micron. These dust particles are elutriated from the reactor and represent an economic loss. Dust generation increases with higher silane concentration, increased reactor temperature, a larger reactor bubble holdup, higher gas velocity, and lowered bed height. However, without adequate bubble fraction, solid mixing is reduced and the particles can agglomerate when silane is decomposing on the surface. In extreme cases, the whole bed can collapse. Operating conditions of the reactor are reported as 600 to 800°C and 5 to 15 psig. Average particle size is 700 to 1100 micron. Operating velocity is 1.2 to 3.5 minimum fluidization. Under these conditions, the reactor operates in a slugging mode.

Ibrahim and Johnston suggest a two-cycle reactor operation. In the initial cycle, the reactor is operated in a high productivity condition with silane concentrations as high as possible. This operating condition results in appreciable homogeneous decomposition and fines generation. In the final cycle, the silane concentration is reduced for a short time. This condition promotes heterogeneous decomposition. A layer of silicon of possibly no more than 1 micron is deposited on the particles, which cements the loosely adhering silicon dust formed in the initial cycle. It is presumed that product withdrawal occurs immediately after the final cycle.

The silicon produced in this process typically contains 10 to 44 ppmw of hydrogen. This hydrogen level can be further reduced to 2 ppmw by heat treatment in a second fluidized bed reactor.

6   METALLURGICAL PROCESSING

6.1   Aluminum Flouride Synthesis

In 1948, Montedison and Lurgi piloted a fluidized bed for synthesis of AlF_3 from aluminum hydroxide $Al(OH)_3$ and hydrofluoric acid by the reaction

$$Al(OH)_3 + 3HF \rightarrow AlF_3 + 3H_2O \qquad (7)$$

The pilot plant produced 28 ton/day of AlF_3 with a purity of 92 to 94% (Reh, 1971). The process has since been commercialized successfully. Moisture-free aluminum hydroxide of about 45 micron average size is fed into the expanded upper section of a two-zone fluidized bed (Fig. 14.) The hydroxide contacts superheated hydrofluoric acid in a zone of increased solid concentration at the bottom of the lower reactor. Hot flue gas is introduced immediately below the expanded section to provide the endothermic heat of reaction and increase operating velocity to around 140 cm/s, affecting circulation of the solids through the cyclone to the

reactor bottom. In this manner the flue gas provides the necessary heat of reaction without reducing the concentration of the reactants. Reaction temperature is around 530°C. Operating pressure is close to ambient.

6.2   Alumina Calcination

This process represents the first large-scale application of circulating fluidized bed (CFB) reactors in the metallurgical industry. The process was piloted by VAW and Lurgi in a 24 ton/day pilot plant in the early 1960s. A 500 ton/day unit was started in 1970. Over 30 units are currently in operation with capacities of up to 2000 ton/day representing over 50% of world's alumina Al_2O_3 production. The process is depicted in Fig. 15. Wet aluminium hydroxide is dried in a cyclone dryer with hot flue gas from the calciner. After recovery in an electrostatic precipitator, the dry hydroxide is air lifted into a venturi preheater where it contacts hot gas from the calciner. The preheated hydroxide is mixed with circulating solids from the calciner in the recycle cyclone and is fed to the bottom of the calciner via the seal loop. Endothermic heat of reaction is provided by flue gas from a burner. The calciner is operated in a staged combustion mode for NO_X control. Operating temperature can be adjusted between 800 and 1200°C depending on the properties required for the product. Operating pressure is near ambient. The product is cooled by heat exchange with cold air and water in the cooler.

6.3   Ore Roasting

Fluidized beds have been used for the roasting of ores since 1947. Over 550 units have been built to date. Both KTI/Dorr–Oliver and Lurgi offer conventional single or double stage fluidized bed roasters. Roasting applications include iron pyrite and pyrrhotite for sulfuric acid, roasts for arsenic removal and roasts of metal sulfides for recovery of zinc, copper, cobalt, nickel, gold, tin, and molybdenum. Both slurry feed and dry feed systems are used. Refer to Kunii and Levenspiel (1991b) and Bunk (1990) for additional information.

In 1989, Lurgi commercialized a 575 ton/day wet feed slurry circulating fluidized bed refractory gold roaster operating at about 640°C and near ambient pressure (Peinemann, 1990). A second unit of identical design was added in 1990. A third unit with dry feed handling and a capacity of about 2000 ton/day gold

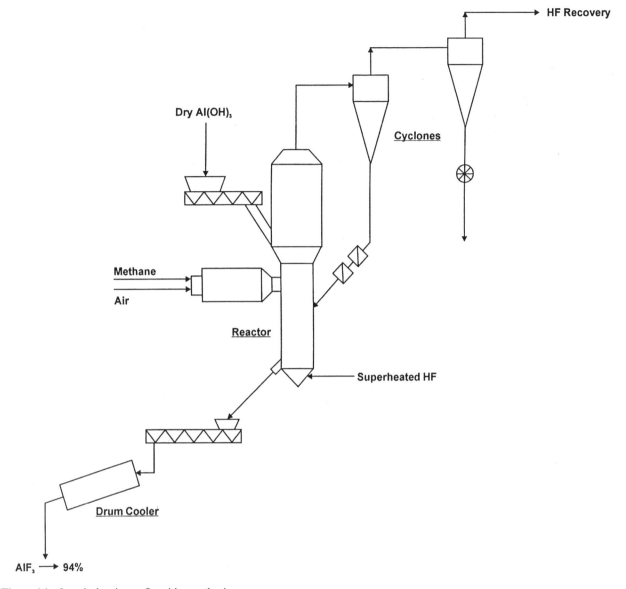

Figure 14 Lurgi aluminum fluoride synthesis.

containing refractory ore was started in 1991. KTI/Dorr–Oliver also has a circulating unit in operation

7 OTHER PROCESSES

7.1 UOP/Hydro Methanol to Olefins

UOP and Norsk Hydro have jointly developed and piloted a fluid bed process for the conversion of methanol to olefins (Fig. 16.) The process uses Union Carbide's (now part of UOP) SAPO-34 catalyst. A two fluidized bed reactor/regenerator system is used,

between which catalyst is circulated. Methanol conversion is essentially 100%. The reaction produces a small quantity of coke that will slowly deactivate the catalyst. To maintain activity, catalyst is circulated to the regenerator where the coke is burnt off with air. The heat of reaction is removed by steam generation. As the reactor severity is increased, the ethylene/propylene ratio and the coke yield increase, while the overall olefin yield drops (Vora et al., 1996). A 0.5 MT/day demonstration unit was operated in the mid-1990s. UOP has announced the successful completion of the demonstration unit and the availability of this process for licensing.

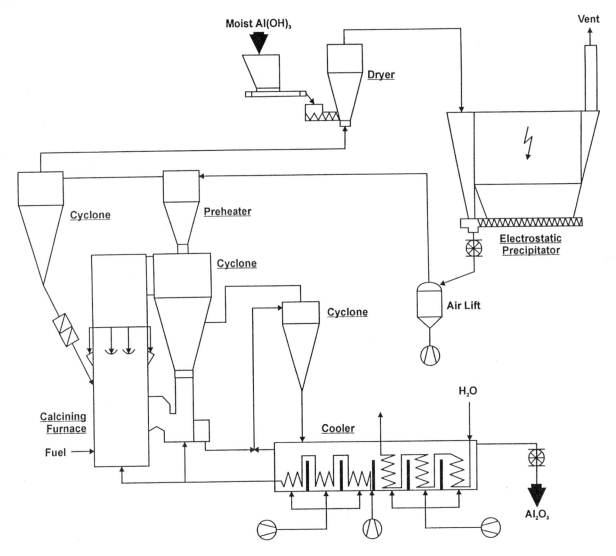

Figure 15 Lurgi alumina calcining.

7.2 Mobil–Badger Technologies

Three fluidized bed processes are available for license from Mobil–Badger (now ExxonMobil–Badger). These are MBR (Mobil benzene reduction), MOG (Mobil olefins to gasoline), and MOI (Mobil olefin interconversion). All three use zeolite catalyst and a dual dense fluidized bed reactor–regenerator system. The MOI process is shown in Fig. 17. The other two processes are conceptually very similar.

The MBR process reduces the benzene content of light reformate, FCC gasoline, or pyrolysis gasoline to below 1 vol% while boosting pool octane up to one point. The zeolite catalyst alkylates benzene with light olefins to form higher octane C8 to C10 aromatics. Single pass benzene conversion is 60 to 70%. However, overall benzene conversion can be increased to 90% by recycling a portion of the MBR reactor effluent. Once-through olefin conversion is greater than 90%.

The MOG process converts olefins, including light olefins contained in dilute streams, to C5+ gasoline pool components with yields of 61 weight percent at high severity and 75 weight percent at low severity. Olefin conversion is over 90%.

The MOI reactor operates at a lower pressure and shorter contact time than the MOG reactor to maximize light olefin production. Operating temperature is

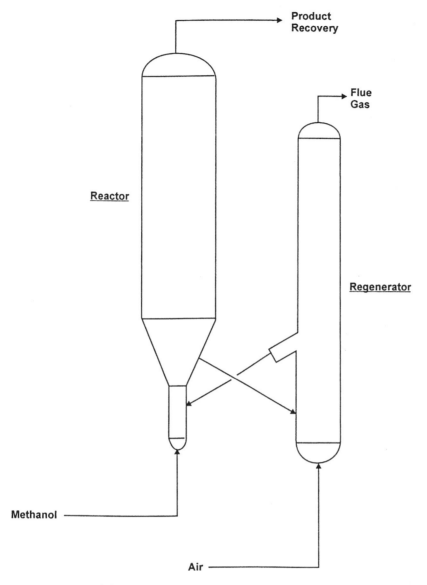

Figure 16   UOP/Hydro methanol to olefin reactor system.

about 538°C. By recycling, up to 40 weight percent of the total feed can be converted to propylene and iso-butylene. The reactor product slate can be changed by adjusting its operating temperature.

7.3   Catalytic Oxidation of Chlorinated Byproducts

The Catoxid fluidized bed process was developed by BF Goodrich (now Geon) in collaboration with Badger. The first unit was started in 1974. At least 4 units have been built.

Chlorinated hydrocarbon liquid streams from the oxychlorination process (this chapter), containing finely divided carbonaceous solids and metallic corrosion products in solution and suspension are pumped into the lower portion of the fluidized bed. Air is used to fluidize the bed and to oxidize the feed. The reaction occurs at below 540°C. Combustion of the feed is essentially complete with no significant breakthrough of elemental chlorine and only minor breakthrough of chlorinated hydrocarbons (Benson, 1979). The heat of combustion is recovered by generating medium to high pressure steam inside coils placed in the bed. The pro-

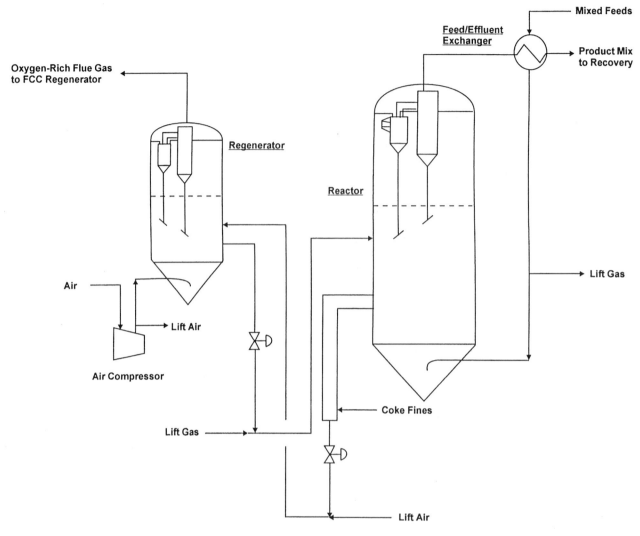

Figure 17 Mobil–Badger MOI reactor system.

cess requires auxiliary fuel only when the feed contains significantly more than 70% chlorine. The auxiliary fuel is required to sustain the correct bed temperature and satisfy the hydrogen–chlorine ratio for hydrogen chloride production.

The combustion gas leaving the Catoxid process contains hydrogen chloride, water, nitrogen, oxygen, and carbon oxides and can be recycled directly back to the oxychlorination reactor, as shown in Fig. 3.

The Catoxid catalyst is robust and is compatible with the oxychlorination reaction. Catalyst makeup is required only for attrition losses. The catoxid catalyst dust moves through the oxychlorination fluid bed reactor and leaves the system with the oxy catalyst fines.

7.4 Isophthalonitrile

GB Biosciences (now part of Zenneca) has been operating a dual fluidized bed process to produce isophthalonitrile by ammoxidation of m-xylene, by the reaction

$$C_8H_{10} + 2NH_3 + 3O_2 \rightarrow C_8H_4N_2 + 6H_2O \qquad (8)$$

The process was commercialized in 1976. A vanadium catalyst with Geldard group A/C characteristics is used. The catalyst is circulated between a reactor operating at 400°C and a reoxidizer operating at 427°C. The reactor operating pressure is not reported but is expected to be 1 to 2 atmosphere. Catalyst losses are kept to a minimum by the use of sintered metal filters (Fig. 18).

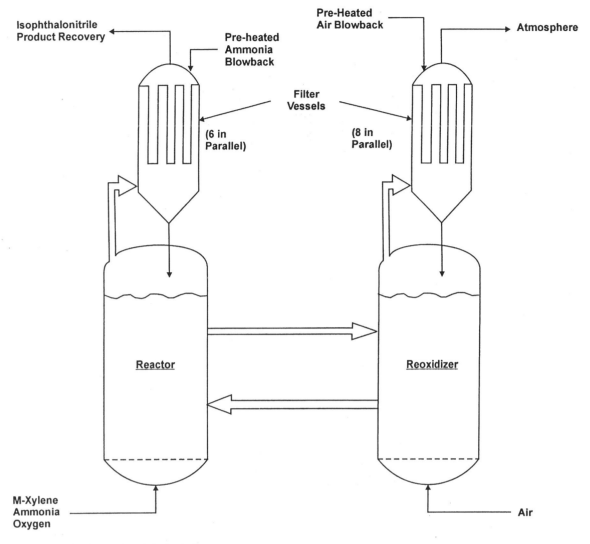

Figure 18 GB Biosciences isophthalonitrile reactor system.

REFERENCES

Benson JS. Catoxid for chlorinated byproducts. Hydrocarbon Processing, pp 107–108, October 1979.

Bolthrunis CO. An industrial perspective on fluid bed reactor models. Chem Eng Progress, pp 51–54, May 1989.

Bunk S. Fluid-bed roasting advances. Randol Gold Forum, pp 173–174, September 1990.

Burdett ID. A continuing success: the UNIPOL process. Chemtech, pp 616–623, October 1992.

Contractor RM. US Patent 4,666,802 to E I DuPont de Nemours and Company, May 26, 1985.

Contractor RM, Bergna HE, Horowitz HS, Blackstone CM, Malone B, Torardi CC, Griffiths B, Chowdry U, Sleight AW. Butane oxidation to maleic anhydride over vanadium phosphate catalysts. Catalysis Today 1:49–58, 1978.

Contractor RM, Pateince GS, Garnett DI, Horowitz HS, Sisler GM, Bergna HE. A new process for n-butane oxidation to maleic anhydride using a circulating fluidized bed reactor. Proceedings of the Fourth International Fluidized Beds Conference, Hidden Valley, 1993, pp 387–391.

Goetsch AD, Say GR, Vargas JM, Eberly PE. US Patent 4,888,131 to Exxon Research and Engineering, 1989.

Graham JJ. Safety features of fluid bed process. Chem Eng Progress 66(9):54, September 1970.

Graham JJ, Way PF. Phthalic anhydride by fluid bed process. Chem Eng Progress 58(1):96–100, 1962.

Ibrahim J, Johnson SW. High purity polysilicon production from silane by fluidized bed process. Proceedings 2nd Symposium on Fluidized Bed Processing of Powder Materials. Soc Chem Eng Japan 1990, pp 26–36.

Jaeger B, Dry ME, Shingles T, Stynberg A. Experience with a new type of reactor for Fischer–Tropsch synthesis. AIChE Spring National Meeting, Orlando, March 1990.

Johnsson JE, Grace JR, Graham JJ. Fluidized-bed reactor model verification on a reactor of industrial scale. AIChE J 33(4):619–627, April 1987.

Jones DH. Development of Sasol's Fischer–Tropsch technology. AIChE Spring National Meeting, Houston, Texas, April 1991.

Kirk–Othmer. Encyclopedia of Chemical Technology. 4th ed. Vol. 24. New York: John Wiley, 1996, pp 860–865.

Kunii D, Levenspiel O. Fluidization Engineering. 2nd ed. Boston Butterworth–Heinemann, 1991a, p 32.

Kunii D, Levenspiel O. Fluidization Engineering. 2nd ed. Boston Butterworth-Heinemann, 1991b, p 51.

Miseralis CD, Way PF, Jones DH. The evolution of fluidized-bed reactors in the phthalic anhydride production. AIChE Spring National Meeting, Houston, Texas, April 1991.

Naworski JS, Velez ES. Oxychlorination of ethylene. In: Leach BE, ed. Applied Industrial Catalysis. Vol. 1. Academic Press, 1983, pp 239–272.

Newton D, Chin JC, Power M. Advances in fluidized bed gas-phase technology. HTI Quarterly, pp 81–85, Autumn 1995.

Ohta M, Yokura M. Fluidized-bed reactor and reaction process using the same. International Patent Application PCT/JP96/00171 to Asahi Kasei Kogyo KKK, January 30, 1996.

Peinemann B. Lurgi's circulating fluid bed roasting process improves gold recovery. Randol Gold Forum, September 1990, pp 169–172.

Pell M, Jordan SP. Effects of fines and velocity on fluid bed reactor performance. AIChE Symp Ser. Vol. 84, No. 262, pp 68–72, 1987.

Reh L. Fluidized bed processing. Chem Eng Prog 167:58–63, 1971.

Shingles T, Jones DH. The development of synthol circulating fluidized bed reactors. Chem SA, pp 179–182, August 1986.

Shingles T, McDonald AF. Commercial experience with synthol CFB reactors. Circulating Fluidized Bed Technology II, Compiegne, 1988, pp 43–50.

Silverman RW, Thompson AH, Stynberg A, Yukawa Y, Shingles T. Development of a dense phase fluidized bed Fischer–Tropsch reactor. Fluidization V. New York: Engineering Foundation, 1986, pp 441–449.

Volk WC, Johnson CA, Stotler HH. Effect of reactor internals on quality of fluidization. Chem Eng Prog 58:44–47, 1962.

Vora BV, Lentz A, Marker TL. UOP/Hydro MTO process. World Petrochemical Conference CMAI, Houston, March 1996.

17

Applications for Coating and Granulation

Gabriel I. Tardos

The City College of the City University of New York, New York, U.S.A.

Paul R. Mort

Procter & Gamble, Cincinnati, Ohio, U.S.A.

1 SUMMARY

We present, in this chapter, applications of fluidized beds to coating and granulation of powders. The process, widely used in industry, is designed to agglomerate fine powders into larger granules and/or to coat large particles or granules with fines. To achieve aggregation and growth and to ensure sticking together of powdery particles, a binder as a solution or a melt is used. The goal of this work is to present a general practical and theoretical framework of binder granulation that takes an agglomeration process from binder selection and testing, to granule formation, growth, and consolidation, and finally to granule deformation and breakup.

For agglomeration and granule growth to take place, a certain amount of binder has to be introduced into the fluid bed granulator: special instrumentation and procedures for binder selection are presented in this chapter. In such granulation processes as detergents and pharmaceutical products, both the powders to be agglomerated and the binders are defined by the formulation, and little liberty is given to alter the chemistry. Binder "selection" in this case is practically reduced to adjusting the properties of the binder using small amounts of additives (mostly surfactants) and

tailoring the binder to exhibit specific behavior. This allows fine-tuning of binder properties that include surface wetting, spreading, adsorption, binder strengthening, and solid bridge strength.

The bulk of the presentation is dedicated to the theory of growth kinetics during granulation and the prediction of critical sizes that delimit different regimes of granulation. Several dimensionless parameters based on energy dissipation principles are presented and examples given about how these parameters, and the critical sizes they define, can be used to predict the outcome of granulation and the scale-up of the process. The above theoretical framework is then tested with experimental data from the literature and with granulation results obtained by the present authors.

A section is dedicated to the interpretation of granulation experiments performed at different scales and how these can be used in scale-up of the process from laboratory and pilot size to full-scale industrial equipment. It is only very recently that criteria for such scale-up have become available mainly from theoretical considerations of granule growth and breakup at the micro-scale, and this new knowledge is briefly presented. Finally, some recent computations are presented in which the process of both agglomeration and growth and deformation and breakup are simulated using numerical techniques. These results are significant because they shed light on the details of the micro-scale process and aid in scale-up.

1 INTRODUCTION

Granulation is a size enlargement operation by which a fine powder is agglomerated into larger granules to generate a specific size and shape, improve flowability and appearance, and, in general, produce a powder with specific properties such as dissolution rates, granule strength, and apparent bulk density. Granulation of powders is practiced at an enormous scale in industries including detergents, foods, agricultural chemicals, and pharmaceutical products. During *binder granulation*, a liquid binder such as a solution or a melt is pumped, poured, or atomized onto an agitated bed of different powders contained in a mixer whose main role is to provide shearing forces in the powder mass. As solvent evaporates from the binder (liquid) or the melt thickens, powder particles stick together and, as interparticle bridges strengthen, larger granules of the original powder are formed. These granules are further consolidated by forces in the mixer and, upon final solidification of the binder or melt, strong agglomerates are left behind.

There are several key product transformations that occur in granulation processes, including binder atomization, fluidization, wetting and spreading of binders on powder surfaces, agglomeration (including nucleation, coalescence, and layering), consolidation, binder solidification, drying, and attrition. In addition, a reactive binder can be used to create solid bridges of the reaction product between particles. In this later case, the kinetics of the chemical reaction is an additional transformation that is key to the overall process result. The interdependence of the many transformations involved in a typical granulation further adds complexity to the process. Given this degree of complexity, one is compelled to identify the key individual transformations and then reduce them to a fundamental basis, i.e., to micro-scale interactions. Fundamental understanding on the micro-scale level provides then a foundation for predicting the interactions on the macro-scale and the effects of process parameters and key material properties.

Binder granulation can be achieved in different types of mixers ranging from rotating drums and pans to high shear mixers and fluidized and spouted beds. During the present analysis, the accent is put on fluid

bed agglomeration and layering or coating, although combined mixers that include both fluidization and mechanical agitation are also considered. Fluidized bed granulation is in some ways different from other types of mixer granulation in that the gas supplied to produce powder agitation through fluidization also causes binder evaporation and cooling (or heating) of the powder. In addition, particle size increase in a fluidized bed is associated with many changes in fluidization characteristics, the most important of which are the mixing properties of the bed. These interacting phenomena make fluid bed granulation by far the most complex while at the same time the most versatile, allowing drying and cooling transformations to be carried out simultaneously with size increase (agglomeration). On the other hand, simultaneous transformations pose difficult challenges to scale-up of industrial applications.

Several excellent reviews of fluid bed granulation were published by Nienow and coworkers (Nienow, 1983; Nienow and Rowe, 1985; Nienow, 1994), in which the operation is described in detail. Heat and mass balances on the system are presented and a granule growth model is proposed. The advantages and disadvantages of the operation are discussed. An overall review of all types of granulation was also presented by Sommer (1988). The present contribution differs from the above works by looking at the micro-scale processes of granule formation, growth, and breakage and by studying the behavior of liquid bridges between agglomerating particles. The micro-scale view is powerful in that it generates an overall picture of binder granulation from such specific phenomena as particle coalescence, layering of fines, deformation and breakage of wet granules, and final product consolidation.

2. THE FLUID BED AS A MIXER/ GRANULATOR

Some typical fluid bed granulation equipment and a few of their main characteristics are given in Fig. 1. The conventional fluid bed granulator (Fig. 1a) can be operated in either "top" (as shown) or "bottom" spray mode. The bed is usually tapered, as shown, to improve solid circulation. The spouted bed granulator (Fig. 1b) is somewhat different in that it has no distributor and that there is a calibrated "gap" between the nozzle and the bottom of the spout that allows discharge of granulated material. It is common industrial practice to combine the two fluid beds presented so far

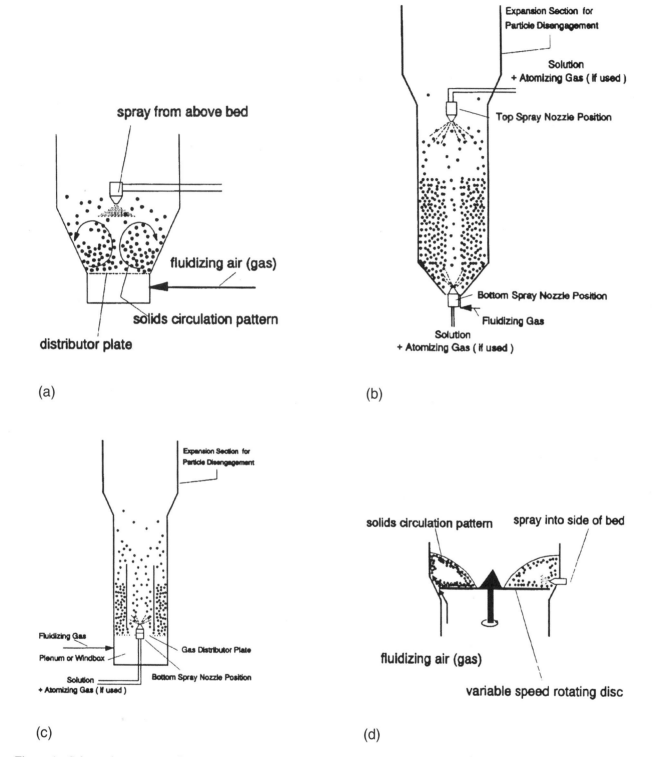

Figure 1 Schematic representation of some industrial granulators. (a) Conventional fluid bed granulator. (b) Conventional spouted bed granulator. (c) Spouted bed with draft tube. (d) High shear rotor granulator.

in large units with the spouted bed forming "units", which when combined yield large granulation areas each with individual spray nozzles. The fluid bed with the "draft tube" (Fig. 1c) is mainly used for coating operations. The tube is used to create strong solid circulation in the bed so that a leaner phase is created inside as particles are blown upward in the tube and a more dense phase at the periphery where drying takes place.

The tendency in present industrial granulation is oriented toward the use of combination equipment such as the high shear–mechanical fluidized bed shown in Fig. 1d. The device has a rotating distributor (that can be flat or made of several interfitting cones) through which air is pumped. Nozzles are placed at the periphery, and a strong solid circulation is achieved due to wall friction on the distributor and on the tapered section of the housing. These devices are very versatile and allow, besides very effective particle growth and coating, also drying, heating, and grinding operations in a very short operation cycle. Several other constructions of "combination" devices are known, and these are exhaustively reviewed by Ormos (1994).

The main purpose of the fluid bed in granulation is to act as a mixer for the granulating powder. This is achieved by creating shear layers in the mix, characterized by an average shear rate, γ°, where particles move relative to each other. The main characteristic of the fluid bed is the relative velocity imparted to the parti-

cles, U_0, that is a strong function of the size of the particles and the gas velocity in the bed and was shown to be given by (Ennis et al.,1991)

$$U_0 \approx a\gamma^\circ = 18 \frac{U_{\mathrm{B}}a}{D_{\mathrm{B}}}\delta^2 \tag{1}$$

where a is the average particle size, U_{B} is the bubble velocity, D_{B} is the bubble diameter, and δ is the dimensionless bubble spacing. The first expression on the RHS applies to combination mixers in general while the last expression on the RHS of Eq. (1) applies to fluidized beds with no rotating parts where shear is induced by the motion of bubbles only.

To obtain certain size granules from a fluid bed granulator, several other unit operations have to be used. These are combined in so called "granulation circuits" where the fluid bed is the central unit with other peripheral equipment also included. An example of such a circuit is given in Fig. 2 (after Dencs and Ormos, 1993). The fluid bed granulator shown is of a very special kind and, as seen, is equipped with grinding rollers activated by a motor; this unit performs granulation and grinding operations simultaneously. In general, however, these two operations are performed separately in different pieces of equipment and therefore a conventional granulation circuit also includes a grinder and some sieving equipment for granule size control. Other peripheral equipment includes tanks and mixers for the preparation and feeding of the binder (Binder solution in Fig. 2),

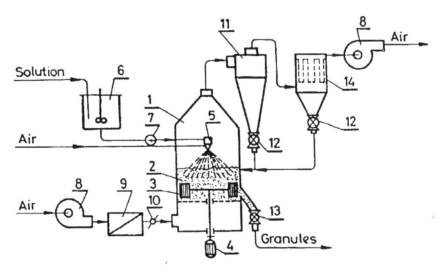

Figure 2 Schematic representation of a fluid bed granulation circuit. Shown in the circuit is a fluid bed grinder–granulator. 1. Fluid bed houseing, 2. Fluidized bed, 3. Grinding roller, 4. Drive, 5. Spray head, 6. Binder solution, 7. Pump, 8. Fluidization air fan, 9. Air heater, 10. Air control valve, 11. Cyclone, 12 and 13. Rotary valves, 14. Filter bag-house. (After Dencs and Ormos, 1993.)

pumps, heaters for air supply, gas cleaning equipment such as cyclones and filters for the separation of fines and granule discharge, and transport devices. More complex circuits have several fluid bed granulators in series in which different operations such as granulation, layering, wetting, drying, heating, and cooling are also performed. For details on such specialty devices, the reader is referred to the review of Ormos (1994) and references within.

3 MICROSCOPIC PHENOMENA

The role of the spray nozzle shown in Figs. 1 and 2 is to introduce binder (liquid) into the mixing powder mass in as uniform a layer as possible. Consequently, we encounter, in the shearing powder mix, individual particles covered by a liquid layer. The process by which these fine powder particles, partially or totally covered by a liquid layer, stick together in a shearing mass of powder to form large granules that possess, in the end, enough strength to survive in the granulator is quite complex (Ennis et al., 1990, 1991; Ennis, 1990; Tardos et al., 1985, 1993, 1997).

Figure 3 depicts some, more common, ways in which agglomerate growth may occur. It must be emphasized that these micro-scale mechanisms depend on the degree of binder dispersion in the powder. Typically, dispersion varies with binder atomization, addition rate, and state of fluidization or shear in the mixer. *Nucleation* is defined as the sticking together of primary particles owing to the presence of a liquid binder on the solid powder surface (Fig. 3a). *Coalescence*, on the other hand, is the process by which two larger agglomerates combine to form a granule (Fig. 3b). In coalescence, the porous granule surface is saturated with binder, and the colliding granules are sufficiently malleable to allow for deformation and bonding (Kristensen et al., 1985; Kristensen, 1991; Kristensen and Schaefer, 1987; Schaefer et al., 1990; Schaefer, 1988; Tardos et al., 1997). Other modes of granule growth are *layering* of a binder-coated granule by small fine particles (Fig. 3c) and the capture of fines by a partially filled binder droplet (Fig. 3d).

The growth mechanisms shown in Fig. 3 depend on the collision and bonding of particles. One can assume that shear forces in a mixer will cause particles to collide at some point along their trajectory. For bonding of particles to occur, it is essential that some binder be present at the point of contact, as depicted in Fig. 4. This can be in the form of pure binder, as shown in the figure, or the solid surface of the particle at the contact

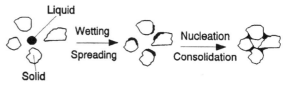

(a) Agglomerate Formation by Nucleation of Particles

(b) Agglomerate Growth by Coalescence

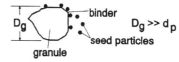

(c) Layering of a Binder Coated Granule

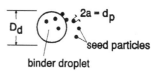

(d) Layering of a Partially Filled Binder Droplet

Figure 3 Granule growth mechanisms (a) Agglomerate formation by nucleation of particles. (b) Agglomerate growth by coalescence. (c) Layering of a binder-coated granule. (d) Layering of partially filled binder droplet.

point can exhibit some "softness", i.e., malleable character, which would allow surfaces to bond upon contact. From the simplified picture of Fig. 4, all mechanisms of Fig. 3 can be easily reconstructed if one allows for variation in size and nature (solid or liquid) of the colliding particles.

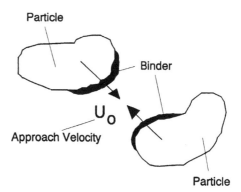

Figure 4 Coalescence of two binder covered particles (granules)

The study of the *kinetics of granule growth and consolidation* in binder-granulation can be undertaken using the picture in Fig. 4 that encompasses, in a simplified way, all mechanisms depicted in Fig. 3. The relative velocity, U_0, between binder-coated particles is generated by the *mixer*, i.e., the fluidized bed, in this case. The magnitude of the relative velocity is mixer dependent and is predicted approximately by Eq. (1). The kinetics of the process essentially reflect the conditions under which two colliding particles will stick together or rebound in the shear field: sticking will yield growth while rebounding will not. The rate of growth will depend on the relative sizes of the individual particles involved. The mechanisms of coalescence (Fig. 3b) and nucleation (Fig. 3a) involve collisions of similar-sized particles and will yield higher growth rates than layering (Fig. 3c) and engulfment (Fig. 3d), which incorporate fines particles into a larger granule. Once formed and grown to a certain size, granules have to survive in the shearing mass; special conditions between the magnitude of shear in the outer shearing mass and the inner strength of the granule have to be satisfied in order for this to happen. Growth is also influenced by granule consolidation that is closely related to the viscosity of the binder after granule formation and the time allowed for deformation of the formed bridges.

Several essential conditions have to be fulfilled a priori in order for granule growth and consolidation to occur. These conditions pertain to the *presence of appropriate amounts of binder on the granular surface at the points of contact* and the binder's physical and flow properties in the bridge; these aspects of granulation will be examined first. Following this, granule growth kinetics and considerations regarding granule consolidation will be presented.

4 CONDITIONS FOR GRANULATION

It is intuitively obvious from the schematic representation in Fig. 4 that a certain amount of binder is required in the powder mass before enough will be present on the surface to ensure stickiness. This critical, minimum amount of binder is an important characteristic of the system and must be determined beforehand. Equally important are (1) the initial distribution of the binder within the bulk powder and (2) the time span over which the binder either spreads on the granular surface and/or penetrates into the pores of the powder. Also critical is the time over which the binder bond increases in strength through evaporation

of solvent, reactive transformation, cooling, or other mechanism of solidification. The critical binder/powder ratio, the characteristic spreading/penetration time, and the binder strengthening rate will be discussed before the theory of growth kinetics is presented.

4.1 Critical Binder (Liquid)/Powder Ratio

Mixer torque rheometry is a useful method of determining critical binder/powder ratios. There has been a large amount of recent work on this subject relative to scale-up of high-shear mixer-agglomerators (Landin et al., 1996a,b; Rowe and Parker, 1994; Hancock et al., 1994). The instrument consists essentially of a mixer in which the powder is slowly agitated and the binder is continuously introduced by a metering pump. The instrument allows the torque of the mixer to be monitored continuously. An important condition to be fulfilled during this procedure is to allow adequate time during binder addition for the liquid to spread and adsorb into the interstices before new binder is introduced. In this way one ensures that the finite spreading and adsorption rate of the liquid does not influence the results.

To illustrate the above procedure, the use of a mixer-torque device to measure torque versus binder content (di-butyl-phthalate, DBT) for a sodium carbonate mixture is shown in Fig. 5. A slight increase in measured torque occurs at $\sim 25\%$ by volume while a sharp increase occurs at $\sim 65\%$ by volume DBT. These two points correspond to the saturation of the capillary-loaded granules and in particular to the

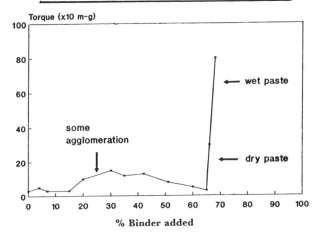

Figure 5 Experimental torque vs. binder/solid ratio for a porous sodium carbonate powder.

crossover to the funicular and the capillary states, respectively. Granulation of the powder shown in Fig. 5 can be performed at the lower binder content of 25–30% binder. Since binder is only filling some voids, growth will occur mostly by nucleation with very little layering and no coalescence because insufficient binder is present on the granule surface. At liquid loading corresponding to the drastic increase in torque (at 65% binder), all mechanisms of growth occur including coalescence: this value is considered for this powder to be the critical binder/solid ratio.

One has to note that the rate of binder addition and the shear rate in the mixer also affect the binder/solid ratio actually achieved during industrial granulation. Decreasing the shear rate and increasing the binder addition rate will result in excessive growth and lumping not because the critical binder/powder ratio was exceeded but because not enough time was given for the liquid to spread and adsorb into pores. The above considerations are presented here as an example of how to determine binder/powder (L/S) ratios. The instrumentation used is not unique, and in general any small size mixer of the medium or high shear type can be used (see also work by Tardos et al., 1993). Alternatively, the binder/powder (solid) ratio can be determined in actual granulation runs by performing granulation with increasing amounts of binder until a wet mass is obtained. It is essential, however, that this determination be done properly, since too high or too low amounts of binder yield undergranulated and dusty or overgranulated and sticky products.

4.2 Binder and Interparticle Bridge Properties

Under normal conditions of granulation, the binder is introduced into the powder bed as a spray of small droplets. These droplets impact on the solid powder surface and deposit on it: if solid–liquid contact angles are such that surface spreading can take place, the binder droplet will flatten and cover an ever-increasing area of the surface. At the same time, however, liquid will penetrate into the surface pores of the granule. Both surface wetting and spreading are necessary, since a nonwetting liquid will either not stick to the surface or be present on or cover a very small area, thereby restricting the number of collisions that yield growth. Penetration into the granule surface pores is also required to give the granules the malleable, plastic property needed for coalescence (Fig. 3b).

To characterize spreading/penetration characteristics of binder/powder systems, sessile drop experiments are performed using commercial goniometers. Binders

are spread onto tablets pressed from the powder to be granulated and the process is observed under a high-magnification microscope. In a more sophisticated version of the instrument, the solid–liquid contact line is digitized by a computer and the liquid volume and spreading area is computed as a function of time (see for details Mazzone et al., 1987; Ennis et al., 1990; Tardos and Gupta, 1996). If spreading, penetration, and liquid bridge formation are found inadequate for a given binder, different binders are chosen or, if this is not an option, binder properties are altered by using additives, surface active agents, etc.

A further condition of successful granulation is the production of granules that gain sufficient strength so that they do not fall apart once they are formed. This is achieved in granulation by choosing an appropriate binder that generates bridges that strengthen in time owing either to solvent evaporation or drying or to cooling of the melt. Strengthening is a result of increased binder viscosity and depends on the binder chemistry and on the heat and mass transfer rates in the granulator. Extensive research showed that, in general, rates of heat and mass transfer in fluid bed granulators are quite high, and so the choice of the appropriate binder chemistry is crucial.

A special instrument was developed (Ennis, 1990; Ennis et al., 1990; Tardos and Gupta, 1996; Tardos et al., 1997) to test binders and measure their strengthening rates. The instrument is essentially a force transducer and an oscillating mechanism activated by an electric motor mounted in the view of a projection microscope. Two powder granules or tablets pressed from the ungranulated material are mounted (glued) onto the transducer and the oscillating arm. A droplet of binder is inserted between the two particles (tablets), and a pendular bridge is formed. The oscillator is put in motion and the bridge strength is recorded by the transducer as a function of time. Both the amplitude and the frequency of oscillation and the bridge temperature can be changed at will. At the same time, pictures of the deforming bridge can be taken, and imperfections in wetting and spreading can also be observed.

The measured force due to the bridge (Gupta and Tardos, 1996; Tardos et al., 1997) increases from an initial value that can be calculated from the knowledge of the liquid surface tension, to a maximum value after which the mostly solidified bridge ruptures. Both the maximum force and the critical bridge rupture time are important for granulation, but it was found that the critical strengthening time had to be within very narrowly defined limits for the granulation to be successful. These limits depend on the granule properties to be achieved: for example, to produce open, high-porosity, easily soluble granules, the binder strengthening time should not exceed one-third to one-half of the total granulation time, while in coating or layering operations it must be very short, of the order of several seconds or a fraction of a minute. The above findings are quite significant, since they show a clear way to design solvent evaporation in the bed when the binder is a solution, or the cooling time when the binder is a melt. One has to stress, however, that the above estimates are probably not very general and may require corrections for different pairs of powders and binders. Since the experimental procedure and the instrument are quite simple and straightforward to use, such measurements should be easy to perform for each specific case.

5   GRANULE GROWTH KINETICS

It will be assumed for the present considerations that sufficient binder is present in the granulator as determined by the binder/powder ratio, and that the binder is appropriately spread on enough granular surfaces to ensure that most random collisions between particles will occur on binder-covered areas. It will also be assumed that the particles are mostly spherical having a characteristic dimension, a. The liquid is characterized by its surface tension γ and its viscosity μ. In those cases where the surface of the colliding particles is itself soft and malleable, the viscosity μ must be taken as that of the surface layer. The relative velocity U_0 is taken to be only the normal component between particles, as seen in Fig. 3, while the tangential component is neglected.

The system of the two approaching particles is shown schematically in Fig. 6. The particles are depicted after the liquid layers on their surfaces have already touched and a liquid bridge has formed. While there may be some additional binder on the free surface of the particles, this is not shown in the figure and is irrelevant for the collision process at hand. As the two particles approach with the relative velocity U_0, the liquid will be squeezed out from the space between them to the point where the two solid surfaces will touch. A solid rebound will occur based on the elasticity of the surface characterized be a coefficient of restitution, e, and the particles will start to move apart. Liquid binder will now be sucked into the interparticle gap up to the point where a liquid bridge will form; upon further movement the bridge will elongate and

finally rupture as the particles separate entirely. When the particles are deformable (malleable) throughout as in the process depicted in Fig. 3b, rebound will not occur, and coalescence will depend on the deformability of the granule–binder system.

It was shown by many researchers (a summary of pertinent work in this area is given in Ennis et al., 1990) that under fairly general conditions the total force, F, induced by a liquid bridge between two solid particles can be calculated from the summation of two effects: a surface tension contribution proportional to the bridge volume or the filling angle, ϕ, and the surface tension of the fluid, γ, and a viscous contribution dominated by the relative velocity and the viscosity of the liquid, μ. The superposition gives accurate results and can be expressed analytically by

$$F = F_{\text{vis}} + F_{\text{cap}} = \frac{3\pi\mu U_0 a^2}{4h} + \pi\gamma a \sin^2[\phi(C_0 + 2)]$$

(2)

where C_0 is the Laplace–Young pressure deficiency due to the curvature of the free surface of the liquid and h is the thickness of the liquid layer (see Fig. 6).

5.1   Conditions of Coalescence

The outcome of the collision of two binder-covered particles is determined by the ratio of the initial kinetic energy of the system and the energy dissipated in the liquid bridge and in the particles (Ennis et al., 1991; Tardos et al., 1997). This can be expressed analytically by the definition of a so-called Stokes number related to particle coalescence, St_{coal}:

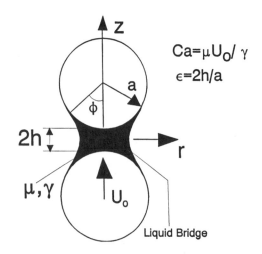

Figure 6   Liquid bridge formed between two moving, spherical particles covered with a liquid layer.

$$St_{coal} = \frac{\text{initial kinetic energy}}{\text{dissipated energy in the bridge}}$$

$$St_{coal} = \frac{2m_p U_0^2}{2F_{vis}h} = \frac{8\rho_p U_0 a}{9\mu} \qquad (3)$$

where ρ_p is the particle density, and m_p is the mass of the particle, F_{vis} is the viscous force given in Eq. (2), and μ is the surface viscosity. This dimensionless number increases in value as particle (granule) size increases during granulation. Simple energy considerations show that if the Stokes number defined above is smaller than a critical value $St_{coal} < St_{coal}^*$, collisions between particles are effective and coalescence occurs, while if $St_{coal} > St_{coal}^*$, particles rebound because the total incoming energy is larger than the one dissipated during collision. One has to note that μ in the above equation can be taken to be the binder viscosity or *an equivalent viscosity of the granular surface.*

An important observation is related to the characteristic particle size, a, which is well defined for two equal particles but, in a more general case, becomes an equivalent size calculated from

$$\frac{2}{a} = \frac{1}{a_1} + \frac{1}{a_2} \qquad (4)$$

where a_1 and a_2 are the sizes of the colliding particles (granules). It is immediately apparent from this equation that during collisions between small and large particles when $a_1 \gg a_2$, the equivalent size is that of the smaller particle, $a \sim a_2$, and so for this case the size of the fine particle is the only relevant dimension to be used in the calculation of the Stokes number. The implication is that for small particles the dimensionless Stokes number is always small, always less then the critical value, St_{coal}^*, and therefore small particles will be preferentially captured by larger ones if some binder is present on the surface in a process called *layered growth* (Fig. 3c).

Trajectory calculations by Tardos and coworkers (Ennis et al., 1991) yield analytical expressions for the critical Stokes number, St_{coal}^*, for simplified cases when either the viscous force or the capillary force in Eq. (2) is dominant. These values only apply for the case when the colliding particles have an internal, solid core to which a restitution coefficient, e, can be assigned. In the more general case, when there is no solid core or when particles are deformable, critical Stokes numbers cannot be calculated analytically and have to be estimated from numerical integration or measured experimentally. Further, in a real process, binder surface tension and viscosity will both act together to dissipate energy and ensure sticking and coalescence, but no simple analytical solution exists for this case. It was also

shown by Ennis and Tardos (Ennis et al., 1991) that conditions based on viscous dissipation are more restrictive then those based on capillary forces, and so the discussion of critical sizes is limited, in this section, to the former. In a subsequent section, computer simulations are presented in which all effects, both capillary and viscous, are taken into account to determine critical values of the Stokes number.

5.2 Prediction of Critical Sizes

To use the above model for actual predictions, it is necessary to assign values to the relative velocity, U_0, between moving particles. Prediction of U_0 at the present level of knowledge is an extremely difficult task since the powder flow field in fluid bed mixers is very complicated and not amenable to simple solutions. This is complicated by the presence of fixed and moving walls such as, for example, draft tubes and rotating propellers. A rough estimate of the relative velocity between particles was given in Eq. (1). Using the above velocities, one can calculate limiting (critical) granule sizes at the point when the process crosses over between granulation regimes associated with coalescence where $St_{coal} = St_{coal}^*$:

$$a_{cr}^{coal} = \left(\frac{9\mu St_{coal}^*}{8\rho_p \gamma^\circ}\right)^{1/2} \sim \frac{A}{[\gamma^\circ]^{1/2}} \qquad (5)$$

Below the critical size, a_{cr}, granule growth is rapid, while above it it slows considerably; this behavior occurs because the coalescence of larger particles ceases at the critical point where $St_{coal} = St_{coal}^*$. Small particles for which the critical Stokes number is still small (since the characteristic dimension is that of the small particle) will continue to stick to large particles or granules in a layering type of mechanism and thus lead to some slow growth. This slow growth process will continue while fines are present and large granules surfaces are covered by binder.

The proportionality condition in Eq. (5), the last expression on the RHS, yields an approximate dependence of critical size on the inverse of the square root of the average shear rate and, since the actual value of the critical Stokes number is not available for most granulations, this is the most information that can be extracted from the above model. To be able to use the model as a predictive tool for detailed calculations, a value of the critical Stokes number has to be estimated from theoretical considerations (see the section on computer simulations) or obtained by direct measurements.

5.3 Wet Granule Deformation and Breakup

It would appear from the above analysis that once granules reach the critical size characteristic of the layering (coating) regime, any further increase in the Stokes number, relative velocity, or shear rate will maintain the size of the granules. This is certainly true for the case in which binder drying and solidification accompanies growth, and granule strength increases appropriately as large granules are formed. For this to happen, however, one has to tailor both the binder characteristics and the heating (or cooling) rate in the agglomerator very carefully to achieve the required properties in the optimal time frame.

We present, in this section, a simplified model that accounts for the behavior of "green," i.e., wet granules. It is assumed that these green agglomerates possess, upon formation, only the strength imparted to them by the liquid bridges that assured coalescence in the first place. They have had, however, no opportunity to strengthen significantly owing to either the *lack of time* or to the fact that the binder did not become more viscous. Such cases of granulation are very common when, for example, oily binders are used that do not evaporate easily or do not solidify, or when the bridge is formed by a slow chemical reaction between the liquid and the powder, i.e., hydration of a powder. One is then left, within the shearing powder mass, with deformable granules that can grow by layering but can also deform and break.

To formalize mathematically the above concepts of granule deformation and breakage, it is useful to define a new dimensionless Stokes number that relates initial kinetic energy in the shearing mass to internal energy resisting deformation, in the form:

$$St_{\text{def}} = \frac{\text{Externally applied kinetic energy}}{\text{Energy required for deformation}}$$

$$St_{\text{def}} = \frac{m_{\text{p}} U_0^2}{2 V_{\text{p}} \tau(\gamma^\circ)} \tag{6}$$

The notations in this equation are similar to those in Eq. (3). In addition, V_{p} is the particle (granule) volume, $m_{\text{p}} = V_{\text{p}} \rho_{\text{p}}$ is the granule mass, and $\tau(\gamma^\circ)$ is some characteristic strength of the granule. In the most general case, this stress can be taken according to the Herschel–Bulkley model

$$\tau(\gamma^\circ) = \tau_{\text{y}} + k \gamma^{\circ n} \tag{7}$$

where τ_{y} is the yield strength, k is an apparent viscosity, and n is the flow index. A "green" unsolidified granule is, from a rheological point of view, a complex system that usually exhibits both a yield strength and some non-Newtonian behavior as shown in Eq. (7). Assuming that the granule is a very concentrated slurry of the binder and the original particles, one can take, as a first approximation, the apparent viscosity being negligible compared with the yield strength. So $\tau(\gamma^\circ) = \tau_{\text{y}}$. Under these conditions, and again taking $U_0 \approx a\gamma^\circ$ as before, one finds

$$St_{\text{def}} = \frac{\rho_{\text{p}} a^2 \gamma^{\circ 2}}{2 \tau_{\text{y}}} \tag{8}$$

The Stokes number defined above increases with increasing particle size, a, and reaches at some point during granulation a critical value, St_{def}^*, above which granules start to deform and break. At the critical point, when $St_{\text{def}} = St_{\text{def}}^*$, a limiting granule size is defined as

$$a_{\text{cr}}^{\text{def}} = \frac{[(2\tau_{\text{y}} St_{\text{def}}^*)/\rho_{\text{p}}]^{1/2}}{\gamma^\circ} \sim \frac{B_0}{\gamma^\circ} \tag{9}$$

This equation is similar in structure to Eq. (5) but predicts an inverse linear dependence of the critical size on the average shear rate.

5.4 Summary of Growth and Deformation Kinetics

The above considerations are summarized in Fig. 7, where both the growth limit characterized by $St_{\text{coal}} = St_{\text{coal}}^*$ and the deformation (breakage) limit characterized by $St_{\text{def}} = St_{\text{def}}^*$ are presented. The ratio of Stokes numbers is presented on the abscissa, while the critical sizes associated with equilibrium points are presented on the ordinate. Both scenarios depicted in the figure are possible: one in which the growth limit is below the breakage limit and another in which the growth limit is above the breakage limit. Arrows in the figure indicate the direction of the drift in the process, while heavy dots represent equilibrium points.

Take for example the situation in Fig. 7a: particles will grow by nucleation and coalescence until they reach the growth limit, $a_{\text{cr}}^{\text{coal}}$. Since this limit is calculated strictly for equal size granules or particles, unequal particles can still grow by layering to yield granules somewhat larger than $a_{\text{cr}}^{\text{coal}}$. Granules that grow beyond $a_{\text{cr}}^{\text{def}}$ become unstable, deform, and eventually break. The process will stabilize at some point between the two limiting sizes and will be characterized by a size distribution that will shift slowly toward larger granules as long as binder and fines are present.

The situation depicted in Fig. 7b is somewhat simpler in that there is only one stable point at the lower critical size. Particles will grow to a size a_{cr}^{def} and will start to deform and break if they grow further. Depending on the balance of growth and breakage forces, this process may yield a narrow size distribution around the critical value. Alternatively, the product distribution may contain excess fines if coalescence and breakage outweigh layering.

Since the critical sizes a_{cr} depend on the average shear rate in different ways [see Eqs. (5) and (9)], it is possible to choose a shear rate that will bring the two curves in Fig. 7 close to each other or even make them overlap. This will yield a granulation with a very narrow size distribution, since small granules will grow and large granules will break and thus in the end yield the critical size. Such an overlap will also generate a situation when $St_{def}^* = St_{coal}^*$; this is important since under these conditions the actual value of the critical Stokes number can be easily measured experimentally (see the section on constant shear experiments).

To put the above considerations in perspective, the simplified granulation model as developed above is presented schematically in Fig. 8. The granulator is assumed to impart a certain known, *constant shear* to the granular media, as indicated by the arrows in the figure. The granules are taken to be of the same size at any given moment, while growth and breakup keep the system in equilibrium as depicted in the figure. The coalescence criterion developed above determines the conditions under which two identical (smaller) granules in the schematic representation to the left coalesce and grow to yield the larger size depicted on the right. Furthermore, the deformation–breakup criterion defines the conditions under which a larger granule shown in the schematic at the right will deform and break to yield the size on the left. This is quite an unrealistic picture of the process, since in reality a wide range of sizes result from both coalescence and breakup. The simplified picture, however, yields some basic understanding of the process, and two semianalytical solutions that can be checked experimentally, and therefore is quite powerful.

Further simplifications, also shown in Fig. 8, are related to the way formed granules are idealized: this is depicted on the right lower part of the figure. Instead of assuming granules to be made of individual smaller particles held together by the binder, we assume instead that granules are spherical and surrounded by liquid as depicted. The premise here is that once granules are

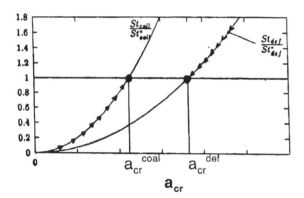

A)Coalescence limit above Deformation limit

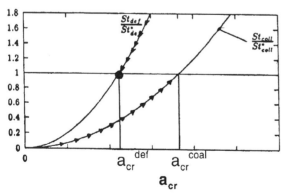

B)Deformation limit above Coalescence limit

Figure 7 Equilibrium Stokes numbers vs. critical granule size. (A) Coalescence limit above deformation limit. (B) Deformation limit above coalescence limit.

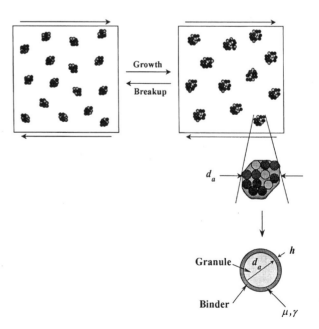

Figure 8 Details of granule growth and breakage.

formed, particles are moved toward the center to form a "Acore," and binder is squeezed toward the surface to insure further sticking. The simplification introduced by this assumption can be relieved somewhat by taking, instead of actual binder properties, the surface tension and viscosity of a binder–particle surface layer.

6   GRANULE CONSOLIDATION

We use the above simplified model to gain some insight into the process by which granules become more dense after formation. This process called "consolidation" can only take place when some bridges inside a formed granule remain viscous and deformable or if the granule itself retains some of its plastic, malleable properties for some time. Under these conditions, external compressive and shear forces will tend to push particles together and hence reduce apparent density. It is clear from the above that the granule has to be in a regime somewhere between the growth limit and the deformation and breakup limit, as depicted in Fig. 7a.

We will now assume that the system subjected to consolidation forces is the one shown in Figs. 4 and 6. The equation describing the motion of the two particles can be written as (Ennis et al., 1991)

$$U = U_0\left[1 - St^{-1}\ln\left(\frac{h}{x}\right)\right] \tag{10}$$

Where U is the instantaneous particle relative velocity, St is the Stokes number (characteristic of either the coalescence or the deformation limit), $2h$ is the initial distance between particles, and $2x$ is the actual separation distance. Solving this equation between two successive consolidation positions during which the particles get closer by a distance Δx, one gets the expression

$$\theta_c = \frac{\Delta x}{2h} = [1 - \exp(-St)] \tag{11}$$

where θ_c is the degree of consolidation. Since the Stokes number is inversely proportional to both binder (granule surface) viscosity and/or the binder yield strength [see Eqs. (3) and (6)], the above expression immediately predicts that if either or both are very large, no consolidation occurs since $\Delta x \sim 0$. Ennis (1990) extended the above correlation to a linear chain of n identical particles that undergo i step-wise consolidations and obtained

$$\theta_c = \frac{1}{(n-1)}\sum_{i=1}^{n-1}\left[1 - \exp\left(\frac{(n-i)St}{n}\right)\right] \tag{12}$$

Both Eq. (11) and Eq. (12) reduce to the consolidation, θ_c, being proportional to the Stokes number ($\theta_c \sim St$) for small values of this parameter ($St \ll 1$). Assuming that the consolidation of a granule takes place homogeneously in three spatial directions and that in each direction the movement is described by, $\theta_c \sim St$, we find a correlation between the bulk density of the granule before, ρ_g and after, ρ_a, consolidation as

$$\frac{\rho_a}{\rho_g} = (1 - \theta_c)^{-3} \approx (1 + 3St) \tag{13}$$

Here we assumed that both the Stokes number and the degree of consolidation are small ($St, \theta_c \ll 1$). The above equation gives a quantitative relation between the density change during consolidation and the shear rate and other parameters of the system through the Stokes number St.

7   EXPERIMENTAL VERIFICATION

This section contains several examples, one from the authors' own experiments, where critical values for the Stokes number were measured experimentally, and others, in which we use the above theoretical framework to match experimental granulation data from the literature.

7.1   Granulation in a Constant Shear Fluidized Bed

This experiment was devised to generate a set of conditions under which the deformation (breakage) limit overlaps the coalescence limit, $St^*_{coal} = St^*_{def}$, as shown in Fig. 7. We use this condition to measure critical values of the Stokes number. This is facilitated by the fact that under these conditions formed granules deform and break immediately after reaching equilibrium; therefore this point can be easily detected experimentally, as mentioned above.

To insure that the shear field in the granulator is constant and uniform, a fluidized bed Couette device was used in which a bed of particles at minimum fluidization conditions was sheared between two concentric rotating rough-walled cylinders (Fig. 9). Fluidized air is supplied in the gap between the stationary outer and the rotating inner cylinder, thereby creating a shear zone in which particles are suspended in the upward moving gas. It was shown by both numerical and analytical calculations (Campbell, 1990; Campbell and Brennen, 1985) that a granular material moving in the rapid granular flow regime, i.e., where inertial interactions between particles over-

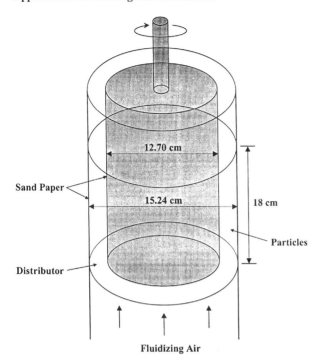

Figure 9 Schematic representation of the fluidized bed Couette device.

whelm frictional effects, indeed exhibits a linear velocity profile, provided that conditions close to nonslip can be generated at the walls. The shear rate for the system can thus be calculated from

$$\gamma^\circ = \frac{V_{\text{wall}}}{\lambda} \tag{14}$$

where V_{wall} is the velocity at the wall and λ, is the gap width.

To simplify the granule growth process in the present experiment, agglomerates were generated by introducing a very viscous binder droplet into a fluidized bed of unsheared particles and by allowing enough time for the granule to form by solids being imbedded into the droplet (i.e., Fig. 3d). The initial granule size could then be calculated from

$$D_g = D_d(\varepsilon_0)^{-1/3} \tag{15}$$

where D_d is the diameter of the binder droplet and ε_0, is the void between the solids; it is assumed in the above equation that the fluid fills all voids. It was found experimentally that this was indeed the case and that due to surface tension, the resulting granule was mostly spherical.

Experiments were performed with the fluidized bed Couette device described above (see also Khan and

Tardos, 1997) using glass particles of a mean size of ~ 300 micrometers so that 80% of the particles lay within the range of 350 to 250 μm. The density of particles was within the range of 2.42 to 2.5 g/cc; under Geldart's classification scheme (Kunii and Levenspiel, 1991), these particles classify as class B. The minimum fluidization velocity of the particles was found experimentally to be about 8 cm/s. The volume fraction or the porosity ε of the bed at minimum fluidization conditions was taken as 0.4 (Kunii and Levenspiel, 1991).

Granules of the above powder using different viscosity Carbowax (PVP) solutions were formed, sheared in the device, and subsequently were let to solidify in a shear-free environment (fluid bed at minimum fluidization conditions). Stokes numbers, St, were calculated using Eq. (3) and taking $U_0 \approx a\gamma^\circ$. A granule deformation parameter defined as $D = (L - B)/(L + B)$, where L and B are the major and minor axes of a deformed agglomerate ($D = 0$ for a sphere), was also determined. The correlation with the Stokes number is given in Fig. 10, where we also superimposed the picture of the granule formed at the given conditions. It was found that for different values of the Stokes number, but otherwise identical conditions, formed granules acquire different shapes. For low values of the Stokes number (high binder viscosity), spherical agglomerates are generated. Increasing the Stokes number (decreasing the binder viscosity) yields regions in which only slightly deformed granules are found and regions in which the granules are totally deformed. A further increase in the value of the Stokes number yields no

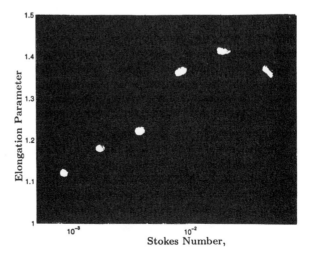

Figure 10 Experimental results. Elongation parameter (D) vs. Stokes number for granules produced in a constant shear Couette device.

agglomerates; binder-covered particles are dispersed into the shearing powder. The important finding is that beyond a *critical value of the Stokes number of about* $St^* = 0.03$, formed granules are so elongated and eroded (see the figure) that their integrity is no longer maintained. This is quite significant, since one would expect a value of the order of unity, $O(1)$, for this critical parameter, and therefore the experiment shows that, at least as a first approximation, the theoretical approach described above is realistic.

Figure 10a shows granules formed in the constant shear granulator from a fine (average size 65 microns) starch powder granulated with water. The pictures were taken with a TV camera placed above the bed and show the upper surface of the granulating powder at six consecutive time steps as depicted in the figure. The powder bed is relatively shallow in the direction perpendicular to the picture (about 3 cm), while the shear is applied in the plane of the picture. One can clearly see significant granule growth during water (binder) addition in pictures A–E; it is also apparent, however, that those granules that form are almost at a constant final size. Subsequent granulation (see pictures D and E) is overwhelmingly controlled by layering of small particles onto large granules. This process continues during wet-massing (while no further binder is added) in picture F until practically no fines are left. This sequence shows quite convincingly that, in a granulator where the deformation–breakage limit is reached, the process follows closely the steps predicted by the theoretical model in Fig. 7. The coalescence–growth phase of granule formation mostly overlaps wetting and spreading of the binder in the shearing mass of powder, while growth by layering immediately follows. The deformation–breakage regime clearly limits growth to yield the final, mostly uniform size distribution observed in Fig. 10a.

7.2 Experimental Data by Watano et al., 1995

Recently reported experiments by the above authors are especially relevant to the present analysis of agglomerate growth and consolidation. These authors published a series of papers on granulation in a Roto-fluidized bed with a top-spray (i.e., a hybrid between devices shown in Figs. 1c and 1d), where shear is induced into the bed by both bubble motion (controlled by the fluidization air velocity) and the motion of the propeller (agitator), which creates shear forces in a plane perpendicular to the gas flow. They measured granule size, size distribution, and bulk density as a function of gas velocity and rotational speed of the propeller (agitator) using three

scales of granulation equipment. In all cases, a pharmaceutical formulation of an initial size of 45 microns was granulated using a polymeric binder. The following sections analyze the data in terms of binder dispersion, growth, and consolidation.

7.2.1 Binder Dispersion

As discussed in the background on microscopic phenomena, the prerequisite for growth by binder granulation is the adequate presence of binder at contact points. Therefore it is important to consider the initial step of binder dispersion and its effect on growth. Inadequate binder dispersion in the agglomeration device can result in product heterogeneity. For example, some particles with ample binder will grow large in size while others remain unagglomerated owing to local scarcity of binder. Binder delivery is measured in terms of flux of binder per surface wetted area of powder. The flux is defined by the binder flow rate relative to the spray coverage area, and the degree of powder fluidization or mixedness that provides fresh powder surface to the spray.

Watano et al. (1995) report three scales of agitated fluid beds that differ in binder spray area relative to batch size (Table 1). The batch size was increased in proportion to the mixer volume, maintaining a fixed powder height/diameter ratio for all scales. In addition, the area of spray coverage was kept in proportion to the cross-sectional area of the vessel. However, the ratio of the spray area to the product bulk volume decreased with scale. This resulted in poorer binder dispersion and broader granule size distribution at larger scales. In addition, the fluidization air velocity was also varied. At higher air velocity, the turnover of powder in the spray zone increased, leading to improved dispersion and narrower size distribution. The combined effect of both variables is shown in Fig. 11 where the geometric standard deviation, (GSD) σ_g is given as a function of relative spray area and gas velocity. As seen, the data is continuously arranged throughout the field of the figure: improved dispersion of binder caused by increase in air flow velocity, increase in spray surface area, or decrease in mass in the mixer yields narrower product size distribution. The data is also continuous in mixer scale as long as the above conditions are equal.

7.2.2 Agglomerate Growth

To compare the results of Watano et al. to the current analysis, we rewrite the result given in Eqs. (7) and (9) but with the assumption that the yield strength, τ_y in

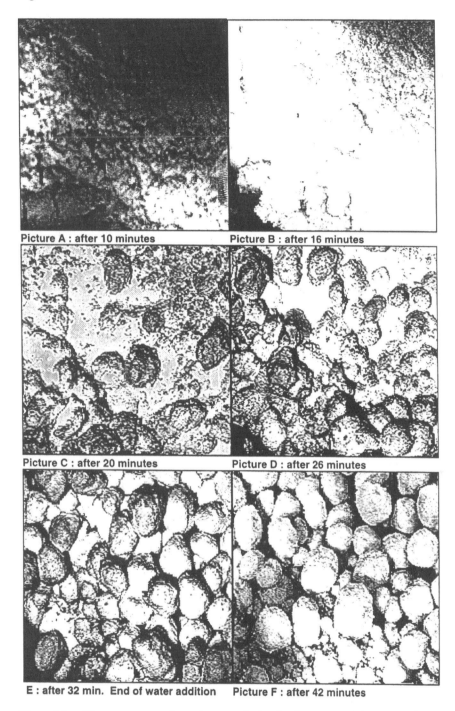

Picture A : after 10 minutes   Picture B : after 16 minutes

Picture C : after 20 minutes   Picture D : after 26 minutes

E : after 32 min. End of water addition   Picture F : after 42 minutes

Figure 10a Time sequence of powder granulation in the constant shear granulator.

Eq. (7) is negligibly small and that the granule's internal strength is given by $\tau(\gamma^{\circ}) = k\gamma^{\circ n}$. Equation (9) can then be rewritten as

$$a_{\mathrm{cr}}^2 \gamma^{\circ 2} = C\gamma^{\circ n} \qquad (17)$$

where C is a constant coefficient that incorporates all parameters in Eq. (9) except the diameter and shear rate. Taking the logarithm of the above expression yields

$$\log(a_{\mathrm{cr}}) = -\left(1 - \frac{n}{2}\right)\log(\gamma^{\circ}) + \log C = m\log(\gamma^{\circ}) + c \qquad (18)$$

Table 1   Scale-Up of an Agitated Fluid Bed Agglomerator

Model*:	NQ-125	NQ-230	NQ-500
Vessel diameter (cm)	12.5	23	50
Cross-sectional area (cm²)	122.7	415.5	1963.5
Spray area (cm²)	38.5	132.7	594.0
Vessel volume ratio	1	6.23	64
Powder feed wt. (kg)	0.36	2.23	22.9
Spray area (cm²)/feed (kg)	106.9	59.5	25.9

Source: Watano et al., 1995.

This expression shows that on a log–log plot, the slope of the curve of granule diameter versus shear rate gives the shear index, n, of the binder through the coefficient m, while the intercept gives a measure of the critical Stokes number through the value of c.

The experimental data shown in Fig. 12 include median granule size as a function of varying tip speed at three independent gas fluidization velocities ($U_f = 0.4$, 0.6, 0.8 m/s) at three machine scales. Equation (18) predicts reduced growth at high propeller tip speed. This effect, however, is counteracted somewhat by increased gas fluidization velocity that cushions particles from the impact of the propeller. Data regression shows that the effect of air fluidization velocity (U_f) relative to the propeller tip speed is approximately to the -2 power. The full set of data in Fig. 12 fit the log–log trend predicted by

$$\log(a_{\mathrm{cr}}) = m\log\left(\frac{U_i}{U_f^2}\right) + c \qquad (19)$$

Here we assumed that the shear rate, γ°, is proportional to the impeller tip speed, U_i, and therefore the quantity U_i/U_f^2 is some modified shear rate. The average value of the slope for all granulators, large and small, is approximately $m = -0.25$, which gives a rheological flow index of $n = 1.5$. Such values are common for concentrated slurries of viscous liquids and small solid particles and characterize a shear thickening behavior. The most important conclusion from the above considerations is that both the slope, m, and the coefficient, c, take approximately the same value for all size granulators operated under different conditions, especially at the high-shear range of the impeller where the power-law term of the viscosity dominates. This implies that the binder viscosity has the same role in both small and large units and that the Stokes number, based on deformation, is the relevant dimensionless parameter. It also implies that the Stokes number may be the appropriate scaling factor when such devices are scaled up.

7.2.3   Agglomerate Consolidation

Experimental support for the expression derived for granule consolidation [Eq. (13)] is found again in the work of Watano et al. (1995). Their data is shown in Fig. 13 where the relative density ratio is given as a function of the impeller tip speed (U_i) for all three

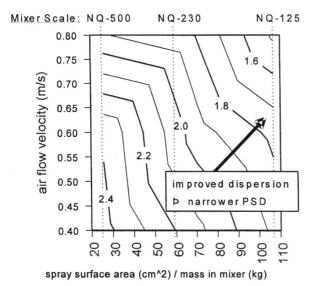

Figure 11   Geometric standard deviation (GSD) of particle size distribution (PSD) as a function of binder dispersion and air flow velocity. (Watano et. al., (1995.)

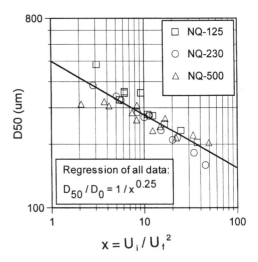

Figure 12   Log–log plot of data by Watano et al. (1995); median particle size (D_{50}) versus impeller tip speed (U_1) and fluidization velocity (U_f). (D_0 is the D_{50} intercept at $x = 1$.)

granulators used at gas fluidization velocities of 0.4, 0.6, and 0.8 m/s. Data regression shows that the effect of air fluidization velocity (U_f) relative to the impeller tip speed is approximately to the -0.5 power. The relative density ratio, $(\rho_a - \rho_g)/\rho_g$, was calculated from the experimental data and by extrapolation for the first point in the figure, assuming that the two densities are equal at zero granulator tip speed. As seen, the bulk density increases with increasing tip speed, i.e., increasing rotation rate or increasing impeller diameter. To provide a comparison with predictions, we combine Eqs. (13), (6), and (7) with $\tau_y = 0$ and $n = 1.5$ (as found from experiments, above) to get

$$\frac{\rho_a - \rho_g}{\rho_g} = K(\gamma^\circ)^{1/2} = K'(U_i)^{1/2} \qquad (20)$$

with K being a parameter that incorporates all variables except the shear rate and where we again assume the propeller tip speed to be proportional to the shear rate.

Watano et al. (1995) found that fluidization gas velocity also has an influence on consolidation, i.e., higher consolidation is obtained at lower gas velocities as shown in the regression in Fig. 13, but this is not captured directly in Eq. (20). A trend, however, can be obtained by combining Eqs. (13), (2), and (4) to yield

$$\frac{\rho_a - \rho_g}{\rho_g} = 48\frac{\rho_p U_B a^2}{D_B \delta^2 \mu} = 34\frac{\rho_p a^2 g^{1/2}}{D_B^{1/2}\delta^2 \mu} \qquad (21)$$

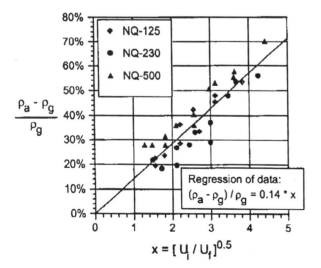

Figure 13 Granule consolidation data by Watano et. al. (1995); relative density increase $[(\rho_a - \rho_g)/\rho_g]$, versus impeller tip speed (U_1) and gas fluidization velocity (U_f).

where we used $U_B = 0.71 (g D_B)^2$. It is likely that fluidization conditions impose a smaller bubble size, D_B, at lower gas velocities, and that would result in larger consolidation; this would explain the result as found by the above authors. Although it is not clear how the dimensionless bubble spacing, δ, changes under these conditions, it appears that a combination of the above two relationships would correctly reflect the trend obtained experimentally,

$$\frac{\rho_a - \rho_g}{\rho_g} = K''\left(\frac{U_i}{U_f}\right)^2 \qquad (22)$$

The regression line in Fig. 13 represents the above correlation with $K'' = 0.14$. It is important to note that correlations (22) and (19) were obtained for all three scales of granulation but that growth (granule size) and consolidation (densification) scale differently with both impeller tip speed and fluidization velocity posing difficult problems in scale-up.

7.3 Experimental Data by Dencs and Ormos (1993, 1994)

These authors, with a long history of eminent granulation research starting as early as 1962, published recently a set of papers reporting on granulations aimed to recover solids from a concentrated aqueous solution. They used a fluidized bed in which shear forces are controlled by bubble motion and by a rotating insert. The insert has two cylindrical rollers mounted on the ends of a rotating arm that keeps the rollers adjacent to the wall of the fluidized bed, as shown in Fig. 2 (a similar construction has the rollers moving parallel to the distributor plate). The gap between the rollers and the wall (distributor) can be adjusted, and by controlling the speed of the rotating arm, strong shear forces can be generated. The device is called "a fluid bed grinder-granulator" by the authors to emphasize the grinding action of the rollers combined with the granulation function of the fluid bed.

The unit can be run in a semicontinuous mode in which the bed is started with seed particles from a previous run and solution is introduced continuously while solids are discharged through a tube close to the distributor. The granulator is operated at some elevated temperature above 100°C so that water from the solution is evaporated and the residual solid material is layered onto the existing solids in the bed. A constant agglomerate diameter is obtained after some time of operation, and from there on a steady state sets in. The data reported in this section is this steady state agglomerate size as a function of different operating variables such as rotation rates of the arm, gap distance between the roller and the wall, gas fluidization velocity, and fluidized bed diameter. Two materials were used: a sucrose solution (700 g/liter) obtained from a carrot processing unit and a zinc sulfate solution (500 g/liter). The data was taken from Dencs and Ormos, (1993, 1994), and an additional large part directly from the authors (Dencs, 1998).

To extract general conclusions from the data, we employ here the same model prediction as used in Eq. (19): $\log(a_{cr}) = m \log(U_i/U_f^2) + c$, where U_i is the impeller tip speed and U_f is the fluidization velocity. In this case however, shear is determined, in addition to the impeller tip speed, U_i, also by the gap distance, r, between the roller and the wall. We rewrite the above relationship to read

$$\log(a_{cr}) = m \log\left(\frac{U_i}{rU_f^2}\right) + c \tag{23}$$

where again $m = -(1 - n/2)$, and c are material constants that determine the non-Newtonian behavior of the solid–water solution and the characteristic Stokes number, respectively, and n is the binder's flow index.

The experimental data is shown in Fig. 14 as the average agglomerate diameter, d_{agg}, as a function of the modified shear rate $\gamma^\circ U_f^2 = U_i/rU_f^2$, as given in Eq. (23). An important conclusion from the figure is that data fit optimization gave the same dependence on the variables as found for the data by Watano et al. The fitted curves to the data obtained for sucrose and zinc sulfate solutions yield a coefficient $m = -0.183$ and $m = -0.15$ (or values of the index $n = 1.63$ and $n = 1.79$, respectively) and $c = 0.37$ and $c = 0.35$, respectively. The values for the shear index, n, are somewhat higher than those obtained during granulation of a pharmaceutical powder by Watano et al. (see Fig. 12, where $n = 1.5$). These values again point to a flow index characteristic of a shear thickening solution in agreement with the general nature of the material.

It is interesting to note that the data show a leveling off as the shear rate is increased, suggesting that at higher shear, characteristic of some additional parameters of the system, the influence of shear on particle size is reduced. This kind of behavior has been observed in previous granulations by several authors and is attributed here to wall slip. In other words, depending on the stickiness (roughness) of the particle (granule) surface, a shear rate is reached by either

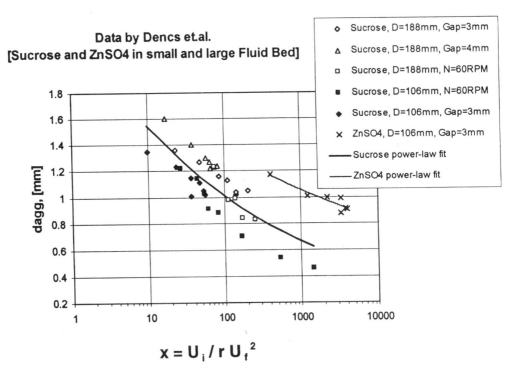

Figure 14   Experimental data by Dencs et al. (1993) and Dencs (1998); agglomerate size versus modified shear rate and fluidization velocity. Data for sucrose and zinc sulfate solutions in small and large fluid beds.

increasing the tip speed of the rotating arm or reducing the gap distance, when particles start to slip significantly. From this point on, further increases in shear rate are not transmitted to the powder, and the agglomerate size is not significantly influenced by further increases in shearing. Further considerations of how the above experimental and theoretical findings are used to scale up industrial granulators can be found in Mort and Tardos (1999).

8 COMPUTER SIMULATION OF GRANULE GROWTH AND BREAKUP

We describe in this section several attempts to simulate granulation in a shearing mass of powder in which a sticky binder is present on a significant fraction of particles. We assume that the shear is constant, that particles are of uniform size and shape (spherical), and that the binder is present on the particle surface in a constant, uniform layer as depicted in Fig. 8. It is further assumed that particles, while spherical, can only move in a two-dimensional space (2D simulation) and that gravity is negligibly small. We follow the concept of *growth by coalescence* if particles are small and of *deformation and breakup* when granules become excessively large, as predicted by the theoretical model presented earlier.

The simulation is based on the discrete element approach widely used in the literature to predict flows of solid-particulate materials. This method in turn is based on the analogy between the character of granular flows with that of the flow of gas molecules and is a generalization of the so called "molecular dynamics simulation" or MDS, employed in physics. Instead of point particles and force potentials, however, rapid granular flow simulations or RGFS use actual expressions for the magnitude of forces to describe interactions between *finite size* particles. During this kind of simulation (RGFG), Newton's equation ($F_{\text{ext}} = m\,dU/dt$) is solved for pairwise interactions between a large number of particles starting from an initial condition, until steady state is achieved. Several of these kinds of calculations have been performed lately, a summary of some results being presented in Khan and Tardos (1997). Interactions between particles in these works are restricted to friction and elastic and sometimes plastic deformations of the surface. It is this restriction that limits the applicability of these models to *dry powder flows*.

During a recent effort (Khan and Tardos, 1997), the above RGFS were extended to include viscous interactions between solid particles that also include the influence of the liquid surface tension; details of the procedure are given in the above paper. The most important addition to the original RGFS simulation is the inclusion of viscous and capillary forces [as given in Eq. (2)] into the force balance used to solve Newton's law ($F_{\text{ext}} = m\,dU/dt$). These forces are added vectorially to supplement friction, elastic, and sometimes plastic forces used in previous simulation. While the addition of viscous forces looks like a simple extension of previous work, it is, in fact, a very serious complication: this is due to the extensive (long) range action of viscous forces and to the fact that they become very large when particles come into close proximity (and become unbounded when particles touch). In addition, simple solutions only apply to very slow movement (low Reynolds number, inertialess flows) of the fluid between the particles.

The simulation was successfully implemented for a maximum of 40,000 particles, of which up to 30% carried the sticky binder; a parallel computer system at the University of Illinois at Urbana–Champaign and a Sun computer cluster at CCNY were used. To limit the complexity of the code and to keep computer running time to acceptable levels (12–14 hours), it was necessary to assume that particles covered with binder (fluid) at the onset of binder introduction were the only ones carrying the binder for the duration of computations. In other words, "sticky" binder-covered particles always remained "sticky" with a constant fluid layer on them, while dry (unsticky) particles remained always dry, and no binder transfer between particles was allowed during a particular run. Additional binder could be introduced at set intervals but only to make more (or less) particles "sticky" which then remained binder covered for the duration of the run.

8.1 Simulation of Granule Growth by Coalescence

The simulation domain with 40,000 particles is presented in Fig. 15. The shear is introduced at a constant rate through the upper and lower cell boundaries as shown. Mirror images of the simulation domain ensure that the influence of solid boundaries is neglected (see Khan and Tardos, 1997). The Stokes [St, defined in Eq. (3)] and capillary (*Ca*, defined in Fig. 6) numbers are the dimensionless parameters of the problem that have to be perceived as dimensionless binder viscosity or shear rate and fluid surface tension, respectively. Increasing binder viscosity and decreasing shear rate result in lower Stokes numbers, while increasing the liquid surface tension results in lower capillary num-

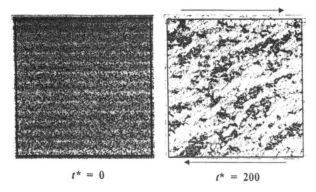

$t^* = 0$ $t^* = 200$

Figure 15 Shear flow of dry and binder covered particles; the 2D simulation domain at dimensionless time $t^* = 0$ and at steady state, $t^* = 200$. Total number of particles, $N = 40,000$ of which 20% (8,000) are covered with binder. Stokes number $St = 0.1$ and capillary number $Ca = 1.0$. Dry particles are shown in gray while binder covered particles (sticky) are depicted in black.

bers. The other important parameter of the system is the ratio of "sticky"/dry particles in the domain (r_B), which is a measure of the total amount of binder present. [It was found during many runs that the binder layer thickness is a secondary parameter if the layer is taken as less then about 15–20% of the particle radius. In this case, the only important consideration is whether the binder is present or the particle is "sticky," while the further influence of the layer thickness can be neglected (for such shallow layers). In practice, creating layers that are more than about 5–10% of the par-

ticle radius is quite difficult, and therefore this is not a serious limitation.]

The simulation domain on the left in Fig. 15 depicts the moment in dimensionless time when the fluid layer is introduced onto the particles at $t^* = 0$; binder-covered, "sticky" particles are shown in black, while dry particles are depicted in gray. Introduction of binder is achieved by randomly assigning stickiness to a percentage of particles (20% or 8,000 in this case) in a domain of dry (unsticky) particles that achieved steady state after some previous running time. The picture at the right in Fig. 15 shows the domain at steady state after a dimensionless time of $t^* = 200$. Lighter areas are devoid of particles, while denser areas depict clusters. The intense black in the picture shows agglomerates, usually surrounded by clusters of dry powder. A pattern recognition routine on a PC was used to generate granule size distributions from pictures similar to the one shown in Fig. 15.

Granule size distributions for values of the Stokes number, $St = 1$, and capillary number, $Ca = 1$, and for a ratio of "sticky"/dry particles (r_B) from 10% to 60%, are shown in Fig. 16. The ratio of binder thickness, h, to the particle diameter, d_p, was taken as $h/d_p = 0.1$. Several other results of this kind, with a more detailed description of the numerical scheme used during computation, are given in Talu et al. (1999). Most of these results were correlated as the average [Sauter mean, $d_{(3,2)}$] granule size (in multiples of initial particle diameters) with the parameters of the system using a general equation of the form

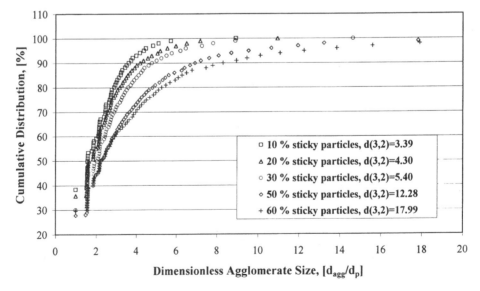

Figure 16 Simulated granulation with increasing amounts of binder. Cumulative particle size distribution vs. relative agglomerate diameter. Stokes number $St = 0.1$ and capillary number $Ca = 1.0$.

$$\frac{d_{\mathrm{agg}}}{d_{\mathrm{p}}} = \left(\frac{St^*}{St}\right)^{\alpha} \tag{24}$$

where the exponent was found to be $\alpha = 0.4$. Values for the critical Stokes number, St^*, were extracted from the computer simulation results. These contain all other parameters of the problem that are not directly incorporated into the Stokes number. They are given for the case of growth by coalescence by

$$St^* = \left(\frac{44 r_{\mathrm{B}}^{1.65}}{Ca^{0.32}}\right)^{2.5} \qquad \text{for } St < 1 \tag{25a}$$

$$St^* = \left[\frac{92 r_{\mathrm{B}}^{1.65}\left(\frac{h}{d_{\mathrm{p}}}\right)^{1/4}}{Ca^{0.32}}\right]^{2.5} \qquad \text{for } St > 1 \tag{25b}$$

It can be easily seen from the above results and from the exponents of different dimensionless quantities in Eqs. (25), that the influence on granule size of binder surface tension (through the Ca number) and binder viscosity (through the St number) are approximately of the same order, while the influence of the amount of binder present (through the binder-covered-to-dry particle ratio, r_{B}) is overwhelming. These conclusions are amply supported by industrial practice where it is well known that only after the amount of binder in the granulator is fixed does the influence of other parameters, such as binder properties, begin to have an influence on granule size.

8.2 Simulation of Granule Deformation and Breakup

A second set of simulations was performed (see also Talu et al., 1999) with all binder covered, "sticky" particles contained in one initial spherical granule as shown in Fig. 17. This was done to simulate the deformation and eventual breakup of a large agglomerate within the shearing mass of dry powder. Figure 17 depicts the simulation domain at steady state after the initially round (spherical) granule was kept in the shear field for a long time (dimensionless time $t^* = 100$). As seen, at a Stokes number of $St = 0.01$ (and $Ca = 10$) the agglomerate does not deform and remains spherical throughout. Increasing the value of this parameter to $St = 0.02$ results in a deformed agglomerate that, however, does not break apart. At larger Stokes numbers (lower binder viscosity or higher shear), the agglomerate breaks into two ($St = 0.03$) and into three parts ($St = 0.04$). The critical value of the Stokes number, when the agglomerate can still be

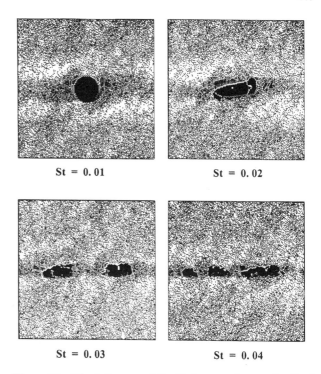

St = 0. 01   St = 0. 02

St = 0. 03   St = 0. 04

Figure 17 The influence of the Stokes number on the final shape of a granule surrounded by dry powder. Binder covered agglomerates in a shearing medium at steady state, $t^* = 100$. Total number of particles in the domain, $N = 10,000$; capillary number $Ca = 10.0$.

considered intact, is somewhere between $St = 0.01$ and 0.02.

Critical values of the Stokes number were calculated for several different initial granule sizes and different values of the capillary number (Ca), and these are shown in Fig. 18. In this figure, the relative agglomerate size $d_{\mathrm{agg}}/d_{\mathrm{p}}$ is given versus the Stokes number for three values of the capillary number $Ca = 10$ (small capillary effects), 1, and 0.5 (large capillary effects). The experimental value for the critical Stokes number, as obtained from the constant shear Couette device (see Fig. 10), is also shown; the fit is quite remarkable even though the value of the capillary number in the experiment is not known (it is assumed to be large, $Ca = 10$ or larger, since the binder is very viscous).

The lines in Fig. 18 were fitted by functions of the general form given in Eq. (24) with an exponent, $a = 1$. Critical Stokes numbers were extracted that, for granule breakup, have the form

$$St^* = \frac{2.25}{Ca} \qquad \text{for } Ca < 10 \tag{26}$$

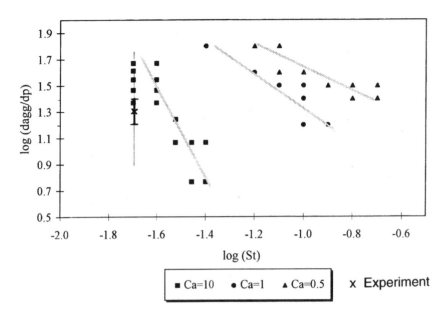

Figure 18 Values of the critical Stokes number, St^*, during granule deformation and break-up. Dimensionless agglomerate size vs. Stokes number for values of the capillary number of $Ca = 10$, 1.0 and 0.5. Experimental result reproduced from Fig. 10.

This last equation yields [with Eq. (24) and $a = 1$] the largest possible agglomerate that will survive in the granulator under given conditions.

CONCLUSIONS

A comprehensive model of granulation was presented, one that takes an agglomeration process from binder selection and testing to granule growth, deformation, and consolidation. An instrument and procedures were described to measure the critical amount of binder that ensures that coalescence and particle growth will occur. It was also shown how the pendular bridge apparatus (PBA), developed earlier by the present authors, can be used to test the bridge-forming characteristics of the binder while at the same time determine binder spreading and penetration rates and the critical time of binder strengthening. These operations can be performed before any granulation runs are carried out, and binder characteristics can be modified to achieve proper wetting, bridge formation characteristics, and binder strengthening (viscosity increase).

A theoretical model of particle/granule growth based on simple principles of energy dissipation in the liquid (binder) bridge between two colliding particles was developed. When the initial energy of the particles is smaller than the dissipated energy, particle coalesce and growth will take place. This concept, ori-

ginally formalized in Ennis et al. (1991) is extended in the present paper to totally deformable particles where the binder viscosity is replaced by an equivalent viscosity of the malleable surface. Furthermore, the concept of energy dissipation is also applied to determine a criterion of "green" particle deformation and breakup and to granule consolidation. A set of dimensionless numbers that characterize granule growth and the deformation process are defined. It is shown that these numbers (Stokes numbers) delimit different granulation regimes, and critical values are found that can be used to calculate the critical granule sizes that are characteristic at the transition from one regime to the other. Several experimental results from the literature are used to illustrate the usefulness of the theoretical approach. A set of very careful experiments performed by the present authors in a specially constructed granulator with constant shear are used to measure critical Stokes numbers. Employing these values, it is possible to predict the outcome of a granulation process provided that the shear in the device is essentially constant and that binder bridge formation is not superimposed on granule growth.

Finally, it was shown that rapid granular flow simulations (RGFS) into which the effect of binder viscosity and surface tension are introduced, readily predict the outcome of granulation. These simulations support the simplified theoretical models of granule growth by coalescence and granule deformation and breakup and give more realistic values for the critical parameters that define different regimes of granulation. Equations were obtained for granule sizes generated in a granulation process as a function of the dimensionless parameters of the problem such as the capillary number, Ca, the Stokes Number, St, and the amount of binder present in the system as defined by the fraction of binder-covered particles in the domain.

Throughout this work, it has been shown that *micromechanical analysis* is a powerful tool in predicting fundamental transformations in the granulation process, including binder dispersion, growth, and deformation. Micromechanical analysis shows the interdependence between key process parameters, such as shear rate, and material properties, such as binder viscosity and agglomerate yield stress. Scale-up of agglomeration in industrial practice is greatly aided by an understanding of the key process transformations and their controlling factors on the *microscale*.

ACKNOWLEDGMENTS

The authors would like to thank Dr. Bela Dencs from the Research Institute for Chemical Engineering at the Pannon University of Agricultural Sciences in Veszprem, Hungary, for the information in Fig. 2 and the data in Fig. 14 and for fruitful discussions on the topic of granulation. The help of Dr. I. M. Khan and Dr. I. Talu from CCNY in preparing the simulation data, and for performing some experiments, is also greatly appreciated

NOTATIONS

a, a_1, a_2	=	particle, granule radius
B	=	width of deformed granule
Ca	=	Capillary number (Fig. 6)
A, B_0, C, K	=	constants
C_0	=	constant in Eq. (2)
$C = \log C$	=	constant in Eq. (18)
$d_{(3,2)}$	=	Sauter mean diameter
d_{agg}	=	agglomerate diameter
d_p	=	particle diameter
D_{50}, d_{50}	=	granule median diameter at 50% of size distribution
$D_0 = D_{50}$ [at $x = 1$]	=	reference granule diameter (in Fig. 12)
D_g	=	granule diameter
D_d	=	binder droplet diameter
$D = (L - B)/(L + B)$	=	deformation parameter
D_B	=	bubble diameter in fluidized bed

e	=	solid core restitution coefficient
F	=	force exerted by liquid bridge
g	=	acceleration due to gravity
h	=	binder layer thickness
k	=	apparent viscosity
L	=	length of deformed granule
$m_p = V_p \rho_p$	=	mass of particle
$m = -(1 - n/2)$	=	constant in Eq. (18)
n	=	flow index [Eq. (7)]
N	=	number of particles in simulation domain
r	=	coordinate, gap width in experimental fluid bed
r_B	=	ratio of binder covered to dry particles in the simulation domain
St	=	Stokes number (dimensionless)
t	=	time
t^*	=	dimensionless time
U	=	particle instantaneous velocity
U_0	=	relative particle velocity
U_B	=	bubble velocity in fluidized bed
U_f	=	gas fluidization velocity
U_i	=	impeller tip speed
V_p	=	volume of particle
V_{wall}	=	granular velocity at the wall of the Couette device
x	=	distance between colliding particles
z	=	coordinate

Greek Letters

α	=	exponent in Eq. (24)
μ	=	binder or granule surface viscosity
γ	=	binder surface tension
γ°	=	shear rate
ϕ	=	filling angle
ε_0	=	void volume in granular bed
ρ_a	=	apparent density of granule
ρ_g	=	granule density before consolidation
ρ_p	=	particle (solid) density
δ	=	bubble spacing in fluidized bed
τ	=	shear stress
τ_y	=	granule yield strength
λ	=	gap width in the Couette device
θ_c	=	degree of consolidation
σ_g	=	geometric standard deviation (GSD)

Subscripts and Superscripts

coal	=	coalescence

def	=	deformation
cr	=	critical
*	=	dimensionless, limiting value
vis	=	due to viscous effects
cap	=	due to capillary effects
wall	=	at the wall
ext	=	exterior

REFERENCES

Adams MJ, Briscoe BJ, Kamjab M. The deformation and flow of highly concentrated dispersions. Adv in Colloid and Interface Sci 44:143, 1993.

Campbell CS. Rapid granular flows. Ann Rev Fluid Mech 22:57, 1990.

Campbell CS, Brennen, CE. Computer simulation of granular shear flows. J Fluid Mech 151:167, 1985.

Dencs B, Ormos Z. A particle size control in a fluidized bed spray-dryer and granulator during the recovery of solids from liquids. Hungarian J Ind Chem Veszprem 21:225–231, 1993.

Dencs B, Ormos Z. A Orlo-gorgos fluidizacios-porlasztos es granulalo berendezes fojadekok szarazanyag-tartalmanak kinyeresere. J Hungarian Chem 7, 1994 (in Hungarian).

Dencs B. Personal communication, April 1998.

Ennis BJ, Tardos GI, Pfeffer R. A micro-level-based classification of granulation regimes and mechanisms. Proceedings of the Second World Congress on Powder Technology, Kyoto, Japan, 1990, p. 409.

Ennis BJ. On the mechanisms of granulation. Ph.D. dissertation, City University of New York, 1990.

Ennis BJ, Li J, Tardos GI, Pfeffer R. The influence of viscosity on the strength of an axially strained pendular liquid bridge. Chem Eng Sci 45:3071–3087, 1990.

Ennis BJ, Tardos GI, Pfeffer R. A micro-level-based characterization of granulation phenomena. Powder Technol 65:257–272, 1991.

Hancock BC, York P, Rowe RC. An assessment of substrate–binder interactions in model wet masses. 1: Mixer torque rheometry. Int J Pharm 102:167–176, 1994.

Khan MI, Tardos GI. Stability of wet agglomerates in granular shear flows. J Fluid Mech 347:347–368, 1997.

Kristensen HG, Holm P, Schaefer T. Mechanical properties of moist agglomerates in relation to granulation mechanisms. Part I: Deformability of moist densified agglomerates. Powder Technol 44:227–237, 1985.

Kristensen HG, Holm P, Schaefer T. Mechanical properties of moist agglomerates in relation to granulation mechanisms. Part II: Effect of particle size distribution. Powder Technol. 44:239–247, 1985.

Kristensen HG, Schaefer T. A review on pharmaceutical wet granulation. Drug Develop Ind Pharm. 13(4,5):803, 1987.

Kristensen HG. Agglomeration of powders. Acta Pharm Soc 25:187–204, 1988.

Kunii D, Levenspiel O. Fluidization Engineering. 2d ed. Boston: Butterworth-Heinemann, 1991.

Landin M, York P, Cliff MJ, Row RC, Wigmore AJ, Scale-up of a pharmaceutical granulation in a fixed bowl mixer–granulator. Int J Pharm 133:127–131, 1996a.

Landin M, York P, Cliff MJ, Row RC, Wigmore AJ. The effect of batch size on scale-up of a pharmaceutical granulation in a fixed bowl mixer–granulator. Int J Pharm 134:243–246, 1996b.

Mazzone DN, Tardos GI, Pfeffer R. The behavior of liquid bridges between relatively moving particles. Powder Technol 51:71, 1987.

Mort PR, Tardos GI. Scale-up of agglomeration processes using transformations. Kona 17, 1999.

Nienow AW. Fluidized bed granulation. In: Stanley-Wood NG, ed. *Enlargement and Composition of Particulate Solids.* London: Butterworths, 1983.

Nienow AW, Rowe PM. Particle growth and coating in gas-fluidized beds. In: Davidson JF, Clift R, Harrison D, eds. *Fluidization.* 2d ed. London: Academic Press, 1985.

Nienow AW. Fluidized bed granulation and coating: application to materials, agriculture, and biotechnology. Proceedings of the First International Particle Technology Forum, Denver, Co., August 17–19, 1994.

Ormos ZD. Granulation and coating. In: Chulia D, Deleuil M, Pourcelot Y, eds. Powder Technology and Pharmaceutical Processes, 1994, pp. 359–376.

Rowe RC, Parker MD. Mixer torque rheometry: an update. Pharm Technol, March, 1994, pp 69–82.

Schaefer T. Equipment for wet granulation. Acta Pharm Soc 25:205–228, 1988.

Schaefer T, Holm P, Kristensen HG. Wet granulation in a laboratory scale high shear mixer. Pharm Ind 52(9):1147, 1990.

Sommer K. Size enlargement. In: Ullman's Encyclopedia of Industrial Chemistry. 5th ed. Vol. B2. VCH Verlag, 1988.

Talu I, Tardos GI, Khan MI. Use of rapid granular flow simulations to predict granulation of fine powders. Powder Technol, in press, 1999.

Tardos GI, Mazzone D, Pfeffer R. Destabilization of fluidized beds due to agglomeration, Part I and Part II. Can J Chem Eng 63:377–383, 384, 1985.

Tardos GI, Garcia J, Ahart R, Ratuiste F. How and why some flow aids promote bulk powder flow. AIChE Symposium Series. Weimer, AW. ed. 89(296), 1993.

Tardos GI, Gupta R. Forces generated in solidifying liquid bridges between two small particles. Powder Technol 87:175–180, 1996.

Tardos GI, Khan MI, Mort PR. Critical parameters and limiting conditions in binder granulation of fine powders. Powder Technol 94:245–258, 1997.

Watano S, Sato Y, Miyanami K, Murakami T. Scale-up of agitation fluidized bed granulation. Chem Pharm Bull 43(7): Part I: 1212, Part II: 1217, Part III: 1224, Part IV: 1227, 1995.

18

Applications for Fluidized Bed Drying

Arun S. Mujumdar

National University of Singapore, Singapore

Sakamon Devahastin

King Mongkut's University of Technology Thonburi, Bangkok, Thailand

1   INTRODUCTION

Fluidized bed dryers (FBDs) have found widespread applications for the drying of particulate or granular solids in the chemical, food, ceramic, pharmaceutical, agriculture, polymer, and waste management industries. More recently, they have also found special applications in the drying of slurries. Suspensions, solutions, dilute pastes, or sludges are atomized into a fluidized bed of inert particles and the dry powder is separated from the exhaust gases.

For drying of powders in the 50 to 2000 μm range, fluidized beds compete successfully with other more traditional dryer types, e.g., rotary, tunnel, conveyor, continuous tray. Among the advantages of the fluidized bed dryers one may cite

High drying rates due to excellent gas–particle contact leading to high heat and mass transfer rates
Smaller flow area
Higher thermal efficiency, especially if part of the thermal energy for drying is supplied by internal heat exchangers
Lower capital and maintenance costs, compared to rotary dryers
Ease of control

However, they also suffer from some limitations such as

High power consumption due to the need to suspend the entire bed in gas phase leading to high pressure drop
Increased gas handling requirement due to extensive recirculation of exhaust gas for high thermal efficiency operation, especially when drying products with extensive internal moisture
High potential of attrition; in some cases granulation or agglomeration
Low flexibility and potential of defluidization if the feed is too wet
Generally not recommended when organic solvents need to be removed during drying.

Table 1 compares the key characteristics of the fluidized bed dryer with those of some of the competing dryers. Note that the flash dryer is applicable only for the removal of surface moisture from smaller size particles, while the fluidized bed is recommended for particles requiring longer drying times, e.g., 10–60 minutes compared to only 10–30 seconds for flash drying. Some of the characteristics listed in Table 1 are subject to modification by changes in the design of the hard-

Table 1 Comparison of Fluidized Bed Dryer with Competing Dryers

Criterion	Rotary	Flash	Conveyor	Fluid bed
Particle size	Large range	Fine particles	500 μm–10 mm	100–2000 μm
Particle size distribution	Flexible	Limited size distribution	Flexible	Limited size distribution
Drying time	Up to 60 min	10–30 s	Up to 120 min	Up to 60 min
Floor area	Large	Large length	Large	Small
Turndown ratio	Large	Small	Small	Small
Attrition	High	High	Low	High
Power consumption	High	Low	Low	Medium
Maintenance	High	Medium	Medium	Medium
Energy efficiency	Medium	Medium	High	High
Ease of control	Low	Medium	High	High
Capacity	High	Medium	Medium	Medium

ware as well as the operating conditions. It is important to point out that for removing organic solvents or handling toxic or flammable solids, the conventional hot air fluidized bed dryer is not a good choice, since there is danger of fire or explosion if flammability limits are exceeded. Also, owing to the inherent need to handle large volumes of air due to the need to achieve fluidization, complete collection of the toxic material is not economically feasible.

2 CLASSIFICATION AND SELECTION CRITERIA

The simplest and most common fluidized bed dryer is a circular cross section vessel in which the particulate material is dried convectively either in the batch mode (for smaller quantities or if the feed is produced in batches) or continuously. Rectangular and other noncircular cross section vessels may be employed if needed, however. For most operations, fluidized beds operate at slight negative pressure to avoid leakage of hot and humid gas into the ambient; it is generally not cost-effective to fabricate fluidized bed dryers as pressure vessels or totally leak-proof.

Figure 1 provides a coarse classification scheme that covers most (but not all) fluidized bed dryers available commercially. Some special variants of the conventional fluidized bed may be needed for special applications. It should be noted at the outset that the well-mixed dryer type (Fig. 2) has the advantage of being able to handle higher moisture content feed which may not be fluidizable under normal conditions. As the wet feed enters a well-mixed fluidized bed of a lower average moisture content, the danger of defluidization or formation of large chunks (or agglomerates) is

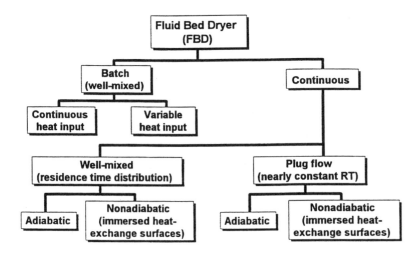

Figure 1 A coarse classification scheme for fluidized bed dryer.

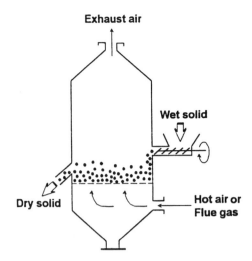

Figure 2 Well-mixed continuous fluidized bed dryer.

reduced. On the negative side, a well-mixed continuous dryer will necessarily lead to a distribution in the moisture content of individual particles; equilibration of moisture during storage yields a product of uniform moisture content. When processing wet feeds that do not fluidize readily, one may use mechanical assists (e.g., agitation or vibration of the bed) to separate the particles or employ solids backmixing. In the latter case, the wet feed is blended in appropriate proportion with dried product so that the mixed feed is fluidizable. This adds to both the capital and operating costs of the drying operation, however.

Further, most fluidized bed dryers are nearly adiabatic contactors, e.g., they are well insulated, and the heat and mass transfer between the drying particles and the hot gas is by convection only. This implies that the product surface temperature (when surface moisture is being removed) will attain the wet-bulb temperature corresponding to the inlet gas temperature and humidity, while the state of the exit gas follows the adiabatic saturation line. When the internal moisture is being removed, the product temperature is above its wet-bulb temperature, and in the limit it attains the fluidizing gas temperature at long drying times.

To improve the thermal efficiency of the dryer by reducing the gas consumption, it is often a good idea to supply a part of the heat for drying by immersing heat exchanger panels or tubes in the bed. Thus lower gas temperatures (and flow rates) can be deployed without extending the drying time excessively, since a part of the heat (30–60 percent) is supplied indirectly by conduction. This is especially important for heat-sensitive materials, which limit the fluidizing gas temperature to

low values. For certain polymer pellet applications, fluidized bed dryers with immersed heat exchange panels (nonadiabatic) are the only cost-effective means for drying. It should be noted, however, that heat transfer considerations make this option most attractive only for fine particles of the order of 100 μm in size.

Table 2 summarizes the diverse variants of the fluidized bed technique; not all the types noted are commercially available, however.

Figure 3 shows the basic steps involved in the selection of a dryer. It is beyond the scope of this presentation to discuss the selection of dryers from a very large pool of possible dryers. The interested reader is asked to refer to Mujumdar and Menon (1995) and Baker (1997) for a more general discussion. Note that while over four hundred dryer types have been reported in the literature, over one hundred are sold commercially. Careful selection of the dryer type is crucial to the success of the final design and optimization of the drying system.

For any dryer, the steps shown in Fig. 3 apply. Unless reliable data exist for the same product it is strongly recommended that laboratory scale tests be carried out to verify that the material can be processed in a fluidized bed. The conditions for fluidization (e.g., minimum fluidization velocity) may be estimated using published correlations, but serious errors may accrue due to the surface wetness of the particulate solid, which may generally behave entirely differently from a surface-dry particle. Even trace amounts of surface wetness may render the material unfluidizable without mechanical energy input. In the case of a fluidized bed dryer with immersed heat exchangers, it is obvious that the material must not stick to the heat exchange surfaces.

While the convective heat and mass transfer rates in wet particle fluidization may be estimated reasonably well with dry particle correlations available readily in the literature, particle-to-immersed-surface heat transfer rates are very sensitive to the presence of surface moisture. Evaporation of the surface moisture may lead to a severalfold increase in the heat transfer rate until the particle is surface-dry, e.g., it reaches its critical moisture content or attains its equilibrium moisture content at the surface.

It is suggested that the laboratory scale tests be carried out in a chamber of at least 100 mm in diameter so that the wall effects are not severe. When possible, pilot scale tests are also recommended owing to the highly nonlinear nature of the hydrodynamics. Especially noteworthy is that the product quality (e.g., physical size, size distributions, structure, texture, color) may vary with the scale of operation depending on the pro-

Table 2 Classification of Fluidized Bed Dryers

Criterion	Type of dryer (application)
Operating pressure	Low-pressure (e.g., for heat-sensitive products)
	Near atmospheric (most common)
	High-pressure (5 bars, steam dryers)
Particulate flow regime	Well-mixed
	Plug flow
	Hybrid (well-mixed followed by plug flow)
Processing mode	Batch
	Continuous
Fluidizing gas flow	Continuous
	Pulsed
Fluidizing gas temperature	Constant
	Time-dependent
Heat supply	Convection or convection/conduction
	Continuous/pulsed
Fluidization action	By gas flow (pneumatic) only
	Downward set flow (jet-zone)
	With mechanical assist, e.g., vibration or agitation for sticky or polydispersed solids
Fluidized material	Particulate solid
	Paste/slurry sprayed onto a bed of inert particles
	Slurry sprayed onto absorbent particles (e.g., silica gel, biomass, etc.)
Fluidizing medium	Air/flue gases/direct combustion products
	Superheated steam (or vapor)
Number of stages	Single
	Multiple

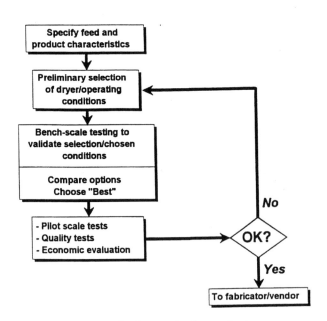

Figure 3 Basic steps in dryer selection.

duct. Attrition and agglomeration rates may differ as well. When the quality of the product is of paramount importance, it is essential to carry out pilot scale tests.

Table 3 summarizes the main criteria that must be specified in the design of a fluidized bed dryer. Detailed design or analysis procedures for the diverse fluidized bed dryer types are beyond the scope of this presentation.

The final selection of the fluidized bed drying system includes the selection of the feed system (including back-mixer, if needed) as well as gas cleaning (using cyclones, bag filters, etc.) and partial exhaust gas recycle for improved thermal efficiency. A typical flow diagram of the whole fluidized bed drying system is shown schematically in Fig. 4. A number of computer codes have been developed for the design of a fluidized bed dryer and its essential ancillaries, e.g., cyclones, fans. An example of such a code is one that is developed by Jumah and Mujumdar (1993). This code also permits estimation of the energy consumption as well as the capital and operating costs of fluidized bed dryers. More sophisticated codes for design and analysis of fluidized bed dryers are available from commercial sources.

Table 3 Design Criteria for a Fluidized Bed Dryer

Criterion/parameter	Main consideration
Feed moisture content, feed rate, product moisture content	Design specifications
Bed area	Calculate in design; given in analysis problem
Maximum air temperature	Depends on heat sensitivity of product
	Maximum product temperature must be specified
Outlet gas temperature and humidity	Exit gas in equilibrium with bed
	Equilibrium moisture content of product must be known
Maximum carryover	Related to attrition
	Quality of product affected

Note that when removing an organic solvent it is normally necessary to use an inert gas (e.g., nitrogen); the solvent is recovered by condensation and the inert gas reheated indirectly and returned. A small "bleed" is necessary to avoid the buildup of impurities in the circulating gas stream. When drying solids that can undergo combustion, care must be taken to avoid conditions susceptible to fire or explosion. Suitably designed rupture disks should be installed and the oxygen level in the drying system (including cyclones and baghouses) must be controlled to be outside the explosion limits. All internal metal parts must be grounded to avoid the potential of buildup of electrostatic charges by the fine dry dust.

Control of fluidized bed dryers is discussed by Jumah et al. (1995). More recently, Liptak (1998) has discussed advanced control strategies for continuous fluidized bed dryers. Attempts have also been reported in the literature on the use of fuzzy logic and artificial neural nets to control fluidized bed drying and granulation.

Finally, some remarks on capital costs of fluidized bed dryers are appropriate. Typically, the equipment cost rises as the 0.60 to 0.70 power of the capacity. The cost of the ancillary equipment, e.g., feeders, blowers, gas cleaning equipment, can exceed the cost of the dryer itself several times, depending on specific applications. It is therefore important to cost the entire drying system rather than just the dryer.

3 BASICS OF DRYING KINETICS

Figure 5 represents a typical textbook drying rate curve obtained from a batch of wet particles fluidized with hot air of fixed temperature, humidity, and flow rate. After an initial transient period (generally a heatup period) if the particles are "very wet," i.e., covered with a liquid film, the drying rate depends entirely on the external heat and mass transfer rates, e.g., a rise in air temperature or air flow rate or a reduction in air

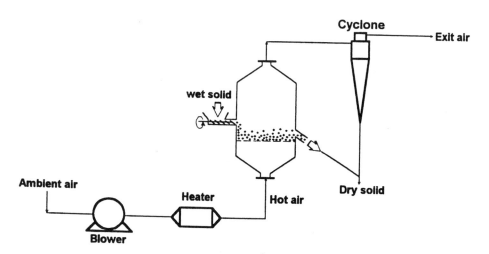

Figure 4 A schematic flow diagram of the whole fluidized bed drying system.

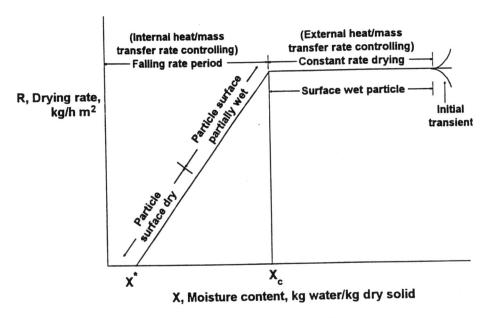

Figure 5 Typical textbook batch drying rate curve under constant drying conditions.

humidity will lead to an increase in the drying rate. This rate continues unchanged as long as the liquid film covers particle surfaces fully, since the drying conditions are held constant. Thus the so-called constant rate drying period exists as long as the drying conditions remain unchanged. Once the particle surfaces become partially dried, the rate of drying is reduced since in the area not covered with liquid film the rate of drying will depend also on the rate at which the internal liquid arrives at the surface prior to vaporization. The moisture content at which this drop in drying rate first occurs is called the critical moisture content. Unfortunately, this critical value is not a material property in that it is also a function of the drying conditions for a given material. It must be determined experimentally. Once the surface film disappears, the surface attains its equilibrium moisture content corresponding to the temperature and humidity conditions it is exposed to. [See Mujumdar and Menon (1995) for details of the definitions and terminology in psychrometry and drying.] Approximate critical moisture content values for some selected materials are given in Table 4.

For a purely convective adiabatic dryer, neglecting sensible heat effects, the surface temperature attained by the drying particles in the constant rate period is the so-called wet-bulb temperature corresponding to the gas conditions they are exposed to. Below the critical moisture content the particle temperature will rise above the wet-bulb temperature. The gas temperature

follows the adiabatic saturation line (in both constant and falling rate periods).

It is useful to summarize the key results of numerous experimental studies on fluidized bed drying kinetics. The following generalizations apply mainly to the conventional hot gas fluidized adiabatic dryer.

3.1 Effect of Bed Height

For materials with high mobility of internal moisture (e.g., iron ore, ion exchange resins, silica gel), most drying takes place close to the distributor plate. Increasing bed height beyond a particular value has

Table 4 Approximate Critical Moisture Contents for Various Materials

Material	Critical moisture content (kg water/kg dry solid)
Salt crystals, rock salt, sand, wool	0.05–0.10
Brick clay, kaolin, crushed sand	0.10–0.20
Pigments, paper, soil, worsted wool fabric	0.20–0.40
Several foods, copper carbonate, sludges	0.40–0.80
Chrome leather, vegetables, fruits, gelatin, gels	> 0.80

no effect on drying rate for the bed. For materials with main resistance to drying within the material, e.g., grains, increase of bed height will decrease the drying rate.

3.2 Effect of Particle Size

The time required to remove a given amount of moisture increases as the square of the particle diameter for a Geldart's type B particle, all other conditions being equal. For type A particles the effect is much smaller. Geldart's classification of powders is discussed in Chapter 3, Bubbling Fluidized Beds.

3.3 Effect of Gas Velocity

In removing surface moisture, gas velocity has a dominant effect. For particles with high internal resistance to moisture transfer, gas velocity has marginal to no effect.

3.4 Effect of Bed Temperature

The effect is complex and depends on the relative significance of external and internal resistances to moisture transfer. Higher external heat fluxes can lead to increased bed temperature, which in turn leads to higher moisture diffusivities and hence drying rates (see Reay and Baker, 1985).

4 FEATURES OF SPECIFIC TYPES OF FLUIDIZED BED DRYERS

Here we summarize the key features of the more commonly used fluidized bed dryers. Some of the more innovative and specialized fluidized bed dryer designs are discussed in the final part of this chapter.

4.1 Batch Fluidized Bed Dryers

Batch fluidized bed dryers are used for low throughput (normally $< 50 \, kg \, h^{-1}$ and good for $< 1000 \, kg \, h^{-1}$), multiproduct applications. Drying air is heated directly or indirectly usually to a fixed temperature. The drying gas flow rate is also usually fixed. However, it is possible to start drying at a higher inlet gas temperature (and flow rate) and lower it once the product moisture content falls below the critical value. Batch fluidized bed dryers are available commercially that can adjust the drying gas conditions automatically to maintain the bed temperature constant throughout the drying

process. Mechanical agitators or vibration may be needed if the material is difficult to fluidize.

4.2 Well-Mixed, Continuous Fluidized Bed Dryers (WMFBDs)

In this type of dryer (Fig. 2), the bed temperature is uniform and is equal to the product and exhaust gas temperatures. However, due to inherent product residence time distribution, product moisture content will span the range from inlet moisture content to a lower value. One advantage of the perfect mixing dryer is that the feed falls into a bed of relatively dry material and so is easy to fluidize.

4.3 Plug-Flow Fluidized Bed Dryers (PFFBDs)

In plug-flow fluidized bed dryers the bed usually has a length-to-width ratio in the range 5:1 to 30:1; the solids flow continuously as a plug through the channel from the inlet to the exit. This ensures approximately equal residence time for all particles, regardless of their size. For nearly monodisperse particles this ensures uniformity of product moisture content. The main operational problems occur at the feed end where wet feedstock must be fluidized directly rather than mixed with drier material as in a well-mixed unit. To handle this problem, several alternative strategies may be employed, e.g.,

Use agitator in feed region
Use backmixing of solids
Use a flash dryer to remove surface moisture prior to fluidized bed drying

At the tail end of drying, thermal efficiency can be poor as little drying takes place while the gas flow rate remains high to maintain fluidization. Zoning of the plenum (so that the drying gas temperature, and velocity to lesser extent, is progressively reduced as the material dries) is helpful to enhance efficiency as well as to reduce thermal degradation of product. A schematic sketch of the plug-flow fluidized bed dryer is shown in Fig. 6.

4.4 Vibrated Fluidized Bed Dryers (VFBDs)

For beds of particles that are difficult to fluidize due to strong polydispersity, particle size, or particle-to-particle adhesive forces (stickiness), it is worth considering a batch or continuous vibrated fluidized bed dryer. An application of nearly vertical sinusoidal mechanical vibration (half-amplitude 3–5 mm; frequency 10–50

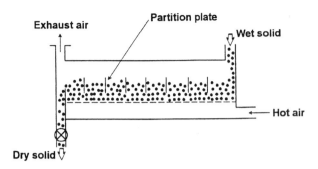

Figure 6 Plug-flow fluidized bed dryer.

Hz) allows "pseudo-fluidization" of the bed with rather low air flow rates. In this case, the requirements of hydrodynamics and heat/mass transfer are effectively decoupled. The gas velocity can be chosen to fit the needs of the drying kinetics; it must be significantly less than the minimum fluidization velocity, since above such a velocity the bed is no longer in continuous contact with the vibrating distributor plate. Furthermore, since the vibrational energy transmitted is attenuated with distance from the plate, the bed depth is limited to about 10–15 cm. Vibrated bed dryers can also be used to reduce attrition by gentle processing. A schematic sketch of this dryer is shown in Fig. 7. Most vibrated fluidized bed dryers are continuous units; the vibration vector is applied at a small angle to the vertical to assist with conveying of the material from the feed end to the exit weir.

4.5 Mechanically Agitated Fluidized Bed Dryers

Several versions of such dryers are in use. For drying of pastes or sludges, one variant uses a cylindrical vessel with a fast spinning agitator at the bottom onto

which the feed drops by gravity for dispersion into an upward spiral of hot drying gas. Other versions use a high rpm chopper that disperses the feed into hot air. More commonly, slowly rotating agitators (or rakes) are used to facilitate fluidization in the feed zone where very wet feed is fed into a continuous plug-flow dryer. Often, the first stage of a plug-flow dryer may be a well-mixed unit designed primarily to act as an efficient solids backmixer to facilitate fluidization in the plug-flow stage.

4.6 Centrifugal Fluidized Bed Dryers

To intensify heat and mass transfer rates for rapid drying of surface-wet particles, a centrifuge-type device may be used so that the drag force due to the fluidizing gas can be balanced with an "artificial gravity" generated by rotating the bed on a vertical axis. The rotating fluidized bed equipment is complex, and the decrease in drying times for most materials is normally not high enough or essential enough to justify the cost and complexity.

4.7 Spouted Bed Dryers (SBDs)

Spouted bed dryers are found suitable for drying of Geldart's type D particles, which are too coarse and dense to fluidize well without channeling. Unlike fluidized beds where the particle motion is random, the movement of particles in spouted beds is a regular recirculatory motion. Both batch and continuous modes of operation are possible. Owing to their limited processing capability per unit floor area and their high power consumption, spouted beds have not yet found major commercial applications. They are more commonly used for roasting (e.g., coffee/cocoa beans,

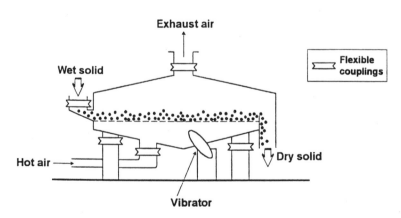

Figure 7 Vibrated fluidized bed dryer.

soya beans). Also, they have found limited applications for small-scale drying of slurries and suspensions sprayed into a spouted bed of inert particles (e.g., Teflon, polyethylene pellets). Another application involving drying is that of the coating of pharmaceutical tablets.

We have not discussed multistage fluidized bed dryers for lack of space. However, in some applications they yield major advantages. A dryer may be a part of a multistage FBD operated at different conditions or as a part of other dryer types, e.g., a flash dryer or spray dryer followed by a FBD.

Table 5 summarizes a selection guide for various fluidized bed dryer types based on their operating char-

acteristics. Suggested dryers for drying of pasty feedstocks are listed in Table 6. For drying granular materials, a suitable dryer may be chosen with the aid of a decision tree illustrated in Fig. 8.

5 DESIGN CONSIDERATIONS

5.1 A Simple Calculation Method for Batch Drying in a Fluidized Bed Dryer

If the amount of moisture to be removed from the particles in the final stages is very small, then the latent heat requirements can be ignored relative to the sensi-

Table 5 Fluidized Bed Dryer Selection Guide (Based on Operating Characteristics)

Selection criterion	WMFBD	PFFBD	VFBD	SBD
Feed Liquid/suspension/slurry	Inert particle bed	No	Inert particle bed	Inert particle bed
Wet particles				
Free-flowing	Yes	Yes	Yes	Yes
Cohesive	No	No	No	No
Particle size				
Small	Yes	Yes	Yes	No
Medium	Yes	Yes	Yes	Yes
Large	No	No	Yes	No
Polydisperse	No	No	Yes	No
Moisture				
Surface	Yes	Yes	Yes	Yes
Internal	Yes	Yes	Yes	Yes
Product specification				
Uniform moisture	No	Yes	Yes	No
Low moisture	No	Yes	Yes	No
Fragile	No	No	No	No
Heat-sensitive	Yes	No	No	No
Drying time				
5–10 min	Yes	No	No	No
10–60 min	Yes	Yes	Yes	Yes
60 min	No	Yes	No	Yes
Throughput				
Low	Yes	Yes	Yes	Yes
Medium	Yes	Yes	Yes	Yes
High	Yes	Yes	Yes	No
Immersed heaters	Yes	No	No	No
Temperature zoning	No	Yes	Yes	No
Flexibility in operation	No	Yes	Yes	No

Table 6 Suggested Fluidized Bed Dryers for Pasty Feedstocks

Feedstock consistency	Dryer suggested
Pumpable slurry, suspension	Fluid bed, spouted bed of inert particles
Soft pastes/sludges	WMFBD/backmix with dry product
Hard pastes	FBD or SBD

ble heat required to bring the product temperature from its wet-bulb value (at the end of the surface drying period in a purely convective dryer) to the final temperature that (from laboratory experiments) the product must reach at the end of drying to achieve the desired moisture content. In such cases, the batch drying time in the final drying period can be estimated reasonably accurately by a simple transient energy balance to determine the time required to heat up the particles from the wet-bulb to the final desired temperature. In most cases, it is also necessary to cool the product down to a lower temperature for storage or packaging. Indeed, the same design equation can be used to estimate the cool down time. The governing equation for the energy balance is (van't Land, 1991)

$$\frac{dT_p}{dt} = \frac{G_g C_{pg}[T_{in} - T_p(t)]}{m_s C_{ps}} \tag{1}$$

Upon integration, this yields the heat-up time, which is the dryer residence time needed for the falling rate period drying:

$$t = \frac{m_s C_{ps}}{G_g C_{pg}} \log\left[\frac{T_{in} - T_p(t=0)}{T_{in} - T_p(t)}\right] \tag{2}$$

In the case when the amount of moisture to be removed is large and the effect of latent heat cannot be neglected, different procedures must be employed. Kunii and Levenspiel (1991) have presented simplified procedures for analysis and design of batch and continuous fluidized bed dryers. As expected, different calculation procedures are required to calculate the drying time for the constant rate and falling rate periods. When the moisture diffusivity is high, the particle size small, and the drying conditions mild, much of the drying will occur in the constant rate period. In practice, fluidized bed dryers are used when a significant portion of the moisture is internal, in which case a simple diffusion model can be used to estimate the drying time for the falling rate period. Interested readers are referred to the design procedures and the example design problems presented by Kunii and Levenspiel (1991).

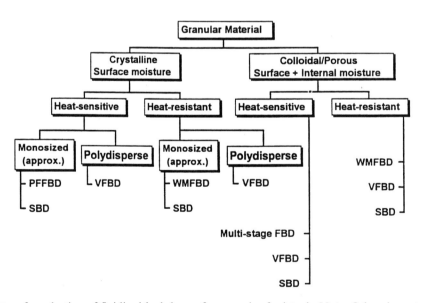

Figure 8 Decision tree for selection of fluidized bed dryers for granular feedstock. Note: Other dryer types are also suited for several of the above applications. If the feedstock contains a solvent, fluidized bed dryers are not recommended in general; indirect or vacuum dryers are preferred instead.

5.2 Predicting the Performance of a Continuous Fluidized Bed Dryer from Batch Drying Data

Provided that a batch drying curve $X(t)$ is available, the performance of a continuously well-mixed fluidized bed dryer can be predicted, at least in principle, by a so-called "integral" model of drying first proposed by Vanecek et al. (1964). The model is based on the following equation for calculating the average outlet moisture content, $\overline{X_0}$, from a continuous dryer:

$$\overline{X_0} = \int_0^\infty E(t)X(t)\,dt \tag{3}$$

where $E(t)$ is a function describing the residence time distribution of the material in the dryer. $X(t)$ is a drying curve function describing how the moisture content varies as the material passes through the dryer.

In a continuously well-mixed dryer, where the residence time distribution is that of a perfectly mixed vessel, Eq. (3) becomes

$$\overline{X_0} = \frac{1}{\tau_m} \int_0^\infty X(t)\exp\left(-\frac{t}{\tau_m}\right) dt \tag{4}$$

where τ_m is the mean residence time of particles in the bed.

Reay and Allen (1982) discussed the difficulty associated with the use of this equation. The batch drying curve $X(t)$ must be obtained at a constant bed temperature corresponding to the temperature selected for the continuous operation. This value is normally not known at the start of the development process, and in any case is extremely difficult to obtain experimentally. The experiment has to be performed with a high initial inlet air temperature, which must then be decreased at the later stage of drying in order to keep the bed temperature constant at the desired value. This is very difficult to achieve even with a sophisticated automatic control system.

The problem was overcome, however, by Reay and Allen (1982) following their derivation of a normalization rule for the effect of bed temperature on batch drying curves for both type A and B materials. They proposed, and verified experimentally, the following steps for constructing a constant bed temperature batch drying curve at any desired bed temperature from a batch drying curve measured at any convenient constant inlet air temperature.

1. Obtain a record of the changing bed temperature T_{ii} during constant inlet air temperature run.

2. Divide the constant inlet air temperature batch drying curve $X(t)$ into increments of length ΔX. For each increment note the time Δt_{ii} required to accomplish that amount of drying.

3. Calculate the time Δt_{ib} required to accomplish the same increment of drying at constant bed temperature by the use of the equation:

$$\frac{\Delta t_{ib}}{\Delta t_{ii}} = \frac{[(p_s - p_i)(X - X_e)]_{T_{ii}}}{[(p_s - p_i)(X - X_e)]_{T_{ib}}} \tag{5}$$

4. Build up the constant bed temperature batch drying curve by increments.

By solving Eq. (4) for a range of mean residence times, a so-called design curve can be constructed and used to extract the mean residence time required to achieve a specified outlet moisture content. The bed area can then be calculated for a given throughput S and bed mass per unit area m_b (Bahu, 1994):

$$A = \frac{S\tau_m}{m_b} \tag{6}$$

The choice of the bed mass per unit area is dictated by the need to break up the wet feed and to ensure rapid dispersion away from the feedpoint.

For a plug-flow fluidized bed dryer, the residence time will deviate from idealized plug-flow because of backmixing. This can be accounted for by employing the axial dispersion number, $B = D\tau_m/L^2$, where D is the particle diffusivity given by Reay (1978):

$$D = \frac{3.71 \times 10^{-4}(U - U_{mf})}{U_{mf}^{1/3}} \tag{7}$$

and L is the bed length. For small deviations from plug-flow, i.e., for small values of $B(< 0.1)$, the residence time function is

$$E(t) = \frac{1}{2\sqrt{\pi B}}\exp\left[-\frac{(I - t/\tau_m)^2}{4B}\right] \tag{8}$$

It should be noted that the validity of Reay's correlation for particle diffusivity has only been confirmed for bed depths up to 0.10 m. There is some evidence in the literature that D may be an order of magnitude larger in much deeper beds (Reay and Baker, 1985). The shallowest practicable beds are thus recommended if the objective is a close approach to plug-flow behavior of solid particles in the bed.

A summary of the basic scale-up of a fluidized bed dryer is illustrated in Fig. 9. For more accurate analysis or optimization, more advanced mathematical models can be found in the literature. Indeed, numerous mathematical models have appeared in the techni-

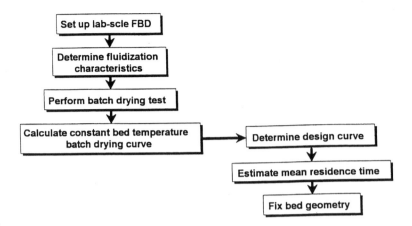

Figure 9 Basic steps in scale-up of fluidized bed dryer.

cal literature on modeling of the heat and mass transfer processes in fluidized bed drying of particles as well as fluidized bed drying of slurries using beds of inert particles. Borde et al. (1997) have presented a simple calculation procedure for the design of a vibrated bed dryer. The review presented by Pakowski et al. (1984) is also very useful to obtain a clear idea about the flow and heat/mass transfer characteristics of vibrated fluidized beds. For the design and analysis of conventional spouted beds as well as certain modified spouted beds the reader is referred to the work of Passos et al. (1987). For the special case of intermittently spouted beds for grain drying, Jumah et al. (1996) present a simple diffusion-based model for the estimation of drying times. Their model may be used only if the entire drying occurs in the falling rate period and the diffusion model adequately describes the falling rate period.

Unfortunately, no degree of sophistication in modeling can predict product quality parameters. Thus pilot testing is essential prior to scale-up to full-scale dryer design.

6 NOVEL FLUIDIZED BED DRYERS

Numerous innovative designs and operational modifications have been proposed in the literature while fewer are readily available commercially.

Tables 7, 8, and 9 summarize the key features of the "innovative" dryer concepts relative to the more conventional FBDs. Many of the ideas presented are new and have not been investigated even at the laboratory stage. It must be noted that not all the novel ideas are necessarily better, but many are worth exploring, since they appear to offer some distinct benefits over the more conventional ones.

6.1 Modified Fluidized Bed Dryers

Depending on special requirements it is possible to develop modified versions of the traditional fluidized bed dryers. Two-stage fluidized bed dryers consisting of several dryers are now in commercial use. Some examples of these combinations are given in Table 10.

Table 7 Fluidized Bed Dryers: Conventional Versus Innovative Concepts

Conventional	Innovative
Convective heat transfer	Convection + conduction (immersed heaters in bed)
Steady gas flow	Pulsed gas flow
Constant gas temperature	Variable gas temperature
Pneumatic fluidization	Mechanically assisted fluidization (vibration/agitation)
Used for drying of particles	Drying pastes, slurries using inert media
Air/combustion gas as drying medium	Superheated steam for fluidization/drying
Air drag resisted by gravity	Centrifugal fluid beds (artificial gravity generated by rotation)
Single stage/multistage fluid beds	Multistage with different dryer types
Simultaneous fluidization of entire bed	Moving fluidization zone (pulsating fluidized bed)

Table 8 Spouted Bed Dryers: Conventional Versus Innovative Concepts

Conventional	Innovative
Pneumatic spouting	Mechanical spouting (screw, vibration)
Single spout	Multiple spouts
Constant gas flow/continuous spouting	Variable gas flow/pulsed gas flow
Constant gas temperature	Variable gas temperature
Drying particles	Drying pastes, slurries using inert media
Spatially fixed spout	Moving spout (rotation, oscillation)
Convective drying	Combined convection and conduction
Axisymmetric	Two-dimensional, annular, hexagonal, etc.

Pulsing the fluidizing gas flow has been demonstrated to enhance drying performance in some cases at the expense of added capital costs. Periodic local fluidization of the bed by moving the distributor zone at regular intervals (so-called pulsated fluidized beds) has the advantage of reduced energy and air consumption when batch drying particulate materials with most of the heat and mass transfer resistances within the particles (Gawrzynski and Glaser, 1996). For larger particles (e.g., Geldart's type D), intermittent spouting of the bed with a rotating spouting jet has been shown by Jumah et al. (1996) to reduce energy consumption at only a marginal increase in drying time for batch drying. Indeed, in this case, it is also possible to introduce another type of intermittency, e.g., periodic heating of the spouting air when drying in the batch mode.

To conserve energy while reducing the drying time in batch fluidized bed drying one may remove the surface moisture using a higher gas temperature initially; the gas temperature can be reduced progressively as the material dries so that it does not exceed its maximum permissible temperature (Devahastin and Mujumdar, 1999). This can be achieved with a fuzzy logic controller or with a model-based control scheme. In plug-flow dryers this idea can be extended by temperature-zoning along the length of the dryer. The final section may generally act as a fluid bed cooler to reduce the product temperature to a safe value for storage or packaging (Bahu, 1997). If the product is highly hygroscopic, and low final moisture contents are desired, the cooling stage may require dehumidified air.

In the following we will review only a few innovative FBD techniques. For further details, see Kudra and Mujumdar (1995).

6.2 Superheated Steam Drying in Fluidized Beds

Although the idea was first published as early as 1898, the potential for superheated steam as a drying medium was not exploited industrially for at least six decades thereafter. Superheated steam drying offers one or more of the following advantages (Mujumdar, 1995):

No fire or explosion hazards
No oxidative damage
Ability to operate at vacuum or high-pressure operating conditions
Ease of recovery of latent heat supplied for evaporation; effectively a multiple effect operation is feasible

Table 9 Vibrated Bed Dryers: Conventional Versus Innovative Concepts

Conventional	Innovative
Constant gas flow	Variable gas flow
Constant gas temperature	Variable gas temperature
Aerated	Nonaerated
Convective drying	Combined conduction/radiation/microwave
Near atmospheric pressure	Vacuum operation
Horizontal trough	Vertical spiral trough
Fixed frequency/amplitude	Variable frequency/amplitude

Table 10   Two-Stage Fluidized Bed Dryers

Type	Remarks
Similar FBD stages stacked one below the other	Reduces floor area/bed depth of each stage; low product moisture content possible
Flash dryer stage preceding FBD stage	Fast removal of surface moisture; reduced stickiness leads to easy fluidization/smaller fluidized bed dryer
Spray dryer stage followed by FBD stage	Significantly reduced spray dryer size
Well-mixed FBD followed by plug-flow FBD stage underneath in same vessel	Ease of fluidization for high moisture content feed; uniform/ low final product moisture content possible

A higher drying rate above a critical "inversion temperature" when removing surface moisture
Better quality product under certain conditions
Closed system operation to minimize air pollution

Offsetting some of these advantages are some limitations, e.g.,

Higher product temperature
Higher capital costs compared to hot air drying
Possibility of air infiltration making heat recovery from exhaust steam difficult by compression or condensation

Among industrial scale applications of superheated steam fluidized bed drying one may cite

Drying of pulverized coal (Faber et al., 1986)
Drying of high-moisture lignite using a fluidized bed with internal heat exchangers to improve energy efficiency and reduce dryer size (Potter et al., 1986)
Drying of pulps, bagasse, sludges, hog fuel, spent grain from breweries, etc., in a pressurized superheated steam dryer (3–5 bars) manufactured by Niro A/S, Denmark (Jensen, 1992)

As may be expected, the net energy consumption of such dryers can be very small. Up to 90 percent fuel saving over conventional dryers has been reported for drying of beet pulp in a high-pressure fluidized bed dryer (Bosse and Valentin, 1988).

6.3   Low-Pressure Fluidized Bed Drying

Owing to the high potential of thermal degradation and fire or explosion hazards if the wet solids contain an organic solvent—common occurrences in the fine chemical and pharmaceutical industries—low-pressure operation offers several advantages (Mujumdar, 1995). Similar problems may arise in coating operations. To alleviate this problem, low-pressure fluidization may

be used to lower the operating temperature, thus reducing the probability of thermal degradation; it may allow operation in conditions out of the flammability limits. Arnaldos et al. (1997) have presented experimental results on the hydrodynamic behavior of vacuum fluidized bed dryers and their applications for drying granular solids and coating of particles. Note that the minimum fluidization velocity increases as the pressure decreases. The flow behavior at lower pressures is quite different from that at atmospheric or high-pressure operation. The fluidization is progressive, and the concept of minimum fluidization velocity loses some of its significance. Experimental measurements of the velocity for complete fluidization are needed for design purposes.

Little work is published on vacuum fluidized bed drying. Arnaldos et al. (1997) report vacuum fluidized bed drying data at 200, 400, and 500 mbar pressures. Silica ($d_p = 975\,\mu m$, $\rho_s = 1650\,kg/m^3$; $\phi_s = 0.7$) and millet ($d_p = 1800\,\mu m$, $\rho_s = 1600\,kg/m^3$; $\phi_s = 0.90$) were used as test materials. Interestingly, the highly porous silica particles dried faster as the pressure was lowered, while the denser millet particles dried slower since much of its resistance to drying is internal. Thus the effect of lower pressure depends on the internal heat and moisture resistance of the particles. Because of their inherently higher capital and operating costs, vacuum fluidized bed dryers have found application mainly in the pharmaceutical industry.

CLOSING REMARKS

Fluidized bed dryers, both conventional and innovative, will continue to find increasing applications in various industries. For further details, the reader is asked to refer to Vanecek et al. (1965), Gupta and Mujumdar (1983), Pakowski et al. (1984), Reay and Baker (1985), Hovmand (1995), Keey (1992), and Mujumdar (1995), which contains an extensive annotated bibliography of the major literature sources on drying. Numerous papers dealing with mathematical models for conventional and modified fluidized bed dryers appear regularly in Drying Technology—An International Journal (Marcel Dekker, NY).

NOTATIONS

A	=	bed cross-sectional area, m^2
B	=	axial dispersion number
C_{pg}	=	specific heat of gas, kJ kg^{-1} K^{-1}
C_{ps}	=	specific heat of solid, kJ kg^{-1} K^{-1}
D	=	particle diffusivity, m^2 s^{-1}
d_p	=	particle diameter, m
G_g	=	gas flow rate, kg s^{-1}
L	=	bed length, m
m_b	=	bed mass per unit area, kg m^{-2}
m_s	=	bed mass, kg
p_i	=	partial pressure of vapor in gas entering bed, kPa
p_s	=	saturation partial pressure of vapor at the bed temperature, kPa
S	=	throughput, kg s^{-1}
T_{ib}	=	bed temperature in a batch drying curve with constant bed temperature, K
T_{ii}	=	bed temperature in a batch drying curve with constant inlet temperature, K
T_{in}	=	inlet gas temperature, K
T_p	=	particle temperature, K
t	=	time, s
X	=	moisture content, kg water/kg dry solid
X_e	=	equilibrium moisture content, kg water/kg dry solid
$\overline{X_0}$	=	mean product moisture content, kg water/kg dry solid
U	=	superficial gas velocity, m s^{-1}
U_{mf}	=	superficial minimum fluidization gas velocity, m s^{-1}

Subscripts

b	=	bed
e	=	equilibrium
g	=	gas
p	=	particle
s	=	solid

Greek Symbols

ϕ_s	=	particle sphericity
ρ_s	=	particle density, kg m^{-3}
τ_m	=	mean particle residence time, s

REFERENCES

Arnaldos J, Martra J, Casal J, Kozanoglu B. Vacuum fluidization processes: drying and coating. Powder Handling and Processing 9:315–319, 1997.

Bahu RE. Fluidised bed dryer scale-up. Drying Technology—An International Journal 12:329–339, 1994.

Bahu RE. Fluidized bed dryers. In: Baker CGJ, ed. Industrial Drying of Foods. London: Blackie Academic and Professional, 1997, pp. 65–89.

Baker CGJ. Dryer selection. In: Baker CGJ, ed. Industrial Drying of Foods. London: Blackie Academic and Professional, 1997, pp. 242–271.

Borde I, Dukhovny M, Elperin T. Heat and mass transfer in a moving vibrofluidized granular bed. Powder Handling and Processing 9:311–314, 1997.

Bosse D, Valentin P. The thermal dehydration of pulp in a large scale steam dryer. Proceedings of the Sixth International Drying Symposium, Versailles, 1988, pp. 337–343.

Devahastin S, Mujumdar AS. Batch drying of grains in a well-mixed dryer—Effect of continuous and stepwise change in drying air temperature. Transactions of the ASAE 42:421–425, 1999.

Faber EF, Heydenrych MD, Seppa RUI, Hicks RE. A techno-economic comparison of air and steam drying. In Mujumdar AS, ed. Drying'86, Vol. 2. Washington: Hemisphere, 1986, pp. 588–594.

Gawrzynski Z, Glaser R. Drying in a pulsed-fluid bed with relocated gas stream. Drying Technology-An International Journal 14:1121–1172, 1996.

Gupta R, Mujumdar AS. Recent developments in fluidized bed drying. In Mujumdar AS, ed. Advances in Drying. Vol. 2. Washington: Hemisphere, 1983, pp. 155–192.

Hovmand S. Fluidized bed drying. In: Mujumdar AS, ed. Handbook of Industrial Drying. 2d ed. New York: Marcel Dekker, 1995, pp. 195–248.

Jensen AS. Pressurized drying in a fluid bed with steam. Proceedings of the Eighth International Drying Symposium, Montreal, 1992, pp. 1593–1601.

Jumah RY, Mujumdar AS. A PC program for preliminary design of a continuous well-mixed fluid bed dryer. Drying Technology—An International Journal 11:831–846, 1993.

Jumah RY, Mujumdar AS, Raghavan GSV. Control of industrial dryers. In: Mujumdar AS, ed. Handbook of Industrial Drying. 2d ed. New York: Marcel Dekker, 1995, pp. 1343–1368.

Jumah RY, Mujumdar AS, Raghavan GSV. A mathematical model for constant and intermittent batch drying of grains in a novel rotating jet spouted bed. Drying Technology—An International Journal 14:765–802, 1996.

Keey RB. Drying of Loose and Particulate Materials. Washington: Hemisphere, 1992.

Kudra T, Mujumdar AS. Special drying techniques and novel dryers. In: Mujumdar AS, ed. Handbook of Industrial Drying. 2d ed. New York: Marcel Dekker, 1995, pp. 1087–1149.

Kunii D, Levenspiel O. Fluidization Engineering. 2d ed. Boston: Butterworth-Heinemann, 1991.

Liptak B. Optimizing dryer performance through better control. Chemical Engineering 105:96–104, 1998.

Mujumdar AS. Superheated steam drying. In: Mujumdar AS, ed. Handbook of Industrial Drying. 2d ed. New York: Marcel Dekker, 1995, pp. 1071–1086.

Mujumdar AS, Menon AS. Drying of solids. In: Mujumdar AS, ed. Handbook of Industrial Drying. 2d ed. New York: Marcel Dekker, 1995, pp. 1–46.

Mujumdar AS, Alterman DS. Drying in the pharmaceutical and biotech fields. In Goldberg E, ed. Handbook of Downstream Processing. London: Blackie Academic and Professional, 1997, pp. 235–260.

Pakowski Z, Mujumdar AS, Strumillo C. Theory and application of vibrated beds and vibrated fluid beds for drying processes. In Mujumdar AS, ed. Advances in Drying. Vol. 3. Washington: Hemisphere, 1984, pp. 245–306.

Passos ML, Mujumdar AS, Raghavan GSV. Spouted bed for drying: principles and design considerations. In

Mujumdar AS, ed. Advances in Drying. Vol. 4. Washington: Hemisphere, 1987, pp. 359–398.

Potter OE, Beeby C. Modeling tube to bed heat transfer in fluidized bed steam drying. In Mujumdar AS, ed. Drying'86. Vol. 2. Washington: Hemisphere, 1986, pp. 595–603.

Reay D. Particle residence time distributions in shallow rectangular fluidised beds. Proceedings of the First International Drying Symposium, Montreal, 1978, pp. 136–144.

Reay D, Allen RWK. Predicting the performance of a continuous well-mixed fluid bed dryer from batch tests. Proceedings of the Third International Drying Symposium, Birmingham, 1982, pp. 130–140.

Reay D, Baker CGJ. Drying. In: Davidson JF, Clift R, Harrison D, eds. Fluidization 2d ed. London: Academic Press, 1985, pp. 529–562.

Vanecek V, Picka J, Najmr S. Some basic information on the drying of granulated NPK fertilizers. International Chemical Engineering 4:93–99, 1964.

Vanecek V, Markvart M, Drbohlav R. Fluidized Bed Drying. London: Leonard Hill, 1965.

van't Land CM Industrial Drying Equipment: Selection and Application. New York: Marcel Dekker, 1991.

19

Circulating Fluidized Beds

John R. Grace and Hsiaotao Bi

University of British Columbia, Vancouver, British Columbia, Canada

Mohammad Golriz

Umeå University, Umeå, Sweden

1 INTRODUCTION

The term circulating fluidized bed, commonly abbreviated CFB, has been in common usage since the mid-1970s, although the origins of the technology date back to the 1940s for catalytic processes (Squires, 1994) and to the 1960s for gas–solid processes (Reh, 1971). The term implies two complementary characteristics for gas–solid systems:

1. A *configuration* where particles, entrained at a considerable flux from a tall main reactor or "riser," are separated efficiently from the carrying fluid, usually external to the reactor, and returned to the bottom of the riser, forming a recirculation loop for the particles. Individual particles circulate around this loop many times before leaving the system, whereas the fluid passes through only once. A typical setup is shown schematically in Fig. 1. As described in Sec. 4 below, many different geometric configurations are utilized in practice.

2. *Operation* at high superficial gas velocity (typically 2–12 m/s) and high particle flux (typically 10–1000 kg/m²s) so that there is no distinct interface in the riser between a dense bed and a dilute region above. Contacting is therefore carried out at gas velocities beyond the bubbling, slugging and turbulent fluidization flow regimes, residing instead in a higher velocity flow regime—fast fluidization, dense suspension upflow, or dilute pneumatic conveying (see Sec. 3 below).

In recent years, the term circulating fluidized bed has also been used for liquid–solid systems and gas–liquid–solid (three-phase) systems. These are treated in Chapters 26 and 27. However, the predominant interest in CFB systems continues to be for gas–solid (two-phase) systems, and we restrict this chapter to this case. Some mention is also made in the literature of internally circulating fluidized beds where the particles circulate around one or more loops within a main reactor space or vessel. This chapter deals principally with gas–solid systems where solids recirculate through an external solids flow system, usually involving one or more cyclones, a standpipe, and a valve or seal, either nonmechanical (e.g., L-valve or loop seal; see Chapter 21) or mechanical (e.g., slide valve).

The key operating variables for commercial operation usually fall into the following ranges:

Superficial gas velocity: 2–12 m/s
Net solids flux through the riser: 10–1000 kg/m²s
Temperature: 20–950°C
Pressure: 100–2000 kPa
Mean particle diameter: 50–500 μm
Overall riser height: 15–40 m

Historical accounts of the development of circulating fluidized beds have been prepared by Reh (1986) and Squires (1994). Advantages of circulating fluidized beds relative to bubbling beds and other types of gas–

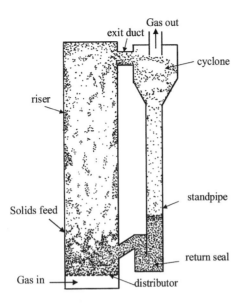

Figure 1 Schematic of typical circulating fluidized bed system.

solid contactors such as packed beds and rotary kilns frequently include

High gas throughputs
Limited backmixing of gas
Long and controllable residence time of particles
Temperature uniformity, without "hot spots"
Flexibility in handling particles of widely differing
    sizes, densities, and shapes
Effective contacting between gas and particles
Lack of bypassing of gas with minimal mass trans-
    fer limitations
Opportunity for separate and complementary
    operation (e.g., catalyst regeneration or particle
    cooling) in the return loop

Disadvantages often include:

Need for very tall vessel: small scale CFB processes
    are therefore seldom viable
Substantial backmixing of solid particles
Internals (e.g., baffles, heat transfer surfaces) not
    viable because of wear/attrition
Wall wastage sometimes a serious problem
Suspension-to-surface heat transfer less favorable
    than for low-velocity fluidization
Lateral gradients can be considerable
Losses of particles due to entrainment.

Extensive reviews of all aspects of circulating fluidized beds up to about 1996 are provided by the books
edited by Kwauk (1994) and Grace et al. (1997). More

limited reviews have been provided by Yoshida and
Mineo (1989) and by Berruti et al. (1995). This chapter
summarizes the most important and useful findings
and approaches to circulating fluidized beds. This is
an active area for research and development with
new results and models emerging each year.
International conferences devoted explicitly to circulating fluidized beds are held every three years, each leading to published proceedings, cited frequently in this
chapter. Many CFB papers also appear in journals and
proceedings of other conferences, e.g., the triannual
Fluidization Conferences under the auspices of the
Engineering   Foundation   and   the   biannual
International Fluidized Bed Combustion conferences.

2 APPLICATIONS

Major commercial applications of circulating fluidized
beds are listed in Table 1, together with key references.
These applications include solid-catalyzed gas reactions and gas–solid reactions, as well as physical operations. While catalytic cracking and solid-fuel
combustion are the predominant catalytic and gas–
solid reaction applications, respectively, there are also
a number of other processes where the unique characteristics of circulating fluidized beds are being
exploited. Further details on the most important applications are presented in Chapters 14 through 16.

3 FLOW REGIMES

3.1 Onset of Fast Fluidization

A circulating fluidized bed (CFB) is operated in the
transport mode, with solids carried over from the top
of the riser separated and returned to the bottom of the
riser via a standpipe and feeding or control device. The
transition from low-velocity fluidization to transport
operation occurs when significant solids entrainment
commences with increasing superficial gas velocity.
At least seven methods have been proposed (Bi et al.,
2000) to quantify the transition. The criteria can be
divided into two groups, one based on solids entrainment and the other on solids concentration profiles.

3.1.1 Transport Velocity, U_{tr}, based on Phase
        Diagrams

A critical solid circulation rate may exist when a sharp
change in pressure drop over the lower part of a riser
takes place when varying the solid circulation flux at a

Table 1 Major Commercial Applications of Circulating Fluidized Beds

Application	Key References	Comments
Fluid catalytic cracking	Avidan, 1997; Sec. 8.1 of this handbook	Hundreds of units worldwide; mainstay of petroleum refining
Fischer–Tropsch synthesis	Shingles & McDonald, 1988; Steynberg et al., 1991; Matsen, 1997	Applied for many years as Synthol process in South Africa
Maleic anhydride	Matsen, 1997; Contractor, 1999	One commercial reactor in Spain
Combustion of coal, biomass, wastes, off-gases	Li and Zhang, 1994; Brereton, 1997; Lee, 1997; Basu, 1999; Plass, 2001	Widespread usage for power generation and boilers in Europe, North America, and Asia
Gasification	Hirschfelder & Vierrath, 1999; Plass, 2001	Commercial units gaining a foothold, especially in Europe
Calcination (e.g., of aluminium trihydrate and carbonates)	Reh, 1971, 1986; Schmidt, 1999	Lurgi units used widely
Catalyst regeneration	Chen et al., 1994	Applied in China
Roasting of ores	Dry & Beeby, 1997; Pienermann et al., 1992	Applied in Australia
Reduction of iron ore	Dry & Beeby, 1997; Husain et al., 1999; Plass, 2001	Lurgi plant in Trinidad
Smelter off-gas treatment	Hiltunen & Moyöhänen, 1992	One plant in Australia supplied by Ahlstrom
Flue gas dry scrubbing of HF, HCl, SO_2, dioxins, mercury, etc.	Graf, 1999; Mayer-Schwinning & Herden, 1999	Commercial units since the 1970s, primarily in Europe

given gas velocity, U_g (see Fig. 2). As U increases beyond a certain point, the sharp change in the pressure gradient disappears. The gas velocity at this critical point, defined as the transport velocity U_{tr}, marks the onset of fast fluidization (Yerushalmi and Cankurt, 1979). An examination of pressure gradient profiles (Bi, 2002) reveals that U_{tr} varies with height. U_{tr} may indicate a transition of axial voidage profiles in the riser.

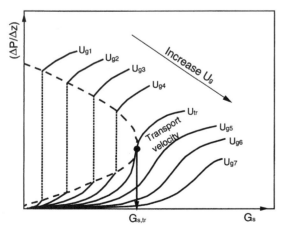

Figure 2 Definition of transport velocity, U_{tr}, of Yerushalmi and Cankurt (1979).

Below this velocity, a distinct interface exists between the top-dilute and bottom-dense regions. Beyond this velocity, the interface becomes relatively diffuse.

To predict U_{tr}, several correlations have been developed, most of the form

$$Re_{tr} = kAr^n \tag{1}$$

Typical k and n values are given in Bi et al. (2000), with $k = 2.28$ and $n = 0.419$ from Bi and Fan (1992). In view of the difficulty of determining U_{tr} experimentally, and the variation of U_{tr} with measurement location and riser geometry, caution needs to be exercised when Eq. (1) is used to predict U_{tr}.

3.1.2 Critical Velocity, U_{se}, Based on Solids Entrainment

In a gas–solids transport system, solids flux and gas velocity are related by

$$G_s = \rho_p(1 - \varepsilon)\left(\frac{U_g}{\varepsilon} - U_{slip}\right) \tag{2}$$

A linear relationship between G_s and U_g in the high-velocity range with ε constant in Fig. 3 suggests that U_{slip} in Eq. (2) approaches a constant value, determined from the intercept and slope of the linear part of the curve. This critical velocity, designated U_{se} (Bi et

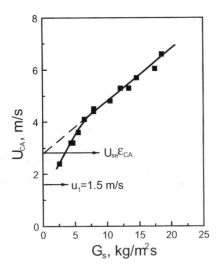

Figure 3 Definition of critical velocity, U_{se}, of Bi et al. (1995).

al., 1995), can be considered a hindered or apparent terminal velocity of bed particles. U_{se} can also be determined in a batch fluidized bed from the emptying time (Bi et al., 1995). A correlation for U_{se} based on data from relatively tall ($H > 5$ m) and large ($D > 0.05$ m) columns (Bi et al., 1995) gives

$$Re_{se} = 1.53 Ar^{0.5} \tag{3}$$

When U_{se} from Eq. (3) < the terminal velocity, u_t, of single particles, U_{se} should be taken $= u_t$.

3.2. Flow Regimes in Gas–Solids Circulating Fluidized Beds

3.2.1 Transition from Pneumatic Transport to Fast Fluidization

Fast fluidization is characterized by a dense region at the bottom of a circulating fluidized bed, leading smoothly (without a sharp interface) into a lean region above (Li and Kwauk, 1980). In contrast, in the dilute-phase flow regime, the pressure gradient, except for an acceleration zone at the bottom, is nearly uniform. Hence the transition from fast fluidization to dilute flow can be characterized by the disappearance of the S-shaped inflection point (Li and Kwauk, 1980), or by the disappearance of nonuniform axial density profiles (Takeuchi et al., 1986).

In conventional pneumatic transport, a minimum in the dP/dz vs. G_s curve at a fixed gas velocity is commonly used to separate dense phase flow from lean phase transport (Leung, 1980). Experimentally, internal solids circulation increases dramatically right after

the gas velocity is reduced to reach the minimum pressure gradient point (Matsumoto and Harakawa, 1987). Radial segregation into a dilute core and dense annulus results from the balance between the force needed to suspend the weight of the particles and wall friction, and the transition velocity is influenced by the unit geometry.

Several correlations have been proposed to predict the transition velocity, U_{CA} (see Bi et al., 1993). The correlations of Bi and Fan (1991),

$$U_{CA} = 21.6\sqrt{gd_p}\left(\frac{G_s}{\rho_g U_{CA}}\right)^{0.542} Ar^{0.105} \tag{4}$$

and Yang (1975),

$$\frac{2gD(\varepsilon_{CA}^{-4.7} - 1)}{(U_{CA}/\varepsilon_{CA} - u_t)^2} = 6.81 \times 10^5 \left(\frac{\rho_g}{\rho_p}\right)^{2.2} \tag{5}$$

show the best agreement with literature data. Improved correlations are needed.

3.2.2 Transition from Fast Fluidization to Dense Suspension Upflow

Stable operation of gas–solids upward transport lines becomes impossible when the blower is unable to provide sufficient pressure head. This condition is referred to as blower-induced instability (Bi et al., 1993).

Another type of instability can occur when an upflow riser is directly coupled with a downcomer that returns entrained particles to the bottom of the riser. A pressure balance between the riser and the downcomer is required to maintain steady operation. If the gas velocity is decreased at a given solids circulation rate, a critical state may be reached at which steady operation at a given solids flux is impossible; instability occurs because solids cannot be fed to the riser at the prescribed rate. Such an instability, referred to as standpipe-induced (Bi et al., 1993), occurs at a lower critical velocity for a higher solids holdup in the riser. The point of instability can be predicted based on an analysis of the pressure balance in the riser–downcomer loop (Bi and Zhu, 1993). To circumvent standpipe-induced instability, the solids inventory in the standpipe needs to be sufficiently high or, alternatively, the riser needs to be uncoupled from the downcomer, e.g., by employing screw feeders.

The third instability, classical choking, occurs on reducing the superficial gas velocity in the riser when slugs start to form right after a dense bed is created in the bottom section of a riser. Stable operation of gas–solids upward slug/plug flow may still be possible in some systems (Konrad, 1986). However, when slug flow develops to such an extent that severe pressure fluctuations cause the whole system to fluctuate dramatically, stable operation is no longer possible. Classical choking only occurs in systems which are capable of slugging. Criteria for classifying systems as slugging or nonslugging have been proposed based on instability analysis of uniform suspension flow (Yousfi and Gau, 1974), stability of slugs (Yang, 1975), and propagation of continuity waves (Smith, 1978). Slug flow is not encountered when the maximum stable bubble size is significantly smaller than the column diameter. The onset of classical choking can be estimated from Yousfi and Gau (1974):

$$U_{CC} = 32\sqrt{gd_p}\left(\frac{G_s}{\rho_g U_{CC}}\right)^{0.28} Re_t^{-0.06} \quad (6)$$

When a riser system has been designed so that blower- and standpipe-induced instabilities and classical choking are all avoided, the flow pattern passes from fast fluidization to dense suspension upflow (Bi and Grace, 1999) with increasing G_s, as shown in Fig. 4. Based on recent experimental observations that solids no longer fall downward near the riser wall when the riser is operated at relatively high G_s and U_g (e.g., Issangya et al., 1998; Karri and Knowlton, 1998) and that gas backmixing decreases after reaching a maximum at high G_s, Grace et al. (1999) proposed that a transition from fast fluidization to dense suspension upflow occurs when the net solids flux near the riser wall changes from downward to upward with increasing G_s at a given U_g. A first correlation (Grace et al., 1999) based on limited experimental data within the range of $500 \geq Ar \geq 6$, $(1 - \varepsilon_{av}) \geq 0.07$, $100 \geq G_s/(\rho_g U_g \geq 7$, $0.305 \geq D \geq 0.051$ m, $27.4 \geq H \geq 6.1$ m, with air at ambient conditions as the gas, gives

$$U_{DSU} = 0.0113 G_s^{1.192} \rho_g^{-1.064} \left[\mu_g g(\rho_\rho - \rho_g)\right]^{-0.064} \quad (7)$$

3.3 Flow Regime and Operating Diagrams

A number of flow regime diagrams have been proposed, beginning with Zenz (1949), who plotted the pressure gradient (dP/dz) vs. superficial gas velocity (Fig. 5). The cocurrent gas–solids flow region spans the flow regimes encountered in the circulating fluidized beds, with the lower limit set by the choking velocity. There is an inoperable region between the low-velocity fluidization and cocurrent upflow, because the transport line is choked as discussed above. Such a regime diagram has been further extended to incorporate more subregions within cocurrent flow (e.g., Drahos et al., 1988). The pressure gradient (dP/dz) is almost proportional to the solids fraction, since friction and acceleration/deceleration terms can usually be neglected. One can alternatively plot solids fraction or bed voidage (e.g., Li and Kwauk, 1980) vs. superficial gas velocity (U_g), normalized superficial gas velocity (U_g/u_t), or slip velocity.

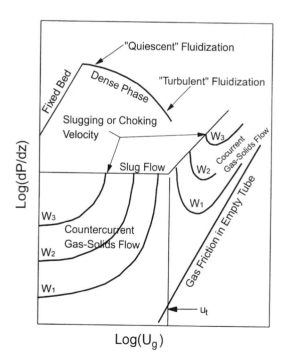

Figure 5 Flow regime diagram based on pressure gradient and superficial gas velocity coordinates. (Adapted from Zenz, 1949.)

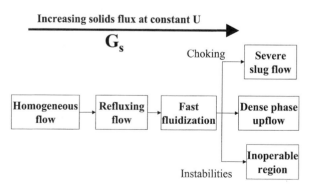

Figure 4 Transitions between dense suspension upflow, fast fluidization, and pneumatic transport with increasing solids circulation flux at a constant gas velocity.

Another set of diagrams, referred to as operating diagrams (e.g., Takeuchi et al., 1986; Yang, 2001) plot G_s or loading ratio $[G_s/(\rho_g U_g)]$ versus U_g or normalized superficial gas velocity. A typical diagram of this type is shown in Fig. 6, with the flow patterns divided into homogeneous dilute-phase flow, core-annular flow, fast fluidization, and dense suspension upflow, with blower and standpipe limitations absent. Bi and Grace (1995) plotted the dimensionless relative velocity between the gas and particles vs. a dimensionless particle diameter based on the Grace (1986a) phase diagram.

It should be noted that severe slugging, blower, and standpipe limitations, as discussed above, are not reflected in almost all of these phase diagrams.

4 HYDRODYNAMICS OF THE FAST FLUIDIZATION FLOW REGIME

4.1 Voidage, Solids Concentration, and Pressure Profiles

4.1.1 Definitions and Measurement Techniques

The voidage is the fraction of volume occupied by the gas in a gas–solid suspension. Voidage is a key variable in circulating fluidized bed risers. Several different voidages are in common usage:

  Instantaneous local voidage in a measuring volume that is small with respect to the equipment but large with respect to individual particles. This is designated ε, or $\varepsilon(x, y, z, t)$ to indicate variation with position and time.
  Local time-average voidage $\bar{\varepsilon}$, $\bar{\varepsilon}(r, z)$, or $\bar{\varepsilon}(x, y, z)$.
  Cross-sectional average voidage at a given height, denoted here by ε_{av} or $\varepsilon_{av}(z)$.
  Overall average voidage for the riser, $[\varepsilon]$, i.e., fraction of entire riser volume occupied by gas.

The overbar signifies a time average, whereas the subscript "av" and square brackets denote spatial averages. Since we are mostly interested in steady flows, time-averaging is often also carried out. For steady axisymmetric flow in a cylindrical riser of radius R and overall height H,

$$\bar{\varepsilon}(r, z) = \lim_{t \to \infty} \frac{1}{t} \int_0^t \varepsilon(r, z)\, dt \qquad (8)$$

$$\varepsilon_{av}(z) = \frac{1}{\pi R^2} \int_0^R \bar{\varepsilon}(r, z) 2\pi r\, dr \qquad (9)$$

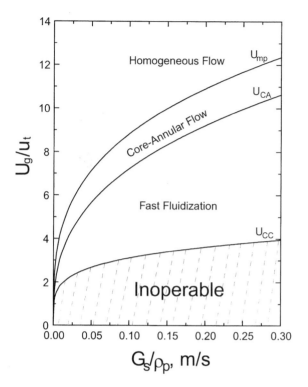

Figure 6 Flow regime diagram based on solids circulation rate and superficial gas velocity. (Adapted from Bi and Grace, 1996.)

In some cases we are interested in the fluctuations themselves, as well as in the time-averaged quantities; we can then write, for example,

$$[\varepsilon] = \frac{1}{H} \int_0^H \varepsilon_{av}(z) \qquad (10)$$

$$\varepsilon(r, z) = \bar{\varepsilon}(r, z) + \varepsilon'(r, z) \qquad (11)$$

where the prime denotes a fluctuating component, i.e., the instantaneous deviation from the time mean. It is also common to report or discuss the volumetric concentration of solid particles, often called the "solids holdup." This terminology is used throughout this chapter and denoted by ε_s, i.e., $\varepsilon_s = 1 - \varepsilon$. Still another way of designating the concentration is to refer to a "suspension density," i.e., $\rho_{susp} = \rho_p(1 - \varepsilon) + \rho_g \varepsilon$. Clearly one can obtain local time-averages and spatial-averages values of ε_s and ρ_{susp} in an analogous manner to obtaining these quantities for ε.

Several techniques have been found to be useful for determining local voidages/solids holdups:

Optical Probes for the measurement of voidages and solids holdups in CFB risers were pioneered by Qin and Liu (1982). In these probes a single fiber or bundle of fibers projects light onto a local region. The solids concentration is inferred from the intensity of either the reflected or the transmitted light. The transmission method is restricted to low solids volume fractions (Werther et al., 1993), while reflective probes can be used over a wide range of particle concentrations after calibration. Advantages of optical probes include insensitivity to electrical and magnetic fields, low signal loss, small size, flexibility of geometry, high sensitivity, rapid response, and low cost. Like any probe they may disturb the flow, but the extent of interference can be minimal if the probe is sufficiently small. Optical fiber probes have also been extended (Johnsson and Johnsson, 2001) to high-temperature systems with optical filtering to separate reflected light from background radiation.

If the particle diameter is larger than the fiber core diameter (type I), single particles are detected. The output signals from the light receiver are then converted to pulses. Averaging over extended periods is required to measure the particle concentration, with only average concentrations determined. On the other hand, if the particle diameter is much smaller than the probe diameter (type II), the reflected light at any instant comes from many particles.

Type II systems are most common. For a type II system, penetration of light and the size of the measuring volume depend on the local volumetric solids concentration (Lischer and Louge, 1992). Zhou (1995) found that penetration length was ~ 7 mm for ε approaching unity, decreasing to ~ 4 mm for a $\varepsilon = 0.6$, and even less near the wall. Reh and Li (1991) and Krol and de Lasa (1998) reduced the measurement volume by having two fiber bundles converging at an angle of 30–$35°$, while Tanner et al. (1994) used a small lens to achieve convergence with a single fiber. For very low solids holdups (0.2 to 0.7%), Sobocinski et al. (1995) used a probe with a rounded (hemispherical) tip.

The accuracy of the measurements relies also on the precision of the calibration. Matsuno et al. (1983) calibrated their probe in a stream of glass beads falling from a vibrating sieve, assuming the velocity to equal the single-particle terminal falling velocity. This method is limited to dilute systems. Matsuno et al. (1983) allowed particles to fall through a stagnant liquid, assuming them to attain their single particle terminal velocity. Hindered settling likely influenced the calibration. Zhang et al. (1991) used the cross-sectional average voidage inferred from static pressure drop measurements. Hartge et al. (1986a) obtained a

linear relationship for quartz particles in a water-fluidized bed for $\varepsilon_s \leq 0.5$; they then assumed linearity and calibrated their probe by comparing the optical signals to two concentrations obtained by γ-ray absorption. Linearity has also been assumed by others (e.g., Kato et al., 1991; Zhou et al., 1994). Werther et al. (1993), using solids–liquid suspensions with FCC and sand particles, suggested an empirical fit of the form $\Delta I = k(1 - \varepsilon)^n$ and found that n was independent of the fluid properties. Nieuwland et al. (1996) obtained similar results. Qin and Liu (1982) derived a power law relationship between the light intensity and the solids holdup for uniform spheres by assuming that the light is reflected from the first two layers of particles only. Lischer and Louge (1992) found a similar power law from in situ calibration using a capacitance probe. A simulation of the output of a single fiber immersed in a suspension of spheres showed the output to be influenced by the particle size and the refractive index of the suspending medium. Therefore calibration with liquids may cause errors. Nonlinear calibration curves were also obtained by Herbert et al. (1994) and Issangya (1998), who placed an optical fiber flush with the inner wall of a tube and recorded the signal as FCC particles fell uniformly.

An iterative procedure for obtaining the true (nonlinear) calibration curve based on a particle dropping test was proposed by Zhang et al. (1998). Nonlinearity and errors can be caused by particles traveling through a "blind region" at the tip of the probe. This can be avoided by adding a thin transparent window (Cui et al., 2001; Liu, 2001).

Capacitance Probes are based on the principle that solids inside a sensing volume alter the effective dielectric constant (permittivity) of the gas/solids mixture. Since fluidized bed particles and gas are typically nonconducting, capacitance is usually easier to measure than electrical resistance or conductivity. Various probe configurations have been used, ranging from parallel plates to needle shapes. The instrumentation is simple, inexpensive, and easy to construct. However, calibration is required for each bed material. Problems include signal drift, e.g. that due to humidity or temperature variations, powder buildup on the probe, and difficulty in delineating the exact measuring volume. Brereton and Grace (1993) found the relationship between the voidage and the signal intensity of a needle probe to be nearly linear. Hage and Werther (1997) applied a water-cooled guarded capacitance probe inside a high-temperature CFB combustor.

X-Ray and Gamma Ray Densitometers have a radioactive source that emits a beam of x- or gamma- rays across the column through its wall to a detector. Photons produce charge pulses (ionization type detector), or light pulses (scintillation type detector) which are then transformed into DC signals. Film exposures can also be taken to give snapshots of the solids distribution (e.g., Weinstein et al., 1986). With calibration, the concentration of particles along a chord can then be determined from the beam attenuation. Weimer et al. (1985) noted that the technique is too slow for time-dependent measurements in fluidized suspensions. The technique gives length-averaged time-averaged densities, rather than localized instantaneous measurements.

Tomographic images and 3-D density maps can be obtained if the radiation source and detector are moved or if several sources and detectors are employed simultaneously to scan the riser along a number of chords (e.g., Galtier et al., 1989; Martin et al., 1992; Simons et al., 1993). Other studies have been intrusive (e.g., Schuurmans, 1980; Wirth et al., 1991), with the radiation source traversing the riser and an external detector. X- and gamma-ray measurements require a large space around the riser and precautions to avoid radiation exposure. In addition, the equipment is expensive. An alternative technique is capacitance tomography (e.g., Malcus and Pugsley, 2001). A system consists of capacitance sensors, a data collection system, and an image reconstruction computer. The capacitance values measured between all pairs of electrodes are fed to a computer, and a cross-sectional image of the component distribution is reconstructed using a suitable algorithm.

4.1.2 Axial Voidage/Solids Holdup Profiles

Capacitance probes and optical fiber probes can also be traversed through risers to obtain local voidages, then Eqs. (8) and (10) are applied to calculate time and spatial averages. However, in practice the great majority of measurements of spatial average voidages have been derived from differential pressure measurements. If friction at the wall and acceleration effects are neglected, the vertical gradient of pressure in the riser acts like a hydrostatic pressure gradient, so that, for gas–solid systems with $\rho_p \gg \rho_g$, the average voidage and solids holdup over the section in question can then be estimated as

$$\frac{dP}{dz} = -g\left[\rho_p(1 - \varepsilon_{av}(z)) + \rho_g\varepsilon_{av}(z)\right] \qquad (12)$$

$$\varepsilon_{av}(z) = 1 + \frac{1}{g\rho_\rho}\frac{dP}{dz} \quad \text{or} \quad \varepsilon_{sav} = -\frac{1}{g\rho_\rho}\frac{dP}{dz}$$

$$(13)$$

By comparing values from this approach with those obtained by other methods, e.g., quick-closing valves (Arena et al., 1986), γ-ray absorption (Hartge et al., 1986a; Azzi et al., 1991), or x-ray computed tomography (Grassler and Wirth, 1999), this approach has been shown to give reasonably accurate measurements. However, corrections may be needed in some cases:

- The neglect of wall friction is reasonable for large columns, but it can lead to significant deviations for D (or equivalent diameter [$= 4 \times$ Area/Perimeter] for noncircular cross sections) $< \sim 0.1$ m.
- The neglect of particle acceleration effects is generally reasonable when the net solids circulation flux, G_s, is less than about 200 kg/m^{2}s. However, significant errors are introduced at the bottom, and, to a lesser extent, at the top of the column, when the acceleration term is ignored at higher G_s. Approaches proposed by Weinstein and Li (1989) and Louge and Chang (1990) can be used to estimate the corrections when accelerational effects are appreciable.

In order to apply Eq. (13), pressure taps should be provided at regular intervals covering the entire height of the riser, beginning just above the gas distributor at the bottom of the riser, with the distance between successive pressure taps small enough that the pressure gradient term, dP/dz, can be estimated with little error. The resulting axial profile of voidage (or, alternatively, solids holdup or pressure profile, from which ε_{av} and ε_{sav} can be readily derived) then provides a "signature" for the riser flow, indicating how the suspension density and solids holdup vary with height. Some representative studies reporting measurements of this nature are listed in Table 2. Most reported measurements have been from small units with air at atmospheric temperature and pressure as the fluidizing medium, but they have covered broad ranges of particle properties, U_g, and G_s.

Typical profiles showing how the solids holdup varies with the net solids circulation rate for constant superficial gas velocity in a column with a smooth top exit appear in Fig. 7. Similar profiles are seen for very different particles (e.g., Jiang et al., 1994). The

Table 2 Representative Data Showing Axial Profiles of Cross-Sectional Mean Voidage or Particle Concentration

Author	Particles	$\bar{d}_p$, μm	U, m/s	G_s, kg/m^{2}s	D, m	H, m	Exit geometry	Other effects studied
Li & Kwauk, 1980	iron, alumina, FCC, pyrite	54–105	0.8–5.6	16–135	0.09	8	smooth	
Rhodes & Geldart, 1986	alumina	64	4	45–115	0.152	6	abrupt	
Bader et al., 1988	catalyst	76	4.3, 9.1	147	0.305	12.2	smooth	
Mori et al., 1992	FCC	54	1.0–1.5	2.4–29	0.05–0.1	1.5–5.5	various	column geometry
Bai et al., 1994	FCC, sand, alumina	70, 321, 633	1.5–3.0	7–128	0.097, 0.15	3.0	smooth	binary mixtures
Brereton & Grace, 1994	sand	148	3.7–9.2	9–89	0.152	9.3	various	exit geometry, secondary air
Karri & Knowlton, 1997	catalyst	76	3.0–8.1	98–586	0.3	13	smooth	pressure
Mastellone & Arena, 1999a	glass, sand, FCC	67–310	3–6	35–144	0.12	5.75	smooth	particle properties
Issangya et al., 1999	FCC	70	4–8	18–425	0.076	6.1	smooth	

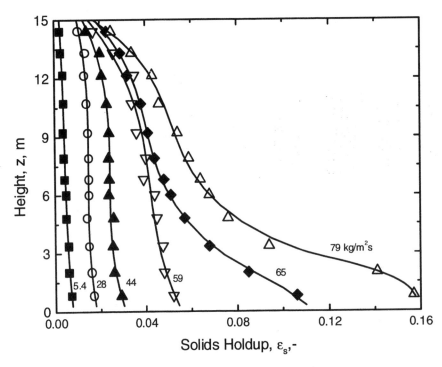

Figure 7 Axial profiles of solids holdup for FCC particles in riser of diameter 0.1 m and height 16 m with a smooth unconstricted exit. $d_p = 67\,\mu$m, $\rho_p = 1500\,$kg/m^3, $U_g = 3.0\,$m/s. (Huang and Zhu, 2001.)

solids holdup is highest near the bottom of the riser and decreases monotonically with height right to the top when there is no constriction or sharp bend at the exit. The holdup is seen to increase at all heights with increasing G_s at constant U_g. Except at very low G_s or very high U_g, where particles tend to be carried right through the riser in pneumatic conveying, the curves tend to be sigmoidal in shape, approaching an asymptotic value at high heights. This shape, first shown by Li and Kwauk (1980), was once considered to be universal. Many risers, however, have constricted top exits that affect the solids concentration profiles significantly. An abrupt or constricted exit acts as an internal separator, with the particles in the CFB riser being large and heavy enough that they do not readily follow the gas as its path varies in direction while negotiating the exit. A significant fraction of the particles arriving at the top (typically 20–80% depending on the exit shape, U_g, and G_s) are therefore internally separated from the gas and sent back down the riser, primarily along the outer walls. With constricted exits, the congregation of particles at the top causes an increase in particle concentration near the exit (e.g., Jin et al., 1988; Brereton and Grace, 1994; Pugsley et al., 1997). The influence of a constricted exit is shown schematically in Fig. 8. Note that as U_g and G_s increase in this

column with a constricted exit, the solids holdup increases throughout, with a particularly large buildup occurring near the top exit. The influence on the shape of voidage profiles of a wide range of variables, includ-

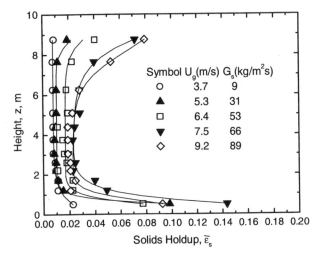

Figure 8 Axial profiles of solids holdup in riser of diameter 152 mm and height 9.3 m with a constricted exit. $d_p = 148\,\mu$m, $\rho_p = 2650\,$kg/m^3, air as fluidizing gas. (Brereton, 1987.)

ing gas velocity, net solids circulation flux, particle diameter, particle density, overall riser height, inlet and outlet geometry, and total inventory of solids, has been well summarized by Bai et al. (1992a).

Voidage profiles can also be influenced by other factors. For example, addition of secondary gas part way up the riser tends to disrupt the downflow of solids along the outer wall, causing the particle concentration to increase locally (e.g., Brereton and Grace, 1994). Similarly, ring baffles along the outer wall of the riser cause particles to be stripped from the descending wall layer, leading to local increases in particle concentration (Jiang et al., 1991; Balasubramanian and Srinivasakannan, 1998; Bu and Zhu, 1999).

Where it is necessary to estimate these profiles in advance or where pressure taps have not been provided, several methods are available for estimating $\varepsilon_{av}(z)$ or $\varepsilon_{sav}(z)$. Bai and Kato (1999) provide excellent summary tables showing previous correlations and sources of data. They then provide useful new correlations based on two cases, one for net circulation fluxes, G_s, less than, and the other for them greater than, the saturation carrying capacity flux, G_s^*, correlated by

$$\frac{G_s^* d_p}{\mu} = 0.125 \mathrm{Fr}_d^{1.85} \mathrm{Ar}^{0.63} \frac{\rho_p - \rho_g}{\rho_g} \qquad (14)$$

where $4.7 < \mathrm{Ar} = \mathrm{Archimedes\ number} = \rho_g(\rho_p - \rho_g) g d_p^3/\mu_g^2 < 1020$, $41 < \mathrm{Fr}_d = \mathrm{Froude\ number} = U_g/(g d_p)^{0.5} < 226$, and $607 < \mathrm{density\ ratio} = (\rho_p - \rho_g)/\rho_g < 3610$. The solids holdups ε_{sd} in the bottom dense zone and ε_s^* at the top exit were correlated by

(i) For $G_s < G_s^*$, profiles are relatively flat with

$$\frac{\varepsilon_{sd}}{\varepsilon_s'} = 1 + 6.14 \times 10^{-3} \left(\frac{\rho_p U_g}{G_s}\right)^{-0.23} \left(\frac{\rho_p - \rho_g}{\rho_g}\right)^{1.21}$$

$$\left[\frac{U_g}{\sqrt{gD}}\right]^{-0.383} \qquad (15)$$

$$\varepsilon_s^* = 4.04(\varepsilon_s')^{1.214} \qquad (16)$$

(ii) For $G_s \geq G_s^*$, there are S-shaped profiles with the limiting holdups given by

$$\frac{\varepsilon_{sd}}{\varepsilon_s'} = 1 + 0.103 \left(\frac{\rho_p U_g}{G_s}\right)^{1.13} \left(\frac{\rho_p - \rho_g}{\rho_g}\right)^{-0.013} \qquad (17)$$

$$\frac{\varepsilon_s^*}{\varepsilon_s'} = 1 + 0.208 \sqrt{\frac{\rho_p U_g}{G_s}} \left(\frac{\rho_p - \rho_g}{\rho_g}\right)^{-0.082} \qquad (18)$$

where $\varepsilon_s' = G_s/[\rho_p(U_g - u_t)]$ is the solids holdup expected for the ideal case where all particles travel with a velocity equal to U_g minus the terminal setting velocity, u_t. Huang and Zhu (2001) found that the distance required to achieve fully developed voidage or solids holdup profiles in a riser of small D increases with increasing G_s and decreases with increasing U_g. Models that predict axial voidage profiles in addition to other hydrodynamic properties are covered in Sec. 4.6 below.

4.1.3 Radial/Lateral Voidage/Solids Holdup Profiles

Measurements based on a variety of probes (e.g., capacitance and optical), as well as x-rays, γ-rays, and tomographic techniques, have shown that particle concentrations near the wall tend to be much greater than in the interior of CFB risers. Table 3 summarizes some major sources of experimental data. An alternative tabulation, also listing the experimental techniques employed, was provided by Xu et al. (1999). Typical results appear in Fig. 9. Note that in this figure and some other measured profiles, the shape is like the letter W (or an M, if the voidage, rather than the solids holdup, is plotted). There is slight asymmetry in this particular profile, due to the side top exit (as discussed below). Except in the bottom zone (Rhodes et al., 1998) or at very high fluxes where a different flow

Table 3 Representative Data Showing Radial Profiles of Local Voidage in CFB Units

Author	Particles	$\bar{d}_p$, μm	U_g, m/s	G_s, kg/m^{2}s	D, m	H, m	z, m
Hartge et al., 1986b	sand	56	3.9	100	0.050	3.3	0.6–2.2
Weinstein et al., 1986	sand	56	5.0	75	0.40	7.8	0.65–6.0
Tung et al., 1988	FCC	54	1.4–3.7	5–91	0.09	10	2.25
Bader et al., 1988	catalyst	76	3.7	98	0.305	12.2	4.0, 9.1
Harris & Davidson, 1994	FCC	60	4.4	26–52	0.140	5.1	4
Jiang et al., 1994	polymer	325	2.4	7–21	0.102	6.3	~0.6
Tanner et al., 1994	glass	110	2.5–6.5	30–107	0.411	8.5	~3–6
Wei et al., 1998	FCC	54	2.3–4.6	40–180	0.186	8	2.3–6.3
Grassler & Wirth, 1999	glass	60	2–8	30–600	0.19	15	4.4–11.6

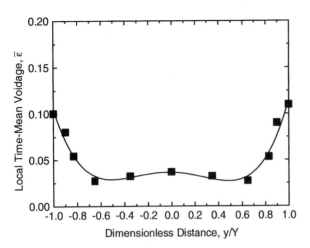

Figure 9 Lateral profile of local time mean solids holdup along a midplane of 146 mm × 146 mm, 9.14 m tall riser at $z = 7.06$ m for $U_g = 5.5$ m/s, $G_s = 40$ kg/m²s. (Zhou et al., 1995a.)

regime can be achieved (see Secs. 3 and 5), particles travel predominantly downward near the wall. The observed pattern of particles being carried upwards in the core of the riser and descending at the outer wall has given rise to core–annulus models. Nevertheless, the variation of local time-mean voidage with radial position is continuous, not subject to a step change, as might be implied by the "core–annulus" designation.

Many groups (e.g., Zhang et al., 1991; Rhodes et al., 1992; Beaud et al., 1996) have found that time mean radial voidage profiles are broadly self-similar. Hence the radial variation can be expressed, at least to a first approximation, as functions of the cross-sectional average voidage, which in turn can be estimated using the experimental methods or correlations discussed above or the predictive methods covered in Sec. 4.6 below. Of the various correlations, one that gives good predictions over a very broad range of suspension densities and net solids fluxes (up to 425 kg/m²s), i.e., for dense suspension upflow as well as fast fluidization, is that of Issangya et al. (2001):

$$\bar{\varepsilon}(r, z) = \varepsilon_{mf} + [\varepsilon_{av}(z) - \varepsilon_{mf}]\varepsilon_{av}^{-1.5+2.1\phi^{3.1}+5.0\phi^{8.8}} \quad (19)$$

where $\phi = r/R$ is the dimensionless radius and R is the column radius. Several alternative forms of correlation (e.g., Zhang et al., 1991; Godfroy et al., 1999b) have also been suggested. Xu et al. (1999) gave separate correlations for the time-mean voidage at the wall,

$$\varepsilon_R(z) = \varepsilon_{mf} + 4(1 - \varepsilon_{mf})[\varepsilon_{av}(z) - 0.75] \quad (20)$$

and on the axis of the riser,

$$\varepsilon_0(z) = 1 - 0.42[1 - \varepsilon_{av}(z)] \quad (21)$$

Both of these equations are only intended for $\varepsilon_{av}(z) \geq 0.75$. Results of Schuurmans (1980) confirm that patterns of radial variation are similar (maximum at center line, minimum at wall) in industrial equipment as in the smaller units in which most experimental measurements have been obtained.

Voidage fluctuations have been recorded in several studies. While early models often assumed that the clusters at the wall have voidages of ε_{mf}, experimental findings show a wide variety of wall voidages. Brereton and Grace (1993) introduced an "intermittency index," defined as the ratio of the standard deviation of voidage at a given point divided by that which would be obtained if a flow having the same time mean voidage was achieved by alternating voids and clusters of voidage ε_{mf}. This index typically has values between ~ 0.1 and 0.6 (Brereton and Grace, 1993; Johnsson et al., 1997). It tends to decrease with increasing height, indicating development of a core–annulus structure with increasing height. Radially it reaches a maximum at a small distance from the outer wall.

4.1.4 Angular Variations and Asymmetries

Flows in CFB risers of circular cross section are usually assumed to be axisymmetric. However, particle feeding and recirculation near the bottom of the riser, as well as particle efflux from near the top, are generally assymmetric, leading to angular variations, especially near the top and bottom of the riser. Some data showing asymmetric distributions and/or angular variations have been reported (e.g., Rhodes and Laussmann, 1992b; Martin et al., 1992; van der Meer et al., 1997; Srivastawa et al., 1998). The presence of the exit on one side leads to significant asymmetry of particle voidage (Zhou et al., 1994), particle velocity (Zhou et al., 1995a) and solids flux (van der Meer et al., 1997; Brobecker et al., 2001) near the top of the risers owing to the lateral motion of gas and particles as they approach the exit. Similarly reentry of circulated solids from one side can lead to marked asymmetry at the bottom (Sun et al., 1999).

4.2 Particle Velocities

4.2.1 Measurement Techniques

It is more difficult to measure local particle velocities than their local concentrations. *Photographic and video techniques* can provide information on particle velocity, direction of motion, and acceleration. However, small, fast-moving particles are difficult to photograph. Backlighting can be used for contrast in two-dimensional risers (Arena et al., 1989) as well as circular or square columns (Rhodes et al., 1991a). Photography is difficult in regions where particle concentrations are high, e.g., the bottom region of CFB risers. Zheng et al. (1992) employed microcomputer-controlled multicolor stroboscopic photography to determine particle velocity, particle acceleration, and directions of motion adjacent to the wall. Boroscope optical fiber probes have been applied (Takeuchi and Hirama, 1991; Li et al., 1991) to study the flow structure in the central region of the riser. However, the viewing area tends to be small. Wider three-dimensional images have been obtained using a laser sheet technique (Kuroki and Horio, 1994), but this is only suitable for dilute suspensions.

When two *optical probes* are separated by a small distance ΔL, the cross-correlation function of the two signals f_1 and f_2 registered by probes 1 and 2 over time t_{tot} is given by

$$\phi_{12}(\tau) = \frac{1}{t_{\text{tot}}} \int_0^{\text{tot}} f_1(t) f_2(t + \tau) \, dt \qquad (22)$$

The mean particle velocity is estimated as $u_p = \Delta L / \tau_{\max}$, where $\tau_{\max}$ is the delay time at which the $\phi_{12}(\tau)$ reaches a maximum. To determine the velocity, one also needs the effective separation distance between the two probes. This generally differs from the actual separation distance and needs to be obtained via calibration. Different probes have been used to measure local time mean particle velocities in CFB risers. Hartge et al. (1988) used a pair of 1 mm diameter fibers separated by 4.4 mm. Horio et al. (1988, 1992) employed a probe with two detection fibers and a single emission fiber in between. To capture individual particle motion, Ohki and Shirai (1976) recommended that the cross-correlation is improved if the diameter of the optical fiber is the same as the particle diameter, with the separation distance between the receiving fibers of similar magnitude.

The delay time or time interval, τ, can also be obtained using peak detectors (Zhou et al., 1995a). To avoid false signals, a five-fiber probe was used by Zhou et al. (1995a). The velocity was accepted only if the difference between two independent values was within a certain tolerance. This probe eliminates errors when light received by the two photomultipliers is from different particles, but only measures velocities of particles traveling verti-

cally, rather than the vertical component of all particle velocities.

In *laser Doppler anemometry*, a laser beam is split into two beams of equal intensity, which are then focused at the measuring point, forming a fringe pattern in the intersection region. When a particle passes normal to the fringes, the intensity of its reflected light varies with a frequency proportional to the particle velocity and inversely proportional to the fringe spacing. Both forward (Arastoopour and Yang, 1992) and backward (Yang et al., 1992) scattering can be used, but forward scattering is only suitable under very dilute flow conditions (e.g., $\varepsilon_s < 2\%$), whereas backscattering provides data at the wall, or one can use an intrusive probe (Wang et al., 1993; Yang et al., 1992).

Particle velocities can also be detected by *tracking the trajectory of a single tagged particle* over extended time (Godfroy et al., 1999a). However, it is currently very difficult to track particles smaller than about 0.5 mm in diameter, so that the technique has not yet contributed significantly to CFB research.

4.2.2 Experimental Results

Some major studies that have reported data related to particle velocities in CFB risers are summarized in Table 4. Time mean particle velocities in fast fluidized beds are upwards in the core of the column, with a magnitude similar to the superficial gas velocity at the column axis. The average velocity then falls as the sensor moves outwards toward the wall, becoming negative in a layer adjacent to the wall. The downward particle velocity right at the wall is generally of the order of 1 m/s. Available data for the velocity at the wall have been correlated (Griffith and Louge, 1998) by

$$v_{\text{pw}} = 36\sqrt{g d_p} \qquad (23)$$

Such core–annulus patterns have been observed frequently in small-scale risers, but also in much larger equipment, even in a 14.7×11.5 m cross section combustor operated by Electricité de France (Caloz et al., 1999).

Typical lateral profiles of time mean local particle velocity, both upward and downward, along a midplane of a column of square cross section are shown in Fig. 10. The superficial gas velocity is seen to have little influence. Ascending particles are dominant in the center of the column, whereas there are more descending than ascending particles near the wall (i.e., as y/Y approaches -1). The magnitudes of the velocities of rising particles at the axis of the column are similar

Table 4 Representative Data Showing Radial Profiles of Particle Vertical Flux or Velocity

Author	Particles	$\bar{d}_p$ μm	U_g, m/s	G_s, kg/m^{2}s	D, m	H, m	z, m
Monceaux et al., 1986	FCC	59	3.2	31–68	0.144	8	$\sim$ 4.8
Bader et al., 1988	catalyst	76	4.6	147	0.305	12.2	9.1
Hartge et al., 1988	FCC, ash	85, 120	2.9, 3.7	49, 30	0.4	8.4	0.9–4.7
Rhodes et al., 1989	alumina	70	3–4	30–40	0.152	6	1.7–4.9
Azzi et al., 1991	FCC	75	6.2	150	0.19–0.95	11.7	3.1–16
Herb et al., 1992	FCC, sand	68, 276	2.4–6	20–44	0.05, 0.15	2.7, 10.8	1.5, 2.1, 5.5
Yang et al., 1992	FCC	54	1.5–6.5	6–25	0.140	11	3.3, 6.6
deDiego et al., 1995	sand, coal	400–650	4–5.5	28–98	0.10	4	2.1–3.2
Wang et al., 1996*	sand	120	2.5*	up to 26	0.161	6.2	1.3, 4.4
Wei et al., 1997	FCC	54	1.9–10.1	37–236	0.186	8.5	1.8, 3.9
Nicolai & Reh, 1997	glass	40–300	1.5–9.6	NA	0.41	8.5	4.0–4.2
Corenella & Deng, 1998	sand	200	3.4–5.4	10–30	0.114	2.7	0.7, 2.5
Issangya et al., 1998	FCC	70	4.5–7.5	38–325	0.076	6.1	2.8
Mastellone & Arena, 1999a, 1999b	FCC, glass, sand	70–310	6	35, 55	0.120	5.75	1.2–4.2
Caloz et al., 1999#	glass	62 for 0.41 083 m units	1.7–3.3, $\sim$ 5.5	17–98	0.41, 0.83, 14.7 $\times$ 11.5	8.5, 11.2, 35	2.6–21.5

* Includes measurements up to 550°C. # Includes measurements under full combustion conditions.

to the superficial gas velocity, while the magnitudes of downward velocities are significantly lower.

There are substantial velocity fluctuations, especially near the wall, with the time mean fluctuating component typically reaching a maximum near the edge of the wall layer (Yang et al., 1992; Caloz et al., 1999). Flows near the wall undergo reversals, i.e., the particle velocity there is sometimes upward and sometimes downward. The fraction of individual particles

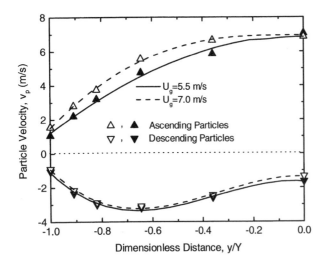

Figure 10 Lateral particle velocity profiles based on optical fiber probe for midplane of 146 mm square column at $z = 6.2$ m at $G_s = 40$ kg/m^{2}s. (Zhou et al. 1995a.)

that are ascending is close to 100% toward the axis of the column, while near the wall, typically 10–40% of particles travel upward, with the rest descending (Zhou et al., 1995a). For rectangular columns, the magnitude of the downwards mean particle velocity and the fraction of particles that are descending when averaged over time are greater in the corners than elsewhere along the outer wall.

Zhou et al. (1995b) inferred horizontal components of velocity for particles crossing the axis of a riser of square cross section. The measured lateral components increased to a maximum of $\sim$ 40% of the superficial gas velocity, U_g, in the middle of the riser and then decreased toward the top.

4.3 Particle Flux

4.3.1 Net Solids Circulation Flux

Most methods of determining the net solids circulation flux, G_s, in CFB systems assume that the cyclones are 100% efficient so that the solids flow can be inferred from the downward flow in the standpipe/downcomer. If plug flow is assumed, the solids flux in the downcomer can be estimated by timing the descent of identifiable particles along the downcomer wall (Burkell et al., 1988). If the standpipe wall is not perfectly smooth, this method underestimates the solids flux because particles at the column surface are retarded by friction and travel more slowly than those in the interior. Another popular method is to close a perforated butterfly valve in the downcomer, with small holes to allow gas to pass. The solids circulation rate is then calculated from the time to accumulate a known volume of solids (Herb et al., 1992), from the time to reach a given pressure drop across the bed that collects above the valve (Brereton, 1987; Li et al., 1991), or from the rate of descent of the moving packed bed level in the downcomer below the valve (Weinstein et al., 1986). Alternatively, the solids flow can be diverted to a measuring cylinder for a given period of time before being switched back to the downcomer (Bai et al., 1987). A drawback of these methods is that they disturb the pressure balance in the system when particles are temporarily diverted.

Patience and Chaouki (1991) correlated the solids flux to the pressure drop over the horizontal section of the exit pipe connecting the riser to the separator at a given gas velocity. Ultrasound can also be employed in the downcomer to measure the solids circulation rate (Tallon and Davies, 1998).

None of the above methods, with the possible exception of the exit pressure drop method (where calibration may be impossible), is likely to be feasible for large-scale commercial CFB systems. Hence industrial CFB systems usually operate without G_s being known.

4.3.2 Local Vertical Flux Measurement Techniques

Time mean local solids fluxes in CFB risers have often been measured using solids sampling probes. Ideally, the suction velocity through the sampling tube matches the local gas velocity at the sampling location to provide isokinetic sampling. However, higher suction velocities may be required to prevent blockage. Therefore nonisokinetic sampling probes have often been employed. Monceaux et al. (1986) found that the solids collected first increased with suction velocity and then stayed constant over a wide range and finally increased again. Bader et al. (1988) and Werther et al. (1993) observed no significant influence of suction gas velocity. Azzi et al. (1991) found that the particle flux remained nearly constant if the suction velocity was within 1.5 m/s of U_g. Suction velocities equal to the superficial gas velocity were reported (Aguillon et al., 1996) to give satisfactory results. Herb et al. (1992) observed that measured solids fluxes in the core region were not strongly dependent on the suction velocity near the isokinetic point. Rhodes (1990) and Miller and Gidaspow (1992) found that although the upward and downward mass fluxes were sensitive to the sampling gas velocity, the net solids mass flux

(obtained by difference) was virtually independent of the suction velocity, provided that it was sufficient to prevent blockage. However, suction velocity has a significant effect when sampling in the wall region. To obtain the true flux near the wall, the suction velocity should be just enough to keep the probe free of blockage (Bierl et al., 1980; Harris et al., 1994). The true flux can be obtained by extrapolating the solids flux to zero suction velocity (Bierl et al., 1980). de Diego et al. (1995), however, found that both the downward and upward solids fluxes in the dense outer region of the riser were greatly affected by the suction velocity. A sensitivity study by Herb et al. (1992) showed that the downward component of the flux was sensitive to the suction velocity near the wall. Liu (2001) found that the sampling method gives reasonable results when the local flow is predominantly in one direction, but significant errors for locally reversing flows. To overcome the sampling problem at the wall, Bolton and Davidson (1988) used 5 mm ID "scoops" to collect particles falling near the wall. Water-cooled sampling probes can be used in high temperature CFB risers to obtain local time mean solids fluxes (Couturier et al., 1991; Johnsson et al., 1997).

Liu (2001) constructed and calibrated an optical fiber probe that can simultaneously measure local particle holdup and local particle velocity. From these products, multiplied by the particle density, one obtains the local instantaneous particle flux. Integration of the local time average fluxes across the cross section led to overall net fluxes that were in good agreement with G_s, the net flux determined independently by trapping recirculating solids in the return loop. The optical fiber technique confirmed that the fluctuations of particle velocity and voidage are strongly correlated, as noted in the next section, so that it is incorrect to assume that one can obtain the time mean flux from the product of particle density, local time mean velocity, and local time mean solids holdup.

4.3.3 Radial Profiles of Solids Flux

Studies that report local particle flux data are summarized in Table 4. For the fast fluidization flow regime, radial profiles of local particle flux are generally similar in shape to the radial profiles of particle velocity described above—upward near the center line, dropping off with increasing r, and negative near the wall. One such experimental profile, shown in Figure 11, is seen to be symmetrical around the axis of the column. Note that the flux is downward in this case only in a thin region adjacent to the outer wall of the column.

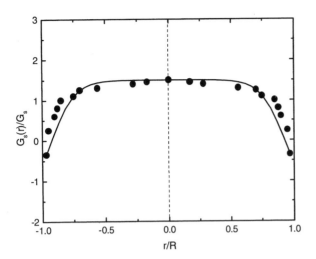

Figure 11 Radial profile of solids flux determined by Bodelin et al. (1994) for $U_g = 5.4$ m/s, $G_s = 43.4$ kg/m^{2}s, $D = 144$ mm, $H = 10$ m, $d_p = 180\,\mu$m, $\rho_p = 2650$ kg/m^3 compared with predictions of Pugsley and Berruti (1996).

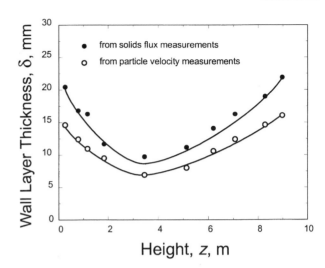

Figure 12 Wall layer thickness based on both where the time mean solids flux is 0 and where the time mean particle velocity is 0 versus height along the mid-plane of 146 mm square cross-section column. $U_g = 5.5$ m/s, $G_s = 40$ kg/m^{2}s, $d_p = 213\,\mu$m, $\rho_p = 2640$ kg/m^3.

Radial profiles of upward particle momentum tend to be similar in shape to the solids flux profiles (e.g., Azzi et al., 1991).

4.3.4 Wall or Annular Layer Thickness

The concept of a core–annulus structure has led many groups to report the thickness of the outer annular wall layer. This thickness is then used in core–annulus reactor models (Sec. 9 below) and in heat transfer models (Sec. 7). Wall layer thicknesses have been based on radial profiles of either particle velocity or solids flux, with the radial position at which the time mean value is 0 taken to define the boundary. However, as shown by Bi et al. (1996) and Fig. 12, the location where the time mean velocity is 0 differs from that where the time mean flux is 0. This arises because fluctuations of local instantaneous voidage are strongly correlated with fluctuations of local instantaneous particle velocity. The most meaningful wall layer thickness is based on the point at which the time mean particle flux is 0. This thickness is well correlated (Bi et al., 1996) by

$$\frac{\delta(z)}{D} = 0.5\left[1 - \sqrt{1.34 - 1.30[1 - \varepsilon_{av}(z)]^{0.2} + [1 - \varepsilon_{av}(z)]^{1.4}}\right]$$

$$0.80 \leq \varepsilon_{av} \leq 0.9985$$

$$(24)$$

If the column cross section is square or rectangular rather than circular, the wall layer tends to be distorted as shown in Fig. 13 (Zhou et al., 1995a), with a greater thickness in the corners.

The annular wall layer thickness may shrink or grow with height. With a constricted exit, the thickness passes through a minimum part way up the column (as in Fig. 12), indicating a net transfer of solids inward from the wall region to the core in the upper part of the riser, while there is a net outward transfer in the lower part of the riser as solids descend along the outer wall.

Several groups (e.g., Rhodes et al., 1992; Horio and Kuroki, 1994; Wei et al., 1995a; Lim et al., 1997) have studied the sizes, shapes, motion, and breakup of individual clusters and streamers. Streamers tend to be U-

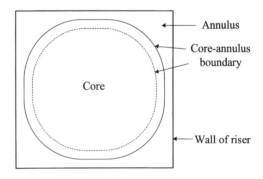

Figure 13 Distorted wall layer boundaries at two heights for riser of square 146 mm × 146 mm × 9.14 m tall column based on particle velocity profiles obtained by Zhou et al. (1995a) for $U_g = 5.5$ m/s and $G_s = 40$ kg/m^{2}s. Solid line: $z = 5.1$ m; broken line: $z = 6.2$ m.

shaped, as shown in Fig. 14, and to become thicker as they descend until they become unstable owing to the stripping of particles by the rapid inner core flow. Idealized streamer shapes assumed by Lim et al. (1997) in modeling streamers are also indicated in the figure.

4.3.5 Lateral Flux

Horizontal particle motion is caused by interparticle collisions and gas turbulence. Unlike diffusion processes, the net transfer is from a more dilute region (core) to a denser one (annulus) (Brereton and Grace, 1993; Davidson, 2000). Particles transferred outward toward the wall tend to be captured by the descending wall layer, while inward transfer occurs as particles are stripped from unstable streamers descending along the outer wall.

The lateral (horizontal) flux of particles has been determined by sampling in a similar manner to the vertical flux sampling method described above, except that the opening of the sampling tube is now vertical, directed toward the near wall to measure the inward flux, or away to determine the outward flux. The net cross-flow is the difference between the inward and the outward flux, and it should be zero if the flow is fully developed. Net horizontal fluxes have been found to be one to two orders of magnitude less than G_s (Zhou et al., 1995b; Jiang and Fan, 1999). The inward and outward fluxes are relatively small in the center of the riser, increase to a maximum near the wall, and

decrease again at the outside (Jiang and Fan, 1999). The net horizontal flux has been found (Zhou et al., 1995b) to be inward near the top of the riser and outward in the lower part, consistent with the appearance of a minimum wall layer thickness in the middle of the riser, as in Sec. 4.3.4 above. There is some evidence (Werther, 1994) that lateral fluxes decrease as the reactor scale increases.

4.4 Gas Velocities

Local gas velocities are very difficult to measure because of the presence of the particles and high levels of turbulence. Experimental techniques include isokinetic sampling (van Bruegel et al., 1969; Harris and Davidson, 1992), injecting a tracer at one level and then cross-correlating the signals from two nearby downstream levels (Horio et al., 1992; Martin et al., 1992), Pitot tubes (Yang et al., 1994), and laser Doppler anemometry (Yang et al., 1992).

Experimental local time mean gas velocities tend to increase monotonically as one traverses inward from the wall of the column, reaching a maximum that is typically 1.5 to 2 times the superficial gas velocity at the column axis, with the profiles being more nonuniform with increasing G_s. Like particle fluctuating velocities, gas fluctuating rms velocities are of order 1 m/s and appear to reach a maximum at a small distance inside the wall, falling to relatively low values near the axis (Yang et al., 1992).

4.5 Geometric and Operating Variable Effects

The influence of riser geometry on hydrodynamics was reviewed by Grace (1997). Only a brief summary is therefore given here.

4.5.1 Influence of Cross Section Shape: Circular vs. Rectangular

While most CFB hydrodynamic studies have been carried out in risers of circular cross section, some have been conducted in risers of rectangular (e.g. Saberi et al., 1998) and square (e.g., Leckner et al., 1991; Zhou et al., 1994) cross section, geometries that are especially relevant to CFB combustors. Some work (e.g., Yerushalmi et al., 1978; Arena et al., 1992) has also been performed in thin ("two-dimensional") risers. Phenomena in rectangular columns are qualitatively similar to those in risers of circular cross section, with the usual dilute upward moving core and dense downward flow in an outer annular region.

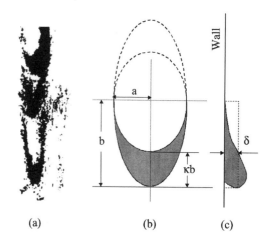

Figure 14 Streamer shapes (a) as viewed at the wall of a column of 146 mm square cross-section; (b) as idealized in simple cluster model; (c) as idealized in plane normal to wall. (Lim et al., 1997.)

Relationships developed for circular risers may then be applied as a first approximation, with the diameter, D, replaced by an equivalent (hydraulic) diameter $= 4 \times$ cross sectional area $\div$ perimeter. However, the corners of rectangular units are very prone to downflow (Andersson and Leckner, 1992; Zhou et al., 1995a; van der Meer et al., 1997), with higher downward velocities and particle concentrations there than elsewhere along the walls. Moreover, the wall layer becomes distorted as shown in Fig. 13. Fabre et al. (1997) found quite different mass flux profiles along two transverse horizontal sections when the length was much greater than the width.

4.5.2 Influence of Wall Shape and Roughness

Refractory walls in CFB reactors tend to be rough. Studies with a roughened wall (Zhou et al., 1996a) demonstrate that wall roughness plays a significant role close to the wall, the primary effect being to increase the voidage there. Studies with membrane walls (see Sec. 7 below) show (Leckner et al. 1991; Zhou et al., 1996b) that particle streamers travel faster and further in the protected troughs along the flat fins between adjacent tubes than along corresponding flat walls. Local voidages also tend to be lower along the fin than on the crests of the tubes.

4.5.3 Influence of Bottom Configuration

Unlike low-velocity fluidized beds, where the distributor is often regarded as the most critical element in achieving a successful design, there is much greater latitude when specifying the distributor and bottom geometry for CFB risers. In fact, it is possible to operate without any distributor plate whatsoever, so long as the solids can be "blown over" at the end of each run to be stored in the return system (i.e., the standpipe and storage vessel, if any). Nevertheless, the inlet configuration plays a significant role, and it would be wrong to consider the design of this region unimportant.

CFB systems have various configurations for recirculating and reinjecting solids. The feed rate may be regulated by a mechanical (e.g., slide valve) or nonmechanical valve (see Chapter 11). L-valves, H-valves, L-valves, V-valves and loop seals are common in CFB combustors, while J-valves and slide valves are popular in fluid catalytic cracking risers. Nonmechanical valves tend to give good flow control for Geldart group B and D particles, while the flow of aeratable group A particles usually needs to be regulated by a slide valve or other mechanical valve. Some common bottom geometries are illustrated in Fig. 15. Cheng et al. (1998) classify inlets according to the degree of restriction of the solids circulation and the configuration of the open area. A strong restriction at the entry helps stabilize the flow, but adds to the system overall pressure drop and hence to the operating cost. Weak restrictions cause the riser to be directly coupled with the entire return system and with the windbox (if any), whereas a strong restriction (high distributor pressure drop) allows the riser to operate somewhat independently. Hartge and Werther (1998) provide useful information on different nozzle types and their influence on erosion. To reduce erosion they recommend nozzles that have a section where there is downward gas flow with a minimum exit velocity.

The geometry and orientation of the entering gas can directly affect axial voidage profiles over a considerable height beginning in the lower part of the riser (Weinstein et al., 1996; Cheng et al., 1998; Kostazos et al., 1998). The bottom zone may behave as a vigorously bubbling bed or as a turbulent fluidized bed. It may be clearly demarcated from the region above, merge smoothly into the upper part of the riser, or show a transitional "splash zone" between a lower dense bed and the riser above. Addition of secondary gas hastens the transition from a denser bottom zone to a more dilute upper region. For low-pressure drop distributors, as in the Chalmers University CFB boiler, surging may occur, with pressure oscillations in the windbox directly affecting the lower zone of the riser, being a source of turbulence (Sternéus et al., 1999) and causing "exploding bubbles" (e.g., Johnsson et al., 1992; Schouten et al., 1999). This effect can be cor-

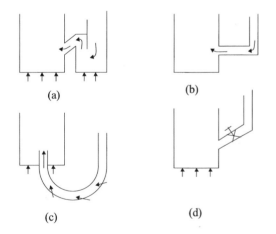

Figure 15 Common CFB riser bottom configurations: (a) with loop seal; (b) with L-valve; (c) with J-valve; (d) with slide valve.

rected by increasing the distributor pressure drop, or by installing horizontal tubes in the bottom zone (Johnsson et al., 2001), thereby converting the bottom region to the turbulent fluidization flow regime.

The height of the bottom zone tends to increase with increasing net solids circulation flux, G_s, and to decrease with increasing superficial gas velocity, U_g (Schlichthaerle and Werther, 1999; Sun et al., 1999). The overall solids holdup in the bottom zone is typically 0.2 to 0.3. The tendency for local solids holdups to be greater near the wall than in the core begins from the very bottom of the riser. Characteristic frequencies tend to be lower in the entry zone than at higher elevations (Louge et al., 1990). If particles are already ascending when they re-enter the riser, e.g., owing to a J-valve, then the bottom zone tends to be less dense and shorter than if the solids enter sideways (e.g., via an L-valve) or obliquely downwards (e.g., via a loop seal) into the lower section of the riser.

The hydrodynamics at the bottom of the riser can also be influenced by whether the lower section of the column is tapered (cross-sectional area increasing with height, Grace, 1997). Tapering leads to increased superficial gas velocity at the bottom that would normally be expected to decrease the solids holdup, but this does not always occur, probably because the tapering also decreases the area available for entraining solids.

4.5.4 Addition of Secondary Gas

Secondary air is commonly injected as jets part way up CFB combustion risers to assist in reducing NO_x emissions. The secondary air jets can strongly influence axial voidage profiles, while the level at which they are injected affects the depth of the dense region at the bottom (e.g., Naruse et al., 1991; Brereton and Grace, 1994; Cho et al., 1996). The jets disrupt the downward flow of solids when they reach the level of the secondary gas injection (Brereton and Grace, 1994; Kim, 1999).

When the secondary air jets cause swirl (tangential motion), the overall voidage can be reduced, creating a more homogeneous bed (Wang and Gibbs, 1991). Jets tend to isolate the region above them from that below, reducing downward flux along the wall (Brereton and Grace, 1994). Near the injection level, the solids holdup tends to increase because of the centrifugal effect of the swirl (Cho et al., 1996). Strong swirl was found to influence riser hydrodynamics for a distance of about D upstream and 2–3 column diameters downstream of the level where the swirl gas was injected (Ran et al., 1999).

4.5.5 Exit Effects

The flow at the top of a riser can be affected profoundly by the geometry of the exit port through which the gas and solids pass to the primary cyclone or other separator. Some configurations employed in practice are illustrated schematically in Fig. 16. Important factors include

The asymmetry of the exit: An exit at one side tends to cause asymmetry, as noted above.

The degree of constriction: A severe reduction in the cross-sectional area leads to a higher exit pressure drop, affecting the overall pressure balance, also reducing the efflux of solids from the riser for a given gas flowrate.

The superficial gas velocity: Exit effects are more important as U_g increases.

The tortuosity of the flow path: The exit acts like a crude gas–solid separator. Any increase in curvature of the gas streamlines as the gas finds its way into the exit makes it more difficult for the particles to remain entrained and increases internal refluxing of solids in the riser.

For horizontal exits, solids may build up in the exit duct and "dump" periodically back into the main riser (van der Meer et al., 1997).

Given these factors, the exit design should be central and smoothly tapered if one wishes to minimize back-mixing of gas and solids and prevent reflection from the top. However, if the designer wishes to increase solids holdup in the riser to provide maximum resi-

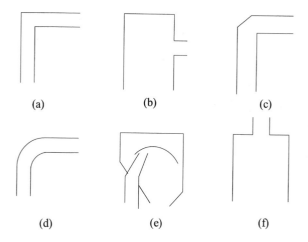

Figure 16 CFB riser top exit configurations: (a) sharp right-angle; (b) exit at side; (c) tapered right angle; (d) long-radius bend; (e) nozzle and low pressure drop impingement separator [gas exit not shown]; (f) top exit with constriction.

dence time, it is desirable to locate the exit on one side and to make its area significantly less than the cross-sectional area of the riser.

Observed effects of exit geometry follow from the above considerations. More constricted exits lead to increased pressure gradients and higher solids holdup in the top region (Jin et al., 1988; Brereton and Grace, 1994; Pugsley et al., 1997). In some cases (e.g., Brereton and Grace, 1994; Pugley et al., 1997), the influence of the exit may extend downward a considerable distance, whereas in other cases it appears to be confined to the top of the riser. Empirical models for predicting the influence of the exit have been proposed by Pugsley (1998) and Gupta and Berruti (2000).

4.5.6 Influence of Temperature and Pressure

Many circulating fluidized beds operate at elevated temperatures and pressures, yet the vast majority of experimental information has been derived from columns operated at ambient conditions. Fang et al. (1999) carried out experiments at high temperature, but in a very small riser at modest gas velocities and solids fluxes, and found a small decrease in overall pressure drop, due presumably to the decreased gas density, although carbon dioxide was used at high temperature and air at room temperature. Increasing temperature leads to a reduction in gas turbulence because of the increased kinematic viscosity of the gas. This is likely to influence particle motion and mixing for relatively small particles (e.g., most catalysts), but less so for the larger denser particles employed in CFB combustors (Senior and Grace, 1998). Measurements inside relatively large CFB combustors (e.g., Leckner et al., 1991; Lin et al., 1999) demonstrate that key hydrodynamic features at low temperature, like down-flowing wall layers and a dilute core, are also applicable at high temperatures.

A study of CFB hydrodynamics at pressures up to 690 kPa (Karri and Knowlton, 1997) showed that the solids holdup tended to decrease with increasing system pressure. This was attributed to decreased wall layer thickness and increased voidage close to the wall as the pressure increased. Wirth and Gruber (1997) found that pressurization caused the solids to be distributed more uniformly over the full height of a CFB riser operated at pressures up to 50 bar.

4.5.7 Internals

Various internals have been tested in laboratory CFB units. A venturi contraction section has been used in some FCC units to promote particle acceleration and mixing (Avidan, 1997a). Ring type and other similar baffles have been tested to improve gas–solids contacting. Some baffles increase the solids holdup in the riser (e.g., Jiang et al., 1991, van der Ham et al., 1993), while others appear to reduce solids holdup (Gan et al., 1990; Zheng et al., 1992).

4.5.8 Geometric Variants

Circulating fluidized beds can involve multiple loops. For example, in a Total resid cracking process with two stages of catalyst regeneration (Dean et al., 1982), the first riser serves as the cracking reactor. After stripping, catalyst particles flow downward to the first-stage regenerator, and then through a second riser to the second-stage regenerator before returning to the cracking riser. The first regenerator acts as the downcomer of the transport riser, while the second serves as the downcomer for the cracking riser. A similar dual-loop circulating fluidized bed has been tested (Issangya et al., 1998; Liu, 2001) to investigate hydrodynamics and mixing at high solids fluxes (up to 600 kg/m^{2}s) and high solids holdup (see Sec. 5.2). Other dual-loop CFB reactors include the EXXON fluid coking process (Matsen, 1996), N-shaped circulating fluidized beds (Qi et al., 1992; Zheng et al., 1989), the Circored iron reduction process (Plass, 2001), and Ferco's CFB gasification system (Weeks and Rohrer, 1997).

A circulating fluidized bed can also be constructed in a concentric configuration as an "internally circulating fluidized bed." The riser is usually located in the interior with the downcomer as an outer annulus (e.g., Fusey el al., 1986). Such geometries simplify construction, facilitate direct heat exchange between the riser and the downcomer, and simplify the design of high-pressure systems.

Other variants of the circulating fluidized bed include the downer reactor (Zhu et al., 1995; Jin et al., 2002), quick-contact (QC) reactor (Gartside, 1989), short-contact-time fluidized bed reactor (SCTFR) (Graham et al., 1991), and entrained flow reactor (EFR) (Raghunathan et al., 1993). As shown in Fig. 17, there is no "fluidized bed" in a downer reactor, with the gas traveling downward, cocurrent with raining particles. The distinct features of the downer reactor include short gas–solids contact times (< 1 s) and reduced gas and solids. Downer reactors are thus promising for very fast reactions, with the intermediate as the desired product, where plug flow and short contact times are essential to prevent over-reaction and ensure good selectivity.

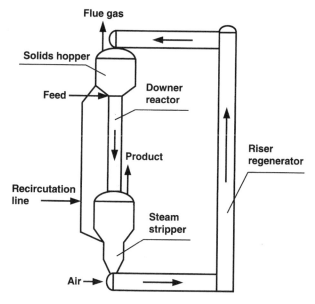

Figure 17 Typical downer geometry.

4.6 Hydrodynamic Models

Models describing the hydrodynamics of CFB systems are of three types: empirical, mechanistic, and based on computational fluid dynamics (CFD).

4.6.1 Empirical Correlations

Given the complexity of multiphase flows and the number of contributing factors, empirical methods are common. Several of these, e.g., the approach adopted by Bai and Kato (1999) for predicting axial voidage profiles, are covered above. However, because of the number of variables involved and their limited ranges of variation in experimental studies, many correlations are only useful for small columns and over very limited ranges of conditions. More comprehensive models are generally needed to describe a broader range of variables over extended ranges of conditions.

4.6.2 Mechanistic Models

Mechanistic models are generally preferred to empirical approaches. Ideally they incorporate the key physical factors affecting the flow, leaving out factors of lesser importance. Mechanistic models for CFB risers are often useful, but many suffer from weaknesses, such as

It is frequently difficult to set the boundary condition at the bottom, i.e., at $z = 0$.

Most approaches assume a sigmoidal voidage profile and/or perfect core–annular flow; as noted above, these are oversimplified.

Axial symmetry is usually assumed, making it impossible to deal adequately with asymmetries such as side feed ports, side reentry of solids, and asymmetric top exits.

The models do not readily allow for the introduction of secondary or tertiary air (or other gases).

However, mechanistic approaches provide rational approaches for predicting key hydrodynamic features. Two useful models will be outlined briefly.

1. Core–annulus model of Senior and Brereton (1992). This model assumes that above an entry region, taken as the volume below any secondary air inlets, particles fall as dense wall streamers of semielliptical cross-sectional shape and voidage of 0.6 at a velocity of 1.1 m/s; a uniform dilute suspension is conveyed upwards in the interior. Conservation of mass and volume are applied to the two phases. The relative velocity between the gas and the particles in the core is set equal to the single particle terminal settling velocity, u_t. The minimum streamer thickness is taken to be 8 mm. Exchange of particles from the streamers to the core is assumed to occur in a manner analogous to the entrainment of droplets in gas–liquid annular flow. Exchange from the core to the streamers is related to lateral particle motion and collisions, with a mass transfer coefficient of 0.2 m/s adopted as the best overall estimate. The model adopts an empirically fitted "top reflection factor," a function of riser exit geometry and gas and particle velocities, to represent the fraction of particles reaching the top that are internally reflected back down the column. The cross-sectional area occupied by the streamers and core varies with height.

With fitted reflection factors, this model does an excellent job of predicting axial density profiles in the upper part of a range of different columns with secondary air addition (Senior and Brereton, 1992), not only for a small-scale pilot plant CFB combustor but also for a prototype boiler of cross-sectional area 0.43 m^2 operated at high temperature (845°C).

2. Core–annulus model of Pugsley and Berruti (1996) and Gupta and Berruti (1998, 2000). This model again assumes a core–annulus structure, a bottom entrance region, and a more dilute upper region. Flow in the upper region in this case is assumed to be fully developed with average voidage

$$\varepsilon_{av} = \frac{U_g \rho_p}{G_s \psi + U_g \rho_p} \qquad (25)$$

ψ is a so-called slip factor correlated for smooth exits and group A particles by

$$\psi = 12.2\mathrm{Fr}_t^{2.01} + 7.65\,\mathrm{Re}_p^{-0.57} \qquad (26)$$

with $\mathrm{Fr}_t = u_t/(gD)^{0.5}$ and $\mathrm{Re}_p = \rho_g U_g d_p/\mu_g$. For abrupt exits, this ψ is multiplied by $(1 + \eta)$, where

$$\eta = 22.5 N_{Gs}^{1.49} N_{Gst}^{-1.36} N_{He}^{-1.32} N_\alpha^{4.06} \qquad (27)$$

with $N_{Gs} = G_s/(\rho_p U_g)$, $N_{Gst} = G_s/(\rho_p u_t)$, $N_{De} = D_e/D$, $N_{He} = (H - H_e)/H$, and $N_\alpha = (\alpha - 45°)/45°$. D_e is the exit duct diameter and H_e the top extension height. Angle α, shown in Fig. 18, is taken as $45°$ for a smoothly curving exit and $90°$ or more for an abrupt exit. For group B powders, Eq. (26) is replaced by a correlation due to Patience et al. (1992):

$$\psi = 1.0 + \frac{5.6}{\mathrm{Fr}} + 0.47\mathrm{Fr}_t^{0.41} \qquad (28)$$

where $\mathrm{Fr} = U_g/(gD)^{0.5}$, while Eq. (27) is replaced by

$$\eta = 0.00091 N_{Gs}^{-2.43} N_{Gst}^{2.02} N_{De}^{-1.54} N_{He}^{-2.80} N_\alpha \qquad (29)$$

Once the average voidage in the fully developed region is estimated, the average upwards particle velocity in the fully developed region is calculated as

$$v_p = \frac{G_s}{\rho_p(1 - \varepsilon_{av})} \qquad (30)$$

In the acceleration zone, Pugsley and Berruti (1996) assumed a core/annulus structure, with the core radius the same as in the fully developed region. A differential force balance was written for a single particle, with the voidage in the annular region assumed to vary linearly with height.

The model gives reasonable predictions for such properties as solids holdup, particle velocity, and solids

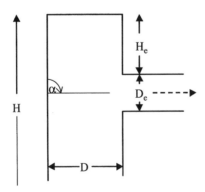

Figure 18 Parameters and dimensionless groups used to characterize riser exit in mechanistic model of Gupta and Berruti (2000).

flux for simple riser geometries over a wide range of fast fluidization conditions. Agreement with a radial profile of local solids flux is illustrated in Fig. 11.

4.6.3 Computational Fluid Dynamic (CFD) Models

CFD codes have had great success in recent decades in simulating single-phase flows. Similar approaches have been attempted by a number of research groups for multiphase flows, but success has been limited for the dense flow conditions encountered in circulating fluidized bed risers. There are several possible formulations of the equations and a number of possible approaches for solving them. Extensive reviews have been provided by Gidaspow (1994), Sinclair (1997), and Kashiwa and Yang (2002). Recent studies include Mathieson et al. (2000) and Zhang and VanderHeyden (2001)

Unfortunately CFD codes, despite the extensive effort and computational power deployed, have in general not matched experimental data any better than the mechanistic models in "challenge problem exercises" conducted at two recent (1995 and 2001) International Fluidization Conferences. While numerical (CFD) approaches will no doubt make major future contributions, there are currently no codes capable of giving consistently reliable predictions for concentrated and turbulent suspension flows of the kind found in CFB risers.

4.7 Scale-Up Considerations

Circulating fluidized beds cover an enormous range of sizes, from centimeters to tens of meters in lateral dimension. While most data have been obtained in laboratory units of ~ 100 to 200 mm diameter, a number of measurements in large-scale commercial units have been reported, as summarized in Table 5. Most of these measurements are limited to the wall region, owing to the difficulty of extending probes into the interior of large high-temperature reactors. The overall flow patterns appear to be qualitatively similar in small and large diameter vessels. For example, one still finds a fast fluidization flow regime with particles traveling predominantly upward in a dilute central core and downward in an outer wall layer. However, some significant quantitative differences arise:

Because a larger riser has less surface area per unit volume, there is greater coverage of the walls by downflowing streamers for the same U_g and G_s. The thickness of the wall layer is approximately proportional to the riser hydraulic diameter to

Table 5 Sources of Hydrodynamic Data from Large-Scale CFB Units

Reference	Cross section × Height (m)	Aspects covered
Couturier et al., 1991	3.15 × 3.15 × 23.8	Pressure profile, solids flux profile
Leckner et al., 1991	1.7 × 1.4 × 13.5	Solids flux near wall
Werther, 1994	various	Pressure profiles, solids flux, wall layer thickness
Zhang et al., 1995	1.7 × 1.4 × 13.5	Particle flux profiles, wall layer thickness
Lafranechère & Jestin, 1995	9.65 eq. dia. × 36.9	Pressure profiles, solids flux profiles, wall layer thickness
Hage & Werther, 1997	5.1 × 5.1 × 28.0	Local particle concentration, local particle velocity
Zhang et al., 1997	1.7 × 1.4 × 13.5	Particle momentum, concentration and velocity
Caloz et al., 1999	14.7 × 11.5 × 35	Particle velocity, fluctuations
Lin et al., 1999	4 × 6.2 × 22.8	Recirculation rate, particle velocities
Schouten et al., 1999	1.2 × 0.8 × 9.0	Bottom bed hydrodynamics
Sternéus et al., 1999	1.7 × 1.4 × 13.5	Turbulence
Johnsson & Johnsson, 2001	1.7 × 1.4 × 13.5	Local particle concentration, probability density, intermittency index
Johnsson et al., 2002	21.1 × 9.9 × 42.5	Local particle momentum, local particle concentration

the power of 0.6 (Werther, 1994), not to the power of 1.

- Large-scale risers, at least those used for combustion processes, commonly operate under very dilute conditions at the top (e.g., see Werther, 1994 and Sec. 5.1 below).
- Turbulence is likely to play a lesser overall role as D increases (Senior and Grace, 1998).

Because of experimental limitations, it is uncommon in experimental CFB studies to vary the riser diameter, or even the H/D ratio, while keeping all other variables fixed.

There have been a number of efforts to introduce a sufficient set of dimensionless groups which, when matched and coupled with geometric similarity, assure dynamic similarity between small- and large-scale units. This work is well summarized by van der Meer et al. (1999). For full hydrodynamic scaling, they recommend that at least five dimensionless groups be matched:

$$\mathrm{Fr_d} = \frac{U_g}{\sqrt{gd_p}} \qquad \mathrm{Re_p} = \frac{\rho_g d_p U_g}{\mu_g} \qquad \frac{G_s}{\rho_p U_g}$$
$$\frac{\rho_p}{\rho_g} \qquad \frac{D}{d_p} \qquad (31)$$

with d_p being the Sauter mean particle diameter. In place of the third group above, Kehlenbeck et al. (2001) suggest a dimensionless mass turnover, involving the solids inventory.

For $\mathrm{Re_p} = \rho_g d_p U_g/\mu_g < 4$, the above list may be reduced to three groups (Glicksman et al., 1993; van der Meer et al., 1999), the Froude number ($\mathrm{Fr_d}$) and dimensionless solids flux $G_s/(\rho_p U_g)$ from the above list, plus either U_g/U_{mf} or U_g/u_t, e.g.,

$$\mathrm{Fr_d} = \frac{U_g}{\sqrt{gd_p}} \qquad \frac{G_s}{\rho_p U_g} \qquad \frac{U_g}{u_t} \qquad (\mathrm{Re_p} < 4) \quad (32)$$

For $\mathrm{Re_p} = \rho_g d_p U_g/\mu_g > 1000$, $\mathrm{Re_p}$ may be dropped from those listed in Eq. (31) above, leaving a subset of four groups to match. D/d_p can be replaced in this four-group set by U_g/U_{mf} or U_g/u_t (Horio et al., 1989; Glicksman et al., 1993; van der Meer et al., 1999), yielding, for example,

$$\mathrm{Fr_d} = \frac{U_g}{\sqrt{gd_p}} \qquad \frac{G_s}{\rho_p U_g} \qquad \frac{\rho_p}{\rho_g}$$
$$\frac{U_g}{u_t} \qquad (\mathrm{Re_p} > 1000) \tag{33}$$

A number of other possible variables are ignored in arriving at the above lists, the most notable being the particle shape, the particle size distribution, the coefficients of restitution and friction for the particles themselves and between the particles and the wall, and any property variations due to temperature or concentration gradients. Advanced fluid dynamic models (e.g., Sinclair and Jackson, 1989; Senior and Grace, 1998) and experimental results (Chang and Louge, 1992) suggest that interparticle collisions and other interactions are of considerable importance for the conditions encountered in CFB risers. The omission of the coefficients of restitution and friction and the particle size distribution, all of which relate to particle–particle interactions, may explain why experimental attempts (e.g., Glicksman et al., 1991) to verify scaling based on any of the above lists (or similar sets of groups) have generally produced disappointing agreement. Nevertheless, the above sets of groups give a reasonable starting point for experimental attempts to achieve hydrodynamic scale modeling. For example, this approach has been used to model the effects of temperature and pressure on CFB hydrodynamics (Chang and Louge, 1992; Louge et al., 1999; Younis et al., 1999).

Scaling may also be based on hydrodynamic models (e.g., Winter et al., 1999). For example, the mechanistic models covered in Sec. 4.5.2 incorporate large-scale data in empirical correlations and mechanistic models used to predict hydrodynamics of CFB risers covering a wide range of sizes.

5 HYDRODYNAMICS OF OTHER RELEVANT FLOW REGIMES

5.1 Pneumatic Transport

CFB systems normally operate at high enough net solids fluxes that type A choking (Bi et al., 1993) occurs, leading to fast fluidization conditions. However, the upper region of some commercial scale CFB combustors (e.g., Werther, 1994) is so dilute that the dilute pneumatic transport flow regime may be of interest. When this occurs, there is no appreciable wall layer and particles travel upward over the entire riser cross section.

The vast majority of work on pneumatic transport has been in pipes and ducts of much smaller diameter and higher H/D ratio than the riser reactors considered in this chapter. CFB research has usually only touched on such dilute conditions (generally $\varepsilon_{sav} < 1\%$) when necessary, e.g., in order to be able to obtain laser sheet images. Correlations and models developed for fast fluidization conditions are unlikely to give accurate predictions when pneumatic transport conditions prevail.

5.2 High-Density Circulating Fluidized Beds

Early studies of CFB hydrodynamics were dominated by modest net particle fluxes (G_s from ~ 10 to $100 \ kg/m^2s$ and solids holdups in the upper part

of the riser well below 10%). While these conditions are relevant to CFB combustors, they do not match those encountered in FCC and other catalytic CFB reactors, where the net solids fluxes can be an order of magnitude higher and solids holdups are commonly 10%. Several recent studies have been dedicated to flow in high-density circulating fluidized bed (HDCFB) systems. These investigations have almost all been performed with FCC (group A) solids, though one study with sand (group B) particles (Karri and Knowlton, 1998) indicates that the results also apply, at least in broad terms, to larger denser particles.

The most notable difference in behaviour as G_s is increased beyond ~ 200–$300 \ kg/m^2s$ at relatively high U_g is the gradual disappearance of net solids downflow at the wall (e.g., Issangya et al., 2000; Karri and Knowlton, 1998, 1999; Liu, 2001). This leads to a different flow regime (dense suspension upflow, as discussed in Sec. 3) and a more homogeneous flow structure. Typical axial solids holdup profiles appear in Fig. 19. Note that at these high values of G_s, the profiles tend to reach relatively constant holdups of the order of 15%, indicating an approach to fully developed flow. Pärssinen et al. (2001) found that there were four zones in axial profiles for a much taller riser.

Despite the lack of time-mean downflow at the wall in dense suspension upflow, the flow is still subject to pronounced radial gradients (Bai et al., 1999; Grassler and Wirth, 1999; Karri and Knowlton, 1999; Liu, 2001), with significantly lower voidages (approaching ε_{mf} on a time mean basis) at the wall, while the particle velocity profile becomes more nonuniform with increasing G_s. The correlation of Issangya et al. (2001) for the radial distribution of voidage (Eq. 19 above) also works well for dense suspension upflow. Intermittency index values are in the same range and again show a maximum near the outer wall (Issangya et al., 2000).

The local time mean vertical component of particle velocity at the center line in dense suspension upflow can approach (Pärssinen et al., 2001) or even exceed (Liu, 2001) $2U_g$. Profiles tend to be very steep, as shown in Fig. 20. As can be seen in the same figure, the local standard deviation of particle velocity is much flatter, with local values exceeding the local time mean particle velocity near the outer wall. This means that there are momentary reversals in particle direction near the wall, even when the time mean direction is upwards.

At the center of the column all particles travel upward, whereas there is a substantial downward flux

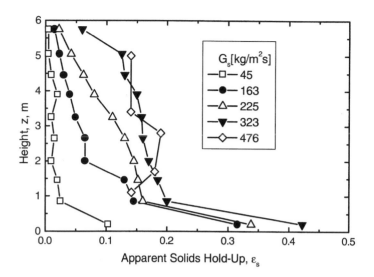

Figure 19 Axial profile of solids holdup for a range of net circulation fluxes in a riser of diameter 76 mm with a smooth unconstricted exit. $d_p = 70\,\mu m$, $\rho_p = 1600\,kg/m^3$, $U_g = 8\,m/s$. Highest fluxes correspond to dense suspension upflow. Note that the entrance region was different for the highest solids flux case. (Issangya et al., 1999; Liu, 2001.)

toward the outer wall, partially offsetting the upward flux there. Typically the local net flux is 1.5–2.5G_s at the axis of the column (Issangya et al., 1998; Karri and Knowlton, 1999; Liu, 2001) falling to nearly 0 at the wall. Results from a column of diameter 0.3 m at high U_g in Fig. 21 indicate that the radial net flux profiles can reverse, showing a minimum at the axis of the column.

6 MIXING

6.1 Gas Mixing

Vertical and horizontal gas mixing are of considerable importance in predicting conversions and selectivities in CFB reactors, especially for catalytic reactions. Gas mixing depends on the flow regime, flow pattern, particle properties, and riser configuration.

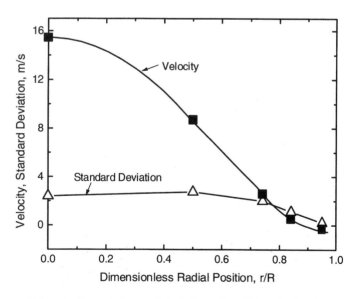

Figure 20 Local time mean particle velocity and standard deviation of particle velocity at $z = 4.2$ m in 76 mm diameter, 6.4 m diameter column operated at $U_g = 6$ m/s, $G_s = 417$ kg/m^{2}s. (Liu, 2001.)

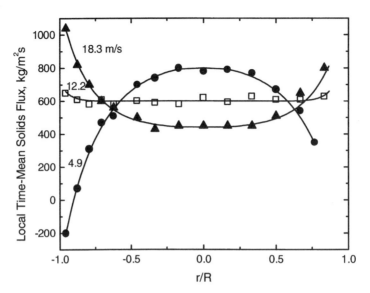

Figure 21 Radial profiles of local solids flux reported by Karri and Knowlton (1999) for $G_s = 586$ kg/m²s, $z = 3.7$ m in a 0.3 m diameter, 13 m long column. Numbers shown on figure are superficial gas velocities.

With the whole riser treated as a closed vessel with well-mixed tracer injection at the inlet and sampling at the outlet, the RTD curve can be obtained from a pulse or a step-change test (Schügerl, 1967; Brereton et al., 1988; Liu et al., 1999). An effective dispersion coefficient, D_{ge}, can then be derived by fitting. Both vertical and horizontal mixing/dispersion of gases contribute to gas dispersion. In steady state tracer experiments, a steady flow of tracer gas is introduced into the riser at a point, and the tracer concentration is measured either downstream or upstream. Ideally, the injection velocity should equal the gas velocity in the riser (Bader et al., 1988). Both adsorbing and nonadsorbing tracers can be used. When an adsorbing tracer is selected, the equilibrium adsorption on the particle surface needs to be identified (Kunii and Levenspiel, 1991). Based on the tracer concentration measured upstream of the injection point, a one-dimensional backmixing coefficient, D_{gb}, can be derived (Kunii and Levenspiel, 1991). The radial dispersion coefficient, D_{gr}, is obtained by analyzing radial concentration profiles measured downstream of the injection point (Bader et al., 1988). To characterize the mixing in the riser fully, all three coefficients need to be studied.

6.1.1 Gas RTD and Effective Gas Dispersion

The overall mixing behavior of the riser can be characterized by the residence time distribution (RTD) curve or an effective dispersion coefficient. To obtain

the RTD of the entire riser, tracer is introduced below the gas distributor, with samples taken at the riser exit to ensure well-mixed injection and sampling. Transient one-dimensional dispersion can be described by

$$\frac{\partial \overline{C}}{\partial t} + U_g \frac{\partial \overline{C}}{\partial z} - D_{ge} \frac{\partial^2 \overline{C}}{\partial z^2} = 0 \qquad (34)$$

With closed–closed boundary conditions, Eq. (34) can be solved. With relatively low dispersion, the variance and the effective dispersion coefficient are related (Levenspiel, 1998) by

$$\frac{\delta_t^2}{\tau^2} = 2\frac{D_{ge}\varepsilon_{av}}{U_g H} - 2\left(\frac{D_{ge}\varepsilon_{av}}{U_g H}\right)^2 \left[1 - \exp\left(-\frac{U_g H}{D_{ge}\varepsilon_{av}}\right)\right] \qquad (35)$$

The effective dispersion coefficient D_{ge} was reported to be in the range 1 to 12.7 m²/s for FCC (Dry and White, 1989; Liu et al., 1999) and (group B) sand (Brereton et al., 1988; Wang et al., 1992). D_{ge} increases with increasing G_s at relatively low solids circulation rates, then tends to level off (Dry and White, 1989) or to decrease after reaching a maximum at $G_s \approx 200$ kg/m²s (Liu et al., 1999). Liu et al. (1999) related the decrease of D_{ge} for $G_s > 200$ kg/m²s to the transition from fast fluidization to dense suspension upflow (see Sec. 5).

Figure 22 shows that the Peclet number, Pe_{ge}, based on the effective dispersion coefficient, decreases with increasing solids flux, especially for $G_s < 100$ kg/m²s.

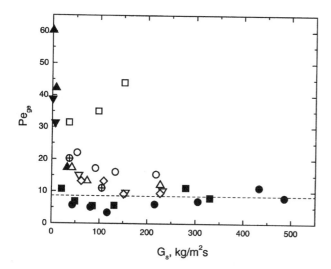

Figure 22 Axial gas Peclet number as a function of solids circulation flux at constant superficial gas velocities. Dry and White (1989): △ $U_g = 2.0$ m/s, ⊕ $U_g = 2.5$ m/s, ○ $U_g = 3.0$ m/s, ▽ $U_g = 4.0$ m/s, ◇ $U_g = 5.0$ m/s, □ $U_g = 8.0$ m/s. Liu et al. (1999): ▲ $U_g = 2.0$ m/s, ▼ $U_g = 2.5$ m/s, ■ $U_g = 4.7$ m/s, ● $U_g = 7.3$ m/s.

At high solids flux, however, Pe_{ge} appears to be no longer sensitive to G_s, with its value of order 10.

Gas dispersion is influenced by the entrance and exit structure of the riser, as well as its length and diameter. Brereton et al. (1988) showed that D_{ge} tends to be higher for an abrupt exit than for a smooth one at a given U_g and G_s. On the other hand, D_{ge} from the smooth exit was higher than D_{ge} from the abrupt exit for a given apparent suspension density.

Effective dispersion is affected by both axial and radial dispersion, as well as by Taylor dispersion due to the radial profile of axial gas velocities. Schügerl (1967) related the effective dispersion coefficient, D_{ge}, axial dispersion coefficient, D_{gd}, radial dispersion coefficient, D_{gr}, and the radial velocity profile by

$$D_{ge} = D_{gd} + M \frac{U_g^2 D^2}{D_{gr}} \tag{36}$$

where M is a constant which characterizes the nonuniformity of the axial velocity profile.

6.1.2 Radial Gas Dispersion

A radial gas dispersion coefficient can be obtained from the steady-state tracer method by fitting measured radial profiles of tracer concentration downstream of the injection point to a two-dimensional

differential equation (Yang et al., 1983; Bader et al., 1988; Kruse et al., 1995),

$$\frac{\partial C}{\partial t} = D_{gd} \frac{\partial^2 C}{\partial z^2} + D_{gr} \frac{1}{\varepsilon r} \frac{\partial}{\partial r} \left(\varepsilon r \frac{\partial C}{\partial r} \right) - u_g(r) \frac{\partial C}{\partial z} \tag{37}$$

D_{gd} and D_{gr} are both assumed to be independent of position, and $u_g(r)$ is the local interstitial gas velocity. If the axial dispersion is much smaller than the radial dispersion, the axial dispersion term can be dropped (van Zoonen, 1962; Yang et al., 1983; Martin et al., 1992; Gayan et al., 1997; Namkung and Kim, 2000). Otherwise, one needs to fit both the axial and the radial dispersion coefficients, which can involve much higher uncertainties (Bader et al., 1988). The radial velocity profile, $u_g(r)$, and local voidage, ε, are required to fit the radial dispersion coefficient, but most studies have assumed a flat radial velocity profile in fitting Eq. (37) to tracer concentration profiles.

Most researchers have reported that the radial dispersion coefficient decreases with increasing G_s at a given U_g and relatively low G_s, and increases with increasing G_s at relatively high solids fluxes (Zheng et al., 1992). Similarly, D_{gr} decreases with increasing solids concentration at low solids concentration owing to suppression of turbulence (van Zoonen, 1962; Adams, 1988; Zheng et al., 1992; Koenigsdorff and Werther, 1995), while increasing with a further increase in solids concentration, probably because of solids downflow in the annulus region (Zheng et al., 1992; Win et al., 1994).

At a given solids circulation rate, D_{gr} decreases with increasing U_g, although Adams (1988) and Zheng et al. (1992) observed that D_{gr} increases with increasing gas velocity at relatively high G_s. When data are plotted as Pe_{gr} vs. Re_p, as in Fig. 23, Pe_{gr} increases with increasing Re_p at given G_s. Pe_{gr} is found to decrease with increasing G_s except at very low solids fluxes (Amos et al., 1993), where Pe_{gr} is observed to increase with increasing G_s.

The radial gas dispersion coefficient associated with the lean phase can be obtained separately (Werther et al., 1992; Koenigsdorff and Werther, 1995; Mastellone and Arena, 1999b). It has been found to increase with increasing G_s (Mastellone and Arena, 1999b) or to be insensitive to G_s, and it tends to increase with increasing U_g (Werther et al., 1992; Koenigsdorff and Werther, 1995).

Zheng et al. (1992) found that D_{gr} increased with increasing ρ_p at a given gas velocity and solids flux for group B particles. Mastellone and Arena (1999b), using group A particles, found that D_{gr} decreased

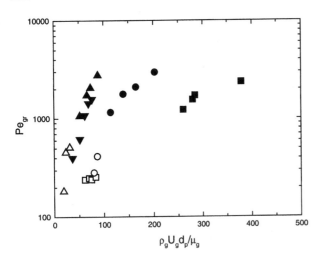

Figure 23 Radial gas Peclet number as a function of Reynolds number. Gayan et al., 1997: ■ $d_p = 710\,\mu m$, $G_s = 30\,kg/m^2s$; ● $d_p = 380\,\mu m$, $G_s = 30\,kg/m^2s$, Adams (1988): □ $d_p = 250\,\mu m$, $G_s = 30\,kg/m^2s$, ○ $d_p = 250\,\mu m$, $G_s = 44\,kg/m^2s$, Bader et al. (1988): △ $d_p = 76\,\mu m$, $G_s = 98\text{--}177\,kg/m^2s$, Yang et al. (1983): ▲ $d_p = 220\,\mu m$, $G_s = 57\,kg/m^2s$, ▼ $d_p = 220\,\mu m$, $G_s = 74\,kg/m^2s$)

with increasing particle density. Gayan et al. (1997) reported that D_{gr} increased when particle size was increased from 380 to $710\,\mu m$ at given U_g and G_s, whereas Mastellone and Arena (1999b) reported a reduction in D_{gr} when d_p increased from 89 to $310\,\mu m$.

Correlations for the radial dispersion coefficient are listed in Table 6. The equation of Yang et al. (1983), based on their own experimental data in a 0.135 m diameter column, underestimates most data obtained in relatively large columns. The Koenigsdorff and Werther (1995) equation can only be applied to very dilute conditions ($\varepsilon_{av} \approx 0.99$). The equation of Amos et al. (1993) predicts that Pe_{gr} increases with increasing G_s over a wide range but is inconsistent with most experimental data. The equations of Gayan et al. (1997) and Namkung and Kim (2000) are recommended until more reliable correlations are available.

6.1.3 Gas Backmixing

Gas backmixing is obtained from steady state tracer injection tests in which tracer is injected continuously from one location and its concentration is monitored upstream of the injection point. A one-dimensional vertical dispersion process with a nonadsorbing tracer can be represented by

$$D_{gb} \frac{d^2\overline{C}}{dz^2} - \frac{U_g}{\varepsilon_{av}} \frac{d\overline{C}}{dz} = 0 \tag{38}$$

where $\overline{C}$ is the cross-sectional average tracer concentration. With boundary conditions

$$\overline{C} = \overline{C}_i \quad \text{at} \quad z = 0 \tag{39}$$

and

Table 6 Correlations for Estimating Gas Radial Dispersion Coefficient

Source	Equation	Comment
van Zoonen (1962)	$Pe_{gr} = 500 \sim 1000$	For small columns only.
Yang et al. (1983)	$D_{gr} = 43.5 \dfrac{U_g}{[(U_g/\varepsilon_{av}) - (G_s/\rho_p[1-\varepsilon_{av}])]} \dfrac{1-\varepsilon_{av}}{\varepsilon_{av}} + 0.7$	No column diameter effect. For small columns only.
Amos et al. (1993)	$Pe_{gr} = 6.46\varepsilon_{av}\left[\dfrac{(G_s + \rho_g U_g)D}{\mu_g}\right]^{0.297}$	That Pe_{gr} increases with G_s is inconsistent with other experimental data.
Koenigsdorff and Werther (1995)	$Pe_{gr} = 150 + 56,000(1 - \varepsilon_{av})$	For dilute flow only with $(1 - \varepsilon_{av}) < 0.01$.
Gayan et al. (1997)	$D_{gr} = 1.4 \times 10^{-6}\left[\dfrac{(G_s + \rho_g U_g)D}{\mu_g}\right]^{-1.14}(G_s + 1)\left(\dfrac{d_p}{\rho_p}\right)^{-1.3} D^{1.85}$	That D_{gr} decreases with increasing G_s at high solids flux is inconsistent with experimental data.
Namkung and Kim (2000)	$\dfrac{Ud_p}{D_{gr}} = 153.1\left(\dfrac{d_p}{D}\right)^{0.96}\left(\dfrac{U_g}{U_{sl}}\right)^{3.73}$	$U_{sl} = \dfrac{U_g}{\varepsilon_{av}} - \dfrac{G_s}{\rho_p(1-\varepsilon_{av})}$

$$\overline{C} = 0 \quad \text{at} \quad z = -\infty \tag{40}$$

the solution for equation (35) is

$$\frac{\overline{C}}{\overline{C_i}} = \exp\left[-\frac{U_g(z_i - z)}{\varepsilon_{av} D_{gb}}\right] \tag{41}$$

The axial dispersion coefficient is obtained by plotting $\ln(\overline{C}/\overline{C_i})$ vs. the distance, z, from the source.

Experimentally it is difficult to generate a plane source of tracer gas. A point source of injection has thus been used in all experiments. When tracer is injected at the axis, little tracer is detected upstream (Li and Weinstein, 1989; Namkung and Kim, 1998). On the other hand, high tracer concentration and significant radial variation were detected when tracer was injected into the annulus. These findings suggest that gas backmixing is significant in the annulus region but negligible in the dilute core region.

To determine a one-dimensional average backmixing coefficient based on equation (41), an average tracer concentration is required. With axial symmetry,

$$\overline{C} = \frac{2}{R^2(U_g/\varepsilon_{av})} \int_0^R C(r) \cdot u_g(r) \cdot r \cdot dr \tag{42}$$

D_{gb} generally increases with increasing G_s at given U_g. Li and Weinstein (1989) found that D_{gb} increased with increasing U_g at constant G_s, while Namkung and Kim (1998) found the opposite trend, possibly because their data were obtained at relatively low fluxes ($G_s < 40\,\text{kg/m}^2\text{s}$), while Li and Weinstein (1989) reached relatively high solids fluxes (up to $200\,\text{kg/m}^2\text{s}$).

6.1.4 Axial Gas Dispersion

Axial gas dispersion can be studied with pulse or step-change injection (van Zoonen, 1962; Li and Wu, 1991; Luo and Yang, 1990; Martin et al., 1992; Bai et al., 1992b). Transient one-dimensional dispersion can be described (Li and Wu, 1991) by

$$\frac{\partial \overline{C}}{\partial t} + U_g \frac{\partial \overline{C}}{\partial z} - D_{gd} \frac{\partial^2 \overline{C}}{\partial z^2} = 0 \tag{43}$$

For pulse injection with open-open boundary conditions, the solution of equation (43) is

$$\overline{C} = \frac{1}{2\sqrt{\pi\theta(\varepsilon_{av}D_{gd}/U_gL)}} \exp\left[-\frac{(1-\theta)^2}{4\theta(\varepsilon_{av}D_{gd}/U_gL)}\right] \tag{44}$$

where L is the distance between the injection and the sampling point for single point sampling or the distance between the two sampling locations for double-point sampling. The variance of the RTD curve is

related to the axial gas dispersion coefficient for relatively low dispersion by

$$\frac{\delta_t^2}{\tau^2} = 2\frac{D_{gd}\varepsilon_{av}}{U_gL} + 8\left(\frac{D_{gd}\varepsilon_{av}}{U_gL}\right)^2 \tag{45}$$

In all reported pulse tests, the tracer was injected on the axis, with the sampling port downstream on the axis. The tracer concentration then reflects the local response on the axis, not the response from the entire cross section. The dispersion coefficient, D_{gd}, obtained in this way is the same as the one-dimensional dispersion coefficient defined by equation (37) only for risers of small diameter because of the strong radial variation of tracer concentrations downstream of the injection point.

D_{gd} generally increases with increasing G_s at given gas velocities for $G_s < 200\,\text{kg/m}^2\text{s}$. For $G_s > 200\,\text{kg/m}^2\text{s}$, van Zoonen (1962) reported a decrease of D_{gd} with increasing solids flux. van Zoonen (1962) and Luo and Yang (1990) found that D_{gd} increases with increasing U_g at constant G_s for $G_s > 50\,\text{kg/m}^2\text{s}$. Li and Wu (1991) and Bai et al. (1992) reported the opposite trend for $G_s < 50\,\text{kg/m}^2\text{s}$.

Luo and Yang (1990) found that D_{gd} decreases with increasing particle size at given U_g and G_s. The effect of column diameter is unclear. Li and Wu (1991) correlated their data by

$$D_{gd} = 0.195\varepsilon_{av}^{-4.11} \tag{46}$$

which ignores any effect of particle properties. On the other hand, Luo and Yang (1990) correlated their D_{gd} data as functions of operating parameters and gas and particle properties as

$$\frac{D_{gd}\rho_g}{\mu_g} = 0.11\left(\frac{\rho_g U_g d_p}{\mu_g}\right)^{1.6}\left(\frac{G_s}{\rho_p u_t}\right)^{0.62}\left(\frac{\rho_p}{\rho_g}\right)^{1.2} \tag{47}$$

6.1.5 Interphase Mass Transfer

Brereton et al. (1988) and Namkung and Kim (1998) considered a core–annulus structure with exchange between the two regions. With one-dimensional flow assumed in both regions, axial dispersion in the annulus and plug flow in the core, one can write

$$\frac{\partial C_c}{\partial t} + u_{gc}\frac{\partial C_c}{\partial z} + \frac{2K_{ca}}{R_c}(C_c - C_a) = 0 \tag{48}$$

$$\frac{\partial C_a}{\partial t} - D_{gza}\frac{\partial^2 C_a}{\partial z^2} - \frac{2K_{ca}R_c}{R^2 - R_c^2}(C_c - C_a) = 0 \tag{49}$$

for the core and annulus regions, respectively. Both the axial dispersion coefficient in the annulus region, D_{gza}, and the interphase exchange coefficient, K_{ca}, are then fitted to concentration profiles. Values of K_{ca} have been reported to range from 0.015 to 0.3 m/s and to increase with increasing G_s. The effect of U_g on K_{ca}, however, is uncertain. Werther et al. (1992) found that K_{ca} increases with U_g, but White et al. (1992) and Namkung and Kim (1998) reported a decrease of K_{ca} with increasing U_g.

6.2 Solids Mixing

Solids mixing can influence such factors as gas–solids contacting, heat transfer, solids conversion, and deactivation of catalyst particles. Solids mixing can be measured by several experimental techniques. Particle dispersion and residence time distributions (RTD) in the riser have been commonly studied by injecting a pulse of tracer particles and then sampling downstream of the injection point or at the riser exit. The tracer particles may be radioactive (Patience et al., 1991), magnetic (Avidan and Yerushalmi, 1985), fluorescent (Kojima et al., 1989), heated (Westphalen and Glicksman, 1995), phosphorescent (Wei et al., 1995b), salt or impregnated with salt (Bader et al., 1988). On-line detection systems can be adopted for most of these tracers but not salt. Tracer particles of the same size, density, and shape as the bed material should be used wherever possible.

6.2.1 Axial Dispersion of Particles

RTD curves of solid tracer particles have been analyzed to obtain an axial dispersion coefficient assuming axially dispersed one-dimensional flow (Smolders and Baeyens, 2000b),

$$\frac{\partial C}{\partial t} = D_{pz}\frac{\partial^2 C}{\partial z^2} - v_p\frac{\partial C}{\partial z} \tag{50}$$

where C is the concentration of tracer particles at time t. The flow is considered fully developed, so that particle velocity v_p is independent of z. With open–open boundary conditions, a solution is

$$E(\theta) = \frac{1}{2\sqrt{\pi\theta/Pe_{pz}}}\exp\left[-\frac{Pe_{pz}(1-\theta)^2}{4\theta}\right] \tag{51}$$

with

$$\theta = \frac{t}{\tau} \quad \text{and} \quad Pe_{pz} = \frac{v_pL}{D_{pz}} \tag{52}$$

where τ is the mean residence time, $E(\theta)$ is the dimensionless RTD function, and L is the distance between the tracer injection and detection points. The mean

residence time, τ, can be estimated from the first moment of the normalized RTD (Patience et al., 1991), the overall solids circulation rate (Zhang et al., 1993), or the 50% value of the cumulative solids residence time distribution of the tracer (Rhodes et al., 1991b). In many studies, v_p in the definition of Pe_{pz} above is replaced by U_g (Rhodes et al., 1991b; Wei et al., 1995b; Smolders and Baeyens, 2000b). This practice will be followed here.

Figure 24 plots the axial Peclet number as a function of solids circulation flux from several sources. Pe_{pz} generally ranges from 1 to 10 for risers with abrupt exits. For a smooth exit (Diguet, 1996), Pe_{pz} was higher due to less particle reflection from the exit and thus less downflow of particles near the wall. Pe_{pz} tends to decrease with increasing solids circulation rate, consistent with the finding that the solids downflow in the annulus region increases with increasing solids circulation. Some researchers (Rhodes et al., 1991b; Wei et al., 1995b; Godfroy et al., 1999a), however, report that Pe_{pz} increases with increasing G_s. At constant G_s, Pe_{pz} tends to increase slightly (Wei et al., 1995b; Diguet, 1996; Smolders and Baeyens, 2000b) or remain constant (Rhodes et al., 1991b; Patience et al., 1991) with increasing U_g, although the axial solids dispersion coefficient in all cases increased with increasing U_g. Using three types of sand of different mean sizes, Patience et al. (1991) found that Pe_{pz} was higher for larger than for smaller particles at the same U_g and G_s.

The axial gas dispersion tends to increase with column diameter, as shown above. If both gas and solids

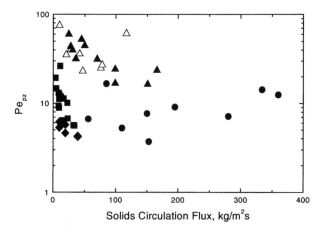

Figure 24 Effect of solids circulation flux on axial solids dispersion. Data: ■ sand, $d_p = 90\,\mu m$ (Smolder and Baeyens, 2000b), △ (Diguet, 1996); ◆ FCC, $d_p = 54\,\mu m$ (Du et al., 1999), ● FCC, $d_p = 56\,\mu m$ (van Zoonen, 1962), ▲ sand, $d_p = 275\,\mu m$ (Patience et al., 1991).

dispersion are mainly due to downflow of particles next to the wall, axial solids dispersion is expected to follow the same trend, i.e., to increase with increasing D. Limited data (Werther and Hirschberg, 1997) indicate that the axial particle dispersion coefficient in the upper fully developed dilute region tends to increase with D. However, Rhodes et al. (1991b) found that axial dispersion decreased with increasing D. Caution thus needs to be exercised in scaling up laboratory data.

Several correlations are available to predict the solids axial dispersion. The Rhodes et al. (1991b) correlation and the Wei et al. (1995b) correlation predict that Pe_{pz} increases with increasing solids flux or solids concentration, while the Smolder and Baeyens (2000) correlation,

$$\mathrm{Pe}_{pz} = 2.4 + \frac{18.3}{(G_s/U_g)^{0.8}} \tag{53}$$

predicts the opposite trend. Equation (53) is recommended, although the effect of D is unclear.

6.2.2   Radial Dispersion of Particles

Radial particle dispersion in CFB risers can be studied by measuring radial concentration profiles of tracer particles injected at a single point upstream of the measurement location (van Zoonen, 1962; Wei et al., 1995b). A two-dimensional dispersion model for fully developed axisymmetric flow with constant dispersion axial and radial coefficients, D_{pz} and D_{pr}, gives

$$\frac{\partial C}{\partial t} = D_{pz}\frac{\partial^2 C}{\partial z^2} + \frac{D_{pr}}{r}\frac{\partial}{\partial r}\left(r\frac{\partial C}{\partial r}\right) - \frac{\partial(v_p C)}{\partial z} \tag{54}$$

Analytical solutions of equation (54) for pulse injection of solid tracers can be obtained with proper boundary conditions by assuming that v_p is independent of radial and axial location (van Zoonen, 1962; Wei et al., 1995b; Patience and Chaouki, 1995). If velocity and solids concentration profiles are nonuniform, equation (54) must be solved numerically (Koenigsdorff and Werther, 1995), and the axial and radial solids dispersion coefficients are then obtained by fitting.

Wei et al. (1995b) and Koenigsdorff and Werther (1995) reported that the radial dispersion coefficient decreased with increasing solids concentration at given gas velocities (see Fig. 25), and that the radial dispersion Peclet number is generally higher than for single-phase turbulent flow. van Zoonen (1962) observed that radial dispersion is more pronounced in the lower entrance region and decreases with height. This is probably associated with developing flow in the

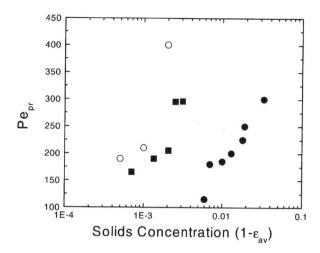

Figure 25   Radial solids dispersion coefficient as a function of solids concentration and gas velocity in CFB risers. Data: ● FCC, $d_p = 54\,\mu\mathrm{m}$, $U_g = 5.7$ m/s (Wei et al., 1995b); ○ SiC, $d_p = 60\,\mu\mathrm{m}$, $U_g = 4.0$ m/s (Koenigsdorff and Werther, 1995); ■ $U_g = 3.0$ m/s (Koenigsdorff and Werther, 1995).

entrance region. At constant solids concentration or constant solids flux, D_{pr} increased with increasing U_g, probably due to increased local turbulence (Wei et al., 1995b; Koenigsdorff and Werther, 1995).

The correlations of Wei et al. (1995b) and Koenigsdorff and Werther (1995), respectively,

$$\mathrm{Pe}_{pr} = 225.7(1 - \varepsilon_{av})^{0.29}\mathrm{Re}_p^{0.3} \tag{55}$$

and

$$\mathrm{Pe}_{pr} = 150 + 5.6 \times 10^4(1 - \varepsilon_{av}) \tag{56}$$

can be used to estimate radial solids dispersions until more accurate correlations are available.

6.2.3   Interphase Solids Exchange Between Core and Annulus

Although unimodal solids RTD curves have usually been reported (Chesonis et al., 1990; Rhodes et al., 1991b; Bai et al., 1992b; Smolders and Baeyens, 2000b; Harris, 2002), bimodal curves have been found for relatively dense conditions (Helmrich et al., 1986; Kojima et al., 1989; Ambler et al., 1990; Wei et al., 1995b). The bimodal distribution is likely associated with the internal circulation of particles and cannot be predicted by a one-dimensional dispersion model. Core–annulus models can account for the bimodal solids RTD distributions, with a core/annulus solids exchange coefficient characterizing cross-flow of particles. An unsteady solids tracer balance relative to the core and annular regions, assuming plug flow and neglecting axial dispersion, yields (Ambler et al., 1990)

$$\frac{\partial(G_c f_c)}{\partial z} + (I_{ca}f_c - I_{ac}f_a) = -\frac{G_c}{v_{pc}}\frac{\partial f_c}{\partial t} \quad (57)$$

$$\frac{\partial(G_a f_a)}{\partial z} + (I_{ca}f_c - I_{ac}f_a) = \frac{G_a}{v_{pa}}\frac{\partial f_a}{\partial t} \quad (58)$$

for the core and annular regions, respectively, where G_c and G_a are the upward and downward solids fluxes in the core and annular regions, respectively. I_{ca} refers to the mass of solids transferred from core to annulus and I_{ac} from annulus to the core, with (I_{ca}-I_{ac}) representing the net lateral flow from the core to the annulus. f_c and f_a represent the mass fraction of tracer in the core and annulus, respectively, while v_{pc} and v_{pa} are average particle velocities in the core and annulus.

Values of I_{ac} ranged from 0.1 to 10 kg/m³s for 90 and 106 μm sand particles (Ambler et al., 1990; Smolders and Baeyens, 2000b), and increased as G_s increased (Smolders and Baeyens, 2000b). Converted into radial solids fluxes, these interchange fluxes (0.025 to 0.25 kg/m²s) fall into the same range of radial solids fluxes (0.1 to 1.5 kg/m²s) as those measured by sampling methods (Zhou et al., 1995b), implying that lateral solids mixing is mainly caused by particle cross flow.

7 HEAT TRANSFER

7.1 Introduction

Favorable heat transfer is a major reason for the success of CFB reactors in such applications as combustion and calcination. CFB heat transfer has previously been reviewed by Grace (1986b), Glicksman (1988), Leckner (1990), Yu and Jin (1994), Basu and Nag (1996), Glicksman (1997), and Molerus and Wirth (1997). Temperature gradients tend to be small within the riser due to vigorous internal mixing of particles, while the gas and particle temperatures are locally nearly equal, except at the bottom (Grace, 1986; Watanabe et al., 1991). Because of wear caused by particle impacts, heat transfer surfaces are predominantly vertical, and only this case is considered here. Many CFB combustion systems are also equipped with external fluidized bed heat exchangers, cooling solids returning to the bottom of the riser while they are being recirculated in the external loop. However, these operate as bubbling beds exchangers, and the reader is referred to Chapter 3. In this section we consider the

case where the bulk solids are hotter than the wall, as in CFB combustion. It is straightforward to change to the case where the transfer is in the opposite direction.

Suspension-to-wall heat transfer is controlled by the flow of the solids and gas near the wall. Most studies have employed small-scale columns at atmospheric temperature and pressure. For simplicity, the experimental risers have generally been idealized in that they have been (Grace, 1996):

Smooth-walled, with roughness elements much smaller than the diameter of the particles
Circular in cross section, i.e., without corners or protruding surfaces
Vertical, i.e., with no tapered or oblique sections
Supplied with air only through a flat multiorifice distributor at the bottom of the riser.

Most boilers operate at 800 to 900°C, but few experimental investigations have been carried out at elevated temperatures. Industrial CFB combustion systems have risers that are usually rectangular in cross section, manufactured from membrane water walls. Surfaces may be rough, especially where refractory is present. The lower section is commonly tapered or constricted. In addition to primary gas from the base, secondary gas may be injected through nozzles at higher levels. Given these factors, data obtained in laboratory CFBs are not always representative of industrial-scale CFB reactors.

Heat transfer models are usually written in terms of either clusters or dense wall layers, based on the hydrodynamics of fast fluidization. For cluster models (Fig. 26), heat can be transferred between the suspension and wall by (1) transient conduction to particle clusters arriving at the wall from the bulk, supplemented by radiation; (2) convection and radiation from the dispersed phase (gas containing a small fraction of solid material). The various components are usually assumed to be additive, ignoring interaction between the convective and radiation components.

7.2 Convection

Any portion of the wall comes into intermittent contact with clusters of particles, interspersed with periods, usually brief, where that portion of the wall is in contact with relatively dilute phase. Heat transfer from the descending clusters is commonly described by unsteady, one-dimensional conduction normal to the wall. Because this transfer relies on renewal of fresh particles from the bulk, this heat transfer mechanism is often referred to as *particle convection*, even though

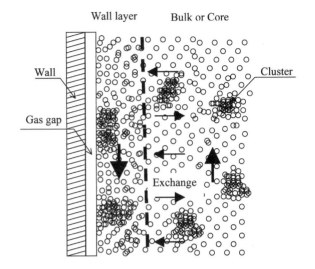

Figure 26 Conceptual view of clusters of particles and gas close to wall of CFB riser.

the transfer between the cluster and the wall is primarily by conduction through the intervening gaseous medium.

7.2.1 Influence of Different Parameters on Convective Heat Transfer

Suspension Density: The cross-sectional average suspension bulk density, $\rho_{susp} = \rho_p(1 - \varepsilon_{av}) + \rho_g\varepsilon_{av}$, is the most important determinant of heat transfer to

the wall. Experimental data from large-scale high-temperature units are plotted in Fig. 27, with corresponding operating conditions summarized in Table 7. The wall-to-suspension heat transfer coefficient increases strongly with increasing ρ_{susp}, primarily because of the particle convection term.

Heat Transfer Surface: The walls of CFB combustion chambers consist of "membrane waterwalls," i.e., parallel tubes connected longitudinally by fins (Fig. 28). The protected channel between tubes is a region of augmented particle downflow and greater particle concentration than the exposed tube surfaces (Wu et al., 1991; Golriz, 1994; Lockhart et al., 1995; Zhou et al., 1996b). Particles descend faster and travel further along the fins on average before being stripped off and returning to the core (Wu et al., 1991; Zhou et al., 1996b). These findings help explain the distribution of local heat transfer coefficients around membrane surfaces (Andersson and Leckner, 1992; Golriz, 1994; Lockhart et al., 1995).

The conductivity and thickness of the membrane water wall surface also play a role (Bowen et al., 1991), because heat transferred to the fin must be conducted to the tube surface and then into the coolant (usually boiling water). In practical systems, the heat transfer coefficient on the coolant side is almost always much higher than on the suspension side, so that the overall heat transfer coefficient is barely affected by the coolant flow rate. A model that accounts for the coolant-side resistance, as well as the suspension-side and

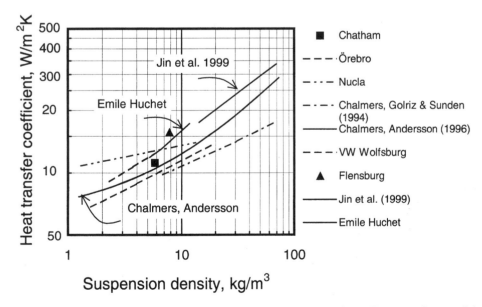

Figure 27 Heat transfer coefficient vs. suspension density for large-scale CFB risers. For operating conditions see Table 7.

Table 7 Experimental Conditions for Large-Scale CFB Heat Transfer Studies

Unit, Capacity, Reference	Size, m	U_g, m/s	T_b, °C	d_p, μm	T_w, °C
Chatham, 72 MW$_{th}$, Couturier et al. (1993)	$4 \times 4 \times 23$	6.4	875	200	500
Örebro, 165 MW$_{th}$, Andersson et al. (1996)	$4.6 \times 12 \times 33.5$	3–5	700–860	280	360
Emile Huchet, 125 MW$_e$, Jestin et al. (1992)	$8.6 \times 11 \times 33$	NA	850	140	340
Nucla, 110 MW$_e$, Nucla Report (1991)	$6.9 \times 7.4 \times 34$	2.6–5.1	774–913	150	330
Chalmers, 12 MW$_{th}$, Leckner (1991)	$1.4 \times 1.7 \times 13.5$	1.8–6.1	640–880	220–440	210
VW Wolfsburg, 145 MW$_{th}$, Blumel et al. (1992)	$7.6 \times 5.2 \times 31$	6.2	850	NS	340
Flensburg, 109 MW$_{th}$, Werdermann & Werther (1994)	$5.1 \times 5.1 \times 28$	6.3	855	220	340
Jianjiang, 50 MW$_{th}$, Jin et al., (1999)	$3.0 \times 6.0 \times 20$	5.1	920	400	290

wall conduction resistances, has been developed by Xie et al. (2003).

Descending Particles: To investigate the effect of the descending wall layer of particles on the heat transfer coefficient at the membrane water wall, Golriz (1994) installed an angled deflector 0.2 m into the combustion chamber of a 12 MW$_{th}$ combustor 0.5, 1.0, 1.5, and 2.0 m above a heat flux meter located 4.0 m above the distributor. The heat transfer coefficient increased and the character of the signal changed with the obstacle diverting clusters and strands away from the membrane wall, exposing the surface to hotter particles. The measured heat transfer coefficient increased by 55% with the deflector 0.5 m above the heat flux meter. The enhancement of the heat transfer decreased as the distance between the meter and deflector increased, disappearing at a separation of 2 m. At 10 m above the distributor plate where the suspension was more dilute, heat transfer was unaffected by a deflector 1.0 m above the heat flux meter. Hyre and Glicksman (1995) found similar augmentation by deflecting descending particles from a plane wall. Cao et al. (1994) increased the heat transfer by 100% by placing a large horizontal ring in a CFB.

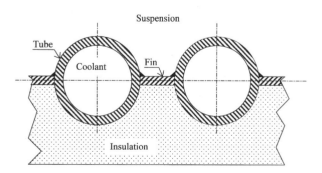

Figure 28 Plan view of section of membrane wall.

Vertical Length of Heat Transfer Surface: Short heat transfer surfaces lead to overestimates of heat transfer to the wall (Wu et al., 1991; Nag and Moral, 1990) because clusters travel along the heat transfer surface over considerable distances, causing the heat transfer coefficient to decrease as temperature equilibration occurs between the cluster and the surface.

A number of researchers have measured descending cluster velocities in the wall layer for a variety of flow conditions and different wall geometries, with most data between 0.5 and 2.0 m/s. The data have been well summarized by Griffith and Louge (1998) and are correlated by equation (22).

The contact length for clusters, L_c, can be 1.5 m or more along the fins of membrane water walls (Wu et al., 1989; Golriz and Leckner, 1992). L_c is much shorter for smooth walls (Wu et al., 1991; Burki et al., 1993; Glicksman, 1997). The experimental results suggest that the cluster residence time at the wall is ~ 0.1 to 1 s for smooth walls and ~ 0.5 to 3 s for sheltered regions, including the fins of membrane walls. However, for short heat transfer surfaces, e.g., 0.01 m long, L_c is only of order 0.01 s. Given the influence of length of heat transfer surface, considerable caution is required when applying data from short surfaces to much longer surfaces (Grace, 1990, 1996).

Bulk Suspension Temperature: The heat transfer coefficient increases with bulk temperature (e.g., see Wu et al., 1989; Basu, 1990; Golriz and Sunden, 1994a,b), due to both the higher gas thermal conductivity and the increased radiation. At temperatures > 500°C, especially for relatively dilute beds (suspension densities < 15 kg/m^3), where radiation tends to be the dominant mode of transfer, the increase in heat transfer with temperature can be very significant. Figure 29 shows the local heat transfer coefficient along a 1.5 m long heat transfer surface for two bulk temperatures (Wu et al., 1989). The 30–40% in-

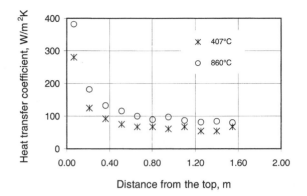

Figure 29 Heat transfer coefficient vs. distance from top of heat transfer surface for two bulk temperatures. (Wu et al., 1989a.)

crease in heat transfer coefficient with increased temperature is predominantly due to increased radiation (see below).

Particle Size: The convective heat transfer coefficient decreases with increasing particle diameter (Strömberg, 1982; Fraley et al., 1983; Basu, 1990) for small heat transfer sections where clusters have brief residence times on the heat transfer surface. The influence of d_p diminishes with longer surfaces, since the contact times on the surface are then much higher than the thermal time constant for individual particles (Wu et al., 1989). Andersson (1996) found that the vertical distribution of heat transfer coefficients in a 12 MW_{th} CFB boiler are independent of d_p for constant U_g/u_t.

Superficial Gas Velocity: Experimental evidence (e.g., Wu et al., 1987; Furchi et al., 1988) indicates little influence of gas velocity for a constant suspension density. This occurs because gas convection is much less important than particle convection for the particle sizes and flow conditions of interest.

Pressure: For a given suspension density, the heat transfer coefficient increases with pressure (Xianglin et al., 1990; Mattmann, 1991). This effect is smaller for smaller particles (Shen et al., 1991).

7.2.2 Convective Heat Transfer Models

Cluster Renewal Models: Most mechanistic models for heat transfer in CFBs are extensions of the model of Mickley and Fairbanks (1955). Descending clusters and strands in the vicinity of the wall surface are modeled as homogeneous semi-infinite media, separated from the wall by a thin gas layer. Transient transfer of heat occurs from the moving clusters to the wall, with the transfer rate depending on the replacement frequency and thermal properties of the clusters.

Emulsion Models: To simulate the core–annulus structure, the cross section in emulsion models is divided into an inner dilute core region where particles are transported upwards, and a denser annular region where particles descend along the wall, as in Fig. 26, but without the clusters. The thickness of the solid layer along the vertical heat transfer surfaces is often approximated as uniform. However, for membrane wall heat transfer surfaces, the annulus layer tends to be thicker at the fin than at the tube crest (Grace, 1990; Golriz 1992).

Descending particles are usually assumed to enter the heat transfer zone at the temperature of the core. Heat transfer from the particles in the wall layer results in a thermal boundary layer. An example of the horizontal temperature profile normal to the fin and crest of a membrane tube wall and normal to the refractory wall 11.0 m above the primary air distributor plate in a 12 MW_{th} CFB boiler (Golriz, 1992) appears in Fig. 30. The thermal boundary layer thickness is smaller at the summit of the tube than at the fin.

The emulsion model is similar to the cluster renewal model, but instead of clusters, the heat transfer surface is assumed to be fully covered by a uniform emulsion layer of voidage significantly higher than ε_{mf}. The predictions from this model are in reasonable agreement with experimental findings (Luan et al., 1999).

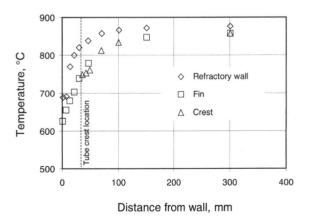

Figure 30 Temperature profile at fin, crest of tube of membrane tube wall, and refractory wall from Golriz (1992). U_g = 3.6 m/s, d_p = 270 μm, T_b = 860°C. Dashed line indicates location of tube crest.

7.2.3 Empirical Correlations

Heat transfer coefficients measured in small and large CFB risers are approximately proportional to the square root of suspension density, despite differences in operating conditions. Hence, in general form,

$$h_{\text{tot}} = a\rho_{\text{susp}}^n + b \qquad (59)$$

Values of a, b, and n from the literature appear in Table 8 with corresponding ranges of conditions.

7.3 Radiation

7.3.1 Some Experimental Investigations

Relatively few studies have been carried out at temperatures high enough ($> \sim 600°C$) that radiation plays a significant role. Basu and Konuche (1988) and Wu et al. (1989) found that the radiative heat transfer coefficient increased with increasing temperature, while the contribution of radiation to the total heat transfer increased with decreasing suspension density. In different studies, the radiative component has been found to contribute more than 50% of the total heat transfer at low particle loadings (e.g., Basu and Konuche, 1988), whereas the contribution is ~ 15 to 40% for higher suspension densities (e.g., Han and Cho, 1999; Luan et al., 1999).

7.3.2 Radiative Heat Transfer Models

Packet Models: Models featuring packets constitute the most common type of model. These usually assume that the radiative component is the sum of radiation from clusters and from the dispersed phase. The radiative heat transfer between clusters and the parallel portion of the surface covered by them is generally based on a gray body approximation. This type of model is employed below.

Non uniform Emulsion Models: Luan et al. (1999) treated the wall layer as a nongray, absorbing, emitting, and anisotropically scattering medium. The radiative heat flux at any distance, x, from the wall in the emulsion layer is estimated by integrating the local spectral radiation intensity. Convective and radiative heat transfer are taken as additive; particles are spherical and uniform in size. A stationary thin surface gas layer is taken with its thickness estimated from the correlation of Lints and Glicksman (1994). The solid concentration is assumed to be uniform in the core and non-uniform in the annulus region. The emulsion layer thickness is estimated from Eq. (24), while its voidage distribution is estimated from the correlation of Zhang et al. (1994). For long surfaces the temperature is assumed to vary only in the emulsion layer and to be given (Golriz, 1995) by

$$\frac{T - T_{\text{w}}}{T_{\text{b}} - T_{\text{w}}} = 1 - \left[-0.023\text{Re}_{\text{p}} + 0.094\left(\frac{T_{\text{b}}}{T_{\text{w}}}\right) + 0.294\left(\frac{z}{H}\right) \right]$$
$$\exp\left[-0.0054\left(\frac{x}{d_{\text{p}}}\right) \right]$$

$$(60)$$

For short surfaces, temperature variation in the emulsion layer is ignored. Temperature gradients in the core are ignored in all cases. The gas and particle temperatures are equal locally. The heat transfer surface is a gray, diffuse surface. The emulsion is taken as a nongray, absorbing, emitting, and scattering medium. Radiation is neglected in the vertical direction.

Table 8 Summary of Empirical Correlations of Heat Transfer in CFB Risers[*]

Investigators	Correlation	Suspension density, kg/m^3	Bulk temperature °C
Basu and Nag (1996)	$h_{\text{ov}} = 40\rho_{\text{sus}}^{0.5}$	$5 < \rho_{\text{sus}} < 20\,\text{kg/m}$	$750°C < T_{\text{b}} < 850$
Andersson & Leckner (1992)	$h_{\text{ov}} = 30\rho_{\text{sus}}^{0.5}$	5–80	750–895
Golriz & Sunden (1994b)	$h_{\text{ov}} = 88 + 9.45\rho_{\text{sus}}^{0.5}$	7–70	800–850
Divilio & Boyd (1994)	$h_{\text{c}} = 23.2\rho_{\text{sus}}^{0.55}$	5–500	Cold condition
Andersson (1996)	$h_{\text{ov}} = 70\rho_{\text{sus}}^{0.085}$	$> 2\,\text{kg/m}^3$	637–883
	$h_{\text{ov}} = 58\rho_{\text{sus}}^{0.36}$	$\leq 2\,\text{kg/m}^3$	

[*] For more details see Golriz and Grace (2002).

The radiative heat flux predicted by the model for a riser of 0.152 m square cross section indicates a rapid decrease with increasing distance from the surface, suggesting that only particles close to the wall contribute significantly to the radiative flux. The predictions of the model are in reasonable agreement with experimental results for the conditions investigated.

Flamant et al. (1996) employed a similar method. The predicted radiative heat flux for a 12 MW_{th} 13.5 m high CFB boiler of 1.7 m × 1.4 m cross section indicates that the radiative component contributes 25 to 50% of the total heat exchange for suspension temperatures between 790 and 830°C.

7.4 Recommended Method for Estimating Total Heat Transfer

Here we outline a practical engineering method (Golriz and Grace, 2002) for estimating average suspension-to-wall heat transfer in large CFB risers operating in the fast fluidization flow regime. Thin packets of emulsion, well-mixed in the horizontal direction, descend along the outside, separated from the outer wall by a thin gap, δ_g. Six resistances shown in Fig. 31 are evaluated. The suspension bulk temperature is T_b, while the wall temperature is T_w. R_1 and R_2 in parallel correspond to the portions of the wall that are uncovered at any instant. The other four resistances apply to portions covered by emulsion. R_3 and R_4 denote bulk-to-emulsion resistances, while R_5 and R_6 represent gas gap resistances. R_2, R_4, and R_6 are radiation resistances and need not be considered for T_b and $T_w < \sim 500°C$. The following steps are recommended:

1. Find the gas density (ρ_g), gas thermal conductivity (k_g), gas specific heat (c_{pg}), particle density (ρ_p), particle thermal conductivity (k_p), particle specific heat (c_{pp}), and particle emissivity (e_p) at temperature $(T_b + T_w)/2$ and estimate the wall emissivity (e_w) at the wall temperature, T_w.

2. Estimate the cross-sectional average suspension voidage, ε_{av}, and wall layer thickness, $\delta(z)$, for the height interval of interest, e.g., using the relationships in Sec. 4.1.2 and Eq. (24) above.

3. Estimate the thickness of the gas film at the wall from the Lints and Glicksman (1993) correlation:

$$\delta_g = 0.0282 d_p (1 - \varepsilon_{av})^{-0.59} \qquad (61)$$

4. Estimate the fractional wall coverage by clusters. For this purpose we recommend an equation (Golriz and Grace, 2002) fitted to the data plotted by Glicksman (1997) that includes the influence of column diameter, D, while also approaching the limiting value of 1 in a rational manners as D increases:

$$f = 1 - \exp\left\{-25,000(1 - \varepsilon_{av})\left[1 - \frac{2}{e^{0.5D} + e^{-0.5D}}\right]\right\} \qquad (62)$$

5. The gas convection coefficient (h_{cg}) is evaluated using a standard steady state correlation for internal

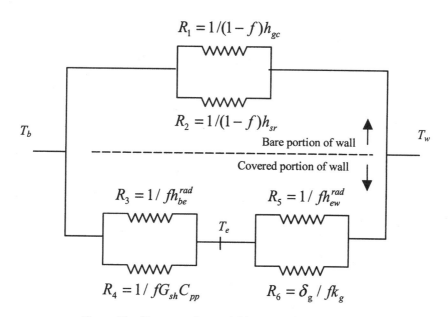

Figure 31 Heat transfer model in network analogy form.

heat transfer for gas flowing alone at a velocity equal to its superficial velocity, U_g, augmented by 15% to account for some particles in the vicinity of the wall during the times when clusters are absent (see Glicksman, 1997) and also for the locally transient nature of the exchange.

6. Estimate the emissivity of the suspension and clusters using the equation of Brewster (1986):

$$e_{\text{susp}} = e_c = \sqrt{\frac{1.5e_p}{(1-e_p)}\left\{\frac{1.5e_p}{(1-e_p)}+2\right\}} - \frac{1.5e_p}{(1-e_p)} \tag{63}$$

A typical value of e_p for silica sand is 0.6 (Flamant et al., 1994).

7. Calculate the radiation heat transfer coefficient corresponding to the fraction of time $(1-f)$ when there is no cluster at a given location on the heat transfer surface. It is assumed that the riser is large enough and the suspension dense enough that only suspension is seen by the bare portions of the wall during these intervals. We can then write

$$h_{sr} = \frac{\sigma\left(T_b^4 - T_w^4\right)}{(T_b - T_w)\left(1/e_{\text{susp}} + 1/e_w - 1\right)} \tag{64}$$

For very small risers or very dilute conditions where exposed portions of the wall can see the opposite wall or side walls, an electric network analogy can be employed (Fang et al., 1995).

8. Estimate the lateral particle flux from a correlation developed by Golriz and Grace (2002):

$$G_{sh} = 0.0225 \ln(\rho_{\text{susp}}) + 0.1093 \tag{65}$$

where the cross-sectional average suspension density, ρ_{susp}, is in units of kg/m^3. The remaining steps depend on the size of the unit and the vertical length of the heat transfer surface.

For large units (e.g., $D > 1$ m) and heat transfer surfaces of height 1.5 m or more,

9a. Evaluate the emulsion temperature as

$$T_e = \frac{T_b R_{56} + T_w R_{34}}{R_{34} + R_{56}} \quad \text{with } R_{34} = \frac{R_3 R_4}{R_3 + R_4}$$

$$\text{and} \quad R_{56} = \frac{R_5 R_6}{R_5 + R_6} \tag{66}$$

10a. Calculate the bulk-to-emulsion and emulsion-to-wall heat transfer coefficients as

$$h_{be}^{\text{rad}} = \frac{2\sigma(T_b^4 - T_e^4)}{(2/e_{\text{sus}} - 1)(T_b - T_e)} \tag{67}$$

$$h_{ew}^{\text{rad}} = \frac{4\sigma(T_e^4 - T_w^4)}{(1/e_{\text{sus}} + 1/e_w - 1)(T_e - T_w)} \tag{68}$$

The coefficient of 2 in equation (67) allows for the radiation shielding effect of the diffuse boundary between the bulk suspension and the emulsion layer facing the wall across the thin gas gap.

11a. We can now estimate the total heat transfer coefficient as

$$h_{\text{tot}} = (h_{gc} + h_{sr})(1-f) + \frac{f}{\dfrac{1}{G_{sh}C_{pp} + h_{be}^{\text{rad}}} + \dfrac{1}{(k_g/\delta_g) + h_{ew}^{\text{rad}}}} \tag{69}$$

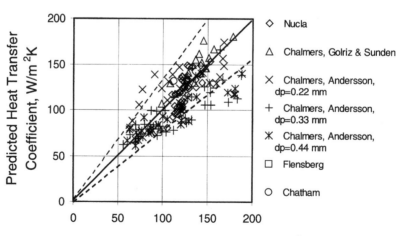

Figure 32 Predicted heat transfer vs. experimental data. Dashed lines indicate ±25% deviations.

As shown in Fig. 32 (see also Golriz and Grace, 2002), this approach gives good predictions for available heat transfer data from large commercial units.

For cases where the surface is shorter than about 1.5 m or where there are appreciable changes in properties such as ρ_{susp} or $\delta(z)$ over the heat exchange surface, both T_e and h_{tot} vary with height. To find local and average heat transfer coefficients, it is then necessary to integrate over the height. For example, a heat balance over a slice of height dz of emulsion (covered portion) for a riser of circular cross section with $\rho_g C_{pg} \ll \rho_p C_{pp}$ yields

$$u_c \frac{D^2 - D_w^2}{4} \rho_p C_{pp}(1 - \varepsilon)\frac{dT_e}{dz} = D\left(\frac{k_g}{\delta_g} + h_{ew}^{rad}\right)(T_e - T_w)$$

$$-D_w\left(G_{sh}C_{pp} + h_{be}^{rad}\right)(T_b - T_e) \tag{70}$$

where $D_w = D - 2\delta(z)$, with $\delta(z)$ given by Eq. (24). Beginning at the top of the heat exchange surface where the boundary condition $T_e = T_b$ is imposed, one can integrate this equation *downwards* to the bottom of the surface. Numerical integration is required for small risers, not only because of the nonlinearity in T_e of the radiation terms in Eqs. (67) and (68), but also because of variations of ρ_{susp} and $\delta(z)$ with z. For the special case where all radiation terms are unimportant and where there is negligible variation of other terms with height, Eq. (70) can be integrated to give

$$T_e = T_{e\infty} + (T_b - T_{e\infty})e^{-z/z_{th}} \tag{71}$$

where

$$T_{e\infty} = \frac{D(k_g/\delta_g)T_w + D_w G_{sh}C_{pp}T_b}{D(k_g/\delta_g) + D_w G_{sh}C_{pp}} \tag{72}$$

and

$$z_{th} = \frac{\rho_p C_{pp}(1 - \varepsilon_e)v_{pw}(D^2 - D_w^2)}{4(Dk_g/\delta_g + D_w G_{sh}C_{pp})} \tag{73}$$

The voidage in the emulsion layer is estimated from an equation proposed by Glicksman (1997):

$$\varepsilon_e = 1 - 1.23(1 - \varepsilon_{av})^{0.54} \tag{74}$$

while the particle velocity of descent in the wall layer, v_{pw}, is calculated from Eq. (23). Once T_e is evaluated, the local time mean heat flux to the wall at the corresponding level is evaluated from

$$q'(z) = (h_{gc} + h_{sr})(1 - f)(T_b - T_w)$$

$$+ \left[(k_g/\delta_g) + h_{ew}^{rad}\right]f(T_e - T_w) \tag{75}$$

The local flux may then be integrated over the height to give the total heat transfer coefficient.

This approach is rational and reasonably accurate except at the leading edge, but it makes a number of assumptions. More comprehensive models are available (e.g., Xie et al., 2003) that adopt more rigorous approaches, but these require considerably more computational effort and are beyond the scope of this chapter.

8 MASS TRANSFER

Mass transfer between gas and particles affects gas–solids contact efficiencies in CFB risers. The mass transfer from a single particle to the suspension in CFB risers has been studied based on the sublimation of naphthalene spheres (Halder and Basu, 1988; Li et al., 1998), dehydration of 2-propanol (Masai et al., 1985), adsorption of CCl_4, naphthalene, H_2S, and NO (Kwauk et al., 1986; van der Ham et al., 1991, 1993; Vollert and Werther, 1994), and heat transfer between a heat pulse and suspension (Dry et al., 1987). For one-dimensional steady-state plug flow of the gas, a mass balance of the adsorbed species in a differential volume element of the reactor (Kwauk et al., 1986; Vollert and Werther, 1994) yields

$$U_g \frac{dC}{dz} + a(1 - \varepsilon_{av})k_m(C - C_s) = 0 \tag{76}$$

where C_s is the concentration of the tracer on particle surface and a is the surface area per unit volume of the particles $(= 6/d_p$ for spherical particles). If both ε_{av} and k_m are assumed to be constant over a short distance dz, Eq. (77) can be integrated to obtain

$$k_m = -\frac{d_p U_g}{6(1 - \varepsilon_{av})z}\ln\left(\frac{C - C_s}{C_0 - C_s}\right) \tag{77}$$

Typical axial profiles of the film–mass transfer coefficient (Kwauk et al., 1986) show that k_m initially increases with height as voidage increases, suggesting improved gas/solids contacting as the dense bed gives way to a more dilute region. This k_m is much lower than for single particles (Venderbosch et al., 1998), as estimated using the well-known Ranz–Marshall equation, and is generally of the same order or even below that in gas–solids bubbling fluidized beds (Halder and Basu, 1988; Vollert and Werther, 1994). A particle surface efficiency, defined as the effective particle surface-to-volume ratio when the film–mass transfer coefficient is calculated based on the Ranz–Marshall correlation, was only 0.01 to 0.13% for FCC particles and decreased with increasing solids concentration (Dry et al., 1987).

Relatively low gas–solids contact efficiency has also been obtained in CFB risers based on the ozone decomposition reaction (Jiang et al., 1991; Kagawa et al., 1991; Sun and Grace, 1992; Ouyang et al., 1993). The low contact efficiency must be mainly attributed to the formation of particle clusters and the core–annulus flow structure in the riser.

To investigate mass transfer between clusters and gas/dilute suspension, Li et al. (1998) constructed artificial clusters by attaching small naphthalene balls to metal wires and weaving them into different shapes. The mass transfer coefficient, k_m, based on a single cluster, was in good agreement with the Ranz–Marshall equation for single particles, and k_m increased with decreasing cluster voidage; k_m was found to be a function of cluster shape with higher values for paraboloid and spherical clusters than for a cylindrical shape. A nonuniform radial distribution of clusters in the riser also led to a lower mass transfer coefficient. For a given number of particles in the riser, the gas–solids mass transfer rate increased as the fraction of dispersed particles increased. Zethræus (1996) found that a simple heterogeneous model that accounts for particle clusters in the bottom dense region and core–annulus flow struc-ture in the upper dilute region can reasonably predict the gas–particle contact efficiency in CFB risers.

9 CIRCULATING FLUIDIZED BEDS AS CHEMICAL REACTORS

9.1 Introduction and Key Considerations

As noted in Sec. 2, most applications of CFB technology involve chemical processes where the principal reactions require solid particles, either as catalyst or as reactant, and a gas. The major factors affecting the performance of CFB reactors are related to the findings covered above:

There are pronounced radial (or lateral) gradients in suspension density, with a greater concentration of particles near the wall than near the axis of the riser.

Because of this greater particle concentration near the wall, concentrations of gaseous reactants tend to be lower at the wall than in the interior of the riser as shown in the examples of Fig. 33.

For the fast fluidization flow regime, particles travel downward near the wall, engendering axial dispersion of both solids and gas.

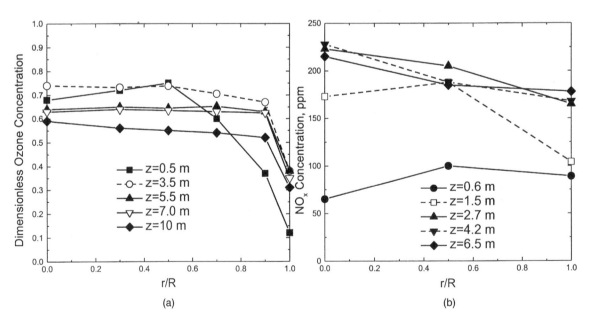

Figure 33 Nonuniform radial profiles of concentration determined experimentally in CFB reactors: (a) concentration of ozone undergoing decomposition in 0.25 m dia × 10.5 m tall riser (Ouyang et al., 1995a); (b) concentration of NO_x in 0.15 × 0.15 m × 7.3 m CFB pilot scale combustor (Brereton et al., 1995).

The riser is subject to significant axial gradients in solids holdup, with the extent of variation dependent on such factors as superficial gas velocity (U_g), net solids circulation flux (G_s), and riser overall height/diameter (H/D) or height/width (H/W) ratio.

The region at the bottom of the riser may operate as a turbulent or bubbling bed, with a gradual transition to fast fluidization, dense suspension upflow, or dilute pneumatic conveying.

For some processes, e.g., CFB combustion, extra gas is introduced at nozzles on the walls well above the bottom primary distributor.

The exit geometry at the top of the riser influences solids reflection and holdup in the upper part of the reactor, affecting solids flow patterns, overall solids holdup, and hence mixing.

Some relevant considerations with respect to the solids recirculation system are

The efficiency of cyclones (or other gas–solids separation devices where applicable) may greatly affect the overall process performance.

Additional reaction (desired or undesirable) may occur in the separators and in the return loop.

Table 9 Examples of Reactions Considered in CFB Reactor Studies and Dimensions of Columns Used in the Experimental Studies

Reaction	M = model, E = Expt'l	References	Riser dimensions
A. Catalytic Reactions			
Claus reaction	M	Puchyr et al., 1996, 1997	NA
Fischer–Tropsch process	E	Shingles & McDonald, 1988	0.1 m id × 13 m, then scaled up by factors of 500 & 2.5
Fluid catalytic cracking	M, E	Fligner et al., 1994	0.3 m id × 12 m
	M	Gao et al., 1999	NA
Methanol-to-olefins	M	Schoenfelder et al., 1994	NA
	M	Gupta et al., 1999	NA
Maleic anhydride from n-butane	M	Pugsley et al., 1992	NA
	E, M	Golbig and Werther, 1996	21 mm i.d.
Oxidative coupling of CH_4	M	Pugsley et al., 1996	NA
Ozone decomposition	E	Jiang et al., 1991	0.10 m id × 6.3 m
	M, E	Ouyang & Potter, 1993, 1995a,b	0.25 m id × 10.9 m
	M, E	Schoenfelder et al., 1996	0.4 m id × 15.6 m
	M, E	Mao et al., 1999	0.25 m id × 10.5 m
	E	Bolland & Nicolai, 1999	411 mm id × 8.5 m
Partial oxidation of methane to syngas	M	Pugsley & Malcus, 1997	NA
Steam methane reforming	M	Matsukata et al., 1995	NA
B. Gas-Solid Reactions			
Calcination	E	Legros et al., 1992	0.15 m sq. × 7.3 m
Combustion	M, E	Li et al., 1995	0.20 m square
	M, E	Talukdar & Basu, 1995a,b	0.20 m sq. × 6.5 m
			0.51 m sq. × 14.3 m
			3.96 m sq. × 21.4 m
	M, E	Mattisson & Lyngfelt, 1998	2.5 m^2 × 13.5 m
			11 m^2 × 17 m
			56 m^2 × 33.5 m
	M, E	Hartge et al., 1999	1.6 m sq. × 13.5 m
	M, E	Torii et al., 1999	0.4 m sq. × 20 m
Gasification	M, E	Jennen et al., 1999	0.3 m id × 8 m
Roasting	E	Luckos & den Hoed, 2001	150 mm id × 6.0 m

NA = not applicable.

Table 9 lists reactions/processes featured in CFB reactor studies or in the development of reactor models, together with relevant references. The ozone decomposition reaction is of no commercial interest, but it is convenient for tests since it is essentially first order, irreversible, and able to proceed at room temperature, and since the concentration of ozone can be readily analyzed at low partial pressures. The other reactions are all of commercial interest.

Understanding the hydrodynamics, mixing, and transfer processes in CFB systems is essential to devising successful reactor models. In preparing models, choices must be made. Overly simple models fail to capture critical aspects of reactor behavior, whereas fully comprehensive models that attempt to include every facet that could affect the performance are likely to be reaction-specific, time-consuming, and difficult to apply. Models commonly contain a series of unproven assumptions and/or empirical constants. Models of intermediate complexity that are mechanistic, capturing the major relevant features, are often the most useful compromises for engineering design, optimization, and control. In the longer term, more sophisticated models, e.g., those based on multiphase computational fluid dynamic (CFD) codes, are likely to have increasing impact, but (as noted above) they have not yet been proven to give reliable predictions. Here we consider models in order of increasing sophistication, abbreviating and extending the review of Grace and Lim (1997).

9.2 One-Dimensional Models

One-dimensional models constitute the simplest type of CFB reactor model. Radial (or lateral) gradients

are completely ignored, with the gas and particle composition, as well as all hydrodynamic variables, treated as invariant across the entire cross section at any level. Some such models also assume that the hydrodynamics are uniform in the axial direction, while others account in some manner for the usual decrease in suspension density with height and/or the presence of a relatively dense zone, possibly subject to a different flow regime, at the bottom. Key gas mixing and hydrodynamic assumptions in one-dimensional models are summarized in Table 10. Gas mixing assumptions cover a wide spectrum from plug flow to relatively well mixed.

One-dimensional models are overly simplified except to provide rough estimates. In practice, the pronounced lateral gradients in CFB risers have important consequences for reactions. For example, high particle concentrations in the outer region can lead to elevated local conversions of gas species toward the wall. When there is downflow of solids at the wall (i.e., fast fluidization), concentration gradients can be pronounced, and substantial backmixing occurs adjacent to the wall. In CFB reactors with secondary gas addition above the bottom of the reactor, gradients associated with the wall jets can persist over considerable heights, possibly even to the top of the reactor, with lateral dispersion too slow to obliterate plumes originating from the secondary jets. Lateral temperature gradients can also be significant.

Models that ignore axial gradients are further removed from reality and are likely to require correction factors (e.g., Ouyang et al., 1993) to fit experimental data. Models that account for axial variations in hydrodynamic variables have found a variety of ways of doing so, e.g.,

Table 10 One-Dimensional Steady-State CFB Reactor Models

Authors	Axial dispersion[*]	Axial hydrodynamic gradients
Hastaoglu et al., 1988	PF	None
Gianetto et al., 1990	PF, ADPF	None
Ouyang et al., 1993, 1995b	PF, PM	None
Marmo et al., 1996	PF, ADPF& (PM + PF)	Solids holdup decays exponentially
Lee & Hyppanen, 1989	PF	$\varepsilon_{av} = f(z)$
Pagliolico et al., 1992	PF	$\varepsilon_{av} = f(z)$
Weiss & Fett, 1986	TIS	$\varepsilon_{av} = f(z)$
Muir et al., 1997	TIS	$\varepsilon_{av} = f(z)$ (each cell uniform)
Arena et al., 1995	TIS + PF	Two separate zones
Zhang et al., 1991	BB + TIS	Two separate zones
Jiang et al., 1991	TIS + PF	5 compartments corresponding to regions bounded by ring baffles

[*] ADPF = axially dispersed plug flow; BB = bubbling bed; PF = plug flow; PM = perfect mixing; TIS = well-mixed tanks in series.

Making voidage a function of z, using empirical evidence, as covered in Sec. 4

Introducing well-mixed compartments in series, each compartment having a different solids concentration, in order to follow the concentration profile in an approximate stepwise manner

Assuming flow regime transitions at one or more levels in the riser, with different mixing representations above and below the transition level(s)

While accounting for axial gradients is helpful, neglecting lateral or radial gradients seriously compromises the ability of one-dimensional models to represent CFB reactors.

9.3 Core–Annulus Models

In core–annulus models, the relatively dense outer region that is subject to solids downflow on a time mean basis, is treated as the annulus region, while the more dilute inner dilute-upflow region comprises the core. Individually, the core and annulus are each usually assumed to be one-dimensional, with radially uniform voidages (ε_a and ε_c for the annulus and core, respectively). This type of model was first introduced for gas mixing (Brereton et al., 1988) and has been extended to core–annulus reactor models in a number of studies, as summarized in Table 11.

Some of these models again ignore axial gradients. For an nth-order solid-catalyzed gas phase reaction

with all of the gas assumed to pass through the core and isothermal conditions, mole balances lead (Kagawa et al., 1991; Marmo et al., 1996) to the governing equations

$$\text{Annulus:} \quad \frac{2K_{ca}R_c(C_a - C_c)}{R^2 - R_c^2} + k_n(1 - \varepsilon_a)\,C_a^n = 0$$

(78)

$$\text{Core:} \quad \frac{U_g R^2}{R_c^2}\frac{dC_c}{dz} + \frac{2K_{ca}(C_a - C_c)}{R_c} + k_n(1 - \varepsilon_c)\,C_c^n = 0$$

(79)

K_{ca} is a core-to-annulus (interregion) mass transfer coefficient, with all other mass transfer resistances (e.g., from the bulk to the particle surfaces within each region) neglected. The final term in each of these equations accounts for reaction, with k_n being the nth-order rate constant. The relevant boundary condition is $C_c = C_0$ at $z = 0$, where C_0 is the inlet concentration of the reacting species. For a first-order reaction, it is straightforward to derive an analytical solution,

$$\text{Conversion} = \frac{C_0 - C_{cH}}{C_0} = 1 -$$

$$\exp\left\{-k_1^*\left[\frac{(1 - \varepsilon_a)(1 - \phi_c^2)K'}{k_1^*(1 - \varepsilon_a)(1 - \phi_c) + K'} + (1 - \varepsilon_c)\phi_c^2\right]\right\}$$

(80)

Table 11 Key Features of Some Core–Annulus CFB Reactor Models

| Authors | Axial dispersion | | Axial gradients | Interregion mass transf. | Core radius/ column radius |
	Core	Annulus			
Kagawa et al., 1991	PF	Stagnant	None	Fitted	0.85
Bi et al., 1992	PF	Stagnant	Allows separately for fine & coarse particles	Rapid	Varies with height
Marmo et al., 1996	PF	Stagnant	PF + CA: $\varepsilon_{av} = f(z)$	Fitted	Presumably fitted
Patience & Chaouki, 1993	PF	Stagnant	None	Fitted	Fitted
Werther et al., 1992	PF	Intermittent up and down	None	$ka = 0.23\,\text{s}^{-1}$ by fitting	0.85
Ouyang & Potter, 1994	PF	PF rel. to solids	None	Fitted	NS
Talukdar et al., 1994	PF	PF	TIS (turbt. & FF zones)	Varied, then fitted	NS
Schoenfelder et al., 1994	PF rel. to solids	PF rel. to solids	4 zones in series	From gas mixing expts	Fitted
Kunii & Levenspiel, 1998	PF	Stagnant	CA + 1-D decay zone	To be fitted	To be fitted
Puchyr et al., 1997	PF	PF rel. to descending solids	Ignored	Modified Higbie penetration theory	Werther correlation

CA = core/annulus; FF = fast fluidization; NS = not specified; PF = plug flow; PM = perfect mixing; TIS = tanks-in-series.

where $\phi_c = R_c/R$ (= dimensionless core radius), $k_1^* = k_1 H/U_g$ (= dimensionless first-order rate constant or Damkohler number), and $K' = 2K_{ca}HR_c/(U_gR^2)$ (= dimensionless core–annulus interregion mass transfer coefficient). Equations (78) to (80) are formally similar to the corresponding equations for the two-phase bubbling bed reactor model (Grace, 1986c), with the dense phase of that model replaced by the CFB annulus, and the bubbles by the core region. It is therefore straightforward to apply analytical solutions for cases provided for the earlier model—half-order and consecutive first-order reactions—to the CFB two-region model. For more complex kinetics, numerical solutions are required.

Conversions are plotted in Fig. 34 for four dimensionless interregion mass transfer coefficients ($K' = 0.01$, 0.1, 1.0, and 10) and two average riser voidages ($\varepsilon_{av} = 0.90$ and 0.96). In each case the corresponding regional voidages are obtained from approximate relationships given by Kagawa et al. (1991), i.e., $\varepsilon_a = 2\varepsilon_{av} - 1$ and $\varepsilon_c = 0.4 + 0.6\varepsilon_{av}$; an overall balance on solids holdup then requires $\phi_c = 0.714$. As expected, the conversion increases with increasing kinetic rate constant and with increasing interregion mass transfer. Increasing the column height, decreasing the superficial gas velocity, or decreasing the voidage of either region is also predicted to improve the conver-

sion. Kagawa et al. (1991) found that $K_{ca} = 0.001$ m/s gave the best fit to experimental concentration profiles. Other workers (e.g., White et al., 1992; Zhao, 1992; Patience and Chaouki, 1993; Schlichtaerle et al., 2001) have assigned larger values of K_{ca}, typical values being of the order of 0.01 to 0.1 m/s. An analogy with gas–liquid annular flow (Senior and Brereton, 1992) gives coefficients of the order of 0.01 m/s. Use of a modified Higbie penetration theory leads to similar values. The interregion mass transfer coefficient is important in core–annulus models, but there are no reliable methods for estimating the coefficient. A value of 0.02–0.05 m/s appears to be reasonable for most purposes.

As indicated in Table 11, models that use a core-annulus structure in combination with axial variation in hydrodynamic properties differ widely in their assumptions. Some (e.g., Marmo et al., 1996) impose a separate region at the bottom to account for turbulent or bubbling fluidization in the bottom zone. Others insert one or more additional zones, e.g., to account for smoother transitions from the distributor region to a core–annular structure, or to account for exit effects. Some of the models from the literature featuring multiple zones are shown schematically in Fig. 35.

There are several ways of allowing for axial voidage variations. In some cases, e.g., Pugsley and Berruti, 1996; Gupta and Berruti, 1998; Gupta et al., 1999, a

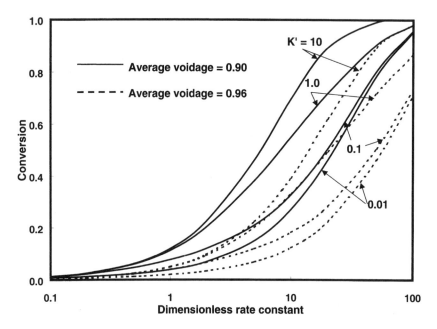

Figure 34 Conversion vs. dimensionless rate constant from simple one-dimensional, two-region, axially uniform model for first-order chemical reaction with different overall voidages of 0.90 and 0.96 and different values of the dimensionless core–annulus interregion mass transfer coefficient.

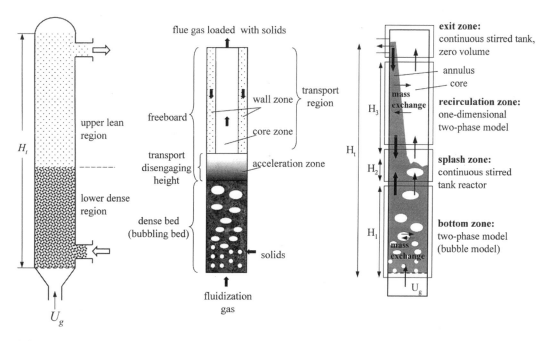

Figure 35 Schematic representation of several multizone models: (a) core–annulus + freeboard approach of Kunii and Levenspiel (1997); (b) four-zone approach of Neidel et al. (1995); (c) five-zone approach of Schoenfelder et al. (1994).

series of equations are written that combine hydrodynamic balances with empirical correlations to give methods for predicting the evolution of voidages and flows as a function of the axial coordinate, z. This approach also incorporates correlations for a slip factor, discussed in Sec. 4.6.2 above.

9.4 More Sophisticated Models

The core–annulus models treated in the previous section improve on the one-dimensional models covered in the preceding section by making some allowance for the difference in behavior between the relatively dense wall region and the dilute core of fast fluidized beds. However, the hydrodynamics are represented in a relatively crude manner. As illustrated in Fig. 33, experimental results show that reactant concentration varies continuously across the entire cross section of the riser, rather than there being a sharp discontinuity at a core–annulus boundary. Hence models are needed that provide for continuous variation across the riser, or at least a greater number of intervals in the lateral direction. Such models include the following:

The model of Werther et al. (1992) still assumes a core–annulus structure, but allows for radial dispersion in the central core, while disregarding gradients in the annular wall zone.

Amos et al. (1993) included radial dispersion in the core zone, but the core occupied the entire cross section, with the solid particles recycling in a separate region beyond the walls of the vessel.

A cluster/gas model (Fligner et al., 1994) with two phases—a cluster phase containing all the particles as spherical clusters, each of voidage ε_{mf}, and a gas phase devoid of particles. Mass transfer is assumed to be controlled by the resistance at the outer surface of the clusters.

Kruse et al. (1995) extended the model of Shoenfelder et al. (1994) to include terms for radial dispersion in both the core and the annulus, with the radial dispersion coefficient assumed to be identical in both zones. These models were further extended by Shoenfelder et al. (1996a) to provide for continuous variations throughout the entire cross section.

Ju (1995) developed a Monte Carlo model for a CFB combustor with the riser divided into 40 cells, 20 in the core and 20 in the annulus. Particles are introduced and tracked one by one, with particles in the core only able to move upwards or sideways subject to the laws of chance, while those in the annulus can only move downwards or sideways. Some particles reaching the top are assumed to be reflected back down the riser. Particles devolatilize and then undergo combustion, allowing the heat release pattern to be approximated by tracking as few as 100 particles.

The probabilistic model of Abba et al. (2002) extends the generalized bubbling turbulent model of Thompson et al. (1999) to cover the fast fluidization flow regime, allowing for smooth transition from bubbling through turbulent fluidization to fast fluidization. A core–annulus structure is assumed at the fast fluidization terminus, with hydrodynamic measurements of regime transitions providing estimates of the relative probabilities of each separate flow regime.

In addition, there are several CFB reactor models that are unrelated to core–annulus models. These are especially appropriate when the solids concentration and gas velocity are high enough that the dense suspension upflow regime is reached, since, in the absence of downflow at the wall, there is no reason to be bound by a core–annulus representation. Among these models are the following:

Cell-based models (e.g., Zhang et al., 1991; Hyppanen et al., 1993; Muir et al., 1997) in which mass and thermal balances allow changes in concentration and energy to be tracked as a function of height and time. Several of these models also consider the recirculation loop in which reactants and energy are recirculated and reintroduced near the bottom of the riser.

Several attempts (e.g., Gao et al., 1999) have been made to extend CFD (computational fluid dynamic) models (see Sec. 4.6.3 above) to include chemical reaction terms.

Note that some of the cited models are intended to provide dynamic predictions that can be used for control purposes or to simulate startup, shutdown, or the response to upsets of CFB reactors.

Virtually all of the above models assume radial symmetry in risers of circular cross section. However, there are various sources of asymmetry in practice:

The solids return system almost invariably returns solids to one side at the base of the reactor.

Similarly the solids feed system is generally asymmetric, feeding solids to one side of the reactor, or, for large units, to a limited number of discrete feed positions on the periphery of the reactor.

The draw-off system at the exit of the reactor is again usually located at one side.

The cross section of many CFB reactors, e.g., of atmospheric pressure combustors and calciners, are rectangular. Corners are regions of additional solids downflow, as discussed in Sec. 4.5.1.

Secondary gas injection at some distance above the bottom of the reactor, commonly found in CFB combustors, causes nonsymmetric jets and plumes that require some distance to dissipate.

9.5 Recommendations

While multiphase CFD models will no doubt make valuable future contributions to predicting the performance of CFB reactors, they are currently unable to make reliable predictions. Simpler reactor models are currently as reliable, while also being much easier to use. For steady-state modeling of reactors operating in the fast fluidization flow regime, core–annulus models coupled with mechanistic hydrodynamic relationships, as in the approach of Gupta et al. (1999), provide reasonable starting points. When the gas velocity is low enough that the flow regime has not fully achieved fast fluidization, then the probabilistic approach pioneered by Abba et al. (2003) provides a useful tool. When radial dispersion and gradients are important, the approach developed by Shoenfelder et al. (1996a) is recommended. This model also appears to provide a rational method for dealing with cases where the flow regime is dense suspension upflow rather than fast fluidization.

NOMENCLATURE

A_c	=	cross-sectional area of column, m^2
Ar	=	Archimedes number, $\rho_g(\rho_p - \rho_g)d_p^3 g/\mu_g^2$
a	=	core–annulus interfacial area per unit volume, m^2/m^3
a, b	=	constants in Eq. (59), $W/m^{2-3n}\,K\,kg^n$ and $W/m^2\,K$, respectively
C	=	tracer concentration, mol/m^3 for gas and kg/m^3 for solids
C_0	=	tracer concentration at injection point, mol/m^3 for gas and kg/m^3 for solids
$\overline{C}$	=	cross-sectional average tracer concentration, mol/m^3
C_c	=	tracer gas or reactant concentration in annulus region, mol/m^3
C_a	=	tracer gas or reactant concentration in core region, mol/m^3
$\overline{C_i}$	=	initial tracer concentration, mol/m^3
C_{pc}	=	specific heat of cluster, $J/kg\,K$
C_{pg}	=	specific heat of gas, $J/kg\,K$
C_{pp}	=	specific heat of particles, $J/kg\,K$

C_s	=	tracer concentration on surface of tracer particles, mol/m^3
D	=	column diameter, equivalent diameter, or width of riser, m
D_{gb}	=	axial gas backmixing coefficient, m^2/s
D_{gd}	=	axial gas dispersion coefficient, m^2/s
D_{ge}	=	effective gas dispersion coefficient, m^2/s
D_{gr}	=	radial gas dispersion coefficient, m^2/s
D_{pr}	=	radial solids dispersion coefficient, m^2/s
D_{pz}	=	axial solids dispersion coefficient, m^2/s
d_p	=	mean particle diameter, m
d_p^*	=	dimensionless particle diameter, $(= \text{Ar}^{1/3})$
$E(\theta)$	=	dimensionless residence time distribution function
e_c	=	emissivity of cluster
e_p	=	emissivity of particles
e_{susp}	=	emissivity of bulk suspension
e_w	=	emissivity of wall surface
Fo	=	Fourier number $= t_c/R_w^2 \Gamma_c$
f	=	fraction of total time during which wall surface is covered by clusters
f_a	=	mass fraction of tracer in the annulus
f_c	=	mass fraction of tracer in the core
g	=	acceleration due to gravity, m/s^2
G_a	=	downward solids flux in annular region, kg/m^{2}s
G_c	=	upward solids flux in core region, kg/m^{2}s
G_s	=	solids net circulation flux or solids entrainment flux, kg/m^{2}s
$G_{s,max}$	=	maximum solids circulation flux, kg/m^{2}s
$G_{s,CA}$	=	saturation carrying capacity, kg/m^{2}s
H	=	total height of riser, m
h_{cc}	=	heat transfer coefficient due to cluster convection, W/m^2 K
h_{cg}	=	heat transfer coefficient due to gas convection when local surface is not covered by a cluster, W/m^2 K
h_{rc}	=	heat transfer coefficient due to cluster radiation, W/m^2 K
h_{rs}	=	heat transfer coefficient due to radiation to dispersed suspension, W/m^2 K
h_{tot}	=	total heat transfer coefficient, W/m^2 K
I_{ac}	=	mass transfer flux from annulus to core per unit length, kg/m^{3}s
I_{ca}	=	mass transfer flux from core to annulus per unit length, kg/m^{3}s
K'	=	dimensionless interregion mass transfer coefficient $= 2K_{ca}HR_c/(UR^2)$
K_{ca}	=	core-to-annulus interregion mass transfer coefficient, m/s
k	=	thermal conductivity, W/m K
k_c	=	effective thermal conductivity of cluster, W/m K
k_g	=	thermal conductivity of gas, W/m K
k_1	=	first-order order rate constant, s^{-1}

k_1^*	=	dimensionless first-order rate constant $= k_1 H/U$
k_m	=	mass transfer coefficient, m/s
k_p	=	thermal conductivity of particles, W/m K
L	=	vertical distance between injection and sampling points, m
L_c	=	average length of travel of cluster along wall, m
M	=	radial gas velocity nonuniformity index
m	=	mass of particles, kg
n	=	order of reaction or exponent in Eq. (59)
P	=	pressure, Pa
Pe_{ge}	=	axial gas Peclet number, $(= U_gL/D_{ge})$
Pe_{gr}	=	radial gas Peclet number, $(= U_gD/D_{gr})$
Pe_{pr}	=	radial particle Peclet number $(= U_gD/D_{pr})$
Pe_{pz}	=	axial particle Peclet number $(= U_gL/D_{pz})$
R	=	radius or hydraulic radius of column, m
r	=	radial coordinate, m
R_c	=	radius of dilute core region, m
Re_p	=	particle Reynolds number, $(= \rho_g U_g d_p/\mu_g)$
R_w	=	thermal gas film conduction resistance $= \delta_g/k_g$, K/W
T	=	emptying time, s
T_b	=	bulk suspension temperature, K
T_c	=	cluster temperature, K
T_w	=	wall temperature, K
t	=	time, s
t_c	=	residence time of cluster at wall, s
U_{CA}	=	type A (accumulative) choking velocity, m/s
U_{CC}	=	type C (classical) choking velocity, m/s
U_{DSU}	=	transition from fast fluidization to dense suspension upflow, m/s
U_g	=	superficial gas velocity, m/s
U_{mf}	=	minimum fluidization velocity, m/s
U_{mp}	=	superficial gas velocity corresponding to minimum pressure drop, m/s
U_{se}	=	onset velocity of fast fluidization at which significant solids entrainment occurs, m/s
U_{slip}	=	slip velocity between gas and particles, m/s
U_{tr}	=	transport velocity, m/s
U^*	=	dimensionless superficial velocity, $(= \text{Re}_g/\text{Ar}^{1/3})$
u_c	=	cluster velocity, m/s
u_g	=	interstitial gas velocity, m/s
u_{gc}	=	average gas velocity in core region, m/s
u_t	=	terminal settling velocity of individual particles, m/s
v_p	=	particle axial velocity, m/s
v_{pa}	=	average particle velocity in the annulus, m/s
v_{pc}	=	average particle velocity in the core, m/s
v_{pw}	=	downwards particle velocity at wall, m/s
W_0	=	total solids inventory in a fluidized bed, kg
x	=	distance from wall, m
z	=	vertical coordinate measured from bottom of riser, m
z_i	=	axial location of tracer injection, m

Greek letters

Γ_c = $\rho_c C_{pc} k_c$, $W^2\,s/m^2 K^4$

δ_g = thickness of gas film between cluster and wall surface, m

ϕ_c = dimensionless core radius = R/R_c

ε = voidage

$[\varepsilon]$ = overall average voidage for the riser as a whole

$\bar{\varepsilon}$ = local time average voidage

$\varepsilon_a, \varepsilon_c$ = voidage of annulus, core

ε_{av} = cross-sectional average voidage at a given height

ε_{cl} = cluster voidage

ε_{CA} = voidage at U_{CA}

ε_c = emulsion layer voidage

ε_{mf} = bed voidage at minimum fluidization

ε_s = solids volume fraction

ε_{sav} = cross-sectional average solids fraction

ε_{sd} = solids volume fraction is dense bottom region

ε_s^* = solids volume fraction at top exit

μ_g = gas viscosity, $kg/m \cdot s$

ρ_b = average bulk bed density, kg/m^3

ρ_c = density of cluster, kg/m^3

ρ_g = gas density, kg/m^3

ρ_p = particle density, kg/m^3

ρ_{susp} = suspension density = $\rho_p(1-\varepsilon) + \rho_g\varepsilon$, kg/m^3

θ = dimensionless time, ($= t/\tau$ or t/t_m)

τ = mean residence time, s

σ = Stefan–Boltzmann constant, $5.67 \times 10^{-8}\,Wm^2 K^4$

σ_t = variance of tracer concentration, s^2

REFERENCES

Abba IA, Grace JR, Bi HT, Thompson ML. An integrated approach to fluidized bed reactor modeling, AIChE J, in press, 2003.

Adams CK. Gas mixing in fast fluidized beds. In: Basu P, Large JF, eds. Circulating Fluidized Bed Technology II. Oxford: Pergamon Press, 1988, pp 299–306.

Ambler PA, Milne BJ, Berruti F, Scott DS. Residence time distribution of solids in a circulating fluidized bed: experimental and modeling studies. Chem Eng Sci 45:2179–2186, 1990.

Amos G, Rhodes MJ, Mineo H. Gas mixing in gas–solids risers. Chem Eng Sci 48:943–949, 1993.

Andersson B-Å. Effect of bed particle size on heat transfer in circulating fluidized bed boilers. Powder Technol 87:239–248, 1996.

Andersson B-Å, Leckner B. Experimental methods of estimating heat transfer in circulating fluidized bed boilers. Int J Heat and Mass Transfer 35:3353–3362, 1992.

Andersson B-Å, Brink K-E, Gustafsson L. Effect of furnace size on CFB wall heat transfer. In: Kwauk M, Li J, eds. Circulating Fluidized Bed Technology V. Beijing: Science Press. 1996, pp 539–544.

Aquillon J, Shakourzadeh K, Guigon P. A new method for local solid concentration measurement in circulating fluidized bed. Powder Technol 86:251–255, 1996.

Arastoopour H, Yang Y. Experimental studies on dilute gas and cohesive particles flow behavior using a laser-Doppler anemometer. In: Potter OE, Nicklin DJ, eds. Fluidization VII. New York: Engineering Foundation, 1992, pp 723–730.

Arena U, Cammarota A, Pistone L. High velocity fluidization behavior of solids in a laboratory scale circulating bed. In: Basu P, ed. Circulating Fluidized Bed Technology. Toronto: Pergamon Press, 1986, pp 119–125.

Arena U, Cammarota A, Marzocchella A, Massimilla L. Solids flow structures in a two-dimensional riser of a circulating fluidized bed. J Chem Eng Japan 22:236–241, 1989.

Arena U, Marzocchella A, Massimilla L, Malandrino A. Hydrodynamics of circulating fluidized beds with risers of different shape and size. Powder Technol 70:237–247, 1992.

Arena U, Chirone R, D'Amore M, Miccio M, Salatino P. Some issues in modelling bubbling and circulating fluidized bed coal combustors. Powder Technol 82:301–316, 1995.

Avidan AA. Fluid catalytic cracking. In: Grace JR, Avidan AA, Knowlton TM, eds. Circulating Fluidized Beds. London: Chapman and Hall, 1997a, pp 466-487.

Avidan AA, Yerushalmi J. Solids mixing in an expanded top fluid bed. AIChE J 31:835–841, 1985.

Azzi M, Turlier P, Large JF, Bernard JR. Use of a momentum probe and gamma-densitometry to study local properties of fast fluidized beds. In: Basu P, Horio M, Hasatani M, eds. Circulating Fluidized Bed Technology III. Oxford: Pergamon Press, 1991, pp 189–194.

Bader R, Findlay J, Knowlton TM. Gas solid flow patterns in a 30.5-cm-diameter circulating fluidized bed. In: Basu P, Large JF, eds. Circulating Fluidized Bed Technology II. Oxford: Pergamon Press, 1988, pp. 123–137.

Bai D, Kato K. Quantitative estimation of solids holdups at dense and dilute regions of circulating fluidized beds. Powder Technol 101:183–190, 1999.

Bai D, Jin Y, Yu ZQ, Yao WH. A study on the performance characteristics of the circulating Fluidized Bed. Chem React Eng and Technol (in Chinese). 3:24–32 1987.

Bai D, Jin Y, Yu ZQ, Zhu JX. The axial distribution of the cross-sectionally averaged voidage in fast fluidized beds. Powder Technol 71:51–58, 1992a.

Bai D, Yi J, Jin Y, Yu ZQ. Residence time distributions of gas and solids in a circulating fluidized bed. In: Potter OE, Nicklin DJ, eds. Fluidization VII, New York: Engineering Foundation, 1992b, pp 195–202.

Bai D, Nakagawa N, Shibuya E, Kinoshita H, Kato K. Axial distribution of solid holdups in binary solids circulating fluidized beds. J Chem Eng Jap 27:271–275, 1994.

Bai D, Issangya AS, Grace JR. Characteristics of gas-fluidized beds in different flow regimes. Ind Eng Chem 38:803–811, 1999.

Balasubramanian N, Srinivasakannan C. Hydrodynamic aspects of a circulating fluidized bed with internals. Ind Eng Chem Res 37:2548–2552, 1998.

Basu P. Heat transfer in high temperature fast fluidized beds. Chem Eng Sci 45: 10, 3123–3136, 1990.

Basu P. Combustion of coal in circulating fluidized bed boilers: a review. Chem Eng Sci 54:5547–5557, 1999.

Basu P, Konuche F. Radiative heat transfer from a fast fluidized bed. In: Basu P, Large JF, eds. Circulating Fluidized Bed Technology II. Oxford: Pergamon Press, 1988, pp 245–254.

Basu P, Nag PK. Heat transfer to walls of a circulating fluidized bed furnace. Chem Eng Sci 51:1–26, 1996.

Beaud F, Louge M. Similarity of radial profiles of solid volume fraction in a circulating fluidized bed. In: Large JF, Laguérie C, eds. Fluidization VIII. New York: Engineering Foundation, 1996, pp 245–253.

Berruti F, Chaouki J, Godfroy L, Pugsley TS, Patience GS. Hydrodynamics of circulating fluidized bed risers: a review. Can J Chem Eng 73:579–602, 1995.

Bi HT. A discussion on the transport velocity. Chem Eng Comm, 189:942–958, 2002.

Bi HT, Fan LS. Regime transitions in gas–solid circulating fluidized beds. Paper #101e, AIChE Annual Meeting, Los Angeles, Nov. 17–22, 1991.

Bi HT, Fan LS. On the existence of turbulent regime in gas–solid fluidization. AIChE J 38:297–301, 1992.

Bi HT, Grace JR. Flow regime maps for gas–solids fluidization and upward transport. Int J Multiphase Flow 21:1229–1236, 1995.

Bi HT, Grace JR. Flow patterns in high-velocity fluidized beds and pneumatic conveying. Can J Chem Eng 77:223–230, 1999.

Bi HT, Zhu JX. Static instability analysis of circulating fluidized beds and the concept of high-density risers. AIChE J 39:1272–1280, 1993.

Bi HT, Jiang P, Jean RH, Fan LS. Coarse-particle effects in a multisolid circulating fluidized bed reactor for catalytic reactions, Chem Eng Sci 47: 3113–3124, 1992.

Bi HT, Grace JR, Zhu JX. On types of choking in pneumatic systems. Int J Multiphase Flow 19:1077–1092, 1993.

Bi HT, Grace JR, Zhu JX. Regime transitions affecting gas–solids suspensions and fluidized beds. Chem Eng Res Des 73:154–161, 1995.

Bi HT, Zhou J, Qin SZ, Grace JR. Annular wall layer thickness in circulating fluidized bed risers. Can J Chem Eng 74:811–814, 1996.

Bi HT, Ellis N, Abba IA, Grace JR. A state-of-the-art review of gas–solids turbulent fluidization. Chem Eng Sci 55:4789–4825, 2000.

Bierl TW, Gajdos LJ, McIver AE, McGovern JJ. Studies in support of recirculating fluidized bed reactors for the processing of coal. DOE Rept. Ex-C-78-01-2449, 1980.

Blumel WP, Kaferstein P, Rummel A, Morl P. Wirbelschichtsysteme. VGB Conference, 1992.

Bolland O, Nicolai R. Describing mass transfer in circulating fluidized beds by ozone decomposition, AIChE Symp Ser 95(321):52–60, 1999.

Bolton LW, Davidson JF. Recirculation of particles in fast fluidized beds. In: Basu P, Large JF, eds. Circulating Fluidized Bed Technology II. Oxford: Pergamon Press, 1988, pp 139–146.

Bowen BD, Fournier M, Grace JR. Heat transfer in membrane waterwalls. Int J Heat Mass Transf 34:1043–1057, 1991.

Brereton CMH. Fluid mechanics of high velocity fluidized beds. PhD diss., University of British Columbia, Vancouver, 1987.

Brereton C. Combustion Performance. In: Grace JR, Avidan AA, Knowlton TM, eds. Circulating Fluidized Beds. Chapman and Hall, London. 1997, pp 369–416.

Brereton CMH, Grace JR. Microstructural aspects of the behavior of circulating fluidized beds. Chem Eng Sci 48:2565–2572, 1993.

Brereton CMH, Grace JR. End effects in circulating fluidized bed hydrodynamics. In: Avidan AA, ed. Circulating Fluidized Bed Technology IV. New York: AIChE, 1994, pp 137–144.

Brereton CHM, Grace JR, Yu J. Axial gas mixing in a circulating fluidized bed. In: Basu P, Large JF, eds. Circulating Fluidized Bed Technology II. Oxford: Pergamon Press, 1988, pp 307–314.

Brereton CMH, Lim CJ, Grace JR, Luckos A, Zhu J. Pitch and coke combustion in a circulating fluidized bed. Fuel 74:1415–1423, 1995.

Brewster MQ. Effective absorptivity and emissivity of particulate media with application to fluidized bed. Trans ASME 108:710–713, 1986.

Brobecker V, Salvaterra A, Ocone R, Geldart D. Solid flux measurements in a circulating fluidized bed riser. In: Kwauk M, Li J, Yang WC, eds. Fluidization X. New York: Engineering Foundation, 2001, pp 269–275.

Bu J, Zhu J. Influence of ring-type internals on axial pressure distribution in circulating fluidized bed. Can J Chem Eng 77:26–34, 1999.

Burkell JJ, Grace JR, Zhao J, Lim CJ. Measurement of solids circulation rates in circulating fluidized beds. In: Basu P, Large JF, eds. Circulating Fluidized Bed Technology II. Oxford: Pergamon Press, 1988, pp 501–509.

Burki V, Hirchberg B, Tuzla K, Chen JC. Thermal development for heat transfer in circulating fluidized beds. AIChE Annual Meeting, St. Louis, 1993.

Caloz Y, Reh L, Cahen C, Evrard R, Piedfer O. Local solids velocities and their fluctuations in CFB units of different sizes. In: Werther J, ed. Circulating Fluidized Bed Technology VI. Frankfurt: DECHEMA, 1999, pp 849–854.

Cao C, Bai D, Jin Y, Yu Z. Mechanism of heat transfer between an immersed vertical surface and suspension in a circulating fluidized bed. In: Kwauk M. Fluidization '94 Science and Technology. Beijing: Science Press, 1994, pp 180–187.

Chang H, Louge M. Fluid dynamic similarity of circulating fluidized beds. Powder Technol 70:259–270, 1992.

Chen J, Cao H, Liu T. Catalyst regeneration in fluid catalytic cracking. In: Kwauk M, ed. Fast Fluidization. Adv Chem Eng 20:389–419, 1994.

Cheng Y, Wei F, Yang G, Jin Y. Inlet and outlet effects on flow patterns in gas–solid risers. Powder Technol 98:151–156, 1998.

Chesonis DC, Klinzing GE, Shah TT, Dassori CG. Hydrodynamics and mixing of solids in a circulating fluidized bed. Ind Eng Chem Res 29:1785–1792, 1990.

Cho YJ, Namkung W, Kim SD, Park S. Effect of secondary air injection on axial solid holdup distribution in a circulating fluidized bed. J Chem Eng Japan 27:158–164, 1996.

Contractor RM. Dupont's CFB technology for maleic anhydride. Chem Eng Sci 54: 5627–5632, 1999.

Coronella CJ, Deng J. A novel method for isokinetic measurement of particle flux within the riser of a circulating fluidized bed. Powder Technol 99:211–219, 1998.

Couturier M, Doucette B, Stevens D, Poolpol S, Razbin V. Temperature, gas concentration and solid mass flux profiles within a large circulating fluidized bed combustor. In: Anthony EJ, ed. Proc 11th Int Conf Fluidized Bed Combustion. New York: ASME, 1991, pp 107–114.

Couturier MF, Steward FR, Poolpol S. Experimental determination of heat transfer coefficients in a 72 MW$_{th}$ circulating fluidized bed boiler. In: Rubow LN, ed. Proc 12th Int Conf on Fluidized Bed Combustion. New York: ASME, 1993, pp 1215–1222.

Cui H, Mostoufi N, Chaouki J. Comparison of measurement techniques of local particle concentration for gas–solid fluidization. In: Kwauk M, Li J, Yang WC, eds. Fluidization X. New York: Engineering Foundation, 2001, pp 779–786.

Davidson JF. Circulating fluidised bed hydrodynamics. Powder Technol 113:249–260, 2000.

de Diego LF, Gayán P, Adánez J. Modelling of the flow structure in circulating fluidized beds. Powder Technol 85:19–27, 1995.

Dean R, Mauleon JL, Letzsch W. TOTAL introduces new FCC process. Oil Gas J 168–176, Oct 11, 1982.

Diguet S. PhD diss., L'Institut National Polytechnique de Toulouse, France, 1996. Data provided in Smolders and Baeyens (2000).

Divilio RJ, Boyd TJ. Practical implication of the effect of solids suspension density on heat transfer in large-scale CFB boilers. In: Avidan AA, ed. Circulating Fluidized Bed Technology IV. New York: AIChE, 1994, pp 334–339.

Drahos J, Cermak J, Guardani R, Schugerl K. Characterization of flow regime transition in a circulating fluidized bed. Powder Technol 56:41–48, 1988.

Dry RJ, Beeby CJ. Applications of CFB technology to gas-solid reactions. In: Grace JR, Avidan AA, Knowlton TM, eds. Circulating Fluidized Beds. London: Chapman and Hall, 1997, pp 441–465.

Dry RJ, White CC. Gas residence-time characteristics in a high-velocity circulating fluidized bed of FCC catalyst. Powder Technol 58:17–23, 1989.

Dry RJ, Christensen IN, White CC. Gas-solids contact efficiency in a high-velocity fluidized bed. Powder Technol 52:243–250, 1987.

Du B, Wei F, Wang F. Effect of particle size on solids mixing in a FCC riser. In: Werther J, ed. Circulating Fluidized Bed Technology VI. Frankfurt: DECHEMA, 1999, pp 405–410.

Fabre A, Molodtsof Y, Koniuta A. Flow structure characterization in a 1 sq m CFB. In: Kwauk M, Li J, eds. Circulating Fluidized Bed Technology V. Beijing: Science Press, 1997, pp 48–53.

Fang Y, Huang J, Zhang J, Wang Y. The study of hydrodynamics of circulating fluidized bed under ambient and high temperature condition. In: Werther J, ed. Circulating Fluidized Bed Technology VI. Frankfurt: DECHEMA, 1999, pp 33–38.

Fang ZH, Grace JR, Lim CJ. Radiative heat transfer in circulating fluidized beds. J Heat Transfer 117: 963–968, 1995.

Flamant G, Lu JD, Variot B. Radiation heat transfer in fluidized beds: a comparison of exact and simplified approaches. J Heat Transfer, 116: 652–659, 1994.

Flamant G, Variot B, Golriz MR, Lu JD. Radiative heat transfer in a pilot scale circulating fluidized bed boiler, In: Kwauk M, Li J, eds. Circulating Fluidized Beds Technology V. Beijing: Science Press, 1996, pp 563–568.

Fligner M, Schipper PH, Sapre AV, Krambeck FJ. Two phase cluster model in riser reactors: impact of radial density distribution on yields. Chem Eng Sci 49:5813–5818, 1994.

Fraley LD, Lin YY, Hsiao KH, Slobakken A. Heat transfer coefficient in a circulating fluidized reactor. ASME paper 83-HT-92, 1983.

Furchi JCL, Goldstein L, Lombardi G, Mohseni M. Experimental local heat transfer in circulating fluidized bed. In: Basu P, Large JF, eds. Circulating Fluidized Bed Technology II. Oxford: Pergamon Press, 1988, pp 13–29.

Fusey I, Lim CJ, Grace JR. Fast fluidization in a concentric circulating bed. In: Basu P, ed. Circulating Fluidized Bed Technology. Oxford: Pergamon Press, 1986, pp 409–416.

Galtier PA, Pontier RJ, Patureaux TE. Near full-scale cold flow model for the R2R catalytic cracking process. In: Grace JR, Shemilt LW, Bergougnou MA, eds. Fluidization VI. New York: Engineering Foundation, 1989, pp 17–24.

Gan N, Jiang DZ, Bai DR, Jin Y, Yu ZQ. Concentration profiles in fast fluidized bed with bluff-body (in Chinese). J Chem Eng Chinese Univ 3:273–277, 1990.

Gao J, Xu C, Lin S, Yang G, Guo Y. Advanced model for turbulent gas–solid flow and reaction in FCC riser reactors. AIChE J 45:1095–1113, 1999.

Gartside RJ. QC—a new reaction system. In: Grace JR, Shemilt LW, Bergougnou MA, eds. Fluidization VI. New York: Engineering Foundation, 1989, pp 25–32.

Gayan P, de Diego LF, Adanez J. Radial gas mixing in a fast fluidized bed. Powder Technol 94:163–171, 1997.

Gianetto A, Pagliolico S, Rovero G, Ruggeri B. Theoretical and practical aspects of circulating fluidized bed reactors for complex chemical systems. Chem Eng Sci 45:2219–2225, 1990.

Gidaspow D. Multiphase Flow and Fluidization: Continuum and Kinetic Theory Descriptions. London: Academic Press, 1994.

Glicksman LR. Circulating fluidized bed heat transfer. In: Basu P, Large JF, eds. Circulating Fluidized Bed Technology II. Oxford: Pergamon Press, 1988, pp 13–29.

Glicksman LR. Heat transfer in circulating fluidized beds. In: Grace JR, Avidan AA, Knowlton TM, eds. Circulating Fluidized Beds. London: Chapman and Hall, 1997, pp 261–311.

Glicksman LR, Westphalen D, Brereton CMH, Grace JR. Verification of the scaling laws for circulating fluidized beds. In: Basu P, Horio M, Hasatani M, eds. Circulating Fluidized Bed Technology III. Oxford: Pergamon Press, 1991, pp 119–124.

Glicksman LR, Hyre M, Woloshun K. Simplified scaling relationships for fluidized beds. Powder Technol 77:177–199, 1993.

Godfroy L, Larachi F, Chaouki J. Position and velocity of a large particle in gas/solid riser using the radioactive particle tracking technique. Can J Chem Eng 77:253–261, 1999a.

Godfroy L, Patience GS, Chaouki J. Radial hydrodynamics in risers. Ind Eng Chem 38:81–89, 1999b.

Golbig KC, Werther J. Selective synthesis of maleic anhydride in a riser-regenerator system. In: Kwauk M, Li J, eds. Circulating Fluidized Bed Technology V. Beijing: Science Press, 1996, pp 394–399.

Golriz MR. Thermal and fluid-dynamic characteristics of circulating fluidized bed boilers, Lic Eng thesis, Chalmers University of Technology, Göteborg, Sweden, 1992.

Golriz MR. Influence of wall geometry on temperature distribution and heat transfer in circulating fluidized bed boilers, In: Avidan AA, ed. Circulating Fluidized Bed Technology IV. New York: AIChE, 1994, pp 693–700.

Golriz MR. An experimental correlation for temperature distribution at the membrane wall of CFB boiler. In: Heinschel KJ, ed. Proc 13th Int Conf Fluidized Bed Combustion. New York: ASME, 1995, pp 499–507.

Golriz MR, Grace JR. Predicting heat transfer in large-scale CFB boilers, In: Grace JR, Zhu J, de Lasa HI, eds. Circulating Fluidized Bed Technology VII. Ottawa: CSChE, 2002, pp 121–128.

Golriz MR, Leckner B. Experimental studies of heat transfer in a circulating fluidized bed boiler. Proc Int Conf Engineering Application Mechanics. Sharif University of Technology, Vol. 3, 1992, pp 167–174.

Golriz MR, Sundén B. A method for temperature measurements in circulating fluidized bed combustors. Exp Thermal Fluid Science 9:274–282, 1994a.

Golriz MR, Sundén B. An experimental investigation of thermal characteristics in a 12 MWth CFB boiler. Exp Heat Transfer J 7:217–233, 1994b.

Grace JR. Contacting modes and behavior classification of gas–solid and other two-phase suspensions. Can J Chem Eng 64:353–363, 1986a.

Grace JR. Heat transfer in circulating fluidized beds. In: Basu P, ed. Circulating Fluidized Bed Technology. Toronto: Pergamon Press, 1986b, pp 63–80.

Grace JR. Modelling and simulation of two-phase fluidized bed reactors. In: de Lasa HI, ed. Chemical Reactor Design and Technology. Dordrecht: Martinus Nijhoff, 1986c, pp 245–289.

Grace JR. Heat transfer in high velocity fluidized beds. In: Hetsroni G, ed. Proc 9th International Heat Transfer Conference, Jerusalem, 1990, pp 329–339.

Grace JR. Influence of riser geometry on particle and fluid dynamics in circulating fluidized beds risers. In: Kwauk M, Li J, eds. Circulating Fluidized Bed Technology. Beijing: Science Press, 1996, pp 16–28.

Grace JR, Lim KS. Reactor modelling for high-velocity fluidized beds. In: Grace JR, Avidan AA, Knowlton TM, eds. Circulating Fluidized Beds. London: Chapman and Hall, 1997, pp 504–524.

Grace JR, Avidan AA, Knowlton TM. eds. Circulating Fluidized Beds. London: Chapman and Hall, 1997.

Grace JR, Issangya AS, Bai DR, Bi HT, Zhu JX. Situating the high-density circulating fluidized beds. AIChE J 45:2108–2116, 1999.

Graf R. Circulating fluidized beds in the flue gas scrubbing—developments, applications and operating experiences in the years 1970 to 2000. In: Werther J, ed. Circulating Fluidized Bed Technology VI. Frankfurt: DECHEMA, 1999, pp 601–607.

Graham RG, Freed BA, Bergougnou MA. Scale-up and commercialization of rapid biomass pyrolysis for fuel and chemical production. In: Klass DL, ed. Energy from Biomass and Wastes XIV. Chicago: Inst Gas Technol, 1991, pp 1091–1104.

Grassler T, Wirth KE. Radial and axial profiles of solids concentration in a high-loaded riser reactor. In: Werther J, ed. Circulating Fluidized Bed Technology VI. Frankfurt: DECHEMA, 1999, pp 65–70.

Griffith AE, Louge MY. The scaling of cluster velocity at the wall of circulating fluidized bed risers. Chem Eng Sci 53:2475–2477, 1998.

Gupta SK, Berruti F. Modeling considerations for large scale high density risers. In: Fan LS, Knowlton TM, eds. Fluidization IX. New York: Engineering Foundation, 1998, pp 205–212.

Gupta SK, Berruti F. Evaluation of the gas–solid suspension density in CFB risers with exit effects. Powder Technol 108:21–31, 2000.

Gupta SK, Pugsley T, Berruti F. A process simulator for circulating fluidized bed chemical reactors. In: Werther J, ed. Circulating Fluidized Bed Technology VI. Frankfurt: DECHEMA, 1999, pp 443–448.

Hage B, Werther J. The guarded capacitance probe—a tool for the measurement of solids flow patterns in laboratory and industrial fluidized bed combustors. Powd Tech 93:235–245, 1997.

Halder PK, Basu P. Mass transfer from a coarse particle to a fast bed of fine solids. AIChE Symp Ser 84(262):58–67, 1988.

Han GY, Cho. Radiative heat transfer in a circulating fluidized bed coal combustor. Powder Technol 102:266–273, 1999.

Harris AT, Thorpe RB, Davidson JF. The measurement of particle residence time distributions in circulating fluidized beds. In: Grace JR, Zhu J, de Lasa HI, eds. Circulating Fluidized Bed Technology VII. Ottawa: CSChE, 2002, pp. 145–152.

Harris BJ, Davidson, JF. Velocity profiles, gas and solids, in fast fluidized beds. In: Potter OE, Nicklin DJ, eds. Fluidization VII. New York: Engineering Foundation, 1992, pp 219–226.

Harris BJ, Davidson JF. A core/annulus deposition model. In: Avidan AA, ed. Circulating Fluidized Bed Technology IV. New York: AIChE, 1994, pp 32–39.

Harris BJ, Davidson JF, Xue Y. Axial and radial variation of flow in circulating fluidized bed risers. In: Avidan AA, ed. Circulating Fluidized Bed Technology IV. New York: AIChE, 1994, pp 103–110.

Hartge EU, Werther J. Gas distributors for circulating fluidized bed combustors. In: Fan LS, Knowlton TM, eds. Fluidization IX. New York: Engineering Foundation, 1998, pp 213–220.

Hartge EU, Li Y, Werther J. Analysis of the local structure of the two phase flow in a fast fluidized bed. In: Basu P, ed. Circulating Fluidized Bed Technology. Toronto: Pergamon, 1986a, pp 153–160.

Hartge EU, Li Y, Werther J. Flow structures in fast fluidized beds. In: Ostergaard K, Sorensen K, eds. Fluidization V. New York: Engineering Foundation, 1986b, pp 345–352.

Hartge EU, Luecke K, Werther J. The role of mixing in the performance of CFB reactors—CFB combustion as an example. In: Werther J, ed. Circulating Fluidized Bed Technology VI. Frankfurt: DECHEMA, 1999, pp 411–416.

Hartge EU, Rensner D, Werther J. Solids concentration and velocity in circulating fluidized beds. In: Basu P, Large JF,

eds. Circulating Fluidized Bed Technology II. Oxford: Pergamon Press, 1988, pp 165–180.

Hastaoglu MA, Berruti F, Hassam MS. A generalized gas–solid reaction model for circulating fluidized beds—an application to wood pyrolysis. In: Basu P, Large JF, eds. Circulating Fluidized Bed Technology II. Oxford: Pergamon Press, 1988, pp 281–288.

Helmrich H, Schurgerl K, Janssen K. Decomposition of $NaHCO_3$ in laboratory and bench scale circulating fluidized bed reactors. In: Basu P, ed. Circulating Fluidized Bed Technology. Oxford: Pergamon Press, 1986, pp 161–166.

Herb B, Dou S, Tuzla K, Chen JC. Solid mass fluxes in circulating fluidized beds. Powder Technol 70:197–205, 1992.

Herbert PM, Gauthier TA, Briens CL, Bergougnou MA. Application of fiber optic reflection probes to the measurement of local particle velocity and concentration in gas–solid flow. Powder Technol 80:243–252, 1994.

Hiltunen M, Myöhänen K. Extremely rapid cooling of hot gas containing sticky components by the means of Fluxflow technology. In: Potter OE, Nicklin DJ, eds. Fluidization VII. New York: Engineering Foundation, 1992, pp 841–848.

Hirschfelder H, Vierrath H. Electricity and syngas from biomass and wastes applying CFB gasification. In: Werther J, ed. Circulating Fluidized Bed Technology VI. Frankfurt: DECHEMA, 1999, pp 459–467.

Horio M, Kuroki H. Three-dimensional flow visualization of dilutely dispersed solids in bubbling and circulating fluidized beds. Chem Eng Sci 49:2413–2421, 1994.

Horio M, Morishita K, Tachibana O, Murata N. Solid distribution and movement in circulating fluidized beds. In: Basu P, Large JF, eds. Circulating Fluidized Bed Technology II. Oxford: Pergamon Press, 1988, pp 147–154.

Horio M, Ishii H, Kobukai Y, Yanmanishi N. A scaling law for circulating fluidized beds. J Chem Eng Japan 22:587–592, 1989.

Horio M, Mori K, Takei Y, Ishii H. Simultaneous gas and solid velocity measurements in turbulent and fast fluidized beds. In: Potter OE, Nicklin DJ, eds. Fluidization VII. New York: Engineering Foundation, 1992, pp 757–762.

Huang W, Zhu J. An experimental investigation on solid acceleration length in the riser of a long circulating fluidized bed. Chin J Chem Eng 9:70–76, 2001.

Husain P, Weber P, Orth A, Eichberger H. Development and experience with fluidized bed based direct reduction processes. In: Werther J, ed. Circulating Fluidized Bed Technology VI. Frankfurt: DECHEMA, 1999, pp 593–598.

Hyppanen T, Lee YY, Kettunen A, Riiali J. Dynamic simulation of a CFB based utility power plant. Proc 12th Intern Fluidized Bed Combustion Conf. New York: ASME, 1993, pp 1121–1127.

Hyre M, Glicksman LR. Experimental investigation of heat transfer enhancements in circulating fluidized beds. Proc 4th ASME/JSME Thermal Eng. Conf, Maui, Hawaii, 1995.

Issangya AS, Bai D, Bi HT, Zhu J, Grace JR. Suspension densities in a high-density circulating fluidized bed riser. Chem Eng Sci 54:5451–5460, 1999.

Issangya AS, Bai DR, Grace JR, Zhu JX. Solids flux profiles in a high density circulating fluidized bed riser. In: Fan LS, Knowlton TM, eds. Fluidization IX. New York: Engineering Foundation, 1998, pp 197–204.

Issangya AS, Grace JR, Bai D, Zhu J. Further measurements of flow dynamics in a high-density circulating fluidized bed riser. Powder Technol 111:104–113, 2000.

Issangya AS, Grace JR, Bai D, Zhu J. Radial voidage variation in CFB risers. Can J Chem Eng 79:279–286, 2001.

Jennen T, Hiller R, Köneke D, Weinspach PM. Modeling of gasification of wood in a circulating fluidized bed. In: Werther J, ed. Circulating Fluidized Bed Technology VI. Frankfurt: DECHEMA, 1999, pp.431–436.

Jestin L, Meyer P, Schmitt G, Morin JX. Transfert de chaleur dans le foyer du lit fluidisé circulant de 125 MW$_e$ de Carlin. Saint-Denis: Elect. de France, 1992.

Jiang P, Fan LS. On the turbulent radial transfer of particles in a CFB riser. In: Werther J, ed. Circulating Fluidized Bed Technology VI. Frankfurt: DECHEMA, 1999, pp 83–88.

Jiang P, Inokuchi K, Jean R-H, Bi HT, Fan LS. Ozone decomposition in a catalytic circulating fluidized bed reactor. In: Basu P, Horio M, Hasatani M, eds. Circulating Fluidized Bed Technology III. Oxford: Pergamon Press, 1991a, pp 557–562.

Jiang P, Bi HT, Jean RH, Fan LS. Baffle effects on performance of catalytic circulating fluidized bed reactors. AIChE J 37:1392–1400, 1991b.

Jiang P, Bi HT, Liang SC, Fan LS. Hydrodynamic behavior of circulating fluidized bed with polymeric particles. AIChE J 40:193–206, 1994.

Jin X, Lu J, Liu Q, Li Y, Xing X, Yue G. Investigation on the heat transfer in a CFB boiler. In: Werther J, ed. Circulating Fluidized Bed Technology VI. Frankfurt: DECHEMA, 1999, pp 355–360.

Jin Y, Yu Z, Qi C, Bai D. The influence of exit structures on the axial distribution of voidage in fast fluidized bed. In: Kwauk M, Kunii D, eds. Fluidization '88 Science and Technology. Beijing: Science Press, 1988, pp 165–173.

Jin Y, Zheng Y, Wei F. State-of-the-art review of downer reactors, In: Grace JR, Zhu J, de Lasa HI, eds. Circulating Fluidized Bed Technology VII. Ottawa: CSChE, 2002, pp 40–60.

Johnsson F, Svensson A, Leckner B. Fluidization regimes in circulating fluidized bed boilers. In: Potter OE, Nicklin DJ, eds. Fluidization VII. New York: Engineering Foundation, 1992, pp 471–478.

Johnsson F, Zhang W, Johnsson H, Leckner B. Optical and momentum probe measurements in a CFB furnace. In:

Kwauk M, Li J, eds. Circulating Fluidized Bed Technology V. Beijing: Science Press, 1997, pp 652–657.

Johnsson F, Sternéus J, He Q, Leckner B. Fluidization regimes in a CFB riser—influence of in-bed tubes. In: Kwauk M, Li J, Yang WC, eds. Fluidization X. New York: Engineering Foundation, 2001, pp 213–220.

Johnsson F, Johansson A, Sternéus J, Leckner B, Hartge EU, Budinger S, Fehr M, Werther J, Sekret R, Bis Z, Nowak W, Noskievic P, Ochodek T, Gadowski J, Jablonski J, Walkowiak R, Kallner P, Engel E, Strömberg L, Hyppänen T, Kettunen A. An investigation of in-furnace processes in a 235 MW$_e$ CFB boiler. In: Grace JR, Zhu J, de Lasa HI, eds. Circulating Fluidized Bed Technology VII. Ottawa: CSChE, 2002, pp 607–614.

Johnsson H, Johnsson F. Measurements of local solids volume-fraction in fluidized bed boilers. Powder Technol 115:13–25, 2001.

Ju DWC. Modelling of steady-state heat release, oxygen profile and temperature profile in circulating fluidized bed combustors. MASc diss., University of British Columbia, Vancouver, Canada, 1995.

Kagawa H, Mineo H, Yamazaki R, Yoshida K. A gas–solid contacting model for fast-fluidized bed. In: Basu P, Horio M, Hasatani M, eds. Circulating Fluidized Bed Technology III. Oxford: Pergamon Press, 1991, pp 551–556.

Karri SBR, Knowlton TM. The effect of pressure on CFB riser hydrodynamics. In: Kwauk M, Li J, eds. Circulating Fluidized Bed Technology V. Beijing: Science Press, 1997, pp 103–109.

Karri SBR, Knowlton TM. Flow direction and size segregation of annulus solids in a riser. In: Fan LS, Knowlton TM, eds. Fluidization IX. New York: Engineering Foundation, 1998, pp 189–196.

Karri SBR, Knowlton TM. A comparison of annulus solids flow direction and radial solids mass flux profiles at low and high mass fluxes in a riser. In: Werther J, ed. Circulating Fluidized Bed Technology VI. Frankfurt: DECHEMA, 1999, pp 71–76.

Kashiwa BA, Yang WC. Computational fluid dynamics for the CFBR: Challenges that lie ahead. In: Grace JR, Zhu J, deLasa H, eds. Circulating Fluidized Bed Technology VII. Ottawa: CSChE, 2002, p. 27–39.

Kato K, Takarada T, Tamura T, Nishino K. Particle hold-up distribution in a circulating fluidized bed. In: Basu P, Horio M, Hasitani M, eds. Circulating Fluidized Bed Technology III. Oxford: Pergamon Press, 1991, pp 145–150.

Kehlenbeck R, Yates J, diFelice R, Hofbauer H, Rauch R. Novel scaling parameter for circulating fluidized beds. AIChE J 47:582–589, 2001.

Kim J. Hydrodynamic behavior of solid transport for a closed loop circulating fluidized bed with secondary air injection. Korean J Chem Eng 16:840–842, 1999.

Koenigsdorff R, Werther J. Gas–solids mixing and flow structure modeling of the upper dilute zone of a circulating fluidized bed. Powder Technol 82:317–329, 1995.

Kojima T, Ishihara KI, Yang GL, Furusawa T. Measurement of solids behavior in a fast fluidized bed. J Chem Eng Japan 22:341–346, 1989.

Konrad K. Dense-phase pneumatic conveying: a review. Powder Technol 49:1–35, 1986.

Kostazos AE, Weinstein H, Graff R. The effect of the location of gas injection on the distribution of gas and catalyst in a riser. In: Fan LS, Knowlton TM, eds. Fluidization IX. New York: Engineering Foundation, 1998, pp 221–228.

Krol S, de Lasa H. CREC-GS-optical probe for particle cluster characterization. In: Fan LS, Knowlton TM, eds. Fluidization IX. New York: Engineering Foundation, 1998, pp 565–572.

Kruse M, Schoenfelder H, Werther J. A two-dimensional model for gas mixing in the upper dilute zone of a circulating fluidized bed. Can J Chem Eng 73:620–634, 1995.

Kunii D, Levenspiel O. Fluidization Engineering. 2d ed. Boston: Butterworth-Heinemann, 1991.

Kunii D, Levenspiel O. Circulating fluidized-bed reactors. Chem Eng Sci 52:2471–2482, 1997.

Kunii D, Levenspiel O. Conversion expressions for FF reactors. In: Fan LS, Knowlton TM, eds. Fluidization IX. New York: Engineering Foundation, 1998, pp 677–684.

Kuroki H, Horio M. The flow structure of a three-dimensional circulating fluidized bed observed by the laser sheet technique. In: Avidan AA, ed. Circulating Fluidized Bed Technology IV. Oxford: Pergamon Press, 1994, pp 77–84.

Kwauk M. ed. Fast Fluidization. Vol. 20, Advances in Chemical Engineering Series. San Diego: Academic Press, 1994.

Kwauk M, Wang ND, Li YC, Chen BY, Shen ZY. Fast fluidization at ICM. In: Basu P, ed. Circulating Fluidized Bed Technology. Oxford: Pergamon Press, 1986, pp 33–62.

Lafanechère L, Jestin L. Study of a circulating fluidized bed furnace behavior in order to scale it up to 600 MW$_e$. In: Heinschel KJ, ed., Proc 13th Int Conf Fluidized Bed Combustion. New York: ASME 2:971–980, 1995.

Leckner B. Heat transfer in circulating fluidized bed boilers. In: Basu P, Horio M, Hasatani M, eds. Circulating Fluidized Bed Technology III. Toronto: Pergamon Press. 1990, pp 27–37.

Leckner B, Golriz MR, Zhang W, Andersson BA, Johnsson F. Boundary layers—first measurements in the 12 MW CFB research plant at Chalmers University. Anthony EJ, ed. Proc 11th Fluidized Bed Comb Conf. New York: ASME, 1991 pp 771–778.

Lee YY, Hyppanen T. Coal combustion model for circulating fluidized bed boilers. Proc 10th Intl Conf Fluidized Bed Combustion. New York: ASME, 1989, pp 753–764.

Lee YY. Design considerations for CFB boilers. In: Grace JR, Avidan AA, Knowlton TM, eds. Circulating Fluidized Beds. London: Chapman and Hall, 1997, pp 417–440.

Legros R, Lim CJ, Brereton CMH, Grace JR. Calcination of lime mud in a circulating fluidized bed. J Pulp Paper Sci 18:71–73, 1992.

Leung LS. Vertical pneumatic conveying: a flow regime diagram and a review of choking versus non-choking systems. Powder Technol 25:185–190, 1980.

Levenspiel, O. Chemical Reaction Engineering 2d ed. New York: John Wiley, 1998.

Levenspiel O, Turner JCR. The interpretation of residence-time experiments. Chem Eng Sci 25:1605–1609, 1970.

Li H, Xia Y, Tung Y, Kwauk M. Micro visualization of two-phase structure in a fast fluidized bed. In: Basu P, Horio M, Hasatani M, eds. Circulating Fluidized Bed Technology III. Oxford: Pergamon Press, 1991, pp 177–182.

Li J, Weinstein H. An experimental comparison of gas back-mixing in fluidized beds across the regime spectrum. Chem Eng Sci 44:1697–1705, 1989.

Li J, Zhang X, Zhu J, Li J. Effects of cluster behavior on gas–solid mass transfer in circulating fluidized beds. In: Fan LS, Knowlton TM, eds. Fluidization IV. New York: Engineering Foundation, 1998, pp 405–412.

Li X, Luo Z, Ni M, Cen K. Modeling sulfur retention in circulating fluidized bed combustors. Chem Eng Sci 50:2235–2242, 1995.

Li Y, Kwauk M. The dynamics of fast fluidization. In: Grace JR, Matsen JM, eds. Fluidization. New York: Plenum Press, 1980, pp 537–544.

Li Y, Wu P. Study on axial gas mixing in a fast fluidized bed. In: Basu P, Horio M, Hasatani M, eds. Circulating Fluidized Bed Technology III. Oxford: Pergamon Press, 1991, pp 581–586.

Li Y, Zhang X. Circulating fluidized bed combustion. In: Kwauk M, ed. Fast Fluidization. Adv Chem Eng 20:333–388, 1994.

Lim KS, Zhou J, Finley C, Grace JR, Lim CJ, Brereton CMH. Cluster descending velocity at the wall of circulating fluidized bed risers. In: Kwauk M, Li J, eds. Circulating Fluidized Bed Technology V. Beijing: Science Press, 1997, pp 218–223.

Lin W, Weinell CF, Hansen PFB, Dam-Johansen K. Hydrodynamics of a commercial scale CFB boiler—study with radioactive particles. Chem Eng Sci 54:5495–5506, 1999.

Lints T, Glicksman LR. The structure of particle clusters near the wall of a circulating fluidized bed. AIChE Symp Ser 89(296):35–52, 1993.

Lischer DJ, Louge M. Optical fiber measurements of particle concentration in dense suspensions: calibration and simulation. Applied Optics 31:5106–5113, 1992.

Liu JZ. Particle and gas dynamics of high density circulating fluidized beds. PhD diss., University of British Columbia, Vancouver, 2001.

Liu JZ, Grace JR, Bi HT, Morikawa H, Zhu JX. Gas dispersion in fast fluidization and dense suspension upflow. Chem Eng Sci 54:5451–5460, 1999.

Lockhart C, Zhu J, Brereton CMH, Lim CJ, Grace JR. Local heat transfer, solid concentration and erosion around membrane tubes in a cold model circulating fluidized bed. Int J Heat Mass Transfer, 38:2403–2410, 1995.

Louge M, Chang H. Pressure and voidage gradients in vertical gas–solid risers. Powder Technol. 60:197–201, 1990.

Louge M, Lischer DJ, Chang H. Measurements of voidage near the wall of a circulating fluidized bed riser. Powder Technol 62:269–276, 1990.

Louge MY, Briscout V, Martin-Letellier S. On the dynamics of pressurized and atmospheric circulating fluidized bed risers. Chem Eng Sci 54:1811–1824, 1999.

Luan W, Lim CJ, Brereton BD, Bowen BD, Grace JR. Experimental and theoretical study of total and radiative heat transfer in circulating fluidized beds. Chem Eng Sci 54:3749–3764, 1999.

Luckos A, den Hoed P. Fluidized-bed roasting of ilmenite. In: Kwauk M, Li J, Yang WC, eds. Fluidization X. New York: Engineering Foundation, 2001, pp 613–620.

Luo G, Yang QL. Axial gas dispersion in fast fluidized beds. Proc 5th National Fluidization Conference, Beijing, 1990, pp 155–158.

Malcus S, Pugsley TS. Investigation of the axial variation of the CFB riser hydrodynamics by means of electrical capacitance tomography (ETC). In: Kwauk M, Li J, Yang WC, eds. Fluidization X. New York: Engineering Foundation, 2001, pp 763–770.

Mao QM, Li XG, Andrews JRG, Potter OE. Catalytic chemical reaction in CFB riser. In: Werther J, ed. Circulating Fluidized Bed Technology VI. Frankfurt: DECHEMA, 1999, pp 521–526.

Marmo L, Manna L, Rovero G. Comparsion among several predictive models for circulating fluidized bed reactors. In: Laguérie C, Large JF, eds. Fluidization VIII. New York: Engineering Foundation, 1996, pp 369–378.

Martin MP, Turlier P, Bernard JR. Gas and solids behavior in cracking circulating fluidized beds. Powder Technol 70:249–258, 1992.

Masai M, Tanaka S, Tomomasa Y, Nishiyama S, Tsuruya S. Reactant-catalyst contact in riser-tube reactor. Chem Eng Comm 34:153–159, 1985.

Mastellone ML, Arena U. The effect of particle size and density on solids distribution along the riser of a circulating fluidized bed. Chem Eng Sci 54:5383–5391, 1999a.

Mastellone ML, Arena U. The influence of particle size and density on the radial gas mixing in the dilute region of the circulating fluidized bed. Can J Chem Eng 77:231–237, 1999b.

Mathiesen V, Solberg T, Hjertager BH. Experimental and computational study of multiphase flow behavior in a circulating fluidized bed. Int J Multiph Flow 26:387–419, 2000.

Matisson T, Lyngfelt A. A sulphur capture model for circulating fluidized-bed boilers. Chem Eng Sci 53:1163–1173, 1998.

Matsen JM. Design and scale-up of CFB catalytic reactors. In: Grace JR, Avidan AA, Knowlton TM, eds. Circulating Fluidized Beds. London: Chapman and Hall, 1997, pp 489–503.

Matsen TM. Mechanisms of choking and entrainment. Powder Technol 32:21–33 1982.

Matsukata M, Matsushita T, Ueyama K. A circulating fluidized bed CH_4 reformer: performance of supported Ni catalysts. Energy and Fuels 9:822–828, 1995.

Matsumoto S, Marakawa M. Statistical analysis of the transition of the flow pattern in vertical pneumatic conveying. Int J Multiphase Flow 13:123–129, 1987.

Matsuno Y, Yamaguchi H, Oka T, Kage H, Higashitani K. The use of optic probes for the measurement of dilute particle concentration: calibration and application to gas-fluidized bed carryover. Powder Technol 36:215–221, 1983.

Mattmann W. Konvektiver Wärmeübergang in vertikalen gas-feststoff-strömungen. Dr Ing diss., Universität Erlagen-Nürnberg, 1991.

Mayer-Schwinning G, Herden H. Use of CFB technology for cleaning off-gases from waste-to-energy plants. In: Werther J, ed. Circulating Fluidized Bed Technology VI. Frankfurt: DECHEMA, 1999, pp 633–637.

Mickley HS, Fairbanks, DF. Mechanism of heat transfer in fluidized beds, AIChE J 1:374–384, 1955.

Miller AL, Gidaspow D. Dense vertical gas–solid flow in a pipe. AIChE J 38:1801–1815, 1992.

Molerus O, Wirth KE. Heat Transfer in Fluidized Beds. London: Chapman and Hall, 1997.

Monceaux L, Azzi M, Molodtsof Y, Large JF. Overall and local characterization of flow regimes in a circulating fluidized bed. In: Basu P, ed. Circulating Fluidized Bed Technology. Toronto: Pergamon Press, 1986, pp 185–191.

Mori S, Kato K, Kobayashi E, Liu D, Hasatani M, Matsuda H, Hattori M, Hirama T, Takeuchi H. Effect of apparatus design on hydrodynamics of circulating fluidized-bed. AIChE Symp Ser 88(289):17–25, 1992.

Muir JR, Brereton CMH, Grace JR, Lim CJ. Dynamic modeling for simulation and control of a circulating fluidized bed combustor. AIChE J 43:1141–1152, 1997.

Nag PK, Moral MNA. Effect of probe size on heat transfer at the wall in circulating fluidized beds. J Inst Energy Res 14:965–974, 1990.

Namkung W, Kim SD. Gas backmixing in a circulating fluidized bed. Powder Technol 99:70–78, 1998.

Namkung W, Kim SD. Radial gas mixing in a circulating fluidized bed. Powder Technol 113:23–29, 2000.

Naruse I, Kumita M, Hattori M, Hasatani M. Characteristics of flow behavior in riser of circulating fluidized bed with internal nozzle. In: Basu P, Horio M, Hasatani M, eds. Circulating Fluidized Bed Technology III. Oxford: Pergamon Press, 1991, pp 195–200.

Neidel W, Gohla M, Borghardt R, Reimer H. Theoretical and experimental investigation of mix-combustion coal/biofuel in circulating fluidized beds. In: Laguérie C, Large JF, eds. Fluidization VIII Preprints, 1995, pp 573–583.

Nicolai R, Reh L. Experimental study on cold-model CFB fluid dynamics. In: Kwauk M, Li J, eds. Circulating Fluidized Bed Technology V. Beijing: Science Press, 1997, pp 54–59.

Nieuwland JJ, Meijer R, Kuipers JAM, van Swaaij WPM. Measurements of solids concentration and axial solids velocity in gas–solid two-phase flows. Powder Technol 87:127–139, 1996.

Nucla Circulating Atmospheric Fluidized Bed Demonstration Project—Final Report, DOE/MC/25137-3046, 1991.

Ohki K, Shirai T. Particle velocity in fluidized bed. In: Keairns, DL ed. Fluidization Technology, Vol. 1, Washington DC: Hemisphere, 1976, pp 95–110.

Ouyang S, Lin J, Potter OE. Ozone decomposition in a 0.254 m diameter circulating fluidized bed reactor. Powder Technology 74:73–78, 1993.

Ouyang S, Lin J, Potter OE. Circulating fluidized bed as a catalytic reactor: experimental study. AIChE J 41:1534–1542, 1995a.

Ouyang S, Li XG, Potter OE. Investigation of ozone decomposition in a circulating fluidized bed on the basis of core–annulus model. In: Laguérie C, Large JF, eds. Fluidization VIII Preprints, 1995b, pp 457–465.

Pagliolico S, Tiprigan M, Rovero G, Gianetto A. Pseudo-homogeneous approach to CFB reactor design. Chem Eng Sci 47:2269–2274, 1992.

Pärssinen JH, Zhu J, Yan AJ, Bu JJ. Flow development in a high-flux and long riser. In: Kwauk M, Li J, Yang WC, eds. Fluidization X. New York: Engineering Foundation, 2001, pp 261–268.

Patience GS, Chaouki J. Solids circulation rate determined by pressure drop measurements. In: Basu P, Horio M, Hasatani M, eds. Circulating Fluidized Bed Technology III. Oxford: Pergamon Press, 1991, pp 627–632.

Patience GS, Chaouki J. Gas phase hydrodynamics in the riser of a circulating fluidized bed. Chem Eng Sci 48:3195–3205, 1993.

Patience GS, Chaouki J. Solids hydrodynamics in the fully developed region of CFB risers. In: Laguérie C, Large JF, eds. Fluidization VIII Preprints, 1995, pp 33–40.

Patience GS, Chaouki J, Kennedy G. Solids residence time distribution in CFB reactors. In: Basu P, Horio M, Hasatani M, eds. Circulating Fluidized Bed Technology III. Oxford: Pergamon Press, 1991, pp 599–604.

Patience GS, Chaouki J, Berruti F, Wong R. Scaling considerations for circulating fluidized bed risers. Powder Technol. 72:31–37, 1992.

Pienemann B, Stockhausen W, McKenzie L. Experience with the circulating fluid bed for gold roasting and alumina calcination. In: Potter OE, Nicklin DJ, eds. Fluidization VII. New York: Engineering Foundation, 1992, pp 921–928.

Plass L. Future R&D directions required in fluidization and fluid–particle systems for sustainable development in chemical industries. In: Kwauk M, Li J, Yang WC, eds.

Fluidization X. New York: Engineering Foundation, 2001, pp 1–14.

Puchyr DMJ, Mehrotra AK, Behie LA, Kalogerakis N. Hydrodynamic and kinetic modelling of circulating fluidized bed reactors applied to a modified Claus plant. Chem Eng Sci 51:5251–5262, 1996.

Puchyr DMJ, Mehrotra AK, Behie LA, Kalogerakis N. Modelling a circulating fluidized bed riser reactor with gas–solids downflow at the wall. Can J Chem Eng, 75:317–326, 1997.

Pugsley TS, Berruti F. A predictive hydrodynamic model for circulating fluidized bed risers. Powder Technol 89:57-69, 1996.

Pugsley T, Lapointe D, Hirschberg B, Werther J. Exit effects in circulating fluidized bed risers. Can J Chem Eng 75:1001–1010, 1997.

Pugsley TS, Malcus S. Partial oxidation of methane in a circulating fluidized-bed catalytic reactor. Ind Eng Chem Res 36:4567–4571, 1997.

Pugsley TS, Patience GS, Berruti F, Chaouki J. Modeling the catalytic oxidation of n-butane to maleic anhydride in a CFB reactor. Ind Eng Chem Res 31:2652–2660, 1992.

Qi CM, Fregeau JR, Farag IH. Hydrodynamics of the combustion loop of N-circulating fluidized bed combustion. AIChE Symp Ser 88(289):26–32, 1992.

Qin S, Liu G. Application of optical fibers to measurement and display of fluidized systems. In: Kwauk M, Kunii D, eds. Fluidization: Science and Technology. Beijing: Science Press, 1982, pp 258–267.

Raghunathan K, Ghosh-Dastidar A, Fan LS. High temperature reactor system for study ultrafast gas–solid reactions. Rev Sci Instrum 64(7):1989–1993, 1993.

Ran X, Wang Z, Wei F, Jin Y. Particle velocity and solids fraction profiles of swirling flow in a riser. In: Werther J, ed. Circulating Fluidized Bed Technology VI. Frankfurt: DECHEMA, 1999, pp 131–136.

Reh L. Fluidized bed processing. Chem Eng Prog 67(2):58–63, 1971.

Reh L. The circulating fluid bed reactor—a key to efficient gas/solid processing. In: Basu P, ed. Circulating Fluidized Bed Technology. Toronto: Pergamon Press, 1986, pp 105–118.

Reh L, Li J. Measurement of voidage in fluidized beds by optical probes. In: Basu P, Horio M, Hasatani M, eds. Circulating Fluidized Bed Technology III. Oxford: Pergamon Press, 1991, pp 163–170.

Rhodes MJ. Modeling the flow structure of upward-flowing gas–solid suspensions. Powder Technol 60:27–38, 1990.

Rhodes MJ, Geldart D. The hydrodynamics of re-circulating fluidized beds. In: Basu P, ed. Circulating Fluidized Bed Technology. Oxford: Pergamon Press, 1986, pp 193–200.

Rhodes MJ, Laussmann P. Characterising non-uniformities in gas–particle flow in the riser of a circulating fluidized bed. Powder Technol 72:277–284, 1992.

Rhodes MJ, Hirama T, Cerutti G, Geldart D. Non-uniformities of solids flow in the risers of circulating fluidized beds. In: Grace JR, Bergougnou MA, Shemilt LW, eds. Fluidization VI. New York: Engineering Foundation, 1989, pp 73–80.

Rhodes MJ, Mineo H, Hirama T. Particle motion at the wall of the 305 mm diameter riser of a cold model circulating fluidized bed. In: Basu P, Horio M, Hasatani M, eds. Circulating Fluidized Bed Technology III. Oxford: Pergamon Press, 1991a, pp 171–176.

Rhodes MJ, Zhou S, Hirama T, Cheng H. Effects of operating conditions on longitudinal solids mixing in a circulating fluidized bed riser. AIChE J 37:1450–1458, 1991b.

Rhodes MJ, Wang XS, Cheng H, Hirama T, Gibbs BM. Similar profiles of solids flux in circulating fluidized-bed risers. Chem Eng Sci 47:1635–1643, 1992a.

Rhodes MJ, Mineo H, Hirama T. Particle motion at the wall of a circulating fluidized bed. Powder Technol 70:207–214, 1992b.

Rhodes MJ, Sollaart M, Wang XS. Flow structures in a fast fluid bed. Powder Technol 99:194–200, 1998.

Saberi B, Shakourzadeh K, Guigon P. Local solid concentration measurement by fiber optics: application to circulating fluidized beds. Trans I Chem Eng 76:748–752, 1998.

Schlichthaerle P, Werther J. Axial pressure profiles and solids concentration distributions in the CFB bottom zone. Chem Eng Sci 54:5485–5493, 1999.

Schlichthaerle P, Hartge EU, Werther J. Interphase mass transfer and gas mixing in the bottom zone of a circulating fluidized bed. In: Kwauk M, Li J, Yang WC, eds. Fluidization X. New York: Engineering Foundation, 2001, pp 549–556.

Schmidt HW. Advanced circulating fluid bed technology for alumina calcination. In: Werther J, ed. Circulating Fluidized Bed Technology VI. Frankfurt: DECHEMA, 1999, pp 587–592.

Schoenfelder H, Werther J, Hinderer J, Keil F. A multi-stage model for the circulating fluidized bed reactor. AIChE Symp Ser 90(301):92–104, 1994.

Schoenfelder H, Kruse M, Werther J. Two-dimensional model for circulating fluidized bed reactors. AIChE J 42:1875–1888, 1996.

Schouten JC, Zijerveld RC, van den Bleek CM. Scale-up of bottom-bed dynamics and axial solids-distribution in circulating fluidized beds of Geldart-B particles. Chem Eng Sci 54:2103–2112, 1999.

Schügerl K. Experimental comparison of mixing processes in two- and three-phase fluidized beds. Drinkenburg AAH, ed. Proceedings Intern Symp on Fluidization. Amsterdam: Netherlands University Press, 1967, pp 782–794.

Schuurmans HJA. Measurements in a commercial catalytic cracking unit. Ind Eng Chem Proc Des Dev 19:267–271, 1980.

Senior RC, Brereton CHM. Modelling of circulating fluidized bed solids flow and distribution. Chem Eng Sci 47:281–296, 1992.

Senior RC, Grace JR. Integrated particle collision and turbulent diffusion model for dilute gas–solid suspensions. Powder Technol 96:48–78, 1998.

Shen X, Zhou N, Xu Y. Experimental study of heat transfer in a pressurized circulating bed. In: Basu P, Horio M, Hasatani M, eds. Circulating Fluidized Bed Technology III. Oxford: Pergamon Press, 1990, pp 451–456.

Shingles T, McDonald AF. Commercial experience with Synthol CFB reactors. In: Basu P, Large JF, eds. Circulating Fluidized Bed Technology II. Oxford: Pergamon Press, 1988, pp 43–50.

Simons SJR, Seville JPK, Clift R, Gilboy WB, Hosseini-Ashrafi ME. Application of gamma-ray tomography to gas fluidized and spouted beds. In: Beck MS, Campogrande E, Morris M, Williams M, Waterfall RC, eds. Tomographic Techniques for Process Design and Operation. Southampton: Computational Mechanics Publications, 1993, pp. 227–238.

Sinclair JL. Hydrodynamic modeling. In: Grace JR, Avidan AA, Knowlton TM, eds. Circulating Fluidized Beds. London: Chapman and Hall, 1997, pp 149–180.

Sinclair JL, Jackson R. Gas-particle flow in a vertical pipe with particle-particle interactions. AIChE J 35:1473-1486, 1989.

Smith TN. Limiting volume fractions in vertical pneumatic transport. Chem Eng Sci 33:745–749, 1978.

Smolders K, Baeyens J. Overall solids movement and solids residence time distribution in a CFB-riser. Chem Eng Sci 55:4101–4116, 2000.

Sobocinski DJ, Young N, de Lasa HI. New fiber-optic method for measuring velocites of strands and solids hold-up in gas–solids downflow reactors. Powder Technol 83:1–11, 1995.

Squires AM. Origins of the fast fluidized bed. In: Kwauk M, ed. Fast Fluidization. Adv Chem Eng 20:1–37, 1994.

Srivastava A, Agrawal K, Sundaresan S, Reddy Karri SB, Knowlton TM. Dynamics of gas-particle flow in circulating fluidized beds. Powder Technol 100:173–182, 1998.

Sternéus J, Johnsson F, Leckner B, Palchonok GI. Gas and solids flow in circulating fluidized beds—discussion on turbulence. Chem Eng Sci 54:5377–5382, 1999.

Steynberg AP, Shingles T, Dry ME, Jager B, Yukawa Y. Sasol commercial scale experience with Synthol FFB and CFB catalytic Fischer–Tropsch reactors. In: Basu P, Horio M, Hasatani M, eds. Circulating Fluidized Bed Technology III. Oxford: Pergamon Press, 1991, pp 527–532.

Strömberg L. Fast fluidized bed combustion of coal. Proc 7th Int Conf Fluidized Bed Combustion. Philadelphia. 1982, pp 1152–1163.

Sun B, Gidaspow D. Computation of circulating fluidized-bed riser flow for Fluidization VIII benchmark test. Ind Eng Chen Res 38:787–792, 1999.

Sun GL, Grace JR. Effect of particle size distribution in different fluidization regimes. AIChE J 38:716–722, 1992.

Sun G, Chao Z, Fan Y, Shi M. Hydrodynamic behavior in the bottom region of a cold FCC riser. In: Werther J, ed. Circulating Fluidized Bed Technology VI. Frankfurt: DECHEMA, 1999, pp 179–184.

Takeuchi H, Hirama T. Flow visualization in the riser of a circulating fluidized bed. In: Basu P, Horio M, Hasatani M, eds. Circulating Fluidized Bed Technology III. Oxford: Pergamon Press, 1991, pp 177–182.

Takeuchi H, Hirama L, Chiba T, Biswas J, Leung LS. A quantitative regime diagram for fast fluidization. Powder Technol 47:195–199, 1986.

Tallon S, Davies CE. Attenuation of sound in beds of particulate materials and application to measurement of flow rate. In: Fan LS, Knowlton TM, eds. Fluidization IX. New York: Engineering Foundation, 1998, pp 597–604.

Talukdar J, Basu P. Modelling of nitric oxide emission from a circulating fluidized bed furnace. In: Laguérie C, Large JF, eds. Fluidization VIII Preprints, Vol. 2, 1995a, pp 829–837.

Talukdar J, Basu P. A simplified model of nitric oxide emission from a circulating fluidized bed combustor. Can J Chem Eng 73:635–643, 1995b.

Talukdar J, Basu P, Joos E. Senstitivity analysis of a performance predictive model for circulating fluidized bed boiler furnaces. In: Avidan AA, ed. Circulating Fluidized Bed Technology IV. New York: AIChE, 1994, pp 450–457.

Tanner H, Li J, Reh L. Radial profiles of slip velocity between gas and solids in circulating fluidized beds. AIChE Symp Ser 90(301):105–113, 1994.

Thompson ML, Bi HT, Grace JR. A generalized bubbling/turbulent fluidized bed reactor model. Chem Eng Sci 54:2175–2185, 1999.

Torii I, Tagashira K, Arakawa Y, Arima K. One-dimensional steady-state simulation model of CFBC. In: Werther J, ed. Circulating Fluidized Bed Technology VI. Frankfurt: DECHEMA, 1999, pp 437–442.

Tung Y, Li Y, Kwauk M. Radial voidage profiles in a fast fluidized bed. In: Kwauk M, Kunii D, eds. Fluidization '88: Science and Technology. Beijing: Science Press, 1988, pp 139–145.

van Bruegel JW, Stein JJM, deVries RJ. Isokinetic sampling in a dense gas–solids stream. Proc Instn Mech Engrs 184:18–23, 1969.

van der Ham A, Prins G, van Swaaij WPM. Hydrodynamics and mass transfer in a regularly packed circulating fluidized bed. In: Basu P, Horio M, Hasatani M, eds. Circulating Fluidized Bed Technology III. Oxford: Pergamon, 1991, pp 605–611.

van der Ham A, Prins GJ, van Swaaij WPM. Hydrodynamics of a pilot-plant scale regularly packed circulating fluidized bed. AIChE Symp Ser 89(296):53–72, 1993.

van der Meer EH, Thorpe RB, Davidson JF. The influence of exit geometry for a circulating fluidised bed with a square cross-sectional riser. In: Kwauk M, Li J, eds. Circulating Fluidized Bed Technology V. Beijing: Science Press, 1997, pp 575–580.

van der Meer EH, Thorpe RB, Davidson JF. Dimensionless groups for practicable similarity of circulating fluidised beds. Chem Eng Sci 54:5369–5376, 1999.

van Zoonen D. Measurements of diffusional phenomena and velocity profiles in a vertical riser. Proc Symp Interaction betw. Fluids and Particles. London: Instn Chem Engrs, 1962, pp 64–71.

Venderbosch RH, Prins W, van Swaaij WPM. Influence of the local catalyst activity on the conversion in a riser reactor. In: Fan LS, Knowlton TM, eds. Fluidization IX. New York: Engineering Foundation, 1998, pp 669–676.

Vollert J, Werther J. Mass transfer and reaction behavior of a circulating fluidized bed reactor. Chem Eng Technol 17:201–209, 1994.

Wang Q, Zhou J, Tu J, Luo Z, Li X. Residence time in circulating fluidized bed. Proc Fluidized Bed Combustion 1:1–6, 1992.

Wang T, Lin ZJ, Zhu CM, Liu DC, Saxena SC. Particle velocity measurements in a circulating fluidized bed. AIChE J 39:1406–1410, 1993.

Wang XS, Gibbs, BM. Hydrodynamics of a circulating fluidized bed with secondary air injection. In: Basu P, Horio M, Hasatani M, eds. Circulating Fluidized Bed Technology III. Oxford: Pergamon Press, 1991, pp 225–230.

Wang XS, Rhodes MJ, Gibbs BM. Solids flux distribution in a CFB riser operating at elevated temperatures. In: Large JF, Laguérie C, eds. Fluidization VIII. New York: Engineering Foundation, 1996, pp 237–244.

Watanabe T, Chen Y, Hasatani M, Xie Y, Naruse I. Gas-solid heat transfer in riser. In: Basu P, Horio M, Hasatani M, eds. Circulating Fluidized Bed Technology III. Oxford: Pergamon Press, 1991, pp 283–288.

Weeks ST, Rohrer JW. Commercial demonstration of biomass gasification. Tappi J 80(5):147–152, 1997.

Wei F, Yang GQ, Jin Y, Yu ZQ. The characteristic of cluster in a high density circulating fluidized bed. Can J Chem Eng 73:650–655, 1995a.

Wei F, Jin Y, Yu ZQ, Chen W, Mori S. Lateral and axial mixing of the dispersed particles in CFB. J Chem Eng Japan 28:506–510, 1995b.

Wei F, Lu FB, Jin Y, Yu ZQ. Mass flux profiles in a high density circulating fluidized bed. Powder Technol 91:189–195, 1997.

Wei F, Lin H, Yi C, Wang Z, Jin Y. Profiles of particle velocity and solids fraction in a high-density riser. Powder Technol 100:183–189, 1998.

Weimer AW, Gyure DC, Clough DE. Application of a gamma-radiation density gauge for determining hydrodynamic properties of fluidized beds. Powder Technol 44:179–194, 1985.

Weinstein H, Li J. An evaluation of the actual density in the acceleration section of vertical risers. Powder Technol 57:77–79, 1989.

Weinstein H, Feindt HJ, Chen L, Graff RA, Pell M, Contractor M, Jordan SP. Riser gas feed nozzle configuration effects on the acceleration and distribution of solids. In: Large JF, Laguérie C, eds. Fluidization VIII. New York: Engineering Foundation, 1996, pp 263–270.

Weinstein H, Shao M, Schnitzlein M. Radial variation in solid density in high velocity fluidization. In: Basu P, ed. Circulating Fluidized Bed Technology. Toronto: Pergamon Press, 1986, pp 201–206.

Weiss V, Fett FN. Modeling the decomposition of sodium bicarbonate in a circulating fluidized bed reactor. In: Basu P, ed. Circulating Fluidized Bed Technology. Toronto: Pergamon Press, 1986, pp 167–172.

Werdermann CC, Werther J. Heat transfer in large-scale circulating fluidized bed combustors of different sizes. In: Avidan AA, ed. Circulating Fluidized Bed Technology IV. New York: AIChE, 1994, pp 428–435.

Werther J. Fluid mechanics of large-scale CFB units. In: Avidan AA, ed. Circulating Fluidized Bed Technology IV. New York: AIChE, 1994, pp 1–14.

Werther J, Hirschberg B. Solids motion and mixing. In: Grace JR, Avidan AA, Knowlton TM, eds. Circulating Fluidized Beds. London: Chapman and Hall, 1997, pp 119–148.

Werther J, Hartge EU, Kruse M. Gas mixing and interphase mass transfer in the circulating fluidized bed. In: Potter OE, Nicklin DJ, eds. Fluidization VII. New York: Engineering Foundation, 1992, pp 257–264.

Werther J, Hartge EU, Rensner D. Measurement techniques for gas–solid fluidized bed reactors. Int Chem Eng 33:18–26, 1993.

Westphalen D, Glicksman L. Lateral solids mixing measurements in circulating fluidized beds. Powder Technol 82:153–168, 1995.

White CC, Dry RJ, Potter OE. Modelling gas-mixing in a 9 cm diameter circulating fluidized bed. In: Potter OE, Nicklin DJ, eds. Fluidization VII. New York: Engineering Foundation, 1992, pp 265–273.

Win KK, Nowak W, Matsuda H, Hasatani M, Kruse M, Werther J. Radial gas mixing in the bottom part of a multi-solid fluidized bed. J Chem Eng Japan 27:696–698, 1994.

Winter F, Mickal V, Hofbauer H, Brunner C, Aichernig C, Liegl J. Scale-up of CFB-fluid dynamics for Group C particles from the laboratory-scale to the industrial-scale. In: Werther J, ed. Circulating Fluidized Bed Technology VI. Frankfurt: DECHEMA, 1999, pp 907–912.

Wirth KE, Gruber U. Fluid mechanics of circulating fluidized beds with small density ratio of solids to fluid. In: Kwauk M, Li J, eds. Circulating Fluidized Bed Technology V. Beijing: Science Press, 1997, pp 78–83.

Wirth KE, Seiter M, Molerus O. Concentration and velocity of solids in areas close to the walls in circulating fluidized systems. VGB Kraftwerkstechnik 10:824–828, 1991.

Wu RL, C.J. Lim J, Chaouki, JR Grace. Heat transfer from a circulating fluidized bed to membrane water-wall surfaces. AIChE J 33:1888–1893, 1987.

Wu RL, Grace JR, Lim CJ, Brereton CMH. Suspension-to-surface heat transfer in a circulating fluidized bed combustor AIChE J 35:1685–1691, 1989.

Wu RL, Lim CJ, Grace JR, Brereton CMH. Instantaneous local heat transfer and hydrodynamics in a circulating fluidized bed. Int J Heat Mass Transfer 34:2019–2027, 1991.

Xie D, Bowen BD, Grace JR, Lim CJ. Two-dimensional model of heat transfer in circulating fluidized beds. Intern J Heat Mass Transf. In press, 2003.

Xu G, Sun G, Nomura K, Li J, Kato K. Two distinctive variational regions of radial particle concentration profiles in circulating fluidized bed risers. Powder Technol 101:91–100, 1999.

Yang GL, Huang Z, Zhao L. Radial gas dispersion in a fast fluidized bed. In: Kunii D, Toei R, eds. Fluidization. Cambridge: Cambridge University Press, 1983, pp 145–152.

Yang H, Gautam M, Mei J. Gas velocity distribution in a circular circulating fluidized bed riser. Powder Technol 78:221–229, 1994.

Yang WC. A mathematical definition of choking phenomenon and a mathematical model for predicting choking velocity and choking voidage. AIChE J 21:1013–1021, 1975.

Yang WC. From pneumatic transport to circulating fluidized beds: the importance of operating maps. In: Kwauk M, Li J, Yang WC, eds. Fluidization X. New York: Engineering Foundation, 2001, pp 333–340.

Yang YL, Jin Y, Yu ZQ, Wang ZW. Investigation on slip velocity distributions in the riser of dilute circulating fluidized bed. Powder Technol 73:67–73, 1992.

Yerushalmi J, Cankurt NT. Further studies of the regimes of fluidization. Powder Technol 24:187–205, 1979.

Yerushalmi J, Cankurt NT, Geldart D, Liss B. Flow regimes in vertical gas–solid contact systems. AIChE Symp Ser 174(176):1–12, 1978.

Yoshida K, Mineo H. High velocity fluidization. In: Doraiswamy LK, Mujumdar AS, eds. Transport in Fluidized Particle Systems. Amsterdam: Elsevier, 1989, pp 241–285.

Younis HF, Glicksman LR, Hyre MR. Measurement and prediction of the hydrodynamics of binary mixtures in a cold model pressurized circulating fluidized bed. In: Werther J, ed. Circulating Fluidized Bed Technology VI. Frankfurt: DECHEMA, 1999, pp 39–45.

Yousfi Y, Gau G. Aerodynamique de l'écoulement vertical de suspensions concentrées gaz-solide—I. Régimes d'écoulement et stabilité aerodynamique. Chem Eng Sci 29:1939–1946, 1974.

Yu Z, Jin Y. Heat and mass transfer. Adv Chem Eng 20: 203–237, 1994.

Zenz FA. Two-phase fluidized-solid flow. Ind Eng Chem 41:2801–2806, 1949.

Zethræus B. A theoretical model for gas–particle contact efficiency in circulating fluid bed risers. Powder Technol 88:133–142, 1996.

Zhang DZ, VanderHeyden WB. High-resolution three-dimensional numerical simulation of a circulating fluidized bed. Powder Technol 116:133–141, 2001.

Zhang H, Johnston PM, Zhu JX, de Lasa HI, Bergougnou MA. A novel calibration procedure for a fiber optic solids concentration probe. Powder Technol 100:260–272, 1998.

Zhang L, Li TD, Zheng, QY, Lu CD. A general dynamic model for circulating fluidized bed combustion with wide particle size distributions. Proc 11th International Fluidized Bed Combustion Conf. New York: ASME, 1991, pp 1289–1294.

Zhang W, Tung Y, Johnsson F. Radial voidage profiles in fast fluidized beds of different diameters. Chem Eng Sci 46:3045–3052, 1991.

Zhang W, Johnsson F, Leckner B. Characteristics of the lateral particle distribution in circulating fluidized bed boilers. In: Avidan AA, ed. Circulating Fluidized Bed Technology IV. New York: AIChE, 1994, pp 266–273.

Zhang W, Johnsson F, Leckner B. Fluid-dynamic boundary layer in CFB boilers. Chem Eng Sci 50:201–210, 1995.

Zhang W, Johnsson F, Leckner B. Momentum probe and sampling probe for measurement of particle flow properties in CFB boilers. Chem Eng Sci 52:497–509, 1997.

Zhang YF, Arastoopour H, Wegerer DA, Lomas DA, Hemler CL. Experimental and theoretical analysis of gas and particle dispersion in large scale CFB. In: Avidan AA, ed. Circulating Fluidized Bed Technology IV. New York: AIChE, 1993, pp 473–478.

Zhao J. Nitric oxide emissions from circulating fluidized bed combustion, PhD diss., University of British Columbia, Vancouver, Canada, 1992.

Zheng CG, Tung YK, Li HZ, Kwauk M. Characteristics of fast fluidized beds with internals. In: Potter OE, Nicklin DJ, eds. Fluidization VII. New York: Engineering Foundation, 1992, pp 275–284.

Zheng QY, Zhu DH, Feng JK, Fan JH. Some aspects of hydrodynamics in N-shaped circulating fluidized bed combustor. Manaker AM, ed. Proc 10th Intern Conf Fluidized Bed Combustion. New York: ASME, 1989, Vol. 2, pp 157–161.

Zheng QY, Xing W, Lou F. Experimental study on radial gas dispersion and its enhancement in circulating fluidized beds. In: Potter OE, Nicklin DJ, eds. Fluidization VII. New York: Engineering Foundation, 1992, pp 285–293.

Zheng Z, Zhu J, Grace JR, Lim CJ, Brereton CMH. Particle motion in circulating and revolving fluidized beds via microcomputer-controlled colour-stroboscopic photography. In: Potter OE, Nicklin DJ, eds. Fluidization VII. New York: Engineering Foundation, 1992, pp 781–789.

Zhou J. Circulating fluidized bed hydrodynamics in a riser of square cross-section. PhD diss., University of British Columbia, Vancouver, 1995.

Zhou J, Grace JR, Qin S, Brereton CHM, Lim CJ, Zhu J. Voidage profiles in a circulating fluidized bed of square cross-section. Chem Eng Sci 49:3217–3226, 1994.

Zhou J, Grace JR, Lim CJ, Brereton CMH. Particle velocity profiles in a circulating fluidized bed of square cross-section. Chem Eng Sci 50:237–244 1995a.

Zhou J, Grace JR, Lim CJ, Brereton CMH, Qin S, Lim KS. Particle cross-flow, lateral momentum flux and lateral velocity in a circulating fluidized bed. Can J Chem Eng 73:612–619, 1995b.

Zhou J, Grace JR, Brereton CMH, Lim CJ. Influence of wall roughness on the hydrodynamics in a circulating fluidized bed. AIChE J 42:1153–1156, 1996a.

Zhou J, Grace JR, Brereton CMH, Lim CJ. Influence of membrane walls on particle dynamics in a circulating fluidized bed. AIChE J 42:3550–3553, 1996b.

Zhu JX, Yu JQ, Jin Y, Grace JR, Issangya AS. Cocurrent downflow circulating fluidized bed (downer) reactors. Can J Chem Eng 73:612–619, 1995.

20

Other Nonconventional Fluidized Beds

Wen-Ching Yang

Siemens Westinghouse Power Corporation, Pittsburgh, Pennsylvania, U.S.A.

1 INTRODUCTION

In a conventional fluidized bed, fluid is passed through a bed of solids via a distributor plate. At a fluid velocity beyond the minimum fluidization or minimum bubbling velocity, visible bubbles appear. The fluid thus passes through the bed in two phases, the bubble and the emulsion phases. The bubble-induced solids mixing and circulation provide the liquidlike behavior of a bed of otherwise immobile solids. The liquidlike behavior of a fluidized bed allows continuous feeding and withdrawal of bed material. The vigorous mixing of solids in the bed gives rise to a uniform bed temperature even for a highly exothermic or endothermic reaction. This leads to easier control and operation. The advantages of a fluidized bed, compared to other modes of contacting such as a packed bed, are numerous, and they are described in detail in Chapter 3, "Bubbling Fluidized Beds," and Chapter 26, "Liquid–Solids Fluidization." Because of the inherent advantages of fluidized beds, they are widely employed in various industries for both physical and chemical operations.

The conventional fluidized beds also possess some serious deficiencies, however. The bubbles that are responsible for many benefits of a fluidized bed represent the fluid bypassing and reduction of fluid–solids contacting. The rapid mixing of solids in the bed leads to nonuniform solids residence time distribution in the bed. The rigorous solids mixing in the bed also leads to attrition of bed material and increases the bed material

loss from elutriation and entrainment. Thus for many industrial applications, the conventional fluidized beds have been modified to overcome those disadvantages. Those modifications, in many ways, alter substantially the operational characteristics of the fluidized beds and also change the design and engineering of the beds. It is the intent of this chapter to document four of the nonconventional fluidized beds in detail: the spouted bed, the recirculating fluidized bed with a draft tube, the jetting fluidized bed, and the centrifugal fluidized bed.

2 SPOUTED BED

The words spouted bed and spouting were first coined by Mathur and Gishler (1955) at the National Research Council of Canada during the development of a technique for drying wheat. The first extensive assimilation of the literature came from the publication of *Spouted Beds* by Mathur and Epstein (1974). A more recent review can be found in Epstein and Grace (1997).

A classical and conventional spouted bed is shown in Fig. 1. The fluid is supplied only through a centrally located jet. If the fluid velocity is high and the bed is low enough, the fluid stream will punch through the bed as a spout as shown in Fig. 1. The spout fluid will entrain solid particles at the spout–annulus interface and form a fountain above the bed. The spout fluid will also leak through the spout–annulus interface into the annulus to provide aeration for the particles in the

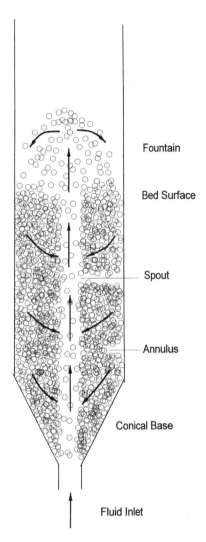

Figure 1 A classical and conventional spouted bed.

velocity and the maximum spoutable bed height requirements.

2.1 Minimum Spouting Velocity

According to Epstein and Grace (1997), the Mathur–Gishler equation shown below remains the simplest equation to estimate the minimum spouting velocity, good to ±15% for cylindrical vessels up to about 0.5 m.

For $D \leq 0.5\,\text{m}$

$$(U_{\text{ms}})_{0.5} = \left(\frac{d_{\text{p}}}{D}\right)\left(\frac{D_{\text{i}}}{D}\right)^{1/3}\sqrt{\frac{2gH(\rho_{\text{p}} - \rho_{\text{f}})}{\rho_{\text{f}}}} \qquad (1)$$

For nonspherical particles, the diameter, d_{p}, to be used in Eq. (1) should be the diameter of a sphere with equal volume. For closely sized near-spherical particles, the volume–surface mean diameter should be employed. For prolate spheroids, the smaller of the two principal dimensions is best used as the particle diameter in Eq (1).

For $D > 0.5\,\text{m}$

$$U_{\text{ms}} = 2.0D(U_{\text{ms}})_{0.5} \qquad D \text{ in meters} \qquad (2)$$

The maximum value of the minimum spouting velocity, U_{m}, occurs when the bed height is at the maximum spoutable bed depth or

$$U_{\text{m}} = \left(\frac{d_{\text{p}}}{D}\right)\left(\frac{D_{\text{i}}}{D}\right)^{1/3}\sqrt{\frac{2gH_{\text{m}}(\rho_{\text{p}} - \rho_{\text{f}})}{\rho_{\text{f}}}} \qquad (3)$$

The U_{m} is closely related to the minimum fluidization velocity, U_{mf}, of the particles by

$$\frac{U_{\text{m}}}{U_{\text{mf}}} = b = 0.9 \qquad \text{to} \qquad 1.5 \qquad (4)$$

A comprehensive review of correlations proposed for the minimum spouting velocity can be found in Mathur and Epstein (1974). A general correlation for the minimum spouting velocity was also suggested by Littman and Morgan (1983).

The minimum spouting velocity has been shown by King and Harrison (1980) to decrease markedly with increasing pressure up to 20 bar. A modified version of the Mathus and Gishler correlation, shown in Eq. (1), was proposed by King and Harrison (1980) as

$$(U_{\text{ms}})_{0.5} = \left[\frac{\rho_{\text{f}}}{\rho_{\text{air}(p=1)}}\right]^{0.2}\left(\frac{d_{\text{p}}}{D}\right)\left(\frac{D_{\text{i}}}{D}\right)^{1/3}\sqrt{\frac{2gH(\rho_{\text{p}} - \rho_{\text{f}})}{\rho_{\text{f}}}} \qquad (5)$$

where $\rho_{\text{air}(p=1)}$ is the density of air at one atmosphere and room temperature.

annulus. The spouted bed is usually constructed as a cylindrical vessel with a conical bottom as shown in Fig. 1 to eliminate the stagnant region. Spouting in a conical vessel has also been employed. Solid particles can be continuously fed into a spouted bed through the concentric jet or into the annulus region and continuously withdrawn from the annulus region, just as in a fluidized bed.

Not all beds are spoutable. For beds with the ratio of nozzle diameter to column diameter D_{i}/D, above a certain critical value, there is no spouting regime. The bed transfers from the fixed bed directly into the fluidized state with increasing gas velocity. To achieve a stable nonpulsating spouted bed, the nozzle-to-particle-diameter ratio, $D_{\text{i}}/d_{\text{p}}$, should be less than 25 or 30. There are also the so-called minimum spouting

For single particle size in a rectangular spouted bed, Anabtawi et al. (1992) showed that the minimum spouting velocity had little difference compared to that in cylindrical beds predictable within the variations from different correlations available for cylindrical beds. A dimensionless correlation for predicting the minimum spouting velocity in a rectangular spouted bed with a mixture of binary particle sizes was recently proposed by Anabtawi (1998).

For a given bed height, the minimum spouting velocity decreases with increasing pressure following the Mathur and Gishler equation shown in Eq. (1). Experimental data obtained by He et al. (1988b) indicated that Mathur and Gishler equation under-predicted the U_{ms} by about 50% for the heavy steel balls and about 39.5% for the large glass beads. The deviation is smaller, about 26%, for the small glass beads. This corresponds to a U_{ms} dependence of gas density of $\rho_f^{-0.36}$ for the steel balls and large glass beads and of $\rho_f^{-0.22}$ for the small glass beads. Thus for correct prediction of the pressure effect on U_{ms}, the exponent on ρ_f in the Mathur and Gishler equation should have different values depending on the particle Reynolds number.

2.2 Maximum Spoutable Bed Height

The maximum spoutable bed height, H_m, can be solved by combining Eqs. (3) and (4) with the Wen and Yu (1966) equation

$$(Re)_{mf} = \frac{d_p U_{mf} \rho_f}{\mu} = 33.7 \left(\sqrt{1 + 35.9 \times 10^{-6} Ar} - 1 \right) \tag{6}$$

to give (Epstein and Grace, 1997)

$$H_m = \frac{D^2}{d_p} \left(\frac{D}{D_i} \right)^{2/3} \frac{568 b^2}{Ar} \\ \times \left(\sqrt{1 + 35.9 \times 10^{-6} Ar} - 1 \right)^2 \tag{7}$$

McNab and Bridgwater (1977) found that Eq. (7) with $b = 1.11$ fitted the experimental data best. For high-temperature applications, take $b = 0.9$ to be conservative (Ye et al., 1992; Li, 1992). More correlations were reviewed in Mathur and Epstein (1974) and Littman et al. (1979).

By differentiating Eq. (7) with respect to Ar, dH_m/dAr, we can find that there is critical particle size at $Ar = 223,000$, calculatable as

$$(d_p)_{crit} = 60.6 \left(\frac{\mu^2}{g(\rho_p - \rho_f)\rho_f} \right)^{1/3} \tag{8}$$

For particles larger than the critical particle shown in Eq. (8), the maximum spoutable bed height, H_m, decreases with increases in particle size, and with particles smaller than the critical particle, H_m increases with increasing d_p. Equation (8) is only good for gas–solids spouting and is not applicable to liquid–solids spouting. For liquid–solids spouting, the maximum spoutable bed height decreases with increases in particle size for all cases. For gas–solids spouting, the critical Archimedes number $Ar = 223,000$ corresponds to a critical Reynolds number of $(Re)_{mf} = 67$, obtainable from the Wen and Yu equation, Eq. (6). Thus for gas–solids spouting, the critical particle diameter is usually in the range of 1.0 to 1.5 mm.

The maximum spoutable bed height, H_m, increases with increasing pressure based on Eq. (7). Thus the region of spoutability is greater at higher pressure. Recent experimental data obtained by He et al. (1998b) provided the evidence for the trend. However, it was found that the McNab and Bridgwater (1977) modification of Eq. (7) over-predicted the H_m up to 36.5% for the steel balls and the smaller glass beads and under-predicted the H_m for the large glass beads by −10.3%.

2.3 Spout Diameter

The average spout diameter has been correlated empirically by McNab (1972) with the equation

$$\overline{D}_s = \frac{2.00 G^{0.49} D^{0.68}}{\rho_b^{0.41}} \tag{9}$$

Equation (9) is a dimensional equation where SI units should be used. Equation (9) is good to ±5.6% at room temperature. He et al. (1998a) employed a fiber-optic probe to measure the spout diameter in a semicylindrical and a full cylindrical spouted bed and found that the presence of the flat front plate in the semicylindrical bed considerably distorted the spout shape. They found that the McNab equation underestimated the spout diameter in a full cylindrical bed by an average of 35.5%. Under pressure, the McNab equation can also introduce error up to 65.5% at a pressure of 343 kPa (He et al., 1998b). Spout diameter tends to increase with increasing pressure.

For elevated temperatures, Wu et al. (1987) suggested the dimensionally consistent semiempirical equation

$$\overline{D_s} = 5.61 \left[\frac{G^{0.433} D^{0.583} \mu^{0.133}}{(\rho_f \rho_b g)^{0.283}} \right] \tag{10}$$

2.4 Voidage Distribution

The voidage at the annulus at minimum spouting, ε_a, can be expected to be close to the voidage at minimum fluidization, ε_{mf}. For narrowly sized spherical particles, this voidage is usually around 0.42. For nonspherical particles, the voidage will be slightly higher. He et al. (1994) found that the voidage in the annulus was somewhat higher than the loose-packed voidage of a packed bed and increased with increasing spouting gas flow.

The voidage in the spout decreases from 1 at the spout inlet almost linearly with height until it reaches around 0.7 at the top of the spout. Models are available for prediction of the voidage distribution for both the annulus and spout (Lim and Mathur, 1978).

2.5 Fluid Flow Distribution

By applying Darcy's law in the annulus of the spouted bed, Mamuro and Hattori (1968) proposed the following equation for the calculation of the superficial fluid velocity, U_a, in the annulus of the spouted bed at height z.

$$\frac{U_a}{U_{mf}} = 1 - \left(1 - \frac{z}{H_m}\right)^3 \tag{11}$$

Equation (11) is expected to apply even for $H \leq H_m$ (Grbravcic et al., 1976) and for annulus Reynolds number one or two orders of magnitude larger than the upper limit of Darcy's law (Epstein et al., 1978).

By continuity at any level of the bed, the gas flow balance can be written as

$$U_{sz}D_s^2 + U_a(D^2 - D_s^2) = UD^2 \tag{12}$$

Thus at any level, the fraction of the total fluid flow passing through the annulus region can be calculated from Eq. (12) if the spout diameter is known. The superficial velocity at minimum spouting, $U = U_{ms}$, can be calculated from Eq. (1). In practice, the operating velocity of a spouted bed is typically 10 to 50% higher than U_{ms}. Under those conditions, the excess gas above that required for minimum spouting can be assumed to pass through the spout as a first approximation.

Typical gas streamlines in the annulus are determined by Epstein and Grace (1997) to be like that shown in Fig. 2. A gas recirculation zone was also

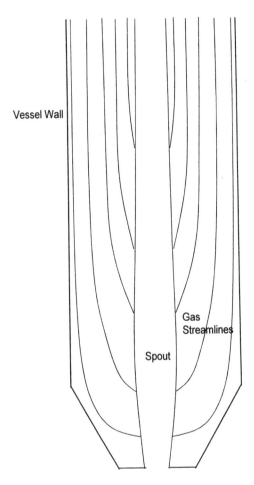

Figure 2 Typical gas streamlines in the annulus of a spouted bed.

observed immediately adjacent to the gas inlet due to the venturi effect above the jet nozzle. The gas flow in the spout is essentially in plug flow and in the annulus, in dispersed plug flow similar to that in a packed bed. Because of the extensive communication of gas between the spout and the annulus regions, the gas residence time distribution deviates substantially from both plug flow and perfect mixing.

2.6 Particle Movement and Fountain Height

Solids continuity dictates that, at any bed level, the particle flowing up in the spout has to be balanced by the particles moving down in the annulus, or

$$W = \rho_p A_s(1 - \varepsilon_s)v_s = \rho_p A_a(1 - \varepsilon_a)v_w \tag{13}$$

Equation (13) neglects the radial variation of particle velocities in the spout and in the annulus. The particle velocity in the annulus is further assumed to equal to the particle velocity observed at the wall, v_w.

He et al. (1994) applied a fiber-optic probe system to measure the vertical particle velocities in the spout, the fountain, and the annulus of a full-column spouted bed. They found that radial profiles of vertical particle velocities in the spout were of near Gaussian distribution rather than parabolic as reported by earlier researchers. In the annulus, vertical particle velocities decreased with decreasing height because of cross-flow of particles from the annulus to the spout. On the contrary, vertical particle velocities increased with decreasing height owing to the reduction of annular cross-sectional area. In the fountain core region, the particles decelerated, attained zero velocity at top of the fountain, and then rained down around the surrounding region. In the radial direction, the particle velocities decreased with increasing radial distance from the axis. The semitheoretical model proposed by Grace and Mathur (1978) as shown in Eq. (14) was found to predict the fountain height quite well.

$$H_F = \varepsilon_{bs}^{1.46} \frac{v_{0\,max}^2}{2g} \frac{\rho_p}{\rho_p - \rho_f} \tag{14}$$

2.7 Pressure Gradient in the Annulus

Experimental data obtained by He et al. (1998b) showed that the longitudinal pressure profile in the annulus was independent of pressure with steel balls and large and small glass beads as bed materials.

2.8 Conical Spouted Beds

Olazar and his associates (1992, 1993a–c, 1995, 1998) studied the design, operation, and performance of a conical spouted bed and found that the conical spouted bed is especially useful for hard-to-handle solids that are irregular in texture or sticky. A conical spouted bed is depicted in Fig. 3. The conical spouted bed exhibits pronounced axial and radial voidage profiles that are quite different from the cylindrical spouted beds.

2.8.1 Minimum Spouting Velocity

$$(Re)_{D_o,ms} = 0.126 Ar^{0.5} \left(\frac{D_b}{D_o}\right)^{1.68} \left[\tan\left(\frac{\theta}{2}\right)\right]^{-0.57} \tag{15}$$

where D_b = upper diameter of the stagnant bed
D_i = diameter of the bed bottom
D_o = diameter of the inlet

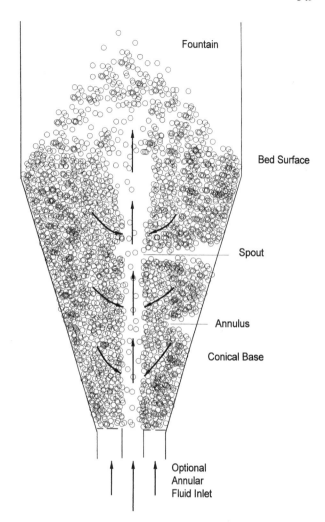

Figure 3 Conical spouted bed.

$(Re)_{Do,ms}$ = Reynolds number at minimum spouting base of D_o
θ = included angle of the cone

2.8.2 Bed Voidage Along the Spout Axis

The bed voidage along the spout axis at $r = 0$ was found to be parabolic and dependent on the system variables employing an optical fiber probe (San Jose et al., 1998). It can be calculated from

$$\varepsilon(0) = 1 - E\left(\frac{z}{H}\right)^2 \tag{16}$$

where E is an empirical parameter varying between 0.3 and 0.6 and is empirically correlated as follows

$$E = 1.20\left(\frac{D_b}{D_o}\right)^{-0.12}\left(\frac{H_o}{D_i}\right)^{-0.97}\left(\frac{U}{U_{ms}}\right)^{-0.7}\theta^{-0.25} \tag{17}$$

where H_o is the height of the stagnant bed

and
$$D_b = D_i + 2H_o \tan\left(\frac{\theta}{2}\right) \qquad (18)$$

2.8.3 Bed Voidage at the Wall

The bed voidage was found by San Jose et al. (1998) to decrease radially toward a minimum at the wall. The bed voidage at the wall can be calculated from the equation

$$\varepsilon(w) = \varepsilon_o \left(1 + \frac{H - z}{H}\right)^{0.5} \qquad (19)$$

where ε_o = loose bed voidage
$\varepsilon(w)$ = bed voidage at the wall
H = height of the developed bed

2.8.4 Bed Voidage Correlation

The general bed voidage correlation can be expressed as

$$\varepsilon = \frac{\varepsilon(0) - \varepsilon(w)}{1 + \exp\left[(r - r_s)/27.81 r_s^{2.41}\right]} + \varepsilon(w) \qquad (20)$$

where r_s is the radius of the spout.

3 RECIRCULATING FLUIDIZED BEDS WITH A DRAFT TUBE

The recirculating fluidized bed with a draft tube concept is illustrated in Fig. 4. This concept was first called a recirculating fluidized bed by Yang and Keairns (1974). Several other names have also been used to describe the same concept: the fluid-lift solids recirculator (Buchanan and Wilson, 1965), the spouted fluid bed with a draft tube (Yang and Keairns, 1983; Hadzismajlovic et al., 1992), the internally circulating fluidized bed (Milne et al., 1992; Lee and Kim, 1992); or simply a circulating fluidized bed (LaNauze, 1976). The addition of a tubular insert, a draft tube, in a spouted fluid bed changes the operational and design characteristics of an ordinary spouted bed. Notably, there is no limitation on the so-called "maximum spoutable bed height." Theoretically, a recirculating fluidized bed with a draft tube can have any bed height desirable. The so-called "minimum spouting velocity" will also be less for a recirculating fluidized bed with a draft tube because the gas in the draft tube is confined and does not leak out along the spout height as in an ordinary spouted bed.

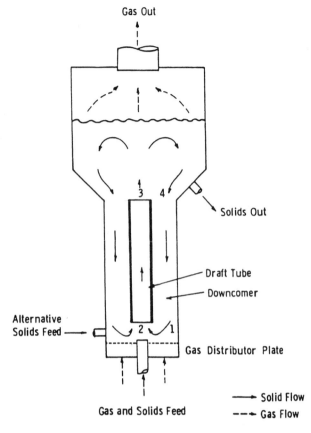

Figure 4 Recirculating fluidized bed with draft tube operated as a pneumatic transport tube.

There is considerably more operation and design flexibility for a recirculating fluidized bed with a draft tube. The downcomer region can be separately aerated. The gas distribution between the draft tube and the downcomer can be adjusted by changing the design parameters at the draft tube inlet. Because the draft tube velocity and the downcomer aeration can be individually adjusted, the solid circulation rate and particle residence time in the bed can be easily controlled. Stable operation over a wide range of operating conditions, solids circulation rates up to 100 metric tons per hour, and a solids loading of 50 (weight of solids/weight of air) in the draft tube have recently been reported by Hadzismajlovic et al. (1992). They used a 95.3 cm diameter bed with a 25 cm diameter draft tube using 3.6 mm polyethylene particles. A detailed discussion of the recirculating fluidized bed with a draft tube is recently published by Yang (1999).

Operating conditions for a recirculating fluidized bed can be flexible as well. The bed height can be lower than the draft tube top or just cover the draft tube top so that a spout can penetrate the bed as in a spouted bed. The bed height can also be substantially higher than the draft tube top, so that a separate fluidized bed exists above the draft tube. Rather than operating the draft tube as a dilute-phase pneumatic transport tube, one can fluidize the solids inside the draft tube at lower velocities to induce the necessary recirculation of the solids. Several studies were conducted in this fashion (Ishida and Shirai, 1975; LaNauze, 1976; LaNauze and Davidson, 1976). The draft tube wall can also be solid or porous, although most of the studies in the literature employ a solid-wall draft tube. Claflin and Fane (1983) reported that a porous draft tube was suitable for applications in thermal disinfestation of wheat where control of particle movement and good gas/solid contacting could be accomplished at a modest pressure drop. The concept can also be employed as a liquid–solids and liquid–gas–solids contacting device (Oguchi and Kubo, 1973). The design and operation of a recirculating fluidized bed with a draft tube are discussed below.

3.1 Draft Tube Operated as a Fluidized Bed

The schematic for this system, where the draft tube is operated as a fluidized bed rather than a dilute phase pneumatic transport tube, is shown in Fig. 5. A mathematical model for the system was developed by LaNauze (1976). The driving force for solids circulation in this case was found to be the density difference between draft tube and downcomer. The solids circulation rate was also found to be affected only by the distance between the distributor and the draft tube and not by the draft tube length or height of bed above it. Because of the lower velocity in the draft tube, the

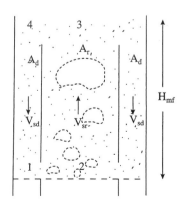

Figure 5 Recirculating fluidized bed with draft tube operated as a fluidized bed.

draft tube diameter tends to be larger compared to when the draft tube is operated in a dilute phase pneumatic transport mode. One disadvantage of operating the draft tube as a fluidized bed is that if the draft tube diameter is too small or the draft tube is too high, the draft tube tends to operate in a slugging bed mode. In fact, the mathematical model developed by LaNauze (1976) described below assumes that the draft tube is a slugging fluidized bed.

The pressure balance for the dense phase in the downcomer in the circulating fluidized system shown in Fig. 5 can be expressed as

$$\Delta P_{1\text{-}4} = \rho_b g H_{mf}(1 - \varepsilon_{bd}) - \frac{\tau_d S_d}{A_d} \tag{21}$$

A similar expression can be written for the pressure balance in the draft tube as

$$-\Delta P_{2\text{-}3} = \rho_b g H_{mf}(1 - \varepsilon_{bf}) + \frac{\tau_r S_r}{A_r} \tag{22}$$

Combining Eqs. (21) and (22), we have

$$\rho_b g H_{mf}(\varepsilon_{br} - \varepsilon_{bd}) = \frac{\tau_d S_d}{A_d} + \frac{\tau_r S_r}{A_r} \tag{23}$$

Experimental evidence indicates that the voidage of the solids flow down the downcomer is close to that of minimum fluidization, thus $\varepsilon_{bd} = 0$.

The bubble voidage in the draft tube, ε_{br}, was calculated on the basis of the velocity of a rising gas slug in a slugging bed relative to its surrounding solids. The total gas superficial velocity in the draft tube, U_{fr}, can be derived as

$$U_{fr} = U_{slug}\varepsilon_{br} + U_{mf} + \frac{V_{sr}\varepsilon_{mf}}{1 - \varepsilon_{mf}} \tag{24}$$

The slug velocity U_{slug}, is defined as the rising velocity of the slug relative to the particle velocity at its nose and can be expressed as

$$U_{slug} = v_p + 0.35\sqrt{gD} \tag{25}$$

and

$$v_p = (U_{fr} - U_{mf}) + V_{sr} \tag{26}$$

Substituting Eq. (25) into (24), we have

$$\varepsilon_{br} = \frac{(U_{fr} - U_{mf}) - V_{sr}\varepsilon_{mf}/(1 - \varepsilon_{mf})}{(U_{fr} - U_{mf}) + V_{sr} + 0.35\sqrt{gD}} \tag{27}$$

The flow rate of particles in the downcomer and the draft tube are related by a mass balance as

$$V_{sr}\rho_p A_r = V_{sd}\rho_p A_d = W_{sd}A_d = W_{sr}A_r \tag{28}$$

By solving Eqs. (24) and (27) simultaneously, the mass flux can be calculated, provided the wall shear stress is known as a function of particle superficial volume flow rate. Botterill and Bessant (1973) have proposed several relationships for shear stress, but these are not general. LaNauze (1976) also proposed a method of measuring this shear stress experimentally.

A similar application of the concept as a slugging lifter of solids was studied by Singh (1978) based on the two-phase theory of fluidization and the properties of slugs.

3.2 Draft Tube Operated as a Pneumatic Transport Tube

Most of the applications for the recirculating fluidized bed with a draft tube operate the draft tube as a dilute phase pneumatic transport tube. Typical experimental pressure drops across the downcomer, ΔP_{1-4}, and the draft tube, ΔP_{2-3}, show that they are essentially similar. Thus successful design of a recirculating fluidized bed with a draft tube requires development of mathematical models for both downcomer and draft tube.

An applicable model is described below for general application.

3.2.1 Downcomer and Draft Tube Pressure Drop

Downcomer Pressure Drop When the downcomer is fluidized, the downcomer pressure drop can be calculated as in an ordinary fluidized bed as

$$\Delta P_{1-4} = L(1 - \varepsilon_d)\rho_p \tag{29}$$

When the downcomer is less than minimally fluidized, the pressure drop can be estimated with a modified Ergun equation substituting gas–solid slip velocities for gas velocities (Yoon and Kunii, 1970), as shown in Eq. (30).

$$\Delta P_{1-4} = \frac{L}{g}\left[150\frac{\mu(U_{gd} + U_{pd})(1 - \varepsilon_d)^2}{d_p^2 \phi_s^2 \varepsilon_d^2} \right.$$
$$\left. +1.75\frac{\rho_f(U_{gd} + U_{pd})^2(1 - \varepsilon_d)}{d_p \phi_s \varepsilon_d} \right] \tag{30}$$

The voidage in the downcomer, ε, can be assumed to be the same as the voidage at minimum fluidization, ε_{mf}, which can be determined in a separate fluidized bed. The agreement between the calculated and the experimental values is usually better than $\pm 10\%$ (Yang and Keairns, 1978a). When the downcomer is not minimally fluidized, the bed voidage depends on the amount of aeration and solid velocity. Use of the

voidage at minimum fluidization is only a first approximation.

Draft Tube Pressure Drop The pressure drop across the draft tube, ΔP_{2-3}, is usually similar to that across the downcomer, ΔP_{1-4}, in magnitude. Thus for a practical design basis, the total pressure drop across the draft tube and across the downcomer can be assumed to be equal. In most operating conditions, the pressure drop at the bottom section of the draft tube has a steep pressure gradient due primarily to the acceleration of solid particles from essentially zero vertical velocity. The acceleration term is especially significant when the solid circulation rate is high or when the draft tube is short.

The model suggested by Yang (1977) for calculating the pressure drop in vertical pneumatic conveying lines can be applied here to estimate the acceleration pressure drop. The acceleration length can be calculated from numerical integration of the equation

$$\Delta L = \int_{U_{pr1}}^{U_{pr2}} \frac{U_{pr}dU_{pr}}{\frac{3}{4}C_{DS}\varepsilon_r^{-4.7}\frac{\rho_f(U_{gr} - U_{pr})^2}{(\rho_p - \rho_f)d_p} - \left(g + \frac{f_p U_{pr}^2}{2D}\right)} \tag{31}$$

The solid friction factor, f_p, can be evaluated with the equation proposed by Yang (1978):

$$f_p = 0.0126\frac{(1 - \varepsilon_r)}{\varepsilon_r^3}\left[(1 - \varepsilon_r)\frac{(Re)_t}{(Re)_p}\right]^{-0.979} \tag{32}$$

The lower limit of integration, U_{pr1}, is derived from

$$W_{sr} = U_{pr}\rho_p(1 - \varepsilon_r) \tag{33}$$

with $\varepsilon_r = 0.5$, and the upper limit, U_{pr2}, by the equation

$$U_{pr} = U_{gr} - U_t\sqrt{\left(1 + \frac{f_p U_{pr}^2}{2gD}\right) \times \varepsilon_r^{4.7}} \tag{34}$$

The total pressure drop in the acceleration region can then be calculated as

$$\Delta P_{2-3} = \int_0^L \rho_p(1 - \varepsilon_r)\,dL + \int_0^L \frac{2f_g\rho_f U_{gr}^2}{gD}\,dL$$
$$+ \int_0^L \frac{f_p\rho_p(1 - \varepsilon_r)U_{pr}^2}{2gD}\,dL$$
$$+ \left[\frac{\rho_p(1 - \varepsilon_r)U_{pr}^2}{g}\right]_{at\cdot L} \tag{35}$$

If the draft tube height is less than the acceleration length, the integration of Eq. (35) is carried out through the whole length of the draft tube. If the draft tube height is larger than the acceleration length, the integration of Eq. (35) is carried out for the total acceleration length, and the extra pressure drop for the rest of the draft tube can then be included to give the total pressure drop in the draft tube. The suggested equations have been applied to actual experimental data satisfactorily (Yang and Keairns, 1978a).

3.2.2 Gas Bypassing Phenomenon

Because of different design and operating parameters, the distribution of the total flow between the draft tube side and the downcomer side can be very different. According to Yang (1999), the important design parameters are the area ratio between the downcomer and the draft tube, the diameter ratio between the draft tube and the draft tube gas supply or the diameter of the solid feeding tube, the distance between the distributor plate and the draft tube inlet, the area ratio of the draft tube gas supply and the concentric solids feeder, and the design of the downcomer gas supply nozzle. In addition to the design parameter, the operating parameters will also affect gas bypassing. The relative strength of the concentric jets of the draft tube gas supply and the solids feeder determines the half-angle of the combined jet, and the jet velocity determines the jet penetration. The jet velocity of the downcomer gas supply nozzles is also important if the jets are horizontal and directed toward the draft tube.

The gas bypassing phenomenon was studied by Stocker et al. (1989) by measuring the differential pressure drops between the draft tube and the downcomer. A more rigorous investigation was conducted by Yang and Keairns (1978a) by injecting gas tracer, carbon dioxide or helium, continuously at different locations and taking gas samples from both the draft tube side and the downcomer side. The actual amounts of gas passing through the draft tube and the downcomer were then obtained by solving mass conservation equations for tracer gas.

Except for the conical distributor plate, no simple gas bypassing relationship exists. No rigorous theoretical model has thus far been proposed. The quantitative gas bypassing information is usually determined experimentally. Qualitatively, the gas bypasses from the draft tube side into the downcomer side for small draft-tube-to-downcomer area ratios and vice versa. Gas bypasses exclusively from the downcomer side to the draft tube side when the distance between

the distributor plate and the draft tube inlet is small. For a conical distributor plate, the angle of the conical plate ($= 45°$ and $60°$) does not seem to affect the gas bypassing characteristics. A more detailed discussion of gas bypassing phenomena is presented in Yang (1999).

3.2.3 Solids Circulation Mechanisms and the Solids Circulation Rate

Both solids circulation mechanisms and the solids circulation rate are important in designing and operating a recirculating fluidized bed with a draft tube. For commercial applications in the area of coating and encapsulation of solid particles, such as in coating of pharmaceutical tablets and in coating seeds for delayed germination and controlling the release rate of fertilizers, the particle residence time and cycle time are important considerations. The performance based on cycle time distribution analysis for coating and granulation was studied by Mann and Crosby (1973, 1975), Mann (1983), and Turton et al. (1999).

Two mechanisms for solids circulation have been observed experimentally (Yang and Keairns, 1978a). High-speed movies (1000 to 1500 frames per second) taken at the inlet and the midsection of the draft tube with a sand bed revealed that solids transport inside the draft tube was not a conventional pneumatic transport, where uniform solid suspension prevailed, but a slugging type transport. The high-speed movies taken at the outlet of a 23 m/s air jet showed that the air jet issuing from the jet nozzle supplying air to the draft tube was composed of bubbles rather than a steady jet. The bubble grew from the mouth of the nozzle until its roof reached the draft tube; then the sudden suction from the draft tube punctured the roof. A continuous stream of dilute solids suspension passed through the roof into the draft tube. Simultaneously, another bubble was initiated. As this bubble grew, it pushed a slug of solids into the draft tube. The high-speed movies taken at the midsection of the draft tube exhibited alternate sections of dilute suspension and solids slug occupying the total cross section of the draft tube.

A steady jet without bubbling can be maintained in a sand bed between the jet nozzle and the draft tube inlet with high jet velocities of the order of 60 m/s and without downcomer aeration. Once the downcomer is aerated, the solids circulation rate increases dramatically and the steady jet becomes a bubbling jet. Apparently, the inward-flowing solids have enough momentum to shear the gas jet periodically into bubbles.

When the polyethylene beads (density $= 907\,\text{kg/m}^3$, average size $= 2800\,\mu\text{m}$) and the hollow epoxy spheres (density $= 210\,\text{kg/m}^3$, average size $= 2800\,\mu\text{m}$) were used as the bed material, a steady jet between the jet nozzle and the draft tube was always observed for all experiments conducted.

Solids circulation rate was found to be strongly affected by the design configuration at the bottom of the draft tube and the downcomer owing to changes in gas bypassing characteristics (Yang and Keairns, 1983). This coupling effect indicates that an understanding of the gas bypassing characteristics is essential. Except for simple cases, the dependency of the gas bypassing characteristics on design and operating parameters are still not amenable to theoretical treatment as discussed earlier. A recent study by Alappat and Rane (1995) on the effects of various design and operational parameters on solids circulation rate essentially affirms the above conclusions.

When only the draft tube gas flow is employed without downcomer aeration, the solid circulation rate depends primarily on the entrainment rate of the jets. Aeration of the downcomer can be provided with greatly increased solid circulation rate. At lower downcomer aeration, the solid circulation rate is essentially similar to that without downcomer aeration. At higher downcomer aeration, however, a substantial increase in solid circulation rate can be realized. Apparently, a minimum aeration in the downcomer is required in order to increase substantially the solid circulation rate.

As expected, the closer the distance between the distributor plate and the draft tube inlet, the lower the solids circulation rate. This is not only because of the physical constriction created by locating the distributor plate too close to the draft tube inlet but also because of the different gas bypassing characteristics observed at different distributor plate locations as discussed earlier. When the distance between the distributor plate and the draft tube inlet becomes large, it can create start-up problems discussed in Yang et al. (1978).

3.3 Design Procedure for a Recirculating Fluidized Bed with a Draft Tube

From the experimental evidence, the design of a recirculating fluidized bed with a draft tube involves the specification of a number of design parameters and an understanding of the coupling effects between the design and the operating variables. A procedure is presented here for the design of a bed to give a specified solids circulation rate. This design procedure assumes that the solids and gas characteristics, feed rates, and operating temperature and pressure are given. The design parameters to be specified include the vessel diameter, draft tube diameter, draft tube height, gas distributor, and distributor position. These parameters can be specified using the solids circulation rate model, experimental data on gas bypassing, and process requirements (e.g., selection of gas velocity in the bed above the draft tube).

Experience indicates that a simple theoretical model to predict gas bypassing that takes into account all the design and operating variables cannot be developed. Empirical correlation, however, can be obtained by conducting experiments with tracer gas injection for a given distributor plate design at different operating conditions and at different distances from the draft tube inlet.

It is also assumed that the total gas flow into the bed is known. When the operating fluidizing velocity is selected for the fluidized bed above the draft tube, the diameter of the vessel is determined. The final design decisions include selection of the draft tube diameter, the distributor plate design, the separation between the draft tube and the distributor plate, and the draft tube height. Selection of the draft tube height may be determined by other considerations, such as solids residence time, though it also affects the solids circulation rate. A gas distributor is selected to be compatible with the process and to maintain the gas velocity in the downcomer near U_{mf}. The draft tube diameter is then selected by using the solids circulation rate model to obtain the desired circulation rate.

The design procedures are thus:

1. Assume a solid circulation rate per unit draft tube area, W_{sr}, and calculate the particle velocity in the downcomer, U_{pd}, from the equation

$$W_{\text{sr}} = U_{\text{pd}}(1 - \varepsilon_{\text{mf}})\rho_p \frac{A_d}{A_r} \qquad (36)$$

2. If the two-phase theory applies, the slip velocity between the gas and the particles in the downcomer must equal the interstitial minimum fluidizing velocity as

$$U_{\text{fd}} + U_{\text{pd}} = \frac{U_{\text{mf}}}{\varepsilon_{\text{mf}}} \qquad (37)$$

where U_{fd} and U_{mf} are positive in the upward direction and U_{pd} is positive in the downward direction.

Calculate U_{fd} from Eq. (37) and the pressure drop across the downcomer, ΔP_{1-4}, from Eq. (30) assuming $\varepsilon_d = \varepsilon_{mf}$.

3. Use trial and error with Eqs. (32), (33), and (34) to evaluate U_{pr}, ε_r, and f_p.

4. Numerically integrate Eq. (31) to obtain the particle acceleration length and Eq. (35) to obtain the pressure drop across the draft tube, ΔP_{2-3}.

5. Compare the pressure drop across the downcomer, ΔP_{1-4}, and that across the draft tube, ΔP_{2-3}. If they are not equal, repeat procedures 1 through 5 until $\Delta P_{1-4} \cong \Delta P_{2-3}$.

3.4 Startup and Shutdown Considerations

Both cold flow experiments and actual pilot-plant experience show that, if operating conditions and design parameters are not selected carefully, start-up (initiating solids circulation) might be a problem (Yang et al., 1978). The primary design parameters that will affect the start-up are the distance between the grid and the draft tube inlet (L_d) and the diameter ratio between the draft tube and the draft tube gas supply nozzle (D/d_D). The maximum allowable distance, L_d, can be determined by applying the jet penetration equation suggested by Yang and Keairns (1978b):

$$\frac{L_j}{d_D} = 6.5\left(\frac{\rho_f}{\rho_p - \rho_f} \cdot \frac{U_j^2}{g d_D}\right)^{1/2} \tag{38}$$

where U_j is the gas velocity issuing from the draft tube gas supply. For high-temperature and high-pressure operations, Eq. (41), to be discussed later, should be used for calculating L_j. Another consideration is that the jet boundary at the end of jet penetration should correspond to the physical boundary of the draft tube inlet. Merry's expression (1975) for jet half-angle can be used for this purpose:

$$\cot(\theta) = 10.4\left[\frac{\rho_f d_D}{\rho_p d_p}\right]^{0.3} \tag{39}$$

or

$$L_d = \frac{(D - d_D)}{2 \cdot \tan(\theta)} \tag{40}$$

The distance between the distributor plate and the draft tube inlet, L_d, selected for the design should be the smaller one of those estimated from Eqs. (38) and (40).

A start-up technique described by Hadzismajlovic et al. (1992) is also worthy of consideration if the draft tube gas supply is retractable. The draft tube gas sup-

ply nozzle can be inserted into the draft tube during start-up and shutdown. This will reduce the difficulty described here during start-up. After start-up, the supply nozzle can be lower to below the draft tube inlet at a predetermined height to provide the normal operation configuration. This will prevent solids from draining into the gas supply nozzle during shutdown. Of course, if the draft tube gas supply nozzle is not movable due to hostile operating conditions, the technique cannot be used. Then the design precautions discussed above during start-up should be followed.

3.5 Multiple Draft Tubes

Studies in the past always concentrate on beds with a single draft tube. A literature survey failed to uncover any reference on the operation of multiple draft tubes. Even in the area of conventional spouted beds, the references on multiple spouted beds are rare. Foong et al. (1975) reported that the multiple spouted bed was inherently unstable owing to pulsation and regression of the spouts. Similar instability was also observed by Peterson (1966), who found that vertical baffles covering at least one-half of the bed height were necessary to stabilize the operation. In an industrial environment where solids are processed in large vessels, multiple draft tubes may be both necessary and beneficial. Exploratory tests in a 2D bed with three draft tubes were reported by Yang and Keairns (1989). The design methodology proposed earlier for beds with a single draft tube is still applicable here for beds with multiple draft tubes.

3.6 Industrial Applications

The application of the recirculating fluidized bed with a draft tube was probably first described by Taskaev and Kozhina (1956) for a coal devolatilizer. Dry coal is introduced into the devolatilizer below the bottom of the draft tube through a coal feeding tube concentric with the draft tube gas supply. The coal feed and recycled char at up to 100 times the coal feed rate are mixed inside the draft tube and carried upward pneumatically in dilute phase at velocities greater than 4.6 m/s. The solids disengage in a fluidized bed above the top of the draft tube and then descend in an annular downcomer surrounding the draft tube as a packed bed at close to minimum fluidization velocity. Gas is introduced at the base of the downcomer at a rate permitting the downward flow of the solids. The recirculating solids effectively prevent agglomeration of the caking coal as it devolatilizes and passes through the plastic stage.

A "seeded coal process" was later developed by Curran et al. (1973) using the same concept to smear the "liquid" raw coal undergoing the plastic transition onto the seed char and the recirculating char during low-temperature pyrolysis. Westinghouse successfully demonstrated a first stage coal devolatilizer with caking coals in a pilot-scale Process Development Unit employing a similar concept where the downcomer was fluidized and the jet issuing from the draft tube was immersed in a fluidized bed above the draft tube (Westinghouse, 1977). The same concept was also proposed for extending fluidized bed combustion technology for steam and power generation (Keairns, et al., 1978). The British Gas Council has also developed the concept for oil and coal gasification (Horsler and Thompson, 1968; Horsler et al., 1969). The development eventually resulted in a large-scale recirculating fluidized bed hydrogenator gasifying heavy hydrocarbon oils (Ohoka and Conway, 1973). McMahon (1972) also described a reactor design for oil gasification using a multiplicity of draft tubes. The Dynacracking process developed by Hydrocarbon Research Inc. in the 1950s (Rakow and Calderon, 1981) for processing heavy crude oil also utilized an internal draft tube. More recently, coal gasification in a recirculating fluidized bed with a draft tube was described by Judd et al. (1984) and a coal–water mixture combustion by Lee and Kim (1992).

Other industrial applications of the concept include that for coating tablets in the pharmaceutical industry (Wurster et al., 1965), for drying of dilute solutions containing solids (Hadzismajlovic, 1989), and for mixing and blending (Decamps et al., 1971/1972; Matweecha, 1973; Solt, 1972; Krambrock, 1976). Both Conair Waeschle Systems and Fuller Company supply commercial blenders based on the concept. The concept was also proposed as a controllable solids feeder to a pneumatic transport tube (Decamps et al., 1971/1972; Silva et al., 1996).

Although most of the experimental data reported here were obtained with large particles, Geldart class B and D powders, it is believed that the concept can equally be applicable for any fine aeratable and free-flowing solids, Geldart's class A powders.

A similar concept has also been used for liquid–solids and liquid–gas–solids contacting devices (Oguchi and Kubo, 1973; Fan et al., 1984) and bioreactors (Chisti, 1989). Bubble columns fitted with draft tubes have also been employed in the chemical process industries as airlift reactors for gas–liquid con-

tacting operations. Examples are the low-waste conversion of ethylene and chlorine to dichloroethane, biological treatment of high-strength municipal and industrial effluent, and bioreactors. Critical aspects of the design and operation of bubble columns with draft tubes have recently been reviewed by Chisti and Moo-Young (1993). Freedman and Davidson (1964) also carried out a fundamental analysis for gas holdup and liquid circulation in a bubble column with a draft tube. Extensive experimentation in a bubble column with a draft tube was conducted by Miyahara et al. (1986), and an in-depth analysis by Siegel et al. (1986). The effects of geometrical design on performance for concentric-tube airlift reactors were studied by Merchuk et al. (1994).

4 JETTING FLUIDIZED BEDS

In a gas fluidized bed, the introduction of gas is usually accomplished through distributors of various designs. Any time the gas is distributed through orifices or nozzles, a jetting region appears immediately above the grid. A large fluctuation of bed density occurs in this zone, indicating extensive mixing and contacting of solids and gas. If the chemical reactions between gas and solids are fast, much of the conversion may occur in this jetting region.

Another type of fluidized bed, where the jetting phenomenon is an important consideration, is the spouted fluid bed, where a large portion of gas goes through a fairly large nozzle. Because of the dominant effect of this jetting action provided by the large nozzle, this type of fluidized bed can be more appropriately called a "jetting fluidized bed," especially when the jet does not penetrate through the bed like that in a spouted bed. A schematic of a typical jetting fluidized bed is shown in Fig. 6. Jetting, bubbling dynamics, and solid circulation are important hydrodynamic phenomena governing the performance and operation of large-scale jetting fluidized beds. They are the focus of our attention in this section. A more extensive discussion can be found in Yang (1999).

4.1 Jet Penetration and Bubble Dynamics

Gas jets in fluidized beds were critically reviewed in Sec. 11 of Chapter 3, "Bubbling Fluidized Beds." Most of the data available now are from jets smaller than 25 mm. The discussion here will emphasize primarily the large jets, up to 0.4 m in diameter, and operation at high temperatures and high pressures.

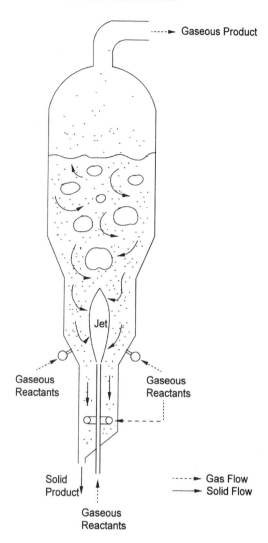

Figure 6 Schematic of a jetting fluidized bed.

The gas jets can also carry solids and are referred to as gas–solid two-phase jets in this discussion.

4.1.1 Momentum Dissipation of a Gas-Solid Two-Phase Jet

Gas velocity profiles in a gas–solid two-phase jet inside a fluidized bed were determined using a pitot tube by Yang and Keairns (1980). The velocity profiles were integrated graphically, and gas entrainment into a jet was found to occur primarily at the base of the jet. A reasonably consistent universal velocity profile can be obtained by plotting $(U_{jr} - U_{jb})/(U_{jm} - U_{jb})$ vs. $r/r_{1/2}$, comparable with the Tollmien solution for a circular homogeneous jet in an infinite medium (Abramovich, 1963; Rajaratnam, 1976).

4.1.2 Jet Penetration and Jet Half Angle

Jet Penetration Jetting phenomena were studied by Yang and Keairns (1978b) in a semicircular column 30 cm in diameter using hollow epoxy spheres ($\rho_p = 210$ kg/m^3) as the bed material and air as the fluidizing medium to simulate the particle/gas density ratio in actual operating conditions at 1520 kPa and 1280°K. A two-phase Froude number, defined as $\rho_f U_j^2/(\rho_p - \rho_f)gd_0$ and derived from both the momentum balance and the dimensionless analysis, was found to correlate jet penetration data well. The correlation was extended to cover the high-pressure jet penetration data of Knowlton and Hirsan (1980) at pressures up to 5300 kPa for fluidized beds of sand ($\rho_p = 2629$ kg/m^3), FMC char ($\rho_p = 1158$ kg/m^3), and siderite ($\rho_p = 3988$ kg/m^3). A subsequent analysis by Yang (1981) indicated that the two-phase Froude number originally suggested could be modified slightly to account for the pressure effect.

$$\frac{L_j}{d_o} = 7.65\left[\frac{1}{R_{cf}} \cdot \frac{\rho_f}{(\rho_p - \rho_f)} \cdot \frac{U_j^2}{gd_o}\right]^{0.472} \tag{41}$$

where

$$R_{cf} = (U_{cf})_p/(U_{cf})_{atm} \tag{42}$$

In the absence of $(U_{cf})_p$ and $(U_{cf})_{atm}$, $(U_{mf})_p$ and $(U_{mf})_{atm}$ can be employed.

The limiting form of Eq. (41) at atmospheric pressure (101 kPa), where the correction factor $R_{cf} = 1$, approaches the correlation originally proposed for atmospheric conditions shown in Eq (38):

Jet Half-Angle The jet half-angle can be calculated from the experimentally measured bubble size and jet penetration depth as follows:

$$\theta = \tan^{-1}\left(\frac{D_B - d_0}{2L_j}\right) \tag{43}$$

Experimentally observed jet half-angles range from 8° to 12° for the experimental data mentioned above. These compare to 10° suggested by Anagbo (1980) for a bubbling jet in liquid.

4.1.3 Bubble Dynamics

To describe the jet adequately, the bubble size generated by the jet needs to be studied. A substantial amount of gas leaks from the bubble to the emulsion phase during the bubble formation stage, particularly when the bed is less than minimally fluidized. A model developed on the basis of this mechanism predicted the experimental bubble diameter well when the experimental bubble frequency was used as an input. The experimentally observed bubble frequency is smaller by a factor of 3 to 5 than that calculated from the Davidson and Harrison model (1963), which assumed no net gas interchange between the bubble and the emulsion phase. This discrepancy is due primarily to the extensive bubble coalescence above the jet nozzle and the assumption that no gas leaks from the bubble phase.

High-speed movies were used to document the phenomena above a 0.4 m diameter jet in a 3 meter diameter transparent semicircular jetting fluidized bed (Yang et al., 1984b). The movies were then analyzed frame by frame to extract information on bubble frequency, bubble diameter, and jet penetration depth. The process of bubble formation is very similar to that described in Kececioglu et al. (1984), but it was much more irregular in the large 3 m bed. A model was developed to describe this phenomenon by assuming that the gas leaks out through the bubble boundary at a superficial velocity equivalent to the superficial minimum fluidization velocity. For a hemispherical bubble in a semicircular bed, the rate of change of bubble volume can be expressed as

$$\frac{dV_B}{dt} = G_j - U_{mf}\frac{\pi D_B^2}{2} \tag{44}$$

where $V_B = \pi D_B^3/12$ for a hemispherical bubble.

Equation (44) can be reduced to show the changes of bubble diameter with respect to time in Eq. (45):

$$\frac{D_B^2 dD_B}{(4G_j - 2\pi U_{mf} D_B^2)} = \frac{dt}{\pi} \tag{45}$$

Integrating Eq. (45) with the boundary condition that $D_B = 0$ at $t = 0$ gives

$$t = \frac{1}{2U_{mf}}\left[\frac{G_j}{\sqrt{2\pi G_j U_{mf}}}\right.$$
$$\left.\ln\left(\frac{2G_j + D_B\sqrt{2\pi G_j U_{mf}}}{2G_j - D_B\sqrt{2\pi G_j U_{mf}}}\right) - D_B\right] \tag{46}$$

The maximum bubble size, where the total gas leakage through the bubble boundary equals the total jet flow, can be obtained from either Eq. (44) or Eq. (45):

$$(D_B)_{max} = \sqrt{\frac{2G_j}{\pi U_{mf}}} \tag{47}$$

The total amount of gas leakage from the bubble at a bubble size D_B is then

$$F = \frac{G_j}{2U_{mf}}\left[\frac{G_j}{\sqrt{2\pi G_j U_{mf}}}\right.$$
$$\left.\ln\left(\frac{2G_j + D_B\sqrt{2\pi G_j U_{mf}}}{2G_j - D_B\sqrt{2\pi G_j U_{mf}}} - D_B\right)\right] - \frac{\pi D_B^3}{12} \tag{48}$$

Equation (46) and the experimental bubble frequency, $n = 1/t$, were used to predict the expected bubble diameter. The predicted bubble diameters are very close to those actually observed. Theoretically, Eqs. (46) and (48) can be solved to obtain both the bubble frequency and the bubble diameter if the total gas leakage at the moment of bubble detachment, F, is known. The bubble growth equations can be similarly derived for a circular jet in a three-dimensional bed. The same experimental observation and conclusions described above for a semicircular bed are expected to hold as well. Bubble frequency from the jet was also studied using a force probe in the same bed. The results were published in Ettehadieh et al. (1988).

The validity of extrapolating the data obtained in a semicircular model to a circular one is also of concern. Not much research has been carried out in this area. Preliminary research results by Whiting and Geldart (1980) indicated that, for coarse, spoutable solids (Geldart's group D powders), semicircular columns could provide information very similar to that from the circular ones.

4.2 Gas Mixing Around the Jetting Region

4.2.1 Gas Mixing Around Single Jets

Gas exchange between the jet and the outside emulsion phase was studied by tracer gas injection and by integration of gas velocity profiles in the jet at various heights above the jet nozzle in a 28.6 cm diameter bed with a 3.5 cm jet using polyethylene beads as bed material (Yang et al., 1984a). The concentration profiles obtained at different elevations were found to be approximately similar if the local tracer concentration is normalized with the maximum tracer concentration at the axis, C/C_m, and plotted against a normalized radial distance, $r/(r_{1/2})_c$, where $(r_{1/2})_c$ is the radial position where the tracer concentration is just half the maximum tracer concentration at the axis. Thus in a permanent flamelike jet in a fluidized bed, not only the velocity profiles in the jet but also the gas concentration profiles are similar.

The gas mixing between the jetting region and the emulsion phase and the gas flow pattern around the jet were determined by solving the tracer gas conservation equation numerically along with the axial velocity profiles in the jet obtained with a pitot tube. It is concluded that the gas mixing in a jetting fluidized bed with a permanent flamelike jet is due primarily to convection and that diffusion plays a negligible role.

The resulting velocity profiles and the flow pattern inside and around the jet indicated that the gas flow direction is predominantly from the emulsion phase into the jet at distances close to the jet nozzle. This flow can be from the aeration flow in the emulsion phase, as in the cases of high jet velocity or high aeration flow, or from the flow recirculated from the upper part of the jet. The entrainment of gas into the jet occurs immediately above the jet nozzle. The extent of this region depends on both the aeration flow outside the jet and the jet velocity. Increases in aeration flow and jet velocity tend to increase the height of this region. Beyond this gas entrainment region, the gas in the jet is then expelled from the jet along the jet height. The gas expelled at the lower part of the jet is recirculated back into the jet, setting up a gas recirculation pattern at the lower part of the jet. The extent of this recirculation pattern increases with increases in jet velocity and with decreases in aeration flow outside the jet.

The axial velocity profiles, calculated on the basis of Tollmien similarity and experimental measurement in Yang and Keairns (1980), were integrated across the jet cross section at different elevations to obtain the total jet flow across the respective jet cross sections. The total jet flows at different jet cross sections are compared with the original jet nozzle flow, as shown in Fig. 7. Up to about 50% of the original jet flow can be entrained from the emulsion phase at the lower part of the jet close to the jet nozzle. This distance can extend up to about 4 times the nozzle diameter. The gas is then expelled from the jet along the jet height.

4.2.2 Gas Mixing Around Concentric Jets

Gas mixing phenomena around a concentric jet were investigated by Yang et al. (1988) in a large semicircular cold flow model, 3 meters in diameter and 10 meters high, with a triple concentric jet nozzle assembly of 25 cm in diameter (see also Yang, 1998). A dividing gas streamline was observed experimentally that prevents the gas mixing between the jetting region and the emulsion phase until at higher bed heights. This dividing gas streamline corresponds roughly to the boundary of down-flowing solids close to the walls, to be discussed later. Several observations were made based on this study. Regardless of the incoming jet flow

rate, the gases that are injected through the concentric jets essentially remain in the core of the reactor and do not fully mix with the gas in the dense solid down-flow region of the bed. Similarly, the gas injected through the conical grid sections is not entrained by the incoming jets. Partial entrainment and mixing of these gases occur at locations where bubble formation and bubble coalescence take place. On the contrary, the mixing among the concentric jets occurs quite fast and is usually completed within the jet penetration length.

4.3 Solids Circulation in Jetting Fluidized Beds

The solids circulation pattern and solids circulation rate are important hydrodynamic characteristics of an operating jetting fluidized bed. They dictate directly the solids mixing and the heat and mass transfer between different regions of the bed.

In many applications the performance of fluidized beds is frequently controlled by the hydrodynamics phenomena occurring in the beds. Applications such as the fluidized bed combustion and gasification of fossil fuels are the cases in point. In those applications, the rates of fuel devolatilization and fines combustion are of the same order of magnitude as the mixing phenomena in a fluidized bed. The mixing and contacting of the gases and solids very often are the controlling factors in reactor performance. This is especially true in large commercial fluidized beds where only a limited number of discrete feed points for reactants is allowed because of economic considerations. Unfortunately, solids mixing in a fluidized bed has not been studied extensively, especially in large commercial fluidized beds, because of experimental difficulties.

4.3.1 Solids Circulation Pattern

Yang et al. (1986) have shown that, based on the traversing force probe responses, three separate axial solids flow patterns can be identified. In the central core of the bed, the solid flow direction is all upward, induced primarily by the action of the jets and the rising bubbles. In the outer regions, close to the vessel walls, the solid flow is all downward. A transition zone, in which the solids move alternately upward and downward, depending on the approach and departure of the large bubbles, was detected in between these two regions.

The solids circulation patterns were investigated with a force probe. Since the force probe is directional, the upward solids movement will produce a positive response from the probe and vice versa, the magnitude

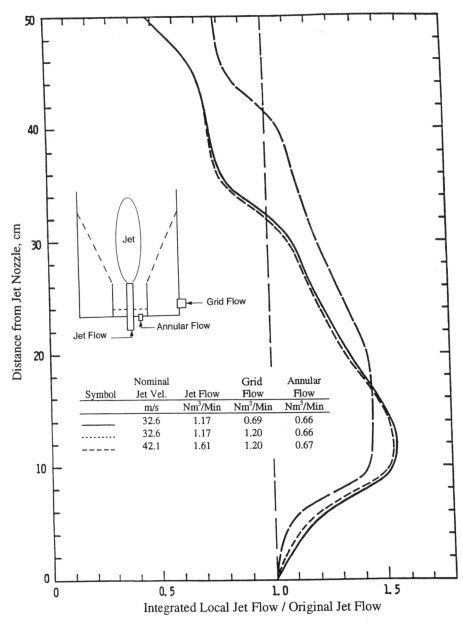

Figure 7 Gas entrainment into a 3.5 cm semicircular jet. (Adapted from Yang, 1998.)

of the response being an indication of the magnitude of the solids circulation rate. The number of major peaks per unit time is closely related to the actual bubble frequency in the bed. The force probe data allow the identification of three major solids flow regions in the 3 m model as shown in Fig. 8. At the central portion of the bed, the solids flow is induced upward primarily by jetting action at the lower bed height and by large bubbles at the higher bed height. At the outer region next to the vessel wall, the solids flow is all downward.

The region has a thickness of approximately 0.25 m. Between these two regions the solids flow is alternatively upward and downward, depending on the approach and departure of large bubbles. No stagnant region was evident anywhere in the bed.

In addition to the three solids circulation regions readily identifiable, the approximate jet penetration depth and bubble size can also be obtained from Fig. 8. The jetting region can be taken to be the maximum average value of jet penetration depth. From the jet

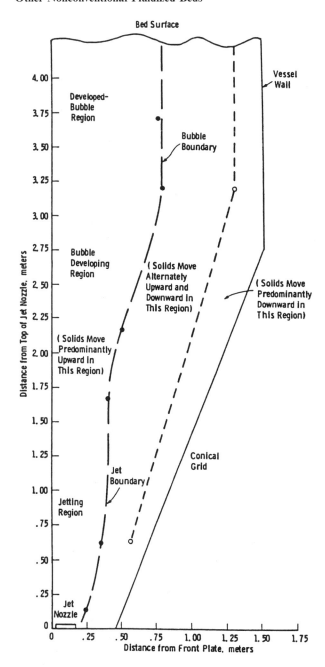

Figure 8 Solids circulating pattern in a jetting fluidized bed.

ing region, and the developed-bubble region. Bubbles were observed to coalesce in the bubble-developing region during analysis of the motion pictures taken through the transparent front plate.

4.3.2 Solids Circulation Rate

The results of the force probe measurement indicate that the solid circulation rates increase with increasing jet flow rates. A simple mechanistic model was developed to correlate the solids circulation data. The model assumes that solids circulation inside the bed is induced primarily by bubble motion. The solids circulation pattern inside the bed can be divided into two major regions radially. In the center of the bed, the particle movement is predominantly upward and is induced by bubbles disengaged from the central jet. This region has a radius similar to the radius of the average bubble size. In the outer region, the particles move primarily downward. In the meantime, the particles in both regions exchange with each other across the neighboring boundary at a constant rate of W_z g/cm^2-s. This mechanistic model is shown schematically in Fig. 9. Material balance in a differential element dz as shown in Fig. 9 gives,

in the bubble street region,

$$K\frac{\partial X_J'}{\partial z} + \pi R_i^2(1 - \varepsilon_{mf})\rho_p\frac{\partial X_J'}{\partial t} + 2\pi R_i W_z \qquad (49)$$
$$(X_J' - X_J) = 0$$

and in the annular region,

$$K\frac{\partial X_J}{\partial z} + \pi(R_0^2 - R_i^2)(1 - \varepsilon_{mf})\rho_p\frac{\partial X_J}{\partial t}$$
$$+ 2\pi R_i W_z(X_J - X_J') = 0 \qquad (50)$$

where

$$K = nV_B f_w(1 - \varepsilon_w)\rho_p \qquad (51)$$

The data do not show any clear dependence on the axial position, z. The axial dependence is thus assumed to be negligible. Equations (49) and (50) are reduced from partial differential equations to ordinary differential equations. If we consider only the annular region, Eq. (50) reduces to

$$\frac{dX_J}{dt} + \frac{2\pi R_i W_z}{\pi(R_0^2 - R_i^2)(1 - \varepsilon_{mf})\rho_p}(X_J - X_J') = 0 \qquad (52)$$

Since both X_J and X_J' are independent of z, the relationship between X_J and X_J' can be approximated by the material balance of the coarse particles injected into the bed to serve as the tracer.

boundary at the end of the jetting region, an initial bubble diameter can be estimated. This value can be taken to be the minimum value of the initial bubble diameter. The diameter of a fully developed bubble can be obtained from the bubble boundary in the developed-bubble region, as shown in Fig. 8. The central region is thus divided further into three separate regions axially: the jetting region, the bubble-develop-

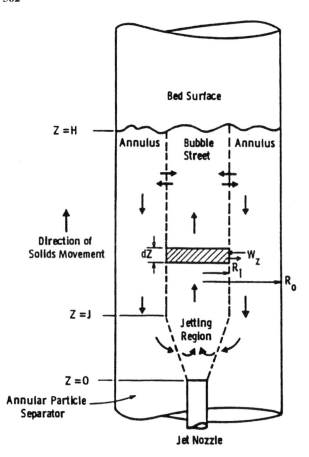

Figure 9 Schematic of a mathematical model for solids circulation in a jetting fluidized bed.

Solving for X_J' we have

$$X_J' = \frac{W_t}{\pi R_i^2 H (1 - \varepsilon_{mf}) \rho_p} - \left[\left(\frac{R_o}{R_i} \right)^2 - 1 \right] \left(\frac{1 - \varepsilon_{mf}}{1 - \varepsilon_i} \right) X_J \tag{53}$$

Substituting Eq. (53) into Eq. (52), after some mathematical manipulation we get

$$\frac{dX_J}{dt} + PX_J - Q = 0 \tag{54}$$

where

$$P = \frac{2 R_i W_z}{(R_o^2 - R_i^2)(1 - \varepsilon_{mf}) \rho_p} + \frac{2 W_z}{R_i (1 - \varepsilon_i) \rho_p} \tag{55}$$

$$Q = \frac{2 W_z W_i}{\pi R_i (R_o^2 - R_i^2) H (1 - \varepsilon_{mf})(1 - \varepsilon_i) \rho_p^2} \tag{56}$$

Equation (54) can be integrated with the boundary condition that $X_J = 0$ at $t = 0$ to give

$$\frac{X_J}{X_J^o} = \frac{1}{t_w P} \{ Pt - [1 - \exp(-Pt)] \} \tag{57}$$

The equilibrium tracer concentration in the bed after complete mixing can be expressed as

$$X_J^0 = \frac{W_t}{\pi H \rho_p [R_i^2 (1 - \varepsilon_i) + (R_o^2 - R_i^2)(1 - \varepsilon_{mf})]} \tag{58}$$

The voidage inside the bubble street, ε_i, can be calculated as

$$\varepsilon_i = \varepsilon_{mf} + f_B (1 - \varepsilon_{mf}) = \varepsilon_{mf} + \frac{n V_B}{\pi R_i^2 U_A} (1 - \varepsilon_{mf}) \tag{59}$$

where f_B is the volumetric fraction of bubbles occupying the bubble street region at any instant; it can be evaluated from

$$f_B = \frac{n V_B}{\pi R_i^2 U_A} \tag{60}$$

If the bubble frequency, bubble diameter, and bubble velocity are known, the solids mixing rate can be calculated.

In correlating the data, the solid exchange rate between the two regions, W_z, was assumed to be constant. Comparison of the calculated and the experimentally observed tracer concentration profiles was good.

The solids mixing study by injection of tracer particles indicated that the axial mixing of solids in the bubble street is apparently very fast. Radial mixing flux depends primarily on the bubble size, bubble velocity, and bubble frequency, which in turn depend on the size of the jet nozzle employed and the operating jet velocity.

4.4 Solid Entrainment Rate into Gas and Gas–Solid Two-Phase Jets

A mathematical model for solid entrainment into a permanent flamelike jet in a fluidized bed was proposed by Yang and Keairns (1982). The model was supplemented by particle velocity data obtained by following movies frame by frame in a motion analyzer. The particle entrainment velocity into the jet was found to increase with increases in distance from the jet nozzle, to increase with increases in jet velocity, and to decrease with increases in solid loading in the gas–solid two-phase jet.

High-speed movies indicated that the entrained particles tended to bounce back to the jet boundary more readily under high solid loading conditions. This may explain why the entrainment rate decreases with increases in solid loading in a two-phase jet. A ready analogy is the relative difficulty in merging into a rush-hour traffic as compared to merging into a light traffic.

A simple model for solid entrainment into a permanent flamelike jet was described in Sec. 11.5.1 of Chapter 3 "Bubbling Fluidized Beds."

4.5 Scale-up Considerations

The development of commercial fluidized bed processors generally requires intermediate stages of testing on physical models simulating commercial equipment. Simulation (or scale-up) criteria derived from fluidized bed momentum-conservation relations may be applied to determine the design and operating conditions for physical models discussed in Chapter 13 "Fluidized Bed Scale-up." Such criteria, while not totally established at this time, have been applied to simulate a pressurized fluidized bed gasifier having a large vertical central jet for fuel feeding, combustion, and gasification (Yang et al., 1995).

Physical models of commercial fluidized bed equipment provide an important source of design information for process development. A physical model of a commercial fluidized bed processor provides a small-scale simulation of the fluid dynamics of a commercial process. While commercial processes will typically operate at conditions making direct observation of bed fluid dynamics difficult (high temperature, high pressure, corrosive environment), a physical model is designed to allow easy observation (room temperature and pressure, nonreactive atmosphere, transparent vessel).

Cold flow studies have several advantages. Operation at ambient temperature allows construction of the experimental units with transparent plastic material that provides full visibility of the unit during operation. In addition, the experimental unit is much easier to instrument because of operating conditions less severe than those of a hot model. The cold model can also be constructed at a lower cost in a shorter time and requires less manpower to operate. Larger experimental units, closer to commercial size, can thus be constructed at a reasonable cost and within an affordable time frame. If the simulation criteria are known, the results of cold flow model studies can then be combined with the kinetic models and the intrinsic rate equations generated from the bench-scale hot models to construct a realistic mathematical model for scale-up.

The need for physical modeling of fluidized bed processors is dictated by the state of the art of fluidized bed scale-up technology. In general, no rational procedure exists for scaling up a new fluidized bed processor concept that precludes the need for physical modeling. Many empirically developed rules of thumb for fluidized bed scale-up exist in specific areas of fluidized bed application that are not generally applicable. Existing mathematical modeling approaches are themselves based heavily on empirical descriptions of fluidized bed fluid dynamics. These bubbling bed models can be applied only where confidence exists for the empirical bubble flow description built into the model. Fluidized bed processors operate over such a broad range of fluid dynamic regimes that this confidence rarely exists for new concepts.

Yang et al. (1995) described the application of this scale-up approach. Comprehensive testing programs were performed on two relatively large-scale simulation units for a period of several years: both a 30 cm diameter (semicircular) Plexiglas cold model and a 3 m diameter (semicircular) Plexiglas cold model operated at atmospheric pressure.

The results are highly significant to the development of fluidized bed technology because they represent a case study of a rational fluidized bed development approach. The extensive data generated are unique in their equipment dimensions, pushing existing models and correlations to new extremes and offering new insights into large-scale equipment behavior. The understanding of the hydrodynamic phenomena developed from the cold flow model studies, and the analytical modeling reported was integrated with parallel studies that investigated coal gasification kinetics, ash agglomeration, char–ash separation, and fines recycle to develop an integrated process design procedure.

4.6 Applications

The primary applications for large-scale jetting fluidized beds are in the area of coal gasification as described by Yang et al. (1995), Kojima et al. (1995), and Tsuji and Uemaki (1994). Smaller scale applications are for fluidized bed coating and granulation discussed in Chapter 17, "Applications for Coating and Granulation."

5 ROTATING FLUIDIZED BEDS

Rotating fluidized beds make use of the centrifugal force, which can reach many times the gravitational force, to increase the minimum fluidizing velocity of the particles and minimize the formation of bubbles even at a very high gas flow rate. This can make the rotating fluidized beds very compact compared to conventional fluidized beds so that they can be utilized in special applications. Pfeffer et al. (1986) described an application using a rotating fluidized bed as a high-efficiency dust filter. Tsutsumi et al. (1994) investigated the reduction of nitrogen oxides with soot emitted from diesel engines using a centrifugal fluidized bed. The fundamental governing equations have been studied by Kao et al. (1987), Chen (1987), Fan et al. (1985), and Chevray et al. (1980). A schematic of a rotating fluidized bed is presented in Fig. 10.

5.1 Minimum Fluidization Velocity

Because both the centrifugal force and the drag force on the particles in the centrifugal fluidized bed vary with radial position, there are three minimum fluidization velocities that can be defined (Qian et al., 1998). The surface minimum fluidization velocity is defined as the point when the inner surface of the bed is fluidized. The pressure drop across the bed can be calculated on the basis of a fixed bed because the bed is essentially a fixed bed except at the surface. The point where the whole bed is fluidized (at the distributor plate) is called the critical minimum fluidization velocity. In this case, the pressure drop can be determined as in a fluidized

bed. Then there is the average minimum fluidization velocity calculated from these two extremes.

When we make use of the Wen and Yu modification of the Ergun equation (1966), the surface minimum fluidization velocity, U_{mfi}, can be derived as

$$\frac{U_{\mathrm{mfi}}\rho_f d_{\mathrm{p}}}{\mu}\frac{r_{\mathrm{o}}}{r_{\mathrm{i}}} = \left[(33.7)^2 + 0.0408\frac{\rho_{\mathrm{f}}(\rho_{\mathrm{p}} - \rho_{\mathrm{f}})d_{\mathrm{p}}^3\omega^2 r_{\mathrm{i}}}{\mu^2}\right]^{0.5} - 33.7 \tag{61}$$

Similarly, the critical minimum fluidization velocity, U_{mfc}, can be expressed as

$$\frac{U_{\mathrm{mfc}}\rho_f d_{\mathrm{p}}}{\mu} = \left[(33.7)^2 + 0.0408\frac{\rho_{\mathrm{f}}(\rho_{\mathrm{p}} - \rho_{\mathrm{f}})d_{\mathrm{p}}^3\omega^2 r_{\mathrm{o}}}{\mu^2}\right]^{0.5} - 33.7 \tag{62}$$

The average minimum fluidization velocity can then be calculated from

$$\frac{U_{\mathrm{mf}}\rho_f d_{\mathrm{p}}}{\mu} = \left[\left(33.7\frac{C_2}{C_1}\right)^2 + 0.0408\frac{\rho_{\mathrm{f}}(\rho_{\mathrm{p}} - \rho_{\mathrm{f}})d_{\mathrm{p}}^3\omega^2}{\mu^2}\frac{C_3}{C_1}\right]^{0.5} - 33.7\frac{C_2}{C_1} \tag{63}$$

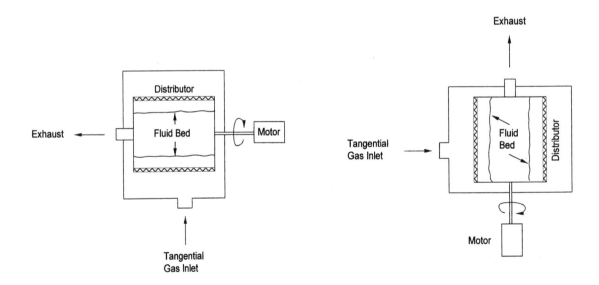

Horizontal Rotating Fluidized Bed   **Vertical Rotating Fluidized Bed**

Figure 10 Schematic of horizontal and vertical rotating fluidized beds.

where

$$C_1 = r_o^2 \left(\frac{1}{r_i} - \frac{1}{r_o} \right)$$

$$C_2 = r_o \ln\left(\frac{r_o}{r_1} \right) \qquad (64)$$

$$C_3 = \frac{(r_o^2 - r_i^2)}{2}$$

5.2 Pressure Drop

For $U_g \leq U_{mfi}$, the bed is a packed bed and the pressure drop can be calculated from the packed bed equation

$$\frac{dP}{dr} = \alpha\left(\frac{U_g r_o}{r} \right) + \beta\left(\frac{U_g r_o}{r} \right)^2 \qquad (65)$$

or

$$\Delta P = \alpha U_g r_o \ln\left(\frac{r_o}{r_i} \right) + \beta U_g^2 r_o^2 \left(\frac{1}{r_i} - \frac{1}{r_o} \right) \qquad (66)$$

where

$$\alpha = \frac{1650(1-\varepsilon)\mu}{d_p^2} \qquad \beta = \frac{24.5(1-\varepsilon)\rho_f}{d_p} \qquad (67)$$

When $U_g \geq U_{mfc}$, the bed is completely fluidized and the pressure drop can be calculated from the fluidized bed equation.

$$\frac{dP}{dr} = (\rho_p - \rho_f)(1-\varepsilon)r\omega^2 \qquad (68)$$

or

$$\Delta P = (\rho_p - \rho_f)(1-\varepsilon)\omega^2 \frac{(r_o^2 - r_i^2)}{2} \qquad (69)$$

With $U_{mfi} < U_g < U_{mfc}$, the bed is partially fluidized, and the pressure drop can be calculated by summing the pressure drop across the fluidized bed and the packed bed:

$$\Delta P = (\rho_p - \rho_f)(1-\varepsilon)\omega^2 \frac{(r_{pf}^2 - r_i^2)}{2} + \alpha U_g r_o \ln\left(\frac{r_o}{r_{pf}} \right)$$
$$+ \beta U_g^2 r_o^2 \left(\frac{1}{r_{pf}} - \frac{1}{r_o} \right) \qquad (70)$$

Qian et al. (1998) found that when a sintered metal distributor plate was used for gas distribution, the experimental pressure drop could be predicted fairly well with the theoretical equation mentioned above. However, when a distributor with slotted openings

was used, the experimental pressure drop is just about 70% of that calculated from the theoretical equation. Apparently, with a slotted distributor plate, the bed was not completely fluidized. The bed above the webs was not fluidized. Chen (1987) also reported two different types of pressure drop in the literature. One shows a pressure drop curve similar to that of a conventional fluidized bed, while the other one exhibits a maximum pressure drop in the fluidized region.

NOTATION

A_a	=	cross-sectional area of annulus at any level
A_d	=	cross-sectional area of the downcomer
A_r	=	cross-sectional area of the draft tube
Ar	=	Archimedes number, $\rho_f(\rho_p - \rho_f)d_p^3 g/\mu^2$
A_s	=	cross-sectional area of spout at any level
C	=	tracer gas concentration
C_{DS}	=	drag coefficient of a single particle
C_m	=	maximum tracer gas concentration at the jet axis
D	=	draft tube diameter
	=	fluidized bed or spouted bed diameter
D_b	=	upper diameter of the stagnant bed in conical spouted bed
D_B	=	bubble diameter
$(D_B)_{max}$	=	maximum bubble diameter
d_D	=	diameter of draft tube gas supply
D_i	=	diameter of spout nozzle in spouted beds
	=	diameter of the bed bottom in conical spouted bed
d_o	=	diameter of jet nozzle
D_o	=	diameter of the inlet in conical spouted bed
d_p	=	mean solid particle diameter
$(d_p)_{crit}$	=	critical particle size in spouted bed
D_s	=	spout diameter
$\overline{D}_s$	=	average spout diameter
F	=	total amount of gas leakage during bubble formation from a jet
f_B	=	volumetric fraction of bubble in bubble street
f_p	=	solid friction factor
f_w	=	wake fraction of the bubble
g	=	gravitational acceleration
G	=	superficial mass flux of spouting fluid
G_j	=	total gas flow rate through the jet
G_r	=	total gas flow rate in the draft tube
H	=	bed height of a fluidized bed or a spouted bed
	=	height of the developed bed in the conical spouted bed
H_F	=	fountain height, m
H_m	=	maximum spoutable bed height
H_{mf}	=	bed height at minimum fluidization

H_o	=	height of stagnant bed in conical spouted bed
L	=	height of draft tube or height of downcomer
L_d	=	distance between the distributor plate and the draft tube inlet
L_j	=	jet penetration length
ΔL	=	distance required for acceleration of particles
n	=	bubble frequency
ΔP	=	pressure drop
$\Delta P_{1\text{-}2}$	=	pressure drop between 1 and 2 (see Fig. 5)
$\Delta P_{1\text{-}4}$	=	pressure drop between 1 and 4 (see Fig. 5)
$\Delta P_{2\text{-}3}$	=	pressure drop between 2 and 3 (see Fig. 5)
$\Delta P_{3\text{-}4}$	=	pressure drop between 3 and 4 (see Fig. 5)
r	=	radial distance from the jet axis or from the spout axis
r_i	=	radius of inner surface of granule bed
r_o	=	radius of rotating fluidized bed
r_{pf}	=	radius of interface of fluidized and packed beds
r_s	=	radius of the spout
$r_{1/2}$	=	radial distance where gas velocity is one-half the maximum gas velocity at the jet axis
$(r_{1/2})_c$	=	radial distance where tracer gas concentration is one-half the maximum at the jet axis
R_i	=	radius of bubble street, $R_i = D_B/2$
R_o	=	bed radius
$(Re)_{Do,ms}$	=	Reynolds number at minimum spouting based on D_o
$(Re)_{mf}$	=	Reynolds number at minimum fluidization, $= d_p U_{mf} \rho_f/\mu$
$(Re)_p$	=	Reynolds number based on the slip velocity, $= d_p(U_{gr} - U_{pr})\rho_f/\mu$
$(Re)_t$	=	Reynolds number based on the terminal velocity, $= d_p U_t \rho_f/\mu$
S_d	=	total wall area in the downcomer
S_r	=	total wall area in the draft tube
t	=	time
t_w	=	total time required to inject all tracer particles
U	=	superficial fluid velocity
U_a	=	superficial fluid velocity in annulus of a spouted bed at any level
U_A	=	absolute bubble velocity
$(U_{cf})_{atm}$	=	complete fluidization velocity at atmospheric pressure
$(U_{cf})_p$	=	complete fluidization velocity at pressure P
U_{fr}	=	superficial fluid velocity in the draft tube
U_g	=	superficial gas velocity
U_{gd}	=	superficial gas velocity in the downcomer
U_{gr}	=	superficial gas velocity in the draft tube
U_j	=	superficial jet nozzle velocity
U_{jb}	=	gas velocity at jet boundary

U_{jm}	=	maximum gas velocity at the jet axis
U_{jr}	=	gas velocity at radial distance r from the jet axis
U_m	=	maximum value of the minimum spouting velocity at the maximum spoutable bed height
U_{mf}	=	superficial minimum fluidization velocity
U_{mfc}	=	critical minimum fluidization velocity
U_{mfi}	=	surface minimum fluidization velocity
$(U_{mf})_{atm}$	=	superficial minimum fluidization velocity at atmospheric pressure
$U_{mf})_p$	=	superficial minimum fluidization velocity at pressure P
U_{ms}	=	minimum spouting velocity
$(U_{ms})_{0.5}$	=	minimum spouting velocity for vessel diameter less than 0.5 meters
U_{pd}	=	solid particle downward velocity in the downcomer
U_{pt}	=	solid particle velocity in the draft tube
U_{slug}	=	rising velocity of the gas slug relative to the particle velocity at its nose
U_{sz}	=	superficial upward fluid velocity in spout or fountain core at any level
U_t	=	terminal velocity of a single solid particle
V_B	=	volume of a gas bubble
$V_{0\,max}$	=	particle velocity along axis at bed surface
v_s	=	local upward particle velocity in spout or fountain core at any level
V_{sd}	=	net upward superficial volumetric flow rate of particles in the downcomer
V_{sr}	=	net upward superficial volumetric flow rate of particles in the draft tube
v_w	=	downward particle velocity at column wall
W	=	mass downflow rate of solids in annulus at any level = mass upflow rate of solids in spout at same level
W_{sd}	=	mass flux of particles in the downcomer, $W_{sd} = V_{sd}\rho_p$
W_{sr}	=	mass flux of particles in the draft tube, $W_{sr} = V_{sr}\rho_p$
W_t	=	cumulative weight of tracer particles injected after time t
W_z	=	radial solids mixing flux
X_j^o	=	tracer particle weight fraction in the bed after complete mixing
X_j, X_j'	=	tracer particle weight fractions in annulus and in bubble street, respectively
z	=	axial coordinate

Greek Letters

ε	=	bed voidage
$\varepsilon(0)$	=	bed voidage at $r = 0$
$\varepsilon(w)$	=	bed voidage at the wall in conical spouted bed

ε_a	=	annulus voidage at minimum spouting
ε_{bd}	=	bubble voidage in the downcomer
ε_{br}	=	bubble voidage in the draft tube
ε_{bs}	=	spout voidage at bed surface
ε_d	=	voidage in the downcomer
ε_i	=	voidage in bubble street
ε_{mf}	=	voidage at minimum fluidization
ε_r	=	voidage in the draft tube
ε_o	=	loose bed voidage in conical spouted bed
ε_s	=	voidage in spout or fountain core at any level
ε_w	=	voidage in bubble wake
μ	=	viscosity of the fluid
$\rho_{air(p=1)}$	=	density of air at one atmosphere and room temperature
ρ_b	=	bed density or bulk density
ρ_f	=	density of the fluid
ρ_p	=	solid particle density
ϕ_s	=	sphericity of the solid particle
ω	=	rotating speed
θ	=	jet half angle
	=	included angle of the cone in conical spouted bed
τ_d	=	particle–wall shear stress in the downcomer
τ_r	=	particle–wall shear stress in the draft tube

REFERENCES

Abramovich GN. The Theory of Turbulent Jets. Cambridge, MA: M.I.T. Press, 1963.

Alappat BJ, Rane VC. Studies on the effects of various design and operational parameters on solid circulation rate in a recirculating fluidized bed. Can J Chem Eng 73:248–252, 1995.

Anabtawi MZ. Minimum spouting velocity for binary mixture of particles in rectangular spouted beds. Can J Chem Eng 76:132–136, 1998.

Anabtawi MZ, Uysal BZ, Jumah RY. Flow characteristics in a rectangular spout-fluid bed. Powder Technol 69:205–211, 1992.

Anagbo PE. Derivation of jet cone angle from bubble theory. Chem Eng Sci 35:1494–1495, 1980.

Botterill JSM, Bessant DJ. The flow properties of fluidized solids. Powder Technol 8:213–222, 1973.

Buchanan RH, Wilson B. The fluid-lift solids recirculator. Mech Chem Eng Trans (Australia) 117–124, May 1965.

Chen YM. Fundamentals of a centrifugal fluidized bed. AIChE J 33:722–728, 1987.

Chevray R, Chan YNI, Hill FB. Dynamics of bubbles and entrained particles in the rotating fluidized beds. AIChE 26:390–398, 1980.

Chisti MY. Airlift Bioreactors. New York: Elsevier, 1989.

Chisti MY, Moo-Young M. Improve the performance of air-lift reactors. Chem Eng Progr 38–45, June 1993.

Claflin JK, Fane AG. Spouting with a porous draft-tube. Can J Chem Eng 61:356–363, 1983.

Curran GP, Pasek B, Pell M, Gorin E. Pretreatment of bituminous coals for pressure gasification. Proceedings of Fluidized Bed Combustion Symposium, American Chemical Society Meeting, Chicago, August 1973.

Davidson JF, Harrison D. Fluidized Particles. Cambridge: Cambridge University Press, 1963.

Decamps F, Dumont G, Goossens W. Vertical pneumatic conveyor with a fluidized bed as mixing zone. Power Technol 5:299–306, 1971/1972.

Epstein N, Grace JR. Spouting of particulate solids. In: Fayed ME, Otten L, eds. Handbook of Powder Science and Technology. New York: Chapman and Hall, 1997, pp 532–567.

Epstein N, Lim CJ, Mathur KB. Data and models for flow distribution and pressure drop in spouted beds. Can J Chem Eng 56:436–447, 1978.

Ettehadieh B, Yang WC, Haldipur GB. Motion of solids, jetting and bubbling dynamics in a large jetting fluidized bed. Powder Technol 54:243–254, 1988.

Fan LS, Hwang SJ, Matsuura A. Some remarks on hydrodynamic behavior of a draft tube gas–liquid–solid fluidized bed. AIChE Symp Ser 80(234):91, 1984.

Fan LT, Chang CC, Yu YS, Takahash T, Tanaka Z. Incipient fluidization condition for a centrifugal fluidization bed. AIChE J 31:999–1009, 1985.

Foong SK, Barton RK, Ratcliffe JS. Characteristics of multiple spouted beds. Mech Chem Eng Trans (Australia) MCII(1,2):7–12, 1975.

Freedman W, Davidson JF. Hold-up and liquid circulation in bubble columns. Trans Inst Chem Eng 47:T251–T262, 1969.

Grace JR, Mathur KB. Height and structure of the fountain region above spouted beds. Can J Chem Eng 56:533–537, 1978.

Grbavcic ZB, Vukovic DV, Zdanski FK, Littman H. Fluid flow pattern, minimum spouting velocity and pressure drop in spouted beds. Can J Chem Eng 54:33–42, 1976.

Hadzismajlovic DzE, Povrenovic DS, Grbavcic ZB, Vukovic DV. A spout-fluid bed dryer for dilute solutions containing solids. In: Grace JR, Shemilt LW, Bergougnou MA, eds. Fluidization VI. New York: Engineering Foundation, 1989, pp 277–284.

Hadzismajlovic DzE, Grbavcic ZB, Povrenovic DS, Vukovic DV, Garic RV. The hydrodynamic behavior of a 0.95 m diameter spout-fluid bed with a draft tube. In: Potter OE, Nicklin DJ, eds. Fluidization VII. New York: Engineering Foundation, 1992, pp 337–344.

He YL, Lim CJ, Grace JR, Zhu JX, Qin SZ. Measurements of voidage profiles in spouted beds. Can J Chem Eng 72:229–234, 1994a.

He YL, Qin SZ, Lim CJ, Grace JR. Particle velocity profiles and solid flow patterns in spouted beds. Can J Chem Eng 72:561–568, 1994b.

He YL, Lim CJ, Qin SZ, Grace JR. Spout diameters in full and half spouted beds. Can J Chem Eng 76:702–706, 1998a.

He YL, Lim CH, Grace JR. Hydrodynamics of pressurized spouted beds. Can J Chem Eng 76:696–701, 1998b.

Horsler AG, Thompson BH. Fluidization in the development of gas making process. Proceedings of Tripartite Chem Engr Conf. Montreal, 1968, p 51.

Horsler AG, Lacey JA, Thompson BH. High pressure fluidized beds. Chem Eng Progr 65:59–64, October 1969.

Ishida M, Shirai T. Circulation of solid particles within the fluidized bed with a draft tube. J Chem Eng (Japan) 8:477–481, 1975.

Judd MR, Masson H, Meihack W. Solid circulation and gasification experiments in a fluidized bed with a draft tube. In: Kunii D, Toei R, eds. Fluidization. New York: Engineering Foundation, 1984, pp 663–670.

Kao J, Pfeffer R, Tardos GI. On partial fluidization in rotating fluidized beds. AIChE J 33:858–861, 1987.

Keairns DL, Yang WC, Newby RA, Hamm JR, Archer DH. Circulating bed boiler concepts for steam and power generation. Proceedings of 13th Intersociety Energy Conversion Eng Conf. Warrendal: Society of Automotive Engineers, 1978, pp 540–547.

Kececioglu I, Yang WC, Keairns DL. Fate of solids fed pneumatically through a jet into a fluidized bed. AIChE J 30:99–110, 1984.

King DF, Harrison D. The minimum spouting velocity of a spouted bed at elevated pressure. Powder Technol 26:103–107, 1980.

Knowlton TM, Hirsan I. The effect of pressure on jet penetration in semi-cylindrical gas-fluidized beds. In: Grace JR, Matsen JM, eds. Fluidization. New York: Plenum Press, 1980, pp 315–324.

Kojima T, Yoshitake H, Kimura T, Matsukata M, Uemiya S. Contribution of local reactions in the grid zone to the performance of a jetting fluidized bed gasifier of coal char. Energy Fuels 9:379–383, 1995.

Krambrock W. Mixing and homogenizing of granular bulk material in a pneumatic mixer unit. Powder Technol 15:199–206, 1976.

LaNauze RD. A circulating fluidized bed. Powder Technol 15:117–127, 1976.

LaNauze RD, Davidson JF. The flow of fluidized solids. In: Keairns DL, ed. Fluidization Technology. Vol. 2. Washington: Hemisphere, 1976, pp 113–124.

Lee WJ, Kim SD. Hydrodynamics and CWM combustion characteristics in an internally circulating fluidized bed combustor. In: Potter OE, Nicklin DJ, eds. Fluidization VII. New York: Engineering Foundation, 1992, pp 479–486.

Li Y. Spouted bed hydrodynamics at temperatures up to 580 °C, M.A.Sc. thesis, University of British Columbia, Vancouver, Canada, 1992.

Lim CJ, Mathur KB. Modeling of particle movement in spouted beds. In: Davidson JF, Keairns DL, eds.

Fluidization. Cambridge: Cambridge University Press, 1978, pp 104–109.

Littman H, Morgan III MH. A general correlation for the minimum spouting velocity. Can J Chem Eng 61:369–373, 1983.

Littman H, Morgan III MH, Vukovic DV, Zdanski FK, Grbavcic ZB. Prediction of the maximum spoutable height and the average spout to inlet tube diameter ratio in spouted beds of spherical particles. Can J Chem Eng 57: 684–687, 1979.

Mamuro T, Hattori H. Flow pattern of fluid in spouted beds. J Chem Eng Japan 1:1–5, 1968.

Mann U. Analysis of spouted-bed coating and granulation. 1. Batch operation. Ind Eng Chem Process Des Dev 22:288–292, 1983.

Mann U, Crosby EJ. Cycle time distribution in continuous systems. Chem Eng Sci 28:623–627, 1973.

Mann U, Crosby EJ. Cycle time distribution measurements in spouted beds. Can J Chem Eng 53:579–581, 1975.

Mathur KB, Epstein N. Spouted Beds. New York: Academic Press, 1974.

Mathur KB, Gishler PE. A technique for contacting gases with coarse solid particles. AIChE J 1: 157–164, 1955.

Matweecha DM. Blending apparatus. U.S. Patents 3,648,985 (1972); 3,729,175 (1973).

McMahon JF. Fluidized bed reactor. U.S. Patent 3,825,477 (1972).

McNab GS. Prediction of spout diameter. Brit Chem Eng Proc Tech 17:532, 1972.

Merchuk JC, Ladwa N, Cameron A, Bulmer M, Pickett A. Concentric-tube airlift reactors: effects of geometrical design on performance. AIChE J 40:1105–1117, 1994.

Merry JMD. Penetration of vertical jets into fluidized beds. AIChE J 21:507–510, 1975.

Merry JMD. Fluid and particle entrainment into vertical jets in fluidized beds. AIChE J 22:315–323, 1976.

Milne BJ, Berruti F, Behie LA. Solids circulation in an internally circulating fluidized bed (ICFB) reactor. In: Potter OE, Nicklin DJ, eds. Fluidization VII. New York: Engineering Foundation, 1992, pp 235–242.

Miyahara T, Hamaguchi M, Sukeda Y, Takahashi T. Size of bubbles and liquid circulation in a bubble column with a draft tube and sieve plate. Can J Chem Eng 64:718–725, 1986.

Oguchi U, Kubo J. Liquid–solid particles or liquid–gas–solid particle contacting method. U.S. Patent 3,754,993 (1973).

Ohoka I, Conway HL. Progress in the gasification of heavy hydrocarbon oils in a recirculating fluidized bed hydrogenator. Proceedings of the 5th Synthetic Pipeline Gas Symposium, Chicago, 1973.

Olazar M, San Jose MJ, Aguayo AT, Arandes JM, Bilbao J. Stable operation conditions for gas–solid contact regimes in conical spouted beds. Ind Eng Chem Res 31:1784–1792, 1992.

Olazar M, San Jose MJ, Aguayo AT, Arandes JM, Bilbao J. Pressure drop in conical spouted beds. Chem Eng J 51:53–60, 1993a.

Olazar M, San Jose MJ, Aguayo AT, Arandes JM, Bilbao J. Design factors of conical spouted beds and jet spouted beds. Ind Eng Chem Res 32:1245–1250, 1993b.

Olazar M, San Jose MJ, Penas FJ, Aguayo AT, Bilbao J. The stability and hydrodynamics of conical spouted beds with binary mixtures. Ind Eng Chem Res 32:2826–2834, 1993c.

Olazar M, San Jose MJ Llamosas R, Alvarez S, Bilbao J. Study of local properties in conical spouted beds using an optical fiber probe. Ind Eng Chem Res 34:4033–4039, 1995.

Peterson WS. Multiple spouted beds. Can. Patent No. 739,660 (1966).

Pfeffer R, Tardos GI, Gal E. The use of a rotating fluidized bed as a high efficiency dust filter. In: Ostergaard K, Sorengsen A, eds. Fluidization V. New York: Engineering Foundation, 1986, pp 667–674.

Qian GH, Bagyi I, Pfeffer R, Shaw H, Stevens JG. A parametric study of a horizontal rotating fluidized bed using slotted and sintered metal cylindrical gas distributors. Powder Technol 100:190–199, 1998.

Rajaratnam N. Turbulent Jets. New York: Elsevier, 1976.

Rakow MS, Calderon M. The dynacracking process—an update. Chem Eng Prog 77:31–36, February 1981.

San Jose MJ, Olazar M, Alvarez S, Bilbao J. Local bed voidage in conical spouted beds. Ind Eng Chem Res 37:2553–2558, 1998.

Siegel MH, Merchuk JC, Schugerl K. Air-lift reactor analysis: interrelationships between riser, downcomer, and gas–liquid separator behavior, including gas recirculation effects. AIChE J 32:1585–1596, 1986.

Silva EMV, Ferreira MC, Freire JT. Mean voidage measurements and fluid dynamics analysis in a circulating fluidized bed with a spouted bed type solids feeding system. Proceedings of 2nd European Thermal-Sciences and 14th UIT National Heat Transfer Conference, Rome, 1996.

Singh B. Theory of slugging lifters. Powder Technol 21:81–89, 1978.

Solt PE. Airlift blending apparatus, U.S. Patent 3,647,188 (1972).

Stocker RK, Eng JH, Svrcek WY, Behie LA. Gas residence time distribution studies in a spouted bed with a draft tube. In: Grace JR, Shemilt LW, Bergougnou MA, eds. Fluidization VI. New York: Engineering Foundation, 1989, pp 269–276.

Taskaev ND, Kozhina ME. Semicoking of Kok-Yangak coal in a circulating bed. Trudy Akad Nauk Kirgiz S.S.R. 7:109, 1956.

Tsuji T, Uemaki O. Coal gasification in a jet-spouted bed. Can J Chem Eng 72:504–510, 1994.

Tsutsumi A, Demura M, Yoshida K. Reduction of nitrogen oxides with soot emitted from diesel engines by a centrifugal fluidized bed. AIChE Symposium Serier 90(301):152, 1994.

Turton R, Tardos GI, Ennis BJ. Fluidized bed coating and granulation. In: Yang WC, ed. Fluidization, Solids Handling, and Processing. Westwood: Noyes, 1999, pp 331–434.

Wen CY, Yu YH. A generalized method for predicting the minimum fluidization velocity. AIChE J 12:610–612, 1966.

Westinghouse Electric Corporation. Advanced coal gasification system for electric power generation. Quarterly Progress Report, NTIS No. FE-1514-61, January 1977.

Whiting KJ, Geldart D. A comparison of cylindrical and semi-cylindrical spouted beds of coarse particles. Chem Eng Sci 35:1499–1501, 1980.

Wu SWM, Lim CJ, Epstein N. Hydrodynamics of spouted beds at elevated temperatures. Chem Eng Commun 62:251–268, 1987.

Wurster DE, Lindlof JA. Particle coating apparatus. U.S. Patent 3,241,520 (1966).

Wurster DE, Lindlof JA, Battista JR. Process for preparing agglomerates. U.S. Patent 3,207,824 (1965).

Yang WC. A unified theory on dilute phase pneumatic transport. J Powder Bulk Solids Tech 1:89–95, 1977.

Yang WC. A correlation for solid friction factor in vertical pneumatic conveying lines. AIChE J 24:548–552, 1978.

Yang WC. Jet penetration in a pressurized fluidized bed. I&EC Fundamentals 20:297–300, 1981.

Yang WC. Comparison of jetting phenomena in 30-cm and 3-m diameter semicircular fluidized beds. Powder Technol 100:147–160, 1998.

Yang WC. Engineering and applications of recirculating and jetting fluidized bed. In: Yang WC, ed. Fluidization, Solids Handling, and Processing. Westwood: Noyes, 1999, pp 236–330.

Yang WC, Keairns DL. Recirculating fluidized bed reactor data utilizing a two-dimensional cold model. AIChE Symposium Series 70(141):27–40, 1974.

Yang WC, Keairns DL. Estimating the acceleration pressure drop and the particle acceleration length in vertical and horizontal pneumatic transport lines. Proceedings of the Pneumotransport 3, Bedford (England): BHRA Fluid Engineering, 1977, pp D7-89–D7-98.

Yang WC, Keairns DL. Design of recirculating fluidized beds for commercial applications. AIChE Symposium Series 74(176):218–228, 1978a.

Yang WC, Keairns DL. Design and operating parameters for a fluidized bed agglomerating combustor/gasifier. In: Davidson JF, Keairns DL, eds. Fluidization. Cambridge: Cambridge University Press, 1978b, pp 208–214.

Yang WC, Keairns DL. Momentum dissipation of and gas entrainment into a gas–solid two-phase jet in a fluidized bed. In: Grace JR, Matsen JM, eds. Fluidization. New York: Plenum Press, 1980, pp 305–314.

Yang WC, Keairns DL. Solid entrainment rate into gas and gas–solid, two-phase jets in a fluidized bed. Powder Technol 33:89–94, 1982.

Yang WC, Keairns DL. Studies on the solid circulation rate and gas bypassing in spouted fluid-bed with a draft tube. Can J Chem Eng 61:349–355, 1983.

Yang WC, Keairns DL. Operational characteristics of a 2-D bed with three draft tubes. In: Grace JR, Shemilt LW, Bergougnou MA, eds. Fluidization. New York: Engineering Foundation, 1989, pp 285–292.

Yang WC, Margaritis PJ, Keairns DL. Simulation and modeling of startup and shutdown in a pilot-scale recirculating fluidized bed coal devolatilizer. AIChE Symposium Series 74(176):87–100, 1978.

Yang WC, Keairns DL, McLain DK. Gas mixing in a jetting fluidized bed. AIChE Symposium Series 80(234):32–41, 1984a.

Yang WC, Revay D, Anderson RG, Chelen EJ, Keairns DL, Cicero DC. Fluidization phenomena in a large-scale cold-flow model. In: Kunii D, Toei R, eds. Fluidization. New York: Engineering Foundation, 1984b, pp 77–84.

Yang WC, Ettehadieh B, Haldipur GB. Solids circulation pattern and particles mixing in a large jetting fluidized bed. AIChE J 32:1994–2001, 1986.

Yang WC, Ettehadieh B, Anestis TC, Kettering RE, Haldipur GB, Holmgren JD. Gas mixing in a large jetting fluidized bed. In: Kwauk M, Kunii D, eds. Fluidization'88: Science and Technology, Beijing: Science Press, 1988, pp 1–16.

Yang WC, Newby RA, Keairns DL. Large scale fluidized bed physical model: methodology and results. Powder Technol 82:331–346, 1995.

Ye B, Lim CJ, Grace JR. Hydrodynamics of spouted and spout-fluidized beds at high temperature. Can J Chem Eng 70: 840–847, 1992.

Yoon SM, Kunii D. Gas flow and pressure drop through moving beds. I&EC Process Des Develop 9:559–565, 1970.

21

Standpipes and Nonmechanical Valves

T. M. Knowlton

Particulate Solid Research, Inc., Chicago, Illinois, U.S.A.

1 INTRODUCTION

Plants that process solids are known to have more problems achieving design capacity than plants handling only liquids and gases. Merrow (1985) listed the major sources of problems in operating solids processing plants in his study on the reasons for their poor performance. He found that one of the primary trouble spots in most solids processing plants was in getting the solids to flow smoothly and consistently. Thus in many instances the key to successful process operation is how well the solids transport systems have been designed. Preventing problems from occurring by good design of the elements of a transportation system can speed plant start-up and minimize downtime.

This chapter discusses two important elements of a solids transport system: standpipes and nonmechanical valves. Although they are very simple in configuration, trying to design and operate these devices without a basic understanding of their principles of operating can lead to much frustration and wasted time. Describing how standpipes and nonmechanical valves operate is the purpose of this chapter.

2 STANDPIPES

A standpipe is essentially a length of pipe through which solids flow. Solids can flow through a standpipe in either dilute or dense-phase flow. Standpipes can be vertical, angled, or a mixture of angled and vertical

pipes called a hybrid standpipe. The standpipe was invented by a research group working at the Jersey Standard Company in the 1940s (Campbell et. al., 1948) trying to develop an FCC unit to produce high-octane aviation gasoline in World War II.

The purpose of a standpipe is to transfer solids from a region of lower pressure to a region of higher pressure. This is schematically shown in Fig. 1, where solids are being transferred from a fluidized bed at pressure P_1 to another fluidized bed operating at P_2, which is higher than P_1.

Solids can be transferred by gravity from a low pressure to a higher pressure if gas flows upward *relative* to the downward flowing solids. This relative gas–solids flow will then generate the sealing pressure drop required for the system. The direction of the actual gas flow in the standpipe relative to the standpipe wall can be either up or down and still have the relative gas–solids velocity, v_r, directed upwards. This is sometimes difficult to understand, but it can be explained with the aid of Fig. 2, and the definition of the relative velocity. The relative gas–solids velocity, v_r, is defined as

$$v_r = |v_s - v_g| \qquad (1)$$

where v_s is the solids velocity, and v_g is the interstitial gas velocity ($v_g = U/\varepsilon$).

It is generally easier to visualize what is occurring in a solids transfer system by mentally traveling along with the solids. Therefore the positive reference direction for determining v_r in this chapter will be the direc-

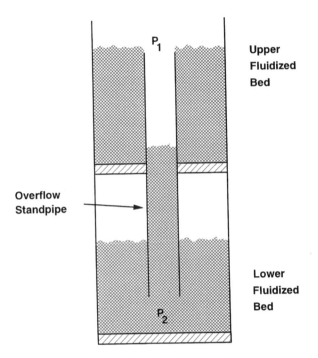

Figure 1 Overflow standpipe.

tion in which the solids are flowing. For standpipe flow, this direction is downward.

In Fig. 2 solids are being transferred downward in a standpipe from pressure P_1 to a higher pressure P_2. Solids velocities in Fig. 2 are denoted by the length of the bold arrows, gas velocities by the length of the dashed arrows, and the relative gas–solids velocity by the length of the thin-lined arrows.

For Case 1 in Fig. 2, solids are flowing downward, and gas is flowing upward relative to the standpipe wall. The relative velocity is directed upward and is equal to the sum of the solids velocity and the gas velocity, i.e.,

$$v_r = v_s - (-v_g) = v_s + v_g \qquad (2)$$

For Case 2, solids are flowing down the standpipe relative to the standpipe wall. Gas is also flowing down the standpipe relative to the standpipe wall, but at a velocity less than that of the solids. For this case, the relative velocity is also directed upward and is equal to the difference between the solids velocity and the gas velocity, i.e.,

$$v_r = v_s - v_g \qquad (3)$$

In both cases, if one were riding down the standpipe with the solids, the gas would appear to be moving upward relative to your reference point.

The gas flowing upward relative to the solids generates a frictional pressure drop. The relationship between the pressure drop per unit length ($\Delta P/Lg$) and the relative velocity for a particular material is determined by the fluidization curve for that material. Normally, this fluidization curve is generated in a fluidization column where the solids are not flowing. However, the relationship also applies for solids flowing in a standpipe.

Nearly all standpipe transfer systems use either Geldart group A or Geldart group B solids. The fluidization curve for Geldart group B solids differs from that for group A solids. For both types of solids, as the relative gas velocity through the bed increases from zero, the $\Delta P/(Lg)$ through the bed increases linearly with v_r. This region is called the packed bed region. At some v_r, the ΔP generated by the gas flowing through the solids is equal to the weight of the solids per unit area, and the solids become fluidized. The relative velocity at this point is termed the interstitial minimum

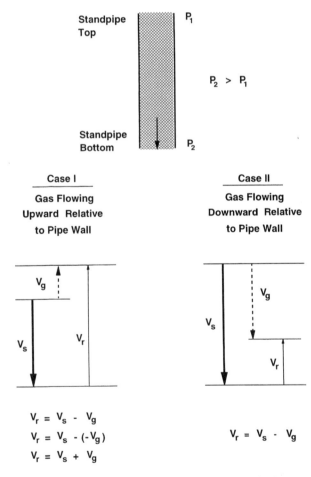

Figure 2 Concept of relative gas–solids velocity.

fluidization velocity, v_{mf}, or U_{mf}/ε_{mf}. The $\Delta P/(Lg)$ at v_{mf} is designated as $\Delta P/(Lg)_{mf}$ and is often referred to as the fluidized bed "density" at minimum fluidization, because $\Delta P/(Lg)$ has the units of density. For more detailed discussion on this subject, see Sec. 1.3 in Chapter 3 "Bubbling Fluidized Beds."

Increases in v_r above v_{mf} do not lead to further increases in $\Delta P/Lg$. For Geldart group B materials, nearly all of any gas flow in excess of that required at v_{mf} goes into the formation of bubbles. Therefore, as v_r increases beyond v_{mf}, $\Delta P/Lg$ remains almost constant and then begins to decrease as the bubble volume in the bed increases.

For Geldart group A materials, as v_r is increased above v_{mf} the solids expand without bubble generation over a certain velocity range. Because of this bubbleless expansion, $\Delta P/L$ decreases over this velocity range. The velocity where bubbles begin to form in group A materials is called the minimum bubbling velocity, v_{mb}. Typical fluidization curves for Geldart group A and Geldart group B materials are shown in Fig. 3.

Standpipes generally operate in three basic flow regimes: packed bed flow, fluidized bed flow, and a dilute-phase flow called streaming flow.

1. *Packed Bed Flow*. In packed bed flow v_r is less than v_{mf}, and the voidage in the standpipe is more or less constant. As v_r is increased, $\Delta P/L_g$ increases more or less linearly in packed bed flow. When a standpipe is operating in the moving packed bed flow regime, a flow condition is sometimes reached that causes the solids to stop momentarily and then start again. This often causes the standpipe to vibrate, and a loud chattering noise can often be heard. This type of flow is called stick–slip flow. It should be avoided, but no method presently exists to predict when it will occur.

2. *Fluidized Bed Flow*. In fluidized bed flow, v_r is equal to or greater than v_{mf}. The voidage in the standpipe can (and generally does) change along the length of the standpipe, and $\Delta P/Lg$ does not change with increasing v_r. There are two kinds of fluidized-bed flow, bubbling and nonbubbling. When a group B solid is fluidized, it always operates in the bubbling fluidized bed mode because bubbles are formed at all relative velocities above v_{mf}. However, for group A solids, there is an operating window corresponding to a relative velocity between v_{mf} and v_{mb}, where the solids are fluidized but no bubbles are formed in the standpipe. A standpipe operating with group A solids and with a relative velocity above v_{mb} operates in the bubbling fluidized bed mode.

Bubbles, especially large bubbles, are undesirable in a standpipe. If a standpipe is operating in the bubbling fluidized bed mode so that the solids velocity, v_s, is less than the bubble rise velocity, u_b, then bubbles will rise and grow by coalescence. The bubbles rising against the downflowing solids hinder and limit the solids flow rate (Knowlton and Hirsan, 1978; Eleftheriades and Judd, 1978) because the solids flow area is reduced by the presence of bubbles. The larger the bubbles are, the greater the hindrance to solids flow.

When the solids velocity in the standpipe is greater than the bubble rise velocity, the solids will carry the bubbles down the standpipe relative to the standpipe wall. It is also possible for bubbles to coalesce and hinder solids flow when they are traveling downward. In this case, the small bubbles are carried downward faster than the larger bubbles. When they catch up to a larger bubble, they coalesce, which results in even larger bubbles and thus hindrance to the solids flow.

Bubbles also reduce the $\Delta P/Lg$ or "density" of the solids in the standpipe. Thus a standpipe operating in the bubbling regime will require a longer length to seal

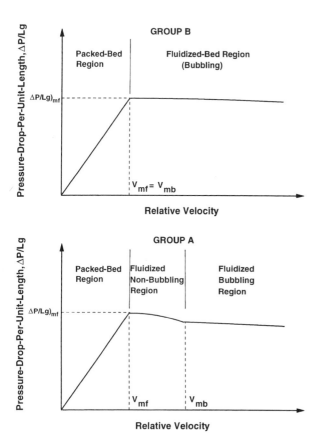

Figure 3   Fluidization curves—group A and group B solids.

the same differential pressure than a standpipe in which the same solids are slightly above minimum fluidizing conditions.

Therefore for optimum fluidized-bed standpipe operation, for group B solids, v_r should be maintained just slightly above v_{mf}; for group A solids, v_r should be maintained in a range just slightly below v_{mb} to just slightly above v_{mb}.

The relative velocity range where it is best to operate standpipes with both group A and group B solids is shown schematically in Fig. 4. In both cases, operation in these areas will either reduce the formation of bubbles or allow standpipe operation with only small bubbles.

Because it is difficult to prevent small bubbles from occurring in standpipes, small-diameter standpipes are more difficult to operate than large-diameter stand-

pipes. In small-diameter standpipes, even small bubbles have a diameter that is a large fraction of the diameter of the standpipe, and these bubbles can hinder solids flow in small standpipes. In large standpipes, this same small bubble diameter will not significantly affect solids flow (Fig. 5)

3. *Streaming Flow.* Underflow standpipes (especially cyclone diplegs) sometimes operate in a dilute-phase streaming flow characterized by high voidages. A substantial amount of gas can be carried down the standpipe when operating in this mode (Geldart and Broodryk, 1991). Excessive gas flow down a dipleg is usually undesirable. It can be minimized by reducing the mass flux in the dipleg and/or increasing the pressure drop that the dipleg has to seal. This usually means increasing the immersion of the dipleg into the fluidized bed.

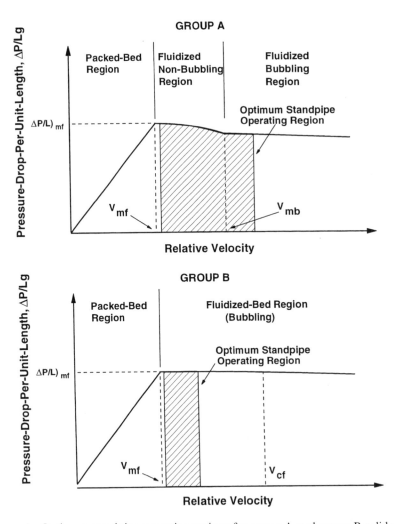

Figure 4 Optimum standpipe operating regions for group A and group B solids.

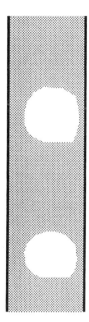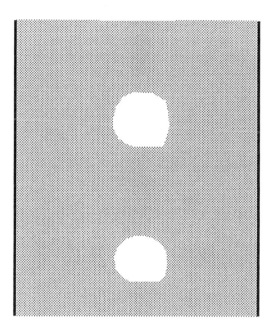

Figure 5 Small and large bubbles in standpipes.

There are two basic types of standpipe configurations, the overflow standpipe (Fig. 6A) and the underflow standpipe (Fig. 6B). The overflow standpipe is so named because the solids overflow from the top of the fluidized bed into the standpipe, and there is *no bed of solids above the standpipe*. In the underflow standpipe, the solids are introduced into the standpipe from the underside, or bottom, of the bed or hopper, and *a bed of solids is present above the standpipe*. With this definition, a cyclone dipleg is classified as an overflow standpipe *because there is no bed of solids* above the entrance to the dipleg.

With two types of standpipe configurations and the two typical standpipe flow regimes (fluidized and packed bed), there are four different types of standpipes:

1. An underflow packed bed standpipe
2. An underflow fluidized bed standpipe
3. An overflow fluidized bed standpipe
4. An overflow packed bed standpipe

All of these standpipe are used extensively in industry except for the overflow packed bed standpipe. It is possible for this type of standpipe to operate, but it is much harder to operate and control than the others. Therefore, it is not used.

Each type of standpipe can be constrained (solids flow limited) at either the top or the bottom. However, top-constrained standpipes are relatively rare, and nearly all standpipes used in fluidized bed standpipe systems are constrained at the bottom. The standpipes discussed in this chapter will all be bottom-constrained standpipes.

In any gas–solids flow system, a pressure drop loop can be defined so that the sum of the pressure drop

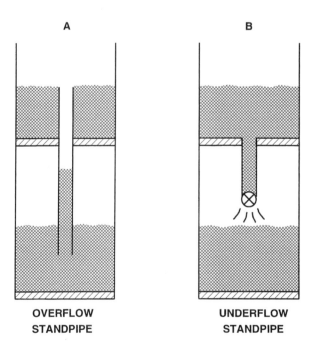

Figure 6 Overflow and underflow standpipes.

components around the loop is zero. In many (but not all) pressure drop loops, the standpipe is the *dependent* part of the loop. This means that the pressure drop across the standpipe will automatically adjust to balance the pressure drop produced by the other *independent* components (which do not automatically adjust their pressure drops) in the loop. How the standpipe pressure drop adjustment is made is different for overflow and underflow standpipes.

Consider the overflow standpipe system shown in Fig. 7A. Solids are being transferred at a low rate from the upper fluidized bed to the lower fluidized bed against the pressure differential P_2-P_1. This pressure drop consists of the pressure drop in the lower fluidized bed from the standpipe exit to the top of the bed (ΔP_{lb}), the pressure drop across the distributor supporting the upper fluidized bed (ΔP_d), and the pressure drop across the upper fluidized bed (ΔP_{ub}), i.e.,

$$P_2\text{-}P_1 = \Delta P_{lb} + \Delta P_d + \Delta P_{ub} \qquad (4)$$

This pressure drop must be balanced by the pressure drop generated in the overflow standpipe. If the stand-

pipe is operating in the fluidized bed mode at a $\Delta P/Lg$ equal to $(\Delta P/Lg)_{mf}$, the solids height in the standpipe, H_{sp}, will adjust so that the pressure buildup generated in the standpipe, ΔP_{sp}, will equal the product of $(\Delta P/Lg)_{mf}$ and H_{sp}, i.e.,

$$\Delta P_{sp} = P_2\text{-}P_1 = \Delta P_{lb} + \Delta P_d + \Delta P_{ub}$$
$$= \left(\frac{\Delta P}{Lg}\right)_{mf} (H_{sp}) \qquad (5)$$

If the gas flow rate through the two beds is increased, ΔP_d increases, while the pressure drops across the two fluidized beds essentially remain constant. Therefore P_2-P_1 increases to P_2'-P_1. The pressure drop across the overflow standpipe will also increase to P_2'-P_1. The standpipe pressure drop increases because the height of fluidized solids in the standpipe increases from H_{sp} to H_{sp}', i.e.,

$$\Delta P_{sp} = P_2'\text{-}P_1 = \left(\frac{\Delta P}{Lg}\right)_{mf} (H_{sp}') \qquad (6)$$

This is shown schematically in Fig. 7B, and the change is also reflected in the pressure diagram. If the increase in the pressure drop across the distributor is so much that H_{sp} must increase to a value greater than the standpipe height available to seal the pressure differential, the standpipe will not operate.

As indicated above, the most common perception of an overflow standpipe is that it consists of a fluidized dense phase at the bottom (the height of which is proportional to the pressure drop across the dipleg), and a dilute phase in the upper part (which does not contribute significantly to the sealing pressure drop in the standpipe). However, recent testing has shown that this picture is not necessarily correct for overflow standpipes operating at appreciable solids mass fluxes. Geldart and Broodryk (1991) reported that at high mass fluxes in an overflow standpipe, there is little or no difference between the densities in the top and bottom sections of the standpipe. This is illustrated in Fig. 8 which shows data obtained in a dipleg at PSRI (1995). As seen in this figure, the typical dense phase at the bottom of the overflow standpipe (cyclone dipleg) only occurs at very low solid mass fluxes. At medium to high solid mass fluxes, the standpipe density was found to be relatively evenly distributed throughout the entire length of the standpipe.

When a cyclone dipleg is operating with a dense phase at the bottom and a dilute phase at the top, the pressure drop through the dilute-phase section at the top is generally assumed to be zero when calculating the required dipleg length. This is not the case,

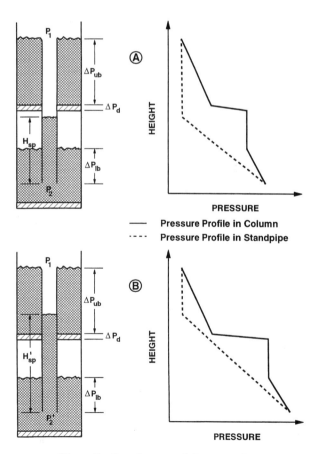

Figure 7 Overflow standpipe operation.

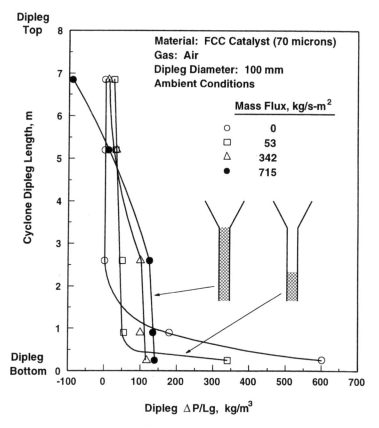

Figure 8 Cyclone dipleg pressure drop profiles.

however, and at high mass fluxes through the dipleg, this dilute-phase pressure drop can be large enough that with a sufficiently long dipleg and a low pressure drop across the dipleg, the dilute phase alone can generate the required sealing pressure drop. This situation is shown schematically in Fig. 9.

When a dipleg operates in streaming flow, substantial amounts of gas can be dragged down the dipleg by the flowing solids. The amount of gas being transferred may be very large (as much as 1/3 of the gas entering the cyclone can be transferred down the dipleg under certain conditions; PSRI, 1995). In many cases, it is not desirable to have so much gas being transferred down the standpipe with the solids. By increasing the pressure drop across the standpipe (increasing the $\Delta P/Lg$ across the standpipe) or reducing the solids mass flux through the standpipe, the amount of gas being transferred down the standpipe can be substantially reduced. Practically, this means that the gas flow down the standpipe or dipleg can be decreased by immersing the standpipe further in the fluidized bed (increasing $\Delta P/Lg$), increasing the diameter of the

standpipe (reducing the mass flux), or reducing the length of the standpipe, if possible (increasing $\Delta P/Lg$).

Wirth (1995) conducted a study of a standpipe operating between a cyclone and a loop seal at the bottom of a standpipe. He also found that if the solids mass flux in the standpipe increased above a certain value, the standpipe lost its dense-phase seal at the bottom. This increased mass flux was accompanied by increasing amounts of gas flowing through the standpipe. If the solids mass flux was increased further beyond a certain threshold value, an even greater flow of gas was observed in the standpipe, with the result that even the upward-flowing part of the loop seal became dilute. Wirth found that this situation could be controlled by decreasing the solids flux through the standpipe. This could be accomplished in two ways, by decreasing the rate of solids flowing around the unit, or more practically, by increasing the diameter of the standpipe. Increasing the diameter of the standpipe decreases the solids mass flux in the downcomer and therefore the dilute-phase density in the standpipe. When the dilute-phase density is decreased below a

WHY DO CYCLONE DIPLEGS OFTEN
OPERATE IN STREAMING FLOW?

- The Total Dipleg Pressure Drop, ΔP_{dip}, is:

$$\Delta P_{dip} = \Delta P_{dll} + \Delta P_{dens}$$

- As the Flux (G) Through the Dipleg is Increased at Constant ΔP_{dip}, ΔP_{dll} Increases

- Therefore, ΔP_{dens} Has to Decrease

- This Trend Continues Until $\Delta P_{dens} = \Delta P_{dll}$

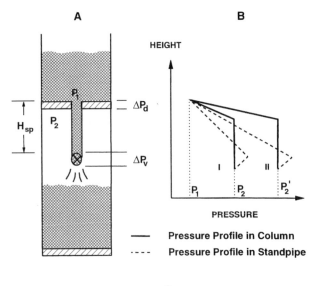

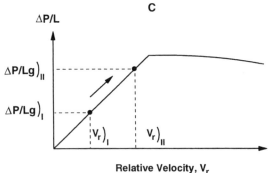

Figure 10 Underflow standpipe operation.

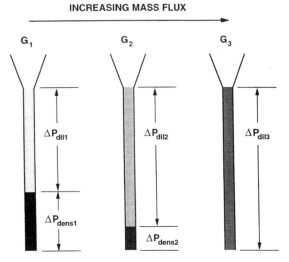

Figure 9 How diplegs operate in streaming flow.

critical value, a dense phase appears at the bottom of the standpipe.

Streaming flow is also more likely to occur with long standpipes. Therefore to prevent this type of situation from occurring and causing poor loop seal operation in combustors, it is recommended that a large-diameter downcomer be used.

In Fig. 10A, solids are being transferred through an underflow standpipe (operating in the packed bed mode) from the upper fluidized bed to the freeboard of the lower fluidized bed against the differential pressure $P_2\text{-}P_1$. The differential pressure $P_2\text{-}P_1$ consists of the pressure drop across the gas distributor of the upper fluidized bed ΔP_d. However, there is also a pressure drop across the solids flow control valve ΔP_v. Therefore the standpipe pressure drop ΔP_{sp} must equal the sum of ΔP_d and ΔP_v, i.e.,

$$\Delta P_{sp} = \left(\frac{\Delta P}{Lg}\right) H_{sp} = \Delta P_d + \Delta P_v = P_2\text{-}P_1 + \Delta P_v$$

$$(7)$$

Thus for this packed bed underflow standpipe case, the standpipe must generate a pressure drop greater than $P_2\text{-}P_1$. This is shown as Case I in the pressure diagram of Fig. 10B.

If the gas flow rate through the column is increased, ΔP_d will increase. If ΔP_v remains constant, then ΔP_{sp} must also increase to balance the pressure drop loop. This is shown as Case II in the pressure diagram of Fig. 10B. Unlike the overflow standpipe case, the solids level in the standpipe cannot rise to increase the pressure drop in the standpipe. However, the $\Delta P/Lg$ in the standpipe must increase in order to balance the pressure drop around the loop. This occurs in a packed bed standpipe because of an increase in v_r in the standpipe. This can be visualized with the aid of Fig. 10C.

For Case I, the pressure drop in the bed was satisfied by having the standpipe operate at point I on the $\Delta P/Lg$ versus v_r curve, as shown in Fig. 10C. When the pressure drop across the distributor increased, the v_r in the standpipe adjusted to generate a higher

$\Delta P/Lg$, $(\Delta P/Lg)_{\text{II}}$, to balance the higher pressure drop.

If the pressure drop across the distributor is increased so that the product of $(\Delta P/Lg)_{\text{mf}}$ (the maximum $\Delta P/Lg$ possible in the standpipe) and the standpipe length, H_{sp}, is less than the sum of ΔP_v and ΔP_d, then the underflow standpipe will not seal.

As stated above, the standpipe is not always the component that adjusts to balance the change in pressure drop in a loop (i.e., it is not always the *dependent* part of the pressure drop loop). In fluidized catalytic crackers, fluidized underflow standpipes are used that do not adjust for changes in pressure drop. This type of standpipe is one of the most widely used standpipes in industry. How does this type of fluidized underflow standpipe operate? Unlike the underflow standpipe in the previous example, the standpipe pressure drop will not adjust to balance pressure drop changes, because the standpipe is operating in the fluidized bed mode. In the fluidized mode, changes in the relative velocity will not cause changes in the pressure drop in the standpipe as in an underflow nonfluidized standpipe. Also, because it is an underflow standpipe and operating full of solids, changes in the bed height to balance the pressure drop are not possible. Therefore this type of

standpipe is designed so that its length is long enough to generate more pressure, or "head," than required. The excess pressure generated by the standpipe is then "burned up" across the slide valve in order to balance pressure drop changes in the other loop components.

With the fluidized underflow standpipe, aeration gas is added to the standpipe to maintain the solids in a fluidized state as they flow down the standpipe. As the solids flow down the fluidized underflow standpipe from a low pressure to a higher pressure, the gas in the standpipe is compressed, which causes the solids to move closer together. When the standpipe is operating at low pressures, the percentage change in gas density from the top of the standpipe to the bottom can be significant. If aeration is not added to the standpipe to prevent this, the solids can defluidize near the bottom of the standpipe (Fig. 11A). Defluidization of solids in the standpipe results in less pressure buildup in the standpipe and a reduction in the solids flow rate around the loop.

To maintain the solids in a fluidized underflow standpipe in a fluidized state, aeration gas is added to the standpipe. Adding the correct amount of gas uniformly (every 2 to 3 meters) in a commercial fluidized underflow standpipe will prevent defluidization

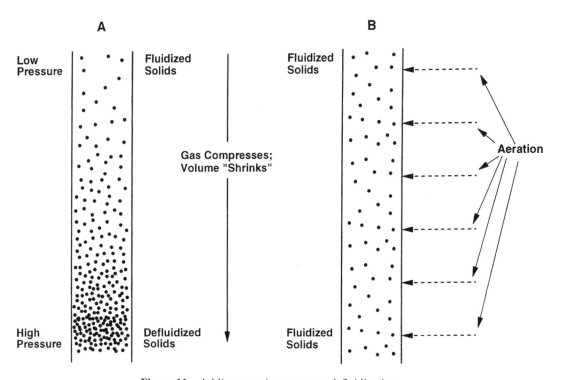

Figure 11 Adding aeration prevents defluidization.

at the bottom of the standpipe (Fig. 11B). If the material flowing in the standpipe is a Geldart group A material, it is required that the aeration be added uniformly along the standpipe. If the aeration is added at only one location (i.e., at the bottom of the standpipe), a large bubble will form in the standpipe at the aeration point (Fig. 12A). If the bubble is large enough, it can restrict the flow of solids down the standpipe. The large bubble forms because it is difficult for the aeration gas to permeate through the very fine solids moving through the standpipe. Therefore it requires a significant area for the gas to dissipate through the very fine particles at the same rate that it is being added through the aeration tap. If the aeration gas is added at several locations, then the bubble size is significantly reduced, and standpipe operation is significantly improved (Fig. 12B). Note in Figs. 12A and 12B that the bubbles are shown extending downward from the aeration point in the direction of flow of the solids. This occurs because the momentum of the solids is much greater than the buoyancy force of the bubble and elongates the aeration bubble in the direction of flow.

For Geldart group B solids, it is often unnecessary to add aeration at several locations along the standpipe to maintain the standpipe in fluidized flow. Adding aeration at the bottom of the standpipe operating with group B solids is sometimes sufficient. This is because the gas can permeate through the larger group B particles much easier than through the group A particles (group A particles have a significantly larger surface area and produce more drag for the same gas flow conditions).

The amount of aeration required to maintain solids in a fluidized state throughout the standpipe was presented by Karri and Knowlton (1993) as

$$Q = 1000 \left[\frac{P_{\mathrm{b}}}{P_{\mathrm{t}}} \left(\frac{1}{\rho_{\mathrm{mf}}} - \frac{1}{\rho_{\mathrm{sk}}} \right) - \left(\frac{1}{\rho_{\mathrm{t}}} - \frac{1}{\rho_{\mathrm{sk}}} \right) \right] \quad (8)$$

In a commercial fluidized underflow standpipe, the amount of aeration theoretically required is added in equal increments via aeration taps located approximately 2 to 3 meters apart. Care should be taken not to overaerate the standpipe. If this occurs, large bubbles are generated in the standpipe that hinder solids flowing down the standpipe. Thus standpipes can be overaerated as well as underaerated.

As indicated above, it is detrimental to have bubbles in standpipes. For fluidized bed underflow standpipes with the standpipe entrance in the fluidized bed, bubbles can be sucked down the standpipe at its entrance if nothing is done to prevent this from occurring. This is especially true when the bed consists of Geldart group

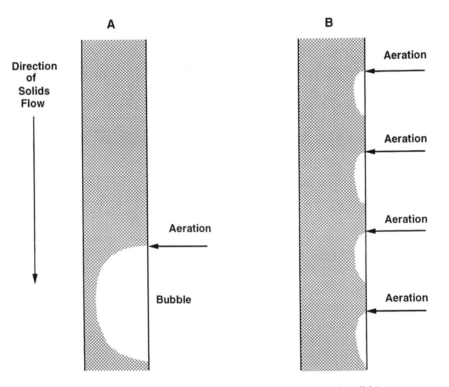

Figure 12 Single and multiple aeration points (group A solids).

A solids. When solids flow from a fluidized bed into the top of an underflow fluidized bed standpipe, the solids are accelerated from a low velocity near 0 m/s in the bed to as much as 2 m/s in the standpipe. This sudden increase in solids velocity can carry bubbles with the solids down into the standpipe and degrade standpipe operation. To prevent this, a cone is often added to the top of the standpipe to minimize the solids velocity at the standpipe entrance and minimize bubble "carry-under." Experience has shown that the diameter of the standpipe inlet cone should be at least four times the area of the standpipe (King, 1992). Standpipe inlet cones typically have an included angle of from 25 to 35 degrees.

In many fluidized beds, a sparger type of gas distributor is used to fluidize the bed. The sparger consists of a pipe with nozzles in it inserted into the bottom of the fluidized bed. Solids flow down through the distributor and into the standpipe. Another technique to prevent bubbles from entering the standpipe can be used with sparger grids. Instead of having the standpipe entrance in the bed, the standpipe entrance is located below the sparger grid. As the solids flow between the sparger grid and the standpipe entrance, the bubbles dissipate and do not enter the standpipe. Generally, an aeration ring is added around the standpipe to ensure that the solids are fluidized as they enter the standpipe.

If Geldart group A solids are defluidized when they enter the standpipe, they are extremely difficult to refluidize (Ross, 1997). Therefore it is necessary to prevent this defluidization. It is somewhat puzzling why it is so difficult to refluidize group A solids. In a laboratory fluidized bed, these solids are relatively easy to fluidize. The reason for the difference is that in the standpipe the solids are flowing at a relatively high solids velocity and have a high momentum. Adding gas to flowing solids streams gives a different result from in a laboratory bed in which the net solids velocity is zero. It is more difficult for gas to permeate through a moving solids stream possessing significant momentum. The gas tends to segregate and cannot permeate the moving solids mass easily. This leads to problems with gas distribution in flowing fluidized solids systems (Rall et al., 1999).

2.1 Geldart Group C Solids Flow in Standpipes

Geldart group C particles typically have an average particle size of less than about 30 microns. These particles are also called cohesive particles because interparticle forces are greater than body forces for these

materials, which causes the particles to clump together. Flour is a typical Geldart group C material. The relative strength of the interparticle forces in these small materials causes the particles to bridge and to stop flowing in standpipes.

However, there is technique that allows group C particles to flow through standpipes. If gas is pulsed into the bottom of a standpipe operating with group C particles, the particles can be made to flow. The theory behind this technique is that the pressure aeration pulse is transmitted through out the standpipe and breaks the solids bridge in the standpipe. Before the bridge can form again, another pulse of aeration is added. This technique has been used with good results in standpipes operating with 10 micron oil shale (Knowlton, 1991). Typically, the pulsing sequence will be a pulse duration of from 0.1 to 0.5 seconds followed by a no-pulse period of from 3 to 10 seconds. This sequence is repeated (by using a timer and a solenoid valve) as long as necessary. For very long standpipes, it may be necessary to add a second pulsing point.

In a 30 cm diameter underflow standpipe with a slide valve at the bottom to control the solids flow rate, it was found that this type of pulsing was necessary to start group C solids flowing in the standpipe and to keep them flowing at low solids velocities. After the solids had reached a relatively high velocity in the standpipe, it was found that the pulsing could be shut off and the solids would flow well.

2.2 Standpipes in Recirculating Solid Systems

Many chemical processes recirculate solids. Catalytic systems recirculate catalyst in a reaction/regeneration cycle. First the catalyst is used to supply heat or a reactant to the process; it is then transferred to a separate vessel to regenerate the catalyst, and then it is returned to the reactor. Circulating fluidized bed combustors recirculate fuel and ash around a loop to burn the fuel completely. A system with a cyclone collecting entrained solids above a fluidized bed and returning the solids to the bed via the cyclone dipleg is also a recirculating solid system. All of these recirculation systems employ standpipes.

There are two basic types of solids recirculation systems, automatic and controlled. In the automatic system, the solids are recirculated around the loop at their "natural" recirculation rate without being controlled. In the controlled recirculation system, a valve is used to control the solids flow rate. In the automatic system, overflow standpipes are almost always employed. In the controlled recirculation systems, the standpipe is an underflow standpipe. It can be either a fluidized or a nonfluidized underflow standpipe.

2.3 Automatic Solid Recirculation Systems

One of the simplest types of automatic solids recirculation systems is the cyclone/dipleg system utilized to collect entrained solids and return them to the bed. This type of system is shown in Fig. 13 for a fluidized bed containing both primary and secondary internal cyclones. External cyclone return systems are also automatic recycle systems. All solids entering the primary and secondary cyclones are returned to the bed automatically via the primary and secondary diplegs.

The pressure balance around the system for returning solids to the bed via the secondary cyclone dipleg is

$$\Delta P_{\text{fluid bed}} + \Delta P_{\text{cy}1} + \Delta P_{\text{cy}2} = \Delta P_{\text{dip}2} = \rho_{\text{dip}2} H_{\text{dip}2} g \tag{9}$$

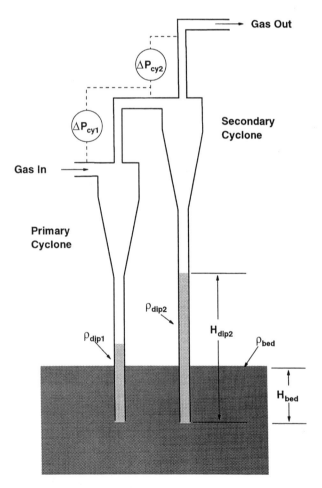

Figure 13   Fluid bed cyclone recirculation system.

For solids to be returned to the bed in the secondary cyclone dipleg, the secondary dipleg length, $H_{\text{dip}2}$, should be at least

$$H_{\text{dip}2} = \frac{\Delta_{\text{fluid bed}} + \Delta P_{\text{cy}1} + \Delta P_{\text{cy}2}}{\rho_{\text{dip}2} g} \tag{10}$$

As shown above, if the primary or secondary cyclone pressure drop increases, the immersion of the secondary dipleg in the fluidized bed increases, or the solids density in the secondary cyclone dipleg decreases, the solids level in the overflow standpipe (secondary cyclone dipleg) increases. Conversely, if the primary or secondary cyclone pressure drop decreases, the immersion of the secondary cyclone dipleg in the fluidized bed decreases, or the solids density in the secondary cyclone dipleg increases, the solids level in the secondary cyclone dipleg decreases to adjust.

2.4 Controlled Solid Recirculation Systems

In a controlled solids recirculation system, a valve is used to control the solids flow rate in the system. In recirculation systems that are controlled, the standpipe is almost always an underflow (either fluidized or nonfluidized) standpipe. A typical type of controlled recirculation system is that used by FCC units in petroleum refineries. This type of recirculation loop (or something very similar) is used by other catalytic and noncatalytic processes as well.

The primary standpipes in FCC units are underflow fluidized bed standpipes, not overflow fluidized bed standpipes. Standpipes in FCC units can be either completely vertical, completely angled, or a combination of vertical and angled sections. Nearly all FCC units incorporate two standpipes in their loop systems. The solids flow rate around the system is controlled by a slide valve or a cone valve in each standpipe.

A typical FCC riser/recirculation system developed by UOP (called a side-by-side unit) is shown in Fig. 14. A pressure drop balance around this unit gives

$$\Delta P_{\text{regen}} + \Delta P_{\text{sp}1} - \Delta P_{\text{sv}1} - \Delta P_{\text{riser}}$$
$$+ \Delta P_{\text{stripper}} + \Delta P_{\text{sp}2} + \Delta P_{\text{sv}2} = 0 \tag{11}$$

In this unit, hot catalyst is introduced into a dilute-phase riser where it is contacted with crude oil. The hot solids vaporize the oil and catalytically crack it into lower molecular weight hydrocarbons in the riser. The catalyst reactivity is significantly reduced by carbon deposition during this step. Therefore the

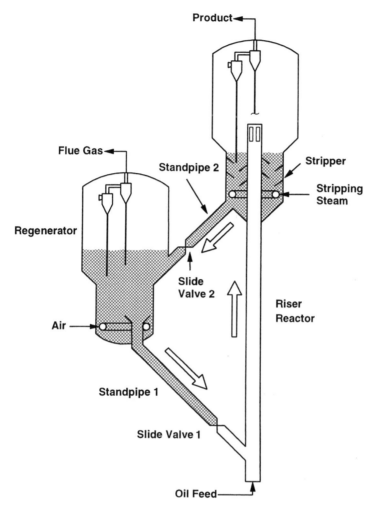

Figure 14 Schematic drawing of side-by-side UOP catalytic cracking unit.

catalyst is transferred to a regenerator where the carbon is burned off of the catalyst. This heats up the catalyst and also restores its reactivity. However, before the catalyst is returned to the regenerator via an angled standpipe/slide valve combination, hydrocarbon vapors are removed from the catalyst in a steam stripper. From the regenerator, the catalyst is transferred down another angled standpipe/slide valve combination and is injected into the bottom of the riser.

In most FCC units, the solids flow rate around the system is controlled by a slide valve. FCC units designed by the Kellogg Brown and Root Company use another type of valve called a cone, or plug valve, to control the solids flow rate (Wrench et al., 1985).

Solids flow in standpipes in FCC units such as those shown in Fig. 14 are generally controlled in the follow-

ing manner. In standpipe 1, slide valve 1 controls the temperature at the outlet of the riser by varying the flow of regenerated solids to the riser. The slide valve in standpipe 2 controls the fluidized solids level in the stripper (King, 1992). Generally, slide valves in FCC standpipes are operated with a pressure drop between about 15 kPa (the minimum required for good control of the catalyst) and 100 kPa. Higher pressure drops result in excessive valve wear. Good valve design will result in the valve operating between 25 and 75% open, with the valve port area about 25 to 50% of the standpipe open area. A typical design is to for the valve to operate in the middle of the ranges shown above (i.e., at about 50 kPa pressure drop and 50% open) as reported by King (1992).

Underflow fluidized standpipes in FCC units are operated in a vertical configuration, a completely angled configuration, or a hybrid configuration in

Vertical   **Hybrid**   **Angled**

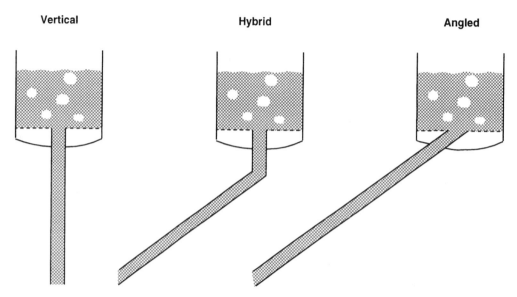

Figure 15 Vertical, angled, and hybrid fluidized underflow standpipes.

which both vertical and angled sections are present (Fig. 15). Angling a standpipe is a convenient way to transfer solids between two points that are separated horizontally as well as vertically. However, it has been found (Karri and Knowlton, 1993; Yaslik, 1993) that long angled underflow fluidized standpipes do not perform as well as vertical standpipes.

Sauer et. al. (1984) and Karri and Knowlton (1993) studied hybrid angled standpipe operation using transparent standpipes to allow visual observation of the flow. Both found the gas and solids separated in the standpipe, with the gas bubbles flow-

ing up along the upper portion of the standpipe while the solids flowed down along the bottom portion of the standpipe (Fig. 16). The pressure buildup in the hybrid standpipe was lower than that in the vertical standpipe, and Karri and Knowlton (1993) reported that the maximum solids mass flux possible in a hybrid angled underflow fluidized standpipe was less than that attainable in a vertical underflow fluidized standpipe (Fig. 17). The principal reason for this is that the rising bubbles in the angled section of the standpipe become relatively large at a low solids flow rate (and low aeration rate). At a certain solids

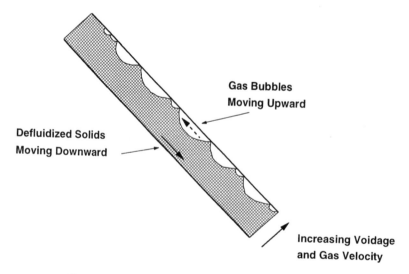

Figure 16 Gas and solids flow in an angled standpipe.

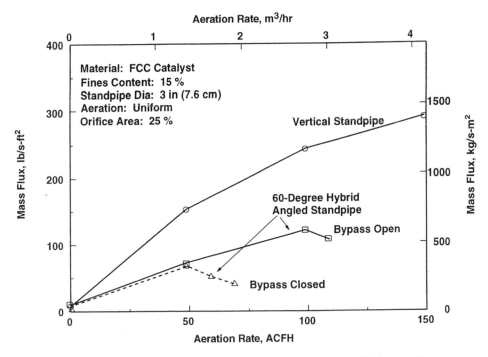

Figure 17 Comparison of flow through a vertical and hybrid standpipe.

mass flux, the bubbles become large enough to bridge across the vertical section at the top of the standpipe, hindering the solids flow. When this occurs, the maximum solids flow rate in the hybrid angled standpipe has been achieved.

Karri et. al. (1995) showed that the solids flow rate through a hybrid angled standpipe can be increased if a bypass line (Fig. 18) is added between the top of the angled section of the standpipe and the freeboard of the bed above it. The effect of adding the bypass line at the top of the standpipe can be seen in Fig. 17. The bypass line allows the bubble gas from the angled section to bypass the vertical section of the pipe so that large bubbles are not formed there. Thus the solids flow rate can be increased. Karri et. al. (1995) reported that if the bypass was used, the solids flow rate could be increased to such a value that the solids velocity in the hybrid standpipe became greater than the bubble rise velocity, and the bubbles were carried down the standpipe with the solids. When the bubbles were being carried down the standpipe by the solids, the bypass line could then be closed and the standpipe would operate without slugging in the vertical section.

Even though vertical standpipes can transfer solids more efficiently than hybrid angled standpipes, true angled standpipes (those containing no vertical section) are commonly operated satisfactorily in large

FCC units with Geldart group A catalyst. However, these standpipes are relatively short and are designed so that the mass flux through them is not too high, so that they can be operated satisfactorily. Yaslik (1993) found that a long angled standpipe had a limited solids circulation rate relative to vertical standpipes. Thus when operating a hybrid angled standpipe or a true angled standpipe it is essential (1) to keep the solids mass flux through the standpipe below a value that will lead to slugging, and (2) to keep the line as short as possible so that the large gas slugs will not have as great a length in which to form.

2.5 Nonmechanical Solids Flow Devices

A nonmechanical solids flow device is one that uses only aeration gas in conjunction with its geometrical shape to cause particulate solids to flow through it. Nonmechanical solids flow control devices have several advantages over mechanical solids flow devices:

1. *They have no moving mechanical parts*, which are subject to wear and/or seizure. This feature is especially beneficial when operating at elevated temperatures and pressures.

2. *They are inexpensive* because they are constructed of ordinary pipe and fittings.

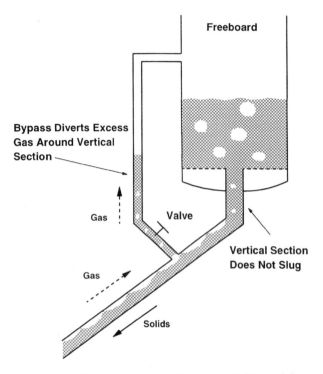

Figure 18 Bypass to increase flow in a hybrid standpipe.

3. *They can be fabricated in-house*, which avoids the long delivery times often associated with the purchase or replacement of large mechanical valves.

Nonmechanical devices can be operated in two different modes:

1. In the valve mode to control the flow rate of particulate solids
2. In an "automatic" solids flow-through mode

There is often confusion as to how these modes differ, and what kind of nonmechanical devices should be used for a particular application. Each mode of operation is discussed below.

2.6 Nonmechanical Valve Mode Operation

In the valve mode of operation, the solids flow rate through the nonmechanical device is controlled by the amount of aeration gas added to it. The most common types of nonmechanical valves are the L-valve and the J-valve. These devices are shown schematically in Fig. 19. The primary difference between these devices are their shapes and the directions in which they discharge solids. Both devices operate on the same principle. It is harder to fabricate a smooth 180-degree bend for a typical J-valve. Therefore the J-valve can be approximated and configured more simply by the geometry shown in Fig. 19C.

The most common nonmechanical valve is the L-valve, because it is easiest to construct, and also because it is slightly more efficient than the J-valve (Knowlton et al., 1981). Because the principle of operation of nonmechanical valves is the same, nonmechanical valve operation is presented here primarily through a discussion of the characteristics of the L-valve.

Solids flow through a nonmechanical valve because of drag forces on the particles produced by the aeration gas. When aeration gas is added to a nonmechanical valve, gas flows downward through the particles and around the constricting bend. This relative gas–solids flow produces a frictional drag force on the particles in the direction of flow. When this drag force exceeds the force required to overcome the resistance

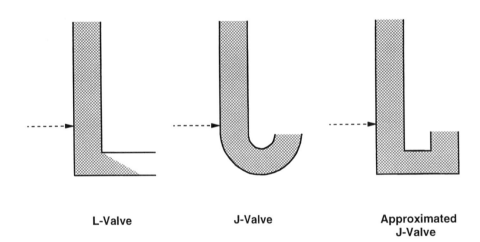

L-Valve   J-Valve   Approximated
                                                                J-Valve

Figure 19 The most common nonmechanical valves.

to solids flow around the bend, the solids flow through the valve.

The actual gas flow that causes the solids to flow around the L-valve is not just the amount of aeration gas added to the valve. If gas is traveling down the moving packed bed standpipe (which occurs in most cases) with the solids, the amount of gas that flows around the L-valve bend, Q_T, is the sum of the standpipe gas flow, Q_{sp}, and the aeration gas flow, Q_A, as shown in Fig. 20A. If the gas is flowing up the standpipe (which would be the case when the standpipe is operating with low solids flow rates and/or large solids), then the amount of gas flowing around the bed, Q_T, is the difference between the aeration gas flow and the gas flowing up the standpipe, as shown in Fig. 20B.

When aeration is added to a nonmechanical valve, solids do not begin to flow immediately. The initial aeration gas added is not enough to produce the frictional force required to start solids flow. Above the threshold amount of gas required to initiate solids flow, additional aeration gas added to the valve causes the solids flow rate to increase, and reducing the amount of aeration to the valve causes the solids flow rate to decrease. In general, there is little hysteresis in the aeration-vs.-solids flow rate curve for a nonmechanical valve.

Nonmechanical valves work best with materials having average particle sizes between 100 and 5000 microns. These materials are in Geldart groups B and D. Materials with average particle sizes greater than about 2000 microns require substantial amounts of gas to generate the drag forces required to make the solids flow around the constricting bend. This is because larger solids have less surface area available for the generation of the drag forces required to produce flow through a nonmechanical valve. These larger materials work best in nonmechanical valves if there are smaller particles mixed in with the larger ones. The smaller particles fill the void spaces between the larger particles and decrease the voidage of the solids mixture, thereby increasing the drag on the entire mass of solids when aeration is added, and causing the solids to move through the constricting bend.

In general, Geldart group A materials (with average particle sizes from approximately 30 to 100 microns) do not work in L-valves. Group A materials retain air in their interstices and remain fluidized for a substantial period of time when they are added to the standpipe attached to the nonmechanical valve. Because they remain fluidized, they flow through the constricting bend like water, and the L-valve cannot control the solids. Although most group A materials cannot be used with nonmechanical devices operating in the control mode, materials at the upper end of the group A classification that contain few fines (particles smaller than 44 microns) can and have been controlled in nonmechanical valves.

Geldart group C materials (with average particle sizes less than about 30 microns) have interparticle forces that are large relative to body forces and are very cohesive (flour is a typical example). These materials do not flow well in any type of pipe, and do not flow well in L-valves.

Nonmechanical valves are used extensively in CFBC systems where Geldart group B solids are used. They are not used in FCC circulating systems where Geldart group A solids are used.

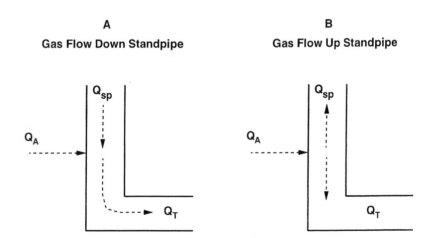

Figure 20 Gas flow around L-valve bend.

A nonmechanical device operating in the valve mode is always located at the bottom of an underflow standpipe operating in moving packed bed flow. The standpipe is usually fed by a hopper, which can either be fluidized or nonfluidized. Knowlton and Hirsan (1978) and Knowlton et al. (1978) have shown that the operation of a nonmechanical valve is dependent upon the pressure balance and the geometry of the system.

Consider the circulating fluidized bed/L-valve return system shown in Fig. 21. The high-pressure point in such a recycle loop is at the L-valve aeration point. The low-pressure common point is at the bottom of the cyclone. The pressure drop balance around the recycle loop is such that

$$\Delta P_{\text{L-valve}} + \Delta P_{\text{CFB}} + \Delta P_{\text{cy}} = \Delta P_{\text{sp}} + \Delta P_{\text{surge hopper}}$$

$$(12)$$

The surge hopper pressure drop is generally negligible, so that the above equation may be written

$$\Delta P_{\text{sp}} \cong \Delta P_{\text{L-valve}} + \Delta P_{\text{CFB}} + \Delta P_{\text{cy}} \qquad (13)$$

The moving packed bed standpipe is the dependent part of the pressure drop loop in that its pressure

drop adjusts to exactly balance the pressure drop produced by the sum of the pressure drops on the independent side of the loop. However, there is a maximum pressure drop per unit length ($\Delta P/Lg$) that the moving packed bed standpipe can develop. This maximum value is the fluidized bed pressure drop per unit length, $\Delta P/Lg)_{\text{mf}}$, for the material.

The independent pressure drop can be increased by increasing any or all of the pressure drops across the CFBC, the cyclone, or the L-valve. For a constant gas velocity in the riser, as the solids flow rate into the bed is increased, the independent part of the pressure drop loop increases. The moving packed bed standpipe pressure drop then increases to balance this increase. It does this by automatically increasing the relative gas–solids velocity in the standpipe. Further increases in the solids flow rate can occur until the $\Delta P/Lg$ in the moving packed bed standpipe reaches the limiting value of $\Delta P/Lg)_{\text{mf}}$. Because of its reduced capacity to absorb pressure drop, a short standpipe reaches its maximum $\Delta P/Lg$ at a lower solids flow rate than a longer standpipe. Thus the maximum solids flow rate through a nonmechanical valve depends upon the length of standpipe above it.

As indicated above during the discussion of how an underflow packed bed standpipe operates, the pressure drop in such a standpipe is generated by the relative velocity, v_r, between the gas and solids. When v_r reaches the value necessary for minimum fluidization of the solids, a transition from packed bed to fluidized bed flow occurs. Any further increase in v_r results in the formation of bubbles in the standpipe. These bubbles hinder the flow of solids through the standpipe and cause a decrease in the solids flow rate.

To determine the minimum standpipe length required for a particular solids flow rate, it is necessary to estimate the pressure drop on the independent side of the pressure drop loop at the solid flow rate required. The minimum length of standpipe necessary, L_{min}, is

$$L_{\text{min}} = \frac{\Delta P_{\text{independent}}}{\Delta P/Lg)_{\text{mf}}} \qquad (14)$$

The actual length of standpipe selected for an L-valve design should be greater than L_{min} to allow for the possibility of future increases in the solids flow requirements and to act as a safety factor. Standpipe lengths are typically designed to be 1.2 to 2 times L_{min}, depending on the length of the standpipe.

To determine the diameter of an L-valve, it is necessary to select the linear solids velocity desired in the standpipe. Nearly all L-valves operate over a linear

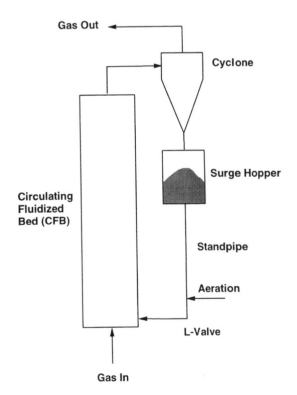

Figure 21 CFB/L-valve loop configuration.

particle velocity range of between 0 to 0.3 m/s in the standpipe. Velocities greater than about 0.3 m/s can result in stick–slip flow in the standpipe. Although an L-valve may theoretically be designed for any linear solids velocity in the standpipe above it, a value near 0.15 m/s is usually selected to allow for substantial increases or decreases in the solids flow rate.

It is desirable to add aeration to an L-valve as low in the standpipe as possible. This results in maximizing standpipe length and the minimizing L-valve pressure drop, both of which increase the maximum solids flow rate through the L-valve. However, if aeration is added too low in the standpipe of the L-valve, gas bypassing results and solids flow control can be insensitive and/or ineffective.

Knowlton and Hirsan (1978) found that aeration was most effective if it was added at a length-to-diameter (L/D) ratio of 1.5 or more above the centerline of the horizontal section of the L-valve. Aeration added at the centerline of the horizontal section or at the bottom of the centerline of the standpipe was found to bypass directly to the top of the horizontal section. Thus it was not being efficiently utilized to drag the solids through the constricting bend. The aeration tap locations and the solids flow rate versus aeration rate curves for each aeration tap location investigated in their study are shown in Fig. 22. In summary, L-valve aeration should be added above

the constricting bend. To assure good operation (i.e., to prevent bypassing), it should be added at an L/D of greater than 1.5 above the centerline of the horizontal section of the L-valve.

In order for an L-valve to operate properly, the L-valve horizontal section length must be kept between a minimum and maximum length. The minimum length must be greater than the horizontal length, H_{min}, to which the solids flow owing to their angle of repose. For design purposes, the minimum horizontal length should be between 1.5 to 2 times H_{min}. In general, the shorter the L-valve, the better the operation of the device.

If the L-valve is operating properly, solids flow through it in small pulses at a relatively high frequency, and for all practical purposes, the solids flow is steady. If the L-valve horizontal section is too long, solids flow can become intermittent (slugging, stopping, and then surging again). In L-valves with long horizontal sections, a slug of solids builds up to such a size that it blocks the pipe, and solids flow stops momentarily. Gas pressure then builds up behind the slug until it becomes so great that the slug collapses and the solids surge momentarily through the L-valve. Another slug then builds up after the gas pressure is released, and the pattern repeats. This cycle does not generally occur if the L-valve horizontal length is less than an L/D ratio of about 8 to 10. Adding additional

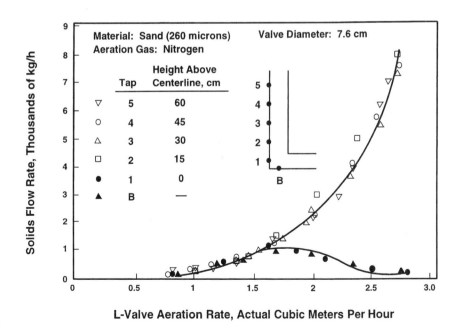

Figure 22 Effect of aeration location on L-valve operation.

gas to the horizontal section prevents slug formation in long L-valves and results in smoother solids flow. The extra gas does not affect control of the valve.

The extra gas for the horizontal section can be added in several ways. Adding aeration taps to the bottom of the L-valve works well except that the taps tend to become plugged if the gas to the taps is shut off. Two other methods have been adopted by various L-valve operators. In one, gas is added to the horizontal section via taps installed along the top of the L-valve section. This prevents plugging of the taps when gas to them is shut off. However, this gas is not as effective as gas added to the bottom, because some of it bypasses along the top of the L-valve. Another method is to insert a tube with holes drilled along its bottom into the lower part of the horizontal section. This tube can be inserted and withdrawn as required. For hot L-valves, this tube may warp and move to the top of the pipe. Therefore, it should be prevented from moving to the top of the section by means of a loose clamp that can allow for longitudinal expansion.

Knowlton and Hirsan (1978) have determined the effect of varying geometrical and particle parameters on the operation of the L-valve. The effects of these parameters are summarized here. L-valve aeration requirements increase with

1. Increasing vertical section diameter
2. Increasing particle size
3. Increasing particle density

L-valve pressure drop increases with

1. Increasing solids flow rate
2. Increasing particle density
3. Decreasing horizontal section diameter

The aeration rate, solids flow rate, and pressure drop relationships in an L-valve have been estimated by Geldart and Jones (1991) and Yang and Knowlton (1993). The Yang and Knowlton method is somewhat involved. It relates L-valve flow to a variable solids area (A_0). This technique first requires estimating or measuring the bulk density at minimum fluidization, the terminal velocity of the average particle size of solids flowing through the L-valve, and the L-valve pressure drop. From these values and the L-valve diameter and horizontal length, the relationship between Q_T, Q_{sp}, and A_0, and W_s (the solids flow rate through the L-valve) is determined by solving the following four equations by trial and error:

1. The Jones and Davidson (1965) equation relating valve pressure drop to solids flow rate and valve open area, i.e.,

$$\Delta P_{L\,valve} = \frac{1}{2\rho_p(1 - \varepsilon_{mf})}\left[\frac{W_s}{C_D A_0}\right]^2 \tag{15}$$

2. The relationship between the amount of aeration added to the L-valve (Q_A), the amount of gas flowing down the standpipe (Q_{sp}), and the amount actually flowing around the L-valve bend (Q_T):

$$Q_T = Q_a + Q_{sp} \tag{16}$$

3. An equation estimating the amount of gas flowing down the standpipe (this equation assumes no slip between the solids and gas in the standpipe):

$$Q_T = \frac{W_s \varepsilon_{mf}}{\rho_p(1 - \varepsilon_{mf})} \tag{17}$$

4. The empirical equation relating Q_T and the variable L-valve area (A_0):

$$\frac{Q_T}{(\pi/4)D_v^2 L_v} = 1.9 + 7.64\frac{U_t A_0}{(\pi/4)D_v^2 L_v} \tag{18}$$

This technique relates L-valve aeration and solids flow rate to within about ±40%. The procedure was developed from data on L-valves ranging in size from 50 to 150 mm in diameter.

If possible, it is recommended that the basic pressure drop as a function of aeration relationship be obtained from a small L-valve test unit. Alternatively, these parameters may be extrapolated from the data presented by Knowlton and Hirsan (1978).

People are often surprised that L-valve aeration gas requirements at high temperatures are small when compared to aeration requirements at low temperature. The reason for this is that the viscosity of a gas increases at high temperatures. Therefore, the drag on the particles for the same external aeration rate increases. This results in an increase in the solids flow rate through the L-valve. Because the viscosity of a gas can increase by a factor of 2 to 2.5 as the temperature is increased, the amount of gas required to flow a certain solids flow rate at high temperature can be as little as half of that at low temperatures for the same aeration rate.

In addition, if particles flowing through the L-valve are Geldart group B particles that lie near the AB boundary, increasing system temperature can cause the particles to cross to the A side of the boundary (Grace, 1986). If this occurs, the L-valve may then experience flow problems at high temperatures not observed at low temperature.

2.7 Automatic Solids Flow Devices

Nonmechanical devices can also be used to pass solids through them automatically. In the automatic mode, they serve as a simple flowthrough device without controlling the solids. If the solids flow rate to the device is changed, the device automatically adjusts to accommodate the changed flow rate. These devices are primarily used to assist in sealing pressure (in conjunction with a standpipe) and to reroute the solids where desired. There are several types of these nonmechanical devices. The most popular are the straight cyclone dipleg, a loop seal (also often called a J-valve, siphon seal, or fluoseal), and an L-valve (in this case "valve" is a misnomer since in this mode it is not controlling the solids flow).

In circulating fluidized bed systems, the most frequent application of automatic nonmechanical devices is to recycle collected cyclone "fines" back to the CFB. The cyclone discharge is at a lower pressure than the desired return point in the CFB, so the solids must be transferred against a pressure gradient. In the past, lockhoppers or rotary valves were sometimes used to transfer these solids—generally with less than satisfactory results. The best and simplest way to return the fines is to use a nonmechanical device. The simplest nonmechanical device used to recycle cyclone fines back to a reactor is a straight cyclone dipleg. However, the dipleg must be immersed in a fluidized dense-phase bed (or fitted with a trickle valve or another control device at its discharge end if it is not immersed in the bed) in order to perform properly. Since cyclones in CFBs are generally external to the bed and located some distance from the CFB, a vertical cyclone dipleg cannot be used. Angled diplegs tend to slug and perform poorly for essentially the same reason as angled standpipes (see above). Therefore, loop seals, seal pots, and L-valves are employed to return the solids to the bed. A V-valve, another type of solids return device, can also be used in this manner. A discussion of the various types of automatic nonmechanical devices used is given below.

2.7.1 Seal Pot

A seal pot is essentially an external fluidized bed into which the cyclone fines discharge via a straight dipleg (Fig. 23A). The solids and gas from the cyclone and the fluidizing gas for the seal pot are discharged via a

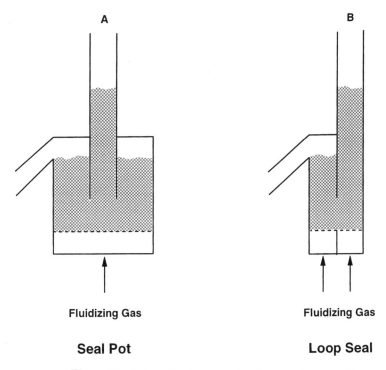

A   **B**

Fluidizing Gas   **Fluidizing Gas**

Seal Pot   **Loop Seal**

Figure 23   Schematic drawing of seal pot and loop seal.

downwardly angled overflow line to the desired return point. Seal pots have been used for many years and are reliable devices. However, one must be sure that the fludizing gas velocity is high enough that the lateral solids transfer rate in the fluidized bed is designed to be greater than the solids flow rate through the dipleg for good operation. This is generally accomplished by increasing the fluidizing gas velocity in the seal pot until satisfactory operation is achieved.

With a seal pot, the solids in the dipleg rise to a height necessary to seal the pressure drop around the recycle solids transfer loop. In general, seal pot transfer problems are associated more with the standpipe than the seal pot itself. These problems depend to a large extent on the particle size of the solids. Geldart groups A and B particles can be transferred with relative ease. Cohesive group C particles are difficult to transfer unless auxiliary means (vibration, pulsing) are used to assist the nonmechanical device. For low solids flow rates, it is recommended that a straight horizontal opening be used at the dipleg exit. Experience has shown that diplegs with mitered bends require a relatively high solids mass flux through them to operate properly.

Aerating standpipes is sometimes necessary in order to achieve optimum solids flow through them. However, too much aeration is just as detrimental as too little aeration. Too much aeration leads to bubble formation and slugging. Often seal pots operate satisfactorily with no additional aeration added to the standpipe, especially with group B solids. However, it is usually wise to add aeration taps into the standpipe discharging into the seal pot. They will be there if needed and do not have to be used if they are not. An aeration point at the bottom of the standpipe and aeration points approximately every 2 to 3 m of solid seal height should be sufficient. It is also best to use a separate aeration control rotameter for each aeration point, especially in research units. In a commercial unit, restriction orifices are generally sufficient to ensure equal flows to each aeration location.

2.7.2 Loop Seal

The loop seal (Fig. 23B) is essentially a variation of the seal pot. Like the seal pot, it is composed of a standpipe and a fluidized bed section. However, the solids from the standpipe enter the fluidized bed from the side. This allows the fluidized bed portion of the device to be smaller in diameter, resulting in a smaller transfer device, lower fluidization gas requirements, and a more efficient operation.

The height of the vertical flow portion of the loop seal can be increased to "insulate" the operation of the loop seal from the pressure fluctuations in the bed to which it is discharging. Increasing the height of the vertical portion of the loop seal also helps prevent blowing the seal leg of the loop seal if a pressure upset occurs. However, a large pressure upset will still blow the seal and cause other operational problems as well.

It is essential to fluidize the upflow section of the loop seal in order for Geldart group B solids to flow smoothly through the loop seal. For Geldart group A solids, little or no fluidization gas may be required to be added to the upflow section, because they may not defluidize when flowing around the bed. With both types of solids, the minimum amount of aeration gas necessary to produce smooth, steady flow is what should be used. Too much gas results in slugging and unsteady flow. It is also recommended that aeration at the base of the loop seal be separated so that the amount of aeration added below the bottom of the dipleg and the amount added for the upflow section can be varied independently of each other.

As noted above, Wirth (1995) found that using a cyclone dipleg too small in diameter caused an excessive amount of gas to be carried down the dipleg. At high mass fluxes, the amount of gas flow was so great that it caused the upward-flowing part of the loop seal to become dilute. This problem was alleviated by increasing the dipleg diameter to decrease the mass flux in the dipleg. Because the amount of gas carried down the dipleg is proportional to the solids mass flux in the downcomer, increasing the dipleg diameter decreased the solids mass flux for the same solids flow rate, and also decreased the amount of gas carried down the dipleg.

2.7.3 N-Valve

A novel type of nonmechanical device called the N-valve was used by Hirama et. al. (1986). However, this type of device is not recommended because gas channeling back from the angled section can cause large bubbles in the standpipe just as was found with the hybrid standpipes discussed earlier.

2.7.4 V-Valve

The V-valve (Fig. 24) is a close relative of the loop seal and also acts as an automatic flow device at the base of standpipes operating in the overflow mode (Li et. al., 1982). The V-valve consists of an angled diverging section connected to a standpipe A circular aperture in

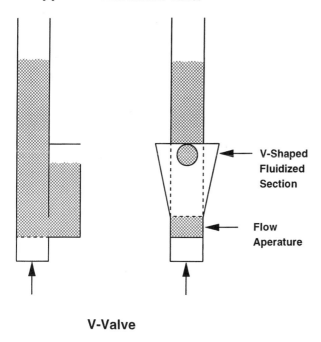

V-Valve

Figure 24   Schematic drawing of V-valve.

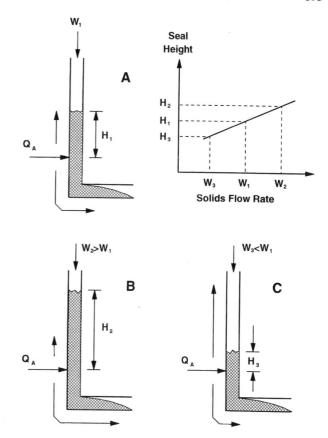

Figure 25   Depiction of automatic L-valve operation.

the standpipe allows the solids to flow from the standpipe into the diverging section. The included angle of the diverging section is usually small—between 5 and 10 degrees. Greater angles cause uneven distribution of the gas and solids.

As with the loop seal, the V-valve will not operate if the solids in the upflow (diverging) section are not fluidized. When the solids in the diverging section are fluidized, solids flow from the standpipe, through the aperture and diverging section, into the fluidized bed. As with the loop seal, the upflow section of the V-valve prevents pressure surges from causing the dipleg to blow.

2.7.5   L-valve

The L-valve can also operate automatically at the bottom of an overflow standpipe. Its operation is somewhat different than that of the other nonmechanical devices. Chan et al. (1988) showed how the L-valve operates in the automatic mode. This operation is described below. Consider the L-valve shown in Fig. 25A, in which solids are flowing through the L-valve at rate W_1 while aeration is being added to the L-valve at a constant rate Q_A. The solids above the aeration point are flowing in the fluidized bed mode and are at an equilibrium height, H_1. For Geldart group B solids, the particles below the aeration point are flowing in the moving packed bed

mode. The fraction of aeration gas required to produce solids mass flow rate W_1 flow around the elbow with the solids, while the remaining portion of the aerating gas flows up the standpipe.

If the solids flow rate into the standpipe above the L-valve increases from W_1 to W_2 (Fig. 25B), the solids level in the standpipe initially rises because solids are being fed to the L-valve faster than they are being discharged. The increased height of solids in the standpipe causes the pressure at the aeration point to increase relative to the bed. If a constant aeration flow to the aeration point is maintained by a critical orifice or a control valve, a greater fraction of the aeration gas flows around the L-valve elbow, causing an increase in solids flow through the L-valve. If enough aeration gas is being added to the L-valve, the system reaches equilibrium at a point where a larger fraction of the aeration gas is flowing around the L-valve bend to cause solids to flow at rate W_2. The height of solids in the standpipe above the L-valve reaches equilibrium at increased height, H_2, and the system is again in balance. If the solids flow rate decreases from W_1 to W_3 (Fig. 25C), the solids level in the standpipe above the L-valve falls, decreasing the pressure at the aeration point relative to the bed. This results in less gas flowing around the L-valve bend. The reduced flow of gas through the L-valve will result in a lower solids flow rate. Balance is reached when solids are flowing through the L-valve at rate W_3, and the solids in the standpipe are at equilibrium height, H_3.

In the description above, gas in the standpipe was assumed to be traveling upward relative to the standpipe wall. It is also possible for gas in the standpipe to travel downward relative to the standpipe wall. In fact, gas travels downward with the solids for most standpipe operation.

The L-valve may not work in the automatic mode if it is used to discharge Geldart group A solids into a dilute-phase environment. This is because group A solids may not deaerate and will flush through the L-valve and not form a pressure seal. However, Geldart group A solids will work in an automatic L-valve if it is discharging into a dense bed. An automatic L-valve works well with Geldart group B solids when discharging into both dense-phase and dilute-phase media.

The L-valve will not operate automatically over an infinite range of solid flow rates. At some increase in the solids flow rate in the example given above, not enough gas is available to fluidize the solids in the standpipe, and they will defluidize. This causes the standpipe to fill with defluidized solids to a level which balances the pressure drop loop. The automatic L-valve can also work if the solids in the standpipe are in the packed bed mode. However, because the $\Delta P/Lg$ in a packed bed can vary over a wide range, the height of solids required for pressure sealing in the standpipe can also vary. This makes it difficult to control the height of solids in the standpipe, and they can easily back up into the cyclone, resulting in significant loss of material. To prevent this from happening, more aeration must be added to the L-valve.

Although the automatic L-valve is geometrically simpler and less expensive to install than the loop seal, for the reason given above, the automatic L-valve can be more difficult to operate when using group B materials. Loop seal operation requires fewer adjustments, and is often preferred over the L-valve for automatic operation with group B materials. For group A materials, either automatic device can be used.

As noted above, it is generally best to design a nonmechanical device by obtaining the required data in a cold-flow test unit using the actual solids

to be transferred. If this is not possible, solids of similar size and density can be used. Nonmechanical valve or automatic nonmechanical device design has not yet reached a stage where it can be done analytically with great confidence. Cold modeling in a unit with 76 to 100 mm diameter transfer lines is relatively inexpensive and minimizes operational problems when incorporating the device into a pilot plant or commercial system.

2.7.6 Cyclone Diplegs and Trickle Valves

As noted above, cyclone diplegs are really overflow standpipes. It is important that they be designed correctly. Poor operation of cyclone diplegs usually results in poor collection efficiency for the cyclone to which it is attached.

Diplegs attached to first-stage cyclones generally give few operational problems because the solids flux rate down them is high. Typical design fluxes for a first-stage cyclone dipleg in FCC units are 350 to 750 kg/s-m². Second stage cyclone diplegs are the ones that tend to have flow problems because the solids in them are finer (sometimes approaching cohesive group C size), and because the solids mass flux through the dipleg is low. Both factors result in more sluggish solids flow, which can sometimes lead to blockage. The minimum dipleg diameter recommended for all commercial cyclone diplegs is about 100 mm. This diameter is large enough to prevent most bridging. Obviously, smaller research units will have smaller diameter diplegs, with the associated plugging problems.

Cyclone diplegs are often designed with a mitered bend at its end (Fig. 26A). This type of bend generally offers no problem for a first-stage dipleg because of the high mass fluxes through it, but it can cause problems in a second-stage dipleg. When a mitered bend is used, the bend causes a restriction in the flow path. If the solids mass flux is not high enough, the dipleg can plug. It is much better to use a straight end dipleg (Fig. 26B) instead of a mitered bend if the mass flux through the dipleg is low (less than 100 kg/s-m²).

During start-up, fluidizing gas preferentially flows up the dipleg of a cyclone until a bed is established to seal it. This occurs because gas does not have to undergo the cyclone entrance and cyclone barrel pressure loss when passing up the dipleg. Many processes operate with high gas flows in the unit during start-up. First stage diplegs generally have enough solids flow through them to allow a seal to be established in spite of the upward gas flow in the dipleg. However, the solids flow through a second stage dipleg is generally

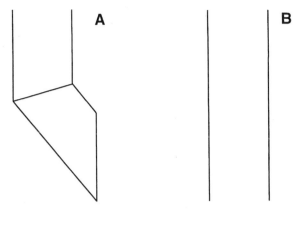

Mitered Dipleg   **Open-Ended Dipleg**

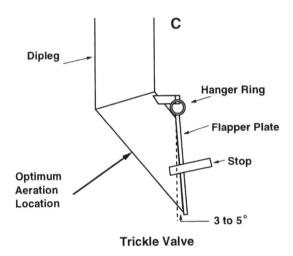

Trickle Valve

Figure 26 Dipleg end configurations.

too small (and the gas flow even greater than through the first-stage dipleg) to establish a seal unless a device to prevent this gas flow is attached to the end of the dipleg. A device called a trickle valve (Fig. 26C) is generally used to prevent this bypass gas flow. A trickle valve is essentially a loosely hung plate that hangs vertically in front of the dipleg discharge opening. Trickle valves allow the establishment of a solids seal upon start-up but can cause problems during operation. Typical problems are plugging (if the trickle valve is not designed correctly) or erratic pulsating flow. It is preferable to avoid trickle valves and to modify the start-up procedure to allow a lower gas flow rate through the unit until the second stage cyclone dipleg is sealed. However, this is not always possible, and trickle valves are often a necessary evil.

In order for the trickle valve to function properly and not bind during operation, it is attached to the dipleg by loose hanger rings. The dipleg opening is generally inclined about 3 to 5 degrees from the vertical so that the flapper plate exerts a positive closing force equal to the moment of its weight about the pivot point of the valve.

Geldart and Kerdoncuff (1992) and Bristow and Shingles (1989) studied trickle valve operation in cold flow models with 100 mm diameter diplegs. Four types of solids flow through the trickle valve were found to occur: (1) constant trickling, (2) dumping, (3) trickling–dumping, and (4) flooding. Constant trickling is the most desirable type of flow because it is smoother and because dipleg blockage is least likely to occur. Constant trickling occurs when solids discharge continually through the valve maintaining a constant height of solids in the dipleg. Flow in this regime was found to occur at high solids velocities in the dipleg (between about 2 and 25 cm/s), high aeration rates equivalent to about 3 cm/s in the dipleg and at the location shown in Fig. 26C (other aeration locations were not nearly as effective), high cyclone differential pressures, and high trickle valve opening torques. All of these parameters tend to increase the amount of gas aeration or leakage rate into the dipleg. Bristow and Shingles (1989) also found that gas leakage through the trickle valve and/or aeration was absolutely necessary for satisfactory trickle valve operation.

The dumping mode of trickle valve operation occurs at low solids velocities, low aeration and/or leakage rates, low cyclone pressure drops, and low valve torques. These are the opposite of what causes the more desirable constant trickling flow. This mode of flow causes more frequent blockage of solids and may cause excessive wear in the upper part of the dipleg discharge.

Trickling–dumping is the regime intermediate between constant trickling and dumping. Material continuously trickles out of the valve while the solids periodically dump when the level in the dipleg reaches the critical height to cause the valve to open.

Flooding occurs when the solids flow into the dipleg is greater than the solids discharge through the valve. This causes solids to back up into the cyclone. Flooding is generally caused by extremely high solids flow rates or excessive leakage or aeration.

Trickle valve diplegs in commercial units often exhibit erosion in the upper part of the dipleg and/or in the upper part of the flapper. This is because gas leakage is preferentially through the upper part of the trickle valve (Fig. 27A). The relatively high leakage

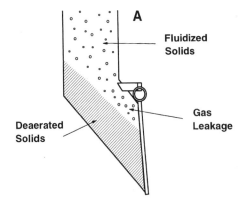

Gas Leakage Through Trickle Valve

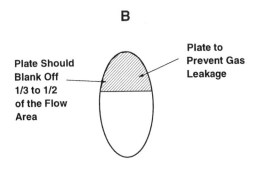

Plate Location for Erosion Prevention

Figure 27   Trickle valve leakage and erosion prevention.

gas velocities (up to 60 cm/s in the dipleg, but much higher in the space between the flapper and the dipleg) carry solids with them and eventually erode the dipleg and flapper (a "horseshoe" pattern occurs on the flapper) in this area. This can be prevented by adding a plate across the upper part of the dipleg (Fig. 27B). This reduces the leakage rate and prevents erosion. Bristow and Shingles (1989) found that a plate covering at least one-third of the dipleg area and up to one-half of the area was effective in reducing the gas leakage rate into the dipleg. However, covering less of the dipleg opening than this amount was ineffective, while covering too much of the opening resulted in blockage of the solids. Although covering the upper part of the dipleg opening can prevent erosion, erosion probably occurs because the trickle valve is operating in the dumping mode. It is much better to modify the dipleg/trickle valve operation to cause constant trickling flow to occur than to install the plate restrictions.

NOMENCLATURE

A_0	=	L-valve variable flow area, m^2
C_D	=	orifice coefficient
D	=	pipe diameter, m
D_v	=	L-valve diameter, m
g	=	gravitational constant, 9.81 m/s^2
H_{dip2}	=	secondary cyclone dipleg length, m
H_{dip2}	=	secondary cyclone dipleg length, m
H_{min}	=	minimum length of L-valve, m
H_{sp}	=	standpipe height, m
H_{sp}'	=	changed standpipe height, m
L	=	length, m
L_{min}	=	minimum standpipe length required to seal, m
L_{sp}	=	standpipe length, m
P	=	pressure, Pa
P'	=	changed pressure, Pa
P_b	=	pressure at bottom of standpipe, Pa
P_t	=	pressure at top of standpipe, Pa
ΔP	=	pressure drop, Pa
ΔP_H	=	higher packed bed pressure drop, Pa
ΔP_{CFB}	=	CFB pressure drop, Pa
ΔP_{cy}	=	cyclone pressure drop, Pa
ΔP_{cy1}	=	primary cyclone pressure drop, Pa
ΔP_{cy2}	=	secondary cyclone pressure drop, Pa
ΔP_d	=	distributor pressure drop, Pa
ΔP_{dip2}	=	pressure drop across secondary cyclone dipleg, Pa/m
$\Delta P_{fluid\ bed}$	=	fluidized bed pressure drop, Pa
$\Delta P_{independent}$	=	independent part of pressure drop loop, Pa
ΔP_{lb}	=	lower fluidized bed pressure drop, Pa
ΔP_{ls}	=	loop seal pressure drop, Pa
$\Delta P_{L\text{-valve}}$	=	L-valve pressure drop, Pa
ΔP_{regen}	=	regenerator pressure drop, Pa
ΔP_{riser}	=	riser pressure drop, Pa
ΔP_{sp}	=	standpipe pressure drop, Pa
ΔP_{sp1}	=	first standpipe pressure drop, Pa
ΔP_{sp2}	=	second standpipe pressure drop, Pa
$\Delta P_{stripper}$	=	stripper pressure drop, Pa
ΔP_{surge}	=	surge vessel pressure drop, Pa
$\Delta P_{surge\ hopper}$	=	pressure drop across surge hopper, Pa/m
ΔP_{sv1}	=	first slide valve pressure drop, Pa
ΔP_{sv2}	=	second slide valve pressure drop, Pa
ΔP_{ub}	=	upper fluidized bed pressure drop, Pa
ΔP_v	=	valve pressure drop, Pa
$\Delta P/Lg$	=	pressure drop per unit length, Pa/m
$\Delta P/Lg)_{mf}$	=	pressure drop per unit length at minimum fluidization, Pa/m
Q	=	aeration required to prevent defluidization, m^3/s
Q	=	volumetric flow rate of gas, m^3/s
Q_A	=	aeration added to L-valve, m^3/s
Q_{sp}	=	amount of gas flowing down standpipe, m^3/s

Q_T	=	total L-valve aeration, m^3/s
U	=	superficial gas velocity, m/s
U_{mf}	=	superficial gas velocity at minimum fluidization, m/s
v_g	=	interstitial gas velocity, Ug/ε, m/s
v_{mb}	=	minimum bubbling velocity, m/s
v_{mf}	=	interstitial minimum fluidization velocity, U_{mf}/ε
v_r	=	relative gas/solid velocity, m/s
v_s	=	solids velocity, m/s
W_s	=	solids flow rate through L-valve, kg/s
ε	=	voidage
ε_{mf}	=	voidage at minimum fluidization
ρ_{dip2}	=	secondary cyclone dipleg density, kg/m^3
ρ_{mf}	=	fluidized-bed density at minimum fluidization, kg/m^3
ρ_p	=	particle density, kg/m^3
ρ_{sk}	=	skeletal density, kg/m^3
ρ_t	=	fluidized bed density at top of standpipe, kg/m^3

REFERENCES

Bristow TC, Shingles T. Cyclone dipleg and trickle valve operation. In: Grace JR, Shemilt LW, Bergougnou MA, eds. Fluidization VI, 1989, pp 161–168.

Campbell DL, Martin HZ, Tyson CW. U.S. Patent 2,451,803, 1948.

Chan I, Findlay J, Knowlton TM. Operation of a nonmechanical L-valve in the automatic mode. Paper presented at the Fine Particle Society Meeting, Santa Clara, CA, 1988.

Eleftheriades CM, Judd MR. The design of downcomers joining gas-fluidized beds in multistage systems. Powder Technology 21:217–225. 1978.

Geldart D, Broodryk N. Studies on the behaviour of cyclone diplegs. Presented at the Annual Meeting of the AIChE, Los Angeles, CA, November 18–21, 1991.

Geldart D, Jones P. The behaviour of L-valves with granular powders. Powder Technology 67:163–172, 1991.

Geldart D, Kerdoncuff A. The behaviour of secondary and tertiary cyclone diplegs. Presented at the AIChE Annual Meeting, Miami Beach, 1992.

Grace JR. Contacting modes and behaviour classification of gas–solid and other two-phase suspensions. Can J Chem Eng 64:353–364, 1986.

Hirama T, Takeuchi H, Horio M. Nitric oxide emission from circulating fluidized bed coal combustion. Proceedings of the 9th International Fluidized Bed Combustion Conference, Boston, pp 898–910, 1986.

Jones DRM, Davidson JF. Rheol Acta, 4:180–191, 1965.

Karri SBR, Knowlton TM. Comparison of group A solids flow in hybrid angled and vertical standpipes. In: Avidan Ar, ed. Circulating Fluidized Bed Technology IV, 1993, pp 253–259.

Karri SBR, Knowlton TM, Litchfield J. Increasing solids flow rates through a hybrid angles standpipe using a bypass line. Proceedings of the Eighth International Conference on Fluidization, Tours, France, 1995.

King D. Fluidized catalytic crackers, an engineering review. In: Potter, OE, Nicklin, DJ, eds. Fluidization VII, 1991, pp 15–26.

Knowlton TM, Aquino MRY. A comparison of several lift-line feeders. Paper presented at the Second Congress of Chemical Engineering, Montreal, Canada, 1981.

Knowlton TM, Findlay JG. Determination of flow-regime boundaries for cohesive particles. Final Report for the US Department of Engergy, Contract No.: DE-AC22-88PC88951, October, 1992.

Knowlton TM, Hirsan I. L-valves characterized for solids flow. Hydrocarbon Processing 57:149–161, 1978.

Knowlton TM, Leung LS, Hirsan I. The effect of aeration tap location on the performance of a J-valve. In: Davidson, JF Keairns, DL, eds. Fluidization II, 1978, p 128–136.

Li X, Liu D, Kwauk M. Pneumatically controlled multistage fluidized beds—II. Proceedings of the Joint Meeting of Chemical Engineers, SIESC and AIChE. Beijing: Chem Industry Press, 1982, pp 382–394.

Merrow EW. Linking R & D to Problems Experienced in Solids Processing. Chemical Engineering Progress, pp 14–22, May 1985.

PSRI Technical Report RR-181. An investigation of the effect of solids mass flux on the axial density profile and the amount of gas entrained in a submerged cyclone dipleg. Chicago, IL, 1995.

Rall R, Pell M. Use of gas–liquid contacting internals for fluidized bed stripping applications. In: Werther J, ed. Proceedings of the Sixth Internation Conference on Circulating Fluidized Beds. Würzburg, Germany, August 22–27, 1999.

Ross J. Personal communication, 1997.

Sauer RA., Chan IH, Knowlton TM. The effects of system and geometrical parameters on the flow of Class B solids in overflow standpipes. AIChE Symposium Series 234, 80, pp 1–10, 1984

Wirth KE. Fluid mechanics of the downcomer in circulating fluidized beds. Proceedings of the Eighth International Conference on Fluidization, Tours, France, 1995.

Wrench RE, Wilson JE, Guglietta G. Design features for improved cat cracker operations. Presented at the first South American Ketjen Catalyst Seminar, Rio de Janeiro, Brazil, 1985.

Yang W-C, Knowlton TM. L-valve equations. Powder Technology 77:49–57, 1993.

Yaslik AD. Circulation difficulties in long angled standpipes. In: Avidan, A ed. Circulating Fluidized Bed Technology IV, 1993, pp 484–492.

22

Cyclone Separators

T. M. Knowlton

Particulate Solid Research, Inc., Chicago, Illinois, U.S.A.

1 INTRODUCTION

There are three primary reasons for separating gases from solids in solids processing plants: (1) to minimize emissions for environmental purposes, (2) to protect other processing equipment (turbines, etc.) from part-icle-laden streams, and (3) to stop unwanted gas–solids reactions from occurring. The primary separator in most fluidized bed processes is a cyclone. The efficiency of separation achieved by the cyclone depends upon the nature of the process. If the process stream is dry and the particles are relatively large and free flowing, collection is relatively easy. If it is not, or there is spe-cial chemistry or extreme temperature involved, parti-culate collection can be challenging. It is the purpose of this chapter to describe the basic operation of cyclone separators and how different conditions affect their performance.

2 CYCLONES

A cyclone is a device that separates particulate solids from a fluid stream by a radial centrifugal force exerted on the particles. This force separates the solids from the gas by driving the solids to the cyclone wall, where they slide to the bottom solids outlet and are collected. Cyclones are used with almost all fluidized beds to remove solids from exit gas streams.

Particles entering a cyclone are subjected to a sub-stantial centrifugal separating force. the force exerted

in the particles relative to the gravitational force is proportional to $(U_i)^2 g r_o$, where g is the gravitational acceleration. For a cyclone with an inlet width that is 20% of the diameter of the cyclone (40% of the radius), the number of g's that the particles experience (assuming the particles are all at the center line of the cyclone inlet) is

$$\frac{U_i^2}{g 0.8 r_o} = \frac{U_i^2}{9.81(0.4 D_b)} \qquad (1)$$

For a cyclone with a barrel diameter of 1 meter and operating with an inlet gas velocity of 20 m/s, over 100 g's will be exerted on the entering particles.

Cyclones can be very small or very large. The smallest cyclones range from approximately 1 to 2 cm in diameter and the largest up to about 10 m in diameter. The number of cyclones used for a single fluidized bed can vary from one to up to 22 sets of first-stage and second-stage cyclones (44 cyclones total).

The advantages of cyclones are that they (1) have no moving parts, (2) are relatively inexpensive to con-struct, (3) have relatively low pressure drops, and (4) have maintenance costs that are low. The limitation of typical cyclones is that they have reduced collection efficiencies for particles approximately 10 microns in diameter and smaller. Special cyclones can be operated and designed that can collect particles down to 3 microns in size with efficiencies approaching 90%. This type of collection system usually consists of many very small cyclones in parallel and is often called a multiclone.

Cyclones have been used to remove particulates from gas streams since the middle of the 19th century (Rietema and Verver, 1961). Early researchers (Rosin et al., 1932; Alexander, 1949; Stairmand, 1951; Ter Linden, 1949; Iinoya, 1953; Lapple, 1951; van Tongren, 1936) conducted the first experiments designed to understand the operation of this mechanically simple but operationally complex device. Many of these experiments (Ter Linden, 1949; Lapple, 1951; Stairmand, 1951, etc.) resulted in cyclone designs based on relative cyclone dimensions. More recently, Zenz (1975) developed an empirical cyclone design procedure that has achieved popular acceptance in the United States.

In Germany, Barth (1956) developed a theoretical model of cyclone operation, which was expanded by Dietz (1981). Muschelknautz (1970a, 1970b) continued the work of Barth and developed a cyclone design technique that is used more in Europe than in the remainder of the world.

2.1 Cyclone Types

A cyclone designed to separate solids from a gas must have a gas–solids inlet that will cause an axial rotation giving rise to a centrifugal force. It must also have an axial gas discharge port and a solids discharge port for the collected solids.

There are two primary types of cyclones: the reverse-flow cyclone and the uniflow cyclone. The reverse-flow (Fig. 1) is by far the most common. It is called a reverse-flow cyclone because the gas–solids mixture enters the cyclone tangentially at its periphery, spirals around the barrel, and then the gas reverses flow and exits through a gas outlet tube (also called the vortex finder or the vortex tube) at the top of the cyclone. The solids spiral down around the barrel of the cyclone at an angle of approximately 15 degrees and enter the cyclone cone attached to the bottom of the barrel. The solids exit the reverse-flow cyclone at the bottom of the cyclone cone.

In the uniflow cyclone (Fig. 2) the gas–solids mixture enters the cyclone and spirals around the barrel, and then the gas exits the cyclone at the bottom in the center of the uniflow cyclone. The solids also exit the uniflow cyclone at the bottom, but along the wall of the cyclone. A small amount of gas (generally 1 to 3% of the inlet gas flow) is also withdrawn with the solids to improve collection efficiency. Gauthier and co-workers (Gauthier et al., 1990a, 1990b, 1992) exten-

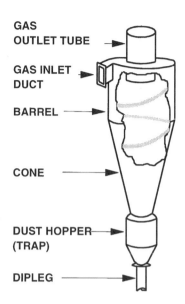

Figure 1 Typical reverse flow cyclone.

sively studied uniflow cyclones, and reported on various aspects of their operation. One application of this type of cyclone is as a short-contact time reactor. Solids spend less time in a uniflow cyclone than in a reverse-flow cyclone.

Over the decades that cyclones have been used, many different reverse-flow cyclone geometries have been tried to improve efficiency, prevent particle attrition, prevent erosion of the cyclone wall, or prevent particle buildup on the cyclone surfaces. However, there are a few basic types that have emerged as the most popular over the years. Some of these cyclone types are shown in Fig. 3. The cyclones shown in this figure are the tangential inlet cyclone, the volute inlet cyclone, and the axial inlet cyclone. This last type of cyclone uses axial swirl vanes to impel the gas–solids mixture into rotary centrifugal motion.

In some cyclones, the roof of the cyclone is angled downward in a helical fashion. In some cyclones with standard roofs, the solids that enter the cyclone do not move downward far enough by the time they traverse one revolution and then collide with the incoming solids. This leads to particle bouncing and reduced cyclone efficiency as well as slightly higher pressure drops. To circumvent this unwanted particle interaction at the inlet, cyclone roofs are sometimes angled downward to force the solids downward and below the incoming solids. In some applications, helical roofs are used to prevent the buildup of sticky material of the cyclone roof by forcing their incoming solids to rub against the top of the cyclone.

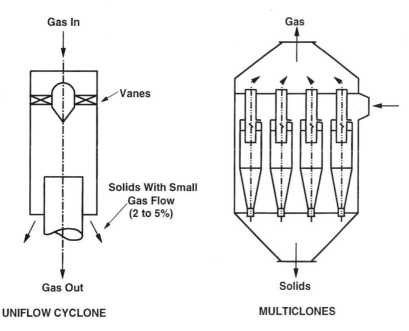

Figure 2 Schematic drawing of uniflow cyclone and multiclones.

2.2 Flow Patterns in Cyclones

In a tangential-inlet reverse-flow cyclone, the cyclone inlet translates the linear inlet gas flow into a rotating vortex flow. As shown in Fig. 1, the gas–solids mixture enters an annulus region between the outer wall of the cyclone and the outer wall of the gas outlet tube. As the gas–solids mixture spirals downwards, it sets up a vortex with an axial direction downward toward the solids outlet.

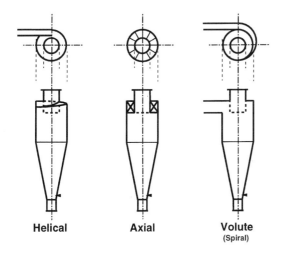

Figure 3 Schematic drawing of tangential, volute, and axial inlet cyclones.

The rotating vortex in the main body of the cyclone below the gas outlet tube sets up a tangential velocity that varies with the radius of the cyclone. As the radius, r_w, decreases from the wall toward the center of the cyclone, the tangential velocity increases to a maximum that occurs at a radius slightly less than the outer radius of the gas outlet tube, r_o, as shown in Fig. 4. At radii much smaller than that of the gas outlet tube, the tangential gas velocity decreases to a much lower value at the center.

The inner vortex (often called the core of the vortex) rotates at a much higher velocity than the outer vortex. In the absence of solids, the radius of this inner vortex has been measured to be 0.4 to $0.8r_o$. With axial-inlet cyclones, the inner core vortex is aligned with the axis of the gas outlet tube. With tangential or volute cyclone inlets, however, the vortex is not exactly aligned with the axis. The nonsymmetric entry of the tangential or volute inlet causes the axis of the vortex to be slightly eccentric from the axis of the cyclone. This means that the bottom of the vortex is displaced some distance away from the axis, and can pluck off and reentrain dust from the solids sliding down the cyclone cone if the vortex gets too close to the wall of the cyclone cone.

At the bottom of the vortex, there is substantial turbulence as the gas flow reverses and flows up the middle of the cyclone into the gas outlet tube. As indicated above, if this region is too close to the wall of the

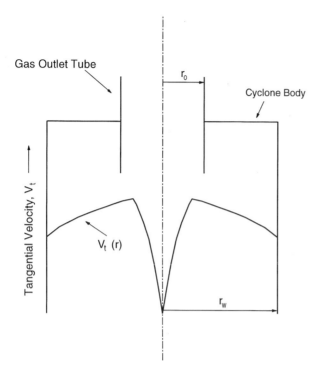

Figure 4 Typical radial tangential velocity profile in a cyclone.

cone, substantial reentrainment of the separated solids can occur, so it is very important that cyclone design take this into account.

Hugi and Reh (1998) have recently reported that (at high solids loadings) enhanced cyclone efficiency occurs when the solids form a coherent, stable strand at the entrance to a cyclone. The formation of such a strand is dependent upon several factors. They reported that they obtained a higher cyclone efficiency for smaller ($d_{p,50} = 40$ microns) solids than for larger solids ($d_{p,50}$ 125 microns). This is not what theory would predict. However, they also found that the smaller particles formed coherent, stable strands more readily than the larger particles, which explained the reason for the apparent discrepancy.

Muschelknautz et al. (1996) also proposed a mechanistic model of cyclone operation. In this model, the gas can carry only a maximum amount of solids (called the critical loading). At any solids loading in excess of this critical loading, the solids are immediately separated from the gas at the inlet to the cyclone, as indicated in Fig. 5. The solids remaining in the gas are then separated in the cyclone barrel and in the inner vortex below the gas outlet tube as if the cyclone were operating at a lower solids loading.

3 SERIES AND PARALLEL CYCLONE ARRANGEMENTS

3.1 Series

A single cyclone can sometimes give sufficient gas–solids separation for a particular process or application. However, solids collection efficiency can usually be enhanced by placing cyclones in series. Cyclones in series are typically necessary for most processes to minimize particulate emissions or to minimize the loss of expensive solid reactant or catalyst. Two cyclones in series are most common, but sometimes three cyclones in series are used. Series cyclones can be very efficient. In fluidized catalytic cracking (FCC) regenerators, two stages of cyclones can give efficiencies of up to and even greater than 99.999%.

Typically, first-stage cyclones will have an inlet gas velocity less than that of second-stage cyclones. The lower inlet velocity of first-stage cyclones results in lower particle attrition rates and lower wall erosion rates. After most of the solids are collected in the first stage, a higher velocity is generally used in second-stage cyclones to increase the centrifugal force on the solids and increase collection efficiency. Erosion rates are generally low in the second stage because of the vastly reduced flux of solids into the second stage cyclone.

Sometimes it is better to use a single high-efficiency cyclone than operate with two cyclones in series. This situation can occur when the first-stage cyclone is very efficient and the solids flowing to the second stage cyclone are very small and cohesive (Geldart group C solids). Cohesive solids are difficult to transfer in cyclone diplegs and may cause the solids to bridge momentarily or not flow at all. Second-stage diplegs operating with trickle or flapper valves and cohesive solids are particularly sensitive to bridging. The inability to flow solids down the second-stage diplegs causes the solids to pass straight through from the inlet to the outlet, thus destroying cyclone efficiency.

In addition, flow of gas up the dipleg of second-stage cyclone diplegs can also be a major problem. If this gas flow is excessive during normal operation, the diplegs will not operate well, and erosion of the cyclone cone and dipleg trickle valve flapper plate may result. This problem occurs because the solids flux in a second-stage dipleg is very low (of the order of 10 to 20 kg/s-m²). For a dipleg density of approximately 800 kg/m³, the velocity of the solids in the dipleg (calculated as the mass flux through the dipleg divided by the suspension density in the dipleg)

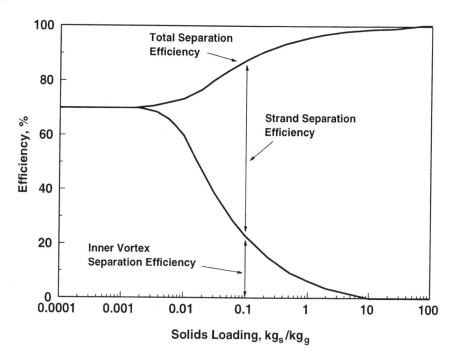

Figure 5 Muschelknautz critical loading plot.

is only 0.025 m/s for a flux in the dipleg of 20 kg/s-m². Gas can easily flow up the dipleg against this very low velocity. Conversely, in first stage cyclone diplegs where the solids flux may be as high as 750 kg/s-m², the solids velocity is so high (approximately 0.94 m/s at a dipleg density of 800 kg/m³) that gas is carried down the dipleg with the solids.

Because of these recurring problems with second-stage cyclone diplegs, companies sometimes choose to use a single high-efficiency cyclone in place of two cyclone stages. Even though the overall efficiency using a single cyclone is slightly less than using two cyclone stages, they find that this cyclone arrangement optimizes the operation of the process by minimizing upsets and reducing downtime using a single high-efficiency cyclone. It is also a much simpler system as well and saves on initial capital costs.

3.2 Parallel

Several small cyclones are placed in parallel when it is not possible to fit a single large cyclone into the available height, or when extremely high centrifugal forces are required. It is difficult to ensure equal distribution of gas and solids into parallel cyclones. This can lead to cyclone inefficiencies and increased wear on the cyclones taking the bulk of the solids flow.

Increasing the pressure drop across the cyclones will improve the solids distribution but is not always sufficient to ensure equal solids flow into each cyclone.

Typically, when the number of parallel cyclones is small (less than 6 to 8), each cyclone will have an individual inlet and outlet duct. However, when the number of cyclones exceeds this number, the cyclones generally have common inlet and outlet plenums and a common collection bin. Parallel arrangements of this type frequently result in the operating problems of equalizing the gas and solid flow rates into each cyclone, and preventing backflow from the common chamber into one or more cyclones and reentrainment of solids from the common collection hopper back into one or more cyclone outlets. To prevent these two problems, a device is often placed at the outlet of each parallel cyclone. This device (generally proprietary) reduces the area available for flow and acts as a "check valve" to prevent backflow and dust reentrainment into the cyclone.

Placing a large number of cyclones in parallel in a common bin can result in distribution problems because it will be easier for the gas and solids to flow through the closest cyclone than one located some distance away from the inlet. Multiple inlets to the common vessel reduce this problem, but result in increased complexity and cost.

Because of problems in the operation of small parallel cyclones, it is generally found that they result in a collection efficiency less than or (if designed well) equal to that of a large, single cyclone processing the same amount of gas and solids.

In the FCC and similar processes there are many units operating with two cyclone stages in series, but with several (2 to 18) cyclones in each stage in parallel. These cyclones can have very high efficiencies even though they operate in parallel. However, each first-stage cyclone generally has a dipleg that is immersed in the bed (there is no common collection chamber). Each second stage cyclone has a dipleg that usually has a trickle valve on its discharge. The trickle valve effectively is a check valve that prevents backflow of most of the gas through the dipleg on startup. In addition, the pressure drop through the cyclone stages is generally sufficient to prevent maldistribution.

There has been very little work in the literature on the operation of parallel cyclones. However, what work does exist indicates that too low a pressure drop results in poor gas–solids distribution, and that too many parallel cyclones in a common hopper are not recommended.

Broodryk and Shingles (1995) studied parallel cyclone operation with cyclones in the freeboard above a turbulent fluidized bed. They found that too low an overall pressure drop (caused by inlet gas velocities of less than about 15 m/s) resulted in preferential flow of gas through two of their three cyclones. They also found that poor trickle valve design and/or operation could induce an imbalance in the gas flow into the cyclones. When preferential gas flow occurred, it resulted in significant solids carryover and reduced cyclone efficiency.

Koffman (1953) studied the efficiency of several different small-diameter (38 and 51 mm diameter) parallel cyclone configurations and reported the relative efficiencies shown in Table 1. Koffman's data show that placing cyclones in a common hopper reduced the collection efficiency, and that adding more cyclones to a common hopper also caused a reduction in efficiency. Smellie (1942), using three 200 mm diameter cyclones, and Whiton (1941), using 9 to 20 cyclones 610 mm in diameter, also reported poorer cyclone performance with parallel cyclone operation. However, O'Mara (1950) reported very little difference between single cyclone and parallel cyclone efficiencies These results indicate that parallel cyclones can be designed and operated with the same efficiencies as a single cyclone, but that it takes good and careful design of the cyclones to accomplish this.

Table 1 Efficiencies of Parallel Cyclones

Configuration	Efficiency, %
Individual cyclone	96.0
Seven cyclones in parallel (individual hoppers)	95.3
Seven cyclones in parallel (common hopper)	94.1
Fourteen cyclones in parallel (common hopper)	92.2

4 INTERNAL AND EXTERNAL CYCLONES

Cyclones can either be placed in the freeboard above the fluidized bed or be located outside of the fluidized-bed vessel. There are advantages and disadvantages to each type of placement, and the optimum type of placement depends on what is best for a particular process.

Internal cyclones have the advantages that they require no inlet piping (their inlets can be open to the freeboard), require no high-pressure shell, and have straight cyclone diplegs. Internal cyclones are generally smaller in diameter than external cyclones because their size is limited by the headspace available in the freeboard above the fluidized bed. These size limitations result in using several smaller cyclones in parallel instead of one large cyclone. In addition, it is difficult to aerate second-stage cyclone diplegs (generally an advantageous technique) when internal cyclones are used. Aerating secondary cyclone diplegs can improve the operation of the dipleg significantly.

The advantages of external cyclones are that they can be much larger than internal cyclones, that they are more accessible than internal cyclones, and that their diplegs can be aerated more easily. The disadvantages of external cyclones are that they require a pressure shell and that external cyclone diplegs generally require a section with an angled or a horizontal pipe to return the solids to the bed. The angled or horizontal dipleg sections can result in poor dipleg operation if not designed correctly.

5 CYCLONE INLET DESIGN

The design of the cyclone inlet can greatly affect cyclone performance. It is generally desired to have the width of the inlet (L_w) be as narrow as possible so that the entering solids will be as close as possible to the cyclone wall where they can be collected.

However, narrow inlet widths require that the height of the inlet (H) be very long in order to give an inlet area required for the desired inlet gas velocities. Therefore a balance between narrow inlet widths and the length of the inlet height has to be made. Typically, low-loading cyclones (cyclones with inlet loadings less than approximately 2 to 5 kg/m^3) have height-to-width ratios (H/L_w) of between 2.5 and 3.5. For high-loading cyclones, this inlet aspect ratio can be increased to as high as 7 or so with the correct design. Such high inlet aspect ratios require that the cyclone barrel length increase substantially.

A common cyclone inlet is a rectangular tangential inlet with a constant area along its length. This type of inlet is satisfactory for many cyclones, especially those operating at low solids loading. However, a better type of inlet is one in which the inner wall of the inlet is angled toward the outer cyclone wall at the cyclone inlet (Fig. 6). This induces solids momentum toward the outer wall of the cyclone. The bottom wall of the inlet is angled downward so that the area decrease along the inlet flow path is not too rapid and acceleration is controlled. In addition, the entire inlet can be angled slightly downward to give enhanced efficiencies. This type of inlet is superior to the constant area tangential inlet, especially for higher solids loadings (greater than 2 to 5 kg/m^3).

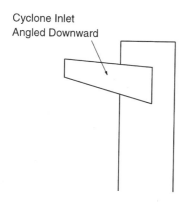

Cyclone Inlet
Angled Downward

Inner Wall of Inlet
Angled to Direct
Solids Toward Wall

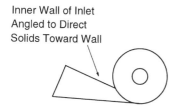

Figure 6 Angled cyclone inlet.

Hugi and Reh (1998) report that continuous acceleration of the solids throughout the inlet is desired for improved efficiency, and that the angled inlet described above achieves this. If the momentum of the solids is sufficient and the solids are continuously accelerating along the length of the inlet, they will form the stable, coherent strand important for high collection efficiencies.

Another superior inlet for high solids loadings is the volute cyclone inlet. At high inlet loadings in a tangential cyclone inlet, the gas–solids stream expands rapidly from its minimum width at the point of contact. This rapid expansion disturbs the laminar gas flow around the gas outlet tube and causes flow separation around the tube. This results in lowered cyclone efficiency. But when a volute inlet is used, the expanding solids stream is further from the gas outlet tube and enters at an angle so that the solids do no induce as much flow separation or asymmetrical flow around the gas outlet tube. This cyclone efficiency is not affected to as great a degree.

The nature of the gas–solids flow in the inlet ducting to the cyclone can affect cyclone efficiency significantly. If the solids in the inlet salt out on the bottom and result in dune formation and the resulting unsteady or pulsing flow, cyclone efficiency is adversely affected. To minimize the possibility of this occurring, it is recommended that the inlet line to the cyclone operate above the saltation velocity (Gauthier et al. 1990b), which will prevent the solids from operating in the dune or pulsing flow regime. If this is not possible, then the inlet line can be angled downward (approximately 15 to 20 degrees) to let gravity assist in the flow of the solids. Keeping the inlet line as short as possible can also minimize any pulsing of the solids flow.

6 EFFECT OF SOLIDS LOADING

Cyclones can collect solids over a wide range of loadings. Traditionally, solids loadings have been reported as either kg of solids per m^3 of gas (kg/m^3), or as kg of solids per kg of gas (kg$_s$/kg$_g$). However, loading based on mass is probably not the best way to report solid loadings for cyclones. This is because the volume of solids processed by a cyclone at the same mass loading can vary greatly, depending on the density of the solids. For example, many polymers have a bulk density of approximately 400 kg/m^3 and iron ore has a bulk density of approximately 2400 kg/m^3. This is a ratio of 6. Therefore, a cyclone operating with polymer would have to process six times the volume of solids that a cyclone operating with iron ore would process at the same mass loading. If the cyclone operating with the polymer were to be designed to operate at high loadings on a mass basis, it would probably plug. In addition, the diplegs below the cyclone operating with the polymer may experience operational problems because of the high volumetric loading. However, because almost all solids have been and are still reported on a mass basis, they will continue to be reported on this basis in this chapter.

At ambient conditions, cyclones have been operated at solid loadings as low as $0.02 \, kg/m^3$ ($0.0125 \, kg_s/kg_g$) and as high as $64 \, kg/m^3$ ($50 \, kg_s/kg_g$) or more. This is a factor of 3200. In general, cyclone efficiency increases with increasing solids loading. This is because at higher loadings, very fine particles are trapped in the interstices of the larger particles and this entrapment increases the collection efficiency of the small particles. Even though collection efficiencies are increased with increased loading, cyclone loss rates are also increased as loading is increased. This is because the cyclone efficiency increase is almost always less than the increase in the solids loading.

When solids are added to a cyclone the overall cyclone pressure drop decreases (Yuu et al., 1978). This effect is significant, as it often results in a pressure drop reduction of greater than 50% when solids are added. The reason for this decrease is that when solids are present at the wall of the cyclone, the increased drag on the gas due to the solids at the wall causes the cyclone tangential velocity to decrease. This results in a decrease in cyclone pressure drop. This condition occurs for low and medium solids loading. At higher solids loading the cyclone pressure drop begins to increase with loading as the pressure drop due to solids acceleration becomes greater than the pressure drop due to the reduction in the tangential velocity. The decrease in cyclone pressure drop at lower loadings and the subsequent increase in cyclone pressure drop at higher loadings are illustrated in Fig. 7.

7 GAS OUTLET TUBE

To prevent solids from the cyclone inlet from bypassing directly into the outlet of the cyclone, a tube the same diameter as the gas outlet is extended into the cyclone to a level equal to or below the bottom of the solids inlet. This prevents solids from bypassing directly into the outlet of the cyclone. This tube is called a gas outlet tube (it is also called a vortex finder or a vortex tube). The gas outlet tube does increase the efficiency of a low-loading cyclone relative to a cyclone that does not have a gas outlet tube. However, many

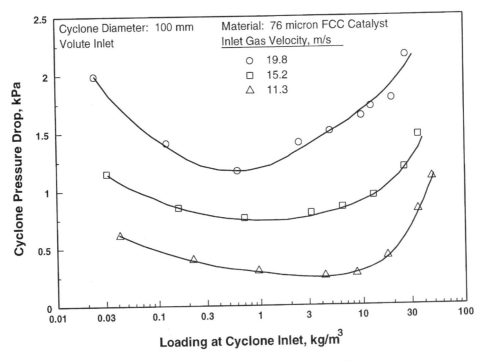

Figure 7 Cyclone pressure drop versus loading.

cyclone designers use a longer gas outlet tube than is required. For low loadings, the length of the gas outlet tube does not have to be greater than the height of the solids inlet. For high loadings, the optimum gas outlet tube length has been shown to be less than the height of the solids inlet. The optimum gas outlet tube length depends upon the gas velocity and the solids loading into the cyclone. At very high loadings, the optimum gas outlet tube length has been reported by Lewnard et al. (1993) to be approximately half that of the cyclone inlet height, or $0.5H$, as shown in Fig. 8.

Decreasing the diameter of the gas outlet tube generally increases the efficiency of a cyclone. This is due to increasing the length and rotating velocity of the vortex below the gas outlet tube. However, if the cyclone is not long enough to contain the increased vortex length and it extends too close to the solids flowing on the wall of the cone, then cyclone efficiency will decrease as the vortex plucks off solids from the wall of the cone, so it is imperative that the cyclone have a sufficient length before reducing the diameter of the gas outlet tube.

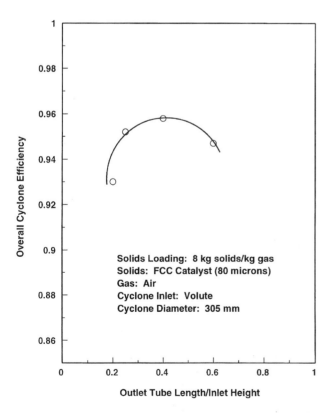

Figure 8 Effect of outlet tube penetration on cyclone efficiency. (From Lewnard et al., 1994.)

Hugi and Reh (1998) operated cyclones at loadings ranging from 10 to $50\,\mathrm{kg_s/kg_g}$ (13 to $64\,\mathrm{kg/m^3}$). They report that if the roof of the cyclone is extended, no gas outlet tube is required, and that this type of cyclone is superior to a cyclone in which the roof is not extended and a gas outlet tube is used. Muschelknautz et al. (1999) agreed with this assessment, but found that even higher efficiencies were possible if the cyclone roof was raised and a vortex tube was used in the cyclone.

Muschelknautz et al. (1999) have also reported that if the gas outlet tube for volute-entry cyclones is moved off-center, a cyclone's efficiency can be improved significantly The reason for this improvement appears to be that the solids entry for a tangential or volute entry cyclone is at the side and is not symmetrical. This causes the vortex to be slightly off-center. If the gas outlet tube is moved slightly, it eliminates eddy formation and improves cyclone efficiency.

8 INLET GAS VELOCITY

The inlet gas velocity for cyclones can range from a low of approximately 10 m/s to a high of over 40 m/s. The gas velocity used depends on the application of the cyclone. Often cyclones are used to separate a fragile, expensive material such as a catalyst from a gas stream. High inlet gas velocities would result in excessive attrition of the catalyst, so for this type of cyclone operation the inlet velocities are reduced to minimize the attrition of the expensive solids. In this case, the gas velocity in the outlet tube is designed to be much higher than the velocity in the inlet. Most of the solids are collected in the barrel of the cyclone, even at the relatively low inlet gas velocity, and most of the remaining solids are separated in the high-velocity vortex below the gas outlet tube.

In second-stage cyclones, the inlet gas velocity is generally higher than the inlet gas velocity of the first stage. In the second-stage cyclone, the loading is low, and an increased gas velocity leading to a higher centrifugal force is required for high separation efficiencies. Also the loading in the second-stage cyclone is so low that erosion and attrition will be relatively small, and so high velocities can be utilized.

High cyclone inlet gas velocities also result in very high cyclone pressure drops. Because the pressure drop across the cyclone is proportional to the square of the inlet gas velocity (i.e., $\Delta P \propto \rho_g g U^2$), lowering the inlet gas velocity can significantly reduce the pressure drop across the cyclone. This will lead to reduced energy expenditures and will also reduce the length of cyclone diplegs required for sealing the cyclones.

9 CYCLONE LENGTH

As described above, the cyclone length should be great enough to contain the vortex below the gas outlet tube. It is generally advisable to have the cyclone somewhat longer than required so that modifications to the gas outlet tube can be made if required. Either the barrel or the cone can be increased in length to contain the vortex. However, cyclone barrels can be made too long. If the barrel is too long, the rotating spiral of solids along the wall can lose its momentum. When this happens, the solids along the wall can be reentrained into the rotating gas in the barrel and cyclone efficiency will be reduced.

Hoffman et al. (2001) studied the effect of cyclone length on cyclone efficiency and showed that the efficiency of a cyclone increases with length. However, they also found that after a certain length, cyclone efficiency decreased. They reported that cyclone efficiency suddenly decreased after a certain cyclone length, which in their cyclone was at a length-to-diameter ratio of 5.65 (although many researchers employ this length-to-diameter ratio as a correlating parameter to make the length parameter dimensionless, it is likely that it is the actual length of the cyclone that is important). Hoffman et al. (2001) stated that the probable reason for the sudden decrease in cyclone efficiency was the central vortex touching and turning on the cyclone cone. When this occurred, the efficiency collapsed and caused increased solids reentrainment.

Hoffman et al. (2001) also reported that cyclone pressure drop decreased with increasing cyclone length. This probably occurs for the same reason that cyclone pressure drop decreases with increasing cyclone loading. For long cyclones, the increased length of the cyclone wall results in a longer path for the gas to travel. This causes more resistance to the flow of the gas in the cyclone (much as a longer pipe produces more resistance to gas flow than a shorter pipe) that results in reducing the tangential velocity in the cyclone and therefore the cyclone pressure drop. Hoffman et al. (2001) were able to predict this effect with various models.

10 CYCLONE WALL ROUGHNESS

Rough cyclone walls reduce cyclone efficiency and decrease the pressure drop in a cyclone. Iinoya (1953) added coarse sandpaper to the wall of a cyclone and found that cyclone pressure drop decreased. The coarser the sandpaper, the lower the pressure drop. Yuu et al. (1978) also found that sandpaper glued on the wall reduced cyclone pressure drop and attributed this to a reduction in the tangential velocity due to the higher drag on the rough walls.

Ionoya (1953) added a 1.2 mm diamond mesh liner to a cyclone and found that the liner significantly reduced cyclone efficiency. When the mesh liner covered 87% of the cyclone inner wall, cyclone efficiency was decreased by 9.5% relative to the cyclone's efficiency without the diamond mesh liner.

Large weld beads, etc., can also reduce cyclone efficiency. If the solids flow along the wall of a cyclone encounter a large protuberance such as a weld bead, the weld bead acts as a type of "ski jump" and causes the solids to be deflected further into the center of the cyclone where they can be thrown into the vortex and carried out of the cyclone. In small pilot or research cyclones this is especially common, because the distance between the wall of the cyclone and the vortex tube is very small. Because of their detrimental effect on cyclone efficiency, weld beads should be ground off to make the cyclone inner wall smooth.

In high-temperature processes, cyclones are often lined with refractory both to minimize heat loss and to protect the metal surfaces from abrasion. These refractor surfaces are not as smooth as metal, but after a few days of operation, the refractory becomes smoother because of the abrasive action of the solids.

With very small laboratory or pilot cyclones, some solids (large polymer beads, spherical particles, etc.) can sometimes bounce off of the cyclone wall immediately across from the cyclone inlet and be deflected into the vortex. Very large particles can be found in the gas outlet stream of the cyclone with these very small cyclones and with particles that bounce. To increase cyclone efficiency with these types of solids, the cyclone barrel diameter can be increased. This increases the distance between the cyclone vortex and the wall and prevents most of the solids from bouncing back into the vortex.

11 CYCLONE DIAMETER

In theory, a smaller diameter cyclone should be able to collect smaller particles because it can develop a higher centrifugal force. However, using smaller cyclones generally means that many have to be used in parallel to accommodate large gas flows. The problem with parallel cyclones (as indicated above) is that it is difficult to get even distribution of solids into all of the cyclones. If maldistribution occurs, this can cause inefficiencies that can negate the natural advantage of the smaller cyclones.

Cyclone diameters can be very large. Perhaps the largest cyclones are those used in circulating fluidized bed combustors, where cyclone diameters approach 10 meters. Large diameter cyclones also result in very long cyclones, and so these large-diameter long-length cyclones are really not feasible as internal cyclones in fluidized beds (they make the vessel too tall).

When using small cyclones in pilot or demonstration plants, particle bouncing or rebounding from the wall can be a problem. Because the cyclones are small, particles can bounce off the cyclone wall into the vortex and be carried out the cyclone exit. This is especially true for plastic particles that can bounce large distances after striking a wall. Therefore, for very small cyclones, it is recommended that the cyclone diameters be increased over the "typical" diameter dimensions to counter this effect. As an example, increasing the diameter of a 150 mm diameter laboratory cyclone to 300 mm in diameter should solve the most egregious bouncing problems. For particles that do not bounce as much as plastic, smaller diameter increases can be made.

12 CYCLONE DIMENSIONS AND DESIGN

Using a force balance (obtained by equating the time that it takes for particles of a particular size to traverse a width equal to the width of the cyclone inlet to the time that the gas resides in the cyclone), Rosin et al. (1932) developed a relation between the fractional efficiency (the efficiency of collection of a particular particle size of a cyclone to system geometrical and system parameters. The relation developed,

$$d_{p,th} \cong \sqrt{\frac{9\mu L_w}{\pi N_s U_i (\rho_p - \rho_g)}} \tag{2}$$

relates the theoretical size that can be collected by a cyclone to the width of the cyclone inlet, L_w, the viscosity of the gas, μ, effective number of spiral paths taken by the gas, N_s, the inlet gas velocity U_i, the particle density, ρ_p, and the gas density, ρ_g. Efficient cyclones collect particles of smaller particle size, so cyclone efficiency, E_o is proportional to $1/d_{p,th}$, or

$$E_o \propto \frac{1}{d_{p,th}} \tag{3}$$

Therefore, cyclone efficiency is increased by increasing the effective number of spiral paths taken by the gas (this increases the residence time of the gas in the cyclone), the inlet gas velocity (this increases the centrifugal force on the particles), and the particle density (which also increases the centrifugal force on a particle). Cyclone efficiency is decreased by larger inlet widths (it takes longer for the solids to get to the wall for greater inlet widths) and increasing gas viscosity (it takes longer for the particles to get to the wall through a more viscous gas). Note that the equation predicts that gas density will not have a large effect on cyclone efficiency. This is because of the term $(\rho_p\text{-}\rho_g)$. Because ρ_p is so much larger than ρ_g, this relation predicts that gas density should have no major effect on cyclone efficiency.

One equation that has been used extensively in cyclone design procedures is that of Lapple (1940). This equation replaces the smallest theoretically collected particle size with a particle size, $D_{p,th}$, collected by the cyclone with a theoretical efficiency of 50%, thus

$$D_{p,th} = \sqrt{\frac{9\mu L_w}{\pi N_s U (\rho_p - \rho_g)}} \tag{4}$$

According to Eq. (4) (which applies at relative low solid loadings in the cyclone), the higher the inlet gas velocity, the greater the cyclone efficiency. In practice, however, this is not the case. In a real cyclone, efficiency increases up to a limiting value of gas velocity and then begins to decrease. The reason for the decrease in cyclone efficiency at high gas velocities is that the inner vortex becomes longer as gas velocity increases and causes the pressure drop of the gas entering the gas outlet tube to increase. The increased pressure drop across the gas outlet tube causes the inner vortex length to increase. At some vortex length, the vortex approaches the side of the cyclone cone. When this happens, the vortex will begin to pluck off the solids flowing on the side of the cone, and cyclone efficiency will begin to decrease. If the vortex length is too long, so that it impinges against the side of the cone, severe erosion of the cyclone will result.

The cyclone design procedure developed by Zenz (1975) utilizes Eq. (4) as the basis of the design procedure. The procedure consists of several steps:

1. Calculate $D_{p,th}$ from Eq. (4).
2. Estimate the effective number of effective spiral gas paths in the cyclone, N_s.

3. Partition the particle size entering the cyclone into at least 10 different cuts, to give particle sizes of the cuts (d_{pi}) from d_{pl} to d_{pn}.
4. Calculate the ratio of $d_{pi}/D_{p,th}$ for each cut and determine the low-loading cyclone collection efficiency for each cut size.
5. Determine the overall cyclone collection efficiency (E_o) for very low loading cyclones (low loading is defined to be 1 grain/ft^3 (0.0023 kg/m^3) by the relation

$$E_o = \sum_{i=1}^{n} x_i E_{oi} \qquad (5)$$

where:

x_i = the weight fraction of each individual
cut size

E_{oi} = the collection efficiency of the cyclone
for each individual cut size, %.

6. Modify the low-loading cyclone collection efficiency for the effect of loading.
7. Determine cyclone pressure drop.

A detailed description of this approach is described below.

12.1.1 Estimation of N_s

The number of effective spiral paths taken by the gas in the outer vortex of the cyclone can be estimated by the curve in Fig. 9. The number of effective cyclone spirals cannot be determined in most cyclones because they are made of metal. However, N_s has been experimentally determined by back-calculation from observed cyclone efficiencies The gas velocity used to determine N_s in Eq. (3) should be the *maximum* of the inlet gas velocity, U_i, or the outlet gas velocity, U_o.

The curve to predict N_s is valid only for very dilute (low loading) cyclones where individual particles can be assumed to be present in the cyclone (Zenz, 2001). The assumption is generally valid for cyclone loadings of no greater than approximately 1 grain/ft^3 (0.002 kg/m^3).

12.1.2 Partition Cuts

When dividing the overall particle size distribution entering the cyclone into several cuts, it is advisable to use a minimum of at least 10 cuts. It is also better not to divide the distribution into equal size cuts. Instead, it is better to bias the cuts so that the small end of the particle size distribution has more cuts than the large end. This is because small particles are more difficult to collect than large particles, and it is the small particles that will determine the collection efficiency of the cyclone. For each d_{pi} in the partition, there will also be a corresponding weight fraction x_i, for each cut.

12.1.3 Calculate $d_{pi}/D_{p,th}$ and Determine E_{oi}

After $D_{p,th}$ has been calculated and the particle size distribution has been partitioned, the ratio of

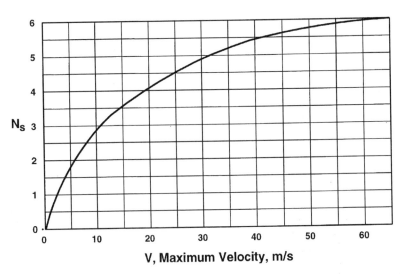

Figure 9   The effect of gas velocity on the number of solid spirals in a cyclone. For maximum velocity, use greater of inlet or outlet velocity.

$d_{pi}/D_{p,th}$ is calculated for each of the cuts in the biased, partitioned particle size distribution. From the values of $D_{pi}/D_{p,th}$, the low-loading collection efficiency (E_{oi}) for each of the cuts can then be determined from Fig. 10. The overall low-loading cyclone collection efficiency is then calculated from the individual cut collection efficiencies by using Eq. (5), i.e.,

$$E_o = \sum_{i=1}^{n} x_i E_{oi}$$

12.1.4 Determine the Effect of Loading on Cyclone Efficiency

The effect of solids loading can be calculated by using the curves shown in Fig. 11. These curves are empirical curves obtained from a considerable amount of data. The curves are plotted on a graph with solids loading on the abscissa and cyclone efficiency on the ordinate. To use the curves, first find the curve corresponding to the low-loading cyclone efficiency calculated above. This efficiency value will be at the lowest loading on the plot. Follow this curve to the actual loading of he cyclone, and read off the actual efficiency of the cyclone when the low-loading curve intersects with the actual cyclone loading.

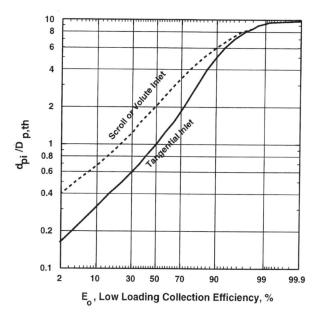

Figure 10 Cyclone efficiency versus $d_{pi}/D_{p,th}$.

12.1.5 Calculate Cyclone Pressure Drop

Cyclone pressure drop is calculated by summing several individual types of pressure-drop terms. The individual pressure drops are listed and described below:

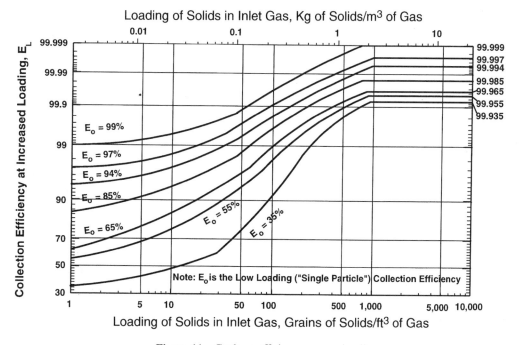

Figure 11 Cyclone efficiency versus loading.

1. Contraction Pressure Drop. The contraction pressure drop applies to the pressure drop produced by gas contracting from a large area to a small area. For internal cyclones, the most common application of the contraction pressure drop is for the contraction from the large area of the freeboard of a fluidized bed to the smaller area of the cyclone inlet. For external cyclones, the contraction pressure drop does not usually apply from the freeboard of the bed to the cyclone inlet. For external cyclones, the contraction pressure drop will usually have to be applied twice, from the area of the freeboard of the bed to the area of the exit gas line, and from the area of the exit gas line to the area of the cyclone inlet. The contraction pressure drop is given by

$$\Delta P_{(f-i)g} = 0.5\rho_g \left(U_i^2 - U_f^2 + K_{fi} U_i^2 \right) \qquad (6)$$

In Eq. (6) K_{fi} is the contraction coefficient due to gas flowing from a larger area to a smaller area. It is a function of the ratio of the smaller area (A_s) to the larger area (A_1). The contraction coefficient in the pressure-drop equation for an internal cyclone is a function of the ratio of the inlet area, A_i, to the area in the freeboard, A_f. Values of this coefficient for various area ratios are shown in Table 2.

For external cyclones, people often measure the pressure drop from a point immediately before the cyclone inlet to a point immediately above the cyclone in the gas outlet tube. This measurement does not include the contraction pressure drop as gas flows from the freeboard of the bed into the exit gas line from the fluidized bed. For dipleg seal height calculations, this pressure drop should be taken into account. It is better to measure the total cyclone pressure drop as the difference between the pressure in the freeboard and the pressure in the gas outlet tube.

Table 2 Values of the Contraction Coefficient for Various Small and Large Area Ratios

Ratio of the smaller to the larger area	K_{fi} or K_o
0.0	0.50
0.1	0.47
0.2	0.43
0.3	0.395
0.4	0.35

2. Acceleration of Solids Pressure Drop. The solids entering the gas with the cyclone must be accelerated from a very low gas velocity in the freeboard of the fluidized bed to the inlet gas velocity of the cyclone. This term is generally the largest of the individual cyclone pressure drop terms listed here, and is given by

$$\Delta P_{(f-i)p} = L U_{pi} \left(U_{pi} - U_{pf} \right) \qquad (7)$$

For small particles having a low terminal velocity, U_t, U_{pi} is usually taken as U_i, and U_{pf} as U_f. In general, this approximation is used for particle sizes less than about 100 microns. Hugi and Reh (1998) recently measured the solid and gas velocities in the inlet of a cyclone and reported that there is generally significant slip between them, even for small particle sizes (40 microns). This probably occurs because of strand or agglomerate formation in the cyclone inlet. However, if no information is available regarding this slip, using the general guidelines given above is probably the best way to proceed.

3. Barrel Friction Pressure Drop. The pressure drop due to solids flowing along the wall of the cyclone barrel is

$$\Delta P_{bf} = \frac{2 f \rho_g U_i^2 \pi D_b N_s}{d_{hi}} \qquad (8)$$

where f is the Fanning friction factor (generally ranging between 0.003 and 0.008) and d_{hi} is the hydraulic diameter of the cyclone inlet given by the relation

$$d_{hi} = \frac{4(\text{Inlet area})}{\text{Inlet perimeter}} \qquad (9)$$

The Reynolds number for the barrel pressure drop friction factor should be based on this diameter. This Reynolds number is given by

$$\text{Re}_{hi} = \frac{d_{hi} U_i \rho_g}{\mu} \qquad (10)$$

4. Gas Reversal Pressure Drop. The gas reversal pressure drop is due to the gas reversing its direction and its subsequent acceleration in the gas vortex in the cyclone. It is given by

$$\Delta P_r = \frac{\rho_g U_i^2}{2} \qquad (11)$$

5. Outlet Exit Contraction Pressure Drop. The outlet contraction pressure drop occurs because of the gas contracting from the area of the cyclone barrel to the area of the gas outlet tube. It is estimated by

$$\Delta P_o = 0.5\rho_g \left(U_o^2 - U_b^2 + K_o U_o^2 \right) \qquad (12)$$

K_o is obtained from Table 2 for vales of the ratio of the outlet area to the inlet area calculated from $(D_o/D_b)^2$.

When calculating dipleg seal heights, it is not necessary to take into account the exit contraction pressure drop in the calculations. This is because the pressure at the top of the dipleg is actually the pressure drop inside the cyclone, not the pressure in the gas outlet tube. This is not really an issue in most calculations because the exit contraction pressure drop is usually very low. Also, it is conservative to include this pressure drop in the calculation because it will result in slightly longer dipleg lengths.

The sum of these five pressure drop terms estimates the total cyclone pressure drop. The cyclone pressure drop estimated by these five terms assumes that the cyclone pressure drop is at a high enough loading so that the cyclone pressure-drop reduction due to loading is at its minimum, as shown in Fig. 7. If the cyclone loading is lower than this (on the left-hand side of the curve in Fig. 7), then a correction or an adjustment for this higher pressure drop at low loadings must be made.

To determine if the solids loading is sufficiently low so that the pressure drop must be adjusted, one must use Fig. 12. To use Fig. 12, first calculate the solids mass flux (in kg/s-m^2) at the inlet of the cyclone. If the value of the mass flux at the cyclone inlet is less than approximately 100 kg/s-m^2, then the calculated pressure drop must be adjusted. The amount of the adjustment is determined from the value on the ordinate of the figure. The ordinate is a ratio of the true cyclone

pressure drop to the calculated pressure drop. If the value of the calculated inlet mass flux is so low that the ratio is significantly greater than one, then the calculated cyclone pressure drop must be multiplied by the value on the ordinate. As an example, assume that the value of the inlet solids flux is 10 kg/s-m^2. The value of the ordinate for this inlet mass flux is approximately 1.2. Thus the calculated pressure drop from the procedure outlined above should be multiplied by 1.2 to obtain the "real" pressure drop.

The cyclone design procedure that has just been described is a tool that can be used to estimate cyclone efficiency. As with most calculation procedures, there is always a difference between what is measured and what is calculated. As Zenz (2001) notes, the primary difference between the measured efficiency and the calculated efficiency using this method is due to the correct estimation of the actual particle size entering the cyclone. Agglomeration can occur in fine dusts to such an extent that the actual cyclone efficiency can be much greater than calculated.

13 VORTEX LENGTH

The spinning gas from the outer annulus flows into the inner rotating core (the vortex) over a certain length of the cyclone. The length of this vortex, L_v, is taken to be the length within the cyclone where the vortex gas can exert a positive driving force (Bryant et al., 1983). This

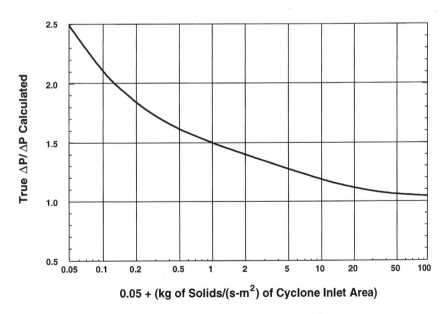

Figure 12 Cyclone pressure drop correction curve.

length is called the vortex length of the cyclone. If the cyclone is so short that the bottom of the vortex length is close to and/or touches the cone, solids will be plucked from the wall of the cone and entrained into the exiting gas in the vortex, so it is of great interest to be able to estimate the length of the spinning vortex.

Alexander (1949) developed an empirical equation to predict L_v:

$$L_v = 2.3 D_o \left(\frac{D_B^2}{A_i}\right)^{1/3} \qquad (13)$$

As Bryant et al. (1983) note, this equation seems incorrect, because as D_o gets smaller, it would be expected that L_v would get larger, not smaller as Eq. (13) predicts. Bryant et al. (1983) then developed an equation for the vortex length that seems more directionally correct:

$$\frac{L_v}{D_B} = \frac{2}{(A_o/A_i)^{1/2}} \qquad (14)$$

This equation is plotted in Fig. 13. For most cyclones, the minimum practical value of A_o/A_i is approximately 0.4. At values much lower than this, the practical outlet diameter becomes too small for most cyclones.

In the same study, Bryant et al. (1983) conducted tests with a cylindrical cyclone with no cone that was

fitted with a sealed piston filled with solids. When the piston was raised in the cylinder the solids were suddenly entrained from the piston at a critical height below the gas outlet tube of approximately $1.6D_b$. When the piston was lowered, the vortex remained attached to the bed surface until a second critical distance was reached. This distance corresponded to a value of approximately $2D_b$. Systematic experiments have not yet been conducted to determine whether the vortex length is actually a function of the diameter of the cyclone. To investigate this, experiments on different diameter cyclones would have to be conducted.

Bryant et al. (1983) also reported that in tests with sticky particles that adhered to the wall of the cyclone, particles adhered to the cyclone cone at a distance greater than $1.6D_b$ below the gas outlet tube; but above this length the cyclone was clean. This result indicates that it would be better to have short cyclones when operating with "sticky" cyclones.

Hoffman et al. (1995) studied the cyclone vortex length and found that separation efficiency was less for cyclones with their natural vortex ending in the cone instead of the barrel. They also reported that the vortex length (swirl intensity) decreased with increasing solids loading and was a strong function of the cyclone length. Akiyama and Marui (1989), in

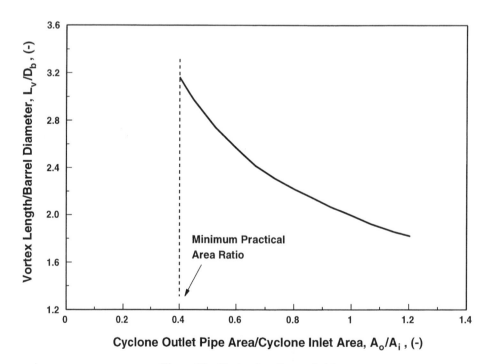

Figure 13 Vortex length vs. A_o/A_i.

their investigation of uniflow cyclones, also reported that the vortex swirl intensity decreased with increasing cyclone length.

13   VORTEX STABILIZERS

As described above, the vortex of a cyclone will precess (or wobble) about the center axis of the cyclone. This motion can bring the vortex into close proximity to the wall of the cone of the cyclone and pluck of and reentrain the collected solids flowing down along the wall of the cone. Sometimes an inverted cone or a similar device is added to the bottom of the cyclone in the vicinity of the cone and dipleg to stabilize and "fix" the vortex. If placed correctly, the vortex will attach to the cone and the vortex movement will be stabilized, thus minimizing the efficiency loss due to plucking the solids off the wall of the cyclone.

14   EFFECT OF PRESSURE AND TEMPERATURE

Most processes operate at high temperatures and/or high pressures, so it is important to know how cyclones operate at these conditions. Efficient cyclones are able to collect very small particle sizes, so that cyclone efficiency is proportional to $1/D_{\mathrm{p,th}}$, i.e.,

$$E_{\mathrm{o}} \propto \frac{1}{D_{\mathrm{p,th}}} = \sqrt{\frac{\pi N_{\mathrm{s}} U (\rho_{\mathrm{p}} - \rho_{\mathrm{g}})}{9 \mu L_{\mathrm{w}}}} \tag{15}$$

The effects of temperature and pressure manifest themselves in the way that they affect gas density and gas viscosity. From Eq. (15), it can be seen that cyclone efficiency is theoretically related to gas density and gas viscosity as

$$E_{\mathrm{o}} \propto \sqrt{\frac{\rho_{\mathrm{p}} - \rho_{\mathrm{g}}}{\mu}} \tag{16}$$

As pressure is increased, gas density will increase. However, the term $\rho_{\mathrm{p}} - \rho_{\mathrm{g}}$ does not change with increases in gas density because particle density is so much greater than the gas density (typically about 2000 kg/m^3 versus approximately 20 kg/m^3 at high pressure) that it dominates this term. Therefore, it would be expected that gas density would have little or no effect on cyclone efficiency. Conversely, cyclone efficiency would be expected to decrease with system temperature because gas viscosity increases with increasing temperature.

Knowlton and Bachovchin (1978) studied the effect of pressure on cyclone performance and found little change in overall cyclone efficiency with pressure over a pressure range from 0 to 55 barg. However, fractional efficiency curves for the same study (Fig. 14) showed that cyclone efficiency decreased with pressure for particle sizes less than about 20 to 25 microns. For particle

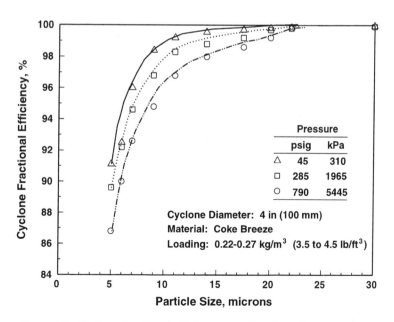

Figure 14　Cyclone fractional efficiency as a function of system pressure.

sizes greater than about 25 microns, there was no effect of pressure on cyclone efficiency. Karri (1998) also showed that cyclone efficiency decreased with pressure, but only for particle sizes less than about 15 microns.

The effect of temperature on cyclone efficiency was studied by both Parker et al. (1981) and Patterson and Munz (1989). Both studies showed that cyclone efficiency decreased with increasing gas viscosity. As with the studies at high pressure, Patterson and Munz (1989) reported that only the collection efficiency of particles less than about 10 microns was reduced because of operation at high temperature (Fig. 15).

NOMENCLATURE

A_f	=	Freeboard area above fluidized beds, m^2
A_i	=	Cyclone inlet area, m^2
A_l	=	Larger of the flow areas in gas contraction flow, m^2
A_o	=	Area of cyclone gas outlet, tube, m^2
A_s	=	Smaller of the flow areas in gas contraction flow, m^2
D_b	=	Cyclone barrel diameter, m
$D_{p,th}$	=	Theoretical particle size collected by a cyclone at 50% efficiency, μm
D_o	=	Gas outlet tube diameter, m
d_{hi}	=	Hydraulic diameter of cyclone inlet, m

$d_{p,50}$	=	Median particle size, μm
$d_{p,th}$	=	Theoretical particle size capable of being collected by a cyclone, μm
E_o	=	Cyclone efficiency, %
E_{oi}	=	Low-loading cyclone collection efficiency, %
f	=	Fanning friction factor
H	=	Cyclone inlet height,
K_{fi}	=	Contraction coefficient for flow from the freeboard to the cyclone inlet
K_o	=	Contraction coefficient for flow from the cyclone barrel to the cyclone outlet
L	=	Solids loading, kg of solids/m^3 of gas
L_v	=	Vortex length below gas outlet tube, m
L_w	=	Cyclone inlet width, m
N_s	=	Effective number of solid spirals in a cyclone
Re_{hi}	=	Reynolds number at the cyclone inlet based on d_{hi}
r_o	=	Radius of cyclone, m
U	=	Superficial gas velocity, m/s
U_f	=	Gas velocity in freeboard of fluidized bed, m/s
U_i	=	Cyclone inlet gas velocity, m/s
U_o	=	Gas velocity in cyclone outlet tube, m/s
Up_i	=	Particle velocity at solids inlet, m/s
U_{pf}	=	Particle velocity in freeboard of fluidized bed, m/s
U_t	=	Particle terminal velocity, m/s
x_i	=	Weight fraction of cut i in particle size distribution
ΔP	=	Pressure drop, cm H_2O

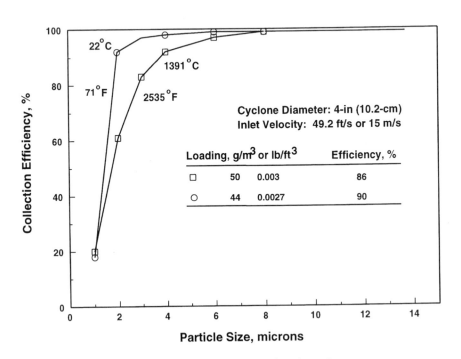

Figure 15 Cyclone fractional efficiency as a function of system temperature.

$\Delta P_{(f-i)g}$ = Gas contraction pressure drop from freeboard to cyclone inlet, cm H_2O

$\Delta P_{(f-i)g}$ = Solids acceleration pressure drop, cm H_2O

P_{bf} = Cyclone barrel friction pressure drop, cm H_2O

ΔP_o = Cyclone exit pressure drop, cm H_2O

μ = Gas viscosity, kg/(m-s)

ρ_p = Particle density, kg/m^3

ρ_g = Gas density, kg/m^3

REFERENCES

Akiyama T, Marui T. Dust collection efficiency of a straight-through cyclone: effects of duct length, guide vanes and nozzle angle for secondary rotational flow. Powder Technol 58:181, 1989.

Alexander RM. Fundamentals of cyclone design and operation. Proceedings of the Australian Institute of Mining and Metallurgy 152, 153:203, 1949.

Barth W. Berechnung und auslegung von zyklonabscheidern aufgrund neuerer untersuchungen. BWK 8(1):1–9, 1956.

Barth W, Leineweber L. Beurteilung und auslegung von zyklonabscheidern. Staub 24(2):41–53, 1964.

Broodryk NJ, Shingles T. Aspects of cyclone operation in industrial chemical reactors. Preprints of Fluidization VIII. Tours, France, May 14–19, 1995, p 1083.

Bryant HS, Silverman RW, Zenz FA. How dust in gas affects cyclone pressure drop. Hydrocarbon Processing June, pp 87–90, 1983.

Dietz PW. Collection efficiency of cyclone separators. AIChE J 27:888, 1981.

Gauthier TA, Briens CL, Bergougnou A, Galtier P. Uniflow cyclone efficiency study. Powder Technol 62:217–225, 1990a.

Gauthier TA, Briens CL, Galtier P, Bergougnou MA. Gas–solid separation in a uniflow cyclone at high solids loadings: effect of acceleration line. In: Basu P, Horio M, Hasatani M, eds. Circulating Fluidized Bed Technology III. Oxford: Pergamon Press, 1990b, pp 639–644.

Gauthier TA, Briens CL, Crousle O, Galtier P, Bergougnou MA. Study of the gas circulation patterns in a uniflow cyclone. Can J Chem Eng 70:209–215, 1992.

Hoffman AC, de Groot M, Peng W, Dries HWA, Kater J. Advantages and risks in increasing cyclone separator length. AIChE J 47(11):2452–2460, 2001.

Hoffman AC, de Jonge R, Arends H, Hanrats C. Evidence of the "natural vortex length" and its effect on the separation efficiency of gas cyclones. Filt Sep 32:799, 1995.

Hugi E, Reh L. Design of cyclones with high solids entrance loads. Chem Eng Technol 21(9):716–719, 1998.

Iinoya K. Memoirs of the faculty of engineering. Nagoya University 5(2): September, 1953.

Kaliski H. Vergleich von fliehkraftentstaubern. Freiberger Forschungsheft A93:78–100, 1958.

Karri SBR. The effect of pressure on cyclone performance. PSRI Research Report No. 74, June 1, 1998.

Knowlton TM, Bachovchin D. Coal processing Technol 4:122–127, 1978.

Koffman JL. Gas and Oil Power 48:89–94, 1953.

Lapple CE. Ind Eng Chem 32:605–617, 1940.

Lapple CE. Dust and mist collection. Air Pollution Abatement Manual, Manufacturing Chemists Association, Washington, D.C., 1951.

Lewnard JJ, Herb BE, Tsao TR, Zenz JA. Effect of design and operating parameters on cyclone performance for circulating fluidized bed boilers. In: Avidan AA, ed. Circulating Fluidized Bed Technology IV. New York: AIChE, 1993, pp 525–531.

Muschelknautz E. Auslegung von zyklonabscheidern in der technischen praxis. Staub-Reinhaltung der Luft 30(5):187–195, 1970a.

Muschelknautz E. Design of cyclone separators in the engineering practice. Staub-Reinhaltung der Luft 30(5):187–195, 1970b.

Muschelknautz E, Krambrock W. Pressure drop and separation efficiency in cyclones. VDI Heat Atlas, 6, and first English edition, Ch. Lj, 1–8, 1996.

Muschelknautz U, Muschelknautz E. Improvements in cyclones in cfb power plants and quantitative estimations of their effects on the boilers solids inventory. In: Werther J, ed. Circulating Fluidized Bed Technology VI. Wurzburg, Germany, 1999, pp 761–767.

O'Mara R. Combustion 21:38–43, 1950.

Parker R, Jain R, Calvert S, Drehmel D, Abbott, J Envir Sci Technol 15(4):451, 1981.

Patterson P, Munz R. Can J Chem Eng 67:321, 1989.

Rentschler W. Abscheidung und druckverlust des gaszyklons in abhängigkeit von der staubbeladung. VDI-Fortschrittbericht, Reihe 3, Nr. 242, 1991.

Rietema K, Verver CG. Cyclones in Industry. New York: American Elsevier, 1961.

Rosin P, Rammler E, Intelmann W. Principles and limits of cyclone dust removal. Zeitschrift Verein Deutscher Ingenieure 76:433, 1932.

Smellie J. Iron Coal Trades Review 144:169, 1942.

Stairmand CJ. Trans Inst Chem Eng 29:356, 1951.

Ter Linden AJ. Inst Mech Engrs 160:233, 1949.

van Tongren H. Mech Eng 58: February, p 127, 1936.

Whiton LC Jr. Trans Am Soc Mech Engrs 213–218, 1941.

Yuu S, Tomosada J, Yuji T, Yoshida K. The reduction of pressure drop due to dust loading in a conventional cyclone. Chem Eng Sci 33:1573–1580, 1978.

Zenz FA. Cyclone separators. In: Chapter 11, Manual on Disposal of Refinery Wastes; Volume on Atmospheric Emissions, API Publication 931. Washington, DC: Pet. Inst. Refining, 1975.

Zenz FA. Cyclone design tips. Chem Eng January, p 60, 2001.

23

Dilute-Phase Pneumatic Conveying

George E. Klinzing

University of Pittsburgh, Pittsburgh, Pennsylvania, U.S.A.

1 INTRODUCTION

Pneumatic conveying occurs in almost every industrial application that uses powders and granular materials. Most of this has been handled by dilute-phase conveying. Particle sizes ranging from a few microns to 3 inches in diameter have been successfully conveyed, and the size of the lines employed for conveying go from $\frac{1}{4}$-inch to 24 inches in diameter. Generally, if a material is sticky or moist, other methods of transport such as belt conveyors are preferred over dilute-phase pneumatic conveying. One guideline in material handling with dilute-phase pneumatic conveying is to avoid abrasive materials; however, this guideline often will be violated to solve a particular conveying problem. Abrasive materials usage causes high maintenance cost for pneumatic conveying systems. Pneumatic conveying has been applied to such diverse situations such as the conveying of ice over 1 kilometer in length to cool gold mines, under reduced gravity conditions to explore the possibility of using the technology on the moon, and even for the conveying of live animals.

Pneumatic conveying protects the product from the environment and the environment from the product because of its enclosed nature. There have been dilute-phase pneumatic conveying applications that extend up to 3 miles in distance, while other uses call for a distance of a few meters.

Dilute-phase pneumatic conveying is not an energy-efficient way of conveying, since generally considerable horsepower is needed to provide the motive air or gas. Dilute-phase pneumatic conveying, however, is easy and convenient to put into operation, even if often it does not work at optimum conditions, mostly owing to poor understanding of the overall concept of the process.

There are five components to a dilute pneumatic conveying system:

1. Conveying line
2. Air/gas mover
3. Feeder
4. Collector
5. Controls

Basically, dilute-phase pneumatic conveying systems can be broken down into three categories based on the physical principle used for conveying.

1. Pressure system
2. Vacuum system
3. Pressure/vacuum system

The modes of pneumatic transport are often classified as dilute, strand (two-phase), and dense phase flows. This discussion will concentrate on dilute transport technology.

2 PARTICLE CHARACTER AND BEHAVIOR

2.1 Size Analysis

In order to convey particles, it is first necessary to characterize the material according to size and size distribution. In the past, the literature was filled with studies that reported that one conveyed sand or a particular seed with little mention of the size and size distribution. The average particle size used in most calculations is the d_{50}, which represents the diameter at which 50% of the particles are less in size. For particles that are formed by a breaking or fracture process, the log normal distribution is most commonly used to represent the size distribution. Particles formed by a condensation process can be described by the regular normal distribution. There are many different types of particle size standards, and depending on the type of operation, one sizing standard or measurement technique can be preferred over another. Microscopes give length mean averages, while light measuring devices give volume or surface mean averages. The common sieving operation provides one with the weight mean. The weight mean average is the one often employed in most of the discussion in pneumatic conveying. The calculations to be employed use the weight mean average for the size of the particle. In general, unless one has a particularly unique or odd size distribution, this average size will be acceptable as an average particle size in the calculations. For a more detailed discussion of size analysis, see Chapter 1, "Particle Characterization and Dynamics."

2.2 Shape Analysis

In addition to knowing the size and distribution of the particles, the shapes of the particles are essential for understanding pneumatic conveying differences. Particles can have various shapes: rounded, angular, and fibrous. The latter is the most difficult to address in choosing a correct particle size for conveying analysis.

The shape of the particle can be given by the traditional sphericity factor, fractal analysis, or by Fourier transform representations. The latter is a bit involved, requiring several coefficients for complex definition. Fractal analysis is receiving more attention of late in representing the shape of the particles handled. Chapter 1, "Particle Characterization and Dynamics," presents a more complete description of shape analysis.

After defining the size and shape, one is now ready to address the overall force balance of the particle–gas system. A macroapproach analysis will be presented in this treatment of dilute-phase transport. Using a kinetic approach with individual particle behaviors is computationally intense but has shown some success in depicting the actual physics of the process. The continuum approach is another procedure that can be computationally intensive but productive. There has been much discussion over the proper force terms and viscosity presentations and turbulence models in these analyses.

2.3 Force Balance

The macroapproach uses Newton's second law initially for a single particle system and then expands to a multiparticle system. Some of the concepts of a single-phase friction factor will be employed in the multiparticle system analysis. In addition to the acceleration and drag forces, the gravity and electrostatic forces should be considered in gas–solid analysis. The variation of the drag coefficient with the kind of particle and condition of flow is important in the analysis. The terminal velocity of the particle is often used as another way to characterize particles, and the larger the value of the terminal velocity, the greater the size and/or density of the particle.

Multiparticle systems must consider the same terms as the single particle analysis with particle–particle and particle–wall interaction essential for a complete description of the system. The pressure behavior is also of the utmost importance because of the nature of gaseous environments. In treating multiparticle systems, the drag term has often been corrected to include a modification to the drag on a single particle. One correction employs the voidage term, which comes from the classic Richardson and Zaki (1954) analysis on settling of multiparticle systems (voidage$^{-4.7}$ where the voidage is defined below). Pneumatic conveying analysis must concentrate on the most important factors in its analysis, the pressure drop loss or the energy loss, the product flow rate, and the volumetric gas flow rate. These factors influence the operational cost of the overall system over long periods of time.

There are several approaches to developing predictive models for pressure drop and energy losses in dilute-phase pneumatic conveying. The German literature, for instance, uses an expression proposed by Barth (1954) that includes a lift term in the horizontal analysis, while the U.S. approach is prone to lump this effect into the frictional term.

If one reviews the basic pressure loss equation,

$$\frac{\Delta P}{L} = \rho \varepsilon g + \rho_p (1 - \varepsilon)g + f_s \rho_p (1 - \varepsilon)\frac{u_p^2}{2D} + \frac{2f_g \rho \varepsilon u_g^2}{2D}$$

(1)

Pressure loss	gas gravity	solid gravity	solids friction effect	gas friction effect

there are a number of parameters that are required to begin to calculate some values:

1. Voidage
2. Particle velocity
3. Friction factors

In order to complete the picture, these parameters will be discussed and guidelines and suggestions will be given on appropriate values and calculations.

2.4 Voidage

The voidage is the amount of void space present in a pipe carrying solids. The voidage in the pipe can be expressed as

$$\varepsilon = 1 - \frac{4W_s}{\pi u_p \rho_p}$$

(2)

For a packed bed, voidage values of 0.4 to 0.6 are common, while in dilute pneumatic conveying, values of 0.98 and above can be found. One should also realize that in pneumatic conveying, especially in the horizontal direction, there is a gradation of the concentration of particles from top to bottom, smaller to larger in value. In most of the calculations presented here, a uniform voidage will be assumed. The superficial gas velocity in a conveying line is the velocity when no particles are present, while the fluid velocity when particles are present is larger by the factor of 1/voidage.

2.5 Particle Velocity

For determining the particle velocity, some guidelines are available. For particles less than 40 micrometers in size, the particle size can be given by the difference between the superficial gas velocity and the terminal velocity:

$$u_p = u_g - u_t$$

(3)

Actually, in our studies we have found that this expression is good up to about 500 micrometers in particle size. Several other studies have given models for the

particle velocities. In order to be sure that actual measurement of the particle velocity is carried out, the Hinkle–IGT (1978) model can be related to several systems and particle parameters:

$$u_p = u_g(1 - 0.68d^{0.92}\rho_p^{0.5}\rho_g^{-0.2}D^{-0.54})$$

(4)

Yang (1977) has developed an implicit equation to predict the particle velocity that is described as

$$u_p = u_g - u_t\left[\left(1 + \frac{2f_s}{gD}u_p^2\right)\varepsilon^{-4.7}\right]^{0.5}$$

(5)

$$f_p = 4f_s$$

(6)

2.6 Friction Factors

The frictional representation in pneumatic conveying assumes that the gas can be treated by itself for representing the gas friction and the solid frictional term for the solid friction. This linear representation has been recently challenged by Weber (1991), who provides some success with a nonlinear combination of these two terms. For the single-phase frictional term, the Koo equation (1932) as shown below is suggested, although the Blasius (1913) could just as well be used.

$$f_g = 0.0014 + 0.125(\text{Re})^{0.32}$$

(7)

For dilute-phase transport, the solid frictional term has been suggested by many investigators. We have found that the expression of Konno and Saito (1969) has the widest applicability; it even spills over into dense phase analysis:

$$f_s = 0.0285(gD)^{0.5}u_p^{-1}$$

(8)

Another frictional representation that has been receiving some attention has been developed by Michaelides (1987):

$$f_s = \frac{K}{4}\frac{\sqrt{gD}}{u_p}$$

(9)

where K depends on the type of particle conveyed. For coal it is 0.058.

The effect of different pipe wall materials has been probed by Rizk (1973) and others. This effect, however, appears to be minor and even of little consequence when a fine powder coats the conveying line wall. For inclined flow Klinzing et al. (1989) has given a frictional representation that covers all kinds of inclined flows from horizontal to vertical:

$$f_s = \frac{D}{2u_p^2}\frac{(3/4)\left(C_D \varepsilon^{-4.7}\rho_g\{u_f - u_p\}^2\right)}{(\rho_p - \rho_s)d} - g\sin\theta \quad (10)$$

where θ is the angle of inclination.

2.7 Acceleration

The analysis so far has assumed that the flows are steady and nonaccelerating. When entering a conveying line, going around a bend or connection, and exiting a line, acceleration effects come into play. The acceleration can be handled with the basic Newton analysis as well as through empiricism. A combination of both of these has been suggested as one approach for analysis. It should also be noted that acceleration effects are more dominant in horizontal conveying than in vertical conveying. A model that has proven useful for acceleration is

$$L_{accel} = 27.66\left(\frac{d}{D}\right)^{0.953}\mu^{-0.0912}\left(\frac{\rho}{\rho_p}\right)^{-0.924} \quad (11)$$

2.8 Saltation

Saltation is a phenomenon that occurs in horizontal flow: particles fall out of the suspension that is conveying the material and deposit on the bottom surface of the pipe. Many studies have been explored to determine the saltation condition, since it is a parameter essential in designing conveying systems. The most commonly used correlation is that of Rizk (1973), which was developed for plastic pellets and shows no effect of particle density or pipe size:

$$\mu = \frac{1}{10^\delta}\left(\frac{u_g}{\sqrt{gd}}\right)^\chi \quad (12)$$

where

$$\delta = 1.44d + 1.96$$
$$\chi = 1.1d + 2.5$$

Matsumoto et al. (1977) have a useful correlation for saltation that breaks the analysis into two parts, one for larger particles and one for smaller ones.

The saltation velocity has been used often as the design parameter for the velocity at the feed point for conveying. In addition, the pickup velocity has also been employed for this purpose.

2.9 Pickup Velocity

The pickup velocity in pneumatic conveying has two different definitions. One is the velocity required to pick up particles from the bottom of the pipe where they are at rest because of a flow condition or blockage. The other definition is the velocity at the feed point of the solids. Mostly in this discussion attention will be paid to the former, although the latter will be mentioned sometimes. A rule of thumb is the relationship that indicates that the saltation velocity is roughly half the pickup velocity. Thus designing on the pickup velocity for the conveying velocity is more conservative. In studying the pickup velocity of particles, one finds that as the particle size decreases from values of 1000 micrometers, the pickup velocity will decrease until a minimum point is reached and it begins to rise with further size decrease. The smaller particles have larger surface and interparticle forces acting on them. For larger particles, recent work by Cabrejos and Klinzing (1994) gives what appears to provide reliable saltation and pickup velocities;

$$\frac{u_s}{\sqrt{gD}} = \frac{u_{s0}}{\sqrt{gD}} + 0.0022\left[\frac{\rho_s}{\rho}\right]^{1.25}\mu^{0.5} \quad (13)$$

$$\frac{u_{gpu}}{\sqrt{gD}} = \left(0.9428Re_p^{0.175}\right)\left(\frac{D}{d}\right)^{0.25}\left(\frac{\rho}{\rho_s}\right)^{0.75} \quad (14)$$

where $Re_p = du_g\rho/\eta$, $25 < Re_p < 5000$; $8 < D/d < 1340$, $700 < \rho/\rho_s < 4240$.

A phase diagram showing the interconnectiveness of these two parameters was also developed and is shown in Fig. 1.

2.10 Compressibility

Up until this point we have not discussed the obvious issue of compressibility of the conveying gas in dilute-phase pneumatic conveying. For short systems (< 300 ft) and low-pressure systems, the change in the gas density is relatively small, and an average value can be employed over the system without encountering much error. For longer distance conveying and high-pressure systems, compressibility is of paramount importance in the analysis. It should also be noted that in most conveying situations the temperature of the gas is relatively constant so isothermal conditions can be applied (Gas temperature = Product temperature).

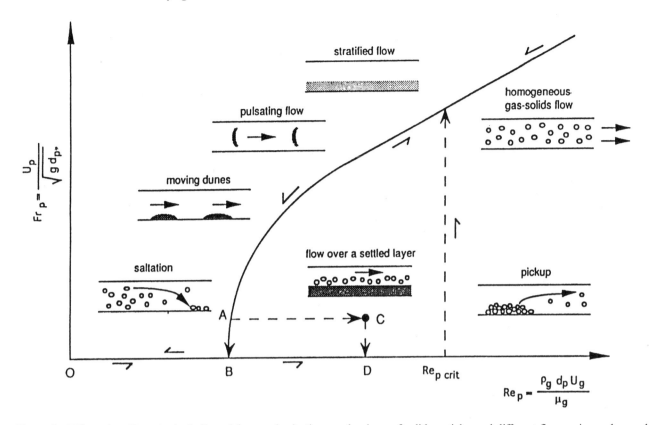

Figure 1 Bifurcation diagram, including pickup and saltation mechanisms of solid particles and different flow regimes observed in fully developed flow of gas–solids suspensions in horizontal pipelines.

2.11 Bends

Most conveying systems have bends or connections. If one has a system that is relatively long (> 300 ft) and has few bends, the presence of the bends has a minor effect on the overall system pressure loss. If a system has several bends and is of relatively short length (< 300 ft), the bend pressure loss is significant. Again, for the analysis of bend pressure loss, the linear combination of the effects of the gas and the effects of the solids is applied. The classic work of Ito (1959, 1960) for single-phase flow is used to establish the base for the pressure loss in bends. The work of Schuchart (1969), although performed on a limited material, has been employed with success to determine the solids contribution to the bend pressure loss.

$$\left(\frac{\Delta P_{\text{bend}}}{\Delta P}\right)_{\text{solids}} = 210\left(\frac{2R_{\text{B}}}{D}\right)^{-1.15} \tag{15}$$

The design suggestion of using the traditional pressure loss factors for single-phase flow in two-phase flow has

also met with some success in analyzing the pressure loss around bends. One factor that is important in bend analysis in pneumatic conveying is to avoid having three bends in quick succession in the design. This arrangement will most often lead to blockages and unsteady operation.

2.12 Choking Conditions

Choking conditions for vertical flow is likened to the saltation condition in horizontal flow. The most complete analysis of choking to date has been carried out by Yang (1975), who correlated a vast amount of existing literature data. These two expressions must be solved simultaneously:

$$\frac{2gD\left(\varepsilon_{\text{choking}}^{-4.7} - 1\right)}{\left(u_{\text{choking}} - u_{\text{t}}\right)^2} = 0.01$$

$$u_{\text{p}}\left(1 - \varepsilon_{\text{choking}}\right) = \left(u_{\text{choking}} - u_{\text{t}}\right)\left(1 - \varepsilon_{\text{choking}}\right) \tag{16}$$

One should note that if a system has both vertical and horizontal legs, then saltation will occur sooner than choking in a system. Only when the pipe diameter is increased in the vertical flow is there concern of choking problems.

With this presentation on pneumatic conveying, the basic design consideration and analysis is reviewed. Further analysis of such systems is possible. One theme that has received much attention is the area of flow classification with several different phase diagram approaches.

The strategy for determining the energy required for transport is given in Table 1.

3   FLOW CLASSIFICATIONS

The flow classification of Geldart (1973) for fluidization has been applied also to pneumatic conveying. Detailed discussion of Geldsart's classification of powders is presented in Chapter 3, "Gas–Solids Fluidization." The four materials A, B, C, and D have a significant relationship with pneumatic conveying ease and ability. The C type powders, which are cohesive in nature, are definitely the most difficult to move, although the piston type flows can be easily formed with these materials. The free-flowing A powders work well in dilute-phase flow. These materials cause few problems in the coupling of the feeder unit with the conveying line. The B materials generally do not move well in the more heavily loaded systems as

Table 1   Design Strategy for Dilute-Phase Transport.

---
1. Select the criteria for the transport velocity: saltation velocity, pickup velocity, specified velocity.
2. Determine the length of each horizontal and vertical section.
3. Determine the number of bends.
4. Determine the pressure drop for each section, horizontal, vertical, and bends.
5. Calculate the acceleration pressure loss for the entrance of the particles into the system.
6. Calculate the over-pressure loss by summing the component parts.
7. Decide if the system should have a stepping pipeline arrangement if the over-distance is greater than 300 feet.
8. Select the blower, feeder, and collector for the system.

---

strand flow/two-phase flow. The D materials have been moved over long distances in dilute-phase conveying in some mining applications.

The classic phase diagram in pneumatic conveying is that proposed by Zenz (1949) having the pressure drop versus gas velocity plotted at constant solids flow rates. A cross plot of this data can also be used to explain phase flow behaviors. The Zenz diagram points clearly to the unsteady regimes and shows that a minimum in pressure drop is achieved at a particular conveying velocity pointing to the ideal operational condition. Using the Zenz diagram one can follow the flow from the beginning to the end of the pipeline, since as the flow continues, the gas velocity increases and the density decreases. One operating condition is not present (note Fig. 2). If one plots the product of the density and gas velocity instead of the gas velocity only, a single operating point for the whole system is achieved and a more comprehensive picture of the system is presented. In pneumatic conveying the concept of the clustering of particles is often mentioned although not truly observed yet. Some indirect measurements also point to the presence of this phenomenon. Clustering in vertical flows has been studied more than in horizontal flows.

Our research team has looked at phase relationships in pneumatic conveying. With the suggestion given by Matsen (1982), the thermodynamic analogy of van der Waals' equation to pneumatic conveying has been used with a certain degree of success. In this analogy the temperature maps as the gas flux, the pressure as the particle flux, and 1/volume as (1-voidage). A division between dilute and dense phase has been clearly demonstrated. Figure 3 depicts this phase behavior with actual experimental points.

Flow pattern identification has recently been probed in our research group by observing the pressure fluctuations seen by a pressure transducer placed on the wall of the conveying line. This concept comes from the use of audio vibrations by Solt (personal communications, 1988) to diagnose the behavior of pneumatic conveying systems. By analyzing the pressure fluctuations with statistical properties, it has been shown that distinct fingerprints can be generated by the different flow conditions. Rapid determination of the flow conditions can permit one to employ this information as a control device for conveying systems. In a similar analysis, fuzzy logic has been employed to develop a phase diagram that has regions of varying degrees of stability of the flow condition. The boundaries dividing the regions are the most unstable. Fuzzy logic can be employed in such diagrams to establish the flow regimes.

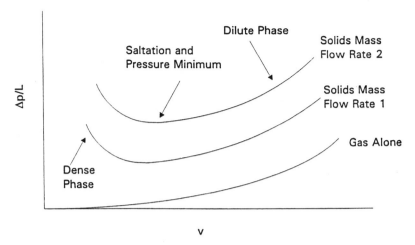

Figure 2 General Zenz-type diagram of pressure drop per unit length versus velocity for varying flowrates. Note dense, dilute, and saltation regimes.

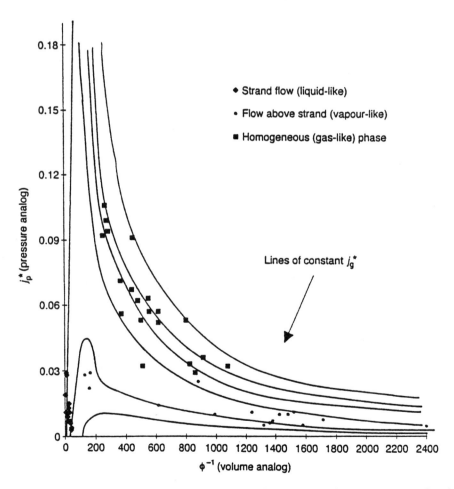

Figure 3 Phase diagram-type plot of reduced pressure analog versus the volume analog at constant reduced gas flow analog. Note similarities of strand flow to liquid phase, flow above strand flow to vapor phase, and homogeneous flow to gas phase.

In order to complete the design of a dilute-phase pneumatic conveying system, other components of the systems must be determined.

4 ESSENTIAL COMPONENTS OF DILUTE-PHASE PNEUMATIC CONVEYING SYSTEMS

4.1 Air Movers

There are three different kinds of air movers that are commonly used for dilute-phase pneumatic conveying:

1. Fans
2. Blowers
3. Compressors

If the power supply is reliable and the flow is relatively dilute, fans are reliable units. If there is an upset in the conveying, fans cannot provide the energy to reentrain the solids from the bottom of the pipe, and the system must be cleaned out before restarting. Of course, fans are the most inexpensive of the three air movers. The blower is the workhorse of pneumatic conveying, and its perfection about the early 1950s gave a boost to pneumatic conveying installations. Blowers can be used for a variety of flow conditions and provide sufficient energy to restart a flow that has been interrupted. The blower produces air that can be quite high in temperature that can also heat the associated piping. If one is conveying plastic particles, limits on the exit pressure may be necessary to reduce the temperature of the conveying air so that the plastic particles do not soften and melt. In order not to exceed the auto-ignition temperature of the oil in the blower, 232°C (450°F), a control should be set to shut down the blower when this condition is approached. When the need is for pressure over 2 bars, the blower cannot respond, so compressors are called into play. These units, however, are expensive and require very good filtration of the input air to function well. In a dusty environment, this restriction can be severe. Plant air that comes from a central compressor system can be used for conveying. Often practitioners think that this air is free and do not factor this cost into their operation, since the compressors are not designed solely for the conveying system.

4.2 Feeders

There are many varieties of feeders for pneumatic conveying systems. For vacuum systems, one can use a simple orifice device for free-flowing material. A nozzle arrangement can be used in a vacuum system much as a vacuum cleaner. This nozzle can be employed for small projects such as the cleanup of a powder spill to large operations such as the unloading of grain from the hold of a ship. In pressure systems, one can use the fluid head pressure as in an airlift operation as well as a fluidized feed arrangement. The venturi feeder can also be used in pressure systems. The most common feeder for pneumatic transport is the rotary feeder. The design of these rotary feeders varies. The rotary feeder usually operates between 10 and 30 rpm. The throat of the rotary feeder should not restrict the flow of solids. Leakage of air through the rotary feeder requires one to design for venting of the feeder.

For pressure systems one often finds lock hopper or gate lock feeders. These units operate similar to the blow tanks.

The screw feeder is another common type of feeder for pneumatic conveying. This type of feeder can often be used as measuring devices. Screw feeders can operate on a wide range of materials from powders up to $\frac{1}{2}$ inch diameter materials. In general, moderately abrasive materials can be tolerated by screw feeders.

Pressure tanks can serve as feeders for pneumatic conveying in a batch or continuous operation. Pressure tanks can be used for both dilute and dense-phase conveying operations.

4.3 Diverter Valves

Diverter valves are units commonly employed in pneumatic conveying to divide flows or redirect flows in various units. One constant concern with diverter valves in dilute-phase conveying is the erosion that may take place at this high impact point.

4.4 Filters

Filters or other collection devices are required for all conveying systems. The reserve-jet air flow systems for cleaning the filter are very common units. For lower pressure, the air-to-filter cloth ratio is usually approximately 1.1 to 0.3 CMM/m^2 (3.5 to 1.0 CFM/sq.ft). For higher pressure reserve flow systems, a higher air-to-cloth ratio is employed, 2.1 to 0.3 CMM/m^2 (7.0 to 1.0 CFM/sq. ft). For more detailed discussions, see Chapters 22 and 28, "Gas–Solids and Liquid–Solids Separation."

4.5 Cyclones

Cyclone collectors are simple units with a complex flow field that even today baffles our understanding. The cyclones work best for the collection of coarse materials. Cyclones are a source of high wear and breakage of the material conveyed. Friable materials should not be employed in cyclone arrangements. See Chapter 22 for more details.

4.6 Scrubbers

When wet environments are used for collection, the resultant product is a slurry. Disposal and handling of this slurry must be dealt with. Often all the scrubber does is to transfer the handling problem from a dry phase to a wet phase.

4.7 Piping Systems Components

The material of construction has a minimal effect on the overall dilute-phase conveying characteristics of pneumatic conveying. Some manufacturers have had success in treating the inside of a pipe so as to create a roughened surface for plastic pellet conveying.

The layout of the piping systems has many important factors in pneumatic conveying. One should keep the flow path as the most direct between two points. Bends should be eliminated as much as possible. Care should be taken in the design that the distance after a feed point before the first bend is inserted in a minimum of 3 meters (10 feet) when two-phase conditions are present. If the flow is dilute or dense, this distance is not crucial. The two-phase condition tends to cause a sloshing of the solids in the bend in an unsteady condition. This sloshing behavior causes plugging and other upsets in the operation of the pneumatic conveying systems. As noted before, one should at all costs avoid more than two bends in quick succession.

Sloping lines are generally not a good idea for pneumatic conveying. When the flow is dilute, sloping will have minor effects on the transport operation. When two-phase flow is present, the slopes of 30 to 85 degrees are dangerous because of the recycling of the solids, which causes instabilities and erosion. When one has a downward flow, generally one does not have any concerns in operation. For the two-phase flow regime, one should have horizontal runs of 12 meters (40 feet) after reaching the bottom to have the phases separate.

Couplings in pneumatic transport are important hardware units. Considerable vibrations can be placed on the system from the feeder or air supply as well as the condition of flow. The piping should be kept aligned and have electrical conductivity achieved by a conductive strap over nonconductive sections such as gaskets. Welded connections are good but difficult to maintain since repairs require breaking the welds. For high-pressure systems, flanges can be used, but alignment of the piping is imperative, since flanged connections are not always properly aligned. Clamps can be used for low-pressure systems but may provide sources of leakage or line separation, and they require electrical grounding.

Bends are crucial hardware elements in all pneumatic conveying systems. Historically the practice has been to use long-radius bends for all turns. Long radius is defined as a radius of 12 pipe diameters or more. These bends have been a great source of wear problems that usually occur on the outside of the bend primarily and secondarily on the inside of the bend. Wear back additions are common in these types of bends. Short-radius bends have a radius of 2.25 pipe diameters. One finds the wear rates of these bends to be higher than those of long radius. The dead end T-bend (barrel bends, etc.) has been found to be a successful wear resistant element in pneumatic conveying. Some secondary wear is experienced about one meter (3 feet) after the bend. In using these bends, sticky and very cohesive materials are not recommended. In general, the pressure loss around the bends is similar in nature for all bends with an equivalent straight pipe loss of approximately 4.6 meters (15 feet).

5 TROUBLESHOOTING

The topic of troubleshooting is extensive and can grow as additional information is obtained. An artificial intelligence (AI) program has been developed in our laboratory for troubleshooting (Dhodapkar and Klinzing, 1993), in which we assimilate the experience of a number of experts and literature citations; yet this program is not complete. This program does provide a beginning for the operations person to attempt to resolve problems in the conveying system as they arise. The technique of troubleshooting a system depends on whether it is brand new or has been working for some time and develops problems.

For a system that used to meet capacity, one can look to worn equipment such as an airlock that has increased leakage or blocked or plugged aeration media. Plugged venting systems, changes in the solids characteristics, such as size distributions, and material feed blocked by foreign material are often-encountered problems.

In order to increase the inadequate capacity of a pneumatic transport system, a number of actions can be taken. For example, consider modifying the air supply quantity:

1. Stream flow system (reduce the air volume to a minimum steady conveying).
2. Two-phase flow system (increase the air volume to reduce slugging).
3. Dense phase system (increase the air volume).

Modifying the air supply pressure: increased pressure may be required.

Change line configurations:

1. Shorten the line length.
2. Eliminate the bends.
3. Increase the line diameter for stepped line at the end of the conveying system.

If one experiences excessive line wear, one can

1. T-bends can be used.
2. The material-to-air ratio can be increased.
3. Step pipe diameters through the system can be carried out to maintain a constant velocity throughout the system.

When material breakage is present, one can

1. Reduce the conveying velocity
2. Eliminate bends, cyclones, and impact points
3. Reduce the velocity before the terminal point of the system.

6 LONG DISTANCE CONVEYING

Conveying of solids over long distances up to a few thousand feet encounters different phenomena owing to the gas expansion of that distance. The gas expansion caused the gas and solid velocities to increase and thus affect the attrition of the particles and the erosion of the pipeline, both of which increase to the nth power with velocity, where n ranges from 2.5 to 5.0. To slow the particles down, one can step the pipeline to larger diameters to decrease the velocities. Another technique developed in our laboratories is to use a flow economizer, which effectively takes gas from the system in a prescribed fashion to reduce the gas velocity. Figure 4 shows a schematic of the flow economizer tested in our laboratory. This unit withdraws a prescribed amount of air at critical points in the transport line.

7 APPLICATIONS OF COMPUTER ANALYSIS

In order to design a dilute-phase pneumatic conveying system, one needs to have a mechanism to select and calculate a number of factors and components. The first question is, What type of system would be appropriate for the conveying operation? Should one use a vacuum or pressure system? Is the dilute-phase the mode most appropriate for transport? Using an expert system approach, our laboratory has developed a program entitled NUSELECT (Pneumatic Conveying Consultants) to help answer these types of questions. After selecting the system, one needs to choose the appropriate feeder to inject the material into the pipeline for transport. Again, we have addressed this ques-

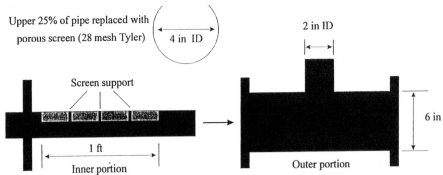

Figure 4 Diagram of the flow enhancer.

tion by developing an expert system program entitled FEEDER for this task. Calculations presented above should provide the basis for predicting the energy needed for transport to help complete the design. Most often a cyclone or reverse jet filter is used for collecting the product after transport. When a system is in operation, more than likely there will be occasions when the system is not functioning properly and troubleshooting the operation is needed. The expert system PANACEA provides the basis for beginning troubleshooting before an expert in the field is called in for analysis.

NOMENCLATURE

C_D	=	drag coefficient
D	=	diameter of the pipe
f_g	=	gas friction factor
f_s	=	solids friction factor
g	=	gravity constant
L	=	length
ΔP	=	pressure drop
u_g	=	gas velocity
u_p	=	particle velocity
W_s	=	solids flow rate

Greek

ε	=	voidage
θ	=	angle of inclination of the pipe
ρ_g	=	gas density
ρ_p	=	particle density
μ	=	loading

REFERENCES

Barth W. Stromungstechnische probleme der verfarhrenstechnik. Chem Ing Tech 20, No. 1:29–32, 1954.

Blasius H. Grenzschichten in flüssigkeiten mit kleiner reibung. Mitt Forschungsarb 131:1–40, 1913.

Cabrejos FJ, Klinzing GE. Pickup and saltation mechanisms of solid particles in horizontal pneumatic transport. Powder Tech 79:173–186, 1994.

Dhodakpar SV, Klinzing GE. In: Roco, M, ed. Expert Systems in Solids Handling. Butterworth-Heinemann, 1993, pp 743–777.

Geldart D. Types of gas fluidization. Powder Tech 7:285-292, 1973.

Institute of Gas Technology. Preparation of a coal conversion system technical data book. Project 8979, ERDA Contract No. EX 76-C-01-2286. October, 1978.

Ito H. Friction factors for turbulent flow in curved pipes. Trans ASME J Basic Engr 81D:123, 1959.

Ito H. Pressure losses in smooth pipe bends. Trans ASME J Basic Engr 82D:131, 1960.

Klinzing GE, Rohatgi ND, Myler CA, Dhodapkar SV, Zaltash A. Pneumatic transport of solids in an inclined pipe. Can J Chem Engr 67:237–244, 1989.

Klinzing GE, Rizk F, Marcus RD, Leung LS. Pneumatic Conveying of Solids. 2d ed. London: Chapman and Hall, 1997.

Konno H, Saito SJ. Pneumatic conveying of solids through straight pipes. Chem Eng Japan 2:211, 1969.

Koo EC, Drew TB, McAdams WH. Friction factors for clean round pipes. Trans AIChE 28:56, 1932.

Matsen JM. Mechanism of choking and entrainment. Powder Tech 32:21–23, 1982.

Matsumoto S, Kikuta M, Maeda S. Effect of particle size on the minimum transport velocity for horizontal pneumatic conveying of solids. J Chem Eng Japan 10, No. 2:273, 1977.

Michaelides EE. Motion of particles in gases: average velocity and pressure loss. Trans ASME 109:172–178, 1987.

Richardson JF, Zaki WN. Sedimentation and fluidization. Part I. Trans Instn Chem Engrs 32:35, 1954.

Rizk F. Dr.-Ing. Dissertation, Technische Hochschule Karlsruhe, 1973.

Schuchart P. Pressure drop across bend in pneumatic transport. Chem Eng Technol 41:1251, 1969.

Weber M. Friction of the air and the air/solid mixture in pneumatic conveying. Bulk Solids Handling 11:99, 1991.

Yang W-C. A mathematical definition of choking phenomena and a mathematical model for predicting choking velocity and choking voidage. AIChE J 21, No. 5:1013–1015, 1975.

Yang W-C. A unified theory on dilute pneumatic transport. J Powder Bulk Solids Handling 1, No. 1:89–95, 1977.

Zenz FA. Two-phase fluid-solids flow. Ind Eng Chem 41:2801–2806, 1949.

24

Electrostatics in Pneumatic Conveying

George E. Klinzing

University of Pittsburgh, Pittsburgh, Pennsylvania, U.S.A.

1 INTRODUCTION

Electrostatics or, more aptly, triboelectric effects in solids processing is always encountered except in the region of the world where the humidity of the ambient is well over 75% most of the time. Electrostatic effects are seen when one walks across a deep-piled rug during winter and then reaches for a metal doorknob: a familiar startling snap most likely occurs. Of course in most cases such an incident does not trigger an explosion or fire. In solids processing, the danger of an explosion and fire is very serious, and such events have had a dramatic effect in several industries. An internet article dealt with the explosion that occurred in a flour mill in the 1700s in Italy (http://www.chemeng.ed.ac.uk/~emju49/SP2001/webpage/intro/intro.html). The report given at the time was quite detailed, and led the reader to the conclusion that electrostatics was the culprit.

Three requirements are needed for a dust explosion: fuel, oxidizing agent, and ignition source. The ignition source must provide a minimum spark energy for ignition, a certain duration of the spark, and a spatial requirement of the spark. It is particularly noteworthy that often the carrier gas in pneumatic conveying is changed to nitrogen in order to reduce the explosion hazard for powder materials. The use of nitrogen will increase the cost of operation. Sometimes the reduction of the percentage of oxygen is all that is needed to reduce the hazard. For example, in the case of some coals one can reduce the oxygen content of a gas stream to 14% as opposed to the 21% normally present in air. One must be careful in the transport of material that has its own oxidizer present, such as smokeless gunpowder. Using nitrogen as the conveyer has no effect on reducing the hazard of pneumatically conveying this material.

Another potentially dangerous situation in pneumatic conveying with electrostatics present exists when there is a flammable vapor present in the gas stream, possibly due to the polymer that is being conveyed. Likewise, when polymer-lined vessels are being employed for material storage, the charge on the material will not be dissipated through this lining, and charge accumulation can build to dangerous levels.

One can liken the charging of materials to a series similar to the electromotive series employed in chemical analysis. For example, one finds that glass and nylon charge positively, while polyethylene and teflon charge negatively. In addition, one finds that bipolar charging can occur with some materials as they are conveyed pneumatically.

2 PNEUMATIC CONVEYING AND CHARGING

Although not all electrostatic discharges will cause an explosion, the resulting damage to electronic equipment, like the discomfort felt by operators who experience the discharge of electrostatics accumulated in a conveying line, makes the elimination of all electrostatic buildup a must. In addition, it is well known that electrostatic charging will increase the pressure requirements for conveying a material (Marcus et al., 1990). Estimating this additional energy requirement is difficult.

Airborne dusts of most polymers are explosive, and many of them are also electrically insulating and triboelectrically active (Jones and King, 1991). Minor electrostatic discharges or drains are beneficial in that they help to dissipate accumulated charge in the system, whereas discharges of large accumulated charges are dangerous and can cause ignition. Of all the solids processing operations, the most dangerous with respect to the handling of fine powders is pneumatic conveying. Jones (1994) has commented on a number of electrostatic hazards in powder handling and cites pneumatic transport as being the process generating the most charge with 10 to 10^2 μC/kg of material flowing.

The particles in this type of transport have thousands of opportunities to interact with both themselves and the surrounding piping and equipment. Contacting between surfaces of different work functions is the driver for electrostatic generation. The work potential is an inherent property of the material and the condition of its surface. Polymers can obtain a wide range of charging from positive to negative. The more the charges are separated in a conveying system, the greater the potential to develop significant voltages and associated currents.

Investigators have studied the use of electrostatic interactions or electrification of solid particles in measuring various two-phase flow parameters in pneumatic transport, namely, the average particle velocity and the mass flow rate (Gajewski et al.,1990; Gajewski et al., 1991; Klinzing et al., 1987). The effect of electrification on system pressure drop, choking, and saltation velocities, flow patterns, pressure fluctuations, and particle velocity was also investigated in the past (Myler et al., 1985; Ally and Klinzing, 1983, 1985; Smeltzer et al., 1982a,b). It was found that at constant loading, greater electrostatic effects were seen for small particles over large particles because of the high particle number density and thus increased interactions for the smaller particles. Bipolar charging was detected for polyethylene powder with fine particles charging negatively and the coarse particles positively (Cartwright et al., 1985). The material electrification obtained experimentally has also been compared to theoretical values (Gajewski, 1989). Nifuku et al. (1989) found that the electrostatic charge increases, peaks, and decreases with increasing powder concentrations or transport velocities; charge generation was greatly influenced by the powder flow pattern in the pipe.

Higher loadings at low relative humidities also produced increased electrostatic effects. Little or no electrostatic charging was generated at ratios of kg of water per kg of solids, which were greater than 0.1. Smeltzer et al. (1982) and Nieh and Nguyen (1988), experimenting with a 2 inch copper pipe and glass beads, found that when the system moisture content exceeded a relative humidity of 76%, the charge on the glass beads became neutralized. The charging exhibited itself as an increase energy requirement for flow, which could under certain conditions be represented by an increase in pressure drop of up to 70% (Ally and Klinzing, 1983).

Boshung and Glor (1980) showed that the particle size and the material are the two most important parameters associated with charging in pneumatic transport systems.

In solids processing one should be on guard for fine particles with a high specific surface area of nonconductive material in an atmosphere of low relative humidity flowing in nonconductive or nongrounded containments. These factors all add up to a troublesome situation.

Adhesion of particles to surfaces due to electrostatics causes problems by coating piping walls. In these cases the conveyed materials contact the wall coating rather than the material of construction of the pipe and can lead to charge buildup. In addition, coating of sensors by the adhesion of particles by electrostatics will impede the process of measuring properties of the gas–solid flow systems. It is well known that powder and granular materials generate charging more readily than large diameter particles.

Not all electrostatic charges should be viewed with suspicion in the handling of powders. One can engineer and manage the charging to provide avenues to measure efficiently the solid flows and separate powders of different properties.

3 USEFULNESS OF ELECTROSTATICS

After first having to deal with the hazards of electrostatic generation in pneumatic conveying and their adverse effects on electronics, these phenomena can be channeled into a useful tool for measurement. The charges generated in the conveying operation can be measured at two distinct points, and these signals can be cross-correlated to provide a measure of the particle velocities. In fact, one can use a coil as an antenna outside a nonconducting pipe to pick up the signals generated (Myler, 1989). This provides a nonintrusive technique for measuring the particle velocity. Another phenomenon that is associated with electrostatics is the magnetic field that it generates associated with the flowing of charged particles or current in the pipe. A polarized light is known to change the degree of polarization as it goes through a magnetic field. Channeling a polarized light through a fiber optic cable wrapped around the conveying pipe will expose this light to the magnetic field, whose orientation can be easily measured. The larger the flow rate of solids, the greater the magnetic field generated. This concept has been used to develop a flow meter in our laboratory (Rader et al., 1995).

Masuda and Matsusaka (1994) have developed a flow meter to measure powder flow rates using electrostatic measurements. The method depends on the impact of the particles with the inside of the tube wall where a current is generated. The initial charge on the particles can be eliminated by using two metallic pipes whose inner surfaces are coated with different materials. The pipe is 6 mm in diameter and carries fly ash. The powder flow rate and the mean particle-charge value could be measured simultaneously.

4 PRECAUTIONS

Electrostatics and electronic equipment such as computers and A/D converters do not mix. One should be very careful, when a system is shown to have electrostatic tendencies and the measuring devices employed have a computer linkage, that the computer should be isolated from electrostatic charging effects. This isolation is sometimes very challenging, since the charging effects can travel on the surface of even nonconducting surfaces to interfere or damage the electronics. Changing the electric signal to a light signal is sometimes a solution to avoid large discharges interfering with and damaging computer operations. Sometimes such conversions will modify the original signal, so knowledge of this modification is essential before applying this procedure.

One of the most dangerous procedures when dealing with electrostatic charging is covering a charged non-conducting surface with a metal foil and shield. This may appear as a possible solution to the grounding of the charge generated, but in effect, what has happened in applying the foil is that one is constructing a large capacitor that can store energy and discharge this

energy over a very short period of time, causing or triggering an explosion.

Another condition that is sometimes ignored until problems occur is proper grounding of surfaces. If one constructs a transport system from metal and if in time, certain regions that would normally serve as a grounding path become interrupted by paint or rust, the system can become nonconducting in nature and interfere with the flow of solids.

The use of nonconducting gaskets in transport line connection also has the potential to create discharges. One often places a conducting strap across the flanges to ensure continuity of current flow to a ground. It is good practice in pneumatic conveying to ensure that all vessels and units are grounded and that this grounded path is not interrupted by corrosive or painted surfaces. In general it is good practice to pay attention to proper housekeeping in the handling of powders and granular material to reduce the potential for explosive limits to be exceeded.

The most convenient method of eliminating electrostatics in a system is by humidifying the conveying gas and ensuring that all possible areas of charge buildup are grounded. Another method is through the addition of a small amount of antistatic agent to the solids inventory. Antistatic agents will operate either to reduce the generation of the charge or to increase the charge leakage (Cross, 1987). Unfortunately, this addition may cause contamination of the material to be conveyed, and it would have to be separated after the collector to obtain the pure material again. A quaternary ammonium salt called Larostat® 519 was used in a 0.15–0.20 wt% mixture with the particles to be transported. The low percentage did not interfere with the characteristics of the gas–solid flows. Experimental tests confirmed and verified this result. The quaternary ammonium salt is a nonpolymeric cationic compound and is considered a hydrophilic additive. This additive increases the surface conductivity by increasing the water adsorption at low relative humidities. Dahn (1992) evaluated possible sources of dust explosions in large-scale conveying systems and recommended ways to minimize them.

More recently the waxing of the interior walls of a nonconductive pipe with a conductive wax similar to that employed in clean room maintenance has proved effective in reducing charging potentials.

A very practical book on electrostatic effects has been prepared by Jones and King (1991). This work gives many procedures for reducing electrostatics in a plant.

5   SOME EXPERIMENTAL OBSERVATIONS OF ELECTROSTATICS IN PNEUMATIC CONVEYING

In order to portray the behavior of electrostatics and their effects on pneumatic conveying, some experiments conducted in our laboratories will be presented. These results will give the flavor of the behavior of this phenomenon and the challenges in dealing with such a system.

5.1   Horizontal System

The conveying air employed was conditioned for humidity to study the conveying characteristics in the absence of electrostatic charging. The pressure drop versus gas velocity plot shows no unusual behavior. However, in the absence of humidification, the conveying characteristics of PVC particles (137μm in diameter) are entirely different. The pressure drop increases by about 100% as the particles start adhering to the wall. Pressure drop is a measure of power consumption in pneumatic conveying systems and therefore a prime concern. The undesired accumulation of particles inside the pipe also leads to unsteady flow conditions. It should be noted that the conveying line, except for the glass viewing section, is made of copper pipe that has been well grounded. It appears that the charged particles do not dissipate the charge completely to the grounded pipe during the short duration of particle–wall collision.

The nature of interparticle forces changes owing to electrostatic charging. These particles become more cohesive and coat the inner surface of the pipe. It can be surmised that the charge on these particles must be renewed constantly by other particles. If the solids were shut off during conveying, or both gas and solids were stopped, the coating of particles on the sight-glass disappeared. No visual observations could be made about the existence of particles on the inner surface of the copper tube. It is unlikely that the particles could be getting charged because of the glass section, since the deposition on the wall could be seen on the entire length of the glass section.

The electrostatic charge on the particles changed the flow patterns usually observed under similar conditions. Figure 1 (Dhodapkar, 1991) shows the range of flow patterns observed under conditions of low humidity (less than 65%). From the experimental data it may be concluded that the appearance and disappearance of these flow patterns are independent of the solids flow rate and depend mainly on the gas velocity.

From Fig. 1 it can be seen that at low gas velocities (Pattern 5) and large gas velocities (Pattern 1), the pipe essentially remains clean at the top. At lower gas velocities a settled layer may be seen that is quite expected. At high velocities a homogeneous flow is observed with no particles sticking to the wall. A possible explanation is that the particles are at equilibrium owing to electrostatic forces and fluid shear forces. It was found that at gas velocities higher than 18 m/s, the fluid shear is greater than the electrostatic forces and therefore no particles can be seen sticking to the wall. In addition, at gas velocities less than 4 m/s, no particles can be seen adhering to the wall because of lack of electrostatic charging.

At intermediate velocities one may observe numerous ring-shaped flow patterns that appear to move in the direction opposite to the direction of the main flow. In all these cases the particles coat the inner surface of

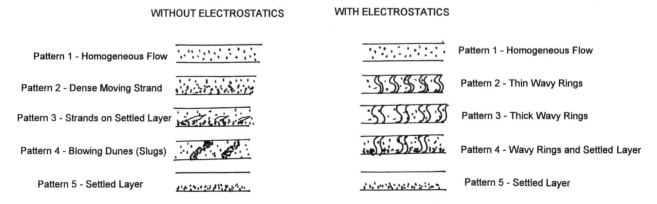

Figure 1   Flow patterns for PVC with electrostatic charging.

the wall, defying gravity. Comparison of Figs. 1 and 2 (Klinzing et al., 1994) indicates that the unique maxima observed in Fig. 2 correspond to the formation of ring-shaped structures on the pipe wall. The intensity of pressure fluctuations were also found to have a maximum (see Fig. 3) at a superficial gas velocity of 9.3 m/s. This is a remarkable departure from the usual behavior.

5.2 Vertical System

Electrostatics effects were experienced several times in the vertical system until the addition of multiple grounding points and Larostat® 519. Electrostatics effects were generated by the high number of particle-particle and particle–wall interactions between the 0.0254 m Lucite tube and the glass particles. The nitrogen gas used as the transport medium had zero relative humidity. Prior to eliminating the electrostatics, interesting data was acquired for 97 μm and 545 μm glass beads.

5.3 Electrostatics with Small Glass Beads

Initially, tests were conducted on 97 μm glass beads at atmospheric pressure with the results of pressure drop versus gas velocity showing the classical Zenz-type diagram. However, upon repeating the set of experiments, electrostatics developed in the system. Electrostatic discharges were evident in the system by visual and audible observations. The classical state diagram, which was found in the first set of experiments, was altered and a double minimum was found. Figures 4 and 5 show the results of the effects of electrostatics on the pressure drop. The minimum $\Delta P/L$ for the second

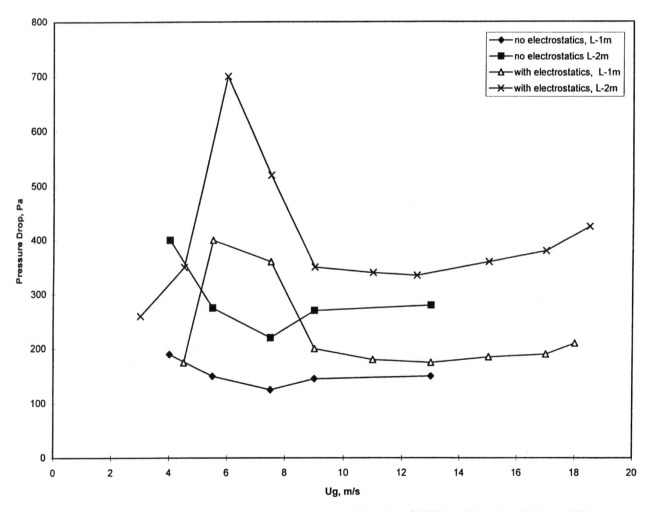

Figure 2 Effect of electrostatics charging on the pressure drop behavior of PVC particles in 0.0254 m (I.D.) system—$W_s = 1.14$ kg/min.

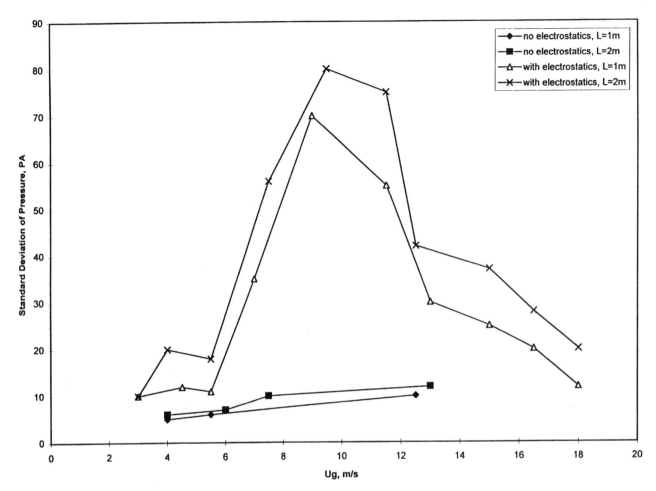

Figure 3 Effect of electrostatics charging on the standard deviation of the pressure drop for PVC particles in 0.0254 m (I.D.) system $W_s = 1.14$ kg/min.

set of experiments is shifted to the right, thereby changing the previous minimum in the gas velocity that occurred without electrostatics. Operating at what was thought of as the previous minimum would have increased the pressure drop significantly, which could result in reduced conveying capacity. Three different solids flow rates were investigated at several gas velocities, with the higher solids flow rates having the most pronounced results.

The solids were mainly in a homogeneous transport state during the initial tests, except at the lowest gas velocity. During the second series of tests, the flow patterns changed from a homogeneous flow at the start of the experiments at low gas velocities to a more clustered or packet flow at the intermediate test velocities. The particles formed clusters for gas velocities in the range of 4.5–6.0 m/s for the particular set of conditions investigated.

The first minimum occurring at the lower gas velocity of the double minimum increased with increasing solids flow rate (Figs. 4 and 5). This increase in pressure drop, as the gas velocity was in the range of 4.5–7.6 m/s, occurred owing to the clustering of the particles caused by the electrostatic generation of the glass beads with the Lucite tubing. This clustering caused the pressure drop in the system to be equivalent to that of larger particles. The greater the charging of the particles, the larger and more defined the clusters were. This was also evident by the tracking of the $\Delta P/L$ vs. time for a given U_g and W_s. The pressure drop would continue to increase until a discharge, of the system, as evident by sight and sound, occurs. At this point the pressure drop would fall back to a certain level and continue to increase again with time until there is another discharge. A certain residual charge resided in the particles and on the pipe.

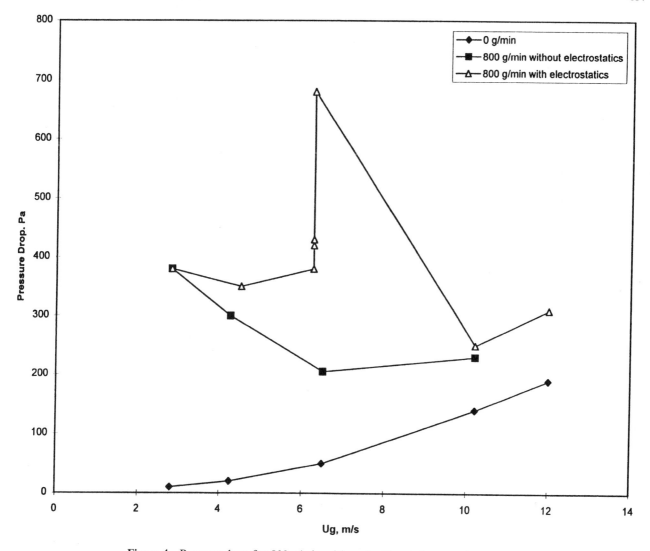

Figure 4   Pressure drop for 800 g/min with and without electrostatics present.

Operating at a gas velocity for what seemed to be a minimum for the pressure drop in the presence of electrostatics could result in the pressure drop being increased substantially. This would lead to undersized equipment for the transport of the material and would result in less material transported than the system was actually designed for.

The double minimum condition has also been reported by Zaltash (1987) and Klinzing et al. (1986). Klinzing suggests that the smaller particles behave as a cluster at the dilute-phase regime, and the clusters tend to break up as the choking regime is approached, resulting in the particles behaving as smaller particles and thus causing an oscillation in the pressure drop. This described flow phenomenon was observed in the

experiments. Figure 6 shows the flow patterns observed with electrostatic charging.

5.4   Electrostatics with Larger Glass Beads

Electrostatics was also observed when $545\,\mu$m glass beads were used as the transport solid. This time the pressure drop was not observed or investigated as a function of gas velocity, but rather pressure fluctuations were observed with time. Electrostatic discharges were observed with arcs discharging from a few centimeters to a meter in length accompanied by a powerful "snapping" noise.

The pressure drop was seen to decrease suddenly upon discharge, rise sharply prior to another discharge

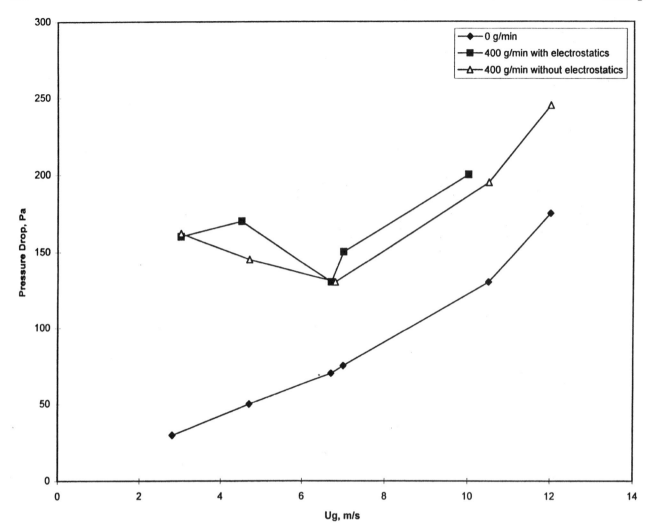

Figure 5 Pressure drop for 400 g/min with and without electrostatics present.

of the solids, and fall back to the original level (pressure drop) after discharge. Table 1 shows a cyclic example of this for the 545 μm glass beads at a gas velocity of 6.6 m/s and a solids feed rate of 1100 g/min. Standard deviation of the pressure drop was also found to increase with the increased charging of the system.

The rise and fall of the pressure signal is similar to that of a voltage vs. time plot observed during the charging and discharging of a capacitor (Floyd, 1987). Figure 7 shows the similarity of the curves. In effect, the transport of the solids causes the particles to become charged, and the charge accumulates on the solids until the solids reach their dielectric breakdown and discharge across an air (or nitrogen) gap. The collector could be considered a large capa-

citor, since the solids were isolated from any conducting medium due to the Lucite walls, and therefore it contained a charge. The pressure drop decreases just prior to the charging portion of the curve. Since the particles are large (545 μm) in comparison with the 9 7 μm glass beads, it is hypothesized that the particles are weakly held together or attracted by electrostatic charges, causing a higher pressure drop than if they were individual entities. The dip in the pressure drop curve prior to the charging and discharging portion of the pressure drop curve could be due to the particles separating (behaving as individual particles) and possibly losing some of the charge. Then upon losing the charge or becoming separated, there is an increasing force that causes the particles to reattract or cluster together.

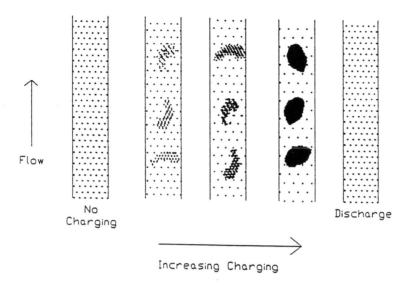

Figure 6 Particle flow patterns observed for vertical flow with electrostatic charging.

5.5 Summary of Testing

Several interesting observations were made and recorded for two different conveying systems that were electrostatically charged. The usual Zenz-type diagram is modified to show two minima in the curve when the systems are charged, rather than a single minimum. Flow patterns will also change from their usual appearance and become more wavy or take the form of rings for horizontal conveying, and form clusters in the case of vertical conveying.

By observing the pressure fluctuations (for the vertical system) of a charged system, the pressure drop versus time curves exhibit the appearance of a capacitor charging and discharging. Visual observations in the vertical column showed that the flow patterns of the particles become clustered until a discharge occurred in the system, from which the particles regain their individuality and the flow becomes more uniform again. Particles appeared to cluster in the gas range (4.5–7.6 m/s) for vertical flow. The particle packets or clusters tend to break up as the choking regime is approached, so that the particles achieve their true identity. As the higher gas velocity is approached, the particle clusters tend to break up owing to the turbulence of the flow. This behavior causes the pressure drop to oscillate with gas velocity over this range.

NOMENCLATURE

t_i = Time, s
U_g = Gas Velocity, m/s
U_t = Terminal velocity of particle, m/s
ΔP = Differential pressure drop, Pa
$\Delta P/L$ = Pressure drop per unit length, Pa/m
σ = Standard deviation, Pa

Table 1 Transient of the Pressure Drop

	ΔP, Pa	Time
Initial reading (taken arbitrarily during	610	t_1
the charging cycle	701	t_2
	720	t_3
	discharge occurred in system	
Readings after discharge	595	t_4

Times between readings, i.e., $t_{i+1} - t_i = \Delta t$, are approximately the same

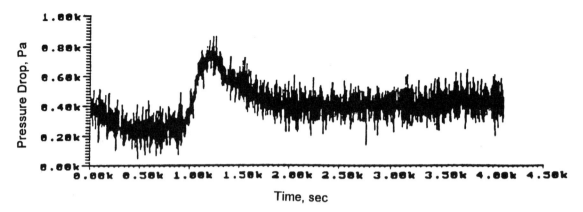

Figure 7 Pressure drop time series trace.

REFERENCES

Ally MR, Klinzing GE. Electrostatic effects in gas-solid pneumatic transport with loadings to 100. J Powder and Bulk Solids Tech 7(3):13–20, 1983.

Ally MR, Klinzing GE. Inter-relation of electrostatic charging and pressure drops in pneumatic transport. Powder Tech 44(1):85–88, 1985.

Boshung P, Glor M. Methods for investigating the electrostatic behavior of powders. J Electrostatics 8:2205–2219, 1980.

Cartwright P, Singh S, Bailey AG, Rose LJ. Electrostatic charging characteristics of polyethylene powder during pneumatic conveying. IEEE Transactions on Industry Applications IA–21(2):541–546, 1985.

Cross JA. Electrostatics: Principles, Problems and Applications. Bristol, U.K.: IOP, 1987, pp 376–382.

Dahn JC. Electrostatic hazards of pneumatic conveying of powders. Plant/Operations Progress 11(3):201–204, 1992.

Dhodapkar SV. Flow pattern classification in gas–solid suspensions. Ph.D. diss., University of Pittsburgh, Pittsburgh, PA, 1991.

Floyd TL. Electronics Fundamentals: Circuits, Devices and Applications. Columbus, OH: Merrill, 1987, pp 334–339.

Gajewski A. Measuring the charging tendency of polystyrene particles in pneumatic conveyance. Journal of Electrostatics 23:55–66, 1989.

Gajewski JB, Glod B, Kala W. Electrostatic method for measuring the two-phase pipe flow parameters. IEEE Industry Applications Society Annual Meeting, Seattle, WA, Oct. 1990.

Gajewski JB, Glod B, Kala W. Electrostatic flow meter for measuring the two-phase flow parameters in pneumatic transport. Results of preliminary tests. Proceedings of the 8th International Conference on Electrostatics, Oxford, U.K., 1991, pp 159–164.

Jones TB. NEPTIS-3. Electrostatic Problems in Powder Technology. Nov. Kyoto: Nisshin Engr, 1994, p 79.

Jones TB, King JL. Powder Handling and Electrostatics: Understanding and Preventing Hazards. Chelsa, MI: Lewis, 1991, pp 1–103.

Klinzing GE, Rohatgi ND, Zaltash A, Myler CA. Pneumatic transport: a review and generalized phase diagram approach to pneumatic transport. XIV Encontro Sobre Escoamento Em Meios Porosos, Octubro 1986, p 14.

Klinzing GE, Zaltash A, Myler C. Particle velocity measurements through electrostatic field fluctuations using external probes. Particulate Science and Technology 5(1):95–104, 1987.

Klinzing GE, Dhodapkar SV, Plasynski SI. NEPTIS-3. Electrostatic Measurement of Powder Flow Rate in Gas–Solids Pipe Flow. Nov. Nisshin Engr, 1994, p 21.

Marcus RD, Leung LS, Klinzing GE, Rizk F. Pneumatic Conveying of Solids: A Theoretical and Practical Approach. London, U.K.: Chapman and Hall, 1990, p 67.

Masuda H, Matsusaka S. NEPTIS-3. Electrostatic Problems in Powder Technology. Nov. Nisshin Engr, 1994, p 30.

Myler CA. Use of a thermodynamic analogy for pneumatic transport in horizontal pipes. Ph.D. diss., University of Pittsburgh, Pittsburgh, PA, 1989.

Myler CA, Zaltash A, Klinzing GE. Gas-solid transport in a 0.0508-m pipe at various inclinations with and without electrostatics, Part I, II. J Powder and Bulk Solids 10:1–17, 1986.

Nieh S, Nguyen T. Effects of humidity, conveying velocity, and particle size on electrostatic charges of glass beads in a gaseous suspension flow. Journal of Electrostatics 21:99–114, 1988.

Nifuku M, Ishikawa T, Sasaki, T. Static electrification phenomena in pneumatic transportation of coal. Journal of Electrostatics 23:45–54, 1989.

Rader J, Prakash A, Klinzing GE. AIChE Symposium Series on Fluidization 91(308):154–163, 1995.

Smeltzer EE, Weaver ML, Klinzing GE. Individual electrostatic particle interaction in pneumatic transport. Powder Tech 33(1):31–42, 1982.

Smeltzer EE, Weaver ML, Klinzing GE. Pressure drop losses due to electrostatic generation in pneumatic transport. Industrial and Engineering Chemistry, Process Design and Development 21(3):390–394, 1982b.

Zaltash A. Application of thermodynamic approach to pneumatic transport at pipe orientations above the horizontal. Ph.D. diss., University of Pittsburgh, Pittsburgh, PA, 1987.

25

Instrumentation and Measurements

Masayuki Horio, Rafal P. Kobylecki*, and Mayumi Tsukada

Tokyo University of Agriculture and Technology, Tokyo, Japan

1 INTRODUCTION

Instrumentation is an important issue in fluidization engineering, particularly because of the multiphase, dynamic, and nonlinear nature of fluidized beds. In commercial plants, instrumentation should be conducted basically for process control, but detection of unusual behavior and prevention of unwanted losses are important as well. For bench and pilot plants instrumentation should be different compared to that for commercial plants, since maximum information output should be aimed at for safe design and scale-up.

After over 60 years of research and experience on fluidization since the first FCC erection, there is an accumulation of a variety of detection and data evaluation methods. At the same time, recent progress in information technologies and in sensor technologies give us more possibilities of innovation in instrumentation and monitoring. New instrumentation techniques are required also by the tighter economic demands for higher efficiency and lower risk as well as by new applications. Depending on the process conditions, we have to evaluate the appropriate detection principles and data processing logics. Figure 1 shows the variety of measurement demands in fluidization engineering, including both research and development (R&D) and commercial operations. In R&D we have much wider

**Current affiliation*: Czestochowa Technical University, Czestochowa, Poland.

choices, including very expensive and sophisticated systems. As is often said, and as is shown in Fig. 1, in the scale-up chemical engineers are encouraged to design their full-scale concept first to find the risky factors in their design. Then they should design scale-down facilities and measurement systems to obtain data for sure scale-up. In the beginning they have to think small in instrumentation, but along with the size increase their instrumentation concepts also have to be scaled up. Figure 1 also shows that even in large-scale or commercial stages, it is recommended to keep laboratory facilities for plant backup, particularly for particle characterization.

The following are the items to be monitored or measured either in commercial processes or in experimental test rigs: (1) temperature, (2) gas pressure, (3) voidage/bed density/bed height/solids inventory, (4) solids mass flow rate/solids mass flux/solids velocity, (5) gas flow rate/gas velocity, (6) chemical composition of gas and solids, (7) particle characterization/particle size distribution (PSD), (8) bed structure and flow regime characterization, including bubble and cluster size. Items 1 and 2 are the most fundamental in process monitoring. Table 1 shows the interrelationships among sensors and parameters to monitor a fluidized bed.

Among all the sensors and probes, pressure sensors can provide the most basic information not only on pressure itself but also on voidage, bed height, and flow regime or quality of fluidization. Pressure monitoring is essential in most cases to maintain an appropriate loop pressure pattern or a macroscopic pressure

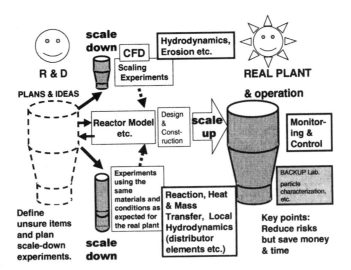

Figure 1 Measurement demands for scale-up and process improvement.

distribution in the plant where particles are usually circulated between columns pneumatically or by gravity flow. In many cases the reaction conditions of each column are quite different. The gas flow for each column is usually separated by the so-called material sealing, where pressure balance takes a major role. Additionally, the state of solids and their distribution in the column can be detected from pressure gradient and pressure fluctuation. As long as the solids are fluidized, their weight is suspended by fluid, and the solids holdup can be detected from the static pressure gradient. In many cases, when a bed is defluidized (e.g., by sintering), the bed pressure drop suddenly decreases. On the other hand, the pressure fluctuations indicate good or poor bubbling as well as the flow regime transitions.

To obtain information of particular interest in the relevant column, it is required to choose appropriate sensing as well as data processing systems and to organize proper measurements to obtain necessary information with a high accuracy. Accordingly, in the present chapter it is intended to provide information to engineers, first on available sensors and probes as well as relevant data necessary to use them, second on data processing methods, and third on experimental methods, particularly for fluidization measurements and monitoring. In Sec. 2, a variety of sensors, probes, and tracers are introduced. In Sec. 3, some information on data processing and visual imaging is provided. In Sec. 4, fluidized bed diagnostic techniques are reviewed.

For further information readers can see Cheremisinoff and Cheremisinoff (1984),

Cheremisinoff (1986), Grace and Baeyens (1986), which all concentrate on bubbling bed measurements, and Werther et al. (1990, 1993), Soo et al. (1994), Yates and Simons (1994), Nieuwland et al. (1996), Louge (1997), Chen (1999) or Werther (1999), although many of them are rather limited in scope.

2 DETECTION PRINCIPLES AND DETECTORS

2.1 Temperature Sensors/Probes

2.1.1 Thermocouples

Thermocouples are the most common temperature measurement devices. A detailed description of thermocouples and other temperature measuring instruments and sensors can be found in Holman (1994) or Stephenson et al. (1999). Maximum temperature limit, up to which a thermocouple can be used without damage, depends strongly on its diameter and the surrounding environment. The thicker the thermocouple, the higher the temperature application limit, but on the other hand the longer the response time.

2.1.2 Optical Pyrometer

The optical pyrometer is a noninvasive temperature measurement device utilizing Planck's law to obtain temperature, T[K], from radiant energy flux, $E(\lambda)$[W], of a particular wavelength λ:

Table 1 Interrelationships Among Measurement Techniques or Sensors and Process Parameters of Fluidized Beds Applicable for Bench Scale (B), Pilot Scale (P), and Commercial Scale Tests (C). Case C also Includes Emergency Measurements

	Flow regime & structure	Particle characteristics	Pressure	Temperature	Voidage	Gas Velocity	Solids velocity	Solids mass flux	Gas/solids composition
Pressure sensors	B, P, C	B, P, C	B, P, C		B, P, C	B, P, C			
Thermocouples				B, P, C			B, P, C		
Pyrometers				B, P, C					
Suction probes	B, P	B, P, C							
Gs meters	B, P, C						B, P	B, P, C	B, P, C
Capacitance probes	B, P				B, P		B, P	B, P, C	
Optical fiber probe	B, P, C	B, P, C			B, P, C		B, P, C	B, P	
Laser sheeting	B				B		B	B	B, P
LDV/PIV	B, P	B, P			B, P	B, P	B, P	B, P	
Tomography	B				B	B, P	B, P		
Solid/gas tracers	B	B			B	B, P	B, P		
Camera observation	B, P, C	B, P, C		B, P, C (IR camera)	B, P, C				
Gas sampling & injection	B, P, C					B, P			B, P, C

$$E(\lambda) = \frac{C_1 \varepsilon(\lambda)}{\pi \lambda^5 (e^{C_2/\lambda T} - 1)} \tag{1}$$

where $\varepsilon(\lambda)$ is emissivity of an object, $C_1 = 3.74 \times 10^{-16}\,\mathrm{Wm^2}$, and $C_2 = 0.0114\,\mathrm{mK}$.

For shorter wavelengths and for lower temperatures, the above equation is simplified to the formula called Wien's law:

$$E(\lambda) = \frac{C_1 \varepsilon(\lambda)}{\pi \lambda^5} \cdot \exp\left(-\frac{C_2}{\lambda T}\right) \tag{2}$$

The schematic of an optical pyrometer is shown in Fig. 2. The radiation of a target body, whose temperature is measured, is viewed through the lens and filters. The aim of the filters is to absorb the radiation and reduce its intensity. A lamp is placed in the optical path of the incoming radiation. By adjusting the lamp current, the color of the filament is changed so that it matches that of the incoming radiation. Temperature calibration is made in terms of the lamp heating current. Additionally, the filter is installed in the eyepiece to ensure that comparisons are made for essentially monochromatic radiation, thus eliminating uncertainties resulting from variation of radiation properties with wavelength. Based on a similar principle, Ross et al. (1981) determined the burning char temperature by comparing its monochromatic photograph with that of a calibrated filament.

To avoid the uncertainties of temperature measurement associated with the uncertainty of emissivity, a two-color pyrometer is recommended, which analyzes the body radiation properties for two wavelengths. In the two-color or ratio pyrometer the radiation is detected at two separate wavelengths for which emissivities of the surface can be considered nearly the same. Thus the ratio of the thermal radiation sensor output calculated for two wavelengths by Wien's law is obtained as

$$X \equiv \frac{E_1}{E_2} = \frac{\varepsilon(\lambda_1)\lambda_1^{-5} e^{C_2/\lambda_1 T}}{\varepsilon(\lambda_2)\lambda_2^{-5} e^{C_2/\lambda_2 T}} \tag{3}$$

Since $\varepsilon(\lambda_1) \approx \varepsilon(\lambda_2)$, Eq. (3) can be rewritten and solved for T as

$$T \approx C_2\left(\frac{1}{\lambda_2} - \frac{1}{\lambda_1}\right)\left(\ln X \frac{\lambda_1^5}{\lambda_2^5}\right)^{-1} \tag{4}$$

2.1.3 Infrared Camera/Thermometer

Another noninvasive tool to measure temperature is the infrared (IR) thermometer, which usually has an optical response time below 0.1 s. The measurement principle is based on determination of the thermal radiation Q of a target body. The radiation is proportional to the temperature of the body according to the Stefan–Boltzmann law

$$Q = \int E(\lambda)\,d\lambda = \varepsilon \sigma T^4 \tag{5}$$

where σ is the Stefan–Boltzmann constant, $\sigma = 5.67 \times 10^{-8}\,\mathrm{W/m^2 K^4}$.

MgF_2, ZnSe, and Sapphire (Al_2O_3) glasses are used as IR transparent media windows, e.g., MgF_2: wavelength 3–5.4 μm, 95% of IR passing through; ZnSe: 0.5–20 μm, 90%; Al_2O_3: 0.2–4 μm, 85% transparency for IR.

Since the IR thermometer must adapt itself to the temperature of the surroundings, it may not measure the temperature accurately without enough time to adapt. The measurement results are also affected by the angle of optical axis to the surface as shown in Fig. 3. When the angle θ is larger than 40°, an IR camera placed at position A gives temperature values higher than the real ones.

IR cameras have been applied to so-called inverse heat transfer problems, which concern the identification of unknown temperature distribution or thermal

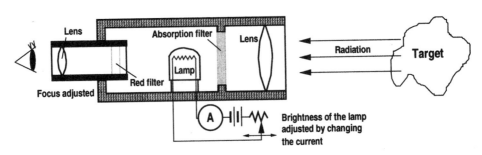

Figure 2 Schematic of the optical pyrometer.

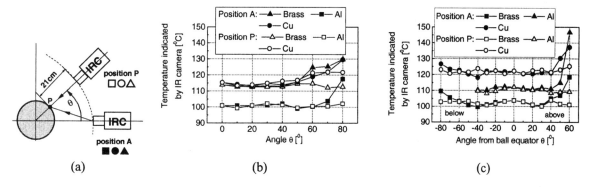

Figure 3 Ball surface temperature versus observation angle, $d_{ball} = 20$ mm. (a) Schematic of measurement; (b) horizontal temperature distribution of the equator; (c) vertical temperature distribution. (Kobylecki and Horio, 2000.)

resistance distribution inside a body from the observed surface temperature distribution (Pasquetti and Le Niliot, 1991; Le Niliot and Gallet, 1998).

2.2 Heat Flux Probes

The overall heat flux in fluidized beds consists of a radiative and a convective component. Assuming, for simplicity, that the probe surface and the suspension flow are gray, parallel, and separated by an opaque gas, one may express the total bed-to-wall heat flux as

$$q_{total} = q_{rad} + q_{conv}$$
$$= \frac{\sigma(T_{bed}^4 - T_{wall}^4)}{(1/\varepsilon_{bed} + 1/\varepsilon_{wall} - 1)} + h(T_{bed} - T_{wall}) \quad (6)$$

where h is the heat transfer coefficient, T is temperature, ε is emissivity, σ is the Stefan–Boltzmann constant, and q is the heat flux.

Since heat transfer in fluidized beds is dominated by surface renewal of bed materials, the bed-to-wall heat transfer coefficient for a vertical surface is much larger at its leading edge. Accordingly, the heat transfer coefficient determined by small probes can be abnormally large. Furthermore, depending on gas and solids motion, surface renewal rate, and the time fraction of coverage by bed material, the heat transfer coefficient varies for different segments of the surface, particularly for horizontal tubes. The requirement to determine the instantaneous heat transfer coefficient is the low thermal inertia of the probe, such as that used e.g., by Mickley et al. (1961) or Ma and Zhu (1999) and shown in Fig. 4a–b, respectively.

In some experiments it is necessary to separate the radiative and convective heat fluxes. This can be done by using two probes of different emissivities (e.g.,

Botterill et al., 1984) or by separating the convective component from the total heat flux by using quartz windows (Basu and Knouche, 1988) or ZnSe windows (Ozhaynak et al., 1984). Quartz windows have rather narrow transmittance bands for infrared radiation (roughly 0.15–3.5 μm wavelengths compared to 0.5–20 μm for ZnSe) but are resistant to thermal shocks. To measure simultaneously the radiative and total heat flux, Luan et al. (1999) proposed the multifunctional probe shown in Fig. 4c. They determined the radiative component of heat flux both by using a ZnSe window to separate the two fluxes and by estimating the radiative flux from comparison of the measurement from two probes of different emissivities. More on heat flux measurement can be found in Mickley and Fairbanks (1955), Botterill (1975), Mathur and Saxena (1987), Leckner and Andersson (1992), Goedicke and Reh (1992), Werdermann and Werther (1993), Couturier et al. (1993), and Molerus and Wirth (1997).

2.3 Gas/Solid Pressure and Force Sensors/Probes

In most cases the pressure is detected using diaphragm sensors classified in terms of sensing element as metal strain gauge, piezoelectric semiconductor, electric capacitance, reluctance, and LVDT (linear variable-differential transformers) sensors. The structure of a silicon semiconductor sensor, where the pressure is detected by a silicon diaphragm, is shown in Fig. 5. In most cases these sensors can be used at temperatures lower than 70°C.

2.3.1 Gas Pressure

Gas pressure sensing is the most basic issue in fluidization and solids transport. The key points of the sensing system are the volumetric capacity of its tubing and the

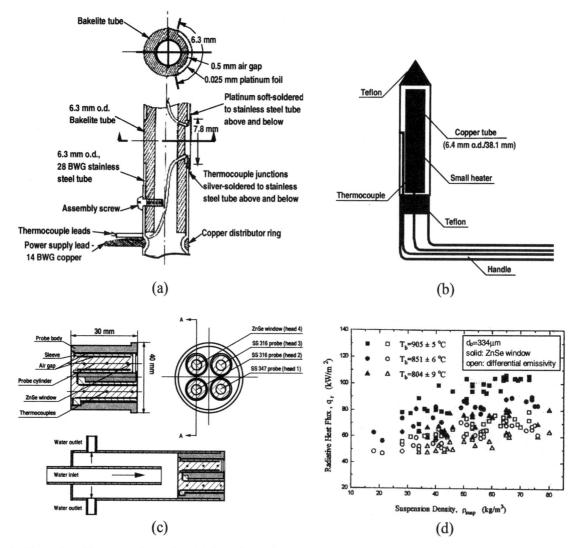

Figure 4 Examples of heat transfer probes. (a) Low thermal capacity instantaneous type of Mickley et al. (1961); (b) miniature heat transfer probe of Ma and Zhu (1999); (c) probe to measure radiative and total heat flux. (Luan et al., 1999.); (d) radiative heat flux vs. suspension density. (Luan et al. 1999.) [(a): Reproduced with permission of the American Institute of Chemical Engineers. Copyright © 1986 AIChE. All right reserved. (b)–(d): with permission of Elsevier Science]

ability to prevent the tube from solids plugging. Figure 6 shows a simplified pressure measurement system consisting of a leading tube of diameter d_{pp}, length L and volume V at the end where the pressure transducer is placed.

If laminar friction resistance is assumed for the flow in the tube, the pressure–amplitude ratio is given by (Holman, 1994)

$$\left|\frac{p}{p_0}\right| = \frac{1}{\left(\left[1-(\omega/\omega_n)^2\right]^2 + 4h^2(\omega/\omega_n)^2\right)^{0.5}} \quad (7)$$

where h is the damping ratio p is the amplitude of pressure signal impressed on the sensor, p_0 is the pressure at the tube inlet, ω is frequency, and ω_n is the resonance frequency.

The resonance frequency, ω_n, damping ratio, h, and phase angle of the signal, ϕ, are given by

$$\omega_n = \sqrt{\frac{3\pi r^2 c^2}{4LV}} \quad (8)$$

$$h = \frac{2\mu}{\rho_f c r^3}\sqrt{\frac{3LV}{\pi}} \quad (9)$$

Instrumentation and Measurements

649

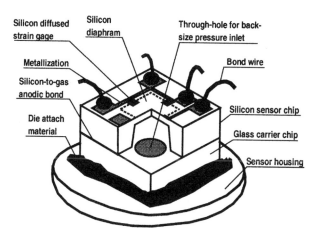

Figure 5 Silicon semiconductor piezo-resistive pressure sensor. (Chau et al., 1999, with permission of CRC Press.)

$$\phi = \tan^{-1}\left[\frac{-2h(\omega/\omega_n)}{1 - (\omega/\omega_n)^2}\right] \qquad (10)$$

where $r = d_{pp}/2$ is the tube radius, c is sound velocity in the fluid, and μ and ρ_f are the viscosity and density of the fluid, respectively.

Figure 7 shows examples of pressure fluctuations for two tubes of various diameters. If the tube diameter is too narrow, the probe gives quite poor response.

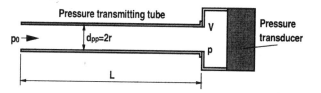

Figure 6 Schematic of a pressure transmitting system. (Holman, 1994.)

2.3.2 Manometer

Manometers are classical and still useful gas pressure measurement devices used also to calibrate semiconductor pressure sensors. Applying the manometers as pressure measurement devices, caution should be taken to prevent the overflowing as well as the spontaneous oscillation of the liquids. To obtain pressure distribution in one glance, the manometers can be assembled and connected to a common-reservoir.

2.3.3 Solid Pressure

To measure the particle pressure separately from the gas pressure, the chamber of the sensor covered by the diaphragm should have one or more bypass tubings to make the gas pressure inside equal to the outside, so

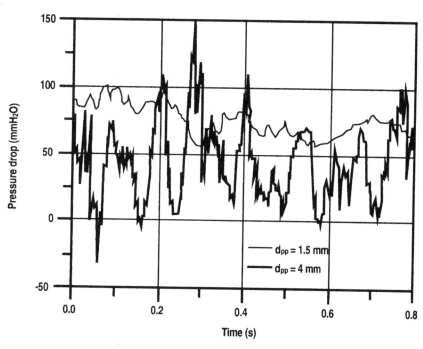

Figure 7 Effect of tube diameter, d_{pp}, on pressure fluctuations. Particles: 68 μm FCC, purge gas velocity: 1 m/s, pressure port length: 2 m dead volume, 450 mm^3. (Xie and Geldart, 1997.) (With permission of Elsevier Science)

that the diaphragm measures only the pressure from solids. The most important design issues are the size of the probe and its dynamic response in the circumstance of fluctuating gas pressure. Depending on the volume inside the diaphragm and the gas flow resistance in the connection tubings, there can be substantial deviation of the dynamic solid pressure data, both in amplitude and phase, as is shown in Fig. 8. Fortunately, the major frequency of gas pressure fluctuation in fluidized beds is up to 10 Hz (even in turbulent conditions), and it should not be difficult to adjust the probe characteristics at the proper range.

2.3.4 Solid Force Sensors

To determine the bulk solid flow in bubbling fluidized beds, force sensors have been used in lab scale and pilot scale units. Examples can be found in Miyauchi and Morooka (1969), Tomita and Adachi (1973), Mann and Crosby (1977), Raso et al. (1983), Fasching and Smith (1993), and Ito et al. (1999).

2.3.5 Fluidized Bed Viscometer

Figure 9 shows two typical probes to measure viscosity: a rotating drum and a rotating bar. It shows also some measurement results obtained with using a rotating bar. In case the internals are heavily placed in the bed, the solid-to-internal friction may affect solids circulation. In such cases, solid viscosity also has to be taken into account in the scaling experiments, as discussed e.g., by Naruse et al. (1996).

2.4 Light Sources and Optical Sensors

2.4.1 Fundamentals of Light Measurement

The main parameters to describe light-related phenomena are classified as radiant quantities and luminous quantities; the latter are psychophysical parameters. Table 2 shows their definitions and units.

Total optical energy input Φ_i consists of three components: reflection Φ_r, transmission Φ_t, and Φ_a adsorption:

$$\Phi_i = \Phi_r + \Phi_t + \Phi_a \tag{11}$$

The commonly used parameters reflectance and transmittance are defined as

$$\text{Reflectance} = \frac{\Phi_r}{\Phi_i} \tag{12}$$

$$\text{Transmittance} = \frac{\Phi_t}{\Phi_i} \tag{13}$$

Light emitted to a suspension is scattered according either to Mie's or to Rayleigh's rule. Mie scattering occurs for particles much larger than the wavelength. In Mie scattering, the spectra of scattered and emitted light are similar. Rayleigh scattering occurs for particles much smaller than the wavelength, and the spectrum of scattered light shifts to the blue side. The directions of scattered light are different for these two scatterings, as shown in Fig. 10.

Owing to the light scatter, the relative decrease of light transmission through the suspension is proportional to the suspension volumetric concentration, ε_p, and the travel path length, dl. Thus for the intensity of light, I, one can write

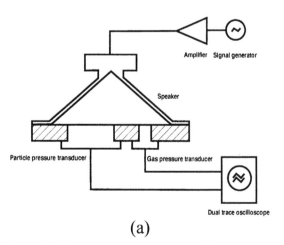

(a)

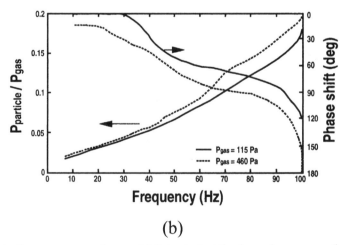

(b)

Figure 8 Frequency response of a particle pressure sensor to the gas pressure change. (a) Setup to test the dynamic response; (b) gas pressure (measured by a pressure transducer) and the phase shift. (Campbell and Wang, 1990.) (With permission of Institute of Physics)

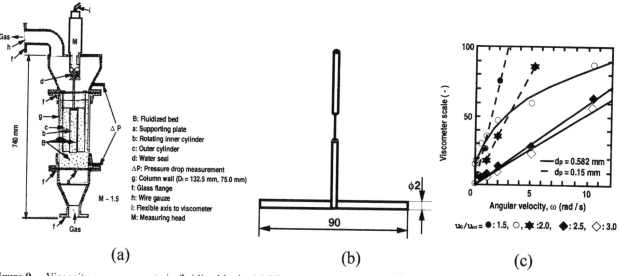

Figure 9 Viscosity measurements in fluidized beds. (a) Viscosity measurement with rotating drum. (Schugerl et al., 1961.) (With permission of Elsevier Science) (b) T-type rotating bar to measure viscosity. (Naruse et al., 1996.) (c) Viscosity measured with a bar viscosimeter. (Naruse et al., 1996.) $D_t = 105$ mm; bed material: SiO_2; $d_p = 582\,\mu$m and $150\,\mu$m.

$$\frac{dI}{dl} = -\alpha I \varepsilon_p \tag{14}$$

For a dilute and monodisperse suspensions, Eq. (14) is integrated and the solution, known as the Lambert-Beer law, is obtained as

$$\frac{I}{I_0} = \exp(-\alpha \varepsilon_p l) \tag{15}$$

where α is the absorbance coefficient and I_0 is the intensity of the light source. Determining α by calibration we can calculate ε_p from the observed I/I_0.

2.4.2 Light Sources

Light sources that have been applied to fluidized bed measurement are lamps (halogen, mercury, metal halide), LEDs (light emitting diodes), and lasers of

Table 2 Radiant and Luminous Quantities

Radiant quantities				Luminous quantities			
Parameter		Definition	Unit	Parameter		Definition	Unit
Radiant energy	Q_e	Total energy emitted by a body (as electro-magnetic wave or particle)	J	Quantity of light	Q_v	$\int \Phi_v dt$	lm · s
Radiant flux	Φ_e	dQ_e/dt	W	Luminous flux	Φ_v	$K_m \int \Phi_{e\lambda}(\lambda)V(\lambda)d\lambda$	lm = cd · sr
Radiant exitance	M_e	$d\Phi_e/dA$	W/m^2	Luminous exitance	M_v	$d\Phi_v/dA$	lm/m^2
Irradiance	E_e	$d\Phi_e/dA$	W/m^2	Illuminance	E_v	$d\Phi_v/dA$	lx = lm/m^2
Radiant	I_e	$d\Phi_e/d\Omega$	W/sr	Luminous intensity	I_v	$d\Phi_v/d\Omega$	cd
Radiance	L_e	$\dfrac{d^2\Phi_e}{dAd\Omega\cos\theta}$	W/(m^{2}sr)	Luminance	L_v	$\dfrac{d^2\Phi_v}{dAd\Omega\cos\theta}$	cd/m^2

dA is infinitesimally small area, $d\Omega$ is an infinitesimally small solid angle, θ is the angle between a normal direction to the surface and a given direction, and $V(\lambda)$ is the spectral luminous efficiency.

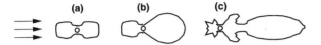

Figure 10 Distributions of scattering angle. (a) Mie scattering; (b) intermediate; (c) Rayleigh scattering.

various types (gas, solid, semiconductor) as listed in Table 3.

The ranges of output wavelengths of major commercial LEDs and lasers applied to fluidized bed measurement are shown in Table 4. An example of an emission spectrum of an LED in comparison with those of various lamps is presented in Fig. 11. The most commonly used gas lasers are He-Ne, CO_2, and Ar ion lasers. In the case of solid-state lasers the light is emitted by atoms, which are fixed within a glassy material or crystal (e.g., Nd YAG laser, where the laser medium is an yttrium-aluminum-garnet matrix with trivalent neodymium ions present as impurities). Semiconductor lasers made usually of AlGaInP, GaAlAs, InGaAsP, or lead salt materials use the special properties of a p-n junction. More precise data can be found e.g., in Weber (1999).

If the phase matching condition is satisfied, the OPO (optical parametric oscillator) laser can continuously emit light of various wavelengths within the range indicated by the equation

$$\lambda_s + \lambda_i = \lambda_p \tag{16}$$

where λ_s, λ_i, and λ_p are the wavelengths of signal, idler, and pumping waves, respectively.

The configuration and wavelength of an OPO laser are shown in Fig. 12. This laser emits light of wavelengths λ_s and λ_i. The laser is pumped with the phase matching angle θ by the light of wavelength λ_p filtered by a nonlinear crystal.

2.4.3 Photodetectors

The types and various applications of photodetectors are shown in Fig. 13. They are classified into two major categories: point sensors and image sensors. The structures and characteristics of the typical point sensors, i.e., CdS cells, photodiodes, and photomultipliers, are shown in Fig. 14. The characteristics of CdS cells and photodiodes are sensitive to temperature and require compensation circuits.

2.4.4 Optical Fibers

Optical fibers are flexible fibrous light transfer media used in optical systems. The fibers are made of plastics or quartz glass. Quartz fibers are much more expensive than plastic ones and not so flexible. However, they are temperature resistant (up to roughly 1000°C) and transmit light more efficiently, particularly in the UV range. Whatever the material, the optical fiber has a center core with a refractive index ρ_1, surrounded by a layer of another material (called cladding) with a refractive index ρ_2. ρ_2 must be lower than ρ_1 in order to keep the light inside the core area. Plastic core is usually made of PS (polystyrene) or PMMC (polymethylmethacrylate). Plastic cladding is usually made of teflon or silicone. Figure 15 shows the transmission

Table 3 Light Sources Applied to Fluidized Bed Measurement

Type		Application	Reference
Lamp	Halogen	Visible light source for optical fiber probe	
		Lighting for picturing	
		Light source for space filtering	Kamiwano and Saito (1984)
	Mercury	UV light source for O_3 detection	Akagi and Furusaki (1983)
		UV light source for fluorescent particles	Morooka et al. (1989)
	Xenon	Lighting for optical fiber imaging	Hatano and Kudo (1994)
	Metal halide	Lighting for optical fiber microscope	Commercially available
Light emitting diode (LED)		Visible light source for optical fiber probe to dip into the bed	Yamazaki et al. (1988)
Laser	Gas	Light source for optical fiber probe	Horio et al. (1988)
		Sheet lighting for visualization	Horio and Kuroki (1994)
		IR laser for contact-free heating (CO_2 laser)	Yamada et al. (1996)
	Semiconductor	Visible light source for optical fiber probe	Werther and Hage (1996)
	Tunable laser	Laser-induced fluorescence (LIF)	

Table 4 Major Commercial LEDs and Lasers

LED			Lasers				
Color	Material	Peak wavelength (nm)	Type	Material	Wavelength	Power	Operation
Red	GaP/GaP	700	Gas	He–Ne	632.8 nm (most cases)	0.1–100 mW	Continuous
	GaA1As/GaAs	660					
	InGaAlP/GaAs	644					
	GaAsP/GaP	635		Argon ion	488 nm, 515.4 nm, and others	100 mW– 20 W	Continuous
	InGaAlP/GaAs	623		CO_2	$10.6\,\mu m$, $9.4\,\mu m$, and others	1W–15 kW > 1 MW	Continuous Pulse
Orange	InGaAlP/GaAs	620					
	InGaAlP/GaAs	612		Excimer		1–50 W	Pulse
	GaAsP/GaP	610					
Yellow	InGaAlP/GaP	590	Solid	Nd: YAG	1064 nm	> 2 kW	Continuous and pulse
	GaAsP/GaP	587					
	GaP/GaP	570					
Green Yellow	InGaAlP/GaAs	574		Nd:YAG (OPO)	300–950 nm		Continuous and pulse
	GaP/GaP	565					
Green	InGaAlP/GaAs	562	Liquid	Organic dye*	250–1100 nm		Continuous and pulse
	GaP/GaP	555					
Blue	SiC/SiC	470					
	GaN/sapphire	450	Semiconductor	GaInP/GaAs	670–680 nm		Continuous
Infrared	GaAs	910					
	GaAs:Si	940					

*Tunable laser, excited by Excimer, Nd: YAG or N_2 gas laser, etc.
Sources: LEDs: Noguchi, 1999. Lasers: Rearranged from Muraoka and Maeda, 1995.

characteristics of PMMA fibers and quartz fibers versus the wavelength.

2.4.5 Optical Fiber Probe

Types of optical fiber probes are presented in Table 5, and some examples of probes are shown in Fig. 16. In the case of the reflection-type optical fiber probe, the volume from which the probe can detect the information depends on fiber configuration. The intensity distribution of reflected light as a function of the distance between a fiber tip and a flat surface is shown in Fig. 17. The size of the core of the optical fiber should be selected depending on the maximum diameter of particles being investigated. If individual particle passages are to be detected, the fiber core diameter should be less than or equal to the particle diameter. On the other hand, if only the passage of solids or the solids concentration is to be detected, the fiber diameter of a nonbundle-type probe or the diameter of a fiber bundle probe should be five to ten times larger than the particle diameter.

The main difficulty in using optical fiber probes is associated with their proper calibration. For solids velocity measurements Ohki and Shirai (1976) calibrated their probe (cf. Table 5) by dipping its top, about 2 mm, into a thin bed of the same material placed on a rotating disk (a music record disk player was used) and obtained a cross-correlation of the signals received from two fibers. The cross-correlation can be successfully applied for the analysis of velocity data unless the velocity fluctuates too often.

To measure solids concentration, the calibration of the probe is more difficult, because it requires a homogeneous suspension. In many cases, the suspension tends to contain agglomerates or clusters, which may seriously affect the output. Various ways to cali-

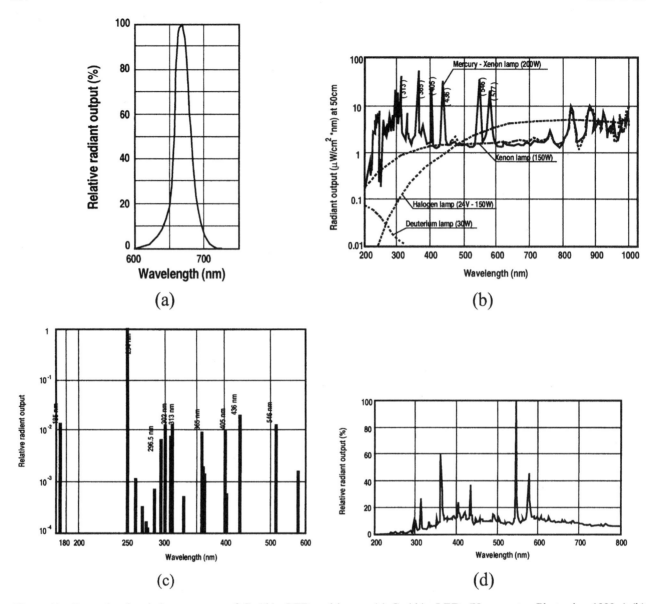

Figure 11 Example of emission spectrum of GaAlAs LED and lamps. (a) GaAlAs LED. (Hamamatsu Photonics, 1999c.) (b) Various lamps. (Hamamatsu Photonics, 2000.) (c) Low-pressure mercury lamp. (Hamamatsu Photonics, 2000.) (d) Metal halide lamp (150 W). (Hamamatsu Photonics, 2000.)

brate the probes can be found, e.g., in Matsuno et al. (1983), Hartge et al. (1986), Lischer and Louge (1992), Amos et al. (1996), or Zhang et al. (1998). Matsuno et al. (1983) calibrated their optical fiber probe using free-falling glass beads assuming that the particle velocity equaled the terminal velocity u_t. For solids concentration up to 50%, Hartge et al. (1986) confirmed a linear relationship between particle volume concentration and light reflection intensity for the liquid–solid fluidized bed. Assuming that a

similar relationship exists for gas–solid systems, they compared the cross section average concentration from the probe with the γ-ray adsorption or pressure gradient data to obtain the calibration coefficient for quartz sand particles. Lischer and Louge (1992) inserted an optical fiber into the center of a capacitance wall probe and calibrated the optical probe by comparing the two signals. Herbert et al. (1994) applied a method similar to that of Matsuno et al. (1983) to FCC particles but used particle velocity

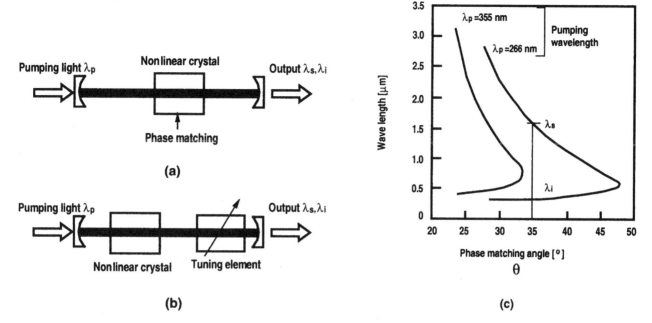

Figure 12 OPO laser configuration and wavelength. (a) Wideband OPO; (b) narrow band OPO; (c) effect of phase matching angle on idler wavelength of BBO crystal OPO. (Muraoka and Maeda, 1995.) (With permission of Sangyotosho)

determined by optical probe measurement. They found a clear effect of an antielectrostatic additive (Larostat 519, PPG/Mazer Chemicals) in dispersing particles, as shown in Fig. 18 (in their original paper Figs. 4a and 4b should be reversed). They obtained a power-law relationship between solids volume fraction and light reflection. Zhang et al. (1998) also reported similar probe calibration.

Rensner and Werther (1993) determined the effective measuring volume of a single fiber optical probe (d = 0.6 mm) for FCC and quartz sand as a function of particle concentration. In both cases (i.e., quartz and FCC), the 50% transmission length was less than 1 mm for the solids volume fraction $\varepsilon_p = 0.002$ and less than 0.1 mm for $\varepsilon_p = 0.2$. Concerning the calibration of particle concentration probes Amos et al. (1996) did detailed analysis and experiments.

2.4.6 LDV—Laser Doppler Velocimetry

Laser Doppler velocimetry (LDV) enables measuring the local fluid velocities from the frequency of light reflection from fine tracer particles (called seed particles) in the fluid. LDV can also be used to detect the velocities of coarse particles or, with seeding, both gas and particle velocities. The wavelength of laser light reflected from a moving particle differs from that of

a stationary source. This phenomenon is called the Doppler effect, and its frequency shift $\Delta \omega_i$ is expressed as

$$\Delta \omega_i = (k_s - k_i) v_p \tag{17}$$

where v_p is particle velocity and k_s and k_i are the wave number vectors of emitted and scattered light, respectively.

Since in many cases the velocity v_p is not sufficiently large to make a large shift, the interference of scattered lights from two laser beams at an angle ϕ is usually utilized. For such a situation let k_1 and k_2 denote the wave number vectors for the two reflected lights, which result in two different Doppler shifts, $\Delta \omega_1$ and $\Delta \omega_2$. Then, subtracting $\Delta \omega_2$ from $\Delta \omega_1$, we obtain the frequency of the interference, f_D, called the Doppler frequency, which can be expressed as a function of the angle ϕ by

$$f_D = \frac{\Delta \omega_2 - \Delta \omega_1}{2\pi} = \frac{2 v_{px}}{\lambda} \sin \frac{\phi}{2} \tag{18}$$

where v_{px} is the x-component of particle velocity (cf. Fig. 19).

When the light reflection from a particle is detected, the signal starts beating with a frequency f_D, and the component of particle velocity can be determined from Eq. (18). There are several modes of light emitting and

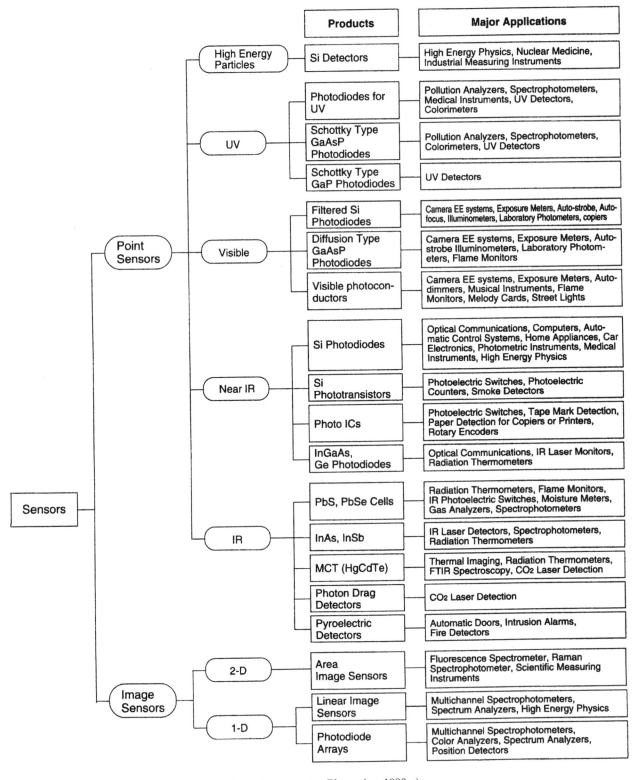

Figure 13 Photodetectors and their applications. (Hamamatsu Photonics, 1999a.)

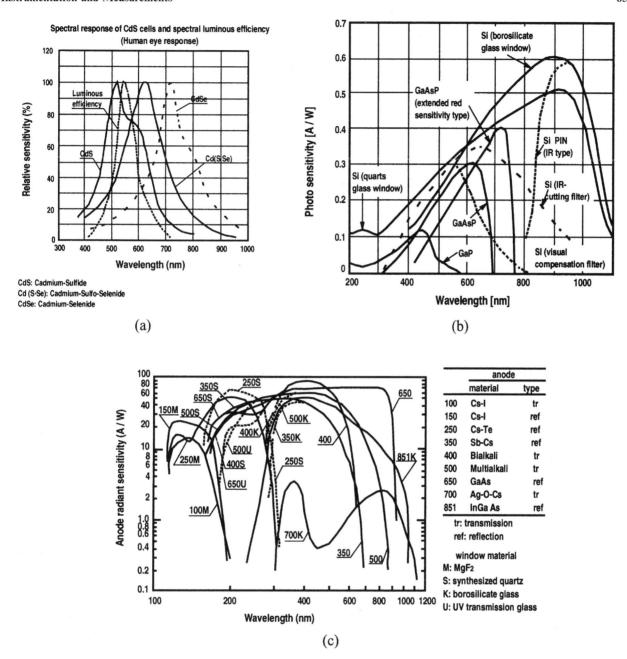

Figure 14 Structures and characteristics of typical point sensors. (a) Spectral responses of photoconductive sensors. (Hamamatsu Photonics, 1999b.) *Note*: Luminous efficiency equals $V(\lambda)$ of Table 2. (b) Spectral responses of photodiodes. (Hamamatsu Photonics, 1998b.) (c) Spectral response of a photomultiplier. (Hamamatsu Photonics, 1998a.)

receiving systems, among which a dual-beam forward-scattering mode is the most popular and commercially available. It is popular because the signal can be easily detected and processed, and high-frequency resolution is not required. In the dual-beam forward-scattering method a laser beam is split in two. The two beams create interference fringes as illu-

strated in Fig. 19. The distance δ between the two bright peaks is expressed by

$$\delta = \frac{\lambda}{2\sin(\phi/2)} \tag{19}$$

where λ is laser wavelength and ϕ is the angle between the beams.

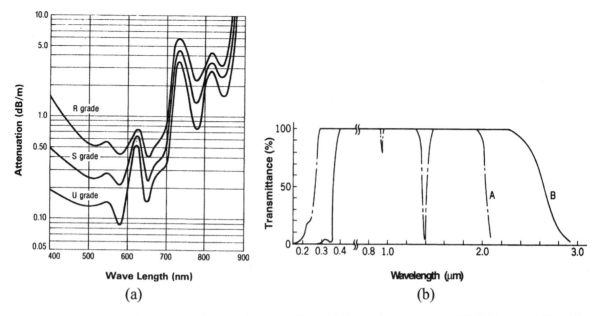

Figure 15 Spectral attenuation and transmittance of optical fibers. (a) Spectral attenuation of PMMA optical fiber (diameter: 0.5–3 mm, room temperature, 65% RH). (Tolay Industries, 1991.) (b) Spectral transmittance of a 1 m long quartz optical fiber. A: SiO_2 core; B: GeO_2/SiO_2 core. (Fujikura Ltd., 1988.)

Table 5 Types of Optical Fiber Probes in the Literature

A. Single Tip

Structure	Type	Fiber diameter	Target parameter/s	Reference
Reflection type				
a	Single fiber	$d_f \gg d_p$	ϵ_p Bubble fraction	Hartge et al. (1986, 1988) Tsukada et al. (1994)
b	Two-fiber	$d_f > d_p$	ϵ_p Bubble fraction	Matsuno et al. (1983) Horio et al. (1985)
		$d_f \sim d_p$	Solid loading and velocity	Cocco et al. (1995)
c	Three-fiber	$d_f \sim d_p$	v_p (one-dimensional) Bubble detection	Ohki and Shirai (1976) Horio et al. (1980a) Hernandez et al. (1998)
d	Multifiber	$d_f \sim d_p$	v_p (two-dimensional)	Patrose and Caram (1982) Ishida et al. (1980)
e	Space filter	$d_f \sim d_p$	v_p (one or two dimensional)	Fielder et al. (1977)
f	Fiber bundle	$d_f \sim d_p$	Particle passage	Morooka et al. (1989)
			v_p and ϵ_p	Nowak et al. (1991) Nieuwland et al. (1996) Tayebi et al. (1999)
		Bore scope	Cluster image Particle image	Li et al. (1991) Hatano et al. (1996)

Table 5 continued

Structure	Type	Fiber diameter	Target parameter/s	Reference
Transmission type				
g	Two in line	$d_f \gg d_p$	Gas concentration Tracer gas passage	Akagi and Furusaki (1983) Horio et al. (1992a)
		Bore scope	Images of clusters and strands	Takeuchi and Hirama (1991) Takeuchi et al. (1998)
h	Two parallel	$d_f \gg d_p$	Bubble passage	Yasui and Johanson (1958)* Kai and Furusaki (1985) Nakajima et al. (1991)
i	Curved fiber type	$d_f \gg d_p$	Bubble passage	Ohki and Shirai (1976) Ji et al. (2000)
		Bore scope	Particle image	Hatano et al. (1998)
Light receiving only				
j	Single fiber	$d_f \sim d_p$	Burning char temperature of pyrometer	LaNauze et al. (1987) Linjewile et al. (1994) Joutsenoja et al. (1999)

* Quartz rods and prisms are used instead of fiber probes.

B. Combined type

Probe shape	Type	Target parameter	References
k	Pair	Velocity (u_b, v_p, v_{cl}, u_g)	Ohki and Shirai (1976) Horio et al. (1988)
		Correlation	Greon et al. (1997)
l	Multi or array	Distribution of bubbles Bubble shape Jet shape	Yamazaki et al. (1988) Lord et al. (1982) Khattab and Ishida (1986)
m	Two parameter measurement	Gas and solid simultaneous detection	Horio et al. (1992a) Hatano et al. (1993)

d_f: fiber diameter, d_p: particle diameter; ⇑: flow direction, →: light direction

If a particle passes through the interference region with velocity component v_{px}, we have a beat of scattered light with the following frequency f:

$$f = \frac{v_{px}}{\delta} = \left(\frac{2v_{px}}{\lambda}\right) \sin\frac{\phi}{2} \qquad (20)$$

This expression is the same as Eq. (18) obtained from Doppler shifts. Equation (20) is usually called the fringe model for the beat frequency in LDV.

A typical signal obtained from a particle passing through the measuring volume and having the diameter less than the fringe distance δ is shown in Fig. 19b and in Fig. 20b (where it is shown as signal A).

In the measurement of solid–fluid two-phase flow we need to discriminate the signals from coarse particles and from fines (i.e., tracer seed). As schematically shown in Fig. 19b, a fine (seed) particle whose diameter is smaller than δ, passing through the interference region, emits a signal described as gas signal. On the other hand, a coarse particle emits the signal with a large offset (described by sand particle signal). Accordingly, in the fluid–solid two-phase flow we can separate the signals from fluid (i.e., from seed) and from coarse particles.

Figure 20 shows a schematic of an LDV system applied to two-phase flows and the signal processing scheme (Tsuji and Morikawa, 1982). The signal has two components, called pedestal and burst. Pedestal is the d.c. component and is obtained by low-pass filtering of the original signal (cf. signal D in Fig. 20b). Burst (cf. signal H in Fig. 20b) is the a.c. component and is obtained by subtracting the pedestal from the original signal. Figure 20c shows an example of a signal from LDV.

In Fig. 20 the signal B indicates the scatter frequency, based on which coarse and fine particles can be discriminated. The discrimination can be made in the following manner: The signals having $L = 1$ and $M = 1$ come from the coarse particles, while those having $L = 1$ and $M = 0$ from the fines. The cases when $L = 0$ are rejected as ambiguous signals, but those with $L = 0$ and $M = 1$ (i.e., from the fines) are accepted because the fine particles usually do not make a large pedestal. The threshold values T_1 and T_2 for pedestal and T_d for burst to obtain logical discriminations should be adjusted carefully by examining the relationship between the threshold values and the number of samples detected.

The seed particles required to measure fluid velocity by LDV must be sufficiently small so that the slip velocity between them and the fluid can be neglected. Additional requirements for the seed particles are large scattering area, uniform size (at least half of δ), no agglomeration, easy to feed, and nontoxic, and they must be "tolerated" in the measurement field (Nakajima, 1995). Spraying water, teflon powder, silicone oil, or other components, as well as chemical reactions, have been used as a source of seed particles. TiO_2, formed from the reaction between $TiCl_4$ and moisture in the air has often been applied. However, the unwanted by-product of this reaction is HCl, which is very corrosive. Tsuji and Morikawa (1982) used NH_4Cl of mean diameter $0.6\,\mu$m.

Recently, the phase Doppler velocimetry (PDV) is becoming a popular technique to measure particle velocity and size. Doppler signals from a PDV are received by an array of photomultipliers, and the phase shifts between the signals from different photomultipliers are analyzed to determine particle size and velocity simultaneously. Figure 21 illustrates the principle of PDV. The particle velocity and diameter are calculated (cf. Bachalo and Houser, 1984, van den Moortel et al., 1997) as

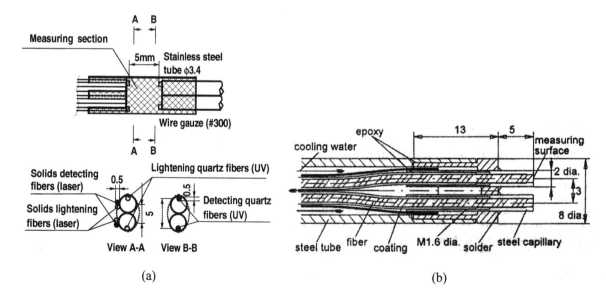

(a)   (b)

Figure 16 Examples of optical fiber probes. (a) Optical fiber probe for simultaneous measurements of solid and gas velocities (solid: light reflection measurement; gas: transmission measurement with O_3 tracer). (Horio et al., 1992a.) (b) Water cooled optical fiber probe. (Werther and Hage, 1996.) (With permission of United Engineering Foundation)

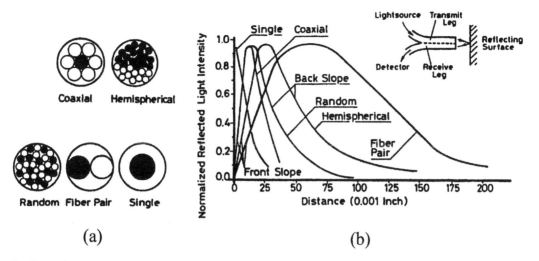

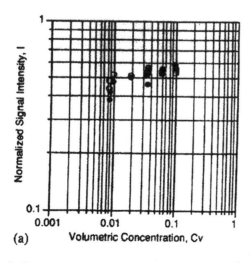

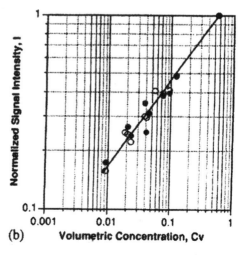

Figure 17 Configuration and characteristics of reflection type optical fiber probes. (a) Fiber configurations. (b) The intensity of reflected light as a function of the distance to a flat target. (Krohn, 1986. Cf. also Reh and Li, 1991.) (With permission of International Society for Optical Engineering)

$$v_p = f_D \cdot \frac{\lambda}{2 \sin(\phi/2)} \tag{21}$$

$$d_p = H \cdot \frac{f \delta \Phi}{2\pi \, \Delta l} \tag{22}$$

Notations in Eqs. (21) and (22) are the same as those for LDV except for f, Φ, Δl, and H, which are the focal length of the transmitter lens, the phase shift between two photomultipliers, the space between detectors (say e.g., #1 and #2 or #1 and #3), and the optical constant, respectively.

The advantage of LDV is that the setup can detect reversal flows for particles, bubbles, and droplets in a single and a multiphase flow system. LDV also enables to measure fluctuating velocity, size, and concentration of suspended particles. However, it is more suitable to apply to the risers of rectangular cross section than to circular ones. The LDV's limitation in determining particle concentration (roughly up to 1%) can be improved greatly by combining it with an optical fiber probe. Wei et al. (1998) reported measurements of particle velocities for solid concentration

Figure 18 Calibration curves for FCC particles. (a) Without Larostat. (b) With 0.5 wt% of Larostat. (Herbert et al., 1994.) (With permission of Elsevier Science)

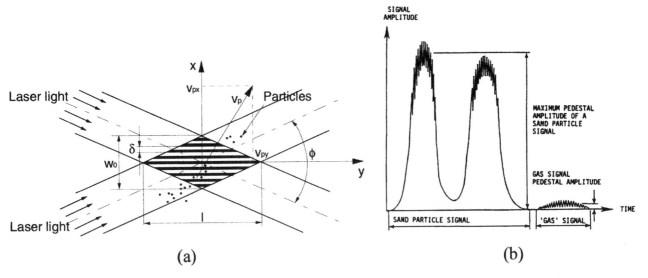

Figure 19 Fringes and output signals of a dual-beam forward-scattering LDV. (a) Interference fringes and a measurement volume. (b) Typical signal from freeboard of a fluidized bed. (Levy, 1986.) (Reproduced with permission of the American Institute of Chemical Engineers. Copyright © 1986 AIChE. All rights reserved.)

up to 0.21. More information on LDV and its applications can be found e.g., in Watrasiewicz and Rudd (1976), Drain (1980), Arastopour and Yang (1992), van de Wall and Soo (1994), and Onofri and Tadrist (1999).

2.4.7 Space Filters

Space filtering is a method of determining particle velocity with the spatial light distribution of a gratelike structure including interference fringes. Dual beam mode LDV is one type of space filter; the others include a reticle type filter and a filter with gratinglike detectors.

In the case of reticle-type filters, the illuminated particle passes through the measurement volume and reflects light to the grating, thus being a source of a periodical signal (Fiedler et al., 1997). From the frequency f of this signal, the particle velocity v_p is determined as

$$v_p = \frac{fg}{\beta} \qquad (23)$$

where g is the grating constant and β is the image scale, i.e., the ratio of image size at grating to that at the measuring point.

An example of a light-reflection-type space filter system is shown in Fig. 22a. The light from halogen lamps is focused by a lens and guided by an optical image fiber bundle into the measuring space (bundle diameter 3 mm, length 0.3 m; the bundle consists of 71,000 optical glass fibers, 10 μm diameter each). The light reflected from the moving particles returns through the same fiber and is detected by two linear image sensors via a lens and a half mirror.

The particle passage through the measuring space is shown in Fig. 22b,c. Each of the wave patterns (I) through (IV) in Fig. 22c corresponds to each state (I) through (IV) in Fig. 22b. In Fig. 22c curve (I) shows the background timing signal to distinguish the output signal of image sensor 1 from sensor 2. When a particle passes sensor 1, a line signal of n_1 bits is detected and it "moves" from the n_1 bit to the n_2 bit within time interval Δt. Thus the radial and vertical components of particle velocity, v_r and v_z, are obtained as

$$v_r = \frac{X_r}{\Delta t} \qquad (24)$$

$$v_z = \frac{W}{\Delta t} \qquad (25)$$

where X_r is the distance between the bit positions n_1 and n_2 and W is the distance between the two image sensors.

A CCD (charge coupled detector) type space filter, shown in Fig. 23, was used by Fiedler et al. (1997) to measure sand particle velocity in a CFB riser. The authors were able to measure the velocities between 0.12 and 40 m/s, at local solid volume concentration 0.1–4%. They reported that particles finer than 5 μm

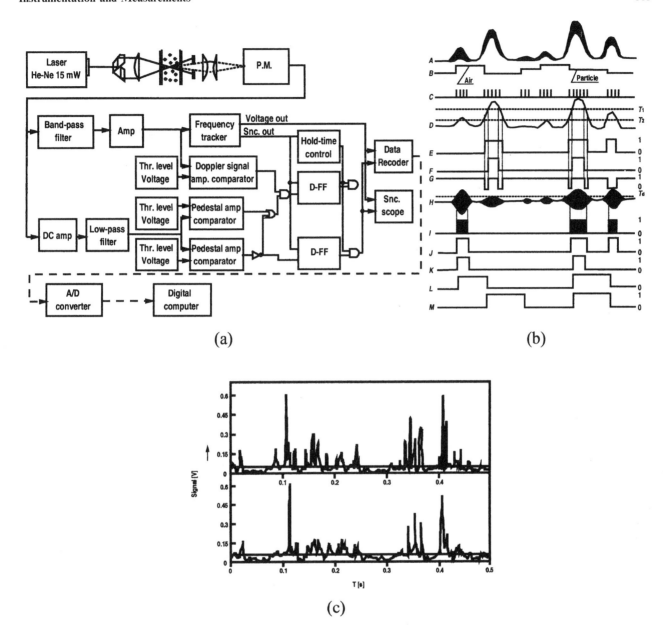

(a)

(b)

(c)

Figure 20 Processing scheme for gas–solid two-phase flow measurement (Tsuji and Morikawa, 1982.) and an example of experimental data. (Lehner et al., 1999.) (a) Optical arrangement and signal processing system. (b) Time chart of the signals. (c) Experimental data obtained with LDV. Description of the signals of case (b): **A**: raw signal; **B**: frequency tracker signal of **A**; **C**: synchronized pulse, which the tracer gives at every signal processing; **D**: Pedestal component of **A** after low-pass filtering; **E**: square wave corresponding to the pedestal amplitude larger than the threshold T_1; **F**: square wave corresponding to the pedestal amplitude smaller than the threshold T_2; **G**: derived from **E** and **F** through an exclusive NOR gate; **H**: alternative signal of **A** (Doppler amplitude, burst signal); **I**: signal produced by comparator; **J**: square wave corresponding to the envelope of Doppler amplitude larger than threshold T_d; **K**: logical product **G** and **J** by AND gate; **L**: signal obtained from **K** and **C** through the D-flip-flop circuit; **M**: signal obtained from **F** and **C** through the D-flip-flop circuit. The main characteristics of the optical system: 15 mW He–Ne laser of wavelength 632.8 nm used as light source, $\phi = 25°$, $\delta = 1.2\,\mu\text{m}$, $w_0 = 76.2\,\mu\text{m}$, $l = 250\,\mu\text{m}$, 41 fringes in the scattering volume, frequency-to-velocity conversion factor 0.6838 ms^{-1} MHz^{-1}. (a,b: Reproduced with permission of Cambridge University Press; c: with permission of DECHEMA e.v.)

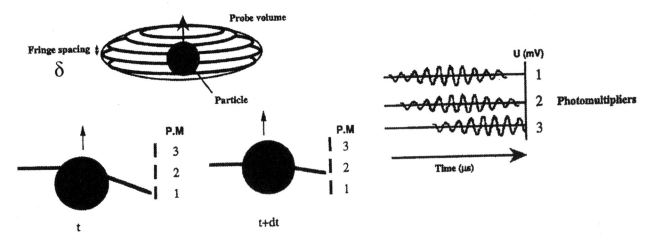

Figure 21 Doppler signals corresponding to a single particle passage received by different photomultipliers. (van den Moortel et al., 1997.) (With permission of Elsevier Science)

adhered to the optical window, but this did not decrease the accuracy of measurement: it only caused an increase in the d.c. level by scattering and reflection.

2.5 Capacitance and Conductivity Sensors/Probes

A capacitance sensor provides information on solids concentration in the target space between its two electrodes. When a voltage E is applied between the electrodes, an equilibrium charge Q is accumulated in them. The charge can be determined by integrating the measured electric current between the electrodes over time. Capacitance is the ratio Q/E. For a set of two parallel plates of area A, separated by a homogenous dielectric of thickness d, the capacitance is given by

$$C = \frac{kA}{d} \qquad (26)$$

where k is the dielectric constant or permittivity of the media between the electrodes [F/m]. k is a function of frequency and varies with temperature. Usually it is expressed in terms of the relative dielectric constant K of the medium as

$$K = \frac{k}{k_{\text{vacuum}}} [-] \qquad (27)$$

In gas–solids systems the volume fraction of solids is inferred from the measurements of the effective dielectric constant (or effective permittivity), expressed by

$$K_{\text{eff}} = \varepsilon K_{\text{f}} + (1 - \varepsilon) K_{\text{s}} \qquad (28)$$

where ε is the voidage, K_{f} is dielectric constant of the fluid, and K_{s} is the dielectric constant of solids (i.e., the ratio of the permittivity of the system to the permittivity of free space).

Some data of the relative dielectric constant, K, are shown in Table 6.

An example of a capacitance probe and its output signals are shown in Fig. 24. To eliminate the influence of "external" capacitance of various parts of the probe (e.g., cables) that can have much larger capacitance than that of the probe itself, a third conductor called a guard is necessary, as discussed by Acree Riley and Louge (1989). More on capacitance measurements can be found in Bottcher (1952), Bakker and Heertjes (1958), Meredith and Tobias (1960), Lanneau (1960), Werther and Molerus (1973), Yoshida et al. (1982), Brereton and Stromberg (1986), Louge and Opie (1990), and Brereton and Grace (1993). Capacitance tomography measurements are described in Sec. 3.3.

2.6 Acoustic sensors

2.6.1 Microphones

Acoustic emission from fluidized beds also provides information on hydrodynamic conditions. Win et al. (1995) used a high-sensitivity microphone to detect the carryover of coarse particles from a multisolid circulating fluidized bed consisting of porous alumina (coarse particles) and glass beads (the fines) They detected acoustic signals, which the particles generated on the cyclone surface. Since coarse particles had larger mass, they generated sounds of magnitude

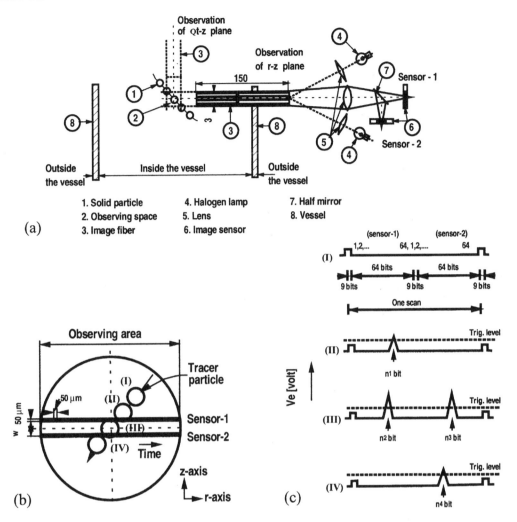

(a)

1. Solid particle
2. Observing space
3. Image fiber
4. Halogen lamp
5. Lens
6. Image sensor
7. Half mirror
8. Vessel

Figure 22 Schematic of a light reflection type space filter (a), tracer particle passage (b), and wave pattern of output signals from light reflection type space filter in a gas–liquid system (c). (Kamiwano and Saito, 1984.) (Reproduced with permission of the American Institute of Chemical Engineers. Copyright © 1984 AIChE. All rights reserved.)

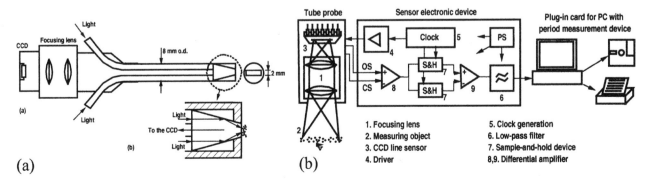

1. Focusing lens
2. Measuring object
3. CCD line sensor
4. Driver
5. Clock generation
6. Low-pass filter
7. Sample-and-hold device
8,9. Differential amplifier

Figure 23 CCD type space filters. Grating constant: 64 μm; image scale (i.e., the size at grating divided by the size at measuring point): 0.0621; sampling frequency: 156 kHz. (a) CCD type space filter probe. (b) Block diagram for CCD type space filter. (Fiedler et al., 1997.) With permission of Elsevier Science

much larger than the fines. Dahlin et al. (1995) applied a knife-edge acoustic sensor connected to a sensitive transducer to detect particles of various sizes in a gas cleanup system.

2.6.2 Acceleration sensors

Sensors of this type were used by Fukayama et al. (2000), who monitored the changes in fluidization conditions inside a 350 MW$_e$ AFBC by determining the acoustic signals collected through in-bed boiler tubes. The signals were detected by the acceleration sensors placed onto the tubes outside the fluidized bed. The monitoring has been performed for 1.5 years. More details can be found in Fukayama et al. (2000).

2.7 Electrostatic Sensors/Probes

Electric charge can be measured by a Faraday cage as shown in Fig. 25a. Fasso et al. (1982) used this cage to study electrostatic charges and concentration of particles in the freeboard of a fluidized bed of glass beads. The gas–solid suspension was sucked into the cage, and the solids charge was determined. The measuring cage was put into another Faraday cage to shield against any external charge. Examples of the electrostatic probes are shown in Fig. 25b,c. Soo et al. (1964) applied the probe shown in Fig. 25b to measure local solids concentration in a gas–solid flow. The authors were able to measure instantaneous concentration, but no impact direction could be identified. Boland and Geldart (1971) used the probe

Table 6 Relative Dielectric Constants of Some Common Materials

Material	Temperature, °C	Frequency, Hz	K
Air	0	$< 3 * 10^6$	1.001
Water	20	Low	80.37
Alumina	20	10^6	4.5–8.5
Nylon 66	25	10^6	3.33
Neoprene rubber	24	10^6	6.26
Polyethylene	23	10^6	2.26
Potassium nitrate	20	10^6	5.6
Silicon rubber	25	10^6	3.1
Vycor glass	20	10^6	3.85
Pyrex glass	20	10^6	4–6

shown in Fig. 25c to determine static electrification in a 2D bubbling fluidized bed. They described a method of visualizing particle motion based on electrostatic charging.

2.8 Gas Sensors

To determine oxygen concentration in a fluidized bed combustor, zirconia sensors (commonly known as λ-probes) have been used. The zirconia sensor acts as a kind of battery, whose output voltage depends on oxygen concentration (the voltage is low at oxidizing and high at reducing conditions). The schematic of the sensor and an example of its ouput signal are shown in Fig. 26.

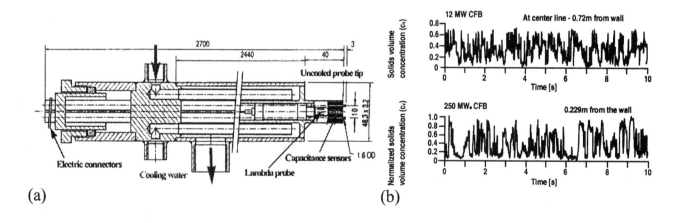

(a)   (b)

Figure 24 An example of a capacitance probe. (a) High-temperature two-needle capacitance probe of Wiesendorf et al. (1999) (With permission of American Society of Mechanical Engineers). (b) Output signals from a 12 mW (Chalmers) and a 250 MW (Gardanne) CFB boiler obtained with the probe shown in (a). (Johnsson et al., 1999.) With permission of DECHEMA e.v.

2.9 Humidity Sensors

The main types of instruments measuring humidity in gases are various types of hygrometers (gravimetric, mechanical, condensation, infrared absorbance detector, electric sensor, thermal conductivity, Al_2O_3/silicon, P_2O_5) and psychrometers.

2.10 Radioactive Sensors

Radioactive particles can be applied as tracers, used to study solids mixing, particle trajectory, or RTD (residence time distribution) of gas or solids within a fluidized bed, as well as to measure the suspension density. Inorganic salt scintillators, i.e., crystals of inorganic salts containing trace quantities of activators to enhance the emission have been used to detect the radiation. A popular type is the sodium iodide scintillator activated with thallium (NaI(Tl)). However, this material is hygroscopic and must be encapsulated. Other commonly used scintillators contain cesium iodide activated with thallium or sodium, CsI(Tl), CsI(Na).

Examples of some radioactive isotopes used for fluidized bed tracking are shown in Table 7. As discussed e.g., by Seville et al. (1995) and Benton and Parker (1996), the scattering or absorption of the emitted radiation may lead to incorrect determination of tracer position. However, since the scattered rays have lower energy than the "correct" ones, by removing the scatter data one may avoid this problem. Application of radioactive particles as tracers is relatively easy and inexpensive, even for large fluidized bed plants. However, it may sometimes disturb the flow and definitely requires strict safety regulations during the mea-

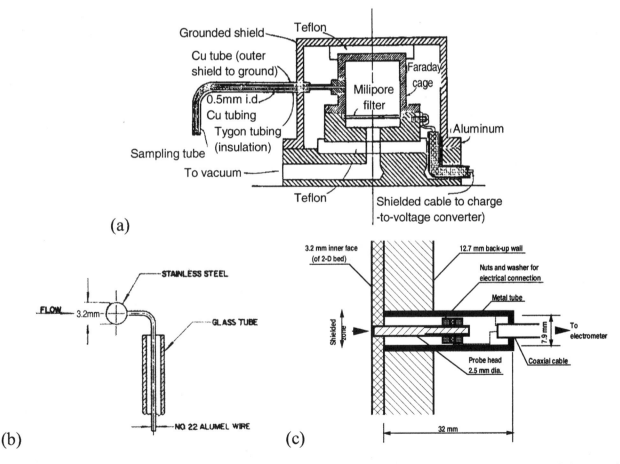

Figure 25 Faraday cage used by Fasso et al. (1982) (With permission of Elsevier Science) (a) and examples of electrostatic probes of Soo et al. (1964) (Reprinted with permission from Ind. Eng. Chem. Fundam., 1964, 3, 98–106 Copyright © 1964 American Chemical Society) (b) and Boland and Geldart (1971) (With permission of Elsevier Science) (c).

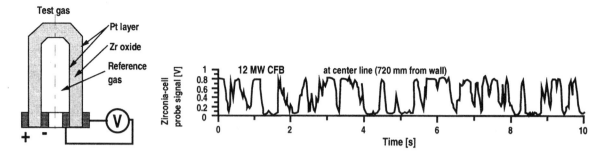

Figure 26 Schematic of a zirconia sensor and an example of the output signal (measurement in a 12 MW CFB boiler). (Johnsson et al., 1999.) With permission of DECHEMA e.v.

surements (particularly for the particles having long half-life). More on radioactive particle tracking can be found in Seville et al. (1995), Stein et al. (1997), and Larachi et al. (1997). More on radioactive tracers is also given in Sec. 2.11.

2.11   Tracers

2.11.1   Solid Tracers

To obtain information on solids movement and mixing, the solid tracers are fed into a fluidized bed and detected either in situ, as in the case of isotope tracers, or by sampling and analysis. It is assumed that the tracers behave like the solids in the flow. Some examples of commonly used tracers are shown in Table 8.

In choosing the radioactive tracer one should take into consideration the following aspects:

1.   The purity of radionuclide; to assure that γ-rays are emitted from disintegration of the selected radionuclide and not from some of its impurities.

2.   The activity of the source; high counting rate can be measured more accurately than a low one.
3.   The energy of γ-rays: according to the Lambert–Beer law, the attenuation is an exponential function of the linear attenuation coefficient and the thickness of the material between the source and the detector.
4.   The half-life: it should be at least an order of magnitude longer than the duration of the tracking study. Otherwise, the loss of source activity may be significant during the experiments. On the other hand, too long a half-life time is not recommended, because it may bring about difficulties in the test equipment after the experiment (e.g., owing to the residual activity of the tracer).

2.11.2   Gas Tracers

By injection of a tracer gas into the flow (continuous or batchwise form) and by measuring its concentration and velocity one obtains information on the residence time distribution (RTD), mixing or mass transfer. Examples of applications of gas tracers to fluidized beds are presented in Table 9. Figure 27 presents a model of gas flow in the fluidized bed and an example of recorded data. Gas tracking is also a useful way to measure mass transfer in fluidized beds, as reported by Ebert et al. (1993) and de Kok et al. (1986).

2.12   Suction Probes for Solids Sampling

There are two kinds of solids suction: isokinetic and nonisokinetic. In isokinetic sampling, the gas suction velocity is adjusted to equal the local gas velocity. This method was mainly applied to aerosol or other

Table 7   Half-Life of Some Positron Emitting Isotopes

Nuclide	Half-life	Nuclide	Half-life
Rb^{82}	78 s	Zn^{62}/Cu^{62}	9.2 h
O^{15}	122 s	Ga^{66}	9.7 h
N^{13}	10 min	Cu^{64}	12.7 h
C^{11}	20.3 min	I^{124}	4.2 days
Ga^{68}	68 min	Ge^{68}/Ga^{68}	271 days
F^{18}	110 min	Na^{22}	2.6 years
Ti^{45}	3.1 h		

Source: Benton and Parker, 1996.

Table 8 Various Types of Particles Used as Tracers in Fluidized Beds

Tracer type	Tracer	Remarks	Author(s)
Chemical	NaCl tracer	FCC reactor, the tracer detected by electric conductivity of sample–water solution	Bader et al. (1988)
	Limestone	Diffusivity in a large-scale FB boiler	Ito et al. (1999)
	Phosphor	Effect of tracer size on solids mixing in the 0.14 m i.d., 10.4 m high FCC riser	Du et al. (1999)
Color	Colored particles	Semicircular fluidized bed, video recording	Yang and Keairns (1982)
Color	Colored particles	Video recording of particle velocity in 2D fluidized bed	Gbavcic et al. (1990)
Fracture	Spheres of different strengths	Kinetic forces acting on particles in a fluidized bed determined from analysis of tracer particles	Kono et al. (1987)
Fluorescent	Paint pigment attached to coal particles	Particle velocity in the cocurrent downward flow of coal and air	Brewster and Seader (1980)
	Fluorescent	Tracer detected by LDV, 319×176 mm cross-section FB	Hamdullahpur et al. (1987)
	FCC impregnated with a fluorescent material	Particles illuminated by UV light, reflected light detected by optical fibers	Kojima et al. (1989) Nowak et al. (1991)
Magnetic	Ferromagnetic	Solids mixing in a bed of FCC, $D_t = 0.15$ m i.d. Bed operated in a bubbling, slugging, and turbulent regime	Avidan and Yerushalmi (1985)
	Sand coated with ferrite	Solids mixing studied by magnetic separation	Horio et al. (1986a)
	Char impregnated with magnetic powder	Measurement of the flow of dark colored particles	Yamaki et al. (1994)
Thermal	Hot particles	Temperature measurements	Westphalen and Glicksman (1995)
Radioactive	Co^{60}	Observation of particle motion in a FB	Kondulov et al. (1964)
	Na^{22}	Studies on particle motion in a 0.03 m i.d. gas-solid fluidized bed. Scintillation detector placed below the fluidized column	Borlai et al. (1967)
	Na^{24}	Particle motion in a FB of glass beads Scintillation counter placed above the bed	Van Velzen et al. (1974)
	Radioactive Na_2CO_3	RTD measurement in CFBs of various sizes	Helmrich et al. (1986)
	Ga^{68}	Particle tracking in a CFB	Ambler et al. (1990)
	Radioactive SiO_2	Studies on RTD in a 82.8 mm i.d. CFB	Patience et al. (1991)
	Sc^{46} oxide	Studies on particle motion and solids mixing in a 3D liquid–solid fluidized bed	Larachi et al. (1994) Larachi et al. (1995)
	Radioactive tracers of various sizes and densities	Measurement of local particle velocities in a cold CFB	Weinell et al. (1995)
	F^{18}	Particle tracking in a 0.15 m i.d. sand bed	Seville et al. (1995)
	Au^{198}	Studies on the hydrodynamics of a CFB tracer	Godfroy et al. (1996)
	Positron emission	Tracer velocity measured 250 times per second	Stein et al. (1997)
	Irradiated Cu	Particle tracking, 3D resolution below 15 mm	Stellema et al. (1998)
	Mn^{56} and cordierite	Studies on particles tracking in the 80 MW_{th} CFB boiler. 16 NaI(T1) detectors used, data collected with the frequency of 20 Hz	Lin et al. (1999a)
	Ca^{137} tracer combined with optical fiber set	Solid concentration axial profile and instantaneous local values measured in a 0.4 m i.d. 15.6 m CFB riser, G_s up to 50 kg/m^{2}s	Schlichthaerle and Werther (1999)

dilute two-phase flow sampling, as well as to fluidized beds (e.g., van Breugel et al., 1969; Nguyen et al., 1989).

In case of a CFB, the amount of sampled solids is not sensitive to the gas suction velocity because of frequent ups and downs of solids in the riser and because of high inertia of particles, as reported e.g., by Monceaux et al. (1986), van Breugel et al. (1969), and Herb et al. (1992). There is a certain suction velocity window where the suction velocity does not affect the solid mass flux, as shown in Fig. 28, and the nonisokinetic sampling is more practical.

Examples of application of solids suction are listed in Table 10. The measurement setup consists usually of two parts: the suction part, immersed in the system, and the measurement section, located outside. More on suction probes and their equipment can be found in Dry (1987), Rhodes and Laussmann (1992a, 1992b), Leckner et al. (1991), Kruse and Werther (1995), and Schoenfelder et al. (1996). A schematic of a solids suction probe is shown in Fig. 29.

In the case of measurement in a CFB, because of the rapid solids up and down motion, the orientation of the suction tube is also important. At each measure-

Table 9 Various Gas Tracers

Author(s)	Tracer type	Comments
Cankurt and Yerushalmi (1978)	CH_4	Radial gas diffusivity estimated from radial profiles of tracer concentration.
Adams (1988)	CH_4	Studies on gas tracking in a CFB riser
Khattab et al. (1988)	O_3	Studies on gas velocity in fluidized beds. The tracer was detected by optical fibers through which UV light was transmitted
Horio et al. (1992a)	O_3	Optical fiber probe applied to measure particle and gas velocities simultaneously; ozone injection upstream of two optical probes, tracer gas velocity measured using UV light, optical fiber and photomultiplier
Ye et al. (1999)	O_3	Ozone decomposition within a CFB. Cast steel used as ozone decomposing catalyst
Sauer and Wallen (1999)	CO_2	Studies on gas mixing in a 2D ABFBC
Yang et al. (1984)	He	Concentration profiles within the bed measured by continuous gas injection
Bader et al. (1988)	He	Continuous injection of gas into an FCC bed
Li and Weinstein (1989)	He	Continuous He injection applied to study gas backmixing in a 0.152 m i.d. CFB
Martin et al. (1992)	He	Gas radial velocity profiles and dispersions in a CFB studied by gas chromatograph analysis of concentration of injected He
Shen et al. (1992)	He	Studies on hydrodynamics and mass transfer in a 2D FB, thermal conductivity detector used to measure tracer concentration
Rivault et al. (1995)	He	Tracer injection into the leg of primary cyclone; tracer gas sampled by a displaceable probe equipped with a filter and analyzed chromatographically
Lin et al. (1999b)	He	Studies on gas RTD in a high density FCC unit. Solids mass flux up to 430 kg/m^2s
Krambeck et al. (1987)	SF_6	Studies on tracer adsorption in the FCC bed. Sulphur hexafluoride injected below the bed. Tracer concentration determined 1.8 m above the bed
Dry et al. (1995)	Warm gas injection, liquid N_2 injection	Studies on gas–solid contact efficiency in two CFBs of different scale, 102 mm i.d. and 600 mm i.d
Li and Wu (1991)	H_2 pulses	Studies on axial gas mixing and RTD in the CFB containing FCC particles ($d_p = 58\,\mu m$)
White and Dry (1989)	Ar pulses	Studies on gas residence time, tracer detected by two rapid response mass spectrometers, reactor 0.09 m i.d. CFB

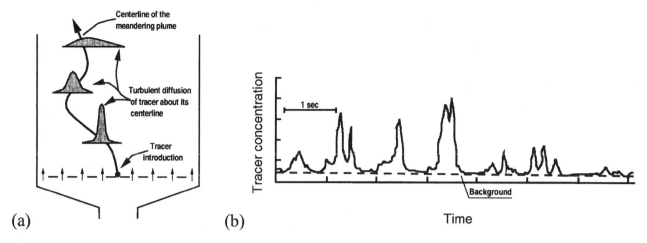

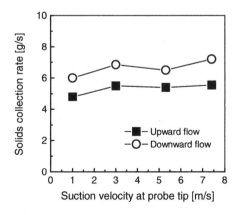

Figure 27 Path of meandering gas in a fluidized bed (a) and the variation of tracer concentration (b). (From Jovanovic et al., 1980.)

ment point the suction should be conducted at both up and down positions of the tube in order to measure the local upward and downward solids mass fluxes (one should remember that even in the core of the riser the solids downflux always exists, as reported e.g., by Herb et al., 1992).

2.13 Gas Sampling

Gas sampling is a basic tool providing information about the local/total gas composition and concentration. Examples of gas sampling studies are listed in Table 11.

In order to separate the gas and solid phase during fluidized bed sampling, a ceramic porous plate at the

Figure 28 Effect of gas suction velocity on solids sampling. (Leckner et al., 1991.) (With permission of American Society of Mechanical Engineers) $U_o = 3.4$ m/s, $G_s = 13$ kg/m²s.

inlet of the sampling system is recommended (e.g., Werther et al., 1990). Naruse et al. (1994) applied the probe shown in Fig. 30a,b and were able to sample the gas separately from the bubble phase and from the dense phase. The bubble passage was detected from pressure drop between two sensors, as shown in Fig. 30c.

3 DATA PROCESSING AND VISUALIZATION SYSTEMS

3.1 Data Processing

Signal processing after the measurements is required: (1) to reduce noises and (2) to obtain characteristic values such as mean values, characteristic frequencies, and phase delay between signals. More information can be found in, e.g., Kawada and Minami (1994) and Edgar et al. (1997). Table 12 shows the major algorithms for data processing applied to the fluidized beds.

Some examples of data processing are shown in Figs. 31–33. The simplest window function is unity for all data (irrespective to the time they were taken) and gives the arithmetic moving average. Figure 31 presents the ammonia IR spectrum processed by a polynomial equation as a window function. The number of averaged samples $2m + 1$ should be chosen appropriately to obtain the necessary information out of the data set.

In the case of stationary signals, the averaging enhances the S/N ratio. Such averaging is used in ordinary chemical analyzers, e.g., FTIR, ICP, EDS, EPMA. If the number of averaging is too large, the

Table 10 Examples of Recent Applications of Solids Suction Probes

Author(s)	Remarks
Verloop et al. (1993)	Dust sampling from a PFBC
Wang et al. (1995a)	Nonisokinetic solids sampling from a 16.1 cm i.d. CFB riser operated at various temperatures
Wang et al. (1995b)	
Nowak et al. (1995)	Solids suction to measure local solids mass fluxes and PSD in a multisolid FB
Mattison and Lyngfelt (1995)	Application of a water-cooled sampling probe to measure the amount of CaS in a 12 MW CFB boiler
Kozinski et al. (1995)	Solids sampling from a CFB sludge waste combustor to detect solids PSD and composition.
Dahlin et al. (1995)	Application of a particle sampling probe to evaluate hot gas cleanup system
Junfu et al. (1999)	Application of a water-cooled solids mass flux meter to measure local G_s at the furnace outlet of a 75 t/h CFB boiler
Brunier et al. (1999)	13 mm i.d. suction probe successfully applied to a CFB boiler

averaging process becomes time-consuming. However, by that averaging the S/N ratio can be improved without knowing the frequency of noises. Figure 32 shows a schematic of the change of the signal vs. the number of averaging.

Filtering in the frequency domain is the most popular method of signal denoising, where a conventional resistor–capacitor circuit is used as an analog filter to eliminate the noises, i.e., signal components having frequencies much higher than that of the main signal.

A schematic of signal processing by low-pass filtering is shown in Fig. 33.

3.1.1 Autocorrelation

The autocorrelation function, defined in Table 12, becomes periodic for periodic signals. Figure 34 shows examples of the autocorrelation functions and power spectra for a bubbling bed and a slugging bed.

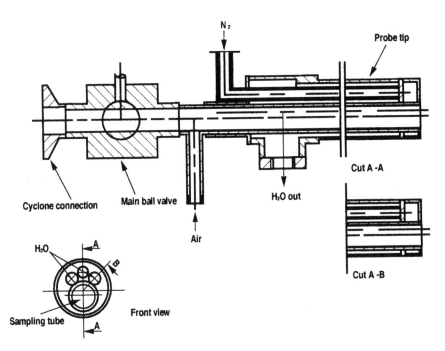

Figure 29 High temperature solids sampling probe of Mattisson and Lyngfelt (1995). (With permission of American Society of Mechanical Engineers)

Table 11 Examples of Gas Sampling in Fluidized Bed Systems

Author(s)	Remarks
Swift et al. (1975)	Flue gas sampling to measure the concentration of Hg, F, and the solids fraction
Yang et al. (1984)	Continuous He injection into a FB. Gas concentration profiles measured
Bader et al. (1988)	Detection of He injected into an FCC riser
Adams (1988)	Application of CH_4 tracer to a CFB riser
Atkinson and Clark (1988)	Detection of bubbles in a FB. Bubble passage determined using a dual static pressure probe
Naruse et al. (1994)	Gas sampling probe, shown in Fig. 30c, applied to sample gas from bubble and
Naruse et al. (1995)	emulsion phases separately. Reactor: 0.1 m i.d. 0.7 m FBC; bed material: sand, fuel: coal
Kassman et al. (1995)	Gas sampling system applied to measure the concentration of NH_3 and HCN inside a 12 mW CFB boiler
Hansen et al. (1995)	Particle probe, λ-probe, gas probe and alkali metal sampling probe used to monitor a 20 MW_e CFB combustor
Mann et al. (1995)	Alkali-sampling probe applied to characterize sorbents for a PFBC.
Hayrinen et al. (1999)	Alkali concentration measurements in a 10 MW PCFB using laser fluorescence and plasmas spectroscopy sensors
Kassman et al. (1999)	Gas sampling from a 12 MW CFB boiler
Lin et al. (1999c)	Simultaneous measurement of N_2O and NO_x concentration, as well as particle temperature in the vicinity of a burning char particle

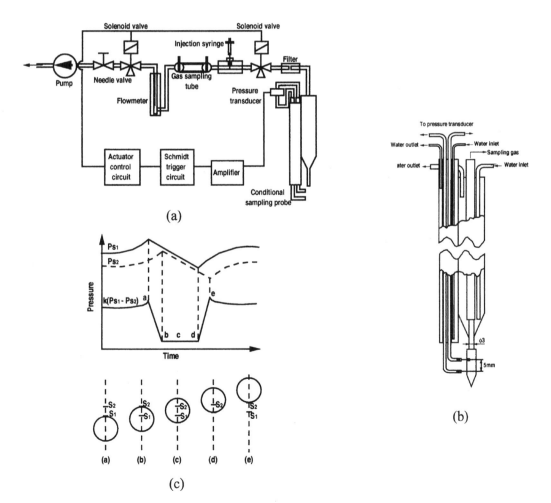

Figure 30 Gas sampling system (a), probe details (b), and bubble passage determined from pressure drop (c). (Naruse et al., 1994.) (With permission of Combustion Institute)

Table 12 Data Processing Algorithm and Application for Fluidization

Processing	Definition/explanation	Examples Signal	Remarks	Reference
Moving average	$y_i = \dfrac{\sum_{j=-m}^{m} w(j)x(i+j)}{\sum_{j=-m}^{m} w(j)}$ $(i = m+1, \cdots n-m)$ $w(j)$: window function			
Arithmetic moving average	$y_i = \dfrac{1}{2m+1}\sum_{j=-m}^{m} x(i+j)$	Pressure	S/N improvement, equivalent to simple average for random noise	Hatano et al. (1999)
Filtering in frequency region	$x(t) \xrightarrow{\text{Fourier}} X(\omega)W(\omega) \xrightarrow{\text{Fourier inverse}} y(t)$ $X(\omega) = \int_{-\infty}^{\infty} x(t)e^{-j\omega t} dt$: Fourier transform $W(\omega)$: filter function		Effective when noise frequency is much higher than the signals	
Low-pass filter	$W(\omega) = \begin{cases} 1(\omega \le \omega_1) \\ 0(\omega > \omega_1) \end{cases}$ $\begin{array}{l}(\omega_1 : \text{cut}-\text{off} \\ \text{frequency})\end{array}$		S/N improvement, elimination of signal from individual particle	
Averaging	$y = \dfrac{1}{N}\sum_{i=1}^{N} x_i$	Pressure	S/N improvement, for signal only repeatable measurement; frequency response information for signal and noise is not necessary	
Root mean square (RMS)	$x_{\mathrm{rms}} = \sqrt{\dfrac{\int_{x_1}^{x_2}(\bar{x}-x)^2 dx}{x_2 - x_1}}$	Pressure fluctuation	Phase transition	Yerushalmi and Cankurt (1979)
Probability distribution function (PDF)	$p(x) = \dfrac{1}{\Delta x}\lim_{T\to\infty}\sum_{i=1}^{n}\dfrac{\Delta t_1}{T}$	Light LDV Light	Determination of threshold value to distinguish: Bubble and emulsion phases Gas and particle signals in pneumatic transport Cluster and lean phase	Horio et al. (1985) Tsuji and Morikawa (1982) Horio et al. (1988)
Autocorrelation function	$R(\tau) = \lim_{T\to\infty}\int_0^T x(t)x(t+\tau)\,dt$	Pressure Pressure Light Light Heat transfer	Slugging condition Bubbling condition Bubbling regularity Freeboard turbulence Dynamics of CFB	Broadhurst and Becker (1976) Fan et al. (1981) Greon et al. (1997) Horio et al. (1980a) Li et al. (1993)
Cross-correlation function	$R_{\mathrm{xy}}(\tau) = \lim_{T\to\infty}\int_0^T x(t)y(t+\tau)$	Pressure Pressure Light	Bubbling condition Pressure wave velocity Particle velocity	Fan et al. (1981) Roy and Davidson (1990) Horio et al. (1980a)

Table 12 continued

Processing	Definition/explanation	Examples		
		Signal	Remarks	Reference
Power spectrum	$S_x(\omega) = \lim\limits_{T\to\infty} \dfrac{1}{T}\|X_T(\omega)\|^2$	Pressure	Bubbling condition	Fan et al. (1981)
		Pressure	Solid structure in a CFB	van den Schaaf et al. (1999a)
		Pressure	Phase transition	Horio and Morishita (1988)
		Heat transfer	Dynamics of a CFB	Li et al. (1993)
		Pressure (Maximum entropy method, MEM)	Analysis of long range force	Mehrabi et al. (1997)
Fractal analysis	Self similarity, evaluated by fractal dimension	Image	Cluster shape	Ito et al. (1994)
		Pressure	Chaos analysis	Fan et al. (1990) Ross-Pence (1997)
		Solid momentum probe signal	Chaos analysis	Bai et al. (1996)
Chaos analysis	Evaluated by Kolmogorov entropy, correlation dimension, Lyapunov exponent, presence of the chaotic attractor, Hurst coefficient, etc.	Pressure, light, γ-ray porosity, hot wire prove signal, etc.	Quantification of the dynamics and phase transition in the BFB, CFB, mixing chamber, etc.	Schouten and van den Bleek (1992), Marcocchella et al. (1997), Ji et al. (2000), Huilin et al. (1995) Briens et al. (1997)
Wavelet transform	Analysis of time-varying signal simultaneously from both time and frequency perspective $W_x(\tau, a) = \dfrac{1}{\sqrt{a}}\displaystyle\int_{-\infty}^{\infty} x(t)\Psi((t-\tau)/a)\, dt$	cf. Table 15.13	cf. Table 15.13	cf. Table 15.13

3.1.2 Cross-correlation

Cross-correlation, defined in Table 12, provides information about the statistical mean delay time between two signals if there is any correlation between them. Figure 35 shows an example of a bed pressure fluctuation signal in a bubbling fluidized bed. From cross-correlation shown in Fig. 35b the obtained mean delay time is 0.12 s.

The degree of interaction between two signals can also be determined from cross-power spectrum, coherence (van den Schaaf et al., 1999b), or cross-covariance function (Greon et al., 1997). In case of wide solid velocity fluctuations with periodic ups and downs, as in the case of particles or clusters, the cross-correlation cannot give the right solid velocity; a peak search algorithm with a time lag determination for every peak of the original signal is required to obtain the proper results.

3.1.3 Power Spectrum

The dominant frequency of a fluctuating signal is obtained from a Fourier transform power spectrum. Figure 36 shows an example of a pressure fluctuation

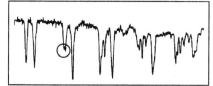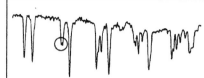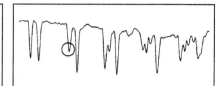

Figure 31 The moving average with second and third polynomial adaptive window. Left: original; center: 7 points averaging; right: 21 points averaging. (Minami, 1986.)

power spectrum in which different dominant frequencies are observed.

3.1.4 Maximum Entropy Method

The maximum entropy method (MEM) is developed to obtain the maximum spectrum information from the limited number of data. It enables us to estimate the power spectrum without an FFT (fast Fourier transform) using a distinct Fourier transform (DFT). The main problems in the FFT method are the so-called spectrum leaks from other frequencies, i.e., in addition to the true range of frequencies, the power spectrum also contains components at other unwanted frequencies, which leads to errors in spectral analysis. To demonstrate the spectrum leaks associated with FFT, suppose that an original continuous signal is the one shown in Fig. 37a. Its Fourier transform power spectrum is shown in Fig. 37b and has one sharp peak. From the limited number of data (c) the FFT is obtained as (d), which is still similar. However, from another set of limited data but half a period longer than the data (c), we obtain a spectrum with a few small peaks. This is the spectrum leak. To cope with this, a windowed Fourier transform shown in (g) with a lenslike window has to be applied to improve the spectrum to (h).

As already mentioned the MEM gives a frequency-leak-free spectrum without FFT. Furthermore, for

signals having sharp spectrum peaks, the MEM gives a sharper spectrum than FFT, as shown in Fig. 38.

3.1.5 Wavelet Transform

By a wavelet transform it becomes possible to detect unusual as well as periodical signals. Several kinds of wavelet transforms have been proposed, including continuous and discrete transforms. Examples are summarized in Table 13.

Gabor wavelet transform, illustrated in Fig. 39a, is one of the continuous transforms, in which the following wavelet, $\Psi((t - \tau)/a)$, is used:

$$\Psi\left(\frac{t - \tau}{a}\right) = \frac{1}{\sqrt{2\pi}\sigma} \cdot \exp\left[-0.5\left(\frac{t - \tau}{a\sigma}\right)^2\right] \cdot \exp\left[\frac{-j\omega_0(t - \tau)}{a}\right] \quad (29)$$

The wavelet transform of a signal $x(t)$, $W_x(\tau, a)$, is defined by

$$W_x(\tau, a) = \frac{1}{\sqrt{a}} \cdot \int_{-\infty}^{\infty} x(t)\Psi\left(\frac{t - \tau}{a}\right) dt \quad (30)$$

where a is a scale defined by using the representative frequency of the signal ω_0, as

$$a \equiv \frac{\omega_0}{\omega} \quad (31)$$

$\Psi(t^*)$ satisfies the following relationships:

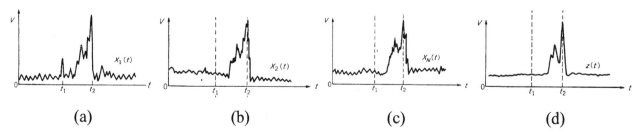

Figure 32 Effect of averaging on the shape of signal: (a) first, (b) second, (c) Nth, (d) final. (Minami, 1986.)

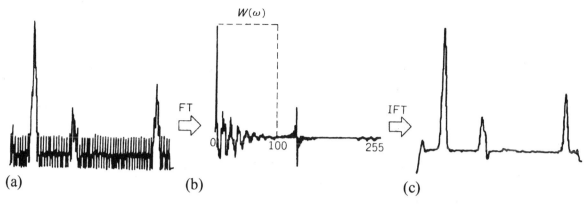

Figure 33 Elimination of high-frequency noises by low-pass filtering. (a) Original signal, (b) Real part of Fourier transform, (c) Processed signal. (Minami, 1986.)

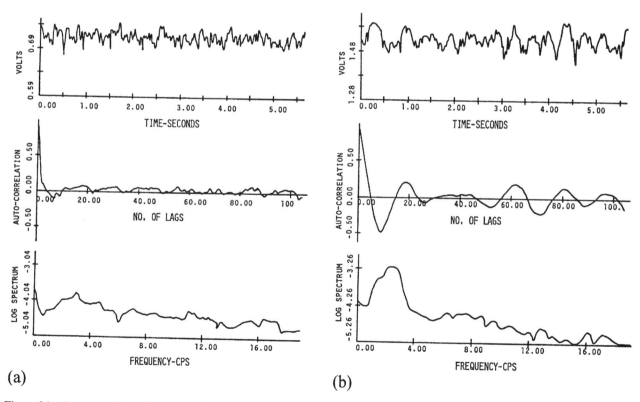

Figure 34 Autocorrelation function and power spectrum. Upper row: original pressure signal; middle row: autocorrelation (no. of lags $20 = 0.45$ s; sampling interval: 22.5 ms; data points: 1000; $D_t = 0.1$ m, $u_0 = 0.1$ m/s; bed material: silica sand, $d_p = 0.183$ mm); bottom row: power spectrum. (a) Smoothly fluidized bubbling bed ($L/D_t = 1$); (b) slugging bed ($L/D_t = 2$). (Broadhurst and Becker, 1976.)

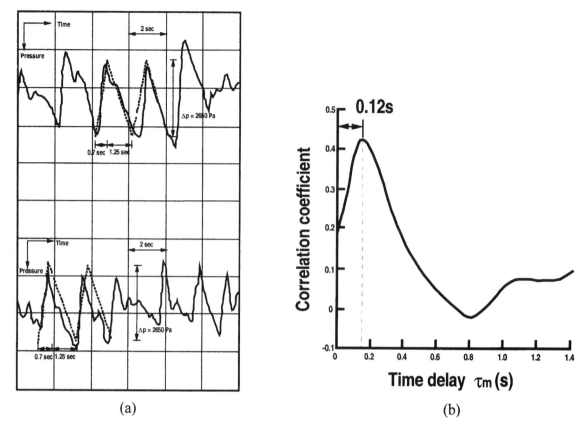

(a)

(b)

Figure 35 Pressure fluctuation signal from a slugging fluidized bed and their cross-correlation function. (Roy et al., 1990.) $A_t = 0.12 \times 0.12\,\mathrm{m}^2$; $L_{mf} = 1.6\,\mathrm{m}$; silica sand bed, $d_p = 240\,\mu\mathrm{m}$. (a) Original signal from pressure probes (upper row: 1.5 m above the distributor; lower row: 0.07 m above the distributor. --- is the theoretical signal from Kehoe and Davidson, 1973). (b) Cross-correlation function. (With permission of Elsevier Science)

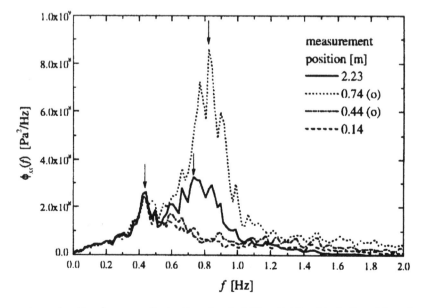

Figure 36 Power spectrum density of pressure fluctuations in a bubbling fluidized bed of sand (van den Schaaf et al., 1999a). D_t: 0.80 m; u_0: 0.44 m/s; d_p: 390 μm; settled bed height = 2.19 m. (With permission of American Society of Mechanical Engineers)

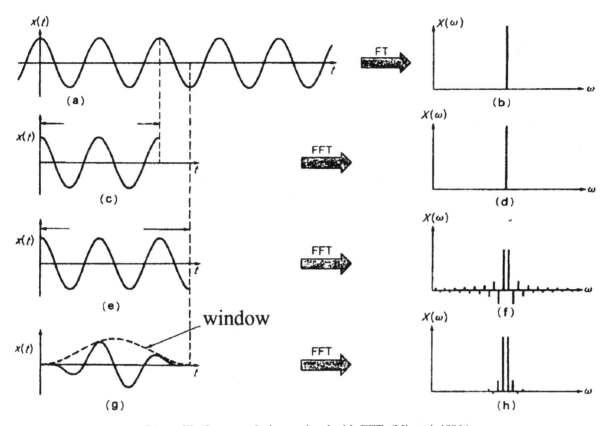

Figure 37 Spectrum leak associated with FFT. (Minami, 1986.)

$$\int_{-\infty}^{\infty} \Psi(t^*)\, dt^* = 0 \qquad (32\text{-}1)$$

$$\int_{-\infty}^{\infty} |\Psi(t^*)|^2\, dt^* = 1 \qquad (32\text{-}2)$$

The second exponential function in Eq. (29) is used in ordinary Fourier transforms, where no window is applied. In the Gabor wavelet transform a window around $t = \tau$ is superposed by the first part of Eq. (29) (i.e., by the error function). On the other hand, in the classical Gabor transform, not the Gabor wavelet transform, a windowed Fourier transform with a fixed width window is applied, as illustrated in Fig. 39b.

3.2 Visual Observation

3.2.1 Photo and Video Imaging

For visual observation through a column wall of a cold test rig, particle adhesion to the wall by static electricity should be reduced. By putting charge absorbers into the bed (Ilias et al., 1988; Chang and Louge, 1992), by coat-ing the riser surface by a conductive material (Myler et al., 1986), or by covering the inside wall with cellophane tape (Horio and Kuroki, 1994; Tsukada et al., 1997), the adhesion trouble can be avoided.

Observation of a high-temperature fluidized bed can be easily done in a lab scale gold mirror furnace consisting of a quartz tube, a gold-coated Pylex tube, and a coil heater between them. If temperature difference is created in the bed by some reactions it can be well observed from the outside as, e.g., done by Horio et al. (1986c). To observe the inside of a hot fluidized bed, an endoscopic system can be used (Zevenhoven et al., 1999b). More information can be found in Horio et al. (1980b), Yang and Keairns (1982), Prins (1987), Yang and Chitester (1988), Takeuchi and Hirama (1991), Zou et al. (1994), and Hull and Agarwal (1995).

3.2.2 Laser Imaging

At low solids concentration the laser sheet imaging shown in Fig. 40a is useful in observing fluidized beds. However, the original images on the light source

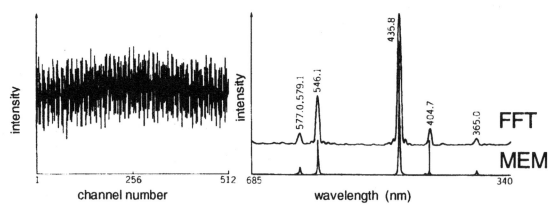

Figure 38 Spectrum of mercury lamp obtained by FFT and MEM. (Kawata et al., 1983.)

side are brighter than those on the other side, owing to the expansion of laser beam. This can be corrected in the following manner: suppose L and L' are the gray scales of a pixel on the original and modified picture, respectively, which are located at a distance r from the rod lens. $L' = Lr_0/r$ gives the corrected scale, where r_0 is the distance of a reference pixel in the image (cf. Fig. 40b).

Table 13 Application of Wavelet Transform to Fluidized Bed Measurement

Type of wavelet transform	Measurement	Derived information	Reference
Orthonormal wavelet transform	Pressure fluctuation	Decomposition of the signals into long-term and short-term correlation components; one represented by the self similarity Hurst's parameter of fractional Brownian motion originated by bubble, and another represented by the intensity parameter of Gaussian white noise originated by gas jetting, small bubble formation, and turbulence of fluidized beds, etc.	Hee et al. (1997)
Morlet wavelet	Heat flux	Self-similar bifurcation and trifurcation phenomena	Ross and Pence (1997)
Discrete wavelet transform based on Mallet and Zhang	Optical fiber probe signal	Phase separation (identification of the transition from the dense phase to the dilute phase)	Ren and Li (1998)
Discrete analog of wavelet transform (orthogonal wavelet basis functions by dilating and translating in discrete steps	Pressure fluctuation	Signal denoising	Roy et al. (1999)
Mallat's pyramidal algorithm used for computing one-dimensional orthogonal wavelet transform	Pressure fluctuation	Signal filtered by a wavelet; peak frequency and value corresponded to the bubble frequency and diameter, respectively	Lu and Li (1999)
Gabor wavelet	Signal from an acceleration sensor	Continuous transform with a Gabor wavelet [cf. Eq. (29)] applied to a commercial scale AFBC	Fukayama et al. (2000)

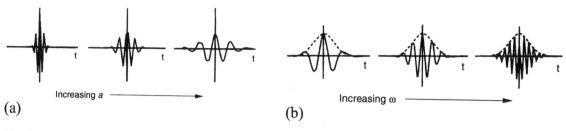

Figure 39 Examples of Gabor wavelet and Gabor function. (a) Gabor wavelet (dilated window); (b) Gabor function (fixed window width). (Sheng, 1996.) (With permission of CRC Press)

At higher solid fluxes, the core of the riser can be observed using a hood or an endoscope as shown in Fig. 40c,d. Examples of laser sheet images are shown in Fig. 40e.

3.2.3 Particle Image Velocimetry (PIV)

If particle images are distinct in a time series of visualized images, one can track the particle and determine its velocity and trajectory. This is the PIV technique. Rix et al. (1996) measured particle movement in the freeboard using PIV. Reese and Fan (1997) tracked local velocities of bubbles and particles in gas–liquid and gas–liquid–solid fluidized beds using a continuous mode argon ion laser as a light source, and a CCD camera for image recording. Assuming linear movement during the time interval between consecutive images, particles and bubbles were tracked and their velocities determined. More information on this technique can be found in Chen and Fan (1992) and in Fan (1995). An experimental setup used by Chen and Fan (1992) is shown in Fig. 41.

3.3 Computer Tomography (CT) and Image Reconstruction

Tomographic visualization is based on numerical solution of a set of linear algebraic equations corresponding to the transmission intensity of high-energy electromagnetic radiation such as x- or γ-rays or other beams (visible light, positrons, neutrons, ultrasound, magnetic resonance), and/or field parameters such as impedance, inductivity, or capacitance.

3.3.1 Image Reconstruction Principle

As for an example, let us suppose that for the construction of a tomographic image we use light of initial intensity I_0. According to the Lambert–Beer law, the intensity of light after transmission through an object of thickness L can be written as

$$I_L = I_0 e^{-\mu\rho L} \tag{33}$$

where μ is the mass attenuation coefficient, ρ is the medium density, and accordingly $\mu\rho$ is the linear attenuation coefficient of the medium the beam passes through. L is the thickness, and the term $\mu\rho L$ is the absorptance of the medium.

Now, let us introduce cells from 1 to m in the objective cross section as shown in Fig. 42a. The cross section is scanned by number of prefix beams i $\sim$ n, and the average passage length for cell j was I_{ij} (for those cells where beam i does not pass $I_{ij} = 0$). The first medium fraction in cell j expressed by ε_j is the main unknown variable. Then the total absorbance I_{Li}/I_0 can be written as

$$\frac{I_{Li}}{I_0} = \exp\left(\sum_j \{\mu_1\rho_1\varepsilon_j + \mu_2\rho_2(1-\varepsilon_j)\}l_{ij}\right) \tag{34}$$

By rearranging the above expression we obtain

$$\sum_j (\mu_1\rho_1 - \mu_2\rho_2)l_{ij}\varepsilon_j = \ln\left(\frac{I_{Li}}{I_0}\right) - \mu_2\rho_2 L_i \tag{35}$$

Thus we have a set of linear algebraic equations to obtain ε_j:

$$(a_{ij})(\varepsilon_j) = (b_i) \tag{36}$$

where

$$a_{ij} = (\mu_1\rho_1 - \mu_2\rho_2)l_{ij} \tag{37}$$

$$b_i = \ln\left(\frac{I_{Li}}{I_0}\right) - \mu_2\rho_2 L_i \tag{38}$$

Now let us study a more general case in which the distribution of attenuation coefficients, described by a

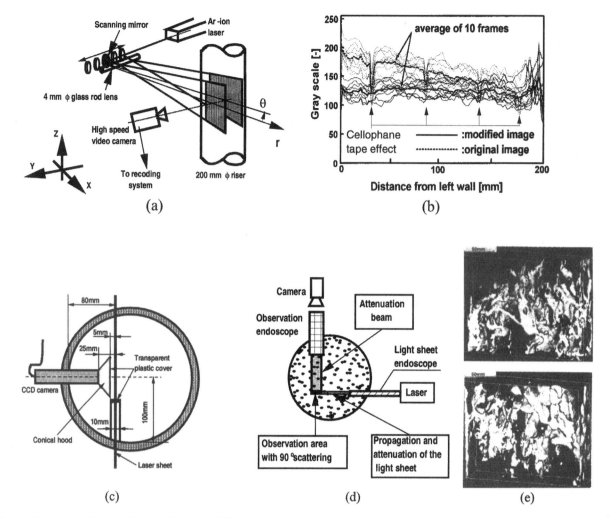

Figure 40 Laser sheet technique, image modification, and laser sheet picturing in dense suspension. (a) Scanning laser sheet imaging (Tsukada et al., 1997); (b) horizontal profiles of gray scales of the images before and after image modification ($r_0 = 0.5$ m, cellophane tape correction by interpolation) (Tsukada et al., 1997); (c) hood system of Kuroki and Horio (1994); (d) endoscopic system of Werther and Rudnick (1996, courtesy of Prof. Werther); (e) 3D image obtained from scanning laser sheet images (Horio and Ito, 1997)—upper: turbulent bed ($u_0 = 0.58$ m/s; $G_s = 0.045$ kg/m^{2}s), and lower: fast bed ($u_0 = 1.1$ m/s, $G_s = 0.087$ kg/m^{2}s).

function $f(x, y)$, is projected through a line L as defined in Fig. 42b. The measured value $P(l, \theta)$ (θ is the projection angle and l is the length, as in Fig. 42b) can be written

$$P(l, \theta) = \int_L f(x, y) \, ds \tag{39}$$

When $f(x, y)$ is a continuous function with a continuous first derivative, Radon (1917) reported that $f(x, y)$ can be reconstructed from an infinite set of line integrals $P(l, \theta)$ by

$$f(x, y) = \frac{1}{4\pi^2} \int_0^{2\pi} \int_{-\infty}^{\infty} \frac{-l}{l - x \cos \theta - y \sin \theta} \cdot \frac{\partial P(l, \theta)}{\partial l} \, dl \, d\theta \tag{40}$$

where $(l - x \cos \theta - y \sin \theta)$ is the perpendicular distance of the point (x, y) from the line l.

In reality the reconstruction is always done as an approximation of Radon's solution for a certain number of projections. As shown in Fig. 42b, for each angle of projection θ, the value of $P(l, \theta)$ depends on the

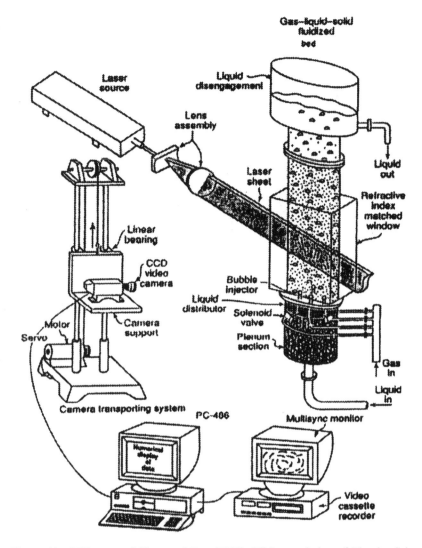

Figure 41 PIV setup of Chen and Fan (1992). With permission of Elsevier Science.

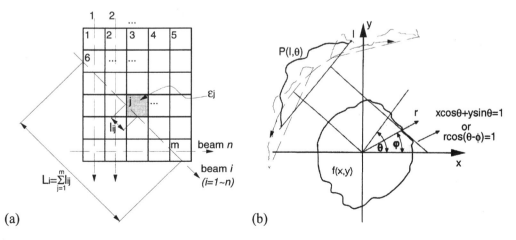

Figure 42 An algebraic approach to image reconstruction in CT (a) and the geometry for CT image reconstruction (b). (Adopted from Kumar and Dudukovic, 1997.) With permission of Elsevier Science.

distance on line l. The back-projected function, $b(x, y)$, for a certain θ is

$$b(x, y) = \int P(l, \theta) \delta(x \cos \theta + y \sin \theta - l) \, dl \qquad (41)$$

where δ is the Dirac delta function.

The complete back-projection, being an approximation of the function $f(x, y)$, is the sum of $b(x, y)$ corresponding to all angles θ, at which the projections were made:

$$\begin{aligned} f(x, y)_{\text{approx}} &= \int_0^{\pi} b(x, y) \, d\theta \\ &= \int_0^{\pi} \int P(l, \theta) \, \delta(x \cos \theta + y \sin \theta - l) \, dl \, d\theta \end{aligned}$$

$$(42)$$

Normalization of $f(x, y)_{\text{approx}}$ is achieved by multiplying by $(-1/2\pi)$.

The quality of image reconstruction depends on the number of projections and detectors, the speed of measurement and of the data acquisition system, and the homogeneity of the measurement volume. More on image reconstruction can be found, e.g., in Kak and Slaney (1988); Kalender (2000); and Kumar and Dudukovic (1997).

The first reported application of tomography was probably that of Grohse (1955), who investigated the variation in density of a bed of silicon powder as a function of fluidizing velocity. At almost the same time Bartholomev and Casagrande (1957) used Co^{60} as γ-ray source to monitor a $20.4''$ i.d. FCC unit and detected the γ-rays by a Geiger–Muller tube.

In capacitance tomography, the measurement setup consists of an array of electrodes installed around the external surface of the monitored system, and measurements of capacitance between various pairs of electrodes are performed, so that data is received on the capacitance in various directions within the system. Then, from the algorithm described above, the volume fraction of solids or gas within the system can be calculated. Examples of various sources for fluidized bed CTs are presented in Table 14. An example setup and results are shown in Fig. 43.

Positron emission tomography (PET) uses radioactive tracers that decay via the emission of a positron. It results in the production of two back-to-back γ-rays, from which the tracer position can be determined. More information can be found, e.g., in Hawkesworth et al. (1991). Another noninvasive

method is nuclear magnetic resonance (NMR) imaging. The measurements are conducted based on the paramagnetic properties of atom nuclei. During the experiments, external signals (i.e., radio frequency pulses and magnetic field gradient pulses) interact with the nucleus spin system positioned in a static magnetic field. NMR imaging enables us to observe dense flows, but it can only be performed with water-containing systems. Details can be found, e.g., in Pangrle et al. (1992), Shattuck et al. (1997), and Sederman et al. (2001).

An advantage of tomography is its capability of monitoring a relatively large volume at the same moment. However, safety considerations concerning particularly the emission of x-rays or γ-rays during tomographic scanning have to be kept in mind. Using x-rays instead of γ-rays enables adjusting the energy of the emitted radiation by changing the voltage of cathode of an x-ray generator. A disadvantage of the tomographic imaging is the difficulty of transforming and analyzing the huge amount of data from the scanner, but faster computers should solve this difficulty.

3.3.2 Neutron Imaging

The interiors of fluidized bed systems can also be observed by application of the neutron imaging technique. This method can provide fast monitoring of relatively large cross-sectional areas at high resolution. The disadvantage of neutron imaging is the necessity of taking special precautions concerning the application of neutrons and the necessity of using a special neutron source. By this method Onodera et al. (1998) observed the flow pattern around the tube and bubble movement within a fluidized bed. Tasdemir et al. (1999) used a panoramic camera and image segmentation to study mixing in a $0.2 \times 0.2 \, \text{m}^2$ cross-section fluidized bed consisting of three types of particles of various colors. Ozawa et al. (1998) applied neutron radiography visualization to a rectangular $30 \times 60 \times 100 \, \text{cm}$ fluidized bed. Since they used sand as the bed solids (sand is not able to capture the neutrons itself), 0.3% of specially prepared sand (i.e., coated with cadmium) was added to capture the neutrons. More on neutron imaging can be found, e.g., in Fredd et al. (1997) and Umekawa et al. (1999a).

3.4 On-Line Particle Size Distribution Measurement

Techniques for in-situ determination of particle size and shape have been developed in the field of granula-

Table 14 Examples of Fluidized Bed CTs

Author(s)	Source Type	Remarks
Weimer et al. (1985)	γ-ray	Monitoring the voidage along the height of a FB using Cs^{137} as radiation source
Seville et al. (1986)	γ-ray	Studies on voidage distribution above the gas distributor in a 0.146 m i.d. fluidized bed of sand
Seo and Gidaspow (1987)	γ-ray, x-ray	Voidage measurement in a FB using simultaneous densitometry
Weinstein et al. (1984)	x-ray	Application of tomography in a CFB, using an optical densitometer to analyze the images
Weinstein et al. (1992)	x-ray	Studies on the distribution of solid fraction in a fast fluidized bed system (d = 0.15 m)
Contractor et al. (1992)	x-ray	Observation of images from fluidized beds using x-ray source and phototransistors
Kantzas (1994)	x-ray	Images of density distribution and gas holdup obtained for fluidized bed of glass beads and polyethylene fluidized by nitrogen
Holoboff et al. (1995)	x-ray	CT scanner used to study variability of gas voidage in a polyolefin/air fluidized bed of 0.1 m i.d
Durand et al. (1995)	x-ray	Imaging technique combined with video camera to study the hydrodynamics of a polyethylene FB
Fiorentino and Newton (1998)	x-ray	Identifying of scale-up issues for predicting large-scale BP reactor performance
Grassler and Wirth (1999)	x-ray	Application of a 60 kV CT scanner to a cold 0.19 m i.d., 15 m high CFB, results obtained with 0.2 mm resolution
Kai et al. (2000)	x-ray	Dynamic imaging of cross-sectional voidage distribution in every 4 ms by a fast scanning system consisting of 18 x-ray sources and 122 detectors
Huang et al. (1989)	Capacitance	Studies on voidage distribution in the fluidized bed
Halow and Nicoletti (1992)	Capacitance	Studies on voidage distribution in the fluidized bed, bed diameter 0.15 m, sensing electrodes installed round the riser in four rings, each containing 32 electrodes
Dyakowski et al. (1997)	Capacitance	Application to CFB, good agreement with comparison of voidage calculated from pressure transducers
Rhodes and Wang (1999)	Capacitance	Studies on distribution of solids volume fraction in a 0.09 m i.d riser, four sets of sensors applied, each sensor contained 12 electrodes

tion (e.g., Tanino et al., 1993). There are three different methods of monitoring the granule size distribution. In the easiest method, the granules are sampled by compressed air and captured on the adhesive tape. Then the particle image on the adhesive tape is taken with a CCD camera, and finally the sampling tube is cleaned by air, the spent part of the adhesive tape is wound up, and new samples are sucked into the tube. In another system, laser scattering of sampling frequency up to 50 Hz is applied for size determination. Since the light scattered by the particles in the measurement volume has a scatter angle unique for the particle diameters, the particle size can be determined by a coaxial multi-circular detector. Another type of instrument provides real images taken by a CCD camera in front of the probe. The measurement system consists of illumination, purging, and telescope devices, and the image processing for determining the mean particle size can be done relatively fast. More information is given, e.g., in Fluidization Handbook (1999).

4 FLUIDIZED BED DIAGNOSTICS

Combined with the proper knowledge of fluidization systems, the instrumentation and measurement techniques can provide us with much meaningful information. Since information in other chapters should be

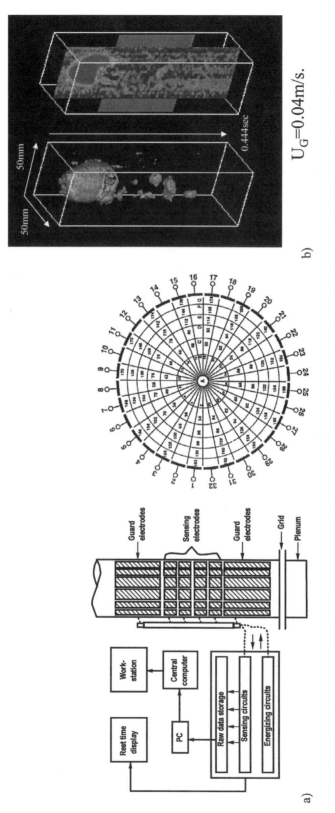

50mm

50mm

0.444sec

$U_G=0.04m/s.$

b)

a)

Guard
electrodes

Sensing
electrodes

Guard
electrodes

Grid

Plenum

Work-
station

Central
computer

Rest time
display

PC

Raw data storage

Sensing circuits

Energizing circuits

Figure 43 CT scanning system of Halow et al. (1993) (a) (with permission of Elsevier Science) and pseudo 3D image of bubbles and their interface structure (Kai et al., 1999.) (b)

referred to for this purpose, only fundamental remarks are presented in this section.

4.1 Diagnostics of Bubbling Bed

4.1.1 Bed Diagnostics by Pressure Measurements

The relationship between bed pressure drop Δp and superficial gas velocity, as well as the pressure distribution along the bed determined by multiple pressure sensors on the column wall, can provide fruitful information on the fluidizing conditions. As illustrated in Fig. 44, the longitudinal pressure distribution curve can indicate: (1) the bed surface location and fluidized solids holdup, (2) segregation of bed materials, and/or (3) the height of the not completely fluidized region above the distributor. The Δp vs. u_0 relationship can of course also indicate (1) the minimum fluidization velocity of the solids in the bed, (2) coexistence of defluidized coarse/heavy particles, and (3) existence of channels or rat holes.

Pressure fluctuation can provide information on bubbling conditions and slugging. For slugging, the frequency f_s, of pressure fluctuation is the frequency of slugs passing by the pressure probe. Suppose slug length is approximately equal to the column diameter, we can write

$$f_s = \frac{(u_0 - u_{mf})(\pi D_t^2/4)}{V_s} = \frac{u_0 - u_{mf}}{D_t} \qquad (43)$$

where D_t is column diameter and V_s is slug volume.

On the other hand, the major frequency f_n of spontaneous oscillation of the system composed of the plenum chamber capacity, the distributor resistance, and the bed inertia, particularly when the distributor resistance is small, is given by (Moritomi et al., 1980)

$$\omega_n = 2\pi f_n = \sqrt{(P_D A_t)/(\rho_p (1 - \varepsilon_{mf}) L_{mf} V_d)} \qquad \text{rad/s}$$

$$(44)$$

where A_t, L_{mf}, P_D, V_d, ε_{mf}, and ρ_p are bed cross-sectional area, bed height at the minimum fluidization condition, time averaged plenum chamber pressure, plenum chamber volume, bed voidage at the minimum fluidization condition, and particle density, respectively.

Pressure fluctuation can be also used to monitor the gas–solid contact mode and to adjust the particle size distribution. Figure 45 is a good example in a catalytic reactor (Ikeda, 1972).

Bed pressure drop can be measured either directly by a pressure tap placed right above the gas distributor such as a vertical pressure probe, a wall pressure tap and/or a distributor buried type, or indirectly from the pressure of the plenum chamber by subtracting the distributor pressure drop measured separately. Pressure taps on the freeboard wall are also necessary in case the gas outlet pressure drop cannot be negligible. To avoid gas dynamic pressure the hole of a pressure tap must face parallel to the main gas flow. To prevent particle entering, one needs to use some filtering material such as a wire gauze (Figs. 46a and 46c) or to slant the pressure tap tube (Fig. 46b). Figure 46d shows immersion type probes, which can be traversed to measure the pressure distribution.

4.1.2 Defluidization Velocity

Defluidization is caused by the accumulation of large or heavy particles in the bed owing to the increasing size of the solids because of agglomeration, coating, or the feed of a large size fraction. Defluidization is serious in commercial-scale operations (e.g., Anthony, 1995; Souto et al., 1996; Skrifvars et al., 1997) and its symptom has to be detected as early as possible. The defluidization can be detected by monitoring the pressure drop and its fluctuation in the bed. When defluidization takes places, the pressure drop suddenly decreases, as reported by Siegell (1984). However, much before that, the accumulation of large particles at the bottom of the bed can be detected by the decrease of the pressure drop and the longitudinal pressure gradient. The temperature of defluidization depends strongly on gas velocity, as reported, e.g., by Siegell (1984).

4.1.3 Bubble Fraction and Visible Bubble Flow Distribution

The local bubble fraction can be determined by either an optical or a capacitance probe. Suppose that a probe output as shown in Fig. 47 is obtained. By introducing a threshold value, which has to be carefully determined by the real eye observation and/or material balance, the time can be divided into bubble passage periods ΔT_{bi} ($i = 1, 2, \ldots$) and emulsion phase passage periods ΔT_{ei}. The sum of bubble passage periods divided by the total observation period $(\Delta T_{bi} + \Delta T_{ei})$ corresponds to the bubble fraction ε_b:

$$\varepsilon_b = \frac{\sum \Delta T_{bi}}{\sum (\Delta T_{bi} + \Delta T_{ei})} \qquad (45)$$

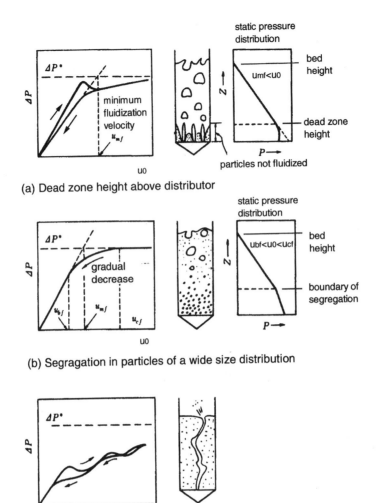

(a) Dead zone height above distributor

(b) Segragation in particles of a wide size distribution

(c) Channeling

Figure 44 Determination of fluidization regime from pressure data. ΔP^*: theoretical pressure drop = (bed weight-buoyancy force)/(total cross section).

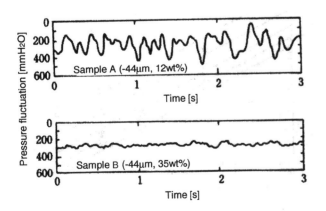

Figure 45 Pressure fluctuations versus time. (Ikeda, 1972.)

The local visible bubble velocity u_{vb} is given by multiplying the voidage by bubble velocity u_b measured by the two needle probes or by a borescope or other image sensors so that

$$u_{vb} = \overline{u_b \varepsilon_b} = \frac{\sum u_{bi} \Delta T_{bi}}{\sum (\Delta T_{bi} + \Delta T_{ei})} \quad \text{(local time average)} \tag{46}$$

By traversing the point probe across the bed one can obtain the bubble flow distribution.

4.1.4 Bubble Size Calculation from Pierced Length

The length of the part of a bubble pierced by a probe l_{bi} (pierced bubble length) is:

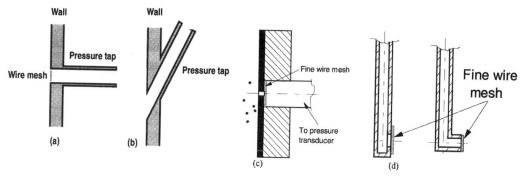

Figure 46 Various types and arrangements of pressure measurement tubes.

$$l_{bi} = u_{bi} \Delta T_i \tag{47}$$

However, l_{bi} does not immediately give the bubble diameter. To obtain the real bubble diameter, D_b, from a series of data for l_{bi}, let us first assume (as a simplest case) that the bubbles are spherical and uniform in size and that they are uniformly distributed radially across the bed. The probability, $\phi(D_b, l_b) \, dl_b$, of having a pierced bubble length within the interval of $l_b \sim l_b + dl_b$ for bubbles of diameter D_b is $8r|dr|/D_b^2$, where the radius, r, is defined by $r^2 = (D_b^2 - l_b^2)/4$ (accordingly we have $r \, dr = -l_b \, dl_b/4$). Then, since $8r|dr|/D_b^2 = 2l_b dl_b/D_b^2$, and since l_b is less than D_b, the probability distribution density function for bubble size $\phi(D_b, l_b)$ is given by

$$\phi(D_b, l_b) = \frac{2l_b}{D_b^2} \quad \text{for} \quad 0 < l_b < D_b \tag{48-1}$$

and

$$\phi(D_b, l_b) = 0 \quad \text{for} \quad l_b < 0 \quad \text{or} \quad D_b < l_b \tag{48-2}$$

Since $\phi(D_b, l_b)$ has to satisfy the normalization condition, we can write

$$\int_0^{D_b} \phi(D_b, l_b) dl_b = 1 \tag{49}$$

Thus the time averaged pierced length is given by

$$\overline{l_b} = \int_0^{D_b} \phi(D_b, l_b) l_b dl_b = \frac{2}{3} D_b \tag{50}$$

Finally by rearranging the above expression, the bubble size is given by

$$D_b = \frac{3}{2} \left(\frac{\sum_{i=1}^N l_{bi}^3}{N} \right)^{1/3} \tag{51}$$

where N is the number of bubble samples.

For a more general case, where the bubble size is distributed (Tsutsui and Miyauchi, 1979), let us assume a distribution density function for bubble diameter, $\Phi(D_b)$. In this case, $\phi(D_b, l_b)$ introduced above in Eq. (48) becomes a conditional probability to get a pierced length l_b from a bubble of size D_b. What we need is the probability density distribution

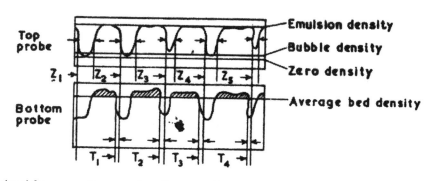

Figure 47 Type of signal from capacitance probes. (Lanneau 1960.) With permission of the Institution of Chemical Engineers.

of pierced length l_b for the case where D_b is distributed, $\phi^*(l_b)$. $\phi^*(l_b)$ can be related to $\Phi(D_b)$ and $\phi(D_b, l_b)$ by

$$\phi^*(l_b)\,dl_b = \int_{l_b}^{\infty} \phi(D_b, l_b)\,dl_b \Phi(D_b)\,dD_b \tag{52}$$

Dividing both sides by dl_b, we obtain

$$\phi^*(l_b) = \int_{l_b}^{\infty} \phi(D_b, l_b)\Phi(D_b)dD_b \tag{53}$$

Differentiating it by l_b, we obtain

$$\frac{d\phi^*}{dl_b} = -\phi(l_b, l_b)\Phi(l_b) \tag{54}$$

Finally, substituting the above expression into the normalization condition

$$\int_0^{\infty} \Phi(D_b)\,dD_b = 1 \tag{55}$$

we obtain

$$\Phi(D_b) = \frac{\left(\dfrac{1}{\phi(D_b, D_b)}\dfrac{d\phi^*(D_b)}{dD_b}\right)}{\int_0^{\infty} \dfrac{1}{\phi(l_b, l_b)}\left(\dfrac{d\phi^*}{dl_b}\right)dl_b} \tag{56}$$

where $\phi^*(l_b)$ is measured and $\phi(D_b, l_b)$ can be calculated for a bubble of particular shape. Then the average bubble size is given by

$$\overline{D_b} = \left(\int_0^{\infty} \Phi(D_b)D_b^3 dD_b\right)^{1/3} \tag{57}$$

If bubbles are not spherical, the same procedure can be applied by changing $\phi(D_b, l_b)$. Cluster size measurements can be done on the same principle, but they have a more complicated structure.

4.1.5 Solid Circulation and Mixing

Solid tracers and probes discussed in Secs. 2.11, 12 can be applied to determine solid circulation and mixing. If real solids sampling is conducted, bed sectionizing is a useful tool—after a certain fluidization period the gas is cut off with a sequencer and the partitioning device is dropped, so that it can quickly reach the bed bottom. Then the bed particles are sampled section by section by a vacuum cleaner (Horio et al., 1986a).

4.1.6 Gas Sampling from the Bed

To sample gas from fluidized beds it is important to know if the sample is taken from the bubble phase,

emulsion phase, and/or cluster or lean phase. There are elaborated probes with an pressure sensor to detect the phase from which the samples are taken, as discussed in Sec. 2.13. Another option is the effective removal of solids from the gas–solid suspension by using a small cyclone. Ceramic or metal filters can also be applied. However, some precaution must be taken in order not to accumulate too much chemically active solid on the filter.

4.2 Diagnostics of Circulating Fluidized Bed

4.2.1 Solid Holdup Distribution

Solid holdup in a CFB can be determined from the time-averaged pressure gradient, $\partial p/\partial z$, based on Eq. (58) applicable except for the acceleration region (the latter should not be too high in circulating fluidized beds, cf. Horio, 1997):

$$\frac{\partial p}{\partial z} = -\rho_p(1-\varepsilon)g \tag{58}$$

where z (m) is height, while g (m/s^2), ρ_p (kg/m^3), and ε are gravity acceleration, particle density, and bed voidage, respectively.

4.2.2 Solid Circulation Rate

Overall solids mass flux is one of the key parameters to determine the regime of circulating fluidization. It also has a great influence on the pressure profile along the riser, as well as voidage and mixing. Table 15 summarizes the previous methods of measuring overall solids mass flux.

Solids circulation rate can be measured directly at the downcomer by devices such as those shown in Fig. 48. Then the overall solid mass flux is obtained by

$$G_s = \frac{w_s}{A} \qquad \text{kg/m}^2\text{s} \tag{59}$$

where w_s is the solids mass flow rate and A is riser cross section.

The solids circulation rate can also be determined from the observed local time-averaged solids velocity, v_p, and solids concentration, $(1-\varepsilon)$, by the following integration over the bed cross section:

$$W_s = \iint_A \rho_p(1-\varepsilon)v_p\,dA \tag{60}$$

If the local pressure is measured in parallel with overall solids circulation, the cross-sectional average solids concentration $(1-\varepsilon)$ can be determined knowing the value of G_s, particle density ρ_p, and velocity v_p. Otherwise, local solids flux can be determined by sucking solids from the riser as described in detail in Sec. 2.12. More on the techniques for measuring solid mass fluxes can be found, e.g., in Burkell et al. (1988).

4.3 Diagnostics of Particle Characteristics

Although particle characterization is not an ordinary instrumentation issue, some supporting laboratory scale equipment helps even the commercial operation by defining the realistic nature of the particles in each occasion. Thus it is recommended to have a small, say 5 cm in diameter, cold test unit with a flowmeter and pressure sensors.

4.3.1 Minimum Fluidization Velocity

Minimum fluidization velocity is determined from bed pressure drop vs. gas velocity relationship for decreasing gas velocity, as already described in Fig. 44a. It should be recognized that the sample must be taken carefully so that it represents the real solids in the bed. In many cases the bed material is already quite different from the one fed to the bed by reaction, attrition, or agglomeration. Bed pressure drop has to be measured for decreasing velocity from a completely fluidized condition. When bed pressure drop is measured for increasing gas velocity, pressure drop vs. gas velocity curves show higher pressure because of changes of particle interaction and bed structure and do not give a reproducible and generally applicable value of minimum fluidization velocity.

Table 15 Various Solids Mass Flux Measurement Methods

Method	Author(s)	Remarks
Installation of a butterfly or slide valve(s) in the downcomer	Lasch et al. (1988) Bader et al. (1988) Hartge et al. (1988) Beaud and Louge (1995)	Measurements based on periodical opening/closing of the valve(s) Difficult to apply to hot FB due to valve requirements and its damage possibility at high temperatures
Disc valve	Ye et al. (1999)	Measurements of solids mass flux in a CFB of cast steel
Three-way valve technique	Arena et al. (1988, 1993)	Measurements based on switching over the three-way valve (placed outside the hot reactor) and accumulating solids in a vessel through the certain time period. No continuous measurements possible
Two ring sensors installed in the CFB recirculating line	Dybeck et al. (1995)	Measurement method based on effect of electrostatic induction of charged particles on a surrounding metallic ring
	Qi and Farag (1995)	Measurements of solids fluxes in a 0.14 m i.d. CFB riser of glass beads and FCC. The solids were collected in each probe for a given time period and then weighed. Local particle concentration was measured using the capacitance probe
Study on five ways of solids mass flux measurement systems	Burkell et al. (1988)	Study on application of closing valve in the return leg, observation of an identifiable particle, using a device to record the force imparted by returning solids form the cyclone, measuring the pressure drop across the constriction in the return loop, and estimating solids mass flow from the heat balance on a calorimetric section in the standpipe
Semicontinuous solids mass flux meter	Horio et al. (1992b)	Application of load cell. The solids were accumulated in the device through a certain time period. Difficult application to hot units due to load cell requirements
	Patience et al. (1990)	Idea based on measurements of pressure drop and gas velocity. Assuming that gas–solid slip velocity is related to the terminal velocity of individual particles, the mass flux was calculated.
	Davies and Harries (1992)	Idea based on application of weighing chamber and an electronic balance
	Kobylecki and Horio (2002)	Scoop-like device measuring continuously the mass fluxes of various solids including group C

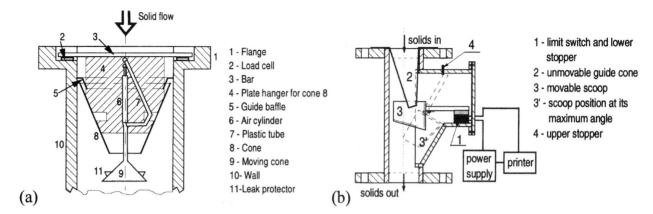

Figure 48 Solids mass flow meters: (a) Horio et al. (1992b), (b) Kobylecki and Horio (2002).

Whether a bed is completely fluidized or not can be examined from the bed pressure drop at the velocity slightly higher than the minimum fluidization velocity. If it is, the pressure drop should equal the bed weight divided by the bed cross section when gas density is sufficiently low that the buoyancy force acting on particles can be neglected (cf. ΔP^* in Fig. 44).

If the particles have a wide size distribution, the pressure drop vs. gas velocity curve shows a rather vague transition from a fluidized condition to the fixed bed condition, as already shown in Fig. 44b. From this figure one can obtain gas velocities corresponding to the beginning of fluidization u_{bf} and to the completion of fluidization u_{cf}.

4.3.2 Agglomerating Behavior of Cohesive Powders

As shown in Fig. 49, channel formation and its collapse and transition to the agglomerating fluidization can be detected from the cyclic response of the bed pressure drop. However, it should be noted here that in the cohesive powder fluidization the wall effects are quite important, and the scale effect related to them cannot be neglected.

Very fine particles are fluidized forming agglomerates. Minimum fluidization velocity of such agglomerates $u_{a,mf}$ can be measured in the same manner as in the case of noncohesive particles, i.e., from the decreasing velocity period (period after #7 in Fig. 49). A particle recycling device, such as a cyclone or a filter, is necessary for fine powder experiments. From $u_{a,mf}$, the apparent agglomerate size d_a can be determined from minimum fluidization velocity correlations such as Wen and Yu's (1966) by using the particle density separately determined by mercury porosimetry or other methods.

The agglomerate size of Geldart's group C powders can be roughly determined by sieve analysis. In such cases, Ro-tap sieving may not be advisable because its action also helps agglomeration or attrition. If direct determination of agglomerate size is critical for operation, as, e.g., in spray granulation, the in situ particle size determination methods already introduced in Sec. 3.4 should be helpful.

4.3.3 Expansion Characteristics

High bed expansion is an important feature of powders suitable for good gas–solid contacting. To quantify the expansion characteristics, the bed contraction measurement test is recommended as a standard method in the laboratory. In this test the gas supply to the bed is stopped after a certain fluidization period and the change of bed height is recorded. During the measurements some attention has to be paid to the effects of plenum chamber volume and the distributor resistance on the result. Figure 50 shows the observed bed contraction curves from which the emulsion phase voidage can be obtained.

For Geldart's group A powders, the bed expands homogeneously even above the minimum fluidization velocity, u_{mf}, until the minimum bubbling velocity u_{mb}. The u_{mb} is another good index of high emulsion phase expansion, as shown in Fig. 51. This figure also illustrates how u_{mb} is determined.

4.3.4 Regime Transition Velocities

In addition to u_{mf} and u_{mb}, other regime transition velocities such as u_c, u_k, and u_{tr} should be important for high-velocity operation. Figure 52 shows the pressure fluctuation response to gas velocity change, from which u_c and u_k are determined. These characteristic

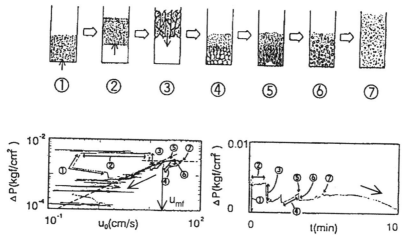

Figure 49 Pressure drop response and fluidization behavior of Al_2O_3 powder (5.0 μm) bed. (Nishii et al., 1993.)

transition velocities should be determined with a column where slugging can be avoided. Furthermore, to avoid criticism as for the early studies (Geldart and Rhodes, 1986), the axial solid holdup should be always monitored, so that the dense bed is kept in the column, preventing the pressure tap from exposure to the lean phase, which causes the decrease in pressure fluctuation at higher gas velocities and gives fictitious values for u_c and u_k. For this issue readers are recommended to check Horio (1997, pp 34–36). A careful evaluation of the regime transition shows that u_c, u_k, and u_{tr} are independent and unique characteristic velocities.

5 SUMMARY

Instrumentation and measurement techniques for fluidized bed processes are reviewed, starting from

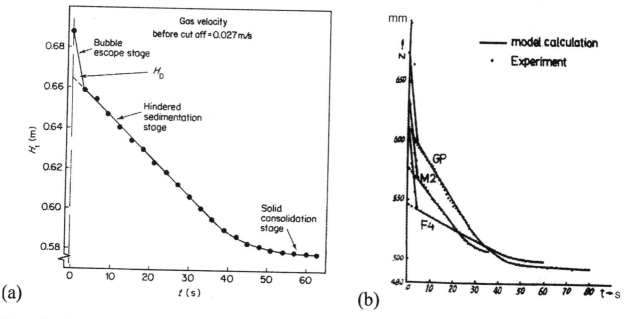

Figure 50 Fluidized bed collapse of group A powder (glass ballotini, 26 μm), reported by Geldart (1986) (© John Wiley & Sons. Reproduced with permission) (a), and collapsing of fluidized bed reported by Tung and Kwauk (1982) (b).

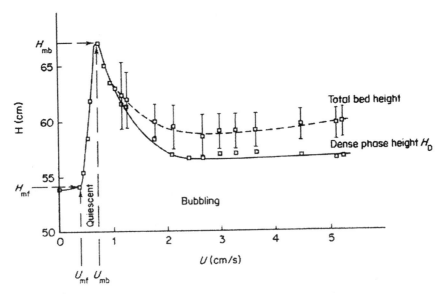

Figure 51 Expansion curve for group A powder (60 μm ballotini). (Abrahamsen and Geldart, 1980.) With permission of Elsevier Science.

fundamental knowledge on sensor elements, probe configuration, and tracers to more sophisticated image analysis systems. Data to evaluate sensing systems are included as much as possible for readers' convenience. In the last part of the chapter a practical approach to the diagnostics of fluidization systems as well as powders are discussed.

ACKNOWLEDGMENTS

The authors thank Ms. Malgorzata Kobylecka for her support in preparing the manuscript.

REFERENCES

Abrahamsen AR, Geldart D. Behaviour of gas-fluidized beds of fine powders, Part II. Power Technol 26:47–55, 1980.

Acree Riley C, Louge M. Quantitative capacitance measurements of voidage in dense gas–solid flows. Particulate Sci Technol 7:51–59, 1989.

Adams C. Gas mixing in fast fluidized beds. In: Basu P, Large J, eds. Circulating Fluidized Bed Technology II. Oxford: Pergamon, 1988, pp 299–306.

Akagi H, Furusaki S. Direct measurement of gas concentration in bubbles in a fluidized bed. Prep of the Meeting of Soc Chem Eng Japan. Shinshu, Japan, 1983, pp 46–47.

Ambler P, Milne B, Berruti F, Scott D. Residence time distribution of solids in a circulating fluidized bed: experimental and modeling studies. Chem Eng Sci 45:2179–2186, 1990.

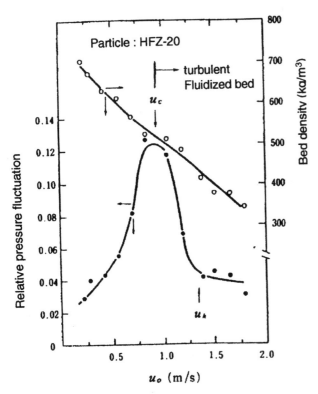

Figure 52 Pressure fluctuation vs. gas velocity. (Yerushalmi and Cankurt, 1979.) (With permission of Elsevier Science)

Amos G, Rhodes M, Benkreira H. Calculation of optic fibers calibration curves for the measurement of solids volume fractions in multiphase flows. Powder Technol 88:107–121, 1996.

Anthony E. Fluidized bed combustion of alternative solid fuels: status, successes and problems of the technology. Prog Energy Combust Sci 21:239–268, 1995.

Arastopour H, Yang Y. Experimental studies on dilute gas and cohesive particle flow behavior using a laser–Doppler anemometer. In: Potter O, Nicklin D, eds. Fluidization VII. New York: Engineering Foundation, 1992, pp 723–730.

Arena U, Cammarota A, Massimilla L, Pirozzi D. The hydrodynamic behavior of two circulating fluidized bed units of different sizes. In: Basu P, Large J, eds. Circulating Fluidized Bed Technology II. Oxford: Pergamon, 1988, pp 223–230.

Arena U, Cammarota A, Marzocchella A, Massimilla L. Hydrodynamics of a circulating fluidized bed with secondary air injection. Proc 12th FBC Conf San Diego, 1993, pp 899–905.

Atkinson C, Clark N. Gas sampling from fluidized beds: a novel probe system. Powder Technol 54:59–70, 1988.

Avidan A, Yerushalmi J. Solids mixing in an expanded top fluid bed. AIChE J 31:835–841, 1985.

Bachalo W, Houser M. Phase Doppler spray analyzer for simultaneous measurements of drop size and velocity distributions. Optical Engineering 23(5):583–590, 1984.

Bader R, Findlay J, Knowlton T. Gas/solid flow patterns in a 30.5 cm diameter circulating fluidized bed. In: Basu P, Large J, eds. Circulating Fluidized Bed Technology II. Oxford: Pergamon, 1988, pp 123–138.

Bai D, Shibuya E, Masuda Y, Nakawaga N, Kato K. Flow structure in a fast fluidized bed. Chem Eng Sci 51:957–966, 1996.

Bakker P, Heertjes P. Porosity measurements in a fluidised bed. British Chem Eng 3(May):240–246, 1958.

Bartholomew RN, Casagrande RM. Measuring solids concentration in fluidized systems by gamma-ray absorption. Ind Eng Chem 49(3):428–431, 1957.

Basu P, Knouche F. Radiative heat transfer from a fast fluidized bed. In: Basu P, Large J, eds. Circulating Fluidized Bed Technology II. Oxford: Pergamon, 1988, pp 245–253.

Beaud F, Louge M. Similarity of radial profiles of solid volume fraction in a circulating fluidized bed. Prep. of Fluidization VIII, Tours, France, 1995, pp 97–104.

Benton D, Parker D. Non-medical applications of positron emission tomography. In: Chaouki J, Larachi F, Dudukovic M, eds. Non-Invasive Monitoring of Multiphase Flows. New York: Elsevier, 1996, pp 161–184.

Boland D, Geldart D. Electrostatic charging in gas fluidized beds. Powder Technol 5:289–297, 1971/72.

Borlai O, Hodany L, Blickle T. Proc Int Symp Fluidization. Drinkenburg A, ed. Amsterdam: University Press, 1967, pp 433–441.

Bottcher C. Theory of Electric Polarization. New York: Elsevier, 1952.

Botterill J. Fluid-Bed Heat Transfer. New York: Academic Press, 1975.

Botterill J, Teoman Y, Yuregir K. Factors affecting heat transfer between gas-fluidized beds and immersed surfaces. Powder Technol 39:177–189, 1984.

Brereton C, Grace J. Microstructural aspects of the behavior of circulating fluidized beds. Chem Eng Sci 48:2565–2572, 1993.

Brereton C, Stromberg L. Some aspects of the fluid dynamic behavior of fast fluidized beds. In: Basu P, ed. Circulating Fluidized Bed Technology. Oxford: Pergamon, 1986, pp 133–144.

Brewster B, Seader J. Nonradioactive tagging method for measuring particle velocity in pneumatic transport. AIChE J 26:325–330, 1980.

Briens C, Mirgain C, Bergougnou M, Pozo M, Loutaty R. Evaluation of gas–solids mixing chamber through cross correlation and Hurst's analysis. AIChE J 43:1469–1479, 1997.

Broadhust T, Becker H. Measurement and spectral analysis of pressure fluctuations in slugging beds. In: Kearins D, ed. Fluidization Technology. Vol. 1. Washington: Hemispher Publishing Co., 1976, pp 63–85.

Brunier E, Yemmou M, Evrard R. Flux particles measurements in industrial CFB. In: Werther J, ed. Proc 6th Conference on CFB. Wurzburg, Germany, 1999, pp 825–830.

Burkell J, Grace J, Zhao J, Lim C. Measurement of solids circulation rates in circulating fluidized beds. In: Basu P, Large J, eds. Circulating Fluidized Bed Technology II. Oxford: Pergamon, 1988, pp 501–509.

Campbell C, Wang D. A particle pressure transducer suitable for use in gas-fluidized beds. Meas Sci Technol 1:1275–1279, 1990.

Cankurt N, Yerushalmi J. Gas backmixing in high velocity fluidized beds. In: Fluidization. Cambridge: Cambridge University Press, 1978, pp 387–392.

Chang H, Louge M. Fluid dynamic similarity of circulating fluidized beds. Powder Technol 70:259–270, 1992.

Chau K, Goehner R, Drubetsky E, Brady H, Bayles W Jr, Pedersen P. Pressure and sound measurement. In: Webster J, ed. The Measurement, Instrumentation, and Sensors Handbook. New York: CRC Press LLC, 1999, Chap. 26.

Chen JC. Experiments that address phenomenological issues of fast fluidization. Chem Eng Sci 54:5529–5539, 1999.

Chen R, Fan LS. Particle image velocimetry for characterizing the flow structure in three-dimensional gas–liquid–solid fluidized beds. Chem Eng Sci 47(13/14):3615–3622, 1992.

Cheremisinoff N. Review of experimental methods for studying the hydrodynamics of gas–solid fluidized beds. Ind Eng Chem Process Des Dev 25:329–351, 1986.

Cheremisinoff N, Cheremisinoff P. Hydrodynamics of Gas-Solids Fluidization. New York: Gulf, 1984.

Cocco R, Cleveland J, Harner R, Chrisman R. Simultaneous in-situ determination of particle loadings and velocities in a gaseous medium. AIChE Symp Ser 91(308):147–153, 1995.

Contractor R, Pell M, Weinstein H, Feindt H. The rate of solid loss in a circulating fluid bed following a loss of circulation accident. In: Potter O, Nicklin D, eds. Fluidization VII. New York: Engineering Foundation, 1992, pp 243–248.

Couturier M, Steward F, Poolpol S. Experimental determination of heat transfer coefficients in a 72 MWth circulating fluidized bed boiler. Proc 12th FBC Conf, San Diego, CA, USA, 1993, pp 1215–1222.

Dahlin R, Pontius D, Haq Z, Vimalchand P, Brown R, Wheeldon J. Plans for hot gas cleanup testing at the power systems development facility. Proc 13th FBC Conf, Orlando, FL, USA, 1995, pp 449–459.

Davies C, Harris B. A device for measuring solids flow-rates: characteristics and application in a circulating fluidized bed. In: Potter O, Nicklin D, eds. Fluidization VII. New York: Engineering Foundation, 1992, pp 741–748.

De Kok J, Stark N, van Swaaij W. The influence of solids specific interfacial area on gassolid mass transfer in gas fluidised beds. In: Ostergaard K, Sorensen A, eds. Fluidization V. New York: Engineering Foundation, 1986, pp 433–440.

Drain L. The Laser Doppler Technique. New York: John Wiley, 1980.

Dry R. Radial particle size segregation in fast fluidized beds. Powder Technol 52:7–16, 1987.

Dry R, White R, Close R, Joyce T. The effect of CFB scale-up on gas–solid contact efficiency. Prep of the Fluidization VIII, Tours, France, 1995, pp 25–32.

Du B, Wei F, Wang Z. Effect of particle size on solids mixing in a FCC riser. In: Werther J, ed. Proc. 6th Conference on CFB. Wurzburg, Germany, 1999, pp 405–410.

Durand D, Llinas R, Newton D. Optimisation of the bubbling in a polyethylene fluid bed process through modelling of experimental test rig data. Prep Fluidization VIII, 1995, pp 923–935.

Durst F, Melling A, Whitelaw J. Principles and Practice of Laser-Doppler Anemometry, 2d ed. New York: Academic Press, 1981.

Dyakowski T, Edwards R, Xie C, Williams R. Application of capacitance tomography to gas–solid flows. Chem Eng Sci 52(13), 1997.

Dybeck K, Nagel R, Schoenfelder H, Werther J, Singer H. The D ring sensor—a novel method for continuously measuring the solid circulation rate in CFB reactors. Prep Fluidization VIII, Tours, France, 1995, pp 9–16.

Ebert TA, Glicksman LR, Lints M. Determination of particle and gas convective heat transfer components in a circulating fluidized bed. Chem Eng Sci 48:2179–2188, 1993.

Edgar T, Smith C, Shinskey G, Gassman G, Schafbuch P, McAvoy T, Seborg D. Section 8: process control. In: Perry R, Green D, Maloney J, eds. Perry's Chemical Engineers' Handbook, 7th ed. New York: McGraw-Hill, Inc., 1997.

Fan LS. Bubble dynamics in liquid–solid suspensions. AIChE Symp Ser 91(308):1–43, 1995.

Fan LT, Ho T, Hiraoka S, Walawender W. Pressure fluctuations in a fluidized bed. AIChE J 27:388–396, 1981.

Fan LT, Neogi D, Yashima M, Nassar R. Stochastic analysis of a three-phase fluidized bed: fractal approach. AIChE J 36(10):1529–1535, 1990.

Fasching G, Smith N. A particle impact probe for hydrodynamic studies in fluidized beds. Powder Technol 75:21–27, 1993.

Fasso L, Chao B, Soo S. Measurement of electrostatic charges and concentration of particles in the freeboard of a fluidized bed. Powder Technol 33:211–221, 1982.

Fiedler O, Werther J, Labahan N, Kumpart J, Christofori K. Measurement of local velocities and velocity distributions in gas–solid flows by means of the spatial filter method. Powder Technol 94:51–57, 1997.

Fiorentino M, Newton D. Application of x-ray imaging to fluidized bed scale-up. In: Fan LS, Knowlton T, eds. Fluidization IX. New York: Engineering Foundation, 1998, pp 589–597.

Fredd C, Fogler H, Lindsay J. Neutron transmission tomography applied to reactive dissolution through percolating porous media. In: Chaouki J, Larachi F, Dudukovic M, eds. Noninvasive Monitoring of Multiphase Flows. New York: Elsevier Science, 1997, pp 185–211.

Fujikura Ltd., Optical Fiber Communication System, Catalog No. T1160-8, 1988, p 51.

Fukayama Y, Horio M, Matsumoto K, Shimohira K. Int Meas Confid IVI IMEKO World Congress, Vienna, Austria, 2000.

Gbavcic Z, Vukovic D, Zdanski F. Tracer particle movement in a two-dimensional water-fluidized bed. Powder Technol 62:199–201, 1990.

Geldart D. Single particles, fixed and quiescent beds. In: Geldart D, ed. Gas Fluidization Technology. New York: John Wiley, 1986, pp 11–32.

Geldart D, Rhodes M. From minimum fluidization to pneumatic transport—a critical review of the hydrodynamics. In: Basu P, ed. Circulating Fluidized Bed Technology. Oxford: Pergamon Press, 1986, pp 21–31.

Godfroy L, Larachi F, Kennedy G, Chaouki J. Simulation measurement of 3-D position and velocity of a single radioactive particle in a CFB riser at high velocity. Proc CFB 5, Beijing, China, 1996, pp 633–638.

Goedicke F, Reh L. Particle induced heat transfer between walls and gas–solid fluidized beds. AIChE Symp Ser 89(296):123–136, 1992.

Grace J, Baeyens J. Instrumentation and experimental techniques. In: Geldart D, ed. Gas Fluidization Technology. New York: John Wiley, 1986, pp 415–462.

Grassler T, Wirth KE. X-ray computed tomography—a non-intrusive technique to characterize local solids distribution in a circulating fluidized bed (CFB). In: Werther J, ed. Proc 6th Conference on CFB. Wurzburg, Germany, 1999, pp 885–890.

Greon J, Berben J, Mudde R, van den Akker H. Local bubble properties in 40 and 80 cm diameter fluidized beds. AIChE Symp Ser 93(317):147–151, 1997.

Grohse E. Analysis of gas fluidized solid systems by x-ray absorption. AIChE J 1:358–365, 1955.

Halow J, Nicoletti P. Observations of fluidized bed coalescence using capacitance imaging. Powder Technol. 69(3):255–277, 1992.

Halow J, Fasching G, Nicoletti P, Spenik L. Observations of a fluidized bed using capacitance imaging. Chem Eng Sci 48:643–659, 1993.

Halow J. Electrical capacitance imaging of fluidized beds. In: Chaouki J, Larachi F, Dudukovic M, eds. Noninvasive Monitoring of Multiphase Flow. New York: Elsevier Science, 1997, pp 263–307.

Hamamatsu Photonics K.K. CdS Photoconductive Cells. Catalog No. KCDS 0001E03, 1999b.

Hamamatsu Photonics K.K. Light Sources. Catalog No. TLSO 0002J04, 2000.

Hamamatsu Photonics K.K. Optosemiconductors. Condensed Catalog, Catalog No. KOTH 0001E03, 1999a.

Hamamatsu Photonics K.K. Photodiodes. Catalog No. KPD 0001J07, 1998b.

Hamamatsu Photonics K.K. Photomultiplier Tubes and Related Products. Catalog No. TPMO 0003J02, 1998a.

Hamamatsu Photonics K.K. Solid State Emitters. Catalog No. KLED 0001E06, 1999c.

Hamdullahpur F, Pegg M, MacKay G. A laser-fluorescence technique for turbulent two-phase flow measurements. Int J Mult Flow 13(3):379–385, 1987.

Hansen P, Bank L, Dam-Johansen K. In-situ measurements in fluidized bed combustors: design and evaluation of four sample probes. Proc 13th Conf FBC. Orlando, FL, USA, 1995, pp 391–398.

Hartge EU, Li Y, Werther J. In: Basu P, ed. Circulating Fluidized Bed Technology. Oxford: Pergamon Press, 1986, pp 153–160.

Hartge EU, Rensner D, Werther J. Solids concentration and velocity patterns in circulating fluidized beds. In: Basu P, Large J, eds. Circulating Fluidized Bed Technology II. Oxford: Pergamon, 1988, pp 165–180.

Hatano H, Kudo N. Microscope visualization of solid particles in circulating fluidized beds. Powder Technol 78:115–119, 1994.

Hatano H, Ogasawara M, Horio M, Hartge EU, Werther J. Local flow of gas and solids in a circulating fluidized bed. Proc 6th SCEJ Symp Fluidization, 1993, pp 107–114.

Hatano H, Matsuda S, Takeuchi H, Pyatenko A, Tsuchiya K. Local interactive patterns of dispersed and swarm particles in a circulating fluidized bed riser. Ind Eng Chem Res 35:4360–4365, 1996.

Hatano H, Takeuchi H, Sakurai S, Masuyama T, Tsuchiya K. Motion of individual FCC particles and swarms in a circulating fluidized bed riser analyzed via high-speed imaging. AIChE Symp Ser 94(318):31–36, 1998.

Hatano, H, Matsuda S, Tsuchiya K. Continuous measurement of velocity and direction of particles using the lag-time method. AIChE Symp Ser 95(321):95–99, 1999.

Hawkesworth M, Parker D, Fowles P, Crilly J, Jeffries N, Jonkers G. Nucl Instr Meth A310:423–434, 1991.

Hayrinen V, Hernberg R, Oikari R, Gottwald U, Monkhouse P, Davidsson K, Lonn B, Engvall K, Petersson J, Lehtonen P, Kuivalainen R. On line determination of alkali concentrations in pressurized fluidized-bed coal combustion by combined measurement techniques. Werther J, ed. Proc 6th Conf on CFB. Wurzburg, Germany, 1999, pp 873–878.

He Z, Zhang WD, Hei KM, Chen BC. Modeling pressure fluctuation via correlation structure in a gas–solids fluidized bed. AIChE J 43(7):1914–1920, 1997.

Helmrich H, Schugerl K, Janssen K. Decomposition of $NaHCO_3$ in laboratory and bench scale circulating fluidized bed reactors. In: Basu P, ed. Circulating Fluidized Bed Technology. New York: Pergamon, 1986, pp 161–166.

Herb B, Dou S, Tuzla K, Chen J. Solid mass fluxes in circulating fluidized beds. Powder Technol 70:197–205, 1992.

Herbert P, Gauthier T, Briens C, Bergougnou M. Application of fiber optic reflection probes to the measurement of local particle velocity and concentration in gas–solid flow. Powder Technol 80:243–252, 1994.

Hernandez D, Lu J, Gauthier D, Flamant G. A new optical probe working in high temperature fluidized beds. Proc 9th Conf Fluidization. New York: Engineering Foundation, 1998, pp 557–564.

Holman J. Experimental methods in engineering, 6th ed. New York: McGraw-Hill, 1994.

Holoboff J, Kantzas A, Kalogerakis N. Utilization of computer assisted tomography in the determination of the spatial variability of voidage in a gas–solid fluidized bed. Prep 8th Int Conf Fluidization. Tours, France, 1995, pp 295–302.

Horio M. Hydrodynamics. In: Grace J, Avidan A, Knowlton T, eds. Circulating Fluidized Beds. New York: Chapman & Hall, 1997, pp 21–85.

Horio M, Ito M. Prediction of cluster size in circulating fluidized beds. J Chem Eng Japan 30(4):691–697, 1997.

Horio M, Kuroki H. Three-dimensional flow visualization of dilutely dispersed solids in bubbling and circulating fluidized beds. Chem Eng Sci 49(15):2413–2421, 1994.

Horio M, Mori S, eds. Fluidization Handbook, Tokyo: Baifukan, 1999, pp 219–221.

Horio M, Morishita K. Flow regimes of high velocity fluidization. Japanese J Multiphase Flow 2:118–136, 1988.

Horio M, Taki A, Hsieh Y, Muchi I. Elutriation and particle transport through the freeboard of a gas–solid fluidized bed. In: Grace J, Matsen J, eds. Fluidization. New York: Plenum Press, 1980a, pp 509–518.

Horio M, Kiyota H, Muchi I. Particle movement on a perforated plate distributor of fluidized bed. J Chem Eng Japan 13(2):137–142, 1980b.

Horio, M, Hayashi H, Morishita K. A novel in-bed heat transfer tube capable of decreasing hw linearly with load turndown. Proc 8th Int Conf Fluidized Bed Combustion. US DOE/METC, 1985, pp 655–663.

Horio M, Takada M, Ishida M, Tanaka N. The similarity rule of fluidization and its application to solid mixing and circulation control. In: Ostergaard K, Sorensen A, eds. Fluidization V. New York: United Engineering Foundation, 1986a, pp 151–158.

Horio M, Nonaka A, Hoshiba M, Morishita K, Kobukai Y, Naito J, Tachibana O, Watanabe K, Yoshida N. Coal combustion in a transparent fluidized bed. In: Basu P, ed. Circulating Fluidized Bed Technology. Oxford: Pergamon Press, 1986b, pp 255–262.

Horio M, Morishita K, Tachibana O, Murata N. Solid distribution and movement in circulating fluidized beds. In: Basu P, Large J eds. Circulating Fluidized Bed Technology II. Oxford: Pergamon Press, 1988, pp 147–154.

Horio M, Mori K, Takei Y, Ishii H. In: Potter O, Nicklin D, eds. Proc Fluidization VII. New York: United Engineering Foundation, 1992a, pp 757–762.

Horio M, Ishii H, Nishimuro M. On the nature of turbulent and fast fluidized beds. Powder Technol 70:229–236, 1992b.

Huang S, Plaskowski A, Xie C, Beck M. Tomographic imaging of two-component flow using capacitance sensors. J Phys E Sci Instrum 22, 1989.

Huilin L, Gidaspow D, Bouillard J. Dimension measurements of hydrodynamic attractors in circulating fluidized beds. AIChE Symp Ser 91(308):103–111, 1995.

Hull A, Agarwal P. Bubble behaviour in a tube bank: modeling and experimental investigation using digital image analysis. Prep Fluidization VIII. Tours, France, 1995, pp 343–350.

Ikeda Y. Recent issues in fluidized bed reactor developments. Chem Eng 10:41–45, 1972 (in Japanese).

Ilias S, Ying S, Mathur G, Govind R. Studies on a swirling circulating fluidized bed. In: Basu P, Large J, eds. Circulating Fluidized Bed Technology II. Oxford: Pergamon Press, 1988, pp 537–546.

Ishida M, Shirai T, Nishiwaki A. Measurement of the velocity and direction of flow of solid particles in a fluidized bed. Powder Technol 27, 1980.

Ito M, Tsukada M, Horio M. Three dimensional meso-scale structure of gas–solid suspensions in circulating fluidized beds, determined by the scanning laser sheet image analysis. Proc 1st Particle Technology Forum, Vol. 1, 1994, pp 428–433.

Ito O, Kawabe R, Miyamoto T, Orita H, Mizumoto M, Miyadera H. Direct measurement of particle motion in

a large-scale FBC boiler model. Proc 15th Conf Fluidized Bed Combustion, Savannah, Georgia, 1999.

Ji H, Ohara H, Kuramoto K, Tsutsumi A, Yoshida K, Hirama T. Nonlinear dynamics of gas–solid circulating fluidized bed system. Chem Eng Sci 55:403–410, 2000.

Johnsson F, Sterneus J, Leckner B, Wiesendorf V, Hartge EU, Werther J, Monat D, Briad P. Fluid dynamics of the bottom zone of CFB combustor. Werther J, ed. Proc 6th Conf CFB. Wurzburg, 1999a, pp 191–196.

Joutsenoja T, Heino P, Hernberg R. Pyrometric temperature and size measurements of burning coal particles in a fluidized bed combustion reactor. Combust Flame 118:707–717, 1999.

Jovanovic G, Catipovic N, Fitzgerald T, Levenspiel O. The mixing of tracer gas in fluidized beds of large particles in fluidization. In: Grace J, Matsen J, eds. Fluidization III. New York: Plenum Press, 1980, pp 325–332.

Junfu L, Jiansheng Z, Guangxi Y, Qing L, Xiaoma L, Jiling L, Xiaoxing Z, Yi L, Long Y, Xudong L, Zhengyi L. The progress of the water cooled separator CFB boiler in China. Proc 15th Conf Fluidized Bed Combustion, Savannah, Georgia, 1999.

Kai T, Furusaki S. Behavior of fluidized beds of small particles at elevated temperature. J Chem Eng Japan 18:113–118, 1985.

Kai T, Takahashi T, Misawa M, Tiseanu I, Ichikawa N, Takada N. Observation of rising bubbles in a fluidized catalyst bed using a fast X-ray CT scanner. Proc, 5th SCEJ Symp. on Fluidization, Tsukuba, 1999, pp. 435–440.

Kai T, Misawa M, Takeuchi T, Tiseanu I, Ichikawa N, Takeda N. Application of fast x-ray CT scanner to visualization of bubbles in fluidized bed. J Chem Eng Japan 33:906–909, 2000.

Kak A, Slaney M. Principles of Computerized Tomographic Imaging. Institute of Electrical and Electronics Engineers, 1988.

Kalender W. Computed Tomography. Publicis MCD Verlag, 2000.

Kamiwano M, Saito F. Measurement method of flow velocity of liquid and irregular solid particles using an image sensor with an image fiber. AIChE Symp Ser 80(241):122–128, 1984.

Kantzas A. Computation of holdups in fluidized and trickle beds by computer-assisted tomography. AIChE J 40(7):1254–1261, 1994.

Kassman H, Abul-Milh M, Amand LE. Measurement of NH3 and HCN concentrations in a CFB boiler, a comparison between a conventional absorption and FTIR technique. Proc 13th FBC Conf. Orlando, FL, USA, 1995, pp 1447–1454.

Kassman H, Karlsson M, Amand LE. Influence of air-staging on the concentration profiles of NH3 and HCN in the combustion chamber of a CFB boiler burning coal. Proc 15th Conf Fluidized Bed Combustion, Savannah, Georgia, USA, 1999.

Kawata S, Minami S. Image data processing for scientific measurement. CQ Publishing, Tokyo, 1994.

Kawata S, Minami K, Minami S. Supersolution of Fourier transform spectroscopy data by the maximum entropy method. Appl. Optics 22:3593–3598, 1983.

Kehoe P, Davidson J. Pressure fluctuation in slugging fluidized beds. AIChE Symp Ser 69(128):34–40, 1973.

Khattab I, Ishida M. Visualization of gas voids behavior above the distributor in a three-dimensional gas fluidized bed. J Chem Eng Japan 19(5):367–374, 1986.

Khattab I, Kuroda C, Ishida M. Radial and vertical distributions of the interstitial gas velocity in a fluidized bed. J Chem Eng Japan 21(3):282–287, 1988.

Kobylecki R, Horio M. On the microscopic aspect of inter-particle heat transfer. Xu D, Mori S, eds. Proc 7th China–Japan Symposium on Fluidization. Xi'an: Xi'an Publishing House, China, 2000, pp 59–64.

Kobylecki R, Horio M. A simple solid mass flow meter for circulating fluidized beds. J Chem Eng Japan 35:456–467, 2002.

Kojima T, Ishihara K, Yang G, Furusawa T. Measurement of solids behavior in a fast fluidized bed. J Chem Eng Japan 22:341–346, 1989.

Kondulov N, Kornilaev A, Skachko I, Akhromenkov A, Kruglov A. An investigation of the parameters of moving particle in a fluidized bed by a radioisotopic method. Int Chem Eng 4, 1964.

Kono H, Soltani-Ahmadi A, Suzuki M. Kinetic forces of solid particles in coarse particle fluidized beds. Powder Technol 52:4-58, 1987.

Kozinski J, Rink K, Lighty JS. Combustion of sludge waste in FBC, distribution of metals and particle sizes. Proc 13th FBC Conf, Orlando, FL, USA, 1995, pp 139–147.

Krambeck F, Avidan A, Lee C, Lo M. Predicting fluid bed reactor efficiency using adsorbing gas tracers. AIChE J 33:1727–1734, 1987.

Krohn D. Intensity modulated fiber optic sensors overview. In: SPIE, Vol. 718, Fiber Optic and Laser Sensors IV, 1986, pp 2–11.

Kruse M, Werther J. 2D gas and solids flow prediction in circulating fluidized beds based on suction probe and pressure profile measurements. Chem Eng Proc 34:185–203, 1995.

Kumar S, Dudukovic M. Computer assisted gamma and x-ray tomography: applications to multiphase flow systems. In: Chaouki J, Larachi F, Dudukovic M, eds. Non-Invasive Monitoring of Multiphase Flows. New York: Elsevier, 1997, pp 47–103.

LaNauze R, Jung K, Dent D, Joyce T, Tait T, Burgess J. Measurement of the temperature of burning particles in fluidized beds. 9th Int Conf FBC. ASME, Boston, MA, USA, 1987, pp 707–712.

Lanneau K. Gas-solids contacting in fluidized beds. Trans Instn Chem Engrs 38:125–143, 1960.

Larachi F, Kennedy G, Chaouki J. A -ray detection system for 3-D particle tracking in multiphase reactors. Nuclear Instrum Methods Phys Res 338, 1994.

Larachi F, Lord E, Chaouki J, Chavarie C, Behie L. Phenomenological study of solids mixing in a binary liquid fluidized bed. Prep Fluidization VIII, Tours, France, 1995, pp 385–392.

Larachi F, Chaouki J, Kennedy G, Dudukovic M. Radioactive particle tracking in multiphase reactors: principles and applications. In: Chaouki J, Larachi F, Dudukovic M, eds. Non-Invasive Monitoring of Multiphase Flows. New York: Elsevier Science, 1997, pp 335–397.

Lasch B, Caram H, Chen J, Korenberg J. Solid circulation in cyclonic fluidized bed combustors. In: Basu P, Large J, eds. Circulating Fluidized Bed Technology II. Oxford: Pergamon Press, 1988, pp 527–535.

Leckner B, Andersson B. Characteristic features of heat transfer in circulating fluidized bed. Powder Technol 70:303–314, 1992.

Leckner B, Golriz M, Zhang W, Andersson B, Johnsson F. Boundary layers—first measurements in the 12 MW CFB research plant at Chalmers University. Anthony E, ed. Proc 11th Int Conf FBC. ASME, New York, USA, 1991, pp 771–776.

Lehner P, Richtberg M, Wirth KE. Measurement of solids velocities in a co-current gas/solids flow with a laser Doppler anemometer and a capacitance probe system. Werther J, ed. Proc 6th Conf CFB. Wurzburg, Germany, 1999, pp 843–848.

Le Niliot C, Gallet P. Infrared thermography applied to the resolution of inverse heat conduction problems: recovery of heat line sources and boundary conditions. Rev Gen Therm 37:629–643, 1998.

Levy Y. LDA technique for measurements in freeboards of fluidized beds. AIChE J 32:1579–1583, 1986.

Li J, Weinstein H. An experimental comparison of gas back-mixing in fluidized beds across the regime spectrum. Chem Eng Sci 44:1697–1705, 1989.

Li Y, Wu P. A study on axial gas mixing in a fast fluidized bed. In: Basu P, Horio M, Hasatani M, eds. Circulating Fluidized Bed Technology III. Oxford: Pergamon Press, 1991, pp 581–586.

Li H, Xia Y, Tung Y, Kwauk M. Microvisualization of two-phase structure in a fast fluidized bed. In: Basu P, Horio M, Hasatani M, eds. Circulating Fluidized Bed Technology III. Oxford: Pergamon Press, 1991, pp 183–188.

Li H, Qian R, Huang W, Bi K. An investigation on instantaneous local heat transfer coefficients in high-temperature fluidized beds II. statistical analysis. Int J Heat Mass Transfer 35:4397–4406, 1993.

Lin J, Grace J, Bi H, Zhu J. Gas mixing in a high density CFB riser. Werther J, ed. Proc 6th Conf CFB. Werther J, ed. Wurzburg, Germany, 1999b, pp 399–404.

Lin M, Kulasekaran S, Ignowski J, Linjewile T, Agarwal P. Emissions of N_2O in fluidized bed combustion of coal. Proc 15th Conf Fluidized Bed Combustion, Savannah, Georgia, USA, 1999c.

Lin W, Weinell C, Hansen P, Dam-Johansen K. Hydrodynamics of a commercial scale CFB boiler— study with radioactive tracer particles. Werther J, ed. Proc 6th Conf CFB. Wurzburg, Germany, 1999a, pp 861–866.

Linjewile T, Hull A, Agarwal P. Optical probe measurements of the temperature of burning particles in fluidized beds. Fuel 73:1880–1886, 1994.

Lischer D, Louge M. Optical fiber measurements of particle concentration in dense suspensions: calibration and simulation. Applied Optics 31:5106–5113, 1992.

Lord W, McAndrew G, Sakagami M, Valenzuela J, Glicksman L. Measurement of bubble properties in fluidized beds. Proc 7th Int Conf Fluidized Bed Combustion. US DOE/EPRI/, 1982, pp 76–88.

Louge M. Measurement techniques. In: Grace J, Avidan A, Knowlton T, eds. Circulating Fluidized Beds. New York: Blackie Academic and Profesional, 1997, pp 312–358.

Louge M, Opie M. Measurements of the effective dielectric permittivity of suspensions. Powder Technol 62:85–94, 1990.

Lu X, Li H. Wavelet analysis of pressure fluctuation signals in a bubbling fluidized bed. Chem Eng Sci 75:113–119, 1999.

Luan W, Lim C, Brereton C, Bowen B, Grace J. Experimental and theoretical study of total and radiative heat transfer in circulating fluidized beds. Chem Eng Sci 54:3749–3764, 1999.

Ma Y, Zhu J. Experimental study of heat transfer in a co-current downflow fluidized bed (downer). Chem Eng Sci 54:41–50, 1999.

Mann U, Crosby E. Flow measurement of coarse particles in pneumatic conveyers. Ind Eng Chem Proc Des Dev 16:9–13, 1977.

Mann M, Swanson M, Yagla S. Characterization of alkali and sulfur sorbents for pressurized fluidized bed combustion. Proc 13th FBC Conf, Orlando, FL, USA, 1995, pp 333–340.

Martin M, Turlier P, Bernard J, Wild G. Gas and solid behavior in cracking circulating fluidized beds. Powder Technol 70:249–258, 1992.

Marzocchella A, Zijerveld R, Schouten J, van den Bleek C. Chaotic behavior of gas–solid flow in the riser of a laboratory-scale circulating fluidized bed. AIChE J 43:1458–1468, 1997.

Mathur A, Saxena S. Total and radiative heat transfer to an immersed surface in a gas-fluidized bed. AIChE J 33:1124–1135, 1987.

Matsuno Y, Yamaguchi H, Oka T, Kage H, Higashitani K. The use of optic fiber probes for the measurement of dilute particle concentrations: calibration and application

to gas-fluidized bed carryover. Powder Technol 36:215–221, 1983.

Mattisson T, Lyngfelt A. The presence of CaS in the combustion chamber of a 12 MW circulating fluidized bed boiler. Proc 13th FBC Conf, ASME, New York, 1995, pp 819–829.

Mehrabi A, Rassamdana H, Sahimi M. Characterization of long-range correlations in complex distributions and profiles. Physical Review E 56:712–722, 1997.

Meredith R, Tobias C. Resistance of potential flow through a cubic array of spheres. J Appl Phys 31(7):1270–1273, 1960.

Mickley H, Fairbanks D. Mechanism of heat transfer to fluidized beds. AIChE J 1:374–384, 1955.

Mickley H, Fairbanks D, Hawthorn R. The relation between the transfer coefficient and thermal fluctuations in fluidized bed heat transfer. Chem Eng Prog Symp Ser 57(32):51–60, 1961.

Minami S (ed.). Wave Data Processing for Scientific Measurement. Tokyo: CQ Publishing, 1986.

Miyauchi T, Morooka S. Circulating flow and its effects on chemical reaction in fluid bed contactor. Kagaku Kogaku 33:369–376, 1969.

Molerus O, Wirth KE. Heat Transfer in Fluidized Beds. New York: Chapman and Hall, 1997.

Monceaux L, Azzi M, Molodtsof Y, Large J. Overall and local characterization of flow regimes in a circulating fluidized bed. In: Basu P, ed. Circulating Fluidized Bed Technology. Toronto: Pergamon, 1986, pp 185–192.

Moritomi H, Mori S, Araki K, Moriyama A. Periodic pressure fluctuation in a gaseous fluidized bed. Kagaku Kogaku Ronbunshu 6:392–396, 1980.

Morooka S, Kusakabe K, Ohnishi N, Gujima F, Matsuyama H. Measurement of local fines movement in a fluidized bed of coarse particles by a fluorescent tracer technique. Powder Technol 58:271–277, 1989.

Muraoka K, Maeda M. Laser Application for Gas Plasma and Gas Measurement. Tokyo: Sangyotosho, 1995.

Myler C, Zaltash A, Klinzing G. Gas-solid transport in a 0.0508 m pipe at various inclinations with and without electrostatics i: particle velocity and pressure drop. J Powder Bulk Solid Technol 10:5–12, 1986.

Nakajima K. Laser Doppler velocimeter. Kikai-no-Kenkyu 47:65–71, 1995.

Nakajima M, Harada M, Asai M, Harada R, Jimbo G. Bubble fraction and voidage in an emulsion phase in the transition to a turbulent fluidized bed. In: Basu P, Horio M, Hasatani M, eds. Circulating Fluidized Bed Technology III. Oxford: Pergamon Press, 1991, pp 79–84.

Naruse I, Ohtake K, Koizumi K, Kuramoto K, Lu G. Gas exchange between the bubble and emulsion phases during bubbling fluidized bed coal combustion elucidated by conditional gas sampling. 25th Symposium on Combustion. Comb Inst 1994, pp 545–552.

Naruse I, Kuramoto K, Lu GQ, Ohtake K. Influence of bubble and emulsion phase on N_2O formation in bubbling fluidized bed coal combustion by conditional sampling. Prep Fluidization VIII, Tours, France, 1995, pp 499–506.

Naruse K, Kashima N, Hosoda S, Toyoda S, Oshita T, Tsukada M, Horio M, Shinozaki S, Sugiyama T. On the effect of fluidized bed viscosity on the solid circulating rate across a tube bank. J Soc Powder Technol Japan 33:468–475, 1996.

Nguyen T, Nguyen A, Nieh S. An improved isokinetic sampling probe for measuring local gas velocity and particle mass flux of gas–solid suspension flows. Powder Technol 59:183–189, 1989.

Nieuwland J, Meijer R, Kuipers J, van Swaaij W. Measurements of solids concentration and axial solids velocity in gas–solid two-phase flows. Powder Technol 87:127–139, 1996.

Nishii K, Shimizu Y, Horio M. Characterization of cohesive powder behavior in fluidized beds. Proc 6th Int Symp Agglomeration. Nagoya, Japan, 1993, pp 740–745.

Noguchi M. How to Use Visual LED Lamps. Transistor Technology, Supplemental Issue, Hardware Design Series No. 8, CQ Pub, 1999, pp 8–16.

Nowak W, Mineo H, Yamazaki R, Yoshida K. Behavior of particles in a circulating fluidized bed of a mixture of two different sized particles. In: Basu P, Horio M, Hasatani M, eds. Circulating Fluidized Bed Technology III. Oxford: Pergamon Press 1991, pp 219–224.

Nowak W, Win K, Matsuda H, Hasatani M, Kruse M, Werther J. Local solids fluxes and particle size distributions in a multi-solid fluidized bed. Prep Fluidization VIII, Tours, France, 1995, pp 209–216.

Ohki K, Shirai T. Particle velocity in fluidized bed. In: Keairns D, ed. Fluidization Technology. Washington: Hemisphere, 1976, pp 95–110.

Onodera T, Okura Y, Umekawa H, Ozawa M, Takenaka N, Matsubayashi M. Large particles and a bubble movement in fluidized-bed. Proc Specialists' Meeting Neutron Radiography Techniques for Application and Utilization, Nov 1998.

Onofri F, Tadrist L. Experimental analysis of spatio-temporal instabilities of the dilute gas–solids flow in a CFB. Werther J, ed. Proc 6th Conf CFB. Wurzburg, Germany, 1999, pp 837–842.

Ozawa M, Umekawa H, Matsuda T, Takenaka N, Matsubayashi M. Flow pattern and heat transfer in tube banks of a simulated fluidized bed heat exchangers. JSME Intern J, Ser B 41(3):720–726, 1998.

Ozawa M, Umekawa H, Takenaka N, Matsubayashi M. Quantitative flow visualization of fluidized-bed under normal and down-flow-mode operations by neutron radiography. Experiments in Fluids, 1999.

Ozhaynak T, Chen J, Frankenfield T. An experimental investigation of radiation heat transfer in a high temperature fluidized bed. In: Kunii D, Toei E, eds.

Fluidization IV. New York: United Engineering Foundation, 1984, pp 371–378.

Pangrle B, Walsh E, Moore S, DiBiasio D. Magnetic resonance imaging of laminar flow in porous tube and shell systems. Chem Eng Sci 47(3):517–526 1992.

Pasquetti R, Le Niliot C. Boundary element approach for inverse heat conduction problems: application to a bidimensional transient numerical experiment. Numerical Heat Transfer, Part B 20:169–189, 1991.

Patience G, Chaouki J, Grandjean B. Solids flow metering from pressure drop measurement in circulating fluidized beds. Powder Technol 61:95–99, 1990.

Patience G, Chaouki J, Kennedy G. Solids residence time distribution in a circulating fluidized bed reactors. In: Basu P, Horio M, Hasatani M, eds. Circulating Fluidized Bed Technology III. Oxford: Pergamon Press, 1991, pp 599–604.

Patrose B, Caram HS. Optical fiber probe transit anemometer for particle velocity measurements in fluidized beds. AIChE J 28:604–609, 1982.

Prins W. Fluidized bed combustion of a single carbon particle. Professor's thesis, Twente University, The Netherlands, 1987.

Qi C, Farag I. Experimental measurement of lateral particle flux over the riser cross-section. In: Large JF, Laguerie C, eds. Fluidization VIII. New York: Engineering Foundation, 1995, pp 65–72.

Radon J. Ueber die bestimmung von funktionen durch ihre integralwerte langs gewisser manningfaltigkeiten. Berichte Saechsische Akad Wiss 69, 1917.

Raso G, Tirabasso G, Donsi G. An impact probe for local analysis of gas–solid flows. Powder Technol 34:151–159, 1983.

Reese J, Fan LS. Particle image velocimetry: application for the characterization of the flow structure in three phase fluidized beds. In: Chaouki J, Larachi F, Dudukovic M, eds. Non-Invasive Monitoring of Multiphase Flows. New York: Elsevier Science, 1997.

Reh L, Li J. Measurement of voidage in fluidized beds by optical probes. In: Basu P, Horio M, Hasatani M, eds. Circulating Fluidized Bed Technology III. New York: Pergamon Press, 1991, pp 163–170.

Ren J, Li J. Wavelet analysis of dynamic behavior in fluidized beds. In: Fan LS, Knowlton T, eds. Fluidization IX. New York: United Engineering Foundation, 1998, pp 629–636.

Rensner D, Werther J. Estimation of the effective measuring volume of single-fibre reflection probes for solid volume concentration measurements. Part Part Syst Charact 10:48–55, 1993.

Rhodes M, Laussmann P. A simple nonisokinetic sampling probe for dense suspensions. Powder Technol 70:141–151, 1992a.

Rhodes M, Laussmann P. Characterizing nonuniformities in gas–particle flow in the riser of a circulating fluidized bed. Powder Technol 72:277–284, 1992b.

702

Rhodes M, Wang X. Studying fast fluidization using electrical capacitance tomography. Werther J, ed. Proc 6th Conf CFB. Wurzburg, Germany, 1999, pp 891–896.

Rivault P, Nguyen C, Laguerie C, Bernard J. Countercurrent stripping dense circulating beds effect of the baffles. Prep of the Fluidization VIII, Tours, France, 1995, pp 491–498.

Rix S, Glass D, Greated C. Preliminary studies of elutriation from gas-fluidised beds using particle image velocimetry. Chem Eng Sci 51(13):3479–3489, 1996.

Ross K, Pence D. Wavelet and fractal analysis of fluidized bed heat transfer, HTD 351. Proc ASME Heat Transfer Div 1:245–253, 1997.

Ross I, Patel M, Davidson J. The temperature of burning carbon particles in fluidised beds. Trans Inst Chem Eng 59:83–88, 1981.

Roy R, Davidson J, Tuponogov V. The velocity of sound in fluidized beds. Chem Eng Sci 45:3233–3245, 1990.

Roy M, Kumar V, Kulkarni B, Sanderson J, Rodes M, van der Stappen M. Simple denoising algorithm using wavelet transform. AIChE J 45:2461–2466, 1999.

Sauer S, Wallen V. Mixing in experiments and modelling of ABFBC. Proc 15th Conf Fluidized Bed Combustion, Savannah, Georgia, USA, 1999.

Schlichthaerle P, Werther J. Axial pressure profiles and solids concentration distributions in the CFB bottom zone. Werther J, ed. Proc 6th Conf CFB. Wurzburg, Germany, 1999, pp 185–190.

Schoenfelder H, Kruse M, Werther J. Two dimensional model for circulating fluidized-bed reactors. AIChE J 40:1875–1888, 1996.

Schouten J, van den Bleek C. Chaotic hydrodynamics of fluidization: consequences for scaling and modeling of fluid bed reactors. AIChE Symp Ser 88(289):70–84, 1992.

Schugerl K, Merz M, Fetting F. Rheological characteristics of gas–solid fluidized bed system. Chem Eng Sci 15:1–38, 1961.

Schuurmans H. Measurements in a commercial catalytic cracker unit. Ind Eng Chem Process Des Dev 19:267–271, 1980.

Sederman A, Alexander P, Gladden L. Structure of packed beds probed by magnetic resonance imaging. Powder Technol 117:255–269, 2001.

Seo Y, Gidaspow D. An x-ray-γ-ray method of measurement of binary solids concentrations and voids in fluidized beds. Ind Eng Chem Res 26:1622–1628, 1987.

Seville J, Morgan J, Clift R. Tomographic determination of the voidage structure of gas fluidized beds in the jet region. In: Ostregaard K, Sorensen A, eds. Fluidization V. New York: United Engineering Foundation, 1986, pp 87–94.

Seville J, Simons S, Broadbent C, Martin T, Parker D, Beynon T. Particle velocities in gas–fluidized beds. Prep Fluidization VIII, Tours, France, 1995, pp 319–326.

Shattuck M, Behringer R, Johnson G, Georgiadis J. Convection and flow in porous media. Visualization by magnetic resonance imaging. J Fluid Mech 332:215–245, 1997.

Shen Z, Briens C, Kwauk M, Bergougnou M. Improved thermal conductivity detector technique for helium measurements in a fluidized bed. Powder Technol 69:249–253, 1992.

Sheng Y. Wavelet transform. In: Poularikas A, ed. The Transforms and Applications Handbook. New York: CRC Press, 1996, pp 747–827.

Siegel R, Howel J. Thermal Radiation Heat Transfer. Tokyo: McGraw-Hill Kogakusha, 1972.

Siegell J. Hight-temperature defluidization. Powder Technol 38:13–22, 1984.

Skrifvars B, Sfiris G, Backman R, Widegren-Dafgard K, Hupa M. Energy and Fuels 11:843–848, 1997.

Soo S, Trezek G, Dimick R, Hohnstreiter G. Concentration and mass flow distributions in a gas–solid suspension. I&EC Fundam 3(2):98–106, 1964.

Soo S, Slaughter M, Plumpe J. Instrumentation for flow properties of gas–solid suspensions and recent advances. Part Sci Tech 12:1–12, 1994.

Souto M, Rodriguez J, Conde-Pmpido R, Guitian F, Gonzalez J, Perez J. Fuel 75(6):675–680, 1996.

Stein M, Martin T, Seville J, McNeil P, Parker D. Positron emission particle tracking: particle velocities in gas fluidized beds, mixers and other applications. In: Chaouki J, Larachi F, Dudukovic M, eds. Non-Invasive Monitoring of Multiphase Flows. New York: Elsevier Science, 1997, pp 309–334.

Stellema C, Mudde R, Kolar Z, Gerritsen, de Goeij J, Van den Bleek C. Single particle tracking in interconnected fluidized beds. In: Fan LS, Knowlton T, eds. Fluidization IX. New York: United Engineering Foundation, 1998, pp 581–588.

Stephenson R, Moulin A, Welland M, Burns J, Sapoff M, Reed R, Frank R, Fraden J, Nicholas J, Pavese F, Stasiek J, Madaj T, Mikielewicz J, Culshaw B. Temperature measurement. In Webster J, ed. The Measurement, Instrumentation and Sensors Handbook. New York: CRC Press LLC, 1999.

Swift W, Vogel G, Panek A, Jonke A. Trace-element mass balances around a bench-scale combustor. Proc 4th FBC Conf, McLean, VA, USA, 1975, pp 525–543.

Takeuchi H, Hirama T. Flow visualization in the riser of a circulating fluidized bed. In: Basu P, Horio M, Hasatani M, eds. Circulating Fluidized Bed Technology III. Oxford: Pergamon Press, 1991, pp 177–182.

Takeuchi H, Pyatenko A, Hatano H. Gross behavior of parabolic strands in a riser. In: Fan LS, Knowlton TM, eds. Fluidization IX. United Engineering Foundation, 1998, pp 173–180.

Tanino T, Yawata Y, Otani S, Inoue H, Sato K, Takeda T, Mizuta T. Particle size control of anti-acid drug granules in fluidized-bed granulation. Proc 6th Int Symp Agglomeration, Nagoya, Japan, 1993, pp 548–553.

Tasdemir A, Sozmen N, Vural H. Determination of mixing quality in fluidized beds by image processing technique. Proc 15th Conf Fluidized Bed Combustion, Savannah, Georgia, USA, 1999.

Tayebi D, Svendsen H, Grislingas A, Mejdell T, Johannessen K. Dynamics of fluidized-bed reactors, development and application of a new multi-fiber optical probe. Chem Eng Sci 54:2113–2122, 1999.

Tolay Industries, Inc., Polymer Optical Fiber, Catalog No. PE0103-10, 1991.

Tomita M, Adachi T. The effect of bed diameter on the behavior of bubbles in gas–solid fluidized beds. J Chem Eng Japan 6:196–201, 1973.

Tsuji Y, Morikawa Y. LDV measurements of an air-solid two-phase flow in a horizontal pipe. J Fluid Mech 120:385–409, 1982.

Tsukada M, Nakanishi D, Horio M. Effect of "transport velocity" in a circulating fluidized bed. In: Avidan A, ed. Circulating Fluidized Bed Technology IV. New York: AIChE, 1994, pp 209–215.

Tsukada M, Ito M, Kamiya H, Horio M. Three-dimension imaging of particle clusters in dilute gas–solid suspension flow. Can J Chem Eng 75:466–470, 1997.

Tsutsui T, Miyauchi T. Fluidity and its influence on behavior in a fluidized bed with fine particles. Kagaku Kogaku Ronbunshu 5:40–46, 1979.

Tung Y, Kwauk M. Dynamics of collapsing fluidized beds. In: Kwauk M, Kunii D, eds. Fluidization Science and Technology. Beijing: Science Press and Gordon and Breach Science Publishers, 1982, pp 155–166.

Umekawa H, Ozawa M, Onodera T, Okura Y, Takenaka N, Matsubayashi M. Visualization of large-particle movement in fluidized-bed by neutron radiography. Proc 6th World Conf Neutron Radiography. Osaka: Gordon and Breach Science Publishers, 1999.

Van Breugel J, Stein J, Vriens R. Isokinetic sampling in a dense gas–solid stream. Proc Instn Mech Engrs 184(113C):18–23, 1969/1970.

Van de Wall R, Soo S. Measurement of particle cloud density and velocity using laser devices. Powder Technol 81:269–278, 1994.

Van den Moortel T, Santini R, Tadrist L, Panataloni J. Experimental study of the particle flow in a circulating fluidized bed using a phase Doppler particle analyser: a new postprocessing data algorithm. Int J Multiphase Flow 23(6):1189–1209, 1997.

Van den Schaaf J, Schouten JC, Johnsson F, van den Bleek CM. Multiple modes of bed mass oscillation in gas–solids fluidized beds. Proc 15th Int Conf Fluidized Bed Combustion. ASME, 1999a.

Van den Schaaf J, Johanson F, Schouten JC, van den Bleek CM. Fourier analysis of nonlinear pressure fluctuations in gas–solid flow in CFB risers—observing solids structures and gas/particle turbulence. Chem Eng Sci 54:5541–5546, 1999b.

Van Velzen D, Flamm H, Langenkamp H, Casile A. Can J Chem Eng 52:156–161, 1974.

Verloop W, Boersma D, Van den Akker H, Hein K. The fluid dynamics of particles in the freeboard of a pressurized fluidized bed combustor. Proc 12th FBC Conf, San Diego, CA, USA, 1993, pp 53–62.

Wang X, Rhodes M, Gibbs B. Solids flux distribution in a CFB riser operating at elevated temperatures. Prep. of the Fluidization VIII, Tours, France, 1995a, pp 41–48.

Wang X, Gibbs B, Rhodes M. Flow structure in a CFB riser—effect of operating temperature. Proc 13th FBC Conf, Orlando, FL, USA, 1995b, pp 663–670.

Watrasiewicz B, Rudd M. Laser Doppler Measurements. Boston: Butterworths, 1976.

Weber M. Handbook of Laser Wavelength. Boston: CRC Press, 1999.

Wei F, Lin H, Cheng Y, Wang Z, Jin Y. Profiles of particle velocity and solids fraction in a high-density riser. Powder Technol 100:183–189, 1998.

Weimer A, Gyure DC, Clough D. Application of a gamma-radiation density gauge for determining hydrodynamic properties of fluidized beds. Powder Technol 44:179–194, 1985.

Weinell C, Dam-Johansen K, Johnsson J. Local up- and downward particle velocities in circulating fluidized beds. Prep Fluidization VIII, Tours, France, 1995, pp 73–80.

Weinstein H, Shao M, Wasserzug L. Radial solid density variations in a fast fluidized bed. AIChE Symp Ser 80(241):117–121, 1984.

Weinstein H, Feindt H, Chen L, Graff R. The measurement of turbulence quantities in a high velocity fluidized bed. In: Potter O, Nicklin D, eds. Fluidization VII. New York: United Engineering Foundation, 1992, pp 305–312.

Wen CY, Yu YH. A generalized method for predicting the minimum fluidization velocity. AIChE J 12:610–612, 1966.

Werdermann C, Werther J. Solids flow pattern and heat transfer in an industrial-scale fluidized bed heat exchanger. Proc 12th FBC Conf, San Diego, CA, USA, 1993, pp 985–990.

Werther J. Measurement techniques in fluidized beds. Powder Technol 102(1):15–36, 1999.

Werther J, Hage B. A fibre-optical sensor for high-temperature application in fluidized bed combustion. In: Large J, Laguerie C, eds. Fluidization VIII. New York: United Engineering Foundation, 1996, pp 577–584.

Werther J, Molerus O. The local structure of gas-fluidized beds. I. A statistically based measuring system. Int J Multiphase Flow 1:103–122, 1973.

Werther J, Rudnick C. Modeling the fluid mechanics of a circulating fluidized bed based on a local flow structure analysis. In: Werther J, Markl H, eds. In-situ Measuring Techniques and Dynamic Modeling of Multiphase Flow Systems SFB 238 Progress Report 1994–1996, Verlag des SFB 238, Hamburg, 1996.

Werther J, Hartge EU, Rensner D. Messtechniken für gas/feststoff-wirbelschichtreaktoren. Chem Ing Tech 62(8):605–613, 1990.

Werther J, Hartge EU, Kruse M. Radial gas mixing in the upper dilute core of a circulating fluidized bed. Powder Technol 70:293–301, 1992.

Werther J, Hartge EU, Rensner D. Measurement techniques for gas–solid fluidized bed reactors. Int Chem Eng 33(1):18–27, 1993.

Westphalen D, Glicksman L. Lateral solid mixing measurements in circulating fluidized beds. Powder Technol 82(2):153–167, 1995.

White C, Dry R. Transmission characteristics of gas in a circulating fluidised bed. Powder Technol 57:89–94, 1989.

Wiesendorf V, Hartge EU, Werther J, Johnsson F, Sterneus J, Leckner B, Montat D, Briand P. The CFB boiler in Gardanne—an experimental investigation of its bottom zone. Proc 15th Conf Fluidized Bed Combustion, Savannah, Georgia, USA, ASME, New York, 1999. FBC99-0151

Win KK, Nowak W, Matsuda H, Hasatani M, Bis Z, Krzywanski J, Gajewski W. Transport velocity of coarse particles in multisolid fluidized bed. J Chem Eng Japan 28(5):535–540, 1995.

Xie H, Geldart D. The time response of pressure probes. Powder Technol 90:149–151, 1997.

Yamada J, Kurosaki Y, Morikawa T. Radiation emitted from fluidized particles adjacent to a heat transfer surface in a fluidized bed. Kikai-Gakkai-Ronbunshu B 62:234–240, 1996.

Yamaki T, Kusakabe K, Morooka S. A probe for detecting the movement of dark-colored particles. Powder Technol 78:189–190, 1994.

Yamazaki M, Fukuta K, Li YH, Tokumoto J. Distribution of porosity of emulsion phase and its effect on conversion in a fluidized bed. J Chem Eng Japan 21(1):47–56, 1988.

Yang WC, Chitester D. Transition between bubbling and turbulent fluidization at elevated pressure. AIChE Symp Ser 84 (262):10–21, 1988.

Yang WC, Keairns D. Solid entrainment rate into gas and gas-solid two phase jets in a fluidized bed. Powder Technol 33:89–94, 1982.

Yang G, Huang Z, Zhao L. Radial gas dispersion in a fast fluidized bed. In: Kunii D, Toei E, eds. Fluidization IV. New York: United Engineering Foundation, 1984, pp 145–153.

Yasui G, Johanson L. Characteristics of gas pockets in fluidized beds. AIChE J 4(4):445–452, 1958.

Yates J, Simons S. Experimental methods in fluidization research. Int J Multiphase Flow 20(suppl):297–330, 1994.

Ye H, Kikuchi R, Nicolai R, Reh L. Characterization of ozone decomposition in a semiindustrial CFB reactor using neutral networks: an experimental study. Werther J, ed. Proc 6th Conf CFB. Wurzburg, Germany, 1999, pp 879–884.

Yerushalmi J, Cankurt N. Further studies of the regimes of fluidization. Powder Technol 24:187–205, 1979.

Yoshida K, Sakane J, Shimizu F. A new probe for measuring fluidized bed characteristics at high temperatures. Ind Eng Chem Fundam 21:83–85, 1982.

Zevenhoven R, Kohlmann J, Laukkanen T, Tuominen M, Blomster AM. Near-wall particle velocity and concentration measurements in circulating fluidised beds in relation to heat transfer. Proc 15th Conf Fluidized Bed Combustion, Savannah, Georgia, USA, 1999.

Zhang H, Johnston P, Zhu J, de Lasa H, Bergougnou M. A novel calibration procedure for a fiber optic solids concentration probe. Powder Technol 100(2/3):200–272, 1998.

Zou B, Li H, Xia Y, Ma X. Cluster structure in a circulating fluidized bed. Powder Technol 78:173–178, 1994.

26

Liquid–Solids Fluidization

Norman Epstein

University of British Columbia, Vancouver, British Columbia, Canada

1 ORIGINS

In seeking the origins of fluidization in general (Brötz, 1952; Leva, 1959; Grace, 1992; Kwauk, 1996), and of liquid fluidization in particular (Di Felice, 1995), it is common to point to the great tome of Agricola (1556), in which a hand jigging operation for ore dressing is described and illustrated. Jigging involves sorting of particles (classification by density) by rapidly alternating surges of liquid (usually water) through a screen supporting the particles. The upward surges represent a series of short-lived fluidizations which are each followed by hindered settling, the short duration of which prevents the development of a significant drag to resist the motion of the particles. This accelerating motion is then primarily determined by the normalized buoyed density of the particles, $(\rho_p - \rho)/\rho_p$, acting under the influence of gravity with little sensitivity to their shape and even size (Brown et al., 1950). In modern jigging operations the screen may be fixed and liquid pulsation effected mechanically, or the screen itself may be given a reciprocating motion; in Agricola's time, the same effect was created by hand oscillation of the screen. The resemblance between jigging and what we now call fluidization, i.e., the continuous upward (or downward) flow of fluid relative to a mobile swarm of particles, is thus small.

Liquid fluidization in the latter sense was used for both sorting (classification by density) and sizing (classification by size) for more than a century (Richards, 1893) and was commonly referred to as "teetering" in

the mineral dressing literature (Gaudin, 1939; Richards and Locke, 1940; Taggart, 1953). A liquid-fluidized bed was referred to as a "teeter column" or as being in the "teeter condition" (Hancock, 1936). It was only after the term fluidization was coined to describe mobile swarms of particles contacted upwardly by gases (ca. 1940) that the same term began to be used for analogous liquid–solid contacting, and the term teeter bed was replaced by liquid-fluidized bed.

2 PARTICULATE FLUIDIZATION

2.1 Introduction

Let us start with the most conventional of liquid-fluidized beds shown schematically in Fig. 1. The cylindrical vertical column is commonly circular in cross section, though it may also be rectangular, e.g., square. The function of the calming or homogenizing section, usually located upstream of the distributor, is to produce, in conjunction with the distributor, as radially uniform a liquid velocity distribution in the fluidized bed as possible, thus eliminating or at least minimizing any tendency toward channeling or bulk circulation ("gulf streaming"). The system shown in Fig. 1 is a semibatch (or semicontinuous) operation, i.e., continuous flow of liquid for an unchanging batch of the fluidized solids, which in this case are bottom-restrained. We shall see later that many other

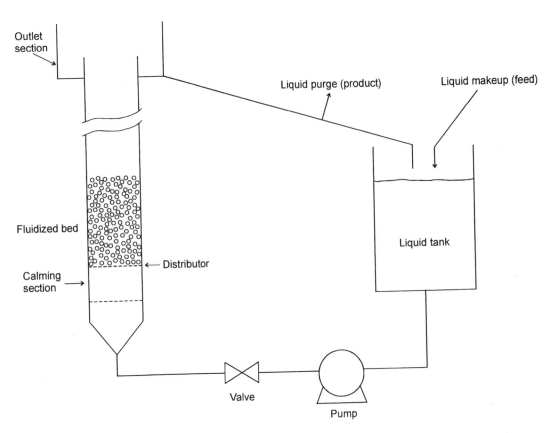

Figure 1 Semibatch liquid fluidization loop. (After Couderc. 1985.)

operating modes are possible under the rubric of liquid–solid fluidization.

We shall also first consider nonporous fluidized particles that are all uniform in density, size, and shape. In their pioneer study of individual particle motion in such a liquid-fluidized bed, Handley et al. (1966) found that, in the absence of gulf streaming, "uniform fluidization" resulted and was "characterized by a homogeneous random motion of the particles and by a constant mean fluid-velocity and momentum distribution throughout the bed," but with larger particle velocity fluctuations (viz. axial/radial ≈ 2.3, as later verified by Carlos and Richardson, 1968a; Latif and Richardson, 1972; Kmiec, 1978) and displacements in the vertical direction than in the horizontal, i.e., $D_{ia} \gg D_{ir}$. We shall therefore begin by assuming, after Couderc (1985), that "liquid–solid fluidization results in stable operation and beds which are homogeneous, with a spatially uniformly distributed concentration of solid particles." This idealized assumption of "particulate fluidization" will be subsequently subjected to significant qualification.

2.2 Buoyancy and Drag

Consider now a representative single particle of volume V and density ρ_p in such a fluidized swarm of monodispersed particles. Macroscopically steady-state fluidization requires that downward gravitational force = upward buoyant force + upward drag, i.e.,

$$V\rho_p g = B + F_D \tag{1}$$

There is a major disagreement in the literature on the correct expression to use for the buoyancy B. The conventional formula for B advocated, for example, by Clift et al. (1987), Jean and Fan (1992), and Clift (1993), is simply the buoyant force under static (no-flow) conditions,

$$B = V\rho g \tag{2}$$

The counterargument, presented among others by Gibilaro et al. (1984, 1987) and Astarita (1993), posits that

$$B = V\rho_B g \tag{3}$$

where ρ_B is the fluidized bed suspension density given by

$$\rho_B = \rho_p(1 - \varepsilon) + \rho\varepsilon \qquad (4)$$

The grounds for this position is that under fluidized bed conditions, the total pressure gradient, $-dP/dz$, which develops to support the suspension, is given by

$$-\frac{dP}{dz} = \rho_B g \qquad (5)$$

so that the effective or dynamic buoyancy in the fluidized state is considerably greater than under static conditions. Combination of Eqs. (1) and (2) gives

$$F_D = V(\rho_p - \rho)g \qquad (6)$$

while combination of Eqs. (1), (3), and (4) results in

$$F_D = V(\rho_p - \rho)\varepsilon g \qquad (7)$$

Note that, by virtue of the greater value in Eq. (1) of B given by Eq. (3) than by Eq. (2), F_D via Eq. (7) is less than F_D via Eq. (6) by a factor of $\varepsilon(\varepsilon_{mf} < \varepsilon < 1)$. Advocates of Eq. (6) argue that it is the drag force F_D that causes the dissipation of mechanical energy and that this dissipation must include the energy required to convert the original static buoyancy of the particle in the unfluidized state to the subsequent dynamic buoyancy in the fluidized state. Partisans on both sides of the argument agree that the frictional pressure gradient, $-dp_f/dz$, of the fluidized bed must be given by the specific weight of the suspension corrected for hydrostatic head, i.e.,

$$-\frac{dp_f}{dz} = \rho_B g - \rho g \qquad (8a)$$

$$= [\rho_p(1 - \varepsilon) + \rho\varepsilon]g - \rho g \qquad (8b)$$

$$= (1 - \varepsilon)(\rho_p - \rho)g \qquad (8c)$$

Since V is the volume of a representative particle in a homogeneously fluidized bed, the representative volume of liquid associated with this particle (in a unit cell so circumscribed that the liquid-to-solid ratio is the same as in the bed as a whole) is $V\varepsilon/(1 - \varepsilon)$ and the correspondingly representative volume of bed is simply $V/(1-\varepsilon)\,[= \{V\varepsilon/(1 - \varepsilon)\} + V]$.

To get from Eq. (6) to Eq. (8), one writes that

$$-\frac{dp_f}{dz} = \frac{\text{drag force on representative particle}}{\text{volume of bed associated with}}$$
$$\text{representative particle}$$

$$= \frac{F_D}{V/(1 - \varepsilon)}$$

$$= (1 - \varepsilon)(\rho_p - \rho)g$$

However, to get from Eq. (7) to Eq. (8), one must write that

$$-\frac{dp_f}{dz} = \frac{\text{drag force on representative particle}}{\text{volume of liquid associated with}}$$
$$\text{representative particle}$$

$$= \frac{F_D}{V\varepsilon/(1 - \varepsilon)}$$

$$= (1 - \varepsilon)(\rho_p - \rho)g$$

If, as is commonly assumed, $-dp_f/dz$ represents frictional force per bed volume (particles plus liquid) rather than simply frictional force per volume of liquid in the bed, then that is an additional argument in favor of Eqs. (2) and (6) over (3) and (7).

We will here adopt the buoyancy convention incorporated in Eqs. (2) and (6) rather than (3) and (7). However, since the latter convention has made significant inroads into the literature, the reader is referred to Table 1 of Khan and Richardson (1990) for useful drag coefficient relationships based on the above alternate conventions, as well as on alternate definitions of the characteristic liquid velocity, and to conversions by Jean and Fan (1992) of several equations incorporating buoyancy as defined by Eq. (3) [derived by Foscolo et al. (1983, 1989), Foscolo and Gibilaro (1984, 1987) and Gibilaro et al. (1985a, 1986)] to the corresponding equations based on buoyancy as defined by Eq. (2).

2.2 Hydrodynamic Representation

After the column of Fig. 1 has been filled with liquid and the monodispersed particles introduced as a fixed packed bed, but before any velocity is imparted to the liquid, the total pressure drop, $-\Delta P$, across height H, measured from immediately above the distributor to a plane well above the fixed or any subsequently fluidized bed, is given by

$$-\Delta P = P_1 - P_2 = \rho g H \qquad (9)$$

where P_1 and P_2 are each determined independently, e.g., by a pressure gauge or transducer. The dynamic pressure drop, $-\Delta p$, due to any motion of the liquid, as measured for example by a differential manometer, is then zero. Once flow is imparted to the liquid, the total pressure drop increases and, at steady conditions, is given by

$$-\Delta P = -\Delta p + \rho g H$$

so that

$$-\Delta p = -\Delta P - \rho g H \qquad (10)$$

Assuming the cross-sectional area of the column is unchanging with bed level, e.g., the bed is untapered, then, neglecting minor bed entrance, bed exit, and wall friction effects, the pressure drop, $-\Delta p$, due to the liquid motion is equivalent to the frictional pressure drop, $-\Delta p_f$, due to flow through the bed of particles. A typical plot of $-\Delta p_f$ as a function of liquid velocity is shown in Fig. 2.

AB corresponds to an immobile fixed bed, the pressure drop across which increases as the velocity increases. From B to C, across the blunt maximum caused by interlocking of particles followed by their rearrangement, the bed, if it has been densely packed, loosens up with each incremental velocity increase; finally, at and beyond C, it continues to expand in the mobile fluidized state, during which the pressure drop remains constant. The constant frictional pressure drop is then given by

$$-\Delta p_f = L(1 - \varepsilon)(\rho_p - \rho)g \qquad (11)$$

which is simply the integral form of Eq. (8). As D is approached, the bed level disappears; beyond D, all particles are carried out of the column by the ascending liquid, and in the ideal case where wall effects, including radial gradients of longitudinal velocity, are absent, the liquid velocity at D is U_0, the free settling terminal velocity of the particles. If the flow is reversed

before the bed disappears, the path described is DCE because when defluidization occurs, the bed settles at its random loose (Oman and Watson, 1944; Eastwood et al., 1969) or minimum fluidization (Leva, 1959) voidage, ε_{mf}. The point C is easier to determine and to reproduce by the velocity-decreasing rather than the velocity-increasing route, although once the packed bed is in the fully expanded or "random loose" condition, the same path is followed (i.e., no hysteresis) on both increasing and decreasing the liquid velocity. The superficial velocity at point C is the incipient or minimum fluidization velocity, U_{mf}.

Plots corresponding to Fig. 2 of bed height L, voidage ε, and liquid–solids bulk density, each as a function of superficial velocity, are shown in Figs. 3, 4, and 5, respectively. For a total mass M of particles, the volumetric particle concentration in either a fixed or a fluidized bed is given by

$$1 - \varepsilon = \frac{M}{\rho_p A L} \qquad (12)$$

Therefore, for a fluidized bed, the product of bed height and particle concentration is constant and can be obtained from either Eq. (11) or Eq. (12):

$$L(1 - \varepsilon) = \frac{-\Delta p_f}{(\rho_p - \rho)g} = \frac{M}{\rho_p A} = \text{constant} \qquad (13)$$

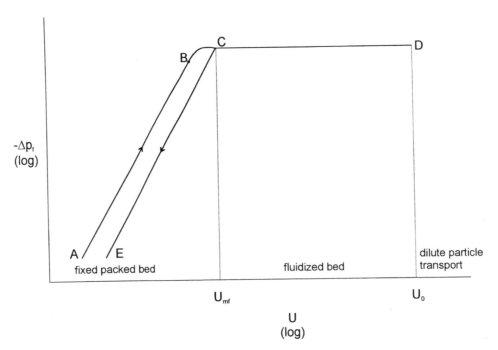

Figure 2 Frictional pressure drop as a function of liquid superficial velocity for monodispersed particles.

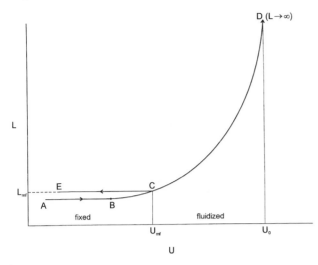

Figure 3 Bed height as a function of liquid superficial velocity.

Thus as L increases, ε increases accordingly, as shown in Figs. 3 and 4. The minimum fluidization voidage, ε_{mf}, is typically about 0.4 for monosize smooth spheres in the absence of significant wall effect ($d_p/D_c \leq 0.01$) but varies considerably with particle geometry (Leva, 1959; Eastwood et al., 1969). Empirically, the slope of log ε vs. log U for liquid fluidization is commonly constant for fixed particle and liquid properties up to $\varepsilon = \varepsilon_c$, where in most instances, $0.8 \leq \varepsilon_c \leq 0.9$ (Wilhelm and Kwauk, 1948; Garside and Al-Dibouni, 1977; Riba and Couderc, 1977; Chong et al., 1979). The characteristics of this slope ($= 1/n$)

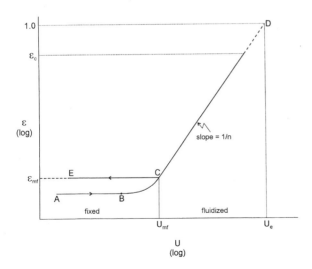

Figure 4 Bed voidage as a function of liquid superficial velocity.

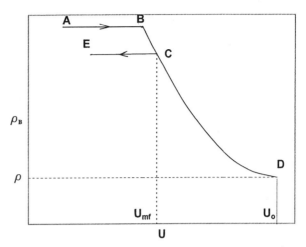

Figure 5 Liquid–solids bulk density as a function of liquid superficial velocity.

will be examined in Sec. 4. The value of the fluidized bed (CD) suspension density in Fig. 5 is given by Eq. (4) as well as by

$$\rho_B = \frac{-\Delta P}{Lg} \tag{14}$$

which is obtained by integrating Eq. (5) from $z = 0$ to $z = L$.

3 MINIMUM FLUIDIZATION

Let us now address the problem of determining the minimum fluidization velocity, U_{mf}, for hard, nonporous particles of uniform size, shape, and density, knowing ε_{mf}, the minimum fluidization voidage. This problem is essentially the same for liquid fluidization as it is for gas fluidization. At the point of incipient fluidization, we will equate the frictional pressure gradient as given by Eq. (8) at $\varepsilon = \varepsilon_{mf}$ with that for a single phase Newtonian fluid flow through the corresponding "loose" packed bed, as given by the widely accepted Ergun (1952) equation:

$$\frac{-dp_f}{dz} = (1 - \varepsilon_{mf})(\rho_p - \rho)g$$
$$= \frac{150 U_{mf}\mu(1 - \varepsilon_{mf})^2}{\phi^2 d_p^2 \varepsilon_{mf}^3} + \frac{1.75 U_{mf}^2 \rho(1 - \varepsilon_{mf})}{\phi d_p \varepsilon_{mf}^3} \tag{15}$$

in which the product of sphericity ϕ and equivolume sphere diameter d_p has been substituted for the equivalent surface-volume sphere diameter d_{sv}. Algebraic manipulation and rearrangement of this equation results in

$$\text{Re}_{mf}^2 + \frac{150}{1.75} \cdot \frac{1 - \varepsilon_{mf}}{\phi} \cdot \text{Re}_{mf} - \frac{\phi \varepsilon_{mf}^3 \text{Ar}}{1.75} = 0 \qquad (16)$$

where $\text{Re}_{mf} = d_p U_{mf} \rho / \mu$ and $\text{Ar} = d_p^3 \rho(\rho_p - \rho)g/\mu^2$. The physically realistic solution of quadratic Eq. (16) is

$$\text{Re}_{mf} = (C_1^2 + C_2 \text{Ar})^{1/2} - C_1 \qquad (17)$$

where

$$C_1 = \frac{150(1 - \varepsilon_{mf})}{2\phi(1.75)} = 42.86 \frac{(1 - \varepsilon_{mf})}{\phi} \qquad (18)$$

and

$$C_2 = \frac{\phi \varepsilon_{mf}^3}{1.75} = 0.5714 \phi \varepsilon_{mf}^3 \qquad (19)$$

Thus knowing d_p, ρ_p, ρ, and μ, knowledge of both ε_{mf} and ϕ is also required in order to solve Eq. (17) for Re_{mf} and hence U_{mf}. Except in the case of monosized smooth spheres for which $\phi = 1$ and $\varepsilon_{mf} \approx 0.4$, the values of ϕ and ε_{mf} are more often than not unavailable, though ε_{mf} is relatively easy to measure (Eastwood et al., 1969). Starting with Wen and Yu (1966), based on assumptions of ϕ and ε_{mf} or on equations relating ϕ and ε_{mf}, or based more directly on empirical fits to experimental data on U_{mf}, many investigators have proposed different combinations of

values for the parameters C_1 and C_2. Several of these are listed in Table 1.

There is evidence that a fluidized bed at a given porosity offers somewhat less resistance to fluid flow than the corresponding fixed bed at the same porosity (Happel and Epstein, 1954; Happel and Brenner, 1957; Richardson and Meikle, 1961a; Barnea and Mednick, 1975), owing to the greater degree of freedom to arrange themselves possessed by the mobile fluidized than by the immobile fixed particles, though this effect is undoubtedly less pronounced at $\varepsilon = \varepsilon_{mf}$ than at higher voidages. Furthermore, the constants 150 and 1.75 in the Ergun equation are not universally accepted, e.g., Carman (1937), based on considerable experimental data, proposed 180 instead of 150 for the viscous flow constant. Therefore the empirical fit basis above is, for the specific conditions fitted, more reliable than the others. The values of C_1 and C_2 in Table 1 have in most cases been subjected by their proponents to this criterion and are simultaneously based, in some cases, on assumptions about ε_{mf} and ϕ.

Since the differences in results for Re_{mf} obtained in applying the various combinations of C_1 and C_2 in Table 1 can be significant (Bin, 1986), some discrimination is warranted. If good estimates of both ε_{mf} and ϕ can be made (cf. Leva, 1959), then Eq. (17) should be used in conjunction with Eqs. (18) and (19). If ε_{mf} is

Table 1 Parameters C_1 and C_2 in Eq. (17)

Investigation	C_1	C_2	Applicability claims
Wen and Yu, 1966	33.7	0.0408	Various particle shapes
Bourgeois and Grenier, 1968	25.46	0.0382	Spheres only
Richardson, 1971	25.7	0.0365	Spheres only
Saxena and Vogel, 1977	25.28	0.0571	Various particle shapes
Babu et al., 1978	25.25	0.0651	Various particle shapes
Grace, 1982	27.2	0.0408	Various particle shapes
Foscolo et al., 1983	25.74	$\varepsilon_{mf}^{4.8}/0.336$	Spheres
Thonglimp et al., 1984	31.6	0.0425	Various particle shapes
Chitester et al., 1984 Lin and Fan, 1997	28.7	0.0494	High pressure; spheres only for liquid fluidization
Nakamura et al., 1985	33.95	0.0465	High temp. & press., spheres only
Lucas et al., 1986	29.5	0.0357	"Round" granular particles, $0.8 \leq \phi \leq 1$
	32.1	0.0571	"Sharp" jagged particles, $0.5 \leq \phi < 0.8$
	25.2	0.0672	"Other" particles, e.g., rings, $0.1 < \phi < 0.5$
Chen, 1987	$33.67\phi^{0.1}$	$0.0408/\phi^{0.45}$	All particle shapes, but must estimate $\phi = d_{sv}/d_p = d_p^2/d_s^2$
Chyang and Huang, 1988	33.3	0.0333	Granular particles
Tannous et al., 1994	25.83	0.043	Many particle shapes

unknown, but a reasonable estimate of ϕ is available, then Eq. (17) should be applied with C_1 and C_2 evaluated from the expressions of Lucas et al. (1986) given in Table 1. Alternately, the previous procedure can be used, after measuring ε_{mf} (as described immediately below) or estimating it from one of the simpler and more creditable of several proposed empirical approximations in the literature for relating ε_{mf} to ϕ, such as that of Chen (1987),

$$\varepsilon_{mf} = \frac{0.415}{\phi^{0.483}} \tag{20}$$

for $0.2 < \phi \leq 1$ and $d_p > 0.05\,\text{mm}$ (Wen and Yu, 1966). Except for spheres ($\phi = 1$), Eq. (20) shows ε_{mf} to be somewhat smaller than the corresponding "random loose" voidages of Brownell et al. (1950), but larger than $\varepsilon_{mf}(\phi)$ from empirical equations proposed by others (Wen and Yu, 1966; Limas-Ballesteros, 1980, as stated by Couderc, 1985). If neither U_{mf} nor ϕ are known, then Eq. (17) should be used with the values of C_1 and C_2 given for the three classes of particles listed beside Lucas et al. (1986) in Table 1.

Measurement of ε_{mf} is best effected by fluidizing the particles involved with a liquid, e.g., water, in the same column for which the subsequent fluidization operation will be carried out and then rapidly shutting off the liquid flow, the volume of particles (M/ρ_p) divided by the volume of collapsed bed (AL_{mf}) representing $1 - \varepsilon_{mf}$. If that is not feasible, a simpler procedure for reproducing the random loosed voidage, which at least for granular particles ($\phi > 0.8$) is a good approximation of ε_{mf}, is to invert a covered cylindrical vessel partially filled with the particles and then quickly righting it. If this vessel is smaller in diameter D_c than the eventual fluidizing column, a correction for wall effect may· be required if $D_c/d_p < 100$ (Eastwood et al., 1969). Enough particles should be used by either method so that $L_{mf} \geq 2D_c$.

A second route to U_{mf} is by putting $U = U_{mf}$ (or $Re = Re_{mf}$) and $\varepsilon = \varepsilon_{mf}$ in one of the many fluidized bed expansion equations relating ε to U (or Re) that have been proposed in the literature, a subject discussed below. This could provide a reasonable second estimate of U_{mf} but, because the expansion equations are usually correlations based on the whole range of ε from ε_{mf} to ε_c or even to unity, they are likely to be less accurate at the ε_{mf} extremity than an equation such as Eq. (17) which is tailored to this extremity.

A third alternative for U_{mf} is the use of empirical equations other than Eq. (17), which have usually been developed for more restrictive conditions. Thus, for liquid fluidization of *spheres*, based on an empirical

fit for a wide variety of solid–liquid combinations over the range of $Re_{mf} = 10$–1000, Riba et al. (1978) have proposed

$$Re_{mf} = 0.0154 Ar^{0.66} \left(\frac{\rho_p - \rho}{\rho}\right)^{0.04} \tag{21}$$

The fact that this equation does not require any specification of ϕ or ε_{mf} is distinctly in its favor, since equations such as (17)–(19) above are very sensitive to ε_{mf}, and even for spheres ($\phi = 1$) an exact specification of ε_{mf} can be elusive (Eastwood et al., 1969). The density ratio, $(\rho_p - \rho)/\rho$, can safely be eliminated from this equation since, for all the solid–liquid combinations investigated, the factor $\overline{(\rho_p - \rho/\rho)}^{0.04}$ was never greater than the scatter of the experimental data. (The index on Ar could then be rounded off to 2/3). Equations for U_{mf} by all three routes are listed in Table 1.6 of Couderc (1985).

Both ε_{mf} and hence also U_{mf} (see Fig. 4) can be significantly reduced by pulsing the fluidizing liquid (El-Temtamy and Epstein, 1986). On the other hand, neither is affected significantly by changes in solids wettability (Mitra and Epstein, 1978). For non-Newtonian liquids that can be represented by the power-law model, an approach analogous to Eq. (17) has been successful in predicting U_{mf} for liquid–solid fluidization of monosize spheres (Miura and Kawase, 1997, 1998).

For solids of *mixed sizes*, assuming any hysteresis effects have been eliminated by first fluidizing the solids and then gradually reducing the liquid velocity to zero, there is no longer a sharp transition from a fixed to a fluidized bed as in Fig. 2. Instead, on raising the velocity, there is a gradual transition from the fully fixed bed to the fully fluidized bed (Couderc, 1985), as illustrated in Fig. 6. The initial deviation from the fully fixed bed line occurs at U_{bf}, the velocity for beginning fluidization; the fully fluidized region starts at U_{tf}, the velocity for total fluidization; and the intersection of the extrapolated fixed bed line with the extrapolated fluidized bed line happens at U_{mfa}, the apparent minimum fluidization velocity (Obata et al., 1982; Casal and Puigjaner, 1983). U_{bf} and U_{tf} can be calculated approximately from Eq. (17), with d_p taken as the smallest and the largest particle size, respectively, while U_{mfa} may be estimated from the same equation, with d_p evaluated as the Sauter or reciprocal mean diameter (Jean and Fan, 1998),

$$d_p = \frac{1}{\displaystyle\sum_{i=1}^{N} (v_i/d_{pi})} \tag{22}$$

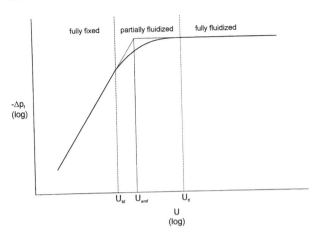

Figure 6 Frictional pressure drop vs. liquid superficial velocity for particles of mixed sizes, shapes, and/or densities. (After Couderc, 1985.)

four transition velocities between an immobile packed bed and full fluidization: (1) the minimum velocity of partial fluidization, U_{mpf}, corresponding to the point at which the pressure drop is a maximum as the velocity is increased; (2) the minimum velocity of full fluidization, U_{mff}, above which the bed expands at a constant pressure drop; (3) the maximum velocity of partial defluidization, U_{mpd}, below which the bed is no longer totally fluidized as the velocity is decreased; (4) the maximum velocity of full defluidization, U_{mfd}, below which a fixed packed bed is fully restored on reducing the velocity. In general, $U_{mfd} < U_{mpf} < U_{mff} = U_{mpd}$, and equations for each of these velocities have been developed by Peng and Fan based on integration of Eq. (15) using the conical boundary conditions appropriate to each.

which, for particles of fixed shape, is the equivalent of the surface-volume mean diameter. Figure 6 would also apply to particles of mixed shapes and densities (as well as mixed sizes), for which estimation of U_{bf} and U_{tf} would require that the particle properties in Eq. (17) be those of the particles with the lowest and the highest individual values of U_{mf}, respectively, while estimation of U_{amf} would require that ρ_p in Eq. (17) be the mean particle density,

$$\rho_p = \frac{\sum_{i=1}^{N} M_i}{\sum_{i=1}^{N} (M_i/\rho_{pi})} \qquad (23)$$

In the case of an upwardly diverging *tapered* or *conical* bed (e.g., $\theta = 5°–30°$) of uniform particles, the bottom of the bed is fluidized before the top because of its higher liquid velocity (Kwauk, 1992). The result is that, on increasing the velocity through the initially fixed packed bed, a very much higher and sharper peak pressure drop is reached than in Fig. 2 before the pressure drop falls again and full fluidization is achieved (Peng and Fan, 1997). The plot of frictional pressure drop vs. liquid velocity then resembles that of a gas-spouted bed (Mathur and Epstein, 1974), as do the corresponding fluid–particle mechanics, and, as in the case of both fluidization and spouting, reducing the velocity produces a hysteresis effect whereby there is no longer (or at most barely) a discernible maximum in the plot of $-\Delta p_f$ vs. U. Peng and Fan (1997) identified

4 FLUIDIZED BED EXPANSION

4.1 Introduction

Crucial to the design of a fluidized bed is a quantitative knowledge of the bed expansion as a function of the liquid superficial velocity, i.e., of either L or ε [the two being interrelated by Eq. (13)] as a function of U. The chaotic behavior of a fluidized bed, in contrast, say, to viscous flow through an ordered (e.g., simple cubic) array of immobilized spheres, renders measurements from the former inherently less reproducible than from the latter.

The equations that have been developed for predicting the frictional pressure drop due to flow through a fixed, including an expanded, packed bed (of which there are many in the literature) yield a relationship between $-dp_f/dz$, ε, and U that, when combined with Eq. (8) to eliminate $-dp_f/dz$, give an equation relating ε to U. Because, as mentioned earlier, particles in a fluidized bed, in contrast to those in a rigid array, are relatively free to move around and arrange themselves so as to minimize or at least decrease the resistance to flow (Richardson and Meikle, 1961), therefore for a given value of ε, U for a fluidized bed will usually be larger than U for a fixed bed under the same pressure gradient. Consequently, the use of a packed bed equation will in most cases underestimate U for a given ε, or overestimate ε and L for a given U (Happel and Epstein, 1951). If, for example, we remove the *minimum* fluidization (subscript mf) restriction from Eqs. (15)–(19), then Eq. (17) becomes

$$Re = \left(\frac{42.86^2(1-\varepsilon)^2}{\phi^2} + 0.5714\phi\varepsilon^3 Ar \right)^{1/2}$$
$$- \frac{42.86(1-\varepsilon)}{\phi} \qquad (24)$$

For a bed of smooth spheres ($\phi = 1$) with $d_p = 0.5$ mm and $\rho_p = 2000$ kg/m^3, contacted by water at 20°C, $Ar = 1225$. If the bed is immobilized at $\varepsilon = 0.6$, then by Eq. (24), $Re = 3.95$. If, however, in the absence of wall effect ($k = 1$), Eqs. (34), (33), and (32) below are used to compute n, Re_0, and Re, respectively, then for the same spheres fluidized by the same liquid to a voidage of 0.6, $n = 3.09$, $Re_0 = 27.9$, and $Re = 5.76$. Thus the required fluidization velocity is in this case almost 50% greater than that predicted by Eq. (24). (It should be noted that for voidages greater than about 75%, the Ergun (1952) equation, from which Eq. (24) originates, starts progressively to break down even as a predictor of pressure drop through an immobilized bed, and it fails entirely to approach the correct limit as ε approaches unity. Several other immobilized bed equations in the literature have been proposed to remedy this deficiency.)

The relative freedom of particle movement characteristic of fluidization is shared also by the initial constant rate stage of sedimentation, of which there is an enormous literature, including a large number of studies in which the two operations are jointly or interchangeably correlated. Assuming an equivalent distribution of particles and its fluctuations, it is easily shown that the liquid superficial velocity required to fluidize a bed of particles to a given voidage is equal (though opposite in sign) to the initial hindered settling velocity at the same voidage of these particles relative to the walls of a sedimentation vessel with an impervious bottom. In the former case, the relative velocity between the moving liquid and the "stationary" particles is $U/\varepsilon - 0 = U/\varepsilon$. In the latter case, the downward motion of the particles at velocity V_p and volumetric flux $V_p(1-\varepsilon)$ results in an upward movement of the displaced liquid at a linear velocity equal to $-V_p(1-\varepsilon)/\varepsilon$. The relative velocity between liquid and particles is therefore $-V_p(1-\varepsilon)/\varepsilon - V_p = -V_p/\varepsilon$. Assuming, after Mertes and Rhodes (1955) and Lapidus and Elgin (1957), that for monodisperse fluid–particle systems, the relative or "slip" velocity is a unique function of voidage (or vice versa) irrespective of the flow direction of either fluid or particles, it follows that $U = -V_p$. Thus plots of $|V_p|$ vs. ε have frequently been interpreted as equivalent to those of U vs. ε. Some caution should be exercised in assuming this

equivalence, as there is some evidence that the design of the distributor (which is absent in sedimentation), through its effect on the uniformity of the liquid flow, may exert some influence on the expansion behavior of a liquid fluidized bed (Adler and Happel, 1962; Jean and Fan, 1989); in addition, the upward moving liquid in the fluidization column is more likely to develop a boundary layer induced velocity profile radially (Neužil and Hrdina, 1965) than the backflowing liquid in the sedimentation vessel, especially at low particle concentrations (high voidages). Sedimentation following agitation of the suspension and/or vertical rotation of the sedimentation vessel may give rise to particle settling velocities somewhat different from sedimentation initiated by first fluidizing the particles with superficial liquid velocity U and then allowing them to settle without liquid flow (Jean and Fan, 1989), $|V_p|$ for the latter then being closest to U. In general, the differences between the velocity–voidage relationships for sedimentation as opposed to fluidization can be considered second-order discrepancies, especially when compared with the larger differences between either of these operations and flow through an immobilized packed bed, so that for most engineering purposes, results for constant rate hindered settling can be applied with little or no modification to liquid fluidization.

4.2 Theoretical Models

Analytical solutions to the bed expansion problem, without resort to empiricism, are almost entirely limited to spheres in creeping (i.e., viscous or Stokes) flow, $Re_0 < 0.2$. Many of these are discussed by Happel and Brenner (1965) and later by Jean and Fan (1989).

It is instructive, after Saffman (1973), to compare the expansion characteristics that are predicted for spheres in a regular periodic array, in a random array in which the particles are held rigidly, and in a random array of freely moving particles. The first case corresponds to a geometrically ordered immobilized packing, the second to a randomly immobilized packing, and only the third resembles a "homogeneously" fluidized bed. Unfortunately, the rigorous statistical solution of the Navier–Stokes equation in creeping flow (Happel and Brenner, 1965) for this third case, by Batchelor (1972), is limited to very dilute suspensions (e.g., $c = 1 - \varepsilon < 0.05$). This solution is

$$\frac{U}{U_0} = 1 - 6.55c \qquad (25)$$

As an example of a regular periodic array, also dilute, we can cite the analytical solution of Hasimoto (1959) for a simple cubic lattice:

$$\frac{U}{U_0} = 1 - 1.76c^{1/3} \qquad (26)$$

For both a face-centered cubic and a body-centered cubic array, the coefficient of $c^{1/3}$ in Eq. (26) changes only slightly—to 1.79 (Hasimoto, 1959).

Rigid random arrays have generally been simulated by cell models that have not been limited to dilute suspensions. An early example of a cell model is that of Brinkman (1947), who considered flow past a single sphere in a porous medium of permeability κ. The flow is described by an equation that collapses to Darcy's (1856) law (in its post-Darcy form, which includes viscosity) for low values of κ and to the creeping flow version of the Navier–Stokes equation for high values of κ. His solution is

$$\frac{U}{U_0} = 1 + 0.75c\left(1 - \sqrt{(8/c) - 3}\right) \qquad (27)$$

which, though it breaks down for $\varepsilon \le 1/3$ (i.e., $c \ge 2/3$, for which $U/U_0 \le 0$), nevertheless gives reasonable answers in the fluidization region, $0.4 < \varepsilon < 1$, and even agreement with the lower values of U/U_0 reported in the literature for liquid-fluidized beds at low Re and $\varepsilon = 0.5$–1 (Verschoor, 1951). To render Eq. (27) comparable to Eqs. (25) and (26), we rewrite it for the case of $c \to 0$, for which it collapses to

$$\frac{U}{U_0} \approx 1 - \left(1.5\sqrt{2}\right)c^{1/2} \approx 1 - 2.12c^{1/2} \qquad (27a)$$

A more rigorous cell model, inasmuch as it involves a solution of the Navier–Stokes equation for creeping flow rather than a modification of that equation as in the case of Brinkman, is that of Happel (1958). In this case the basic cell is that of a single sphere surrounded by a concentric spherical envelope of fluid, the volume of which bears the same ratio to the volume of the cell as bed voidage does to unity. The crucial feature of this model is that the outer surface of the fluid envelope is frictionless (zero shear stress), so that it is often referred to as the "free surface model." The solution is

$$\frac{U}{U_0} = \frac{3 - 4.5c^{1/3} + 4.5c^{5/3} - 3c^2}{3 + 2c^{5/3}} \qquad (28)$$

which at low concentrations reduces to

$$\frac{U}{U_0} \approx 1 - 1.5c^{1/3} \qquad (28a)$$

A comparable cell model is that of Kuwabara (1959), the only difference being that the spherical cell surface is in this case assumed to be at zero vorticity rather than at zero shear stress. The coefficient of $c^{1/3}$ in Eq. (28a) for dilute suspensions becomes 1.8 in Kuwabara's solution, instead of 1.5.

It is notable that in the equations for both the ordered arrangement of spheres and the cell models, $1 - (U/U_0)$ for dilute suspension is directly proportional to $c^{1/3}$ (or $c^{1/2}$ in the case of Brinkman's model) rather than to c as in Eq. (25), a fact that would appear to render those equations inadequate for fluidization or sedimentation purposes (Batchelor, 1972; Saffman, 1973). If we compare the actual values of $1 - (U/U_0)$ predicted by the above equations at $c = 0.01$, we find that $1 - (U/U_0) = 0.0655$, 0.379, 0.212, and 0.323 from Eqs. (25), (26), (27a), and (28a), respectively, which, combined with experimental findings that values of $1 - (U/U_0)$ for uncharged settling spheres at $c = 0.01$ are even lower than 0.0655 (Buscall et al., 1982; Tackie et al., 1983), confirms this inadequacy, especially for dilute suspensions.

Two attempts have been independently made to modify the free surface model of Happel (1958) so that it includes the additional degrees of freedom of movement characteristic of fluidization or sedimentation as opposed to a fixed array of spheres. Jean and Fan (1989) incorporated a vertical line-up of particle pairs into doublets to account for possible alignment (which reduces flow resistance) in a fluidized bed, and they also incorporated a normal distribution function of ε (= 0.4–1.0), with the standard deviation as a system parameter, to account for voidage nonuniformities. The standard deviation varied with distributor design and liquid velocity, and by nonlinear regression of experimental results at low Re_0 was found to reach a maximum near an average bed voidage of 0.75, which agrees roughly with Al-Salim's (Eisenklam, 1967) findings at moderate Re_0 and with Trupp's (1968) findings at high Re_0 (in the Newton region, $Re_0 > 500$) that the root-mean-square of the local voidage fluctuations (and their frequency too, for Al-Salim's data) are a maximum at an average $\varepsilon \approx 0.70$. The values of U/U_0 thus predicted by Jean and Fan fall into closer line with experimental values at low Re_0 than does the unmodified Eq. (28). More recently Smith (1998) modified Eq. (28) by allowing the quantity of liquid in each spherical cell to vary randomly, subject only to the constraints that the minimum local voidage = 0.2595, corresponding to a rhombohedral close packing (which is the densest possible for monosize spheres), and that the overall voidage be maintained at the specified value (between 0.5 and 1). Improved agreement with empirical results was again obtained and, unlike the solution of Jean and Fan (1989) or even Eq. (25) above, the present solution predicts sufficient particle clustering at low particle concentrations that U/U_0 exceeds unity at $c \leq 0.03$, approaching 1.07 as c approaches zero. Values of U/U_0 exceeding unity to as high as 1.07 have been observed between $c = 0.001$ and $c = 0.03$ by Kaye and Boardman (1962), who explained their sedimentation results by the assumption of cluster formation, which at low particle concentrations overrides the small return flow effect associated with hindered settling. For even smaller concentrations ($c < 0.001$), the values of U/U_0 measured by Kaye and Boardman, unlike the theoretical values of Smith, reduce as expected to unity. The minor defect in Smith's theory is inconsequential to fluidization practice, in which volumetric particle concentrations are invariably larger than 0.001.

The zero vorticity cell model has been extended numerically to intermediate Reynolds numbers for spheres (LeClair and Hamielec, 1968), and both the zero vorticity and the free-surface cell models (unmodified) to spheroids at low Reynolds numbers (Epstein and Masliyah, 1972). However, these extensions are more applicable to immobilized packed beds than to fluidized beds, and only in the absence of turbulence, which for unexpanded, fixed packed beds of spheres develops at Re > 110–150 (Jolls and Hanratty, 1966).

Grbavčić et al. (1991) have developed an "intuitive" variational method of predicting U for any given value of ε that is not restricted to any given flow regime. The method, which has been applied successfully to water fluidization of glass spheres over the range of $Re_0 = 180$–920 (Grbavčić et al., 1991), is succinctly summarized by Jean and Fan (1998), as

$$U = U_{mf}\left[\frac{\varepsilon^3(1-\varepsilon)\alpha'_{mf}}{\varepsilon^3_{mf}(1-\varepsilon_{mf})\alpha'}\right]^{1/2} \qquad (29)$$

where

$$\frac{\alpha'}{\alpha'_{mf}} = (1-B_2) + \frac{1}{\lambda_B}\left\{1 - \left[\lambda_B\left(\frac{\varepsilon-\varepsilon_{mf}}{1-\varepsilon_{mf}}\right) + B_1\right]^2\right\}^{1/2} \qquad (29a)$$

$$B_1 = \left[1 + \left(\frac{U^2_{mf}}{U^2\varepsilon^3_{mf}}\right)^2\right]^{-1/2} \qquad (29b)$$

$$B_2 = \frac{(1-B_1^2)^{1/2}}{(1-B_1^2)^{1/2} - B_1} \qquad (29c)$$

and

$$\lambda_B = (1-B_1^2)^{1/2} - B_1 \qquad (29d)$$

This method, which requires prior knowledge of ε_{mf}, U_{mf}, and U_0, predicts ε all the way from ε_{mf} to unity for particles of any shape. ε_{mf} and U_{mf} can be determined as discussed in Sec. 3, while equations for U_0 will be given below. Outside the creeping flow region, this method is the only serious "theoretical" competitor to the empirical methods discussed below for predicting bed expansion.

4.3 Empirical Equations

Perhaps the simplest and certainly the most widely used, if not always the most accurate, of the empirical equations proposed in the literature for predicting the expansion of a liquid fluidized bed is that of Richardson and Zaki (1954), which was anticipated by several prior investigators (Hancock, 1937/38; Lewis et al., 1949; Jottrand, 1952; Lewis and Bowerman, 1952). In its most primitive form this equation is simply

$$\frac{d\log U}{d\log \varepsilon} = n = \text{constant} \qquad (30)$$

i.e., a plot of log U vs. log ε, or of U vs. ε on log–log coordinates, for any given combination of liquid and monodispersed solids, can be well approximated by a straight line of slope n, at least from ε equal to or somewhat greater than ε_{mf} to $\varepsilon = \varepsilon_c$, where $\varepsilon_c \cong 0.850 \pm 0.15$ (see Figs. 4 and 7), i.e., for relatively concentrated suspensions. Commonly, Eq. (30) has been integrated with the boundary condition $U = U_0$ at $\varepsilon = 1$, so that

$$\frac{U}{U_0} = \varepsilon^n = (1-c)^n \qquad (31)$$

which is the most commonly written form of the Richardson–Zaki equation. Unfortunately, plots of log U vs. log ε for concentrated suspensions, when linearly extrapolated to log $\varepsilon = 1$, often yield intercepts equal to log U_e, where $U_e \leq U_0$ (Fig. 7). Therefore Eq. (31) must be modified to

$$\frac{U}{U_0}\left(= \frac{Re}{Re_0}\right) = k\varepsilon^n \qquad (32)$$

where $k = U_e/U_0 \leq 1$.

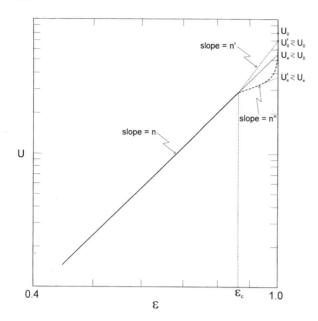

Figure 7 Expansion characteristics of a stable liquid fluidized bed. Concentrated suspensions follow bold solid line of slope n, which *may* at $\varepsilon = \varepsilon_c$ change to either slope $n' > n$ all the way to $\varepsilon = 1$ (bold dotted line), or slope $n'' < n$ to $\varepsilon < 1$ (bold dashed line).

To use Eq. (32) in order to predict U for a given value of ε (or vice versa) requires knowledge of U_0, n, k, and preferably also ε_c. For smooth spheres ($\phi = 1$), to which we shall initially restrict this discussion, the free-settling velocity U_0 is most conveniently calculated from the empirical correlation of Turton and Clark (1987),

$$\mathrm{Re}_0 = \frac{d_\mathrm{p} U_0 \rho}{\mu} = \mathrm{Ar}^{1/3}\left[\left(\frac{18}{\mathrm{Ar}^{2/3}}\right)^{0.824} + \left(\frac{0.321}{\mathrm{Ar}^{1/3}}\right)^{0.412}\right]^{-1.214} \quad (33)$$

which applies up to $\mathrm{Re}_0 = 260,000$ (Haider and Levenspiel, 1989).

The index n for spheres was originally correlated by five empirical equations relating this index to Re_0 and the wall effect ratio, $d_\mathrm{p}/D_\mathrm{c}$ (Richardson and Zaki, 1954). However, the wall effect on n has not been confirmed by subsequent investigators (Chong et al., 1979; Di Felice and Parodi, 1996), and several investigators have proposed a single equation relating n to either Re_0 (Wallis, 1969; Garside and Al-Dibouni, 1977; Rowe, 1987) or Ar (Khan and Richardson, 1989), the advantage of the latter being that Ar, unlike Re_0,

can be calculated directly without further ado if the particle and liquid properties are known. The equation of Khan and Richardson, which has been carefully adjusted to the available data, is

$$\frac{4.8 - n}{n - 2.4} = 0.043\mathrm{Ar}^{0.57} \quad (34)$$

This equation, as well as two of the others, reflect the fact that the upper limit of n, for spheres in the Stokes region ($\mathrm{Re}_0 < 0.2$, Ar < 4), is somewhere between 4.6 and 4.9, while the lower limit, for spheres in the Newton region ($\mathrm{Re}_0 > 500$, Ar $> 85,000$), falls between 2.3 and 2.4. This much variation in n indicates that, whereas U_0 exceeds U_mf in the Stokes regime by about two orders of magnitude, it does so by only one order of magnitude in the Newton regime. The equation of Garside and Al-Dibouni yields values of n about 0.3 higher than Eq. (34) for all values of Ar and Re_0, possibly due to the inclusion, in their determinations of n, of data points for which ε exceeds ε_c and which showed a slope n' exceeding n, as illustrated in Fig. 7.

Greater uncertainty is associated with the prediction of k. In the original paper by Richardson and Zaki (1954), it was concluded that $k = 1$ for sedimentation, while for fluidization k was simply a wall effect factor given by

$$k = 10^{-d_\mathrm{p}/D_\mathrm{c}} \quad (35)$$

and independent of Re_0. Subsequently, however, Khan and Richardson (1989) have dropped any distinction between sedimentation and fluidization as far as k is concerned, and have proposed, on the basis of a regression analysis of experimental data covering the range $\mathrm{Re}_0 = 0.01$–7000 and $d_\mathrm{p}/D_\mathrm{c} = 0.001$–0.2, that

$$k = 1 - 1.15\left(\frac{d_\mathrm{p}}{D_\mathrm{c}}\right)^{0.6} \quad (36)$$

Equation (36), about which there is considerable data scatter (coefficient of correlation = 0.867), shows a much larger wall effect than Eq. (35). To complicate the issue still further, Rapagnà et al. (1989) have found experimental values of k that are in some cases even lower than those given by Eq. (36) over the range $\mathrm{Re}_0 = 50$–1000, for which they proposed the following correlation, ignoring $d_\mathrm{p}/D_\mathrm{c}$, again with considerable data scatter:

$$k = 1.09 - 0.11 \log \mathrm{Re}_0 \quad (37)$$

On the other hand, based on the work of Fan et al. (1985), in the absence of wall effect, $k \approx 1$ for $\mathrm{Re}_0 < 35$ (Di Felice, 1995), which agrees qualitatively with the trend displayed by Eq. (37). In fact, the data of Rapagnà et al. for $Re_0 \leq 100$ are better correlated by the equation

$$k = \left(\frac{7}{Re_0}\right) + 0.8 \tag{38}$$

which extrapolates to $k = 1$ at $Re_0 = 35$. It is therefore recommended that, for $Re_0 = 100–1000$, k be taken as the lower of the values given by Eqs. (36) and (37); for $Re_0 = 35–100$, as the lower of Eqs. (36) and (38); and for $Re_0 < 35$, as the value given by Eq. (36), i.e., unity in the absence of wall effect. For $Re_0 > 1000$, it is not known whether the trend of k with Re_0 shown by Eq. (37) continues, but Eq. (36) is still presumably applicable to account approximately for any wall effect.

It is possible in many cases to reduce the uncertainty associated with an assumed value of k by integrating Eq. (30) with the boundary condition $U = U_{mf}$ at $\varepsilon = \varepsilon_{mf}$, so that

$$\frac{U}{U_{mf}} = \left(\frac{\varepsilon}{\varepsilon_{mf}}\right)^n \tag{39}$$

Dividing Eq. (31) by Eq. (38) in order to solve for k, we find

$$k = \frac{U_{mf}}{U_0 \varepsilon_{mf}^n} \tag{40}$$

If ε_{mf} and U_{mf} are known or determined by the methods described in the previous section, and n is determined from Eq. (34), then k can be estimated from Eq. (40). Agreement, say, within 5% of a previous estimate should bolster confidence in that estimate. Disagreement, however, even assuming that ε_{mf}, U_{mf}, and n are correct, could be due to an unsatisfactory original estimate of k, but it could also be due to the fact that the primary straight line relationship between $\log U$ and $\log \varepsilon$ sometimes starts at a value of ε somewhat in excess of ε_{mf} (Wilhelm and Kwauk, 1948; Richardson and Zaki, 1954).

The critical voidage, ε_c, above which there is a change of slope in the plot of $\log U$ vs. $\log \varepsilon$ for concentrated suspensions, varies between 0.85 and unity when the slope change is positive, e.g., at $Re_0 = 50–1000$, according to Rapagnà et al. (1989), and between 0.85 (or even 0.9, Chong et al., 1979) and 0.7 when the change of slope is negative, which occurs at $Ar \leq 1600$ ($Re_0 \leq 35$) according to Fan et al. (1985). Most liquid fluidization applications occur at $\varepsilon < \varepsilon_c$.

Referring to Fig. 7 it can be seen that, discounting those situations in which the slope of $\log U$ vs. $\log \varepsilon$ remains unchanged all the way to $U = U_e = U_0'$ (i.e., $\varepsilon_c \approx 1$), the region in which ε exceeds ε_c bifurcates on

log–log coordinates into two routes. In the first route, exemplified by the data of Rapagnà et al. (1989), the dilute suspension data follow a straight line of slope $n' > n$, which these investigators correlated very roughly by the equation

$$n' = 6.4 - 0.61 \log Re_0 \tag{41}$$

Thus in the absence of wall effects,

$$\frac{U}{U_0} = \varepsilon^{n'} = (1 - c)^{n'} \tag{42}$$

which for dilute solutions is well approximated by

$$\frac{U}{U_0} \cong 1 - n'c \tag{42a}$$

Equation (42a) is of the same form as Eq. (25) above derived by Batchelor (1972) for dilute suspensions in viscous flow ($Re_0 \leq 0.2$). Note that as Re_0 decreases, the value of n' in Eq. (41) appears to be approaching the 6.55 of Eq. (25), and a value of $n' \approx 6.5$ was actually found by Di Felice and Parodi (1996) in some of their viscous sedimentation experiments. In the presence of a measurable wall effect ($d_p/D_c > 0.01$), the intercept of the dilute suspension line is U_0', which for viscous flow (Di Felice and Parodi, 1996) is related to U_0 by either of the two empirical equations proposed by Francis (1933) to describe the settling velocity of a ball falling axially through a liquid in a cylindrical vessel. The simpler of the two equations, which give almost identical results up to $d_p/D_c = 0.3$, is

$$\frac{U_0'}{U_0} = \left(1 - \frac{d_p}{D_c}\right)^{2.25} \tag{43}$$

Equation (43) yields answers very close to those derived theoretically by Ladenburg (1907) and Faxén (Emersleben and Faxén, 1923) up to $d_p/D_c = 0.1$ and 0.2, respectively, and agrees fortuitously also with the value of $k(= U_e/U_0)$ given by Eq. (35) up to $d_p/D_c = 0.2$. The agreement of U_0' by these equations with the dilute suspension intercept at $\varepsilon = 1$ has been demonstrated by Garside and Al-Dibouni (1977) for d_p/D up to 0.06 and Re_0 up to 1200, i.e., well beyond viscous flow. The corresponding equation of Munroe (1988–89), obtained by fluidizing single spheres in a cylindrical vessel in the range of $Re_0 = 150–3400$, shows a much smaller wall effect. An empirical equation for U_0'/U_0 proposed by Di Felice (1996), which spans the whole range of Re_0 from the Stokes to the Newtonian region, including the data of Fidleris and Whitmore (1961) in the intermediate region, is

$$\frac{U_0'}{U_0} = \left[\frac{1 - (d_p/D_c)}{1 - (0.33 d_p/D_c)} \right]^a \tag{44}$$

where the single parameter a is given by

$$\frac{3.3 - a}{a - 0.85} = 0.1 \, \text{Re}_0 \tag{44a}$$

A more accurate two-parameter model was subsequently developed by Kehlenbeck and Di Felice (1999).

The other dilute suspension route illustrated in Fig. 7 involves an initial slope decrease to $n'' < n$, which usually occurs at $\text{Re}_0 \leq 35$ (Chong et al., 1979; Fan et al., 1985; Di Felice, 1995). The slope n'' has been correlated with Ar by Fan et al. (1985), as has the corresponding intercept $U_e'(< U_e)$, obtained by linear *extrapolation* to $\varepsilon = 1$. The final termination of the actual (as opposed to the extrapolated) data points for the very dilute suspensions is at $\varepsilon = 1$, $U = U_0'$ (Beňa et al., 1963). It should be noted that the rather abrupt changes of slope on log–log coordinates illustrated in Fig. 7 do not occur when the same data are plotted on arithmetic coordinates, a fact that is important for the variational method of Grbavčič et al. (1991) summarized earlier.

Liquid fluidized beds of nonspherical, rough or very fine particles have been shown by many investigators (Jottrand, 1952; Lewis and Bowerman, 1952; Richardson and Zaki, 1954; Whitmore, 1957; Richardson and Meikle, 1961a; Fouda and Capes, 1977; Chong et al., 1979; Cleasby and Fan, 1981) to expand according to Eq. (32), with k depending not only on d_p/D_c and possibly on Re_0 but also on sphericity ϕ (Dhamarajah and Cleasby, 1986). The dependence of k on d_p/D_c is again better represented by Eq. (36) than by Eq. (35) (Dhamarajah and Cleasby, 1986). The value of n in Eq. (32) is always greater than that given by Eq. (34). This positive deviation increases with decreasing sphericity, decreasing particle size, and decreasing Re_0. Thus at $\text{Re}_0 < 0.1$, n for particles with $d_p > 280 \, \mu\text{m}$ increased from 4.8 to 5.8 as ϕ decreased from unity to 0.7 (Chong et al., 1979) and increased from 6.9 to 9.5 as d_p of methacrylate powder decreased from 194 to 65 μm (Whitmore, 1977); while n decreased from about 3.5–5.5 (over a wide sphericity spread) at $\text{Re}_0 = 10$ (Ar $\approx$ 310) to about 2.4–2.6 (over a similarly wide spread in ϕ) at $\text{Re}_0 = 1000$ (Ar $\approx$ 330,000) (Dhamarajah and Cleasby, 1986). The last observation indicates that the particle shape effect on n decreases markedly as Re_0 increases. However, no reliable quantitative correlations between n and ϕ, d_p, or Re_0 have yet been developed. Chianese et al. (1992) scored a

modicum of success in predicting their experimental values of n (5.41–4.90 for $\text{Re}_0 = 6.6$–31.0) for water fluidization of sodium perborate crystals ($\phi = 0.52$) by means of Eq. (40), assuming $k = 1$, evaluating ε_{mf} from one of the two equations relating ε_{mf} to ϕ proposed by Wen and Yu (1966) [though Eq. (20) above gives even better agreement with their experimentally measured values of $\varepsilon_{mf}(\approx 0.56)$], U_0 by means of the isolated sphere correlations of Clift et al. (1978) assuming equivolume spheres, and U_{mf} by means of Eq. (21) above (despite its proposed restriction to spheres and to a range of Re_{mf} considerably higher than those obtained in this study), using the surface-volume diameter $d_{sv} = \phi d_p$ as d_p in the Reynolds and Archimedes numbers. A more recent and convenient empirical equation for determination of U_0 for nonspherical isometric particles is that of Haider and Levenspiel (1989),

$$\text{Re}_0 = \frac{d_p U_0 \rho}{\mu} = \text{Ar}^{1/3} \left(\frac{18}{\text{Ar}^{2/3}} + \frac{2.335 - 1.744\phi}{\text{Ar}^{1/6}} \right)^{-1} \tag{45}$$

which is recommended for $\text{Re}_0 < 25,000$ and $0.5 \leq \Phi < 1$.

Another approach to nonspherical particles and/or rough spheres, originating with Steinour (1944) and adopted by many other sedimentation workers, assumes that such particles immobilize liquid around their surface irregularities and thereby behave as smooth spheres with an effective volumetric concentration Kc, where the hydrodynamic volume factor K is the volume of the liquid envelope plus solid divided by the solid volume. The effective voidage then becomes $1 - Kc$, the effective particle density $\rho_{eff} = [\rho_p + \rho(K - 1)]/K$, and the effective particle diameter $d_{eff} = d_p K^{1/3}$. By this means, Gasparyan and Ikaryan (1962) attempted to reconcile their hindered settling results for nonspherical particles with the original empirical equations of Richardson and Zaki (1954) for spheres. The same approach was later applied to liquid fluidization by Fouda and Capes (1977), who determined n and k for their experiments from the original empirical equations of Richardson and Zaki (1954) for spheres, U_0 from tables provided by Heywood (1962), and K iteratively by fitting the equation

$$\frac{U}{U_0} = k(1 - Kc)^n \tag{46}$$

to their experimental data. They found values of K that ranged from 1.2 for crushed silica to values sometimes exceeding 3 for flat mica particles. Subsequently they discovered that all their results (for crushed silica, crushed steel, aluminum squares, and mica) could be correlated by (Fouda and Capes, 1979; Epstein, 1979)

$$K(1 - \varepsilon_b) = 0.603 \pm 0.016 \text{ (standard deviation)} \tag{47}$$

where ε_b is the voidage of the bed after the liquid flow has been shut off and the bed has settled. Assuming that this settled bed voidage is equivalent to ε_{mf} (Eastwood et al., 1969), then the combination of Eqs. (47) and (20), with a slight correction to render the result compatible with smooth spheres (for which $\phi = 1$ and $K = 1$), results in

$$K = \frac{0.603}{1 - 0.415/\phi^{0.483}} - 0.03 \tag{48}$$

Equation (46), in conjunction with Eq. (48) and the previous equations recommended for n, k, and U_0 [assuming spheres of diameter d_{eff} and density ρ_{eff}, or alternately U_0 by Eq. (45)], is thus a coherent method for dealing with the expansion of nonagglomerating nonspherical particles, albeit one that requires further testing, qualification and elaboration. Note that a high value of n in Eq. (32) is accompanied by a high value of K in Eq. (46), in which the value of n is that from Eq. (34) for unflocculated spheres, i.e., the lowest possible for the given Ar or Re_0.

The increase of n as d_p decreases and hence as specific surface increases could in part be due to the electrokinetic Dorn effect, which retards sedimentation in ionic liquids (Tackie et al., 1983; Masliyah, 1994), and is most certainly associated with the flocculation of colloidal particles (Dixon, 1977). Equation (46) is again applicable to colloidal suspensions, with K now denoting the volume of the flocs divided by the volume of the particles contained by the flocs, and U_0 the terminal free settling velocity of an average floc rather than of the particles within the floc (Michaels and Bolger, 1962). The values of K in Eq. (46) and of n in Eq. (32) are now typically an order of magnitude greater than for larger nonagglomerated particles (Dixon, 1977).

Another approach to bed expansion is based on a postulate by Wen and Yu (1966), anticipated by the drag force measurements of Rowe and Henwood (1961), Rowe (1961), and Richardson and Meikle (1961b) on immobilized beds of spheres at various voidages and Reynolds numbers. This postulate states that the drag force F_D on a given sphere in a fluidized swarm of monosized spheres to the drag force F_{DS} on a single isolated sphere past which the same fluid flows at the same superficial velocity, U, is independent of the Reynolds number, $Re = d_p U \rho/\mu$ and depends only on the voidage, ε. Thus

$$\frac{F_D}{F_{DS}} = f(\varepsilon) = \varepsilon^{-\beta} \tag{49}$$

To arrive at an expression for $f(\varepsilon)$, we first consider the Stokes regime, for which

$$\frac{F_{DS}}{F_{D0}} = \frac{3\pi\mu U d_p}{3\pi\mu U_0 d_p} = \frac{U}{U_0} \tag{50}$$

where F_{D0} is the drag force exerted on an isolated sphere at its terminal settling velocity U_0. But from Eq. (34), in the Stokes regime (Ar < 4, Re_0 < 0.2), $n = 4.8$. Assuming $k = 1$ in Eq. (32), it follows that

$$\frac{F_{DS}}{F_{D0}} = \varepsilon^{4.8} \tag{51}$$

for creeping flow. Now consider the Newton regime, for which

$$\frac{F_{DS}}{F_{D0}} = \frac{0.44\pi d_p^2 \rho U^2/8}{0.44\pi d_p^2 \rho U_0^2/8} = \frac{U^2}{U_0^2} \tag{52}$$

From Eq. (34), in the Newton regime (85,000 < Ar < 1.4×10^{10}, 500 < Re_0 < 200,000), $n = 2.4$. Therefore, again assuming $k = 1$ in Eq. (32), it follows that

$$F_{DS}/F_{D0} = (\varepsilon^{2.4})^2 = \varepsilon^{4.8} \tag{53}$$

for the high Reynolds number regime. Since

$$F_D = V(\rho_p - \rho)g \tag{6}$$

it follows that $F_D = F_{D0}$, the drag force at the terminal velocity being balanced by the same gravitational force modified by the same buoyancy as given by Eq. (6). Therefore

$$F_D/F_{DS} = f(\varepsilon) = \varepsilon^{-4.8} \tag{54}$$

at both ends of the Reynolds number spectrum. Wen and Yu (1966) assumed that the same voidage function, which they took as $\varepsilon^{-4.7}$ instead of $\varepsilon^{-4.8}$, would also apply to intermediate values of Re_0. Therefore

$$\frac{F_D}{F_{DS}} = \frac{C_D}{C_{DS}} = \varepsilon^{-4.7} \tag{55}$$

But for Re < 1000 (Schiller and Naumann, 1933),

$$C_{DS} = \left(\frac{24}{Re}\right)(1 + 0.15\,Re^{0.687}) \tag{56}$$

Substituting Eq. (56) into Eq. (55) and rearranging,

$$C_D\,Re^2\varepsilon^{4.7} = 24Re + 3.6Re^{1.687} \tag{57}$$

In a gravitational field characteristic of fluidization, for which F_D is given by Eq. (6), in which $V = \pi d_p^3/6$,

$$C_D\,\mathrm{Re}^2 = \frac{(\pi d_p^3/6)(\rho_p - \rho)g}{\pi d_p^2 \rho U^2/8}\left(\frac{d_p U \rho}{\mu}\right)^2$$
$$= \frac{4d_p^3(\rho_p - \rho)\rho g}{3\mu^2} = \frac{4\mathrm{Ar}}{3} \tag{58}$$

Eq. (57) is therefore equivalent to

$$\mathrm{Ar}\varepsilon^{4.7} = 18\mathrm{Re} + 2.7\mathrm{Re}^{1.687} \tag{59}$$

which is explicit in ε if Ar and Re are known, and can be solved for Re by rapid iteration if Ar and ε are specified.

It turns out that Eq. (59) applies well at both low and high Reynolds numbers, but underpredicts Re (and hence U) at intermediate values of Re, particularly at the lower values of ε (Khan and Richardson, 1990; Di Felice, 1995). It shares this characteristic with an equation similar in form obtained by equating the relationship for pressure gradient proposed later by Foscolo et al. (1983) with the pressure gradient as given by Eq. (8), viz.,

$$\mathrm{Ar}\varepsilon^{4.8} = 17.3\mathrm{Re} + 0.336\mathrm{Re}^2 \tag{60}$$

and with another empirical equation similar in form proposed still later by Hartman et al. (1992):

$$\mathrm{Ar}\varepsilon^{4.73} = 20.4\mathrm{Re} + 1.44\mathrm{Re}^{1.80} \tag{61}$$

To remedy the deficiency of Eq. (60) at intermediate Reynolds numbers, the equation was modified empirically by Gibilaro et al. (1986) by raising both the viscous and the inertial terms on the right-hand side to a common exponent and raising the sum of these terms to the reciprocal of this exponent, which is expressed as a function of voidage. The modified equations are

$$\mathrm{Ar}\varepsilon^{4.8} = [(17.3\mathrm{Re})^\alpha + (0.336\mathrm{Re}^2)^\alpha]^{1/\alpha} \tag{62}$$

$$\alpha = 2.55 - 2.1[\tanh(20\varepsilon - 8)^{0.33}]^3 \tag{63}$$

Equation (59) can be modified more simply by rewriting it as

$$\mathrm{Ar}\varepsilon^\beta = 18\mathrm{Re} + 2.7\mathrm{Re}^{1.687} \tag{64}$$

since β has been empirically, though roughly, fitted by Di Felice (1994) as

$$\beta = 4.7 - 0.65e^{-(1.5 - \log \mathrm{Re})^2/2} \tag{65}$$

[In this equation for β, the 3.7 given by Di Felice, who couples his result with F_D as given by Eq. (7), is here changed to 4.7, since we have adopted the convention for F_D incorporated in Eq. (6)]. Knowing Ar (from

fluid and particle properties) and specifying U and hence Re, β can first be determined from Eq. (65) and then ε from Eq. (64). If Ar and ε are specified, then Re can be determined iteratively by simultaneously satisfying both Eqs. (64) and (65).

Equations (59)–(65) are all applicable to monosized spherical particles. Additional empirical equations or graphical methods (e.g., Barnea and Mizrahi, 1973) for spheres relating Ar, ε, and Re over various ranges of Ar or Re, both of the same form as the above equations and some of slightly different form, as well as some, such as the "logistic" equation of Garside and Al-Dibouni (1977), with the primary variables assembled in different dimensionless groups, are tabulated by Couderc (1985), Khan and Richardson (1989), and Jamialahmadi and Müller-Steinhagen (1999). Included also in the same or in different (Couderc, 1985) tables are relationships for nonspherical particles, some of which show no shape influence in the actual equations, a notable exception being the equation of Limas-Ballesteros et al. (1982a) for relatively high Reynolds numbers. Many of these equations are critically discussed by Di Felice (1995), who interchanges the definitions of Ar $[= d_p^3(\rho_p - \rho)\rho g/\mu^2]$ and Ga $(= d_p^3 \rho^2 g/\mu^2)$, following in this respect other authors (e.g., Wen and Yu, 1966), some of whom perform this interchange inconsistently (e.g., Khan and Richardson, 1989). It should also be noted that GaMv $[= \mathrm{Ga}(\rho_p - \rho)/\rho]$ in Couderc's tables, or ArM in Di Felice's review, is equivalent to Ar.

For both upwardly diverging and downwardly diverging *tapered beds* with an included angle of 5°, Di Felice et al. (1991a) found that liquid deceleration or acceleration effects could be neglected, so that the voidage for monodisperse solids at any level could then be estimated by applying the expansion equations above, e.g., Eqs. (32)–(34) and (36)–(38) for spheres, at the liquid velocity prevailing at that level. For an upwardly diverging bed with an included angle of 2.7°, Scott et al. (1978) correctly predicted the bed expansion and the decreasing frictional pressure drop with increasing U by combining the force balance given by Eq. (11) with Eq. (59) for ε, both applied in small increments of bed height. Other investigators (Kolár, 1963; Koloini and Farkas, 1973; Maruyama et al., 1984; Webster and Perona, 1988, 1990) of upwardly diverging beds, however, found this procedure somewhat inaccurate even for cone angles as low as 0.5° (and as high as 25.1°), usually overpredicting the degree of expansion that actually occurs, and they have recommended empirical equations for specific cases, e.g., empirical determination of k and n in Eq.

(32) for a given cone angle and liquid–particle system. The moderate bed contraction often observed compared to that for the equivalent cylindrical column can be explained by the additional bulk mixing generated in a tapered bed (Webster and Perona, 1988, 1990), a contraction effect of bulk mixing discussed below in Sec. 9. The bulk mixing and consequent bed contraction were probably attenuated in the study of Di Felice et al. (1991a) by their use of a ball distributor, described later.

When particles of *mixed size, shape, and/or density* are fluidized by a liquid, they may, as will be discussed later, segregate completely according to species (a given size, shape, and density), segregate incompletely (i.e., mix partially), or mix completely. Irrespective of the degree of segregation or mixing, however, the overall bed expands as if it were simply the sum of the N individual species, each acting independently of the other (Lewis and Bowerman, 1953; Richardson and Zaki, 1954; Hoffman et al., 1960; Cleasby and Bauman, 1976; Epstein et al., 1981; Yu and Shi, 1985), so that

$$1 - \varepsilon = \left[\sum_{i=1}^{N} \frac{c_i/c_t}{1 - \varepsilon_i} \right]^{-1} \qquad (66)$$

where ε_i is the voidage when species i is fluidized alone at the same superficial liquid velocity as the mixture, and $c_i/c_t (= v_i)$ is the volume fraction of fluid-free solids which is species i. This experimentally verified "serial model" works best in the absence of bulk circulation within the fluidized bed (Epstein et al., 1981) but is in contradiction to theories (Gibilaro et al., 1985b; Chiba, 1988) that predict some bed contraction when particle mixing occurs, and to a few experiments with fully mixed binary particle species where some bed contraction was measured (Chiba, 1988). The experimental difficulty of detecting the predicted contraction probably arises from the smallness of this effect, illustrated in the next section by the small difference between the binary line ABD and the closest monocomponent lines below them in Fig. 9.

5 PARTICLE MIXING AND SEGREGATION

5.1 Deductions from Pressure Gradient Measurements

When two or more species of particles are fluidized by a liquid in the same column, they may mix completely, mix partially (i.e., segregate partially) or segregate completely, depending on their relative sizes, shapes,

and densities. A common situation, e.g., in crystallization, is that of particles having uniform shape and density but differing in size. A constant axial pressure gradient throughout the bed then denotes perfect mixing of the particles, whereas variation of this gradient signifies segregation, the larger the variation the larger the degree of segregation (Neužil, 1964; Scarlett and Blogg, 1967). For multisized solids in a liquid fluidized bed at superficial velocity U, the voidage at height z above the distributor may be obtained from the pressure gradient at that level by means of Eq. (8), rewritten as

$$\varepsilon = 1 - \frac{-dp_f/dz}{g(\rho_p - \rho)} \qquad (67)$$

By means of the bed expansion equations developed above, e.g., Eqs. (32)–(34) and (36)–(38), we can relate this value of ε for the given ρ_p, ρ, μ, and U to a given value of d_p, namely $d_p(z)$; and in general,

$$\varepsilon = f(d_p) \qquad (68)$$

The mass dm_z of particles in the bed over a differential height dz is given by

$$dm_z = \rho_p A(1 - \varepsilon)dz \qquad (69)$$

Therefore the mass m_z of particles between the distributor and the height z is given by

$$m_z = \int_0^{m_z} dm_z = \rho_p A \int_0^z (1 - \varepsilon) \, dz \qquad (70)$$

Substituting for ε according to Eq. (67),

$$m_z = \frac{\rho_p A}{g(\rho_p - \rho)} \int_0^z \left(-\frac{dp_f}{dz} \right) dz \qquad (71)$$

The total mass M of particles in the bed is then given by

$$M = \frac{\rho_p A}{g(\rho_p - \rho)} \int_0^L \left(-\frac{dp_f}{dz} \right) dz \qquad (72)$$

The cumulative fraction of particles larger than $d_p(z)$, all of which are presumed to be located below level z, becomes

$$\frac{m_z}{M} = \frac{\int_0^z (-dp_f/dz) \, dz}{\int_0^L (-dp_f/dz) \, dz} = \frac{\int_0^z [1 - f(d_p)] \, dz}{\int_0^L [1 - f(d_p)] \, dz} \qquad (73)$$

The cumulative particle size distribution of the solids can then be generated both as m_z/M vs. z and as m_z/M vs. d_p (the latter of which will coincide with the size distribution of the solids as determined by other means, e.g., screen analysis, only if perfect stratification by size occurs). The corresponding plot of d_p vs. z has been dubbed the "perfect classification model" by Al-Dibouni and Garside (1979). Even where some particle mixing occurs, this model gives a good representation of $\overline{d_p}$ vs. z, where $\overline{d_p}$ is either the mass average (Garside and Al-Dibouni, 1974) or the mass median (Al-Dibouni and Garside, 1979) local particle diameter at level z.

For the particular case of a binary solid mixture, correct estimate of a local mean particle diameter allows one to determine the local composition of the solids. Thus Di Felice et al. (1987) employed Eqs. (62) and (63) to determine the local mean diameter as a function of bed level in the transition zone between the fully segregated large particles at the bottom and the fully segregated small particles at the top (Fig. 8c), using the axial voidage profiles that Juma and Richardson (1983) obtained for their two binaries (from the measured pressure profiles) by means of Eq. (67). Assuming, after Wen and Yu (1966), that the computed diameters represented Sauter means, they were thus able to calculate the concentrations of large and small particles locally, and their answers agreed reasonably well with the measured local concentrations of the two solid species. A more approximate estimate of these local concentrations was made by Epstein and Pruden (1999) from the local voidages, applying locally the serial model previously described (in Sec. 4) for the overall bed.

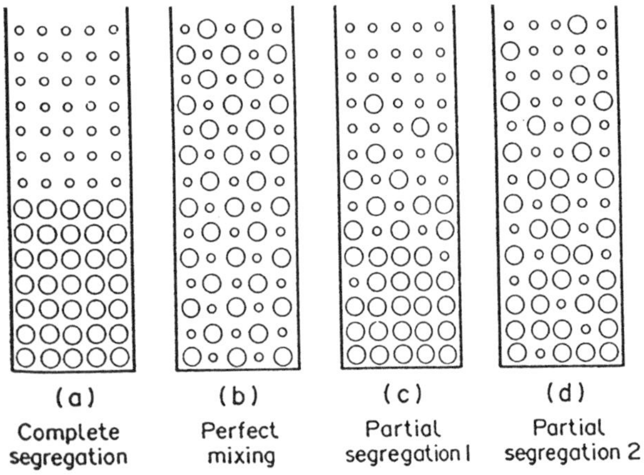

Figure 8 Mixing and segregation patterns for liquid fluidization of binary-sized particles of equal density. (From Gibilaro et al., 1985.)

In mineral processing, in addition to size, the particle density and possibly the shape may also vary. Using the measured variation of $(-dp_f/dz)$ with z, the procedure of Di Felice et al. (1987), just described, can also be applied to determine the changing composition with bed level of two spherical particle species differing in both diameter and density, by means of an additional equation for the volume-average density,

$$\rho_p = \frac{c_1\rho_{p1} + c_2\rho_{p2}}{c_1 + c_2} \tag{74}$$

used in Eq. (67) to related ε to the measured frictional pressure gradient. Thus Eqs. (67), (62), (63), and (74) are solved iteratively along with the equation for the Sauter mean diameter,

$$d_p = \frac{c_1 + c_2}{c_1/d_{p1} + c_2/d_{p2}} \tag{75}$$

to obtain ε, α, ρ_p, d_p, c_1, and c_2, where $c_1 + c_2 = 1 - \varepsilon$ (Gibilaro et al., 1986a; Di Felice et al., 1987).

5.2 Binary Mixtures—Overview

To simplify the analytical description of the phenomena involved in particle mixing and segregation, let us continue to focus on a binary mixture of particles, i.e., two species (components). In the case of a size difference alone, the larger particles will always separate to some degree below the smaller, unless the size difference is small enough and/or complicating factors such as bulk circulation or hydrodynamic instability are large enough to mix the two particle species completely. Under no circumstances, however, will the larger particles segregate above the smaller (in conventional as opposed to e.g., inverse fluidization). If, in addition, there is a shape difference, it is that species with the larger magnitude of the product ϕd_p that will normally segregate below. If there is also a particle density difference, then if that difference is in the same direction as that of ϕd_p, the segregation will be accentuated; if, however, the density difference is in the opposite direction, then depending on the magnitude of this difference, the segregation may be attenuated throughout the entire voidage range as the liquid velocity is increased, it may be reversed over the entire voidage range, or it may be reversed at the lower voidages and attenuated at the higher. The last situation is the interesting bed (or layer) inversion phenomenon, which will be discussed later. First we will discuss the one-way segregation phenomenon, as exemplified by a mixture of particles of a given material having two different sizes.

5.3 Binary Mixtures of Equal Density Particles

For simplicity, let us consider spheres of density ρ_p and of diameters d_L and d_S. where $d_L > d_S$, fluidized by a liquid having a superficial velocity, U, between U_{mfL} and U_{0S}. In the absence of the smaller particles, the fluidized bed suspension density will be given by

$$\rho_{BL} = \rho_p(1 - \varepsilon_L) + \rho\varepsilon_L \qquad (76)$$

while in the absence of the larger particles, the suspension density is given by

$$\rho_{BS} = \rho_p(1 - \varepsilon_S) + \rho\varepsilon_S \qquad (77)$$

Subtracting Eq. (77) from Eq. (76), we obtain

$$\rho_{BL} - \rho_{BS} = (\rho_p - \rho)(1 - \varepsilon_L) - (\rho_p - \rho)(1 - \varepsilon_S) \qquad (78)$$

and the dimensionless density ratio, $\gamma = (\rho_{BL} - \rho_{BS})/(\rho_p - \rho)$, is then given by

$$\gamma = \varepsilon_S - \varepsilon_L \qquad (79)$$

where ε_S and ε_L are the voidages displayed by the respective monocomponent beds when they are each separately fluidized by the same liquid at the same superficial velocity, U, as the binary bed. These voidages can be separately measured or they can be predicted by the bed expansion equations elaborated earlier. Assuming no major disturbances due to bulk circulation or instability effects, the value of γ can function as a crude qualitative measure of the degree of particle segregation as follows (Epstein and Pruden, 1999): $\gamma < 0.015 \pm 0.005$: little segregation, good particle mixing (Fig. 8b). $0.015 \pm 0.005 < \gamma < 0.045 \pm 0.015$: partial segregation with no interface between layers (Fig. 8d). $0.045 \pm 0.015 < \gamma \leq 0.1$: segregation with fuzzy interface and some intermixing (Fig. 8c). $\gamma > 0.1$: clean-cut segregation with sharp interface and little intermixing (Fig. 8a). The last criterion is consistent with the generalization of Di Felice (1995) that "complete segregation is to be expected when the size ratio is greater than about two." The fuzzy interface that occurs with the third criterion (Fig. 8c) is a transition zone, the depth of which extends over the whole bed length for the second criterion (Fig. 8d).

The implications of Eq. (79) can be made manifest by writing the Richardson–Zaki (1954) equation, Eq. (32) above, for the two monocomponent beds represented by Eq. (79), so that

$$U = k_S U_{0S}\varepsilon_S^{n_S} = k_L U_{0L}\varepsilon_L^{n_L} \qquad (80)$$

But in general (Pruden and Epstein, 1964),

$$U_0 \propto d_p^{(3-m)/m} = a'd_p^{(3-m)/m} \qquad (81)$$

where the exponent m varies from 1 in the Stokes regime to 2 in the Newton regime and the coefficient a' varies correspondingly from $(\rho_p - \rho)g/18\mu$ to $[(\rho_p - \rho)g/0.33\rho]^{1/2}$. Therefore Eq. (80) becomes

$$k_S a'_S d_S^{(3-m_S)/m_S}\varepsilon_S^{n_S} = k_L a'_L d_L^{(3-m_L)/m_L}\varepsilon_L^{n_L} \qquad (82)$$

If the size difference between the two species of particles is sufficiently small that $k_S \approx k_L$, $n_S \approx n_L$, $a'_S \approx a'_L$, and $m_S \approx m_L$, then Eq. (82) can be simplified and rearranged to

$$\varepsilon_S = \varepsilon_L \left(\frac{d_L}{d_S}\right)^{(3-m)/mn} \qquad (83)$$

Therefore from Eq. (79),

$$\gamma = \varepsilon_L \left[\left(\frac{d_L}{d_S}\right)^{(3-m)/mn} - 1\right] \qquad (84)$$

Equation (84) shows that the degree of segregation (or stratification) by size increases with an increase in

1. d_L/d_S, i.e., in size difference, as would be expected.
2. ε_L, i.e., in bed expansion, as has been frequently observed (Wakeman and Stopp, 1976; Juma and Richardson, 1979; Epstein and Pruden, 1999), though not by Garside and Al-Dibouni (1974), who found this trend to be interrupted by strong mixing in the voidage range of $\varepsilon = 0.65$–0.80.
3. $(3 - m)/mn$, which is roughly equivalent to $(3 - \beta/n)/\beta$, where n is given by Eq. (34) and β by Eq. (65), and varies from 0.43 to 0.22 as one moves from the Stokes to the Newton regime. Thus as the extent first of inertial effects and then of turbulence increases, the degree of segregation decreases (Epstein and Pruden, 1999), even without complicating effects such as bulk circulation and hydrodynamic instability.

Another approach to the problem of binary particle stratification is based on the more traditional concept of axial dispersion according to Fick's law of diffusion in competition with segregation, as formulated by Kennedy and Bretton (1966). Assuming semibatch liquid fluidization that results in partial segregation of the type represented by Fig. 8c or 8d, then at any bed level within the partially mixed region, the volumetric particle mixing flux for species i must equal the particle segregation flux for that species, i.e.,

$$\frac{D_i dc_i}{dz} = c_i U_{pi} \qquad (85)$$

where D_i is the axial dispersion coefficient (or dispersivity) of species i and U_{pi} the segregation (or classification) velocity of that species through a swarm of species j. Kennedy and Bretton proposed that the segregation velocity U_{pi} is the difference between the local interstitial (or "slip") velocity of the fluid in the bed and the interstitial velocity that would be required to maintain particles exclusively of that size in suspension at the voidage existing in the local mixed-particle environment, so that

$$U_{pi} = \frac{U}{\varepsilon} - \frac{U_i}{\varepsilon} \qquad (86)$$

For a monocomponent system or for the fully segregated regions of either a partially (Fig. 8c) or fully (Fig. 8a) segregated bed, $U_{pi} = 0$, i.e., $U = U_i$, assuming i signifies the particles in the segregated region involved. Substituting for U_i according to Eq. (32), the result is

$$\frac{D_i dc_i}{dz} = c_i \left[\left(\frac{U}{\varepsilon} \right) - k_i U_{0i} \varepsilon^{n_i - 1} \right]_{i = L, S} \qquad (87)$$

For the large particles, both sides of this equation are negative, while for the small particles, both are positive. The exact solution of Eq. (87) presents problems because of the difficulty in satisfactorily defining the relevant boundary conditions (Richardson and Afiatin, 1997). However, a number of techniques have been proposed to arrive at solutions from which D_i can be determined from measured axial concentration profiles of c_L, c_S, and/or ε ($= 1 - c_L - c_S$).

Thus Al-Dibouni and Garside (1979), for partially segregated beds with no transition zone (Fig. 8d), rewrote and integrated Eq. (87) as

$$\int_0^z \left(\frac{U}{\varepsilon} - k_i U_{0i} \varepsilon^{n_i - 1} \right) dz = D_i \int_{c_{i0}}^{c_i} \frac{dc_i}{c_i} = D_i \ln \frac{c_i}{c_{i0}} \qquad (88)$$

and determined D_i for each species from the best straight line through the moderately scattered plot of the left-hand-side of Eq. (88) vs. $\log c_i / c_{i0}$, the slope of which equals $2.303 D_i$. Juma and Richardson (1983), looking at transition zones in partially segregated beds (Fig. 8c), owing to measurements of pressure gradients and hence voidages as a function of bed level, determined that the square-bracketed portion of Eq. (87), viz. U_{pi}, could be represented as

$$U_{pi} = a_i + b_i z, \qquad i = L, S \qquad (89)$$

and they determined the values of a_L, b_L, and a_S, b_S, accordingly. Eq. (89) could then be integrated as

$$D_L \int_{1-\varepsilon_L}^{c_L} \frac{dc_L}{c_L} = D_L \ln \frac{c_L(z)}{1 - \varepsilon_L} = \int_0^z (a_L + b_L z)\, dz$$

$$= a_L z + \frac{b_L z^2}{2} \qquad (90)$$

$$D_S \int_{c_S}^{1-\varepsilon_S} \frac{dc_S}{c_S} = D_S \ln \frac{1 - \varepsilon_S}{c_S(z)} = \int_z^H (a_S + b_S z)\, dz$$

$$= a_S(H - z) + \frac{b_S(H^2 - z^2)}{2} \qquad (91)$$

where the voidage at $z = 0$ (the bottom of the transition zone) and below is ε_L and the voidage at $z = H$ (the top of the transition zone) and above is ε_S. D_L and D_S best-fitted to the experimental measurements according to Eqs. (90) and (91) were then obtained and adjusted to best-match the profile of $c_L + c_S$ ($= 1 - \varepsilon$) vs. z.

In contrast to the procedures and the empirical results of the above two investigations, Gibilaro et al. (1985b) assumed that $D_L = D_S$, so that Eq. (87) applied to the small particles could be divided by the same equation applied to the large particles, whence

$$\frac{dc_S}{dc_L} = \frac{c_S(U - k_S U_{0S} \varepsilon^{n_S})}{c_L(U - k_L U_{0L} \varepsilon^{n_L})} \qquad (92)$$

where $\varepsilon = 1 - c_L - c_S$. A boundary condition for this equation is $c_L = 0$, $\varepsilon = \varepsilon_S$, $c_S = 1 - \varepsilon_S$, $U = k_S U_{0S} \varepsilon_S^{n_S}$ at $z = +\infty$, where ε_S is the monocomponent voidage at the given superficial velocity U (or alternatively $c_S = 0$, $\varepsilon = \varepsilon_L$, $c_L = 1 - \varepsilon_L$, $U = k_L U_{0L} \varepsilon_L^{n_L}$ at $z = -\infty$). By L'Hôspital's rule,

$$\left. \frac{dc_S}{dc_L} \right|_{c_L \to 0} = \frac{(1 - \varepsilon_S) n_S k_S U_{0S} \varepsilon_S^{n_S - 1}}{U - k_L U_{0L} \varepsilon_S^{n_L}} \qquad (93)$$

which yields the initial gradient for the downward marching solution of Eq. (92) as c_S vs. c_L, and thence to an integration of Eq. (87) in parallel with Eq. (92) from $c_L = 0$ to $c_L = 1 - \varepsilon_L$, thus generating curves of c_L vs. z and the corresponding c_S vs. z, for any assumed value of D_i. Thus, by this method, curves are generated without advance knowledge of the voidage or concentration profiles. The final matching of the results to the experimental concentration profiles requires, in addition to the appropriate choice of D_i, satisfaction of the integral conditions stipulating the total volume fluidized of each solid component:

$$V_i = A \int_0^L c_i\, dz \qquad i = L, S \qquad (94)$$

a requirement that is equivalent to the specification of Kennedy and Bretton (1966) and Al-Dibouni and Garside (1979) that the overall average concentration of each particle species be satisfied. The results can represent all cases shown in Fig. 8. A more involved method, by Patwardhan and Tien (1984), solves Eq. (85) to generate both the voidage profiles and the particle concentration profiles, but even with D_i as an adjustable parameter, the success for the concentration profiles was only moderate.

Van der Meer et al. (1984) and, subsequently, both Di Felice et al. (1987) and Asif and Petersen (1993) assumed that U_{pi} in Eq. (85) varied linearly with c_i from zero in a monocomponent bed of i particles in which $\varepsilon = \varepsilon_i$, $c_i = 1 - \varepsilon_i$ to U_{pi0} in a monocomponent bed of j particles in which $\varepsilon = \varepsilon_j$ $c_i = 0$, so that

$$U_{pi} = U_{pi0}\left(1 - \frac{c_i}{1 - \varepsilon_i}\right) \tag{95}$$

Substituting this equation into Eq. (85), taking $i = L$ and the upward direction as positive,

$$D_L \frac{dc_L}{dz} = c_L U_{pL0}\left(1 - \frac{c_L}{1 - \varepsilon_L}\right) \tag{96}$$

that is,

$$\frac{dx_L}{x_L(1 - x_L)} = \frac{U_{pL0}}{D_L} dz \tag{97}$$

where $x_L = c_L/(1 - \varepsilon_L)$. Integrating Eq. (97) between the limits $x_L = 0.5$ [i.e., $c_L = 0.5(1 - \varepsilon_L)$], $z - \bar{z}_L$, and x_L, z, the result is

$$\ln \frac{x_L}{1 - x_L} = \frac{U_{pL0}(z - \bar{z}_L)}{D_L} \tag{98}$$

from which a plot of $z - \bar{z}_L$ vs. $\ln x_L/(1 - x_L)$ yields a slope equal to D_L/U_{pL0}, where U_{pL0} is taken as the measured or predicted velocity of a single large particle moving through a bed of the smaller particles fluidized by the same superficial velocity U as the binary under investigation (Martin et al., 1981; Prudhoe and Whitmore, 1964; Grbavčić and Vuković, 1991; Van der Wielen et al., 1996; Di Felice, 1998; Mostoufi and Chaouki, 1999). From Eq. (98) it follows that

$$x_L = \frac{1}{1 + e^{-Pe_L(Z - \bar{Z}_L)}} \tag{99}$$

where $Pe_L = LU_{pL0}/D_L$ and $Z - \bar{Z}_L = (z - \bar{z}_L)/L$. Similarly it can be shown that

$$x_S = \frac{1}{1 + e^{-Pe_S(Z - \bar{Z}_S)}} \tag{100}$$

where $Pe_S = LU_{pS0}/D_S$ and $Z - \bar{Z}_S = (z - \bar{z}_S)/L$. But, in accord with Di Felice et al. (1987), c_L and c_S are linearly related, so that

$$\frac{c_S}{1 - \varepsilon_S} = 1 - \frac{c_L}{1 - \varepsilon_L} \tag{101}$$

Equation (101), first stated as such by Van Duijn and Rietema (1982), is simply the serial model, Eq. (66), applied locally to a binary. Thus

$$x_S = 1 - x_L = \frac{e^{-Pe_L(Z - \bar{Z}_L)}}{1 + e^{-Pe_L(Z - \bar{Z}_L)}} = \frac{1}{1 + e^{+Pe_L(Z - \bar{Z}_L)}} \tag{102}$$

Comparing Eqs. (102) and (100), $Pe_L = -Pe_S$ and $\bar{Z}_L = \bar{Z}_S$. Asif and Petersen (1993) accordingly use $Pe_i = LU_{pi0}/D_i$ and $\bar{Z}/L$ as the adjustable parameters.

Methods of determining D_i by the Fickian approach above have been supplemented by stochastic or random walk methods (Yutani et al., 1982; Yutani and Fan, 1985; Dorgelo et al., 1985; Kang et al., 1990), and a plethora of D_i values have already been reported in the literature. These vary from as low as 0.24 cm^2/s to as high as 515 cm^2/s (Asif and Petersen, 1993), a range which brackets the smaller spread in values of D_{ia} (0.5–33 cm^2/s) obtained for monocomponent self-dispersion, i.e., by observing the transient mixing of a single species of particles initially divided into two layers, one colored and the other uncolored (Brötz, 1952; Kennedy and Bretton, 1966; Carlos and Richardson, 1968b). Attempts to correlate D_i with system properties and operating conditions have met with limited success. The most recently proposed dimensionally homogeneous correlation, for a large number of binary systems of spheres from the literature covering the ranges Re = 0.5–1000 and ε = 0.50–0.95, by Asif and Petersen (1993), is equivalent to

$$D_i = \frac{(7.9 \pm 1.1)d_p^2(\rho_p - \rho)g}{\varepsilon \rho U}\left(\frac{U - U_{mf}}{U_0}\right)^{2.14 \pm 0.05}$$
$$\pm 16.4\% \tag{103}$$

in which U_{mf} presumably applies to the larger particles, so that both are fully fluidized. When $D_i = D_s$, this equation thus gives recognition to the properties of both particle species, directly via d_p and ρ_p and indirectly via U_0 for the smaller particles, and indirectly via U_{mf} for the larger. When, however, $D_i = D_L$, only the properties of the larger particles are manifested in Eq. (103), in violation of experiments (e.g., Juma and Richardson, 1983) that indicate that both species play a role in the determination of D_i.

The data on D_i in the literature, including those that form the basis of Eq. (103), show an important divergence. Many, e.g., those of Juma and Richardson (1983), Van der Meer et al. (1984), and Dorgelo et al. (1985), show no maximum in D_i with respect to U or ε, while others, e.g., those of Al-Dibouni and Garside (1979), Yutani et al. (1982), and Kang et al. (1990), show a definite maximum at $\varepsilon \cong 0.7$, 0.7–0.8, and 0.6–0.7, respectively. Equation (103) shows no maximum D_i within the fluidization voidage range, although at least two of the data sets (Yutani et al., 1982; Kang et al., 1990) on which it is based do so. Interestingly, Handley et al. (1966) found that the turbulent particle velocity components and the turbulent fluid velocity fluctuations (which were about one-half the former) both rise to a maximum, at $\varepsilon \le 0.70$ and $\varepsilon \le 0.65$, respectively. Trupp (1968) then showed that the root mean square voidage fluctuation in a liquid fluidized bed of heavy spheres in the Newton region ($\mathrm{Re}_0 > 500$) occurred at $\varepsilon = 0.7$, and following his reasoning and subsequently that of Al-Dibouni and Garside (1979), the collision frequency of such a bed should be expected to be proportional to $U(1 - \varepsilon)$, since average particle velocities were shown to vary directly as U. Invoking Eq. (32) for U, the collision frequency is proportional to $kU_0\varepsilon^n(1 - \varepsilon)$, which shows a maximum with respect to ε at $\varepsilon = n/(n + 1)$. In the Newton region, $n = 2.4$, so that the maximum occurs at $\varepsilon = 0.706$. At lower Reynolds numbers, n is higher, and therefore the critical voidage can be expected to be correspondingly higher. Trupp's results were vindicated by Bordet et al. (1972) and were consistent with the subsequent observations of Kmiec (1978).

The divergence between data that show a maximum D_i within the fluidization region and those that do not is a puzzle yet to be solved. The extreme sensitivity of these data to gulf-streaming (Al-Dibouni and Garside, 1979) and to the slightest deviation of the fluidization column from verticality (Van der Meer et al., 1984) are undoubtedly among the factors contributing to this puzzle, as well as to the unsatisfactory data scatter and state of data correlation in the literature. Uncertainties arising from the various experimental methods of measuring the extent of particle mixing are also contributing factors (Di Felice, 1995).

5.4 Binary Mixtures of Equal Size Particles

Although the case of multisized particles of fixed density is far more common in industrial situations than that of particles of uniform size having different

densities, it will nevertheless be instructive briefly to consider spheres of diameter d_p and of density ρ_{ph} and ρ_{pl}, where $\rho_{ph} > \rho_{pl}$, fluidized by a liquid having superficial velocity, U, between U_{mfh} and U_{0l}. In the absence of the lighter particles, the fluidized suspension density will be given by

$$\rho_{Bh} = \rho_{ph}(1 - \varepsilon_h) + \rho\varepsilon_h \quad (104)$$

while in the absence of the heavier particles, the suspension density is given by

$$\rho_{Bl} = \rho_{pl}(1 - \varepsilon_l) + \rho\varepsilon_l \quad (105)$$

Subtracting Eq. (105) from Eq. (104),

$$\rho_{Bh} - \rho_{Bl} = (\rho_{ph} - \rho)(1 - \varepsilon_h) - (\rho_{pl} - \rho)(1 - \varepsilon_l) \quad (106)$$

which is equivalent to

$$\gamma_h = \frac{\rho_{Bh} - \rho_{Bl}}{\rho_{ph} - \rho} = 1 - \varepsilon_h - \frac{\rho_{pl} - \rho}{\rho_{ph} - \rho}(1 - \varepsilon_l) \quad (107)$$

where the nondimensionalized bulk density difference γ_h and Eq. (107) are now assumed to be qualitatively indicative of the degree of segregation by particle density in the same way as γ and Eq. (79) are indicative of the degree of segregation by particle size. Substituting the generalization (Pruden and Epstein, 1964)

$$U_0 \propto (\rho_p - \rho)^{1/m} = b(\rho_p - \rho)^{1/m} \quad (108)$$

into Eq. (32) for each particle species, and making the assumption that the density difference between the two species of particles is small enough that the indices m and n, and the coefficients k and b, are essentially the same for each of the two particle species, then since U is given,

$$(\rho_{ph} - \rho)^{1/m}\varepsilon_h^n = (\rho_{pl} - \rho)^{1/m}\varepsilon_l^n \quad (109)$$

or

$$\varepsilon_h = \varepsilon_l\left(\frac{\rho_{pl} - \rho}{\rho_{ph} - \rho}\right)^{1/mn} \quad (110)$$

By means of Eqs. (83) and (79) for a constant density binary of different particle sizes, and Eqs. (110) and (107) for a fixed size binary of different particle densities, Table 2 indicates what happens to the nondimensionalized bulk density difference in each case as one moves from a very dilute system (e.g., $\varepsilon_S = 1 = \varepsilon_l$) to a very dense system (e.g. $\varepsilon_S = 0.5 = \varepsilon_l$). It is seen that, irrespective of flow regime, γ decreases as ε decreases, while γ_h does the reverse. This tendency by γ was already rationalized in the discussion (item 2) of Eq. (84). Substituting Eq. (110) into Eq. (107), we obtain

Table 2 Nondimensionalized Bulk Density Differences of the Two Particle Species in Two Types of Binaries

	$\rho_p = $ constant $d_L/d_S = \sqrt{2}$			$d_p = $ constant $(\rho_{ph} - \rho)/(\rho_{pl} - \rho) = \sqrt{2}$		
	ε_S	ε_L	γ	ε_l	ε_h	γ_h
		Eq. (83)	(Eq. (79)		(Eq. (110)	Eq. (107)
Stokes regime:	1.0	0.866	0.134	1.0	0.930	0.070
$m = 1, n = 4.8$	0.5	0.433	0.067	0.5	0.465	0.181
Newton regime:	1.0	0.930	0.070	1.0	0.930	0.070
$m = 2, n = 2.4$	0.5	0.465	0.035	0.5	0.465	0.181

$$\gamma_h = 1 - \frac{\rho_{pl} - \rho}{\rho_{ph} - \rho} - \varepsilon_l \left[\left(\frac{\rho_{pl} - \rho}{\rho_{ph} - \rho} \right)^{1/mn} - \frac{\rho_{pl} - \rho}{\rho_{ph} - \rho} \right]$$

$$(111)$$

Since mn equals 4.8 both in the Stokes and the Newton regimes and is in any case always in excess of unity, it follows that the expression in square brackets is always positive, so that γ_h (in contrast to γ) always increases as the voidage decreases. Thus we can explain the empirical rule, long known in the ore dressing industry (Gaudin, 1939), that classification by size, i.e., sizing, is best performed under dilute ("free settling") conditions, while classification by density, i.e., sorting, prevails under concentrated ("hindered settling" or "teeter bed") conditions. We shall see below how this rule manifests itself in the bed inversion phenomenon.

Note also in Table 2 [or in Eqs. (84) and (111)] that, in contrast to segregation by size, which is attenuated as one moves from the Stokes to the Newton regime, i.e., as Re_0 increases, there is no such effect in the case of segregation by density.

Axial dispersivities for particle species of different density, whether or not the sizes are the same, can be determined by methods similar to those for fixed density, mixed size binaries, but their experimental determination has received much less attention in the literature. Equation (103), to the extent that it is applicable at all, is restricted to binaries of equal density particles.

5.5 Binary Mixtures of Particles Differing in Size and Density: Bed Inversion

Consider a binary mixture of spheres of diameters d_{lL} and d_{hS} ($< d_{lL}$) and densities ρ_{plL} and ρ_{phS} ($> \rho_{plL}$),

respectively. For the larger, less dense particles (species lL) subjected alone to liquid fluidization,

$$\rho_{BlL} = \rho_{plL}(1 - \varepsilon) + \rho\varepsilon \qquad (112)$$

The voidage ε can be expressed in terms of the superficial liquid velocity U, explicitly by means of the Richardson–Zaki relation, Eq. (32), or by Eq. (64) with (65), and implicitly (requiring iteration) by means of Eq. (62) with (63), applied to the given solid–liquid system. A plot of ρ_{BlL} vs. U from U_{mflL} to U_{0lL} can thus be generated. Similarly, for the smaller, denser particles (species hS) subjected alone to fluidization by the same liquid at the same temperature and pressure,

$$\rho_{BhS} = \rho_{phS}(1 - \varepsilon) + \rho\varepsilon \qquad (113)$$

Again, ε can by the same means as for Eq. (112) be expressed in terms of U, so that a plot of ρ_{BhS} vs. U from U_{mfhS} to U_{0hS} can then be obtained. If the two curves do not intersect between U_{mflL} and U_{0hS}, then no bed inversion will occur, and the component that has a consistently higher bulk density will always tend to segregate toward the bottom of the bed. The reduced bulk density difference, $|\Delta\rho_B|/(\rho_{plL} - \rho)$, then, provides an indication of the degree of segregation in accord with $\gamma = (\rho_{BL} - \rho_{BS})/(\rho_p - \rho)$ for pure sizing (Epstein and Pruden, 1999). If, however, the two curves do intersect between $U = U_{mflL}$ ($> U_{mfhS}$) and $U = U_{0hS}$ ($< U_{0lL}$), as in Fig. 9, then bed inversion will occur.

The intriguing subject of bed inversion has received much attention in the literature, both experimental (Hancock, 1936; Cleasby and Woods, 1975; Van Duijn and Rietema, 1982; Moritomi et al., 1982; Epstein and LeClair, 1985; Gibilaro et al., 1986a; Jean and Fan, 1986; Di Felice et al., 1988; Qian et al., 1993; Funamizu and Takakuwa, 1996) and theore-

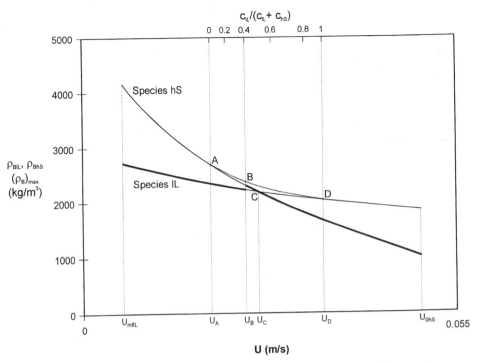

Figure 9 The bed inversion phenomenon of a binary as manifested by the bulk densities of the two layers as a function of liquid velocity. The dotted line ABD merging with the species hS line to the left of A and the species lL line to the right of D represents the bottom layer. Assuming $v_{IL} = 0.4$, the two thickened solid lines represent the top layer below and above the inversion velocity, U_B, respectively, while B represents conditions at the singular inversion point. ρ_{BIL} and ρ_{BhS} are the bulk densities of the monocomponent species lL and species hS beds, respectively. $(\rho_B)_{max}$ refers to the bulk densities of the bottom layer along the dotted line ABD, and $c_{lL}/(c_{lL} + c_{hS})$ are the corresponding compositions of the binary solids in this layer. (After Epstein and Pruden, 1999.)

tical (Van Duijn and Rietema, 1982; Patwardhan and Tien, 1985; Epstein and LeClair, 1985; Moritomi et al., 1986; Gibilaro et al., 1986a; Jean and Fan, 1986; Syamlal and O'Brien, 1988; Funamizu and Takakuwa, 1996; Richardson and Afiatin, 1997; Asif, 1997, 1998; Epstein and Pruden, 1999; Howley and Glasser, 1999). The earlier work incorrectly assumed that the inversion velocity was entirely independent of the overall solids composition of the binary and that therefore inversion occurred at the intersection point C on Fig. 9 (Hancock, 1936; Van Duijn and Rietema, 1982; Epstein and LeClair, 1985). According to this view, an idealized inversion phenomenon, i.e., one not subject to flow disturbances that induce mixing, would show only the three segregation patterns (a), (c), and (e) of Fig. 10. In the light of further careful experimentation (e.g., Moritomi et al., 1982; Jean and Fan, 1986), backed by theory (Gibilaro et al., 1986a), it was found that the relative proportions of the two solids species in the binary influenced the inversion velocity, and that this fluid velocity was preceded and

followed, respectively, by the two additional segregation patterns (b) and (d) of Fig. 10.

To gain further insight into this phenomenon, apply Eq. (62) with (63) or, alternately, (64) with (65) to various binary mixtures of the two solids species fluidized by the given liquid, assuming after Wen and Yu (1966) and Gibilaro et al. (1986a) that d_p is the Sauter mean particle diameter given by Eq. (75), evaluating ρ_p as the volume-average density by Eq. (74), and noting that

$$\varepsilon = 1 - c_1 - c_2 = 1 - c_{lL} - c_{hS} \tag{114}$$

Assuming knowledge of ρ, μ, d_{lL}, ρ_{lL}, d_{hS}, and ρ_{hS}, then for any given value of U, Eq. (62) or Eq. (64) becomes a relationship between c_{lL} and c_{hS}. Thus plots of c_{hS} vs. c_{lL} can be generated with U as a parameter. From the families of c_{hS} and c_{lL} curves, one can determine plots of ρ_B vs. c_{lL} (or ρ_B vs. c_{hS}) by substituting Eqs. (74) and (114) into Eq. (4), so that

$$\rho_B = c_{lL}\rho_{plL} + c_{hS}\rho_{phS} + (1 - c_{lL} - c_{hS})\rho \tag{115}$$

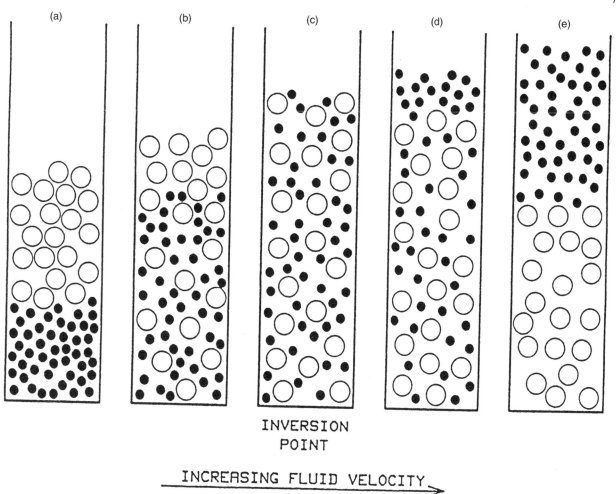

INVERSION
POINT

INCREASING FLUID VELOCITY

Figure 10 Bed inversion phenomenon. Blackened circles represent higher density, smaller spheres (species hS); open circles represent lower density, larger spheres (species lL). (Based on experimental observations of Moritomi et al., 1982, and Jean and Fan, 1986, after Di Felice, 1995.)

For the lower values of U between U_{mflL} and U_{0hS} the plots of ρ_B vs. c_{lL} rise monotonically over the entire range of c_{lL}, while for the higher values of U between the same limits, they fall monotonically. However, for a range of intermediate values of U, the plots of ρ_B vs. c_{lL} show maxima (Gibilaro et al., 1986a), the locus of which is represented by curve ABD on Fig. 9. (For consistency the monocomponent species lines are best generated by the same equation, either (62) or (64), as used to generate ABD.) This curve represents the range of possible inversion velocities and the corresponding bulk densities over which the bottom layer is a binary mixture of solids, the solids composition, $c_{lL}/(c_{lL} + c_{hS})$, of which is given by the nonlinear scale at the top of Fig. 9. It also represents the range of possible inversion velocities and the corresponding

bulk densities of the mixed bed at the inversion point, depending on the overall solids composition, v_{lL}. For example, assuming an overall solids composition, $c_{lL}/(c_{lL} + c_{hS}) = v_{lL} = 0.4$, given by point B on Fig. 9, then on increasing the liquid superficial velocity from U_{mflL} to U_{0hS}, we can observe five stages to the inversion process:

1. From $U = U_{mflL}$ to $U = U_A$, the bottom layer contains pure solids species hS and the top pure species lL, both layers expanding with increasing U.

2. From $U = U_A$ to $U = U_B$, the expanding bottom layer composition is a mixture of the two species, with $c_{lL}/(c_{lL} + c_{hS})$ increasing from 0 to 0.4 and ρ_B decreasing from ρ_{BA} to ρ_{BB} as U increases, while the contracting top layer continues to contain pure lL.

3. At $U = U_B$, the inversion velocity, the entire bed attains a uniform composition equal to that of the overall solids composition, $x_{IL} = 0.4$, and a bulk density equal to ρ_{BB}.

4. From $U = U_B$ to $U = U_D$, the contracting bottom layer solids composition continues to be a mixture of the two species, with $c_{IL}/(c_{IL} + c_{hS})$ increasing from 0.4 to unity and ρ_B decreasing from ρ_{BB} to ρ_{BD} as U increases, while the expanding top layer now contains pure species hS.

5. Finally, from $U = U_D$ to $U = U_{0hS}$, the bottom layer now contains pure species lL and the top layer continues with pure hS, both layers again now expanding with increasing U.

Deviations from this idealized "complete segregation model" due to particle dispersion, e.g., the penetration by the smaller, denser particles of the upper zone in Fig. 10a, have been addressed by Di Felice (1993) and by Asif (1997).

Jean and Fan (1986) have adopted another approach to predicting the bed inversion velocity. Recall from Eqs. (85)–(87), as applied to the present situation, that the segregation velocity of species hS with respect to species lL is given by

$$U_{phS} = \left(\frac{U}{\varepsilon}\right) - k_{hS} U_{0hS} \varepsilon^{n_{hS}-1} \tag{116}$$

The inversion condition is then taken as

$$\frac{d(U_{phS})}{dU} = 0 \tag{117}$$

Accurate agreement with the experimentally measured value of inversion velocity was thus obtained when the corresponding voidage in Eq. (116) was measured; moderately good prediction of this velocity could be effected by formulating this voidage in terms of the previously discussed serial model, Eq. (66), applied to the binary solids mixture:

$$\frac{1}{1-\varepsilon} = \frac{v_{lL}}{1-\varepsilon_{lL}} + \frac{1-v_{lL}}{1-\varepsilon_{hH}} \tag{118}$$

with the monocomponent voidages, ε_{lL} and ε_{hH}, each expressed as functions of U by means of Eq. (32) and their respective values of both $kU_0 (= U_e)$ and n.

6 INVERSE FLUIDIZATION

Solids having density smaller than that of a given liquid may be fluidized downward by that liquid, the distributor of which is located at the top rather than the bottom of the fluidizing column. Since gravity now

acts in the same direction as the drag, while buoyancy acts in the opposite direction, Eq. (8) must be modified to

$$-\frac{dp_f}{dz} = (1-\varepsilon)(\rho - \rho_p)g = (1-\varepsilon)|\rho_p - \rho|g \tag{119}$$

The first mention of such inverse fluidization occurs in the appendix to a thesis by Page (1970), who summarizes the results of a contemporaneous report by Field and Riley (1970), the original of which has since been destroyed. These results and more detailed, subsequent hydrodynamic studies (Fan et al., 1982; Garnier et al., 1990; Karamanev and Nikolov, 1992b) show that both minimum fluidization velocity and bed expansion can, with minor qualifications, be predicted by the same methods as for conventional fluidization. For free rising single spheres in water with $\rho_p < 300$ kg/m^3, the Newton regime starts at $Re_0 \cong 130$ because of the additional turbulence then generated by wake shedding, which produces a spiral particle trajectory and an increase in the constant drag coefficient, C_D, from Newton's theoretical value of 0.5 and the actual value of about 0.44 to a measured value of 0.95 (Karamanev and Nikolov, 1992a). Thus, for $\rho_p < 300$ kg/m^3 and $Re_0 > 130$ (i.e., $Ar > 12,000$),

$$F_{D0} = 0.95 \left(\frac{\pi}{4}\right) d_p^2 \left(\frac{1}{2}\right) \rho U_0^2 = \left(\frac{\pi}{6}\right) d_p^3 |\rho_p - \rho| g \tag{120}$$

whence

$$U_0 = 1.18 \sqrt{\frac{d_p |\rho_p - \rho| g}{\rho}} \tag{121}$$

which gives values of U_0 about 32% lower than the usual equation for the Newton regime [with 0.44 instead of 0.95 in Eq. (120) and 1.74 instead of 1.18 in Eq. (121)]. Therefore for very light spheres ($\rho_p < 130$ kg/m^3), provided $Ar > 12,000$, U_0 in the Richardson–Zaki relation, Eq. (32), should be obtained from Eq. (121) rather than from Eq. (33). The minimum velocity for inverse fluidization of such spheres is slightly lower than predicted by Eq. (17), owing presumably to the higher drag exerted on them (Karamanev and Nikolov, 1992b).

In operating an inverse liquid fluidized bed, extra measures must be taken to rid the liquid of undesired bubbles since, unlike in conventional fluidization, the larger bubbles tend to move in a direction (upward) opposite to that of the exit-bound liquid (Karamanev and Nikolov, 1992b). Inverse liquid fluidization, in which the particles are coated with biofilm, has thus far found its principal application as a vehicle for biochemical reactions (Chavarie and Karamanev, 1986; Nikolov and Karamanev, 1990), with particular emphasis on ferrous ion oxidation (Karamanev and Nikolov, 1988) and wastewater treatment (Karamanev and Nikolov, 1996).

7   VERTICAL MOVING FLUIDIZED SYSTEMS (CONTINUOUS FLUIDIZATION)

7.1   Monodisperse Solids

Continuing with our assumption of relatively homogeneous particulate fluidization, but switching from semibatch operation, i.e., from an unchanging batch of bottom- or top-restrained (the latter in inverse fluidization) solids to a continuous vertical flow of both solids and liquid, we consider first the case of monosized constant density solids. Studies by Mertes and Rhodes (1955) and by Lapidus and Elgin (1957) have shown that, irrespective of the flow direction of either liquid or solids, the relative or "slip" velocity, U_r, between the liquid and the solids in the nonaccelerating region of a given liquid–solids system is always the same unique function of the voidage, so that in general,

$$U_r = \frac{U}{\varepsilon} - \frac{U_p}{1 - \varepsilon} = f'(\varepsilon) \tag{122}$$

in which the upward direction is taken as positive. Equation (122) has been verified, i.e., $f'(\varepsilon)$ has been shown to be the same for cocurrent upward flow (Struve et al., 1958), countercurrent flow (Price et al., 1959), and cocurrent downward flow (Quinn et. al., 1961), as well as for both unfed, i.e., semibatch, fluidization ($U_p = 0$) and constant-rate bottom-restrained sedimentation ($U = 0$) (Mertes and Rhodes, 1955). The Richardson–Zaki (1954) equation, Eq. (32), can therefore be generalized as

$$\frac{U}{\varepsilon} - \frac{U_p}{1 - \varepsilon} = kU_0\varepsilon^{n-1} \tag{123}$$

and all the other relationships between U and ε for semibatch fluidization can similarly be applied to moving fluidized systems by substituting $U_r\varepsilon$ for U in these equations or plots, with U_r given by Eq. (122). The useful graphical plot of Barnea and Mizrahi (1973) is already expressed in terms of U_r, based as it is on data for sedimentation, semibatch fluidization, and various modes of fully continuous operation.

In the case of countercurrent operation (Lapidus and Elgin, 1957; Price et al., 1959), flooding, i.e., backup of solids, will occur when

$$\left. \frac{\partial U}{\partial (1 - \varepsilon)} \right|_{U_p} = 0 \tag{124}$$

Applying this criterion to Eq. (123), we arrive at

$$U_p = -nkU_0\varepsilon^{n-1}(1 - \varepsilon)^2 \tag{125}$$

and

$$U = kU_0\varepsilon^n[1 - n(1 - \varepsilon)] \tag{126}$$

Thus, for a given solids flux U_p, flooding will occur when the in situ voidage falls to that given by Eq. (125), which occurs when the liquid flux U rises to that given by Eq. (126); and if $|U_p|$ is increased, the flooding will occur at a lower value of ε and a correspondingly lower value of U.

7.2   Multidisperse Solids

For solids containing more than one particle size or particle density, the situation is not nearly as clearcut, except in the case of dilute ($\varepsilon > 0.94$) suspensions in creeping flow, for which the theory of Batchelor (1982) and Wen (Batchelor and Wen, 1982) has been verified for sedimentation of bidisperse submicron spheres by Al-Naafa and Selim (1992). For more concentrated suspensions, e.g., $0.06 < \varepsilon < 0.50$, Finkelstein et al. (1971) found that, in both countercurrent and cocurrent downward flow, both species of binary-sized solids moved at the same rate, provided that the solids residence time was short (< 1 minute) and the size ratio did not exceed 3. For more common situations, however, in which N solids species have left their initial acceleration zone, Richardson and Shabi (1960) have suggested that the relative or slip velocity, U_{ri}, of each particle species (e.g., each narrow cut of sieved solids of fixed density) depends on the total surrounding voidage in a manner similar to that of the same species acting monodispersely in the same liquid, viz.,

$$U_{ri} = \frac{U}{\varepsilon} - \frac{U_{p,i}}{c_i} = k_iU_{0i}\varepsilon^{n_i-1}, \qquad i = 1, 2, 3, \dots, N \tag{127}$$

Lockett and Al-Habbooby (1973) provided some experimental backing for Eq. (127), for both countercurrent flow of three sized (i.e., ρ_p = constant) binaries and one sized ternary, and for batch sedimentation of several sized binaries, but only when the voidage exceeded 0.6, below which the inapplicability of Eq. (127) was attributed to interactions between particle species. Subsequently they claimed Eq. (127) to be applicable also to particles of different densities

(Lockett and Al-Habbooby, 1974), but the small difference in density of the equal-sized binary particle species on which they based this claim, and experiments by later workers, render such a conclusion generally inadmissible. These experiments and some theory addressed themselves both to the problem of species interactions and that of density differences among species.

Mirza and Richardson (1979), based on batch sedimentation experiments with several sized binaries down to $\varepsilon \cong 0.5$, concluded that the species interaction effect could be accounted for by the factor $\varepsilon^{0.4}$ applied to the right-hand side of Eq. (127), so that

$$U_{ri} = k_{0i}U_{0i}\varepsilon^{n_i-1}\varepsilon^{0.4} = k_{0i}U_{0i}\varepsilon^{n_i-0.6} \quad (128)$$

At the same time, Masliyah (1979) addressed the density difference problem. For a monodisperse system, i.e., a single particle species, both Eq. (123) and Eq. (127) are equivalent to

$$U_r = kU_0\varepsilon^{n-1} \quad (129)$$

Since from Eq. (4),

$$\varepsilon = \frac{\rho_p - \rho_B}{\rho_p - \rho} \quad (130)$$

it follows that

$$U_r = kU_0\varepsilon^{n-2}\frac{\rho_p - \rho_B}{\rho_p - \rho} \quad (131)$$

By using the form of the governing momentum equation given by Wallis (1969), Masliyah showed analytically that Eq. (131) can be generalized for each species i of a multispecies particle system to

$$U_{ri} = k_iU_{0i}\varepsilon^{n_i-2}(\rho_{pi} - \rho_B)/(\rho_{pi} - \rho) \quad (132)$$

where for N particle species,

$$\rho_B = (\rho_p)_{avg}(1 - \varepsilon) + \rho\varepsilon \quad (133a)$$

$$= \sum_{i=1}^{N}\rho_{pi}c_i + \rho\left[1 - \sum_{i=1}^{N}c_i\right] \quad (133b)$$

Subsequently, Patwardhan and Tien (1985) adopted Masliyah's model but assigned an effective voidage, ε_{ei}, to each particle species to account for the different environment experienced by that species in the presence of the others than in their absence. Thus Eq. (132) becomes

$$U_{ri} = k_iU_{0i}\varepsilon_{ei}^{n_i-2}\frac{\rho_{pi} - \rho_B}{\rho_{pi} - \rho} \quad (134)$$

where, by a heuristic cell model approach, ε_{ei} was formulated as

$$\varepsilon_{ei} = 1 - \left[1 + \left(\frac{d_{avg}}{d_i}\right)\{(1 - \varepsilon)^{-1/3} - 1\}\right]^{-3} \quad (135)$$

with

$$d_{avg} = \frac{\sum_{i=1}^{N}c_id_i}{\sum_{i=1}^{N}c_i} = \frac{\sum_{i=1}^{N}c_id_i}{(1 - \varepsilon)} \quad (136)$$

Note that for a monodisperse system, $\varepsilon_{ei} = \varepsilon$ and Eq. (134) reverts back to Eq. (131). Both Eqs. (132) and (134) were able to map features of the binary batch sedimentation data of Richardson and Meikle (1961a) for spheres of different sizes and densities having the same value of U_0.

A different approach was adopted by Selim et al. (1983a,b), who considered the buoyancy effect induced by the smaller (or less dense) particles on the motion of the larger. Thus particles i are treated as moving in a fluid of viscosity equal to that of the pure liquid and density equal to that of a mixture of the pure fluid and of all particles smaller than d_{pi} (or less dense than ρ_{pi}). For example, for a ternary mixture of spheres having diameters d_1, d_2, and d_3 and densities ρ_{p1}, ρ_{p2}, and ρ_{p3}, respectively, such that the relative ranking of U_0 in the given liquid is $U_{01} > U_{02} > U_{03}$,

$$U_{r1} = k_1U_{01}'\varepsilon^{n_1-1} \quad (137)$$

where U_{01}' is the terminal free settling velocity of the fastest moving particles in a fluid of density equal to

$$\rho_{2,3} = \frac{c_2\rho_{p2} + c_3\rho_{p3} + \varepsilon\rho}{1 - c_1} \quad (138)$$

and

$$U_{r2} = k_2U_{02}'\varepsilon^{n_2-1} \quad (139)$$

where U_{02}' is the terminal free settling velocity of the intermediate-velocity particles in a fluid of density equal to

$$\rho_3 = \frac{c_3\rho_{p3} + \varepsilon\rho}{1 - c_1 - c_2} \quad (140)$$

while

$$U_{r3} = k_3U_{03}\varepsilon^{n_3-1} \quad (141)$$

Thus only the smallest particles will behave as predicted by Eq. (127). This model was found to agree well with both multisized sedimentation data ($0.5 < \varepsilon < 0.85$) and the binary-sized countercurrent results of Lockett and Al-Habbooby (1973).

All five models discussed above have been subjected to comparison with data on gravity separation of bidisperse suspensions in which light and heavy particles of the same size are vertically separated from each other in a liquid of intermediate density (Law et al., 1987), and with data on sedimentation of polydisperse concentrated suspensions (Al-Naafá and Selim, 1989). The unequivocally inferior model that emerges from a combination of these comparisons is that of Lockett and Al-Habbooby (1973), while the equivocally superior one is that of Selim et al. (1983a,b).

The model of Selim et al. (1983a,b) can be thought of as intermediate between the "overall voidage model" (Di Felice, 1995) of Lockett and Al-Habbooby (1973) and the pseudofluid model of Di Felice et al. (1991a). In the latter model, the application of which has thus far been limited to binaries, if we assume that $U_{01} > U_{02}$, then

$$U_{r1} = \frac{U + U_{p2}}{\varepsilon + c_2} - \frac{U_{p1}}{c_1} = k_1 U_{01}''(1 - c_1)^{n_1 - 1} \quad (142)$$

That is, the larger and/or heavier particles are assumed to move through a pseudofluid having the properties, and flowing with the superficial velocity (or volumetric flux), of the combined pure liquid and smaller and/or lighter particles. The free settling terminal velocity, U_{01}'', is thus based on a pseudofluid density of

$$\rho_2 = \frac{c_2 \rho_{p2} + \varepsilon \rho}{1 - c_1} \quad (143)$$

as in the model of Selim et al., while the pseudofluid viscosity on which U_{01}'' is based can be estimated from one of several empirical or semiempirical equations (e.g., Trawinski, 1953; Hetzler and Williams, 1969; Saxton et al., 1970; Rigby et al., 1970) for the effective viscosity of fluidized beds, or from one of several theoretical or semitheoretical equations (cf. Table 8 of Barnea and Mizrahi, 1973) for the apparent viscosity of suspensions. Di Felice (1998) has most recently recommended the semitheoretical equation of Thomas (1965), which applies over the entire voidage range of 0.4–1 and, for the liquid suspension of the smaller and/or lighter particles, is given by

$$\mu_2 = \mu\left[1 + 2.5c_2' + 10.05(c_2')^2 + 0.0027e^{16c_2'}\right] \quad (144)$$

where $c_2' = c_2/(c_2 + \varepsilon)$. The pseudofluid model has scored its greatest success with single foreign spheres ($c_1 = 0$) settling in a fluidized bed of smaller and/or lighter spheres (Di Felice, 1998), though it has also yielded good predictions for the larger solid species of a sized binary in both batch and continuous sedi-

mentation and fluidization (Di Felice et al., 1989, 1991b). It has, however, not yet succeeded in satisfactorily modeling the movement of the smaller species (Di Felice, 1995).

A subtle mutation of Masliyah's (1979) model, arrived at quite independently, has recently been proposed by Asif (1997) and by Galvin et al. (1999). If Eq. (130) is substituted into Eq. (131), the result is

$$U_r = kU_0\left(\frac{\rho_p - \rho_B}{\rho_p - \rho}\right)^{n-1} \quad (145)$$

By more heuristic approaches than that of Masliyah, both Asif and Galvin et al. generalized this empirical equation as applying to each species of a multispecies system, so that

$$U_{ri} = k_i U_{0i}\left(\frac{\rho_{pi} - \rho_B}{\rho_{pi} - \rho}\right)^{n_i - 1} \quad (146)$$

which is *not* equivalent to Eq. (132) except in the case of a monodisperse suspension. Substituting for ρ_B in Eq. (145) its value given by Eq. (8a) results in

$$U_{ri} = k_i U_{0i}\left[1 - \frac{(-dp_f/dz)}{(\rho_{pi} - \rho)g}\right]^{n_i - 1} \quad (147)$$

from which U_{ri} can be determined for each species by measurement of frictional pressure gradient, assuming knowledge of $k_i U_{0i} (= U_{ei})$ and n_i for each. Reasonable agreement with Eq. (147) was obtained by Galvin et al. for measured values of U_{ri} of each particle species in a liquid fluidized bed operating with a continuous feed consisting of three species of particles differing in size, shape, and density, separating into overflow and raked underflow streams (a "teetered bed separator"). Galvin et al. disclaim any possibility of Eq. (146) or (147) applying to situations in which one of the particle species is less dense than the suspension (i.e., $\rho_{pi} < \rho_B$), and it is likely that the same limitation applies to Eqs. (132) and (134) in the case of countercurrent flow of upward liquid and denser solids.

Externally circulating liquid-fluidized beds and the simplified one-dimensional hydrodynamics thereof made a brief appearance in the literature a decade ago with respect to particle mixing (Di Felice et al., 1989) and fermentation (Pirozzi et al., 1989), but more detailed experimental investigations of their hydrodynamic complexities have not been reported on until quite recently (Liang et al., 1996, 1997; Liang and Zhu, 1997; Zheng et al., 1999). Nonuniformities in flow structure appear to be much attenuated when compared with externally circulating gas-fluidized beds.

8 LIQUID DISPERSION

8.1 Axial Dispersion

The dispersion of liquid in the axial or longitudinal direction within a particulately liquid-fluidized bed is qualitatively and quantitatively intermediate between that of liquid flow through a packed bed and liquid flow through an otherwise empty tube. Thus, generally speaking, the axial dispersion coefficient, D_a, increases in magnitude as the fluidizing liquid velocity increases, encompassing for different conditions values below $1 \, \text{cm}^2/\text{s}$ to values as high as $100 \, \text{cm}^2/\text{s}$, and the corresponding axial Peclet number, $\text{Pe}_a (= d_p U / \varepsilon D_a)$, decreases from values of 0.4–2 for packed beds to 1.0–0.01 or even lower (Scott et al., 1993) for liquid-fluidized beds. Axial liquid dispersion in a fluidized bed above the entry region is contributed to by particle motion, fluid turbulence, particle wake-shedding (Letan and Elgin, 1972), and voidage waves (Kramers et al., 1962) but only minimally by longitudinal velocity gradients in the radial (or transverse) direction, especially if these are abated by uniform liquid distribution of the inlet liquid and by sufficiently high values of D_c/d_p (e.g., > 50), assuming $\varepsilon < 0.9$.

The experimental determination of axial dispersion coefficients has been effected mainly by measurement of the outlet response to a controlled tracer input. As with particle dispersion, liquid axial dispersivities, when plotted against liquid velocity or voidage, in some cases rise monotonically from minimum fluidization to particle transport (e.g., Bruinzeel et al., 1962; Chung and Wen, 1968; Tang and Fan, 1990) and in some cases show a maximum, usually at $\varepsilon = 0.7 \pm 0.1$ (e.g., Cairns and Prausnitz, 1960a; Mehta et al., 1976; Kikuchi et al., 1984). Usually this maximum is followed by a minimum (Kramers et al., 1962) and sometimes even a tendency towards a second maximum at $\varepsilon \approx 0.9$ (Mehta et al., 1976). Even in the absence of a maximum, Kramers et al. have noted a sharp rise in the rate of increase of D_a with ε at $\varepsilon \approx 0.7$ for water fluidization of 1 mm glass spheres. They attributed this rise to the fact that above this voidage the velocity of horizontal voidage disturbances (Slis et al., 1959), $(1 - \varepsilon) dU/d\varepsilon = n(1 - \varepsilon) U_e \varepsilon^{n-1}$, subsequently referred to as continuity waves (Wallis, 1969), exceeds the interstitial liquid velocity, $U/\varepsilon = U_e \varepsilon^{n-1}$ (where n = Richardson–Zaki index), so that eddies produced

by individual particles are then supplemented by eddies produced by these voidage fluctuations. Note that when

$$n(1 - \varepsilon) U_e \varepsilon^{n-1} = U_e \varepsilon^{n-1} \tag{148}$$

the corresponding voidage is

$$\varepsilon = \frac{n - 1}{n} \tag{149}$$

so that for nonflocculating spheres in Newtonian liquids, the critical voidage will vary from 3.8/4.8 ($= 0.79$) in the Stokes region to 1.4/2.4 ($= 0.58$) in the Newton region. In the early study of Cairns and Prausnitz (1960a), water fluidization of both 1.3 and 3.0 mm lead spheres showed maxima (at $\varepsilon = 0.6$–0.7), which suggests that the transition from particulate to aggregative fluidization (see Sec. 9) characteristic of the latter is an additional factor contributing to the maxima.

An empirical equation for D_a of liquid fluidized beds, correlating a large range of experimental data [$d_p = 0.5$–14.3 mm, $\rho_p = 1280$–11,300 kg/m^3, $\mu = 0.00056$–0.0619 kg/(m)(s), and rheological index = 0.858–1.0 for some non-Newtonian power law liquids (Wen and Fan, 1973)] with a standard deviation of 8%, is given by (Krishnaswamy et al., 1978)

$$\left[1 - \left(\frac{2 D_a \varepsilon}{UL} \right)^{0.5} \right] \varepsilon^{0.25} = 0.74 \tag{150}$$

The presence of bed height L [$= L_{mf}(1 - \varepsilon_{mf})/(1 - \varepsilon)$] in the correlation is, however, at odds with the results of Wen and Fan (1973), which explicitly show no dependence of D_a on L_{mf}. Furthermore, data on D_a limited to the lower Reynolds number region, obtained subsequently by Kikuchi et al. (1984), were poorly predicted by Eq. (150) and by seven other empirical equations in the literature. For $0.15 < \text{Re} < 100$, $0.41 < \varepsilon < 0.93$, and $8 \times 10^{-5} < e_d < 0.44 \, \text{m}^2/\text{s}^3$, these investigators recommended

$$\frac{D_a}{\nu} = 500 e_d^{0.43} \exp \left[-20.5 (0.75 - \varepsilon)^2 \right] \pm 38\% \tag{151}$$

where e_d, the energy dissipation rate per unit mass of liquid, is given by

$$e_d = \frac{(\rho_p - \rho)(1 - \varepsilon) g U}{\varepsilon \rho} = \frac{(\rho_p - \rho)(1 - \varepsilon) g U_e \varepsilon^{n-1}}{\rho} \tag{152}$$

Note that the square-bracketed term in Eq. (151) is a maximum at $\varepsilon = 0.75$ while e_d, like the continuity wave velocity given by the left-hand side of Eq.

(148), and other related hydrodynamic phenomena that are proportional to $(1 - \varepsilon)\varepsilon^{n-1}$, including the bulk turbulent intensity (Joshi, 1983; Kang and Kim, 1986), all show a maximum with respect to voidage at $\varepsilon = (n - 1)/n$, which is thus considered a crucial turning point for these phenomena (Kang and Kim, 1988).

8.2 Radial Dispersion

Radial dispersivities of liquid in liquid-fluidized beds are typically one or two orders of magnitude smaller than the corresponding axial dispersivities, so that the radial Peclet number, $Pe_r (= d_p U/\varepsilon D_r)$, usually falls within the range of 1–10 (Couderc, 1985; Jean and Fan, 1998), D_r increasing with d_p at any given value of the interstitial velocity, U/ε (Kang and Kim, 1986), and with ρ_p at any given value of ε (Cairns and Prausnitz, 1960b). Two studies of radial dispersion have shown minima in Pe_r at $\varepsilon \approx 0.7$ (Hanratty et al., 1956; Cairns and Prausnitz, 1960b), which Hanratty et al. attributed to a measured maximum in the scale of turbulence at the same voidage, while Cairns and Prausnitz observed, "the most active local particle motion occurs in the range of $\varepsilon = 0.7$." A third study (Kang and Kim, 1986) showed similar minima in Pe_r at $\varepsilon \approx (n - 1)/n$, which they attributed to the maxima in e_d and bulk intensity of turbulence detailed in the paragraph above. The experimental values of n for this study were such that both $(n - 1)/n$ and the measured critical values of ε were closer to 0.6 than to 0.7.

9 NONHOMOGENEITIES AND INSTABILITY

A frequently encountered but underreported disruption of an otherwise relatively "homogeneous" liquid-fluidized bed occurs when nonuniformity of the liquid velocity profile above the distributor gives rise to bulk circulation of the solids. This effect can also manifest itself as a convective instability associated with a uniform distributor having an insufficient pressure drop compared to that of the fluidized bed (Medlin et al., 1974; Medlin and Jackson, 1975; Agarwal et al., 1980). Most commonly it occurs with upflow of solids at the center of the column and downflow near the walls (e.g., Latif and Richardson, 1972). The net effect, usually unnoticed and therefore underreported, is that the bed expansion or overall voidage is smaller than in the absence of the bulk circulation, i.e., smaller than would be predicted by, e.g., Eq. (32), for the same superficial velocity (Epstein et al., 1981).

As shown roughly by Hiby (1967), and more rigorously by Masliyah (1989), this phenomenon can be explained by the incremental increase of voidage in the solids upflow region ("channeling") and the corresponding incremental decrease in the downflow region, the result of which is a net decrease of total drag on the fluidized solids. Masliyah demonstrated experimentally that the same contraction effect can be brought about by mechanical stirring of a liquid-fluidized bed.

An arguably ubiquitous nonhomogeneity, at least for $U/U_{mf} > 1.5 \pm 0.3$ (Ham et al., 1990), which has received much theoretical and experimental attention in the literature, e.g., by Jackson (1963, 1985), Anderson and Jackson (1968, 1969), and Homsy and coworkers (e.g., El-Kaissy and Homsy, 1976; Didwania and Homsy, 1981; Ham et al., 1990), are the horizontal striations or bands of low particle concentration, dubbed "parvoids" by Hassett (1961a,b), which rise wavelike through a liquid fluidized bed and become more pronounced as they rise. The growth rate of these voidage waves, which may sometimes be accompanied by short-lived bubblelike formations (Cairns and Prausnitz, 1960b; Hassett, 1961a), tend to increase with increasing voidage for a given system, and to increase with increasing bubble size for a given voidage. This type of disturbance, which is associated with liquid fluidization of relatively low-density solids, e.g., water fluidization of 1 mm glass beads (Jackson, 1994), is relatively innocuous, i.e., it is minimally disruptive of the operating characteristics of a nondisturbed bed and its effects are incorporated in the previously written empirical equations, e.g., in the equations for predicting bed expansion. Jackson (1994) has pointed out that this type of instability should not be confused with the disruptively persistent bubbling and consequent large-scale particle circulation (Jean and Fan, 1998) characteristic of aggregative fluidization, which is a property of most gas-fluidized beds but for liquid fluidization is usually restricted to relatively high density or large particles, e.g., water fluidization of 3 mm lead spheres (Cairns and Prausnitz, 1960b). For aggregative fluidization, most of the previously written empirical equations, including those for predicting bed expansion, no longer apply. These two types of instability are not explicitly distinguished from each other in Table 1 of Di Felice (1995), in which are recorded the nonhomogeneous behavior observed in 26 different studies of liquid fluidization from the literature.

A straightforward quantitative criterion which purports to distinguish between particulate and aggregative fluidization of both gas and liquid fluidized beds has been deftly developed by Foscolo and Gibilaro (1984, 1987) and Gibilaro et al. (1986b). The sources of the instability are the previously mentioned voidage disturbances or continuity waves, the kinematic velocity of which is given by

$$U_\varepsilon = k U_0 n (1 - \varepsilon) \varepsilon^{n-1} \qquad (153)$$

After Wallis (1969), the countervailing stabilizing force is provided by the particle concentration gradients that give rise to the so-called elastic waves, the dynamic velocity of which is derived as

$$U_E = \sqrt{\frac{3.2 g d (1 - \varepsilon)(\rho_p - \rho)}{\rho_p}} \qquad (154)$$

On equating U_ε with U_E and rearranging the result, one arrives at

$$\left[\frac{g d (\rho_p - \rho)}{k^2 U_0^2 \rho_p} \right]^{0.5} - 0.56 n (1 - \varepsilon)^{0.5} \varepsilon^{n-1}$$
$$= \begin{cases} +ve, & \text{stable (particulate)} \\ 0, & \text{stability limit} \\ -ve, & \text{unstable (aggregative)} \end{cases} \qquad (155)$$

The first term in Eq. (155) is equivalent to $(\mathrm{Ar}\rho/\rho_p)^{0.5}/k\mathrm{Re}_0$. Since in general (Epstein and LeClair, 1985),

$$U_0 = \frac{c_m (\rho_p - \rho)^{1/m} g^{1/m} d^{(3-m)/m}}{\rho^{(m-1)/m} \mu^{(2-m)/m}} \qquad (156)$$

where m varies from 1 in the Stokes region to 2 in the Newton region, and c_m varies correspondingly from $1/18\,(=0.0556)$ to $[4/(3)(0.44)]^{1/2}\,(=1.74)$, Eq. (155) at the stability limit can be rewritten as

$$\left[\frac{\rho^{(2m-2)/m} \mu^{(4-2m)/m}}{k^2 c_m^2 g^{(2-m)/m} d^{(6-3m)/m} (\rho_p - \rho)^{(2-m)/m} \rho_p} \right]^{1/2}$$
$$- 0.56 n (1 - \varepsilon)^{0.5} \varepsilon^{n-1} = 0 \qquad (157)$$

Thus aggregative fluidization is favored by high values of ρ_p and d and low values of μ and ρ (and the reverse in each case for particulate fluidization), which agrees qualitatively with experimental observations. Quantitatively, three situations arise (Gibilaro et al., 1986b): (1) The left-hand side (LHS) of Eq. (155) is positive for all values of ε, in which case particulate fluidization prevails from minimum fluidization to particle transport, e.g., for 0.1 mm copper spheres fluidized by water. (2) The LHS of Eq. (155) is zero at $\varepsilon = \varepsilon_{b1}$ and $\varepsilon = \varepsilon_{b2}$, i.e., positive for ε up to ε_{b1}, negative between ε_{b1} and ε_{b2}, and positive again at $\varepsilon > \varepsilon_{b2}$,

where $\varepsilon_{mf} < \varepsilon_{b1} < \varepsilon_{b2} < 1$. Thus particulate fluidization prevails from ε_{mf} to the minimum bubbling voidage ε_{b1}, aggregative fluidization from ε_{b1} to the maximum bubbling voidage ε_{b2}, and particulate fluidization again in the very dilute phase regime from ε_{b2} to a voidage of unity, e.g., for 0.5 mm copper spheres fluidized by water, for which $\varepsilon_{b1} \approx 0.5$ and $\varepsilon_{b2} \approx 0.95$. (3) The LHS of Eq. (155) is zero for $\varepsilon = \varepsilon_{b1} < \varepsilon_{mf}$ and $\varepsilon = \varepsilon_{b2} < 1$ and negative between ε_{b1} and ε_{b2}, in which case vigorous bubbling starts as soon as the bed is fluidized and ceases only at ε_{b2}, the maximum bubbling voidage, above which dilute phase particulate fluidization prevails, e.g., for 2 mm copper spheres fluidized by water (Chen et al., 1999), for which $\varepsilon_{b2} \approx 0.97$.

Gibilaro et al. (1990) have subsequently corrected Eq. (155) for the added or virtual mass effect due to the fluid acceleration that accompanies particle acceleration. The correction involves adding $\rho/2$ to both ρ_p and ρ wherever they occur in the equation, as well as multiplying the second term in the equation by a factor exceeding unity that approaches unity as ρ/ρ_p approaches zero. However, since this correction only becomes important as ρ/ρ_p approaches unity, i.e., under conditions when the LHS of Eq. (155) will in most cases be positive whether or not the correction is applied (and will hence indicate particulate fluidization, in agreement with experiment), and given the experimental uncertainty involved in pinpointing regime transitions, this added complication can be safely dispensed with, at least for purposes of predicting aggregative vs. particulate fluidization.

The predictions of Eq. (155) have reasonably well matched the relatively few available regime transition data on liquid fluidization (Gibilaro et al., 1986b), as well as the much more numerous data on gas fluidization (Foscolo and Gibilaro, 1984, 1987; Chen et al., 1999; these references also include some liquid fluidization comparisons). However, Batchelor (1988) has questioned the basis of Eq. (154) and has proposed other stabilizing mechanisms, while Jackson (1994) has also criticized Foscolo and Gibilaro for not properly distinguishing between the stability of a uniform bed against small perturbations and the phenomenon of aggregative fluidization. The alternative stability criterion developed by Batchelor (1988) is similar in form to Eq. (155), but it contains two quantities that are more difficult to evaluate. Until the challenging task of solving this problem by two-dimensional nonlinear stability analysis is accomplished (Jackson, 1994), Eq. (155) can serve as a useful semitheoretical or semiempirical guideline.

10 DISTRIBUTOR DESIGN

The simplest configuration for a liquid-fluidized bed distributor, common in the early mineral dressing literature (Richards and Locke, 1940; Taggart, 1953) on teeter-bed classifiers, is a tapered tube with a small included angle (e.g., $20 \pm 10°$) which diverges smoothly into the cylindrical column above, and into which the fluidizing liquid is introduced either axially (through an optional solids-retaining screen) or concentrically. This type of distributor is not often used now, even for laboratory purposes, as it will usually violate two conditions that are considered important for many practical applications, namely, that the frictional pressure drop, $-\Delta p_d$, across the distributor be "sufficiently large" with respect to the frictional pressure drop, $-\Delta p_f$, across the fluidized bed, and that the longitudinal liquid velocity distribution above the distributor be uniform in the lateral direction.

A liquid distributor in current practice consists most commonly of a plate with evenly spaced perforations, but other distributors used include a sintered metal or glass plate, or several layers of metal gauze, or any number of other uniformly (or relatively so) porous materials; and each is normally preceded by a calming section (Fig. 1), the function of which is to equalize the liquid flow as much as is feasible before it gets to the distributor. This section, the bottom of which is a wire screen and the length of which can vary from less than to several times the column diameter, is usually filled from bottom to top with rings, saddles, spheres, or other immobilized packings, or alternately with tubular or planar flow straighteners. Bascoul et al. (1988) found that by only partially filling their calming section with the same particles as used in the fluidized bed above, the calming section, which then acted as a fluidized rather than a fixed bed, produced greater uniformity of flow from the distributor and thus much reduced channeling. The action of this distributor was then similar to that of an upwardly diverging conical bed of polydisperse solids, in which the well-agitated condition of the fluidized coarser solids in the higher velocity bottom zone serves to disperse the fluidizing liquid as it enters the upper zone of finer solids, thus performing the function of a normal liquid distributor for the latter, the lower liquid velocity of which provides the additional advantage of preventing excessive entrainment (Kwauk, 1992). The "ball distributor", developed by Hiquily et al. (1979) for gas fluidization, though anticipated by Adler and Happel (1962) and Grimmett and Brown (1962) for liquid fluidization, consists only of a screen or porous plate topped by 50–150 mm of spheres, e.g., 2–3 mm lead shot, which are much heavier than the particles to be fluidized immediately above them. The simplicity and efficacy of this type of distributor has resulted in its wide adoption for liquid fluidization (Epstein et al., 1981; Rapagnà et al., 1989; Masliyah, 1989; Di Felice, 1993).

The requirement of a minimum ratio of distributor to fluidized-bed frictional pressure drop, coupled with a modicum of distributor uniformity, arises from the need to suppress the convective instabilities that give rise to large-scale particle circulation, usually upward in the central regions and downward near the walls. This effect in a two-dimensional column has been studied by Jackson (1985) and coworkers (Medlin et al., 1974; Medlin and Jackson, 1975; Agarwal et al., 1980), who found both theoretically and experimentally that the value of $\Delta p_d / \Delta p_f$ required for convective stability decreases with increasing bed depth for a given voidage but increases with bed width W up to some limiting value of W above which no further increase of $\Delta p_d / \Delta p_f$ is required. This limiting value of W increases with bed depth. The analysis and experimental results, which were limited to water fluidization of 0.8 mm glass beads from $\varepsilon_{mf} = 0.42$ to $\varepsilon = 0.48$, indicate in their most conservative interpretation that $\Delta p_d / \Delta p_f$ should exceed 0.2 to assure convective stability for large-diameter (e.g., 1 m) beds, while lower values of this ratio will be sufficient for smaller bed diameters, especially with deep beds. However, Latif and Richardson (1972), working at much higher voidages ($\varepsilon = 0.55–0.95$, $U/U_{mf} = 2.3–8.6$) with 6.2 mm soda glass spheres fluidized by dimethyl phthalate ($\rho = 1183$ kg/m³, $\mu = 0.0105$ Pa·s) in a 102 mm diameter cylindrical column that showed large-scale circulation patterns that occupied an increasing volume of the bed from the lower part upward as the voidage was increased, reported that their multilayered metal gauze distributor delivered a uniform flow and that "under all the experimental conditions the pressure drop across the distributor was considerably greater than that across the fluidized bed." These results imply an even higher $\Delta p_d / \Delta p_f$ requirement at high voidages than the conservative value recommended above based on studies at ε close to ε_{mf}. It should be noted, though, on the reasonable assumption of Darcy's law for flow through a porous distributor, that Δp_f is essentially invariant with velocity, so that a distributor that satisfies the criterion $\Delta p_d / \Delta p_f = 0.2$ at minimum fluidization will yield considerably higher values of $\Delta p_d / \Delta p_f$ at higher voidages, e.g., at $U = 5U_{mf}$, $\Delta p_d / \Delta p_f$ would equal unity.

Even if the above large-scale convective circulation is avoided by sufficiency of Δp_d, significant inhomogeneities in flow and particle distribution can result from small imperfections in construction of liquid distributors of various designs (Volpicelli et al., 1966; Agarwal et al., 1980; Asif et al., 1992). Typically one may find, close to the distributor, high-velocity, high-voidage channeling streams interspersed with relatively stagnant, low-voidage regions. Particles with densities very close to that of the fluidizing liquid, e.g., water-fluidized polystyrene ($\rho_p = 1050 \, \text{kg/m}^3$, $d_p = 3$ mm) particles, have been found to be most vulnerable to the design of a perforated plate distributor (Asif et al., 1991), but even for somewhat higher density particles, e.g., polypropylene ($\rho_p = 1610 \, \text{kg/m}^3$, $d_p = 3$ mm), the liquid axial dispersivity near the distributor tends to be much greater than higher up in the fluidized bed, unless the hole density of the perforated plate, for a given fractional free area F, is sufficiently high, e.g., 1.3 (2 mm diameter) holes/cm^2 for $F = 0.04$ (Asif et al., 1991, 1992). For water fluidization of 2 and 3 mm glass ($\rho_p = 2460 \, \text{kg/m}^3$) spheres, however, this vulnerability declines to the point that distributor effects are found to be negligible over a wide range of hole density (Asif et al., 1992), presumably owing to the stabilizing effect of the higher density particles. According to Briens et al. (1997), channeling in liquid fluidized beds can arise not only from imperfections or poor design of the liquid distributor but also when the particles to be fluidized are very angular, e.g., biobone particles. These investigators found that, while channeling can be gradually reduced by increasing the liquid velocity from U_{mf} upward at the expense of increased pumping cost, particle attrition, and particle entrainment, a method of reducing channeling in their 0.1 m diameter bed without incurring these undesirable consequences was simply to introduce secondary liquid through horizontal jets issuing from a sparger located 90 mm above their perforated plate distributor.

11   MASS (AND HEAT) TRANSFER BETWEEN PARTICLES AND LIQUID

Steady state heat transfer and mass transfer (of solute A) between the surface of the solid particles and the liquid in a vertical fluidized bed can be represented differentially by the energy balance,

$$\frac{\rho c_p U dT}{dz} = \frac{(\lambda + \lambda_{\text{eddy}})d^2 T}{dz^2} + h_p S_v (T_s - T) \quad (158)$$

and the corresponding solute balance,

$$\frac{U dC_A}{dz} = \frac{D_a d^2 C_A}{dz^2} + k_p S_v (C_{AS} - C_A) \quad (159)$$

respectively, where $S_v = 6(1 - \varepsilon)/\phi d_p = S/AL$, $T = $ bulk temperature, and $C_A = $ bulk concentration of A at the bed level under scrutiny, T_S and C_{AS} are the corresponding surface temperature and concentration (e.g., the saturation concentration in the case of a dissolving solute), and D_a, discussed earlier, includes both the molecular diffusivity D_A and the longitudinal eddy diffusivity of solute A in the liquid. As it turns out, the heat transfer coefficient in a typical liquid-fluidized bed is high enough that thermal equilibrium between the liquid and the surface of the solids is reached within a very few particle layers of the distributor. This makes it very difficult (e.g., Holman et al., 1965), but at the same time of minor practical importance, to make reliable measurements of the particle-liquid heat transfer coefficient h_p. Because the molecular diffusivity of mass, D_A, of solute A in a liquid is some two or three orders of magnitude smaller than the molecular diffusivity of heat, $\lambda/\rho c_p$, (i.e., Sc $\gg$ Pr), particle liquid mass transfer is much slower and therefore much easier to measure. We shall therefore focus on mass transfer.

If we assume plug flow of the liquid, $D_a = 0$ and Eq. (159) can be rearranged to

$$\frac{dC_A}{C_{AS} - C_A} = \left(\frac{k_p S}{UAL}\right) dz \quad (160)$$

which, when integrated between the limits $z = 0$, $C_A = C_{A1}$ and $z = L$, $C_A = C_{A2}$, results in

$$\ln\left(\frac{C_{AS} - C_{A1}}{C_{AS} - C_{A2}}\right) = k_p S/UA \quad (161)$$

and is equivalent to

$$m_A = Q(C_{A2} - C_{A1}) = k_p S(\Delta C_A)_{\text{l.m.}} \quad (162)$$

For the opposite extreme of perfect mixing, $dC_A/dz = 0$ and $C_A = C_{A2}$ for all values of z except $z = 0$, so that

$$m_A = Q(C_{A2} - C_{A1}) = k_p S(C_{AS} - C_{A2}) \quad (163)$$

Since $(\Delta C_A)_{\text{l.m.}} > (C_{AS} - C_{A2})$, and since the real situation falls between the two, it follows that determination of k_p assuming plug flow will tend to underestimate k_p while its determination assuming perfect mixing will tend to overestimate k_p. Arters and Fan (1986) have solved Eq. (159) for finite D_a, assuming appropriate boundary conditions, and for two-phase particle–liquid fluidization, k_p, based on Eq. (162), is much closer to the answer by this solution than k_p by Eq. (163). Measured axial solute profiles are also much closer to those for plug flow than those for perfect mixing. All this provides justification for applying the plug flow model to the determination of k_p, a procedure used by most investigators of particle–liquid mass transfer. Values of k_p obtained from empirical correlations generated by these investigators should then be applied using Eq. (162).

The most frequently applied method in the literature for measuring the total mass transfer rate, m_A, between a cluster of suspended particles and the surrounding liquid is by fluidizing a total mass M of slightly soluble particles (e.g., benzoic in water), which slowly dissolve into the liquid, so that over a finite time period, Δt, m_A can be determined as $\Delta M / \Delta t$, and the slightly altered particle dimensions averaged from before and after the run. This procedure in the case of slow dissolution is often more accurate than determining m_A as $Q(C_{A2} - C_{A1})$ because of the large error associated with measuring the small change in C_A. Besides solids dissolution methods, used especially by investigators from Toulouse (e.g., Couderc et al., 1972; Damronglerd et al., 1975; Laguerie and Angelino, 1975; Vanadurongwan et al., 1976; Tournié et al., 1979), and also by many others (e.g., McCune and Wilhelm, 1949; Evans and Gerald, 1953; Fan et al., 1960), other methods include adsorption (e.g., Ganho et al., 1975), ion exchange (e.g., Rahman and Streat, 1981) and crystallization under mass-transfer-controlled (e.g., low Re) conditions (Laguerie and Angelino, 1975). Some investigators (e.g., Riba and Couderc, 1980) use an electrochemical method that, like the earlier heat and mass transfer experiments of Rowe and Claxton (1965) for a fixed array of spheres, focuses on the transfer from a single fixed particle within the fixed or fluidized array of particles. This method, as representative of what happens when all the particles are involved in transferring mass, must be treated with some caution, particularly for dense suspensions at low Reynolds numbers, as it neglects entirely any interaction of the diffusional boundary layers from the various particles. For heat transfer, focus on a single particle is even less representative of what happens when all the particles are absorbing or emitting heat because of the thicker and therefore more interactive thermal boundary layers. The early notion that for a fixed or fluidized bed of spheres, Nu_p ($= h_p d/\lambda$) or Sh_p ($= k_p d/D_A$) cannot decline to less than 2 as Re approaches zero (as in the case of an isolated sphere) has been laid to rest by Cornish (1965) and by Nelson and Galloway (1975), who showed that for dense assemblages of particles both Nu and Sh decline to zero as Re approaches zero.

The early correlations of low flux particle–liquid mass transfer in fluidized beds were patterned after those for fixed beds, in which, for spheres in the absence of free convection,

$$Sh_p = fctn(Re, \varepsilon, Sc) \tag{164}$$

(For forced-convection heat transfer, Nu_p replaces Sh_p and Pr replaces Sc, but such a conventional analogy between heat and mass transfer in liquid–fluidized beds has been questioned by Briens et al., 1993.) However, in fluidized beds, unlike fixed beds, ε is not an independent variable but is related to Re and the properties of both liquid and particles as incorporated in the Archimedes number, Ar, and exemplified by equations such as (61), (62), or (64), whereby

$$\varepsilon = fctn(Re, Ar) \tag{165}$$

Combining Eqs. (164) and (165), we obtain

$$Sh_p = fctn(Re, Ar, Sc) \tag{166}$$

Correlation for spheres then commonly takes the exponential form

$$Sh_p = K' Re^a Ar^b Sc^c \tag{167}$$

Actually, $Ar = Ga \cdot (\rho_p - \rho)/\rho$, and the Toulouse group that initiated correlation along these lines always fitted their data to

$$Sh_p = K' Re^a Ga^b \left(\frac{\rho_p - \rho}{\rho} \right)^{b'} Sc^c \tag{168}$$

but even for their spread of $(\rho_p - \rho)/\rho = 0.27$–1.14, the largest difference between b and b' in any such correlation developed by this group was only 0.023 (Tournié et al., 1979), while the one study that extended the upper range of $(\rho_p - \rho)/\rho$ to 7.07 found that $b - b' = 0$ (Nikov and Delmas, 1987). Thus Eq. (167) takes precedence over Eq. (168) (Epstein, 1992).

The index on Re in Eq. (167) has been generally found to be zero (Nikov and Karamanev, 1991, 1992) or very close to zero, e.g., +0.004 by Tournié et al. (1979) for Re = 1.6–1320 and −0.07 by Riba and Couderc (1980) for Re ≈ 150–25,000 and $\varepsilon = 0.45$–0.90. Within the experimental error it is conveniently taken as zero, which means that the effect of any increase in liquid superficial velocity for a given bed is compensated for by a corresponding increase in voidage. We are then left with an equation similar in form to that recommended by Calderbank (1967) for mass transfer into liquids from both solid spheres and small gas bubbles, namely,

$$Sh_p = K_0' Ar^b Sc^c \tag{169}$$

for which Calderbank proposed and rationalized $b = 1/3$, which is in remarkable agreement with values of $b = 0.323$, 0.33, 0.333, and 0.33 found experimentally by Tournié et al. (1979), Riba and Couderc (1980), Nikov and Delmas (1987), and Nikov and Karamanev (1991), respectively, the last for inverse fluidization. The range of Ar encompassed by these studies was 670–1.1×10^8. Since both Sh_p and $Ar^{1/3}$ are directly proportional to d, it follows that k_p is independent of particle size over this range, a result reported explicitly by more than one author (e.g., Riba and Couderc, 1980; Panier et al., 1980; Nikov and Delmas, 1987).

There is more disagreement on the value in Eq. (169) of the exponent c, which can vary from 1/3 assuming boundary layer theory to ½ assuming penetration theory, but the empirical values of which usually come closer to the former than to the latter. The lower the value assigned to c, the higher the value of the dimensionless constant K_0'. Thus Tournié et al. (1979) for $Sc = 305$–1595 found $c = 0.400$ and $K_0' = 0.245$, modified slightly to $K_0' = 0.228$ by Arters and Fan (1986); Riba and Couderc (1980) found $c = 0.34$ and $K_0' = 0.267$ for $Sc = 550$–7700; while Nikov and Delmas (1987) for $Sc = 860$–$19,900$ and Nikov and Karamanev (1991) for $Sc = 938$–2181 both found $c = 1/3$ but $K = 0.34$ and 0.28, respectively, in substantial agreement with Calderbank (1967), who proposed $c = 1/3$ and $K_0' = 0.31 \pm 0.03$. Thus experimental uncertainties at this stage prompt us to write

$$Sh_p = (0.23 \rightarrow 0.31)Ar^{1/3}Sc^{0.40 \rightarrow 0.33} \qquad (170)$$

In Eq. (170), application of the upper and lower values of Sc along with the corresponding lower and upper values of the coefficient K_0' should give values of Sh_p that bracket reality.

Some investigators (e.g., Kikuchi et al., 1983) have proposed a dimensionless correlation for Sh in which the key dimensional variable is e_d, the energy dissipation term given by Eq. (152). The underlying theory has been criticized by Arters and Fan (1986) on the grounds that it assumes all energy input to the system is dissipated as turbulence when in fact viscous dissipation is far from negligible, and the actual correlation of literature data by Kikuchi et al. showed considerable scatter.

Equation (170) applies to *low flux* mass transfer between liquid and fluidized *spheres*, *all* of which and the *total surface* of which are contributing to the mass transfer. For nonspherical particles, Limas-Ballesteros et al. (1982b) found that with both Sh_p and Ar based

on the equivolume sphere diameter, d_p, Sh_p was modified by the factor $\phi^{1.35}$ over the sphericity range 0.7–1. For spheres with a surface area partly active and partly inert, Panier et al. (1980) found electrochemically that, for different ratios and geometries of the active and inert parts, Sh_p (in which k_p is based only on the active area) increased as the fraction ϕ_A of active area decreased, the modification factor being $\phi_A^{-0.25}$ over the range $\phi_A = 0.057$–0.98. A qualitatively similar mass transfer intensification effect occurs when inert particles of higher density and smaller diameter are mixed with active particles; the higher the fraction of inert particles the greater the observed increase in Sh_p (Yang and Renken, 1998). Finally, when the solute flux due to mass transfer from or to the surface of the particles is not completely overshadowed by the mainstream mass flux (or mass velocity), e.g., for dissolution of very soluble solids, Sh_p by Eq. (170) must be corrected owing to distortion of the mainstream velocity and concentration profiles, rapid change in the area of the solid–liquid interface, and sharp physical property variations with concentration near the interface. Any analogy, conventional or otherwise, between heat and mass transfer breaks down in these circumstances. Chhun and Couderc (1980) found that for dissolution experiments with B' = surface mass flux/mainstream mass flux = 0.013–0.674,

$$\frac{Sh_p}{Sh_{p,low\ flux}} = \left(1 + B'\right)^{-0.56} \qquad (171)$$

which signifies a reduction of Sh_p as B' increases. The reverse effect, viz. an increase in Sh_p, occurs when the mass transfer is from the liquid to the solid surface (Bird et al., 1960), e.g., as in adsorption.

12 HEAT AND MASS TRANSFER BETWEEN SUBMERGED SURFACES AND LIQUID

12.1 Heat Transfer

Heat transfer between a liquid fluidized bed and a submerged surface in the bed, most commonly the bed wall itself, has been the subject of many experimental studies, the conditions, scope, and results of which have been summarized by Haid et al. (1994) in their Tables 1 and 2. A typical set of results is shown in Fig. 11, where it is seen that the increase of the heat transfer coefficient, h, with superficial liquid velocity, U, that occurs in the fixed packed condition continues in somewhat attenuated form when the bed is fluidized, eventually achieving a maximum and then declining to

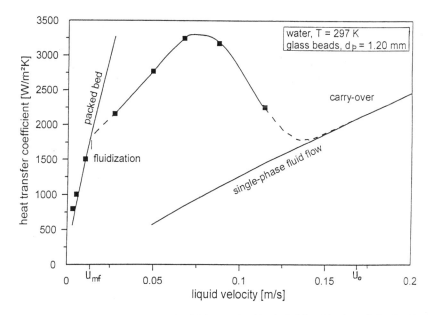

Figure 11 Heat transfer coefficient between wall and liquid for packed bed, fluidized bed, and single-phase flow as a function of superficial liquid velocity. (After Haid et al., 1994; data points from Kato et al., 1981.)

merge with the single-phase fluid flow line at $U \approx U_0$. The increase in h for a liquid-fluidized bed relative to single-phase liquid flow at the same value of U can be by as much as a factor of seven (Richardson et al., 1976; Klaren and Halberg, 1980; Bremford et al., 1996).

Using 2665 data points from 43 publications, different weighting procedures and different numbers of dimensionless parameters, Haid et al. (1994) and Haid (1997) arrived at six empirical equations, of which the one with the lowest standard deviation (11.5%) and the smallest average relative error (32.0%), considerably smaller than that obtained for the same data using 38 other proposed correlations from the literature, was

$$\mathrm{Nu}\left(=\frac{hd_p}{\lambda}\right) = 0.0413\mathrm{Re}^{0.79}\mathrm{Pr}^{0.67}(1-\varepsilon)^{0.12}\varepsilon^{-1.66}$$

$$\times \left(\frac{d_p}{D_h}\right)^{0.10}\left(\frac{\rho_p - \rho}{\rho}\right)^{0.04} \quad (172)$$

where D_h is the hydraulic diameter of the column (or annulus, in the case of a vertically immersed cylindrical heater or cooler). The range of parameters (some interrelated, e.g., ε depends upon Re for any given value of Ar) covered by Eq. (172) was $\mathrm{Re} = 0.020$–9400, $\mathrm{Pr} = 1.65$–7700, $\varepsilon = 0.40$–0.99, $\mathrm{Ar} = 3.85$–$67,000,000$ and $d_p/D_h = 0.0013$–0.210. If Eq. (32) is combined with Eq. (172), then

$$\mathrm{Nu} \propto \varepsilon^{0.79n-1.66}(1-\varepsilon)^{0.12} \quad (173)$$

from which it follows that Nu is a maximum at $\varepsilon = (0.79n - 1.66)/(0.79n - 1.54)$. Thus for spheres in the Stokes regime ($\mathrm{Re}_0 < 0.2$), $n = 4.8$ and $\varepsilon_{max} = 0.95$, while in the Newton regime ($\mathrm{Re}_0 > 500$), $n = 2.4$ and $\varepsilon_{max} = 0.66$. Experimentally measured values of ε_{max} generally fall between 0.62 and 0.82, with the largest cluster in the vicinity of 0.73 (Jamalahmadi et al., 1997). Note that Eq. (172) is inapplicable at $\varepsilon = 1$.

Combination of Eqs. (172) and (32) is also equivalent to

$$h \propto d_p^{-1}\mathrm{Re}_0^{0.72}d_p^{0.17} \quad (174)$$

For spheres in the Stokes regime, $\mathrm{Re}_0 = \mathrm{Ar}/18 \propto d_p^3$, whence from Eq. (174) it follows that $h \propto d_p^{1.33}$, while in the Newton regime, $\mathrm{Re}_0 = \sqrt{3\mathrm{Ar}} \propto d_p^{3/2}$, whence $h \propto d_p^{0.25}$. Thus the larger the particles the larger the value of h, but the smaller the influence of d_p on h. Both effects are in agreement with experimental results (Haid et al., 1994; Jamialahmadi et al., 1996). Equation (172) also implicitly indicates that, at any given bed voidage, h increases with particle density, primarily because the value of Re required for expansion to the given voidage will increase with particle density and secondarily (and very slightly) through the term $(\overline{\rho_p - \rho}/\rho)^{0.04}$. This effect too is in agreement with experiment (Jamialahmadi et al., 1995).

An attempt at a more mechanistic, rather than purely empirical, approach to the heat transfer problem in turbulent flow has been made by Jamialahmadi et al. (1995, 1996, 1997), who divided the total heat transfer surface area into two parts—A_p, the portion affected by particle contact at any instant, and A_c, the remaining portion. From A_p (assuming a bed being heated), heat is transferred by transient conduction to the adjacent liquid layer, which is then transported in the wake of departing particles into the liquid bulk and replaced by cooler liquid. This turbulence promoting disruption of both the viscous sublayer and the thermal boundary layer is the principal source of heat transfer enhancement to single-phase liquid flow. Some heat, but a much smaller fraction than in the case of gas fluidization, is also assumed to be transferred via conduction into and transport of particles, i.e., by particle convection. From A_c, heat is transferred in parallel by normal single-phase forced convection, depending upon the type and configuration of surface involved. Crucial to this model is knowledge of the area ratio $A_p/(A_p + A_c)$, which in turn depends on the fraction of the total number of fluidized particles that are in contact with the heat transfer surface at any instant, n_p/N_p. At this juncture, Jamialahmadi et al. were forced to resort to empiricism, i.e., n_p/N_p had to be expressed as exponential functions of d_p/D_h, $1 - \varepsilon$, and $\varepsilon - \varepsilon_{mf}$, with the coefficient and three exponents evaluated from the experimental data, which required assigning different values of each for wall-to-bed, vertically immersed plate, vertically immersed cylindrical and horizontally immersed cylindrical heating; and with different values again for aggregative than for particulate fluidization. They thus generated a total of $4 \times 4 \times 2 = 32$ empirical constants, plus another 24 for improved prediction of h_{max} (Jamialahmadi et al., 1997). The criterion recommended for ε_{max}, corresponding to maximum heat transfer, was that the ratio U_ε/U_E, which is obtained on dividing Eq. (153) by Eq. (154) and is proportional to $(1 - \varepsilon)^{0.5}\varepsilon^{n-1}$, be maximized, the result being

$$\varepsilon_{max} = \frac{n - 1}{n - 0.5} \qquad (175)$$

Equation (175) yields values at $n = 2.4$–4.8 of $\varepsilon_{max} = 0.74$–0.88, whereas Eq. (149), obtained either from Eq. (148) or by simply maximizing U_ε given by Eq. (153), yields $\varepsilon_{max} = (n - 1)/n$, from which $\varepsilon_{max} = 0.58$–0.79, in better agreement with the experimentally observed ranges of the wall-to-bed maximum for both heat and mass transfer (Kang and Kim, 1988).

One factor taken into account by the equations of Jamialahmadi et al. (1995, 1996, 1997) but not by Eq. (172) is particle convection, the heat transfer coefficient of which is written as directly proportional to $\sqrt{\lambda_p \rho_p c_{pp} f}$, where f is the collision frequency of the particles. The proportionality constant for cylindrical particles is assigned twice the value for that of equal sized spheres to account for the somewhat higher value of h measured for the former than for the latter (Richardson et al., 1976), an effect implicitly accounted for in Eq. (172) by use of the equivolume sphere diameter to characterize the particle size of cylinders. The particle conductivity, λ_p, and heat capacity, c_{pp}, are not however accounted for in Eq. (172). As it turns out, though, a 391-fold increase in $\lambda_p \rho_p c_{pp}$ as between lead glass and copper-coated aluminum spheres of equal diameter (1.98 mm) and density (2890 kg/m^3) fluidized by water showed only a small gain in h for the latter, and only at $\varepsilon < \varepsilon_{max}$ (≈ 0.75), the gain increasing to a maximum of 18% as ε decreased to 0.45 (Wasmund and Smith, 1967; Haid et al., 1994); while a 255-fold increase in $\lambda_p \rho_p c_{pp}$ as between copper and lead spheres of equal diameter (= 4.5 mm) fluidized by water produced a maximum net gain in h for the copper at $\varepsilon \leq \varepsilon_{max}$ (≈ 0.72) of only 14%, both experimentally and as fitted by Jamialahmadi et al. (1995). Given that the column diameter for the latter runs was only 23.8 mm, so that the inordinately high wall effect ratio, $d_p/D_c = 0.19$, could have given rise to an unrepresentatively high degree of particle convection, Molerus and Wirth (1997) have questioned the generality of this result and have recommended conservatively that, until further evidence proves otherwise, the thermal properties of the particles can be ignored in calculating the heat transfer coefficient for liquid-fluidized beds, as in Eq. (172).

12.2 Mass Transfer

Turning to mass transfer between fixed submerged surfaces and liquid-fluidized beds, it has been noted by several authors (King and Smith, 1967; Briens et al., 1993; Schmidt et al., 1999) that the conventional analogy between heat and mass transfer is inapplicable. Thus, substituting Sc for Pr and Sh ($= k_m d_p/D_A$) for Nu in Eq. (172) results in overprediction of Sh by factors varying from about 5 to 2 as measured Sh rises from 2 to 30 (Schmidt et al., 1999). Schmidt et al. attribute this result to the presence of particle convection in heat transfer and its absence in mass transfer, but in the light of the discussion immediately above, factors of 2–5 cannot be accounted for by this explanation. King and Smith (1967) have argued that the analogy breaks down because for mass transfer the entire resistance to transfer is close to the submerged surface, while for heat transfer there is also an appreciable resistance in the bed, but this observation is merely a consequence of the fact that Sc often (as in water) exceeds Pr by orders of magnitude. This fact in and of itself should not preclude the applicability of, for example, the Chilton–Colburn (1934) analogy, which is applicable to heat and mass transfer from a pipe surface to flowing liquids despite the same discrepancy between Sc and Pr, i.e., between a thin mass transfer boundary layer and the corresponding thicker heat transfer boundary layer. A more credible explanation for the apparent breakdown of the analogy is provided by Briens et al. (1993), who invoke the film-penetration model of Toor and Marchello (1958) in conjunction with the large difference between Sc and Pr. However, application of this model depends on knowledge of two parameters, the film thickness and the fractional rate of surface renewal, which requires a measurement of Sh and Nu for every operating condition. Though theoretically sound, the method does not lend itself at present to quantitative prediction of Sh for fixed immersed surfaces.

Most investigators have correlated their mass transfer results in the Beek (1971) form of the Chilton–Colburn (1934) equation,

$$j_m \varepsilon = \frac{k_m}{U/\varepsilon} Sc^{2/3} = a'' \left(\frac{Re}{1-\varepsilon} \right)^{-b''} \tag{176a}$$

which is equivalent to

$$Sh = a'' Re^{1-b''} (1-\varepsilon)^{b''} \varepsilon^{-1} Sc^{1/3} \tag{176b}$$

Values of a'' and b'' from different studies, mainly using the electrochemical technique of Lin et al. (1951), have been tabulated in several papers (Storck and Coeuret, 1980; Lee et al., 1997; Schmidt et al., 1999). The coefficient a'' in the case of some studies (e.g., Jottrand and Grunchard, 1962), is itself a function of ε and/or Sc, but it is more usually a constant. A plot by Lee et al. (1997) of ten correlations from the literature as $j_m \varepsilon$ vs. Re shows a fourfold variation of $j_m \varepsilon$ at $Re/(1-\varepsilon) = 1$ (i.e., a'' varies from 0.3 to 1.2), but a much smaller variation ($j_m \varepsilon = 0.023$–0.031) at $Re/(1-\varepsilon) = 1000$. The data of Lee et al. for mass transfer from an axially immersed cylindrical surface in an aqueously fluidized bed of glass beads ($d_p = 0.30$–0.77 mm) at Sc = 1542, $\varepsilon = 0.55$–0.80, and $Re/(1-\varepsilon) = 4.2$–157 showed mass transfer enhancement factors of 2–7 relative to single-phase

flow at the same superficial velocity and yielded a representative correlation, well within the extremities of the others, namely Eq. (176a) with $a'' = 0.44$ and $b'' = 0.394$. If one combines Eq. (32) with this result, then

$$Sh \propto \varepsilon^{0.606n-1} (1-\varepsilon)^{0.394} \tag{177}$$

from which it follows that Sh is a maximum at $\varepsilon = (n-1.65)/(n-1)$. Thus ε_{max} for spheres varies from 0.83 in the Stokes regime ($n = 4.8$) to 0.54 in the Newton regime ($n = 2.4$). Experimentally, values of ε_{max} that have been reported vary from 0.58 to 0.75 (Schmidt et al., 1999), a range about 0.05 lower than that for the corresponding heat transfer process. As ε approaches unity, Eq. (176) breaks down.

Most reported mass transfer studies are for a single or a relatively narrow range of Sc, so that the 1/3 power of Sc inherent in the Chilton–Colburn equation is assumed rather than tested. Recently the 1/3 power was verified by an extensive series of runs encompassing Sc = 151–7021, for oxygen transfer from an axially immersed cylindrical membrane in a water-fluidized bed of spheres with $d_p = 0.325$–3 mm, $\rho_p = 2500$–11,343 kg/m^3, $Re/(1-\varepsilon) = 0.9$–1652, Ar = 32–481,365, and $\varepsilon = 0.4$–1. The correlating equation for Sh, which contains a minor term to describe molecular diffusion not caused by the fluidized bed itself and a major term to account for turbulence and fluidization, is (Schmidt et al., 1999)

$$Sh = 0.14 Re^{1/3} Sc^{1/3} + 0.13(1-\varepsilon)(\varepsilon - \varepsilon_{mf})$$
$$\times \left(\frac{Sc}{Re} \right)^{1/3} Ar^{2/3} \tag{178}$$

The correlation index (or "coefficient of determination") R^2 for this equation was 0.98, as compared to $R^2 = 0.8$ for the same data correlated by Eq. (176a). Neglecting the first term on the right-hand side of Eq. (178) relative to the second, assuming $\varepsilon_{mf} = 0.4$, and again invoking Eq. (32), we find that

$$Sh \propto (1-\varepsilon)(\varepsilon - 0.4)\varepsilon^{-n/3} \tag{179}$$

from which it follows that for $n = 4.8$–2.4, $\varepsilon_{max} = 0.59$–0.65, somewhat narrower than, but within, the range of reported measurements.

13 APPLICATIONS (as abridged by Wen-Ching Yang)

Some applications of liquid-fluidized beds, such as particle classification, are over a century old, while many, such as fluidized-bed electrolysis and bioreactors, are of more recent vintage. None have achieved the industrial and commercial prominence of gas fluidization applications exemplified by catalytic cracking or coal combustion. The brief discussion below will be limited to identifying the salient features of some applications, or broad categories thereof, which have been reported on in literature accessible to this author, and few specific examples will be cited. The reader seeking more information on a given application should consult the literature on the given unit operation or process industry involved.

Particle classification by size (sizing), density (sorting), or even shape (shaping?) depends on the segregation characteristics of liquid-fluidized beds described in Sec. 5. For particles of fixed known density and shape, axial particle size distribution at equilibrium in semibatch fluidization can be estimated from the axial pressure profile, using the perfect classification model discussed in that section. For narrowly sized solids of varying density, particle density measurements from successive bottommost suspension samples in conjunction with frictional pressure drops measured before and after the samplings will yield the axial density distribution of the solids, by appropriate use of Eqs. (11) and (12) (Galvin and Pratten, 1999). The particle stratification in a rising stream of liquid, which occurs once equilibrium is achieved within a semibatch fluidized bed, can be maintained intact for continuous separation of the particles involved via a discharging underflow and overflow, provided the feed rate is kept sufficiently low. An annularly compartmentalized unit of some elegance for effecting continuous classification has been patented by Delachanal (1963). Traditionally, liquid-fluidized or teeter bed classifiers were devoted mainly to mineral separations, but recently they have extended their domain to other materials (Galvin et al., 1998, 1999). In the actual separations, density always dominated over size.

Sorting of particles, particularly of bidisperse mineral mixtures, is more cleanly effected by *sink-and-float separation*. This is most simply done by using a liquid nonsolvent intermediate in density between that of the two particle species. With overflow and underflow discharge streams, and a continuous feed suspension introduced at an intermediate position, the sharpness of separation decreases as the feed rate, the feed solids concentration, and the underflow/overflow ratio are increased (Nasr-El-Din et al., 1988, 1990). In the absence of an acceptable liquid, a homogeneous suspension, viz., a water-fluidized bed of narrowly cut fine sand (e.g., $-325 + 400$ mesh), could

serve instead. By adjusting the upward liquid flow carefully, the required suspension density can be achieved. Coarse coal particles can thus be effectively separated into a clean fraction of low ash content that floats and denser fractions of higher ash content that sink (Needham and Lynch, 1945).

Backwashing of downflow granular filters (or fixed bed ion exchange columns) by water fluidization of the filter medium or media (or ion exchange resin) is a well entrenched industrial procedure, especially for municipal-water sand filters, where what is removed in the backwash is primarily the filtered solids. The voidage at which maximum removal occurs is about 0.65–0.70, and since typically $n = 3.1$–3.4 for the sand used as filter media, the voidage range is close to that predicted by Eq. (149), which we have already seen as generating the critical or optimum voidage for several other hydrodynamic or hydrodynamically related phenomena. However, since the curve of removal vs. ε is quite flat near the maximum, more practical bed expansions of 40–50% (rather than 100%, corresponding to $\varepsilon = 0.7$) will produce almost as much filter cleaning as operating at the optimum (Amitharajah, 1978). Improved cleaning will normally be effected by an upward air scour before or during the water fluidization, and this is especially essential to the satisfactory functioning of wastewater filters, which receive heavier, more variable, and stickier suspended-solids loads than potable water filters (Cleasby et al., 1975; Cleasby and Lorence, 1978). Wastewater is therefore not recommended as the backwash liquid. In the case of petroleum refinery wastewaters, the backwash includes much oil in addition to the suspended solids (Brody and Lumpkins, 1977; Kempling and Eng, 1977), and the stickiness of other wastewater deposits is also probably contributed to by nonaqueous, immiscible liquid contaminants. Dual-media filters generally perform somewhat better than single-medium, and no worse than triple-media (Cleasby and Lorence, 1978). Typically, in a dual-media filter, anthracite coal sits on top of denser silica sand in a coal:silica size ratio between 2:1 and 4:1 (Cleasby and Sejkora, 1975) or even 6:1 (Brody and Lumpkins, 1977), and in the case of a triple-media filter a third component of even greater density and smaller size, e.g., garnet sand (Cleasby and Woods, 1975), lies below the silica sand.

In-situ fluidized washing of soils is based on the same principle as filter backwashing, but for soil washing a downward water jet that spreads out and then reverses direction has been proposed (Niven and Khalili, 1998a). As in the case of granular filter backwashing, the use of an air scour, i.e., air–water fluidization, greatly increases the cleaning efficiency (Niven and Khalili, 1998b).

Another long-standing and industrially well-established process involving liquid fluidization is that of seeded *crystal growth* in a bed fluidized by a moderately supersaturated (metastable) solution of the solute to be precipitated. Because stratification by size of the growing crystals accompanies the process, the units involved are usually called classifying crystallizers, of which the most common is the Krystal (or Oslo or Jeremiassen) crystallizer (Svanoe, 1940; Bamforth, 1965; Perry et al., 1984). Intensive study of the operating characteristics and design of such crystallizers has been undertaken by Mullin (1993) and coworkers (e.g., Mullin and Garside, 1967; Mullin and Nyvlt, 1970; Garside et al., 1972), as well as by many others (e.g., Bransom, 1960), most recently by Tai et al. (1999). Most studies involve the crystallization of inorganic salts from aqueous solution, but the fluidized-bed crystallization of acetylsalicylic acid (aspirin) from absolute alcohol has also been reported (Glasby and Ridgway, 1968). Crystal growth rates in lean beds ($\varepsilon > 0.98$) have been found to be substantially the same as those in the industrially more common dense beds ($\varepsilon \leq 0.8$), and in both types of beds, the crystal growth rate is almost always significantly smaller than the corresponding mass-transfer controlled dissolution rate (Garside et al., 1972; Phillips and Epstein, 1974; Jira-Arune and Laguerie, 1979; Tai et al., 1987). Fluidized-bed crystallizers can be operated in both the batch mode and the continuous mode, with bottom discharge of the enlarged crystal product. Models to describe the behavior of both batch (Shiau et al., 1999) and continuous (Frances et al., 1994) fluidized-bed crystallizers have been formulated.

Leaching, i.e., physical or chemical dissolution of a soluble component embedded in the inert matrix of a granular solid, and *washing* for removal of the leaching or other residual solution within the intraparticle pores and the interparticle voids of the leached or other granules, can be accomplished quite effectively by liquid fluidization of the solids with the required solvent. The principal applications, actual or potential, are for extraction of mineral or metal values from ores, and of vegetable oils from seeds. A thorough discussion of fluidized bed leaching/washing is provided by Kwauk (1991/92). In the leacher/washer illustrated in Fig. 12, the slurry feed is hydraulically distributed into an enlarged settling head, where much excess liquor is removed. Solid particles then fall countercurrently against a rising stream of liquor into the leaching/

washing region, in which a reasonably well-defined interface usually separates an upper dilute-phase zone from a lower dense-phase zone. The leaching solvent or washing liquid is sparged in at the bottom of the dense-phase region, below which the solid slurry is compressed, densified, and discharged. Compared to more conventional leaching/washing equipment (Treybal, 1980), such fluidized leachers/washers are deemed by Kwauk (1991/92) to have the advantages of complete hydraulic operation with no mechanical parts, continuous (though imperfect) countercurrent contacting in a single column, low solvent-to-solids ratios, low space requirements, and ease of automation. Other more complicated fluidized leaching/washing setups, some with countercurrent staging, are also

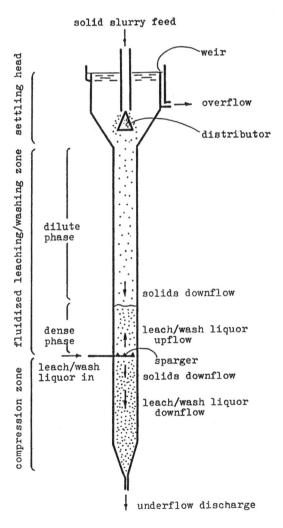

Figure 12 A fluidized leacher/washer. (From Kwauk, 1991/92.)

described by Kwauk, as well as by Slater (1969), who included pulsed fluidization in his review of countercurrent liquid–solids contactors.

Adsorption and *ion exchange* are both operations that are conventionally carried out batchwise in fixed packed beds but can gain certain advantages when performed in the fluidized state. Though physical adsorption involves only two resistances in series (external mass transfer of solute to particle macrosurface and internal diffusion within particle pores to microsurface), while ion exchange involves five (external mass transfer and internal diffusion of ions inwardly, exchange of ions, internal diffusion, and external mass transfer of exchanged ions outwardly), the two operations commonly employ similar types of equipment that are subject to similar design procedures (Treybal, 1980). Adsorption from liquids is mainly applied industrially to the decontamination of water, aqueous solutions, and petroleum products, while ion exchange is principally, though not exclusively, used for water purification and metal recovery from hydrometallurgical leach liquors. Though both operations must usually be coupled with regeneration (elution) of the adsorbent or ion exchange resin, and often with washing as well, it is the original loading (adsorption, exhaustion) process that is most subject to improvement by liquid fluidization and that has received most attention industrially and academically in this respect. Since the voidage is greater for the fluidized bed, the same superficial velocity (upflow in fluidization, downflow in fixed bed) for a given batch of solids will result in a lower pressure drop (Himsley and Farkas, 1977). The advantage of continuous operation can be achieved by adopting a setup similar to that of Fig. 12 (Slater and Lucas, 1976; Koloini and Zumer, 1979). If the loaded adsorbent or ion exchange particles are denser than the unloaded ones, as in the case of copper being exchanged with hydrogen ions (Selke and Bliss, 1951), the loaded particles move to the bottom of the bed, thus enhancing the countercurrency of the operation. Further enhancement in that direction is provided by staging, which can have the desirable result of reducing the required solids inventory appreciably (Slater, 1982). Detailed reviews of the various staging methods proposed and of those adopted industrially, especially for uranium recovery from its leach liquor, have been published by Slater (1969), Streat (1980), and Slater (1981). The simplest and most recently revived proposal for a multistage fluidized bed adsorber is a sieve-plate column without downcomers or pulsation or controlled cycling; see Grünewald and Schmidt-Traub (1999). The authors report that stable operation requires, among other conditions, that the lowest plate have a smaller free hole area than the others.

Flocculation in order to achieve *clarification* of turbid liquids is a process similar to adsorption that can be effected by the liquid fluidization of some seeded flocs, which then enhance subsequent flocculation of the suspended colloidal particles that cause the turbidity (Svarovsky, 1990). One continuous process of this kind involves liquid-fluidized microsand coated (activated) with an alginate flocculant that serves to flocculate and retain aluminum or ferric salts from incoming suspensions and is then externally washed to remove the floc from the sand and then recycled (Sibony, 1981).

Electrolysis in a fluidized bed both with inert and with electrically conductive particles (extended electrodes) had among its earliest proponents the team of Le Goff et al. (1969). Electrolytic recovery of metals from dilute streams with consequent purification of these streams can be much enhanced and rendered economical by *inert particle fluidized bed electrolysis*. Inert glass beads, typically 0.6 mm in diameter, are continuously fluidized in the electrolytic cell by the dilute solution, and the overflow stream is recirculated by a pump through the liquid distributor at the bottom of the cell. Planar metallic mesh electrodes, anodes interdispersed with cathodes, are vertically immersed within the full depth of the fluidized bed with mesh apertures much greater than the bead size, so that the fluidized beads can move freely through and around the electrodes. Since for electrolysis of dilute aqueous solutions (< 5 kg metallic ions/m^3) ion transport becomes mass transfer controlled, and since the mass transfer enhancement factor between the immersed electrodes and the liquid owing to the presence of the fluidized beads can, as discussed in Sec. 12, be as much as 7, ion transport is augmented accordingly. To achieve maximum mass transfer enhancement, the operation is carried out at bed expansions beyond ε_{mf} of 50–100%, i.e., at $\varepsilon \approx 0.6$–0.7, which is in accord with the findings summarized in Sec. 12. Reports by Lopez-Cacicedo (1981) and Boyanov et al. (1988) recommend connecting several electrolytic cells in series as a cascade.

Fluidized bed electrodes, since the initial reports by Goodridge and coworkers (e.g., Backhurst et al., 1969; Goodridge et al., 1971), have received much more attention (see Goodridge and Wright, 1983 and Salas-Morales et al., 1997, for incomplete but useful reference lists). A fluidized bed electrode (FBE) consists of a bed of electrically conducting (metallic or metal-coated, depending on desired particle density, which influences fluidization velocity) particles fluidized by electrolyte flow, to which DC current is fed by one or more connecting rods or plates known as current feeders, and which are often separated from one or more immersed counterelectrodes by a diaphragm of porous or ionically conductive material that is desirably long lasting. Provided the bed expansion from the static condition is small, e.g., 5–25%, some electrical contact is maintained between the particles and the current feeder and between the particles themselves, so that the surfaces of the particles then act as a large extension of the feeder surface, thereby greatly increasing the current density based on the counterelectrode surface area. Though the many trials and proposed applications of FBEs include their use for fuel cells, organic and inorganic synthesis, industrial wastewater treatment, and electrowinning (or electrodeposition) of metals, the qualified successes of this technology have been mainly in the area of electrowinning, especially from *dilute* solutions, with the concomitant reduction of metallic components in the solutions involved. Figure 13 is a *schematic* view of a continuous fluidized bed electrowinning cell, in the cathode compartment of which metal particles are fluidized by the process stream or wastewater to be treated. Dimensions and configurations of *actual* cells differ greatly, while the metals that have been subjected

to FBE deposition are also various (Van der Heiden et al., 1978). In fact, as Salas-Morales et al. (1997) have most recently reaffirmed, the intermittency of contact of the FBE particles with each other and with the current feeder render them periodically subject to chemical attack by a strongly acid electrolyte, even when the bed expansion is small, so that the FBE has been in the main unsatisfactory for electrowinning from *concentrated* electrolytes. These investigators have reviewed the prior literature on, and have therefore recommended—as a device which performs between the extremes of a fixed bed electrode and a FBE—a spouted bed electrode, in which the 98% of the particles that at any moment reside within the annulus as a moving packed bed are in good electrical contact with each other and with the current feeder. A similar recommendation has been made by Hadžismajlović et al. (1996) and by Dweik et al. (1996).

Liquid-fluidized bed heat exchangers (FBHX), in addition to enhancing the clean wall-to-liquid heat transfer coefficient as discussed in Sec. 12, have the additional important attribute of vigorously combating scaling and other types of fouling of the heat transfer surface without the use of chemical additives. They do so because the bed particles incessantly scour the surface Kim and Lee (1997), act as alternative deposition sites for whatever precipitation does occur, and scour each other thoroughly so that, for example, all the precipitated calcium sulfate from saline water leaves with the exit liquor and can be filtered out (Hatch et al., 1966; Meijer et al., 1980). Although tests have been performed with the liquid-fluidized bed on the unbaffled shell side of a shell-and-tube heat exchanger, both with the tubes horizontal and with the tubes vertical (Cole and Allen, 1978), preference has in most practical cases been given to locating the fluidized bed inside the tubes, which must therefore be vertically oriented with a single upward pass (Klaren and Bailie, 1988) and with proper care taken to design for even distribution of both the liquid and the fluidized bed particles among the tubes (Rautenbach and Kollbach, 1986). The main required modifications to a conventional shell-and-tube exchanger are a larger inlet chamber with a distribution system and a larger outlet chamber to act as freeboard for separation of the liquid from the top of the fluidized bed, and these can sometimes be retrofitted to a conventional exchanger (Kollbach et al., 1987; Klaren and Sullivan, 1999). The particles can be kept in a stationary fluidized condition or they can be circulated by means of an internal downcomer (Klaren and Bailie, 1988) or externally (Klaren and Sullivan, 1999). The

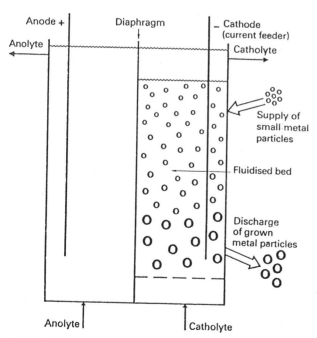

Figure 13 Schematic of a fluidized bed electrowinning cell. (From Van der Heiden et al., 1978.)

development of a multistage flash/fluidized bed exchanger (MSF/FBE), in which the fluidized brine feed acts to condense the stagewise flashed steam, resulted in lower heat consumption, greater flexibility with respect to brine loads and temperatures, and reduction in flash chamber volume by a factor of 6 relative to a conventional MSF exchanger (Veenman, 1976, 1977). In general, as summarized by Kollbach et al. (1987) and others, FBHX units will usually maintain either totally clean surfaces or reduce fouling sufficiently that operation can be continued without cleaning (Müller-Steinhagen et al., 1994); and wall-to-bed heat transfer coefficients are enhanced by factors of 2 to 7 at superficial velocities below 0.5 m/s, maximum enhancement commonly occurring at $\varepsilon \approx 0.7$ (Klaren and Halberg, 1980; Rautenbach and Kollbach, 1986). FBHX technology has made inroads into seawater desalination, geothermal energy utilization, wastewater evaporation, and pulp and paper production, and it has been proposed for lube oil dewaxing (Kollbach et al., 1987; Klaren and Bailie, 1989; Klaren and Sullivan, 1999).

Thermal energy storage by encapsulating phase change material into hollow spheres to be thermally cycled by water fluidization has been demonstrated experimentally (Sozen et al., 1988). Specifically, 96% Glauber's salt ($Na_2SO_4 \cdot 10H_2O$), which undergoes an endothermic phase change to anhydrous Na_2SO_4 (plus H_2O) at 32.4°C, and can be reversibly regenerated exothermically at the same temperature, was injected along with 4% borax (a nucleation catalyst) into thousands of 25 mm o.d. hollow polypropylene spheres. These spheres were then cyclically fluidized at $U/U_{mf} = 1.2$–2.6 in a 0.34 m i.d. column by hot (inlet temperature ≈ 39°C) and cold (inlet temperature ≈ 15°C) water, each for intervals exceeding 1 hour. Good heat transfer resulted because of the large capsule–water surface area engaged, but more importantly, an unchanging heat storage efficiency of about 60% was obtained, even after 96 cycles. Although heat recovery efficiencies up to 83% for stoichiometric Glauber's salt were subsequently obtained using the same capsules in a rotating drum (Sozen et al., 1988), the greater simplicity, higher heat transfer surface area per unit volume, and lower costs associated with fluidization continue to endow this technique with some advantages.

Fluidized bed bioreactors have received considerable attention during the past three decades, and the literature on this subject has mushroomed. Much of the effort in this respect has been devoted to wastewater treatment, which involves biodegradation of waste chemicals (e.g., organics, ammonia, nitrates), but there is now also a considerable literature on fluidized bed fermentation, the object of which is biosynthesis of useful products (e.g., alcohol). A broad collection of papers with useful discussions on various aspects of fluidized bed biological treatment of potable water and wastewater has been edited by Cooper and Atkinson (1981). An extensive review of both aerobic wastewater treatment and fermentation in fluidized beds has been provided by Fan (1989). There is an excellent update mainly on wastewater treatment by Wright and Raper (1996) and a significant textbook entry by Grady Jr et al. (1999). Anaerobic bioreactors, though their fluid feed is a liquid, nevertheless commonly produce in situ a gaseous product (Parkin and Speece, 1984), so that, like aerobic reactors in which air enters with the liquid feed, they are actually three-phase systems, but they can be treated as two-phase liquid-fluidized beds when the gaseous product is hydrodynamically negligible. What most characterizes fluidized-bed bioreactors is their biocatalytic use of immobilized enzymes or microbial cells, which are attached to the surface of biologically inert nonporous particles, or entrapped within the matrices of porous particles or gels, or encapsulated within a semipermeable barrier such as a membrane, or self-aggregated flocs (Karel et al., 1985). The first of these four immobilization methods, referred to as the attached growth technique for microbial cells, and involving growth of a biofilm (layers of cells and excreted slime) on the surface of each support or carrier particle during the course of the fluidization, is the one most used in fluidized-bed wastewater treatment. The combination of inert carrier particle (or "biomass support particle") and attached biofilm is usually referred to as a bioparticle. The carrier particles can be inorganic, e.g., sand, glass, alumina, and many others, or organic, e.g., activated carbon, coal, polyethylene, and many others; they are often spherical or cylindrical and are usually though not always moderately narrow cuts of sizes which for different applications have varied from 0.1 to 6 mm. For wastewater treatment, the most commonly used biocarriers are sand and activated carbon, the latter of which can simultaneously act as an adsorbent for organic contaminants (Andrews and Tien, 1981). The growth of biofilm on the carrier particles changes their size, effective density, and surface properties—and thereby their fluidization characteristics, such as U_{mf}, U_0, and n. For spherical carrier particles, some investigators find that the values of these parameters given by the equations in Sec. 4 apply reasonably well to the resulting bioparticles, if the total equivolume sphere diameter and weighted density of the latter at any given time are used in the equations, while others find that empirical modifications must be made to the accepted equations for their prediction assuming rigid spheres. Since the biocoated particles may be neither smooth nor rigid, their drag coefficients are then higher than for smooth rigid spheres, so that U_{mf} and U_0 are correspondingly lower, i.e., bed expansion for a given value of U is higher; and the measured values of n are considerably higher (Thomas and Yates, 1985). Empirical determination of U_{mf}, U_0 (or U_e), and n as a function of bioparticle size is recommended in the event that knowledge of bed expansion characteristics is crucial to the bioreactor design. In the case of bioflocs or of particle matrices that retain some of their porosity after cell entrapment (Atkinson et al., 1979), it has been shown theoretically for permeable spheres in the Stokes regime ($Re_0 < 0.2$; Neale et al., 1973) and experimentally both in the Stokes regime (Matsumoto and Suganuma, 1977) and at higher Reynolds numbers (Masliyah and Polikar, 1980; Webb et al., 1983) that the drag on such particles is lower than for impermeable spheres of the same size and bulk density, U_0 correspondingly higher, and the bed expansion for a given superficial velocity correspondingly lower. Fluidized bed bioreactors, schematized in Fig. 14, come in many configurations, including conventional upward fluidization with con-

stant cross section, inverse (e.g., Nikolov and Karamanev, 1987), tapered (e.g., Scott et al., 1978; Allen et al., 1979), zig-zag (Nakamura et al., 1979), baffled (Parkin and Speece, 1984), with internal circulation (e.g., via draft tube), and with external circulation (Zhu et al., 1999). Wright and Raper (1996) list twelve advantages of fluidized bed attached growth bioreactors as compared to more traditional suspended growth bioreactors in which the microorganisms are not anchored or immobilized. Of greater relevance are the advantages that liquid-fluidized beds have over other attached growth bioreactors (e.g., fixed beds). These include (Allen et al., 1979; Grady Jr et al., 2000) freedom from plugging, easy passage of insoluble foreign material or unwanted microorganisms, lower pressure drop, superior mass and heat transfer characteristics, better control of biofilm thickness, larger surface areas for biofilm development, and easy circulation or removal of bioparticles for excess biomass separation (Fig. 14). Separation of the excess biomass is commonly effected outside the reactor either hydrodynamically or mechanically (Cooper et al., 1981), or inside the reactor by scouring with air fed exclusively to the draft tube (Nikolov and Karamamev, 1987; Wright and Raper, 1996). The overgrowth of biofilms can lead to upward elutriation of bioparticles from a conventional fluidized bed or downward elutriation from an inverse fluidized bed,

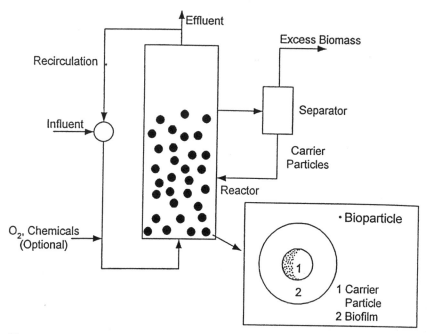

Figure 14 Schematic diagram of a fluidized bed bioreactor. (After Grady Jr et al., 1999.)

a problem that is sometimes dealt with by the use of a retaining grid at the top of the conventional bed, giving rise to semifluidization and clogging (Fan, 1989), but is more adeptly handled by having some type of expanded freeboard section at the top of the conventionally operated column (Wright and Raper, 1996) or at the bottom of the inverse bed (Nikolov and Karamanev, 1987). The elutriation problem is much attenuated or even eliminated by use of a tapered fluidized bed (Scott et al., 1978).

ACKNOWLEDGMENTS

Thanks are due to John Grace, Arturo Macchi, Dusko Posarac, and Yunbi Zhang for help in locating references, and to the Natural Sciences and Engineering Research Council of Canada for continuing research support.

NOMENCLATURE

a	=	exponent in Eq. (44), dimensionless
a, b, c	=	exponents in Eq. (167), dimensionless
a_i	=	constant for particle species i in Eq. (89), $\mathrm{m \cdot s^{-1}}$
a'	=	coefficient in Eq. (81), $\mathrm{m^{(2m-3)/m} \cdot s^{-1}}$
a''	=	constant in Eq. (176), dimensionless
A	=	cross-sectional area of column, $\mathrm{m^2}$
A_c	=	portion of submerged surface not affected by particle contact, $\mathrm{m^2}$
A_p	=	portion of submerged surface affected by particle contact, $\mathrm{m^2}$
Ar	=	Archimedes number $= d_p^3 (\rho_p - \rho)\rho g/\mu^2$, dimensionless
b	=	coefficient in Eq. (108), $\mathrm{m^{(m+3)/m} \cdot kg^{-1/m} \cdot s^{-1}}$
b_i	=	constant for particle species i in Eq. (89), $\mathrm{s^{-1}}$
b'	=	exponent in Eq. (168), dimensionless
b''	=	exponent in Eq. (176), dimensionless
B	=	buoyant force on single particle in swarm of particles, N
B_1	=	parameter defined by Eq. (29b), dimensionless
B_2	=	parameter defined by Eq. (29c), dimensionless
B'	=	surface mass flux/mainstream mass flux, dimensionless
c	=	volumetric particle concentration of monodisperse solids, $\mathrm{m^3 \cdot m^{-3}}$
c_i	=	local or overall volumetric concentration in liquid of particle species i, $\mathrm{m^3 \cdot m^{-3}}$

c_{i0}	=	local volumetric concentration in liquid of particle species i at $z = 0$, $\mathrm{m^3 \cdot m^{-3}}$
c_m	=	coefficient in Eq. (156), dimensionless
c_p	=	specific heat capacity of liquid, $\mathrm{J \cdot m^{-1} \cdot K^{-1}}$
c_{pp}	=	specific heat capacity of solid particles, $\mathrm{J \cdot m \cdot K^{-1}}$
c_t	=	local or overall volumetric concentration in liquid of particle species i, $\mathrm{m^3 \cdot m^{-3}}$
c^*	=	saturation concentration in Fig. 12, wt%
$c^* - c$	=	bulk liquid undersaturation for crystal dissolution points in Fig. 12, wt%
$c - c^*$	=	bulk liquid supersaturation for crystal growth points in Fig. 12, wt%
c_2'	=	volumetric concentration of smaller particles in pseudofluid through which larger particles move, $\mathrm{m^3 \cdot m^{-3}}$
C_A	=	bulk concentration of solute A in liquid solution at given bed level, $\mathrm{kg \cdot m^{-3}}$
C_{AS}	=	concentration of solute A in liquid solution at particle surfaces, $\mathrm{kg \cdot m^{-3}}$
C_{A1}	=	value of C_A at $z = 0$, $\mathrm{kg \cdot m^{-3}}$
C_{A2}	=	value of C_A at $z = L$, $\mathrm{kg \cdot m^{-3}}$
$(C_A)_{l.m.}$	=	logarithmic mean of $C_{AS}-C_{A1}$ and $C_{AS}-C_{A2}$, $\mathrm{kg \cdot m^{-3}}$
C_D	=	drag coefficient of particle in bed fluidized at superficial velocity U, dimensionless
C_{DS}	=	drag coefficient of isolated particle at velocity U, dimensionless
C_{D0}	=	drag coefficient of isolated particle at velocity U_0, dimensionless
C_1, C_2	=	constants in Eq. (17), dimensionless
d	=	diameter of spherical particle, mm or m
d_{avg}	=	average diameter of spherical particles as given by Eq. (136), mm or m
d_{eff}	=	effective particle diameter $= d_p K^{1/3}$, mm or m
d_i	=	spherical particle diameter of species i particles, mm or m
d_p	=	diameter of sphere having same volume as particle, mm or m
d_{pi}	=	equivolume sphere diameter of species i particles, mm or m
$\bar{d}_p$	=	local mass average or mass median particle diameter, mm or m
d_s	=	diameter of sphere having same surface area as particle, mm or m
d_{sv}	=	diameter of sphere having same surface-to-volume ratio as particle, mm or m
D_a	=	axial dispersion coefficient of liquid, $\mathrm{cm \cdot s^{-1}}$ or $\mathrm{m \cdot s^{-1}}$
D_A	=	molecular diffusivity of solute A in liquid, $\mathrm{cm^2 \cdot s^{-1}}$ or $\mathrm{m^2 \cdot s^{-1}}$
D_c	=	column diameter, m
D_h	=	hydraulic diameter $= 4 \times$ cross-sectional area/wetted perimeter, m

D_i	=	axial dispersion coefficient of particle species i with respect to particle species j, $\mathrm{cm^2 \cdot s^{-1}}$ or $\mathrm{m^2 \cdot s^{-1}}$	k_p	=	mass transfer coefficient between particles and liquid in fluidized bed, $\mathrm{m \cdot s^{-1}}$
D_{ia}	=	axial self-dispersivity of particle species i, $\mathrm{cm^2 \cdot s^{-1}}$ or $\mathrm{m^2 \cdot s^{-1}}$	K	=	volume of particle and "immobilized" liquid/volume of particle, dimensionless
D_{ir}	=	radial self-dispersivity of particle species i, $\mathrm{cm^2 \cdot s^{-1}}$ or $\mathrm{m^2 \cdot s^{-1}}$	K'	=	constant in Eq. (167), dimensionless
D_L	=	axial dispersivity of larger particles with respect to smaller particles, $\mathrm{cm^2 \cdot s^{-1}}$ or $\mathrm{m^2 \cdot s^{-1}}$	K'_0	=	constant in Eq. (169), dimensionless
			L	=	fluidized bed depth, m
D_r	=	radial dispersion coefficient of liquid, $\mathrm{cm^2 \cdot s^{-1}}$ or $\mathrm{m^2 \cdot s^{-1}}$	L_mf	=	fluidized bed depth at minimum fluidization, m
D_S	=	axial dispersivity of smaller particles with respect to larger particles, $\mathrm{cm^2 \cdot s^{-1}}$ or $\mathrm{m^2 \cdot s^{-1}}$	m	=	regime-dependent exponent in Eqs. (81), (108), and (156)
			m	=	$2 +$ slope of $\log C_{D0}$ vs. $\log \mathrm{Re}_0$ at Re_0 of monodisperse solids–liquid system under investigation, dimensionless
e	=	base of natural logarithms ≈ 2.718, dimensionless			
e_d	=	rate of energy dissipation per unit mass of liquid, $\mathrm{m^2 \cdot s^{-3}}$	m_A	=	mass transfer rate of solute A between particles and liquid, $\mathrm{kg \cdot s^{-1}}$
f	=	collision frequency of particles, $\mathrm{s^{-1}}$	m_z	=	mass of particles between distributor and height z, kg
$f(d_\mathrm{p})$	=	function of particle diameter that yields voidage, dimensionless	M	=	total mass of particles, kg
$f(\varepsilon)$	=	function of voidage defined by Eq. (49), dimensionless	M_i	=	mass of particle species i, kg
			n	=	Richardson–Zaki expansion index defined by Eq. (32), dimensionless
$f'(\varepsilon)$	=	function of voidage that yields slip velocity, $\mathrm{m \cdot s^{-1}}$	n_i	=	Richardson–Zaki expansion index for particle species i alone, dimensionless
F	=	fractional free area of perforated plate distributor, dimensionless	n_p	=	number of particles in contact with heat transfer surface at any instant
F_D	=	drag force on single particle in particle swarm, N	n'	=	expansion index exceeding n for $\varepsilon > \varepsilon_\mathrm{c}$, dimensionless
F_DS	=	drag force on isolated spherical particle at velocity U, N	n''	=	expansion index less than n for $\varepsilon > \varepsilon_\mathrm{c}$, dimensionless
F_D0	=	drag force on isolated spherical particle at velocity U_0, N	N	=	total number of different particle species
g	=	acceleration of gravity, $\mathrm{m \cdot s^{-2}}$	N_p	=	total number of fluidized particles
Ga	=	Galileo number $= d_\mathrm{p}^3 \rho^2 g / \mu^2$, dimensionless	Nu	=	Nusselt number for heat transfer between submerged surface and liquid in fluidized bed $= h d_\mathrm{p} / \lambda$, dimensionless
h	=	heat transfer coefficient between submerged surface and liquid in fluidized bed, $\mathrm{W \cdot m^{-2} \cdot K^{-1}}$			
			Nu_p	=	Nusselt number for heat transfer between particles and liquid in fluidized bed $= h_\mathrm{p} d_\mathrm{p} / \lambda$, dimensionless
h_max	=	maximum value of h with respect to U or ε, $\mathrm{W \cdot m^{-2} \cdot K^{-1}}$			
			$-\Delta p$	=	pressure drop across bed due to liquid motion, Pa
h_p	=	heat transfer coefficient between particles and liquid in fluidized bed, $\mathrm{W \cdot m^{-2} \cdot K^{-1}}$	$-\Delta p_\mathrm{d}$	=	pressure drop across distributor, Pa
			$-\Delta p_\mathrm{f}$	=	frictional pressure drop across bed, Pa
H	=	distance between pressure taps in Eqs. (9) and (10), or thickness of transition zone in Eqs. (89)–(91), m	$-dp_\mathrm{f}/dz$	=	frictional pressure gradient, $\mathrm{Pa \cdot m^{-1}}$
			P_1	=	total pressure immediately above distributor, Pa
j_m	=	Chilton–Colburn mass-transfer factor $= k_\mathrm{m} / U \mathrm{Sc}^{2/3}$, dimensionless	P_2	=	total pressure at plane above bed, Pa
			$-\Delta P$	=	total pressure drop $P_1 - P_2$, Pa
k	=	wall-effect factor U_e/U_0, dimensionless	$-\Delta P/\Delta z$	=	total pressure gradient, $\mathrm{Pa \cdot m^{-1}}$
k_i	=	wall-effect factor for particle species i alone, dimensionless	Pe_a	=	axial Peclet number of liquid $= d_\mathrm{p} U / \varepsilon D_\mathrm{a}$, dimensionless
k_m	=	mass transfer coefficient between submerged surface and liquid in fluidized bed, $\mathrm{m \cdot s^{-1}}$	Pe_i	=	Peclet number of single species i particle in bed of species $j = L U_{\mathrm{p}i0}/D_i$, dimensionless
			Pe_L	=	Peclet number of single larger particle in bed of smaller particles $= L U_{\mathrm{pL0}}/D_\mathrm{L}$, dimensionless

Pe_r	=	radial Peclet number of liquid = $d_p U / \varepsilon D_r$
Pe_S	=	Peclet number of single smaller particle in bed of larger particles = $L U_{pS0}/D_S$, dimensionless
Pr	=	Prandtl number = $c_p \mu / \lambda$, dimensionless
Q	=	volumetric flow rate of liquid, $m^3 \cdot s^{-1}$
R^2	=	coefficient of determination, dimensionless
Re	=	particle Reynolds number = $d_p U \rho / \mu$, dimensionless
Re_{mf}	=	particle Reynolds number at minimum fluidization = $d_p U_{mf} \rho / \mu$, dimensionless
Re_0	=	free-settling terminal particle Reynolds number = $d_p U_0 \rho / \mu$, dimensionless
S	=	total surface area of particles, m^2
S_v	=	surface area of particles/volume of bed, $m^2 \cdot m^{-3}$
Sc	=	Schmidt number = $\mu / \rho D_A$, dimensionless
Sh	=	Sherwood number for mass transfer between submerged surface and liquid in fluidized bed = $k_m d_p / D_A$, dimensionless
Sh_p	=	Sherwood number for mass transfer between particles and liquid in fluidized bed = $k_p d_p / D_A$, dimensionless
Δt	=	time interval, s
T	=	bulk temperature, K
T_s	=	particle surface temperature, K
U	=	superficial liquid velocity, $m \cdot s^{-1}$
U_{amf}	=	apparent minimum fluidization velocity of multidisperse solids, $m \cdot s^{-1}$
$U_A, U_B,$ U_C, U_D	=	values of U shown in Figure 8, $m \cdot s^{-1}$
U_{bf}	=	beginning fluidization velocity of multidisperse solids, $m \cdot s^{-1}$
U_e	=	value of U when linear plot of $\log U$ vs. $\log \varepsilon$ at $\varepsilon < \varepsilon_c$ is extrapolated to $\varepsilon = 1$, $m \cdot s^{-1}$
U_{ei}	=	value of U_e for particle species i alone, $m \cdot s^{-1}$
U_e'	=	value of U when linear plot of $\log U$ vs. $\log \varepsilon$ at $\varepsilon > \varepsilon_c$ with slope n'' is extrapolated to $\varepsilon = 1$, $m \cdot s^{-1}$
U_E	=	dynamic velocity of elastic waves, $m \cdot s^{-1}$
U_i	=	superficial velocity required to fluidize monodisperse particle species i to same voidage as exists in multispecies system, $m \cdot s^{-1}$
U_{mf}	=	minimum fluidization velocity (superficial), $m \cdot s^{-1}$
U_{mfd}	=	maximum velocity of full defluidization of tapered bed, $m \cdot s^{-1}$
U_{mff}	=	minimum velocity of full fluidization of tapered bed, $m \cdot s^{-1}$
U_{mpd}	=	maximum velocity of partial defluidization of tapered bed, $m \cdot s^{-1}$
U_{mpf}	=	minimum velocity of partial fluidization of tapered bed, $m \cdot s^{-1}$
U_p	=	superficial velocity of solids, $m \cdot s^{-1}$
U_{phS}	=	segregation velocity of denser smaller particles through less dense larger particles, $m \cdot s^{-1}$
U_{pi}	=	segregation velocity of species i particles through species j, $m \cdot s^{-1}$
$U_{p,i}$	=	superficial velocity of species i particles, $m \cdot s^{-1}$
U_{pi0}	=	segregation velocity of single species i particle through swarm of species j, $m \cdot s^{-1}$
U_{pL0}	=	segregation velocity of single larger particle through swarm of smaller particles, $m \cdot s^{-1}$
U_{pS0}	=	segregation velocity of single smaller particle through swarm of larger particles, $m \cdot s^{-1}$
U_r	=	relative or slip velocity between liquid and solids, $m \cdot s^{-1}$
U_{ri}	=	relative velocity between liquid and particle species i, $m \cdot s^{-1}$
U_{tf}	=	minimum velocity for total fluidization of multidisperse solids, $m \cdot s^{-1}$
U_0	=	terminal free settling velocity of particles, $m \cdot s^{-1}$
U_0'	=	terminal settling velocity of single particle along axis of finite diameter cylindrical column, $m \cdot s^{-1}$
U_{0i}	=	terminal free settling velocity of species i particles, $m \cdot s^{-1}$
U_{01}'	=	terminal free settling velocity of fastest moving particles of a ternary particle mixture according to model of Selim et al. (1983a,b), $m \cdot s^{-1}$
U_{02}'	=	terminal free settling velocity of intermediate velocity particles of a ternary particle mixture according to model of Selim et al. (1983a,b), $m \cdot s^{-1}$
U_{01}''	=	terminal free settling velocity of faster moving particles of a binary particle mixture according to pseudofluid model, $m \cdot s^{-1}$
U_ε	=	kinematic velocity of continuity waves, $m \cdot s^{-1}$
v_i	=	volume fraction of species i particles in solids mixture, dimensionless
V	=	volume of representative particle in fluidized bed, m^3
V_i	=	total volume of species i particles in fluidized bed, m^3
V_p	=	constant rate particle sedimentation velocity in downward direction, $m \cdot s^{-1}$
W	=	width of two-dimensional bed, m
x_L	=	$c_L/(1 - \varepsilon_L)$, dimensionless
x_S	=	$c_S/(1 - \varepsilon_S)$, dimensionless
z	=	vertical distance above distributor, or vertical distance above bottom of binary transition zone, m
$\bar{z}_L$	=	value of z when $x_L = 0.5$, m

$\bar{z}_S$ = value of z when $x_S = 0.5$, m
Z = z/L, dimensionless
$\bar{Z}_L$ = $\bar{z}_L/L$, dimensionless
$\bar{Z}_S$ = $\bar{z}_S/L$, dimensionless

Greek Letters

α = exponent in Eq. (62), dimensionless
α'/α'_{mf} = parameter given by Eq. (29a), dimensionless
β = exponent in Eq. (49), dimensionless
γ = sizing density ratio $= (\rho_{BL} - \rho_{BS})/(\rho_p - \rho)$, dimensionless
γ_h = sorting density ratio $= (\rho_{Bh} - \rho_{B1})/(\rho_{ph} - \rho)$, dimensionless
ε = local or overall fractional void volume, i.e., voidage, dimensionless
ε_b = settled bed voidage after sedimentation, dimensionless
ε_{b1} = minimum bubbling voidage, dimensionless
ε_{b2} = maximum bubbling voidage, dimensionless
ε_c = critical bed voidage at which slope of $\log U$ vs. $\log \varepsilon$ changes, dimensionless
ε_{ei} = effective voidage associated with particle species i, dimensionless
ε_i = voidage when particle species i is fluidized alone at same superficial liquid velocity as for multispecies system, dimensionless
ε_{max} = voidage at which various hydrodynamic and hydrodynamically related phenomena maximize, dimensionless
ε_{mf} = voidage at minimum fluidization, dimensionless
θ = included angle of conical or tapered bed, degrees
κ = permeability of porous medium, m^2
λ = thermal conductivity of liquid, W $\cdot$ m^{-1} $\cdot$ K^{-1}
λ_B = parameter defined by Eq. (29d), dimensionless
λ_{eddy} = eddy conductivity of heat, W $\cdot$ m^{-1} $\cdot$ K^{-1}
λ_p = thermal conductivity of particles, W $\cdot$ m^{-1} $\cdot$ K^{-1}
μ = liquid viscosity, Pa $\cdot$ s
μ_2 = pseudofluid viscosity for determining motion of larger particles through suspension of smaller particles, Pa $\cdot$ s
ρ = liquid density, kg $\cdot$ m^{-3}
ρ_B = bulk density of liquid-fluidized bed, kg $\cdot$ m^{-3}
$\rho_{BA}, \rho_{BB},$
ρ_{BD} = bulk densities, $(\rho_B)_{max}$, of bottom layer at points A, B, and D, respectively, in Fig. 8, kg $\cdot$ m^{-3}

$(\rho_B)_{max}$ = maximum bulk density of binary particle mixture over range of possible bed inversion velocities, i.e., bulk density of bottom layer, kg $\cdot$ m^{-3}
ρ_{eff} = effective particle density $= [\rho_p + \rho(K - 1)]/K$, kg $\cdot$ m^{-3}
ρ_p = particle density, kg $\cdot$ m^{-3}
$\bar{\rho}_p$ = mean density of heterogeneous particle, kg $\cdot$ m^{-3}
ρ_{pi} = density of species i particle, kg $\cdot$ m^{-3}
ρ_2 = pseudofluid density of binary particle system including only the slower moving particles, kg $\cdot$ m^{-3}
$\rho_{2,3}$ = pseudofluid density of ternary particle system excluding the fastest moving particles, kg $\cdot$ m^{-3}
ρ_3 = pseudofluid density of ternary particle system including only the slowest moving particles, kg $\cdot$ m^{-3}
$\sum$ = summation
ϕ = particle sphericity, dimensionless
ϕ_A = fraction of total particle surface area that is mass-transfer active, dimensionless

Subscripts

h = heavier particles of monosize binary mixture
hS = heavier (i.e., denser) and smaller particles of binary mixture
i = particle species i, which is representative of all other particle species
j = particle species j
l = lighter particles of monosize binary mixture
lL = lighter (i.e., less dense) and larger particles of binary mixture
L = larger particles of fixed density binary mixture
mf = minimum fluidization
S = smaller particles of fixed density binary mixture
1, 2, 3 = particle species 1, 2, and 3, respectively

REFERENCES

Adler IL, Happel J. The fluidization of smooth spheres in liquid media. Chem Eng Prog Symp Ser 58(38):98–105, 1962.

Agarwal GP, Hudson JL, Jackson R. Fluid mechanical description of fluidized beds. Experimental investigation of convective instabilities in bounded beds. Ind Eng Chem Fundam 19:59–66, 1980.

Agricola G. De Re Metallica. Translated from the first Latin edition of 1556 by H. C. Hoover and L. H. Hoover. New York: Dover, 1950, pp. 310–311.

Al-Dibouni MR, Garside J. Particle mixing and classification in liquid fluidised beds. Trans IChemE 57:94–103, 1979.

Allen BR, Coughlin RW, Charles M. Fluidized-bed enzyme reactors. Ann NY Acad Sci 326:105–117, 1979.

Al-Naafá MA, Selim MS. Sedimentation of polydisperse concentrated suspensions. Can J Chem Eng 67:253–264, 1989.

Al-Naafá MA, Selim SM. Sedimentation of monodisperse and bidisperse hard-sphere colloidal suspensions. AIChE J 38:1618–1630, 1992.

Amitharajah A. Optimum backwashing of sand filters. J Environ Engng Div ASCE 104:917–932, 1978.

Anderson TB, Jackson R. Fluid mechanical description of fluidized beds: stability of the state of uniform fluidization. Ind Eng Chem Fundam 7:12–21, 1968.

Anderson TB, Jackson R. A fluid mechanical description of fluidized beds: comparison of theory and experiment. Ind Eng Chem Fundam 8:137–144, 1969.

Andrews GF, Tien C. Bacterial film growth in [sic] adsorbent surfaces. AIChE J 27:396–403, 1981.

Arters DC, Fan L-S. Solid–liquid mass transfer in a gas-liquid-solid fluidized bed. Chem Eng Sci 41:107–115, 1986.

Asif M. Modeling of multisolid liquid fluidized bed. Chem Eng Technol 10:485–490, 1997.

Asif M. Generalized Richardson-Zaki correlation for liquid fluidization of binary solids. Chem Eng Technol 21:77–82, 1998.

Asif M, Petersen JN. Particle dispersion in a binary solid-liquid fluidized bed. AIChE J 39:1465–1471, 1993.

Asif M, Kalogerakis N, Behie LA. Distributor effects in liquid fluidized beds of low-density particles. AIChE J 37:1825–1832, 1991.

Asif M, Kalogerakis N, Behie LA. Hydrodynamics of liquid fluidized beds including the distributor region. Chem Eng Sci 47:4155–4166, 1992.

Astarita G. Forces acting on particles in a fluidised bed. Chem Eng Sci 48:3438–3440, 1993.

Atkinson B, Black GM, Lewis PJS, Pinches A. Biological particles of given size, shape, and density for use in biological reactors. Biotechnol Bioeng 21:193–200, 1979.

Babu SP, Shah B, Talwalkar A. Fluidization correlations for coal gasification materials—minimum fluidization velocity and fluidized bed expansion ratio. AIChE Symp Ser 74(176):176–186, 1978.

Backhurst JR, Coulson JM, Goodridge F, Plimley RE, Fleischmann M. A preliminary investigation of fluidized bed electrodes. J Electrochem Soc 116:1600–1607, 1969.

Bamforth AW. Industrial Crystallization. London: Leonard Hill, 1965, pp 49–55.

Barnea E, Mednick RL. Correlation for minimum fluidisation velocity. Trans Instn Chem Engrs 53:278–281, 1975.

Barnea F, Mizrahi J. A generalized approach to the fluid dynamics of particulate systems. Part I. General correlation for fluidization and sedimentation in solid multiparticle systems. Chem Eng J 5:171–189, 1973.

Bascoul A, Delmas H, Couderc JP. Caractéristiques hydrodynamiques de la fluidisation liquide–solide: influence du distributeur. Chem Eng J 37:11–24, 1988.

Batchelor GK. Sedimentation in a dilute dispersion of spheres. J Fluid Mech 52(2):245–268, 1972.

Batchelor GK. Sedimentation in a dilute polydisperse system of interacting spheres. Part 1. General theory. J Fluid Mech 119:379–408, 1982.

Batchelor GK. A new theory of the instability of a uniform fluidized bed. J Fluid Mech 193:75–110, 1988.

Batchelor GK, Wen CS. Sedimentation in a dilute polydisperse system of interacting spheres. Part 2. Numerical results. J Fluid Mech 124:495–528, 1982.

Beek WI. Mass transfer in fluidized beds. In: Davidson JF, Harrison D, eds. Fluidization. London: Academic Press, 1971, pp 431–470.

Beňa J, Ilavský J, Kossaczký E, Neužil L. Changes of the flow character in a fluidized bed. Collection Czechoslov Chem Commun 28:293–309, 1963.

Biń AK. Minimum fluidization velocity at elevated temperature and pressure. Can J Chem Eng 64:854–857, 1986.

Bird RB, Stewart WE, Lightfoot EH. Transport Phenomena. New York: John Wiley, 1960, p 663.

Bordet J, Coeuret F, Le Goff P, Vergues F. Étude par conductance électrique des fluctuations de porosité locale dans les lits fluidisés liquide–solide. Powder Technology 6:253–261, 1972.

Bourgeois P, Grenier P. The ratio of terminal velocity to minimum fluidizing velocity for spherical particles. Can J Chem Eng 46:325–334, 1968.

Boyanov BS, Donaldson JD, Grimes SM. Removal of copper and cadmium from metallurgical leach solutions by fluidised bed electrolysis. J Chem Technol Biotech 41:317–328, 1988.

Bransom SH. Factors in the design of continuous crystallizers. Brit Chem Eng 5:838–844, 1960.

Bremford DJ, Müller-Steinhagen H, Duffy GG. Heat transfer to black liquor in a fluidized bed. In: Celata GP, Di Marco P, Mariani A, eds. Proc 2d European Thermal-Sciences and 14th UIT National Heat Transfer Conference. Edizioni ETS, 1996, pp 1545–1551.

Briens CL, Del Pozo M, Chiu K, Wild G. Modeling of particle-liquid heat and mass transfer in multiphase systems with the film-penetration model. Chem Eng Sci 48:973–979, 1993.

Briens LA, Briens CL, Margaritis A, Cooke SL, Bergougnou MA. Characterization of channeling in multiphase systems. Application to a liquid fluidized bed of angular biobone particles. Powder Technology 91:1–9, 1997.

Brinkman HC. A calculation of the viscous force exerted by a flowing fluid on a dense swarm of particles. Appl Sci Res A1:27–34, 1947.

Brody MA, Lumkins RJ. Performance of dual-media filters—1. Chem Eng Prog 73(4), 83–86, 1977.

Brötz W. Grundlagen der wirbelschichtverfahren. Chem–Ing–Technik 24(2):60–81, 1952.

Brown GG, Foust AS, Kata DL, Schneidewind R, White RR, Wood WP, Brown GM, Brownell LE, Martin JJ, Williams GB, Banchero JT, York JL. Unit Operations. New York: John Wiley, 1950, pp 91–94.

Brownell LE, Dombrowski HS, Dickey CA. Pressure drop through porous media part IV—new data and revised correlation. Chem Eng Prog 46:415–422, 1950.

Bruinzeel C, Reman GH, van der Laan ETR. Eddy diffusion in particulately fluidized beds: model experiments for the design of a large-scale unit. In: Proc Symp on Interaction Between Fluids and Particles. London: Instn Chem Engrs, 1962, pp 120–126.

Buscall R, Goodwin JW, Ottewill RH, Tadros ThF. The settling of particles through Newtonian and non-Newtonian media. J Colloid Interface Sci 85:78–86, 1982.

Cairns EJ, Prausnitz JM. Longitudinal mixing in fluidization. AIChE J 6:400–405, 1960a.

Cairns EJ, Prausnitz JM. Macroscopic mixing in fluidization. AIChE J 6:554–560, 1960b.

Calderbank PH. Gas absorption from bubbles. Chemical Engineer (London) 45(211):CE209-CE253, 1967.

Carlos CR, Richardson JF. Solids movement in liquid fluidised beds—I. Particle velocity distribution. Chem Eng Sci 23:813–824, 1968a.

Carlos CR, Richardson JF. Solids movement in liquid fluidised beds—II. Measurement of axial mixing coefficients. Chem Eng Sci 23:825–831, 1968b.

Carman PC. Fluid flow through granular beds. Trans Instn Chem Engrs (London) 15:150–166, 1937.

Casal J, Puigjaner L. Segregation and apparent minimum fluidization velocity in particulate fluidization. Chem Eng Commun 23:125–136, 1983.

Chavarie C, Karamanev DG. Use of inverse fluidization in biofilm reactors. Proc Internat Conf Bioreactor Fluid Dynamics, BHRA, Cambridge, England, 1986, pp 181–190.

Chen JJJ. Comments on "Improved equation for the calculation of minimum fluidization velocity." Ind Eng Chem Res 26:633–634, 1987.

Chen Z, Gibilaro LG, Foscolo PU. Two-dimensional voidage waves in fluidized beds. Ind Eng Chem Res 38:610–620, 1999.

Chhun T, Couderc JP. High flux mass transfer in liquid fluidization. Chem Eng Sci 35:1707–1715, 1980.

Chianese A, Frances C, Di Berardino F, Bruno L. On the behaviour of a liquid fluidized bed of monosized sodium perborate crystals. Chem Eng J 50:87–94, 1992.

Chiba T. Bed contraction of liquid-fluidised binary solid particles at complete mixing. In: Yoshida K, Morooka S, eds. Proc Asian Conf on Fluidized Bed and Three-Phase Reactors. Tokyo: Tokyo University, 1988, pp 385–392.

Chilton TH, Colburn AP. Mass transfer (absorption) coefficients—prediction from data on heat transfer and fluid friction. Ind Eng Chem 26: 1183–1187, 1934.

Chitester DC, Kornosky RM, Fan L-S, Danko JP. Characteristics of fluidization at high pressure. Chem Eng Sci 39:253–261, 1984.

Chong YS, Ratkowsky DA, Epstein N. Effect of particle shape on hindered settling in creeping flow. Powder Technology 23:55–66, 1979.

Chung SF, Wen CY. Longitudinal dispersion of liquid flowing through fixed and fluidized beds. AIChE J 14:857–866, 1968.

Chyang CS, Huang WC. Characteristics of large particle fluidization. J Chin Inst Chem Engrs 19(2):81–89, 1988.

Cleasby JL, Baumann AR. Backwash of granular filters used in waste water filtration. Final Report on EPA Project R 802140, Engineering Research Institute, Iowa State University, 1976.

Cleasby JL, Fan KS. Predicting fluidization and expansion of filter media. J Environ Engng Div ASCE 107:455–471, 1981.

Cleasby JL, Lorence JL. Effectiveness of backwashing for wastewater filters. J Environ Engng Div ASCE 104:749–765, 1978.

Cleasby JL, Sejkora GD. Effect of media intermixing on dual media filtration. J Environ Engng Div ASCE 101:503–516, 1975.

Cleasby JL, Woods CF. Intermixing of dual media and multimedia granular filters. J Am Water Works Assoc 67:197–203, April 1975.

Cleasby JL, Stangl EW, Rice GA. Developments in backwashing of granular filters. J Environ Engng Div ASCE 101:713–727, 1975.

Clift R. An Occamist review of fluidized bed modelling. AIChE Symp Series 89(296):1–17, 1993.

Clift R, Grace JR, Weber ME. Bubbles, Drops, and Particles. New York: Academic Press, 1978, p. 114.

Clift R, Seville JPK, Moore SC, Chavarie C. Comments on buoyancy in fluidized beds. Chem Eng Sci 42:191–194, 1987.

Cole LT, Allen CA. Liquid-fluidized-bed heat exchanger flow distribution models. Proc Intersoc Energy Convers Eng Conf 13:1129–1134, 1978.

Cooper PF, Atkinson B, eds. Biological Fluidised Bed Treatment of Water and Wastewater. Chichester: Ellis Horwood, 1981.

Cooper PF, Wheeldon DHV, Ingram-Tedd PE, Harrington DW. Sand/biomass separation with production of a concentrated sludge. In: Cooper PF, Atkinson B, eds. Biological Fluidised Bed Treatment of Water and Wastewater. Chichester: Ellis Horwood, 1981, pp 361–367.

Cornish ARH. Note on minimum possible rate of heat transfer from a sphere when other spheres are adjacent to it. Trans Instn Chem Engrs 43:T332-T333, 1965.

Couderc J-P. Incipient fluidization and particulate systems. In: Davidson JF, Clift R, Harrison D, eds. Fluidization. 2d ed. London: Academic Press, 1985, pp 1–46.

Couderc JP, Gibert H, Angelino H. Transfert de matière par diffusion en fluidisation liquide. Chem Eng Sci 27:11–20, 1972.

Damronglerd S, Couderc JP, Angelino H. Mass transfer in particulate fluidisation. Trans Instn Chem Engrs 53:175–180, 1975.

Darcy HPG. Les Fontaines Publiques de la Ville de Dijon. Paris: Victor Dalmont, 1856.

Delachanal M. Apparatus and process for sorting solid particle material in a fluidized liquid medium. U.S. Patent 3,075,643, Jan. 29, 1963.

Dhamarajah AH, Cleasby JL. Predicting the expansion behavior of filter media. J. Am Water Works Assoc 78:66–76, December 1986.

Didwania AK, Homsy GM. Flow regimes and flow transitions in liquid fluidized beds. Internat J Multiphase Flow 7:563–580, 1981.

Di Felice, R. Mixing in segregated, binary-solid liquid fluidized beds. Chem Eng Sci 48:881–888, 1993.

Di Felice R. The voidage function for fluid-particle interaction systems. Int J Multiphase Flow 20:153–159, 1994.

Di Felice R. Hydrodynamics of liquid fluidisation. Chem Eng Sci 50:1213–1245, 1995.

Di Felice R. A relationship for the wall effect on the settling velocity of a sphere at any flow regime. Intern J Multiphase Flow 22:527–533, 1996.

Di Felice R. The applicability of the pseudo-fluid model to the settling velocity of a foreign particle in a suspension. Chem Eng Sci 53:371–375, 1998.

Di Felice R., Parodi E. Wall effects on the sedimentation velocity of suspensions in viscous flow. AIChE J 42:927–931, 1996.

Di Felice R, Gibilaro LG, Waldram SP, Foscolo SP. Mixing and segregation in binary-solid liquid fluidised bed. Chem Eng Sci 42:639–652, 1987.

Di Felice R, Gibilaro LG, Foscolo PU. On the inversion of binary-solid liquid fluidised beds. Chem Eng Sci 43:979–981, 1988.

Di Felice R, Gibilaro LG, Rapagnà S, Foscolo, PU. Particle mixing in a circulating liquid fluidized bed. AIChE Symp Series 85(270):32–36, 1989.

Di Felice R, Foscolo PU, Gibilaro LG, Wallis GB, Carta R. Expansion characteristics of tapered fluidized beds. AIChE J 37:1668–1672, 1991a.

Di Felice R, Foscolo PU, Gibilaro LG, Rapagnà S. The interaction of particles with a fluid-particle pseudo-fluid. Chem Eng Sci 46:1873–1877, 1991b.

Dixon DC. Comments on "Hindered settling theories", Powder Technology 17:147–149, 1977.

Dorgelo EAH, van der Meer AP, Wesselingh JA. Measurement of the axial dispersion of particles in a liquid fluidized bed applying a random walk model. Chem Eng Sci 40:2105–2111, 1985.

Dweik BM, Liu CC, Savinell RF. Hydrodynamic modelling of the liquid–solid behaviour of the circulating particulate bed electrode. J Appl Electrochem 26:1093–1102, 1996.

Eastwood, J, Matzen EJP, Young MJ, Epstein N. Random loose porosity of packed beds. Brit Chem Eng 14:1542–1545, 1969.

Eisenklam P. Discussion. In: Drinkenburg AAH, ed. Proc Internat Conf on Fluidization. Amsterdam: Netherlands Univ. Press, 1967, pp 549–551.

El-Kaissy MM, Homsy GM. Instability waves and the origin of bubbles in fluidized beds—Part 1: experiments. Internat J Multiphase Flow 2:379–395, 1976.

El-Temtamy SA, Epstein N. Effect of low velocity liquid pulstions on some hydrodynamic characteristics of liquid and gas–liquid fluidized beds. AIChE J 32:509–512, 1986.

Emersleben O, Faxén H. The motion of a rigid sphere along the axis of a tube filled with viscous fluid. Arkiv Mat Astron Fysik 17(27):1–28, 1923.

Epstein N. Hydrodynamic particle volume and settled bed volume. Can J Chem Eng 57:383, 1979.

Epstein N. Letter to the editor. AIChE J 38:637–638, 1992.

Epstein N, LeClair BP. Liquid fluidization of binary particle mixtures—II. Bed inversion. Chem Eng Sci 40:1517–1526, 1985.

Epstein N, Masliyah JH. Creeping flow through clusters of spheroids and elliptical cylinders. Chem Eng J 3:169–175, 1972.

Epstein N, Pruden BB. Liquid fluidisation of binary particle mixtures—III. Stratification by size and related topics. Chem Eng Sci 54:401–415, 1999.

Epstein N, LeClair BP, Pruden BB. Liquid fluidization of binary particle mixtures—I. Overall bed expansion. Chem Eng Sci 36:1803–1809, 1981.

Ergun S. Fluid flow through packed columns. Chem Eng Prog 48:89–94, 1952.

Evans GC, Gerald CF. Mass transfer from benzoic acid granules to water in fixed and fluidized beds at low Reynolds numbers. Chem Eng Prog 49:135–140, 1953.

Fan L-S. Gas–Liquid–Solid Fluidization Engineering. Boston: Butterworths, 1989, pp 453–588, 415–420.

Fan L-S, Muroyama K, Chern S-H. Hydrodynamic characteristics of inverse fluidization in liquid–solid and gas–liquid–solid systems. Chem Eng J 24:143–150, 1982.

Fan L-T, Yang Y-C, Wen C-Y. Mass transfer in semifluidized beds for solid-liquid system. AIChE J 6:482–487, 1960.

Fan Z, Xuanyu Z, Lichang X. Fluidization of uniformly sized spheres. In: Kwauk M, Kunii D, Jiansheng Z, Hasatani M, eds. Fluidization '85. Science and Technology Conference Papers. Second China–Japan Symposium. Amsterdam: Elsevier, 1985, pp 283–294.

Fidleris V, Whitmore RL. Experimental determination of the wall effect for spheres falling axially in cylindrical vessels. Brit J Appl Phys 12:490–494, 1961.

Field CJ, Riley PJ. Tripos Part II Research Report, Cambridge University, 1970.

Finkelstein E, Letan R, Elgin JC. Mechanics of vertical moving fluidized systems with mixed particle sizes. AIChE J 17:867–872, 1971.

Foscolo PU, Gibilaro LG. A fully predictive criterion for the transition between particulate and aggregative fluidization. Chem Eng Sci 39:1667–1675, 1984.

Foscolo PU, Gibilaro LG. Fluid dynamic stability of fluidised suspensions: the particle bed model. Chem Eng Sci 42:1489–1500, 1987.

Foscolo PU, Gibilaro LG, Waldram SP. A unified model for particulate expansion of fluidised beds and flow in fixed porous media. Chem Eng Sci 38:1251–1260, 1983.

Foscolo PU, Di Felice R, Gibilaro LG. The pressure field in an unsteady-state fluidized bed. AIChE J 35:1921–1926, 1989.

Fouda AE, Capes CE. Hydrodynamic particle volume and fluidized bed expansion. Can J Chem Eng 55:386–391, 1977.

Fouda AE, Capes CE. Hydrodynamic particle volume and fluidized bed expansion. Can J Chem Eng 57:120–121, 1979.

Frances C, Biscans B, Laguerie C. Modelling of a continuous fluidized-bed crystallizer. Chem Eng Sci 49:3269–3276, 1994.

Francis AW. Wall effect in falling ball method for viscosity. Physics 4:403–406, 1933.

Funamizu N, Takakuwa T. A minimal potential energy model for predicting stratification pattern in binary and ternary solid–liquid fluidized beds. Chem Eng Sci 51:341–351, 1996.

Galvin KP, Pratten SJ. Application of fluidization to obtain washability data. Minerals Engineering 12:1051–1058, 1999.

Galvin KP, Pratten S, Nguyen-Tran-Lam G. Differential settling in a teeter bed separator. Presented at Third World Congress on Particle Technology. Brighton, England: I Chem E, 1998, Paper 228, pp 1–11.

Galvin KP, Pratten S, Nguyen-Tran-Lam G. A generalized empirical description for particle slip velocities in liquid fluidized beds. Chem Eng Sci 54:1045–1052, 1999.

Ganho R, Gibert H, Angelino H. Cinétique de l'adsorption du phenol en couche fluidisée de charbon actif. Chem Eng Sci 30:1231–1238, 1975.

Garnier A, Chavarie C, André G, Klvana D. The inverse fluidization airlift bioreactor, part I: hydrodynamic studies. Chem Eng Comm 98:31–45, 1990.

Garside J, Al-Dibouni M. Behaviour of liquid fluidized beds containing a wide size distribution of solids. In: Angelino H, Couderc JP, Gibert H, Laguerie C, eds. Fluidization and Its Applications. Toulouse: Cepadues-Éditions, 1974, pp 53–62.

Garside J, Al-Dibouni MR. Velocity–voidage relationships for fluidization and sedimentation in solid-liquid systems. Ind Eng Chem Process Des Dev 16:206–214, 1977.

Garside J, Gaska G, Mullin JW. Crystal growth rate studies with potassium sulphate in a fluidized bed crystallizer. J Cryst Growth 13/14:510–516, 1972.

Gasparyan AM, Ikaryan NS. Hindered settling of particles V. Izv Akad Nauk Arm SSR Ser Tekhn Nauk 15(4):53–64, 1962.

Gaudin AM. Principles of Mineral Dressing. 1st ed. New York: McGraw-Hill, 1939, p 205.

Gibilaro LG, Waldram SP, Foscolo PU. Authors' reply to comments by N. Epstein. Chem Eng Sci 39:1819–1820, 1984.

Gibilaro LG, Di Felice R, Waldram SP, Foscolo PU. Generalized friction factor and drag coefficient correlations for fluid–particle interactions. Chem Eng Sci 40:1817–1823, 1985a.

Gibilaro LG, Hossain I, Waldram SP. On the Kennedy and Bretton model for mixing and segregation in liquid fluidized beds. Chem Eng Sci 40:2333–2338, 1985b.

Gibilaro LG, Di Felice R, Waldram SP, Foscolo PU. A predictive model for the equilibrium composition and inversion of binary-solid liquid fluidised beds. Chem Eng Sci 41:379–387, 1986a.

Gibilaro LG, Hossain I, Foscolo PU. Aggregate behaviour of liquid fluidised beds. Can J Chem Eng 64:931–938, 1986b.

Gibilaro LG, Di Felice R, Waldram SP, Foscolo PU. Authors' reply to Clift et al. Chem Eng Sci 42:194–196, 1987.

Gibilaro LG, Di Felice R, Foscolo PU. Added mass effects in fluidized beds: application of the Geurst–Wallis analysis of inertial coupling in two-phase flow. Chem Eng Sci 45:1561–1565, 1990.

Glasby J, Ridgway K. The crystallization of aspirin from ethanol. J Pharm Pharmac 20(suppl.):94S–103S, 1968.

Goodridge F, Wright AR. Porous flow-through and fluidized bed electrodes. In: Yeager E, Bockris JO'M, Conway BE, Sarangapani S, eds. Comprehensive Treatise of Electrochemistry. Vol. 6. Electrodics: Transport. New York: Plenum Press, 1983, pp 393–440.

Goodridge F, Holden DI, Murray HD, Plimley RF. Fluidized-bed electrodes. Trans Instn Chem Engrs 49:128–141, 1971.

Grace JR. Fluidized bed hydrodynamics. In: Hetsroni G, ed. Handbook of Multiphase Systems. New York: Hemisphere and McGraw-Hill, 1982, p 8–6.

Grace JR. Agricola aground: characterization and interpretation of fluidization phenomena. AIChE Symp Series 88(289):1–16, 1992.

Grady Jr CPL, Daigger GT, Lim HC. Biological Wastewater Treatment. New York: Marcel Dekker, 1999, pp 614–615, 621–622, 653–654, 809–840, 951, 956–957, 971, 975–978.

Grbavčić ZB, Vuković DV. Single-particle settling velocity through liquid fluidized beds. Powder Technology 66:293–295, 1991.

Grbavčić ZB, Garić RV, Hadjismalović DzE, Jovanović S, Vuković DV, Littman H, Morgan III MH. Variational model for prediction of the fluid–particle interphase drag coefficient and particulate expansion of fluidized and sedimenting beds. Powder Technology 68:199–211, 1991.

Grimmet ES, Brown BP. A pulsed column for countercurrent liquid–solids flows. Ind Eng Chem 54:24–28, 1962.

Grünewald M, Schmidt-Traub H. Adsorption in multistage fluidized bed column. Chem Eng Technol 22:206–209, 1999.

Hadžismajlović DE, Popov KI, Pavlović MG. The visualization of the electrochemical behaviour of metal particles in spouted, fluidized and packed beds. Powder Technology 86:145–148, 1996.

Haid M. Correlations for the prediction of heat transfer to liquid–solid fluidized beds. Chemical Engineering and Processing 36:143–147, 1997.

Haid M, Martin H, Müller-Steinhagen H. Heat transfer to liquid–solid fluidized beds. Chemical Engineering and Processing 33:211–225, 1994.

Haider A, Levenspiel O. Drag coefficient and terminal velocity of spherical and nonspherical particles. Powder Technology 58:63–70, 1989.

Ham JM, Thomas S, Guazzelli E, Homsy GM, Anselmet M-C. An experimental study of the stability of liquid-fluidized beds. Internat J Multiphase Flow 16:171–185, 1990.

Hancock RT. The teeter condition. Mining Magazine (London) 55:90–94, 1936.

Hancock RT. The law of motion of particles in a fluid. Trans Inst Min Engrs (London) 94:114–121, 1937/38.

Handley D, Doraisamy A, Butcher KL, Franklin NL. A study of the fluid and particle mechanics in liquid-fluidized beds. Trans Instn Chem Engrs 44:T260-T273, 1966.

Hanratty TJ, Latinen G, Wilhelm RH. Turbulent diffusion in particulately fluidized beds of particles. AIChE J 2:372–380, 1956.

Happel J. Viscous flow in multiparticle systems: slow motion of fluids relative to beds of spherical particles. AIChE J 4:197–201, 1958.

Happel J, Brenner H. Viscous flow in multiparticle systems: motion of spheres and a fluid in a cylindrical tube. AIChE J 3:506–513, 1957.

Happel J, Brenner H. Low Reynolds Number Hydrodynamics. Englewood Cliffs, NJ: Prentice-Hall, 1965.

Happel J, Epstein N. Viscous flow in multiparticle systems: cubic assemblages of uniform spheres. Ind Eng Chem 46:1187–1194, 1954.

Hartman M, Trnka D, Havlin V. A relationship to estimate the porosity in liquid–solid fluidized beds. Chem Eng Sci 47:3162–3166, 1992.

Hasimoto H. On the periodic fundamental solutions of the Stokes equations and their application to viscous flow past a cubic array of spheres. J Fluid Mech 5:317–328, 1959.

Hassett NJ. Flow patterns in particle beds. Nature 189:997–998, 1961a.

Hassett NJ. The mechanism of fluidization. Brit Chem Engng 19:777–780, 1961b.

Hatch LP, Weth GG, Wachtel SJ. Scale control in the high temperature distillation of saline waters by means of fluidized-bed heat exchangers. Desalination 1:156–164, 1966.

Hetzler R, Williams MC. Fluidized bed viscosity and expansion correlated with glass forming liquid model. Ind Eng Chem Fundam 8:668–677, 1969.

Heywood H. Uniform and non-uniform motion of particles in fluids. In: Proc Symp on Interaction Between Fluids and Particles. London: Instn Chem Engrs, 1962, pp 1–8.

Hiby JW. Particle forces in homogeneous beds. In: Drinkenburg AAH, ed. Proc Internat Conf on Fluidization. Amsterdam: Netherlands Univ. Press, 1967, pp 21–30.

Himsley A, Farkas EJ. Applications of fluidized bed ion exchange techniques to extractive hydrometallurgy. Presented at 27th Canadian Chemical Engineering Conference. Calgary, Alberta, 1977, Paper 10-2, pp 1–12.

Hiquily N, Couderc JP, Angelino H. Analyse des conditions de fonctionnement du distributeur à billes. Powder Technology 22:59–69, 1979.

Hoffman RF, Lapidus L, Elgin JC. The mechanics of vertical moving fluidized systems: IV. Application to batch-fluidized systems with mixed particle sizes. AIChE J 6:321–324, 1960.

Holman JP, Moore W, Wong VM. Particle-to-fluid heat transfer in water-fluidized systems. Ind Eng Chem Fundam 4:21–31, 1965.

Howley MA, Glasser BJ. Hydrodynamics of a binary-particle fluidized-bed: segregation and mixing. In: Glicksman LR. Fluidization and Fluid–Particle Systems, AIChE 1999 Annual Meeting, pp 152–158.

Jackson R. The mechanics of fluidized beds – Part I: The stability of the state of uniform fluidisation. Trans Instn Chem Engrs 41:13–21, 1963.

Jackson R. Hydrodynamic stability of fluid-particle systems. In: Davidson JF, Clift R, Harrison D, eds. Fluidization. 2d ed. London: Academic Press, 1985, pp 47–72.

Jackson R. Progress toward a mechanics of dense suspensions of solid particles. AIChE Symp Series 90(301):1–30, 1994.

Jamialahmadi M, Müller-Steinhagen H. Hydrodynamics and heat transfer of liquid fluidized bed systems. Chem Eng Communications 179:35–79, 2000.

Jamialahmadi M, Malayeri MR, Müller-Steinhagen H. Prediction of heat transfer to liquid–solid fluidized beds. Can J Chem Eng 73:444–455, 1995.

Jamialahmadi M, Malayeri MR, Müller-Steinhagen H. A unified correlation for the prediction of heat transfer coefficients in liquid/solid fluidized systems. Trans ASME J Heat Transfer 118:952–959, 1996.

Jamialahmadi M, Malayeri MR, Müller-Steinhagen H. Prediction of optimum operating conditions of liquid fluidized bed systems. Can J Chem Eng 75:327–332, 1997.

Jean R-H, Fan L-S. On the criteria of solids layer inversion in a liquid–solid fluidized bed containing a binary mixture of particles. Chem Eng Sci 41:2811–2821, 1986.

Jean R-H, Fan L-S. A fluid mechanic-based model for sedimentation and fluidization at low Reynolds number. Chem Eng Sci 44:353–362, 1989.

Jean R-H, Fan L-S. On the model equations of Gibilaro and Foscolo with corrected buoyancy force. Powder Technology 72:201–205, 1992.

Jean R-H, Fan L-S. Multiphase flow: liquid/solid fluidized bed systems. In: Johnson RW, ed. Handbook of Fluid Dynamics. Boca Raton, FL: CRC Press, 1998, pp 19-1–19-25.

Jira-Arune N, Laguerie C. Croissance des cristaux de chlorure de potassium en couche fluidisée: comparaison avec la dissolution. Chem Eng J 18:47–57, 1979.

Jolls KR, Hanratty TJ. Transition to turbulence for flow through a dumped bed of spheres. Chem Eng Sci 21:1185–1190, 1966.

Joshi JB. Solid-liquid fluidised beds: some design aspects. Chem Eng Res Des 61:143–161, 1983.

Jottrand R. An experimental study of the mechanism of fluidization. J Appl Chem (suppl. issue 1) 2:17–26, 1952.

Jottrand R, Grunchard F. Transfert de matière dans les lits fluidisés liquides. In: Symp on Interaction Between Fluids and Particles. London: Instn Chem Engrs, 1962, pp 211–216.

Juma AKA, Richardson JF. Particle segregation in liquid–solid fluidised beds. Chem Eng Sci 34:137–143, 1979.

Juma AKA, Richardson JF. Segregation and mixing in liquid fluidized beds. Chem. Eng. Sci. 38:955–967, 1983.

Kang Y, Kim SD. Radial dispersion characteristics of two- and three-phase fluidized beds. Ind Eng Chem Process Des Dev 25: 717–722, 1986.

Kang Y, Kim SD. Solid flow transition in liquid and three-phase fluidized beds. Particulate Science and Technology 6:133–144, 1988.

Kang Y, Nah JB, Min BT, Kim SD. Dispersion and fluctuation of fluidized particles in a liquid–solid fluidized bed. Chem Eng Comm 97:197–208, 1990.

Karamanev DG, Nikolov LN. Influence of some physico-chemical parameters on bacterial activity of biofilm: ferrous iron oxidation by *thiobacillus ferrooxidans*. Biotechology and Bioengineering 31:295–299, 1988.

Karamanev DG, Nikolov LN. Free rising spheres do not obey Newton's law for free settling. AIChE J 38:1843–1846, 1992a.

Karamanev DG, Nikolov LN. Bed expansion of liquid–solid inverse fluidization. AIChE J 38:1916–1922, 1992b.

Karamanev DG, Nikolov LN. Application of inverse fluidization in wastewater treatment: from laboratory to full-scale bioreactors. Environmental Progress 15(3):194–196, 1996.

Karel SF, Libicki SB, Robertson CR. The immobilization of whole cells: engineering principles. Chem Eng Sci 40:1321–1354, 1985.

Kato Y, Uchida K, Kago T, Morooka S. Liquid hold-up and heat transfer coefficient between bed and wall in liquid-

solid and gas–liquid–solid fluidized beds. Powder Technology 28:173–179, 1981.

Kaye BH, Boardman RP. Cluster formation in dilute suspensions. In: Proc Symp on Interaction Between Fluids and Particles. London: Instn Chem Engrs, 1962, pp 17–21.

Kehlenbeck R, Di Felice R. Empirical relationships for the terminal settling velocity of spheres in cylindrical columns. Chem Eng Technol 21:303–308, 1999.

Kempling JC, Eng J. Performance of dual-media filters—2. Chem Eng Prog 73(4):87–91, 1977.

Kennedy SC, Bretton RH. Axial dispersion of spheres fluidized with liquids. AIChE J 12:24–30, 1966.

Khan AR, Richardson JF. Fluid-particle interactions and flow characteristics of fluidized beds and settling suspensions of spherical particles. Chem Eng Commun 78:111–130, 1989.

Khan AR, Richardson JF. Pressure gradient and friction factor for sedimentation and fluidisation of uniform spheres in liquids. Chem Eng Sci 45:255–265, 1990.

Kikuchi K, Sugawara T, Ohashi H. Correlation of mass transfer coefficient between particles and liquid in liquid fluidized beds, J Chem Eng Japan 16:426–428, 1983.

Kikuchi K, Konno H, Kakutani S, Sugawara T, Ohashi H. Axial dispersion of liquid in liquid fluidized beds in the low Reynolds number region. J Chem Eng Japan 17:362–367, 1984.

Kim N-H, Lee Y-P. A study on the pressure loss, heat transfer enhancement and fouling control in liquid fluidized bed heat exchangers. In: Panchal CB, ed. Fouling Mitigation of Industrial Heat-Exchange Equipment. New York: Begell House, 1997, pp 421–433.

King DH, Smith JW. Wall mass transfer in liquid-fluidized beds. Can J Chem Eng 45:329–333, 1967.

Klaren DG, Bailie RE. Fluid bed heat exchangers solve fouling problems. Amer Papermaker, Pacific Ed., Feb 1988, pp 32–33.

Klaren DG, Bailie RE. Consider nonfouling fluidized bed exchangers. Hydrocarbon Processing 68(7):48–50, 1989.

Klaren DG, Halberg N. Development of a multistage flash/fluidized bed evaporator. Chem Eng Prog 43(7):41–43, 1980.

Klaren DG, Sullivan DW. Nonfouling heat exchanger performance in severe fouling services: principle, industrial applications and operating installations. In: Preprints, Internat Conf Petroleum Phase Behavior and Fouling. Houston: AIChE Spring Meeting, March 1999, pp 359–366.

Kmiec A. Particle distributions and dynamics of particle movement in solid–liquid fluidized beds. Chem Eng J 15:1–12, 1978.

Kolár V. Fluidization of solid particles by liquid in conical vessels. Collection Czechoslov Chem Commun 28:1224–1231, 1963.

Kollbach J, Dahm W, Rautenbach R. Continuous cleaning of heat exchanger with recirculating fluidized bed. Heat Transfer Engineering 8(4):26–32, 1987.

Koloini T, Farkas EJ. Fixed bed pressure drop and liquid fluidization in tapered or conical vessels. Can J Chem Eng 51:499–502, 1973.

Koloini T, Zumer M. Ion exchange with irreversible reaction in deep fluidized beds. Can J Chem Eng 57:183–190, 1979.

Kramers H, Westermann MD, de Groot JH, Dupont FAA. The longitudinal dispersion of liquid in a fluidized bed. In: Proc Symp on Interaction Between Fluids and Particles. London: Instn Chem Engrs, 1962, pp 114–119.

Krishnaswamy PR, Ganapathy R, Shemilt LW. Correlating parameters for axial dispersion in liquid fluidized systems. Can J Chem Eng 56:550–553, 1978.

Kuwabara S. The forces experienced by randomly distributed parallel circular cylinders or spheres in a viscous flow at small Reynolds numbers. J Phys Soc Japan 14:527–532, 1959.

Kwauk M. Particulate fluidization: an overview. Adv Chem Eng 17:207–360, 1991–92.

Kwauk M. Fluidization: Idealized and Bubbleless, with Applications. New York: Science Press and Ellis Horwood, 1992.

Kwauk M. Comment. Chem Eng Sci 51(24): iii–iv, 1996.

Ladenburg R. On the influence of the walls on the motion of a sphere in a viscous fluid. Annalen der Physik 23:447–458, 1907.

Laguerie C, Angelino H. Comparaison entre la dissolutien et la croissance de cristaux d'acide citrique monohydrate en lit fluidisé. Chem Eng J 10:41–48, 1975.

Lapidus L, Elgin JC. Mechanics of vertical-moving fluidized systems. AIChE J 3: 63–68, 1957.

Latif BAJ, Richardson JF. Circulation patterns and velocity distributions for particles in a liquid fluidised bed. Chem Eng Sci 27:1933–1949, 1972.

Law H-S, Masliyah JH, MacTaggart RS, Nandakumar K. Gravity separation of bidisperse suspensions: light and heavy particle species. Chem Eng Sci 42:1527–1538, 1987.

LeClair BP, Hamielec AE. Viscous flow through particle assemblages at intermediate Reynolds numbers: steady state solutions for flow through assemblages of spheres. Ind Eng Chem Fundam 7:542–549, 1968.

Lee J-K, Chun H-S, Shemilt LW. Wall-to liquid mass transfer in liquid–solid and gas–liquid–solid fluidized beds. J Chem Eng Japan 30:246–252, 1997.

Le Goff P, Vergnes F, Coeuret F, Bordet J. Applications of fluidized beds in electrochemistry. Ind Eng Chem 61(10):8–17 (1969).

Letan R, Elgin JC. Fluid mixing in particulate fluidized beds. Chem Eng J 3:136–144, 1972.

Leva M. Fluidization. New York: McGraw-Hill, 1959.

Lewis EW, Bowerman EW. Fluidization of solid particles in liquids. Chem Eng Prog 48:603–610, 1952.

Lewis WK, Gilliland ER, Bauer WC. Characteristics of fluidized particles. Ind Eng Chem 41:1104–1117, 1949.

Liang W-G, Zhu J-X. A core–annulus model for the radial flow structure in a liquid–solid circulating fluidized bed (LSCFB). Chem Eng J 68:51–62, 1997.

Liang W-G, Zhu J-X, Jin Y, Yu Z-Q, Wang Z-W, Zhou J. Radial nonuniformity of flow structure in a liquid–solid circulating fluidized bed. Chem Eng Sci 51:2001–2010, 1996.

Liang W, Zhang S, Zhu J-X, Jin Y, Yu Z, Wang Z. Flow characteristics of the liquid–solid circulating fluidized bed. Powder Technology 90:95–102, 1997.

Limas-Ballesteros R. Fluidisation de particules non sphériques. Transfert de matière entre le solide et le liquide. Thèse de Docteur Ingénieur, I.N.P. Toulouse, France, 1980.

Limas-Ballesteros R, Riba JP, Couderc JP. Expansion de couches de particules non sphériques fluidisées par un liquide. Entropie 106:37–45, 1982a.

Limas-Ballesteros R, Riba JP, Couderc JP. Dissolution of non-spherical particles in solid–liquid fluidization. Chem Eng Sci 37:1639–1644, 1982b.

Lin T-J, Fan L-S. Characteristics of high pressure liquid–solid fluidization. AIChE J 43:45–57, 1997.

Lin CS, Denton EB, Gaskill HS, Putnam GL. Diffusion-controlled electrode reactions. Ind Eng Chem 43:2136–2143, 1951.

Lockett MJ, Al-Habbooby HM. Differential settling by size of two particle species in a liquid. Trans Instn Chem Engrs 51:281–292, 1973.

Lockett MJ, Al-Habbooby HM. Relative particle velocities in two-species settling. Powder Technology 10:67–71, 1974.

Lopez-Cacicedo CL. The electrolytic recovery of metals from diluent effluent streams. J Separ Proc Technol 2(1):34–39, 1981.

Lucas A, Arnaldos J, Casal J, Puigjaner L. Improved equation for the calculation of minimum fluidization velocity. Ind Eng Chem Process Des Dev 25:426–429, 1986.

Martin BLA, Kolar Z, Wesselingh JA. The falling velocity of a sphere in a swarm of different spheres. Trans IChemE 59:100–104, 1981.

Maruyama T, Maeda H, Mizushina T. Liquid fluidization in tapered vessels. J Chem Eng Japan 17:132–139, 1984.

Masliyah JH. Hindered settling in a multi-species particle system. Chem Eng Sci 34:1166–1168, 1979.

Masliyah JH. Effect of stirring on a liquid fluidized bed. AOSTRA J Res 5:49–60, 1989.

Masliyah JH. Electrokinetic Transport Phenomena. Edmonton: AOSTRA, 1994, pp 165–171.

Masliyah JH, Polikar M. Terminal velocity of porous spheres. Can J Chem Eng 58:299–302, 1980.

Mathur KB, Epstein N. Spouted Beds. New York: Academic Press, 1974.

Matsumoto K, Suganuma A. Settling velocity of a permeable model floc. Chem Eng Sci 32:445–447, 1977.

McCune LK, Wilhelm RH. Mass and momentum transfer in a solid–liquid system. Ind Eng Chem 41:1124–1134, 1949.

Medlin J, Jackson R. Fluid mechanical description of fluidized beds. The effect of distributor thickness on convective instabilities. Ind Eng Chem Fundam 14:315–321, 1975.

Medlin J, Wong H-W, Jackson R. Fluid mechanical description of fluidized beds. Convective instabilities in bounded beds. Ind Eng Chem Fundam 13:247–259, 1974.

Mehta SC, Shemilt LW. Frequency response of liquid fluidized systems. Part II: Effect of liquid viscosity. Can J Chem Eng 54:43–51, 1976.

Meijer JAM, Van Rosmalen GM, Veenman AW, Van Dissel CM. Scale inhibition of calcium sulfate by a fluidized bed. Desalination 34:217–232, 1980.

Mertes TS, Rhodes HB. Liquid–particle behavior. Chem Eng Prog 51:429–432, 517–522, 1955.

Michaels AS, Bolger JC. Settling rates and sediment volumes of flocculated kaolin suspensions. Ind Eng Chem Fundamentals 54:24–33, 1962.

Mirza S, Richardson JF. Sedimentation of suspensions of particles of two or more sizes. Chem Eng Sci 34:447–454, 1979.

Mitra AK, Epstein N. Effect of solids wettability on liquid fluidization. Can J Chem Eng 56:520–522, 1978.

Miura H, Kawase Y. Hydrodynamics and mass transfer in three-phase fluidized beds with non-Newtonian fluids. Chem Eng Sci 52:4095–4104, 1997.

Miura H, Kawase Y. Minimum liquid fluidization velocity in two- and three-phase fluidized beds with non-Newtonian fluids. Powder Technology 97:124–128, 1998.

Molerus O, Wirth K-E. Heat Transfer in Fluidized Beds. 1st ed. London: Chapman and Hall, 1997, p 167.

Moritomi H, Iwase T, Chiba T. A comprehensive interpretation of solid layer inversion in liquid fluidised beds. Chem Eng Sci 37:1751–1757, 1982.

Moritomi H, Yamagishi T, Chiba T. Prediction of complete mixing of liquid-fluidized binary solid particles. Chem Eng Sci 41:297–305, 1986.

Mostoufi N, Chaouki J. Prediction of effective drag coefficient in fluidized beds. Chem Eng Sci 54: 851–858, 1999.

Müller-Steinhagen H, Jamialahmadi M, Robson B. Mitigation of scale formation during the Bayer process—measurements at the Alcoa of Australia refinery in Kwinana. In: Mannweiler U, ed. Light Metals 1994. Minerals, Metals and Materials Society, 1994, pp 121–127.

Mullin JW. Crystallization. 3d ed. Oxford: Butterworth-Heinemann, 1993.

Mullin JW, Garside J. The crystallization of aluminium potassium sulphate: a study in the assessment of crystallizer design data. Part II: Growth in a fluidized bed crystallizer. Trans Instn Chem Engrs 45:T291–T295, 1967.

Mullin JW, Nyvlt J. Design of classifying crystallizer. Trans Instn Chem Engrs 48:T7–T14, 1970.

Munroe HS. The English vs. the continental system of jigging—is close sizing advantageous? Trans Am Inst Min Engrs 17:637–659, 1888/89.

Nakamura K, Kumagai H, Yano T. Performance of a zig-zag fluidized bed as an immobilized enzyme reactor. In:

Linko P, Larinkari J, eds. Food Process Engineering, Vol 2. Enzyme Engineering in Food Processing. London: Applied Science, 1979, pp 186–191.

Nakamura M, Hamada Y, Toyama S, Fouda AE, Capes CE. An experimental investigation of minimum fluidization velocity at elevated temperatures and pressures. Can J Chem Eng 63: 8–13, 1985.

Nasr-El-Din H, Masliyah JH, Nandakumar K, Law D H-S. Continuous gravity separation of a bidisperse suspension in a vertical column. Chem Eng Sci 43:3225–3234, 1988.

Nasr-El-Din H, Masliyah JH, Nandakumar K. Continuous gravity separation of concentrated bidisperse suspension in a vertical column. Chem Eng Sci 45: 849–857, 1990.

Neale G, Epstein N, Nader W. Creeping flow relative to permeable spheres. Chem Eng Sci 28:1865–1874, 1973.

Needham LW, Lynch S. The use of suspensions as heavy liquids. Trans Instn Chem Engrs 23:93–105, 1945.

Nelson PA, Galloway RT. Particle-to-fluid heat and mass transfer in dense systems of fine particles. Chem Eng Sci 30:1–6, 1975.

Neužil L. Fluidization of polydisperse materials. I. Theoretical considerations on the study of the segregation of a mixture of particles of equal density. Collection Czechoslov Chem Commun 29:571–578, 1964.

Neužil L, Hrdina M. Effect of walls on velocity profile of fluid in a fluidized bed. Collection Czechoslov Chem Commun 30:3063–3070, 1965.

Nikolov L, Karamanev D. Experimental study of the inverse fluidized bed biofilm reactor. Can J Chem Eng 65:214–217, 1987.

Nikolov LN, Karamanev, DG. The inverse fluidized bed biofilm reactor: a new laboratory scale apparatus for biofilm research. J. Fermentation and Bioengineering 69(4):265–267, 1990.

Nikov I, Delmas H. Solid–liquid mass transfer in three-phase fixed and fluidized beds. Chem Eng Sci 42:1089–1093, 1987.

Nikov I, Karamanev D. Liquid–solid mass transfer in inverse fluidized bed. AIChE J 37:781–784, 1991.

Nikov I, Karamanev D. Reply to letter. AIChE J 38:638–639, 1992.

Niven RK, Khalili, N. In situ fluidisation by a single internal vertical jet. J Hydraulic Res 36(2):199–228, 1998a.

Niven RK, Khalili N. In situ multiphase fluidization ("upflow washing") for the remediation of hydrocarbon contaminated sands. Can Geotech J 35:938–960, 1998b.

Obata E, Watanabe H, Endo N. Measurement of size and size distribution of particles by fluidization. J Chem Eng Japan 15:23–28, 1982.

Oman AO, Watson KM. Pressure drops in granular beds. National Petroleum News 36:R-795–801, Nov. 1, 1944.

Page RE. Some aspects of three-phase fluidization. Ph.D. thesis, Cambridge University, 1970, p 138.

Panier P, Riba JP, Couderc JP. Étude de l'extraction liquide-solide en couche fluidisée par voie electrochimique. Chem Eng J 20:157–165, 1980.

Parkin GF, Speece RE. Anaerobic biological waste treatment. Chem Eng Prog 80(12):55–58, 1984.

Patwardhan VS, Tien C. Distribution of solid particles in liquid fluidized beds. Can J Chem Eng 62:46–54, 1984.

Patwardhan VS, Tien C. Sedimentation and liquid fluidization of solid particles of different sizes and densities. Chem Eng Sci 40:1051–1060, 1985.

Peng Y, Fan LT. Hydrodynamic characteristics of fluidization in liquid–solid tapered beds. Chem Eng Sci 52:2277–2290, 1997.

Perry RH, Green DW, Maloney JO. Perry's Chemical Engineers' Handbook. 6th ed. New York: McGraw-Hill, 1984, pp 19.37–19.38.

Phillips VR, Epstein N. Growth of nickel sulfate in a laboratory-scale fluidized-bed crystallizer. AIChE J 20:678–687, 1974.

Pirozzi D, Gianfreda L, Greco G (Jr). Development of a circulating fluidized bed fermentor: the hydrodynamic model for the system. AIChE Symp Series 85(270):101–110, 1989.

Price BG, Lapidus L, Elgin JC. Mechanics of vertical moving fluidized systems II. Application to countercurrent operation. AIChE J 5:93–97, 1959.

Pruden BB, Epstein N. Stratification by size in particulate fluidization and in hindered settling. Chem Eng Sci 19:696–700, 1964.

Prudhoe J, Whitmore RL. Terminal velocity of spheres in fluidized beds. Brit Chem Eng 9:371–375, 1964.

Qian D, Wang Z, Chen G. On the inversion phenomena in a liquid–solid fluidized bed. Chemical Reaction Engineering and Technology 9:485–491, 1993.

Quinn JA, Lapidus L, Elgin JC. The mechanics of moving vertical fluidized systems V. Concurrent cogravity flow. AIChE J 7:261–263, 1961.

Rahman K, Streat M. Mass transfer in liquid fluidized beds of ion exchange particles. Chem Eng Sci 36:293–300, 1981.

Rapagnà S, Di Felice R, Gibilaro LG, Foscolo PU. Steady-state expansion characteristics of monosize spheres fluidised by liquids. Chem Eng Commun 79:131–140, 1989.

Rautenbach R, Kollbach J. New developments in fluidized bed heat transfer for preventing fouling. Swiss Chem 8(5):47–55, 1986.

Riba JP, Couderc JP. Expansion de couches fluidisées par des liquides. Can J Chem Eng 55:118–121, 1977.

Riba JP, Couderc JP. Transfer de matière autour d'une sphère immergée dans une couche fluidisée par un liquide. Intern J Heat Mass Transfer 23:909–917, 1980.

Riba JP, Routie R, Couderc JP. Conditions minimales de mise en fluidisation par un liquide. Can J Chem Eng 56:26–30, 1978.

Richards RH. Close sizing before jigging. Trans Am Inst Min Engrs 24:409–486, 1893.

Richards RH, Locke CE. Textbook of Ore Dressing. 3d ed. New York: McGraw-Hill, 1940, pp 136–143, 154–157.

Richardson JF. Incipient fluidization and particulate systems. In: Davidson JF, Harrison D, eds. Fluidization. London: Academic Press, 1971, pp 25–64.

Richardson JF, Afiatin E. Fluidisation and sedimentation of binary mixtures of particles in a liquid. Proc 9th International Conference on Transport and Sedimentation of Solid Particles. Cracow, Poland: Zeszyty Naukowe Akademii Rolniczej we Wroclawiu Nr 315, 1997, pp 487–502.

Richardson JF, Meikle RA. Sedimentation and fluidisation – Part III: The sedimentation of uniform fine particles and two-component mixtures of solids. Trans Instn Chem Engrs 39:348–356, 1961a.

Richardson JF, Meikle RA. Sedimentation and fluidisation – Part IV: Drag force on individual particles in an assemblage. Trans Instn Chem Engrs 39:357–362, 1961b.

Richardson JF, Shabi FA. The determination of concentration distribution in a sedimenting suspension using radioactive solids. Trans Instn Chem Engrs 38:33–42, 1960.

Richardson JF, Zaki WN. Sedimentation and fluidisation: Part I. Trans Instn Chem Engrs 32:35–53, 1954.

Richardson JF, Romani MN, Shakiri KJ. Heat transfer from immersed surfaces in liquid fluidized beds. Chem Eng Sci 31:619–624, 1976.

Rigby GR, Van Blockland GP, Park WH, Capes CE. Properties of bubbles in three-phase fluidized beds as measured by an electroresistivity probe. Chem Eng Sci 25:1729–1741, 1970.

Rowe PN. Drag forces in a hydraulic model of a fluidised bed—Part II. Trans Instn Chem Engrs 39:175–180, 1961.

Rowe PN. A convenient empirical equation for estimation of the Richardson–Zaki exponent. Chem Eng Sci 42:2795–2796, 1987.

Rowe PN, Claxton KT. Heat and mass transfer from a single sphere to fluid flowing through an array. Trans Instn Chem Engrs 43:T321–T331, 1965.

Rowe PN, Henwood GA. Drag forces in a hydraulic model of a fluidised bed—Part I. Trans Instn Chem Engrs 39:43–54, 1961.

Saffman PG. On the settling speed of free and fixed suspensions. Studies in Appl. Math L11(2):115–127, June 1973.

Salas-Morales JC, Evans JW, Newman OMG, Adcock PA. Spouted bed electrowinning of zinc: Part 1. Laboratory-scale electrowinning experiments. Metall Mat Trans B 28:59–68, 1997.

Saxena SC, Vogel GJ. The measurement of incipient fluidisation velocities in a bed of coarse dolomite at temperature and pressure. Trans IChemE 55:184–189, 1977.

Saxton JA, Fitton JB, Vermeulen T. Cell model theory of homogeneous fluidization: density and viscosity behaviour. AIChE J 16:120–130, 1970.

Scarlett B, Blogg MJ. The motion and distribution of particles in a liquid-fluidized bed. In: Drinkenburg AAH, ed. Proc Internat Conf on Fluidization. Amsterdam: Netherlands Univ. Press, 1967, pp 82–89.

Schiller L, Naumann A. Über die grundlegenden berechnungen bei der schwerkragtaufbereitung. Z Ver Deutsch Ing 77:318–320, 1933.

Schmidt S, Buchs J, Born C, Biselli M. A new correlation for the wall-to-fluid mass transfer in liquid–solid fluidized beds. Chem Eng Sci 54: 829–839, 1999.

Scott CD, Hancher CW, Shumate II SE. A tapered fluidized bed as a bioreactor. Enzyme Eng 3:255–261, 1978.

Scott TC, Cosgrove JM, Asif M, Petersen JN. Hydrodynamic studies of an advanced fluidized-bed bioreactor for direct interaction with coal. Fuel 72:1701–1704, 1993.

Selim MS, Kothari AC, Turian RM. Sedimentation of binary suspensions. AIChE Symp Series (Fluidization and Fluid Particle Systems: Theories and Applications, Knowlton TM, ed.) 79(222):103–108, 1983a.

Selim MS, Kothari AC, Turian RM. Sedimentation of multisized particles in concentrated suspensions. AIChE J 29:1029–1038, 1983b.

Selke WA, Bliss H. Continuous countercurrent ion exchange. Chem Eng Prog 47:529–533, 1951.

Shiau L-D, Cheung S-H, Liu Y-C. Modelling of a fluidizedbed crystallizer in a batch mode. Chem Eng Sci 54: 865–871, 1999.

Sibony MJ. Clarification with microsand seeding. A state of the art. Water Res 15:1281–1290, 1981.

Slater MJ. A review of continuous counter-current contactors for liquids and particulate solids. Brit Chem Eng 14:41–46, 1969.

Slater MJ. Recent industrial-scale applications of continuous resin ion exchange systems. J Separ Proc Tech 2(3):2–12, 1981.

Slater MJ. The relative sizes of fixed bed and continuous counter-current flow ion exchange equipment. Trans IChemE 60:54–58, 1982.

Slater MJ, Lucas BH. Flow patterns and mass transfer rates in fluidized bed ion exchange equipment. Can J Chem Eng 54:264–268, 1976.

Slis PL, Willemse Th W, Kramers H. The response of the level of a liquid fluidized bed to a sudden change in the fluidizing velocity. Appl Sci Res A8:209–218, 1959.

Smith TN. A model of settling velocity. Chem Eng Sci 53:315–323, 1998.

Sozen ZZ, Grace JR, Pinder KL. Thermal energy storage by agitated capsules of phase change material. Ind Eng Chem Res 27:679–691, 1988.

Steinour HH. Rate of sedimentation: suspensions of uniform-size angular particles. Ind Eng Chem 36: 840–847, 1944.

Storck A, Coeuret F. Mass transfer between a flowing liquid and a wall or an immersed surface in fixed and fluidized beds. Chem Eng J 20:149–156, 1980.

Streat M. Recent developments in continuous ion exchange. J Separ Proc Technol 1(3):10–18, 1980.

Struve DL, Lapidus L, Elgin JC. The mechanics of moving vertical fluidized systems III. Application to

concurrent countergravity flow. Can J Chem Eng 36:141–152, 1958.

Svanoe H. "Krystal" classifying crystallizer. Ind Eng Chem 32:636–639, 1940.

Svarovsky L. Solid–Liquid Separation. 3d ed. London: Butterworth, 1990, p 130.

Syamlal M, O'Brien TJ. Simulation of granular layer inversion in liquid fluidized beds. Chem Eng Sci 14:473–481, 1988.

Tackie E, Bowen BD, Epstein, N. Hindered settling of uncharged and charged submicrometer spheres. In: Pfeffer R, ed. Fourth International Conference on Physicochemical Hydrodynamics. Annals of the New York Academy of Sciences 104:366–367, 1983.

Taggart AF. Handbook of Mineral Dressing, Vol. I. New York: John Wiley, 1953, pp 8–47.

Tai CY, Chen C-Y, Wu J-F. Crystal dissolution and growth in a lean fluidized-bed crystallizer. Chem Eng Commun 56:329–340, 1987.

Tai CY, Chien W-C, Chen C-Y. Crystal growth kinetics of calcite in a dense fluidized-bed crystallizer. AIChE J 45:1605–1614, 1999.

Tang W-T, Fan L-S. Axial liquid mixing in liquid–solid and gas–liquid–solid fluidized beds containing low density particles. Chem Eng Sci 45:543–551, 1990.

Tannous K, Hemati M, Laguerie C. Charactéristiques au minimum de fluidisation et expansion des couches fluidisées de particules de la catégorie D de Geldart. Powder Technology 80:55–72, 1994.

Thomas DG. Transport characteristics of suspensions: VIII. A note on the viscosity of Newtonian suspensions of uniform spherical particles. J Colloid Sci 20:267–277, 1965.

Thomas CR, Yates JG. Expansion index for biological fluidised beds. Chem Eng Res Des 63:67–70, 1985.

Thonglimp V, Hiquily N, Laguerie C. Vitesse minimale de fluidisation et expansion des couches fluidisées par un gaz. Powder Technology 38:233–253 (1984).

Toor HL, Marchello JM. Film-penetration model for heat and mass transfer. AIChE J 4:97–101, 1958.

Tournié P, Laguerie C, Couderc JP. Correlations for mass transfer between fluidized spheres and a liquid. Chem Eng Sci 34:1247–1255, 1979.

Trawinski H. Effective zahigkeit und inhomogenität von wirbelschichten. Chemie-Ing.-Tech. 25:229–238, 1953.

Treybal RE. Mass-Transfer Operations. 3d ed. New York: McGraw-Hill, 1980.

Trupp AC. Dynamics of liquid fluidised beds of spheres. I Chem E Symp Ser 30:182–189, 1968.

Turton R, Clark NN. An explicit relationship to predict spherical particle terminal velocity. Powder Technology 53:127–129, 1987.

Vanadurongwan V, Laguerie C, Couderc JP. Influence des propriétés physiques sur le transfert de matière en fluidisation liquide. Chem Eng J 12:29–31, 1976.

Van der Heiden G, Raats MS, Boon HF. Fluidised bed electrolysis for removal or recovery of metals from dilute solutions. Chemistry and Industry 13:465–468, 1978.

Van der Meer AP, Blanchard CMRJP, Wesselingh JA. Mixing of particles in liquid fluidized beds. Chem Eng Res Des 62:214–222, 1984.

Van der Wielen LAM, Van Dam MHH, Luyben KChAM. On the relative motion of a particle in a swarm of different particles. Chem Eng Sci 51:995–1008, 1996.

Van Duijn G, Rietema K. Segregation of liquid-fluidized beds. Chem Eng Sci 37:727–733, 1982.

Veenman AW. The MSF/FBE: An improved multi-stage flash distillation process. Desalination 19:1–14, 1976.

Veenman AW. Heat transfer in a multi-stage flash/fluidized bed evaporator (MSF/FBE). Desalination 22:55–76, 1977.

Verschoor H. Experimental data on the viscous force exerted by a flowing fluid on a dense swarm of particles. Appl Sci Res A2:155–161, 1951.

Volpicelli G, Massimilla L, Zenz FA. Nonhomogeneities in solid–liquid fluidization. Chem Eng Prog Symp Ser 62(67):42–50, 1966.

Wakeman RJ, Stopp BW. Fluidisation and segregation of binary particle mixtures. Powder Technology 13:261–268, 1976.

Wallis GB. One-Dimensional Two-Phase Flow. New York: McGraw-Hill, 1969.

Wasmund B, Smith JW. Wall to fluid heat transfer in liquid fluidized beds, Part 2. Can J Chem Eng 45:156–165, 1967.

Webb C, Black GM, Atkinson B. Liquid fluidisation of highly porous particles. Chem Eng Res Des 61:125–134, 1983.

Webster GH, Perona JJ. Liquid mixing in a tapered fluidized bed. AIChE J 34:1398–1402, 1988.

Webster GH, Perona JJ. The effect of taper angle on the hydrodynamics of a tapered liquid–solid fluidized bed. AIChE Symp Ser 86(276):104–112, 1990.

Wen CY, Fan L-S. Axial dispersion of non-Newtonian liquids in fluidized beds. Chem Eng Sci 28:1768–1772, 1973.

Wen CH, Yu YH. Mechanics of fluidization. Chem Eng Prog Symp Ser 62(62):100–111, 1966.

Whitmore RL. The relationship of the viscosity to the settling rate of slurries. J Inst Fuel 30:238–242, 1957.

Wilhelm RH, Kwauk M. Fluidization of solid particles. Chem Eng Prog 44(3):201–218, 1948.

Wright PC, Raper JA. A review of some parameters involved in fluidized bed bioreactors. Chem Eng Technol 19:50–64, 1996.

Yang J, Renken A. Intensification of mass transfer in liquid fluidized beds with inert particles. Chemical Engineering and Processing 37:537–544, 1998.

Yu H, Shi Y. Characteristics of fluidization of binary particle mixtures. In: Kwauk M, Kunii D, Jiansheng Z, Hasatani M, eds. Fluidization '85 Science and Technology Conference Papers. Second China-Japan Symposium. Amsterdam: Elsevier, 1985, pp 250–261.

Yutani N, Fan LT. Mixing of randomly moving particles in liquid–solid fluidized beds. Powder Technology 42:145–152, 1985.

Yutani N, Ototake N, Too JR, Fan LT. Estimation of the particle diffusivity in a liquid–solids fluidized bed based on a stochastic mode. Chem Eng Sci 37:1079–1085, 1982.

Zheng Y, Zhu J-X, Wen J, Martin SA, Bassi AS, Margaritis A. The axial hydrodynamic behavior in a liquid–solid circulating fluidized bed. Can J Chem Eng 77:284–290, 1999.

Zhu J-X, Zheng Y, Karamanev DG, Bassi AS. (Gas–)liquid–solid circulating fluidized beds and their potential applications to bioreactor engineering. Can J Chem Eng 78:82–94, 2000.

Manuscript received February 28, 2000.

27

Gas–Liquid–Solid Three-Phase Fluidization

Liang-Shih Fan and Guoqiang Yang

The Ohio State University, Columbus, Ohio, U.S.A.

1 INTRODUCTION

Gas–liquid–solid fluidization systems have been applied extensively in industry for physical (e.g., sand filter cleaning and granular material drying), chemical (e.g., hydrogen peroxide production and methanol synthesis), petrochemical (e.g., resid hydrotreating and hydrotreating of tar sands), and biochemical (e.g., treatment of lactose wastewater and bioleaching of metals from ores) processing (Shah, 1979; L'Homme, 1979; Ramachandran and Chaudhari, 1983; Fan, 1989). The interest in the application of three-phase fluidization systems has promoted continued research and development efforts in these systems. Examples are biological operation for human viral vaccine production using microcarrier cultures of animal cells (Kalogerakis and Behie, 1995), ethanol fermentation using immobilized cells in a multistage fluidized bed bioreactor (Tzeng et al., 1991), cleaning and desulfurization of high sulfur coal by selective flocculation and bioleaching (Attia et al., 1991), hydrocarbon production in fossil fuel processing using mild hydrocracking of shale oil (Souza et al., 1992), production of paraffin waxes using the SASOL slurry bed process (slurry bubble column) for Fischer–Tropsch synthesis (Inga, 1994), and production of light hydrocarbon diesel using the Texaco T-Star process (three-phase fluidized bed) for vacuum gas oil hydrotreating/hydrocracking (Johns et al., 1993). Most of these processes with considerable commercial interest are conducted under high pressure and high temperature, for example, methanol synthesis (at $P = 5.5\,\text{MPa}$ and $T = 2\ 60°C$), resid hydrotreating (at $P = 5.5–21\,\text{MPa}$ and $T = 300–425°C$), Fischer–Tropsch synthesis (at $P = 1.5–5.0\,\text{MPa}$ and $T = 250°C$), and benzene hydrogenation (at $P = 5.0\,\text{MPa}$ and $T = 180°C$) (Fox, 1990; Jager and Espinoza, 1995; Saxena, 1995; Mill et al., 1996; Peng et al., 1999). Fundamental study of flow characteristics under high-pressure and high-temperature conditions is crucial for the development of these industrial processes, and much progress has been made recently regarding high-pressure and high-temperature systems with relevance to industrial processes.

The fundamental characteristics of three-phase fluidization including bubble characteristics, hydrodynamics, and heat and mass transfer properties along with many industrial processes have been extensively reported in Fan's (1989) book as well as its companion book on bubble wake dynamics (Fan and Tsuchiya, 1990). As both books are widely referenced in the field of three-phase fluidization, this chapter is presented mainly as an update to these two books. The chapter will cover the continued research progress made over the past ten years on the fundamentals of three-phase fluidization. Major findings on fluidization and bubble dynamics under ambient conditions and the relevant literature reported earlier will be covered. Furthermore, new research on the high-pressure and high-temperature three-phase fluidization will be highlighted as well as computational fluid dynamics.

2 BUBBLE DYNAMICS

In gas–liquid–solid fluidization systems, bubble dynamics plays a key role in dictating the transport phenomena and ultimately affects the overall rates of reactions. It has been recognized that the bubble wake, when it is present, is the dominant factor governing the system hydrodynamics. In general, consideration of the flow associated with the bubble wake near the bubble rear, whether laminar or turbulent, is essential to characterize the complete behavior of the rising bubble including its motion. Conversely, examining the shape, rise velocity, and motion of a bubble can provide an indirect understanding of the dynamics of the liquid–solid flow around the bubble. Temperature and pressure are the key operating variables dictating the physical properties of the gas and liquid phases and hence the bubble behavior. The liquid–solid suspension through which the gas bubbles rise may be characterized as a pseudohomogeneous medium under certain operating conditions. The most marked effects are flow instability induced by bubble wake, and the resulting large-scale vortical motion and local solids concentration gradient. Interactions between neighboring bubbles are usually recognized in terms of bubble coalescence and/or breakup. Direct causes of these phenomena, however, lie in the interactions between the bubbles and the surrounding flow.

In the following, several fundamental aspects of the dynamics of single and multibubbles in liquid–solid suspensions are discussed, which are of paramount importance to the transport processes in the three-phase fluidization system. Specifically, four subjects are covered: (1) plenum bubble behavior, (2) bubble rise characteristics, (3) bubble coalescence, and (4) bubble breakup. The effects of pressure and temperature on these phenomena are illustrated.

2.1 Plenum Bubble Behavior

Bubble behavior in the plenum region is important for understanding bubble characteristics in multiphase fluidization systems, particularly under low gas velocities. In this section, the behavior of bubble formation from a single orifice will be emphasized.

There are two typical mechanical arrangements for bubble formation from a single orifice, that is, the orifice connected or not connected to a gas chamber, as shown in Fig. 1. For bubble formation from a single orifice without a gas chamber, the gas flow rate through the orifice is always constant, which is referred to as constant flow conditions. The phenomenon of

bubble formation from a single orifice connected to a gas chamber varies with gas injection conditions, which are characterized by the dimensionless capacitance number N_c defined as $4V_c g \rho_l / \pi D_o^2 P$ (Kumar and Kuloor, 1970; Tsuge and Hibino, 1983). When N_c is smaller than 1, the gas flow rate through the orifice is almost constant during the bubble formation process, similar to the first mechanical arrangement. When N_c is larger than 1, the gas flow rate through the orifice is not constant, and it is dependent on the pressure dif-

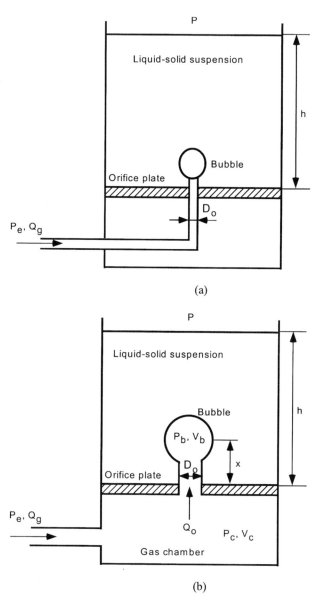

Figure 1 Typical mechanical arrangements for single bubble formation: (a) single orifice without a gas chamber; (b) single orifice connected to a gas chamber.

ference between the gas chamber and the bubble. Such bubble formation conditions are characterized as variable flow conditions by Yang et al. (2000a) or as constant pressure and intermediate conditions by Tsuge and Hibino (1983).

Numerous experimental and modeling studies have been conducted over the past decades on bubble formation from a single orifice or nozzle submerged in liquids, mostly under ambient conditions (Kupferberg and Jameson, 1969; Kumar and Kuloor, 1970; Azbel, 1981; Lin et al., 1994; Ruzicka et al., 1997). Among various factors that affect the bubble formation, the wettability of the orifice surface is one of the important factors that affect the initial size of the bubble formed on the orifice. Lin et al. (1994) found that initial bubble size increases significantly with the contact angle between the bubble and the orifice surface when the contact angle exceeds the threshold value of 45°. A few studies were conducted at elevated pressures (La Nauze and Harris, 1974; Idogawa et al., 1987; Tsuge et al., 1992; Wilkinson and van Dierendonck, 1994). These studies indicated that an increase in gas density reduces the size of bubbles formed from the orifice.

Bubble formation in liquids with the presence of particles, as in slurry bubble columns and three-phase fluidized bed systems, is different from that in pure liquids. The experimental data of Massimilla et al. (1961) in an air–water–glass beads three-phase fluidized bed revealed that the bubbles formed from a single nozzle in the fluidized bed are larger than those in water, and the initial bubble size increases with the solids concentration. Yoo et al. (1997) investigated bubble formation in pressurized liquid–solid suspensions. They used aqueous glycerol solution and 0.1 mm polystyrene beads as the liquid and solid phases, respectively. The densities of the liquid and the particles were identical, and thus the particles were neutrally buoyant in the liquid. The results indicated that initial bubble size decreases inversely with pressure under otherwise constant conditions, that is, gas flow rate, temperature, solids concentration, orifice diameter, and gas chamber volume. Their results also showed that the particle effect on the initial bubble size is insignificant. The difference in the finding regarding the particle effect on the initial bubble size between Massimilla et al. (1961) and Yoo et al. (1997) is possibly due to the difference in particle density.

Bubble formation in a hydrocarbon liquid and liquid–solid suspension with significant density difference between the liquid and solid phases was investigated by Luo et al. (1998a) and Yang et al. (2000a) under various gas injection conditions. A mechanistic

model was developed to predict the initial bubble size in liquid–solid suspensions at high-pressure conditions. The model considers various forces induced by the particles and is an extension of the two-stage spherical bubble formation model developed by Ramakrishnan et al. (1969) for liquids. In the two-stage spherical bubble formation model, bubbles are assumed to be formed in two stages, the expansion stage and the detachment stage. The bubble expands with its base attached to the nozzle during the first stage. In the detachment stage, the bubble base moves away from the nozzle, although the bubble remains connected with the nozzle through the neck. The shape of the bubble is assumed to remain spherical during the entire bubble formation process. It is also assumed in this model that a liquid film always exists around the bubble. During the expansion and detachment stages, particles collide with the bubble and stay on the liquid film. The particles and the liquid surrounding the bubble are set in motion as the bubble grows and rises.

The volume of the bubble at the end of the first stage and during the second stage can be described by considering a balance of all the forces acting on the bubble if the instantaneous gas flow rate, Q_o, or the instantaneous gas velocity, u_o, through the orifice, is known. The forces induced by the liquid include the upward forces (effective buoyancy force, F_B, and gas momentum force, F_M), and the downward resistance (liquid drag, F_D, surface tension force, F_σ, bubble inertial force, $F_{I,g}$, and Basset force, F_{BA}) as shown in Fig. 2. It is assumed that the particles affect the bubble formation process only through two additional downward forces on the bubble, that is, the particle–bubble collision force, F_C, and the suspension inertial force, $F_{I,m}$. The suspension inertial force is due to the acceleration of the liquid and particles surrounding the bubble. Therefore the overall force balance on the bubble in this model can be written as

$$F_B + F_M = F_D + F_\sigma + F_{BA} + F_{I,g} + F_C + F_{I,m} \quad (1)$$

The expansion stage and the detachment stage follow the same force balance equation, although the expression for the same force in the two stages may be different owing to different bubble moving velocities in the two stages. The expressions for all the forces under two stages are listed in Table 1. The particle–bubble collision force is merely the rate of momentum change of the particles colliding with the bubble surface. The suspension inertial force is derived from the suspension flow field around an accelerating bubble, obtained from a particle image velocimetry (PIV) sys-

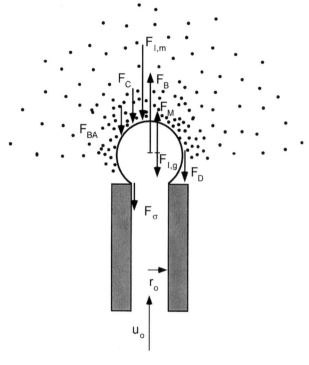

Figure 2 The balance of all the forces acting on a growing bubble. (From Luo et al., 1998a.)

tem. The final expression of the suspension inertial force is

$$F_{\mathrm{I,m}} = \frac{d\left(\iiint \rho_{\mathrm{m}} u_{\mathrm{m}} \delta V\right)}{dt} = \zeta \frac{d}{dt}\left[\rho_{\mathrm{m}}\left(\frac{4}{3}\pi r_{\mathrm{b}}^3\right)u_{\mathrm{b}}\right] \quad (2)$$

The coefficient ζ in Eq. (2) is equal to 3.86 for bubbles formed in liquid–solid suspensions based on PIV measurements (Luo et al., 1998a) and to 11/16 for bubbles formed in liquids, corresponding to the added mass in inviscid liquids (Milne-Thomson, 1955).

For bubble formation from a single orifice without a gas chamber, the motion equation of the rising bubble itself is sufficient to predict the initial bubble size. For the case in which the orifice is connected to a gas chamber, the gas flow rate through the orifice is not constant and depends on pressure fluctuations in both the chamber and the bubble. In order to simulate bubble formation under such conditions, the pressure fluctuations in the gas chamber and in the bubble must be considered to account for the time-variant orifice gas flow rate as illustrated below.

The instantaneous gas flow rate through the orifice depends on the pressure difference in the gas chamber, P_{c}, and inside the bubble, P_{b}, as well as the flow resis-

Table 1 Expressions for the Forces Involved in the Bubble Formation Process

Forces	Expansion stage	Detachment stage
F_{B}	$\dfrac{\pi}{6}d_{\mathrm{b}}^3(\rho_{\mathrm{l}}-\rho_{\mathrm{g}})g$	Same as expansion stage
F_{M}	$\dfrac{\pi}{4}D_{\mathrm{o}}^2\rho_{\mathrm{g}}u_{\mathrm{o}}^2$	Same as expansion stage
F_{D}	$C_{\mathrm{D}}\left(\dfrac{\pi}{4}d_{\mathrm{b}}^2\right)\dfrac{\rho_{\mathrm{l}}u_{\mathrm{b}}^2}{2}\left(C_{\mathrm{D}}=\dfrac{24}{\mathrm{Re}}\right)$	Same as expansion stage
F_{σ}	$\pi D_{\mathrm{o}}\sigma\cos\gamma$	Same as expansion stage
$F_{\mathrm{I,g}}$	$\dfrac{d}{dt}\left[\rho_{\mathrm{g}}\left(\dfrac{\pi}{6}d_{\mathrm{b}}^3\right)u_{\mathrm{b}}\right]$	Same as expansion stage
F_{BA}	Not applicable	$\dfrac{3}{2}d_{\mathrm{b}}^2\sqrt{\pi\rho_{\mathrm{l}}\mu_{\mathrm{l}}}\displaystyle\int_0^t\dfrac{du/dt}{\sqrt{t-\tau}}\,d\tau$
F_{C}	$\dfrac{\pi}{4}D_{\mathrm{o}}^2(1+e)\varepsilon_{\mathrm{s}}\rho_{\mathrm{s}}u_{\mathrm{e}}^2$	$\dfrac{\pi}{4}d_{\mathrm{b}}^2\varepsilon_{\mathrm{s}}\rho_{s}u^2$
$F_{\mathrm{I,m}}$	$\dfrac{d\left(\iiint \rho_{\mathrm{m}}u_{\mathrm{m}}\delta V\right)}{dt}=\zeta\dfrac{d}{dt}\left[\rho_{\mathrm{m}}\left(\dfrac{\pi}{6}d_{\mathrm{b}}^3\right)u_{\mathrm{b}}\right]$	Same as expansion stage

Source: Luo et al. (1998a).

tance of the orifice, which can be described by the orifice equation as given in Eq. (3). The pressure in the gas chamber is evaluated by applying the first law of thermodynamics and choosing the gas chamber as an open system as given in Eq. (4) (Wilkinson and van Dierendonck, 1994). The pressure inside the bubble is derived from the pressure balance at the bubble–liquid interface governed by a modified Rayleigh equation (Pinczewski, 1981). In order to simulate bubble formation in liquid–solid suspensions, the effect of particles on the pressure balance at the bubble–liquid interface must be considered. The inertial term in the modified Rayleigh equation is modified by considering the contribution of the suspension inertial force, as given in Eq. (5) for two stages (Yang et al., 2000a).

$$\Delta P = |P_c - P_b| = \left(\frac{Q_o}{k_o}\right)^2 \tag{3}$$

$$\frac{dP_c}{dt} = \frac{\gamma}{V_c}\left(P_e Q_g - P_c Q_o\right) \tag{4}$$

$$P_b - P_0 = \zeta\rho_m\left[\frac{r_b}{3}\frac{d^2 r_b}{dt^2} + \left(\frac{dr_b}{dt}\right)^2\right]$$
$$+ \frac{2\sigma}{r_b} + \frac{4\mu_l}{r_b} - \frac{1}{4}\rho_g\left(\frac{Q_o}{\frac{1}{4}\pi D_o^2}\right)^2 \tag{5a}$$

for the expansion stage and

$$P_b - P_0 = \zeta\rho_m\left[\frac{r_b}{3}\frac{d^2 x}{dt^2} + \frac{dr_b}{dt}\frac{dx}{dt}\right]$$
$$+ \frac{2\sigma}{r_b} + \frac{4\mu_l}{r_b}\frac{dr_b}{dt} - \frac{1}{4}\rho_g\left(\frac{Q_o}{1/4\pi D_o^2}\right)^2 \tag{5b}$$

for the detachment stage. P_0 is the hydrostatic pressure at the bubble surface. The four terms on the right-hand side of Eq. (5) represent the contributions of inertial, surface tension, viscous, and gas momentum forces, respectively. Combining Eqs. (1) and (3) through (5), and solving these coupled ordinary differential equations simultaneously, the dynamic behavior of bubble formation can be obtained. If a certain bubble detachment criterion is used, the initial bubble size can be estimated.

The effect of pressure on the initial bubble size under various bubble formation conditions is shown in Fig. 3. The solid lines in the figures represent the model predictions. Under constant flow conditions ($N_c \leq 1$), the pressure effect on the initial bubble size is not significant; however, under variable flow conditions ($N_c > 1$), pressure has a significant effect on the

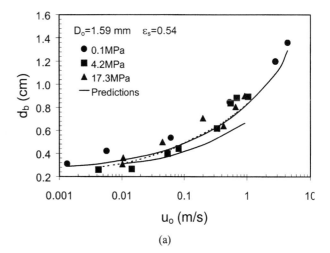

(a)

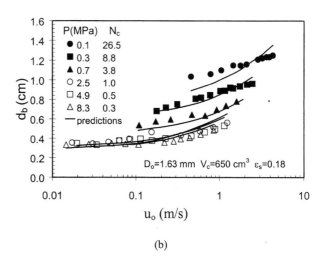

(b)

Figure 3 Initial bubble size in liquid–solids suspensions as a function of pressure and gas velocity for bubble formation under various conditions: (a) single orifice without a gas chamber. (From Luo et al., 1998a.) (b) Single orifice connected to a gas chamber. (From Yang et al., 2000a.)

initial bubble size. The initial bubble size decreases significantly with an increase in pressure when $N_c > 1$. The different pressure effects can be explained based on the model simulation results. The simulated bubble formation time and instantaneous gas flow rate through the orifice are shown in Fig. 4. Under variable flow conditions, both the bubble formation time and the orifice gas flow rate decrease with increasing pressure, resulting in a significant reduction in the initial bubble size. On the other hand, for bubble formation under constant flow conditions, the gas flow rate through the orifice is almost constant during the formation process, and the bubble formation time slightly

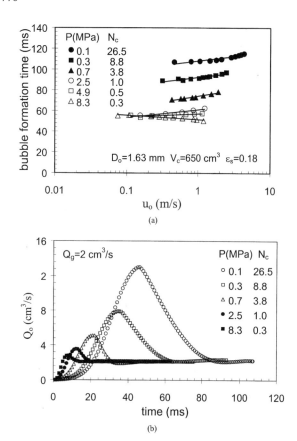

(a)

(b)

Figure 4 Simulated (a) bubble formation time and (b) instantaneous gas flow rate through the orifice during the bubble formation process in the liquid–solid suspension ($D_o = 1.63$ mm, $V_c = 650$ cm^3, $\varepsilon_s = 0.18$). (From Yang et al., 2000a.)

changes with pressure, causing insignificant pressure effect.

Based on the model calculation, a comparison of the magnitudes of various forces acting on the bubble at the end of formation process in a slurry system is shown in Fig. 5. As shown in the figure, the suspension inertial force is the dominant downward force and the buoyancy force is the major upward force. At low pressures, the effect of gas momentum on bubble formation is negligible; however, at high pressures, the gas momentum force could play an important role in providing an upward force. It is noted that the Basset force is also important in dictating the bubble formation process, especially at low pressures due to the large bubble size and fast bubble acceleration. When the pressure increases, both buoyancy and suspension inertial forces decrease significantly owing to reduced bubble size, and all the forces are important for determining the size of the bubble at high pressures owing to similar magnitudes.

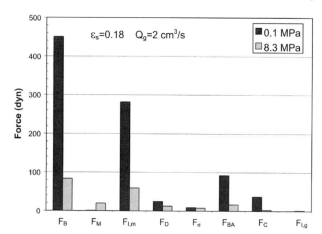

Figure 5 Comparison of various forces acting on a bubble formed in the liquid–solid suspension at different pressures ($D_o = 1.63$ mm, $V_c = 650$ cm^3, $\varepsilon_s = 0.18$, $Q_g = 2$ cm^3/s). (From Yang et al., 2000a.)

2.2 Bubble Rise Characteristics

The bubble rise velocity is an important parameter in characterizing the hydrodynamics and transport phenomena of bubbles in liquids and liquid–solid suspensions. The rise velocity of a single gas bubble depends on its size: for small bubbles, the rise velocity also strongly depends on liquid properties such as surface tension and viscosity; for large bubbles, the rise velocity is insensitive to liquid properties (Fan, 1989). Under limited conditions, the rise velocities of single bubbles in liquid–solid suspensions were found to be similar to those in highly viscous liquids (Massimilla et al., 1961; Darton and Harrison, 1974). Liquid–solid suspensions can thus be characterized as Newtonian homogeneous media, but they often exhibit non-Newtonian or heterogeneous behavior (Tsuchiya et al., 1997). Differences in fluidizing media, pressure, and temperature lead to different bubble rise characteristics.

This section focuses on the bubble rise characteristics in liquids and liquid–solid suspensions at elevated pressure and temperature. A nonwater based liquid medium is employed to illustrate the effect of liquid properties on the bubble rise velocity. For a bubble rising in a liquid–solid suspension, the bubble rise velocity is discussed in light of both the apparent homogeneous (or effective) properties of the suspension and the recently evolved numerical prediction based on a computational model for gas–liquid–solid fluidization systems. Computational modeling will be covered later in Sec. 4.

2.2.1 Single Bubble Rise Velocity in Liquids

In most applications of three-phase fluidization systems, the liquid phase is a hydrocarbon based medium. The physical properties of hydrocarbon liquids vary dramatically with pressure and temperature. Lin and Fan (1997) and Lin et al. (1998) developed various in-situ techniques to measure the physical properties of a hydrocarbon liquid (e.g., Paratherm NF heat transfer fluid) at elevated pressures and temperatures. The hydrostatic weighing method, the falling ball technique, and the emerging bubble technique were used to measure in-situ liquid density, viscosity, and surface tension, respectively. Based on their measurements, the liquid and interfacial properties change significantly with pressure and temperature. For example, at room temperature, as the pressure increases from 0.1 to 20 MPa, the liquid density of Paratherm NF heat transfer fluid increases by approximately 5%, the liquid viscosity increases by 65%, and the surface tension decreases by 25% (Lin and Fan, 1997; Lin et al., 1998). Therefore it is important to consider these changes in processing data, developing models, and conducting numerical simulations, particularly the variations of liquid viscosity and surface tension.

Since the bubble rise velocity depends on liquid properties, it is expected that pressure and temperature would also affect the bubble rise characteristics. Krishna et al. (1994) studied the pressure effect on the bubble rise velocity and found that the single bubble rise velocity does not depend on the gas density over the range of 0.1 to 30 kg/m³. The conclusion is limited to a narrow range of pressures. Lin et al. (1998) measured the rise velocity of single bubbles of known sizes in the Paratherm NF heat transfer fluid at various pressures and temperatures. The pressure ranges from 0.1 to 19.4 MPa. Figure 6 shows the relationship between the bubble rise velocity and the bubble size for two different temperatures. The bubble size in the figure is represented by the equivalent spherical diameter, d_b. For a given bubble size, u_b tends to decrease with increasing pressure at both temperatures. The effects of pressure and temperature, or more directly, the effects of physical properties of the gas and liquid phases on the variation of u_b with d_b, could be represented or predicted most generally by the Fan–Tsuchiya equation (Fan and Tsuchiya, 1990) among three predictive equations. The other two are the modified Mendelson's wave-analogy equation (Mendelson, 1967) by Maneri (1995) and a correlation proposed by Tomiyama et al. (1995).

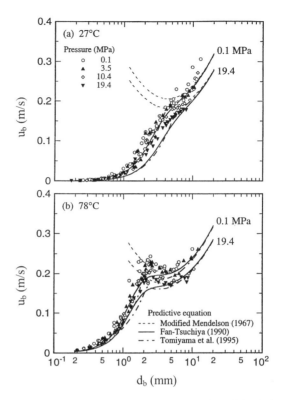

Figure 6 Effect of pressure on rise velocity of single bubbles at different temperatures: (a) 27°C and (b) 78°C. (From Lin et al., 1998.)

The Fan–Tsuchiya equation, generalized for high-pressure systems, can be written in a dimensionless form:

$$u_b' = u_b \left(\frac{\rho_l}{\sigma g}\right)^{1/4} = \left\{ \left[\frac{Mo^{-1/4}}{K_b}\left(\frac{\Delta\rho}{\rho_l}\right)^{5/4} d_b'^2\right]^{-n} + \left[\frac{2c}{d_b'} + \left(\frac{\Delta\rho}{\rho_l}\right)\frac{d_b'}{2}\right]^{-n/2} \right\}^{-1/n} \quad (6)$$

where the dimensionless bubble diameter is given by

$$d_b' = d_b\left(\frac{\rho_l g}{\sigma}\right)^{1/2} \quad (7)$$

Three empirical parameters, n, c, and K_b, in Eq. (6) reflect three specific factors governing the rate of bubble rise. They relate to the contamination level of the liquid phase, to the varying dynamic effects of the surface tension, and to the viscous nature of the surrounding medium, respectively. The suggested values of these parameters are

$$n = \begin{cases} 0.8 & \text{for contaminated liquids} \\ 1.6 & \text{for purified liquids} \end{cases} \quad (8a)$$

$$c = \begin{cases} 1.2 & \text{for monocomponent liquids} \\ 1.4 & \text{for multicomponent liquids} \end{cases} \quad (8b)$$

$$K_b = \max(K_{b0} Mo^{-0.038}, 12) \quad (8c)$$

where

$$K_{b0} = \begin{cases} 14.7 & \text{for aqueous solutions} \\ 10.2 & \text{for organic solvents/mixtures} \end{cases}$$
$$(8d)$$

The modified Mendelson's equation is a special form of the Fan–Tsuchiya equation where the viscous term, that is, the first term on the right side of Eq. (6), is omitted. Just as general as the Fan–Tsuchiya equation for bubbles in liquids, the correlation by Tomiyama et al. (1995) given in terms of drag coefficient,

$$C_D = \frac{4}{3} \frac{g\Delta\rho d_b}{\rho_l u_b^2} \quad (9)$$

consists of three equations under different system purities:

$$C_D = \max\left\{ \min\left[\frac{16}{Re}(1 + 0.15 Re^{0.687}), \frac{48}{Re} \right], \frac{8}{3}\frac{Eo}{Eo+4} \right\} \quad (10a)$$

for purified systems,

$$C_D = \max\left\{ \min\left[\frac{24}{Re}(1 + 0.15 Re^{0.687}), \frac{72}{Re} \right], \frac{8}{3}\frac{Eo}{Eo+4} \right\} \quad (10b)$$

for partially contaminated systems, and

$$C_D = \max\left[\frac{24}{Re}(1 + 0.15 Re^{0.687}), \frac{8}{3}\frac{Eo}{Eo+4} \right] \quad (10c)$$

for sufficiently contaminated systems. It is noted that u_b can be obtained explicitly from the Fan–Tsuchiya equation [Eq. (6)] for a given d_b as well as gas and liquid physical properties, while it can only be obtained implicitly from Tomiyama's correlation [Eq. (10)].

Based on the comparison of three predictive equations as shown in Fig. 6, the modified Mendelson equation, which is valid only under the inviscid condition, has limited predictive capability at the low temperature (Fig. 6a), suggesting that the viscous force predominates in the bubble rise process. On the other hand, at the high temperature (Fig. 6b), there is a strong agreement over the bubble size range of $d_b > 2$ mm including the sharp breakpoint/peak. This indicates that the liquid used tends to behave as a pure inviscid liquid. Note that over the pressure range from 0.1 to 19.4 MPa, the liquid viscosity varies from 29 to 48 mPa · s at 27°C, whereas it is almost constant within a range from 4.7 to 5.2 mPa · s at 78°C (Lin et al., 1998).

The Fan–Tsuchiya equation applied for the given liquid, a pure ($n = 1.6$), multicomponent ($c = 1.4$), and organic solvent ($K_{b0} = 10.2$), demonstrates good overall predictive capability except for the sharp peak existing under the high temperature (Fig. 6b). The equation by Tomiyama et al. (1995) also has good general applicability, especially around the peak point occurring near $d_b = 2$ mm at 78°C; however, it tends to underestimate the bubble rise velocity over the rest of the bubble size range.

The consistent difference in u_b prevailing between 0.1 and 19.4 MPa for $d_b > 2$ mm is due to the significant increase in gas density (as large as a 200-fold increase with pressure from 0.1 to 19.4 MPa). The density effect is accounted for in Fan–Tsuchiya equation in terms of $\Delta\rho/\rho_l$ or in Tomiyama's equation in terms of both $\Delta\rho/\rho_l$ and Eo. As can be seen from the equations and figure, the density difference between the continuous liquid phase and the dispersed gas phase plays an important role in determining bubble rise velocity, especially for large bubbles.

The Re–Eo relationship is often utilized in representing the general rise characteristics of single bubbles in liquids (Clift et al., 1978; Bhaga and Weber, 1981). Figure 7 shows the Re–Eo relationship at different Mo values. The thin background lines signify the general, quantitative trend for the rise velocity of single bubbles in purified Newtonian liquids under ambient conditions, plotted with constant intervals of log Mo. The figure shows the general agreement in correlation predictions. The experimental results under four conditions are also plotted in the figure (Lin et al., 1998). By employing accurate values for the physical properties of the liquid phase and the gas density at given pressures and temperatures, the experimental results can be successfully represented over the entire Eo range by the Fan–Tsuchiya equation. This equation was also proven to represent single bubble rise velocity in various liquids under ambient conditions (Tsuchiya et al., 1997). So it is clear that much more is known now than before about single bubbles rising in liquids at high pressure and high temperature.

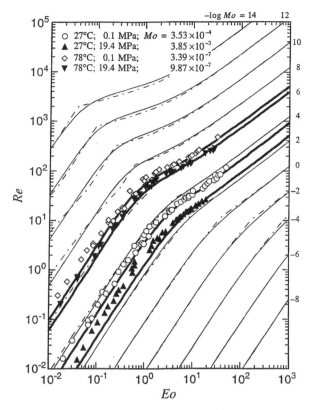

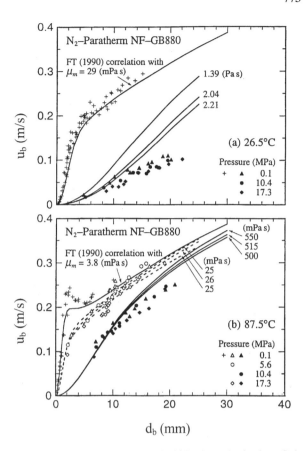

Figure 7 The rise characteristics of single bubbles in liquids represented by the Re–Eo relationship (symbols represent experimental results in Paratherm NF heat transfer fluid under various pressure and temperature conditions; ——— and — — represent predictions by the Fan–Tsuchiya equation and correlation of Tomiyama et al. (1995) at regular intervals of Mo values, respectively; ——— represents predictions by the Fan–Tsuchiya equation at measured Mo values for comparison with experimental results. (From Lin et al., 1998.)

Figure 8 Effect of pressure on bubble rise velocity in a fluidized bed of 0.88 mm glass beads at (a) 26.5°C and (b) 87.5°C (solids holdups for +, open, and filled symbols are 0, 0.384, and 0.545, respectively). (From Luo et al., 1997b.)

2.2.2 Single Bubble Rise Velocity in Liquid–Solid Suspensions

Much progress has also been made regarding single bubble rise characteristics in liquid–solid suspensions at high pressure and high temperature. This section covers the effect of suspension properties on single bubble rise velocity. Figure 8 shows the effect of pressure on the bubble rise velocity in a fluidized bed with Paratherm NF heat transfer fluid and 0.88 mm glass beads at two different temperatures (Luo et al., 1997b). At both temperatures, the bubble rise velocity decreases with an increase in pressure for a given solids holdup. The extent of the reduction is as high as 50% from 0.1 to 17.3 MPa. A more drastic reduction in u_b, however, arises from the addition of solid particles. While the particle effect is small at a low solids holdup ($\varepsilon_s < 0.4$), the effect is appreciable at a high solids holdup ($\varepsilon_s = 0.545$), especially for high liquid viscosity (Fig. 8a). A comparison of bubble rise velocity at 26.5°C and 87.5°C, for the same solids holdup, indicates that the viscosity effect appears to be significant.

The bubble rise velocity in a fluidized bed containing smaller particles (0.21 mm glass beads) is shown in Fig. 9 (Luo et al., 1997b). The effect of particle size on the bubble rise velocity can be seen from Figs. 8 and 9. While the extent of decrease in bubble rise velocity with increase in pressure is comparable between large particles (0.88 mm glass beads) and small particles (0.21 mm glass beads), the extent of decrease in bubble rise velocity with increasing solids holdup is much smaller for the smaller particles. This difference in the sensitivity of u_b reduction to solids holdup variation is clearly seen for the high solids holdup cases.

In the presence of solid particles, as a first approximation, it can be assumed that the particles modify

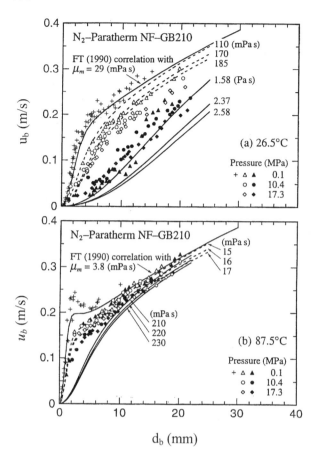

Figure 9 Effect of pressure on bubble rise velocity in a fluidized bed of 0.21 mm glass beads at (a) 26.5°C and (b) 87.5°C (solids holdups for +, open, and filled symbols are 0, 0.381, and 0.555, respectively). (From Luo et al., 1997b.)

only homogeneous properties of the surrounding medium. Based on this homogeneous approach, the Fan–Tsuchiya equation, Eq. (6), can be extended to liquid–solid suspensions by replacing the liquid properties, ρ_l and μ_l, by the effective properties of the liquid–solid suspension, ρ_m and μ_m, respectively (Tsuchiya et al., 1997). The effective density can be estimated by

$$\rho_m = \rho_l(1 - \varepsilon_s) + \rho_s \varepsilon_s \tag{11}$$

The effective viscosity of liquid–solid suspensions is estimated by

$$\frac{\mu_m}{\mu_l} = \exp\left[\frac{K\varepsilon_s}{1 - (\varepsilon_s/\varepsilon_{sc})}\right] \tag{12}$$

with two parameters correlated by Luo et al. (1997b):

$$K = \frac{3.1 - 1.4 \tanh[0.3(10 - 10^2 u_t)]}{\phi} \tag{13a}$$

and

$$\varepsilon_{sc} = \left\{1.3 - 0.1 \tanh[0.5(10 - 10^2 u_t)]\right\}\varepsilon_{s0} \tag{13b}$$

where u_t is in units of m/s. The ranges of applicability of Eqs. (12) and (13) are $840 < \rho_l < 1000\,\mathrm{kg/m^3}$; $1 < \mu_l < 47\,\mathrm{mPa \cdot s}$; $19 < \sigma < 73\,\mathrm{mN/m}$; $0 < \varepsilon_s < 0.95\varepsilon_{s0}$; $7.9 \times 10^{-4} < u_t < 0.26\,\mathrm{m/s}$; $0.88 < \phi \leq 1$; and $0.56 < \varepsilon_{s0} < 0.61$.

The Fan–Tsuchiya equation with constant values of μ_m estimated from Eq. (12) predicts reasonably well the general trend of bubble rise velocity variation in liquid–solid suspensions as shown in Figs. 8 and 9. However, a detailed match between the experimental results and predictions appears to be difficult to attain by assigning a constant value of μ_m for each condition. A more elaborate analysis is required to account for the effect of bubble size on interactions of the bubble with the surrounding medium (non-Newtonian approach) or with individual particles (heterogeneous approach).

The effect of particle wettability on the bubble rise velocity was examined by Tsutsumi et al. (1991a,b). Nonwettable particles have a larger contact angle than wettable particles. That is to say that nonwettable particles favor contact between bubbles and solids. Tsutsumi et al. (1991a,b) observed that particle–particle aggregates and particle–bubble aggregates are formed at low gas velocities with nonwettable particles. Consequently, the bubble rise velocity is smaller than that for wettable particles. Further, the liquid velocity for minimum fluidization is lower and the liquid velocity for transition to the transport regime is higher for wettable particles than for nonwettable particles. On the other hand, a negligible effect of particle wettability was observed on the rise velocity of large bubbles ($d_b > 15\,\mathrm{mm}$). For large bubbles with circular cap shapes, the attachment of particles to the bubbles occurs only at the bubble base and was not observed on the bubble roof owing to the fluid shear effects caused by the fast rising bubbles.

2.3 Bubble Coalescence

The coalescence of bubbles to form larger bubbles is another quantifiable area of study in three-phase fluidization, which is important for determining the actual bubble size in the system. Much work has been accomplished to understand how bubble coalescence occurs. For gas–liquid systems, the experimental results available in the literature indicate that an increase of pressure retards the bubble coalescence (Sagert and Quinn,

1977, 1978). There are three steps in the bubble coalescence process (Vrij, 1966; Chaudhari and Hoffmann, 1994): (1) approach of two bubbles to form a thin liquid film between them; (2) thinning of the film by the drainage of the liquid under the influence of gravity and suction due to capillary forces; and (3) rupture of the film at a critical thickness. The second step is the rate-controlling step in the coalescence process, and the bubble coalescence rate can be approximated by the film thinning rate (Vrij, 1966). The film thinning velocity can be expressed as (Sagert and Quinn, 1977, 1978)

$$-\frac{dl}{dt} = \frac{32l^3\sigma}{3\phi R_d^2 \mu_l d_b} \tag{14}$$

where the parameter ϕ is a measure of the surface drag or velocity gradient at the surface due to the adsorbed layer of the gas.

It is known that surface tension decreases and liquid viscosity increases with increasing pressure. In addition, ϕ increases with pressure. As seen from Eq. (14), all of these variations contribute to the reduction of the film thinning velocity, and hence the bubble coalescence rate, as pressure increases. As a result, the time required for two bubbles to coalesce is longer and the rate of overall bubble coalescence in the bed is reduced at high pressures. Moreover, the frequency of bubble collision decreases with increasing pressure. An important mechanism for bubble collision is the bubble wake effect (Fan and Tsuchiya, 1990). When the differences in bubble size and bubble rise velocity are small at high pressures, the likelihood of small bubbles being caught and trapped by the wakes of large bubbles decreases. Therefore bubble coalescence is suppressed by the increase in pressure, due to the longer bubble coalescence time and the smaller bubble collision frequency.

The presence of pulp, even at very low pulp consistencies (0.1%) in the column leads to enhanced bubble coalescence and hence a narrowing of the gas velocity for the dispersed bubble regime as the pulp consistency increases (Reese et al., 1996). Bubble coalescence inhibitors such as inorganic salts (e.g., sodium chloride and sodium phosphate dibasic) and organic compounds (e.g., ethanol, n-pentanol, iso-amyl alcohol, and benzoic acid) can be effectively applied to the liquid at concentrations up to 200 ppm to inhibit bubble coalescence behavior in three-phase fluidization (Briens et al., 1999). With the addition of the bubble coalescence inhibitor, the bed hydrodynamics at low gas velocities are significantly different from the case without the inhibitor, and the influence of the gas distributor becomes marked (Nacef et al., 1995).

2.4 Bubble Breakup

Another quantifiable area of bubble dynamics is bubble breakup. Bubble breakup is a key phenomenon to study because it determines the bubble size in the system and hence affects heat and mass transfer between phases. Bubble breakup has been investigated theoretically and experimentally. There are two causes for bubble breakup: bubble–particle collision and bubble instability. In this section, only bubble breakup due to bubble instability will be discussed.

It is known that the variation of bubble size with pressure is the key for understanding pressure effects on hydrodynamics. The upper limit of the bubble size is set by the maximum stable bubble size, D_{max}, above which the bubble is subjected to breakup and hence is unstable. Several mechanisms have been proposed for the bubble breakup phenomenon, and based on these mechanisms theories have been established to predict the maximum bubble size in gas–liquid systems. In this section, the mechanisms of bubble breakup and the theories to predict the maximum bubble size are covered.

Hinze (1955) proposed that bubble breakup is caused by the dynamic pressure and the shear stresses on the bubble surface induced by different liquid flow patterns, e.g., shear flow and turbulence. When the maximum hydrodynamic force in the liquid is larger than the surface tension force, the bubble disintegrates into smaller bubbles. This mechanism can be quantified by the liquid Weber number. When the Weber number is larger than a critical value, the bubble is not stable and disintegrates. This theory was adopted to predict the breakup of bubbles in gas–liquid systems (Walter and Blanch, 1986). Calculations by Lin et al. (1998) showed that the theory underpredicts the maximum bubble size and cannot predict the effect of pressure on the maximum bubble size.

A maximum stable bubble size exists for bubbles rising freely in a stagnant liquid without external stresses, e.g., rapid acceleration, shear stress, and/or turbulence fluctuations (Grace et al., 1978). Rayleigh–Taylor instability has been regarded as the mechanism for bubble breakup under such conditions. A horizontal interface between two stationary fluids is unstable to disturbances with wavelengths exceeding a critical value if the upper fluid has a higher density than the lower one (Bellman and Pennington, 1954):

$$\lambda_c = 2\pi \sqrt{\frac{\sigma}{g(\rho_l - \rho_g)}} \tag{15}$$

Grace et al. (1978) modified the Rayleigh–Taylor instability theory by considering the time available for the disturbance to grow and the time required for the disturbance to grow to an adequate amplitude. Batchelor (1987) pointed out that the observed size of air bubbles in water was considerably larger than that predicted by the model of Grace et al. (1978). Batchelor (1987) further took into account the stabilizing effect of the liquid acceleration along the bubble surface and the nonconstant growth rate of the disturbance. In Batchelor's model, the magnitude of the disturbances is required to predict maximum bubble size; however, the magnitude of the disturbances is not known. The models based on the Rayleigh–Taylor instability predict an almost negligible pressure effect on the maximum bubble size; in fact, Eq. (15) implies that the bubble is more stable when the gas density is higher.

The Kelvin–Helmholtz instability is similar to the Rayleigh–Taylor instability, except that the former allows a relative velocity between the fluids, u_r. Using the same concept of Grace et al. (1978), Kitscha and Kocamustafaogullari (1989) applied the Kelvin–Helmholtz instability theory to model the breakup of large bubbles in liquids. Wilkinson and van Dierendonck (1990) applied the critical wavelength to explain the maximum stable bubble size in high-pressure bubble columns:

$$\lambda_c = \frac{2\pi \sqrt{\dfrac{\sigma}{g(\rho_l - \rho_g)}}}{\dfrac{\rho_l}{\rho_l + \rho_g} \dfrac{\rho_g u_r^2}{2\sqrt{\sigma g(\rho_l - \rho_g)}} + \sqrt{1 + \dfrac{\rho_l^2 \rho_g^2 u_r^4}{4(\rho_l + \rho_g)^2 \sigma g(\rho_l - \rho_g)}}} \tag{16}$$

Disturbances in the liquid with a wavelength larger than the critical wavelength can break up a bubble. Equation (16) indicates that the critical wavelength decreases with an increase in pressure and therefore bubbles are easier to break apart by disturbances at higher pressures. However, the critical wavelength is not equivalent to the maximum stable bubble size, and Eq. (16) alone cannot account for the effect of pressure on the maximum bubble size.

All of the models mentioned above do not account for the internal circulation of the gas. The internal circulation velocity is of the same order of magnitude as the bubble rise velocity. A centrifugal force is induced by this circulation, pointing outward toward the bubble surface. This force can suppress the disturbances at the gas–liquid interface and thereby stabilize the interface. Centrifugal force explains the underestimation of D_{max} by the model of Grace et al. (1978). On the other hand, the centrifugal force can also break apart the bubble, as it increases with an increase in bubble size. The bubble breaks up when the centrifugal force exceeds the surface tension force, especially at high pressures when gas density is high. Levich (1962) assumed the centrifugal force to be equal to the dynamic pressure induced by the gas moving at the bubble rise velocity, that is, $k_f \rho_g u_b^2 / 2 (k_f \approx 0.5)$, and proposed a simple equation to calculate the maximum stable bubble size:

$$D_{max} \approx \frac{3.63\sigma}{u_b^2 \sqrt[3]{\rho_l^2 \rho_g}} \tag{17}$$

Equation (17) severely underpredicts the maximum bubble size in the air–water system, although it shows a significant effect of pressure on the maximum bubble size.

An analytical criterion for the bubble breakup can be derived by considering a single large bubble rising in a stagnant liquid or slurry at a velocity of u_b, without any disturbances on the gas–liquid interface. The bubble is subjected to breakup when its size exceeds the maximum stable bubble size due to the circulation-induced centrifugal force. Schematic of the internal circulation model for bubble breakup is shown in Fig. 10 (Luo et al., 1999). Large bubbles normally assume a spherical cap shape, and in the model, the spherical cap bubble is approximated by an ellipsoidal bubble with the same volume and the same aspect ratio (height to width). The circulation of gas inside the bubble can be described by Hill's vortex (Hill, 1894). To model bubble breakup, it is necessary to evaluate the x-component of the centrifugal force, F_x, induced by the circulation on the entire bubble surface as shown in Fig. 10. A rigorous theoretical derivation from Hill's vortex yields the expression for F_x:

$$F_x = \frac{9\pi \rho_g u_b^2 a^2}{64\sqrt{2}\alpha} \tag{18}$$

The surface tension force is the product of the surface tension and the circumference of the bubble,

$$F_\sigma = \sigma L = \sigma \int_{ellipse} \sqrt{(\delta r_c)^2 + (\delta z)^2} \tag{19}$$

$$= 4\sigma a E(\sqrt{1 - \alpha^2})$$

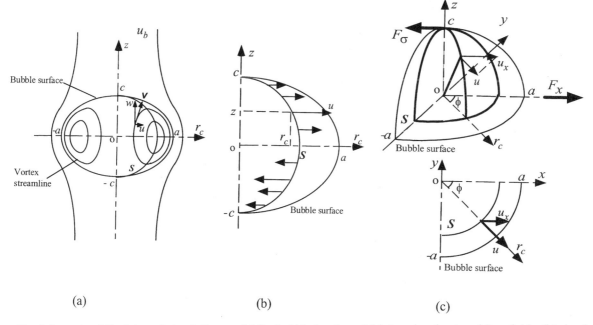

(a)   (b)   (c)

Figure 10 Schematic of the internal circulation model for bubble breakup: (a) internal and external flow fields; (b) circulation velocity on surface S; (c) force balance and 3-D view of surface S and the flow pattern on S. (From Luo et al., 1999.)

Also, the volume equivalent bubble diameter, d_b, is related to a and α by

$$a = \frac{d_b}{\sqrt[3]{8\alpha}} \tag{20}$$

Note that the centrifugal force is affected significantly by the gas density, the aspect ratio of the bubble, the bubble size, and the bubble rise velocity. The bubble is not stable if F_x is larger than F_σ, that is,

$$u_b^2 d_b \geq \frac{8\alpha^{4/3} \mathrm{E}(\sqrt{1-\alpha^2})}{0.312} \frac{\sigma}{\rho_g} \tag{21}$$

When the centrifugal force is larger than the surface tension force, the bubble would be stretched in the x-direction. During the stretching, the aspect ratio, α, becomes smaller while d_b and u_b can be assumed to remain constant. As a result, the centrifugal force increases, the surface tension force decreases, and the bubble stretching becomes an irreversible process. Using the Davies–Taylor equation (Davies and Taylor, 1950) for the bubble rise velocity, the maximum stable bubble size is expressed by

$$D_{\max} \approx 7.16\alpha^{2/3} \mathrm{E}(\sqrt{1-\alpha^2})^{1/2} \sqrt{\frac{\sigma}{g\rho_g}} \tag{22}$$

The simplified forms of Eq. (22) are:

$$D_{\max} \approx 2.53 \sqrt{\frac{\sigma}{g\rho_g}} \quad (\text{for } \alpha = 0.21) \tag{23a}$$

in liquids, and

$$D_{\max} \approx 3.27 \sqrt{\frac{\sigma}{g\rho_g}} \quad (\text{for } \alpha = 0.3) \tag{23b}$$

in liquid–solid suspensions. Further, based on the Davies–Taylor equation, the rise velocity of the maximum stable bubble is

$$u_{\max} = \left(\frac{1.6\sigma g}{\rho_g}\right)^{1/4} \tag{24}$$

The comparison of experimental maximum bubble sizes and the predictions by various instability theories is shown in Fig. 11. The internal circulation model can reasonably predict the observed pressure effect on the maximum bubble size, indicating that the internal circulation model captures the intrinsic physics of bubble breakup at high pressures. The comparison of the predictions by different models further indicates that bubble breakup is governed by the internal circulation mechanism at high pressures over 1.0 MPa, whereas the Rayleigh–Taylor instability or the Kelvin–Helmholtz instability is the dominant mechanism at low pressure.

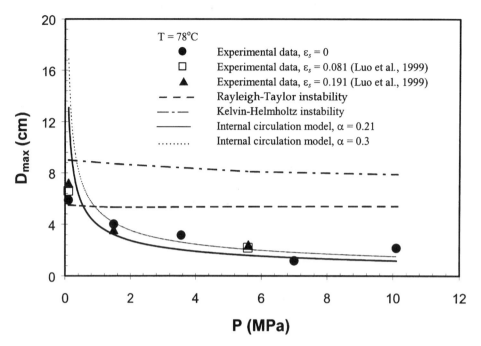

Figure 11 Comparison of maximum stable bubble size obtained experimentally and the predictions by various models. (From Fan et al., 1999.)

3 TRANSPORT PHENOMENA

Bubble dynamics and characteristics discussed above determine the hydrodynamic and heat and mass transfer behaviors in three-phase fluidization systems, which is important for better design and operation of three-phase fluidized beds. In this section, various hydrodynamic variables and transfer properties in three-phase systems are discussed. Specifically, areas discussed in the hydrodynamics section are minimum fluidization, bed contraction and moving packed bed phenomenon, flow regime transition, overall gas holdup and hydrodynamic similarity, and bubble size distribution and the dominant role of larger bubbles. Later in this section, important topics covering transport phenomena will be discussed, which include heat and mass transfer and phase mixing.

3.1 Hydrodynamics

Flow behavior of cocurrent upward gas–liquid flow to a bed of particles is characterized by a complex interaction among gas, liquid, and solid phases. A typical flow regime map for three-phase fluidized beds is shown in Fig. 12 (Zhang et al., 1997). Seven different flow regimes are identified in this map, which vary with gas and liquid velocities. The regime map is operating

condition specific, and the flow regime map shown in Fig. 12 is obtained in an air–water–1.5 mm glass bead three-phase fluidized bed. The fluidized bed is 8.26 cm in diameter and 2 m in height. This map puts the common three-phase fluidization operating conditions represented by the dispersed bubble and coalesced bubble flow regimes in proper perspective.

3.1.1 Minimum Fluidization

The first topic of study for the hydrodynamics of three-phase fluidization is minimum fluidization. For a given gas velocity, the minimum liquid flow rate required to fluidize a bed of particles (U_{lmf}) may generally be determined from the change in the bed dynamic pressure drop behavior that occurs as the bed changes from a fixed bed to a fluidized bed. There are considerable variations on minimum fluidization phenomena among small/light, large/heavy, and mixed particle systems. However, U_{lmf} in general can be evaluated mechanistically by considering an intrinsic condition for minimum fluidization where the total pressure drop over a bed of particles at the fixed state is equal to the total bed weight per unit bed cross-sectional area as formulated by Song et al. (1989). In this evaluation, the pressure drop in the fixed bed can be described by a flow model developed by Chern et al. (1983, 1984).

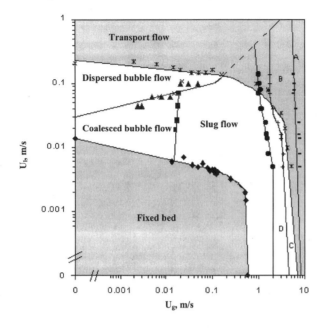

A: Two-phase annular flow; B: Two-phase bridging flow
C: Bridging flow; D: Churn flow.

(a)

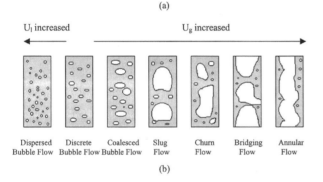

(b)

Figure 12 (a) Flow regime map for an air–water–1.5 mm glass bead three-phase fluidized bed (shaded portions refer to prefluidized and postfluidized regimes); (b) schematic diagram of corresponding flow regimes. (From Zhang et al., 1997.)

The minimum fluidization velocity can also be effectively described based on the gas-perturbed liquid model (Zhang et al., 1998a). The model assumes that the solid particles in the bed are fully supported by the liquid and there is little direct interaction between gas and solid phases. In this model, the gas phase plays a simple role in the bed by occupying space with particles fluidized by the liquid. The minimum fluidization velocity along with minimum liquid velocity for particle transport for three-phase fluidized beds of relatively large particles ($d_p > 1$ mm) (Zhang et al., 1998b) can

be well predicted by this model. Naturally, the model is more suitable to be applied to systems where bubble or wake induced flow is insignificant. For example, bed contraction, which is a typical phenomenon of the bubble wake, cannot be described by this model. The minimum fluidization velocity for non-Newtonian fluid following the power-law model can be obtained by equating the pressure drop in a fixed bed to that in a fluidized bed, similar to the fluidization of Newtonian fluids; the pressure drop in the fixed bed can be evaluated by the free-surface cell model for non-Newtonian fluids (Miura and Kawase, 1998). New approaches in examining the minimum fluidization condition include artificial neural network (Larachi et al., 2000) for predicting the minimum fluidization velocity and V-statistics (Briens et al., 1997) for identifying minimum fluidization of the bed. These approaches would not replace the more mechanistic based and currently used techniques in the quantification of minimum fluidization conditions.

3.1.2 Bed Contraction and Moving Packed Bed Phenomenon

For three-phase fluidization systems, two striking phenomena pertaining to macroscopic hydrodynamic behavior are bed contraction and moving packed bed flow. These two topics will be covered in this section.

3.1.2.1 Bed Contraction. Three-phase fluidized beds using small particles display unique bed expansion characteristics, that is, upon initial introduction of the gas into the liquid–solid fluidized bed, contraction, instead of expansion, of the bed occurs (Massimilla et al., 1959). An increasing gas flow rate causes further contraction up to a critical gas flow rate beyond which the bed expands (Turner, 1964; Oestergaard, 1964). A quantitative elucidation of the bed contraction phenomenon was reported by Stewart and Davidson (1964) and El-Temtamy and Epstein (1979). Basically, bed contraction is caused by the behavior of the bubble wake, which entraps liquid and particles and therefore is associated with large bubble systems. The entrainment of the liquid and particles by the bubble wake reduces the effective amount of liquid in the bed used to fluidize the remaining particles. The bed contraction phenomenon has been extensively studied under ambient fluidization conditions. Bed contraction has also been observed at high pressure (Jiang et al., 1997).

On the basis of the generalized wake model of Bhatia and Epstein (1974), a criterion for the bed contraction was developed (Jean and Fan, 1987). In the generalized wake model, the three-phase fluidized bed is assumed to consist of three regions, that is, the gas bubble region, the wake region, and the liquid–solid fluidized region. Bed contraction will occur when

$$
\psi = \left(\frac{n}{n-1} + k\right)\frac{U_1}{\varepsilon_1} + \frac{xk(U_g/\varepsilon_g)}{n-1}
$$
$$
- \left[(1+k)U_1 + \frac{k}{n-1}\left(\frac{U_g}{\varepsilon_g} - \frac{U_1}{\varepsilon_1}\right)\right]
$$
$$
+ xk\left(U_1 - \frac{U_1}{\varepsilon_1}\right)\left(\frac{n}{n-1}\right) < 0 \qquad (25)
$$

Here, n is the Richardson–Zaki index, k is the ratio of wake size to bubble size, and x is the ratio of solids concentration in the wake region to that in the liquid–solid fluidized region.

3.1.2.2 Moving Packed Bed Phenomenon. Moving packed bed flow is characterized by the motion of solids in piston flow in a three-phase fluidized bed. Moving packed bed flow, which usually occurs during start-up, depends not only on the gas and liquid velocities but also on how they are introduced into the bed. Moving packed bed flow is caused by the surface phenomena involving fine bubbles attached onto particles and the subsequent formation of a fine bubble blanket under the packed solids; a liquid flow would move the entire bed upward. This phenomenon is thus associated with the small bubble system. The moving packed bed flow phenomenon in a three-phase fluidized bed is a known, anomalous event in the resid hydrotreating industry. It was observed in the 1960s in the bench and pilot units during the development and commercialization of the resid hydrotreating process (Fan, 1999). The reactor was typically operated at pressures between 5.5 and 21 MPa and temperatures between 300 and 425°C. In the early 1970s, moving packed bed flow was observed in a commercial three-phase fluidized bed reactor. The occurrence of a moving packed bed in a three-phase fluidized bed could simply be circumvented by utilizing a start-up procedure that involves degassing the bed first and then introducing liquid flow to expand the bed prior to commencing gas flow. Commercial operators of three-phase fluidized bed reactors have long recognized and undertaken a proper start-up procedure of this nature since observing this anomalous event. As the small bubbles can also be generated under ambient conditions using surfactants in an air–water system, the moving packed bed flow phenomenon was reported in open literature first by Saberian-Broudjenni et al. (1984)

and later by Bavarian and Fan (1991a,b) in small columns with small bubbles generated in the same manner.

3.1.3 Flow Regime Transition

Two main flow regimes are commonly identified for three-phase fluidization systems based on the bubble flow behavior: the dispersed (or homogeneous) bubble flow and the coalesced bubble (or churn-turbulent) flow regimes. Knowledge of the transition from the homogeneous bubble flow to the churn-turbulent flow regimes is important for the design and operation of industrial reactors. The transition velocity is defined as the superficial gas velocity at which the transition from homogeneous bubble flow to churn-turbulent flow occurs. The transition velocity depends on gas distributor design, physical properties of the phases, operating conditions, and column size. The flow regime transition has been studied extensively under ambient conditions over the last three decades (Wallis, 1969; Joshi and Lali, 1984; Shnip et al., 1992; Tsuchiya and Nakanishi, 1992; Zahradnik et al., 1997). Most of these studies pointed out a critical role played by the liquid-phase turbulence during the transition and employed phenomenological models to predict the flow transition.

The regime transition is also influenced by the operating pressure, and the pressure effect has been examined by many researchers in gas–liquid systems (Tarmy et al., 1984; Clark, 1990; Krishna et al., 1991, 1994; Wilkinson et al., 1992; Hoefsloot and Krishna, 1993; Reilly et al., 1994; Letzel et al., 1997; Lin et al., 1999). The flow regime transition is normally identified based on the following approaches: instability theory, analysis of fluctuation signals, and the drift flux model. Based on the stability theory of Batchelor (1988) and Lammers and Biesheuvel (1996), the influence of pressure on the stability of bubbly flows in a nitrogen–water system was identified (Letzel et al., 1997). It is found that a higher gas density has a stabilizing effect on the flow and that the gas fraction at the instability point (i.e., transition point) increases with gas density, while the gas velocity at the instability point only slightly increases with gas density. However, the conclusion is limited to a narrow range of operating pressures (0.1 to 1.3 MPa). Using the standard deviation of the pressure fluctuation and the drift flux model, the flow transition from the homogeneous bubble flow regime to the churn-turbulent flow regime in a nitrogen–Paratherm NF heat transfer fluid system was investigated over a wide range of operating conditions (e.g., pressures up to 15.2 MPa and temperatures up to 78°C) (Lin et al., 1999). The effects of pressure and temperature on the transition velocity are clearly shown in Fig. 13a. Increasing pressure or temperature delays the regime transition. The pressure effect on the flow regime transition is mainly due to the variation in bubble characteristics, such as bubble size and bubble size distribution. Bubble size and bubble size distribution are closely associated with factors such as initial bubble size, bubble coalescence rate, and breakup rate. Under high pressure conditions, bubble coalescence is suppressed and bubble breakup is enhanced. Also, the distributor tends to generate smaller bubbles. All these factors contribute to small bubble sizes and narrow

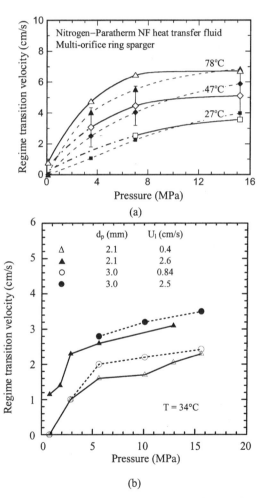

Figure 13 The regime transition velocity (a) in a bubble column. Open symbols are obtained by standard deviation of pressure fluctuation and drift flux model; closed symbols are calculated by the correlation of Wilkinson et al. (1992). (From Lin et al., 1999.) (b) In a three-phase fluidized bed. (From Luo et al., 1997a.)

bubble size distributions and consequently delay the flow regime transition at high pressures.

A correlation was also proposed to estimate the gas holdup and gas velocity at the transition point under high-pressure conditions (Wilkinson et al., 1992). This predictive scheme incorporates the concept of the bimodal bubble size distribution presented by Krishna et al. (1991), that is, the churn-turbulent regime is characterized by a bimodal bubble size distribution, consisting of fast rising large bubbles ($d_b > 5$ cm) and small bubbles (typically, $d_b < 5$ mm). They found that the transition velocity depends on the physical properties of the liquid and can be estimated by the following correlations:

$$\varepsilon_{g,tran} = \frac{U_{g,tran}}{u_{small}} = 0.5\exp(-193\rho_g^{-0.61}\mu_l^{0.5}\sigma^{0.11}) \quad (26)$$

$$u_{small} = 2.25\left(\frac{\sigma}{\mu_l}\right)\left(\frac{\sigma^3\rho_l}{g\mu_l^4}\right)^{-0.273}\left(\frac{\rho_l}{\rho_g}\right)^{0.03} \quad (27)$$

where u_{small} is the rise velocity of small bubbles. As shown in Fig. 13a, the correlation can reasonably predict the transition velocity at a given temperature and pressure when the in-situ physical properties of the fluids are used in the correlation.

The effects of pressure, particle size, and liquid velocity on the regime transition velocity in a three-phase fluidized bed are shown in Fig. 13b (Luo et al., 1997a). The transition velocity is identified by analyzing the drift flux of gas. The drift flux of gas increases with the gas holdup in the dispersed regime; in the coalesced bubble regime, the rate of increase is much larger. As the pressure increases, the transition gas velocity and the gas holdup at the transition point increase. The pressure effect on the regime transition is significant at low pressures, but the effect levels off at a pressure around 6 MPa. The transition velocity increases with liquid velocity and slightly increases with particle size, similar to the regime transition behavior at ambient conditions. The addition of fine particles to the liquid phase promotes bubble coalescence and thus accelerates the transition to the churn-turbulent regime (Clark, 1990).

3.1.4 Overall Gas Holdup and Hydrodynamic Similarity

Gas holdup is a key parameter to characterize the macroscopic hydrodynamics of three-phase fluidization systems. The gas holdup depends on gas and liquid velocities, gas distributor design, column geometry (diameter and height), physical and rheological properties of the gas and liquid, particle concentration, and physical properties of the particles. The gas holdup generally increases with gas velocity, with a larger rate of increase in the dispersed bubble regime than in the churn-turbulent regime. Such distributors as perforated plates, nozzle injectors, and spargers affect the gas holdup only at low gas velocities. The effect of column size on gas holdup is negligible when the column diameter is larger than 0.1 to 0.15 m (Shah et al., 1982). The influence of the column height is insignificant if the height is above 1 to 3 m and the ratio of the column height to the diameter is larger than 5 (Kastaneck et al., 1984). Gas holdup decreases as liquid viscosity and/or gas–liquid surface tension increase; however, the effect of liquid density is not clear. The addition of particles into a bubble column leads to a larger bubble size and thus a decreased gas holdup, especially when the particle concentration is low. The particle size effect on the gas holdup can be ignored in the particle size range of 44 to 254 μm. When highly viscous pseudoplastic liquids such as xanthan solution with an effective viscosity of 300 mPa · s are used, the gas holdup increases with decreasing liquid viscosity and particle size when the liquid velocity is less than 6 cm/s (Zaidi et al., 1990).

Numerous studies have been conducted to investigate the effects of pressure and temperature on the gas holdup in bubble columns (Deckwer et al., 1980; Tarmy et al., 1984; Idogawa et al., 1986; Kojima et al., 1991; Wilkinson et al., 1992; Reilly et al., 1994; Daly et al., 1995; Jiang et al., 1995; Inga, 1997; Letzel et al., 1997; Lin et al., 1998) and three-phase fluidized beds (Luo et al., 1997a). Empirical correlations have been proposed for gas holdup in bubble columns operated at elevated pressure and temperature. It is commonly accepted that elevated pressures lead to a higher gas holdup in both bubble columns and three-phase fluidized beds except in those systems that are operated with porous plate distributors and at low gas velocities. The increased gas holdup is directly related to the smaller bubble size and, to a lesser extent, to the slower bubble rise velocity at higher pressures (Luo et al., 1997b). Figure 14 shows bubbles emerging from the three-phase fluidized bed of Paratherm NF heat transfer fluid and 2.1 mm glass beads over a wide range of operating conditions. As shown in the figure, bubble size is drastically reduced as pressure increases. The most fundamental reason for the bubble size reduction can be attributed to the variation in physical properties of the gas and liquid with pressure.

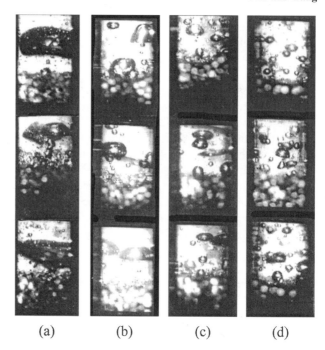

(a)   (b)   (c)   (d)

Figure 14 Visualization of bubbles emerging from the surface of a three-phase fluidized bed at (a) $P = 0.1$ MPa; (b) $P = 3.5$ MPa; (c) $P = 6.8$ MPa; (d) $P = 17.4$ MPa. (From Fan et al., 1999.)

The gas holdup in slurry bubble columns at high pressures was also investigated by some researchers (Deckwer et al., 1980; Kojima et al., 1991; Daly et al., 1995; Inga, 1997; Luo et al., 1999). A dynamic gas disengagement technique has been proven suitable for the measurement of gas holdup in a slurry bubble column under high-pressure conditions (Luo et al., 1999). The pressure effect on the gas holdup in slurry bubble columns is similar to that in bubble columns. Elevated pressures lead to higher gas holdups in a slurry bubble column. The presence of particles reduces the gas holdup at both ambient and elevated pressures as shown in Fig. 15. An empirical correlation was obtained to estimate the gas holdup in high-pressure slurry bubble columns as

$$\frac{\varepsilon_g}{1-\varepsilon_g} = \frac{2.9\left(U_g^4\rho_g/\sigma g\right)^\alpha \left(\rho_g/\rho_m\right)^\beta}{\left[\cosh\left(\mathrm{Mo}_m^{0.054}\right)\right]^{4.1}} \tag{28}$$

where Mo_m is the modified Morton number for the slurry phase, $g\left(\rho_m - \rho_g\right)\left(\xi\mu_l\right)^4/\rho_m^2\sigma^3$, and

$$\alpha = 0.21\,\mathrm{Mo}_m^{0.0079} \tag{29a}$$

$$\beta = 0.096\,\mathrm{Mo}_m^{-0.011} \tag{29b}$$

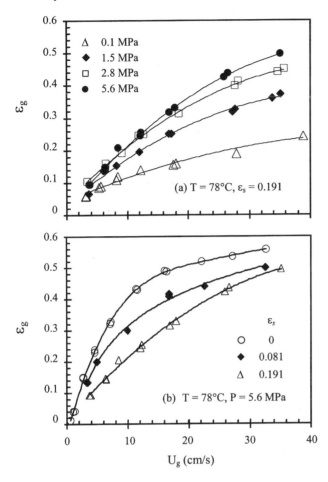

Figure 15 Effects of (a) pressure and (b) solids concentration on the gas holdup in a slurry bubble column (nitrogen-Paratherm NF heat transfer fluid–100 μm glass beads). (From Luo et al., 1999.)

A correction factor ξ accounts for the effect of particles on the slurry viscosity

$$\ln \xi = 4.6\varepsilon_s \{ 5.7\varepsilon_s^{0.58} \sinh[-0.71 \exp(-5.8\varepsilon_s)$$
$$\times \ln(\mathrm{Mo})^{0.22}] + 1 \} \qquad (30)$$

The physical meaning of the dimensionless group of $U_g^4 \rho_g / \sigma g$ in the gas holdup correlation can be shown by substituting the rise velocity of the maximum stable bubble [Eq. (24)] into the group:

$$\frac{U_g^4 \rho_g}{\sigma g} \propto \left(\frac{U_g}{u_{\max}} \right)^4 \qquad (31)$$

Clearly, the dimensionless group represents the contribution of large bubbles to the overall gas holdup, which is the major reason why the correlation is able to cover a wide range of experimental conditions.

For high-pressure bubble columns and slurry bubble columns operated under a wide range of conditions, hydrodynamic similarity requires the following dimensionless groups to be the same: $U_g/u_{\max}$, $\mathrm{Mo_m}$, and ρ_g/ρ_m (Luo et al., 1999). To simulate the hydrodynamics of industrial reactors, cold models could be used and milder pressure and temperature conditions could be chosen, as long as the three groups are similar to those in the industrial reactor. Safoniuk et al. (1999) identified eight key variables that affect hydrodynamics of three-phase fluidization systems, yielding five independent dimensionless groups according to the Buckingham PI theorem,

$$\mathrm{Mo} = \frac{g\Delta\rho\mu_l^4}{\rho_l^2 \sigma^3} \qquad \mathrm{Eo}' = \frac{g\Delta\rho d_p^2}{\sigma}$$
$$\mathrm{Re}' = \frac{\rho_l d_p U_l}{\mu_l} \qquad \beta_d = \frac{\rho_p}{\rho_l} \qquad \beta_U = \frac{U_g}{U_l} \qquad (32)$$

Gas holdup data in the freeboard region of the bed and the extent of the bed expansion from two cold models (one 8.3 cm ID and the other 91.4 cm ID) were used in their analysis, which satisfactorily verify the hydrodynamic similarity. More studies are needed based on these two approaches (Luo et al., 1999; Safoniuk et al., 1999) to arrive at general similarity rules for scale-up applications.

The hydrodynamic properties such as phase holdup for reaction systems that actually generate gas (e.g., bioreaction, Buffiere et al., 1998) and consume gas (e.g., methanol synthesis, Fan, 1989) are different from those obtained from gas injected systems. The gas holdup in the case of the bioreaction system was found to be higher than that in the gas injected system at the same gas flow rate, owing to small bubbles generated in the bioreaction system (Buffiere et al., 1998).

3.1.5 Bubble Size Distribution and Dominant Role of Large Bubbles

The bubble size can be measured by photographic or probe techniques. In multibubble systems, a mean bubble size is usually used to describe the system. The mean bubble size is commonly expressed through the Sauter, or volume-surface, mean. For a group of bubbles with measured diameters, the Sauter mean is

$$d_{vs} = \frac{\sum n_i d_{bi}^3}{\sum n_i d_{bi}^2} \qquad (33)$$

where n_i is the number of bubbles in the class i with its volume equivalent size d_{bi}.

Studies have been conducted to investigate pressure or gas density effects on the mean bubble size and bubble size distribution in bubble columns (Idogawa et al., 1986, 1987; Jiang et al., 1995; Soong et al., 1997; Lin et al., 1998) as well as in three-phase fluidized beds (Jiang et al., 1992, 1997). According to these experimental studies, pressure has a significant effect on the mean bubble diameter. The mean bubble diameter decreases with increasing pressure; however, above a certain pressure, the bubble size reduction is not significant. The effect of pressure on the mean bubble size is due to the change of bubble size distribution with pressure. At atmospheric pressure, the bubble size distribution is broad, while under high pressure, the bubble size distribution becomes narrower, as shown in Fig. 16 (Luo et al., 1999). The bubble size is mainly determined by three factors, that is, bubble formation at the gas distributor, bubble coalescence, and bubble breakup. When the pressure is increased, the bubble size at the distributor is reduced (Luo et al., 1998a),

bubble coalescence is suppressed (Jiang et al., 1995), and large bubbles tend to break up (Luo et al., 1999). The combination of these three factors causes the decrease of mean bubble size with increasing pressure.

The bubble size distribution can normally be approximated by a log-normal distribution with its upper limit at the maximum stable bubble size. The contribution of bubbles of different sizes can be examined by analyzing the relationship between overall gas holdup and bubble size distribution. In slurry bubble columns, the gas holdup can be related to the superficial gas velocity, U_g, and the average bubble rise velocity, $\bar{u}_b$ (based on bubble volume), by a simple equation:

$$U_g = \varepsilon_g \bar{u}_b \tag{34}$$

When the distributions of bubble size and bubble rise velocity are taken into account, $\bar{u}_b$ can be expressed as

$$\bar{u}_b = \frac{\displaystyle\int_{d_{b,min}}^{d_{b,max}} V_b(d_b) f(d_b) u_b(d_b)\, dd_b}{\displaystyle\int_{d_{b,min}}^{d_{b,max}} V_b(d_b) f(d_b)\, dd_b} \tag{35}$$

The outcome of Eq. (35) and the gas holdup strongly depend on the existence of large bubbles, because of their large volume and high rise velocity. An experimental study by Lee et al. (1999) revealed that, in the coalesced bubble regime, more than 70% of the small bubbles are entrained by the wakes of large bubbles and consequently have a velocity close to large bubbles. It is clear that the large bubble behavior characterizes the overall hydrodynamics due to their large volume, their high rise velocity, and their large associated wakes.

3.2 Heat Transfer

In the last several sections, hydrodynamic characteristics of three-phase fluidization systems were covered. Starting from this section, several key transport phenomena will be discussed. This section examines heat transfer characteristics in three-phase fluidization systems; mass transfer will be discussed in a later section.

Comprehensive reviews on the heat transfer behavior in various gas–liquid and gas–liquid–solid systems under ambient conditions are given by Kim and Laurent (1991), and more recently by Kumar et al. (1992, 1993a,b) and Kumar and Fan (1994). There are two types of heat transfer coefficients, that is, heat transfer coefficients between the column wall and the bed (wall-to-bed), and heat transfer coefficients

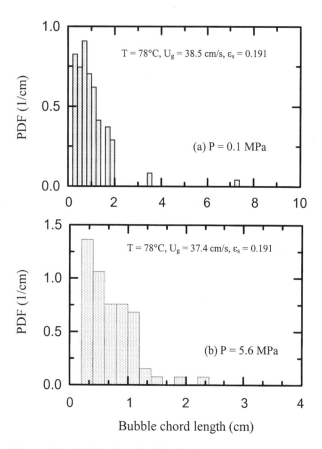

Figure 16 Bubble size distribution in a N_2-Paratherm NF heat transfer fluid–100 μm glass beads slurry system at (a) $P = 0.1$ MPa and (b) $P = 5.6$ MPa. (From Luo et al., 1999.)

between the surface of an immersed heating object and the bed (surface-to-bed). Values of these two types of heat transfer coefficients under ambient conditions have been reported extensively.

3.2.1 Effect of Operating Variables

Heat transfer in a three-phase fluidization system has a strong dependence on the physical properties of the liquid phase and a weak dependence on the gas phase properties. The heat transfer coefficient has been found to increase with an increase in the gas and liquid velocities, the size and density of the particles, the diameter of the column, and the thermal conductivity and heat capacity of the liquid. On the other hand, the heat transfer coefficient decreases with an increase in the liquid viscosity and the diameter of the immersed object (Kim and Laurent, 1991). Among various factors affecting the heat transfer behavior, the effects of phase velocities are most pronounced. The heat transfer coefficient initially increases quite rapidly with gas velocity but becomes less marked at higher gas flow rates, asymptotically approaching a maximum value (Baker et al., 1978). The plot of heat transfer coefficient vs. liquid velocity exhibits a maximum, similar to that observed in liquid–solid fluidized beds. The bed voidage corresponding to the maximum heat transfer rate decreases with increasing particle size, but increases with an increase in the liquid viscosity (Kang et al., 1985). A higher wall-to-bed heat transfer rate is found in a three-phase fluidized bed compared to that in liquid–solid and gas–liquid systems except under conditions of small particles and high gas velocities (Chiu and Ziegler, 1983).

3.2.2 Bubble Wake Effect

An understanding of heat transfer mechanisms is important for developing a heat transfer correlation and model, and studies have been conducted to investigate the mechanism of heat transfer in three-phase fluidization systems. Deckwer proposed a heat transfer mechanism in bubble columns based on Higbie's surface renewal theory combined with Kolmogoroff's theory of isotropic turbulence (Deckwer, 1980). Several other investigators extended this concept to three-phase fluidized beds by modifying the energy dissipation rate to include the increased surface renewal due to the increased turbulence created by the solid particles (Suh et al., 1985; Magiliotou et al., 1988; Suh and Deckwer, 1989). Suh et al. (1985) and Suh and Deckwer (1989) neglected the effect of particle

convective transport on heat transfer, while Magiliotou et al. (1988) proposed that the particles also contribute to heat transfer in both liquid–solid and gas–liquid–solid fluidized beds in conjunction with isotropic fluid microeddies. Being semitheoretical in approach, these studies do not fully account for the inherent mechanism underlying the heat transfer in three-phase fluidized beds.

Heat transfer characteristics in three-phase fluidized beds are intimately associated with bubble motion, bubble size, and phase holdup, which are in turn affected by the hydrodynamic behavior of the system, including wake flow. The effect of the bubble wake on heat transfer can be illustrated by the measurement of the instantaneous local variations in the heat transfer coefficient due to the passage of gas bubbles in liquid and liquid–solid systems (Kumar et al., 1993a,b; Kumar and Fan, 1994). Simultaneous visualization of the flow around the heat transfer probe can also be performed to establish the correspondence between the local hydrodynamic behavior and the instantaneous heat transfer rate. Figure 17 shows a representative example of the time-dependent heat transfer coefficient in a liquid–solid fluidized bed of low-density gel beads with the injection of gas bubbles (Kumar et al., 1993a). Associated photographs of the bubble-wake-induced liquid–solid flow patterns in the vicinity of the heat transfer probe are also shown in the figure. Each photograph corresponds to a specific heat transfer coefficient. In the photographs the bright dots are the gel particles and the bright vertical object is the heat transfer probe located at the center of the column. The injected bubble volume is 3 cm^3.

Photograph A shows the instantaneous flow field where a marked increase in the heat transfer coefficient results as the bubble approaches the heat transfer surface. It is also observed that the instantaneous heat transfer coefficient starts increasing well before the bubble approaches the lower edge of the probe, although the photograph for this case is not shown here. This increase in heat transfer coefficient is attributed to the local turbulence caused by the approaching bubble. The bubble is a spherical cap, and the wake structure appears to be symmetrical about the vertical axis of bubble movement.

Photograph B shows a large vortex in the primary wake. The vortex entrains liquid and particles from around the wake, causing rapid surface renewal at the probe surface, resulting in increased heat transfer rates. In the two-dimensional plane of visualization, the probe surface appears to lie in the vortex structure close to the wake central axis.

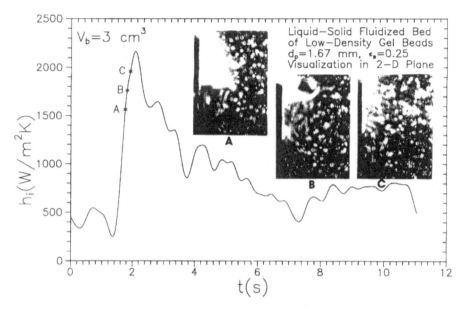

Figure 17 Variation of instantaneous heat transfer rate with bubble passage synchronized with the visualization of flow patterns in the vicinity of the heat transfer probe in a liquid–solid fluidized bed of low-density gel beads. (From Kumar et al., 1993a.)

Photograph C shows the flow pattern near the maximum in the heat transfer signal. The heat transfer surface experiences high shear flow due to the primary-wake-induced upward liquid and solid flow toward the wake central axis, causing enhancement in the heat transfer. The velocity of gel particles near the heat transfer surface is roughly estimated from the streak of particles. The upward particle velocity is in the range of 20 cm/s, which is close to the bubble rise velocity (24 cm/s), confirming that the particle is in the central wake region. Thus the maximum heat transfer rate is obtained along the wake central axis in the primary wake region. Figure 17 clearly demonstrates the importance of bubble wake behavior on the heat transfer characteristics of three-phase fluidized beds.

3.2.3 Pressure Effect

It is known that pressure has a significant effect on the hydrodynamics of three-phase fluidized beds, and it would also affect the heat transfer characteristics, because heat transfer behavior strongly depends on the hydrodynamics of the system. The effect of pressure on heat transfer is mainly through the variations in liquid properties and hydrodynamic parameters, as summarized in Table 2. The overall effect of pressure on heat transfer behavior depends on the outcome of the counteracting effects of each individual factor.

The effects of gas velocity and pressure on the surface-to-bed heat transfer coefficient in a three-phase fluidized bed are shown in Fig. 18 (Luo et al., 1997a). With an increase in pressure, the heat transfer coefficient increases, reaches a maximum at pressures of 6 to 8 MPa, and then decreases. The following correlation can be used to predict the heat transfer coefficient in three-phase fluidized beds at high pressures:

$$h = h' \varepsilon_g^{0.45} \left(\frac{0.396}{U_g^{0.45}} + \frac{0.6768}{u_{pt,0}} \right) \tag{36}$$

where h' is the heat transfer coefficient of a liquid–solid fluidized bed with the same solids holdup, and $u_{pt,0}$ is the particle terminal velocity in the fluidizing liquid at ambient pressure. The units for U_g and $u_{pt,0}$ in the equation are in m/s. The heat transfer coefficient, h', can be calculated by the correlation (Richardson et al., 1976):

$$Nu' = 0.67 \, Re^{0.62} Pr^{0.33} \frac{\varepsilon_s^{0.38}}{1 - \varepsilon_s} \tag{37}$$

The heat transfer behavior between an immersed solid surface and surrounding bulk fluids in a slurry bubble column (nitrogen-Paratherm NF heat transfer fluid—53 μm glass beads) at elevated pressures is shown in Fig. 19 (Yang et al., 2000b). It is found that pressure has a significant effect on the heat transfer characteristics in a slurry bubble column. The heat

Table 2 Variation of Various Parameters with Pressure and Their Effect on Heat Transfer Coefficients in Three-Phase Fluidization Systems

Parameter	Effect of the parameter on heat transfer coefficient	Effect of pressure increase on parametric value	Parametric effect on heat transfer coefficient with increase in pressure
μ_l	−	+	−
ρ_l	+	+ (small)	+ (small)
k_l	+	+ (small)	+ (small)
C_{pl}	+	+ (small)	+ (small)
σ	No direct effect	−	No direct effect
d_b	+	−	−
ε_g	+	+	+

+: Increase; − decrease.
Source: Luo et al. (1997a).

transfer coefficient decreases appreciably with increasing pressure. The variation in heat transfer coefficient with pressure is attributed to the counteracting effects of the increased liquid viscosity, decreased bubble size, and increased gas holdup or frequency of bubble passage over the heating surface as the pressure increases. In slurry bubble columns, it is observed that the bubble size reduces significantly with an increase in pressure, especially under low pressures, which would result in a decrease in the heat transfer coefficient. Therefore the bubble size is the most important factor affecting the

heat transfer rate in slurry bubble columns. The addition of fine particles to the liquid phase enhances heat transfer substantially, and the effect of temperature on the heat transfer behavior is mainly determined by the change in liquid viscosity.

It is noted that the pressure effects on the heat transfer coefficient are different between large-particle and small-particle systems (Luo et al., 1997a; Yang et al., 2000b). Similar observations were also found for hydrodynamics and bubble characteristics between large-particle and small-particle systems. Luewisutthichat et al. (1997) photographically studied bubble characteristics in multiphase flow systems. They found that large-particle systems (i.e., three-phase fluidized beds) exhibit appreciably different

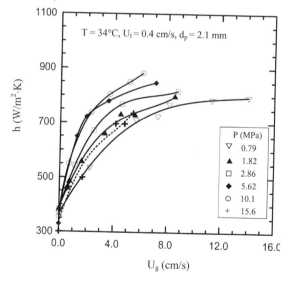

Figure 18 Heat transfer coefficients as a function of gas velocity at different pressures in a three-phase fluidized bed (nitrogen-Paratherm NF heat transfer fluid–2.1 mm glass beads. (From Luo et al., 1997a.)

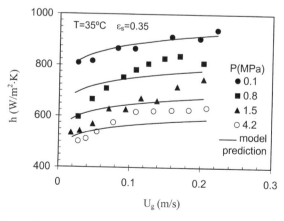

Figure 19 Effect of pressure and gas velocity on heat transfer coefficients in a slurry bubble column (nitrogen-Paratherm NF heat transfer fluid–53 μm glass beads). (From Yang et al., 2000b.)

bubble behavior from small-particle systems (i.e., slurry bubble columns). In small-particle systems, the bubble flow behavior is similar to that in two-phase systems, while large-particle systems exhibit a broad bubble size distribution with a large Sauter mean diameter. Therefore the effect of pressure on heat transfer behavior, which is closely dependent on the hydrodynamics and bubble characteristics, would be different in three-phase fluidized beds and slurry bubble columns. In a three-phase fluidized bed, the variation of bubble size with pressure may become less important in affecting the heat transfer coefficient, while the increase in the gas holdup or the frequency of bubble passage over the heating surface with pressure could be a dominant factor in determining heat transfer behavior. Therefore the heat transfer coefficient would increase with increasing pressure, as observed by Luo et al. (1997a).

Various correlations are also developed to predict the heat transfer coefficient in slurry bubble columns. For systems where the gas holdup is not affected by pressure, the following correlation can be used to estimate the heat transfer coefficient (Deckwer et al., 1980):

$$St_m = 0.1 \left(Re_m Fr\, Pr_m^2 \right)^{-0.25} \tag{38}$$

This correlation is obtained based on the surface renewal model and Kolmogoroff's theory of isotropic turbulence. In the model, the liquid–solid suspension is considered as a homogeneous phase, and consequently an estimation scheme of the physical properties of the suspension from the individual phase is required.

For systems in which pressure has a significant influence on the hydrodynamics, the effect of gas holdup on the heat transfer coefficient needs to be considered, and the following correlation can be used to quantify the heat transfer coefficient in such systems (Yang et al., 2000b):

$$St_m = 0.037 \left[\left(Re_m\, Fr\, Pr_m^{1.87} \right) \left(\frac{\varepsilon_g}{1 - \varepsilon_g} \right) \right]^{-0.22} \tag{39}$$

3.2.4 Heat Transfer Model

The consecutive film and surface renewal model originally developed by Wasan and Ahluwalia (1969) can be used to analyze the heat transfer behavior in slurry bubble columns (Yang et al., 2000b). For fine particles, the liquid–solid suspension can be reasonably treated as a pseudohomogeneous fluid phase. The model assumes that a thin fluid film (liquid or liquid–solid mixture) with a thickness δ exists surrounding the heat-

ing surface; and fluid elements are forced to contact the outer surface of the film due to the passage of bubbles. The fluid elements contact the film for a short time, t_c, and then are replaced by fresh fluid elements. The heat is transferred to the bulk fluid through conduction by the fluid film and unsteady state conduction by the fluid elements. The heat transfer coefficient is then expressed in terms of the physical properties of the fluid, the film thickness, and the contact time between the fluid elements and the film (Wasan and Ahluwalia, 1969):

$$h = \frac{2k_l}{\sqrt{\pi \alpha t_c}} + \frac{k_l \delta}{\alpha t_c} \left[e^{\alpha t_c / \delta^2} \left(1 - \mathrm{erf}\, \frac{\sqrt{\alpha t_c}}{\delta} \right) - 1 \right] \tag{40}$$

Based on Eq. (40), the heat transfer coefficient is a function of film thickness and contact time between the fluid elements and the film. A thinner film and shorter contact time lead to a higher heat transfer rate. On the basis of the border diffusion layer model (Azbel, 1981), the order of magnitude of the film thickness is estimated by (Kumar and Fan, 1994)

$$\delta = \frac{6.14L}{Re_m^{3/4}} Pr_m^{-1/3} \tag{41}$$

where Re_m is equal to $\rho_m L u_b / \mu_m$. Assuming that the contact time between the fluid elements and the film is equal to the contact time between the bubbles and the film, that is, the bubble motion is considered as the driving force for the fluid element replacement, the contact time can be estimated from

$$t_c = \frac{L}{u_b} \tag{42}$$

where u_b is the actual bubble rise velocity in a stream of bubbles (Kumar and Fan, 1994). By considering the pressure effects on the physical properties of the liquid and bubble characteristics such as bubble size and bubble rise velocity, this model can be used to analyze the heat transfer behavior in a high-pressure system. As shown in Fig. 19, this model can reasonably predict the pressure effect on the heat transfer coefficient in slurry bubble columns.

3.3 Mass Transfer

Mass transfer is an important component of transport phenomena in multiphase reactor operation. Two topics will be discussed in this section, i.e., interfacial area and liquid-side mass transfer coefficient. For mass transfer phenomena in gas–liquid or gas–liquid–solid systems, the interfacial area and the liquid-side mass transfer coefficient are considered the most important transport properties. The liquid-side mass transfer coefficient incorporates the effects of the liquid flow field surrounding rising gas bubbles. The interfacial area reflects the system bubble behavior. The most common approach in treating gas–liquid mass transfer is to combine the mass transfer coefficient and interfacial area terms into a single volumetric mass transfer coefficient ($k_l a$) averaged over the entire column height (Fan, 1989).

3.3.1 Interfacial Area

The interfacial area depends on bubble size and the number of bubbles in the system, and can be expressed by the equation

$$a = \frac{6\varepsilon_g}{d_b} \qquad (43)$$

An increase in pressure increases the gas holdup and decreases the bubble size. Therefore a significant increase in the interfacial area at high pressures is expected, which would result in an overall increase of $k_l a$.

Typical results regarding the influence of pressure on the specific gas–liquid interfacial area in a bubble column are shown in Fig. 20 (Oyevaar et al., 1991). The interfacial area in the figure was determined by means of CO_2 absorption into aqueous solutions of diethanolamine (DEA) in a column of 8.1 cm ID. The pressure has a positive effect on the interfacial

area. The study in a large bubble column with a diameter of 15.6 cm also confirmed this conclusion (Stegeman et al., 1996). The positive influence of the operating pressure on the interfacial area is clearly attributed to smaller bubble size and higher gas holdup at high pressures. The increase of the interfacial area with pressure is more pronounced at higher gas velocities.

On the basis of Eq. (43) and the observation that the Sauter mean bubble diameter is proportional to $\rho_g^{-0.11}$, Wilkinson et al. (1992) proposed a procedure to estimate the interfacial area in a bubble column under high pressure as shown in Eq. (44):

$$
\begin{aligned}
\frac{a(\text{high pressure})}{a(\text{atmospheric})} &= \frac{\varepsilon_g(\text{high pressure})}{\varepsilon_g(\text{atmospheric})} \\
&\quad \cdot \left[\frac{\rho_g(\text{high pressure})}{\rho_g(\text{atmospheric})}\right]^{0.11}
\end{aligned} \qquad (44)
$$

Equation (44) can be used to estimate the interfacial area at high-pressure conditions based on the atmospheric data. However, this procedure needs to be further verified.

Studies of the interfacial area in a packed-bed column at elevated pressures indicate the insignificant effect of pressure for three-phase systems (Molga and Westerterp, 1997). The possible explanation for the lack of pressure effect in three-phase fluidized beds is that the column packing controls the bubble coalescence and breakup process. Therefore the variation of bubble size with pressure is relatively smaller with packing than without. A comparison of interfacial areas obtained in bubble columns with packing and without packing indicates higher interfacial areas in the packed-bed bubble column, which implies that packing tends to break up the bubbles.

3.3.2 Liquid-Phase Mass Transfer Coefficient

The second important parameter to characterize mass transfer behavior is the liquid-phase mass transfer coefficient, k_l. The effect of pressure on the liquid-phase mass transfer coefficient is complicated. Since the liquid-phase mass transfer coefficient depends on the gas–liquid system and the pressure affects the physical properties of gas and liquid phases significantly, a change in the liquid-phase mass transfer coefficient at high pressures is also expected. The following illustrates expected relations between the liquid-phase mass transfer coefficient and the physical properties of the liquid phase.

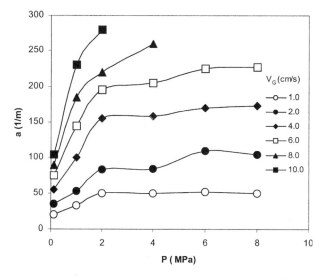

Figure 20 Pressure effect on interfacial area at different gas velocities in a bubble column with perforated plate gas distributor (nitrogen–aqueous diethanolamine solution, $D_c = 8.1$ cm). (From Oyevaar et al., 1991.)

Normally, an increase in pressure results in a decrease in surface tension and an increase in liquid viscosity (Lin et al., 1998). The decrease of surface tension would reduce the liquid flow over the surfaces of rising gas bubbles, resulting in a reduction of bubble rise velocity and a longer contact time between the liquid and bubbles (Chang and Morsi, 1992). Since the mass transfer coefficient is inversely proportional to the square root of the contact time (Higbie, 1935), k_l would decrease as the pressure increases due to the decrease in surface tension. The increase of liquid viscosity with pressure would also decrease the mass transfer coefficient, since k_l is inversely proportional to the liquid viscosity (Calderbank and Moo-Young, 1961). Furthermore, an increase in pressure also reduces the bubble size, and hence the liquid-phase mass transfer coefficient, since k_l is known to be lower for small bubbles (Calderbank and Moo-Young, 1961). Based on the above analysis, the liquid-phase mass transfer coefficient k_l would decrease with increasing pressure. Most experimental data available in the literature are expressed by the volumetric mass transfer coefficient, $k_l a$. Since the interfacial area increases significantly with increasing pressure, it is expected that the variation of interfacial area with pressure will be the predominant factor in determining $k_l a$.

The following are some experimental studies to quantify the pressure effect on the volumetric mass transfer coefficient. Wilkinson et al. (1994) studied the influence of pressure on the volumetric mass transfer coefficient in a bubble column with an inner diameter of 0.158 m. The uncatalyzed oxidation of sodium sulfite was chosen as a model reaction for determining the volumetric mass transfer coefficient. The superficial gas velocity and operating pressure varied in the ranges of 0–0.15 m/s and 0.1–0.4 MPa, respectively. Their experimental results showed that both the interfacial area and the volumetric mass transfer coefficient increase with increasing pressure, especially at high gas velocities. At low superficial gas velocities ($U_g < 0.03$ m/s), the influence of pressure is relatively small. Similar pressure effect on the volumetric mass transfer coefficient was also observed by Letzel (1997) using a dynamic pressure-step method developed by Linek et al. (1989, 1993). Figure 21 shows the effect of pressure on the volumetric mass transfer coefficient in the nitrogen-water system (Letzel, 1997). Based on experimental results, a procedure for the estimation of the liquid volumetric mass transfer coefficient at high pressures is proposed, which is similar to that for predicting the interfacial area (Wilkinson et al., 1992),

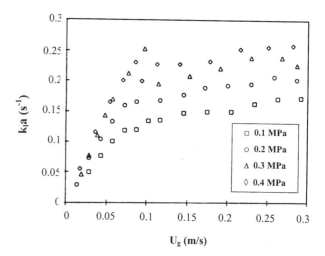

Figure 21 Volumetric mass transfer coefficient, $k_l a$, as a function of superficial gas velocity at different pressures in a nitrogen–water system. (From Letzel, 1997.)

$$\frac{k_l a(\text{high pressure})}{k_l a(\text{atmospheric})} = \left[\frac{\varepsilon_g(\text{high pressure})}{\varepsilon_g(\text{atmospheric})}\right]^n$$

$$\text{with} \quad n = 1.0\text{–}1.2 \quad (45)$$

The limitation of available high-pressure mass transfer studies is that the measuring technique is only suitable for low-pressure conditions (normally less than 1.0 MPa); thus further studies to cover wide ranges of pressure and gas velocity are needed.

Although some work has been conducted to study the mass transfer behavior in bubble columns at elevated pressures, little is known for gas–liquid–solid systems, especially with fine particles. Muroyama et al. (1997) studied mass transfer from an immersed cylinder to the bed and found that a 50 to 75% enhancement of mass transfer coefficients for using fine particles ($d_p = 0.25$–0.5 mm) over coarse particles ($d_p = 1.1$ mm). This finding is of interest to electrode applications of fluidized beds. Using the electrochemical method of Reiss and Hanratty (1963), Nikov and Delmas (1992) studied the local liquid–solid mass transfer coefficients of a particle fixed in the bed. The time-averaged local mass transfer coefficients are related to the shear stress distribution around the particle. They found that local liquid–solid mass transfer coefficients vary with the polar angle measured with respect to the frontal stagnant point. The presence of gas increases the liquid–solid mass transfer coefficient as well as the velocity gradient and the shear force. An increase in the bubble diameter and the number of bubbles around the fixed particle results in an increase in the local mass transfer coefficient as well as the velocity gradient around the particle, particularly in the frontal zone.

3.4 Phase Mixing

Backmixing is a flow pattern that is intermediate between the two ideal cases of plug flow and perfect mixing. Backmixing of individual phases in three-phase fluidization systems is an important parameter for designing reactors and predicting reactor performance. Generally, in three-phase fluidized beds appreciable backmixing occurs in the liquid and solid phases, and relatively little backmixing exists in the gas phase. The extents of backmixing of both the liquid and the solid phase depend heavily on the rising motion of the gas bubbles. In this section, liquid-phase mixing will be emphasized, and recent development in solids mixing will be introduced briefly.

Liquid-phase mixing in three-phase fluidized beds can be described using the dispersion model. A two-dimensional model considers both radial and axial dispersions. Both axial and radial dispersion coefficients are strong functions of operating conditions such as liquid and gas velocities and properties of liquid and solid phases. Evaluations of liquid-phase dispersion coefficients are based on a tracer injection method and subsequent analysis of the mean and the variance of the system response curves.

Typically, there are two ways to inject tracers, steady tracer injection and unsteady tracer injection. It has been verified that both methods lead to the same results (Deckwer et al., 1974). For the steady injection method, a tracer is injected at the exit or some other convenient point, and the axial concentration profile is measured upward of the liquid bulk flow. The dispersion coefficients are then evaluated from this profile. With the unsteady injection method, a variable flow of tracer is injected, usually at the contactor inlet, and samples are normally taken at the exit. Electrolyte, dye, and heat are normally applied as the tracer for both methods, and each of them yields identical dispersion coefficients. Based on the assumptions that the velocities and holdups of individual phases are uniform in the radial and axial directions, and the axial and radial dispersion coefficients, E_{zl} and E_{rl}, are constant throughout the fluidized bed, the two-dimensional unsteady-state dispersion model is expressed by

$$\frac{\partial C}{\partial t} = E_{zl}\frac{\partial^2 C}{\partial z^2} + \frac{E_{rl}}{r}\frac{\partial}{\partial r}\left(r\frac{\partial C}{\partial r}\right) - \frac{U_l}{\varepsilon_l}\frac{\partial C}{\partial z} \qquad (46)$$

Equation (46) can be reduced to the axial dispersion (one-dimensional) model when the radial dispersion is negligible in comparison with the axial dispersion.

The studies regarding the axial liquid-phase mixing in gas–liquid systems at atmospheric conditions are extensive, especially for the air–water system (Fan, 1989). Recently, some work has been conducted to study the liquid mixing under high-pressure conditions. Houzelot et al. (1985) measured the axial dispersion of the liquid phase in a bubble column with a diameter of 5 cm. Their study found an insignificant pressure effect on the axial dispersion of the liquid phase; their study was limited by narrow experimental conditions, that is, very low superficial gas velocity (< 6 mm/s) and pressure (< 3 atm). Under such conditions, the flow is always in the homogeneous bubbling regime and a significant change in liquid-phase mixing within such narrow operating conditions is not expected. Sangnimnuan et al. (1984) experimentally investigated the extent of liquid-phase backmixing under coal hydroliquefaction conditions (e.g., temperature between 164 and 384°C and pressure between 4.5 and 15 MPa) in a very small bubble column reactor (1.9 cm in diameter). They did not describe the effect of pressure on the axial mixing, and their study was limited by the small scale of the reactor.

Holcombe et al. (1983) determined the liquid axial-dispersion coefficient in a 7.8 cm diameter bubble column under pressures in the range of 3.0–7.1 atm. The superficial gas velocity varied up to 0.6 m/s. They used heat as a tracer to measure the thermal dispersion coefficient, which was found to be comparable to the mass dispersion coefficient. The effect of pressure on thermal dispersion coefficients was negligible in the pressure range of their study. Wilkinson et al. (1993) measured the liquid axial dispersion coefficient in a batch-type bubble column of 0.158 m in diameter for the nitrogen–water system at pressures between 0.1 and 1.5 MPa using an electrical conductivity cell. They found that the liquid axial dispersion coefficient actually increases with increasing pressure, especially under high gas velocity conditions (> 0.10 m/s). It is noted that the available theories in the literature to describe liquid mixing at atmospheric pressure cannot explain the pressure effect observed in their study. Tarmy et al. (1984) investigated liquid backmixing in industrial coal liquefaction reactors using radioactive tracers. The operating pressure in their study varied up to 17 MPa, and they found that the measured dispersion coefficients at high pressures were up to 2.5 times smaller than the values predicted by literature correlations, which were developed based on the experimental data under ambient conditions. Studies of gas–liquid dispersion behavior in coal liquefaction reactors using a neutron absorption tracer technique indicated a similar trend of the pressure effect (Onozaki et al., 2000a,b). The axial dispersion coefficients of the liquid phase under coal liquefaction conditions (P = 16.8–18.8 MPa) are smaller than those estimated from literature correlations. These observations imply a decreasing trend for the pressure effect on liquid mixing.

It has been shown that the steady-state thermal dispersion method is suitable for measuring the axial dispersion coefficients of the liquid phase at high pressures in bubble columns, particularly for hydrocarbon liquids in which the electrical conductivity technique is not applicable (Yang and Fan, 2002). In this technique, heat is introduced close to the outlet of the liquid phase and the upstream temperature profile in the liquid is measured. The dispersion coefficient is then determined from the axial temperature profile based on the one-dimensional dispersion model. The effects of pressure and gas velocity on the axial dispersion coefficients in the 5.08 cm and 10.16 cm columns for the nitrogen-Paratherm heat transfer fluid system are shown in Fig. 22 (Yang and Fan, 2002). The axial dispersion coefficient decreases significantly when the pressure is increased from ambient to elevated pressures for both columns, indicating distinct flow behaviors for ambient pressure and elevated pressure. When the pressure is further increased, the decrease rate of the axial dispersion coefficient becomes smaller. The pressure effect is more pronounced at higher gas velocities and in larger columns. A decrease in the liquid mixing at high pressures was also observed by Tarmy et al. (1984) and Onozaki et al. (2000a,b). Other researchers found that an insignificant effect of pressure on liquid mixing in small columns (diameter normally less than 10.0 cm) is possibly due to the narrow operating conditions in their studies, that is, either low gas velocities or a narrow pressure range (Houzelot et al., 1985; Holcombe et al., 1983). As shown in Fig. 22a, in small columns, the variation of the axial dispersion coefficient with pressure is not pronounced at low gas velocities (i.e., in the homogeneous bubbling flow regime) and low pressures.

The effects of gas velocity and pressure on liquid mixing can be explained based on the mixing mechanisms of gross liquid circulation and local turbulent fluctuations. The available theories in the literature describing liquid mixing under atmospheric pressure are based on liquid turbulence induced by rising bubbles (Baird and Rice, 1975), large-scale liquid internal circulation (Joshi, 1980), or a combination of these two

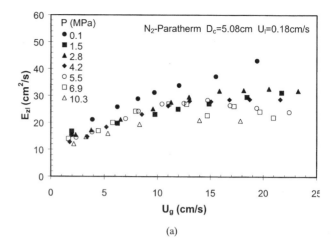

(a)

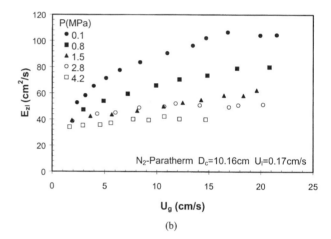

(b)

Figure 22 Pressure effect on liquid mixing in different sizes of bubble columns: (a) $D_c = 5.08$ cm and (b) $D_c = 10.16$ cm. (From Yang and Fan, 2002.)

mechanisms (Degaleesan et al., 1997). Based on the internal circulation model, Joshi (1980) proposed the following equation to predict the average liquid circulation velocity in bubble columns:

$$V_c = 1.31\left[gD_c\left(U_g - \frac{\varepsilon_g}{1 - \varepsilon_g}U_l - \varepsilon_g U_{b\infty}\right)\right]^{1/3} \quad (47)$$

where $U_{b\infty}$ is the terminal bubble rising velocity. The second term on the right-hand side of Eq. (47) is normally negligible compared to the other two terms at low liquid velocity. It can be seen that both the gas holdup and the bubble rising velocity affect the liquid circulation velocity, which can be treated as a measure of the extent of liquid circulation effect on the liquid mixing. It is known that the gas holdup increases and the bubble size and rising velocity decrease with increasing pressure. The counteracting effects of gas hold-up and bubble rise velocity on the liquid circulation velocity may result in unchanged or slightly changed liquid circulation velocity at elevated pressures, as shown in Eq. (47). On the other hand, when the system pressure increases, bubbles become smaller, and the liquid entrainment by the bubble wake and the turbulence induced by the motion of bubbles are reduced. Therefore liquid mixing is reduced at elevated pressures. The combination of variations of liquid-phase turbulent fluctuations and internal liquid circulation gives rise to the overall effect of pressure on the liquid mixing.

Recent study of flow fields and Reynolds stresses in high-pressure bubble columns using laser Doppler velocimetry (LDV) also confirmed that the bubble-induced turbulence of the liquid phase is depressed at higher pressures (Lee et al., 2001). Figure 23 shows the effect of pressure on the profiles of axial liquid velocity and Reynolds normal stress measured by the LDV technique. The ensemble-averaged velocity shows the gross circulation pattern, and as the pressure increases, the magnitudes of the averaged velocity of the liquid phase at the center and in the wall region decrease. The Reynolds stresses decrease significantly with increasing pressure, indicating the smaller extent of liquid-phase fluctuations at high pressures. The fluctuations of the liquid phase are mainly caused by the motion of gas bubbles, and the fluctuations are damped out as the pressure increases because of a narrower bubble size distribution with a smaller mean bubble size.

Recently, research in mixing behaviors of both the liquid and the solid phase in three-phase fluidized beds under ambient conditions has continued. Chen et al. (1995) and Zheng et al. (1995) studied the axial variations of the gas holdup, axial liquid mixing, and gas–liquid mass transfer. They observed a significant increase of the axial dispersion coefficients of the liquid phase and a decrease of the gas–liquid mass transfer coefficients with an increase of the axial distance under certain operating conditions, particularly under high liquid flow rates. They attributed the significant axial variation in part to the coexistence of the dispersed bubble and coalesced bubble regimes along the axial direction. Kim et al. (1992) proposed empirical correlations to account for the axial dispersion coefficient of the liquid phase over a wide range of operating conditions:

$$\mathrm{Pe}_z = \left(\frac{d_p U_1}{E_{zl}}\right) = 11.96 \left(\frac{U_1}{U_1 + U_g}\right)^{1.03} \left(\frac{d_p}{D_c}\right)^{1.66}$$

(48a)

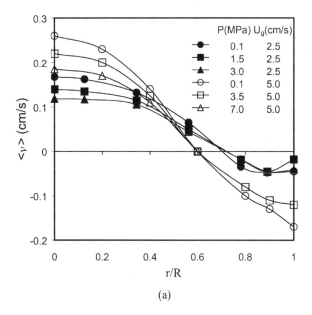

(a)

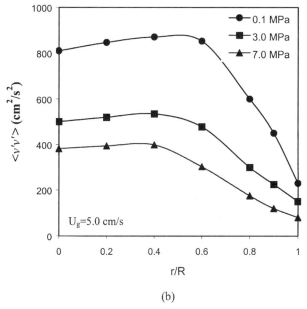

(b)

Figure 23 Pressure effects on the profiles of (a) axial liquid velocity; (b) Reynolds axial normal stress in a bubble column ($D_c = 5.08$ cm, nitrogen–water). (From Lee et al., 2001.)

for the bubble coalescing regime with a range of variables of $0.004 \leq (d_p/D_c) \leq 0.024$ and $0.250 \leq [U_1/(U_1 + U_g)] \leq 0.857$, and

$$\mathrm{Pe}_z = \left(\frac{d_p U_1}{E_{zl}}\right) = 20.72 \left(\frac{U_1}{U_1 + U_g}\right)^{1.03} \left(\frac{d_p}{D_c}\right)^{1.66}$$

(48b)

for the bubble disintegrating regime with a range of variables of $0.004 \leq (d_p/D_c) \leq 0.012$ and $0.143 \leq [U_l/(U_l + U_g)] \leq 0.857$.

The degree of solids mixing is a crucial component for such application as bioreactors. For this type of application, the uniformity of the particle properties is important for determining the overall rate of reaction, and this uniformity of solids is strongly dependent upon the degree of solids mixing. For large bubble systems, the mechanism for solids mixing is primarily through the bubble wake. For small bubble systems or any bubble system with light particles, particle drift effects or random motion of the particles can cause solids mixing. Particle motion can be induced by velocity gradients caused by passing gas bubbles or from local nonuniformities in the flow field. A detailed introduction to solids mixing behavior was given in Fan's 1989 book; this section is an update of recent studies on solids mixing.

The turbulence of the solid phase in three-phase fluidization measured using radioactive particle tracking (RPT) was found to be anisotropic (Cassanello et al., 1995). The same technique was also used to examine the mixing behavior of a binary mixture of solid particles (Cassanello et al., 1996; Kiared et al., 1997). The results indicated that there are two separate regions containing solid particles: the upward flow in the core region established by bubble wakes (wake phase) and the downward flow in the wall region (emulsion phase) with cross-flow between two regions, which could form the basis for a macromixing model to describe the solids residence time distribution and solids mixing in the three-phase fluidized bed. In a more phenomenological approach, axial mixing and segregation of solids particles in a slurry bubble column or a three-phase fluidized bed are commonly described by a sedimentation–dispersion model (Matsumoto et al., 1997; Hidaka et al., 1998). There are various equations employed for the sedimentation–dispersion model and a general form can be given as

$$\frac{\partial C_s}{\partial t} = \frac{\partial}{\partial z}\left[E\frac{\partial C_s}{\partial z} - (V - U)C_s\right] \quad (49)$$

where C_s is the solids concentration in a solid–liquid mixture, E is the solid phase axial dispersion coefficient, and V and U are liquid and solids velocities, respectively.

A theoretical analysis of the sedimentation–dispersion model by Jean et al. (1989) indicated that in a common expression of the sedimentation–dispersion model with U_l (defined as the cross-sectional averaged

linear liquid velocity) as V and u_t (defined as solids settling velocity) as U, under the coalesced bubble flow regime and low liquid velocity or slow slurry flow conditions, E and u_t do not have their alleged physically compatible definitions and should be regarded as purely empirical parameters and as bearing no physical significance. Under the dispersed bubble flow regime and high liquid velocity or fast slurry flow conditions, the physically compatible definitions for E and u_t would hold for this model (Jean et al., 1989). The general prediction for u_t is available in the literature (e.g., Di Felice, 1995). The surface property of particles (e.g., biofilm-coated particles compared to regular spherical particles with the same density and size) may affect the drag coefficient of the particle and hence the value of the terminal settling velocity. Thus sedimentation and dispersion behavior of these two particles would be different in a three-phase fluidized bed.

4 COMPUTATIONAL FLUID DYNAMICS

The earlier studies mainly focused on experimentally examining the macroscopic fluid dynamic behavior of three-phase fluidized beds, and developing empirical correlations. With increasing computer power, the employment of the computational fluid dynamic approach has gained considerable attention. Over the past decade, significant advances have been made in numerical modeling of two-phase flows such as gas–solid and gas–liquid flows. For example, the two-fluid model (e.g., Gidaspow, 1994) and the discrete particle model (e.g., Tsuji et al., 1993) were developed to simulate collision-dominant gas–solid flows. The direct numerical simulation technique based on the front-capturing method (e.g., Boris, 1989), the tracking method, which includes the volume-tracking (e.g., Harlow and Welch, 1965) and the front-tracking (e.g., Unverdi and Tryggvason, 1992) methods, and the lattice Boltzmann method (e.g., Sankaranarayanan et al., 2002) were developed to capture the free surfaces of gas–liquid flows.

An understanding of three-phase flows is still limited because of complicated phenomena underlying interactions between phases, such as the particle–bubble interaction and the liquid interstitial effect during particle–particle collision. Recently, several computational fluid dynamics models were reported to simulate three-phase fluidization behavior (Gidaspow et al., 1994; Grevskott et al., 1996; Mitra-Majumdar et al., 1997). These models are based on the multifluid approach in which both the gas and the solid phase are treated as pseudocontinuous phases. Alternatively, a second-order moment three-phase turbulence model can be used. This approach was demonstrated satisfactorily in the simulation of the gas–liquid flow in a bubble column (Zhou et al., 2002) and is readily extendable to the gas–liquid–solid flow simulation. There is, however, a practical need for a discrete simulation that illustrates the inherent discrete flow characteristics of individual phases such as the bubble wake structure.

The discrete phase simulation method has demonstrated its potential in simulating three-phase flows (Li et al., 1999; Zhang et al., 2000). In this approach, the Eulerian volume-averaged method, the Lagrangian discrete particle method (DPM), and the volume-of-fluid (VOF) volume-tracking method can be employed to describe the motion of liquid, solid particles, and gas bubbles, respectively. A bubble induced force (BIF) model, a continuum surface force (CSF) model, and Newton's third law are applied to account for the couplings of particle–bubble, bubble–liquid, and particle–liquid interactions, respectively. A close distance interaction (CDI) model is included in the particle–particle collision analysis, which considers the liquid interstitial effect between colliding particles (Zhang et al., 1999).

In the following, the computational models for each individual phase will be discussed. Relevant underlying equations will be given, and the assumptions of each model will be explained. Finally, simulated results of several important phenomena in three-phase flows will be demonstrated.

4.1 Liquid-Phase Model

The governing equations for the continuous phase of multiphase flows can be derived from the Navier–Stokes equations for single-phase flows. Considering the existence of dispersed particles, a volume-averaging technique is used to develop a set of partial differential equations to describe the mass and momentum conservation of the liquid phase. The continuity equation for the liquid phase can be given as

$$\frac{\partial \varepsilon_l}{\partial t} + \nabla \cdot (\varepsilon_l \boldsymbol{v}) = 0 \tag{50}$$

The momentum equation for the liquid phase is

$$\rho_l \frac{\partial (\varepsilon_l \boldsymbol{v})}{\partial t} + \rho_l \nabla \cdot (\varepsilon_l \boldsymbol{v}\boldsymbol{v}) = -\nabla p + \nabla \tau + \rho_l \boldsymbol{g} + \boldsymbol{f}_b \tag{51}$$

where $\boldsymbol{v}$ is the liquid velocity vector; ε_l is the liquid holdup; ρ_l is the liquid density; p is the scalar pressure;

τ is the viscous stress tensor; $\boldsymbol{g}$ is gravitational acceleration; $\boldsymbol{f}_b$ is the total volumetric body force acting on the liquid phase excluding the gravitational force, that is, the volumetric particle–fluid interaction force ($\boldsymbol{f}_{pf}$) plus the volumetric bubble–fluid interaction force ($\boldsymbol{f}_{bf}$). Based on Newton's third law of motion, the force acting on a particle from the liquid phase, $\boldsymbol{F}_{fp}$, yields a reaction force on the liquid. Therefore the momentum transfer from particles to the liquid–gas phase is taken into account in Eq. (51) by adding the volumetric particle–fluid interaction force, $\boldsymbol{f}_{pf}$, given below to the body force term,

$$\boldsymbol{f}_{pf} = -\frac{\sum \boldsymbol{F}_{fp}^k}{\Delta V_{ij}^k} \qquad \boldsymbol{x}_p^k \in \Omega_{ij} \tag{52}$$

where the subscript ij defines the location of a computational cell; Ω and ΔV are the domain and volume of this cell, respectively; $\boldsymbol{x}_p^k$ is the location vector of particle k; $\boldsymbol{F}_{fp}$ is the fluid–particle interaction force acting on any individual particles, which includes the drag ($\boldsymbol{F}_D$), buoyancy ($\boldsymbol{F}_B$), added mass ($\boldsymbol{F}_{AM}$), and Basset force ($\boldsymbol{F}_{BA}$):

$$\boldsymbol{F}_{fp} = \boldsymbol{F}_D + \boldsymbol{F}_B + \boldsymbol{F}_{AM} + \boldsymbol{F}_{BA} \tag{53}$$

It is noted that the buoyancy force is also included in Eq. (53) as one of the fluid–particle interaction forces. Due to the small particle size, the Saffman and Magnus forces are ignored in Eq. (53).

The bubble–fluid interaction force, $\boldsymbol{f}_{bf}$, is obtained by using a continuum surface force (CSF) model (Brackbill et al., 1992)

$$\boldsymbol{f}_{bf} = \pm \delta(\boldsymbol{x}, t) \sigma \kappa(\boldsymbol{x}, t) \nabla \alpha(\boldsymbol{x}, t) \tag{54}$$

where the plus sign is used for the liquid phase and the minus sign for the gas phase; $\delta(\boldsymbol{x}, t)$ is the Delta function, which equals 1 at the gas–liquid interface and 0 elsewhere; σ is the surface tension of the interface; $\kappa(\boldsymbol{x}, t)$ is the curvature of the free surface; $\alpha(\boldsymbol{x}, t)$ is the volume fraction of the fluid.

The Newtonian viscous stress tensor is used in Eq. (51) and given as

$$\tau = 2\mu_l \boldsymbol{S} = \mu_l [(\nabla \boldsymbol{v}) + (\nabla \boldsymbol{v})^T] \tag{55}$$

where $\boldsymbol{S}$ is the rate-of-strain tensor and μ_l is the liquid viscosity.

The liquid properties on a particle are obtained by an area-weighted averaging method based on the properties at the four grid points of the computational cell containing the particle. The cell-averaged liquid holdup obtained as the volume fraction of the liquid in the computational cell is only used for solving the volume-averaged equations of the liquid phase. When evaluating the particle drag coefficient in the liquid–solid medium, the liquid holdup is based on particle-centered averaging over a prescribed area.

4.2 Gas-Phase Model

The simulation of the flow inside the gas bubble is important for predicting bubble behaviors under high-pressure conditions, since the effect of gas density on the flow behavior is significant. The model equations for the gas phase are straightforward, as the flow inside the gas bubble is governed by single-phase Navier–Stokes equations. However, modeling the bubble–suspension interface is not a straightforward task. The difficulty of numerical simulation is caused by the discontinuous jump of properties across the interface between the gas bubble and the liquid–solid suspension. To circumvent this problem, a continuous transition method (CTM) can be employed. In this method, the discontinuous characteristics are replaced by a smooth variation of the properties (e.g., density and viscosity) from one phase to another within the finite interface thickness. The continuous transition method can overcome the problem of numerical divergence while simulating the flow field at both sides of the interface where the physical properties of the fluids strongly differ.

The scalar function of fluid volume fraction, $\alpha(x, t)$, solved by the VOF method (Hirt and Nichols, 1981) is used to construct this continuous transition function. A fluid property at the interface is then given by

$$Q = Q_m^* \alpha(x, t) + Q_g^*[1 - \alpha(x, t)] \tag{56}$$

where Q represents a property of the fluid, and Q_m^* and Q_g^* represent the properties of the liquid–solid suspension and the gas bubble, respectively. By definition, $\alpha(x, t) = 1$ in the liquid or liquid–solid mixture, $0 < \alpha(x, t) < 1$ at the free surface, and $\alpha(x, t) = 0$ in the gas bubble. Therefore Q is replaced by Q_m^* when $\alpha(x, t)$ equals 1 in the liquid–solid suspension or Q_g^* when $\alpha(x, t)$ equals 0 inside the gas bubble.

The advection equation for $\alpha(x, t)$ is

$$\frac{\partial \alpha}{\partial t} + (v \cdot \nabla)\alpha = 0 \tag{57}$$

On the gas–liquid free surfaces, the stress boundary condition follows the Laplace equation as

$$p_s = p - p_v = \sigma \kappa \tag{58}$$

where the surface pressure, p_s, is the surface-tension-induced pressure jump across a fluid interface. The continuum surface force (CSF) model converts the surface force into a volume force within free surfaces, as shown in Eq. (54).

4.3 Discrete Particle Method

The motion of a particle in a flow field can be described in Lagrangian coordinates with its origin attached to the center of the moving particle. The motion of a single particle can be described by its acceleration and rotation in a nonuniform flow field. The particle accelerating in the liquid is governed by Newton's second law of motion as

$$m_p \frac{dv_p}{dt} = F_{total} \tag{59}$$

The total force acting on a particle is composed of all applicable forces, including drag (F_D), added mass (F_{AM}), gravity (F_G), buoyancy (F_B), Basset force (F_{BA}), and other forces (ΣF_i):

$$F_{total} = F_D + F_{AM} + F_G + F_B + F_{BA} + \sum_i F_i \tag{60}$$

The drag force acting on a suspended particle is proportional to the relative velocity between the phases:

$$F_D = \tfrac{1}{2}\varepsilon_l^2 C_D' \rho_l A |v - v_p|(v - v_p) \tag{61}$$

where A is the exposed frontal area of the particle in the direction of the incoming flow, and C_D' is the effective drag coefficient, which is a function of the particle Reynolds number, Re_p. For isolated rigid spherical particles the drag coefficient C_D can be estimated by the following equations (Rowe and Henwood, 1961)

$$C_D = \begin{cases} \dfrac{24}{Re_p}(1 + 0.15 Re_p^{0.687}) & Re_p < 1000 \\[2mm] 0.44 & Re_p \geq 1000 \end{cases} \tag{62}$$

where $Re_p = \rho_l \varepsilon_l d_p |v - v_p|/\mu_l$. In the liquid–solid or gas–solid suspensions, the drag force depends strongly on the local phase holdup in the vicinity of the particle under consideration. The effective drag coefficient can be obtained by the product of the drag coefficient for an isolated particle and a correction factor as given by (Wen and Yu, 1966)

$$C_D' = C_D \varepsilon_l^{-4.7} \tag{63}$$

The added mass force accounts for the resistance of the fluid mass that is moving at the same acceleration as the particle. For a spherical particle, the volume of the added mass is equal to one-half of the particle volume, so that

$$F_{AM} = \frac{1}{2} \rho_l V_p \frac{d}{dt} (v - v_p) \qquad (64)$$

The Basset force accounts for the effect of past acceleration. The original formulation of the Basset force is derived from the creeping flow condition. For a particle moving in a liquid with a finite Reynolds number, the modified Basset force is given as (Mei and Adrian, 1992)

$$F_{BA} = 3\pi\mu_l d_p \int_0^t K(t - \tau) \frac{d(v - v_p)}{dt} \, d\tau \qquad (65)$$

$K(t - \tau)$ in Eq. (65) is given as

$$K(t - \tau) = \left\{ \left[\frac{\pi(t - \tau)\nu_l}{r_p^2} \right]^{1/4} \right.$$

$$\left. + \left[\frac{1}{2} \pi \frac{(U + v_p - v)^3}{r_p \nu_l f_H^3(\mathrm{Re})} (t - \tau)^2 \right]^{1/2} \right\}^{-2}$$

$$f_H(\mathrm{Re}) = 0.75 + 0.105\mathrm{Re},$$

$$\mathrm{Re} = \frac{U d_p}{\nu_l} \qquad (66)$$

where ν_l is the kinematic viscosity of the fluid and U is the mean stream velocity.

The sum of the gravity and buoyancy forces has the form

$$F_G + F_B = \rho_p V_p g - \rho_l V_p g = (\rho_p - \rho_l) V_p g \qquad (67)$$

An inertial term, $\rho_l V_p dv/dt$, is added to the buoyancy force, which is important when the fluid is accelerated. When particles approach the gas–liquid interface, which is identified as $0 < \alpha(x, t) < 1$, the surface tension force acts on the particles through the liquid film. Since the size of computational cell is larger than the thickness of the gas–liquid interface film, a bubble-induced force model (BIF) is applied to the particle:

$$F_{bp} = V_p \sigma \kappa(x, t) \nabla \alpha(x, t) \qquad (68)$$

If the particle overcomes this bubble-induced force, the particle will penetrate the bubble surface. The penetrating particle breaks the bubble surface momentarily upon contact. If the penetrating particle is small, the bubble may recover its original shape upon particle penetration (Chen and Fan, 1989a,b). However, if there are several particles colliding with the bubble surface simultaneously, the resulting force may cause the bubble to break.

The general scheme of a stepwise molecular dynamic (MD) simulation (Allen and Tildesley, 1987), based on a predictor–corrector algorithm, is used to compute the particle motion. When particles are moving close to each other, the close-range particle–particle interaction including collision takes place as illustrated in the following.

4.4 Particle–Particle Collision Dynamics

In this section, particle–particle collision dynamics will be discussed. A hard sphere approach is used for the particle–particle collision analysis. In this approach, it is assumed that collisions between spherical particles are binary and quasi-instantaneous, and further, that there is a sequence of collisions during each time step. The equations, which are similar to the stepwise molecular dynamic simulation, are used to locate the minimum flight time of particles before any collision.

4.4.1 Liquid Shear Effect

The liquid shear effect on particle motion is important in liquid–solid systems. Particularly, the liquid shear effect between particles becomes significant when two particles move close to each other in liquid–solid systems, especially when the distance between two particles is less than $0.1 d_p$. Thus the close-distance interaction (CDI) model is used to determine the particle contact velocity just before collision; this model considers the strong damping effect of the liquid film before particle contact. The particle normal contact velocity can be described by (Zhang et al., 1999)

$$\left(1 + \frac{1}{2} \frac{\rho_l}{\rho_p} + \frac{3}{16} \frac{\rho_l}{\rho_p} \frac{r_p^3}{h^3} \right) \frac{dv_p}{dh}$$

$$= \frac{9}{2} \frac{\mu_l f \phi}{r_p^2 \rho_p} \frac{(v_p - v)}{v_p} + \frac{9}{32} \frac{\rho_l}{\rho_p} \frac{r_p^4}{h^4} \frac{(v_p - v)|v_p - v|}{v_p}$$

$$- \left(1 - \frac{\rho_l}{\rho_p} \right) \frac{g}{v_p}$$

$$+ \frac{9\mu_l \int_0^t K(t - \tau)[d(v_p - v)/dt] d\tau}{2r_p^2 \rho_p v_p} \qquad (69)$$

where h is the distance from the center of the approaching particle to the midpoint between the two particles; r_p is the radius of the particle; and f (Schiller and Naumann, 1933) and ϕ (Zhang et al., 1999) are the correction functions and can be expressed as

$$\phi = \exp\left[\left(\frac{\mathrm{Re_p}}{1.7}\right)^{0.44}\left(\frac{\rho_p}{\rho_l}\right)^{0.19}\left(\frac{r_p}{h}\right)^{\mathrm{Re_p^{0.47}}}\right] \qquad (70)$$

$$f = 1 + 0.15\mathrm{Re_p^{0.687}}$$

Using the Runge–Kutta method, Eq. (69) can be solved to determine the particle normal contact velocity just before the collision.

4.4.2 Particle Collision Analysis

When two particles are in contact, a collision analysis can be conducted to obtain the velocities of the particles after the collision. To simplify the analysis, it can be assumed that the tangential traction and the resulting displacements have no effect on the normal collision. For the collision between particles a and b, the normal components after collision can be obtained by solving the equations for the restitution coefficient and the conservation of momentum,

$$\frac{U_a^{N'} - U_b^{N'}}{U_b^N - U_a^N} = e$$

$$m_a U_a^N + m_b U_b^N = m_a U_a^{N'} + m_b U_b^{N'} \qquad (71)$$

where U^N is the normal velocity of the particle (a or b) at the contact point before or after (with superscript $'$) the collision.

From Mindlin's contact theory, there are three kinds of frictional contact during the collision: sliding contact, nonsliding or sticking contact, and torsion of elastic particles in contact (Mindlin, 1949). By neglecting the effect of particle torsion during the collision, the simplified Mindlin's contact theory is applied to obtain the tangential components after the collision. If the incident angle, defined as the ratio of the particle–particle relative velocity in the tangential direction to velocity in the normal direction, is less than the critical angle ($\alpha_{cr} = \tan^{-1}(2f_k)$, where f_k is the friction coefficient), sticking collision occurs:

$$U_a^{T'} = U_b^{T'} \qquad (72)$$

Otherwise, sliding collision occurs, in which

$$(U_a^T - U_b^T) - (U_a^{T'} - U_b^{T'}) = 2f_k(U_a^N - U_b^N) \qquad (73)$$

where U^T is the tangential particle velocity (a or b) at the contact point (Fan and Zhu, 1998).

The conservation of momentum is given as

$$m_a U_a^T + m_b U_b^T = m_a U_a^{T'} + m_b U_b^{T'} \qquad (74)$$

The tangential velocities after the collision can be obtained by solving Eq. (72) or (73) together with Eq. (74).

The collision also induces a change in particle rotation. The angular velocities after the collision are determined by

$$I_a(\omega_a' - \omega_a) = m_a(U_a^{T'} - U_a^T) \cdot r_a$$

$$I_b(\omega_b' - \omega_b) = m_b(U_b^{T'} - U_b^T) \cdot r_b \qquad (75)$$

where ω is the angular velocity of the particle (a or b) and I is the moment of inertia defined by $I = (2/5)m_p r_p^2$.

The tangential velocities of the particle center are given as

$$U_{ac}^{T'} = U_a^{T'} - \omega_a' \cdot r_a$$

$$U_{bc}^{T'} = U_b^{T'} - \omega_b' \cdot r_b \qquad (76)$$

Particle collision is important in simulating bubble behavior in liquid–solid suspensions, especially for high solids holdup conditions. A simulation without considering particle collision leads to inappropriate nonuniformity of particle distribution in the flow field and hence false flow field information. Numerical instability may also occur as a result of inappropriate particle accumulation in a computational cell. In the collision model discussed above, only the binary-particle collision mechanism is considered, which limits the model to low solids holdup conditions (less than 30–40% by volume). For higher solids holdup cases, the multiparticle collision mechanism needs to be considered.

4.5 Simulation Examples

The discrete phase simulation method described in Secs. 4.1 through 4.4 is capable of predicting the flow behavior in gas–liquid–solid three-phase flows. In this section, several simulation examples are given to demonstrate the capability of the computational model. First, the behavior of a bubble rising in a liquid–solid suspension at ambient pressure is simulated and compared to experimental observations. Then the effect of pressure on the bubble rise behavior is discussed, along with the bubble–particle interaction. Finally, a more complicated case, that is, multibubble formation dynamics with bubble–bubble interactions, is illustrated.

The behavior of a bubble rising in a liquid or liquid–solid suspension can be simulated by the computational model. Simulated results and experimental observations of a single bubble rising in a liquid–

solid suspension under ambient conditions are shown in Fig. 24 (Fan et al., 1999). The simulation domain is $30 \times 80 \, \text{mm}^2$ and the computational grid size is $0.15 \times 0.16 \, \text{mm}^2$. One thousand glass beads with a density of 2,500 kg/m^3 and a diameter of 1.0 mm are modeled as the solid phase. An aqueous glycerine solution (80 wt%) with $\rho_l = 1,206 \, \text{kg/m}^3$, $\mu_l = 52.9$ mPa · s, and $\sigma = 62.9$ mN/m is modeled as the liquid phase. A circular bubble with a diameter of 10 mm is initially imposed in the computational domain with its center 15 mm above the bottom. Initially, the particles are randomly positioned in a $30 \times 240 \, \text{mm}^2$ area. Then the simulation is performed for particles settling at a liquid velocity of 5 mm/s. At this stage, the bubble is treated as an obstacle and fixed in the original place. An equilibrium bed height is reached at 80 mm, which gives a three-dimensional equivalent solids holdup of 0.44. After the bed reaches its equilibrium height, the simulation is restarted with bubble tracking and particle movement. The time step of simulation for the liquid and solid phases is 5 μs.

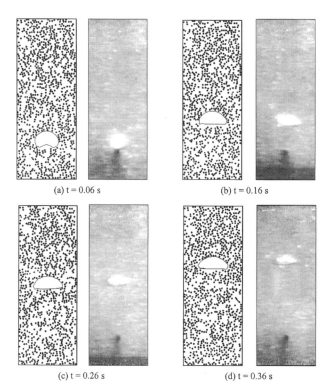

(a) t = 0.06 s   (b) t = 0.16 s

(c) t = 0.26 s   (d) t = 0.36 s

Figure 24 Simulated and experimental results of a bubble rising in a liquid–solid fluidized bed under ambient conditions (nitrogen–80% glycerine solution–1.0 mm glass beads, $d_b = 10.0$ mm, $\varepsilon_s = 0.44$). (From Fan et al., 1999.)

The experimental results in Fig. 24 were obtained in a two-dimensional column with a thickness of 7 mm. The solids holdup, the liquid velocity, and the liquid and solids properties are the same as the simulation conditions. As shown in the figure, the simulated bubble rise velocity and bubble shape generally agree well with experimental results, indicating the validity of the computational model.

The pressure effect on the single bubble rise behavior can be clearly seen from the model simulations. The simulated behavior of a single bubble rising in a liquid–solid suspension at elevated pressures ($P = 17.3$ MPa) is shown in Fig. 25 (Zhang et al., 2000). The size of the simulation domain is the same as in the previous case. Paratherm NF heat transfer fluid, nitrogen, and glass beads with a diameter of 0.88 mm are used as liquid, gas, and solid phases, respectively. The numerical simulation indicates that the bubble rise velocity decreases with an increase in pressure, and it is in good agreement with the experimental data and the prediction by the Fan–Tsuchiya equation. Elevated pressure also causes the bubble shape to become flatter owing to the variation of properties inside the bubble.

The simulation also provides some information about the bubble–particle interaction as shown in Fig. 25. As the bubble rises in the liquid–solid fluidized bed, an interaction between the bubble and particles takes place. In the simulation model, the bubble-particle interaction is accounted for by adding a surface-tension-induced force to the particle motion equation. This force is also added to the source term of the liquid momentum equation for the liquid elements in the interfacial area to account for the particle effect on the interface. The particle movement is determined from the resulting total force acting on the particle. From the simulation results, it is seen that most particles contacting the bubble do not penetrate the bubble; only one or two particles penetrate. Instead, they pass around the bubble surface. When the particles penetrate the bubble, they fall through quickly to the bubble base because of the low viscosity and density of the gas phase.

The last example demonstrated in this section is the simulation of multibubble formation dynamics (Li et al., 2001). Multibubble formation behavior is more complicated, and even difficult to study experimentally. However, the simulation can provide more information about the dynamics of multibubble formation, particularly for cases with interaction between bubbles formed from different orifices. The simulation of two-bubble formation at high pressures ($P = 6.6$ MPa) in

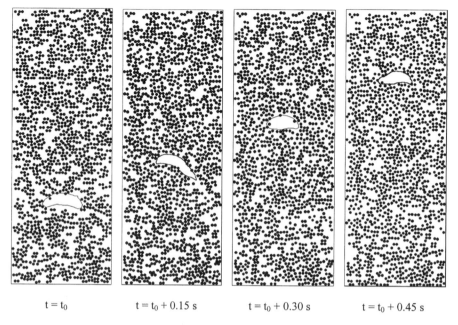

| t = t₀ | t = t₀ + 0.15 s | t = t₀ + 0.30 s | t = t₀ + 0.45 s |

Figure 25 Simulation of a single bubble rising in a liquid–solid suspension at high pressure (nitrogen-Paratherm NF heat transfer fluid–0.88 mm glass beads, $P = 17.3\,\text{MPa}$, $\varepsilon_s = 0.384$, $d_b = 7.5$ mm). (From Zhang et al., 2000.)

Paratherm liquids is shown in Fig. 26. In the simulations, an 80 mm (width) × 50 mm (height) domain is used with two orifices at the bottom, an exit at the top, and two sidewalls as boundary conditions. The liquid initially fills the domain to a certain height. At the beginning of the simulation, gas is injected into the liquid through two orifices. Paratherm NF heat transfer fluid and nitrogen are used as liquid and gas phases, respectively. The gas inlet velocity is 14.0 cm/s. The diameter of each orifice is 6 mm. To study the interaction between bubbles formed from different orifices, the orifice separation distance is set as 20 mm. As can be seen from the figure, due to the closeness of the orifices, the fluctuations of flow field induced by the bubble forming from one orifice have an effect on the bubble forming from the other orifice. This interactive effect yields varied bubble shapes and an alternate detachment pattern of bubbles from two orifices. Initially the bubble from each orifice rises rectilinearly (Fig. 26a); after detachment, however, the bubbles tend to move to the centerline, rise in zigzag fashion, and break up at the free surface (Figs. 26d–f). Based on simulation results, it is found that owing to the closeness of the orifices, two forming bubbles induce a high liquid velocity in the area near the centerline. The high velocity thus results in a lower pressure in this area and draws bubbles toward the centerline. In other words, bubble and bubble wake interaction leads to a zigzag bubble

rising path, and induces a complicated wake flow field, which, in turn, affects the bubble formation behavior.

From the above examples, the computational fluid dynamics simulation is able to capture the dynamic behavior of each individual phase in three-phase flows, and it also demonstrates its great potential in other fields.

5 SUMMARY

Gas–liquid–solid three-phase fluidized bed systems have been widely used in many industrial applications. Successful application of three-phase fluidization systems lies in a comprehensive understanding of bubble dynamics, hydrodynamics, and transport phenomena, particularly under high pressure and temperature conditions. High-pressure and high-temperature operation of three-phase fluidized beds is commonly encountered in most industrial applications of commercial interest. The flow characteristics of reactors at high pressure and temperature are distinct from ambient conditions; for example, elevated pressure leads to higher gas holdup and smaller bubble size in the system, and thus dramatically affects transport phenomena including heat and mass transfer, and phase mixing. The effects of pressure and temperature on fluid dynamics and transport properties are mainly in turn due to

Figure 26 Simulation of two-bubble formation in Paratherm NF heat transfer fluids with bubble–bubble interactions ($P = 6.6$ MPa, $D_o = 6$ mm, $L_{or} = 20$ mm). (From Li et al., 2001.)

variations in bubble characteristics and changes in the physical properties of fluid phases. General flow characteristics in a three-phase fluidized bed are described in this chapter with specific attention given to high-pressure phenomena.

To date, the design of three-phase fluidized beds still relies heavily on experimental observations, empirical correlations, and engineering models. With increasing computer power, the employment of the computational fluid dynamics approach has gained considerable attention in recent years. The computational approach offers viable and attractive options that complement the traditional experimental approach. In this chapter, a discrete phase simulation method is demonstrated in the simulation of three-phase flows. This method has proven to be effective in simulating bubble behavior and fluid dynamics in three-phase fluidization systems. However, many challenges still remain in computation, such as how to incorporate kinetics and transport properties into fluid dynamics calculations, how effectively to simulate the full flow field in large columns, and how to employ properly the closure relationships into turbulence calculations. Furthermore, more experimental studies on the liquid turbulence induced by rising bubbles are required for improving modeling and exploring the mechanisms of various transport phenomena.

NOMENCLATURE

A	=	cross-sectional area of particle
a	=	half-width of bubble; interfacial area
b	=	bubble breadth
C	=	tracer concentration
C_D	=	drag coefficient
C_D'	=	modified drag coefficient
C_{pl}	=	liquid heat capacity
C_s	=	solids concentration in a liquid–solid mixture
c	=	parameter in Fan–Tsuchiya equation reflecting surface tension effect; half-height of ellipsoidal bubble
D_c	=	column diameter
D_{max}	=	maximum stable bubble size
D_o	=	orifice diameter
d_b	=	volume equivalent bubble diameter
d_b'	=	dimensionless bubble diameter
d_p	=	particle diameter
d_{vs}	=	Sauter mean bubble diameter
E	=	solid phase axial dispersion coefficient
$E(\sqrt{1-\alpha^2})$	=	complete second kind elliptic integral
Eo	=	Eötvös number based on bubble diameter
Eo$'$	=	Eötvös number based on particle diameter
E_{rl}	=	radial liquid dispersion coefficient
E_{zl}	=	axial liquid dispersion coefficient
e	=	restitution coefficient
F_{AM}	=	added mass force
F_B	=	effective buoyancy force; buoyancy force
F_{BA}	=	Basset force
F_{bp}	=	bubble–particle interaction force
F_C	=	particle–bubble collision force

F_D	=	liquid drag force
F_{fp}	=	fluid–particle interaction force
F_G	=	gravity force
F_i	=	other forces
$F_{I,g}$	=	bubble inertial force
$F_{I,m}$	=	liquid–solid suspension inertial force
F_M	=	gas momentum force
F_{total}	=	total force
F_x	=	x-component of centrifugal force induced by gas circulation inside bubble
F_σ	=	surface tension force
Fr	=	Froude number
f	=	correction factor for particle collision
f_b	=	volumetric body force
f_{bf}	=	volumetric bubble–fluid interaction force
f_k	=	friction coefficient
f_{pf}	=	volumetric particle–fluid interaction force
$f(d_b)$	=	probability density function of bubble size
g	=	gravitational acceleration
H	=	column height
h	=	level of liquid or liquid–solid suspension in the column; time-averaged heat transfer coefficient; separation distance from the center of the approaching particle to the symmetric plane of two colliding particles
h'	=	heat transfer coefficient in liquid–solid fluidized beds
h_i	=	instantaneous heat transfer coefficient
I	=	moment of inertia
K	=	proportionality constant for calculating the effective viscosity of liquid–solid suspensions
K_b	=	parameter in Fan–Tsuchiya equation reflecting viscous nature of surrounding medium
K_{b0}	=	proportionality constant in Fan–Tsuchiya equation
k	=	ratio of wake size (or volume) to bubble size (or volume) in the generalized wake model
k_l	=	liquid thermal conductivity; liquid-phase mass transfer coefficient
k_o	=	orifice constant
L	=	circumference of the ellipsoidal bubble; length of the heat transfer probe
L_{or}	=	distance between two orifices
l	=	thickness of the liquid film between two coalescing bubbles
Mo	=	Morton number based on liquid properties
Mo$_m$	=	modified Morton number based on slurry properties
m_p	=	particle mass

N_c	=	dimensionless capacitance number
Nu$'$	=	Nusselt number in liquid–solid fluidized beds
n	=	parameter in Fan–Tsuchiya equation reflecting system purity; Richardson–Zaki index in the generalized wake model
n_i	=	number of bubbles
P	=	system pressure
P_b	=	pressure in the bubble
P_c	=	pressure in the gas chamber
P_e	=	pressure at the gas inlet to the chamber
Pe$_z$	=	axial liquid Peclet number
P_0	=	hydrostatic pressure at the bubble surface
Pr	=	Prandtl number based on liquid properties
Pr$_m$	=	Prandtl number based on slurry properties
p	=	scalar pressure
p_s	=	surface pressure
p_v	=	vapor pressure
Q	=	property of fluid
Q_g	=	volumetric gas flow rate into the gas chamber
Q_g^*	=	property of gas
Q_m^*	=	property of liquid–solid suspension
Q_0	=	volumetric gas flow rate through the orifice
R	=	column radius
R_d	=	radius of a contacting circle between two bubbles
Re	=	bubble Reynolds number based on liquid properties; particle Reynolds number based on mean stream velocity
Re$'$	=	Reynolds number based on particle diameter
Re$_m$	=	Reynolds number based on slurry properties
Re$_p$	=	particle Reynolds number
r	=	r-axis in a cylindrical coordinate system
r_b	=	radius of bubble
r_c	=	radius in a cylindrical coordinate system
r_o	=	radius of orifice
r_p	=	radius of particle
S	=	rate-of-strain tensor
St$_m$	=	Stanton number based on slurry properties
T	=	temperature
T	=	matrix transform
t	=	time
t_c	=	contact time between liquid element and film
U	=	solids velocity in the sedimentation–dispersion model; mean stream velocity
U^N	=	normal velocity of particle

U^T	=	tangential velocity of particle
U_b	=	bubble rise velocity relative to the liquid phase
$U_{b\infty}$	=	terminal bubble rising velocity
U_g	=	superficial gas velocity
$U_{g,tran}$	=	regime transition gas velocity
U_l	=	superficial liquid velocity
U_{lmf}	=	liquid minimum fluidization velocity
u	=	rise velocity of bubble base
u_b	=	bubble rise velocity relative to the liquid phase; bubble rise velocity in a stream of bubbles
u_b'	=	dimensionless bubble rise velocity
$\bar{u}_b$	=	average bubble rise velocity
u_e	=	bubble expansion velocity
u_m	=	suspension velocity
u_{max}	=	rise velocity of maximum stable bubble
u_o	=	superficial gas velocity through the orifice
$u_{pt,0}$	=	particle terminal velocity in the fluidizing liquid at the ambient pressure
u_r	=	relative velocity between liquid and gas bubble
u_{small}	=	small bubble rise velocity
u_t	=	particle terminal velocity in liquid
u_x	=	x-component of the circulation velocity of gas inside a bubble
V	=	liquid velocity in the sedimentation–dispersion model
V_b	=	bubble volume
V_c	=	volume of gas chamber; average liquid circulation velocity
V_p	=	particle volume
v	=	liquid velocity vector
v_p	=	particle velocity vector
x	=	distance between bubble center and orifice plate; ratio of solids concentration in the bubble wake region to that in the liquid–solid fluidized region in the generalized wake model
x_p^k	=	location vector of particle k
y	=	lateral displacement from the axis of symmetry of a bubble
z	=	z-axis in a cylindrical coordinate system
$\langle v \rangle$	=	ensemble-average axial liquid velocity
$\langle v'v' \rangle$	=	Reynolds axial normal stress

Greek Letters

α	=	aspect ratio of bubble; thermal diffusivity; volume fraction of fluid
α_{cr}	=	critical angle for sticking collision
β_d	=	ratio of particle density to liquid density
β_U	=	ratio of superficial gas velocity to liquid velocity

δ	=	thickness of fluid film surrounding the heating surface; delta function
ε_g	=	gas holdup
$\varepsilon_{g,tran}$	=	gas holdup at the regime transition point
ε_l	=	liquid holdup
ε_s	=	solids holdup
ε_{sc}	=	critical solids holdup
ε_{s0}	=	solids holdup at incipient fluidization
ϕ	=	particle sphericity; parameter reflecting the surface drag in the equation calculating film thinning velocity; correction factor; azimuthal angle in spherical or cylindrical coordinates
γ	=	contact angle between bubble and orifice surface; heat capacity ratio
κ	=	free surface curvature
λ_c	=	critical wavelength for bubble breakup
μ_g	=	gas viscosity
μ_l	=	liquid viscosity
μ_m	=	effective viscosity of liquid–solid suspension
ν_l	=	liquid kinematic viscosity
ρ_g	=	gas density
ρ_l	=	liquid density
ρ_m	=	density of liquid–solid suspension
ρ_p	=	particle density
ρ_s	=	solids density
σ	=	surface tension
τ	=	viscous stress tensor
Ω	=	computational domain
ω	=	angular velocity of particle
ξ	=	correction factor reflecting particle effects on the slurry viscosity in the gas holdup correlation
ζ	=	coefficient of suspension inertial force
ΔV	=	volume of a computational cell
$\Delta \rho$	=	density difference between liquid and gas phases

REFERENCES

Allen MP, Tildesley DJ. Computer Simulation of Liquids. Oxford: Clarendon Press, 1987.

Attia YA, Elzeky M, Bavarian F, Fan LS. Cleaning and desulfurization of high sulfur coal by selective flocculation and bioleaching in a draft tube fluidized bed reactor. In: Dugan P, Quigly D, Attia YA, eds. Processing and Utilization of High Sulfur Coals IV. Elsevier, 1991, pp 181–201.

Azbel D. Two-Phase Flows in Chemical Engineering. Cambridge, UK: Cambridge University Press, 1981.

Baird M, Rice R. Axial dispersion in large scale unbaffled columns. Chem Eng J 9:171–174, 1975.

Baker CGJ, Armstrong ER, Bergougnou MA. Heat transfer in three-phase fluidized beds. Powder Technol 21:195–204, 1978.

Batchelor GK. The stability of a large gas bubble rising through liquid. J Fluid Mech 184:399–422, 1987.

Batchelor GK. A new theory of the instability of a uniform fluidized bed. J Fluid Mech 193:75–110, 1988.

Bavarian F, Fan LS. Mechanisms of hydraulic transport of a packed bed at the start-up of a three-phase fluidized bed. Chem Eng Sci 46:3081–3087, 1991a.

Bavarian F, Fan LS. Hydraulic transport of a packed bed during the start-up of a three-phase fluidized bed with large gas holdups. Ind Eng Chem Res 30:408–414, 1991b.

Bellman R, Pennington RH. Effect of surface tension and viscosity on Taylor instability. Quarterly of Applied Math 12:151–162, 1954.

Bhaga D, Weber ME. Bubbles in viscous liquids: shapes, wakes and velocities. J Fluid Mech 105:61–85, 1981.

Bhatia VK, Epstein N. Three phase fluidization: a general wake model. In: Angelino H, Couderc JP, Gibert H, Laguerie C, eds. Fluidization and Its Applications. Toulose: Cepadues-editions, 1974, pp 372–392.

Boris JP. New directions in computational fluid dynamics. Annu Rev Fluid Mech 21:345–385, 1989.

Brackbill JU, Kothe DB, Zemach C. A continuum method for modeling surface tension. J Comput Phys 100:335–354, 1992.

Briens LA, Briens CL, Margaritis A, Hay J. Minimum liquid fluidization velocity in gas–liquid–solid fluidized beds of low-density particles. Chem Eng Sci 52:4231–4238, 1997.

Briens CL, Ibrahim YAA, Margaritis A, Bergougnou MA. Effect of coalescence inhibitors on the performance of three-phase inverse fluidized-bed columns. Chem Eng Sci 54:4975–4980, 1999.

Buffiere P, Fonade C, Moletta R. Liquid mixing and phase hold-ups in gas producing fluidized bed bioreactors. Chem Eng Sci 53:617–627, 1998.

Calderbank PH, Moo-Young MB. Continuous phase heat and mass transfer properties of dispersions. Chem Eng Sci 16:39–54, 1961.

Cassanello M, Larachi F, Marie MN, Guy C, Chaouki J. Experimental characterization of the solid phase chaotic dynamics in three-phase fluidization. Ind Eng Chem Res 34:2971–2980, 1995.

Cassanello M, Larachi F, Guy C, Chaouki J. Solids mixing in gas–liquid–solid fluidized beds: experiments and modelling. Chem Eng Sci 51:2011–2020, 1996.

Chang MY, Morsi BI. Mass transfer in a three-phase reactor operating at elevated pressures and temperatures. Chem Eng Sci 47:1779–1790, 1992.

Chaudhari RV, Hoffmann H. Coalescence of gas bubbles in liquids. Reviews in Chemical Engineering 10:131–190, 1994.

Chen YM, Fan LS. Bubble breakage mechanisms due to collision with a particle in a liquid medium. Chem Eng Sci 44:117–132, 1989a.

Chen YM, Fan LS. Bubble breakage due to particle collision in a liquid medium: particle wettability effects. Chem Eng Sci 44:2762–2767, 1989b.

Chen Z, Zheng C, Feng Y, Hofmann H. Distributions of flow regimes and phase hold-ups in three-phase fluidized beds. Chem Eng Sci 50:2153–2159, 1995.

Chern SH, Muroyama K, Fan LS. Hydrodynamics of constrained inverse fluidization and semifluidization in a gas–liquid–solid system. Chem Eng Sci 38:1167–1174, 1983.

Chern SH, Fan LS, Muroyama K. Hydrodynamics of cocurrent gas–liquid–solid semifluidization with a liquid as the continuous phase. AIChE J 30:288–294, 1984.

Chiu TM, Ziegler EN. Heat transfer in three-phase fluidized beds. AIChE J 29:677–685, 1983.

Clark KN. The effect of high pressure and temperature on phase distributions in a bubble column. Chem Eng Sci 45:2301–2307, 1990.

Clift R, Grace JR, Weber ME. Bubbles Drops, and Particles. New York: Academic Press, 1978.

Daly JG, Patel SA, Bukur DB. Measurement of gas holdups in a three-phase bubble column by gamma-ray densitometry. In: Large JF, Laguerie C, eds. Fluidization VIII. New York: Engineering Foundation, 1995, pp 647–655.

Darton RC, Harrison D. The rise of single gas bubbles in liquid fluidised beds. Trans Instn Chem Engrs 52:301–306, 1974.

Davies RM, Taylor G. The mechanics of large bubbles rising through extended liquids and through liquids in tubes. Proc Roy Soc London A200:375–390, 1950.

Deckwer WD. On the mechanism of heat transfer in bubble column reactors. Chem Eng Sci 35:1341–1346, 1980.

Deckwer WD, Burckhart R, Zoll G. Mixing and mass transfer in tall bubble columns. Chem Eng Sci 29:2177–2188, 1974.

Deckwer WD, Louisi Y, Zaidi A, Ralek M. Hydrodynamic properties of the Fischer–Tropsch slurry process. Ind Eng Chem Process Des Dev 19:699–708, 1980.

Degaleesan S, Dudukovic MP, Toseland BA, Bhatt BL. A two-compartment convective-diffusion model for slurry bubble column reactors. Ind Eng Chem Res 36:4670–4680, 1997.

Di Felice R. Hydrodynamics of liquid fluidization. Chem Eng Sci 50:1213–1245, 1995.

El-Temtamy SA, Epstein N. Contraction or expansion of three-phase fluidized beds containing fine/light solids. Can J Chem Eng 57:520–522, 1979.

Fan LS, Gas–Liquid–Solid Fluidization Engineering. Stoneham, MA: Butterworth, 1989.

Fan LS. Moving packed bed phenomenon in three-phase fluidization. Powder Technology 103:300–301, 1999.

Fan LS, Tsuchiya K. Bubble Wake Dynamics in Liquids and Liquid–Solid Suspensions. Stoneham, MA: Butterworth-Heinemann, 1990.

Fan LS, Zhu C. Principles of Gas–Solid Flows. New York: Cambridge Press, 1998.

Fan LS, Yang GQ, Lee DJ, Tsuchiya K, Luo X. Some aspects of high-pressure phenomena of bubbles in liquids and liquid–solid suspensions. Chem Eng Sci 54:4681–4709, 1999.

Fox JM. Fischer–Tropsch reactor selection. Catal Lett 7:281–292, 1990.

Gidaspow D. Multiphase Flow and Fluidization. New York: Academic Press, 1994.

Gidaspow D, Bahary M, Jayaswal UK. Hydrodynamic models for gas–liquid–solid fluidization. Numerical Methods in Multiphase Flows, FED 185, ASME, New York, 1994, pp 117–129.

Grace JR, Wairegi T, Brophy J. Break-up of drops and bubbles in stagnant media. Can J Chem Eng 56:3–8, 1978.

Grevskott S, Sannas BH, Dudukovic MP, Hjarbo KW, Svendsen HF. Liquid circulation, bubble size distributions, and solids movement in two- and three-phase bubble columns. Chem Eng Sci 51:1703–1713, 1996.

Harlow FH, Welch JE. Numerical calculation of time-dependent viscous incompressible flow of fluid with free surface. Phys Fluids 8:2182–2193, 1965.

Hidaka N, Kakoi K, Matsumoto T. Flow behavior of a gas–liquid–solid fluidized bed in a self-circulating operation through an external conduit. Ind Eng Chem Res 37:240–246, 1998.

Higbie R. The rate of absorption of pure gas into a still liquid during short periods of exposure. Trans Am Inst Chem Eng 31:365–375, 1935.

Hill MJM. On a spherical vortex. Phil Trans Roy Soc London 185:213–223, 1894.

Hinze JO. Fundamentals of the hydrodynamic mechanism of splitting in dispersion processes. AIChE J 1:289–295, 1955.

Hirt CW, Nichols BD. Volume of fluid (VOF) method for the dynamics of free boundaries. J Comput Phys 39:201–225, 1981.

Hoefsloot HCJ, Krishna R. Influence of gas density on the stability of homogeneous flow in bubble columns. Ind Eng Chem Res 32:747–750, 1993.

Holcombe NT, Smith DN, Knickle HN, O'Dowd W. Thermal dispersion and heat transfer in nonisothermal bubble columns. Chem Eng Commun 21:135–150, 1983.

Houzelot JL, Thiebaut MF, Charpentier JC, Schiber J. Contribution to the hydrodynamic study of bubble columns. Int Chem Eng 25:645–650, 1985.

Idogawa K, Ikeda K, Fukuda T, Morooka S. Behavior of bubbles of the air-water system in a column under high pressure. Int Chem Eng 26:468–474, 1986.

Idogawa K, Ikeda K, Fukuda T, Morooka S. Formation and flow of gas bubbles in a pressurized bubble column with a single orifice or nozzle gas distributor. Chem Eng Comm 59:201–210, 1987.

Inga JR. Personal communication, 1994.

Inga JR. Scaleup and scaledown of slurry reactors: a new methodology. PhD diss., University of Pittsburgh, PA, 1997.

Jager B, Espinoza R. Advances in low temperature Fischer–Tropsch synthesis. Catal Today 23:17–28, 1995.

Jean RH, Fan LS. Letter to the editor: Bed contraction criterion for three phase fluidization. Can J Chem Eng 65:351–352, 1987.

Jean RH, Tang WT, Fan LS. The sedimentation–dispersion model for slurry bubble columns. AIChE J 35:662–665, 1989.

Jiang P, Arters D, Fan LS. Pressure effects on the hydrodynamic behavior of gas–liquid–solid fluidized beds. Ind Eng Chem Res 31:2322–2327, 1992.

Jiang P, Lin TJ, Luo X, Fan LS. Visualization of high pressure (21 MPa) bubble column: bubble characteristics. Chem Eng Res Des 73:269–274, 1995.

Jiang P, Luo X, Lin TJ, Fan LS. High temperature and high pressure three-phase fluidization–bed expansion phenomena. Powder Technology 90:103–113, 1997.

Johns WF, Clausen GA, Nongbri G, Kaufman H. Texaco T-star process for ebullated bed hydrotreating/hydrocracking. National Petroleum Refiners Association (NPRA) Conference, San Antonio, TX, 1993.

Joshi JB. Axial mixing in multiphase contactors—a unified correlation. Trans Instn Chem Engrs 58:155–165, 1980.

Joshi JB, Lali AM. Velocity–holdup relationship in multiphase contactors—a unified approach. In: Doraiswamy LK, Mashelkar RA, eds. Frontiers in Chemical Reaction Engineering. New York: Wiley Eastern, 1984, pp 314–329.

Kalogerakis N, Behie LA. Oxygenation capabilities of new generation three phase two region bioreactors for microcarrier cultures of animal cells. In: Large JF, Laguerie C, eds. Fluidization VIII. New York: Engineering Foundation, 1995, pp 695–704.

Kang Y, Suh IS, Kim SD. Heat transfer characteristics of three phase fluidized beds. Chem Eng Commun 34:1–13, 1985.

Kastaneck F, Zahradnik J, Kratochvil J, Cermak J. Modeling of large-scale bubble column reactors for nonideal gas–liquid systems. In: Doraiswamy LK, Mashelkar RA, eds. Frontiers in Chemical Reaction Engineering. New York: Wiley Eastern, 1984, pp 330–344.

Kiared K, Larachi F, Guy C, Chaouki J. Trajectory length and residence-time distributions of the solids in three-phase fluidized beds. Chem Eng Sci 52:3931–3939, 1997.

Kim SD, Laurent A. The state of knowledge on heat transfer in three-phase fluidized beds. Int Chem Eng 31:284–302, 1991.

Kim SD, Kim HS, Han JH. Axial dispersion characteristics in three-phase fluidized beds. Chem Eng Sci 47:3419–3426, 1992.

Kitscha J, Kocamustafaogullari G. Breakup criteria for fluid particles. Int J Multiphase Flow 15:573–588, 1989.

Kojima H, Okumura B, Nakamura A. Effect of pressure on gas holdup in a bubble column and a slurry bubble column. J Chem Eng Japan 24:115–117, 1991.

Krishna R, Wilkinson PM, van Dierendonck LL. A model for gas holdup in bubble columns incorporating the influence of gas density on flow regime transitions. Chem Eng Sci 46:2491–2496, 1991.

Krishna R, de Swart JWA, Hennephof DE, Ellenberger J, Hoefsloot CJ. Influence of increased gas density on hydrodynamics of bubble column reactors. AIChE J 40:112–119, 1994.

Kumar S, Fan LS. Heat-transfer characteristics in viscous gas–liquid and gas–liquid–solid systems. AIChE J 40:745–755, 1994.

Kumar R, Kuloor NR. The formation of bubbles and drops. Advances in Chemical Engineering 8:255–368, 1970.

Kumar S, Kusakabe K, Raghunathan K, Fan LS. Mechanism of heat transfer in bubbly liquid and liquid–solid systems: single bubble injection. AIChE J 38:733–741, 1992.

Kumar S, Kusakabe K, Fan LS. Heat transfer in three-phase fluidized beds containing low-density particles. Chem Eng Sci 48:2407–2418, 1993a.

Kumar S, Kusakabe K, Fan LS. Heat transfer in three-phase fluidization and bubble-columns with high gas holdups. AIChE J 39:1399–1405, 1993b.

Kupferberg A, Jameson GJ. Bubble formation at a submerged orifice above a gas chamber of finite volume. Trans Instn Chem Engrs 47:T241–250, 1969.

Lammers JH, Biesheuvel A. Concentration waves and the instability of bubbly flows. J Fluid Mech 328:67–93, 1996.

La Nauze RD, Harris IJ. Gas bubble formation at elevated system pressures. Trans Instn Chem Engrs 52:337–348, 1974.

Larachi F, Iliuta I, Rival O, Grandjean BPA. Prediction of minimum fluidization velocity in three-phase fluidized-bed reactors. Ind Eng Chem Res 39:563–572, 2000.

Lee DJ, Luo X, Fan LS. Gas disengagement technique in a slurry bubble column operated in the coalesced bubble regime. Chem Eng Sci 54:2227–2236, 1999.

Lee DJ, McLain BK, Cui Z, Fan LS. Pressure effect on the flow fields and the Reynolds stresses in a bubble column. Ind Eng Chem Res 40:1442–1447, 2001.

Letzel HM. Hydrodynamics and mass transfer in bubble columns at elevated pressures. PhD diss., Delft University of Technology, The Netherlands, 1997.

Letzel HM, Schouten JC, van den Bleek CM, Krishna R. Influence of elevated pressure on the stability of bubbly flows. Chem Eng Sci 52:3733–3739, 1997.

Levich VG. Physiochemical Hydrodynamics. Englewood Cliffs, NJ: Prentice Hall, 1962.

L'Homme GA. Chemical Engineering of Gas–Liquid–Solid Catalyst Reactions. Liege: CEBEDOC, 1979.

Li Y, Zhang J, Fan LS. Numerical simulation of gas–liquid–solid fluidization systems using a combined CFD-VOF-DPM method: bubble wake behavior. Chem Eng Sci 54:5101–5107, 1999.

Li Y, Yang GQ, Zhang JP, Fan LS. Numerical studies of bubble formation dynamics in gas–liquid–solid fluidization at high pressures. Powder Technology 116:246–260, 2001.

Lin TJ, Fan LS. Characteristics of high-pressure liquid–solid fluidization. AIChE J 43:45–57, 1997.

Lin JN, Banerji SK, Yasuda H. Role of interfacial tension in the formation and the detachment of air bubbles: a single hole on a horizontal plane immersed in water. Langmuir 10:936–942, 1994.

Lin TJ, Tsuchiya K, Fan LS. Bubble flow characteristics in bubble columns at elevated pressure and temperature. AIChE J 44:545–560, 1998.

Lin TJ, Tsuchiya K, Fan LS. On the measurements of regime transition in high pressure bubble columns. Can J Chem Eng 77:370–374, 1999.

Linek V, Benes P, Vacek V. Dynamic pressure method for k_1a measurement in large-scale bioreactors. Biotechnol Bioeng 33:1406–1412, 1989.

Linek V, Benes P, Sinkule J, Moucha T. Non ideal pressure step method for k_1a measurement. Chem Eng Sci 48:1593–1599, 1993.

Luewisutthichat W, Tsutsumi A, Yoshida K. Bubble characteristics in multi-phase flow systems: bubble sizes and size distributions. J Chem Eng Japan 30:461–466, 1997.

Luo X, Jiang P, Fan LS. High pressure three-phase fluidization: hydrodynamics and heat transfer. AIChE J 43:2432–2445, 1997a.

Luo X, Zhang J, Tsuchiya K, Fan LS. On the rise velocity of bubbles in liquid–solid suspensions at elevated pressure and temperature. Chem Eng Sci 52:3693–3699, 1997b.

Luo X, Yang GQ, Lee DJ, Fan LS. Single bubble formation in high pressure liquid–solid suspensions. Powder Technology 100:103–112, 1998a.

Luo X, Lee DJ, Lau R, Yang GQ, Fan LS. Maximum stable bubble size and gas holdup in high-pressure slurry bubble columns. AIChE J 45:665–680, 1999.

Magiliotou M, Chen YM, Fan LS. Bed immersed object heat transfer in a three-phase fluidized bed. AIChE J 34:1043–1047, 1988.

Maneri CC. New look at wave analogy for prediction of bubble terminal velocities. AIChE J 41:481–487, 1995.

Massimilla L, Majuri N, Signorini P. Gas absorption in solid–liquid fluidized systems. Ricerca Sci 29:1934–1940, 1959.

Massimilla L, Solimando A, Squillace E. Gas dispersion in solid–liquid fluidized beds. British Chem Eng 6:232–239, 1961.

Matsumoto T, Hidaka N, Takebayasi Y, Morooka S. Axial mixing and segregation in a gas–liquid–solid three-phase fluidized bed of solid particles of different sizes and densities. Chem Eng Sci 52:3961–3970, 1997.

Mei R, Adrian RJ. Flow past a sphere with an oscillation in the free-stream and unsteady drag at finite Reynolds number. J Fluid Mech 237:323–341, 1992.

Mendelson HD. The prediction of bubble terminal velocities from wave theory. AIChE J 13:250–253, 1967.

Mill PL, Turner JR, Ramachandran PA, Dudukovic MP. The Fischer–Tropsch synthesis in slurry bubble column reactors: analysis of reactor performance using the axial dispersion model. In: Nigam KDP, Schumpe A, eds. Three-Phase Sparged Reactors. Gordon and Breach, 1996, pp 339–386, 679–739.

Milne-Thomson LM. Theoretical Hydrodynamics. 3d ed. London: Macmillan, 1955.

Mindlin RD. Compliance of elastic bodies in contact. J Appl Mech 16:259–268, 1949.

Mitra-Majumdar D, Farouk B, Shah YT. Hydrodynamic modeling of three-phase flows through a vertical column. Chem Eng Sci 52:4485–4497, 1997.

Miura H, Kawase Y. Minimum liquid fluidization velocity in two- and three-phase fluidized beds with non-Newtonian fluids. Powder Technology 97:124–128, 1998.

Molga EJ, Westerterp KR. Gas–liquid interfacial area and holdup in a cocurrent upflow packed bed bubble column reactor at elevated pressure. Ind Eng Chem Res 36:622–631, 1997.

Muroyama K, Yoshikawa T, Takakura S, Yamanaka Y. Mass transfer from an immersed cylinder in three-phase systems with fine suspended particles. Chem Eng Sci 52:3861–3868, 1997.

Nacef S, Poncin S, Wild G. Onset of fluidization and flow regimes in gas–liquid–solid fluidization—influence of the coalescence behavior of the liquid. In: Large JF, Laguerie C, eds. Fluidization VIII. New York: Engineering Foundation, 1995, pp 631–638.

Nikov I, Delmas H. Mechanism of liquid–solid mass transfer and shear stress in three-phase fluidized beds. Chem Eng Sci 47:673–681, 1992.

Oestergaard K. Fluidization. Fluidisation Papers Joint Symp, London, 1964, pp 57–66.

Onozaki M, Namiki Y, Sakai N, Kobayashi M, Nakayama Y, Yamada T, Morooka S. Dynamic simulation of gas–liquid dispersion behavior in coal liquefaction reactors. Chem Eng Sci 55:5099–5113, 2000a.

Onozaki M, Namiki Y, Ishibashi H, Takagi T, Kobayashi M, Morooka S. Steady-state thermal behavior of coal liquefaction reactors based on NEDOL process. Energy Fuels 14:355–363, 2000b.

Oyevaar MH, Bos R, Westerterp KR. Interfacial areas and gas hold-ups in gas–liquid contactors at elevated-pressures from 0.1 to 8.0 MPa. Chem Eng Sci 46:1217–1231, 1991.

Peng XD, Toseland BA, Tijm PJA. Kinetic understanding of chemical synergy under LPDME conditions–once-through applications. Chem Eng Sci 54:2787–2792, 1999.

Pinczewski WV. The formation and growth of bubbles at a submerged orifice. Chem Eng Sci 36:405–411, 1981.

Ramachandran PA, Chaudhari RV. Three-Phase Catalytic Reactors. Gordon and Breach Science, 1983.

Ramakrishnan S, Kumar R, Kuloor NR. Studies in bubble formation—I: Bubble formation under constant flow conditions. Chem Eng Sci 24:731–747, 1969.

Reese J, Jiang P, Fan LS. Bubble characteristics in three-phase systems used for pulp and paper processing. Chem Eng Sci 51:2501–2510, 1996.

Reilly IG, Scott DS, de Bruijn TJW, MacIntyre D. The role of gas phase momentum in determining gas holdup and hydrodynamic flow regimes in bubble column operations. Can J Chem Eng 72:3–12, 1994.

Reiss LP, Hanratty TJ. An experimental study of the unsteady nature of the viscous sublayer. AIChE J 9:154–160, 1963.

Richardson JF, Roman MN, Shakiri KJ. Heat transfer from immersed surfaces in liquid fluidized beds. Chem Eng Sci 31:619–624, 1976.

Rowe PN, Henwood GA. Drag forces in a hydraulic model of a fluidized bed—Part I. Trans Instn Chem Engrs 39:43–54, 1961.

Ruzicka MC, Drahos J, Zahradnik J, Thomas NH. Intermittent transition from bubbling to jetting regime in gas–liquid two phase flows. Int J Multiphase Flow 23:671–682, 1997.

Saberian-Broudjenni M, Wild G, Charpentier JC, Fortin Y, Euzen JP, Patoux R. Contribution to the hydrodynamic study of fluidized gas–liquid–solid reactors. Entropie 120:30–44, 1984.

Safoniuk, M, Grace JR, Hackman L, McKnight CA. Use of dimensional similitude for scale-up of hydrodynamics in three-phase fluidized beds. Chem Eng Sci 54:4961–4966, 1999.

Sagert NH, Quinn MJ. Influence of high-pressure gases on the stability of thin aqueous films. J Colloid Interface Sci 61:279–286, 1977.

Sagert NH, Quinn MJ. Surface viscosities at high pressure gas–liquid interfaces. J Colloid Interface Sci 65:415–422, 1978.

Sangnimnuan A, Prasad GN, Agnew JB. Gas hold-up and backmixing in a bubble-column reactor under coal-hydroliquefaction conditions. Chem Eng Commun 25:193–212, 1984.

Sankaranarayanan K, Shan X, Kevrekidis IG, Sundaresan S. Analysis of drag and virtual mass forces in bubble suspensions using an implicit formulation of the Lattice Boltzmann method, J. Fluid Mech., in press, 2002.

Saxena SC. Bubble column reactors and Fischer–Tropsch synthesis. Catal Rev Sci Eng 37:227–309, 1995.

Schiller L, Naumann A. Über die grundlegenden berechnungen bei der schwerkraftaufbereitung. Ver Deut Ing 77:318–324, 1993.

Shah YT. Gas–Liquid–Solid Reactor Design. McGraw-Hill, 1979.

Shah YT, Kelkar BG, Godbole SP, Deckwer WD. Design parameter estimations for bubble column reactors. AIChE J 28:353–379, 1982.

Shnip AI, Kolhatkar RV, Swamy D, Joshi JB. Criteria for the transition from the homogeneous to the heterogeneous regime in two-dimensional bubble column reactors. Int J Multiphase Flow 18:705–726, 1992.

Song GH, Bavarian F, Fan LS, Buttke RD, Peck LB. Hydrodynamics of three-phase fluidized bed containing cylindrical hydrotreating catalysts. Can J Chem Eng 67:265–275, 1989.

Soong Y, Harke FW, Gamwo IK, Schehl RR, Zarochak MF. Hydrodynamic study in a slurry-bubble-column reactor. Catalysis Today 35:427–434, 1997.

Souza GLM, Afonso JC, Schmal M, Cardoso JN. Mild hydrocracking of an unstable feedstock in a three-phase fluidized-bed reactor: behavior of the process and of the chemical compounds. Ind Eng Chem Res 31:2127–2133, 1992.

Stegeman D, Knop PA, Wijnands AJG, Westerterp KR. Interfacial area and gas holdup in a bubble column reactor at elevated pressures. Ind Eng Chem Res 35:3842–3847, 1996.

Stewart PSB, Davidson JF. Three-phase fluidization: water, particles, and air. Chem Eng Sci 19:319–321, 1964.

Suh IS, Deckwer WD. Unified correlation of heat transfer coefficients in three-phase fluidized beds. Chem Eng Sci 44:1455–1458, 1989.

Suh IS, Jin GT, Kim SD. Heat transfer coefficients in three phase fluidized beds. Int J Multiphase Flow 11:255–259, 1985.

Tarmy B, Chang M, Coulaloglou C, Ponzi P. Hydrodynamic characteristics of three-phase reactors. Chemical Engineer 407:18–23, 1984.

Tomiyama A, Kataoka I, Sakaguchi T. Drag coefficients of bubbles. 1. Drag coefficients of a single bubble in a stagnant liquid. Nippon Kikai Gakkai Ronbunshu B-hen 61:2357–2364, 1995.

Tsuchiya K, Nakanishi O. Gas holdup behavior in a tall bubble column with perforated plate distributors. Chem Eng Sci 47:3347–3354, 1992.

Tsuchiya K, Furumoto A, Fan LS, Zhang J. Suspension viscosity and bubble velocity in liquid–solid fluidized beds. Chem Eng Sci 52:3053–3066, 1997.

Tsuge H, Hibino S. Bubble formation from an orifice submerged in liquids. Chem Eng Comm 22:63–79, 1983.

Tsuge H, Nakajima Y, Terasaka K. Behavior of bubbles formed from a submerged orifice under high system pressure. Chem Eng Sci 47:3273–3280, 1992.

Tsuji Y, Kawaguchi T, Tanaka T. Discrete particle simulation of two-dimensional fluidized bed. Powder Technol 77:79–87, 1993.

Tsutsumi A, Nieh JY, Fan LS. Particle wettability effects on bubble wake dynamics in gas–liquid–solid fluidization. Chem Eng Sci 46:2381–2384, 1991a.

Tsutsumi A, Dastidar AG, Fan LS. Characteristics of gas–liquid–solid fluidization with nonwettable particles. AIChE J 37:951–952, 1991b.

Turner R. Fluidization in the petroleum industry. Fluidisation Papers Joint Symp, London, 1964, pp 47–56.

Tzeng JW, Fan LS, Gan YR, Hu TT. Communication to the editor: Ethanol fermentation using immobilized cells in multistage fluidized bed bioreactor. Biotech Bioeng 38:1253–1258, 1991.

Unverdi SO, Tryggvason G. A front-tracking method for viscous, incompressible, multi-fluid flows. J Comput Phys 100:25–37, 1992.

Vrij A. Possible mechanism for the spontaneous rupture of thin, free liquid films. Discuss Faraday Soc 42:23–33, 1966.

Wallis GB. One-Dimensional Two-Phase Flow. New York: McGraw-Hill, 1969.

Walter JF, Blanch HW. Bubble break-up in gas–liquid bioreactors: break-up in turbulent flows. Chem Eng J 32:B7–B17, 1986.

Wasan DT, Ahluwalia MS. Consecutive film and surface renewal mechanism for heat and mass transfer from a wall. Chem Eng Sci 24:1535–1542, 1969.

Wen CY, Yu YH. Mechanics of fluidization. Chem Eng Progr Symp Ser 62:100–111, 1966.

Wilkinson PM, van Dierendonck LL. Pressure and gas density effects on bubble break-up and gas hold-up in bubble columns. Chem Eng Sci 45:2309–2315, 1990.

Wilkinson PM, van Dierendonck LL. A theoretical model for the influence of gas properties and pressure on single-bubble formation at an orifice. Chem Eng Sci 49:1429–1438, 1994.

Wilkinson PM, Spek AP, van Dierendonck LL. Design parameters estimation for scale-up of high-pressure bubble columns. AIChE J 38:544–554, 1992.

Wilkinson PM, Haringa H, Stokman FPA. Liquid mixing in a bubble column under pressure. Chem Eng Sci 48:1785–1191, 1993.

Wilkinson PM, Haringa H, van Dierendonck LL. Mass transfer and bubble size in a bubble column under pressure. Chem Eng Sci 49:1417–1427, 1994.

Yang GQ, Fan LS. Axial liquid-phase mixing in high-pressure bubble columns. AIChE J, In press, 2002.

Yang GQ, Luo X, Lau R, Fan LS. Bubble formation in high-pressure liquid–solid suspensions with plenum pressure fluctuation. AIChE J 46:2162–2174, 2000a.

Yang GQ, Luo X, Lau R, Fan LS. Heat-transfer characteristics in slurry bubble columns at elevated pressures and temperatures. Ind Eng Chem Res 39:2568–2577, 2000b.

Yoo DH, Tsuge H, Terasaka K, Mizutani K. Behavior of bubble formation in suspended solution for an elevated pressure system. Chem Eng Sci 52:3701–3707, 1997.

Zahradnik J, Fialova M, Ruzicka M, Drahos J, Kastanek F, Thomas NH. Duality of the gas–liquid flow regimes in bubble column reactors. Chem Eng Sci 52:3811–3826, 1997.

Zaidi A, Deckwer WD, Mrani A, Benchekchou B. Hydrodynamics and heat transfer in three-phase fluidized beds with highly viscous pseudoplastic solutions. Chem Eng Sci 45:2235–2238, 1990.

Zhang JP, Grace JR, Epstein N, Lim KS. Flow regime identification in gas–liquid flow and three-phase fluidized beds. Chem Eng Sci 52:3979–3992, 1997.

Zhang JP, Epstein N, Grace JR. Minimum fluidization velocities for gas–liquid–solid three-phase systems. Powder Technology 100:113–118, 1998a.

Zhang JP, Epstein N, Grace JR. Minimum liquid velocity for particle transport from gas–liquid fluidized beds. In: Fan LS, Knowlton TM, eds. Fluidization IX. New York: Engineering Foundation, 1998b, pp 645–652.

Zhang J, Fan LS, Zhu C, Pfeffer R, Qi D. Dynamic behavior of collision of elastic spheres in viscous fluids. Powder Technology 106:98–109, 1999.

Zhang J, Li Y, Fan LS. Numerical studies of bubble and particle dynamics in a three-phase fluidized bed at elevated pressures. Powder Technology 112:46–56, 2000.

Zheng C, Chen Z, Feng Y, Hofmann H. Mass transfer in different flow regimes of three-phase fluidized beds. Chem Eng Sci 50:1571–1578, 1995.

Zhou LX, Yang M, Lian CY, Fan LS, Lee DJ. On the second-order moment turbulence model for simulating a bubble column. Chem Eng Sci 57:3269–3281, 2002.

28

Liquid–Solids Separation

Shiao-Hung Chiang, Daxin He, and Yuru Feng

University of Pittsburgh, Pittsburgh, Pennsylvania, U.S.A.

1 INTRODUCTION

Liquid–solids separation represents a group of unit operations widely used by chemical, mineral, paper, food, biotechnology, water treatment, waste remediation and other activities. Technically, it involves the separation, removal, and collection of solid particulate matter existing in a dispersed or colloidal state in a liquid suspension. The separation is most often performed by employing mechanical forces causing fluid flow through a complex media structure. Thus the design of a liquid–solids separation process requires a combined knowledge of fluid mechanics, particle mechanics, and material properties of the media.

Generally, the separation of solid particulates from liquid suspensions consists of one or more process steps: pretreatment, solids concentration, solids separation, and posttreatment. To implement these steps four major unit operations are commonly used:

1. Filtration (cake filtration, deep-bed filtration, and membrane filtration)
2. Sedimentation (thickening and clarification)
3. Centrifugation (centrifugal filtration and sedimentation)
4. Hydrocyclones (classification and clarification)

These unit operations are integral parts of many industrial processes mainly to serve the purposes of recovering

1. Valuable solids (the liquid being discarded)
2. The liquid (the solids being discarded)
3. Both the solids and the liquid
4. Neither (e.g., to prevent water pollution)

In all cases, the liquid–solids separation plays a crucial role in producing materials to meet the desired product quality and in maintaining environmental protection.

1.1 Liquid–Solids Systems

There are three major classes of liquid–solids systems: aqueous systems, nonaqueous systems, and biological systems. The following provides a brief description of these systems from the standpoint of liquid–solids separation.

1.1.1 Aqueous Systems

Water is the most commonly used liquid in industry and in all human activities. Thus most liquid–solids separation equipment is specifically designed for treating aqueous systems. In water suspensions, the sizes of solid particles are usually very small (in the micron and submicron size range) and surface-active forces often play an important role. Representative examples are the mineral slurries, lyophilic colloids, slimes, and wastewater mixtures. The treatment of aqueous systems involves four stages: flocculation (clarification), sedimentation (settling), consolidation (compression or compaction), and phase separation (filtration or centrifugation).

1.1.2 Nonaqueous Systems

A major problem in the treatment of nonaqueous systems is the removal of colloidal particles to produce acceptable liquid products (e.g., fuel oil). Such a situation is encountered in hydrocarbon production from tar sands and oil shale. The particles (such as oxides, silicates, and clay mineral) suspended in the hydrocarbon liquids originate from a rock matrix. Particle separation problems occur in the solvent extraction of bitumen with nonaqueous media such as toluene. Electrostatic forces (bonding forces) play a predominant role in the physical state of these nonaqueous systems. In many instances, by addition of "antisolvents," and selecting the proper temperature and agitation, these systems can be altered to improve solid separation. Separation of carbon black particles suspended in tetralin using Aerosol OT as a surfactant and by filtration through a bed of sand (deep-bed filtration) is another example of liquid–solids separation in nonaqueous systems.

1.1.3 Biological Systems

Biological treatment is often employed to decompose organic substances in wastewater to remove biological oxygen demand (BOD) or chemical oxygen demand (COD) components using microorganisms under aerobic or anaerobic conditions. As a result of these biological reactions, cells (or biomass) are produced as an aqueous sludge that needs to be separated from the effluent stream. In a microbial fermentation process, fermentation broth contains a complex aqueous mixture of cells, soluble extracellular products, intracellular products, and unconverted substrate or unconvertible components. In order to produce the desired product, a biological system always involves both a bioreactor and a separation/product recovery section. The product recovery section usually consists of solids removal (filtration), primary isolation, purification, and final isolation (crystallization) steps.

1.2 Industrial Applications

As mentioned earlier, liquid–solids separation technology is basic to many manufacturing industries (chemical, mineral, food, beverages, etc.) as well as to pollution abatement and environmental control. It is difficult to find any important engineering enterprises in which liquid–solids separation does not play an important part. Major industrial and commercial applications of four key unit operations for liquid–solids separation are summarized here:

Unit Operation	Applications
Filtration	Suspended, precipitated, and oversize particle removal; process water, wash water, waste oil, boiler feed water cleanup; mineral dewatering; and recovery of valuable products.
Sedimentation	Portable water clarification; municipal treatment; storage pond for toxic waste and mineral processing waste; and pretreatment step for feed to filters and centrifuges.
Centrifugation	Wastewater and sludge thickening; metalworking coolants cleaning; purification of marine fuels; dehydration of tar; and clarification of beer, wine, fruit juice, and varnishes.
Hydrocyclones	Mineral processing and coal cleaning; cooling oils cleaning and clarification; and industrial wastewater treatment.

These unit operations will be discussed separately in the following sections.

2 FILTRATION

The filtration operation involves the separation, removal, and collection of a discrete phase of matter existing in suspension. The undissolved solid particles are separated from the liquid suspension by means of a porous medium (i.e., filter medium). Filtration leads to the formation of a cake containing a relatively low proportion of residual filtrate. Depending upon the mechanism for arrest and accumulation of particles, the filtration operation can generally be classified into three types: cake filtration, deep-bed filtration, and membrane filtration (see Fig. 1).

2.1 Cake Filtration

Most of the liquid filtration operations follow the mechanism of cake filtration. As the filtration proceeds, the particles retained on the filter medium will form a growing cake with porous structure. The small particles that are able to pass through the pores initially will be trapped at a greater depth as they traverse through this porous cake. This cake becomes the true filter medium and hence plays a very important part in the entire filtration operation. The mechanism of flow within the cake and the external conditions imposed on the cake are the basis for modeling a filtration process.

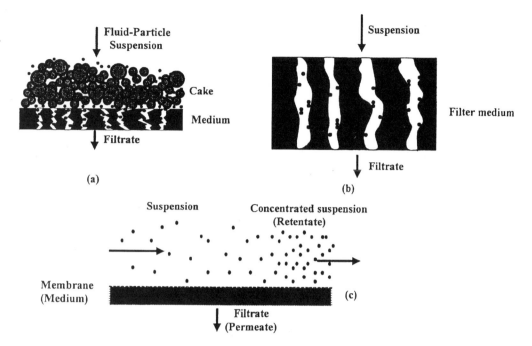

Figure 1 Mechanisms of filtration: (a) cake filtration, (b) deep-bed filtration, (c) cross-flow membrane filtration. (From Chiang and He, 1995.)

Under a pressure gradient, if the particle packing arrangement in the cake can sustain the drag force without deformation, the cake is regarded as incompressible. However, stresses developed in the particulate structure usually lead to deformation and compression with substantial changes in the porosity and permeability as a load (fluid drag) is applied. This kind of filter cake is known as a compressible cake.

2.1.1 Cake Filtration Equation—Two Resistance Model

The structure of the deposited cake resembles the structure of a packed bed. Based on the principles for flow through a packed bed with the consideration of resistances offered by both the filter medium and the filter cake, an equation of cake filtration can be obtained (Akers and Ward, 1977; Cheremisinoff and Azbel, 1989; Tiller et al., 1987a):

$$\frac{dV}{dt} = \frac{A(-\Delta P)}{\mu[(\alpha c V/A) + R_m]} \tag{1}$$

where V is the filtrate volume, t the filtration time, μ the fluid viscosity, c the mass of solid per unit volume of filtrate, A the cross-sectional area of filter, ΔP the pressure drop across the filter, and R_m the resistance of

filter medium. The specific cake resistance, α, can be expressed as

$$\alpha = \frac{1}{k\rho_s(1 - \varepsilon)} \tag{2}$$

where k is the permeability, ρ_s the density of the solid particles, and ε the porosity of the filter cake, defined as the ratio of the volume of voids in the cake to the total volume of filter cake.

Equation (1), known as the two-resistance filtration model, is a simple expression for describing the filtration of incompressible cakes, the specific cake resistance, α, can be regarded as a constant. In the case of compressible cakes, the effect of variation in cake porosity on specific cake resistance must be considered (Tiller and Shirato, 1964). Filtration tests should be performed under different pressure drops to establish an empirical relation between the specific cake resistance, α, and the pressure drop across the filter cake, ΔP_c (McCabe et al., 1993):

$$\alpha = \alpha_0(\Delta P_c)^n \tag{3}$$

where α_0 and n are empirical constants. The constant n represents the compressibility coefficient of the filter cake. For an incompressible cake, n equals zero. The value of n ranges from 0.25 for moderately compressible to 1.0 for supercompressible cakes (Tiller et al., 1987b).

2.1.2 Constant Pressure Filtration

For constant pressure filtration, Eq. (1) can be rearranged to give

$$\frac{dt}{dV} = \frac{\mu \alpha \rho s}{\Delta P (1 - ms) A^2} V + \frac{\mu R_m}{A \, \Delta P} \tag{4a}$$

where

$$\frac{\rho s}{1 - ms} = c \tag{4b}$$

In Eq. (4b), s represents the slurry concentration, ρ the filtrate density, and m the mass ratio of wet cake to dry cake. For incompressible cakes ($\alpha = $ constant), Eq. (4a) may be simplified to

$$\frac{dt}{dV} = K'V + R' \tag{5}$$

where K' and R' are constants.

2.1.3 Constant Rate Filtration

Many industrial filtration processes can be approximated as constant rate filtration. In this case, the linear velocity of filtrate, $u = (V/At)$, remains constant. The relation between the overall pressure drop, ΔP, and the filtration time, t, can be expressed as (McCabe et al., 1993)

$$(\Delta P - \Delta P_m)^{(1-n)} = \alpha_0 \mu c \left(\frac{V}{At}\right)^2 t = K_r t \tag{6}$$

where ΔP_m is the pressure drop across the filter medium and $K_r = \alpha_0 \mu c (V/At)^2$ is a constant.

In industrial filtration operation carried out under variable pressure and variable rate conditions, the method of Tiller must be employed to integrate Eq. (1) for the general case (Tiller, 1958).

2.2 Deep-Bed Filtration

Deep-bed filtration (or depth filtration) is known by various terms like blocking filtration, surface filtration, and clarification. Deep-bed filtration (Rajagopalan and Tien, 1979; Stamatakis and Tien, 1993; Tien and Payatakes, 1979) is normally preferred in treating large quantities of liquids containing low solid concentration (less than 500 mg/liter) with particles size less than 30 μm. In this operation, the particles to be removed are often substantially smaller than the pores of the filter medium and will penetrate a considerable depth

before being captured. Silica sand, anthracite coal, active carbon, and fibers are most commonly employed as filter media.

A rational design of a deep-bed filtration process is based on the rate of clarification (removal of suspended particulate) and the pressure drop (due to medium clogging) required for maintaining a given throughput. The flow of a suspension through a deep-bed of grains (filter media) results in the penetration of the particles into the filter medium where deposition takes place on the grain surfaces at various depths.

2.2.1 Basic Equations for Deep-Bed Filtration

The fundamental equations describing the particle retention behavior in a deep-bed filtration are the continuity equation, the rate equation, and the expression for pressure drop. The removal rate of suspended solids as a function of solid concentration is written as

$$\frac{\partial C}{\partial L} = -\lambda C \tag{7}$$

Where C is the solid concentration in the suspension, L the distance from the top of the bed to the section under consideration, and λ the filter coefficient or impediment modulus. The variation of λ depends on the extent of particle retention and parameters that determine the mode of deposit morphology. A proposed expression for λ can be written in a general functional form:

$$\lambda = \lambda_0 F(\alpha, \sigma) \tag{8}$$

where λ_0 is the filter coefficient of the clean filter bed, α a parameter vector, σ the specific deposit (the volume of deposited matter per unit volume of filter bed), and $F(\alpha, \sigma)$ is a function representing the effect of particle deposition in the filter bed on the filter coefficient, λ. A list of useful expressions for function F may be found in the reference (Tien, 1989).

2.2.2 Pressure Drop in Deep-Bed Filtration

For a clean filter bed, the pressure drop can be calculated using the Carman–Kozeny equation,

$$-\left(\frac{\partial P}{\partial L}\right)_0 = \frac{150}{d_p^2} \mu u \frac{(1 - \varepsilon_0)^2}{\varepsilon_0^3} \tag{9}$$

where d_p is the particle diameter in the packed bed. This can serve as a basis for evaluating the change of pressure drop due to filter bed clogging by using empirical correlations in terms of volume occupied by the deposited particles (Ives and Pienvichitr, 1965; Ives, 1985; Tien, 1989).

2.3 Membrane Filtration

Membrane separation processes have been applied to many industrial production systems for the purpose of clarification, concentration, desalting, waste treatment, or product recovery. Broadly speaking, membrane filtration can be classified as microfiltration, ultrafiltration, nanofiltration, reverse osmosis, and dialysis or electrodialysis. In this section, the discussion will only cover microfiltration and ultrafiltration, both of which are pressure-driven membrane processes.

2.3.1 Basic Concepts

A membrane filtration process is illustrated in Fig. 1c. A suspension flows tangentially to the membrane surface into the filter under an applied pressure. The membrane functions as a filter medium. The components that pass through the membrane constitute the permeate (or filtrate), while the components retained on the membrane surface form the retentate.

The membrane retention is determined almost solely by the size and shape of the particles (including macromolecules) in suspension. For particle sizes larger than the pore size of the membrane, capture of particles is by means of interception. For particles that are smaller than the pore size, the predominant capture mechanism is inertial impaction. As the particle size and/or inertial mass decreases, diffusion is the primary capture mechanism. In addition, increased viscosity of the fluid will greatly diminish the effect of inertial impaction and diffusion. In this case, interception would be the primary collection mechanism.

In a membrane filtration process, the retentate in suspension may build up a high concentration adjacent to the membrane surface forming a dyamic boundary layer (a gel layer). This concentration gradient becomes a driving force to pull the retentate from the boundary layer back to the bulk flow. This phenomenon is referred to as concentration polarization. The accumulation of retentate at the membrane surface will result in a hydraulic resistance that may reduce the permeability of the membrane. This phenomenon is called fouling. Membrane fouling is a common phenomenon observed in the operation of any membrane filtration process, which leads to a reduction in permeate flux and selectivity.

The retention characteristics of a membrane can be described using a retention coefficient (rejection coefficient), which is defined as

$$R = \frac{C_S - C_P}{C_S} \tag{10}$$

and

$$R_0 = \frac{C_b - C_P}{C_b} \tag{11}$$

where R is the true retention coefficient and R_0 the observed retention coefficient. The variables C_S and C_b are the solid concentration in the retentate at the membrane surface and in the bulk flow, respectively, and C_P is the solid concentration in the permeate.

Membrane filters of early design can foul quickly owing to the concentration polarization effect. To control this effect, the configuration of a membrane filter is designed as a cross-flow mode (Belfort, 1986). In cross-flow configuration, the main direction of the suspension flow is perpendicular to the permeate flow. As permeate flow passes through the membrane, a portion of the particles is deposited at the surface of the membrane to form a cake layer. The cake thickness varies with the operating time and so does the rate of permeation. Thus the time-dependent behavior represents a major factor of the cross-flow membrane filtration.

2.3.2 Microfiltration

Microfiltration (MF) is a membrane filtration in which the filter medium is a porous membrane with pore sizes in the range of 0.02–10 μm. It can be utilized to separate materials such as clay, bacteria, and colloid particles. The membrane structures have been produced from the cellulose ester, cellulose nitrate materials, and a variety of polymers. A pressure of about 1–5 atm is applied to the inlet side of suspension flow during the operation. The separation is based on a sieve mechanism. The driving force for filtration is the difference between applied pressure and back pressure (including osmotic pressure, if any). Typical configurations of the cross-flow microfiltration process are illustrated in Fig. 2. The cross-flow membrane modules are tubular (multichannel), plate-and-frame, spiral-wound, and hollow-fiber as shown in Fig. 3. The design data for commercial membrane modules are listed in Table 1.

The microfiltration membranes are known to be highly porous. Thus the separation behavior within these membranes is mainly based on pore size. As the flow paths through the membrane are tortuous with

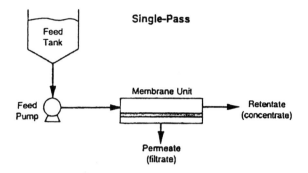

Single-Pass

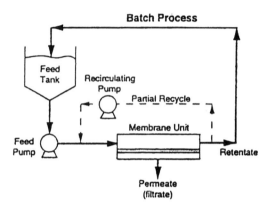

Batch Process

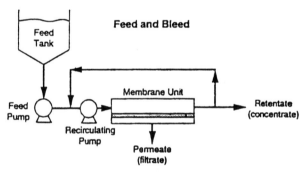

Feed and Bleed

Figure 2 Membrane filtration process configurations. (From Zeman and Zydney, 1996.)

dead-end passages and variable channel dimensions, the Carman–Kozeny equation may be used. Therefore, the flux across the microfiltration membrane (J) can be calculated by

$$ J = \frac{dV}{A dt} = \frac{\varepsilon^3}{K(1-\varepsilon)^2 S_p^2} \frac{\Delta P}{\mu L_m} \qquad (12) $$

where K is the Kozeny coefficient, depending only on the pore configuration, S_p the surface area of a single particle; and L_m the effective membrane thickness. All other parameters are as defined in the previous section. For more detailed information regarding the performance of microfiltration, readers are referred to

Microfiltration and Ultrafiltration by Zeman and Zydney (1996).

2.3.3 Ultrafiltration

Ultrafiltration (UF) is a membrane filtration process used for separating or collecting submicron-size particles (0.001 to 0.02 µm) from a suspension (Cheryan, 1986). It is usually employed to concentrate or fractionate a solution containing macromolecules, colloids, salts, or sugars. The UF membrane functions as a sieve with the pore size of molecular dimensions.

A UF process can be operated in either batch or continuous mode. The determination of membrane surface area (A) for the UF process requires three parameters: (1) flux, J, which is the measure of the membrane productivity; (2) the volume of permeate, V_p, passed through the membrane, and (3) the volume of retentate, V_r, retained on the membrane surface. The average flux, J_{av}, can be estimated from the equation

$$ J_{av} = J_f + 0.33(J_i - J_f) \qquad (13) $$

where J_f is the final flux at the highest concentration and J_i the initial flux. The material balance gives

$$ V_f = V_r + V_p \qquad (14) $$

where V_f is the volume of feed. The volume concentration ratio (VCR) is defined as

$$ VCR = \frac{V_f}{V_r} \qquad (15) $$

Eqs. (13) through (15) can be used to estimate the membrane surface area (Chiang and He, 1995), which can be expressed as

$$ A = \frac{V_f - V_p}{J_{av}} \qquad (16) $$

The configuration of UF is usually designed as polymeric and asymmetric modules for high productivity and resistance to plugging. Membrane modules used in the UF process design are similar to those adopted in MF process design (see previous section).

2.4 Filter Media

After specifying the filter type and the optimum operating conditions, the remaining issue for a filter design is the selection of the most suitable filter medium. In a filtration process, the fundamental role of a filter medium is to separate effectively the particulates from a flowing fluid to provide a sufficiently clean filtrate without clogging and damaging the medium (i.e., low

Table 3 Generalized Summary of Filter Media Based on Rigidity

Main type	Subdivisions	Smallest particle retained, microns*
Solid fabrications	(a) flat wedge-wire screens	100
	(b) wire-wound tubes	10
	(c) stacks of rings	5
Metal sheets	(a) perforated	20
	(b) woven wire	5
Rigid porous media	(a) plastics	5
	(b) sintered metals	10
	(c) ceramics & stoneware	5
	(d) carbon	1
		1
Plastic sheets	(a) woven monofilaments	10
	(b) porous sheets	10
	(c) membranes	<0.1
Woven fabrics	(a) mono- or multifilaments	10
	(b) staple fiber yarns	5
		10
Nonwoven media	(a) polymeric nonwovens (melt blow, spun bonded, etc)	10
	(b) felts & needle felts	
	(c) paper media	5
	cellulose	2
	glass	0.5
	(d) filter sheets	5
		5
Cartridges	(a) yarn wound	3
	(b) bonded beds	
	(c) sheet fabrications	1
		<0.1
Loose media	(a) fibers	0.2
	(b) powders	0.2
Membranes	(a) ceramic	<0.1
	(b) metal	
	(c) polymeric	

*Very rough indication.
Source: Purchas D. Handbook of Filter Media. 1st ed. UK: Elsevier Science, 1996, p 4.

ne key advantage of perlite over
purity.
und wool pulp) is applied to
his filter aid forms a much
with good permeability but
e retentivity than diatomite
ulose is higher than that of
his and other filter aids are
(such as precoat stabiliza-

2.5.2 Filtration with Filter Aids

In general, filter aid filtration should be used only
systems that meet two key requirements. First,
desired product is the filtrate, not the cake. Sec
the filter aid is acceptable in the filter cake or the
cake can be easily repulped and refiltered to remov
filter aid. In addition, the particle-settling rate sh
be less than 0.012 m/min and the particle concent
lower than 0.1 wt% (<0.3% for rotary drum pre

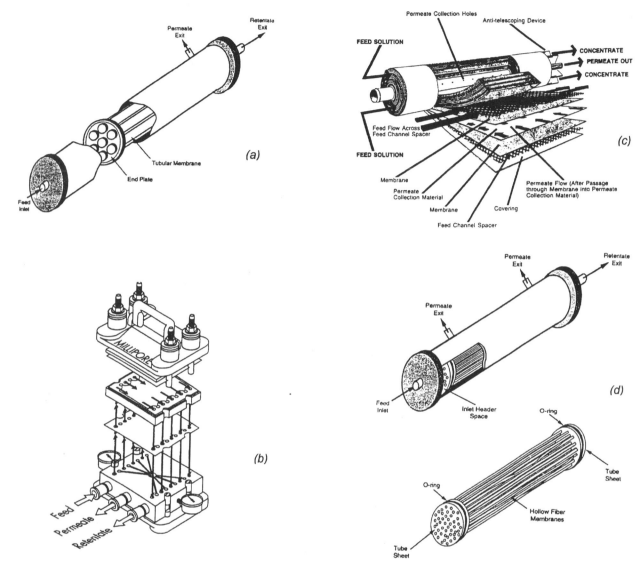

Figure 3 Membrane modules for microfiltration and ultrafiltration. (a) Tubular membrane module. (From Zeman and Zydney, 1996.) (b) Cassette membrane assembly. (From Perry et al., 1997.) (c) Spiral wound membrane module. (From Purchase, 1996.) (d) Hollow fiber membrane module. (From Zeman and Zydney, 1996.)

energy consumption and long operating period). Thus proper selection of the filter medium is often the most crucial step for assuring efficient filtration operation. The major considerations include the permeability of the medium relative to a pure liquid, its retention capacity relative to particulates of known size, and the medium pore size distribution.

The resistance (or permeability) of a filter medium directly affects the capital and operating costs. Most manufacturers also employ the permeability of filter media as a measure of particle retention, which is related to its pore size and porosity. A high permeabil-

ity of the media is, therefore, used as an indication of high porosity and in turn a low particle retentivity. The ideal medium would provide the maximum open (free) area for flow while it meets the required retentivity. A comparison of the free area for commonly used filter media is presented in Table 2.

An alternative classification of filter media is based on the minimum size of the trapped particles (see Table 3). Obviously, this classification provides only a general guideline to the types of media available. In fact, many process parameters (such as the particulate concentration, etc.) affect the retention behavior of

Membrane Modules

Product	Code[a]	Membrane type	Height or i.d. (mm)	L (m)	Membrane area (m²)	V'_w (ms⁻¹)	$Re'_w \times 10^{-3}$	$Re'^d_w \times 10^{-3}$
Ultrafiltration								
	US1	Polysulphone fluoride	2.54	0.37	0.067	$1.89\text{–}4.25 \times 10^{-4}$	21.84	48.0–108.0
	US2	Polyvinylidene copolymer	2.54	0.37	0.067	$1.80\text{–}1.42 \times 10^{-5}$	21.84	30.0–36.0
	US3	VC-AN copolymer	2.54[b]	0.37	0.067	$0.95\text{–}1.18 \times 10^{-5}$	21.84	24.0–30.0
	US4	Polysulphone	5.90[b]	0.22	0.150	$0.69\text{–}6.94 \times 10^{-4}$	6.02	4.0–41.0
Lab Unit 35	US5	Polysulphone	5.19[b]	0.076	0.018	$0.16\text{–}8.10 \times 10^{-5}$	19.30	6.0–42.0
Lab Unit 20	UT1	Cellulose acetate	7.60[b]	3.05	0.204	9.43×10^{-5}	43.26	0.78
HFA	UT2	Polyvinylidene fluoride	11.36–22.7	3.05	0.022	$1.41\text{–}2.82 \times 10^{-4}$	5.39	19.30
HFM	UT3	Polysulphone	3.32–6.48	0.635	0.635	$0.116\text{–}1.62 \times 10^{-4}$	4.87	36.0–72.0
PM	UT4	Modacryl polymer		1.10	0.30	$1.39\text{–}1.74 \times 10^{-5}$	0.083	1.3–18.0
XM	UT5	Polyamidimide		0.60	0.032	$0.116\text{–}6.94 \times 10^{-4}$	33.44	15.0–19.0
BM			2.84–6.02			0.0021×10^{-4}		0.07–4.20
			0.011–0.266			5.65×10^{-4}	3.60	20.0–89.0
Microfiltration								
Dyna-Sep Sampler	MT1	Polypropylene	4.21–7.97	5.50	1.2	$0.556\text{–}2.78 \times 10^{-4}$		10.0–50.0
Microdyne	MT2	Polypropylene	1.0–3.0	1.80		0.56		
		Polypropylene				1.83		

Source: Belfort G. In: Muralidhara HS, ed. Advances in Solid–Liquid Separation. Columbus: Battelle Press, 1986, pp 182–183, Table 2.

[a] UT = ultrafiltration silt. UT = ultrafiltration tube; MT = microfiltration tube; MT = microfiltration.

Liquid–Solids Separation

Table 2 Typical Porosity of Filter Media

Filter media	% free area
Wedge wire screen	5–40
Woven wire	15–20
twill weave	25–50
square	30–40
Perforated metal sheet	45
Porous plastics (molded powder)	25–55
Sintered metal powder	50–60
Crude kieselguhr	80
Membranes	60–95
Paper	70–85
Sintered metal fibers	80–90
Refined filter aids (diatomite, perlite)	93
Plastic, ceramic foam	

Source: Purchas D. Handbook of Filter Media. 1st ed. UK: Elsevier Science, 1996, p 31.

particulates within the filter media. The detailed information regarding the filter media properties and their selection can be found in the Handbook of Filter Media (Purchas, 1996) and the Filters and Filtration Handbook (Dickenson, 1994).

A good filter medium should have the following features:

1. A wide size distribution of particles from the slurry, producing a clean filtrate
2. An economic filtration time, i.e., minimum filtrate flow resistance
3. Easy discharge of the filter cake from the medium
4. Sufficient strength to withstand filtering pressure and mechanical wear
5. Avoidance wedging of particles into its pores and an adequate medium lifetime
6. Low cost

2.5 Filter Aids

In order to improve the filtration characteri... hard-to-filter suspensions, such as those w... filtration rate, rapid medium binding, or ... tory clarity, addition of another parti... material is one of the best alternat... Such solid material is termed a filter ... added to the suspensions build up a ... able, and rigid lattice structure for ret... particles and allowing the liquid to pass th...

As a general rule, a suitable filter aid should be capable of creating a thin layer structure with a maximum pore size over the medium's external surface and producing a prespecified filtrate clarity at an optimum filtration rate.

Filter aids may be applied to the filtration oper... in two ways. The first way is to precoat the filte... ium using a precoat filter aid. The precoat is to ... as the actual filter medium. The function o... application is to prevent the filter medium... ging or fouling as well as to facilitate th... the formed cake at the end of filtratio... way is to pretreat the suspension u... powder with a coarser size distrib... filtration process. Such material is... admix. The functions of body ai... porosity of filter cake and to de... ity, resulting in a decrease in t'... turn an increase in the filtra...

2.5.1 Requirements f...

Filter aid selection... tests. The require... of the selected... follows:

1. Form... por...
2. H...
3. ...
4. ...

from 48–96 kg/m³. T... diatomite is its relative... Cellulosic fiber (gro... cover metallic cloths. ... more compressible cake... displays a smaller partic... or perlite. The cost of cel... diatomite or perlite. Thus ... only applied to special cases... tion or chemical resistance).

Pressure or vacuum cake filters can be operated with both precoat and body aids. Most filter aids are used on a one-time basis, although tests reusing precoat filter aid have been demonstrated in a few pressure filters (Schweitzer, 1997).

2.6 Filtration Equipment

Filtration equipment selection is often complex. First, the suspension properties and process conditions can change tremendously. Secondly, a wide variety of filters are commercially available. In practice, projected results of the selection are worked out most reliably from actual or pilot plant database or analogous application profiles. The basic filtration equations can serve as a useful guide for performing the comparison among the selected filters and evaluating their acceptability. In specifying and selecting filtration equipment, attention should be placed on options that would offer a low cake resistance. This resistance is directly related to the filter capacity. The cost profiles are also important information for making the final decision. Therefore a proper selection of a filter should consider a number of factors including slurry properties, cake characteristics, anticipated capacity, process operation requirements, and cost estimate (Fitch, 1977; Mayer, 1988; Rushton et al., 2000).

2.6.1 Cake Filtration Equipment

Many different types of equipment are being marketed for cake filtration. Only the three most commonly used ones are discussed here.

Rotating Drum Filters. Rotating drum filters most frequently work under vacuum conditions, such as rotary vacuum drum filters with external filtering surfaces (see Fig. 4). The design configuration consists of a cloth-covered hollow drum with a slotted face, the outer circumference containing a shallow tray-shaped compartment. The drum surface is partially submerged in the feed suspension. The filter drum is divided into several operating zones: filtration, first dewatering, cake washing, second dewatering, cake removal, and cloth cleaning. The drum rotates at 10 to 60 revolutions per hour by a variable speed motor. Each zone of the drum is connected to a collection port on the automatic valve. In the course of one revolution, the drum area passes through these zones in succession by means of a control head device. In the filtration zone, the filter cake builds up on the drum outer surface under vacuum conditions (approximately 400 to 160 torr). After

cake dewatering and washing, compressed air and a scraper are used for cake removal and clean-blowing the filter cloth. Fouling by small particles is a frequent problem in the cases of suspended particles with wide size ranges. Rotary drum filters also have a design version used as a pressure filter. Unlike the vacuum drum filter, the pressure drop required for cake formation is controlled between the filter pressure vessel and the filtrate separator. The filter cake can be washed and discharged by the same method as described previously. Pressure drum filters are particularly suitable for processing of foodstuffs, antibiotics, dyestuffs and solvent, and water treatment. The typical applications and performance guide for rotary drum vacuum filters can be found in Dickenson (1994).

Rotary Disc Filters. Rotary disc filters are another kind of rotating filters, as shown in Fig. 5 (Svarovsky, 1990). This type of filter provides a much larger filter area per unit of floor area at lower cost than those of the rotary drum filter. Their applications are in coal preparation, ore dressing, and pulp or paper processing. The rotary disc filter is constructed by a number of discs (up to 12 or more) mounted on a horizontal hollow shaft. Each disc is equipped with interchangeable elements and has an individual slurry compartment. Submergence up to 50% of the filtering surface can be attained by a level control. The cake formed on the emergent sector of the disc is treated and removed by washing and scraper before reentering the trough. The filter area or filtering capacity can be adjusted by the change in the number of discs. Disc filters are available in filtering areas from 0.5 to 300 m². The disc filter is mostly used as a dewatering or thickening device. Owing to the vertical filtering surface, cake washing in the disc filter is not as efficient as in a drum filter. The rotary disc filter is particularly beneficial in cases of limited space and where cakes do not require washing.

Horizontal Filters. To avoid the poor cake pickup and washing inherent in the design of rotary drum and vertical disc filters, one such design, the Nutsch filter, employs a flat filtering plate covered with filtering cloth. This type of filter basically consists of a large false-bottom tank with a loose filter medium. The Nutsch filter takes advantage of the effect of gravity in cake formation and is able to provide uniform washing.

The horizontal belt filter consists of a series of Nutsch filters as a long chain moving along a closed path with the belt speed in the range of 3–30 m/min. A

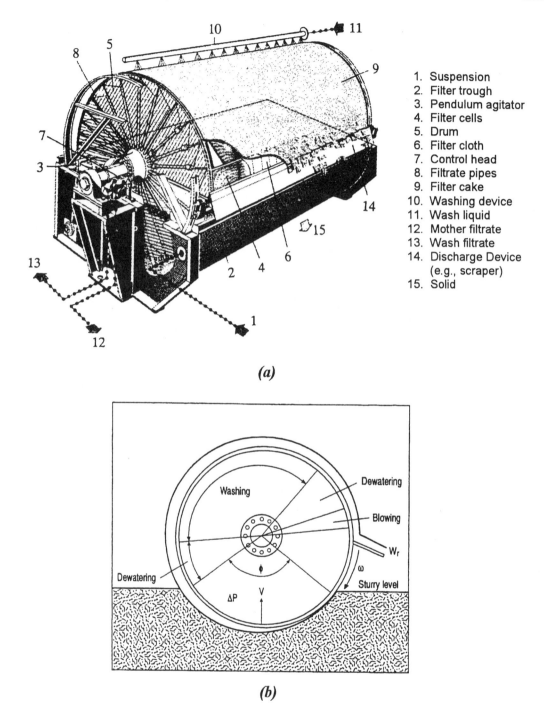

(a)

(b)

Figure 4   Rotary drum vacuum filter. (a) Cut-away view. (From Dickenson, 1994.) (b) Filtration cycle. (From Rushton, 2000.)

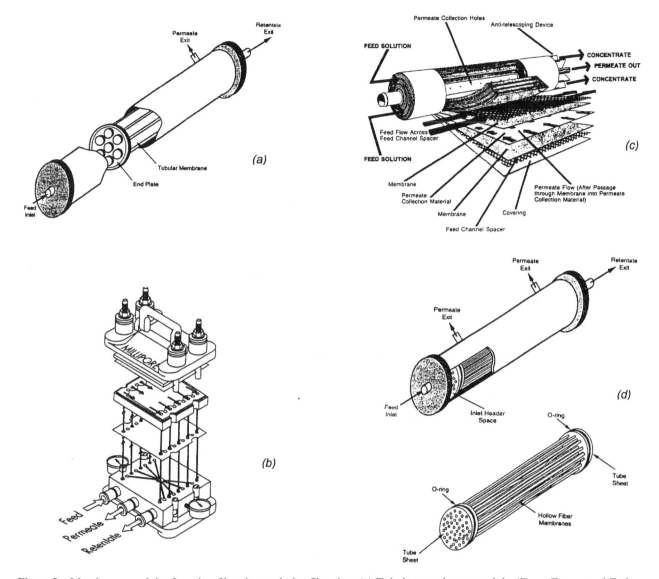

Figure 3 Membrane modules for microfiltration and ultrafiltration. (a) Tubular membrane module. (From Zeman and Zydney, 1996.) (b) Cassette membrane assembly. (From Perry et al., 1997.) (c) Spiral wound membrane module. (From Purchase, 1996.) (d) Hollow fiber membrane module. (From Zeman and Zydney, 1996.)

energy consumption and long operating period). Thus proper selection of the filter medium is often the most crucial step for assuring efficient filtration operation. The major considerations include the permeability of the medium relative to a pure liquid, its retention capacity relative to particulates of known size, and the medium pore size distribution.

The resistance (or permeability) of a filter medium directly affects the capital and operating costs. Most manufacturers also employ the permeability of filter media as a measure of particle retention, which is related to its pore size and porosity. A high permeabil-

ity of the media is, therefore, used as an indication of high porosity and in turn a low particle retentivity. The ideal medium would provide the maximum open (free) area for flow while it meets the required retentivity. A comparison of the free area for commonly used filter media is presented in Table 2.

An alternative classification of filter media is based on the minimum size of the trapped particles (see Table 3). Obviously, this classification provides only a general guideline to the types of media available. In fact, many process parameters (such as the particulate concentration, etc.) affect the retention behavior of

Table 1 Design Data for Commercial Membrane Modules

Configuration	Company Location	Product	Code[a]	Membrane type	U_m (ms^{-1})	Height or i.d. (mm)	L (m)	Membrane area (m^2)	V_w (ms^{-1})	Re[c] ×10^{-3}	Re$_w$[d] ×10^3
Ultrafiltration											
Flat plate	Dorr-Oliver USA	Ioplate	US1	Polysulphone	5.0–12.2	2.54	0.37	0.067	1.89–4.25 × 10^{-4}	21.84	48.0–108.0
			US2	Polyvinylidene fluoride	5.0–12.2	2.54	0.37	0.067	1.80–1.42 × 10^{-4}	21.84	30.0–36.0
			US3	VC-AN copolymer	5.0–12.2	2.54	0.37	0.067	0.95–1.18 × 10^{-4}	21.84	24.0–30.0
Flat plate	DDS Denmark	Lab Unit 35	US4	Polysulphone	0.68–1.36[b]	5.90[b]	0.22	0.150	0.69–6.94 × 10^{-5}	6.02	4.0–41.0
		Lab Unit 20	US5	Polysulphone	0.10–0.20[b]	5.19[b]	0.076	0.018	0.16–8.10 × 10^{-5}	0.78	6.0–42.0
Tube	Abcor 24.0 USA	HFA	UT1	Cellulose acetate	7.60[b]	2.54	3.05	0.204	9.43 × 10^{-5}	19.30	
Tube	Romicon-Amicon USA	HFM	UT2	Polyvinylidene fluoride	11.36–22.7[b]	2.54	3.05	0.204	1.41–2.82 × 10^{-4}	43.26	36.0–72.0
		PM	UT3	Polysulphone	3.32–6.48	1.10	0.635	0.022	0.116–1.62 × 10^{-4}	5.39	1.3–18.0
Tube	Berghof Germany	XM	UT4	Modacryl polymer	2.84–6.02	1.10	0.635	0.0021	1.39–1.74 × 10^{-4}	4.87	15.0–19.0
		BM	UT5	Polyamidimide	0.011–0.266	0.60	0.30	5.65 × 10^{-4}	0.116–6.94 × 10^{-5}	0.083	0.07–4.20
Microfiltration											
Tube	Enka/Membrana Germany	Dyna-Sep Sampler	MT1	Polypropylene	4.21–7.97	5.50	1.83	0.032	0.365–1.62 × 10^{-4}	33.44	20.0–89.0
Tube		Microdyne	MT2	Polypropylene	1.0–3.0	1.80	0.56	1.2	0.556–2.78 × 10^{-4}	3.60	10.0–50.0

[a]US = ultrafiltration slit, UT = ultrafiltration tube, MT = microfiltration tube.
[b]Estimate
[c]$v = 10^{-6}$ m^2/s for water at 25°C
[d]Re$_w$ = (hV_w/v) or (dV_w/v)
Source: Belfort G. In: Muralidhara HS, ed. Advances in Solid–Liquid Separation. Columbus: Battelle Press, 1986, pp 182–183, Table 2.

Table 2 Typical Porosity of Filter Media

Filter media	% free area
Wedge wire screen	5–40
Woven wire	
twill weave	15–20
square	25–50
Perforated metal sheet	30–40
Porous plastics (molded powder)	45
Sintered metal powder	25–55
Crude kieselguhr	50–60
Membranes	80
Paper	60–95
Sintered metal fibers	70–85
Refined filter aids (diatomite, perlite)	80–90
Plastic, ceramic foam	93

Source: Purchas D. Handbook of Filter Media. 1st ed. UK: Elsevier Science, 1996, p 31.

particulates within the filter media. The detailed information regarding the filter media properties and their selection can be found in the Handbook of Filter Media (Purchas, 1996) and the Filters and Filtration Handbook (Dickenson, 1994).

A good filter medium should have the following features:

1. A wide size distribution of particles from the slurry, producing a clean filtrate
2. An economic filtration time, i.e., minimum filtrate flow resistance
3. Easy discharge of the filter cake from the medium
4. Sufficient strength to withstand filtering pressure and mechanical wear
5. Avoidance wedging of particles into its pores and an adequate medium lifetime
6. Low cost

2.5 Filter Aids

In order to improve the filtration characteristics of hard-to-filter suspensions, such as those with slow filtration rate, rapid medium binding, or unsatisfactory clarity, addition of another particulate solid material is one of the best alternative measures. Such solid material is termed a filter aid. Filter aids added to the suspensions build up a porous, permeable, and rigid lattice structure for retaining solid particles and allowing the liquid to pass through.

As a general rule, a suitable filter aid should be capable of creating a thin layer structure with a maximum pore size over the medium's external surface and producing a prespecified filtrate clarity at an optimum filtration rate.

Filter aids may be applied to the filtration operation in two ways. The first way is to precoat the filter medium using a precoat filter aid. The precoat is to behave as the actual filter medium. The function of such an application is to prevent the filter medium from clogging or fouling as well as to facilitate the removal of the formed cake at the end of filtration. The second way is to pretreat the suspension using a filter aid powder with a coarser size distribution prior to the filtration process. Such material is called body aid or admix. The functions of body aid are to increase the porosity of filter cake and to decrease its compressibility, resulting in a decrease in the cake resistance and in turn an increase in the filtration rate.

2.5.1 Requirements for Filter Aid Selection

Filter aid selection should be based on laboratory tests. The requirements for preliminary evaluation of the selected filter aids may be summarized as follows:

1. Form a thin and rigid lattice layer with high porosity
2. Have low specific surface or coarse size
3. Have a narrow fractional size distribution by removing the finer size fractions
4. Create a rapid particle bridging and settling or a uniform filter aid layer
5. Be chemically inert and able to prevent medium cracking and clogging

These requirements are all found in the two most common filter aids. The first one is the diatomaceous silica type filter aid (also called diatomite, kieselguhr or diatomaceous earth), which contains 90% or almost pure silica and particle size mostly smaller than 50 μm. Its bulk density ranges from 128 to 320 kg/m^3. Calcinated diatomaceous additives display their high retention ability with relatively low hydraulic resistance. the relative permeability of the calcinated diatomite increases up to 3–20 times that of natural diatomite. The disadvantage of diatomite is that it may foul filtering liquids by dissolved salts and colloidal clays. The second most common filter aid is expended perlite, which is a glasslike volcanic rock. The porosity of this filter aid is in the range of 0.85–0.9. Its bulk density ranges

Table 3 Generalized Summary of Filter Media Based on Rigidity

Main type	Subdivisions	Smallest particle retained, microns*
Solid fabrications	(a) flat wedge-wire screens	100
	(b) wire-wound tubes	10
	(c) stacks of rings	5
Metal sheets	(a) perforated	20
	(b) woven wire	5
Rigid porous media	(a) plastics	10
	(b) sintered metals	5
	(c) ceramics & stoneware	1
	(d) carbon	1
Plastic sheets	(a) woven monofilaments	10
	(b) porous sheets	10
	(c) membranes	< 0.1
Woven fabrics	(a) mono- or multifilaments	10
	(b) staple fiber yarns	5
Nonwoven media	(a) polymeric nonwovens (melt blow, spun bonded, etc)	10
	(b) felts & needle felts	10
	(c) paper media	
	cellulose	5
	glass	2
	(d) filter sheets	0.5
Cartridges	(a) yarn wound	5
	(b) bonded beds	5
	(c) sheet fabrications	3
Loose media	(a) fibers	1
	(b) powders	< 0.1
Membranes	(a) ceramic	0.2
	(b) metal	0.2
	(c) polymeric	< 0.1

*Very rough indication.
Source: Purchas D. Handbook of Filter Media. 1st ed. UK: Elsevier Science, 1996, p 4.

from 48–96 kg/m^3. The key advantage of perlite over diatomite is its relative purity.

Cellulosic fiber (ground wool pulp) is applied to cover metallic cloths. This filter aid forms a much more compressible cake with good permeability but displays a smaller particle retentivity than diatomite or perlite. The cost of cellulose is higher than that of diatomite or perlite. Thus this and other filter aids are only applied to special cases (such as precoat stabilization or chemical resistance).

2.5.2 Filtration with Filter Aids

In general, filter aid filtration should be used only for systems that meet two key requirements. First, the desired product is the filtrate, not the cake. Second, the filter aid is acceptable in the filter cake or the filter cake can be easily repulped and refiltered to remove the filter aid. In addition, the particle-settling rate should be less than 0.012 m/min and the particle concentration lower than 0.1 wt% (< 0.3% for rotary drum precoat).

Pressure or vacuum cake filters can be operated with both precoat and body aids. Most filter aids are used on a one-time basis, although tests reusing precoat filter aid have been demonstrated in a few pressure filters (Schweitzer, 1997).

2.6 Filtration Equipment

Filtration equipment selection is often complex. First, the suspension properties and process conditions can change tremendously. Secondly, a wide variety of filters are commercially available. In practice, projected results of the selection are worked out most reliably from actual or pilot plant database or analogous application profiles. The basic filtration equations can serve as a useful guide for performing the comparison among the selected filters and evaluating their acceptability. In specifying and selecting filtration equipment, attention should be placed on options that would offer a low cake resistance. This resistance is directly related to the filter capacity. The cost profiles are also important information for making the final decision. Therefore a proper selection of a filter should consider a number of factors including slurry properties, cake characteristics, anticipated capacity, process operation requirements, and cost estimate (Fitch, 1977; Mayer, 1988; Rushton et al., 2000).

2.6.1 Cake Filtration Equipment

Many different types of equipment are being marketed for cake filtration. Only the three most commonly used ones are discussed here.

Rotating Drum Filters. Rotating drum filters most frequently work under vacuum conditions, such as rotary vacuum drum filters with external filtering surfaces (see Fig. 4). The design configuration consists of a cloth-covered hollow drum with a slotted face, the outer circumference containing a shallow tray-shaped compartment. The drum surface is partially submerged in the feed suspension. The filter drum is divided into several operating zones: filtration, first dewatering, cake washing, second dewatering, cake removal, and cloth cleaning. The drum rotates at 10 to 60 revolutions per hour by a variable speed motor. Each zone of the drum is connected to a collection port on the automatic valve. In the course of one revolution, the drum area passes through these zones in succession by means of a control head device. In the filtration zone, the filter cake builds up on the drum outer surface under vacuum conditions (approximately 400 to 160 torr). After

cake dewatering and washing, compressed air and a scraper are used for cake removal and clean-blowing the filter cloth. Fouling by small particles is a frequent problem in the cases of suspended particles with wide size ranges. Rotary drum filters also have a design version used as a pressure filter. Unlike the vacuum drum filter, the pressure drop required for cake formation is controlled between the filter pressure vessel and the filtrate separator. The filter cake can be washed and discharged by the same method as described previously. Pressure drum filters are particularly suitable for processing of foodstuffs, antibiotics, dyestuffs and solvent, and water treatment. The typical applications and performance guide for rotary drum vacuum filters can be found in Dickenson (1994).

Rotary Disc Filters. Rotary disc filters are another kind of rotating filters, as shown in Fig. 5 (Svarovsky, 1990). This type of filter provides a much larger filter area per unit of floor area at lower cost than those of the rotary drum filter. Their applications are in coal preparation, ore dressing, and pulp or paper processing. The rotary disc filter is constructed by a number of discs (up to 12 or more) mounted on a horizontal hollow shaft. Each disc is equipped with interchangeable elements and has an individual slurry compartment. Submergence up to 50% of the filtering surface can be attained by a level control. The cake formed on the emergent sector of the disc is treated and removed by washing and scraper before reentering the trough. The filter area or filtering capacity can be adjusted by the change in the number of discs. Disc filters are available in filtering areas from 0.5 to 300 m². The disc filter is mostly used as a dewatering or thickening device. Owing to the vertical filtering surface, cake washing in the disc filter is not as efficient as in a drum filter. The rotary disc filter is particularly beneficial in cases of limited space and where cakes do not require washing.

Horizontal Filters. To avoid the poor cake pickup and washing inherent in the design of rotary drum and vertical disc filters, one such design, the Nutsch filter, employs a flat filtering plate covered with filtering cloth. This type of filter basically consists of a large false-bottom tank with a loose filter medium. The Nutsch filter takes advantage of the effect of gravity in cake formation and is able to provide uniform washing.

The horizontal belt filter consists of a series of Nutsch filters as a long chain moving along a closed path with the belt speed in the range of 3–30 m/min. A

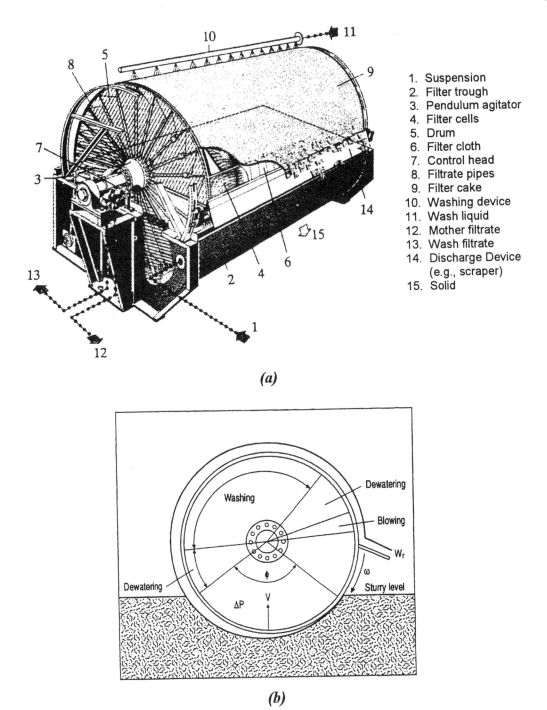

1. Suspension
2. Filter trough
3. Pendulum agitator
4. Filter cells
5. Drum
6. Filter cloth
7. Control head
8. Filtrate pipes
9. Filter cake
10. Washing device
11. Wash liquid
12. Mother filtrate
13. Wash filtrate
14. Discharge Device
    (e.g., scraper)
15. Solid

(a)

(b)

Figure 4   Rotary drum vacuum filter. (a) Cut-away view. (From Dickenson, 1994.) (b) Filtration cycle. (From Rushton, 2000.)

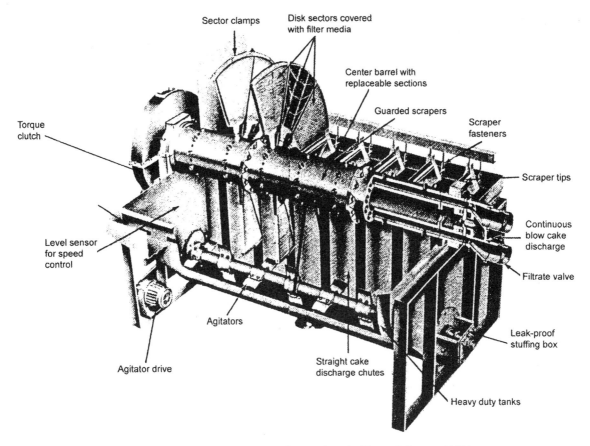

Figure 5 Rotary vacuum disc filter (Eimco). (From Osborne, 1990.)

simple belt vacuum filter is illustrated in Fig. 6. Filtrate permeates through the filter medium and is directed to centrally situated drainage pipes. In this filter, the suspension feed, the wash liquor feed, the cake discharge, and cloth cleaning are continuous. The cake thickness generally ranges from 1 to 25 mm. Cakes up to 100–150 mm thick are possible with some fast-draining materials. The chamber and collector are divided into sections from which filtrate and washing liquid can be discharged. Efficient cake discharge can be accomplished by separating the belt from the filter cloth and directing the latter over a set of discharge rollers. The simplicity in design is an advantage of belt filters. In addition, the countercurrent cake washing and removal of thin cake can be easily achieved in such filters. The shortcomings of this type of filter are large area requirements, inefficient use of the available filter area, and ineffective washing at the belt edges.

Other types of filters, such as the rotating table filter, the candle filter, filter presses, etc., are described in detail in the recent filtration literature (Dickenson, 1994; Schweitzer, 1997; Svarovsky, 1990).

2.6.2 Deep-Bed Filtration Units

Commercial deep-bed filters consist of a cylindrical or rectangular packed bed through which the suspension to be filtered is passed. The common types of deep-bed filter include the slow sand filter, the rapid filter, and direct filtration with a flocculated mixture. Typical deep-bed filters are 0.5–0.3 m in height and 1 m in diameter. Smaller packing material (filter media) provide a greater surface area and result in a more effective capture of suspended solids, but the bed pressure drop and clogging tendency also increase. Often, the design of the deep-bed filter is to employ mixed size media packed in multilayers, as shown in Fig. 7 (Cheremisinoff, 1998).

In water filtration, the dual media filters are usually designed using coarse anthracite coal on the top of fine silica sand. The coarse anthracite layer serves to prevent the formation of surface deposits on the sand bed, resulting in the formation of a compressible cake along the bed depth. Thus the depth removal of particulates throughout the bed would be the key feature.

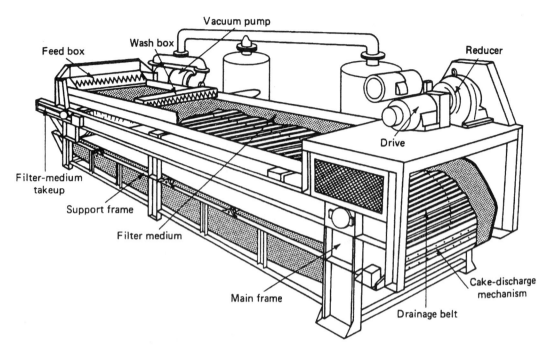

Figure 6   Horizontal belt filter. (From McCabe et al., 1993.)

This arrangement leads to longer operating cycles, particularly in low-pressure systems. For potable or wastewater filtration, a minimum top effective size of 1.1 to 1.7 mm for the coarse medium (anthracite) is recommended. It has been demonstrated that the use of coarser media can reduce the filtration costs, because such a filter can be run at a higher flow rate with a lower pressure drop. Obviously, the use of coarser particles requires deeper bed media to meet the demand of equivalent media surface area for filtrate quality. In some cases, a thin layer of dense solids, such as alumina or gravel, has been used as a third layer situated beneath the sand in the modified design of a dual layer deep-bed filter. However, modern filter design still mostly stems from the design version used in the 1980s. It uses coarse media at the bottom and finer sand media at the top (see Fig. 7a–c). It was reported that good quality water without chemical pretreatment was produced at a rate of 4–5 million gallons per acre per day. A comparison of operating characteristics of various deep-bed filters is presented in Table 4 (Rushton et al., 2000).

NOMENCLATURE

A	=	cross-sectional area of filter
C	=	solid concentration in the suspension
C_b	=	retentate concentration in the bulk flow
C_P	=	permeate concentration
C_S	=	retentate concentration at the membrane surface
c	=	mass of solid per unit volume of filtrate
d_p	=	particle (grain) diameter
$\boldsymbol{F}$	=	function describing the effect of particle deposition on deep-bed filtration coefficient
J	=	flux
J_{av}	=	average flux
J_f	=	final flux at the highest concentration
J_i	=	initial flux
K	=	Kozeny coefficient
K'	=	constant
K_r	=	constant
k	=	permeability
L	=	distance from the top of the bed to the section under study or bed depth
L_m	=	effective membrane thickness
m	=	mass ratio of wet cake to dry cake
n	=	compressibility constant of filter cake
ΔP	=	overall pressure drop across the filter
ΔP_c	=	pressure drop across the filter cake
P_m	=	pressure drop across the filter septum
R	=	true retention coefficient
R_0	=	observed retention coefficient
R_m	=	resistance of filter medium (e.g., septum)
R'	=	constant
S_p	=	surface area of a single particle
s	=	slurry concentration
t	=	filtration time

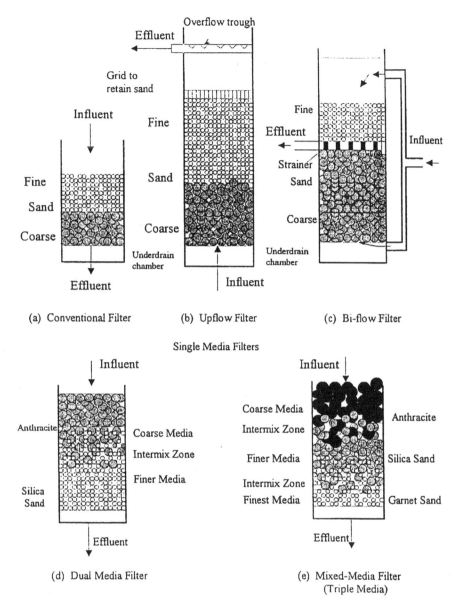

Figure 7 Common deep-bed filter operating configuration. (From Cheremisinoff, 1998.)

u	=	liquid flow rate
V	=	filtrate volume
V_f	=	volume of feed
V_p	=	volume of permeate passed through the membrane
V_r	=	volume of retentate, retained on the membrane surface
VCR	=	volume concentration ratio

Greek Symbols

α	=	specific cake resistance

α_o	=	empirical constant defined
α	=	vector parameter in function F, Eq. (8)
ε	=	porosity of the filter cake
ε_0	=	porosity of clean filter bed
λ	=	deep-bed filter coefficient or impediment modulus
λ_0	=	deep-bed filter coefficient of the clean filter bed
μ	=	fluid viscosity
ρ	=	filtrate density
ρ_s	=	density of the solid particles
σ	=	specific deposit in the filter bed

Table 4 Characteristics of Different Types of Deep-Bed Filters

Characteristic	Continuous upflow	Coarse downflow	Multimedia
Direction of flow	upwards	downwards	downwards
Cleaning	continuous	sequenced	sequenced
Media size, mm	sand, 1–2	sand, 1–3	sand, 0.4–1 anthracite, 0.8–2
Sphericity of media	0.8	0.8–0.9	sand, 0.8 anthracite, 0.7
Media depth, m	1.2	1.5–2	sand, 0.5 anthracite, 0.3
Vessel depth, m	3.5–7	3–5	2–3.5
Surface loading, $m^3/m^2\,h$	8–14	6–18	8–16
Hydraulic stability	low	high	high
Removal efficiency	moderate	low	sand: high anthracite: low
Back-wash cycle			
Water, $m^3/m^2\,d$	19	30	30
Air, m^3/m^2	10–12 at 7.5 bar	120 at 1 bar	50 at 0.5 bar
Time	continuous	2 h/d	1 h/d
Relative capital cost	high	low	moderate
Relative operating cost	high	low	low

Source: Rushton et al., 2000, p 225.

REFERENCES

Akers, RJ, Ward AS. Liquid filtration theory and filtration pretreatment. In: Orr, CC, ed. Filtration: Principles and Practices, Part I. New York: Marcel Dekker, 1977, pp 169–250.

Belfor G. Fluid mechanics and cross-flow membrane filtration. In: Muralidhara HS, ed. Advances in Solid–Liquid Separation. Columbus: Battelle Press, 1986, pp 182–183.

Cheremisinoff NP. Solid/Liquid Separation. 2d ed. Lancaster, PA: Technomic, 1995, p 240.

Cheremisinoff NP. Liquid Filtration. 2d ed. Boston: Butterworth-Heinemann, 1998, p 277.

Cheremisinoff NP, Azbel DS. Liquid Filtration for Process and Pollution Control. New Jersey: SciTech Publishers, 1989, p 59.

Cheryan M, Ultrafiltration Handbook, Lancaster, PA: Technomic, 1986, p 2.

Chiang SH, He DX. Section 62: Fluid/particle separation. In: Dorf RC, ed. Engineering Handbook. Boca Raton, FL: CRC Press, 1995, pp 644–651.

Dickenson C. Filters and Filtration Handbook. 3d ed. UK: Elsevier Science, 1994, p 214.

Fitch B. When to Use Separation Techniques other than Filtration. In: Shoemaker W, ed. AIChE Symposium Series, 171, Vol. 73. New York: AIChE, 1977, pp 107–108.

Ives KJ. Deep bed filters. In: Mathematical Models and Design Methods. In: Rushton A, ed. Solid–Liquid Separation. Boston: Martinus Nijhoff, 1985, pp 90–149.

Ives KJ, Pienvichitr V. Kinetics of the filtration of dilute suspensions. Chem Eng Sci (20):965, 1965.

Jacobs LJ, Penney WR. In: Rousseau RW, ed. Handbook of Separation Process Technology. New York: John Wiley, 1987, p 168.

Mayer E. Solid/liquid separation-selection techniques. Fluid/Particle Separation Journal 1(2):132–139, 1988.

McCabe WL, Smith JC, Harriott P. Unit Operations of Chemical Engineeering. 5th ed. New York: McGraw-Hill, 1993, pp 1010–1027.

Perry RH, Green DW, Maloney JO. Perry's Chemical Engineers' Handbook. 7th ed. New York: McGraw-Hill, 1997, p 22–42.

Purchas D. Handbook of Filter Media. 1st ed. UK: Elsevier Science, 1996, p 4, 31, 415.

Rajagopalan R, Tien C. The theory of deep bed filtration. In: Wakeman RJ, ed. Progress in Filtration and Separation. New York: Elsevier Scientific, 1979, pp 179–269.

Rushton A, Ward AS, Holdich RG. Solid–Liquid Filtration and Separation Technology. 2d ed. Weinheim, Germany: Wiley-VCH, 2000, pp 225, 453.

Schweitzer PA. Handbook of Separation Techniques for Chemical Engineers. 3d ed. New York: McGraw-Hill, 1997, pp 4.3, 4.61, 4.140.

Stamatakis K, Tien C. A simple model of cross-flow filtration based on particle adhesion. AIChEJ 39(8): 1292–1302, 1993.

Svarovsky L. Pressure filtration and vacuum filtration. In: Svarovsky L, ed. Solid–Liquid Separation. 3d ed. London: Butterworths, 1990, pp 402–475.

Tien C. Granular Filtration of Aerosols and Hydrosols. Boston: Butterworths, 1989, pp 22–27.

Tien C, Payatakes AC. Advances in deep bed filtration. AIChE J 25(5):737–759, 1979.

Tiller FM. The role of porosity in filtration, Part 3: Variable pressure–variable rate filtration. AIChE J 4(2):170–175, 1958.

Tiller FM, Shirato M. The role of porosity in filtration, VI. New definition of filtration resistance. AIChE J 10(1):61–68, 1964.

Tiller FM, Alciatore A, Shirato M. Filtration in the chemical process industry. In: Matteson MJ, Orr C, eds. Filtration: Principles and Practices. 2d ed. New York: Marcel Dekker, 1987a, pp 361–474.

Tiller FM, Yeh CS, Leu W. Compressibility of particulate structures in relation to thickening, filtration and expression—A review. Separa Sci Technol 22(2&3):1044, 1987b.

Zeman LJ, Zydney AL. Microfiltration and Ultrafiltration: Principles and Applications. New York: Marcel Dekker, 1996, pp 332, 335, 381.

3 SEDIMENTATION

Sedimentation is the separation of suspended solid particles from a fluid stream by the action of a body force on the settling behavior of the particle. The body force may be either gravitational or centrifugal force. This section covers gravity sedimentation, represented by clarification and thickening. The equipment used for these two operations are called clarifiers and thickeners, respectively. Centrifugal sedimentation will be discussed in the next section.

From a unit operation standpoint, clarification and thickening are essentially based on the same design principles, and each combines features of the other. The key objective of a clarification operation is to remove small quantities of suspended particulates from the liquid stream to produce a clarified effluent or overflow stream. In thickening operation, the goal is to concentrate the dilute suspensions for their further treatment in filters or centrifuges. Key features of these two types of operation are presented in Fig. 8.

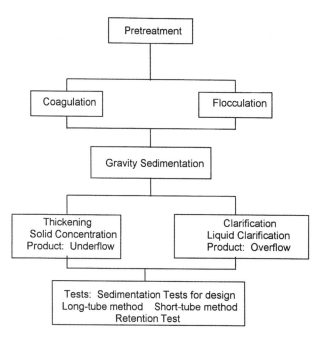

Figure 8 Pretreatment requirements and key features of gravity sedimentation.

3.1 Sedimentation Fundamentals

The settling behavior of suspended particulates in a gravity field is mainly affected by the concentration of the particulate solids and their aggregation status. In a dilute suspension, the settling solid behaves as individual particles, and the process is regarded as dilute sedimentation. This operating regime is called a particulate (or free) settling regime (see Fig. 9). Most clarifier operations fall into this regime. As the solid concentration increases, the suspended particles have more chances to approach each other closely and gradually form an aggregation (or cloud) state.

Once the concentration reaches a level in which the suspended particles settle as a mass, the corresponding sedimentation is known as hindered or zone settling. In this regime, the settling behavior is related more to the solid concentration rather than to particle size. Most thickener sizing calculations are based on this regime. As the solid concentration further increases to a higher level, a settled bed of sediment mass (or settled units) is compressed by the overburden of sediment on top of them. Liquid is expressed from the lower sediment layers and flows upward through the sediment. Their settling behavior is affected not only by hydrodynamic forces but also by the depth of the settling layers. This regime is termed a compression regime. Sedimentation

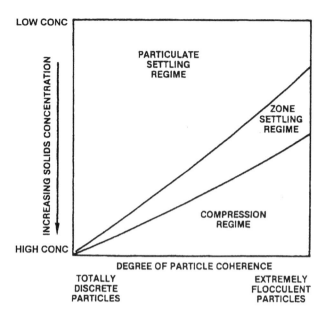

Figure 9 Characteristics of sedimentation processes in a suspension. (From Perry et al., 1997.)

with the addition of flocculant usually falls into this regime. It should be noted that when the suspension becomes concentrated, the separation behavior is more like filtration than settling in certain cases. Under this situation, the development of channels is quite possible. This type of microstructure formation is, however, seldom achievable in industrial settlers due to horizontal shear. A typical thickener is illustrated in Fig. 10. A feed suspension to be clarified or thickened can be operated in any regime. When it becomes concentrated because of sedimentation, the settling particles may initially behave in the particulate settling regime and subsequently in the zone settling and compression regimes. Therefore the design of sedimentation equipment must consider all three regimes.

In dilute sedimentation, solid particulates have no tendency to aggregate with one another. In this case, the detention time has little or no effect on the settling behavior. The sedimentation rate is equal to the particle settling rate, which may be evaluated in terms of the terminal settling velocity of the particle. On the other hand, in the hindered settling regime, clusters of particles (settling units) develop and settle as a sediment mass. The primary feature of this regime is that the settling rate of the suspension is a function of particle concentration. The pertinent equations and related constant are given in Tables 5a and 5b for both free and hindered settling regimes.

3.2 Thickeners

As shown in Fig. 10, there are four zones existing in a thickener: a clarification zone, a feed zone, a transition zone, and a compression zone. As solids thicken, a critical concentration will be reached. This concentration would affect the passage of solids to the underflow, causing a buildup in the compression zone thickness. The thickener design and operation must prevent the solid concentration from reaching the critical value.

3.2.1 Thickener Design

Three key parameters are required for thickener design: the thickener basin area (or unit area), the thickener basin depth, and the torque for the rake. These are discussed here.

Thickener Basin Area (Unit Area). The thickener basin area is determined from the solid flux rates at the critical concentration in the hindered settling regime. By definition, the critical point can be experimentally determined from the solids concentration just prior to the beginning of the compression zone. For a nonflocculated system, the basin area, expressed as the unit area (m^2/ton/d), A_0, can be calculated using the equation (Osborne, 1990)

$$A_o = \frac{1/C - 1/C_u}{u_i} \qquad (17)$$

where C is the test solid concentration; C_u the underflow solids concentration; and u_i the initial settling rate at the test condition.

For a flocculated suspension, the unit area of the thickener is determined by

$$A_o = \frac{t_u}{C_o H_o} \qquad (18)$$

where t_u is the settling time; C_o the test or feed solids concentration; and H_o the initial height of suspension in the test. Scale-up factors used in the thickener design usually vary, but a 1.2 to 1.3 multiplier applied to the unit area calculated from laboratory data is sufficient.

Thickener Basin Depth. In the hindered settling conditions, the pulp depth is unimportant in the determination of thickening rate and can be omitted. As the pulp enters into the compression regime, the pulp depth and the agitation affect the thickening rate. In this case, the compression zone unit volume may be calculated from the equation (Perry et al., 1997)

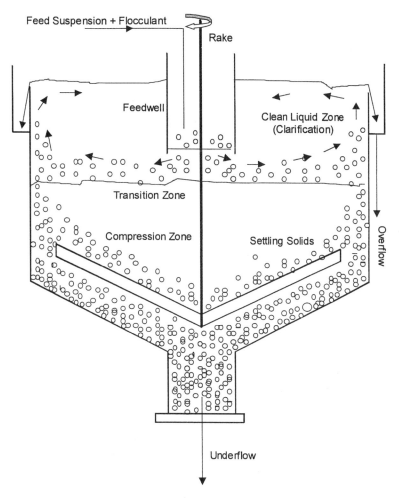

Figure 10 Operating zone in a typical thickener.

$$V_c = \frac{t_c(\rho_s - \rho_l)}{\rho_s(\rho_p - \rho_l)} \qquad (19)$$

where V_c is the unit volume of the compression zone defined as volume per weight of solids per day, t_c the compression time for a required particular underflow concentration, and ρ_s, ρ_l, and ρ_p are the average densities of solids, liquid, and pulp, respectively.

The depth of the compression zone equals to (V_c/A_0). The design thickener depth is calculated as the sum of the depth for the compression zone plus a clear zone of approximately 2 to 3 m allowing for clarification and transition from feed concentration to compression zone concentration. However, a greater depth may be used to attain a better overflow clarity.

Torque Requirement. The torque requirement for the thickener operation is based on the force necessary to drive the rake mechanism through the thick-

ened slurry. All other mechanical parts must also be designed for this same load. Most thickener suppliers base the torque requirements on operating data from experience with similar applications. The maximum torque selection for a thickener may follow the expression written as (Perry et al., 1997)

$$T = k_T D^2 \qquad (20)$$

where T is the torque; D the thickener diameter; and k_T a constant dependent upon the application. The value of k_T can be estimated from the process data shown in Table 6. A much higher value of k_T than the required one would lead to an increase in the unnecessary capital cost, while too small a value of k_T would severely affect the capability of the unit to handle process upsets and produce the desired underflow concentration. Normally, sedimentation units should be operated at the torque capability that is

Table 5a Equations for the Settling Rate Calculations

Operating regime	Equations	Note
Particulate settling	$u_t = \dfrac{K d_p^2 (\rho_s - \rho)}{\mu}$	$\mathrm{Re_p} = \dfrac{d_p u_t \rho}{\mu} < 0.1$ particle size range of 1 to 200 µm
	$u_t = \left[\dfrac{4(\rho_s - \rho)g d_p}{3\rho C_D} \right]^{0.5}$	$\mathrm{Re_p} \geq 0.2$

u_t = terminal settling velocity; d_p = the particle size;
ρ_s = the solid particle density; ρ the liquid density;
μ = the liquid viscosity; K (Kozeny constant) = 0.002 for u_t in
    m/h, d_p in micrometer, and μ in cp;
g = gravitational constant; C_d = drag coefficient

Hindered setting	$u = u_t \varepsilon^n$	non-flocculated suspension (Richardson and Zaki, 1954a,b)
	n is a function of the particle Reynolds number (see Table 5b)	
	$u = u_t [1 - k_\phi \phi_S]^{(4.70 + 17.8 d^*/D)}$	flocculated suspension (Scott, 1984)

$\phi = k_\phi \phi_s$
ϕ = settling units
ϕ_s = volumetric dry concentration
k_ϕ = the factor allowing for the liquid closely associated with the solids
d^* = the mean volume-surface-length diameter relevant to sedimentation

Table 5b Exponent n as a Function of the Particle Reynolds Number and Vessel Diameter (D)

$\mathrm{Re_p} = d_p u_t \rho / \mu$	n for small vessel	n for large vessel
$\mathrm{Re_p} < 0.2$	$4.65 + 19.5\, d_p/D$	4.65
$0.2 < \mathrm{Re_p} < 1$	$(4.35 + 1.75\, d_p/D)\,\mathrm{Re_p}^{-0.03}$	$4.35\,\mathrm{Re_p}^{-0.03}$
$1 < \mathrm{Re_p} < 200$	$(4.45 + 18\, d_p/D)\,\mathrm{Re_p}^{-0.1}$	$4.45\,\mathrm{Re_p}^{-0.1}$
$200 < \mathrm{Re_p} < 500$	$4.45\,\mathrm{Re_p}^{-0.1}$	$4.45\,\mathrm{Re_p}^{-0.1}$
$\mathrm{Re_p} > 500$	2.39	2.39

Source: Rushton et al., 2000, p 109.

not greater than 20% of the design level to avoid shutdown due to inevitable process upsets or overloads. Rake speed requirements depend on the type of solids entering the thickener. Rake speed ranges used are 3 to 8 m/min for slow-settling solids, 8 to 12 m/min for fast settling solids, and 12 to 30 m/min for coarse solids or crystalline materials.

The detailed description of the major mechanical components used for the thickeners, such as tank, feed well, drive support structure, drive and lifting devices, rake structure, underflow withdrawal, and overflow collection systems can be found in the literature (Osborne, 1990; Perry et al., 1997; Schweitzer, 1997).

3.2.2 Thickener Types and Selection

A thickener consists of several basic components: a tank to contain the slurry, a feed well for feed supply (to minimize the turbulence effect), a rotating rake mechanism, an underflow solids withdrawal system, and an overflow launder. Recirculation of the underflow back to the thickener feed line is a common practice, but care must be exercised in the design to avoid

Table 6 A Guide to Torque Specification for Gravity Sedimentation Devices

	Duty classification			
	Light	Standard	Heavy	Extra heavy
Unit area, $m^2/(t/d)$	5	1–5	0.2–1	< 0.2
Underflow, wt%	5	5–25	25–55	> 55
Particle size				
% + 65 mesh	0	0–5	5–15	> 15
% − 200 mesh	100	85–100	50–85	< 50
Solids sp. gr.	1.25–1.5	1.5–2.7	2.7–4.0	> 4.0
k factor	15–75	75–150	150–300	> 300

Source: Schweitzer PA. Handbook of Separation Techniques for Chemical Engineers. 3d ed. New York: McGraw-Hill, 1997, p 4-140, Table 1.

the overload and underflow line plugging. There are two basic types of continuous thickeners: conventional thickeners and high-rate thickeners.

The conventional thickeners can be operated with or without flocculants. They may be divided into three classes based on their specific drive-support configurations: bridge supported, center column supported, and traction driven. The design configuration of bridge-supported thickener is illustrated in Fig. 11. The diameter of this type of thickener can be selected up to 45 m. The key advantages of this thickener include (1) ability to produce a denser and more consistent underflow concentration; (2) use of a simplified lifting; and (3) fewer parts subject to mud accumulation. The center-column-supported thickeners are usually designed for large diameter units (> 20 m). The mechanism is supported by a stationary center column. The traction thickeners are an economical configuration of the center column support. They are mostly adaptable to the larger tanks over 60 m in diameter. For example, the Superthickener or Caisson thickener is a very big center-column-supported unit over 120 m in diameter. In this thickener, the underflow is withdrawn into the column and pumped back to the circumference by the underflow pump. Its maintenance generally is easier than other types of thickeners, but its installed cost may be higher.

High-rate thickeners or high-capacity thickeners are designed to take advantage of maximizing the flocculation efficiency. They have specially designed feed wells and flocculation systems as illustrated in Fig. 12. The feed well design is aimed at providing a good dispersion of the flocculants in the feed and transporting the flocculated feed suspension into the settling zone of the thickener without destruction of the newly formed flocs. The increase in flocculation efficiency will

normally increase the solid settling rate 2–10 times over that obtained from conventional thickeners. Obviously, the required unit area will reduce by the same factor. As a result, the capital cost will be reduced significantly. However, care must be exercised in adding flocculants, since overflocculation will produce a gellike structure that is hard to remove from the thickener.

Numerous combination and component design variations are possible. Selection of the type of thickener should primarily be based on installation and operating costs. The selection of the final design is made after an initial conceptual approach is defined. To specify a thickening system, one should consider the following: (1) the characteristics of feed suspension: flow rate, solid loading, particle-size distribution, and pH value; (2) the model experimental results for underflow concentration and overflow clarity; (3) a tank size estimate; (4) the required power or torque; (5) the capability of the selected rake; (6) the materials of construction; (7) a control scheme; and (8) a cost estimate. Table 7 lists typical design sizing criteria and operating conditions for a number of commercial thickeners and clarifiers. The information presented in the table can be used as a guide for selecting the gravity sedimentation units and preliminary cost estimate.

3.3 Clarifiers

Continuous clarifiers arre generally employed to treat dilute suspensions (industrial process streams and domestic wastewater) with a large percentage of relatively fine (usually smaller than 10 μm) solids. The clarified liquid is the main product. Clarifiers typically are designed for lighter duty operation than thickeners.

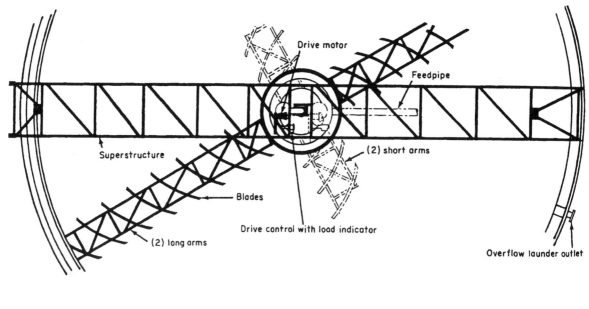

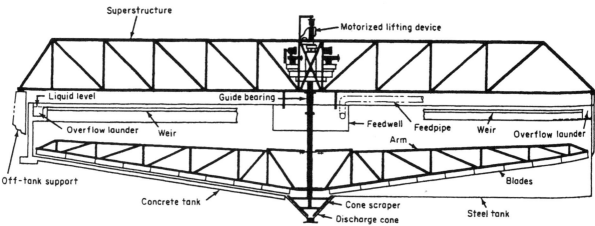

Figure 11 Conventional thickener with bridge-supported mechanism (Eimco). (From Schweitzer, 1997.)

3.3.1 Clarifier Design

As mentioned earlier, clarifiers and thickeners are very similar in their overall designs and layouts. The design considerations of the thickeners can also be used as a basis for clarifier design. The design configuration of clarifiers may have the following features: (1) rapid and good mixing of chemical additives with feed slurry and adjustable recirculator speed with at least a range of 3 : 1; (2) control of the mixer speed (not exceeding 2 m/s) and the scraper tip speed (less than 3 m/min with speed variation of 3 : 1); (3) a discharge system for easy automation and variation of amount discharged; (4) devices for measuring and varying the slurry concentration in the contacting zone; (5) suspension blanket levels at a minimum of 1.6 m below the clean water surface. The design criteria and operating conditions for commercial clarifiers are given in Table 7.

Clarifier design has traditionally been based on the principle of dilute sedimentation. The design calculations of a clarifier include feed well design, sedimentation basin design (providing enough residence time for the separation), and solid flux (G). The other design parameters, such as the torque requirement and clarifier area, are similar to those of the thickeners, which have been discussed in the previous section (3.2.1).

The design configurations of the feed well should provide: (1) a decreasing velocity gradient inside and outside the feed well; (2) gentle mixing to promote ideal conditions for dispersing flocculant or coagulant

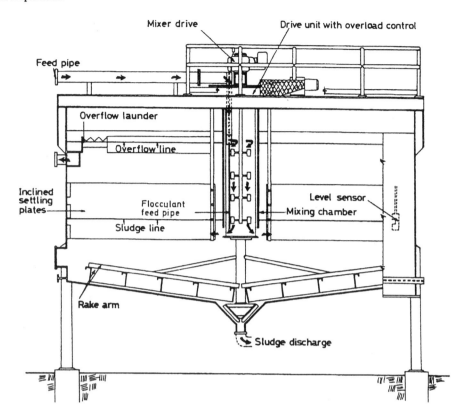

Figure 12 High-capacity thickener (Eimco). (From Osborne, 1990.)

Table 7 Design Criteria and Operating Conditions for Commercial Thickeners and Clarifiers

	Solid %		Unit area, m^2 day/mg	Overflow rate, $m^3/(m^2 h)$
Slurry	Feed	Underflow		
Alumina-hydrate	1–10	20–50	1.2–3	0.07–0.12
Alumina-red mud (primary)	3–4	10–25	2–5	—
Coal refuse	0.5–6	20–40	0.5–1[a]	0.7–1.7
FGD[b] waste	3–12	20–45	0.3–3[a]	—
Magnesium hydroxide (seawater)	1–4	15–20	3–10	0.5–0.8
Metallurgical				
Copper concentrates	14–50	40–75	0.2–2	—
Copper tailings	10–30	45–65	0.4–1	—
Magnetic tailings	2–5	45–60	0.5–1.5	1.2–2.4
Municipal waste				
Primary clarifier	0.02–0.05	0.5–1.5	—	1–1.7
Primary thickening sludge	1–3	5–10	8	—
Phosphate slimes	1–3	5–15	1.2–18[a]	—
De-inking waste	0.01–0.05	4–7	—	1–1.2
Paper-mill waste	0.01–0.05	2–8	—	1.2–2.2
Water softening lime-sludge	5–10	20–40	0.6–2.5	—

[a]High-rate thickeners using required flocculant dosages operate at 10 to 50% of these unit areas.
[b]FGD represents flue-gas desulfurization.
Source: Perry RH, Green DW, Maloney JO. Perry's Chemical Engineers' Handbook. 7th ed. New York: McGraw-Hill, 1997, pp 18–72, Table 18-7.

into the feed stream without creating too much turbulence, while keeping the maximum overflow clarity. The usual operating range for the upflow clarifier is 2 to $5\,m^3/(h\,m^2)$. The retention time required for the floc separation in a sedimentation basin can be estimated using Stokes' law or Newton's law, especially for well-defined homogeneous systems. It should be noted that allowances must be made for deleterious effects due to convection, air-induced surface waves, and inlet and outlet turbulence.

If the clarifier is assumed to operate close to its capacity limit, the solids flux, G, into the clarifier at a steady state can be determined using a simple material balance:

$$G = \frac{C(Q + Q_u)}{A} \tag{21}$$

where C is the solids concentration in feed suspension, Q the effluent flow rate, Q_u the underflow rate, and A the surface area of clarifier.

3.3.2 Clarifier Types and Selection

A clarifier typically consists of a concentric circular compartment for conditioning and settling, a clarified liquid overflow section (effluent launder), and a concentrated underflow discharge system. Figure 13 illustrates the design of an upflow clarifier, which combines the functions of mixing, flocculation, and sedimentation in a single unit. This clarifier is also known as a solid-contact clarifier. In such a clarifier, the impeller is installed underneath the draft tube as a primary mixing device. In another version of the upflow clarifier design (Eimco Process Equipment Co.), the impeller is set

above the draft tube to provide secondary mixing for flocculation–feed contact. Also, this new type of upflow clarifier is equipped with a rake. The operation of the clarifier may be divided into several distinct layers: the clarified water zone, the feed zone (coagulant and raw water addition with primary and secondary mixing), and the compacting zone. In the feed zone, the coagulants and/or flocculants are mixed rapidly with the raw water and the return flow under agitation. With proper coagulation/flocculation a uniform dense floc can be produced. For operating flexibility, it is desirable to control the mixing intensity and the sediment rake speed independently. In the clarified zone, the suspended solid concentration is very low, and the clarified stream will escape over weirs as purified effluent. In the compacting zone, the formed flocs with the treated water pass through the sedimentation basin, which provides a retention time of 1 or 2 hours for the floc particles to settle. When the floc particles are too heavy to circulate up through the draft tube, a modified design using external recirculation of a portion of the thickened underflow is recommended. Solid-contact clarifiers are usually used for clarifying turbid water or slurries that require adding coagulant or flocculant for the removal of bacteria, suspended solids, or color.

Inverted cone clarifiers are another type of solid-contact clarifiers. The mixed feed suspension is introduced from the top of a cone, which has an angle ranging from 60° to 90°. These clarifiers are operated without rakes or agitators. The slurry interface is usually within 1 or 2 m of the overflow level. The suspended solids mixed with the chemical additives are

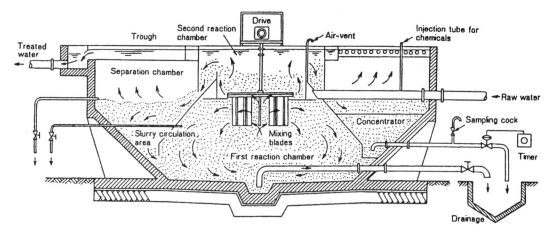

Figure 13 Structure of an upflow clarifier. (From Mukai, 1986).

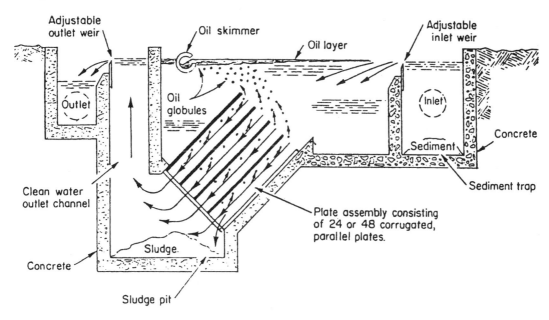

Figure 14 Schematic of a tilted-plate clarifier (Heil Process Equipment Co.). (From Schweitzer, 1997.)

kept in a fluidized state. Typical operating rates are around 5 to $10 \, m^3/(m^2 \, h)$. The concentration of suspended solids in the overflow ranges from 10 to 40 mg/L. The operating and maintenance costs are low compared to other types of solid-contact clarifiers. These clarifiers are well suited to the case where the flow rate varies only slightly.

In recent years, tilted-plate clarifiers (see Fig. 14) have been widely used. This type of clarifier is also known as lamella or tube settler. It contains a multiple plate assembly inclined at 30° to 60° from the horizontal. Clarified liquid and settled solids may flow countercurrently or cocurrently along each channel. Generally, a plate angle of 30° to 40° is a suitable inclination when operating cocurrently, whereas an increase of up to 55° to 60° may be necessary with the countercurrent arrangement. The suspended solids settle only a short distance in the channel before falling into the base. Settled solids are collected in a lower compartment and discharged by pumping. The selection of the plate spacing should be based on two considerations. First, the plate spacing must be large enough to accommodate the opposite flows. Secondly, the channel space arrangement should be able to limit the interference and plugging as well as to provide enough residence time for the solids to settle in a short distance. The tilted-plate clarifiers usually use 10 to 50 plates with spacing from 0.05 to 0.1 m. The channel lengths range from 1 to 3 m with widths of about 1.2 m. Operating capacities vary from 0.5 to $3 \, m^3/(m^2 \, h)$.

The major advantage of the tilted-plate clarifier is its increased capacity per unit of plate area compared to the conventional clarifier. There are two key shortcomings existing in this type of clarifier. First, it produces a varying underflow solids concentration lower than other type of gravity clarifiers. Secondly, it is difficult to clean the scale formed within the channel. The applications of tilted-plate clarifiers include clarification of plating and pickling wastes, paper mill effluent, and tertiary wastewater treatment. For the clarifier selection one can follow the same guidelines as for thickener selection, as discussed in the last section.

NOMENCLATURE

A	=	clarifier surface area
A_o	=	unit area
C	=	solid concentration in the feed
C_o	=	test or feed solid concentration
C_u	=	underflow solids concentration
D	=	thickener diameter
G	=	solid flux into the clarifier
H_o	=	initial height of suspension in the test
k_T	=	constant dependent upon the application of thickener
Q	=	effluent flow rate
Q_u	=	underflow rate
T	=	torque
t_C	=	compression time for a required particular underflow concentration

t_u = settling time
u_i = initial settling rate or hindered settling velocity at solids concentration C_i
V_C = unit volume of the compression zone

Greek Symbols

ρ_l = average density of liquid
ρ_p = average density of suspension or pulp
ρ_s = solid particle density

REFERENCES

Mukai T. Solid–liquid separations of water and wastewater treatment systems in Japan. In: Muralidhara HS, ed. Advances in Solid–Liquid Separation. Columbus: Battelle Press, 1986, p 409.

Osborne DG. Gravity thickening. In: Svarovsky L, ed. Solid–Liquid Separation. 3d ed. London: Butterworths, 1990, pp 132–201.

Perry RH, Green DW, Maloney JO. Perry's Chemical Engineers' Handbook. 7th ed. New York: McGraw-Hill, 1997, pp 18–60 to 18–72.

Richardson JF, Zaki WN. Sedimentation and fluidization: Part 1. Trans Inst Chem Engrs 32:35–53, 1954a.

Richardson JF, Zaki WN. The sedimentation of a suspension of uniform spheres under conditions of viscous flow. Chem Eng Sci 3:65–73, 1954b.

Rushton A, Ward AS, Holdich RG. Solid–Liquid Filtration and Separation Technology. 2d ed. Reinheim, Germany: Wiley-VCH, 2000, p 109.

Schweitzer PA. Handbook of Separation Techniques for Chemical Engineers. 3d ed. New York: McGraw-Hill, 1997, pp 4.140–4.156.

Scott KJ. Hindered settling of a suspension of spheres. Critical evaluation of equations relating settling rate to mean particle diameter and suspension concentration. CSIR Report CENG 497. CSIR, Pretoria, South Africa, 1984.

4 CENTRIFUGATION

Centrifugation can be viewed as an extension of the conventional filtration and gravity sedimentation. In this case, the centrifugal force replaces the pressure force in filtration and the gravitational force in sedimentation, respectively. However, the operating principles remain the same. The deciding factor for separating particles from liquid is the density difference between the solids and the suspending liquid. Centrifuges are available in a wide variety of types and sizes with a centrifugal force ranging from less than $100G$ up to $10,000G$, where G is the gravitational acceleration. They can be broadly divided into two major types: sedimentation centrifuges and filter centrifuges.

4.1 Sedimentation Centrifuges

Centrifugal sedimentation removes solid particles from a solid–liquid suspension by employing centrifugal force to induce setting effects. Owing to its much higher acceleration, centrifugation can extend the range of sedimentation to finer particles and it can also separate emulsions, which might normally be stable in the gravity field. The sedimentation centrifuges are not usually sensitive in their solids handling capacity to feed solid concentration because the liquid does not have to filter through the solids or a filter medium. They are effective for separating particles ranging from 6 mm (1/4 in.) down to submicron sizes. Flocculants are often used to promote agglomeration of particles to accelerate the settling rate of very fine materials.

4.1.1 Principles of Sedimentation Centrifuges

In sedimentation centrifuges, liquid and solid are acted on by two forces: gravity acting downward and centrifugal force acting horizontally. In commercial units, however, the centrifugal force component is normally so large that the gravitational component may be neglected. The separating power of the sedimentation centrifuges is often measured by comparing the centrifugal force (R_c) in the device with the gravity acceleration (G), which is referred to as the relative centrifugal force or the centrifugal number (N_c). The centrifugal number, N_c, typically varies from 200 times gravity to 360,000 times gravity.

The separation happens in a sedimentation centrifuge when the solid particles are removed from the fluid. In order for a particle of a given size to be removed from the fluid a sufficient time should be allowed for the particle to settle and reach the wall of the separator bowl. If it is assumed that the solid particle moves at its terminal velocity at all times, the smallest particle that should just be removed can be calculated. Consider the simplest, tubular type centrifuge, which is shown in Fig. 15 (McCabe et al., 1993). In a tubular centrifuge, the bowl consists of a vertical tube with a large height-to-diameter ratio, which rotates at a high speed about its vertical axis. The feed point is at the bottom, and the liquid discharge is at the top. Assuming that the incoming fluid starts to rotate with the bowl, its angular velocity will soon

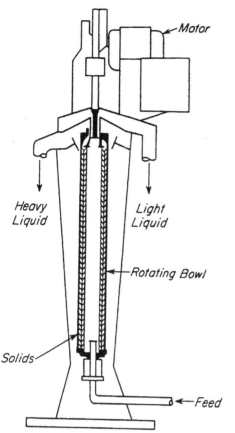

Figure 15 Schematic of a tubular centrifuge. (From McCabe et al., 1993.)

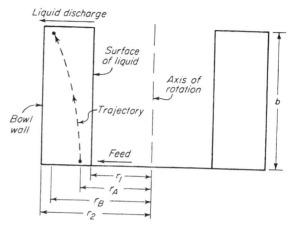

Figure 16 Particle trajectory in sedimenting centrifuge. (From McCabe et al., 1993.)

time. The following equation gives the relationship between the feed rate and the particle cut diameter (McCabe et al., 1993):

$$q_c = \frac{\pi b \omega^2 \Delta\rho D_{pc}^2}{18\mu} \frac{r_2^2 - r_1^2}{\ln[2r_2/(r_2 + r_1)]} \qquad (22)$$

where r_1 is the radius of inner surface of the liquid, r_2 the radius of inner surface of the bowl, ω the angular velocity, $\Delta\rho$ the density difference between solids and liquid, and q_c the feed volumetric flow rate corresponding to the cut diameter D_{pc}. At this feed rate most of the particles having diameters greater than D_{pc} will be removed from the liquid by the centrifuge, and the particles remaining in the liquid will be smaller than D_{pc}. Most of the sedimentation centrifuges are capable of removing solids particle sizes well into the submicron range.

This analysis is an oversimplification, since the flow pattern of the fluid in the centrifuge bowl is much more complicated than that assumed in the Fig. 16. The only way to describe fully the separation performance of a sedimentation centrifuge is by the grade efficiency curve. Knowledge of the grade efficiency curve allows accurate and reliable predictions of total efficiency with different feed solids, subject to the operating characteristics, the state of the dispersion of solids, other variables remaining constant.

4.1.2 Major Types of Sedimentation Centrifuges

A sedimentation centrifuge consists of an imperforate bowl into which a suspension feed is fed. The bowl rotates at high speed. The liquid after separation is removed through a skimming tube or over a weir

become identical with that of the bowl. There is therefore no tangential flow in the bowl, and the fluid only rotates with the bowl and moves upward through the bowl at a constant velocity, carrying solid particles with it. As the solid particles move with the fluid, they are subjected to high centrifugal forces and begin to settle at some position in the fluid, say at a distance r_A from the vertical axis, following trajectory similar to that shown in Fig. 16. Its settling time is limited by the residence time of the fluid in the bowl; at the end of this time let the particle be at a distance r_B from the axis of the rotation. If $r_B < r_2$, the solid particle leaves the centrifuge with the liquid; if $r_B = r_2$, it is deposited on the bowl wall and removed from the liquid.

Ambler (1952) introduced the cut point concept in the sedimentation centrifuge separation. It is defined as the diameter of the particle that just reaches one-half the distance between r_1 and r_2. If a solid particle is to be removed from the fluid, it must travel the distance $(r_2 - r_1)/2$ to the bowl wall in the available residence

while the separated solids either remain in the bowl or are intermittently (or continuously) discharged from the bowl. Typically a sedimentation centrifuge will have the following main components (as shown in Fig. 15):

1. A bowl or rotor in which the centrifugal force is applied to a solid–liquid suspension
2. A means for feeding the suspension into the rotor and a means for discharging the separated components from the bowl either in batch or continuous mode
3. A drive shaft, axial, and thrust bearings
4. A drive mechanism to rotate the shaft and bowl
5. A casing to contain the separated components
6. A frame for support and alignment

Sedimentation centrifuges can be classified by several criteria, including the centrifugal number, N_c, the range of throughputs, the solids concentration in suspension that can be handled, the bowl design, and the solids discharge mechanism. Figure 17 gives a classification of sedimentation centrifuges based on the design of the bowl and of the solids discharge mechanism: tubular, multichamber, imperforate basket, scroll type, and disk centrifugres (Svarovsky, 1985). Due to the design of the bowl structure, the bowl has to be cleaned manually for both tubular and multichamber centrifuges, so they are usually used as liquid classifiers and suitable for relatively low solids concentrations. The disc type and the scroll type centrifuges can be operated continuously so that they are suitable for very high solids concentrations, sometimes up to 50% by volume. The imperforate basket type centri-

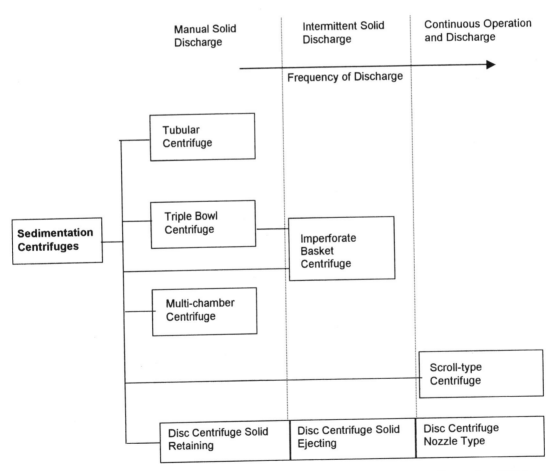

Figure 17 Classification of sedimentation centrifuges. (From Svarovsky, 1985.)

fuge is somewhere in between and suitable for moderate feed solids concentrations.

4.2 Filter Centrifuges

Filter centrifuges separate solid particles and liquid from a solid–liquid suspension by employing pressure resulting from the centrifugal action to force the liquid through the filter medium, leaving the solid particles behind. The density difference between the solids and the liquid, which governs the separation in sedimentation, is no longer a necessary prerequisite.

The common feature of all filter centrifuges is a rotating basket equipped with a filter medium. Solids, which form a porous cake, can be separated from the suspension liquid in a filter centrifuge. Slurry is fed to a rotating basket having a slotted or perforated wall covered with a filter medium such as canvas or metal cloth. The separation process takes place much like the cake filtration (see Sec. 2.1) process, except the pressure gradient is created by the centrifugal action of the rotating basket.

4.2.1 Principles of Filter Centrifuges

The centrifugal filtration process starts with the feed slurry to an empty centrifuge and is followed by the filter cake deposition and the flow of clear filtrate or wash water through the cake. The basic theory of constant pressure filtration can be modified to apply to the filter centrifuges (McCabe et al., 1993).

Based on the two-resistance model (see Sec. 2.1.1), the volumetric flow rate of filtrate, q, is expressed as

$$q = \frac{\rho\omega^2(r_2^2 - r_1^2)}{2\mu(\alpha m_C/\bar{A}_L\bar{A}_a + R_m/A_2)} \tag{23}$$

where A_2 is the area of the filter medium, $\bar{A}_L$ the arithmetic mean cake area, $\bar{A}_a$ the logarithmic mean cake area, R_m the filter medium resistance, α the specific cake resistance, and m_c the total mass of the solids in the filter. The mean areas $\bar{A}_L$ and $\bar{A}_a$ are defined by the equations

$$\bar{A}_a = (r_i + r_2)\pi b \tag{24}$$

$$\bar{A}_L = \frac{2\pi b(r_i - r_2)}{\ln(r_2/r_i)} \tag{25}$$

where b is the hieght of the basket and r_i the inner radius of the cake.

4.2.2 Major Types of Filter Centrifuges

In general, the filter centrifuges can be classified into two groups: batch and continuous type. Several selected filtering centrifuges are listed in Fig. 18.

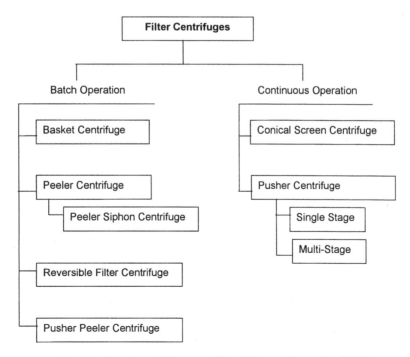

Figure 18 Classification of filter centrifugal filters. (From Alt, 1986.)

More complete information can be found in several references (Perry et al., 1997; Svarovsky, 1990; Zeitsch, 1990).

Batch-Type Filter Centrifuges. Batch-type filter centrifuges are much older in conception than continuous ones, but they are by no means obsolete. The total number of batch type far exceeds that of the continuous type. Their main advantage is high separation efficiency linked to the outstanding purity of the separated solids and liquids. Among the commonly used batch-type filter centrifuges are the three-column basket centrifuges and the peeler centrifuges.

Basket Centrifuges The simplest and most common form of the batch-type filter centrifuge is the basket centrifuge. It consists of a cylindrical basket, or drum, which is suspended on three columns and because of this characteristic it is also known as the three-column centrifuge. The basket centrifuges are normally arranged with a vertical axis of rotation. A schematic of this design is shown in Fig. 19. The basket centrifuge represents the earliest centrifuge used for liquid–solid separations. It remains extensively used throughout the process industries worldwide.

The basket centrifuges are constructed with baskets, as their name suggests. Generally, the mantle surface of the basket is perforated with a large number of holes covered on the inside by one or more coarse screens. The latter serves as a backup for the filter medium that can have the shape of a bag lining the perforated wall of the basket. The feed slurry enters the rotating basket from the top of the basket, before or after the start of the rotation. Liquor (filtrate) drains through the filter medium into the casing and out a discharge pipe. The solid particles deposit against the basket wall and form cake of 5 to 15 cm thick. The discharge of the solids is achieved by stopping the machine and manually removing the solid cake or replacing the bag. Operation cycles of these machines can be varied to achieve the desired performance. The basket provides a good surface for washing the filter cake. Wash liquid may be sprayed through the cake to remove the soluble material. The cake is then spun as long as needed, often at a higher rotating speed than those during the loading, filtering, and washing steps, to provide maximum dryness.

The dimensions of the baskets range from 76 to 122 cm in diameter and 46 to 76 cm deep and turn at speeds from 600 to 1,800 rpm. These machines can usually handle a capacity up to about 12 ft^3 of product per cycle. They are widely employed in the fine chemical and pharmaceutical industries, where products are produced on a batch basis. The characteristics of batch operation allow easy changing of the operation parameters from batch to batch to meet production needs.

In these machines the solid cake is removed by moving a plough into the cake after the basket is slowed down to a preset speed. The plough directs the solids toward the center, where they fall through the bottom openings of the basket. The entire operation can be automated by means of timers and solenoid-operated valves, which control the various parts of the opera-

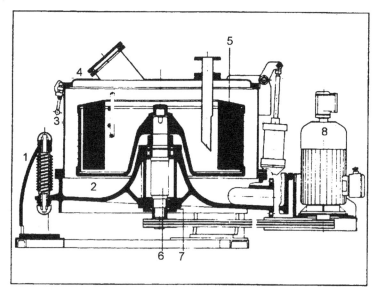

1. Link column
2. Lower housing
3. Upper housing
4. Lid
5. Drum
6. Shaft
7. Bearing
8. Drive

Figure 19 Schematic of a three-column centrifuge. (From Alt, 1986.)

tion: feeding, filtering, washing, spinning, rinsing, and unloading. Any portion of the cycle may be lengthened or shortened as desired.

Peeler Centrifuges. The peeler centrifuge (as shown in Fig. 20) is a variation of the basket centrifuge. It usually rotates at constant speed about a horizontal axis instead of a vertical axis. The filter cake in the peeler centrifuge is unloaded periodically while turning at full speed by a heavy knife: the "peeler." During the unloading, the peeler rises and cuts solid cake out with considerable force through a discharge chute. In general, the peeler centrifuge is fully automated in its operation.

In the three-column basket centrifuge the basket rotates vertically. The effect of the gravity tends to form a nonuniform cake, its lower portion being thicker and containing larger particles than its upper portion. Because of the unevenly distributed solid cake, the washing can cause variations in the purity of the solid product. In such cases, peeler centrifuges are preferable over three-column centrifuges.

Furthermore, the peeler centrifuges have higher productive capacity since they do not require any nonproductive periods of deceleration and acceleration for unloading solids. Usually they are not used for treating feed containing solid particles finer than 150 mesh. Hence these machines have been found particularly attractive where the filtration and drainage periods are relatively short. On the other hand, they are not suitable for handling slow-draining solids, which

would give uneconomically long cycles, or sticky solids, which do not discharge cleanly through the chute. There is considerable breakage or degradation of the particles during high-speed unloading by the peeler. Also, the permeability of the residual heel could be unfavorably affected by plugging it with fractured fines. Consequently, it may require washing to recondition the plugged residual heel.

Continuously Fed Filter Centrifuges. There are several different types of continuously fed filter centrifuges. Among them, the pusher centrifuges and the conical screen filter centrifuges are the most common designs.

Pusher Centrifuges. The pusher centrifuge is so named because of the pushing mechanism employed to transport the solids across the basket. A schematic of a one-stage pusher centrifuge is shown in Fig. 21. The first pusher centrifuge was designed more than a century ago (Alt, 1986). It consists of a rotating perforated rotor with a slotted wall and the circular pusher plate reciprocating with frequencies ranging from 20 to 100 strokes per minute. The plate attached to the feed funnel rotates in the same direction as the rotor. Feed enters the small end of the feed funnel from a stationary pipe at the axis of the rotation of the rotor. It travels toward the large end of the feed funnel. When it spills off the feed funnel onto the rotor wall, it moves in same direction as the wall and very nearly the same rotation speed. The main filtration occurs within

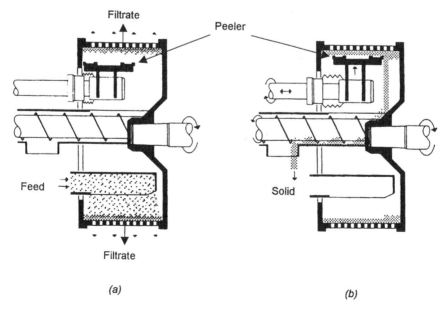

Figure 20   Peeler centrifuge: (a) feed and separation phase, (b) solids discharge phase. (From Jacobs and Penney, 1987.)

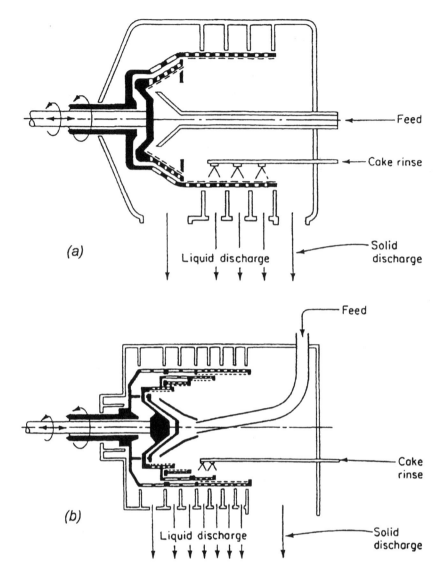

Figure 21 Schematic of (a) single-stage pusher centrifuge, (b) two-stage pusher centrifuge. (From Perry et al., 1997.)

the stroke zone of the pusher plate. A layer of solid cake 2.5 to 7.5 cm thick is formed. This solid layer is moved over the filtering surface by the reciprocating pusher plate. Each stroke of the pusher plate moves the solid layer a few inches toward the lip of the rotor; on the return stroke a space is opened on the filtering surface in which more solid cake can be deposited. When the cake reaches the lip of the rotor, it is thrown into the large casing by the centrifugal acceleration. The liquor passes the filtering screen and leaves the machine by ducts well separated from the solids discharge.

The pusher centrifuges are normally used when the feed can be concentrated above 60% by volume, since the operation capacity of the pusher centrifuge is greatly increased as the solid concentration is increased. They are capable of handling fragile crystals, which may be damaged in other types of centrifuges. The small pusher centrifuges can handle a few grams per second, while the large ones have capacities up to 55,000 kg/h. Based on the number of rotors, pusher centrifuges are divided into two types: single stage pusher centrifuges with only one rotor and multistage pusher centrifuges with two or more rotors. The multistage pusher centrifuge is particularly suitable for particles forming a soft cake or having a high frictional resistance to sliding on the filter medium. A schematic of a two-stage pusher centrifuge is also shown in Fig. 21. The purpose of the multistage design is to keep the length of the rotor short.

The capacity and performance of the pusher centrifuge depend on the particle size and shape of the feed: the coarser the feed particle size, the higher the operation capacity and the lower the moisture. The thickness of the cake is one of the most important factors affecting the operation of the pusher centrifuge. The pusher centrifuge has three significant operating characteristics dependent on the cake thickness: the solid discharge, the filter rate, and the pushing force. Generally, the thickness of the cake on the rotor is proportional to the friction of the cake over the filter medium, the centrifugal force, and the length of the rotor.

Conical Screen Filter Centrifuges The conical screen centrifuge is another very commonly used continuous centrifuge. Instead of using the pusher action to discharge the solids, it uses a helical conveyor that is turning slightly slower or faster than the rotor to discharge the solids. Such a solid discharge mechanism can be used for both cylindrical and conical rotors. The differential speed of the conveyor controls the rate at which the solids move through the drainage zone. A schematic of a conical screen centrifuge is shown in Fig. 22. The conical screen centrifuges may have vertical or horizontal axes of rotation.

The conical screen centrifuges are used mainly for processing coarser particles, for example crystalline salts, coal, and minerals. These centrifuges are suitable for relatively large throughputs, up to 320,000 kg/h. The various types of conical screen centrifuges operate successfully in a wide range of applications. Among them, the sliding filter centrifuges and the

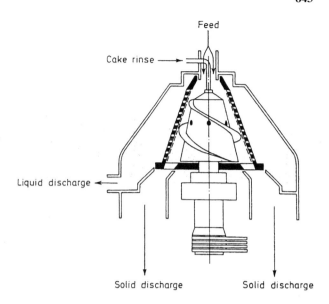

Figure 22 Schematic of a conical conveyor discharge centrifuge. (From Zeitsch, 1990.)

vibrating filter centrifuges are the most commonly used. In some cases, the angle of the conical screen can be set sufficiently large to permit the cake to overcome its friction on the filter medium so that it becomes self-discharging. Compared with the pusher centrifuge, the conical screen centrifuge has a much greater particle breakage.

Table 8 shows the application range of different types of centrifuges. Several important feed characteristics, including the minimum particle size, the maxi-

Table 8 Application Range of Centrifuges

	Tubular	Disc (nozzle type)	Conveyor bowl
A. Sedimentation type centrifuges			
Minimum particle size, micron	0.1	0.25	2
Maximum particle size, micron	200	50	5,000
Allowable concentration of feed solid, %	0.1	2–20	2–60
Condition of cake	Pasty, firm	Fluid	Firm, pasty
Typical solids handling rate, lb/h	0.1–5	10–30,000	100–100,000
	Batch vertical	Conical basket	Pusher
B. Filtering Centrifuges			
Minimum particle size, micron	10	250	40
Maximum particle size, micron	1,000	10,000	5,000
Allowable concentration of feed solid, %	2–10	40–80	15–75
Condition of cake	Pasty	Relatively dry	Relatively dry
Typical solids handling rate, lb/h	0.1–1.0	5–40	0.5–5.0

Source: Cheremisinoff PN. Solid/Liquids Separation. Lancaster PA: Technomic, 1995, p 240, Table 6.3.

mum particle size, the maximum solid concentration, and the solid handling capacity are presented. This list can serve as a useful guideline for an initial selection of suitable centrifuges.

NOMENCLATURE

$\bar{A}_L$	=	arithmetic mean cake area
$\bar{A}_a$	=	logarithmic mean cake area
A_2	=	area of filter medium
b	=	height of the basket in a filter centrifuge
D_{pc}	=	cut diameter
G	=	gravity acceleration
m_c	=	total mass of the solids in the filter
N_c	=	centrifugal number
q_c	=	feed volumetric flow rate
r_1	=	distance from the surface of liquid to the axis of rotation
r_2	=	distance from the wall of the bowl to the axis of rotation
r_A, r_B	=	distance from the particle to the axis of rotation
r_i	=	inner radius of the cake
R_c	=	centrifugal force
R_m	=	filter medium resistance

Greek Symbols

α	=	specific cake resistance
ρ	=	density of liquid (filtrate)
$\Delta\rho$	=	density difference between solids and liquid
μ	=	liquid viscosity
ω	=	angular velocity

REFERENCES

Alt C. Centrifugal separation. In: Muralidhara HS, ed. Advances in Solid–Liquid Separation. Columbus, Ohio: Battelle Press, 1986, pp 107–139.

Ambler CM. The evaluation of centrifuge performance. Chem Eng Progress 48:150–158, 1952.

Jacobs LJ, Penney WR. In: Rousseau RW, ed. Handbook of Separation Process Technology. New York: John Wiley, 1987, p 168.

McCabe WL, Smith JC, Harriott P. Unit Operations of Chemical Engineering. 5th ed. New York: McGraw-Hill, 1993, pp 1011–1028, 1064–1072.

Perry RH, Green DW, Maloney JO. Perry's Chemical Engineers' Handbook. 7th ed. New York: McGraw-Hill, 1997, pp 18–64.

Svarovsky L. Solid–Liquid Separation Processes and Technology. Amsterdam: Elsevier, 1985, pp 72–106.

Svarovsky L. Separation by centrifugal sedimentation. In: Svarovsky L, ed. Solid–Liquid Separation. 3d ed. UK: Butterworths, 1990, p 264.

Zeitsch K. Centrifugal filtration. In: Svarovsky L, ed. Solid–Liquid Separation. 3d ed. UK: Butterworths, 1990, pp 476–532.

5 HYDROCYCLONES

The hydrocyclone is based on the principle of centrifugal force causing the separation of solids from a liquid by the differences in density and particle size. A typical hydrocyclone consists of a cylindrical section and a conical section, as shown in Fig. 23. It does not have any internal rotating parts. An external pump is used to transport the liquid suspension to the hydrocyclone through a tangential inlet at high velocity, which in turn generates the fluid rotation and the necessary centrifugal force. The outlet for the bulk of the liquid is connected to a vortex finder located on the axis of the cylindrical section of the vessel. The underflow, which carries most of the solids, leaves through an opening (apex) at the bottom of the conical section.

The principle and basic design of the hydrocyclone has been known for more than a century (Bretney, 1891), but it did not find significant application in industry until the late 1940s. These separation devices were first used in mining and mineral processing, but in recent years their applications have spread to many other industries, including chemical manufacturing, power generation, and environmental cleanup.

5.1 Separation Efficiencies

The solid–liquid separation in hydrocyclones is never complete, because there is always liquid discharging with the solids through the underflow. The term separation efficiency used for the hydrocyclone is usually defined for measuring the capability of the hydrocyclone of separating the solids from the feed into the underflow. There are a number of different terms for the separation efficiency used in the literature. They include total efficiency, reduced total efficiency, grade efficiency, reduced grade efficiency, and cut size.

5.1.1 Total Efficiency

Total efficiency is defined as the total solids (mass or volume) reported in the underflow as a fraction of the total solid in the feed:

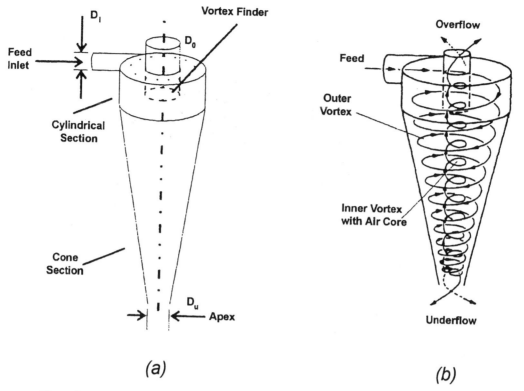

Figure 23 Hydrocyclone: (a) schematic diagram, (b) flow pattern. (From Rushton, 2000.)

$$E_T = \frac{UC_u}{QC_f} \qquad (26)$$

where U is the underflow volumetric flow rate, Q the feed volumetric flow rate, C_u the solid concentration in the underflow, volume, or mass fraction, and C_f the solid concentration in the feed, volume, or mass fraction.

Because an overall mass balance must apply, the total efficiency E_T can be determined by measuring any two of the three streams (feed, underflow, and overflow) for total solid amount, assuming no accumulation of solids in the hydrocyclone.

5.1.2 Reduced Total Efficiency

There are two problems associated with total efficiency. First, if a hydrocyclone delivers both liquid and solid to the underflow and nothing to the overflow, an ideal total efficiency of 1 will result. Second, without any separation, a hydrocyclone, by simply splitting the feed to an overflow and an underflow, will result in certain "guaranteed" total efficiency. In order to overcome these weaknesses, several alternative definitions of efficiency are used. One of the definitions is the reduced total efficiency E_T':

$$E_T' = \frac{E_T - R_f}{1 - R_f} \qquad (27)$$

where R_f is the underflow-to-throughput ratio defined as

$$R_f = \frac{U}{Q} \qquad (28)$$

5.1.3 Grade Efficiency

In place of the total efficiency, a grade efficiency corresponding to a particular particle size is used, since a hydrocyclone is a size-dependent separator. A graphical representation of the relationship between the grade efficiency and the particle size is called the grade efficiency curve, as shown in Fig. 24 (Svarovsky, 1985).

The grade efficiency curve can be determined by measuring the total efficiency and the particle size distribution of any two of the three streams (feed, underflow, and overflow). One of the following equations can be used for calculating the grade efficiency:

Feed and underflow:
$$G(x) = E_T \cdot \frac{dF_u(x)}{dF_f(x)} \quad (29)$$

Feed and overflow:
$$G(x) = 1 - (1 - E_T) \cdot \frac{dF_o(x)}{dF_f(x)} \quad (30)$$

Overflow and underflow:
$$G(x) = 1 + \left(\frac{1}{E_T} - 1\right) \cdot \frac{dF_o(x)}{dF_u(x)} \quad (31)$$

where $F_f(x)$ is the cumulative percentage of solids with particle size x in the feed, $F_o(x)$ the cumulative percentage of solids with particle size x in the overflow, and $F_u(x)$ the cumulative percentage of solids with particle size x in the underflow.

Similar to the reduced total efficiency, a reduced grade efficiency $G'(x)$ is introduced as

$$G'(x) = \frac{G(x) - R_f}{1 - R_f} \quad (32)$$

5.1.4 Cut Size

A term closely related to separation efficiency is the cut size (d_{50}), which is defined as the particle size that has a 50% chance of being separated when it is subject to the action of a hydrocyclone. The majority of the solid particles finer than the cut size in the feed will report to the overflow, while the majority of those coarser will be separated and report to the underflow. The cut size can be determined from the grade efficiency curve (see Fig. 24).

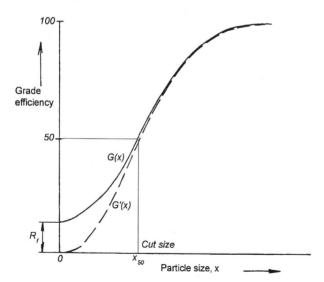

Figure 24 Grade efficiency and cut size in hydrocyclone operation. (From Svarosvsky, 1985.)

5.2 Fundamentals of Hydrocyclone Separation

A full understanding of the hydrocyclone requires a detailed analysis of the flow pattern within its body. A number of reviews on this subject may be found in the literature (Bradley and Pulling, 1959; Fontein, 1951; Kelsall, 1952). Only a brief qualitative description will be presented in this section.

5.2.1 Fluid Flow in Hydrocyclones

The pattern of fluid flow within the hydrocyclone body is best described as a spiral within a spiral with circular symmetry. A schematic view of the spiral flow inside a hydrocyclone is shown in Fig. 23b. The entering fluid flows down the outer regions of the hydrocyclone body. This combined with the rotational motion creates the outer spiral. At the same time, because of the wall effect, some of the downward moving fluid begins to feed across toward the center. The amount of inward motion of fluid increases as the fluid approaches the cone apex, and fluid that flows in this inward stream ultimately reverses its direction and flows upward to the cyclone overflow outlet via the vortex finder. This reversal applies only to the vertical component of velocity, and the spirals still rotate in the same circular direction. In the meantime, the downward flow near the wall carries solid particles to the apex opening (bottom outlet).

In addition to the main flow pattern there exists a secondary flow pattern, short circuit flow, at the top of the hydrocyclone. The short circuit flow is a flow pattern that moves across the cover of the cylindrical section to the base of the vortex finder. It flows along the outer wall of the vortex finder until it combines with the fluid in the overflow created by the main flow pattern. This short circuit flow pattern is due to the wall effect of the cyclone top cover and the outer wall of the vortex finder. The quantity of the short circuit flow can be as much as 15% of the total feed flow.

The central air core is another important flow pattern in the hydrocyclone. The rotation of the fluid in the hydrocyclone creates a low-pressure axial area that results in the formation of a rotating free liquid surface. The central air core is cylindrical in shape and filled with air the whole way through the length of the hydrocyclone. the central air core tends to stabilize the vortex flow pattern within the hydrocyclone.

5.2.2 Particle Motion in Hydrocyclones

Particles (including immiscible liquid droplets) will separate from the suspending liquid if their density is different from that of the suspending liquid owing to the centrifugal effects generated by the spiral flow in the hydrocyclone. Usually it is assumed that particles are heavier than that of the suspending fluid and that they move readily outward.

As discussed earlier, there are two spiral flow patterns existing in the hydrocyclone. Only particles existing in the outer spiral flow will be separated by the centrifugal force. Any particles in the inner spiral flow will pass upward to the overflow outlet. It should be noted that there are two important stages in the process of particle separation. One is the separation of the solids from the main body of the flow into the boundary layer adjacent to the inner wall of the hydrocyclone by centrifugal forces. The other is the removal of the separated solids from the boundary layer by downward fluid flow (not by gravity) to the apex of the cone and out of the hydrocyclone.

5.2.3 Velocity and Pressure Distributions

The velocity of the fluid flow in a hydrocyclone can be resolved into three components: tangential, axial, and radial. The most useful and significant of these three components is the tangential velocity.

The tangential velocity of the fluid in the hydrocyclones increases as the radius decreases, which is expressed by the empirical relationship

$$VR^n = \text{const} \tag{33}$$

where V is the tangential velocity, R the radius, and n the empirical exponent, usually between 0.5 and 0.9. For the outer regions of a free vortex by definition n will be 1. From Eq. (33) one can conclude that as the radius R approaches zero the tangential velocity V will approach infinity. In practice, this cannot happen since Eq. (33) only holds true until small values of the radius are reached, when the tangential velocity starts to fall with a further decrease in radius. The static pressure increases radially outward because of the centrifugal field induced by vortex flow in the hydrocyclone. This static pressure head is primarily determined by the distribution of the tangential velocities within the flow. Therefore the tangential velocity distribution can be estimated from the simple measurements of the radial static pressure distribution.

Since the outer and the inner layers move in opposite vertical directions (i.e., the flow in the outer vortex moving down and the inner vortex moving up), there is a well-defined locus of zero vertical velocity between the two vortices. This locus forms an invisible boundary, which plays an important role in particle separation (see the equilibrium orbit theory discussed in the next section).

The radial velocity is the smallest of the three velocity components. It has been found that the inward radial flow (toward the center of the vessel) is a maximum near to the wall and it diminishes with decrease in radius until it is zero at the air core interface.

5.2.4 Theories of Separation

A number of physical models have been proposed for the separation process in a hydrocyclone (Driessen MG, 1951; Bradley and Pulling, 1959; Fahlstrom, 1960; Kelsall, 1952; Rietema, 1961; and Schubert and Neesse, 1980). Among these, different phenomenological approaches have led to the development of two basic theories: the equilibrium orbit theory and the residence time theory.

Equilibrium Orbit Theory. The general concept that particles of a given size reach an equilibrium radial orbit position in the hydrocyclone forms the basis of equilibrium orbit theory. The fine particles reach equilibrium at small radii where the flow is moving upwards and transports fines to the overflow, while the coarse particles find equilibrium position at large radii where the flow is moving downwards and carries these particles to the underflow outlet (apex). The dividing surface is the locus of zero vertical velocity (LZVV). The size of the particles that find equilibrium radius on LZVV will be the cut size that has an equal chance to finish in either overflow or underflow.

In developing the equilibrium orbit theory, a key assumption made by Bradley and Pulling (1959) is the existence of a "mantel" in the hydrocyclone, which precludes inward radial velocity in the region immediately below the vortex finder. Furthermore, the LZVV is assumed to be in the form of an imaginary cone whose apex coincides with the apex of the hydrocyclone and whose base is at the bottom of the mantel. Based on these assumptions, the equilibrium orbit theory has led to the development of empirical correlations for determining the cut size and pressure drop in hydrocyclone operation.

The major deficiency of the equilibrium orbit theory lies in its lack of consideration of the effect of turbulence flow on particle separation and the residence time of the particles in the hydrocyclone (as not all particles are able to find equilibrium orbits within their residence time). In spite of such weaknesses, it proves to be a reasonable approach for determining the hydrocyclone performance at low solids concentrations (2.0% by volume).

Residence Time Theory. Residence time theory is based on the consideration of whether a particle will reach the cyclone wall in a given residence time. In the development of this theory, the distribution of all particles across the inlet is assumed to be homogeneous (Rietema, 1961). The cut size will be the size of those particles that enter the center of the inlet pipe and just reach the wall within the residence time. Using this theory coupled with extensive experimental test data, Rietema was able to establish a set of empirical correlations and suggest a criterion (a characteristic cyclone number) for "optimum" design of hydrocyclones.

Despite the very different approaches and assumptions, the forms of correlations obtained by the equilibrium orbit theory and residence time theory are similar. For specific hydrocyclone designs, both theories provide their respective empirical equations for determining the cut size and pressure drop in terms of three dimensionless groups, the Stokes number at cut size, Stk_{50}, the Euler number, Eu, and the Reynolds number, Re (see discussions in Sec. 5.4 below):

$$Stk_{50} = \frac{d_{50}^2(\rho_s - \rho)v}{18\mu D} \quad (34)$$

$$Eu = \frac{\Delta p}{v^2\rho/2} \quad (35)$$

$$Re = \frac{Dv\rho}{\mu} \quad (36)$$

where d_{50} is the cut size, ρ_s the solid density, ρ the liquid density, v the superficial velocity of liquid, Δp the pressure drop, D the hydrocyclone diameter, and μ the liquid viscosity.

5.3 Design and Operation Variables

In practice the performance of a hydrocyclone is affected by two groups of variables. These are: (1) the operating variables that are related to the operating conditions but independent of hydrocyclone size and proportions; and (2) the design variables that are dependent on the physical dimensions and proportions of the hydrocyclone.

5.3.1 Operating Variables

The most important operating variable is the pressure drop, Δp, which is proportional to the square of volumetric feed rate, Q (which is directly related to the tangential velocity):

$$\Delta p \propto Q^2 \quad (37)$$

As the operating pressure drop increases, the tangential velocity increases. Consequently, the higher separation efficiency or smaller cut size d_{50} results. Generally, this relationship only holds to be true with the operating pressure up to 2 bar. Beyond that, further increase in the operating pressure has little effect on the separation efficiency.

The second important operating variable is the feed solid concentration. With increasing feed solid concentration, the separation efficiency falls off rapidly owing to its effect on liquid flow pattern and the interaction among solid particles. Therefore hydrocyclones are usually operated with dilute feed solids concentrations (<2% by volume).

The adjustment of the underflow orifice (apex) opening is also a very important operating variable. Correct adjustment of this opening is vital for the best operation of the hydrocyclone. In general, an increase in the underflow orifice size causes an increase in operating capacity, but it tends to reduce the cut size and the underflow solids concentration.

5.3.2 Design Variables

The key design variables are defined as those that are associated with the hydrocyclone dimensions, including cyclone diameter, D, inlet diameter, D_i, outlet (apex) diameter, D_u, and vortex finder diameter, D_o. In addition, the total length of the cyclone, L, the length of the cylindrical section, l, and the cone angle, θ, are also important. They all affect the operating performance of the hydrocyclone.

The cyclone diameter or the diameter of the base of the cone is a primary design variable, and all other dimensions are usually related to it. For a given feed flow rate, the effects of cyclone diameter on the operating performance of the hydrocyclone can be described by proportionalities

$$d_{50} \propto D^x \quad (38)$$

and

$$\Delta P \propto D^y \quad (39)$$

where x ranges from 1.36 to 1.52 and y from -3.6 to -4.1. In other words, at a constant feed flow rate, the larger the cyclone diameter the larger the cut size and the smaller the operating pressure drop. If the feed flow rate is varied to maintain a constant pressure drop, the values of x will then vary from 0.41 to 0.50.

The inlet diameter, D_i, affects both the capacity and the separation efficiency, since it controls the inlet velocity and therefore the tangential velocities inside the cyclone. Based on the theoretical analysis, the cyclone resistance coefficient (in terms of Euler number) increases exponentially with a decrease in D_i:

$$\text{Eu} \propto D_i^{(-n)} \qquad (40)$$

where n is in the range of 1.2 to 4 with an average of about 2.

At constant operating pressure drop, the feed flow rate is proportional to the square root of $1/\text{Eu}$. Therefore the capacity of the cyclone, Q, is roughly proportional to D_i. In addition, the effect of D_i on the cut size d_{50} is significant. According to Bradley (1965), the cut size d_{50} is related to inlet diameter by the relationship

$$d_{50} \propto D_i^z \qquad (41)$$

where z varies from 0.6 to 0.68.

5.4 Design and Scale-Up

A reliable scale-up and performance prediction of the hydrocyclone is limited to the low solids concentration ($<2\%$ by volume). Under dilute conditions the flow pattern in hydrocyclones is unaffected by the presence of solid particles, and the particle–particle interaction is not significant. In contrast to low feed concentration conditions, quantitative prediction of the hydrocyclone performance under higher feed concentrations is uncertain.

Generally, the scale-up of hydrocyclones is based on the concept of cut size d_{50} on the grade efficiency curve, because the shape of the grade efficiency curve remains the same for a family of geometrically similar hydrocyclone designs. Dimensional analysis coupled with theoretical considerations gives two functional relationships for a hydrocyclone (Svarovsky, 1990):

$$\text{Stk}_{50} \cdot \text{Eu} = C_1 \qquad (42)$$

and

$$\text{Eu} = K_p \cdot (\text{Re})^{n_p} \qquad (43)$$

where the constants C_1, K_p, and n_p are empirical performance constants for geometrically similar hydrocyclones.

Table 9 summarizes the dimensions and experimentally determined performance constants for several known hydrocyclone designs (Svarovsky, 1984). In the absence of actual test results, Eqs. (42) and (43), using constants given in Table 9, can serve as useful guides for estimating the performance of an existing hydrocyclone or selecting a proper hydrocyclone size for a given duty.

Table 9 Summary of Several Known Hydrocyclone Designs

Type and size of hydrocyclone	Geometrical proportions					Scale-up constants			Running cost criterion[a]
	D_i/D	D_0/D	l/D	L/D	Cone angle, θ, degrees	Stk_{50}Eu	K_p	n_p	$(\text{Stk}_{50})^{4/3}\text{Eu}$
Rietema's design $D = 0.075\,\text{m}$	0.28	0.34	0.40	5	20	0.0611	316	0.134	2.12
Bradley's design $D = 0.038\,\text{m}$	0.133	0.20	0.33	6.85	9	0.1111	446.5	0.323	2.17
Mozley cyclone $D = 0.022\,\text{m}$	0.154	0.214	0.57	7.43	6	0.1203	6381	0	3.20
Mozley cyclone $D = 0.044\,\text{m}$	0.160	0.25	0.57	7.71	6	0.1508	4451	0	4.88
Mozley cyclone $D = 0.044\,\text{m}$	0.197	0.32	0.57	7.71	6	0.2182	3441	0	8.70
Warman 3″ Model R $D = 0.076\,\text{m}$	0.29	0.20	0.31	4.0	15	0.1079	2.618	0.8	2.07
R W 2515 (AKW) $D = 0.125\,\text{m}$	0.20	0.32	0.8	6.24	15	0.1642	2458	0	6.66

[a] Running cost criterion is directly proportional to the pressure drop.
Source: Svarovsky L, 1984, p 60, Table 5.1.

NOMENCLATURE

C_1	=	empirical constant
C_f	=	solid concentration in the feed
C_u	=	solid concentration in the underflow
d_{50}	=	cut size
D	=	hydrocyclone diameter
D_i	=	inlet pipe diameter
D_o	=	diameter of the vortex finder
D_u	=	underflow (apex) diameter
E_T	=	total efficiency
E_T'	=	reduced total efficiency
Eu	=	Euler number
$F_f(x)$	=	Cumulative percentage of solids with particle size x in the feed
$F_o(x)$	=	Cumulative percentage of solids with particle size x in the overflow
$F_u(x)$	=	Cumulative percentage of solids with particle size x in the underflow
$G(x)$	=	grade efficiency
$G'(x)$	=	reduced grade efficiency
K_p	=	empirical constant
l	=	length of the cylindrical section
L	=	total length of the hydrocyclone
n	=	empirical constant
n_p	=	empirical constant
Δp	=	operating pressure drop
Q	=	feed volumetric flow rate
R	=	radius
Re	=	Reynolds number
R_f	=	underflow to throughput ratio
Stk_{50}	=	Stokes number of cut size
U	=	underflow volumetric flow rate
v	=	superficial velocity
V	=	tangential velocity
x	=	empirical constant
y	=	empirical constant
z	=	empirical constant

Greek Letters

θ	=	cone angle
$\Delta\rho$	=	density difference between solids and liquid
μ	=	liquid viscosity

REFERENCES

Bradley D. The Hydrocyclone. London: Pergamon Press, 1965.

Bradley D, Pulling DJ. Flow patterns in the hydraulic cyclone and their interpretation in terms of performance. Trans Inst Chem Eng 37:34–45, 1959.

Bretney E. Water Purifier. U.S. patent No. 453,105, May 26, 1891.

Driessen MG. Theory of flow in cyclone. Rev Ind Mining, Special Issue 4: pp 449–461, 1951.

Fahlstrom PH. Discussion (on hydrocyclone) Proceedings of International Mining Processing Congress. Institute of Mining and Metallurgy, 1960, pp 632–643.

Fontein DF. Cyclone separator of solid mixtures of different grain size and specific gravity. U.S. patent No. 2,550,341, 1951.

Kelsall DF. A study of the motion of solid particles in a hydraulic cyclone. Trans Inst Chem Eng 30:87–104, 1952.

Rietema K. Performance and design of hydrocyclones. Parts I to IV. Chemical Engineering Science 15:298–325, 1961.

Schubert H, Neesse T. A hydrocyclone separation model in consideration of turbulent multi-phase flow. In: Proc International Conference on Hydrocyclones. Cambridge, BHRA Fluid Engineering, Cranfield, 1–3 October 1980, Paper 3, 23–36.

Svarovsky L. Hydrocyclones. New York: Technomic, 1984, pp 36–60.

Svarovsky L, ed. Solid–Liquid Separation Processes and Technology. Amsterdam: Elsevier, 1985, pp 18–27.

Svarovsky L. Hydrocyclones. In: Svarovsky L, ed. Solid–Liquid Separation, 3d ed. UK: Butterworths, 1990, pp 203–248.

Index

Acrylonitrile, 421–422

Adsorber design (*see* Mass Transfer)

Agglomeration of particles, 148–150 (*see also* Coating and granulation)
    effect of temperature and pressure, 148

Alumina calcination, 438–440

Aluminum flouride synthesis, 438–439

Angle (*see also* Particles)
    of internal friction, 24
    of repose, 24

Application:
    chemical production and processing, 421–444
    coating and granulation, 445–468
    drying, 469–484
    fluid catalytic cracking, 379–396
    gasifiers and combustors, 397–420

Archimedes number, 25, 55, 90

Attrition, 13, 163, 201–239
    attrition index, 13
    in fluidized beds, 201–239
        continuous operation, 228–233
        factors affecting attrition, 202–209
        modes of attrition, 202–209
        sources of attrition, 218–224
    friability tests, 212
    at submerged jets, 163, 219–221
    test apparatus, 15, 213–218
        cyclone, 216–218
        Davidson jet cup, 15, 214
        fluidized bed, 214–216
        by impaction principle, 15, 214

Baffles (*see* Internals in fluidized beds)

Barnea and Mednick correlation, 43

Bed-wall heat transfer (*see* Heat transfer)

Biofluidization (*see* Liquid fluidized beds)

Biomass:
    combustion (*see* Fluidized bed combustor)
    gasification (*see* Fluidized bed gasifier)

Blake correlation, 40

British standard sieve, 7–8

Brownell and Katz correlation, 41

Bubbling fluidized beds, 53–112
    attrition in (*see* Attrition)
    bed expansion, 67
        aggregative beds, 67
    bubble phase, 67–75
        bubble coalescence, 72–75
        bubble formation, 71–72
        bubble in liquid, 68
        bubble rise and growth, 72–75
        Davidson's isolated bubble model, 69–71
        effect of internals on, 171–172, 189–193
    effect of temperature and pressure, 129–141
        applications of dimensional analysis, 147–148
        experimental observations, 135–139
        Group A powders, 133–134
        Group B powders, 134
        Group D powders, 140
    essential elements, 54
    fast fluidization, 172, 174
        flow structure, 172, 174
    flow structure, 172
        effect of internals, 193–194
    heat transfer (*see also* Heat transfer)
        effect of temperature and pressure, 144–146
    industrial applications, 310–312
    jetting phenomena, 78–89

[Bubbling fluidized beds]
    attrition at submerged jets, 163
    concentric jets, 85
    effect of temperature and pressure, 141
    gas entrainment into jets, 85–86
    gas velocity profile, 84
    initial bubble size and frequency, 84–85
    jet half angle, 84
    jet penetration correlations, 82
    jet penetration depth, 81
    jetting regimes, 80
    momentum dissipation, 81
    multiple jets, 86–87
    potential core, 81
    solid entrainment into jets, 85–86
    two-phase jets, 85–86
minimum bubbling velocity, 54
minimum fluidization velocity, 62–64, 129–132
    of binary mixtures, 94
    Carman equation, 62
    effect of temperature and pressure, 129–132
    Ergun equation, 63
    of multicomponent mixtures, 96
    Leva equation, 63
    Rowe equation, 63
    Wen and Yu equation, 63
models, 239–255
    bubble assemblage model, 246–250
    bubbling bed models, 244–246
    comparison of models, 250
    Davidson and Harrison model, 241
    for freeboard region, 251
    for grid region, 251
    Kunii and Levenspiel model, 246–250
    multiple-region models, 250–252
    Partridge and Rowe model, 244
    pseudo-homogeneous models, 239–240
    theoretical studies, 243
    two-phase models, 240–244
regimes, 57–62, 172–175
regimes transitions, 58–61
    between bubbling and slugging, 59
    between bubbling and turbulent, 59
    between particulate and aggregative, 58
    transition to fast fluidization, 61
two-phase theory of fluidization, 64–65
particle mixing, 87–104
    bed turnover time, 91–92
    convective transport, 90–91
    effect of internals, 194–195
    lateral mixing, 90
    mixing index, 97
    solid residence time distribution, 92
particle segregation, 87–104
    analogy to gas-liquid-solid phase equilibrium, 93–94
    classification of flotsam and jetsam, 94
    effect on fluidization phenomena, 99

[Bubbling fluidized beds]
    effect of particle properties, 97–99
    effect of pressure and temperature, 99
    industrial applications, 102–103
    mathematical models, 103–104
    rate of particle segregation, 99–102
    segregation index, 97
scaleup (see Scaleup)
slugging beds, 75–78
    bed expansion, 76
    effect of expanded bed section, 78
    minimum bed height required for slugging, 77
    slug frequency, 76
    slugging criteria, 76
    slug length, 76
    type A and type B, 76
    wall slug, 76, 78
turbulent fluidization, 173
    flow structure, 173
viscosity, 65–66

Calcination (see Chemical production and processing)
Carman and Kozeny correlation, 42
Carman equation, 62
Carryover (see Entrainment)
Catalytic cracking (see FCC)
Catalytic oxidation of chlorinated byproducts, 441–442
Catalytic reactors (see Reactor design)
Centrifugal fluidized beds (see Rotating fluidized beds)
CFB (see Circulating fluidized beds)
Chemical production and processing, 421–444
    chemical synthesis, 421–428
        acrylonitrile, 421–422
        maleic anhydride, 422–425
        oxychlorination of ethylene, 425
        phthalic anhydride, 425–428
    gas to liquids, 428–432
        Exxon process, 430–431
        Sasol process, 428–430
        Syntroleum, 431–432
    heavy oil upgrading, 434–437
        combicracking, 435–437
        flexicoking, 435
        fluid coking, 434–435
        HDH, 437
        H-oil and LC fining, 435
        U-CAN, 437
    metallurgical processing, 438–439
        alumina calcination, 438
        aluminum flouride synthesis, 438
        ore roasting, 438–439
    other processes, 439–443
        catalytic oxidation of chlorinated byproducts, 441–442
        isophthalonitrile, 442–443
        Mobil/Badger technologies, 440–441
        UOP/Hydro methanol to olefins, 439

[Chemical production and processing]
  polymerization of olefins, 432–434
    UNIPOL process, 432–433
  semiconductor silicon, 437–438
    MEMC process, 437–438
Chlorination (see Chemical production and processing)
Choking (see Pneumatic conveying, Circulating fluidized beds)
Circulating fluidized beds, 146, 485–544
  advantages and disadvantages, 485–486
  applications, 486–487 (see also Application)
  CFB as chemical reactors, 524–530
    core-annulus models, 527–529
    key considerations, 524–526
    multizone models, 529–530
    one-dimensional models, 526–527
    type of reactors, 525
  critical velocity, 487–488
    definition, 488
  effect of temperature and pressure, 146
  flow regimes, 486–490
    onset of fast fluidization, 486–488
    operating diagram, 489–490
    transition from fast fluidization to dense suspension upflow, 488–489, 508
    transition from pneumatic transport to fast fluidization, 488, 508
  gas mixing, 509–514
    axial gas dispersion, 513
    backmixing, 512–513
    gas residence time distribution, 510–511
    interphase mass transfer, 513–514
    radial gas dispersion, 511–512
  gas velocities, 501
  geometric/operating variable effect, 501–506
    addition of secondary gas, 503
    bottom configuration, 502–503
    cross-sectional shape, 501–502
    downer geometry, 504–505
    exit effect, 503–504
    internals, 504
    temperature and pressure, 504
    wall shape and roughness, 502
  heat transfer, 516–523 (see also Heat transfer)
    convective heat transfer, 516–520
    convective heat transfer models, 519–520
    estimating total heat transfer, 521–523
    large scale CFB experiments, 518
    radiation heat transfer experiments, 520
    radiation heat transfer models, 520–521
  hydrodynamic models, 505–506
    CFD models, 506
    empirical correlations, 505
    mechanistic models, 505–506
  mass transfer, 523–524 (see also Mass transfer)
  particle flux, 498–501
    lateral flux, 501

[Circulating fluidized beds]
    measurement techniques, 499
    net solids circulation flux, 498–499
    radial profiles, 499–500
    wall/annular layer thickness, 500–501
  particle velocities, 496–498
    experimental results, 497–498
    measurement techniques, 496–497
  scaleup, 506–508 (see also Scaleup)
    data from large units, 507
  solids concentration, 490–496
    annular variations and asymmetries, 496
    axial solids holdup profiles, 492–495
    measurement techniques, 490–492
    radial solids holdup profiles, 495–496
  solids mixing, 514–516
    axial dispersion, 514–515
    exchange between core and annulus, 515–516
    radial dispersion, 515
  transport velocity, 486–487
  typical circulating fluidized bed, 486
  voidage, 490–496 (see also Voidage)
    annular variations and asymmetries, 496
    axial voidage profiles, 492–495
    measurement techniques, 490–492
    radial voidage profiles, 495–496
Circulating systems (see Circulating fluidized beds)
Classification of particles (see Particles)
Coal:
  combustion (see Fluidized bed combustor)
  gasification (see Fluidized bed gasifier)
Coalescence of bubbles (see Bubbling fluidized beds)
Coating and granulation, 445–468
  computer simulation, 463–466
    granule deformation and breakup, 465–466
    granule growth by coalescence, 463–465
  conditions for granulation, 450–452
    critical binder/powder ratio, 450–451
    interparticle bridge properties, 451–452
  experimental verification, 456–463
    in constant shear fluidized beds, 456–458
    data by Dencs and Ormos, 461–463
    data by Watano et al., 458–461
  fluid bed as a mixer/granulator, 446–449
    fluid bed granulation circuit, 448
    industrial fluid bed granulators, 447
  granule consolidation, 456
  granule growth kinetics, 452–456
    conditions of coalescence, 452–453
    prediction of critical size, 453
    summary, 454–456
    wet granule deformation and breakup, 454
  microscopic phenomena, 449–450
    granule growth mechanisms, 449
Combicracking, 435–437
Combustor (see Fluidized bed combustor)
Coulter counter, 11

Cyclones (*see* Cyclone separators)
Cyclone separators, 599–617
   arrangements, 602–604
      internal or external, 604
      parallel, 603–604
      series, 602–603
   cyclone design, 608–614
      diameter, 608
      length, 608
      wall roughness, 609
      Zenz design procedure, 609–613
   effect of pressure and temperature, 615–616
   effect of solids loading, 605–606
   inlet design, 601, 604–605
   inlet gas velocity, 607–608
   outlet design, 606–607
      effect of outlet tube penetration, 607
   types, 600–601
      multiclones, 601
      reverse flow cyclone, 600
      uniflow cyclone, 601
   vortex length, 614–615
   vortex stabilizers, 615

Darcy's law, 39–40
Davidson jet cup, 15
Davidson's isolated bubble model, 69–71
Distributors (*see* Gas distributors)
Downer reactors, 336–339
Drag coefficient, 131
   particle in an array, 43
   single particle, 14–18
      corrections for non-sphericity, 17–18
      empirical expressions, 16–17
      for particles lighter than surrounding fluid, 18
      theoretical expressions, 16
Drying, fluid bed dryers 469–484
   classification of fluid bed dryers, 470–473
      comparison with other dryers, 470
   design considerations, 477–480
      calculation method for batch drying, 477–478
      performance prediction, 479–480
   design criteria for fluid bed dryers, 473
   drying kinetics, 473–475
      effect of bed height, 474–475
      effect of bed temperature, 475
      effect of gas velocity, 475
      effect of particle size, 475
   features of fluid bed dryers, 475–477
      batch dryers, 475
      centrifugal dryers, 476
      mechanically-agitated dryers, 476
      plug-flow dryers, 475
      spouted bed dryers, 476–477
      vibrated dryers, 475
      well-mixed continuous dryers, 475
   novel fluid bed dryers, 480–482

[Drying, fluid bed dryers]
      low-pressure fluid bed drying, 482
      modified fluid bed dryers, 480–481
      superheated-steam drying, 481–482
   scaleup, 480 (*see also* Scaleup)
   selection criteria, 470–473
      decision tree, 478
      selection guide, 477

Elutriation (*see* Entrainment)
Entrainment, 113–128
   effect of attrition, 209–212
   effect of temperature and pressure, 143–144
   elutriation constant, 122–123
      correlations, 122–123
   experimental investigations, 116–120
      effect of bed height, 119
      effect of freeboard geometry, 118
      effect of internals, 119
      effect of particle size, 116–117
      effect of superficial gas velocity, 118
      effect of temperature and pressure, 119–120
   gas-solid flow in freeboard, 115–116
   models, 120–125
      freeboard height below TDH, 124–125
      freeboard height exceeding TDH, 120–121
   particle ejection into freeboard, 114–115
      entrainment flux at bed surface, 125
   particle size distribution, 210–212
      effect of attrition, 210–212
   TDH (transport disengaging height), 121
      correlations, 124
Ergun Equation, 42, 63
Explosion hazard (*see* Pneumatic conveying)
Exxon gas to liquids process, 430–431

Fast fluidization, 142–144, 174–175 (*see also* Circulating
      fluidized beds)
   effect of temperature and pressure, 142–143
   flow structure, 174–175
FBC processes (*see* Fluidized bed combustor)
FCC, 379–396
   fundamentals, 381–385
      catalytic cracking reactions, 381–382
      FCC catalyst, 381
      heat balance, 383–384
      hydrogen balance, 385
      pressure balance and catalyst circulation, 384–385
      thermal cracking reactions, 382–383
   modern FCC design, 385–395
      feed injection system, 385–389
      regenerator, 390–393
      riser/reactor, 389–390
      standpipe and standpipe inlet, 393–395
      stripper, 390
      third-stage separator, 393
   what is, 379–381

Feret diameter, 2, 7
Filtration (*see* Liquid-soilds separation)
Fischer-Tropsch synthesis (*see* Chemical production and
    processing)
Fixed bed, 29–52
    gas velocity distribution, 45–46
    heat transfer, 46–49
        effective radial thermal conductivity, 47
        with immersed surfaces, 47
        Molerus and Wirth correlations, 47
        particle to fluid, 46
        through wall, 46
    mass transfer, 49, 288–289
        effective radial diffusivity, 49
        mass transfer coefficient, 49, 289
        single particle, 49, 288
    packing characteristics, binary mixtures 32–34
        critical ratio of entrance, 34–35
        critical ratio of occupancy, 34–35
        non-spherical particles, 33–34
        spherical particles, 32–34
    packing characteristics, mono-sized spheres 29–32
        random packing, 29–31
        regular packing, 29–31
    packing characteristics, non-spherical particles 35–36
        Hausner ratio and sphericity, relationship 36
    permeability, 44
    pressure drop, 39–43
        Barnea and Mednick correlation, 43
        Blake correlation, 40
        Brownell and Katz correlation, 41
        Carman and Kozeny correlation, 42
        Darcy's law, 39
        Ergun correlation, 42
        general friction factor correlation, 43
        modified Ergun correlation, 43, 709–712
    surface area, 31–32, 289–291
        specific surface area of bed, 31–32
        total particle exterior surface area, 289–291
    voidage, 31, 37–39, 44
        correlations, 37–39
        factors affecting, 36–37
        radial oscillations, 38
        of randomly-packed, mono-sized particles, 36
        rough estimate, 36
        stagnant voidage, 44
        voidage and sphericity, relationship, 37
Flexicoking, 435
Flotsam (*see* Particle segregation)
Flow regime transitions, 58–61
    between bubbling and slugging, 59
    between bubbling and turbulent, 59
    between particulate and aggregative, 58
    transition to fast fluidization, 61
Fluid coking, 434–435
Fluidized bed dryers (*see* Drying)

Fluidization (*see also* Bubbling fluidized beds, Circulating
    fluidized beds)
Fluidization engineering (*see* Reactor design)
Fluidized bed applications (*see* Application)
Fluidized bed combustor, 409–418
    design considerations 412–418
        coal-fueled AFBC, 413–416
        coal-fueled PFBC, 416–418
        biomass-fueled AFBC, 416
    future of, 418
    principles, 409–412
        combustion process schemes, 411
    status of technology, 412
Fluidized bed gasifier, 398–409
    design considerations 406–409
        biomass gasification, 408–409
        coal gasification, 406–408
    future of, 418
    principles, 398–400
        functions and requirements, 398–399
        products and applications, 398
        reaction environment, 399–400
    status of technology, 400–406
Fluidized beds (*see* Bubbling fluidized beds)
Fluidized bed viscometer, 650
Freeboard (*see* Bubbling fluidized beds)

Gas distributor, 155–170 (*see also* Liquid fluidized beds)
    attrition at grid, 161
    attrition at submerged jets, 163
    bubble caps, 155–156
    conical grids, 157
    design criteria, 157–161
        effect of temperature and pressure, 164
        jet penetration, 157–158
        pressure drop, 158–159
    design equations, 159–160
        hole layout, 160
        hole size, 160
    design examples, 166–169
        FCC grid design, 166–167
        polyethylene reactor grid design, 168
        power consumption, 169
    erosion, 161–163
        bed walls and internals, 161–162
        distributor nozzles, 162–163
    mass transfer in grid region (*see* Mass transfer)
    nozzles, 155–156
        sizing, 161
    perforated plates, 155
    pierced sheet grids, 157
    plenum, 164–165
        configurations, 165
        design, 164
    port shrouding, 161
    power consumption, 165–166
    spargers, 156

[Gas distributor]
    design criteria, 160–161
    weeping, 163–164
Gasifier (*see* Fluidized bed gasifier)
Gas-solids separation (*see* Cyclone separators)
Gasification (*see* Fluidized bed gasifier)
Geldart classification of powders, 54–55
General friction factor correlation, 43
German DIN sieve, 7–8
Granulation (*see* Coating and granulation)
Grid, (*see* Gas distributor)

Hardgrove Grindability Index, 13
Hatch-Choate transformation, 23
HDH, 437
Heat transfer, 144–148, 257–286
    in bubbling fluidized beds, 144–146, 258–268
        bed-surface transfer, 261–265
        convective heat transfer, 261–265
        examples, 273–278
        freeboard heat transfer, 266–268
        particle-gas transfer, 259–261
        radiative heat transfer, 265–266
    in fast fluidized beds, 268–273
        bed-surface transfer, 269–273
        convective heat transfer, 270–273
        examples, 278–283
        particle-gas transfer, 269
        radiative heat transfer, 273
Heywood shape factor, 5
H-oil and LC fining, 435
Hydrocyclone (*see* Liquid-solids separation)

Incipient fluidization (*see* Minimum fluidizing velocity)
Industrial applications (*see also* Liquid fluidized beds,
        Nonconventional fluidized beds, Three phase
        fluidized beds)
    chemical production and processing, 421–444
    coating and granulation, 445–468
    drying, 469–484
    fluid catalytic cracking, 379–396
    gasifiers and combustors, 397–420
Instrumentation, 643–704
    acoustic sensors, 666
        acceleration sensor, 666
        microphone, 666
    capacitance and conductivity sensors, 664–666
    circulating fluidized bed diagnostics, 690–691
        solid circulation rate, 690
        solid holdup distribution, 690
    computer tomography (CT) and image reconstruction,
        681–685
        image reconstruction principle, 681–685
        neutron imaging, 685
    data processing, 671–681
        auto-correlation, 676
        cross-correlation, 676

[Instrumentation]
        maximum entropy method, 676
        power spectrum, 676
        wave transformation, 679–680
    electrostatic sensors, 667
    fluidized bed diagnostics, 687–694
        agglomerating of cohesive powders, 692
        bubble fraction, 688–689
        bubble size from pierced length, 689–690
        defluidization velocity, 687–688
        expansion characteristics, 692–693
        gas sampling from bed, 690
        minimum fluidization velocity, 692
        pressure measurement, 687
        regime transition velocities, 694
        solid circulation and mixing, 690
        visible bubble flow distribution, 688–689
    gas sampling, 671, 673
    gas sensors, 667
    gas/solid pressure and force sensors, 647–650
        fluidized bed viscometer, 650
        gas pressure, 647–649
        manometer, 649
        solid force, 650
        solid pressure, 649–650
    heat flux probes, 647
    humidity sensors, 667
    on-line particle size distribution measurement,
        685–687
    optical sensors, 650–664
        fundamentals of light measurement, 650–651
        laser doppler velocimetry (LDV), 655–662
        light sources, 651–652
        optical fiber probes, 653–655, 658–661
        optical fibers, 652–653
        photodetectors, 652
        space filters, 662–664
    radioactive sensors, 668
    suction probes for solids sampling, 670–672
    temperature sensors/probes, 644–647
        infrared camera/thermometer, 646–647
        optical pyrometer, 644–646
        thermocouple, 644
    tracers, 668–670
        gas, 670
        solid, 668–670
    visual observation, 681
        laser imaging, 681
        particle image velocimetry (PIV), 681
        photo and video imaging, 681
Internal friction angle, 24
Internals in fluidized beds, 171–199
    classification, 175–186
        baffles, 175
        inserted bodies, 184–186
        other configurations, 186–189
        packings, 182–184

[Internals in fluidized beds]
    tubes, 180–182
  reactors with internals, 184,187, 188–189
Isophthalonitrile production, 442–443

Jet in fluidized beds, 78–87 (*see also* Jetting fluidized
    beds)
  attrition at submerged jets, 163
  characteristics of jets, 81–84
    gas velocity profile, 84
    initial bubble size and frequency, 84–85
    jet half angle, 84
    jet penetration correlations, 82–83
    momentum dissipation, 81
    penetration depth, 81–84
    potential core, 81
    effect of temperature and pressure, 141–142
  concentric jets, 85
  gas entrainment into jets, 85–86
  jetting regimes, 80–81
  multiple jets, 86–87
  solid entrainment into jets, 85–86
    model, 86
  two-phase jets, 85
Jetsam (*see* Particle segregation)
Jetting fluidized beds, 556–563
  application, 563
  bubble dynamics, 557–558
  gas mixing, 558–559
    around concentric jets, 559
    around single jet, 558–559
  jet penetration, 556–557 (*see also* Jet in fluidized beds)
    gas-solid two-phase jet, 557
    jet half angle, 557
    jet penetration depth, 557
    momentum dissipation, 557
  scaleup, 563 (*see also* Scaleup)
  solids circulation, 559–562
    solids circulation pattern, 559–561
    solids circulation rate, 561–562
  solids entrainment into jets, 562–563

Kunii and Levenspiel model, 246–250 (*see also* Bubbling
    fluidized beds)

Liquid fluidized beds, 705–764
  applications, 743–750
    adsorption and ion exchange, 746
    backwashing of granular filters, 744
    bioreactor, 748
    electrolysis, 746
    flocculation, 746
    fluidized bed electrodes, 746
    heat exchanger, 747
    in-situ fluidized washing of soils, 744
    leaching, 745
    particle classification, 744

[Liquid fluidized beds]
    seeded crystal growth, 745
    thermal energy storage, 748
  bed expansion, 712–721
    empirical equations, 715–721
    theoretical models, 713–715
  distributor design, 737–738
  fluidized systems, continuous 731–734
    monodisperse soids, 731
    multidisperse solids, 731–734
  heat transfer, 738–743
    between particles and liquid, 738–740
    between submerged surface and liquid, 740–742
  inverse fluidization, 730–731
  liquid dispersion, 734–735
    axial dispersion, 734–735
    radial dispersion, 735
  mass transfer, 738–743
    between particles and liquid, 738–740
    between submerged surface and liquid, 742–743
  minimum fluidization velocity, 709–712
  nonhomogeneities and instability, 735–736
  particle mixing and segregation, 722–730
    binary mixtures of equal density particles,
      723–727
    binary mixtures of equal size particles, 727–728
    binary mixtures of particles differing in size and
      density, bed inversion 728–730
    deduction from pressure gradient, 722–723
  particulate fluidization, 705–709
    buoyancy and drag, 706–707
    hydrodynamic representation, 707–709
    mixing and segregation pattern, 722
Liquid-solids separation, 811–850
  centrifugation, 836–844
    filter centrifuges, 839–844
    sedimentation centrifuges, 836–839
  filtration, 812–827
    cake filtration, 812–814
    deep-bed filtration, 814–815
    equipment, 821–826
    filter aids, 819–821
    filter media, 816–819
    membrane filtration, 815
  hydrocyclones, 844–849
    cut size, 846
    design and operation variables, 848–849
    fundamentals, 846–848
    scaleup, 849
    separation efficiencies, 844–846
  sedimentation, 827–836
    clarifiers, 831–835
    fundamentals, 827–828
    thickeners, 828–831
Log-normal size distribution, 23
Loop seal (*see* Nonmechanical valves)
L-valve (*see* Nonmechanical valves)

Maleic anhydride production, 422–425
Martin diameter, 2, 7
Mass transfer, 287–307
  in bubbling fluidized beds, 287–307
    average mass transfer coefficients, 291–297
    bubbling bed approach, 294–298
    bubble-cloud exchange, 295–297
    Chavarie and Grace model, 299–301
    correlation for multiple-bubble beds, 301
    emulsion-cloud exchange, 295–297
    fluidized particles and gas exchange, 291–293
    highly adsorbing particles, 298
    homogeneous bed approach, 287–294
    isolated spheres and bed transfer, 293–294
    Kunii and Levenspiel model, 295–298
    nonporous and nonadsorbing particles, 297
    Partridge and Rowe model, 298–299
    porous or partially adsorbing particles, 298
    single sphere and gas transfer, 288
  comparison of mass transfer coefficient in fixed and
    fluidized beds, 293
  effect of particle adsorption, 302
  example problems, 303–304
  in fixed beds, 288–291
    average mass transfer coefficient, 289–291
    fixed bed particles and gas transfer, 288–291
    homogeneous bed approach, 287–294
  in grid region, 301–302
  relation between mass and heat transfer, 302–303
MEMC high purity silicon process, 437–438
Minimum bubbling velocity (see Bubbling fluidized beds)
Minimum fluidizing conditions (see Bubbling fluidized
  beds)
Minimum fluidizing velocity (see Bubbling fluidized beds)
Mode, median, mean, 22
Models, 239–255
  bubbling fluidized beds, 239–252
    bubble assemblage model, 246–250
    bubbling bed models, 244–246
    comparison of models, 250
    Davidson and Harrison model, 241
    for freeboard region, 251
    for grid region, 251
    Kunii and Levenspiel model, 246–250
    multiple-region models, 250–252
    Partridge and Rowe model, 244
    pseudo-homogeneous models, 239–240
    theoretical studies, 243
    two-phase models, 240–244
  CFB as chemical reactors, 524–530
    core-annulus models, 527–529
    key considerations, 524–526
    multizone models, 529–530
    one-dimensional models, 526–527
    type of reactors, 525
Modified Ergun correlation, 43, 709–712
Multiclones, 601

Newton's law regime, 15
Nonconventional fluidized beds, 545–570
  Jetting fluidized beds, 556–563
  recirculating fluidized beds, 550–556
  rotating fluidized beds, 563–565
  spouted beds, 545–550
Nonmechanical solids flow devices (see Nonmechanical
  valves)
Nonmechanical valves, 571–597
  common nonmechanical valves, 586
  cyclone dipleg, 594–596
  loop seal, 591–592
  L-valve, 587–591, 593–594
    aeration, 589–590
    loop configuration, 588
    mode of operation, 591
  N-valve, 592
  seal pot, 591
  trickle valve, 594–596
  V-valve, 592–593
N-valve (see Nonmechanical valves)

Ore roasting, 438–439
Orifices, discharge from (see Gas distributors)
Oxychlorination of ethylene, 425

Packed beds (see Fixed beds)
Particles (see also Powders)
  angle of internal friction (typical values), 24
  angle of repose (typical values), 24
  characterization techniques, size 6–14
    cascade impaction, 10
    commercial equipment, 12–13
    Coulter counter, 11
    elutriation, 10
    gravity and centrifugal, 8–9
    imaging technique, 7–8
    resistivity and optical zone sensing, 10–11
    sedimentation, 8
    sieve analysis, 7–9
  characterization technique, surface 12
  density, 6
    skeleton density, 6
  diameter, multiparticle system 21–24
    arithmetic mean, 21
    geometric mean, 22
    harmonic mean, 22
    length mean, 21
    statistical characterization, 22
    surface mean, 21
    volume mean, 21
    volume-surface mean, 21
    weight mean, 21
  diameter, single particle 2–9
    drag diameter, 2
    equivalent diameter, 3

[Particles]
Feret diameter, 2, 7
free-falling diameter, 2
Martin diameter, 2, 7
perimeter diameter, 2
projected area diameter, 2
sieve diameter, 2
Stokes diameter, 2, 9
surface diameter, 2
surface-volume diameter, 2
volume diameter, 1
mechanical properties, 13
attrition index, 13
erosiveness index, 14
Hardgrove Grindability Index, 13
shape, 3
circularity, 3
elongation ratio, 5
flatness ratio, 5
Heywood shape factor, 5
operational circularity, 4
sphericity, 3, 290
determination via Ergun equation, 42
nonspherical particles, 4
operational sphericity, 4
regularly shaped solids, 4
typical values, 290
Particle dynamics, 14
single particle, 14–20
corrections for non-sphericity, 17
drag coefficient, 14
empirical expressions, 16–17
intermediate regime, 15
Newton's law regime, 15
for particles lighter than fluid, 18
Stokes free fall velocity, 16
Stokes law regime, 15
theoretical expressions, 16
terminal velocity, 18, 132
by Haider and Levenspiel equation, 18
by polynomial equations, 19
of porous spheres, 20
effect of temperature and pressure, 132
wall effect, 16
Particle mixing, 87–97
bed turnover time, 91
convective transport, 90
lateral mixing, 90
mixing index, 97
solid residence time distribution, 92
Particle segregation, 87–102
analogy to gas-liquid-solid phase equilibrium, 93
classification of flotsam and jetsam, 94
effect on fluidization phenomena, 99
effect of particle properties, 97
effect of pressure and temperature, 99
industrial applications, 102

[Particle segregation]
mathematical models, 103
rate of particle segregation, 99
batch systems, 99
continuous operation, 102
segregation index, 97
Permeability, 44
Phthalic anhydride production, 425–428
Plenum, 164
configurations, 165
design, 164
Pneumatic conveying:
dilute phase, 619–629
bends, 623
choking, 623–624
compressibility, 622
computer analysis, 628–629
design strategy, 624
electrostatics, 631–641
essential components, 619, 626–627
flow classification, 624–626
flow structure, 175
force balance, 620–621
friction factors, 621–622
long distance conveying, 628
particle acceleration, 622
particle velocity, 621
pickup and saltation mechanisms, 623
pickup velocity, 622
saltation, 622
system configurations, 619
troubleshooting, 627–628
voidage, 621
Pneumatic transport (*see* Pneumatic conveying)
Polymerization of olefins, 432–434
Powders:
powder classifications, 54–57
classification by bed collapse, 56–57
classification by Grace, 55
classification by Goossen, 55–56
Geldart classification, 54–55
Molerus' interpretation, 54–55
Pressure effect, on fluidization, 129–154

Reactor design, 309–342
fluidized bed reactors, 309–312
comparison of fluidization properties, 324
flow regime determination, 317, 322
gas-mixing, 322
gas-solid contact, 319–322
general development procedure, 313–315
heat transfer, 322
inter-phase mass transfer, 319–321
particle selection and catalyst development, 315
solid-mixing, 321–322
various configurations, 312
modeling, 333–339

[Reactor design]
  bubbling fluidized bed reactors, 333–334
  riser and downer reactors, 336–339
  turbulent fluidized bed reactors, 334–336
 process requirements, 326–332
  catalytic reactions, 326–327
  effect of moisture content, 331–332
  gas-phase olefin polymerization, 328–332
  gas-solid reactions, 327–328
Reactor-regenerator (*see* FCC)
Recirculating fluidized beds, 550–563
 design procedure, 554–555
 draft tube as a fluidized bed, 551–552
 draft tube as a transport tube, 552–554
  downcomer pressure drop, 552
  draft tube pressure drop, 552–553
  gas bypassing phenomenon, 553
  solids circulation mechanisms, 553–554
  solids circulation rate, 553–554
 industrial applications, 555–556
 multiple draft tubes, 555
 startup and shutdown considerations, 555
Regimes of fluidization (*see* Bubbling fluidized beds,
      Circulating fluidized beds)
Repose angle (*see* Angle)
Reverse flow cyclone, 600
Riser reactors, 336–339
Rosin-Rammler particle size distribution, 24
Rotating fluidized beds, 563–565
 horizontal rotating fluidized beds, 564
 minimum fluidization velocity, 564–565
 pressure drop, 565
 vertical rotating fluidized beds, 564
RTD of solids (*see* Bubbling fluidized beds)

Saltation (*see* Pneumatic conveying)
Saturation carrying capacity (*see* Entrainment)
Sasol gas to liquids process, 428–430
Scaleup, 343
 effect of bed diameter, 345–353
  apparent reaction rate constant, 345
  circulating or fast fluidized beds, 350–352
  CO conversion in Hydrocol reaction, 347
  hydrodynamics, 347–353
  solid diffusivity, 346
 scale models, 353–376
  application, bubbling fluidized beds, 368–370
  application, circulating fluidized beds, 370–374
  design, 359–363
  scaling parameter development, 353–355
  simplified scaling relationships, 355–359
  verification, bubbling and slugging beds, 363–366
  verification, pressurized bubbling beds, 366–367
  verification, spouting beds, 366
Screens, Tyler (*see* Tyler sieve)
Seal pot (*see* Nonmechanical valves)
Sedimentation (*see* Liquid-solids separation)

Semiconductor silicon production, 437–438
Sintering of particles, 148–150
 effect of temperature and pressure, 148–150
Slugging beds, 75–78
 bed expansion, 76–77
  effect of expanded bed section, 78
 minimum bed height required for slugging, 77
 slug frequency, 76–77
 slugging criteria, 76
 slug length, 76–77
 type A and type B slugging beds, 76
 wall slug, 76, 78
Solid circulation loops (*see* Circulating fluidized beds)
Solid circulation systems (*see* Circulating fluidized beds)
Solids (*see* Particles)
Sphericity (*see* Particles)
Spouted beds, 545–550
 conical spouted beds, 549–550
  minimum spouting velocity, 549
  voidage along spout axis, 549–550
  voidage at wall, 550
  voidage correlation, 550
 conventional spouted beds, 546–549
  fluid flow distribution, 548
  fountain height, 548–549
  maximum spoutable bed height, 547
  minimum spouting velocity, 546–547
  particle movement, 548–549
  pressure gradient in annulus, 549
  spout diameter, 547–548
  voidage distribution, 548
Standpipe, 517–597
 aeration, 579–580
 concept of relative gas-solids velocity, 572
 cyclone dipleg, 577–578
 fluidized bed flow, 573–574
 Geldart group C powders in, 581
 optimum operating regions, 574
 packed bed flow, 573
 in recirculating solid systems, 581–585
  automatice systems, 582
  controlled systems, 582–585
 standpipe configurations, 575–576
 streaming flow, 574
 vertical, angled and hybrid standpipes 584, 586
Static charge (*see* Pneumatic conveying)
Static electricity (*see* Pneumatic conveying)
Stokes diameter, 2, 9
Stokes free fall velocity, 16
Stokes law regime, 15
Syntroleum, 431–432

TDH (transport disengaging height), 121–125
 estimation, 124–125
Terminal particle velocity, 18–21, 132
 by Haider and Levenspiel equation, 18–19
 by polynomial equations, 19–20

[Terminal particle velocity]
  of porous spheres, 20–21
  effect of temperature and pressure, 132
Temperature effect, on fluidization, 129–154
Three-phase fluidization (*see* Three-phase fluidized beds)
Three-phase fluidized beds, 765–809
  bubble dynamics, 766–778
    bubble breakup, 775–778
    bubble coalescence, 774–775
    bubble formation, 766–770
    bubble rise characteristics, 770–774
  computational fluid dynamics (CFD), 794–801
    discrete particle method, 796–797
    gas-phase model, 796
    liquid-phase model, 795–796
    particle-particle collision dynamics, 797–798
    simulation examples, 798–801
  heat transfer, 784–788
    effect of bubble wake, 785–786
    effect of operating variables, 785
    effect of pressure, 786–788
    heat transfer models, 788
  hydrodynamics, 778–784
    bed contraction, 779–780
    bubble size distribution, 783–784
    dominant role of large bubbles, 783–784
    flow regime transition, 780–781
    gas holdup, 781–783
    minimum fluidization velocity, 778–779
    moving packed bed phenomenon, 780
    similarity, 781–783
  mass transfer, 788–791
    interfacial area, 789
    liquid phase mass transfer coefficient, 789–791

[Three-phase fluidized beds]
  phase mixing, 791–794
Transport disengaging height (*see* TDH)
Trickle valve (*see* Nonmechanical valves)
Turbulent fluidization, 173–174 (*see also* Bubbling
    fluidized beds)
  flow structure, 173–174
  transition to, 195–197
    effect of internals, 195–197
  scaleup (*see* Scaleup)
Turbulent fluidized bed reactors, 334–336
Two-phase theory of fluidization, 64–65
Tyler sieve, 7–8

U-CAN, 437
Uniflow cyclone, 601
UNIPOL process, 432–433
UOP/Hydro methanol to olefins process, 439
U.S. standard sieve, 7–8

Voidage:
  in bubbling beds (*see* Bubbling fluidized beds)
  in fast fluidized beds (*see* Circulating fluidized beds)
  in fixed beds (*see* Fixed beds)
  in freeboard (*see* Bubbling fluidized beds)
  in turbulent beds (*see* Turbulent fluidized bed
      reactors)
Vortex length, 614–615
Vortex stabilizers, 615
V-valve (*see* Nonmechanical valves)

Wakes, behind bubbles (*see* Bubbling fluidized beds)
Wen and Yu equation, 63
Winkler reactor (*see* Fluidized bed combustor)